Property of
S0-ATS-772

PHOTONIC SWITCHING TECHNOLOGY

IEEE Press
445 Hoes Lane, P.O. Box 1331
Piscataway, NJ 08855-1331

PHOTONIC SWITCHING TECHNOLOGY

Systems and Networks

Edited by

Hussein T. Mouftah
Queen's University, Ontario, Canada

Jaafar M. H. Elmirghani
University of Northumbria, United Kingdom

A Selected Reprint Volume

IEEE Communications Society, *Sponsor*

The Institute of Electrical and Electronics Engineers, Inc., New York

This book may be purchased at a discount from the publisher
when ordered in bulk quantities. Contact:

IEEE Press Marketing
Attn: Special Sales
445 Hoes Lane, P. O. Box 1331
Piscataway, NJ 08855-1331
Fax: (732) 981-9334

For more information on the IEEE Press,
visit the IEEE home page: http://www.ieee.org/

Printed in the United States of America
10 9 8 7 6 5 4 3 2 1

ISBN 0-7803-4707-2

IEEE Order Number: PC5761

Library of Congress Cataloging-in-Publication Data

Photonic switching technology : systems and networks / edited by
Hussein T. Mouftah, Jaafar M. H. Elmirghani.
p. cm.
Includes bibliographical references and index.
ISBN 0-7803-4707-2
1. Optical communications. 2. Telecommunication—Switching
systems. 3. Integrated optics. 4. Switching circuits.
I. Mouftah, Hussein T. II. Elmirghani, Jaafar M. H.
TK5105.59.P58 1998
621.382'7—dc21 98-7873
CIP

Contents

Preface

In the last two decades, there has been an enormous growth in the data rate and bandwidth requirements of users. Broadband services and multimedia traffic have to be transmitted and switched at an ever-increasing rate. To cope with this explosion in data rates and the demand for information, communication, and entertainment, there has been a steady investment and a steady evolution in photonic technologies that have been designed to cope with the increasing demand. In particular, in the last decade we have witnessed the wide exploitation of many of these technologies in a wide range of applications relating to the telecommunications industry. The technological evolution has been remarkable in two major areas—transmission and switching. Photonic technologies have been developed to a state where they have clearly shown orders of magnitude improvement as far as transmission aspects are concerned and as such have widely replaced their electronic counterparts in this area. Photonics, in the switching context, clearly offers a distinct advantage in avoiding the signal degradation and the bottleneck associated with electrooptic and optoelectronic conversions that are necessary if photonics are used in transmission, (which is a fact today) and electronics are used in switching. On the other hand, photonic switching devices are generally diffraction limited and as such are characterized by dimensions in the order of the wavelength of operation which makes them bulky and cost can be an issue.

Therefore, the role of photonic components and technologies in large switched networks will mainly be determined by the enhanced functionality that can be gained by using photonics (e.g., transparency, avoiding electronic bottlenecks, etc.) weighed against cost, size, and versatility of the functions available. Successful demonstrations and exploitation have been witnessed on the component level (switches, routing devices), system level (large photonic switch architectures), and the network level (wavelength-switched networks). The aim of this text is to provide an introduction to photonic switching technologies, photonic switch architectures, and photonic components used in all-optical networks. Our goal is to develop the reader's awareness of many of these emerging technologies and to enable him or her to carry out a judgment on the status and future of photonics and switching.

Chapter 1 presents an overview of the whole area of photonic switching technology for optical switching systems and networks. It starts with a comparative study on the use of electronic and optical switching for broadband networks and guided-wave optical networks. The chapter goes on to give a full review of many currently available switching technologies and architectures. Consideration is given to self-routing switch architectures, guided-wave switches, wavelength-division multiplexing (WDM) switches, high-level optoelectronic integration, space switching fabrics, and multidimensional switches. This is followed by a comprehensive discussion on integrated-optics planar components. The chapter then discusses the possible applications for optical networks based on WDM and how they complete and complement high-speed networks. At the end of this chapter, a comprehensive survey of asynchronous transfer mode (ATM) switch architectures is presented.

One of the first technologies that has been developed for use in photonic switches is introduced in Chapter 2. Electrooptic technologies used to implement directional couplers have formed the basis of the first widely used 2×2 photonic switches. Such switches have later been developed to higher dimensions. The chapter introduces such switches and discusses their application in photonic switching networks. In particular, the use of a directional coupler as a switch is described and the basic modes of operation are highlighted. Some of the system design constraints are discussed and the switching environment best suited to directional couplers is described. Experimental results are also presented and a treatment is given of switch arrays including parameters such as insertion loss, crosstalk, and switching voltage. Potential network exploitation of photonic directional coupler switches are given, including point-to-point and multicast architectures. The architectures presented are also compared to previously known networks.

The chapter continues by outlining some of the limitations of electrooptic switches. Crosstalk in a lossy directional coupler is evaluated. The chapter concludes by describing the integration, on silicon, of many of the components required for photonic networks and switching. The components described include optical splitters, optical switches, and various kinds of filters.

Chapter 3 is devoted to optical amplifier based switches. Switches based on the semiconductor optical amplifier as well as erbium fiber amplifiers are considered. A treatment of semiconductor optical amplifier-based switches is presented and designs for 2×2 and an asymmetric 2×3 element together with several designs for 4×4 switches are given. A device that utilizes semiconductor laser amplifier in a loop mirror (SLALOM) configuration is later described. Different modes of operation of the device are detailed including single-pulse switching and two-pulse switching at rates in the order of 1–100 Gbit/s. Space division switching fabrics based on semiconductor amplifier switches are also discussed. Bounds on the switch size and the number of possible switches are derived based on the amplifier saturation power.

The chapter finally discusses optical switches based on erbium-doped fiber amplifiers (EDFA). It provides an analysis of nonlinear optical switching in an erbium-doped fiber. In particular, it presents a model that relates the physical properties of the amplifier (dopant spectroscopic parameters, fiber parameters, pump power, signal wavelength, and signal loss) to the phase shift that is exploited in interferometric switch arrangements. The model enables the design of switches with low loss, a low drive power and minimum dependence on operating wavelength.

The use of optical memory elements for switching is covered in Chapter 4. The scene is set by a treatment of symmetric self-electrooptic effect devices (S-SEED). The devices are reviewed and their bistable behavior is demonstrated showing how they can be used as both relational and logic devices. It is also demonstrated that sophisticated functionality can be achieved including self-routing and optical packet switching. The discussion of S-SEED devices continues and the results demonstrate the parallel operation of a 32×16 two-dimensional array of SEED devices (512 devices). Bistable operation is illustrated and "static" operation as a random access memory (RAM) is achieved. The operation of large optical switching device arrays is discussed. Bistable characteristics and set-reset operation in all-optical devices are studied reporting detailed results on the set-reset dynamics and illustrating switch-on times in the subnanosecond order. The chapter finishes on a slightly different theme, introducing new optical devices that can perform a number of functions including optical logic operations. The devices are based on nonlinear fiber couplers.

Free-space optical switching is dealt with in Chapter 5. Five free-space switching fabric demonstrators constructed in a research laboratory are described. An overview of the SEED technology, which was the device platform used by the demonstrators, is given. This is then followed by a discussion of the architecture, optics, and optomechanics developed for each of the five demonstrators. The chapter then considers new optical interconnections suitable for three-dimensional combining of lens arrays, along with a new multistage interconnection network with a self-routing function. Two separate lens arrays are fabricated to construct an 8×8 network of this design. Optical interconnections are then further evaluated through a review of the European community's European Strategic Program for Research in Information Technology II Optical Interconnections for VLSI and Electronic Systems (ESPRIT II OLIVES) program under which components for practical demonstrations of optical interconnections for high-performance processors are under development. Subsequently, back plane interconnect systems are dealt with and applications in high-speed digital systems are indicated. At the end of this chapter, a description is given for a reduced switching delay in wavelength-division multiplexed two-dimensional multiple-plane optical interconnections using multiple-wavelength VCSEL arrays.

Photonic switch architectures are considered in Chapter 6. Photonic switching fabrics based on both guided-wave structures and free-space structures are reviewed. The discussion on guided-wave structures extends to fabrics based on space channels, time channels, and wavelength channels and that on free-space channels includes an evaluation of relational and digital free-space switching fabrics. The chapter then reviews the progress in photonic switching toward the realization of photonic packet switches that have the versatility to build complete switching networks. In particular, attention is given to all-optical photonic packet switches in which the data portion of the packet remains all-optical throughout the switching and the routing process. The control portion, however, may or may not be regenerated optoelectronically. The theme of packet switching architectures is then continued by considering photonic fast packet switching architectures. Architectures that exploit wavelength, time, code, and space to achieve the required fast packet switching are described. A tree-structure based architecture is reported for an 8×8 polarization-independent switch matrix. The chapter concludes by introducing a switch architecture that has the potential to achieve high dimensionality. The architecture resembles a collapsed network and utilizes a dedicated path for all input/output port connections, thus eliminating the limitations associated with architectures employing 2×2 switches. The architecture is shown to achieve a fully connected 120×120 dimension.

Chapter 7 is concerned with time-division multiplexing (TDM) optical switch architectures. It starts by evaluating various ATM optical switch architectures. It then describes the work carried out within the Research and Development in Advanced Communications in Europe (RACE) project R2039 asynchronous transfer mode optical switching (ATMOS). Four different switch architectures are discussed all based on a high-speed optical routing matrix with electrical control. The key basic optical components and subsystems are discussed and performance at 10 and 20 Gbit/s is reported. The discussion on ATM photonic switch architecture is extended by describing an ultrafast photonic ATM (ULPHA) switch. The switch is an output buffer type ATM switch based on TDM. Experimental results are reported at 25 Gbit/s for a video distribution switching experiment.

Packet-switched networks can exhibit resource contention. A technique for resolving such a potential contention in all-optical networks is reported and evaluated. The contention resolution by delay lines (CORD) methodology exploits optical delay lines and switching matrices. The discussion on switching in the time domain is continued by an evaluation of a time domain approach for avoiding crosstalk in optical interconnection networks. Tree-type photonic switching networks are introduced and all known tree-type architectures are reviewed while new solutions are proposed. Implementation issues on single and multiple substrates are discussed and the various characteristics of tree-type networks are compared.

Wavelength-division switching is considered in Chapter 8. The chapter discusses tunable filters, tunable/switched sources, wavelength conversion, all-optical WDM packet networks and concludes with performance and limitations issues.

Multiwavelength networks are reviewed and an evaluation is given of the technologies, architectures, and limitations of such networks. Acoustooptic (AO) tunable filters employed as wavelength routing switches are introduced. The design of both hybrid and fully integrated AO switches is reviewed and consideration is given to design methodologies that can be used to reduce crosstalk and that can result in wavelength misadjustment-tolerant switches. Recent progress achieved through the use of tunable/switched Y-laser sources is reviewed. Such devices are shown to exhibit a wide tuning range of more than 50 nm and their application in data and packet switching is described.

In high-capacity dynamic WDM networks, wavelength contention can arise. This can be effectively reduced through wavelength conversion. The technologies available for such wavelength conversion are reviewed. All-optical WDM packet networks are then described with special consideration for networks using self-routing packets in a multihop network. The chapter concludes by evaluating the crosstalk and interference penalty incurred in all-optical networks utilizing static routers. The penalty is shown to strongly depend on the sources used, particularly the linewidth of the laser carriers.

Optical cross-connects are dealt with in Chapter 9. An optical cross-connect system incorporating a new concept, transparent optical multiwavelength networks, for flexible and high-capacity transport networks is described. The key technologies used are space switching devices and wavelength selective devices. Two optical cross-connect demonstrators set up within the European RACE R1003 "OSCAR" program are reviewed. Moreover, multiwavelength cross-connects are reviewed and a class of wavelength interchange devices is proposed and used as building blocks. Three different structures that can be constructed using these devices are described and their blocking performance, complexity, modularity, and wavelength channel spacing are examined. A WDM-based supercomputer interconnect is described where the concept is based on combining the rich interconnect structure of WDM with the fast, low-latency mesh of crossbar switches developed for workstation groups.

The chapter continues by giving attention to optical cross-connects for transport networks. Optical devices suitable for cross-connect node implementation are described and results for experimental demonstrations at data rates up to 10 Gbit/s are presented. The chapter concludes by discussing the wavelength path (WP) concept and how it can be exploited in the design of an optical path cross-connect (OPXC). The OPXC design offers 16 sets of input and output fiber ports with each fiber carrying eight multiwavelength signals each operating at 2.5 Gbit/s. The OPXC thus feature a full capacity of 320 Gbit/s. Experimental demonstrations are given for the transport of data at this system bit rate via a cross-connection node and over 320 km of fiber at high signal integrity.

The last chapter is completely devoted to pilot systems and demonstrators. The chapter starts with an overview of the multiwavelength optical networking program (MONET). The program activities are in three parts: Network architecture and economics, networking demonstrations, and supporting technology. Attention is then focused on the results reported by the Phase I program of the Optical Networks Technology Consortium (ONTC). The program is concerned with a multiwavelength reconfigurable WDM/ATM/SONET network testbed. The architecture of the testbed and its nodes is presented together with a description of the key component technologies. A report is then given on the European RACE COBRA project. The COBRA consortium has performed four field trials to verify the suitability of dense WDM with heterodyne detection for different network applications. The work reports on the demonstrator concepts as well as the components and subsystems developed within the consortium. The work conducted at Bellcore on the multiwavelength network LAMBDANET is summarized in this chapter. The basic network architecture is described and experimental results are reported demonstrating the feasibility of the approach.

The chapter then continues by considering the advanced research projects agency (ARPA) sponsored consortium. A description is given of the technologies, architectures, and applications developed for an ultrafast 100 Gbit/s TDM network. The consortium envisages high-end users operating at bit rates in the order of 10–100 Gbit/s as well as aggregates of lower-speed users. A WDM-based computer communication network is the last demonstrator considered. The network, referred to as STARNET, is a high-speed, high-performance packet-switched computer network based on a physical passive star topology.

This text integrates many of the technologies, architectures, and network applications encountered today in the field of optical switching. It should appeal to professional and practicing engineers working in the communications industry who would like to acquire an appreciation of the photonic technologies that can be used in switching. Furthermore, the book contains a wealth on systems, networks, and application issues which are generally highlighted using some key papers. Special attention has been given to demonstrators and pilot systems, thus giving the text a more comprehensive and practically oriented coverage.

Hussein T. Mouftah
Jaafar M. H. Elmirghani

Chapter 1

Overview

While electronic-based asynchronous transfer mode (ATM) switch architectures have been extensively researched, their throughput is limited by the speed of their electronic circuits. The major speed-limiting factors in high-speed data transfer circuits are the waveform distortion due to electromagnetic induction and the R-C time constants of the electronic circuits. In addition, high power consumption and heat production are problems in high-speed electronic circuits because more power is required for faster operation of logic gate transistors. One terabit per second is thus considered the practical limit of throughput that can be achieved by purely electronic technologies. Optical circuits, on the other hand, are free from these problems because optical signals can be transmitted without distortion, even at system rates higher than 1 Tb/s. Moreover, the low-loss characteristics of optical fiber allow the high bandwidths in the transmission system to be fully utilized. Photonic technologies are becoming increasingly attractive for use in switching systems because the high bandwidth of photonic technology is a key to breaking through the switching throughput limitation of purely electronic circuits.

This chapter presents an overview of the whole area of photonic switching technology for optical switching systems and networks. Paper 1.1 presents comparative studies on the use of electronic and optical switching for broadband networks and guided wave optical networks. In this paper, Thylen et al. discuss generic switching functions in telecommunications networks with a view of the possible role of photonics, and discuss the implementation of the switching functions from a photonics and electronics point of view. Then they apply this reasoning to look at scaling and transparency of all-optical networks based on their current research.

In paper 1.2 J. E. Midwinter examines some of the many ways in which optics can contribute to the network nodes, ranging from all-optical networks through to synchronous digital-optic free-space switches. He has deliberately chosen the title of this paper to be photonics in switching rather than photonic switching since he believes that there is no realistic possibility of optics entirely replacing electronics in switching as it has effectively done in transmission. He concludes that although there is a great potential for photonics in switching, probably the only completely safe prediction is that we will see an increasing penetration of optical interconnects in switching systems and nodes, assuming a broad but flexible interpretation of what constitutes an optical interconnect.

Optical fiber architectures of the future are envisioned to employ extensive splitting of optical signals, as well as multiple wavelength division multiplexing and demultiplexing. These architectures and networks have motivated the development and commercialization of today's planar passive components. Beyond and including these applications, the prospects for new and exciting uses of planar devices are excellent. Nolan et al. present in paper 1.3 a comprehensive discussion on integrated-optics planar components. Components covered include MxN splitters and other integrated passive components, Mach-Zehnder multiplexers and demultiplexers, thermooptic switches, waveguide arrays, planar optical amplifiers, and spatial solitons.

The following two papers discuss the possible applications for optical networks based on wavelength-division multiplexing (WDM) and how they compete and complement current high speed networks. Paper 1.4 first outlines the best-case scenario for this technology and describes the spectrum of proposed optical networks (WDM links, passive optical access networks, broadcast-and-select networks, and wavelength routing networks). Then it focuses on wavelength routing networks and describes their advantages and disadvantages relative to other competing alternatives for very-high-speed networks. P. E. Green, Jr. presents in paper 1.5 an inventory of the current state and future prospects for all-optical networks which are defined as networks in which signal paths between end-user nodes remain entirely in optical form without intervening electronic conversions. His emphasis is on WDM, its applications, and system aspects.

At the end of this chapter, Awdeh and Mouftah present in paper 1.6 a comprehensive survey of ATM switch architectures. They give a descriptive overview of these architectures by classifying them according to the basic structure, into three categories: crossbar-based, disjoint-path-based, and banyan-based. A considerable portion of this paper is devoted to reviewing some necessary background as well as discussing the various buffering strategies employed by ATM switches. Although the paper focuses on electronic switch architectures, many of the switch design basic concepts and in particular the use of unbuffered multistage interconnection networks, are applicable in the optical domain.

Switching Technologies for Future Guided Wave Optical Networks: Potentials and Limitations of Photonics and Electronics

Lars Thylén, Royal Institute of Technology
Gunnar Karlsson, Swedish Institute of Computer Science
Olle Nilsson, Chalmers University of Technology

The field of photonic networks continues to be fueled both by the vision of the broadband electronic or rather photonic highways and by the seemingly inexhaustible possibilities offered by photonics and electronics to bring ever higher performance.

The continued rapid progress in the deployment of fiber optics communications has led to widespread interest in employing photonics to implement switching functions, hitherto reserved for electronics. Significant research effort has been devoted to this, and different specialized concepts have been suggested and partly demonstrated, albeit not in the field. At the same time, mainstream research and development in the area of telecommunications switching is solidly based on electronics.

In this article we first give a short introduction to photonic switching. We then discuss generic switching functions in telecommunications networks with a view of the possible role of photonics, and discuss the implementation of the switching functions from a photonics and electronics point of view.

We then apply this reasoning to look at scaling and transparency of all-optical networks based on our current research.

PHOTONIC SWITCHING

It might be illuminating to see what sort of capacity increases we are talking about in the future broadband network: In a current digital exchange handling 100,000 subscribers, we have a total aggregate capacity of, say, 10 Gb/s. If each subscriber, instead of having access to a narrowband 64 kb/s channel, were upgraded to a broadband capacity of, say. 100 Mb/s, the aggregate capacity of the switch would be around 10 Tb/s, with an ensuing capacity increase in the transport network. Obviously, this is a quantum leap, and fiber optics can cope with this from a transmission point of view. The question asked now is whether photonics also has a role in switching.

The area of photonic switching [l] emerged as a natural outgrowth of the rapid development in fiber optics: due to the anticipated total dominance of optical fiber in the wired network, it was natural to investigate the potential of a deeper penetration of photonics, in addition to sheer point-to-point transmission, where the technology has proven successful in a short time span. Here, it is necessary to consider the unique properties of optics:

- The virtually unlimited bandwidth available in the optical fiber: The carrier frequency is approximately 200 THz, and the available wavelength window, say 200 nm, corresponds to half a billion telephone channels or more than 300,000 high-definition television channels.
- The weak interaction between the information carriers, the photons, in transparent media. This is actually a condition for the success of the optical fiber as a transmission medium, and it leads to some interesting consequences:
  - Optical circuits can be crossed in a plane without "short circuits," but not without crosstalk. This has important consequences for the topologies of photonic integrated circuits and for optical interconnects.
  - The difficulty of "controlling light by light."
- The strong interaction between light and, for example, semiconductors in terms of absorption or amplification and refractive index changes at resonant bandgap wavelengths.

One scenario for the introduction of photonic switching, starting from today's network (Fig. 1, top) is depicted in the middle of Fig. 1, where the switching fabric is optical but still controlled by electronics; this can be labeled *optical interconnect with electronic control.*

This is the "near-term" vision of photonic switching, and the optical switching fabric can, for instance, be optical cross-connects and optical add-drop multiplexers. Photonic switch fabrics for the asynchronous transfer mode (ATM) [2] and synchronous transfer mode (STM) also belong to this category, but are expected to lie further in the future. A potential last stage, fully optical switching, where both the controlling and the controlled signals are optical, is a long-term research item. The reason for the difficulties encountered at this final stage is that fully optical switching requires optical random access memories, logic, and control. At the present time, and for the foreseeable future, purely optical devices for these operations are either far too power-consuming, too slow in relation to electronics, or both.[1] This might also be dictated from fundamental physics [3]. For specialized, nonintegrated

[1] *For a discussion of some fundamental constraints of optical switching, see: R. L. Fork,* Phys. Rev. A*, vol. 26, p. 2049, 1982, and P. W. Smith,* Bell Syst. Tech. J.*, pp. 1975–93, 1982. Further reading: H. M. Gibbs,* Optical Bistability: Controlling Light with Light*, [Academic Press, New York, 1985], H. Haug (ed.).*

Reprinted with permission from *IEEE Communications Magazine,* Vol. 34, No. 2, pp. 106-113, February 1996.

functions, such as demultiplexing of, say, 200 Gb/s bit-streams, all-optical technology such as the nonlinear loop mirror might be viable [4].

It should be noted that the distinction between switching and transmission is becoming somewhat blurred; with the use of digital crossconnects and add-drop multiplexers, the transmission layer assumes some of the roles of the switching layer.

Switching Functions

Multiplexing in optical networks may utilize one or more of the dimensions space, time, and wavelength (frequency). Pure space-division multiplexing means that each communication uses a dedicated physical path (an optical fiber) through the network. There is thus only a single channel per fiber. For most types of communication this use is prodigal; better utilization of the available network resources is obtained if the links are shared among the users of the network. Space-division multiplexing is therefore usually combined with wavelength-division or time-division multiplexing (WDM and TDM, respectively) to give more channels per link, and thus to support a higher number of simultaneous communications.

The space dimension is tacitly assumed as part of all multiplexing modes in the following discussion (for instance, a WDM-based network is both space- and wavelength-divided).

WDM means that the optical bandwidth of a link is split into fixed, nonoverlapping spectral bands. Each such band constitutes a wavelength channel that can be used for a specific bit-rate and transmission technique, independent of the choices for other channels. This is referred to as "bit-rate and code-format transparency."

TDM can be either synchronous or asynchronous (ATDM). TDM divides the transmission on a link in time into frames (often 125 μs long), which in turn are split into slots. A channel consists of one or more slots in every frame that is sent over the link.

Both WDM and TDM yield channels with fixed shares of a link's bandwidth and transmission capacity, respectively (the channels on a link are physical). The allocated amount of resource cannot be exceeded when necessary and is wasted when not used. The amount of bandwidth or capacity per channel is usually determined at the specification and design of the system. For instance, European telephony networks are built with a structure of digital channels of (approximate) width 64 kb/s, 2 Mb/s, 8 Mb/s, 32 Mb/s, and 140 Mb/s (rates E0 up to E4 in the plesiochronous digital hierarchy — PDH), and 155 Mb/s, 622 Mb/s, and 2.5 Gb/s (rates for STM 1, 4, and 16 in the synchronous digital hierarchy — SDH).

ATDM differs in this respect in that it does not mandate a specific amount of transmission capacity per channel (the channels are said to be logical or virtual). The capacity could, in principle, be chosen with infinitely fine bit-rate granularity. ATDM also allows (but does not mandate) statistical allocation of transmission capacity to a channel. This means that the channel has only a probabilistic guarantee to access the link. The likelihood of rejection (blocking) is specified as the quality of service that the user has requested and depends on the total amount of traffic offered to the link by all the users sharing it. (Statistical multiplexing is possible also in the spectral domain through code-division multiple access.)

Any switch has to perform two basic functions irrespective of transfer mode: demultiplexing incoming channels according to a given routing pattern and multiplexing the channels on to the outgoing link. The switch is thus built for a particular multiplexing and routing strategy. We can categorize networks according to the combinations of these functions:

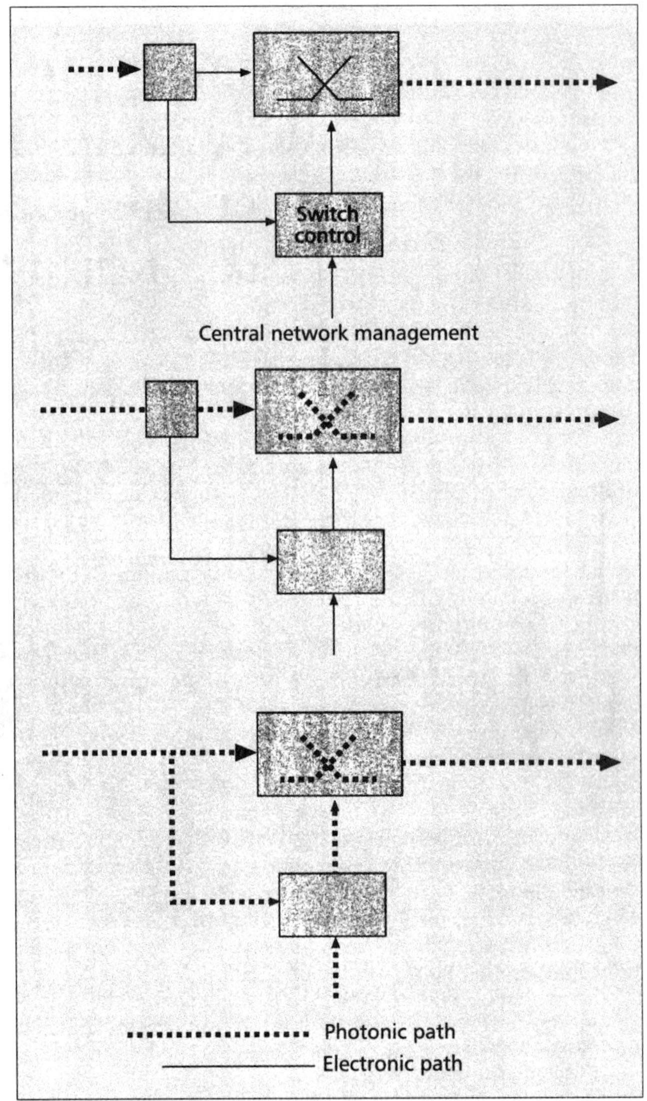

■ **Figure 1.** *Development from today's situation, photonic transmission and electronic control (top), via emerging network scenarios with photonic transmission and switching functions such as crossconnecting controlled by electronics (middle), to a possible future fully photonic switching system (bottom). The realism of the last option is heavily dependent on breakthroughs in, above all, photonics material technology.*

- Circuit-switching: fixed channel allocation (WDM, TDM, or ATDM), fixed route established before transmission
- Virtual circuit-switching: statistical allocation (ATDM), fixed route
- Datagram-switching: statistical allocation (ATDM), route selected per packet during transmission

Note that ATDM is possible in all three modes depending on how the capacity is shared (deterministically or statistically) and whether a connection is established or not. A connection or circuit (we use them synonymously) is a path through a network with a channel allocated on each link it traverses. ATM is virtually circuit-switched, and generic STM is circuit-switched; the internet protocol uses datagram-switching.

There is commonly a distinction made in vocabulary which differentiates between cross-connection, switching, and routing. The first is used when the connection pattern is semi-permanent and the setup is done by the network operator. Switching is based on connections set up by user signaling, and routing is connectionless switching of datagrams (a datagram is a packet routed independent of other packets from the same source). For example, SDH use crossconnects, ATM uses both switches and crossconnects, and the Internet protocol (IP) uses routers. "Switches" in this discussion should be interpreted generally to include all types.

SWITCHING STRUCTURES

A switch can have an internal structure based on space-division interconnection or a shared medium (Fig. 2a and b). In the first case, an input port has dedicated links to all outputs. The link has to provide the same bandwidth or capacity as the port. Each channel has to be routed over the internal link to the explicitly or implicitly addressed output. The demultiplexing of channels thus precedes the multiplexing. The growth in the number of internal links is quadratic. For example, the size limitation of optical space-division switches, as discussed in the next section, appears to be less than 100 input and output ports.

In the second principle, using a shared medium, all inputs share a common link that interconnects all inputs and outputs. It is the responsibility of the output port to receive all the channels destined to it. Multiplexing thus precedes demultiplexing. The shared medium has to provide bandwidth or capacity equal to the aggregate of the inputs. The medium should, for transparency, be shared using the same multiplexing format as the transmission links, and the number of channels over the medium has to equal the total number of incoming channels.

Shared-medium switches using WDM seem ideally suited to photonics, considering the claims of a virtually unlimited optical "ether." A large amount of research has been done on different structures resting on such a concept. However, several factors limit the alleged switching capacity: loss, crosstalk, nonlinearities, and noise.

The two mentioned structures are nonblocking: the switch cannot internally block incoming traffic from reaching a free output. Very large switches cannot, however, be built by these structures due to the growth in number of links, in bandwidth, or in number of distinct channels. Instead, nonblocking switching elements are interconnected into a larger switch fabric, as shown in Fig. 2c.

SWITCHING OF CHANNELS

Pure space switching means that a channel is moved from one link to another, but the same channel is used on both links. A connection can only be established when one and the same channel is available on all links from sender to destination. This means that the user is blocked, even though there might be vacant channels on all links.

The blocking probability could be reduced by allowing a connection to change from one channel on an input link of the switch to another channel on the output link. For WDM this corresponds wavelength conversion (the signal is moved from one spectral band to another). For synchronous TDM, the change of channel is commonly done by a *time-slot interchanger* which can move time-slots within a frame by *buffering* them. In ATDM the channels are logical, and are identified by some address. Switching channels is done simply by a change of address.

The number of channels may be a limiting factor in all-optical networks. This may cause users to be denied service due to an insufficient supply of channels, rather than insufficient amounts of transmission capacity (or bandwidth). If few channels can be provided per link, then more links have to be available to keep the blocking probability low. For example, the number of 2 x 2 space-switches needed in a network with F wavelengths, wavelength conversion, and with M users is at least $M\log_2 M - 2M\log_2 F - 1.44M$, if all permutations of pairwise communication should be possible simultaneously [5]. Multicast would increase that number.

Shortage of channels is not a likely problem with ATDM systems such as ATM. The channels are logical and their number is given by the length of the addresses. For example, ATM allows up to roughly 270 million unique virtual circuits

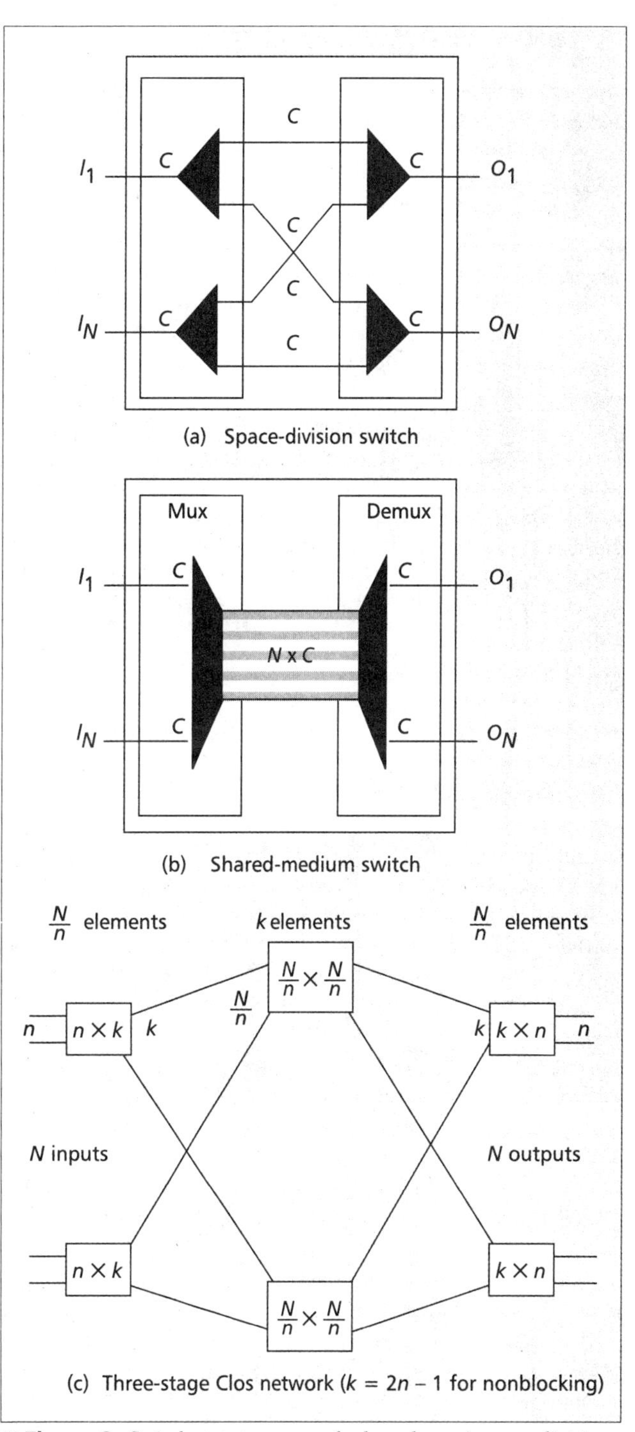

Figure 2. *Switch structures may be based on a) space division; b) shared medium. These two structures, with the dimensions shown, are nonblocking (C can denote both the capacity, or bandwidth, and the number of channels.). Large nonblocking switches can be built by interconnection of switching modules in structures such as Benes, Banyan, and c) Clos networks.*

per link (28 bits of address). There would, of course, be only a small amount of capacity available on the average for each circuit if so many should share a link.

Routing and Multiplexing

The demultiplexing in a switch separates the channels of an input link and routes them to their destined outputs, where they are multiplexed with channels coming from other inputs (or vice versa for a shared-medium switch).

The routing pattern of wavelength channels only changes when a connection is added. Between such instances, the routing is fully independent of time. The routing pattern of a TDM switch is equally static, but each time slot of a frame must be switched according to the existing pattern. ATDM does not have a fixed routing pattern since packets do not come in any predefined order for the outputs. The time dynamics is lower for ATDM than for TDM, as long as the length of a packet exceeds the length of a time slot.

Multiplexing of wavelength channels is just a passive combination of them provided that they are on distinct wavelengths. This can be done passively — giving a loss of 3 $\log_2 N$ dB (for N inputs) — or, preferably, actively by means of a wavelength selective multiplexer to avoid such losses (see the fifth section). The multiplexing of time slots for TDM requires that they do not overlap in time. This requires synchronization of the inputs that send to a particular output. For ATDM the switch has to ensure that packets do not overlap when signals from different inputs are added at a switch output. If the multiplexing is statistical, then there can be more traffic destined to an output link than can go out of it. Packets that cannot be delivered immediately must be buffered at the switch inputs or outputs.

Discussion

Every communication over a network requires a channel per link. As we mentioned, pure space division would then require one set of links for each communication. However, this solution does not scale well with the number of connected users and the number of simultaneous connections that need to be supported. The poor scaling might be ameliorated by using WDM.

The number of wavelength channels that can be accommodated on a fiber is discussed later; it can be calculated from different considerations, one being the limitation posed by Raman scattering, which limits the number of channels to the order of hundreds within the fiber optics gain window, around 4 THz (by employing quantum limited detection this number might be increased). This would mean that the network still has to be built with a high degree of space division and would have to provide wavelength conversion to fully utilize the wavelength dimension.

To further improve the link-sharing and thereby the scaling properties of the network, TDM or ATDM should be used. In particular, ATDM would scale well in the number of channels since the channels are logical. The functions required for all of this are wavelength conversion, synchronization, and buffering. These are straightforward to implement electronically, but less so optically. This is discussed in the next section. The choice of switch structure depends on how well it can support simultaneous switching in time and wavelength.

Photonic Implementations

In the previous section we identified the functions required for a general switching system. The photonic implementation will thus require:

- Temporal multiplexing and demultiplexing with buffering and synchronization
- Wavelength multiplexing and demultiplexing with wavelength conversion
- Space-division switching

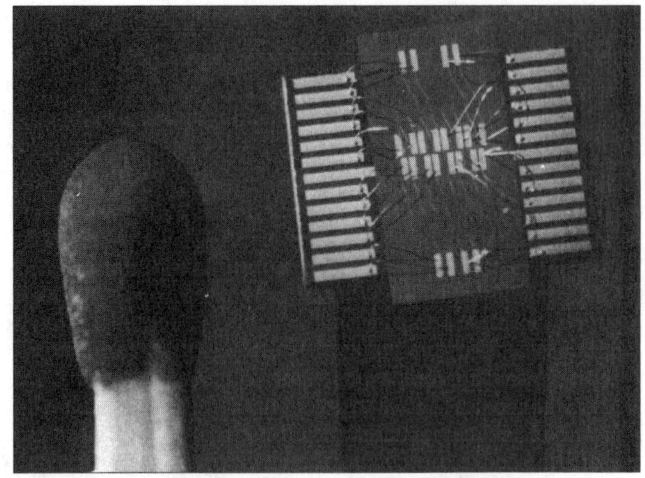

■ **Figure 3.** *Indium phosphide 4 x 4 strictly nonblocking space switch with zero dB insertion loss or gain. Switching is brought about by gating on and off the kernel of 16 integrated optical amplifiers. [Courtesy of Ericsson.]*

Current electronic ATM switches, for instance, employ all of these functions except WDM. The use of this additional degree of freedom is unique to photonics and was utilized in the most elaborate demonstration of an optical ATM system to date [6]. Laser amplifier gate switches (logic), optical filters (i.e., WDM used for space switching), and fiber optics delay lines (buffering) were used to implement all the functions required above. Synchronization, however, had to be done electronically, although there have been proposals for also doing this optically. The long-term prospects of these approaches are not clear at this stage.

Photonic Implementation of Switching Functions

One of the reasons for the slow progress of photonic penetration in switching is the comparatively immature photonics device technology; photonics could be said to be at a state comparable to electronics in the '60s. However, rapid progress in this field is being made, and especially promising is the development in integrated photonics, analogous to integrated electronics. Furthermore, much research effort is devoted to integrated optoelectronics with the vision of combining the advantages of photonics and electronics . Another reason is the current mismatch between the photonic technology and the switch functions required, as noted in the previous section. A review of the status of photonics device technology of relevance to this field is given in [1], and below we discuss the situation in relation to the switching functions identified in the previous section

Temporal (De)Multiplexing — In today's systems this is carried out electronically using different types of clock circuits, involving storage and logic elements based on silicon (Si) (or, in some cases, gallium arsenide — GaAs) transistor technology. To perform these operations, synchronization is required. The necessity of involving storage and logic operations makes synchronous TDM difficult to carry out optically, due to the lack of a viable optical memory. However, multiplexing and demultiplexing of synchronized optical streams can be carried out provided that the pulses are suitably shorter than the bit-slots. Hence, multiplexing can be achieved by passively combining bit-streams in fiber couplers. All-optical demultiplexing of synchronized bit-streams has been reported

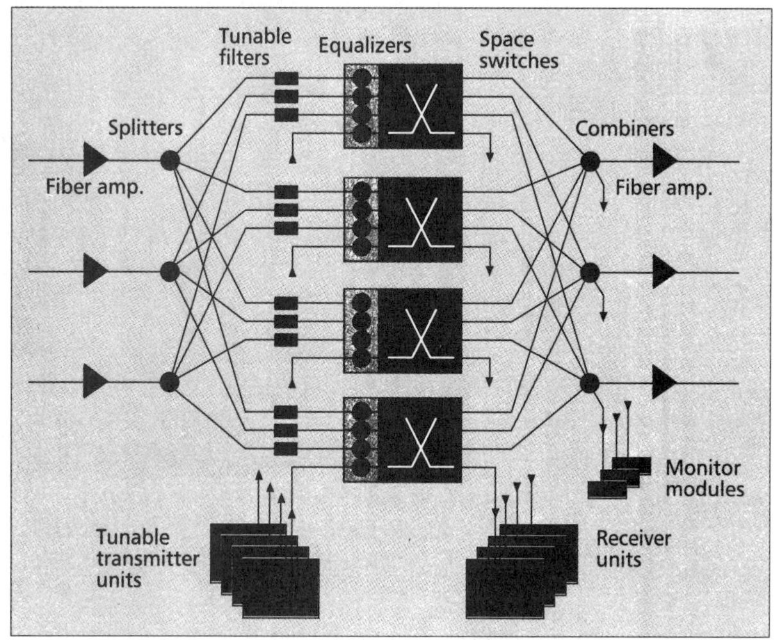

■ **Figure 4.** *An optical crossconnect node for traffic management.*

utilizing a number of techniques (e.g., using nonlinear loop mirrors [4]).

Buffering and Synchronization — This is probably the most difficult area for photonic implementation, and the crucial shortcoming of photonics is the absence of a photonic logic element comparable to the electronic transistor. For buffering, the fiber optics delay line has been the commonly used element; bistable optical elements were considered prime candidates in the '80s for applications such as random access optical memories . However, these devices have difficulty in competing with electronic devices regarding speed, switching energy, and power consumption. A systems study we have carried out, comparing photonic and electronic switching in a conventional time-space-time switch, showed the inadequacy from the performance and economic points of view of current and envisaged future photonic memory elements. The issue of synchronization is analogous to buffering in the sense that delays, albeit shorter, are involved; in [7], a scheme for optical synchronization is proposed.

Wavelength-Division (De)Multiplexing (WDM) — This is implemented by a variety of techniques, still mostly based on gratings. WDM can obviously be used for increasing the capacity in transmission links. The role of WDM in switching with or without wavelength conversion is more unclear and is a subject of research. Nearly all of the optical multiplexing techniques demonstrated today are not reconfigurable because the mapping of wavelength channel to spatial port is static . A prerequisite to perform real switching operations is obviously to be able to either control this mapping or do wavelength conversion. The latter can be done all-optically, or by opto-electronic-opto conversion, involving detection, amplification, and light generation.

It has long been recognized that the wavelength domain of photonics adds a significant new degree of freedom which could allow new network concepts to be developed. However, this has not happened to any appreciable extent; rather, the wavelength domain has been embedded in existing systems concepts. It appears that wavelength-division techniques for switching are most applicable to cross-connect applications (since it avoids the issue of synchronization). The use of active WDM, with opto-electronic wavelength conversion, in a cross-connect type of application has been shown to increase the capacity as well as the flexibility of networks (next section). In [8] the first demonstration of a managed multiwavelength network is described.

Space Switching — Initially, research in photonic switching was dominated by space-division circuit switching (in principle analogous to the old "crossbar" systems) where the device base was the strongest, primarily in the shape of lithium niobate ($LiNbO_3$) switching arrays. Current space-switch device technology is dominated by different types of semiconductor (indium phosphide — InP) switches, some with integrated optical amplification. The original reason for the interest in space switching was the *optical transparency* discussed in the next section. The characteristics of these switches, whether based on gating of optical amplifiers [9] or more conventional refractive index changes (directional coupler, digital switches), can be summarized as follows:

- Speed of rearrangement.
  - Currently reported: subnanoseconds.
  - Achievable: Tens of picoseconds.
- Insertion loss.
  - Currently reported: 0 dB, or gain with the use of optical amplifiers.
- Optical bandwidth (the bandwidth of the signals routed through the switch fabric): approximately 3 THz.
- Single-chip switch size.
  - Currently reported: The current record in switch size is 16 x 16 in $LiNbO_3$ and 8 x 8 in semiconductors. For switches based on integrated amplifiers (InP) exhibiting 0 dB insertion loss, the largest matrix reported is 4 x 4, incorporating 24 semiconductor amplifiers [9] (Fig. 3)
  - Achievable: Probably less than 100 input and 100 output ports

It seems that InP switches (e.g.,based on amplifier gates) will be ready for systems introduction in a few years. These switches are used in the cross-connect system described in the next section. The limitation in the number of ports for a single-chip switch is given by the available physical chip size and the comparatively large length of the switch elements (in the order of hundreds of micrometers). The speed of rearrangement means that, in principle, switching at the bit, byte, or ATM cell level is possible, as long as synchronization issues are resolved.

In view of the limited single-chip switch sizes and the uncertainty about the possibility of making very large ($\gg$ 100 x 100) switch fabrics, such as Clos nets, it appears that photonic space switches will be most suited to the cross-connect application, due to size and synchronization issues. This is the same conclusion as in the wavelength switching case.

A final remark: Whatever technology we develop for guided wave networks, it is required to be *polarization-independent*. Since optical fibers in use today do not preserve polarization, this additional requirement, innocent as it might appear, in most cases entails a major complication.

Network Scaling and Transparency

The creation of an all-optical network comprising trunk and local levels without any electronic intermediation raises a number of questions. The intention of this section is to elucidate some of them.

CROSS-CONNECT

A fair amount of work has been devoted to optical cross-connects, where the switch routes very-high-bandwidth data streams (say 10 Gb/s per port) but is only reconfigured at a rate corresponding to load changes and faults in the overall network. For this, it appears that a photonic implementation is very attractive from a systems point of view. This idea was the basis of the European RACE project Multiwavelength Transport Network (MWTN), which intended to demonstrate that an optical layer could form an extension to the electronic transport layer and offer increased capacity and flexibility by employing multiple space switching. These features were indeed demonstrated using nodes of the type depicted in Fig. 4, which shows a cross-connect node. The node includes an optical layer and an electronic layer. The latter is used for higher-granularity switching or to resolve wavelength contention. Each of the three incoming fibers carries four wavelengths (the same set is used on all fibers: 1.548 μm, 1.552 μm, 1.556 μm, and 1.560 μm). The optical signals are amplified, split by fiber power splitters, and brought to tunable filters, which select wavelengths as prescribed by the management system. The wavelength channels are then routed in the space switch either to output fiber combiners, where they are merged with other wavelength channels for subsequent amplification in line-fiber amplifiers, or to the electronic layer. Conversely, the node can inject traffic from three transmitter units comprising wavelength-tunable lasers. Several issues had to be resolved:

- Power equalization, wavelength identification, and filter tuning to the set of wavelengths were accomplished by using a system of pilot tones.
- Monitoring of the network was done by using power splitters to branch off the signal at selected points.
- Management of the network.

In-depth investigations of the scalability of this specific network were carried out as part of the MWTN project [10]. Scaling pertains to node size (number of ports times number of wavelengths per port) and the number of nodes that can be passed in the optical domain before the signal has to be regenerated. Figure 5 shows the number of nodes that can be passed under the stated conditions, and samples a complex parameter space. As a general conclusion, on the order of tens of nodes can be passed, given that they have the size indicated. Larger nodes will incur proportionally higher losses due to the passive signal splitting and combining used, with a subsequent demand for optical amplification. Since amplification is always associated with noise, the network will be reduced to fewer nodes that can be passed before electric intermediation. In the investigation, the noise from optical amplifiers, gain saturation, and stimulated Raman scattering were the limiting factors considered. A full investigation of scaling of optical networks would in addition have to address crosstalk. Figure 5 illustrates in a concrete way the role of optical amplification: it compensates losses and, hence, increases tremendously the transmission capacity. At the same time, the introduced noise limits the capacity. Crosstalk between wavelength channels is a further limiting factor.

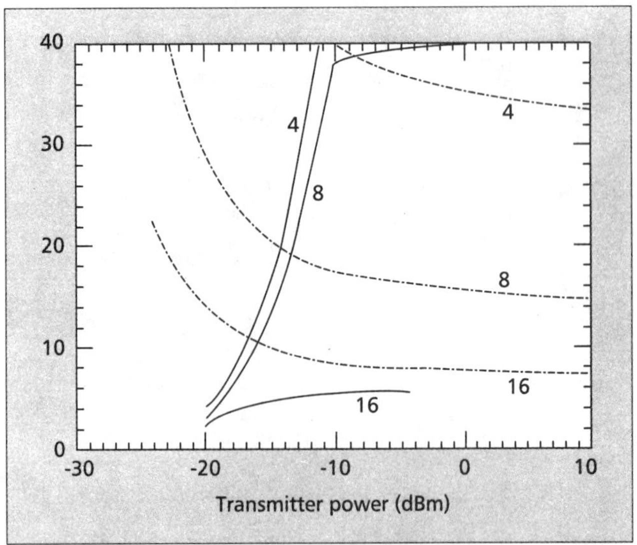

■ **Figure 5.** *Number of nodes in the multiwavelength transport network of [9] that can be passed under the stated conditions as a function of average transmitter power per WDM channel with the number of wavelength channels as parameter at bit rate 2.5 Gb/s per channel and bit error rate 10^{-12}. Channel (de)multiplexing is performed with power splitters, combiners, and tunable filters. Solid lines represent limits from optical noise accumulation; dashed lines, limits from stimulated Raman scattering. This figure samples a complex parameter space. [Courtesy L. Gillner, Ericsson.]*

TRANSPARENCY

The most touted reason for developing all-optical networks is transparency. Given an optical channel, the user may freely decide what coding format and bit rate to choose (independence might be a better word than transparency). This obviously holds for space- and wavelength-divided networks. One may also conceive of a transparent TDM network where possibly a different coding method and bit rate are used for each slot in a frame. Note that ATDM cannot offer code and bit-rate transparency since the address in a packet has to be decoded in every switch.

Transparency in principle may not carry over to practice. A deployed network will certainly set some limits for the coding formats and the bit rate, such as power limitation, accumulation of noise, and dispersion. With TDM, design of the transmitter would have to consider the multiplexing format, the duration of the slots and frames, which might have a bearing on the coding format for bit synchronization. The transmitting equipment is, of course, not transparent to the used multiplexing format (WDM or TDM, number of channels, location of channels in wavelength or time, reference points for those locations).

The channel structure may also be affected whenever a network is upgraded to support more users.

It might be valuable to see what sacrificing transparency might imply. We consider three levels: optical (no electronics in the signal path), coding, and bit-rate transparency (signal propagation and routing independent of coding format and bit rate, respectively).

- By giving up optical transparency, analog bit-rate-transparent electronics may be used in the signal path. It can be used to alleviate dispersion and coherent crosstalk (i.e., coherently enhanced crosstalk due to interference between sources of nominally the same wavelength [11]), and may also provide amplification, wavelength conversion, and access to monitoring signals.
- Clocked electronics, which is not coding-transparent, makes it possible to regenerate the signal, in addition to the aforementioned.
- By finally giving up bit-rate transparency, full electronic switches may be used. ATM switches could, for instance, help avoid blocking due to an insufficient supply of channels in the network (see the "Switching of Channels" section).

The alleged reasons for transparency are that the transmission and switching functions provided by electronics could

There is scant evidence today that all-optical solutions will be more cost-effective than opto-electronic ones. Photonics is currently far less developed than electronics in this respect

become bottlenecks, and that purely optical solutions could be more cost-effective.

The fear of electronic bottleneck might be exaggerated. Communication traffic is still foreseen to be generated and absorbed by electronic equipment. Thus, network electronics can at least match the speed of individual connections. The question is simply how much aggregation of traffic electronics may support, which is intimately tied to the ultimate capacity of electronics. As for speed, this may well be in the terabit-per-second range; however, the integration level at these speeds is an open question. Little aggregation means that a higher degree of space division has to be used, which may result in higher cost for a given traffic volume. However, it is not obvious that photonics offers a solution here. The power consumption of electronics should, in general, be lower than that of optics [3], and it would be difficult to argue that a power-limited system should be all-optical. Furthermore, there is scant evidence today that all-optical solutions will be more cost-effective than opto-electronic ones. Photonics is currently far less developed than electronics in this respect, reflected in the lack of standards for optical components and building practices that could support mass production.

We have, in this discussion, pointed out the limited scaling of current all-optical networks. Also, we have found that cost is the only potential but still unproven advantage of full transparency. One could therefore argue that a viable alternative to all-optical networks could be networks based on fiber optic transmission, optical cross-connects, and electronic switching. This would mean that all-optical networks of limited size would be joined at the transmission level by electronic repeaters. The optical switches would only offer cross-connect functionality (space and wavelength), and full-feature switching would be offered electronically at a small subset of the nodes. Wavelength conversion could be supported by simple analog opto-electronics, the optimal mix of electronics and photonics depending on the system in question. In this scenario, a significantly larger deployment of optoelectronics, preferably in the form of monolithically integrated devices, would be the case.

Concluding Remarks

Compatibility with the transmission medium has been the traditional argument for photonic implementation of switching networks: a transparent network, virtually unaffected by the data rates, coding formats, and number of wavelengths involved, would, of course, imply a considerable enhancement of flexibility, capacity, and reliability in the total network. However, the optical network is analog, noisy (due to optical amplifiers), nonlinear, nonregenerating (today) and sensitive to crosstalk due to coherence effects [11] in contrast to electronic, digital, signal-restoring systems. Hence the scalability of optical networks is inherently limited, as discussed above, and one will most likely deal with "islands" of transparent optical networks interfaced by optoelectronic conversions. At this stage it is not possible to give the details of any trade-off between the optical and electronic domains. We consider it a subject of further research, which will have to involve detailed studies of the limiting factors mentioned above. Such studies should not only aim at long distance, few-channel systems — as has been the case until now — but also at local-area multichannel systems where such issues as signaling and operations and maintenance are important. The detailed systems scenario and trade-offs will have to take the continued rapid development of electronics as well as photonics into account. Thus, we arrive at the fairly truistic statement that photonic switching should be regarded as a complement to electronics rather than a competitor. Maybe this could be formulated in such a way that we have an electronic bottleneck in terms of transmission capacity and a photonic one in channel granularity [12], the latter being due to the current difficulties in generating enough channels in the spatial or wavelength domain by all optical means. In the temporal domain, the channels are currently generated by electronics, driving photonic devices. All optical TDM is a possibility, as an example, [13] gives an interesting example on how to utilize the extreme capacity of the optical medium in the temporal domain, by switching with ultra short pulses. This is an approach which is analogous to today's heavily TDM oriented systems, however, progress in this area will require breakthroughs in nonlinear materials.

The field of photonic networks continues to be a dynamic one, fueled on one side by the vision of the broadband electronic or rather photonic highways and on the other side by the seemingly inexhaustible possibilities offered by photonics and electronics, with physical limits still remote, to bring ever higher performance. Thus, this is a truly interdisciplinary field.

References

[l] L. Thylén"$LiNbO_3$ and Semiconductor Guided Wave Optics in Switching and Interconnects," *Photonic Switching and Interconnects*, ed. A. Marrakchi, [Excel Dekker, 1994].

[2] M. de Prycker, *Asynchronous Transfer Mode: Solution for Broadband ISDN*, 2nd Ed., [Ellis Horwood, New York, 1993].

[3] L. Thylen and T. Palm, "Switch Energies of Electronic and Optical Switches: A Simple Comparative Study," *IEEE J. Quantum Electron.*, vol. 30, 1994, p. 1163.

[4] I. Glesk, J. Sokoloff, and P. Prucnal, "Demonstration of All Optical Demultiplexing of TDM Data at 250 Gb/s," *Elect. Lett.*, vol. 30, no. 4, 1994, pp. 339–40.

[5] R. A. Barry and P. A. Humblet, "On the Number of Wavelengths and Switches in All-Optical Networks," *IEEE Trans. on Commun.*, vol. 42, no. 2/3/4, Feb./Mar./Apr. 1994.

[6] J. Dupraz, "ATM: Current Status and Perspectives of Evolution," *Proc. ECOC '94*, Firenze, Italy, 1994, p. 555.

[7] M. Burzio *et al.*, "Optical Cell Synchronization in an ATM Optical Switch," *Proc. ECOC '94*, Firenze, Italy, 1994, p. 581.

[8] G. Hill *et al.*, "A Transport Network Layer Based on Optical Network Elements," *IEEE J. Lightwave Tech.*, vol. 11, no. 5/6, 1993, pp. 667–79.

[9] M. Gustavsson *et al.*, "Monolithically Integrated 4 x 4 InGaAsP/InP Laser Amplifier Gate Switch Arrays," *Elect. Lett.*, vol. 28, 1992, p. 2223.

[10] L. Gillner and M. Gustavsson, "Scalability of Optical Multi-Wavelength Switching Networks: Power Budget Analysis," submitted for publication.

[11] D. Blumenthal and L. Thylen, "Coherent Crosstalk Induced BER Floors in *N* x *N* Space Photonic Switches," *Proc. IEEE/LEOS Topical Mtg. on Optical Networks and Their Enabling Technologies*, Lake Tahoe, NV, 1994.

[12] D. Blumenthal, "Channel Capacity and Bit-Rate Limitations of WDM Multihop Photonic Switched All Optical Networks," submitted to *J. Lightwave Tech.* special issue on WDM technologies and networks, Apr. 1995.

[13] M. N. Islam, *Ultrafast Fiber Switching Devices and Systems*, [Cambridge Univ. Press, 1992].

Photonics in Switching: the Next 25 Years of Optical Communications?

J.E. MIDWINTER

Indexing terms: Optical communication, Optical switching, Communication systems theory

Abstract: Twenty five years ago, modern optical communications was launched in the now famous paper outlining the theory of single mode fibres by Charles Kao and George Hockham. Almost exactly half that time ago, the first fibre systems carried telephone traffic (BTRL and STL) whereas today, fibre is the universal wideband transmission medium. However, having made almost unlimited bandwidth available at low cost, the problems have now moved firmly to the network nodes, the switches, and much interest centres on whether optics can contribute here. The paper will examine some of the many ways in which this might come about, ranging from 'all optical networks' through to 'synchronous digital-optic free-space' switches.

Introduction

Background

Just 25 years ago, the now famous paper by Kao & Hockham [1] was published in IEE Proceedings outlining the theory of dielectric single mode optical waveguides and linking them to the idea of telecommunications cable transmission using light in place of electrons. It led to collaboration between the Standard Telecommunications Laboratories (now BNR Europe) and the British Post Office Research Laboratories (now BT Laboratories) which led to the establishment of a target attenuation figure of 20 dB/km for a 'real' system [4] and to the demonstration that some optical materials had intrinsic attenuations very much lower than this [5]. In 1970, at the 'Trunk Telecommunications Conference' held at the IEE in London, Corning Glass Works (USA) announced the experimental achievement of fibre having an attenuation of 20 dB/km [6]. The major ingredients thus existed for a successful engineering programme to begin.

By 1976, we found a number of groups starting to plan experimental field systems trials and early in 1977, a fibre cable was laid from BTRL Martlesham through to Ipswich. This was used for experimental 8 and 140 Mbit/s transmission trials as well as for extensive studies of installation practices [2, 7]. Later in the same year a trial 140 Mbit/s system built by STL was installed and operated over a 9 km route from Stevenage to Hitchin [3]. This system included remotely powered optical regenerators and thus more closely resembled a fully engineered system. Together, they demonstrated basic system feasibility and led, in 1979, to British Telecom placing orders for £22 million of traffic carrying fibre systems to be installed on a very wide variety of challenging routes around the UK, a family known as the 'proprietary optical line systems' or 'POLS' [8].

The first papers and most early study concentrated on single mode fibre. In the early 1970s, this changed to multimode and graded-index fibre [see for example References 9 and 10] because these were expected to be easier to splice and launch power into while offering adequate performance and these were used exclusively for the early production systems. However, by the late 1970s, it was becoming apparent that by moving to a longer wavelength of 1300 or 1550 nm much lower attenuation was possible and to exploit this, much lower pulse spreading would be highly desirable. Coupled with the clear evidence that material dispersion was zero at a wavelength close to 1300 nm [11], the move to single mode fibre started and by the early 1980s, we see overwhelming evidence that it was emerging as the preferred choice for long distance transmission [12]. Since then, the fundamental capability of single mode fibre in point-to-point transmission has been explored extensively and is summarised in Fig. 1 which is based upon data from a variety of different sources.

Further major developments have occurred in transmission since then, most notably the development of the erbium doped fibre amplifier (EDFA) which makes possible amplification in a fibre almost identical to the main transmission fibre, across a spectral window of about 30 nm centred near to the 1550 minimum attenuation wavelength [13]. In Fig. 2 we show the gain curve for a typical amplifier superimposed on the transmission spectrum of a typical transmission fibre showing the width of the spectral window available. As a result it is likely to be possible soon to make 'lossless' optical pipes thousands of kilometres long and with a transmission bandwidth of order 30 nm which, in electrical terms, corresponds to a spectral window of about 4000 GHz, roughly equivalent to 13 times the whole EM spectrum from DC to millimetre wavelength. How this capacity might be used remains an area of intense debate.

By comparison with these developments, change has been slow in the telecommunications networks although the same period has seen the widespread introduction of digital transmission and switching, steady growth in the range of services offered (e.g. ISDN 64 kbit/s digital, FAX, Video Conference etc.) and both continuing and

Paper 8366J, delivered before the IEE on the 10th October 1990

J.E. Midwinter is the Pender Professor with the Department of Electrical and Electronic Engineering, University College London, Torrington Place, London WC1E 7JE, United Kingdom intense discussion of broadband services, implying everything from $(n) \times 64$ kbit/s ($n = 2, 3, 4$ etc.) through to switched 2 Mbit/s channels and on up to 140 Mbit/s per customer under the general banner of BB-ISDN, broad band integrated service digital network [14]. With this have emerged ideas of asynchronous time multiplexing,

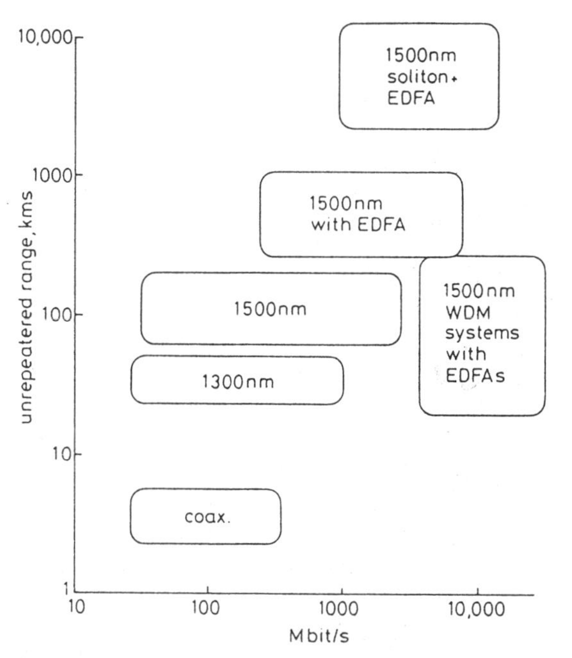

Fig. 1 *Transmission capability of single mode fibre*

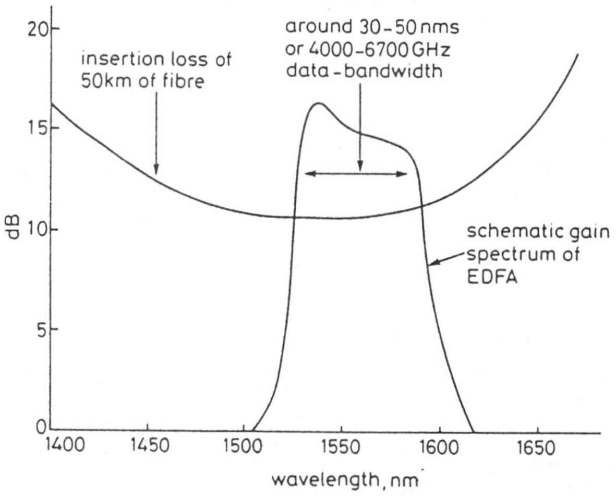

Fig. 2 *Fibre spectral window with approximate window for erbium doped fibre amplifier superimposed*

ATM, whereby the use of a packet like format for transmission could open up the possibility of variable bandwidth transmission and perhaps the ubiquitous IT (information technology) wall socket that would deliver and receive any service or mix of services that the user demanded ranging from telephony to TV, e-mail to whole data file-transfer [15]. All this adds up to a massive problem converging on the network nodes because it envisages rapidly escalating data rates per bearer (say from 140/560 Mbit/s to 4 Gbit/s), rapidly increasing numbers of bearers (say from 1 wavelength bearer per fibre to ten) and from a predominantly circuit switched network to a predominantly packet like ATM network. It is thus natural to ask if optics can do anything to assist the electronics at the switching nodes.

The meaning of 'switching'

We should first note that 'switching' takes on two distinct meanings. There is the switch which is the analogue of the light switch or the reed relay, that changes a connection as in Fig. 3*a*. There are many optical analogues of

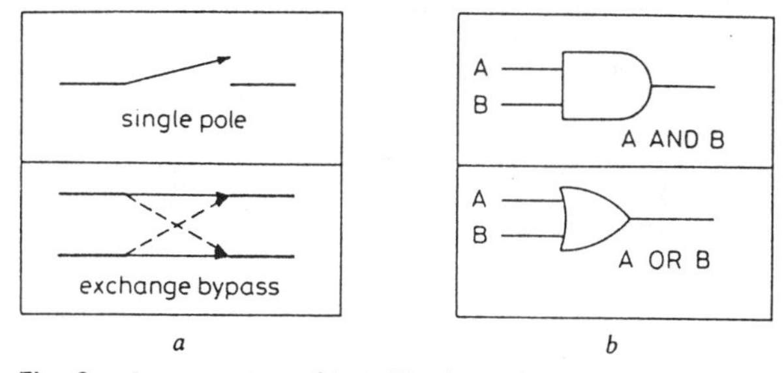

Fig. 3 *Interpretations of 'switch' as logic element and externally controlled routing element*
a Analogue switches
b Logical switches

this. The second meaning comes from digital logic where devices that carry out the logical operations such as A AND B or A OR B are referred to as switches. Evidently, for digital traffic, the switch of Fig. 3*a* can be rendered using logical devices as the circuit of Fig. 3*b*. There are also optical analogues of this latter class of device but it is important to recognise that the two classes have very different properties. For the former, the control and data channels are completely separated whereas in the latter, they are intimately linked. Moreover, the latter implies clocked synchronous data with a very close relationship between the time for a logical operation (such as AND or OR) (and hence the 'switch reconfiguration time') and the basic clock cycle time whereas the former might take a millisecond to reconfigure but pass 10 Gbit/s data. This distinction is not often made in electronic switching circles because of the predominance of electronic logic as the switch medium. However, in optics, we shall see that it may be very important.

Crossbars and multistage networks

The simplest switching matrix to imagine is the $N \times N$ crossbar formed of an $N \times N$ square array of 'wires' with open/close switches placed at the wire crossings. To connect input i to output j one merely has to close the switch at the array location (i, j). Evidently, such a matrix requires N^2 discrete switches and this number rises very rapidly with N. Moreover, the path distance varies dramatically between input and output ports, embracing one crossing from input N to output 1 (the path across one corner of the square) through to $(2N - 1)$ crossings between input 1 and output N (the path right along two sides of the square). In optical terms, this tends to mean insertion losses that vary wildly with connection pattern. Moreover, we shall see that guided-wave optical techniques do not in general lend themselves to very high complexity 'circuits' and hence it is appropriate to consider building up a large network from an assembly of smaller building blocks. The Clos network [16] provides a well known model of such a nonblocking network and is illustrated in Fig. 4, which illustrates a 3 stage $N \times M$ network composed of n input stages, m output stages and $(n + m - 1)$ centre stages, the number required to ensure a nonblocking operation. In many cases, $n = m$ and $N = M$ and moreover, $n \sim \sqrt{(N)}$. Such networks do not avoid the necessity of complex interconnection patterns but they do allow large matrices to be assembled from numbers of smaller building blocks. Moreover, it is not essential that all the elements of such a multistage network operate on the same principle so that the input

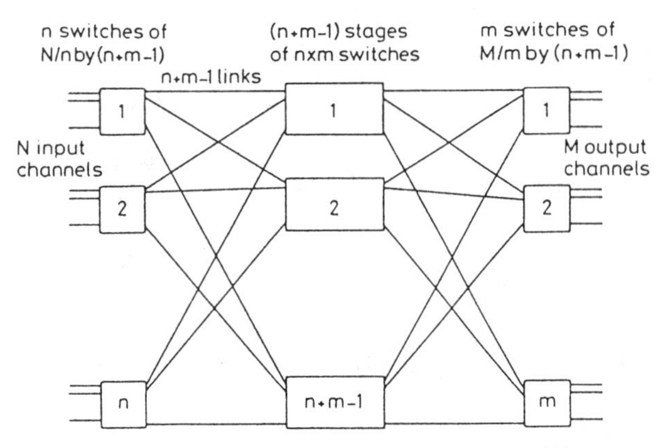

Fig. 4 *Schematic layout of Clos network showing two-fold expansion in centre stages arising from absolute nonblocking requirement*

and output stages might be time or wavelength switches whereas the centre stage might be a space switch etc. Of particular interest for large switches constructed in this manner is the obvious potential for inserting gain stages in the optical path between chosen stages, an approach exploited by the Japanese 128 × 128 lithium niobate switch mentioned in the following [17].

A further large class of switches are the reconfigurably nonblocking multistage networks stemming from the work of Benes [18] where the number of discrete switches required to connect N inputs to N outputs in a nonblocking manner is greatly reduced. However, a penalty is paid in that when changing the connections of one or two circuits through the matrix, it may be necessary to relocate several others as well.

Self-routing switches

Many of these multistage networks offer fascinating self-routing capabilities with the result that packet type data with suitable header addresses can worm their way through a complex network, setting up the switches and their own route as they proceed. A family of switches based on this concept is well known in the USA under the ATT name of Starlite [19] which exploits a combination of a Batcher sorting network followed by a Banyan fanout network. They appear to form a very attractive basis for ultralarge throughput switches using optoelectronics.

The general principle involved is that packets are injected into the switch input ports simultaneously. The N inputs are connected to $N/2$ 2 × 2 switching processors which, apart from having the capability to transport data from inputs A and B to outputs C and D or D and C according to setting, also have sufficient intelligence to read the incoming address headers on the two packets and carry out bit-by-bit comparisons of the headers. Then using various simple algorithms, the on-board logic can set the switch state to exchange or bypass and latch it there until a global reset signal is applied. The reader is referred elsewhere for details of the algorithms involved [19].

Time against space switching

So far we talked about switching in terms of physically connecting 'in space' different circuits envisaged as discrete wires. However, we note that a 140 Mbit/s TDM transmission signal contains 1920 discrete telephone circuits, all time shared and interleaved. If we imagine a box acting as a switch and connecting two such transmission highways together, then we realise that between such time divided highways, the switching operation consists of moving signals from one time slot associated with the signal from Mr. Jones to another time slot associated with the signal to Mr. Smith. This is a time slot interchange switch. Electronically, it takes the general form shown in Fig. 5 whereby successive bytes of signal are

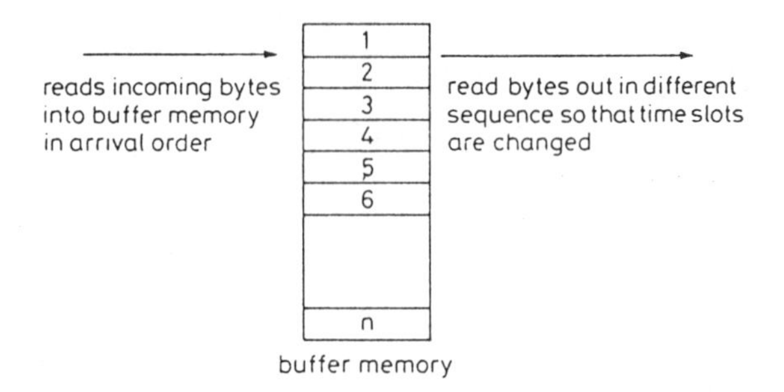

Fig. 5 *Schematic diagram of electronic time-slot-interchange switch*

read into a buffer memory and then read out again in a different sequence. It can be shown that for such a switch, if N is the number of multiplexed channels then approximately N buffer memories are required [20].

Control against data flow

If we consider a TDM telephony stream at 1.2 Gbit/s composed of ISDN 64 kbit/s channels in which the mean holding time per circuit is 1 min, then the number of new circuits to be established every second is only about 256, representing a modest computing/control overhead. However, if we take the same data flow and treat it as composed of 53 byte ATM cells, then neglecting the overhead in the transmission stream, this corresponds to about 2.8 million ATM cell/s, each of which has to be routed independently. The increase in the control load is quite dramatic. From this observation there have arisen many suggestions that it would be attractive if packets in an optical network could self-route themselves to their destinations. These generally fall into two categories, first, techniques for using the header on a packet travelling along a single fibre highway to establish the state of a single switch on the highway such that it is either diverted off the highway into a receiver or allowed to bypass that terminal. Such techniques have obvious applications in ring type LANs. The second and more challenging approach is to consider a self-routing multistage interconnection network in which each node of the network has 'on-board' intelligence that can perform a local routing function. The Batcher-Banyan switch type of switch is an excellent example of this type of network and has been implemented electronically in the ATT 'Starlite' design.

In both cases, it is clear that digital logic is required to decode the header(s) and set the optical switch state prior to the arrival of the packet. Some delay is thus generally envisaged in the optical data path. However, examination of this approach to routing brings one face-to-face with the dilemma of whether to use optical or electronic logic in the control circuit. To date virtually all serious studies have concluded that electronic logic is so superior to optical logic that there is no contest [21]. Hence, one envisages the switching node involving an intimate integration of optical and electronic functions, an approach that has been widely described as based on 'optically connected electronic islands'.

Optical switch implementation

Guided wave optical switches

The first optical switches to attract widespread interest arose from studies of 'integrated optics' using materials such as lithium niobate to form optical 4-port directional couplers switched by means of the electro-optic effect [22]. A typical structure is shown in Fig. 6 where we can

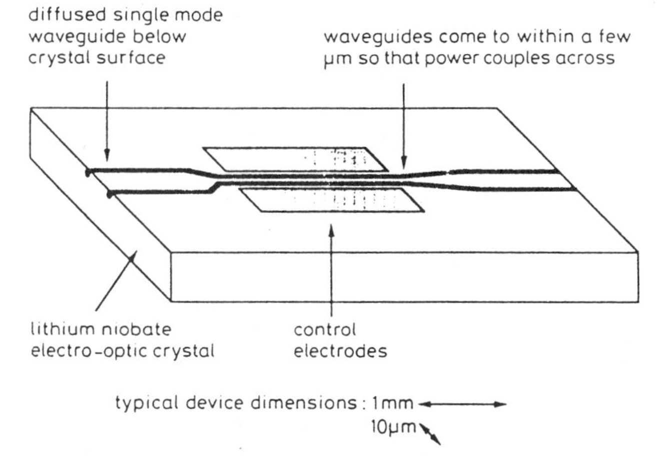

Fig. 6 *Typical guided-wave directional coupler layout and dimensions*

see the two input single mode guides, the interaction region in which power crosses over or not according to the phase relationships of the two waveguides and the two output guides. The control electrodes are placed on the surface of the crystal material and the overall dimensions of the resulting device are typically 1 cm to 1 mm long and perhaps 10–20 μm wide, a length to breadth ratio of 100–1000. The electrodes are also long and typically require 5 to 20 V to induce switching. Hence the device exhibits large capacitance to the electrical drive circuit and the switch-reset time is limited by electrical considerations, typically to around 1 ns although much faster can be achieved.

However, it is important to note that if a square matrix of such devices were required, say with 100 inputs and 100 outputs, then this would imply 10^4 such devices and the electrical drive problem would become horrific. The current state of the art is that switching arrays of various designs and up to sizes of about 8 × 8 have been made on single crystals. These are normally of the multistage rather than crossbar type, which not only need fewer discrete devices but also give similar insertion loss between any pair of input and output ports. Once a route is established through such a switch, the pathway is data-transparent, passing a spectral bandwidth typically of 30 nm or more albeit with attenuation on any desired channel and some crosstalk from scattering between channels. Typical attenuation through a 1 × 16 array is 7.5 dB with interchannel crosstalk at 27 dB. An example is shown in Fig. 7 which shows a 1 : 16 switch mounted in a hybrid package complete with drive electronics, developed by GEC Marconi [23]. The reset time to establish a pathway to any one of the 16 exit ports is typically 3–6 ns. Note that dimensions of this complex package are approximately 1 × 5 × 25 cm.

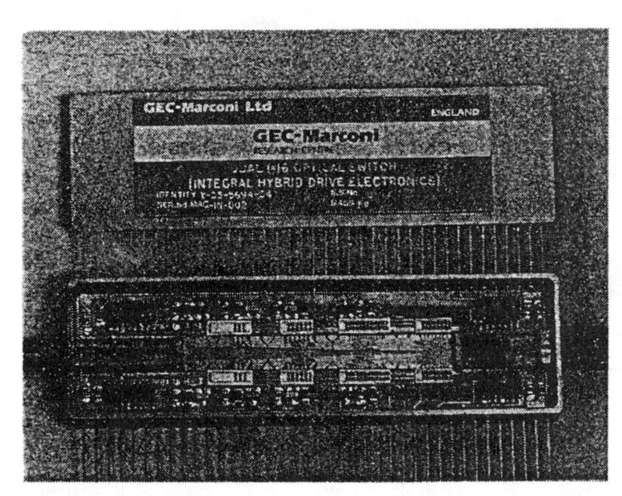

Fig. 7 *GEC Marconi 1 : 16 switch developed for RACE-OSCAR project (Package includes switching chip as well as fibre interfaces and control circuits)*

It is now sobering to examine how larger switches can be constructed from such building blocks. In Fig. 8 we show schematically a 128 × 128 port switch recently reported by NEC from Japan [17]. Using the data quoted in their paper, we have estimated some of the details of the experiment which are summarised in Table 1. This underlines the awesome complexity because the component cost alone would require the total annual research budget of several University Research Departments. Add to this the need to drive thousands of electri-

Table 1: Estimated numbers of components in Japanese 128 × 128 switch demonstrator

Vital statistics for NEC 128 × 128 guided-wave switch [17]
Active component count
number of 4 × 7 switch matrix modules = 176
number of 8 × 8 switch matrix modules = 49
each 4 × 7 stage contains 28 directional couplers
each 8 × 8 stage contains 64 directional couplers
number of 4 × 4 building blocks = 548
total directional coupler count = 8064
number of control lines (bond wires) = 8064 (or 2 × 8064 if separate ground wires)
number of laser amplifiers = 784
Singe mode connection count
connections per butterfly (including amplifiers) = 1568
number of guided-wave lines between modules in 4 × 7 stages = 224
total number of guided-wave connections (including input & outputs)
= (2 × 128) + (4 × 224) + (2 × 1568) = 4288

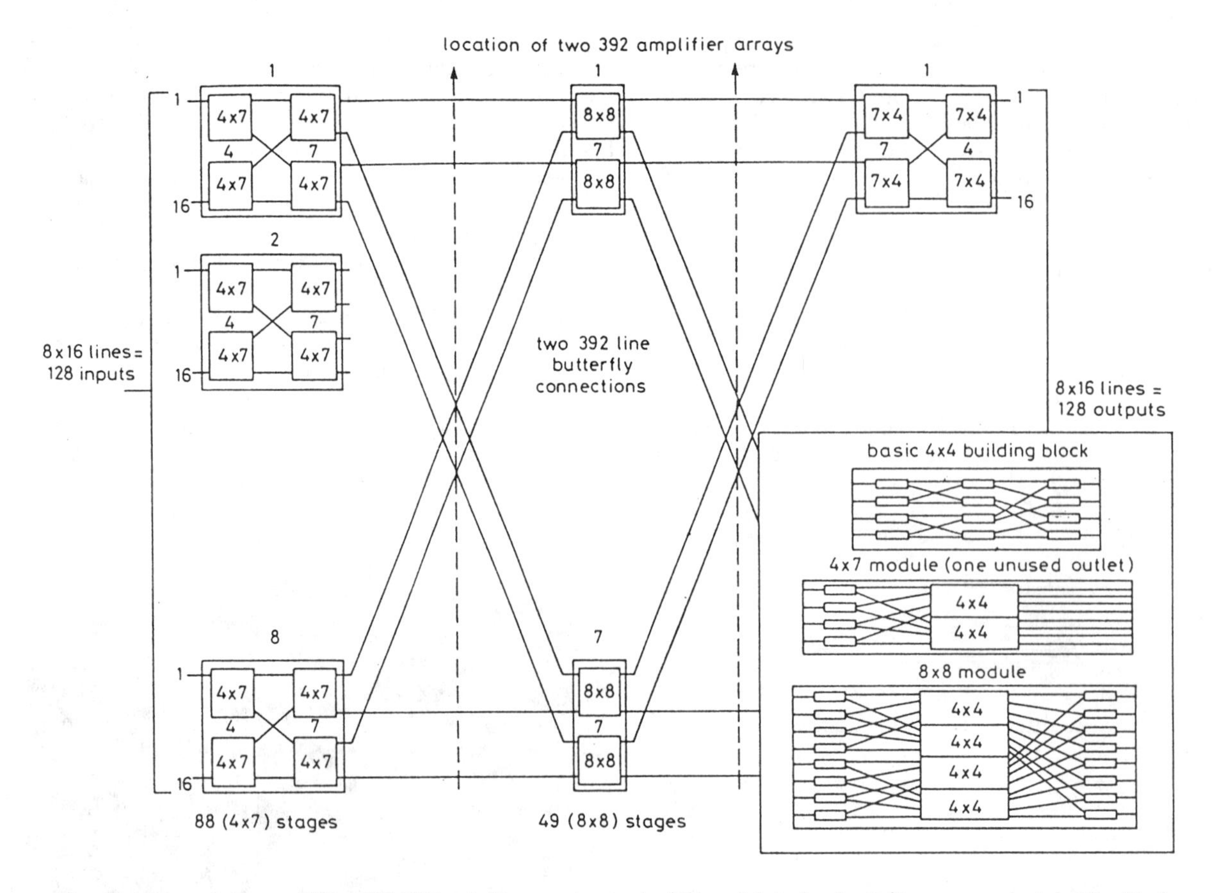

Fig. 8 *Schematic layout of 128 × 128 NEC switching experiment using lithium niobate 4 × 4 switching arrays as basic building blocks*

cal control signals for pathway control and the required insertion of hundreds of semiconductor laser amplifiers to make up for the 48 dB matrix insertion loss and we recognise a truly 'hero' experiment.

A variety of other techniques exist in optics to form guided-wave switching matrices that are broadly compatible with single mode optical fibres. For example, a semiconductor laser chip with antireflection coatings on its mirror-facets and placed between two fibres (Fig. 9A) acts as a good ON/OFF switch according to whether it is electrically energised or not. Linking passive 3 dB power splitters and an array of four such devices offers the circuit of Fig. 9B which operates similarly to the directional-coupler shown above (Fig. 6) except that it can provide lossless fanout delivering one input to two outputs without attenuation [24]. Once again, arrays of such sub-assemblies lead to electrically controlled switching matrices in principle. However, the practical problems in fabricating and controlling such arrays are formidable and remind us that optoelectronics tends to be a discrete component or low level integration technology.

In summary, these technologies lead to small switch arrays with awesome data throughput potential (e.g. BW of 4000 GHz per port) but their route reconfiguration times are set by electrical control problems that are probably more complex than those in digital electronic switches. Hence they provide huge throughput capability but relatively slow reset times and thus do not seem well suited to switching byte-multiplexed traffic and it seems open to doubt as to whether they are useful switching at the ATM cell level either. They do seem well suited to switching large blocks of traffic as might be required in protection or digital cross-connect operation.

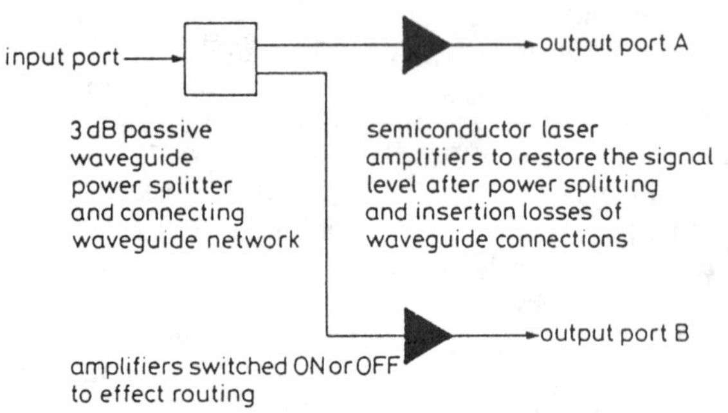

Fig. 9A *Semiconductor laser/amplifier switch schematic layout*

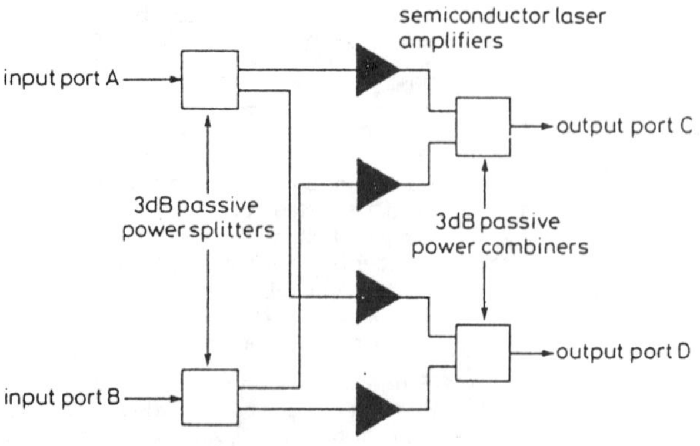

Fig. 9B *Configured as 2 × 2 directional coupler etc.*
Passive insertion loss must be at least 6 dB, compensated for by amplifier again

Wavelength division multiplex switches
Optics brings a completely new dimension to switching through the wavelength domain. We have already seen that typical optical devices and media of relevance offer about 30 nm spectral bandwidth which corresponds to about 4000 GHz electrical spectrum. A wide variety of tunable sources, detectors and/or receiver filters are now slowly becoming available. Typically, these can handle of order 10 wavelength channels per device although many more may be possible, perhaps 100 or eventually 1000. A single $N \times N$ power splitter, splitting the power from each of N input ports equally to each of N output ports inevitably delivers at best $1/N$th of the input power from any one transmitter to any one receiver but instantly makes possible $N \times N$ switched networks provided also that N wavelengths are available. In the absence of amplifiers, such a network is probably limited to about 100 terminals by the 20 + dB insertion loss that would imply. Much interest thus centres upon networks of the general type shown in Fig. 10 which, for generality, shows tunable transmitters and receivers although in practice, probably only one of these would be tuned. A network in which each transmitter operates on a discrete fixed wavelength allows for mixed directed and broadcast communication because more than one receiver can tune to a single transmitter whereas in other cases only a single receiver tunes a given transmitter.

Such networks are characterised by huge throughput capability (i.e. the product of the number of transmitters times the bit rate each is clocked at) because there is no contention involved in such a system. A switch fabricated in this manner has a reconfiguration time set by the sum of the times it takes to establish (logically) the route required, the time to send that information to the desired transmitter or receiver and the time for that element to retune its wavelength, leaving aside time for the resolution of collisions which inevitably occur when two or more sources wish to speak to one receiver. However, for ATM traffic, the control overhead seems likely to be the limiting factor if switching is to be done at the discrete ATM cell level [25]. Typical tuning times for discrete DFB semiconductor laser sources or filters based upon similar technology is measured in nanoseconds. Less obvious is how wavelength calibration would be maintained. However, if we examine a single fibre termination into such a switching subsystem, it is likely to look like that shown in Fig. 11 which highlights several factors.

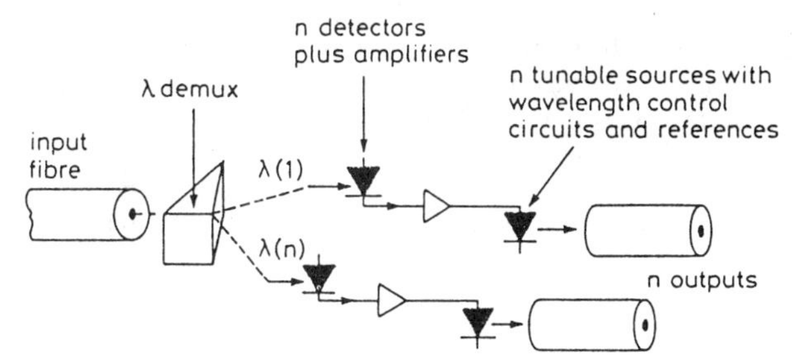

Fig. 11 *WDM network-switch termination indicating urgent need for higher levels of component integration*
n might typically be 10–20

True optical transparency has been lost because optoelectronic conversion is implicit at both entrance and exit interfaces, although there may be no need to restrict operation to a preset synchronous bit rate; i.e. variable bit rate may be allowed. However, if the fibre carries wavelength multiplexed channels then it is fairly clear that the complete interface module should be monolithicly integrated although with today's optoelectronic technology, that would be extremely difficult. The mixed optoelectronic circuit of Fig. 11 is considerably more complex than anything that has been fabricated monolithically to date.

It should also be noted that we have postulated a large $N \times N$ passive power splitter but if that is made up from discrete 2×2 couplers, then it is also a major challenge. For example, a 128×128 power splitter composed of a binary logarithmic array of 2×2 couplers requires $(64 \times 7) = 448$ devices implying 1024 single mode splices. (Expanding to 256×256 inflates these numbers to $128 \times 8 = 1024$ 2×2 couplers with 2304 single mode splices.) Here again the case for some form of component integration seems overwhelming although the requirement is only for passive elements and an appropriate

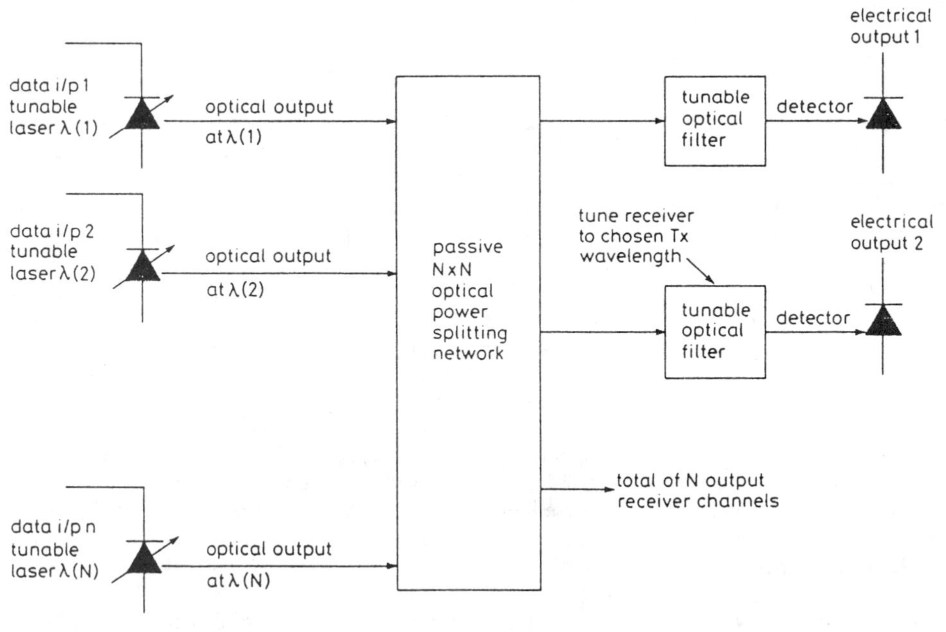

Fig. 10 *Generalised WDM network*

technology may be planar doped silica waveguide on silicon substrates [26].

Time-slot interchanging in optics

Unfortunately, there is no real equivalent in optics to a buffer memory, the nearest being a delay line. Hence an optical implementation of such a switch tends to involve a structure as shown in Fig. 12 where each delay line memory is digitally switchable from 0 to $S - 1$ time slots. Such memories can be constructed in principle using lengths of fibre as the delay line and lithium niobate switches to reconfigure them to provide the desired delay [27]. Note that the speed of light in an optical fibre corresponds to about 20 cm/ns so that an 8 bit delay at 1.2 Gbit/s corresponds to about 1.3 m of fibre, a reasonable coiled length to handle. However, if we then assume that we wish to switch a 32 channel multiplex we need lengths of 0, 1.3, 2.6, 5.2, 10.4, 20.8 and 41.6 m that can be combined in each channel or about 2.4 km in total, still manageable but becoming bulky. Expanding this to handle a 256 channel multiplex at 1.2 Gbit/s, the number expands alarmingly to about 170 km of fibre delay line. We are again struck by the fact that what has proved to be elegant in electronics because of high level integration is exceedingly cumbersome in optics because of the almost complete absence of integration.

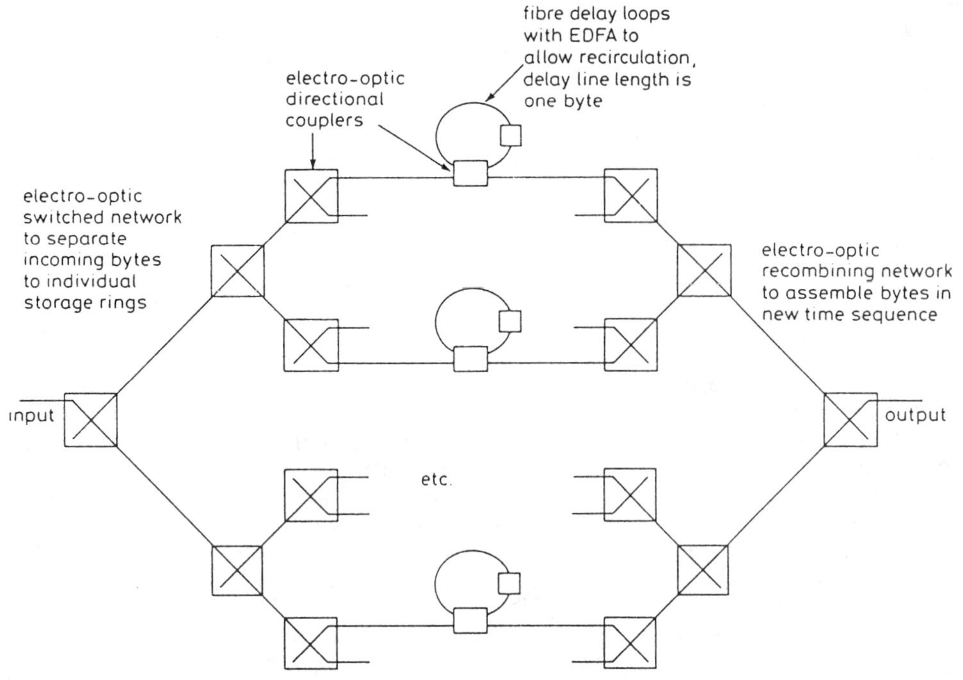

Fig. 12 *Optical delay line time-slot-interchange switch, example of how optics does not easily copy electronics*

High level optoelectronic integration

Introduction

A recurring theme in the discussion so far has been the notion that although optics can reproduce most functions used in electronic switching systems and can often do so with performance advantage at least in terms of data throughput capability, it is very difficult to see how the guided wave approaches readily scale to high levels of complexity because almost all optoelectronics remains a discrete component technology. However, there is one major exception to this statement and for that reason, it has been seen as a potential way forward in switching applications. Two dimensional planar arrays of MQW modulator/detectors and/or surface emitting lasers offer some interesting new possibilities when it comes to chip-level wiring and may open up a new era of both optical interconnect and optoelectronic system.

MQW device arrays in III-V materials SEEDs, modulators and detectors etc.

This new class of optoelectronic devices has been pioneered at ATT Bell Labs [28] and at UCL [29]. The typical device structure now used is shown in Fig. 13 and comprises a PIN diode placed over an epitaxially grown dielectric reflector stack with the intrinsic region of the PIN diode composed of about 50 pairs of quantum well layers, each about 100 Å thick so that the total structure is typically just a few micrometres thick. The UCL designed device, when used as an electrically driven modulator, modulates incident light to better than 20 dB depth for about 5 V drive and with a minimum insertion loss of less than 3 dB. Because the devices can be made very small, they offer an extremely power efficient way of impressing an electrical signal onto an optical interrogation beam. The same device, simply operated in reverse bias, works as a very fast high quantum-efficiency photodetector so that identical devices can take electrical information from a circuit and impress it on an optical carrier beam and somewhere else, return it to electrical form.

Virtually the same devices can be operated as optically triggered logic devices. The R-SEED consists of a suitably designed MQW modulator connected to an electrical bias rail through a suitably chosen resistive load. Such a device shows strong switching behaviour as a function of the incident optical power opening the way to its use in optically activated logic. The example shown in Fig. 14 shows the switching characteristics of a recently reported device, designed and fabricated at UCL [30], which follows earlier designs from ATT Bell Labs, but which offers enormously improved contrast (from 3 dB to better than 15 dB).

The great advantage of all these devices is that they occupy very small amounts of real-estate, typically

10 × 20 μm, relative to guided wave devices, and, being planar in their design and fabrication, can be fabricated in very large arrays using the standard processes of semi

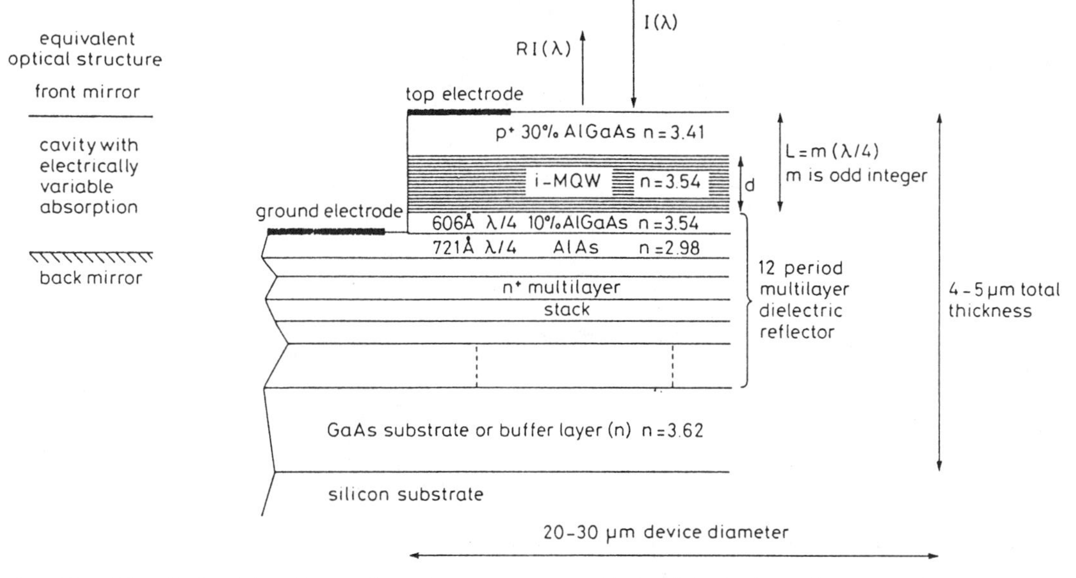

Fig. 13 *PIN MQW modulator/detector/logic element structure suitable for fabrication in very large arrays (e.g. 256 × 256)*

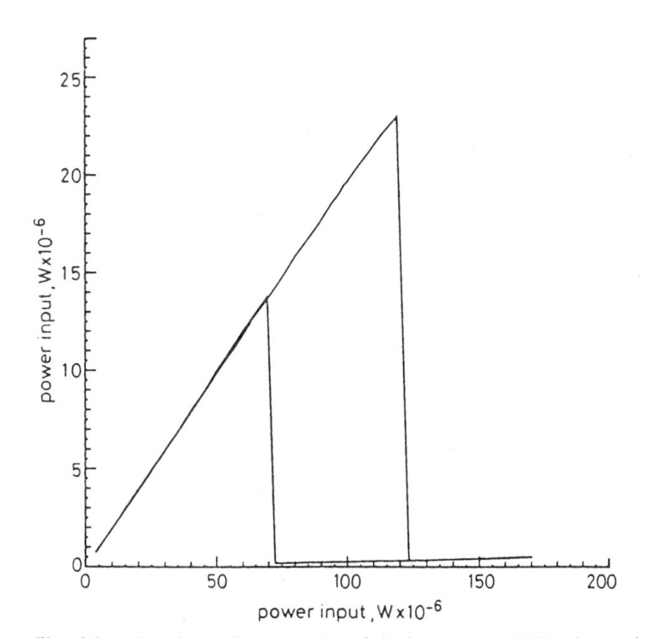

Fig. 14 *Switching characteristic of high contrast UCL designed SEED showing hysterisis loop in input/output characteristic and better than 18 dB contrast*

conductor technology. Of course, this is also true of semiconductor lasers to a degree but they then have to be separated into discrete devices or at best, very small arrays, and mounted individually before they can be used. Using free space imaging techniques, these optical elements can be accessed (interconnected) optically in sizeable numbers in a manner that is virtually unthinkable using guided-wave optics. Moreover, the very fact that the technology of planar integration can be fully exploited offers potential for great cost reduction that seems ruled out in the optical guided-wave technologies. However, we note in passing that although these devices are small by the standards of optical devices, they remain large by electronic standards and we (UCL) believe that this will seriously limit the degree to which they can realistically expect to replace electronic logic, tending to confine their application to optical interconnect and optoelectronic interfacing applications.

Free space interconnect

The interest in 2D arrays of such planar devices is greatly enhanced by the potential offered by imaging optical interconnects. The concept of a simple imaging array interconnect is shown in Fig. 15 whereby a lens connects one, 2D array of 'sources' to a matching 2D array of 'receivers'. Even modest arrays of, say 8 × 8 dimension, clocked at 1 Gbit/s imply an awesome pin-out capability. 'System' studies of such interconnects suggest that they could compete on power efficiency grounds with electrical interconnects (metallisation) over distances as short as a few millimetres so that the concept is gaining ground

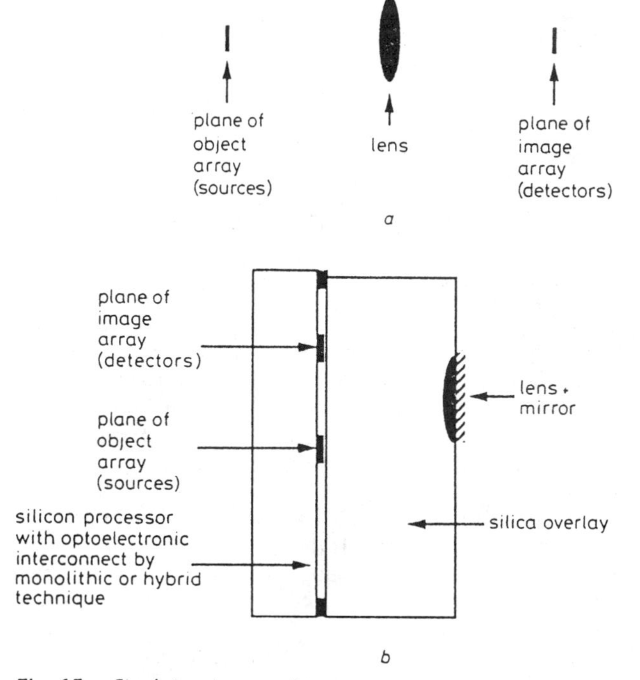

Fig. 15 *Simple imaging array-interconnect*

a Basic concept using simple lens
b Beginnings of packaging such interconnects with electronics

steadily that future ultrahigh data flow processors will make extensive use of optical array interconnects within the wafer and between the wafer and the outside world [31, 32].

Point to point imaging is valuable but is only one 'wiring layout' of interest. In switching matrices, other patterns occur repeatedly such as crossovers and perfect-shuffles (see Fig. 16). The crossover has been imple-

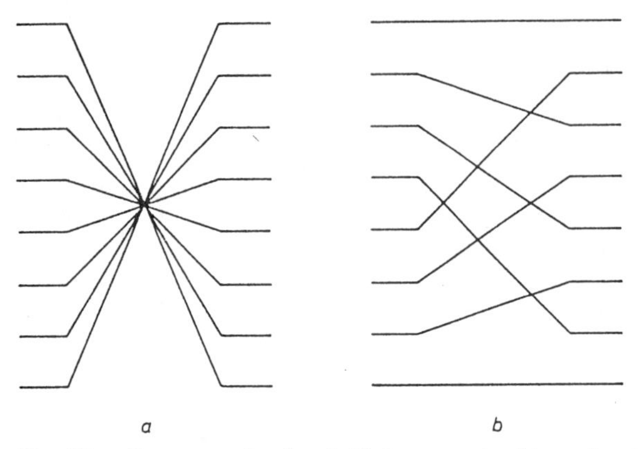

Fig. 16 *Crossover and perfect-shuffle interconnects wiring patterns*
a crossover
b Perfect-shuffle

mented in bulk imaging optics form at ATT-Bell Labs [33] as shown in Fig. 17 using lenses, beam splitters and prisms. Although this looks very cumbersome, it is worth remembering that the single optical assembly is capable of providing a crossover connection between two large arrays, perhaps as large as 128 × 128, and thus replacing some 16000 metal lines. An elegant proposal for imple-

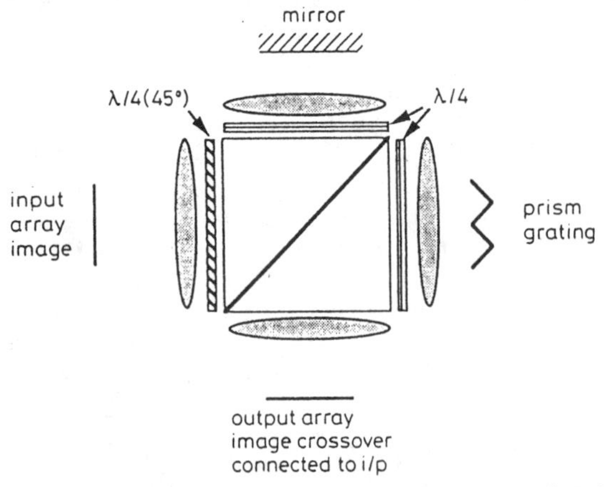

Fig. 17 *Optical crossover as developed by ATT-Bell Labs. using prism mirror array to provide crossover connection and plane mirror an overlay of direct connections*

menting a 2D perfect shuffle interconnect is shown in Fig. 18 again using bulk optical imaging components. It relies upon the observation that the perfect shuffle operation involves cutting an array into two (or 4 for 2D) and then overlaying and interleaving the resulting sectors. In the context of hybrid self-routing switches, this particular layout has been shown to lead to extremely well structured data flow patterns. It then leads to concepts such as that shown schematically in Fig. 19 for a hybrid opto-electronic implementation of a self-routing high speed switch matrix.

The obvious objection to these concepts arises from the packaging problems implicit in them. Bulk optics is usually construed as placed on an 8′ × 4′ air-bearing vibration-isolation table and aligned by vast numbers of micromanipulators whereas complex electronic or optoelectronic components are envisaged in chip carriers with bond wires connecting their outputs to pins and PCBs. The two do not obviously mix and this probably remains the area requiring the greatest study.

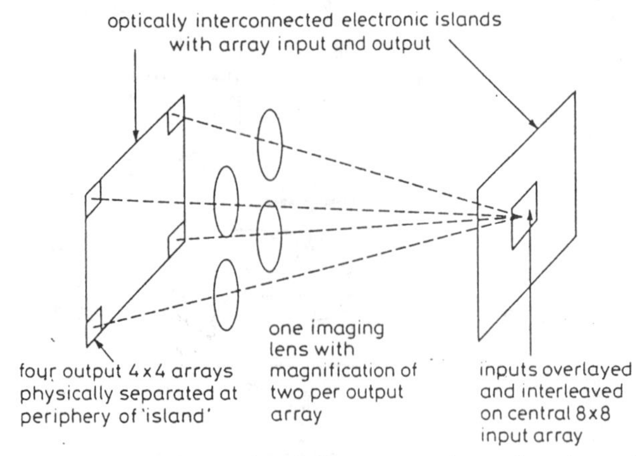

Fig. 18 *Optical 2D perfect-shuffle interconnection configured around 4 microlenses*

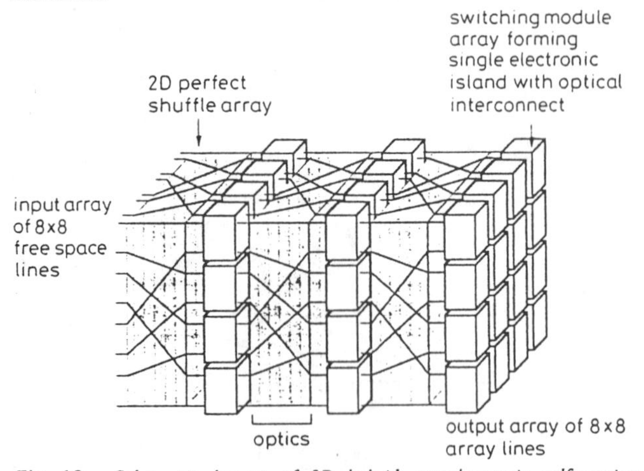

Fig. 19 *Schematic layout of 3D hybrid optoelectronic self-routing switch exploiting free-space optical interconnections with electronic processor islands*

ATT Free-space switching demonstrator

The one serious attempt to tackle some of these problems so far is that of ATT Bell Labs in which a prototype multistage interconnection network was assembled using symmetric SEEDs as the active elements in arrays typically of 32 × 64 and with imaging interconnections and crossovers between them [34]. The layout of the demonstrator is shown photographically in Fig. 20 and sche-

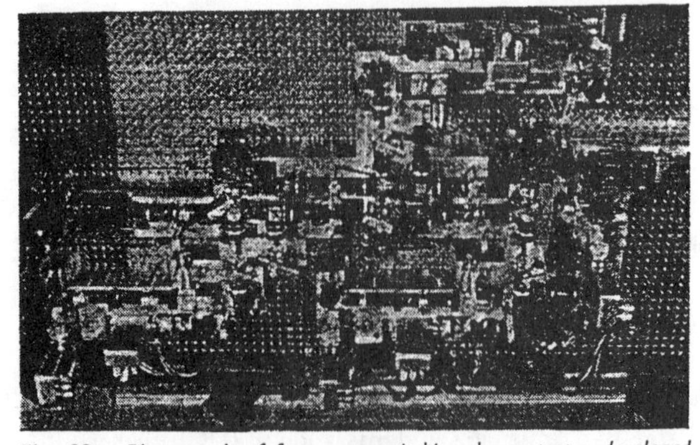

Fig. 20 *Photograph of free-space switching demonstrator developed by ATT Bell Labs*

matically in Fig. 21. Three active stages were built each of which contains of a symmetric-SEED array clocked by a laser 'power' source onto which control and data signals are fed to interact logically to generate an output signal that is clocked into the next stage of the processor. In its operation, this system is very much closer in concept to a digital electronic system than to any other optical system because it relies fundamentally upon binary digital logical signal interactions. Where it differs from an all electronic system and most guided-wave optical systems is that it exploits the third space dimension for communication between processor arrays and it does so uniquely through the use of imaging optical elements. In Fig. 20, we clearly see the lens and prismatic beam splitting optical elements laid out over a considerable area of optical bench. This highlights one of the major unsolved problems of this approach, that the technology is not readily compatible with current electrical packaging technology and that major advances will be required here, probably using microlens and holographic elements before it can be reduced to something much closer to a compact solid 3D assembly.

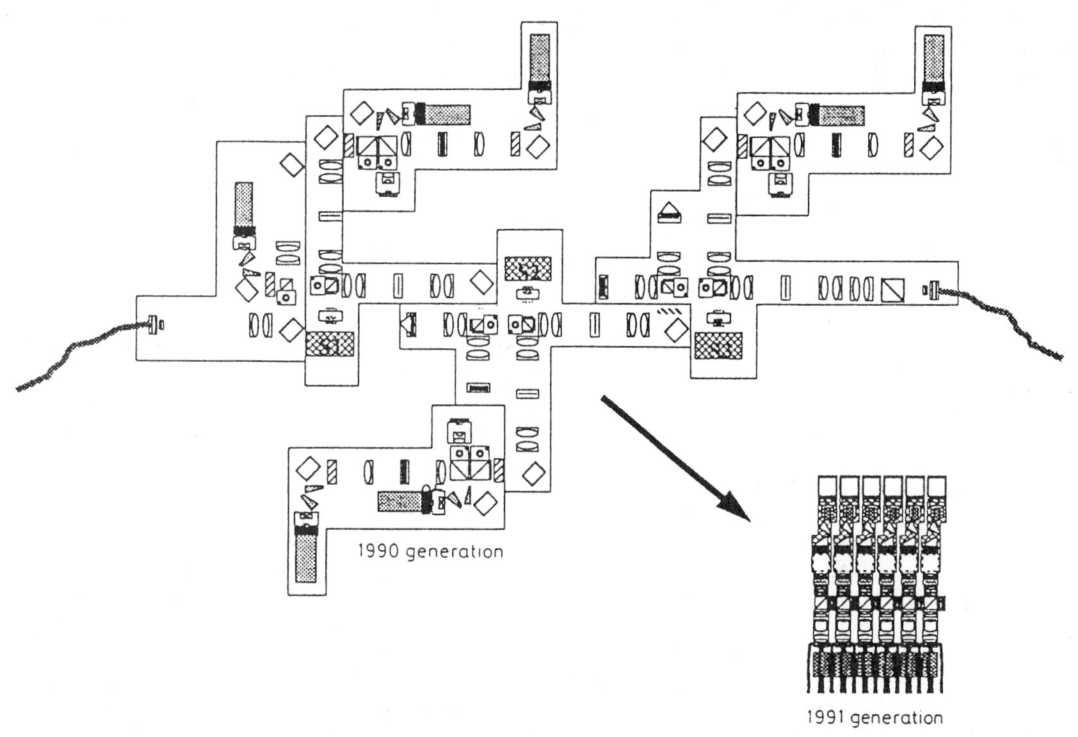

Fig. 21 *Schematic layout of Fig. 20 showing location of 3 SEED arrays (S1, S2 and S3) and connecting optics, crossovers etc.*

Thus the first impression is of an optical system of awesome complexity, and so it is. However, by comparison to the Japanese lithium niobate switch demonstrator already mentioned, we quickly recognise that the complexity is of a very different nature. Instead of vast numbers of discrete or low level integration components, this approach uses much higher levels of integration, fewer active components and in place of huge numbers of guided wave connections, relatively small numbers of carefully aligned imaging interconnections. Moreover, scaling the ATT system to a larger structure, say 256 input ports, while very challenging optically, does not lead to the same awesome complexity extrapolation that the guided-wave approach does. Against that, it must be said that the ATT approach seems to imply fundamentally rigidly clocked synchronous data flow so that ideas of 'optical transparency' have been lost in its implementation. However, so do almost all other proposals to implement optical switching fabrics, if only in terms of their requirement for buffer memory at their front ends. To make informed comparisons thus requires consideration not just of the optoelectronic technology but also the service and network requirements as well as the detailed cost projections for the different approaches. Next to nothing has been published on this at present.

Optical against electrical interconnects

We have highlighted the theme of optical integration so far, and this approach clearly favours the planar imaging route. We have also commented that for very high throughput rapidly reconfiguring switches, there is great attraction in a self-routing approach, a theme that seems not to have been much studied in Europe. Such an approach implies distributed logic throughout the switching fabric and given the extreme improbability of optics competing seriously with electronics in complex logical processing, this implies either an all electronic implementation or a hybrid optoelectronic approach.

This observation has led to much debate as the relative merits of optics and electronics in logic and general processing and a reasonably clear consensus now exists that electronics is always best for logical processing unless the processor is 'communication' limited. If the ability to transport and deliver data to the right place at the right time is a serious limitation, then optics may be able to help although it is still limited by the speed of light, a relatively slow 20 cm/ns in glass and not much better in vacuum ($\times 1.5$). However, this has stimulated intense interest in the 'smart pixel' or 'optically interconnected electronic island' approach to fast processor design and has given a hefty push to attempts to find ways to integrate sophisticated array interconnects with processed silicon electronic circuits. This seems certain to remain a subject of intensive development in the next few years. At present, the optoelectronic interface devices discussed above can be either grown epitaxially onto silicon [35] or mounted by hybrid technology. Alternatively, III-V electronic components can be monolithically integrated with III-V optoelectronics but in every case, many compromises have to be made and it seems certain that we are currently only seeing the beginning of these technologies.

Multidimensional switches

In electronic switching, much interest centres on mixed mode switches using, for example, time-space-time fabrics. Optics brings additional dimensionality to the equation, embracing the wavelength domain as well as time and three space dimensions. Whether such approaches offer a real advantage remains to be proven. So far there have been isolated reports of such approaches [36] but little hard data seems to exist to show how they are superior to a multistage switch constructed using similar technology stages. Each substage seems limited by the same constraints already alluded to and the overall assembly still suffers from the problems of limited device integration unless in the design process, that is overcome. At present, only the free space option seems to promise any significantly different extrapolations in the right direction.

Transparent optical networks—conclusions for the future?

Perhaps the greatest difficulty facing those who are considering which aspect of photonic switching technology to push derives from the confusion of vision one has of where the network is going. Two diverging views exist: we see an essentially linear development, based on an all digital network, with most switching nodes operating as digital logical processing points, clocking messages in whatever form synchronously from one port to another via the appropriate switching fabric. The alternative scenario sees a much more revolutionary future that exploits in a more imaginative way the opportunities now developing in fibre transmission and which seeks to make the network nodes optically (and data) transparent [37]. Such an approach seems axiomatically to block the ATT-demonstrator (Figs 20 and 21) route forward and to favour an optical guided-wave approach. However, neither approach at present appears to have been thought through in the detail necessary to enable one to make affirmative statements for or against either option. It is clear that the two routes lead to widely differing requirements in terms of components and systems and lead to networks with widely differing characteristics. What can we conclude with reasonably certainty?

Optical switches clearly do not simply plug in as replacements for electronic ones. The family of guided-wave optical switches can generally be characterised as well suited to forming small fabrics, offer massive throughput but with route-set-up times that are very long compared to the minimum bit intervals they might allow. This leads to thoughts of transmission formats having guard bands (dead spaces) between the blocks to be switched during which the switch can be reset.

It is clear that if the switch must be bit-synchronised, then at present some form of electronic synchronisation circuit is required probably with electronic buffer memory. This seems to restrict the whole system to the speed of the electronic buffers etc. However, once that is required, then approaches such as the 'Starlite' self-routing designs may become attractive and, with free space optical technology and electronic islands, might be extended to considerably greater throughputs than is currently possible in purely electronic configurations. We can also rephrase this last observation by saying that ultimately, most fast electronics is communication rather than device limited so that an intimate fusion of the two technologies promises major advances. However, we should not underestimate the difficulty of this because at present an acceptable total technology does not exist, even though many extremely promising ingredients do. Optical interconnects are steadily penetrating deeper into electronic systems. Optical backplanes and optical 'printed circuit boards' are now close to market so the optically interconnected electronic island vision is not as far away as it might seem.

Wavelength routing offers an interesting variant to space switching in a closed network but seems to necessarily involve optoelectronic buffer-interfaces at the node interface. And because routing information must be gleaned to set up the switch, electronic intervention in the data stream again seems axiomatic.

More visionary ideas of an all-pervasive optical ether using a truly optically transparent network in which 'all signals at 1552.7597 nm go to Birmingham etc.' are difficult to flesh out in sufficient detail to critically assess but do seem to run into some reasonably fundamental objections. For example, taking the spectrum available in the fibre as 4000 GHz, the number of wideband channels that might be squeezed into this space might be 1000 but seems unlikely to be more and that does not allow for a very large address space without wavelength re-use. Nevertheless, it is clear that the changes already set in motion as a result of EDFAs will stimulate a great deal of rethinking on the likely and potential shape of the telecommunications network of the future and this is one of the many ideas that will start to be evaluated more critically and seriously. Meanwhile we have a ferment of different technologies, ideas and approaches many of which interact back on the network format that we might consider. Probably the only completely safe prediction is that we will see an increasing penetration of optical interconnects in switching systems and nodes and by embracing a broad but flexible interpretation of what constitutes an optical interconnect, we can be sure of being right in our prediction.

References

1 KAO, K.C., and HOCKHAM, G.: *IEE Proc.*, 1966, **113**, p. 1151
2 BERRY, R.W., BRACE, D.J., and RAVENSCROFT, I.A.: *IEEE Trans.*, 1978, **COM-26**, pp. 1020–1027
3 HILL, D.R.: '140 Mbit/s optical fibre demonstration system', *in* SANDBANK, C.P. (Ed.): 'Optical fibre communication systems' (Wiley, Chichester, UK, 1980) Chap. 11
4 ROBERTS, F.F.: (formerly BPO, now deceased) — private communication
5 KAO, K.C., and DAVIES, T.W.: *J. Sci. Instrum.*, 1968, **1**, pp. 1063–68 and WRIGHT, C.K., and KAO, K.C.: *J. Sci. Instrum.*, 1969, **2**, pp. 331–335
6 MAURER, R.D.: Conf. 'Trunk telecommunications'. IEE, Savoy Place, London, Sept. 1970
7 MIDWINTER, J.E., and STERN, J.R.: 'Propagation studies of 40 km of graded index fibre installed in cable in an operational duct route', *IEEE Trans.*, July 1978, **COM-26**, pp. 1015–1020
8 MIDWINTER, J.E.: 'Optical fibre transmission systems: overview of present work', *Post Off. Electr. Eng. J.* (GB), October 1977, **70** Part 3, pp. 144–5

9 GLOGE, D., and MARCATILLI, E.A.J.: *Bell Syst. Tech. J.*, 1973, **52**, p. 1563

10 KECK, D., and OLSHANSKY, R.: *Appl. Opt.*, 1971, **15**, p. 483

11 PAYNE, D., and GAMBLING, W.A.: *Electron. Lett.*, 1975, **11**, pp. 176–177

12 MIDWINTER, J.E.: 'Studies of monomode long wavelength fibre systems at the British Telecom Research Laboratories', *IEEE J. Quantum Electron.*, June 1981, **QE-17**, pp. 911–918

13 PAYNE, D.N.: 'Advances in active fibers'. Invited Review Paper 20.A-3.1, Integrated Optics and Optical Fibre Communication IOOC-89, July 18–21, Kobe Japan, Pub. IEICE Japan 1989

14 NTT Review, May 1991 devoted to BB-ISDN, Network & System Technologies

15 *Int. J. Digi. Analog Cabled Syst.*, 1988, **1** (4), (special issue devoted to ATM)

16 CLOS, C.: 'A study of non-blocking switching networks', *Bell Syst. Tech. J.*, 1953, **32**, pp. 407–424

17 BURKE, C., FUJIWARA, M., YAMAGUCHI, M., NISHIMOTO, H., and HONMOU, H.: 'Studies of a 128 line space division switch using Lithim Niobate switch matrices and optical amplifiers'. Topical Meeting on 'Photonic Switching', Salt Lake City, Utah, March 6–8, 1991, Pub. Optical Society of America, Washington, USA 1991

18 BENES, V.E.: 'Optimal rearrangeable multistage connecting networks', *Bell Syst. Tech. J.*, 1964, **43**, pp. 1641–1656

19 HUANG, A., and KNAUER, S.: 'Starlite, a wideband digital switch'. Proc. IEEE Global Telecomm. Conf., Atlanta, Georgia, USA, pp. 121–125, Pub. IEEE New York, 1984

20 FREEMAN, R.L.: 'Telecommunication system engineering', *in* 'Digital switching & networks' (Wiley, New York, 1989) Chap 10.

21 MIDWINTER, J.E.: 'Digital Optics, Smart Interconnect or Optical Logic?', *Phys. Technol.*, Part I, pp. 101–108, May 1988, Part II, pp. 153–157, July 1988

22 SCHMIDT, R.V., and ALFERNESS, R.C.: 'Directional coupler switches, modulators & filters using alternating $\Delta\beta$ techniques', *IEEE Trans.*, 1979, **CAS-26**, pp. 1099–1108

23 Switch developed for the RACE-OSCAR programme by GEC-Marconi UK

24 HIMENO, A., TERUI, H., and KOBAYASHI, M.: 'Guided wave optical gate matrix switch', *IEEE J. Lightwave Technol.*, 1988, **6**, pp. 30–35

25 BRACKETT, C.A.: 'Capacity of multi-wavelength optical-star packet switches and implications for packet length'. Topical Meeting on 'Photonic Switching', Salt Lake City, Utah, 6–8 March, 1991, Pub. Optical Society of America, Washington, 1991

26 KAWACHI, M.: 'Integrated Micro-optic components based on silica waveguides'. European Conference on Optical Communications, ECOC-87, Helsinki, Finland, 13–17 Sept, 1987

27 THOMPSON, R.A., and GIORDANO, P.P.: 'An experimental photonic time-slot interchanger using optical fibers as re-entrant delay line memories', *IEEE J. Lightwave Technol.*, 1987, **LT-5**, pp. 154–162

28 MILLER, D.A.B. *et al.*: 'The quantum well self electro-optic; bistability, oscillation and self-linearized modulation', *IEEE J. Quantum Electron.*, 1985, **QE-21**, pp. 1462–1476

29 WHITEHEAD, M., RIVERS, A., PARRY, G., ROBERTS, J.S., and BUTTON, C.: 'Low voltage multiple quantum well modulator with on-off ratio better than 100 : 1', *Electron. Lett.*, 1989, **25**, pp. 984–985

30 GRINDLE, R., and MIDWINTER, J.E.: 'High contrast SEED'. Submitted to *Electron. Lett.*

31 FELDMAN, M.R., ESENER, S.C., GUEST, C.C., and LEE, S.H.: 'Comparison between optical and electrical interconnects based on power & speed considerations', *Appl. Opt.*, 1988, **27**, pp. 1742–1751

32 MIDWINTER, J.E.: 'Communications, VLSI, optoelectronics & self routing switches', 'Prestige poster session'. XIII Int. Switching Symp., Stockholm, May 27–June 1, 1990

33 CLOONAN, T.J., HERRON, M.J., TOOLEY, F.A.P., RICHARDS, G.W., McCORMICK, F.B., KERBIS, E., BRUBAKER, J.L., and LENTINE, A.L.: 'An all-optical implementation of a 3D crossover switching network', *IEEE Photonics Technol. Lett.*, 1990, **2**, p. 438

34 McCORMICK, F.B., TOOLEY, F.A.P., CLOONAN, T.J., BRUBAKER, J.L., LENTINE, A.L., HINTERLONG, S.J., and HERRON, M.J.: 'A digital free space photonic switching network demonstration using S-SEEDs'. CLEO 1990, Technical Digest Series, Vol. 7, Post Deadline Paper CPD-1, Pub. Optical Society of America, Washington 1990

35 BARNES, P., ZOUGANELLI, P., RIVERS, A., WHITEHEAD, M., PARRY, G., WOODBRIDGE, K., and ROBERTS, C.: 'A GaAs/AlGaAs MQW modulator using a multilayer stack grown on Si substrate', *Electron. Lett.*, 1989, **25**, p. 995

36 SMITH, D.W., HEALEY, P., and CASSIDY, S.A.: 'Extendible optical interconnection network'. Topical meeting on 'Photonic Switching', Salt Lake City, March 1–3, 1989, Pub. Optical Society of America, Washington, USA

37 COCHRANE, P.: 'Future directions in long haul fibre optic systems', *Br. Telecom Technol. J.*, April 1990, **8**, pp. 5–17

Integrated-Optics Planar Components

Optical fiber architectures of the future are envisioned to employ extensive splitting of optical signals, as well as multiple wavelength division multiplexing and demultiplexing. These architectures and networks have motivated the development and commercialization of today's planar passive components. Beyond and including these applications, the prospects for new and exciting uses of planar devices are excellent.

D. A. Nolan, V. A. Bhagavatula, and C. Lerminiaux

Soon after the demonstration of low-loss transmission in optical fibers, the realization and implementation of long-distance optical fiber links took place. Research into the possibility of manipulating light in integrated glass components was at that time already under way in a number of laboratories. However, applications for these devices were not yet apparent. Meanwhile, the commercialization of the optical fiber technology, including fiber fabrication, cabling, the development of the needed active components such as solid state lasers, the transmitters, receivers and systems themselves took place at rapid paces. Current fiber-optic networks are deployed predominantly in what is known as the trunk or the feeder distribution. Essentially, they are point-to-point transmission systems and as such do not require passive integrated components. Optical fiber architectures of the future, e.g., passive optical networks (PONs) and all optical networks (AONs), however, are envisioned to employ extensive splitting of optical signals, as well as multiple wavelength division multiplexing and demultiplexing. Planar passive components already are providing such functions in what is known as fiber-in-the-loop (FITL) systems. These applications will be discussed in more detail below, but it is important to realize that it is these architectures and networks that have motivated the development and commercialization of today's planar passive components. Both beyond and including these applications, the prospects for new and exciting uses of planar devices are excellent.

Among the first to report on the fabrication of passive planar waveguides were T. Izawa and H. Nakagome of NTT Corporation, Japan. In 1972 they reported that they had achieved low-loss propagation in an ion-exchanged bulk glass. Until that time, waveguides had been formed by using thin film deposition techniques. Within months, Giallorenzi and others at the U.S. Naval Research Laboratory reported their results on the formation of ion-exchange waveguides in glass. From the extent and magnitude of their work, they obviously had been pursuing this approach for some time. After years of research and development, in the mid-'80s, Corning Incorporated and NSG, Japan, commercialized 1 x N multimode splitters based on an ion exchange technology. Soon thereafter single-mode components were introduced and today 2 x 16 splitters [1] for FITL applications are available.

In another approach, M. Kawachi and others at NTT Corp. reported in 1983 that they successfully fabricated silica-based, single-mode waveguides and directional couplers on silicon substrates. They used a flame hydrolysis process, similar to what is used in the fabrication of optical fiber. A few years later, they showed how one can integrate passive devices on one chip in order to obtain multiple wavelength multiplexers and demultiplexers. PIRI Inc. has commercialized this technology and offers a number of passive single-mode components.

Both of these planar technologies will be discussed in more detail below, along with the emerging applications for these components, including PONs and AONs. Components of immediate interest include M x N splitters, and other integrated passive components. Components under investigation in the laboratory include: Mach-Zehnder multiplexers and demultiplexers, thermo-optic switches, waveguide arrays, planar optical amplifiers, and spatial solitons.

Ion Exchange

The formation of optical waveguides in glass by ion exchange has been explored since the early '70s. Since then, a variety of glass compositions and dopants have been used to realize planar integrated optical components [2, 3].

The ion exchange technology modifies the refractive index of the glass by replacing the alkali ions present in the glass host by other alkali ions. This can be done when heating the substrate at a temperature high enough to let the alkali ions be mobile in the host structure formed by the silicate network.

D. A. NOLAN is senior research associate for Guided Wave Optics at Corning Incorporated.

V. A. BHAGAVATULA is a senior research assocociate in Optoelectronics Research at Corning Incorporated.

C. LERMINIAUX is in charge of the Advanced Optics Group at Corning Europe.

Reprinted with permission from *IEEE Communications Magazine*, Vol. 32, No. 7, pp. 62-67, July 1994.

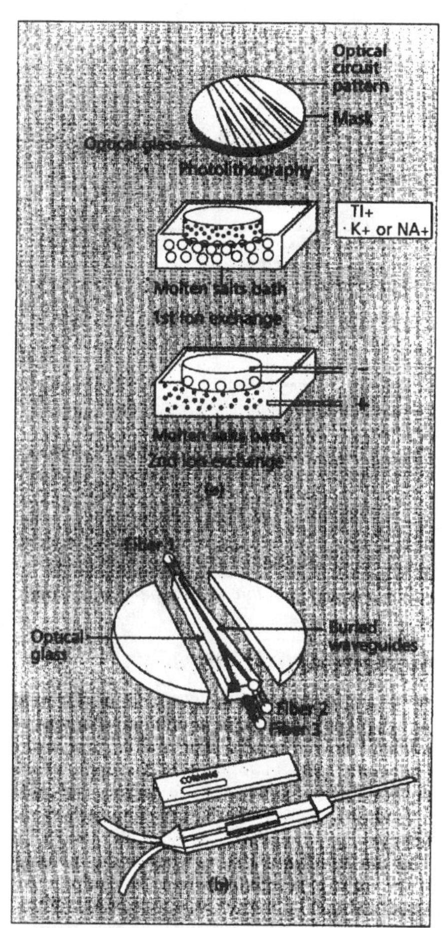

■ **Figure 1.** *Ion exchange process: photolithography, 1st and 2nd ion rxchange, pigtailing, and packaging.*

The choice of the glass substrate and the replacing ion is of great importance. The glass must be of optical quality, i.e., of high purity and homogeneous, to avoid any source of scattering losses. Its index of refraction must be close to that of optical fiber to minimize interconnection losses; it must have low transmission losses (<.1 dB/cm) and finally contain monovalent ions (generally cesium, silver, or potassium). Different companies and universities have developed their own glass (BGG 31 or 35 from IOT, etc.).

The replacing ion must fulfill some conditions:

- Permitting a change of the refractive index high enough to permit light guidance. Silver, thallium, or potassium are the best candidates for this requirement.
- Avoid mechanical stresses that could generate polarization dispersion in the waveguide. Potassium is known to generate much stress.

For these reasons, the commercial devices actually produced are made using thallium (Corning) or silver (IOT).

Nevertheless, in this last case the glass composition has to be carefully designed to avoid any chemical instability of the exchanged ion (reduction in metallic silver).

A typical process for making waveguides (Fig. 1) is as follows:

- Deposition of a metallic mask on the glass wafer and patterning through a lithographic process.
- First ion exchange to modify the refractive index. This usually is done in a molten salt bath, but also can be obtained by dry exchange.
- Removing the metallic mask.
- A second step to bury and round the area where the refractive index has to be modified. This is done either by a second ion exchange step (Corning) or by thermal rediffusion (NSG, IOT).
- Dicing to separate the individual components.
- Pigtailing the input and output fibers. This is made using index-matching adhesives through active or passive alignment. The fibers can be individually micropositioned, to assure the best loss performances, or mass pigtailed, by placing the fibers in blocks.
- Packaging.

Typical loss figures of waveguides obtained in such a way are on the order of 0.3dB fiber-to-fiber, which can be split in 0.1dB/interface for the coupling losses and significantly less than 0.1 dB/cm for propagation losses.

Vapor Deposition/Thin Film Techniques

With their potential for optical integration and mass production, vapor deposition techniques are becoming increasingly important for planar passive component fabrication. Although proposals for the application of this technique date back to the early '70s, significant progress has been made only in the '80s [4]. In this approach, shown schematically in Fig. 2, films of glass (mainly doped silica) are deposited on silica or silicon substrates. Film deposition techniques include flame hydrolysis, plasma-enhanced CVD (PECVD), and sputtering techniques. In the flame hydrolysis process, which is similar to optical fiber blank fabrication, silica or doped silica soot with the required composition is deposited on a flat substrate to the required thickness. Normal dopants include TiO_2, B_2O_3, GeO_2. Next, it is consolidated into clear glass at high temperatures. Unlike flame hydrolysis, the PECVD process deposits a clear film at much lower temperatures (~350° C). With the PECVD process, the higher index core layer can be made by P-doped silica or by silicon oxynitrides (SiON). After depositing a thin film, the optical circuit pattern is transferred from a mask onto the glass film, using lithographic and plasma-etching methods. An advantage of this technique is that the mask pattern sizes transferred are the same size as the waveguides. This one-to-one pattern transfer relaxes the dimensional tolerances on the mask quite significantly. Even with single-mode devices, feature sizes can be as large as 8 to 10 μm, which are well within the capability of the lithography process. This approach is well suited for single-mode technology, and is being studied by NTT, Photonics Integration Research Institute, AT&T, LETI (France), and Corning.

Using this technique, various low-loss single-mode devices have been fabricated. Single-mode waveguides with champion research results of less than 0.1 dB/cm for transmission loss, and

The formation of optical waveguides in glass by ion exchange has been explored since the early '70s.

The trend observed in the recent deployment of optical communications networks is an unquestionable move toward system architectures using fiber-optic splitters or couplers.

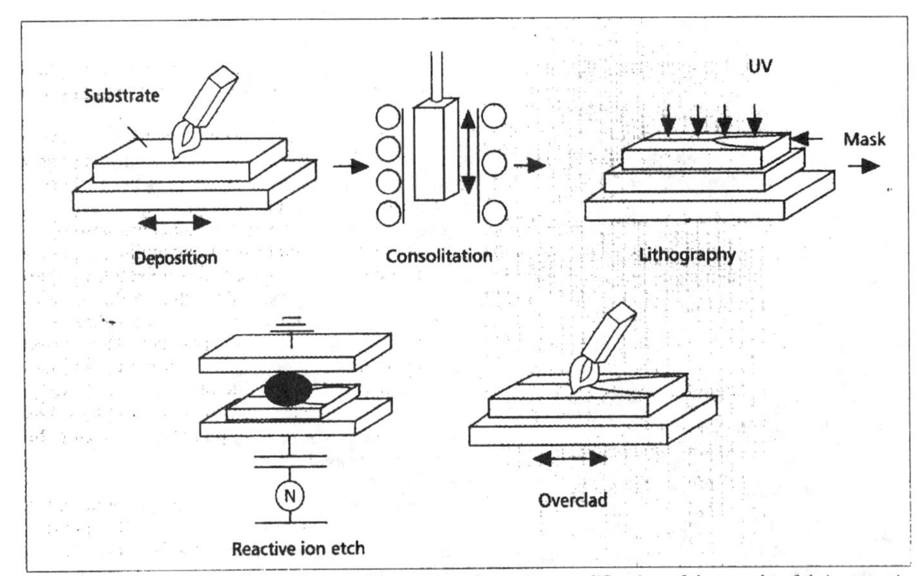

■ **Figure 2.** *Vapor deposition technique. The process shown is a modification of that used to fabricate optical fiber and is known as the flame hydrolysis technique.*

fiber coupling losses of less than 0.05dB/interface, have been achieved by NTT. 1 x *N* and *N* x *N* couplers with excess loss in the range of 1 to 4 dB have been obtained by various groups. In addition, this technology, among others, offers the possibility of integrating a number of passive functions on a single chip, and the possibility of the hybrid integration of both active and passive components on silicon and other active materials. With this approach, passive alignment and multiport pigtailing techniques are being developed for mass production to make low loss, low-cost planar components.

With the possibility of monolithic integration, planar techniques using semiconductors [5] and polymers to fabricate photonic circuits are under investigation. With these techniques, the light-conducting material is a polymer with high light transmission characteristics or a semiconductor like GaAs or InP. The advantage of polymer materials is the simplicity of thin film fabrication and of patterning the optical circuits. However, the attenuation characteristics that can be obtained and the thermal and environmental stability of the devices generally are not as good as glass films. A number of devices, including 1 x *N* and *N* x *N* couplers, have been manufactured by the polymer technique, and a transmission loss of about 1.0 dB/cm has been obtained. This approach is followed mainly by DuPont and NTT. Because of the nonlinear properties of polymer materials, significant future research activity is expected on these devices. With the potential for monolithic opto-electronic integration, much of the work with semiconductor waveguides is limited to active devices, such as modulators, switches, opto-electronic integrated circuits (OEICs), etc. Many of the opto-electronic companies like AT&T, NEC, and Hitachi are actively involved in this area. Semiconductor waveguiding devices have been fabricated with losses as low as 1.0 dB/cm. Future device research will focus on OEICs with very high performance.

System Architectures

The trend observed in the recent deployment of optical communications networks is an unquestionable move toward system architectures using fiber-optic splitters or couplers. To supply a variety of services, the telecommunications and cable TV industries are considering installing single-mode fiber (as near as possible) to the subscriber FTTS. This can be either FTTC or fiber-to-the-building, if the opto-electronic conversion is done in a street cabinet or in the basement of a building. To achieve the access of every home to broadband services on an economic basis, this means sharing the fibers in the feeder part of the local loop and, therefore, an increasing use of passive components.

The main type of architecture being installed now is PON (Fig. 3), where extensive use of 1 x *N*, 2 x *N* splitters and WDMs is made.

The 1 x *N* device will split the signal to many customers; the second input of the 2 x *N* splitters will permit either a multiplexing of different signals or the necessary path redundancy to reach a splitting point; and the WDM will allow a separation of the different signals.

A higher complexity undoubtedly will evolve in future networks.

The fiber bandwidth will be finally used, and dense wavelength division multiplexing will allow the propagation of many channels on the same fiber. The MUNDI RACE project developed by BT, Siemens, GEC, Philipps, DBP, and Corning, partly founded by the EEC, is an example of such evolution. The optimal networks will, therefore, evolve from totally passive toward a more and more all-optical. The next step will be to use the wavelength itself to route the signal as proposed by IBM ("Rainbow") or the EEC ("RACE Project 1036: WTDM Broadband Customer Premises Network").

Optical amplifiers will be needed in the network itself to provide enough power for such evolution. Finally, a pure AON will make use of the light itself to control the signals optically. Solitons will

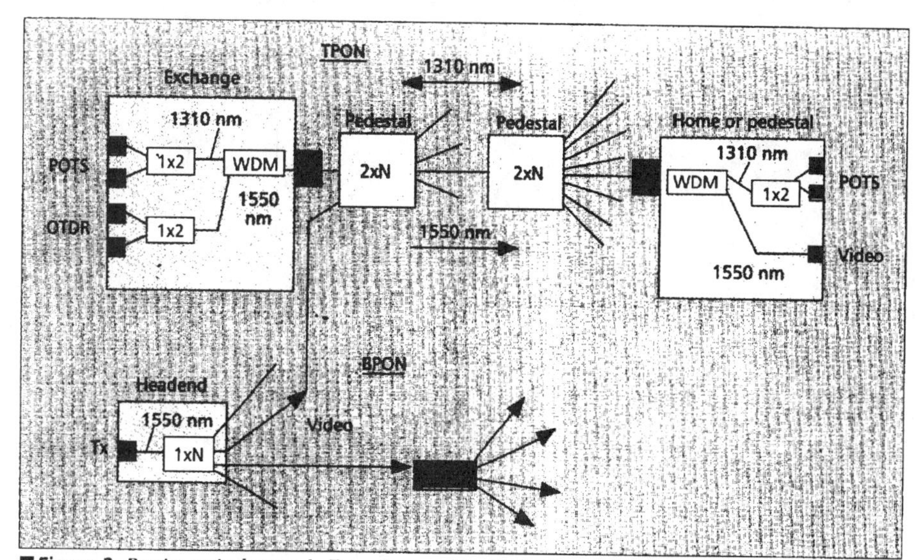

■ **Figure 3.** *Passive optical network. Extensive use of 1* x N, *2* x N, *and WDM passive components.*

be the tool of an optical TDM/WDM technology. Nevertheless, all of these developments will only take place when the components have reached a sufficient level of performance. Reliability is an issue for the tunable components of the AON. Climatically, the FTTS environment is unprotected; operating temperatures range from −40 to +85°C. Flooding has to be considered. Therefore, standardization is needed. For the U.S. market, Bellcore set up a list of "Generic Requirements for Fiber Optic Branching Components" that is mainly used by system installers as a reference [6].

Component Types

Optical communication networks require a variety of passive, active, and hybrid components. With the continual advances in architecture, there is an ever-increasing need for novel components and circuits. One of the basic building blocks of the passive components is the 1 x *N* power splitter shown in Figs. 4(a) and 4(b), which splits a single channel into *N* different output channels. Similarly, this type of coupler can be used to combine *N* inputs into one output. An extension of the 1 x *N* coupler is the *M* x *N* coupler shown in Fig. 4(c). In this case, a signal input on any of the *M* input fibers is split equally to all of the *N* output fibers.

Another important class of devices is the wavelength division multiplexer-demultiplexer, which is used to combine or to separate signals of different wavelengths onto a fiber. Such devices, shown in Fig. 4(d), are used in systems where different wavelength sources are used to transmit over the same fiber, or where one wavelength is used to transmit and another is used to receive. For example, in telecommunications applications, systems are being deployed that transmit both 1310 and 1550 nm signals over the same fiber. Narrowband WDM star networks such as LAMBDANET, proposed by Bellcore, utilize multiple wavelength transmission with separations as small as 2 nm.

Along with power splitting and WDM applications, polarization is an important parameter to fiber systems and ultimately coherent communications. Polar-

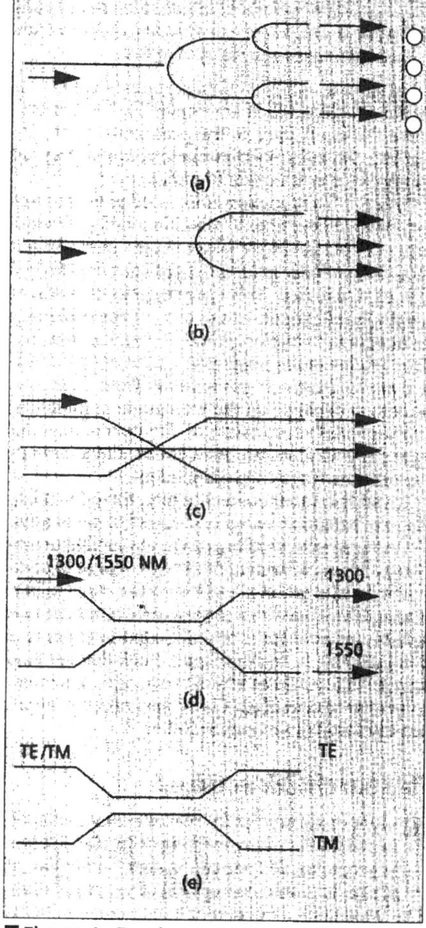

■ **Figure 4.** *Coupler types: a, b) 1* x N *or tree coupler/combine; c)* N x N *coupler; d) wavelength division multiplexer (WDM); and e) polarization splitter.*

A *higher complexity will evolve in future networks. The fiber bandwidth will be used, and dense wavelength division multiplexing will allow the propagation of many channels on the same fiber.*

F*uture opportunities for planar components utilize the integration capability of the planar waveguide processes*

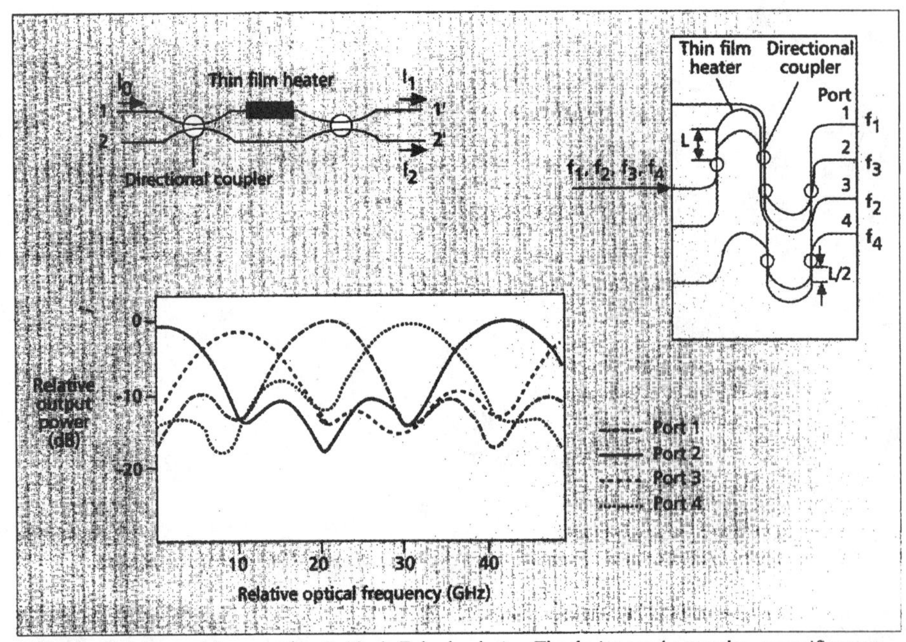

■ **Figure 5.** *Integrated four wavelength Mach-Zehnder device. The device can be tuned to a specific wavelength through a thermo-optic effect.*

ization couplers, which are usually 1 x 2, either preserve the input state of the polarization, or else separate the input polarization into two orthogonal components as shown in Fig. 4(e).

Many of the functions discussed generally are implemented by either Y-junction couplers or proximity couplers. The Y-junction splitter is based on splitting a single guide into two or more additional guides. Generally, this type of coupler is wavelength independent. In the case of the proximity or directional coupler, power is coupled between propagating mode fields and undergoes a periodic transfer from one guide to another. This phenomenon depends on the waveguide parameters, the physical separation of the cores, and the operating wavelength. As a result, these devices typically are wavelength dependent.

In addition to these techniques, these functions also can be implemented in a variety of novel ways. For example, power splitting can be accomplished using planar lenses. Wavelength division can be accomplished using gratings etched on the waveguides. Some of these novel techniques would be much more suitable for multi-function, integrated devices where compact size is an advantage. Such devices also may provide a unique way of coupling light sources to waveguides and fibers. Other types of novel devices are discussed in sections to follow.

Research Directions

Future opportunities for planar components utilize the integration capability of the planar waveguide processes, i.e., one can integrate a number of simple devices onto one optical waveguide circuit. The integration capability of planar components is utilized for applications in the PON architecture. Here, 1 x 2 and 2 x 2 couplers are integrated onto one chip to fabricate 1 x N and 2 x N (N = 4, 8 16, 32) components. Future networks, in particular the AON will use N x N stars, dense wavelength components, and tunable filters. Integrated Mach-Zehnder devices are candidate components for the AON. Fig. 5 shows a schematic of a four-wavelength WDM integrated Mach-Zehnder device. These components have been under investigation for a number of years. In 1988 researchers at NTT's opto-electronics research laboratories in Japan reported in detail on the fabrication of integrated Mach-Zehnder interferometers on silicon substrates. The wavelength selectivity of single Mach-Zehnder devices had at that time been well known, but an important outcome of this research was the successful demonstration of the concatenation of several Mach-Zehnder devices on one circuit in order to fabricate a four-frequency demultiplexer with 10-Ghz spacing (0.08 nm). They also reported that they were able to tune these devices over the range of the four wavelengths by depositing chromium patches on the planar guides to form resistive heaters. The index of the Mach-Zehnder arms change with temperature, and in this way they were able to affect the wavelength selectivity of the devices. Thermo-optic tuning occurs on the order of milliseconds, which is useful for circuit switching, but not packet switching, which occurs on the order of microseconds. Packet switching is fundamental to the AON, and such devices will require an electro-optic tuning mechanism. On the other hand, the ultra-high-speed AON takes advantage of the tremendous speeds of optical phenomena. One can readily envision tuning via nonlinear optic effects in these integrated "passive" components. One disadvantage of the Mach-Zehnder scheme is that the space utilization rapidly increases as the number of channels (and therefore devices) increases. In order to address this issue, NTT has reported on the successful fabrication of arrayed-waveguide gratings. Here, rather than cascading 2 x 2 couplers, a parallel array of guides of different lengths enables one to achieve wavelength interference and therefore selectivity. Using this same technology, Dragone and others at AT&T have fabricated integrated *N* x *N* star couplers and integrated *N* x *N* multiplexers on silicon. The value of *N* is as high as 100 for the star design and as high as 11 for the multiplexers. Their grating designs and input and output optics allow for efficient circuit designs.

The realization of optical amplification in rare-earth doped fibers has initiated studies of rare-earth doping in planar structures. Fiber devices such as optical amplifiers utilize the advantage of long optical paths in order to obtain excellent performance. Planar structures offer the long-term possibility of device integration with amplification. Amplification at 1.06 μms via Nd^{3+} doping and 1.55 μm via Er^{3+} doping has been demonstrated in planar structures. Major efforts are underway at 1.55 μm since it is expected that transmission in the AON will be at 1.55 μm for a number of reasons. An important issue to be resolved is the clustering and interaction of Er^{3+} ions as a function of concentration and composition. Clustering limits the performance of an optical amplifier as it effectively limits the the number of ions available for the amplification process. Nonetheless, progress continues at a significant pace at AT&T, Corning and NTT. Gains approaching 20 dB recently have been demonstrated in erbium-doped glass waveguides.

Ultimately, it is anticipated that the ultra-high-speed AON will be the communication network of the future. In this network, optical phenomena are used to generate and control temporally short soliton pulses. Soliton transmission in fibers, as theoretically predicted by Hasagawa of AT&T, could be commercially used in ultra long distance communication in the next few years [REF?]. Interestingly, the existence of spatial or planar solitons was predicted by Zakharov and Shabat of the former Soviet Union [REF?] before Hasagawa published his theoretical works on the existence of temporal solitons in fiber. In the spatial soliton case, diffraction and self focusing compensate with one another, which results in a beam that can propagate without spreading. Researchers at Bellcore already have demonstrated the existence of spatial bright solitons in glass [9]. Both bright and dark solitons are mathematically predicted to exist. While bright solitons are the result of self focusing as mentioned above, dark solitons are localized depressions and can be considered as the absence of light. Bright solitons will form with intense optical power (hundreds of kilowatts) and a positive nonlinear index. A negative nonlinear index is required for the formation of dark solitons and as such has not been observed in glass but has in fact been demonstrated in gasoline. The prospects of manipulating light with light in a planar device with or without a guide are indeed exciting, but much work needs to be done before such devices can be realized for optical switching.

Ultimately, it is anticipated that the ultra-high-speed AON will be the communication network of the future.

References

[1] A. Beguin *et al.*, "Fabrication and Performance of Low Loss Optical Components Made by Ion Exchange Ion Glass", *J. Lightwave Technol.*, vol. 6, no. 10, Oct. 1988.

[2] T. Findakly, *Opt. Eng.*, vol. 24, p. 244, 1985.

???Y.H. Won, P.C. Jassaud and G.H. Charter, *Appl. Phys. Lett.*, vol. 37, p. 269, 1980.

[3] R. V. Ramaswamy and R. Srivastava, Special IEEE Joint Issue on Integrated Optics, *Journal of Quantum Electronics* and *J. Lightwave Technol.*, vol. 6, p. 984, 1988.

[4] M. Kawachi, "Silica waveguides on silicon and their application to Integrated-optic components," *Opt. Quantum Electron.*, vol. 22, p. 391, 1990.

[5] H. Hayashi, "Long Wavelength Optoelectronic Integrated Circuits," OFC '92, San Jose, Calif., p. 23, 1992.

[6] Bellcore TR-NWT-001209, "Generic Requirements for Fiber Optic Branching Components," Nov. 1, 1992.

[7] N. Takato *et al.*, "Silica-Based Single-Mode Waveguides on Silicon and their Application to Guided-Wave Optical Interferometers", *J. Lightwave Technol.*, vol. 6, 1003-1010, 1988.

[8] C. Dragone, "Efficient *N* x *N* Star Couplers Using Fourier Optics", *J. Lightwave Technol.*, 7, pp. 479-489, 1989.

[9] J. S. Aitchison *et al.*, *Opt. Lett.*, vol. 15, p. 471, 1990.

On the Future of Wavelength Routing Networks

ORI GERSTEL
IBM T.J. WATSON RESEARCH CENTER

Abstract

This article discusses the possible applications for optical networks based on wavelength division multiplexing and how they compete and complement current high-speed networks (SONET, ATM). We first outline the best-case scenario for this technology and describe the spectrum of proposed optical networks (WDM links, passive optical access networks, broadcast-and-select networks, and wavelength routing networks). Then we focus on wavelength routing networks and describe their advantages and disadvantages relative to other competing alternatives for very-high-speed networks. Finally, we analyze the different markets for such networks in the telco and data communications arena.

Wavelength division multiplexing (WDM) is the technology of transmitting multiple data streams independently on a single fiber using different light wavelengths. This technology, in the form of point-to-point multiplexer/demultiplexer systems, has successfully passed its first real-world test, and has moved from laboratories and testbeds into commercially available and deployed systems. It is now time to consider whether this stage is the first step towards the vision of the much researched all-optical networks, or this is the only niche for this technology.

WDM technology clearly has great advantages for the long distance telephone companies, since it enables them to dramatically increase their (currently almost saturated) trunk capacities without going into the painful process of laying more fiber in the ground. As a result, all three large long distance carriers in the United States (AT&T, MCI, and Sprint) have embarked on programs of deploying WDM point-to-point technology on many of their long-haul routes. Similar systems have already been deployed n the enterprise world to connect large data processing centers to their backup sites.

On the other hand, from the research point of view, optical networks based on WDM hold great promises for the future of both tele- and data communication networks, since they currently appear to provide one of the only solutions that overcomes the inherent limitations of electronics and probably the simplest solution in sight to enable high-speed networks to become really high-speed (say, above 10 Gb/s). However, the reasons a given technology becomes successful lie not in its theoretical merits, but in how well it competes with alternative solutions, both potential ones as well as those already in use. Several questions arise in this context: Do we need such high-speed networks at all? Isn't the combination of WDM links and existing high-speed electrical switches (e.g., synchronous optical network, or SONET, and asynchronous transfer mode, or ATM) good enough? When will optical networks be cost-effective enough for commercial use? Will they ever be fast enough concerning setting up connections?

The answer to the first question is clearly in the affirmative: there just never seem to be enough communication resources. We believe that the answers to the other questions are quite favorable for optical networks as well, if not in the immediate future, then in a few years from now. In the rest of this article we attempt to clarify this position.

The article first outlines the optical networking vision: assuming that these networks will become prevalent, how will they change the world of networking? We then present a brief technical overview of the field, and narrow down the discussion to the current commercially viable solutions: WDM links and wavelength routing networks. Next we compare all-optical networking with competing electrical solutions for very-high-speed networking, and outline the potential markets for the technology and the specific needs in each case, after which we conclude the article.

The Vision

The best way to understand what a new technology is all about may well be to put aside all the technical details, all current technological difficulties, and all competing technologies that threaten to nip it in the bud, and to concentrate on the vision (or the dream, as its opponents may call it): Assuming all goes in favor of this technology, how will it change the world?

Looking at the evolution of communication networks and standards, they seem to become more and more complex (and hence less manageable) as time goes by. This trend is due in part to the increase in their sizes and bit rates, but mainly to the diversity of the traffic they carry and services they support. The vision of WDM optical networks offers a change in this course of evolution into much simpler network architectures.

This work was supported in part by grant MDA 972-95-C-0001 from the Defense Advanced Research Projects Agency (DARPA).

Reprinted with permission from *IEEE Network*, Vol. 10, No. 6, pp. 14-20, November/December 1996.

Their transparency, abundance in resources, and passive nature may eliminate the need for sophisticated mechanisms to optimize the utilization, control, and management of integrated networks. The architectural simplicity is achieved through traffic segregation as opposed to the current trend of traffic aggregation.

All that remains is to concentrate on the endpoints, namely, how to design computers that can make use of so much bandwidth. Even more important, what new applications are now enabled by these bit rates?

Let us concentrate on the futuristic scenario depicted in Fig. 1, wherein the fiber infrastructure is extended to the home, and in which we can make efficient use of the thousands of wavelengths that theoretically may be multiplexed into a single fiber. Then our global network would be made of fibers interconnected by optical cross-connects, with optical multiplexers at the endpoints. This entire network may be viewed as a huge, sophisticated piece of glass, almost passive in terms of electrical power. When end user X wishes to communicate with end user Y, X requests the network control entity to establish the connection.[1] The network then assigns a wavelength λ_x to this connection, sets the switches along the path to support it, and informs both X and Y of the existence of this new connection. Here ends the role of the network in the connection, as opposed to conventional networks in which the network takes an active part in the transfer of the data. Now, when user X sends a light encoded signal on wavelength λx, it is optically routed from X to Y, received optically at Y, and converted to the electrical domain, to be processed by Y's application. The endpoints of the connection now have an ultra-high-speed, low-noise pipe between them, equivalent to a private fiber that serves them exclusively. The network, acting as a passive piece of glass, is sensitive neither to the protocol that X and Y choose to use nor to the bit rate. It may even be insensitive to the nature of the data (digital or some analog signal). As a result, there need only be a handful of simple protocols of which the network will be aware (and thus will have to be standardized). The rest is up to the users.

Contrast this scenario with the complexity and extensive monitoring and management required by "traditional" networks, with SONET setting the current record, probably to be eclipsed by broadband integrated services digital network (B-ISDN)/ATM. It is enough to see how much standardization effort is put into other alternatives for high-bandwidth integrated networks to realize the promise that optical networks provide. While ATM standardization bodies, researchers, and implementors struggle with definitions of different traffic classes, congestion control mechanisms, quality of service definitions, and the implications of these features on switch architectures — not to mention policing mechanisms, resource allocation problems, dealing with cell losses, buffer management, and pricing — all these issues become much simpler or trivial when considering optical networks instead.

As is always the case with vision, reality tells a different story, which we outline in the rest of this article.

[1] *Note that this description only fits connection-oriented applications. It also assumes that the number of active connections per end user is rather small. We believe that almost all bandwidth-consuming applications conform with these assumptions. A fixed connection from an end user to a simple, Internet Protocol (IP)-style network will still be necessary to accommodate light-load connectionless traffic with loose performance guarantees.*

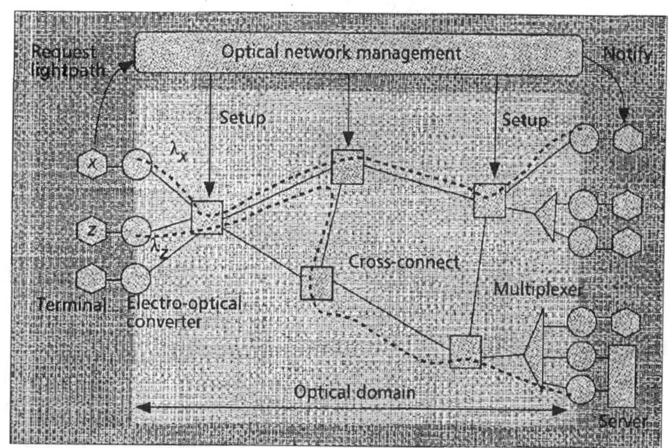

■ Figure 1. *An all-optical network.*

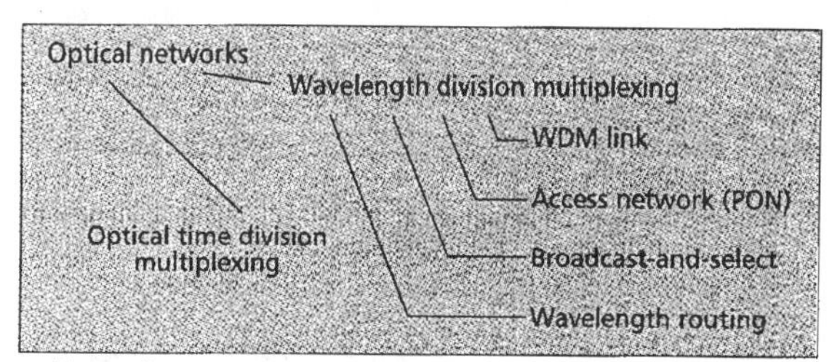

■ Figure 2. *A taxonomy of optical networks.*

The Technology Behind Optical Networks

This section briefly outlines the current architectures of optical networks. It also serves to filter out from the rest of the discussion directions that are beyond the scope of this article. An in-depth discussion of the optical devices used to build these networks, as well as some networking aspects, can be found in [1].

Optical networks can be divided according to the taxonomy tree in Fig. 2. The main distinction between the various network types is based on the multiplexing scheme: whether it is done in the frequency domain (WDM) or the time domain, as in optical time division multiplexing (OTDM). WDM networks may be further split into:

- Point-to-point links — in which both ends of the link have identical equipment to transmit and receive the channels
- Access networks — in which one side of the link gets split among different locations (homes) and requires simpler equipment at the home
- Broadcast-and-select systems—in which the signal is broadcast to multiple endpoints rather than a single endpoint
- More scalable and complex networks, in the form of wavelength routing networks, realized by introducing switching nodes to connect multiple point-to-point links

WDM Links

WDM technology is based on the ability to transmit several light signals on a single fiber using different wavelengths. It turns out that such different light wavelengths do not interfere with each other, and thus can be split apart at the other end of the fiber to form separate channels (Fig. 3 is a schematic drawing of such a system). In this figure only half-duplex channels are depicted. To realize the other direction, a duplicate system is typically used in that direction. In some systems (e.g., IBM 9729 [2]) the very same fiber and grating are used, but on a different set of wavelengths. The WDM link essen-

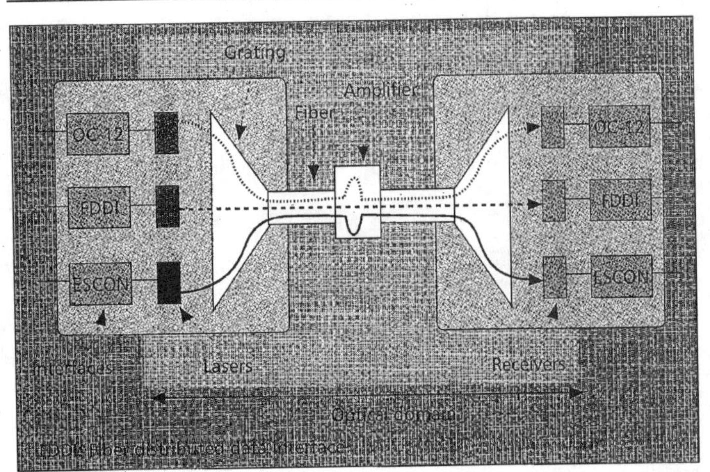

■ Figure 3. *A WDM link.*

tially comprises the following elements (scanning the figure from left to right):

- Different interfaces per port to enable different protocols to communicate over the link.
- An electro-optical converter which includes a laser per channel at different wavelengths.
- An optical multiplexer, typically a piece of glass called a grating.
- Due to attenuation, amplifiers may be needed along the fiber or at the endpoints.
- When the signal gets to the other end of the fiber it is split by an optical demultiplexer, which acts like a prism, to separate wavelength-specific optical signals (the same grating could be used again).
- A wavelength-insensitive receiver converts the signal into electrical form.
- The signal is output via the specific interface of the channel's port.

The pace of improvement for this technology is spectacular: while in 1994 the only commercial product for the telco market multiplexed four OC-48 (2.5 Gb/s) channels to a distance of 550 km, the high-end multiplexer in 1995 has multiplexed 8 such channels, the current record is 16 channels to 600 km, and a 32-channel system to 1200 km has been announced and should be available next year. As for the enterprise market, a multiprotocol lower-speed and shorter-distance system has existed since 1994. As mentioned earlier, such systems are already deployed and are rapidly gaining popularity due to their maturity, the large gain they offer, and the simplicity of integration with legacy equipment.

Broadcast-and-Select Networks

Broadcast-and-select networks are based on a passive star coupler device connected to several stations in a star topology [3, 4]. This device is a piece of glass that splits the signal it receives on any of its ports to all the ports. As a result it offers an optical equivalent of radio systems: each transmitter broadcasts its signal on a different wavelength, and the receivers can tune to receive the desired signal (see Fig. 4 for a schematic drawing of such a system). The main networking challenge in such networks pertains to the coordination of a pair of stations in order to agree and tune their systems to transmit and receive on the same wavelength [5]. One design issue that must be determined before deciding on these protocols is the tunable part of the system. It is possible to either have the transmitters each fixed on a different wavelength and have tunable receivers, have fixed receivers and tunable transmitters, or have tuning abilities in both components. It has been shown in [5] that it is more advantageous to have tunable receivers and fixed transmitters than the other way around. The advantage of these networks is in their simplicity and natural multicasting capability. However, they have severe limitations since they do not enable reuse of wavelengths and are thus not scalable beyond the number of supported wavelengths. Another factor that hinders the scalability of this solution and disables it from spanning long distances is the splitting of the transmitted energy to all the ports. For these reasons the main application for broadcast-and-select is high-speed local and metropolitan area networks. However, the relatively high costs of WDM transmitters and receivers compared to the low costs (less than $1000/port) of other technologies (e.g., ATM and switched Ethernet) do not enable broadcast-and-select networks to be competitive in this arena currently. The few niches that appear to be appropriate for such networks are in broadcast studios and supercomputer centers. Due to these reasons we will ignore broadcast-and-select networks for the rest of the discussion.

Wavelength Routing Networks

A scalable optical network can be constructed by taking several WDM links and connecting them at a node by a switching subsystem. Using such nodes (also called wavelength routers) interconnected by fibers, diverse networks with complex and large topologies may be devised [6, 7]. Each wavelength router makes its routing decision based on the input port and wavelength of a connection going through it. Thus, if a light signal of λ_1 enters a router at a port x it is switched to some output port y. At the other end of the fiber, attached to y, the signal enters another router in which a similar routing decision is made. This process continues until the signal is switched to an output port of the system (Fig. 1). Another optical signal coming into the same router on a different λ_2 will be routed differently. Such an end-to-end connection is called a lightpath, and it provides a high-

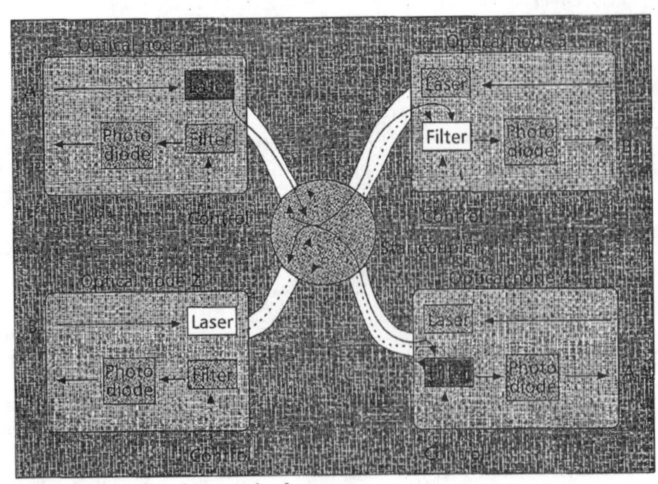

■ Figure 4. *A broadcast-and-select system.*

speed transparent pipe to its end users. At the same time, another lightpath can reuse the same wavelength in some other part of the network, as long as both lightpaths do not use it on the same fiber. Since such "spatial reuse" of wavelengths is supported by wavelength routing networks, they are much more scalable than broadcast-and-select networks. Another important characteristic which enables these networks to span long distances is that the energy invested in a lightpath is not split to irrelevant destinations. There is a large diversity of capabilities that a wavelength router can provide, depending on the components in use and design of the node. Most notably, nodes may provide configurable lightpaths versus fixed routing, full wavelength conversion versus limited conversion versus no conversion at all, fault tolerance in the optical layer versus reliance on higher layers. Nodes may also vary in their scalability to increasing numbers of local or network ports. As for the design of the node itself, current commercial technology enables either of the first two of the following designs. The third design relies on large optical switches and wavelength converters, a technology far from commercially available and therefore a longer-term option.

Electro-Optical Node — Converts the optical signal into the electrical domain, performs the switching in this domain, and regenerates the optical signal at the outputs (Fig. 5a). This design easily enables wavelength conversion and maintains a high-quality signal for multiple hops. On the other hand, it does not support transparency. This design will not be discussed in what follows because it only represents an evolutionary phase toward all-optical networks.

Simple All-Optical Node — Separates the different wavelengths from each input and sends all channels of λ_i to the same switch, which optically switches them to the output ports (Fig. 5b). This design does not allow wavelength conversion, thereby restricting the reuse of wavelengths in the system. This may prove to be a cost-effective solution because it does not require a (costly) transceiver per channel per node.

Full-Conversion All-Optical Node — Enables each wavelength to be converted to any other wavelength. It is based on a large optical switch which takes a channel and switches it to any other channel (on any fiber). Before being multiplexed into the fiber, each channel is converted to the appropriate wavelength by fixed wavelength converters (Fig. 5c).

Other Architectures

Passive optical (access) networks (PONs) enable bidirectional communication between a server (such as a cable TV provider) and a set of customers [8, 9]. The main challenge in these networks is to design such a system based on WDM technology, in which the equipment at the customers' side is as cheap, simple, and durable as possible, but above all, identical for all customers (or else it will create impossible management overheads, since it will be necessary to ensure that no two customers have identical wavelength-specific devices). PONs are thoroughly discussed in another article in this issue [10] and are therefore beyond the scope of the current article. Optical TDM attempts to copy conventional TDM ideas and realize them optically, thereby achieving much higher speeds [11, 12]. Since this technology is in its very first steps, where the most elementary functions (such as realizing all-optical buffers and synchronization) have only been conceptually proven, we estimate that it will not be a potential player in the commercial arena in the next 10 years. For this reason we shall ignore it in the rest of the article.

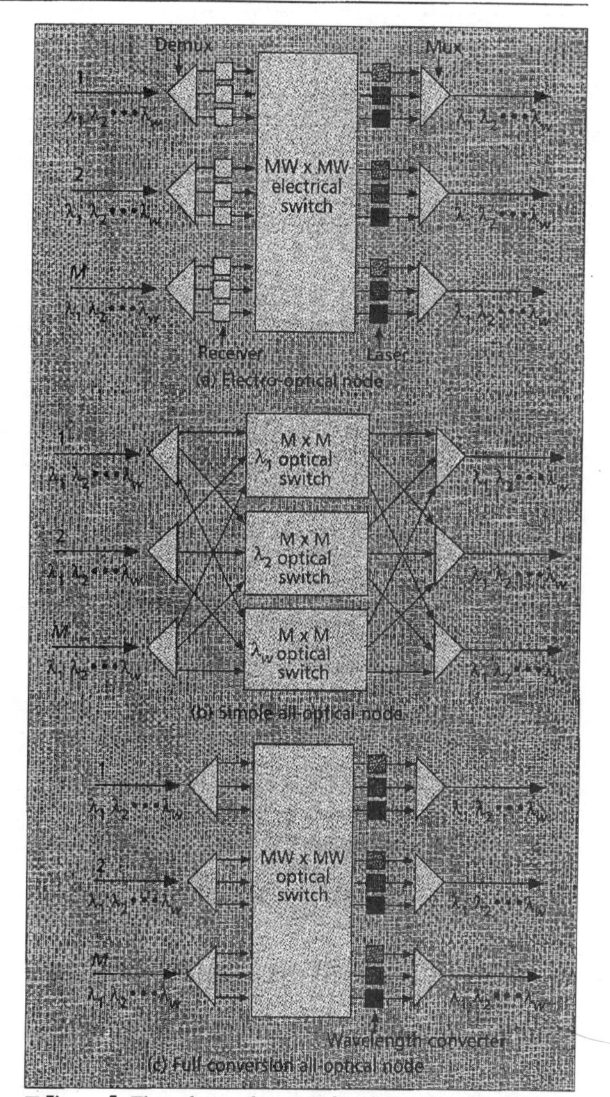

■ Figure 5. *Three designs for a wavelength routing node.*

Competing Technologies

So far we have discussed the technical properties of different optical network architectures and focused the discussion on WDM links and wavelength routing networks. In this section we compare these networks to other high-speed networking solutions which are based on electrical switching. The electrical networking example considered for the purposes of this comparison is SONET/SDH (synchronous digital hierarchy). The amount of processing required for ATM cell switching does not currently enable it to compete as a very-high-speed alternative (2.5 Gb/s and above).

The following solutions are the main alternatives to very high speed networking (see Fig. 6 for a graphical demonstration).

Very-High-Speed TDM Links and Switches

One obvious solution is to push current high-speed TDM networks to even higher speeds. This solution works for fairly high speeds: current SONET links operate at 2.5 Gb/s (OC-48), and could be pushed up to 10 Gb/s (OC-192). However, the technology seems to already be approaching its limits, as dictated by the maximum speed of current electronics; even the current OC-48 nodes are based on expensive gallium arsenide technology, and require very careful nonstandard design. Optical transmitters and receivers are currently limited to 10 Gb/s speeds. Furthermore, in some cases where old fiber is installed in the ground, polarization mode dispersion limits the bit rate to less than 10 Gb/s for long distances.

Parallel Fibers, Parallel Electrical Switches

Another straightforward solution is to use parallel fibers between sites connected by lower-speed electrical switches. This is indeed a good idea in places with a rich optical infrastructure. However, for long distances this is still a costly solution, since each of these fibers requires its own set of optical amplifiers every 80–120 km. Such equipment constitutes a large portion of the fiber cost, needs to be managed, and considerably complicates the system. This solution is not very scalable, as additional parallel fibers will result in linear increases in the cost. In places where there is not enough fiber, the high costs of laying more fiber in the ground and legal complexities involved in getting the "right-of-way" permission from land owners to install it are the major disadvantages of this solution.

WDM Links, Parallel Electrical Switches

Here, the parallel fibers of the previous solution are replaced by separate channels of a single WDM link. These channels are interconnected by lower-speed existing TDM equipment. An important cost advantage of this solution over the previous one is that there is no need for an amplifier per channel, and all the wavelengths are amplified together by a single optical amplifier. A central advantage of combining optical transmission with electrical switching over all-optical networks is that it is based on today's technology, and is thus cheaper, more reliable, and more flexible.

All-Optical Networks

Wavelength routing networks allow the setup of lightpaths which remain in the optical domain across the network. Thus, they enable the creation of configurable higher-level topologies based on traffic analysis, and painless reconfiguration as traffic demands change. In any case, it is clear that they offer an almost unlimited upgrade path for the future, which is not the case with the other solutions. The latter all-optical solution has the following advantages over all the other solutions:

Transparency — Since no electrical processing is involved, wavelength routing networks are not aware of the structure of the data, and can carry diverse protocols and bit coding structures. Electrical solutions carry a single form of traffic and require costly conversion devices from other protocols to the supported standard, which also complicate the management of the network. Another type of transparency supported by wavelength routing networks (although to a lesser extent) is bit rate transparency. Such networks will carry quite a large spectrum of bit rates, up to a maximum rate determined by the design point of the system.

Future-Proofness — A corollary of the above is that all-optical networks will carry most future protocols at many different bit rates without having to replace components of the network. Thus, the investment in this technology is protected against future developments (insofar as anything can be protected against the future).

Reduced Processing — Electrical solutions involve considerably more processing than their all-optical counterparts, a fact

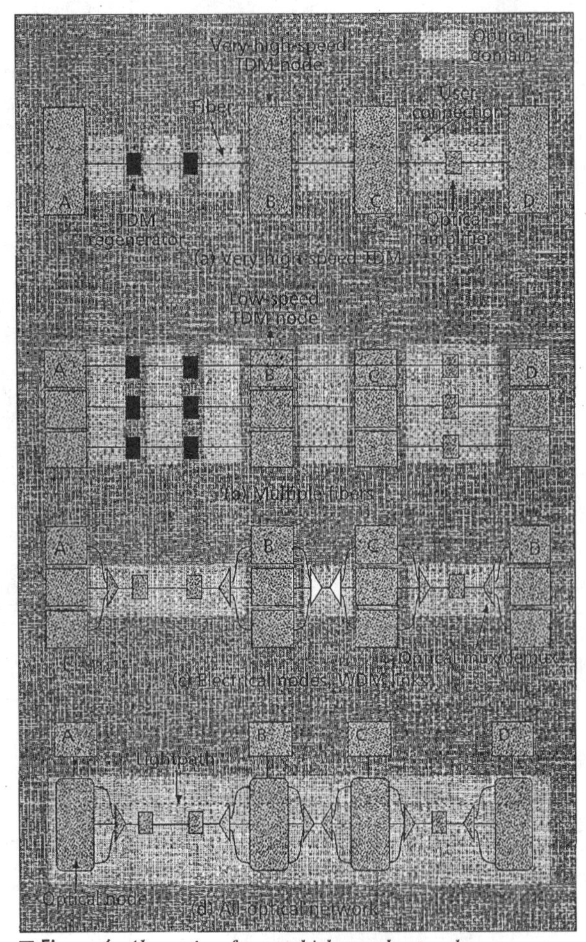

■ Figure 6. *Alternatives for very-high-speed networks.*

which implies more hardware or more expensive hardware. Consider a bit in a connection between nodes A and D in Fig. 6a–c. This bit is converted to the electrical domain, reclocked and processed[2] by each and every node on the path (nodes B and C in this example). In Fig. 6d, however, there is a lightpath from node A to node D, and our bit remains in the optical domain at nodes B and C. Thus, the electrical switches of nodes B and C are not bothered by it and can be made smaller, and less costly.

Reduced Management — Whenever a bit is interpreted, an error may occur. In turn, this event must be detected and reported (especially in the telco world, where network management is much less oblivious to such events). Thus, if bits are interpreted only at the border of a network, much less quality-of-service-related management is necessary (fault management is, however, still necessary).[3]

---

[2] *In the case of ATM this processing is quite complex and includes looking up a routing table, and replacing the routing label in each cell with a new label, but even in the case of SONET the amount of processing is nontrivial, involving checking for errors in the start of header (SOH), recalculating a new SOH, and adjusting the pointers.*

[3] *If more quality-of-service monitoring is necessary on a link by link basis (as opposed to the end-to-end monitoring assumed here), it can still be achieved by monitoring a special supervisory channel. This still implies one monitored channel per fiber, instead of monitoring all the channels.*

The main disadvantages of wavelength routing networks are the following.

Immaturity — At this stage, optical components are not yet mature. Some of them are technically mature — for example, distributed feedback (DFB) lasers but suffer from commercial immaturity as indicated by their high cost, resulting from the lack of mass production (most of them are handmade by order and have to be individually tuned). Other components are technically immature (e.g., optical switches), as indicated by their large physical dimensions and their less reliable nature. However, it seems that this is not an inherent problem and will be resolved in the near future.

System Design Problems — Many design issues for wavelength routing network systems are not yet fully understood and solved. Examples of such problems are the wavelength allocation problem and the dynamic gain equalization problem. There has been quite extensive research on wavelength allocation, but the problem is far from being resolved efficiently, even for simple network topologies. This fact limits the scalability of the network, especially if the number of wavelengths per fiber is low.

A much more severe obstacle for having scalable wavelength routing networks is the physical layer design, particularly variation in the signal quality of individual lightpaths, which is very hard to control. Since optical amplifiers do not amplify all wavelengths by the same amount, and some lightpaths travel many hops while others travel a single hop, the energy of some lightpaths may be very low at their destination, while others have high energy. Thus, it is necessary to equalize the gain, for example, by having adaptive filters or transmitting different energy levels depending on the route of lightpaths. To further complicate the picture, the network has to react to sudden changes in the configuration of lightpaths (e.g., due to link failures). This dynamic gain equalization problem is very complex and far from being well understood or solved.

Markets for the Technology

After understanding the differences between possible solutions for very-high-speed transport, we explore the different markets that this technology may have in the near future and how it fits into the existing frameworks.

Interexchange Market

As mentioned in the introduction, long-haul telephone companies have been the most prominent sector for WDM links so far. The only major market where WDM does not seem to have much of a future is Japan, because of its predominantly installed base of dispersion-shifted fiber which does not support WDM (this fiber is, in fact, very good for conventional high-bit-rate optics and OTDM).

The reason this market is ripe for such a new technology stems from the rapidly shrinking pools of free capacity in existing fibers in the ground on one hand, and from the above-mentioned very high costs of installing long distance fiber on the other. In fact, these costs are so high that they make the relatively high prices of current WDM multiplexers seem reasonable.

Despite the requirement for WDM point-to-point solutions, this market may not need all-optical networks in the near future. The specific features offered by such networks, such as configurability and transparency, seem less attractive here, where single-hop OC-48 SONET lightpaths are all that is necessary at this stage. In the long run, if SONET networks are to be replaced by wavelength routing networks (as may be the case for regional communication providers, see discussion below), a similar need may evolve in this market as well.

Enterprise Market

The main drive for WDM links in the data communications environment has come so far from large financial institutions that wish to protect their valuable data by duplicating it at a geographically remote backup site.[4] The need for wavelength routing networks in such applications is obvious if more than one primary and backup site exist in a large corporation.

The focus of such corporate networks is very different from that of long-haul carriers. While the latter are more interested in high aggregate capacity at one protocol (SONET), the former are more interested in having a larger number of channels, each operating at lower speeds (typically less than 1 Gb/s). Transparency is also a very important issue in this case, since the spectrum of protocols used between the sites is large (fiber distributed data interface, or FDDI, ESCON, fiber channel, ATM, and others). Monitoring and fault localization are central issues for the telcos, while enterprise networks typically have much less stringent requirements.

Another difference between these markets lies in fault tolerance. While SONET networks provide their own backup mechanisms, and thus do not need the optical layer below them to perform fault recovery (which can cause more havoc if not very carefully integrated), such fault tolerance is crucial in the data center backup case, where no such fault tolerance exists. Furthermore, since the telcos are heavily invested in legacy SONET equipment, it will be harder for them to integrate new optical layer fault tolerance into their systems.

In the long run, we expect wavelength routing networks in the data communications sector to provide a low-level, transparent, and configurable infrastructure for more specific technologies, mainly ATM and Transmission Control Protocol (TCP)/IP. Such a layered approach is not redundant, as the low-level optical layer and high-level electrical network play different roles. The main goal of the optical layer is to relieve high-layer nodes from the above-mentioned extra processing by providing high-capacity pipes of fixed bit rate that connect physically remote switching nodes. The goal of the electrical layer is to make efficient use of these pipes by statistically multiplexing lower-bandwidth bitstreams with complex behavior (such as the ATM variable bit rate class) onto them. These two types of connections also operate on different time scales. While ATM virtual connections (or TCP connections), not to mention datagrams, may have short life spans (from seconds down to milliseconds), lightpaths will typically operate on much longer time scales of hours or days, trying to adapt the network to changes in its usage pattern. Therefore, it is sufficient to have low-speed optical switching and configuration management.

A crucial factor in the penetration of optical technology into this sector is its cost. Current components, in particular lasers, receivers, multiplexers, and switches, are just too expensive to justify their deployment in this very competitive market. Therefore, the major research effort which will determine the amount of success of optical networking for data communication is that of reducing the cost of its components.

Regional Communication Providers

This sector may be divided into three subsectors:

- Traditional telephony service (local exchange)
- High-speed pipes for private business-oriented networks

[4] *The 1993 World Trade Center bombing has been one of the triggers for the interest of this community in disaster-recovery via WDM.*

• High-speed access networks to support high-definition television (HDTV), interactive TV, Internet access, and so on.

Regional networks combine requirements from both the long-haul providers and data processing centers. On one hand, such companies aggregate numerous telephone sources into high-bandwidth SONET rings to connect their central offices. On the other hand, they supply high-speed data channels to businesses, which need to run diverse protocol suites on top of them. In fact, while small enterprises may find the installation of a WDM link for disaster recovery purposes too dear, economy-of-scale considerations enable regional providers to offer affordable high-bandwidth pipes to many small organizations. As is the case for the enterprise world, prices are an important issue here, because the providers will be very cautious in investing in such a non-core business.

In the short term, this sector is expected to take the same upgrade path as the long-haul carriers, replacing simple fiber links with WDM links. However, since much of the network complexity lies within the realm of regional providers, they are expected to be the first to suffer from inflating network management and maintenance burdens due to the increasing quantities of SONET equipment required to connect all these channels. Thus, in the long run they may pioneer the effort of replacing SONET-based networks with wavelength routing networks.[5]

A different market which falls under this category is that of providing high-speed access to the home. For such systems optical access networks (PONs) may be an economical solution, which can integrate very well with wavelength routing networks.

Concluding Remarks

Watching the proliferation of wavelength routing network testbeds, it is tempting to conclude that such optical networks are almost ready to become deployed in the near future. On the other hand, while testbeds prove a concept, they are far from robust, reliable, and cost-effective products acceptable to telcos. Taking into account the great efforts that went into the SONET standardization process, the huge recent investments in SONET equipment, and the currently modest demand for very high bandwidth, it seems almost hopeless to expect the telcos to replace SONET with a new and much less mature technology.

Our prediction is that wavelength routing networks will become a commercially viable solution two to four years from now, which will very gradually replace high-speed SONET networking as such, while lower-speed (155 Mb/s up to 2.5 Gb/s) SONET and ATM will be deployed directly on top of this optical layer. We believe that the way to practical optical TDM networking is much longer, and even when such networks become practical it will be quite hard to justify their deployment over the (by then established) competition from the above-mentioned integration between lower-speed electronics and wavelength routing networks.

In the near future (in the next five years), we expect to see much research and development activity in wavelength routing networks, resulting in reasonable solutions to the main open problems (such as dynamic gain equalization and wavelength allocation), more reliable and cost-effective components, and complex optical network experiments and testbeds. Meanwhile, as far as the "real world" is concerned, it is only reasonable to expect extensive deployment of WDM links between existing electrical nodes. During this period, "almost-all-optical" solutions based on electro-optical, high-granularity cross-connects will be introduced. Such networks will not replace SONET, but instead will provide virtual point-to-point links to SONET, leaving most of the networking mechanisms (such as fault recovery) to the SONET layer.

And then again, we may be wrong.

Acknowledgment

I would like to thank Rajiv Ramaswami and Paul Green for teaching me most of what I know in this exciting field. I would also like to thank the referees for their detailed and valuable comments.

References

[1] P. E. Green, *Fiber-Optic Networks*, Upper Saddle River, NJ: Prentice Hall, 1992.

[2] F. J. Janniello *et al.*, "Multi-Protocol Optical Fiber Multiplexer for Remote Computer Interconnection," *OFC '95 Tech. Digest*, 1995.

[3] M. S. Goodman *et al.*, "The LAMBDANET Multiwavelength Network: Architecture, Applications and Demonstrations," *IEEE JSAC*, vol. 8, no. 6, Aug. 1990, pp. 995–1004.

[4] F. J. Janniello, R. Ramaswami, and D. G. Steinberg, "A Prototype Circuit-Switched Multiwavelength Optical Metropolitan-Area Network," *IEEE/OSA J. Lightwave Tech.*, vol. 11, May/June 1993, pp. 777–82.

[5] M. S. Chen, N. R. Dono, and R. Ramaswami, "A Media-Access Protocol for Packet-Switched Wavelength-Division Metropolitan Area Networks," *IEEE JSAC*, vol. 8, no. 6, Aug. 1990, pp. 1048–57.

[6] H. J. Westlake *et al.*, "Reconfigurable Wavelength Routed Optical Networks: A Field Demonstration, *Proc. Euro. Conf. Opt. Commun.*, 1991, pp. 753–56.

[7] G. K. Chang et al., "Experimental Demonstration of a Reconfigurable WDM/ATM/SONET Multiwavelength Network Testbed," *OFC '94 Tech. Digest*, 1994, Postdeadline paper PD9.

[8] N. J. Frigo *et al.*, "A Wavelength-Division-Multiplexed Passive Optical Network with Cost-Shared Components," *IEEE Photon. Tech. Lett.*, vol. 6, no. 11, 1994, pp. 1365–67.

[9] P. P. Iannone *et al.*, "A WDM PON Architecture with Bidirectional Optical Spectral Slicing," *OFC '95 Tech. Digest*, 1995, Paper TuK2.

[10] N. J. Frigo, "Local access optical networks," *IEEE Network Magazine*, Nov. 1996.

[11] D. M. Spirit, A. D. Ellis, and P. E. Barnsley, "Optical Time Division Multiplexing: Systems and Networks," *IEEE Commun. Mag.*, Dec. 1994, pp. 56–62.

[12] R. S. Tucker, Photonic Packet Switching, *OFC '95 Tutorial Book*, 1995, pp. 3–30.

[5] *Note that such a change pertains only to the backbone networks. SONET will still dominate the arena, but in the lower multiplexing levels of 622 Mb/s to 2.5 Gb/s. In fact, even in places where the actual SONET nodes will be replaced by all-optical nodes, the new nodes will probably still carry SONET frames.*

Optical Networking Update

PAUL E. GREEN, JR.

(Invited Paper)

***Abstract*—This paper presents an inventory of the current state and future prospects for networks in which signal paths between end user nodes remain entirely in optical form without intervening electronic conversions. The emphasis is on wavelength-division multiplexing (WDM). The applications and system aspects are stressed relative to details of the supporting technologies. The case to be made for optical networks as the basis of an entire possible future generation of networking is examined, after which the various architectural choices are discussed. Next, the limits on what can be achieved, mostly arising from limitations within the available technology, are treated. After a review of the history of all-optical networking, the paper concludes with speculations about where the applications, the technology and the architectural character of these systems will be going in the years ahead.**

I. Introduction

THE PURPOSE of this paper is to provide a "level set" about purely optical networks, or all-optical networks (AONs) as they are sometimes called [1]–[5]: what they are good for, how far we have gotten with them, and how far we have yet to go. The emphasis will be on the system aspects and on the applications, with secondary emphasis on the technology required to build them. The presentation will attempt to give equal weight to serving the needs of end users (usually at a computer or workstation) and improving the communication providers' infrastructure.

All-optical networks are those in which the path between the using nodes at the ends remains entirely optical from end to end. Such paths are termed *lightpaths*. Each lightpath may be optically amplified or have its wavelength altered along the way, but it is a purely optical path. As the optoelectronic technology to build optical networks has gotten closer to functional and economic feasibility, more and more groups worldwide are studying them as a possible base upon which to build the networks of the future, both within the wide-area backbone and for metropolitan and local area distribution facilities. In view of the potential and recent advances in optical networks, people are beginning to consider that they are a particularly promising candidate for the generation of networking to follow the emerging ATM generation, as sketched in the upper part of Fig. 1 [6].

II. Why Optical Networking?

The motivations for thinking that optical networks might be a particularly promising follow-on arise from the abilities

Manuscript received September 15, 1995; revised November 29, 1995.

The author is with the IBM T. J. Watson Research Center, Yorktowne Heights, NY 10598 USA.

Publisher Item Identifier S 0733-8716(96)03668-2.

197X	198X	199X	200X
Host-centric	Router-centric	Switch-centric	All-optical?
SNA, DNA	TCP/IP	ATM	This issue

Command line	Graphical user interface	World Wide Web, Java, ...

Fig. 1. Recent and projected history. Top: Four networking generations. Bottom: Three generations of the diffusion of computee power to end users.

of fiber to satisy the growing demand for 1) bandwidth per user, 2) protocol transparency, 3) higher path reliability, and 4) simplified operation and management. In all these respects, there is evidence that classical time division (TDM) approaches realized in electronic circuitry of ever-increasing speed, variety and complexity are slowly beginning to prove insufficient. By insufficient is meant either that the older technology cannot do the required job as cheaply or that it cannot do it at all and the optical approach can. We begin this overview by discussing the four benefits to be expected from optical networking.

A. The Growing Demand for Bandwidth

The ever-growing need for more bandwidth per user has historically taken many forms, but it is timely to focus on a particular new example that is likely to have deep and permanent significance. Per user bandwidth demand for the World Wide Web mode of PC usage has suddenly undergone a marked acceleration to a factor of eight per year [7] as huge numbers of users are flocking to it. The Web, or in more fundamental terms, "point-and-click" access to objects independent of location, seems to be best interpreted as a generalization of the two earlier generations of diffusion of computer usage into society at large. The three stages are represented as the bottom row of Fig. 1.

The annual factor of eight growth in bit rate required *by each user* is a faster growth rate than that of almost any other known measure of either user demand or technology performance. This set of data is even more daunting when one appreciates that most Web users encounter response times measured in tens of seconds or minutes rather than the milliseconds or seconds they really want. If sessions are to get shorter, the peak bit rate per user will have to increase.

It is difficult to predict where the next innovation will take us in bandwidth per user. For example, the recent introduction of the Java language has suddenly introduced the prospect of huge further increases in the file sizes downloaded from any

Reprinted with permission from *IEEE Journal on Selected Areas in Communications,* Vol. 14, No. 5, pp. 764-779, June 1996.

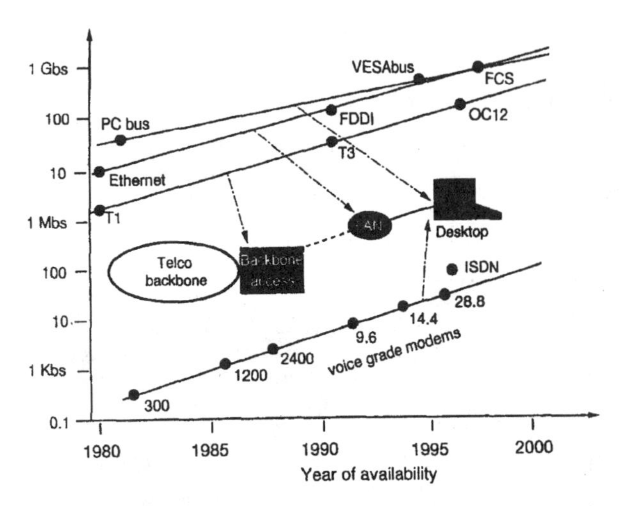

Fig. 2. Evolution of bit rate capabilities of transmission technologies within the local environment and between the local and long haul environments.

server (which can be any node on the network), since "files" now means not only data objects, but all necessary application code to run on these data objects. Some of the visionaries seeking to understand the implications of Java envision downloads replacing local accesses to rotating storage, including those to CD/ROMs.

That a business-as-usual answer to these challenges will be insufficient is shown equally dramatically [8] by Fig. 2, which plots the bit rate performance history of the physical level network technologies available. This figure is disturbing in two ways. The first is the mismatch between the available 1.5 per year increase of available electronic TDM technology and the required factor of eight, and the second is the infamous "last mile" deficiency of a factor of 10^4 in performance between the two high speed environments, local and backbone access. For those fortunate enough to work for large organizations with 10 Mb/s LANs and T-carrier access to the telco backbone, the Web is an entirely pleasurable experience. However, for most of us who want audio and video with our text, the voice grade modem (or the only slightly better ISDN) is a serious roadblock to full participation.

It seems clear that there is only one technology that offers any hope of breaking these bottlenecks on a sustained basis: optical fiber. Copper of any form, even coax, clearly offers only temporary relief if bandwidth per user and number of users per termination (household or office) are to continue to grow indefinitely. Radio, the other enduring technology of the future, has insufficient bandwidth, even as cell sizes and spot beams become quite small. Radio and fiber appear to be dual technologies: radio can go essentially anywhere but cannot do much when it gets there; fiber has severely limited pervasiveness but brings almost unlimited potential to points that it reaches. The total bandwidth of radio on the planet Earth is 25 GHz, whereas each fiber has 25 000 GHz. These figures represent the total accessible bandwidth, not an achievable instantaneous rate for an individual channel; obviously, no individual optical transmission will run at 25 000 GHz, any more than an individual RF system will use all 25 GHz of spectrum.

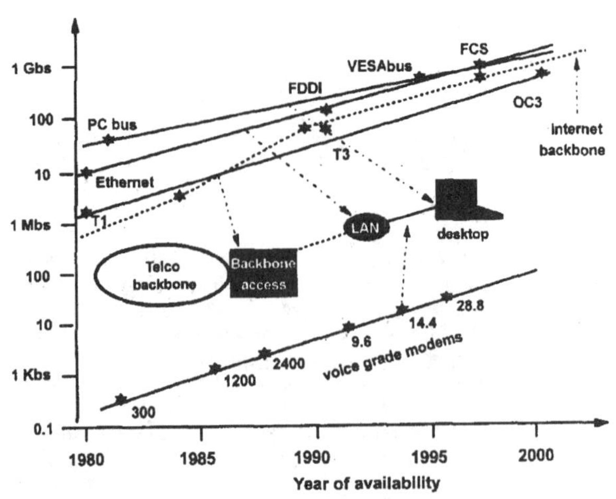

Fig. 3. Evolution of fiber proximity to the best-served 10% of homes and businesses (from [10]). FBB = fiber backbone, DLC = digital loop carrier, FSA = fiber serving area, A/HFC = active (EDFA) hybrid fiber/coax, P/HFC = passive (no EDFA) hybrid fiber/coax, FTTC = fiber to the curb, and FTTH = fiber to the home.

Fortunately, fiber is gradually becoming pervasive. In the United States alone, the installation rate has been 4000 strand miles per day for many years, so that a total installed base of over 10 million strand miles has now accumulated [9]. As Fig. 3 shows [10], the interconnection of telco central offices and cable TV trunk distribution has been more or less completed, and fiber has now started reaching outward toward us users. While bare fiber to the premises is readily available to large businesses today on a rental basis in most U.S. cities, for most users the picture is not so favorable. If one examines fiber availability to the public at large, as in Fig. 3, it is seen that this option is so costly that it is estimated that the most fortunate ten percent of homes and businesses will be reached by their own individual fibers no earlier than 2005 in the U.S. Other countries, notably Japan [11], are thinking of accelerating this process by the heavy use of public funds.

The competition from coax for the last few hundred feet (e.g., as part of hybrid fiber coax sytems) will persist until replacement of coax with fiber becomes cheaper or more in line with premises bandwidth requirements. Furthermore, even if fiber were to reach most homes today, it would probably be used in TDM mode until the demand and the technology costs would favor WDM. Note from Fig. 3 that there is a time window within which optical networking has an opportunity to prove itself to be the preferred option in cost and performance (e.g., with WDM set-top boxes and WDM feature cards for PCs) before fiber to the premises becomes the norm.

B. Error Performance

Unlike radio or copper, fiber has the intrinsic capability to drive the error rate to arbitrarily low levels using link budget improvements alone, instead of the more familiar methods of source and channel coding, equalization, exotic modulation formats, and so forth. With optical networks it is usually sufficient to use simple on-off keying, along with optical

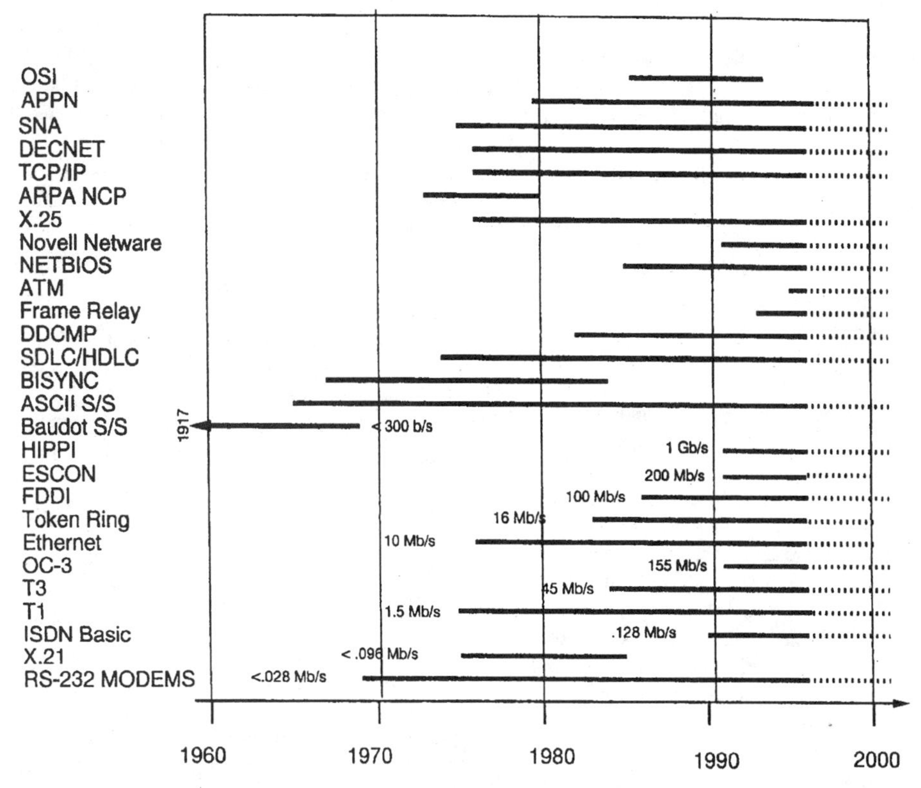

Fig. 4. The proliferation of protocols, as seen by the end user.

amplification where needed. There are two reasons for this, the large bandwidth per channel available, and the fact that moving photons do not interact, whereas moving electrons do, producing things like electrical crosstalk and impulse noise, which in turn produce error rate floors. Because of the large available bandwidth, source coding (compression) is often not required, and the need for channel coding, e.g., using sophisticated modulation formats with long constraint lengths, is obviated by the fact that it is easier to get the same performance improvement by link budget increases alone. The only channel equalization ever likely to be required might be power level control plus simple fixed electronic equalization to compensate for dispersion at long distances and high bit rates [12] and fixed optical equalization of the erbium amplifier passband.

C. Protocol Transparency

Figure 4 shows the steady increase in the variety of protocols installed by information processing users. This listing does not include minor variants, but only those in which end users have major investments today. A similar list made for use inside the common carrier plant would show a much more controlled proliferaton, not only because of the slower pace of technological change, but because of a long tradition of worldwide standardization unmatched by the computer industry.

Every one of the protocols indicated by the dotted lines at the right will be found today to a significant extent in businesses around the world, and the mix varies from installation to installation. One reason the list is so large is that those in the upper part of the figure are required (at all intermediate nodes as well as the end nodes) in order to provide end-to-end service, running on top of those at the bottom of the figure. With the lightpaths of optical networks end-to-end service is provided transparently with no buffering or logic in between.

Protocol variegation is one of the biggest inhibitors to flexibility and growth in the information industry [13]. Innovation is stifled and time to market for new solutions slowed down by the need to recover investments in existing protocol-specific communication solutions, both hardware and software. Attempts to enforce stability and uniformity by dictating comprehensive standards like OSI have failed rather spectacularly, as new innovations have upset these "best laid plans" and customers have proved reluctant to scrap existing facilities, postponing needed upgrades in favor of stability.

Protocol transparency is achieved when the path, all-optical or not, offers the flexibility of requiring the end users to understand each other only, rather than requiring that both obey the networks protocol dictates. Historically, it is clear that the telephone system's great transparency (after the connect phase and before the termination phase) has been one source of its great success as a base for the variety of services that it now delivers to us. It is equally clear that somewhat the same thing underlies the success of the very simple datagram-based IP in allowing the Internet to be built from many disparate technologies and protocols. It remains to be seen whether the cell-based packet switched variant, ATM, will bring similar benefits. The protocol-transparent nature of the lightpaths of optical networking, plus the ten orders of magnitude improvement in both exploitable bandwidth and also in achievable error performance compared to voice grade copper; all these offer society a base upon which to innovate an unprecedented breadth of applications and the protocol stacks to support them.

D. Reliability, Availability, and Maintainability

Electronic networks, of both the computer type and the provider type, have evolved to become structures of great logical complexity with many failure modes and corresponding diagnostic complexity. Fortunately, the reliability of LSI has outdistanced the unreliability factor related to complexity, so that these networks rarely fail (unless such equipment is running at bit rates that are at the edge of what the electronic technology will allow). Still, a nonelectronic approach of competitive cost and greater structural simplicity has advantages.

Optical networks are very simple structurally, each path having essentially no fan-out or parallelism, and no logic or memory along the way, and moreover, such networks make heavy use of passive components. The points of failure, which are typically due to loss of power, are few, and usually associated with optical amplification and electronically actuated switching of optical paths. Such a network is likely to be much more reliable and available than one making the heavy use of powered TDM electronics that forms the basis for today's LANs, MANs, and WANs.

All-optical operation does pose some problems of remote access to diagnostic information. An example is the erbium doped fiber amplifier (EDFA), for which it is desirable to monitor several parameters continuously. The very fact that this device amplifies all the WDM signals "in bulk" without separating them means that some ingenuity is required in providing some sort of channel to communicate the monitored variables to the maintenance facility. Other AON components probably do not have to be monitored at all; for example, it is difficult to imagine what can fail in a passive star coupler.

Needless to say, LANs and MANs, which today rarely number more than 250 nodes, are much easier to maintain than the vast structure of such a WAN as characterizes any telco, where the number of subscribers can be in the millions. The same difference in scale may be expected to exist in managing the availability and maintainability of all-optical LANs, MANs, and even private WANs, compared to AONs of the scale that would apply to the telco environment.

The competition to optical networking, both on cost and usability grounds, comes from such electronic TDM technologies as ATM switches (over $200 per port, typically), local area networks ($200 per port), central office switches (over $1000 per trunk or subscriber port), and PBXs (under $1000 per trunk port or subscriber port). These figures are prices, not manufacturing costs.

III. Architectural Options

A. WDM versus SDM, TDM, and CDMA

To take full advantage of the potential of fiber, the use of wavelength division multiplexing techniques is turning out to be the option of choice, the one being studied by almost all the two dozen or so optical networking research, development and product undertakings around the world. Traditional time division multiplexing, which in its electronic form has served so well throughout the fifty year history of digital communications, is under pressure from the so-called *TDM bottleneck*, the fact that TDM demands that each port handle not only its own bits but also those belonging to many or most of the other ports on the network. This is true independently of topology, being just as true of rings and meshes as it is of busses. The bit rate of the front end electronics then scales as the product of number of ports and per port bit rate. As networks evolve to higher values of this product, it inevitably exceeds that of the fastest available digital technology. WDM and space division, on the other hand, demand that each electrical port handle only its own bits.

Optical physical space-division transmission and switching (SDM), as a second way of avoiding the TDM bottleneck, has been a success for short distance transmission, but a disappointment for switching. The earlier promise of photonic switches (those in which the path is all-optical) has eluded practical realization for all but the smallest switch sizes (say, 8×8). The environment in which optical space division makes particular sense is in the nonswitched short distance "interconnect" application [14], on backplanes and between boxes in the same space, e.g., between the various processors, each other and their peripherals in one computer room or multiprocessor complex.

The outcome of an argument between SDM and WDM seems to be that SDM is preferred for short distances and can therefore employ multimode fiber and short wavelength optoelectronics (capitalizing on the coincidences that gallium arsenide also provides good electronics and that silicon provides good short-wavelength photodetection). WDM at long wavelengths is the preferred solution for distances longer than, say, a fraction of a kilometer, because of the expense of multistrand fiber paths and also because of attenuation and dispersion problems. For any distance, dense WDM mandates single-mode fiber, because, for each of the narrow-band sources that WDM requires, there are not enough different radiated wavelengths to average out the random variations of the received strengths of the different propagating modes supported by multimode fiber (modal noise).

The development of very short pulse technology has sustained a certain interest in all-optical TDM, for very localized networks in which dispersion is kept small, and for larger networks by the use of soliton propagation, in which dispersion effects are cancelled out by nonlinear effects. Compared to WDM networks, optical TDM networks are in their infancy, partly due to the very primitive and expensive nature of the devices required. Perhaps some reason for building AONs on an optical TDM basis will be found, but at present, immediate feasibility seems to lie more with WDM than with TDM. In addition to dispersion effects, the time division approach of transmitting a number of bits per frame equal to the number of bits per node *times the number of nodes supported* has two other disadvantages: it destroys the protocol transparency by

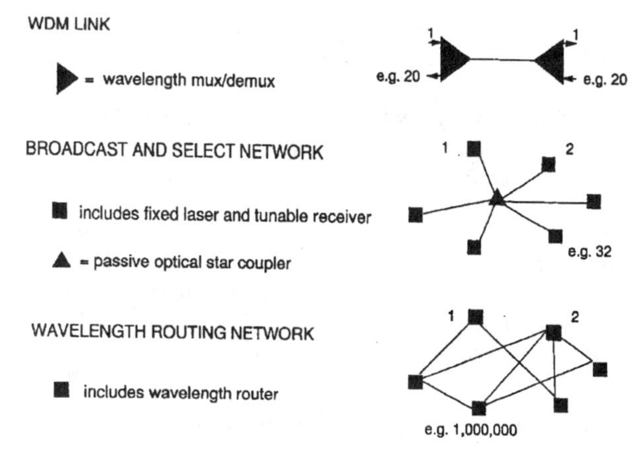

Fig. 5. The three forms of WDM network.

dictating framing format, and it exacerbates synchronization problems, since timing must be more accurate by a factor equal to the number of nodes.

Code division multiaccess (CDMA—also called spread spectrum) exacerbates the problems of dispersion and synchronization even further, especially for direct-sequence CDMA in which each bit time has been broken up into the number of pieces (chips) equal to the "processing gain," (a multiplier of two to three orders of magnitude) in order to provide the desired quasiorthogonality between the time-concurrent signals from the many nodes on the common channel. This is true whether the signals are derived in the time or frequency domain. For frequency hopping CDMA systems, this processing gain is much smaller, but such CDMA systems do not seem to have been proposed for optical networking.

B. Three Kinds of All-Optical Networks

All-optical structures can be divided into three classes on the basis of topology, the simple WDM point-point link (a degenerate form of network—really not a network in the usual sense), the broadcast and select star network (most useful for LANs and MANs) and the highly scalable wavelength routing mesh network (appropriate for WANs). These are illustrated in Fig. 5.

Figure 6(a) is an expanded block diagram of a typical WDM link [44], showing a grating multiplexor at each end, with wavelength-differentiated distributed feedback (DFB) lasers at the transmitting ends of the different wavelength channels and photodetectors at the receiving ends. Among the issues to be faced in building such a WDM link, in addition to the obvious ones of component cost and reliability are the registration between the wavelengths of the lasers and multiplexors and between the two multiplexors, and whether to minimize the fiber usage by bidirectional transmission (as shown in the figure) or to use one fiber in each direction, which then allows amplifiers to be added more easily if the link length is insufficient without them.

While with the link there are only two nodes, each with many input/output ports, the two forms of true network have many nodes but sometimes as few as a one port per node.

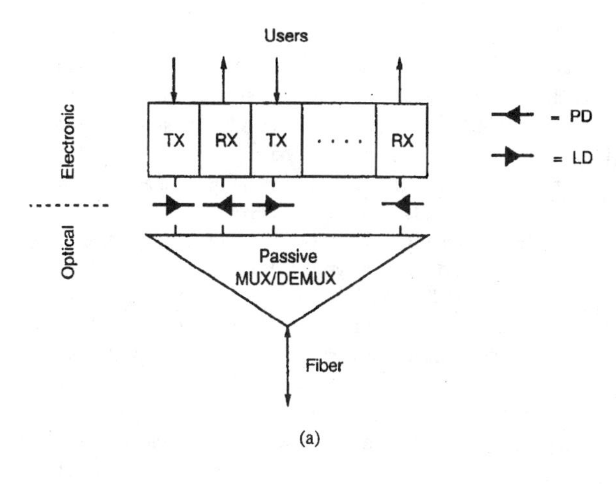

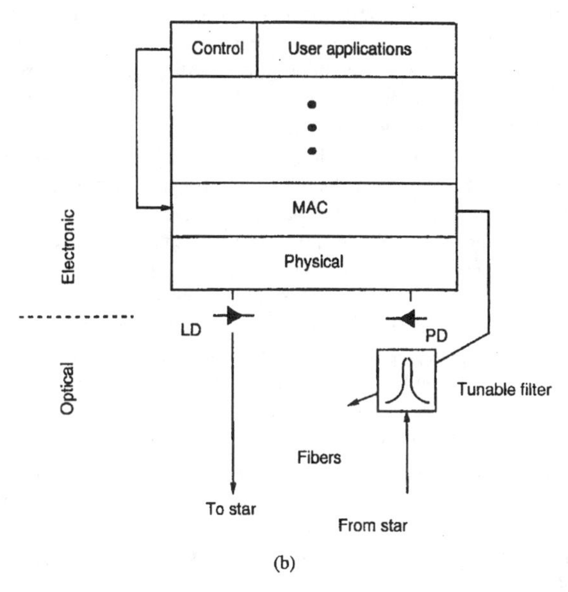

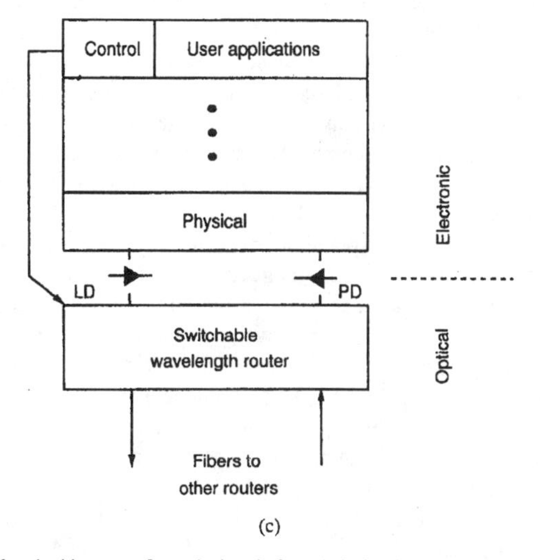

Fig. 6. Architecture of a typical node for: (a) A simple WDM point-to-point link (case of bidirectional transmission), (b) a typical broadcast and select network (tunable receiver case), and (c) a typical wavelength routing network.

The broadcast and select form of network works by assigning a single optical frequency to the transmit side of each port in the network, merging all the transmitted signals at the center of the network in an optical star coupler, and then broadcasting the mix to the receive sides of all ports. In principle, one can place the wavelength tunability required for dynamic access in all the receivers, all the transmitters, or in both transmit and receive sides of each port. Fig. 6(b) shows the structure of a typical broadcast and select node of today that uses the fixed tuned sources and tunable receivers that represent commercially available technology. By means of a suitable media access control (MAC) protocol [15], shown running as an application, when one node wants to talk to another (either by setting up a fixed circuit or by exchanging packets), in the case that only the receivers are tunable, the destination's receiver tunes for each interchange to the source's transmit wavelength and vice versa. The entire inner structure, consisting of fiber strands and star coupler (perhaps with further combining or splitting outside the star), is completely passive, unpowered, and therefore extremely reliable and easy to manage.

The design issues that have to be addressed with broadcast and select networks include not only the access protocol, but the technology for tuning the receiver and making sure that the correspondence is maintained between the tuned receiver wavelength and that of the transmitter.

Aside from high cost, which is currently a problem with any optical network, there are two other things wrong with broadcast and select networks. The power from each transmitter, since it is being broadcast to all receivers, is mostly wasted on receivers that do not use it. Secondly, the maximum number of nodes the network can have is the size of the wavelength pool, the number of resolvable wavelengths (call it N). Today, even though there are 25 000 GHz of fiber capacity waiting to be tapped, the wavelength resolving technology is rather crude, allowing a network of only up to about 100 wavelengths to be built [16].

Clearly, a new form of network that allows only 100 nodes does not constitute a revolution; technology must provide more wavelengths, and also some means must be provided for assuring additional scalability by using each wavelength many places in the network at the same time. A wavelength routing network accomplishes the latter function, and also avoids wastage of transmit power, by channeling the energy transmitted by each node along a restricted route to the receiver instead of letting it spread out over the entire network, as with the broadcast and select architecture. As the name implies, at each intermediate node between the end nodes, light coming in on one port at a given wavelength gets routed out of one and only one port by a *wavelength router* component that has purely optical paths. This component could, in principle be fixed or switchable, and could involve no change in the wavelengths of the optical signals passing through or could enforce a pattern of wavelength shifts.

Static routers provide insufficient flexibility for the inevitable change of traffic pattern, while wavelength-shifting routers seem to be much more difficult to build than those that pass the optical signals through unshifted. Thus the current emphasis is on switchable nonshifting routers. To complete the lightpath between end users, the settings of all the routers along the path need to be coordinated, either from a single central controller or by distributed and coordinated action of a set of controllers, one located at each router.

Figure 6(c) shows the structure of a canonical wavelength routing node in the latter case [17] of noncentralized control. One wavelength, λ_1, is used in this realization to exchange control and management information with the other controllers. In the case shown, the node also runs user applications, and is not just an intermediate node. The controller software is shown also running as an application.

C. Circuit Switching versus Packet Switching

As far as the end user is concerned, there is sometimes a preference for circuit switching and sometimes for packet switching. The former provides protocol transparency during the data transfer interval, and the latter provides concurrency (many apparently simultaneous data flows over the same physical port, by the use of protocol-specific methods of time-slicing). In both cases, very large bit rates are possible without requiring the electronics to handle traffic from extraneous nodes other than the communicating partners.

For wavelength routing networks of large diameter, circuit switching is almost mandated, as an interesting indirect consequence of the unavailability of purely optical technology for doing the intermediate node buffering, header recognition and processing traditionally used for packet switching networks. For short distances, say within a single LAN or small MAN, propagation time between the controllers associated with the wavelength routers at intermediate nodes is so small that lightpaths can be set up and taken down on a packet by packet basis by coordinated action of the controllers along the path without the control delay overhead becoming hopelessly large. For WAN distances, the long propagation time between controllers translates into long setup times for these lightpaths, with a correspondingly intolerable overhead in the case of packet switching. Perhaps some day a practical form of optical buffering and packet header processing will be found, whereupon the same kind of packet switched WANs with many intermediate nodes that one sees today can be done all-optically; today such networks are not practicable for usefully long packet or cell lengths.

Thus, on bit-efficiency grounds, fast (submicrosecond) tuning time seems to be required for packet switching of gigabit or sub-gigabit data streams, in LANs and MANs; any form of circuit switching, including WANs based on wavelength routing, can tolerate tuning times of milliseconds or more, because the propagation time is the controlling factor anyhow. This makes building wavelength routers much easier than if they had to tune as rapidly as sources and receivers for packet switching.

D. Networks of Networks

To build any sort of large national or international infrastructure of the future involving all-optical architectures throughout, selected pieces of the three types we have described will have to be pieced together to form large scalable networks of optical subnetworks and optical links. This, however, is in the distant future, if ever. Meanwhile, hybrid WDM/TDM networks of networks are likely to evolve with ever larger islands of all-optical nature being imbedded in the traditional TDM electronic background. For example, local high bandwidth demands may dictate the use of broadcast and select LANs, which are then knitted together with high speed electronic technology, for example, several supercomputer sites interconnected with SONET long-haul facilities. Another example that is real today is the interconnection of large computer centers, each using electronic TDM throughout, by means of dense WDM links.

Perhaps the most immediate candidate for a mix of the three AON architectures is the combination of broadcast and select LANs and MANs connected together by a backbone formed by wavelength routing. Such networks can be scaled to very large sizes in principle.

The central issue in scalability when the paths must remain all-optical concerns how best to use the N available wavelengths. Fig. 7(a) shows reuse of available wavelengths within an individual wavelength routing network, as already described. When the overall network is divided into subnetworks, perhaps connected in a hierarchy with a "backbone" at the top level, as shown in Fig. 7(b), wavelength reuse is employed differently, and a slightly greater number of connections for a given N results. Traffic from a node in the lowest level uses a wavelength outside the set reserved for local traffic if it wants to send to a node in some higher hierarchical level or to a node in another subnetwork at its level. However, if wavelength converters are available, it is very efficient to use the full wavelength pool for each hierarchical level or interconnecting subnetwork, as shown in Fig. 7(c). While wavelength converters may be of only modest effectiveness in increasing scalability in wavelength routed single networks (as shown in Fig. 3), they are probably the key to truly large all-optical networks of networks.

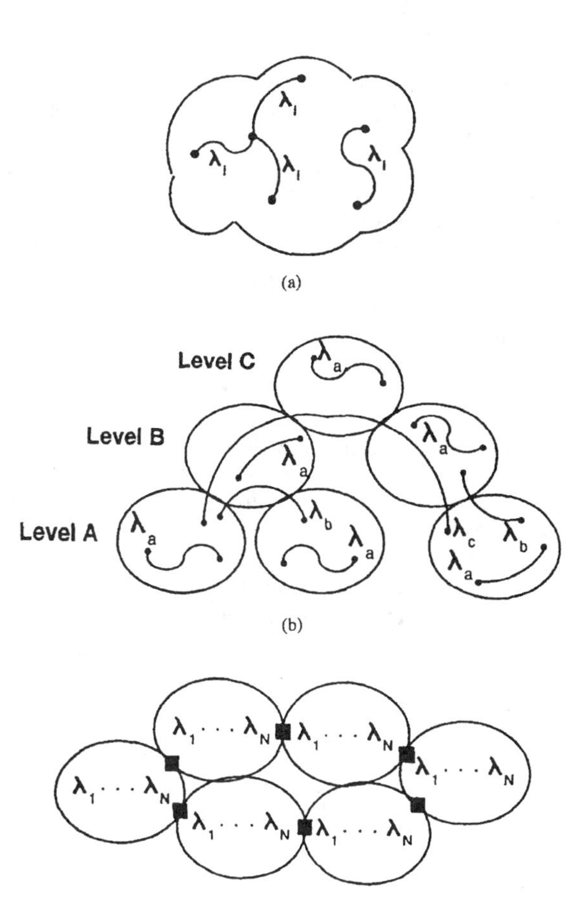

Fig. 7. Wavelength reuse. (a) Reuse of one wavelength λ_i within one network or subnetwork and (b) within a hierarchy of networks. Staying within a level uses wavelengths $\lambda_1 \cdots \lambda_a \cdots \lambda_A$, going from Level A to B (and maybe back to another Level A uses wavelengths $\lambda_{A+1} \cdots \lambda_b \cdots \lambda_B$, and going all the way through level C uses $\lambda_{B+1} \cdots \lambda_c \cdots \lambda_N$, $(A < B < C)$. (c) General network interconnection using wavelength converters (square dots).

E. Subcarrier Multiplexed (SCM) Networks

Point-to-point optical links have for some years been carrying multiple TV distribution information streams by modulating a number of frequency-separated RF subcarriers onto a common laser at a single wavelength [18]. There has been considerable interest in extending this idea to WDM networks in which each wavelength carries several subcarrier signals, thus increasing the number of channels available.

The principal problem that has limited the applicability of such networks is splitting loss. In addition to any splitting loss incurred in a broadcast and select network in sending all wavelengths to all receivers, there is a second splitting loss associated with "power backoff." Call the number of subcarriers per wavelength M. Instead of running the transmitter at full peak power, intermodulation considerations require expensive linear laser diodes to which each of the M subcarriers contributes at best one Mth of the total drive power. Careful consideration [5], [19] of subcarrier WDM networks shows that they are only feasible in conditions of copious link budgets, especially because of this double splitting loss. An extensive discussion of optical networks with SCM is given in [5, chap. 8].

F. Almost All-Optical Networks

For true protocol transparency, a lightpath traversing an intermediate node must remain in optical form, but for most signal formats encountered in practice, it is only the zero-crossings (transition times between zero and one bits) that carry the needed information; as long as the transition times are preserved, transparency is maintained. Such formats include FM, PM, and binary digital data. In such cases, a partially effective alternative to wavelength-broadened EDFAs is to interpose at the intermediate points in the network at each wavelength a simple conversion from optical to electrical form and back again that attempts no more than just to preserve the the zero-crossing instants [20].

Such a shortcut was considered some years ago for a single link, and has been termed 2R (regeneration and reshaping), where "regeneration" means amplification and reshaping means trying to reproduce the square wave with its zero crossings. This is to be compared to 1R (simple amplification) or 3R (regeneration, reshaping, and reclocking). 3R, which is the full treatment widely used in telco repeaters, presupposes a specified bit rate, and is therefore nontransparent, as well as being more expensive. With 2R, no attempt at clock resynchronization is made and amplitude details are not preserved either. At a considerable cost in performance compared to either 3R or optical amplification, this artifice, when applied individually to all the wavelengths traversing a given node, maintains most of the transparency while providing other benefits, such as facilitating wavelength-by-wavelength add-drop, bidirectional transmission, wavelength switching and conversion, escaping the bandwidth limitation of erbium amplifiers, and keeping each optical segment short enough to limit the buildup of effects of dispersion, nonlinearity, coherent crosstalk (see Section IV-B), and amplifier noise.

A minor comment about transparency is in order. While it is possible to design receivers that will respond to almost any bit rate, if these are nonadaptive, then since optimum electrical filter settings are bit rate dependent, performance will be sacrificed at some bit rates. Thus, it should be realized that the transparency offered by 2R remodulation comes at an additional price in link budget.

IV. Limits

A. Capacity Limits

In Fig. 8, the rightmost diagonal line shows how the 25 000 GHz of capacity of a broadcast and select network could, in principle, be divided up between a few extremely fast channels or many more lower bitrate channels. A conservative modulation efficiency of 2.5 Hz per baud is assumed, in other words, no fancy bandwidth conserving source coding is used. The fiber capacity could intrinsically support 10 000 uncompressed HDTV images at 1 Gb/s each, for example, or one thousand OC-192 transmissions at 10 Gb/s each, or any other combinations that add up to 10 000 Gb/s. Of course, today the limit is set by the size of the wavelength pool, 100 wavelengths or less.

When amplification using EDFAs is required (either because of distance attenuation, splitting loss, or component excess loss), this figure is reduced from 25 000 GHz down to 5000 GHz, since the erbium gain spectrum is that much narrower than the low-attenuation passband of the glass fiber. Even this requires careful wavelength flattening (Section IV-E). The digital capacity consequences are shown by the second diagonal line.

With wavelength routing, the number of supportable connections is the available number of wavelengths multiplied by a *wavelength reuse factor* [21], plotted in Fig. 9. This factor grows with the topological connectedness of the network, since the more complex the topology the more ways there are to lay down many paths that are physically edge-disjoint. For example, the figure shows the reuse factor growing with number of nodes for a fixed number of ports per node. In a 1000-node network of nodes with four ports each, the reuse factor averages around fifty, so that with a pool of 100 wavelengths each node could have around five concurrently active lightpaths.

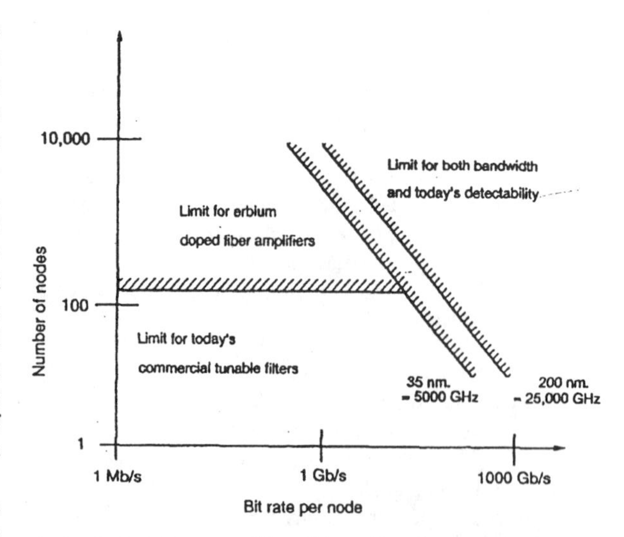

Fig. 8. Intrinsic bit rate capability of fiber and broadcast and select networks.

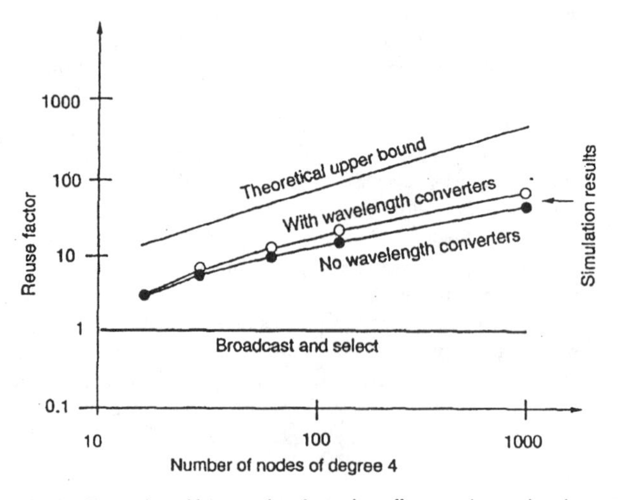

Fig. 9. Factor by which wavelength routing allows each wavelength to be used many places in the network simultaneously. One percent blocking probability and a pool of $N = 10$ wavelengths are assumed.

Scalability to large network sizes requires a large wavelength pool, wavelength routing or, more likely, both.

B. Limits Imposed by Crosstalk

Today, the number of wavelengths that a system can actually use is limited not by the modulation rate (as used in the calculations of the diagonal lines of Fig. 8), but to much smaller capacities by the stability and wavelength resolving ability of the technology. Today the preferred receiver technology is incoherent detection preceded by optical narrowband filtering, since the coherent reception alternative requires a considerable increase in system cost and complexity [22]. Current channel spacings are typically 0.5–1.0 nm, which

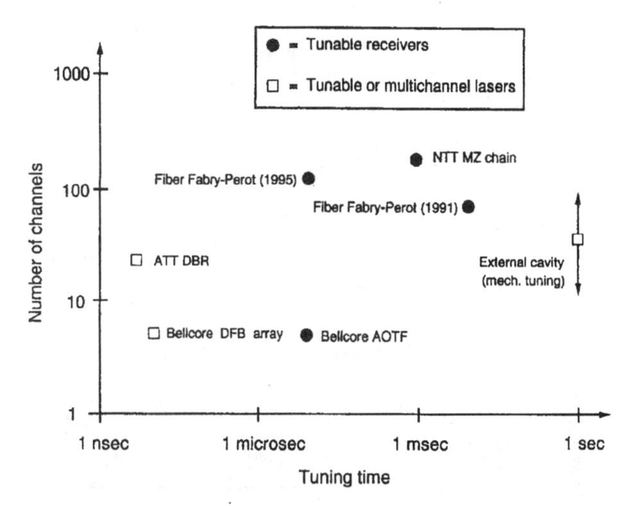

Fig. 10. Number of resolvable wavelengths and tuning speed of currently system-usable tunable transmitters, and receivers.

at 1.5-μm wavelength is 60–120 GHz, a number so large that it will probably be quite a while before the technology evolves to the point that modulation bandwidth affects channel spacing. Laser chirp, however, does have to be considered for current channel spacings and multi-gigabit rates, and must be controlled by careful isolation of the laser diodes, and eventually by external modulation.

Figure 10 shows the number of channels and speed of tunability achieved for several tunable receiver and transmitter technologies that are current candidates for system usage. As of this writing, there are really only three tunable laser solutions that could be considered for actual system service, in this author's opinion: Bell Labs' 24-wavelength two-section DBR laser, as used in the AON testbed [23], DFB arrays such as that used by Bellcore in the ONTC testbed [24], and various external cavity lasers. The former two devices have a tuning time of a few nanoseconds, while the external cavity devices require many milliseconds or a few seconds.

Wavelength selective technology is often not limited in the range that it can tune over but in the narrowness of its passband. For example a Fabry–Perot interferometer can easily tune over its entire free spectral range merely by changing the cavity length by $\lambda/2$, and this FSR can be made to cover even the entire 25 000 GHz (200 nm). However, to keep the crosstalk penalty from other WDM channels below 1 dB, the number of resolvable channels must be no larger than about 60% of the filter's *finesse* [25]. Since today's commercial FPI's have finesses no higher than about 200, this leads to a maximum number of channels of about 120, which is shown as the horizontal limit in Fig. 8.

Other technologies are limited in different ways in the number of channels that can be fabricated. For example, planar arrayed waveguide gratings, or *phased arrays* are limited in the number of waveguide paths that can be lithographed side by side without making them so close together that crosstalk rises to harmful levels. The same goes for the many side by side steps in a planar grating. Current numbers of wavelengths for service-usable phased arrays are running at 64 or less [26]. Mach–Zehnder chains have been made to resolve as many as 128 wavelengths (seven elements in the chain) before problems of control and overall waveguide attenuation become too large [16].

Other devices are limited by physical size. Size is particularly important if the device is fabricated from an expensive material, such as lithium niobate or indium phosphide. An acoustooptic filter or electrooptic filter will resolve a number of channels, for a given crosstalk level, that scales with the length of the device in multiples of the wavelength along the path of the acoustic or electrooptic drive sinusoid, respectively. Current numbers are less than 10 [27]. Long cascades of any kind of wavelength resolving device, such as Mach–Zehnder interferometers, pose obvious size and attenuation limitations. Gratings and phased arrays are size limited, since the number of resolvable wavelengths scales with the number of teeth in the grating, or equivalently, the number of length-differentiated paths in the phased array.

When the signal from an interfering channel is at a wavelength close enough to be passed by the receiver optical filter, but outside the modulation bandwidth of the optical carrier, the interfering signal adds its optical power to that of other noise. This situation is referred to as *incoherent crosstalk*. However, if the interference is within the postdetection bandwidth (as set by the modulation rate), it is much more harmful, appearing somewhat as a spurious modulation sideband at the electrical, not the optical power level. Several situations can produce this *coherent crosstalk*. In any network having topological circuits (e.g., rings or meshes), if a transmitted signal is not adequately removed from the circuit completely at some point around the periphery, then remnants of it continually recirculate, thus delivering multiple, delayed copies to the intended receiver. An even more widely studied case [28] concerns wavelength routing networks in which nonideal switches within the router components allow some of the signal at one wavelength arriving at one port to leak into exit ports other than the intended one, thus contaminating any intended signal at that wavelength at that output port. By expanding the number of switches required and connecting them in a *wavelength dilation* pattern [29], one can suppress the coherent crosstalk from the output, at an obvious cost in complexity and component excess loss.

C. Limits Imposed by Component Drift

Drift is often the determining factor on number of wavelengths in the system. Piezoelectrically tunable FPIs are available with finesses of 10 000 or more [30] but have unusably high drifts with temperature, aging, and other factors. Generally speaking, all WDM receivers today, even when made using such presumably stable elements as fixed silica gratings with multiple wavelength selectable outputs, require some sort of stabilization. With gratings and phased arrays, and today's 0.5–1.0 nm channel spacing, temperature stabilization appears to be sufficient, but receivers using analog tuning, e.g., FPIs, usually require either direct control of the wavelength using some sort of AFC loop [31], perhaps by referencing the wavelength to an external standard [32], or by referencing the cavity length itself [33].

The stability of commercial DFB lasers as sources is much better than that of receivers, since after burn-in the aging effects are minimal, and since the devices are packaged with good thermal control loop capability [34].

It seems a safe bet that as the WDM network art progresses to more wavelengths per system, component drifts will play more of a dominant role.

D. Limits Imposed by Fiber Nonlinearities

The transmit signal power applied to an individual wavelength of a WDM optical link or network must be great enough to provide adequate SNR at the receiver, after suffering the effects of attenuation (distance, splitting and component excess loss), crosstalk, and the admixture of unwanted noises like amplified stimulated emission (ASE) within amplifiers. On the other hand, the transmit power cannot be too great. If the link is too long and the power levels too high, any one of several physical effects may cause the tiny nonlinearities of glass fiber to introduce crosstalk; the original signal at a wavelength is depleted and inserted as crosstalk into other channels [35]. There are also potential limits on total power in the fiber imposed by safety considerations [36]. To date, the latter have proved to be manageable by use of protective circuitry for powering off transmitter light upon either loss of received signal or exposure of the laser radiation outside the contained propagation environment, but nonlinearities are a serious impediment for system growth of geographically large networks.

The two principal nonlinear physical effects that cause the trouble are four-wave mixing—FWM—i.e., third harmonic distortion and stimulated Raman scattering (SRS), the pumping by shorter wavelength signals of stimulated emission in the longer ones. Crosstalk levels due to SRS increase with both number of channels and their spacing. While the relative roles of the various effects is dependent upon the detailed system design, a rough summary is offered by Fig. 11, a classic figure from several years ago [35], still approximately valid. An antidote to FWM is to make sure that there is enough dispersion accumulated along the path to destroy the phase relationships of the intermodulation products. There is no corresponding known fix for Raman scattering, which has not become a problem yet, but which is likely to become so for long links with high bit rate (and therefore high power levels) as the number of wavelengths gets in the hundreds.

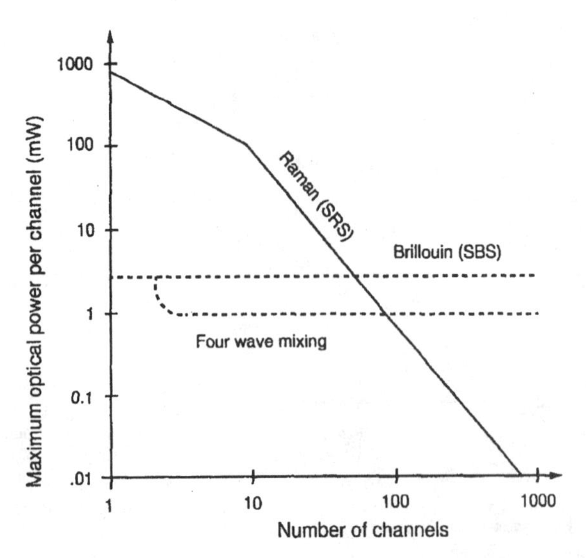

Fig. 11. Summary of the effect of fiber nonlinearities in limiting the number of wavelengths and the power per wavelength in nondispersion shifted fiber of 22 km length. A 1 dB link penalty is assumed, and the channel spacing is 10 GHz (roughly 0.8 nm. at 1.5 μm wavelength) (from [35]).

E. Limits Imposed by Optical Amplifiers

EDFAs [37], [38] have the advantages over semiconductor laser diode amplifiers that multichannel crosstalk is essentially absent (due to the long 11 ms carrier lifetime of the erbium ion). However, they have the disadvantage of a narrower passband (35 nm versus 200 nm), which moreover has significant peaks and valleys.

There is therefore a need to equalize the gain spectrum, particularly if a number of amplifier stages are to be cascaded. There are currently several methods of attacking this problem, and it is too early to tell which one will emerge as the preferred one. A broad review of EDFA wavelength flattening is contained in [39, Sect. B]. The available approaches include equalizing filters (either within the fiber or externally), use of host glasses with flatter spectra (particularly fluoride glass), and methods of decoupling the gain at one wavelength from that at another, for example by causing different portions of the doped section to amplify different wavelengths (spatial hole-burning).

Electronic amplifiers have always had internally produced noise, whether thermionic shot effect, thermal or Johnson noise in the front end resistance or transistor internal noise effects, and erbium amplifiers are no different. The inevitable spontaneous emission in the earlier portions of the amplifier get amplified, and this ASE must be carefully taken into account when cascading EDFAs in a long link or network. (It is small comfort to know that the same thing was necessary forty years ago with long FDM telco links using vacuum tube electronics.)

For modest numbers of channels and the wavelength spacings of 1.0 nm or less that are being used in WDM systems, the dominant effects are four-wave mixing and amplified stimulated emission, so the design of the link focusses on using enough transmitter power to overcome ASE, but not so much that FWM becomes excessive. At least one model is available [40] that takes both these into account for designing long amplifier cascades for WDM service, assuming ideally flat EDFAs, and making a conservative assumption about preservation of relative signal phase when passing through the amplifiers.

V. Brief History of All-Optical Networking

The fact that fiber had a valuable bandwidth potential far in excess of anything that was ever likely to be exploited by conventional means, and that WDM was the natural way to unlock this potential, has been realized for over a quarter of a century. As attempts to act on this insight took shape, the fundamental limits to the achievable that were covered here in the preceding section gradually came more clearly

TABLE I
HISTORICAL HIGHLIGHTS. L = LINK, B = BROADCAST. R = WAVELENGTH ROUTING, AND BERT = BIT-ERROR RATE TEST

Year	Sys. Type	Name	Organization	Ref.	# Wave-Lengths	Source	Mux	Demux	Traffic•	Per Ch. Bitrate (Gb/s)	Dist. (km)	System Type
71	L	–	IBM	43	5	LED	Rod	Filter	BERT	1 MHz	3m	Experiment
85	B	–	Toshiba	47	4	LED	Star	Thinfilm	FM TV	4.5 MHz	5	Experiment
86	B	–	BTL	54	4	DFB	Star	Dichr. gel	Data, TV	0.28	5	Experiment
86	B	Lambdanet	Bellcore	48	18	DFB	Star	Grating	BERT	2	58	Testbed
88	B	--	AT&T	60	3	Ext.-cav.	Star	Coherent	BERT	0.045	--	Experiment
:8, 90	B	--	AT&T	58	4	DBR	Star	FFP	BERT	0.045	–	Experiment; upgraded to 1.2 Gb/s [59]
90	B	--	NTT	16	100	DFB	Star	MZI chain	BERT	0.62	50	Testbed
91	R	–	BTL	52	3	DFB	Grating	Grating	BERT	0.62	45	Experiment
91, 4	B	Rainbow	IBM	50	6	DFB	Star	FFP	Video/FDDI/HIPPI	0.3; 1	25	Testbed; upgraded to 1 Gb/s HIPP [51]
91	B	Teranet	Columbia University	57	3	DFB	Star	FFP	BERT	1	–	Experiment
93	R	MWTN	BTL/Ericsson/Pirelli/etc.	53	4	DFB	Grating	Various	SDH	0.62	–	Proposal
93	B	STARNET	Stanford University	61	2	Nd:YAG	Star	Coherent	BERT	2.5	–	Experiment
93	B	COSNET	Dutch PTT	62	3	DFB	Star	Coherent	Video	0.155	205	Testbed
93	B	COBRA (R1010)	Philips/GMTT/Siemens/etc.	63	10	DFB	Star	Coherent	Video	0.14	10	Testbed
95	B, R	AON	AT&T/MIT/DEC	55	20	Step DBR	Ph. array	FFP	3 classes; Telemedicine	10	100	Testbed
95	R	ONTC	Bellcore/Hughes/NT/etc.	56	4	DFB array	Thinfilm	Thinfilm	Video/ATM	0.155	150	Testbed; also subcarriers
95	L	Model TXT/EM	Pirelli	45	4	DFB	Star	Interf.	SONET	2.4	550	Product
95	L	NLGN	AT&T	46	8	DFB	Ph. array	Ph. array	OC48	2.4	360	Product
95	L	Model 9729	IBM	44	20	DFB	Grating	Grating	ESCON, CL	0.2, 1	50	Product

into view. So far, no "show-stoppers" have been found that cannot be designed around, and serious systems are being built to deal with serious applications. Without trying to cover all the various laboratory prototypes and "hero experiments" that were part of the evolution, we shall attempt here simply to hit the high spots on the road to where the situation stands today. A tabulation of some key parameters of most of the important systems that have actually been built and extensively tested is given in Table I. For further reading, there are several recent books [3], [5] and journal special issues [2], [4], [41], [42] that give rather comprehensive surveys of the many major (and also minor but necessary) steps along the road.

A. Links

It was only recently that the term "WDM" meant anything more than two wavelengths, one in the 1.3-μm fiber passband and the other around 1.5 μm. Operational use of two-wavelength WDM goes back to the 1970's, often with the two being used in counterpropagating directions. Today, the term "WDM" is now usually understood to mean *dense WDM*, that is, tens to hundreds of wavelengths (or more), usually in the low-attenuation 1.5-μm window.

WDM links are the operational precursors to multinode WDM networks, since they are so much easier to build. In 1971, investigators at IBM built a five-wavelength system at several Mb/s per wavelength as a laboratory demonstration of the potential of single link dense WDM [43]. Today, 20-wavelength IBM 9729 systems are in commercial use for computer site interconnection over dark fiber [44]. Four-wavelength WDM links became available from Pirelli in 1995 for the telco backbone application [45], while an eight-wavelength such system was recently announced by AT&T [46]. Even though these commercial systems are expensive, simply because the optoelectronic technology is currently so expensive, they are practical because of the even larger cost of the multiplicity of the single-wavelength fiber transmission paths that they displace.

In the telco world, there is currently a debate about whether to do OC-192 (10 Gb/s) directly in 10 Gb/s electronics or by means of four OC-48s at 2.4 Gb/s each. As matters currently stand, it looks like electronics will win this one, but it is problematic how many more like this it can win before succumbing to the inevitable. In principle, soliton propagation can support very high bitrates, in the tens of Gb/s, but the signal must still be multiplexed, launched, received, dispersion and synchronization dealt with, and demultiplexed at the full bit rate, all of which add to the costs.

The telco community and the end users both use current commercial WDM for largely the same purposes, for productivity increases for a given amount of fiber, and to make the system protocol insensitive, and thereby more "futureproof." In the case of the IBM 9729, the problems solved also include the use of remodulation (Section III-E), to turn multiple WDM links into rings and fixed wavelength routing meshes, albeit at greater expense today than is desirable.

B. Broadcast/Select Networks

Once it became clear that amplification of terahertz optical bandwidths was feasible, this gave great encouragement to those who had been experimenting with AONs, because it told them that such networks would not always be limited to short distances and the small numbers of nodes implied by low splitting loss.

Some years before that, the use of wavelength division had been widely agreed upon and many laboratory experiments had been carried out, using both coherent and incoherent detection. Broadcast and select optical networks offered an attractive way of building LANs and MANs of a few dozen high bit rate nodes, and can be built today from commercially available parts.

The earliest broadcast and select network to be built in the laboratory for any kind of applicability in service usage was apparently the four wavelength analog video distribution system using LEDs built at Toshiba in 1985 [47]. The first system realization that even approached "industrial strength" came with Bellcore's 18-wavelength *Lambdanet*, built in 1987 [48]. Lambdanet used a star coupler to broadcast from the DFB laser of each node a TDM composite of a number of bitstreams to all receivers, each of which received all 18 wavelengths and selected the desired bitstreams not by tuning, but by TDM demultiplexing.

In 1990, NTT Laboratories achieved what still stands as the largest number of wavelength channels to be used in a network prototype [16]. The system was not a true any-any network, but rather a distribution system in which over 100 collocated lasers at different wavelengths could carry different OC-12 bitstreams, so that any tunable receiver could select between them.

By 1991, it was possible to build WDM networks of almost service usable quality, thanks especially to the arrival of commercial tunable filters [33], [49]. IBM field tested two of these, Rainbow-1 [50] in 1991 for 300 Mb/s digital video and 100 Mb/s FDDI traffic between eight of a possible 32 nodes, and Rainbow-2 in 1994 at Los Alamos National Laboratory for 1 Gb/s HIPPI supercomputer interconnection, with TCP/IP protocol offload [51].

Broadcast and select local networks of a similar type form the outermost local (zeroth) level of the AON prototyped by the AON consortium of AT&T Bell Labs, MIT and DEC [23]. In this network, instead of the classical form of star coupler fashioned from many biconical tapered 3 dB couplers, the star coupler was of the more recent planar waveguide type. AON probably also holds the record for distance covered, 50 km.

It is interesting to observe from Table I that all the optical networks built so far, even the broadcast and select LANs and MANs, use circuit switching, doubtless more a reflection of the state of the technology than of intrinsic need. Note from Table I that several of the systems have employed coherent detection. Whether this is robust and economical enough to see commercial service any time soon is questionable, in this author's opinion. Broadcast and select networks having a ring topology have often been proposed, but not built.

C. Wavelength Routing Networks

The components to build broadcast and select networks have been available on the street for four years now, while most interesting forms of optical wavelength routers are still prohibitively expensive. Architecturally, the function of a nonwavelength changing wavelength router is the complex arrangement shown in Fig. 12(a), made out of two grating MUX/DEMUX units for each of P bidirectional ports, plus a number of $P \times P$ photonic switches equal to N, the number of wavelengths in the pool.

While Fig. 12(a) can be thought of as only a schematic representation, this arrangement was implemented quite literally in the first demonstration of wavelength routing networks by British Telecom Laboratories [52]. Later, because of its scalability due to wavelength reuse, wavelength routing became the main focus of ARPA's support of optical networking in the U.S., and of many of the RACE programs in Europe. The + Level 1 (regional) network layer used a fixed wavelength router consisting of a planar waveguide phased array that used more than the single input port that phased array operation implies. For the key switchable wavelength router component of the ONTC consortium [24], an acoustooptic tunable filter with a clever scheme of multiple variable frequency RF drive signals was initally proposed, but eventually replaced by a dielectric thin film multiplexor and opto-mechanical switch. A grating-based version of Fig. 12(a) was used in the recent work of the Multiwavelength Transport Network (MWTN) consortium [53].

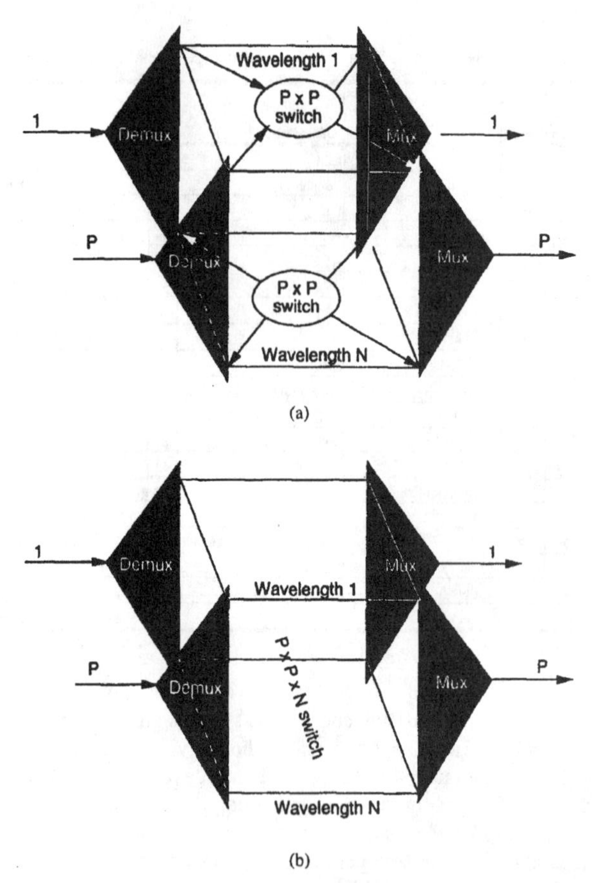

Fig. 12. Abstract architecture of (a) a switchable wavelength router of N wavelengths and P ports, showing separability of the switching fabric by wavelength and (b) a switchable wavelength converter, requiring a constrained $P \times P \times N$ switch.

VI. Future Application Expectations

When one reflects on the richness of the information infrastructure that has been built in the last twenty years on the modest base of telco 22-gauge copper, then as optical networking introduces the potential of ten orders of magnitude improvements in both total accessible bandwidth and in achievable error rate, the possibilities seem limitless. Probably the safest (but not very helpful) thing to say in trying to prognosticate the applications of the future is that in view of such a large discontinuity, in the long run the most important emerging applications will not be any that one would predict today.

Nevertheless, a few comments on the near-term future might be in order. The introductory pages of this paper tried to make the case that AONs will be needed soon, for reasons of capacity, protocol insensitivity, low error rate, reliability and manageability. The preceding historical review of the field has shown that the earliest dense WDM systems to come on line as a functioning part of the overall information infrastructure capitalize on the first two of these properties, greater bandwidth and wider protocol independence, in a manner that is economically superior to alternatives based on TDM or space division.

However, the real application sets that have been tested so far are really quite limited; they are actually niches rather than broad domains. While various laboratory testbeds and even field tests have experimented with video distribution [47], [54], [50], [62], [63], supercomputer interconnection [51], and telemedicine [23], in all these cases, if WDM networking had not been available, some form of TDM could have been found that would do the job just as easily and economically. The real test today of the usefulness of optical networking is whether it provides a solution that is superior to existing alternatives by a large enough margin overcome the burden of unfamiliarity.

As bit rates go up, eventually there will be traffic streams whose rates are so high and whose protocol variety is so pressing that the laws of physics dictate that only optical transmission can handle them. Meanwhile, as optoelectronic costs go down, the laws of economics will say that WDM will be the preferred solution when it is the lower cost solution, no matter what the laws of physics say. For example, the emerging wide use of dense WDM on simple point to point links is sometimes done because it is cheaper than parallel individual fibers, sometimes on the physical basis that there are simply not enough fibers available at almost any cost, and sometimes in order to avoid paying for protocol-specific conversion equipment to match some local bit rate to some transmission facilities bit rate.

Assuming some sort of preexisting fiber infrastructure, installed and already paid for, it is mostly optoelectronic component costs that stand in the way of wider all-optical networking usage. It is the conviction of most people working on these networks that if device costs were two orders of magnitude smaller than they are today, there would be many more niches of applicability and today's niches (e.g., video, telemedicine, and computer interconnection) would grow into broad application areas of significant economic importance. As an example, one could imagine that a $300 500-channel set-top box based on WDM distribution of digital HDTV might have a significant impact. So would a $300 100 Mb/s PC card for connection into the Internet of the future (see Fig. 2).

VII. Future Technology Expectations

How rapidly will we get the two orders of magnitude price reduction? The problem should be soluble in principle. After all, the only building blocks needed to build successful all-optical networking on an existing fiber base are tunable and multichannel sources, tunable and multichannel receivers, EDFAs (or laser diode amplifiers), switchable routers and converters, and passive splitters/combiners.

One of the themes of the opening paragraphs of this paper was that fiber is slowly reaching out to every home and desktop, but that the time scale is slow enough that there is a real opportunity for the optical networking community to put WDM solutions "on the table" as alternatives to classical TDM.

An examination of the reasons for current high optical networking technology costs shows that there are two main culprits, a dependence on bulk optics and a marketplace that has historically been oriented toward low volumes with high permissive pricing structures. The latter is changing as the classical usage of fiber for telco interoffice connections of great cost is being superseded by growing interest in fiber to the premises and fiber in the premises, for TV distribution and computer communication. Another way of saying this is that there are at least four orders of magnitude more TVs and PCs than there are telephone exchanges. The compact disk player is the existence proof of what can be done to reduce optoelectronic costs if high volumes can be achieved.

The dependence of WDM communication on bulk optics, with its attendant expensive materials, manufacturing and testing cost levels is visibly giving way to the same methodology that revolutionized electronics: lithographic fabrication. A good example is passive couplers and splitters. Until two or three years ago, these were usually assembled from fused biconical tapered fiber couplers; today the preferred solution is a small planar waveguide structure in silica on either silicon or silica, with suitable V-grooved or butt-coupled pigtail attachment.

However, harnessing lithography for optical networks as effectively as it has been for LSI will not be without its problems [64]. Since the optical components are analog rather than digital, narrower percent tolerances will have to be maintained. Whereas in microelectronics, chip size has tended to remain constant while packing density increased, with planar waveguide lithography there are fixed limits to packing density, set by waveguide curvature, waveguide size and other parameters.

Because of this last factor, materials whose cost is less sensitive to area than present waveguide materials must be found, and polymers seem to offer particular promise. It is interesting to note that the tolerances required for today's polymer compact disks (with a current manufacturing cost of 35 cents) are no less severe than those for planar couplers, gratings or arrayed waveguide gratings (phased arrays). Among the problems that will have to be solved with polymers are those of making their attenuation and birefringence properties as small and as wavelength independent as with today's silica waveguides.

One of the principal challenges at both the component and system level is to increase the size of the available pool of wavelengths, which should probably be 1000 or more ultimately. Techniques of wavelength regulation and referencing will become necessary with growth of the number of channels that can be fitted within the 200-nm silica attenuation window, to say nothing of the 35-nm EDFA window. What would really be desirable is some sort of all-digital frequency synthesis technology so common with RF equipment, but no such technologies is in sight for frequencies of 10^{15} Hz. In the absence of frequency synthesizers, referencing against atomic lines [32] appears to be the most feasible alternative.

Currently, the most popular wavelength resolution technology seems to be the planar phased array. These are being manufactured with 32 wavelengths, with 64 and possibly 128 in the future. However, phased arrays do no scale in size as gracefully with number of wavelengths as do planar diffraction gratings. For example, three years ago early multichannel receivers composed of arrays of wavelength-differentiated photodiodes used gratings having 70 or more resolvable wavelengths [65], [66], and 128 was even attempted [67]; all these devices were only 2-3 cm on a side. For both pure gratings and arrayed waveguide gratings of only several tens of wavelengths, simple temperature control may be adequate for stabilization.

It is quite possible that, as it becomes necessary to crowd wavelength channels closer together, coherent detection could become attractive again. In principle, passive filters could be built in the optical domain with just as narrow passbands as at IF, either as single resonant structures or by cascading. For example, using the single device of Reference [30] with a free spectral range of 100 nm, achieves an intrinsic but unstabilized bandwidth of only 1 GHz. Yet, in the end it might prove desirable to get the narrow bandwidth at IF using coherent rather than at the optical level using direct detection with filtered preamplification. Either way, it is worth noting that stabilizing either a filter or a local oscillator to 0.1 GHz amounts to stabilization to a part in 2×10^7, not impossible, but difficult to do cheaply.

There has been a long-standing but fluctuating interest in totally optical wavelength converter components, particularly switchable wavelength converters, as sketched in Fig. 12(b). Actually NEC demonstrated an 8×8 laboratory device six years ago based on nonlinear optical components [68]. So far, no field-usable such component has been developed.

VIII. Future Systems Expectations

It seems most unlikely that, given the available technology, architecture, application pressures and pervasiveness of optical fiber, society will allow the "unused 24 999" GHz of bandwidth to languish.

The demands of modern communication focus not only on bandwidth per user but number of users, or to put it another way, the user often sees interconnectability as more important than bit rate. Scalability of optical networking to very large numbers of users is therefore of great concern. Because of this, no matter how much progress is made at the component and receiver architecture level in increasing the size of the wavelength pool, wavelength reuse appears to be a permanent requirement. Individual LANs and even MANs using either WDM packet or circuit switching will continue to make sense, and each seldom involves more than several hundred nodes anyhow, since beyond that they become difficult to administer. But beyond that, wavelength reuse must be employed, either by wavelength routing within the backbone, by wavelength conversion at the boundary between LAN/MAN and backbone, or by use of both. While purely optical wavelength converters are not really practical today, remodulation (Section III-F) can sometimes be used for this function today, at least for binary data.

The injection of optical networking into the midst of the existing communication infrastructure will presumably take place in stages. The only penetration visible today lies in a few situations where the cost levels of the alternatives are very high. As the solutions become cheaper, the domains of applicability will spread, probably beginning with WDM links, then isolated broadcast and select MANs for selected wideband applications such as uncompressed video production or supercomputer interconnection. In the computer networking sector, the requirement of multiple session concurrency in the users' computers, even for small machine sizes, suggests that today's circuit switched broadcast and select LANs and MANs will have to be succeeded by packet switching.

In the long run, the widespread viability of broadcast and select networks probably lies not only in reducing per-network costs, but in using them as the lowest level of hierarchical networks of two or more layers in which the outer layer consists of broadcast and select LANs and MANs and the inner layer is a wavelength routed WAN.

Wavelength routing should then follow, with its general ability to span large distances and large levels of scalability. There is great appeal to the idea of standardizing on a given wavelength routing architecture (at both the optical and control point levels) as a basic layer underlying current protocol stacks, particularly ATM and TCP/IP. Most of the basic pieces of a total such architecture have recently become available [17], [21]. Given such an agreed upon architecture, as bare fiber paths become available, both providers and users can build their own networks on an optical base, thus gaining improved bandwidth capability, less use of the intermediate nodes for processing through traffic, ability to handle foreign protocol streams (once the ATM or TCP/IP message exchange has set up the lightpaths), and other advantages.

While both ATM and TCP/IP are fairly "slim" protocol stacks, in terms of software pathlength (number of instructions required to send or receive a message), improvements will ultimately be necessary. With bandwidth demands growing at the top of the protocol stack from new applications (such as multimedia) and new fiber-based wideband facilities appearing at the bottom of the stack, the layers in the middle will ultimately have to become more efficient. Today's presentation layers will probably be the performance bottleneck long before the lower layers of ATM or TCP/IP have to be dealt with.

IX. Conclusion

The present time seems to be a particularly good one to pause and take stock of where optical networks have come from and where they are going. The technology difficulties are beginning to come under control, there are finally some real products to show for all the research efforts, and the factors that lead to preferring optical solutions are beginning to become urgent, particularly the growth of bandwidth requirements. In the next few years, as new wideband services (such as video on demand) proliferate, as the legislative, regulatory and industry structure turmoil brings in new players with new agendas, the optical option will be reaching maturity and will become a serious candidate for widespread implementation.

If the present trends in extending fiber to the desktop persist in the way that was predicted in Section I, the optical networking community has time to provide solutions that exploit this fiber in the most cost effective way. In this author's opinion, the challenge is not more hero experiments and arcane theories of exotic protocols, but 1) systems and 2) components that work in actual service and are cheap enough to be widely effective for end users. This Special Issue treats the first challenge and the companion issue of the *Journal of Lightwave Technology* [42] the second.

REFERENCES

[1] C. A. Brackett, "Dense wavelength division multiplexing networks: Principles and applications," *IEEE J. Select. Areas Commun.*, vol. 8, no. 6, 1990, pp. 948–964.

[2] N. K. Cheung, K. Nosu, and G. Winzer, Eds., "Special issue on dense wavelength division multiplexing techniques for high capacity and multiple access communications systems" *IEEE J. Select. Areas Commun.*, vol. 8, no. 6, pp. 945–1214, 1990.

[3] P. E. Green, Jr., *Fiber Optic Networks*. Englewood Cliffs: Prentice-Hall, 1993.

[4] M. J. Karol, G. Hill, C. Lin and K. Nosu, Eds., "Special issue on broadband optical networks," *J. Lightwave Technol.*, vol. 11, no. 5/6, pp. 665–1124, 1993.

[5] D. J. G. Mestdagh, *Fundamentals of Multiaccess Optical Fiber Networks*. Norwood, MA: Artech, 1995.

[6] F. P. Corr, Private communication, Mar. 1995.

[7] A. R. Rutkowski, "Collected Internet growth history of number of hosts and packets per month," Private communication, Internet Society, Mar. 26, 1995.

[8] A. G. Fraser, "Banquet talk," in *2nd IEEE Workshop High Performance Commun. Subsyst.*, Sept. 2, 1993.

[9] J. Kraushaar, "Fiber deployment update—End of year 1993," *FCC Common Carrier Bureau*, May 13, 1994.

[10] D. Charlton, "Through a glass dimly," *Corning, Inc.* 1994.

[11] J. West, "Building Japan's information superhighway," Center for Res. Info Tech. Organization, Univ. of California at Irvine, Feb., 1995.

[12] J. H. Winters and S. Kasturia, "Adaptive nonlinear cancellation for high-speed fiber-optic systems," *J. Lightwave Technol.* vol. 10, no. 7, pp. 971–977, 1992.

[13] P. E. Green, Jr., Ed., *Network Interconnection and Protocol Conversion*. New York: IEEE Press, 1988.

[14] A. Husain and J. N. Lee, Eds., "Special issue on optical interconnections for information processing," *J. Lightwave Technol.*, vol. 13, no. 6, pp. 985–1183, 1995.

[15] R. Ramaswami, "Multiwavelength lightwave networks for computer communication," *IEEE Commun. Mag.*, pp. 78–88, Feb. 1993.

[16] H. Toba *et al.*, "100-channel optical FDM transmission/distribution at 622 Mb/s over 50 km using a waveguide frequency selection switch," *Electron. Lett.*, vol. 26, no. 6, pp. 376–377, 1990.

[17] R. Ramaswami and A. Segall, "Distributed network control for wavelength routed optical networks," *INFOCOM-96*, submission.

[18] T. E. Darcy, "Subcarrier multiplexing for multiple access lightwave networks," *J. Lightwave Technol.*, vol. 5, no. 8, 1987.

[19] M. Choy, S. M. Altieri, and K. N. Sivarajan, "A 200 Mb/s packet switched WDM SCM network using fast RF tuning," in *Proc. SPIE*, 1992, pp. 32–34.

[20] P. E. Green, F. J. Janniello, and R. Ramaswami "WDM Protocol-Transparent Distance Extension Using R2 Remodulation," *IEEE J. Select. Areas Commun.*, this issue, pp. 962–967.

[21] R. Ramaswami and K. Sivarajan, "Optimal routing and wavelength assignment in all-optical networks," in *IEEE INFOCOM'94 Conf. Rec.*, 1994.

[22] P. E. Green and R. Ramaswami, "Direct detection lightwave systems: Why pay more?" *IEEE LCS Mag.*, vol. 1, no. 4, pp. 36–49, Nov. 1990.

[23] S. B. Alexander *et al.*, "A precompetitive consortium on wide-band all-optical networks," *J. Lightwave Technol.*, vol 11, no. 5/6, pp. 714–735, 1993.

[24] G. K. Chang *et al.*, "Subcarrier multiplexing and ATM/SONET clear-channel transmission in a reconfigurable multiwavelength all-optical network test bed," in *IEEE/OSA OFC Conf. Rec.*, Feb., 1995, 269–270.

[25] P. A. Humblet and W. M. Hamdy, "Crosstalk analysis and filter optimization of single- and double-cavity Fabry–Perot filters," *IEEE J. Select. Areas Commun.*, vol. 8, no. 6, Aug. 1990.

[26] S. Suzuki, Y. Inoue, and Y. Ohmori, "Polarization-insensitive arrayed-waveguide grating multiplexer with SiO_2-on-SiO_2 structure," *Electron. Lett.*, vol. 30, no. 8, pp. 642–643, 1994.

[27] J. L. Jackel, J. E. Baran, A. d'Alessandro, and D. A. Smith, "A passband flattened acousto optic filter," *IEEE Photon. Technol. Lett.*, vol. 7, no. 3, pp. 318–320, 1995.

[28] E. L. Goldstein and L. Eskildsen, "Scaling limitations in transparent optical networks due to low level crosstalk," *IEEE Photon. Technol. Lett.*, vol. 7, no. 1, 1995, pp. 93–4.

[29] J. Sharony, K.-W. Cheung, and T. Stern, "The wavelength dilation concept in lightwave networks—implementation and system considerations," *J. Lightwave Technol.*, vol. 11, no. 5/6, p. 905, 1993.

[30] "Supercavity—High-finesse tunable Fabry-Perot filter," Newport Corporation, Fountain Valley, CA, 1990.

[31] C. M. Miller and F. J. Janniello, "Passively temperature-compensated fiber Fabry–Perot filter and its application in a wavelength division multiple access network," *Electron. Lett.*, vol. 23, pp. 781–783, 1987.

[32] B. Glance, U. Koren, C. A. Burrus and J. D. Evankow, "Optical frequency synthesizer," *Electron. Lett.*, vol 25, no. 17, pp. 1193–1195, 1989.

[33] H. R. Hicks, N. K. Reay and P. D. Atherton, "The application of capacitance micrometry to the control of Fabry-Perot etalons," *J. Phys. E, Sci. Instrum.*, vol. 17, pp. 49–55, 1984.

[34] R. S. Vodhanel, M. Krain, R. E. Wagner, and W. B. Sessa, "Long-term wavelength drift of the order of -0.01 nm/yr for 15 free-running DFB laser modules," in *OFC-94 Conf. Rec.*, 1994, pp. 103–104.

[35] A. R. Chraplyvy, "Limitations on lightwave communications imposed by optical-fiber nonlinearities," *J. Lightwave Technol.*, vol. 8, no. 10, pp. 1548–1557, 1990.

[36] "Compliance guide for laser products," U.S. Department of Health and Human Services, Publication FDA 86-8260, Sept. 1995.

[37] E. Desurvire, *Erbium-Doped Fiber Amplifiers—Principles and Applications*. New York: Wiley, 1994.

[38] C. R. Giles, M. Newhouse, J. Wright, and K. Hagimoto, Eds., "Special issue on system and network applications of optical amplifiers," *J. Lightwave Technol.*, vol. 13, no. 5, pp. 701–981, 1995.

[39] J.-M. P. Delavaux and J. A. Nagel, "Multi-stage erbium-doped fiber amplifier designs," *J. Lightwave Technol.*, vol. 13, no. 5, pp. 703–720, 1995.

[40] K. Inouye, "A simple expression for optical FDM network scale considering fiber four-wave mixing and optical amplifier noise," *J. Lightwave Technol.*, vol 13, no. 5, pp. 856–861, 1995.

[41] R. Cruz, G. Hill, A. Kellner, R. Ramaswami, G. Sasaki, and Y. Yamabashi, Eds., "Special Issue on optical networks," *IEEE J. Select. Areas Commun.*, vol. 14, 1996.

[42] M. Fujiwara, M. Goodman, M. O'Mahony, O. Tonguz, and A. Willner, Eds., "Special issue on multiple wavelength technologies and networks," *J. Lightwave Technol.*, vol. 14, 1996.

[43] D. J. Stigliani, D. Hanna, and D. Lynch, "Wavelength division multiplexing in light interface technology," IBM Federal Systems Division Report 71-531-001, Mar. 1971.

[44] IBM Model 9729 Optical Wavelength Division Multiplexor, RPQ-8Q1488, IBM, Network Hardware Division, Res. Triangle Park, NC 27709, Nov. 14, 1995; See also: F. J. Janniello, R. A. Neuner, R. Ramaswami, and P. E. Green, "Multiple-protocol optical-fiber multiplexer for remote computer interconnection," *IEEE J. Select. Areas Commun.*, submission.

[45] Pirelli Corporation, Model TXT/EM-XXX transceivers. (See also: F. Diner, P. Jaggi, A. Cavaciuti, and F. Meli, "Network design considerations of n-channel WDM architectures with EDFA's for central office applications," in *Proc. Nat. Fiber Optic Engrs. Conf.*, June 1995, pp. 1207–1214.)

[46] C. Fan, L. Clard, R. Kurtz, and J. Shrimpton, "Planning the Next Generation Lightwave Network (NLGN) for capacity upgrade and service growth," presented at the European Institute R&D in Telecomm, WDM Workshop Operators and Suppliers, Brussels, Aug. 14, 1995.

[47] T. Shibagaki, H. Ibe, and T. Ozeki "Video transmission characteristics in WDM star networks," *J. Lightwave Technol.*, vol. 3, no. 3, pp. 490–495, 1985.

[48] M. S. Goodman, H. Kobrinski, and K. W. Loh, "Application of wavelength division multiplexing to communication network architectures," in *Proc. Int. Commun. Conf.*, 1986.

[49] C. M. Miller and F. J. Janniello, "Passively temperature-compensated fiber Fabry–Perot filter and its application in wavelength division multiple access computer networks," *Electron. Lett.*, vol. 26, no. 25, pp. 2122–2123, 1990.

[50] F. J. Janniello, R. Ramaswami, and D. G. Steinberg, "A prototype circuit-switched multi-wavelength optical metropolitan-area network," *J. Lightwave Technol.*, vol. 11, May/June, pp. 777–782, 1993.

[51] W. E. Hall *et al.*, "The Rainbow-II gigabit optical network," *IEEE J. Select. Areas Commun.*, submission.

[52] H. J. Westlake, P. J. Chidgey, G. R. Hill, P. Granestrand, L. Thylen, G. Grasso, and F. Meli, "Reconfigurable wavelength routed optical networks: A field demonstration," in *Conf. Rec., Eur. Conf. Opt. Commun.*, 1991, pp. 753–756.

[53] G. R. Hill *et al.*, "A transport network layer based on optical network elements," *J. Lightwave Technol.*, vol. 11, no. 5/6, pp. 667–679, 1993.

[54] D. B. Payne and J. R. Stern, "Transparent single-mode fiber optical networks," *J. Lightwave Technol.*, vol. 4, no. 7, pp. 864–869, 1986.

[55] R. A. Barry *et al.*, "All-optical network consortium," *IEEE J. Select. Areas Commun.*, this issue, pp. 999–1013.

[56] G. K. Chang *et al.*, "Subcarrier multiplexing and ATM/SONET clear-channel transmission in a reconfigurable multiwavelength all-optical network test bed," in *Conf. Rec., OFC-95*, 1995.

[57] R. Gidron and A. Temple, "TeraNet: A multihop multichannel ATM lightwave network," in *Conf. Rec., ICC-91*, 1991, pp. 602–608.

[58] I. P. Kaminow, P. P. Iannone, J. Stone, and L. W. Stulz, "FDMA-FSK star network with a tunable optical filter demultiplexor," *J. Lightwave Technol.*, vol. 6, no. 9, pp. 1406–1414, 1988.

[59] A. Willner, I. P. Kaminow, M. Kuznetzov, J. Stone, and L.W. Stulz, "1.2 Bg/s closely-spaced FDMA-FSK direct-detection star network," *J. Lightwave Technol.*, vol. 2, no. 3, pp. 223–226, 1990.

[60] B. S. Glance *et al.*, "Densely spaced FDM coherent star network with optical signals confined to equally spaced frequencies," *J. Lightwave Technol.*, vol. 6, no. 11, pp. 1770–1781, 1988.

[61] T.-K. Chiang, S. K. Agrawal, D. Mayweather, D. Sadot, and L. Kazovsky, "Implementation of STARNET: A WDM computer communication network," *IEEE J. Select. Areas Commun.*, this issue, pp. 824–839.

[62] A. C. Labrujere *et al.*, "COSNET—A coherent optical subscriber network," *J. Lightwave Technol.*, vol. 11, no. 5/6, pp. 865–874, 1993.

[63] G.-d. Khoe *et al.*, "Coherent multicarrier technology for implementation in the customer access," *J. Lightwave Technol.*, vol. 11, no. 5/6, pp. 695–713, 1993.

[64] R. Maerz, *Integrated Optics—Design and Modeling.* Norwood, MA: Artech House, 1995.

[65] C. Cremer, N. Emeis, M. Schier, G. Heise, G. Ebbinghaus, and L. Stoll, "Grating spectrograph integrated with photodiode array in InGaAsP/InGaAs/InP," *IEEE Photon. Technol. Lett.*, vol 4, no. 1, pp. 108–110, 1992.

[66] J. B. D. Soole, A. Scherer, H. P. LeBlanc, R. Bhat, and M. A. Koza, "Spectrometer on a chip: a monolithic WDM component," in *Conf. Rec., OFC-92*, 1992, p. 123.

[67] K. Liu, F. Tong, and S. W. Bond, "Planar grating wavelength demultiplexer," in *Proc. SPIE*, 1993, pp. 278–285.

[68] S. Suzuki *et al.*, "A photonic wavelength-division switching system using tunable laser diode filters," *J. Lightwave Technol.*, vol. 8, pp. 660–666, 1990.

Survey of ATM Switch Architectures ☆

RA'ED Y. AWDEH [a], H.T. MOUFTAH [b,*]

[a] BELL-NORTHERN RESEARCH, P.O. BOX 3511, STATION C, OTTAWA, ONTARIO, CANADA K1Y4H7
[b] DEPT. OF ELECTRICAL AND COMPUTER ENGINEERING, QUEEN'S UNIV., KINGSTON, ONTARIO, CANADA K7L3N6

Accepted 21 July 1994

Abstract

For reasons of economy and flexibility, BISDN (*Broadband Integrated Services Digital Network*) is expected to replace existing application-oriented communication networks. ATM (*Asynchronous Transfer Mode*) is a high-speed packet-switching technique that has emerged as the most promising technology for BISDN. Since early 1980s, a large number of architectures have been proposed for ATM switching. In this paper, we present a descriptive survey of ATM switch architectures, with emphasis on electronic space-division point-to-point switches.

Keywords: ATM switching; Buffering strategies; Fast packet switching; Switch architectures

1. Introduction

Currently, heavy research is being carried out to design an integrated network that supports all services, existing and emerging, in a unified fashion [20], [25], [56], [67], [68], [107], [161], [187], [213]. Flexibility to accommodate volatile changes in service mixes, ease of installation and maintenance, better user access, and more efficient resource utilization, are among the reasons for such a drive. The advances in fiber optics technology have made available huge amounts of transmission bandwidth, and have resulted in the emergence of new applications, specially real-time services, which require much higher bandwidth than possible in existing networks. Fiber is already widely deployed in both public and private networks, which served to reduce its cost dramatically. Also, as the geometries of various VLSI components continue to shrink, both higher speeds and higher levels of circuit integration are made possible [25]. However, the design of a network that can support very high bandwidth services remains a challenge. *Broadband Integrated Services Digital Networks* (BISDNs) are expected to provide diverse services with diverse performance requirements. Three issues are to be satisfactorily resolved before BISDN becomes a reality [181].

- The development of a network protocol capable of supporting the diversity of services expected by BISDN.
- The design of switching nodes which are within the implementation capability of current VLSI technology.
- The development of a control strategy that

☆ This work was performed while the first author was at Queen's University. The views expressed in this paper do not necessarily represent those of BNR.

* Corresponding author. Tel (613) 545-2925, Fax (613) 545-6615, Email mouftahh@qucdn.queensu.ca

guarantees for every user the required quality of service regardless of the network traffic conditions.

The third issue is still an open and even an unclear problem. In BISDN, the performance bottleneck of the network, which was once the channel transmission speed, is shifted to the processing speed at the switching nodes and the propagation delay of the channel [20]. Many flow/congestion control strategies for BISDN have been proposed [20], [108]. Regarding the first issue, ATM (*Asynchronous Transfer Mode*) has been selected as the multiplexing and switching technique for BISDN, and is currently receiving much standardization activity [20], [32], [68], [187]. With regard to the second issue, a large number of ATM switch architectures have been proposed [2], [8], [115], [157], [168], [173], [213], and it has been shown that many of them can be implemented using current technology. Berthold [25] has shown that currently-available digital VLSI technology is more than adequate to the ATM switching needs. Rooholamini et al. [190] and Steffora [200] describe several already commercially-available ATM switches. It is our intention in this paper to survey the different ATM switch proposals. Our survey paper differs from existing surveys in the following points:

- It focuses on *both* the buffering strategy and internal structure of ATM switches in a *clear* and *comprehensive* manner.
- It includes for the first time some recently-proposed categories of ATM switches.
- It covers in detail many more ATM switch proposals (some of which have been proposed only recently) than any existing survey.

2. Definitions

This section provides some background and describes some assumptions related to this survey.

2.1. *Broadband Integrated Services Digital Network (BISDN) [20], [25], [56], [67], [68], [107], [161], [184], [187], [213]*

A major step that has already been accomplished towards BISDN is Narrowband (N) ISDN or simply ISDN. ISDN is a concept that was developed during the 1970s with the advent of digital telephone systems. It has been deployed in many countries in the world, where *digital* communications at the aggregate rate of 144 Kbps (two 64 Kbps circuit switched channels for voice or data, and one 16 Kbps packet switched signalling channel) is made available to the public. Besides extending digital transmission capabilities to existing twisted wire pairs, another new aspect of ISDN is the use of an out-of-band common channel for signalling, which insures the transparency of the information channels. Unfortunately, ISDN has many shortcomings including the following. The number of simultaneous connections to a subscriber is determined by the number of physical channels installed to the subscribers premises, which makes ISDN inflexible with regard to future or even some existing services. Moreover, there is no provision either for a range of services with different bit rates or for services with variable-bit rates. Even for the limited range of narrowband services it offers, ISDN is not really integrated as there are still two different bearer services resulting in two overlay networks.

The main concept of BISDN is the support of a wide range of existing and emerging voice, video, and data applications within the same network. It will provide an integrated access for its users in a flexible and cost-effective manner. The network will include support for: interactive and distributive services, broadband and narrowband rates, bursty and continuous traffic, connection-oriented and connectionless services, and point-to-point and multipoint communications. The general goal of BISDN is to define a user interface and a network that will meet the required quality of service of each of these applications.

2.2. *Synchronous Transfer Mode (STM) [20], [25], [33], [60], [68], [105], [161], [184], [216]*

STM was the first switching and multiplexing technique to be considered for BISDN due to its compatibility with most existing systems, and the desire to preserve the investment in existing equipments while *evolving* to a more flexible network. In the multiplexing system (multiplexing refers to the arbitration of access to a link and should be distinguished from switching), time slots within a periodic structure called *frame*, are allocated to a service for the duration of a connection. This is usually referred to as *time-division multiplexing*. An STM channel is identified by the position of its time slot in a frame. STM provides fixed throughput and constant delay, and thus it is suitable for fixed-rate services. STM switching is known as *circuit switching*, and is widely used in voice telephone networks. The switching function in circuit switches is performed continuously for the duration of the connection. Circuit switching can be performed in space, in time, or in a combination of both.

Rigidly-structured STM is very inefficient in handling the different and variable-bit rates required by the diversity of services which are expected to be supported by BISDN. Even in complex multirate STM which allows the allocation of bandwidths equal to integer multiples of some basic rate, the choice of this basic rate is an uneasy engineering design decision. In general, as the number of channels increases, STM-based approaches grow cumbersome.

2.3. Packet switching [21], [26], [67], [68], [184], [213], [216]

Baran [21] was the first to propose the concept of packet switching in 1964. In packet switching networks, user information is organized into packets (usually of variable length) which contain additional information used inside the network for routing, error detection and correction, flow control, etc. Packets are transmitted from a source to a destination through multiple switches in a store-and-forward manner. Sources transmit their packets as soon as they are available, which is referred to as *statistical-multiplexing*. Switching of packets is done by computers running communication processes. In contrast to circuit switching, the switching function here needs to be performed only when packets are present at the inputs of the switch.

Packet switching networks, such as those based on X.25, were designed at the time only poor-to-medium transmission links were available. In order to offer an acceptable end-to-end performance, complex protocols were therefore necessary to perform flow and error control (e.g., packet re-transmission) on each link of the network. Also, because packets have variable lengths, complex buffer management was required inside the network. All of the above result in large delay and large jitter (the variance of the delay) on this delay, making conventional packet switching unsuitable for high-speed integrated services networks.

2.4. Asynchronous Transfer Mode (ATM) [20], [25], [32], [56], [67], [68], [86], [161], [187], [216]

ATM has been chosen as the transfer mode for BISDN since it overcomes the problems of STM and conventional packet switching altogether. In short, ATM is a high-speed connection-oriented packet-switching technique with minimal functionality in the network. This is why ATM is also known as *fast packet switching*. Turner [216] was among the first who noticed that the disadvantages usually attributed to packet switching are not due to any inherent properties, but are side effects of the conventional implementations. ATM differs from conventional packet switching in many aspects. First, high-level protocol functions such as error control are performed on an end-to-end and not link-by-link basis. This is possible because of the high-quality links which are going to be used in BISDN. Furthermore, ATM uses short fixed-length packets called cells (this is why ATM is also known as *cell switching*). Each cell consists of 53 bytes or octets; 48 for the payload and 5 for the header or label. This choice simplifies the design of switching nodes, and reduces delay and jitter especially for delay-sensitive services such as voice and video. (Also, is was argued that this results in controlled or "fair" handling of multiple traffic streams [187].)

ATM is a connection-oriented transfer mode. In a connection, three phases can be distinguished: set-up, information transfer, and teardown. This paper deals with the second phase only. The first and third phases have to do with

ATM signalling standards [32], [187], and are out of the scope of the current work. We conclude this subsection by pointing out that the term "Asynchronous" in ATM does not necessarily imply asynchronous transmission or switching systems. Rather, it implies aperiodicity; i.e., no source shall own a time slot on a periodic basis. In ATM, the association of a cell with a connection is made explicit through the label of that cell.

2.5. ATM switch

An ATM switch of size N can be regarded as a box with N input ports and N output ports, that routes cells arriving at its input ports to their desired output ports (destinations). For the sake of simplicity, in this paper it is assumed that cells arrive at the input ports in a time-slotted fashion, and that all input lines are synchronized. (In reality, arriving cells at different input ports need to be aligned and synchronized to the local clock.) The minimum slot size is equal to the transmission time of a single cell. Input and output lines are assumed to operate at the same speed. Each arriving cell carries two fields: the ATM header, and the payload (the actual information to be transmitted). Prior to entering the switching fabric, each cell is provided with a local switching header that is used within the switch. A local header may include the following two fields:

- An activity bit (a) to indicate the presence of a cell when $a = 1$.
- An address field which contains the address of the local destination port. Usually, N is assumed to be a power of two, and in this case the address field will be in a binary form $d_1 d_2 ... d_n$, where $n = \log_2 N$ and d_1 is the most significant bit (MSB).

ATM switching fabrics can be classified into time-division fabrics and space-division fabrics [213]. In the former, all cells flow through a single resource that is shared by all input and output ports. This resource may be either a common memory, or a shared medium such as a ring or a bus. The capacity of the entire switching fabric is limited by the bandwidth of this shared resource. Thus, this class of switches do not scale well to large sizes. Another factor that usually limits the

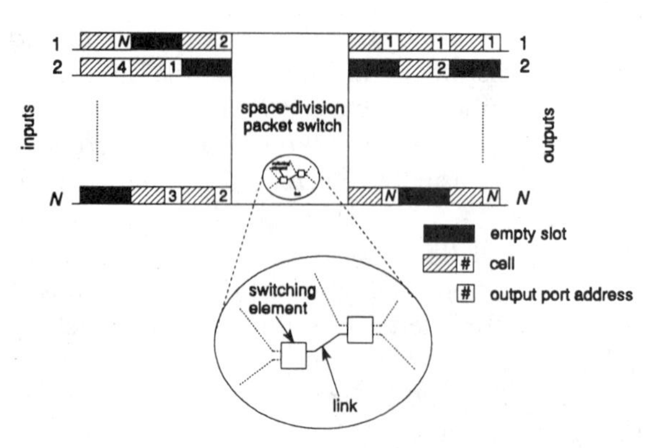

Fig. 1. A generic space-division ATM switch.

maximum size of a common memory architecture, is the centralized control requirement. On the other hand, in a space-division switch multiple of concurrent paths can be established between input and output ports, so that many cells may simultaneously be transmitted across the switching fabric. Space-division ATM switches are usually constructed from identical basic switching blocks, called *switching elements* (SEs), which are interconnected in a specific topology by *links*. The switch directs incoming cells to their destinations, as specified by the information contained in their local headers (see Fig. 1). A problem arises when it is impossible for all required paths to be simultaneously established. This is called *blocking*, and it negatively affects the throughput performance of the particular space-division switch.

With a multiplicity of paths in a space-division switching fabric, a routing function is now required to select a path to the requested output port for each incoming cell. The routing can be centralized or distributed. With the former, a central unit examines the destination addresses of all incoming cells, and sets up the required paths in the switching fabric accordingly. In this case, the capacity of the central processor establishes a physical limit to the performance and scalability of the switch. Distributed-routing, better known as *self-routing*, switches avoid central control bottlenecks by distributing the routing function among several smaller processors. Here, each SE in the fabric makes a very fast local routing decision, simply by examining the destination addresses (or part of them) of the cells at its inlets. In all cases, each cell arrives at its desired destination regardless of its port of entry.

Many ATM switch architectures have been proposed within the past fifteen years or so [2], [8], [115], [157], [168], [173], [199], [213]. Most of the proposed architectures are based on highly-parallel structures, are characterized by distributed control, and perform switching at the hardware level. Since most of the proposed ATM switches are of the space-division type, time-division architectures are not covered by this survey. Interested readers are referred to [213] in which both an excellent discussion and a survey of time-division fast packet switches are given. Before discussing the different approaches of classifying space-division ATM switches, we give some important definitions which are relevant to any space-division packet switch:

- *Internal blocking*. Occurs when two or more cells, destined to different output ports, contend over the same internal link. A switch that does not suffer from internal blocking is called nonblocking. (A nonblocking packet switch is not necessarily nonblocking as a circuit switch.)
- *Output blocking*. Occurs when more than one cell request the same output port within the same time slot. A blocking switch suffers from both internal and output blocking. Nonblocking switches can be further classified into output-blocking and output-nonblocking architectures. An output-nonblocking switch is able to clear all incoming cells in any given time slot to the buffers of the requested output ports before the next time slot.
- *Cell-sequencing*. A switch is said to preserve cell-sequencing if, for each input-output pair, it delivers incoming cells in the order by which they have arrived at the switch.
- *Speed-up*. Speeding-up an ATM switching fabric is sometimes necessary to improve the performance and/or compensate for processing overheads. The speed-up can be implemented in time or in space. In the former, the switch is said to have a speed-up factor of S if the internal switching fabric operates S times faster than the external lines. In the latter implementation, S concurrent paths are provided to each output port.
- *Multicast connections*. BISDN must support multipoint communications in which more than two users are participating in a connection. At the switch level, a multicast or point-to-multipoint connection refers to the situation where an incoming connection requests K output ports ($1 \leq K \leq N$). When $K = 1$, we are back to point-to-point (unicast) connections, while $K = N$ refers to broadcasting.
- *Performance measures*. The performance of an ATM switch is usually evaluated based on three measures: throughput, delay, and cell loss probability. Throughput (TP) is defined as the average number of cells which are successfully delivered by the switch per time slot per input line. Maximum throughput (TP_{max}) is the value of TP under maximum-load conditions. While TP_{max} is an important performance measure, it will not be directly felt by network users. On the other hand, the end-to-end delay, which includes the delay of individual switching nodes, will be experienced by network users. Switch delay (D) is defined as the average time (in time slots) a cell spends from the time it arrives at an input port, till the time it is successfully delivered on its requested output line. D includes the time spent in any input, internal, and/or output buffers. Cell loss probability (P_{loss}) is defined as the fraction of cells lost within the switch. Cell loss might occur as a result of blocking and/or buffer overflows. Because cell re-transmission takes place on an end-to-end basis in ATM networks, and because of the high speeds involved in these networks, P_{loss} is considered as a very important performance measure. Typically, an ATM switch is required to support a high throughput, a small delay, and an extremely low cell loss probability. Finally, an important performance measure that is often ignored in most switch proposals is the jitter, which should be very small.

Space-division ATM switch architectures can be classified according to many different criteria. Some typical classifications are the following:

- *Single-stage versus multistage* [157]. Input and output ports are interconnected through one stage only in single-stage switches; thus, a cell is switched in a single phase. On the other hand, in multistage switches, switching is performed by SEs in consecutive stages.

- *Single-path versus multipath* [168]. In a single path switch, only one path exists for any input-output pair, while in a multipath switch there are more than one path for each input-output pair.
- *Blocking versus nonblocking* [46], [173]. As has been explained in the above definitions.
- *Unicast versus multicast*. As has been explained in the above definitions. While some of the architectures to be surveyed in this paper do support multicasting, the emphasis is on unicast switching.
- *Buffering based classifications* [168], [173], [181]. Based on the location of buffers, ATM switches can be input-buffered, internally-buffered, output-buffered, or any combinations of these. Also, based on the memory sharing policy, ATM switches can have dedicated buffers, shared buffers, or a mixture of both.
- *Structure based classification* [2], [8], [213]. According to their internal structures, ATM switches can be possibly classified into crossbar based, disjoint-path based, and banyan based.

In this paper, while keeping in mind all the above classifications, we first review the buffering based classifications, then we classify ATM switches into the following categories:

- Crossbar based switches.
- Disjoint-path based switches.
  - Paths are disjoint in space.
  - Paths are disjoint in time.
- Banyan based switches.
  - Internal node buffering.
  - Multiple banyans.
  - Dilation.
  - Sorting.
  - Deflection-routing.
  - Load-sharing.
  - Expansion.

2.6. Traffic model

Here, we refer to the traffic as seen by the input ports of the switch. The traffic model is described by two random processes. The first is the process that governs the arrival of cells in each time slot. The second process describes the distribution by which arriving cells choose their destination ports. It is clear that input traffic can follow an infinite number of models. In the following, we describe three of the most frequently-used traffic models for the performance evaluation of ATM switches.

Uniform traffic

In this model, cells arrive at the input ports of the switch according to independent and identically distributed Bernoulli processes, each with parameter p $(0 < p \leq 1)$. In other words, at an input port in a given time slot, a cell arrives with probability p, and there is no arriving cell with probability $1 - p$. Thus, p represents the input load or the arrival rate to each input port of the switch. An incoming cell chooses its destination uniformly among all N output ports, and independently from all other requests; i.e., it chooses a particular output port with probability $1/N$. This traffic model is sometimes referred to as the independent uniform traffic model [213], or simply the random traffic model. Asserting the assumption of uniformity for real-life situations may lead to optimistic evaluation of performance measures. However, a large number of studies on the performance evaluation of ATM switches assume uniform traffic. The main reasons behind this trend are as follows:

- This assumption makes the analytical evaluation of the switch more tractable, specially if the switch is of the blocking type and/or employs a complex buffering strategy.
- A distribution/randomization network can be used at the front end of the switch to randomize incoming traffic.
- It was observed that the traffic arriving at the switching nodes is less bursty than the traffic arriving at user access nodes, due to the inherent smoothing that takes place when bursty cell streams are queued and then released at a given rate (the link service rate) to the network [87]. Furthermore, it was shown that subsequent stages of switching cause the traffic to become even less bursty [69], [87], making the uniform traffic assumption closer to reality.

By definition, any traffic model that is not uniform is called a nonuniform traffic model. In this paper, unless otherwise specified, we assume the traffic to be uniform.

Bursty traffic

Future BISDN is expected to support virtually all existing and emerging services including voice, video and data. Strong correlation may exist among cells originating from the same source, giving rise to bursty traffic. A bursty source generates cells at a peak or a near-peak rate for very short durations, and remains almost inactive in between. Several models have been proposed to describe such *bursty* sources [20], [198]. One popular and simple model is the On/Off model, where the source alternates between a busy (also called on, active, or burst) period, and an idle (also called off or silent) period. The length of the active period (in time slots) is geometrically distributed with parameter a. The probability that the active period lasts for a duration of i time slots is given by

$$B(i) = a(1-a)^{i-1}, i \geq 1. \tag{1}$$

In the above equation, it is assumed that a burst contains at least one time slot. The average length of a burst is given by

$$A = \sum_{i=1}^{\infty} iB(i) = \frac{1}{a}. \tag{2}$$

Similarly, the length of the idle period (in time slots) is geometrically distributed with parameter s. Thus, the probability that the idle period lasts for a duration of i time slots is given by

$$I(i) = s(1-s)^{i}, i \geq 0. \tag{3}$$

The average idle period length is obtained by

$$B = \sum_{i=0}^{\infty} iI(i) = \frac{1-s}{s}. \tag{4}$$

Cells are continuously generated during the active period. Thus, the offered load to an input port is obtained by

$$p = \frac{A}{A+B}. \tag{5}$$

All cells belonging to the *same* burst address the same output port which is chosen uniformly among all N output ports, independently from all other bursts. Finally, uniform traffic results when $A = 1$; in this case, $p = s$.

Hot-spot traffic

Hot-spot traffic refers to a situation where many input ports prefer to communicate with one output port (the hot-spot). This kind of traffic may arise in many real-life applications [231]. It occurs, for example, when many callers compete to call a popular location in a telephone network. Another example is a local area network consisting of many diskless computer systems and a single file server. In the model introduced in [186], a single hot-spot of higher access rate is superimposed on a background of uniform traffic. The cell arrival process is the same as for the uniform traffic case. If we let h be the fraction of cells directed at the hot-spot, then p, the arrival rate to an input port, can be expressed as $p = ph + p(1-h)$. In other words, hp cells are directed at the hot-spot, and $p(1-h)$ cells are uniformly distributed over all output ports. Also, the average number of cells which request the hot-spot per time slot is $phN + p(1-h)$.

3. Buffering strategies

Due to the statistical nature of the traffic, buffering in any packet switch is unavoidable. This is true because even with an output-nonblocking switch (which can clear all incoming cells to the output side of the switch before a new time slot begins), two or more cells may address the same output port within the same time slot. In such a situation, given the assumption that input and output lines operate at the same speed, each output line can serve only one cell per time slot; other cells must be buffered. This is called *output buffering* since the buffers will be physically located at the output side of the switch.

With a switch that is able to deliver a maximum of one cell to each output port in any given time slot, output buffering is not needed. In such switches, buffers can be placed at the input ports (called *input buffering*), within the switching fabric at possible points of contention (called *internal buffering*), or both (called *input-internal buffering*). On the other hand, if the switch can deliver more than one cell (but not all possible cells) to each output port simultaneously, then it also needs output buffering. This yields *input-output buffering*, *internal-output buffering*, or *input-internal-output buffering*. Finally, some switches while being blocking require only output buffering because they employ certain methods to significantly reduce blocking while offering multiple paths to each output port. Such switches can be engineered to make cell loss probability within the switching fabric extremely small.

In this section, we consider a generic nonblocking space-division packet switch, without regard to its internal structure except for the assumption that it has no internal buffers. First, we review and compare two extreme strategies of external buffering, namely, input buffering and output buffering. Then, we review various methods for improving the performance of input buffering switches. Where applicable, we assume that the size of each input buffer is B_{in} cells, and the size of each output buffer is B_{out} cells. Since internal buffering is architecture-dependent, we defer its discussion to the following sections. However, it is noteworthy that generally speaking, internal buffering is not desirable for one or more of the following reasons [139]:

- In multipath architectures, cells may be delivered out-of-sequence.
- It complicates the internal design of the SEs.
- It complicates fault-diagnosis testing if a cut-through mechanism [122] is implemented (this mechanism will be explained later in this paper).
- It limits the maximum length of packets which can use the switch to the maximum internal buffer size.

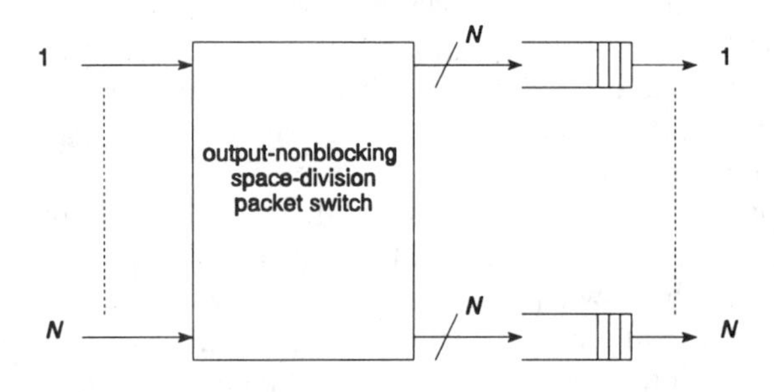

Fig. 2. A generic nonblocking output buffering switch.

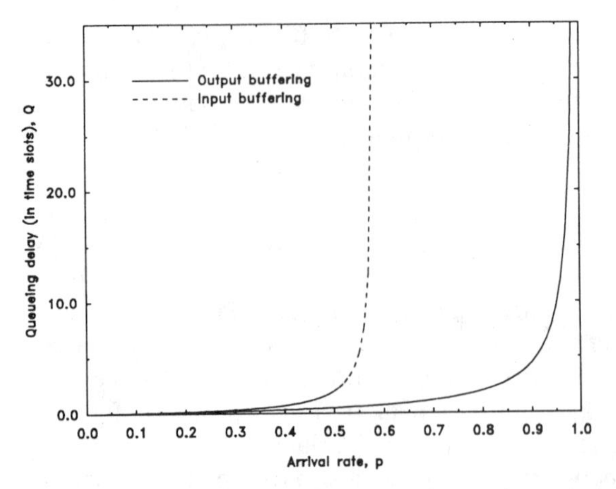

Fig. 3. Throughput-delay performance of pure input and output buffering for nonblocking switches.

Finally, unless otherwise specified, the numerical results presented in this section are for the case where both the switch and buffer sizes are infinite.

3.1. Output buffering

Here, we assume an output-nonblocking switch (Fig. 2) where all arriving cells in a given time slot are cleared to the output side (i.e., are switched) before the beginning of the next time slot, even if all N inputs have cells destined to the same output port. This can be achieved by, for example, speeding-up any switching fabric by a factor of N. However, only one cell can be served by an output line in each time slot and other cells with the same output request have to buffered, if space is available. The system is stable, since the average utilization of an output line is the same as that of an input line. The average queueing delay Q when $N = \infty$ and $B_{out} = \infty$ is the same as that of an M/D/1 queue [75], [101], [120]:

$$Q = \frac{p}{2(1-P)}; \quad 0 \leq p < 1. \qquad (6)$$

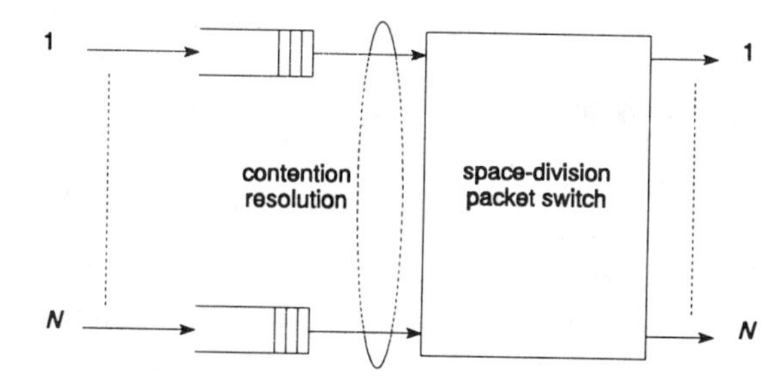

Fig. 4. A generic input buffering switch.

The above equation is plotted in Fig. 3. Studies concerning the performance evaluation of nonblocking output buffering switches under nonuniform traffic models include [31], [48], [76], [153], and [228].

3.2. Input buffering

With an input buffering switch (Fig. 4), an arriving cell enters a *first-in first-out* (FIFO) buffer located at its port of entry, if space is available. In each time slot, the switch resolves contentions prior to switching. If all *head-of-line* (HOL) cells are destined to distinct output ports, then all of them are admitted and switched to their desired output lines. However, if K HOL cells ($1 < K \leq N$) are destined to a particular output port, only one cell is chosen, according to some selection policy, to be switched and other cells wait to participate in the next time slot selection process.

Several selection policies (also called contention resolution mechanisms or HOL arbitration schemes) have been proposed and studied in the literature including the following:

- *Random* selection [14], [42], [65], [97], [101], [104], [118], [120], [149], [150], [151].
- *HOL FIFO* (or *oldest HOL*) selection [17], [18], [34], [42], [65], [85], [97], [112], [154], [182].
- *HOL LIFO* (last-in first-out) selection [42].
- *Global FIFO* (or *earliest arrival*) selection [17].
- *Longest queue* selection [17], [18], [85], [120].
- *Oldest queue* selection [17], [85] where age is defined as the length of time a non-empty queue has been unserved.
- *Cyclic* (or *round-robin*) selection and its variations [16], [17], [30], [99], [153], [155].
- *HOL blocking based* selection [97]: to select a cell that is causing HOL blocking (defined below).
- *Prediction based* selection [99].

It should be emphasized here that while the selection policy among contending HOL cells can be non-FIFO, the service discipline within each input buffer is still assumed to be FIFO.

Input buffering switches with FIFO queueing suffer from the so-called *HOL blocking* [101], [104], [120], [185]: in any given time slot while a cell is waiting its turn for access to an output port, other cells may be blocked behind it despite the fact that their destination ports are possibly idle. As a consequence, the maximum throughput is limited to $(2-\sqrt{2}) \approx 0.586$ [101], [104], [120], regardless of the specific selection policy used. This figure also represents the saturation value of p, above which steady state queue sizes grow without limit, and cells experience infinite waiting times. The following equation [104] describes the relationship between the average queueing delay Q and the arrival rate p (for $N=\infty$, $B_{\text{in}}=\infty$, and random selection):

$$Q = \frac{(2-p)(1-p)}{(2-\sqrt{2}-p)(2+\sqrt{2}-p)} - 1; \quad 0 \leq p < 2-\sqrt{2}. \tag{7}$$

The above equation is plotted in Fig. 3.

It is noteworthy that a maximum throughput of $(1-e^{-1}) \approx 0.632$ can be achieved by eliminating the input buffers and dropping cells upon contention [120], [176]. In general, the throughput with this blocked-loss scheme is given by the following equation [120], [176] (was derived in [202] for $p=1.0$):

$$TP = 1 - \left(1 - \frac{p}{N}\right)^N; \quad 0 \leq p \leq 1. \tag{8}$$

Unfortunately, the above maximum throughput of 63.2% (compared to 58.6% for FIFO input buffering) is achieved at the expense of high cell loss probabilities which are unacceptable for ATM switches. For example, at $p=0.50$ and

$N = \infty$, 21.3% of all incoming cells are dropped, compared to P_{loss} of 10^{-6} for FIFO input buffering with $B_{in} = 20$ and random selection [104].

Despite the fact that with FIFO queueing, the maximum throughput of an input buffering switch is 58.6% regardless of the selection policy, other performance measures are affected by the specific selection policy used when operating below the saturation level (i.e., when $p < 0.586$). For example, the longest queue selection policy results in less queueing delay, compared to the random selection policy [120], and to the HOL FIFO selection policy [18], [154]. Also, both the global FIFO and the longest queue selection policies have better cell loss performance, compared to the oldest queue or the HOL FIFO policies [17], [18]. Priority schemes [114] insure the delivery of delay-sensitive cells by giving them higher priority. Needless to say that the different selection policies have different implementation complexities. Finally, the performance of nonblocking input buffering switches was evaluated under nonuniform traffic models in [149] [150], [151], [153], and [155].

3.3. Comparison

It can be easily seen from Fig. 3 that output buffering significantly outperforms input buffering with regard to the throughput-delay performance under uniform traffic. (Although output buffering has significantly better throughput-delay performance than input buffering under uniform traffic for all switch dimensions, the advantage of output buffering over input buffering decreases as the switch dimensions become more and more asymmetric [154], [155].) Also, output buffering switches lend themselves naturally to multicast and broadcast functionalities. Despite the above, output buffering switches are usually more complex. In addition to the complexity of the switching fabric itself, which must transfer all incoming cells in any given time slot to the output side of the switch before the beginning of the next time slot, output memories are not easy to implement. Each output memory must have a minimum bandwidth of $(N+1)\nu$ bps (where ν is the speed of the external lines), corresponding to a maximum of N write and a single read operations [213]. Reduction in the memory speed is possible by using a bit-sliced organization of the memory [213], or by using output port concentration [230]. On the other hand, memory speed does not constitute a major concern for input buffering switches.

The superiority of output buffering over input buffering is valid under any traffic model, assuming infinite-size buffers. However, in reality buffer sizes are finite. It was shown that for the same values of N, p (uniform traffic), and target P_{loss}, an output buffering switch requires less amount of buffers [68], [153]. However, in an interesting simulation study [153], it has been concluded that it is *not* true that output buffering has better performance than input buffering in all situations. In particular, under bursty traffic, output buffering could have higher P_{loss} than input buffering, for the same buffer size. This has been explained as a result of inherent buffer sharing effects of input buffering under bursty traffic: although buffers are not actually shared, simultaneous cell arrivals with a common destination are automatically distributed across several buffers in input buffering. In the same study, it was also concluded that TP_{max} is not a good performance metric for input buffering under bursty traffic, since it is necessary to operate the switch at loads much lower than TP_{max} in order to obtain small P_{loss}. In general, bursty traffic requires much larger amounts of buffers, compared to uniform traffic, to achieve the same cell loss probability, for any buffering strategy [48], [153], [175], [203].

While complicating the design of a switch, using a shared-memory implementation of the buffers significantly reduces the amount of memory needed to achieve a certain cell loss performance, for any buffering strategy under uniform traffic [75], [78], [101], [110], [111], [160], [175], [181], [213]. In general, this is true under any balanced traffic model [78], [111] in which the overall input load is uniformly distributed across all input ports (input balance) and the destination requests are uniformly distributed across all output ports (output balance). This definition includes besides uniform traffic, the bursty traffic model as defined previously. As an example [175], for an output buffering switch of size (8×8), total buffer of 8000 cells, a 10^{-10} target P_{loss}, and an average burst length of 5 cells, the shared buffering technique allows up to 88% load compared to 45% allowed with dedicated buffering. Unfortunately, the advantage of buffer sharing diminishes under imbalanced traffic [78], such as hot-spot traffic [186]. This is because *favored* ports monopolize the use of the shared buffer, causing performance degradation for the whole switch [101], [213]. Therefore, the appropriate buffer sharing policy must lie somewhere between total separation and full sharing. Several buffer sharing policies have been proposed to avoid performance degradation of a shared buffer under imbalanced load, such as sharing with minimum and/or maximum allocation constraints [44], [78], [119], [160].

3.4. Improved input buffering

The problem of "ordinary" input buffering is HOL blocking resulting from FIFO queueing in each input buffer. Thus, the throughput performance of an input buffering switch is expected to be improved by allowing non-FIFO service. With the w-window mechanism [101], input ports which are not selected to transmit their HOL cells, contend again with their second cells for access to any of the remaining idle output ports, and so on up to w times, in each time slot ($w = 1$ corresponds to FIFO input buffering). This mechanism is sometimes referred to as *bypass queueing* or *look-ahead contention resolution*. Also, the window size w is sometimes called the *scanning range* or the *number of bypass offers*. TP_{max} improves as w increases for a given N. As an example, using simulation it was shown that for $N = 128$ and $w = 8$, $TP_{max} = 88\%$. In the same reference, it was concluded that input buffering even with an infinite window size does not attain the optimal throughput-delay performance of output buffering. However, approximate analytical models show that $TP_{max} = 100\%$ can be achieved using $w = \infty$ [142], [181]. Thomas and Man [212] noticed that the original windowing mechanism of [101] strives to minimize the number of input buffers which are not served at all.

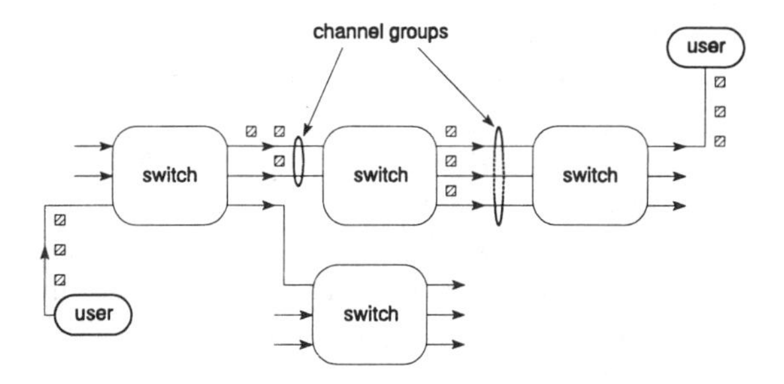

Fig. 5. The concept of channel grouping.

The same reference proposed an alternative approach where the goal is to maximize the number of assigned output ports in each time slot, even by allowing multiple selections from the same input buffer. For $N = 128$ and $w = 8$, a TP_{max} of 94% results. In general, more significant improvements over the basic windowing mechanism is achieved for smaller w and larger N. A similar improved windowing mechanism was also suggested in [218] and [219] in the context of internally-buffered multistage interconnection networks, under the name *parallel bypass queueing*. Window mechanisms were also studied in [35], [142], [192], [193], and [195].

In a network, it is possible to have multiple links interconnecting adjacent switching nodes, and these links can provide cells with multiple paths to a number of downstream switches as shown in Fig. 5. This results in the concept of *output channel grouping* [177], in which the output ports of a switch are partitioned into disjoint sets, with each port in a set offering the same service as the other members of the set. This technique can be used to more efficiently utilize the huge bandwidth of fiber optics, while avoiding the need for ultra-fast electronic switches, and to provide higher reliability by offering alternative paths. It was shown that output channel grouping also results in improving the throughput performance of an input buffering switch as the size of each group grows. As an example, TP_{max} increases from 58.6% to 87.8% for groups of 16 output ports, and to 91.2% for groups of 32 output ports [154], [155], [181]. However, Liew and Lu [154] have shown that with a group size of 1024, a TP_{max} of 98.4% results, which suggests that increasing the output group size above a certain value does not result in any more significant improvements. Studies dealing with output channel grouping have also been reported for cases involving non-uniform partitioning of the output ports [47], [136], and nonuniform traffic models [47], [129], [136], [152], [154], [155], [172]. Output channel grouping is also known as *multichannel bandwidth allocation, output trunk grouping, multiservice switching, output pooling*, or *output link group routing*. In [152], the concept of *input channel grouping* was examined, where the input ports are partitioned into disjoint sets each with a demultiplexer and each demultiplexer skips those input links with full buffers; an incoming cell is lost only if all input buffers within the same input channel group are full. Notice that the above is an implementation of partially-shared input buffering. While output channel grouping relieves traffic output contentions, input channel grouping prevents individual input links from overloading [152]. It has been shown that both techniques have significant impact in improving the performance of an input buffering nonblocking switch, especially in a highly nonuniform traffic environment [152].

Output conflicts and consequently HOL blocking can be reduced by switch expansion [153], [154], [155], [181]; i.e., by having the number of output ports O greater than the number of input ports I. The ratio $E = O/I$ is called the expansion factor, and as it increases, the throughput performance of an input buffering switch improves. As an example, for $E = 8$, the maximum throughput is 93.8%, compared to 58.6% for $E = 1$ [154], [155], [181]. For $E = \infty$, a 100% maximum throughput is achieved [181]. Finally, despite the "similarity" between output channel grouping and switch expansion, the concepts behind them are different. However, these two approaches can be used together; e.g., for $E = 2$ and an output channel group size of 2, $TP_{max} = 96.6\%$ [154], [155].

Using independent output port schedulers has been investigated in [170]. The scheduling algorithm has two phases: request and arbitration. In the request phase, each input buffer sends a request to the appropriate output port scheduler. While these schedulers are located in a centralized contention control unit, each schedules its own transmissions independently of the others. In the arbitration phase, each output port scheduler passes consecutive assigned slot times to requesting input ports. Each acknowledged input port checks its sending table: if the assigned time slot has already been reserved, the cell must participate in the request phase again in the next time slot. $TP_{max} = 65\%$ can be achieved with this algorithm (82% if input ports are organized into groups of 4, where all members of a group coordinate their requests; or if each input queue runs 4 times the speed of the input ports [171]). This scheduling algorithm can be improved further by re-using those time slots which inputs cannot use due to earlier assignments [121]. The modified algorithm achieves a TP_{max} of 92% (95% with 4-member input groups). Matsunaga and Uematsu [158] proposed a similar algorithm but in which there exists full coordination between the outputs in the scheduling process. It has been shown that $TP_{max} = 94\%$ can be achieved. Another scheduling algorithm that resolves conflicts and optimizes performance by maximizing throughput was proposed in [90].

Ali and Youssefi [3] considered having a separate queue for each output port at each input port; i.e., a total of N^2 input queues for an $(N \times N)$ switch. In each time slot, HOL cells in all N input queues corresponding to a particular output link contend for routing. Winners are chosen in such a way as to maximize the throughput of the switch. A neural network implementation of the controller has been suggested to overcome the high complexity required by the optimal selection process. A similar approach was previously suggested in [85], but using a heuristic selection scheme with a number of operations of $O(N)$ in time. Del Re and Fantacci [65] examined the same structure under two other selection policies: random and HOL FIFO. In all cases, the throughput-delay performance approaches that of pure output buffering at the expense of large buffer requirements. Chen and Guerin [41] considered only two queues per input port; one for high-priority traffic, and the other for low-priority traffic. For each input port, the high-priority queue is served first, and upon contention between different input ports, high-priority cells win. The maximum throughput achieved by this approach is 60.6%. Multiple queues per input port were also studied in [220] in the context of internally-buffered multistage interconnection networks.

3.5. Combined input-output buffering

In order to achieve an acceptable performance level while avoiding the complexities associated with pure output buffering, many switch designs employ a combined input-output buffering strategy. This can be achieved by speeding-up a nonblocking switch by a factor (S) greater than 1. By doing so, up to S cells can be switched to each output port in every time slot, which necessitates the use of output buffers. In input-output buffering switches, one of two possible mechanisms can be used: queue loss [42], [182], or backpressure [17], [18], [19], [34], [96], [109], [111], [112], [118], [182], [194]. In the queue loss mechanism, cells can be lost after being switched as a result of output buffer overflows. However, with the backpressure mechanism, cell flow from the input buffers to the output buffers is limited by the current free space in each output buffer jointly with the speed-up factor.

Within the past three years, great interest has been shown in input-output buffering switches built around nonblocking fabrics. An analytical model of such switches has been developed in [182] for $N = \infty$, arbitrary values of B_{in}, B_{out}, and S, and both queue loss and backpressure. The analysis of [118] makes the above assumptions but considers backpressure only. In [194], an approximate model that allows N to be finite for the backpressure mechanism has been reported. The analysis of [42] assumes Poisson arrivals of variable-length packets, $N = \infty$, $B_{in} = \infty$, $B_{out} = \infty$, and arbitrary S. It also estimates the packet loss rate resulting from finite input and output buffers, with queue loss. Assuming that $S = B_{out}$, $N = \infty$, $B_{in} = \infty$, Poisson arrivals, fixed or variable-length packets, and synchronous or asynchronous switch operation, an input-output buffering switch with backpressure has been analyzed in [112]. The packet loss rate resulting from finite B_{in} has also been estimated. Other studies dealing with input-output buffering nonblocking switches under uniform traffic include [34], [96], [109], [110], [142], [174], and [180]. Also, the performance of these switches under nonuniform traffic models has been examined in [18], [42], [99], [111], and [180].

Some of the main observations about input-output buffering nonblocking switches include the following:

- Given a fixed total buffer budget, there exists an optimal placement of buffers among input and output ports to minimize P_{loss}. This is true for the queue loss mechanism [42], [182], as well as for the backpressure mechanism [182].
- The backpressure mechanism requires less $B_{in} + B_{out}$ to achieve a given P_{loss} under a very wide range of load values [182]. Also, the queue loss mechanism requires in general larger B_{out} to achieve a target P_{loss} given a fixed $B_{in} + B_{out}$ [182].
- With backpressure, the effect of B_{out} is more dominant on the performance than that of B_{in} [118].
- Assuming infinite size buffers at input and output ports, fixed-length packets, and $N = \infty$, $TP_{max} = 88.5\%$, 97.6%, and 99.6%, when $S = 2$, 3, and 4, respectively [96], [174]. Buffering cells at the input ports results in a significant reduction in the value of S needed to attain a given P_{loss} at a given load value [174]. Assuming $B_{out} = \infty$, $p = 0.90$, and target $P_{loss} = 10^{-6}$, we need $S = 8$, 4, and 3, if $B_{in} = 0$, 1 (with priority scheme), and 9, respectively [174].
- With $S = B_{out}$, fixed-length packets, Poisson arrivals, and backpressure, TP_{max} is the same for both synchronous and asynchronous modes of switch operation [112]. However, for a given load, asynchronous operation results in a slightly better performance in terms of delay and packet loss probability [112]. On the other hand, if the synchronous mode assumes Bernoulli arrivals, it achieves better delay performance compared to asynchronous operation with Poisson arrivals [109].
- Under Poisson arrivals, a higher TP_{max} is achieved with fixed-length packets compared to that achieved with packets having exponentially-distributed lengths [42]. For the latter

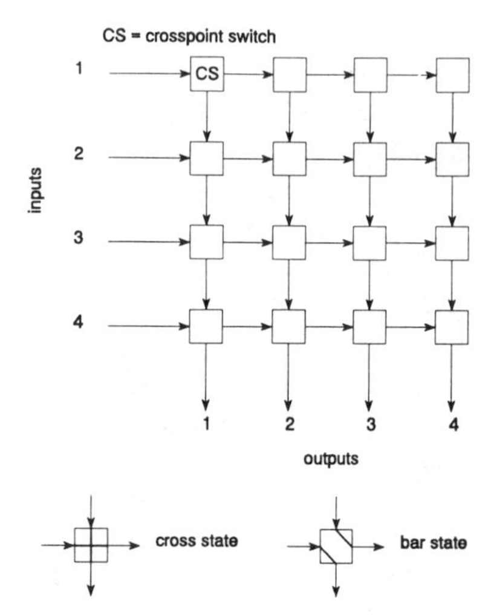

Fig. 6. The crossbar switch.

case, $TP_{max} = 50.0\%, 82.8\%, 96.1\%$, and 99.3%, when $S = 1, 2, 3$, and 4, respectively [42].

4. Crossbar based switches

The term "crossbar" derives from a particular design of a single-stage single-path nonblocking switching fabric, originally developed for circuit switching [25], [33], [55], [184], [213]. Later on, crossbar switches were considered for interconnecting processors and memory modules in multiprocessor systems [27], [84], [159], [176]. Currently, many ATM switch proposals are based on the crossbar switch, or use it as the basic building block. A crossbar switch is schematically shown in Fig. 6 for $N = 4$, where horizontal lines represent the inputs to the switch, and vertical lines represent the outputs. Basically, an $(N \times N)$ crossbar switch consists of a square array of N^2 individually-operated crosspoints, one for each input-output pair. In general, crosspoints could be electromechanical relays or semiconductor switches (transistors, SCRs, or logic gates) [25]. (Clearly, the first option is not suitable for a high-speed packet switching environment.) Each crosspoint has 2 possible states: cross (default) and bar. A connection between input port i and output port j is established by setting the (i, j)th crosspoint switch to the bar state.

Crossbar switches have always been attractive to switch designers because they are nonblocking, simple in architecture, and modular. However, they have the following two main drawbacks [213]:

- Square growth of complexity (i.e., the number of crosspoints is of $O(N^2)$).
- Different input-output pairs may have different transit delays.

Because of the first problem, crossbar switches do not scale well to large sizes. The second drawback introduces a fairness problem when the switch operates in a self-routing mode, and can be solved using artificial time delays.

Despite being nonblocking, crossbar switches suffer from output blocking. Therefore, as in any packet switch, buffers are necessary to reduce cell losses. In the following, we review crossbar based switches by classifying them according to their buffering strategies.

4.1. Input buffering

In this case (Fig. 7), a contention resolution mechanism is needed to select one cell only out of those requesting the same output port. This can be done in one of two ways: centralized or distributed. The former approach requires a central controller that resolves output contentions in a slot-by-slot basis. Such an approach keeps the

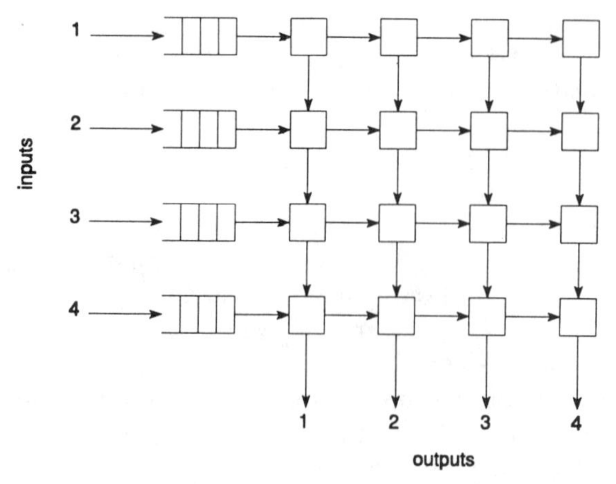

Fig. 7. An input-buffered crossbar switch.

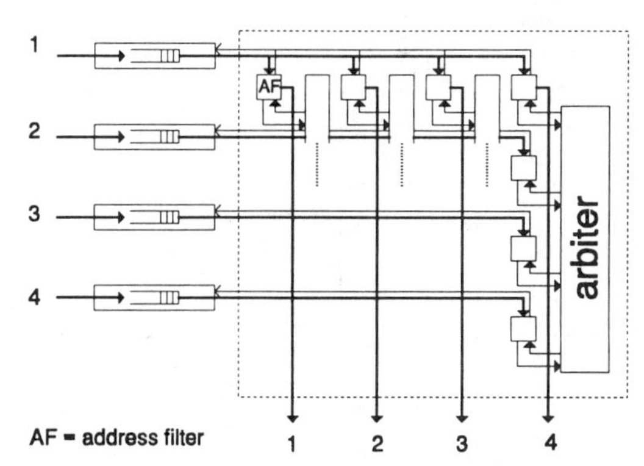

Fig. 8. The arbiter implementation of an input-buffered crossbar switch.

switching fabric itself as simple and as economical as possible. However, the centralized control becomes a major bottleneck and constitutes a second limiting factor (besides the square growth of complexity) to the scalability of the switch. The other approach is to distribute the control over all output ports; i.e., each output port has its own controller, called *arbiter* [2], [213], as shown in Fig. 8. Each arbiter sees all cells which are destined to the corresponding output port, selects one of them according to some rule, and blocks all others by a means of a backpressure signal. Only a single bus is needed per input port to broadcast the cell and its destination port address, and only a single reverse control line is required per input port.

Input-buffered crossbar switches suffer from HOL blocking, and the various methods which have been discussed earlier to improve the performance of an input-buffered nonblocking switch can be used here. In particular, Del Re and Fantacci [64], [65] showed that the arbiter implementation (Fig. 8) can be modified to significantly improve the performance. This is done by splitting each input buffer into N separate queues, one for each output port, and having these N queues share the same input memory. Routing requests jointly with memory locations of new arriving cells are broadcast over the input buses to all arbiters, and each arbiter maintains its own destination queue. A cell is transmitted when its routing request is selected by the arbiter of the desired output port. Interestingly, it was shown that the above approach achieves a smaller P_{loss} for a given load value and switch size, compared to classical output buffering [64], [65].

4.2. *Internal or crosspoint buffering*

Buffers can be placed within the switching fabric at possible locations of contention. For example, in the *Bus Matrix switch* [169], each crosspoint switch is replaced by an address filter (AF) and a FIFO buffer, as shown in Fig. 9 for $N = 4$. While an incoming cell is broadcast to all attached AFs, it can pass only through the one whose address matches its destination address. An arbitration mechanism determines which of the HOL cells in the N buffers connected to an output bus, may have access to the corresponding output line. This architecture has a throughput-delay performance which approaches that of output buffering switches, for large enough buffer sizes [3], [65]. However, the total memory required to achieve a certain performance level is larger than that needed by output buffering, since each output buffer is now distributed over N separate buffers [213]. Also, it is always desirable, from an implementation point of view, to separate buffering and switching functions [213]. A crossbar switch with FIFO buffers at each crosspoint has also been discussed in [189] under the name of the *Butterfly switch*.

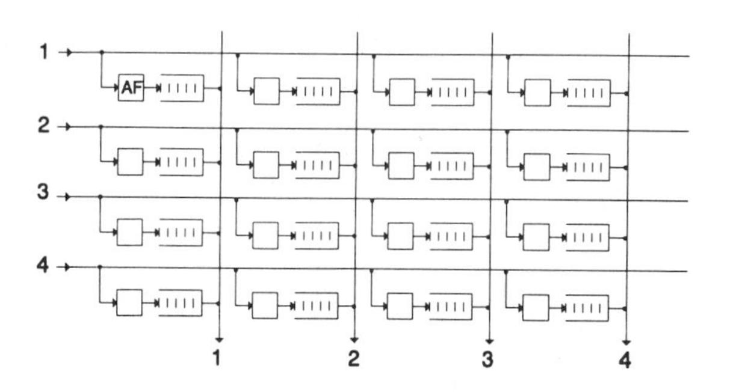

Fig. 9. A crosspoint-buffered crossbar switch.

4.3. Input-internal buffering

To reduce the complexity of crosspoint-buffered crossbar switches [169], [189], which resulted in limiting an actual implementation to a (2×2) matrix [169], and at the same time to improve the performance of pure input buffering, the *Limited Intermediate Buffer* switch (LIB) has been proposed in [97]. Here, a single buffer (called an intermediate buffer) is provided at each crosspoint resulting in a total internal buffer size of N^2 cells, in addition to the FIFO buffers at the input ports (Fig. 10). An HOL cell can advance to the intermediate buffer connected to its desired destination port, only if that buffer is empty. Several policies for selecting a cell to be forwarded to an output line from the corresponding N intermediate buffers were examined, and it was found that the choice of the selection policy significantly influences the performance of the switch. A maximum throughput of 87.5% can be achieved for a (16×16) switch, compared to 60% for pure input buffering [101]. Surprisingly, it was found that the LIB switch has a maximum throughput that increases as the switch size increases. The explanation given in [97] was that the positive effect of the intermediate buffers becomes more dominant than the negative effect of the HOL blocking as N increases. Evaluating the performance of the LIB switch under two-priority traffic classes has been reported in [98].

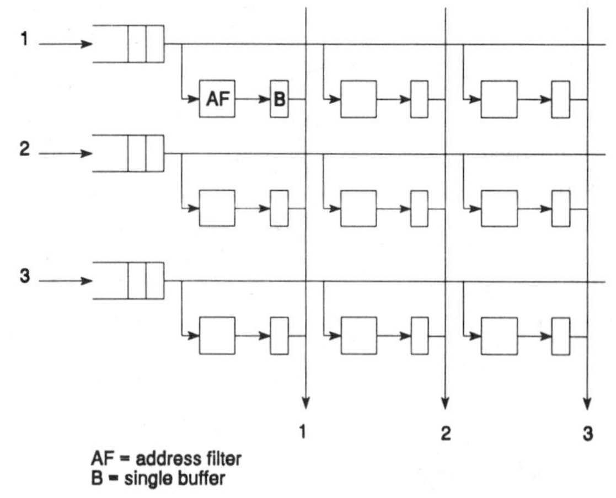

Fig. 10. The Limited Intermediate Buffer switch.

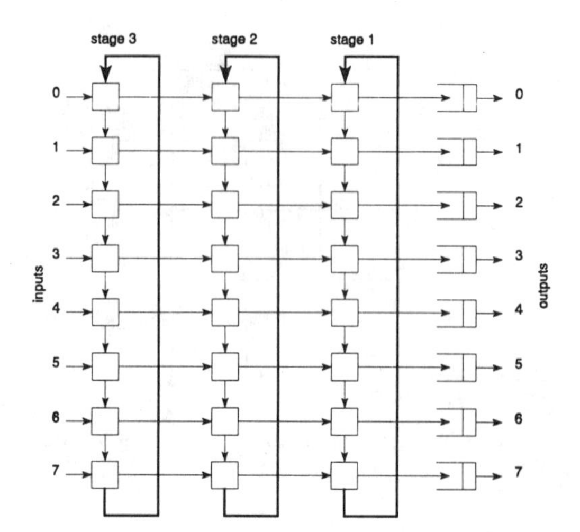

Fig. 11. An (8×8) input-to-output address-difference-driven switch.

5. Disjoint-path based switches

In this section, we review some switching fabrics with the capability of disjointly establishing all possible N^2 paths. Thus, no blocking of any type can occur within the switching fabric itself, even among cells destined to the same output port. However, buffering at the output ports is needed to cope with the possibility of multiple cell arrivals. Being pure output-buffered, these architectures have the best possible throughput-delay performance. It can be seen that the Bus-Matrix [169] and Butterfly [189] switches which have been described previously, also lie under the current category. In disjoint-path based switches, paths can be disjoint either in time or in space; in the following we adopt this classification.

5.1. Paths disjoint in time

Speeding-up *any* switching fabric by a factor of N makes it a disjoint-path based switch in which paths are disjoint in time. A multistage self-routing switch, in which each stage is sped-up

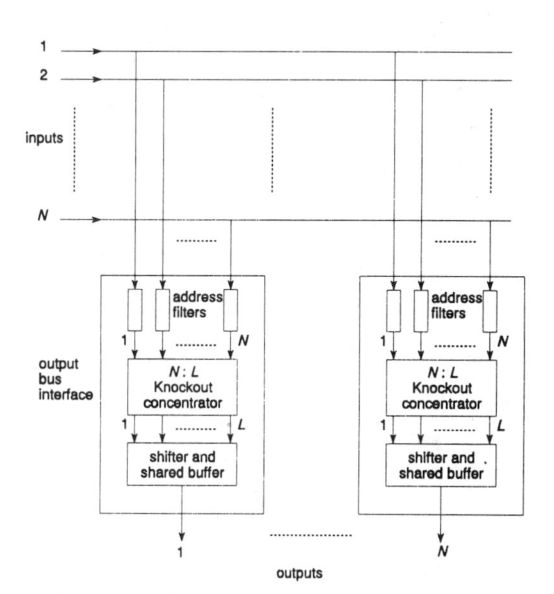

Fig. 12. The Knockout switch.

by a factor of N was proposed in [113]. This switch (Fig. 11) uses the difference between the input and output addresses of a cell as it routing tag. It consists of $n = \log_2 N$ stages, each with N (2×2) SEs. Let Y be the output port number and X be the input port number of an incoming cell, then the difference $F = (Y - X) \bmod N = f_n f_{n-1} \ldots f_1$ is used as the routing header (where f_n is the MSB). If the stages of the switch are numbered from 1 to n starting from the output side, then an SE in stage k examines f_k of an incoming cell: if $f_k = 1$, the cell is shifted by 2^{k-1} downward in the same switching stage and then transferred to the following stage; if $f_k = 0$, the cell is transferred directly to the following stage. Cell loss may occur only as a result of output buffer overflows. This switch is capable of handling multicast connections, in addition to unicast connections. Unfortunately, the requirement to run each stage N times faster than the external lines places a limit on the maximum size of the switch. Also, each output buffer must be capable of receiving up to N cells within a single time slot.

5.2. Paths disjoint in space

The most well-known example of this category is the *Knockout switch* [230] shown in Fig. 12. It exploits a key observation: in any practical switching system, cell loss within the network is unavoidable. This architecture uses a fully-interconnected topology to passively broadcast all incoming cells to all outputs. Each output port has a bus interface that performs several functions: filtering (to discard cells not intended to that particular port), concentration (only L out of N cells which may be destined to that port in a given time slot are admitted and the rest are dropped), and finally buffering (using a FIFO buffer that is shared among all L lines in a clever arrangement). A concentration cell loss probability smaller than 10^{-6} at a load of 0.90 can be achieved with $L = 8$ for an arbitrary large value of N. Larger values of L are required under

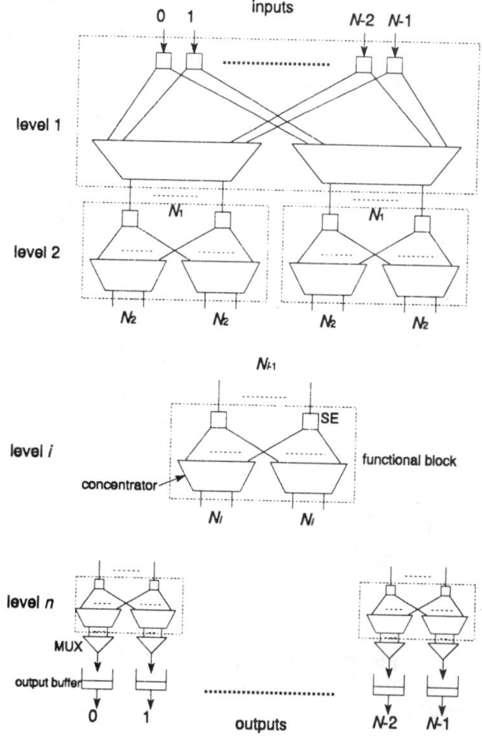

Fig. 13. The Christmas-Tree switch.

imbalanced traffic models [231]. Extending the function of the Knockout switch to variable-length packets was carried out in [79], and to multicasting in [80]. A Knockout switch employing priority-based concentration was studied in [82].

Despite its excellent performance, the Knockout switch is complex for large switch sizes. It requires N^2 distinct physical paths, N^2 address filters, and $N(N \times L)$ Knockout concentrators. Efforts to reduce the complexity of the Knockout switch include the following. The use of a single repeat contention resolution scheme was shown to reduce the required L by a factor of 2 [81]. In [126], it was suggested to build a large switch modularly from smaller Knockout switches in a multistage manner. In [221], it was observed that the high complexity of the Knockout switch is a result of the separation of the distribution and concentration functions. The same reference proposed the *Christmas-Tree switch* (Fig. 13), which is a self-routing output buffering switch architecture that interleaves the distribution and concentration functions in such a way as to achieve high performance with fewer than N^2 paths. While cell filtering and concentration (or contention resolution) functions are performed *centrally* at each output port in the Knockout switch [230], these functions are distributed in small switching

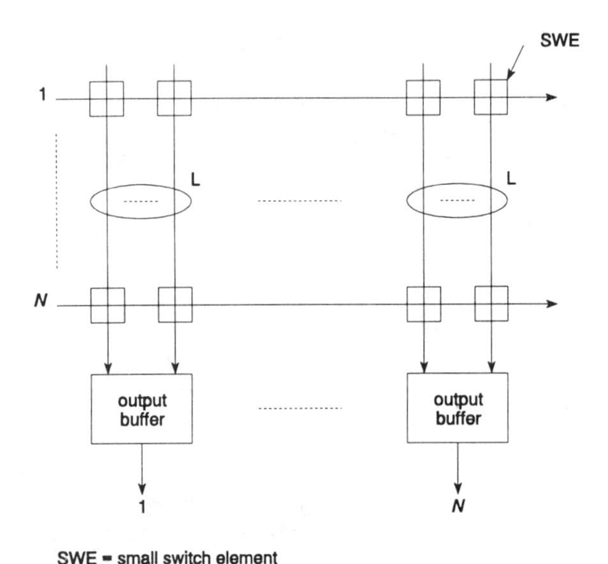

Fig. 14. The distributed Knockout switch.

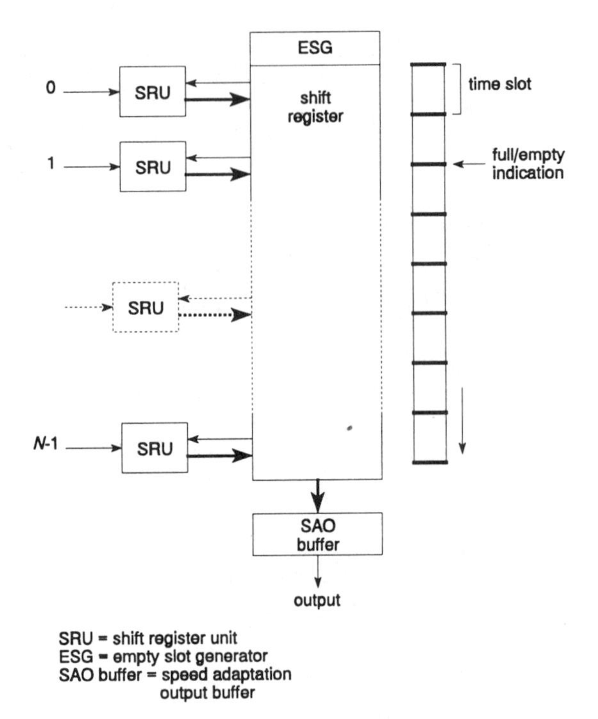

Fig. 15. An output bus interface of the GAUSS switch.

elements (SWEs) located at the intersections of the crossbar lines (Fig. 14) in the *distributed Knockout switch* [39]. The SWEs examine incoming cells from horizontal lines and route them to one of the L vertical lines of each output port. The number of vertical lines is reduced from N^2 to LN, and concentration within each output bus interface is no longer needed, however, at the expense of LN^2 SWEs. It was pointed out in [39] that this switch design has a regular and uniform structure, is modular, and with relaxed synchronization requirements. Several methods for reducing L were also discussed. It should be noted that both the multistage-based Knockout switch [126] and the Christmas-Tree switch [221] do not belong to the disjoint-path based class of switches; they are mentioned here because of their relation to the Knockout switch.

Several disjoint-path switch designs which differ from the Knockout switch in the structure and operation of the output bus interfaces were also proposed, such as the *GAUSS switch* [70], and the *Cylinder switch* [162]. The GAUSS (Grab Any

UnuSed Slot) switch was introduced in [70] and analyzed in [1]. In addition to the knockout principle, it also exploits also the observation that the probability of two cells originating from the same input port and heading for the same output port in successive time slots, gets smaller as N increases. (Notice that this observation may not be valid under bursty traffic.) An output bus interface is called here a GAUSS module and is shown in Fig. 15. It consists of a parallel shift register, shift register units (SRUs) and a speed adaptation output buffer (SAO-buffer). Each input of a GAUSS module has an SRU connected to the parallel shift register. This shift register is used to convey cells arriving at the SRUs to the SAO-buffer. Empty slots are inserted at the top of the shift register and can be filled along their way through the shift register. The speed at which the slots are shifted is higher than the cell arrival rate at the inputs. Each SRU uses an address filter to detect whether a received cell is destined for that output. If so, the SRU seizes an empty slot by means of 'grab any' mechanism. A small buffer is needed in the SRU to take care of the speed difference between the inputs and the shift register. A buffer is also necessary since an SRU may have to wait before it can access the shift register. A cell that arrives at an SRU with a busy buffer is lost. At the end of the shift register the SAO-buffer receives the emitted cells from the SRUs. The ratio of the speed of the shift register to the speed at the inputs is called the speed-up factor L, and is determined by the amount of tolerable cell loss probability. It has been shown that the GAUSS switch has better average cell loss performance compared to the Knockout switch for the same N, L, and load value, which is a result of using buffers in each SRU [1]. However, the GAUSS switch is unfair, since upper input lines are favored over lower ones.

The Cylinder switch [162] solves a basic problem with the Knockout switch, namely the possibility of discarding a cell at a concentrator (i.e., after it has been switched), even though a space is available in the corresponding output buffer. Each output bus interface (called here a ring buffer multiplexer) consists of a buffer ring that is connected to all input lines by input elements

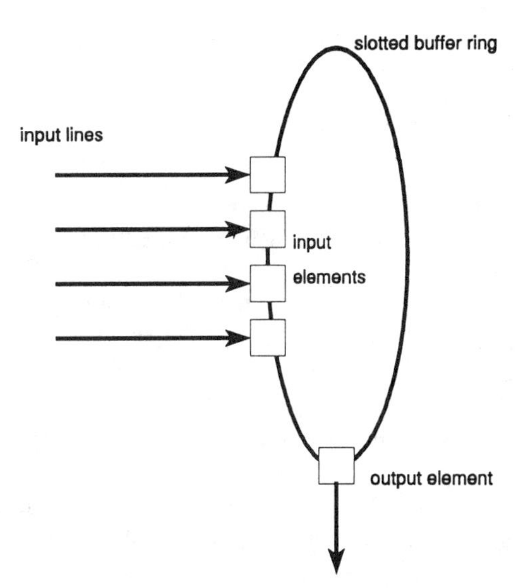

Fig. 16. An output bus interface of the Cylinder switch.

which perform the filtering function, and to the corresponding output line by an output element (Fig. 16). A buffer ring is a set of cell-width parallel shift registers arranged to form a ring (or a cylinder) that is shared between all input lines. Cells rotate in the buffer ring till they are read out by the corresponding output element, one in each time slot. The capacity of the ring is designed to be much greater than that of the input or output lines. Besides the above speed requirement, the switch requires a minimum of N shift registers in each of the N buffer rings.

6. Banyan based switches

The concept of multistage interconnection networks (MINs) was first introduced in the context of circuit switching [24], [33], [46], [55], [84], [105], [159]. The aim was to design a nonblocking switch with less complexity than the crossbar switch. In 1953, Clos [55] showed that the use of a MIN of three or more stages built from relatively small crossbar switches, can significantly reduce the number of crosspoints required for building a large size switch. A 3-stage $(N \times N)$ Clos network is shown in Fig. 17. This network is nonblocking if $m \geq 2s - 1$ [55], where m and s are as defined in

Fig. 17. (Notice that m paths exist for each input-output pair.) Such a network results in crosspoint savings compared to the crossbar switch for all $N \geq 36$ [159]. Despite the above, Clos network requires complex computations to find a connection path compared to the crossbar switch which can establish a connection without path search.

At this point, it is worth mentioning that the definition of the blocking property in a circuit switch is different from that in a packet switch. In the former [33], [46], [84], [105], [159], if any point-to-point connection between unused input-output ports:

- can be established without any disturbance to the existing connections, then the switch is *nonblocking* (or *strictly nonblocking*);
- can be established without any disturbance to the existing connections, provided that the existing connections were set-up according to some routing algorithm, then the switch is *wide-sense nonblocking*;
- can be established if one or more existing connections are rearranged, then the switch is *rearrangeably nonblocking*;
- cannot be established at all, then the switch is *blocking*.

A switch that is nonblocking in a packet switching environment is not necessarily nonblocking as a circuit switch. On the other hand, a circuit switch that is even rearrangeably nonblocking is nonblocking as a packet switch. One known MIN that

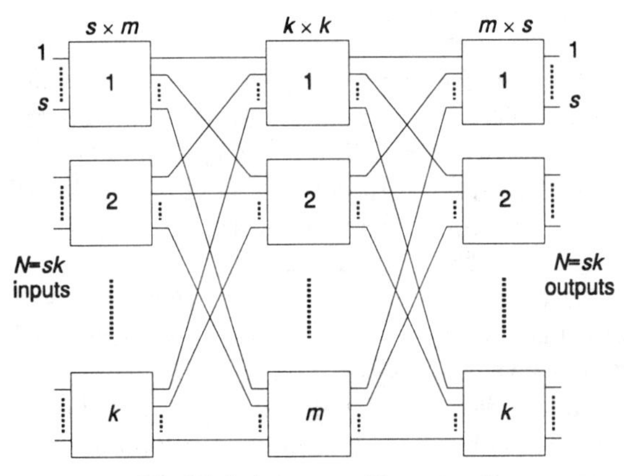

Fig. 17. A three-stage Clos network.

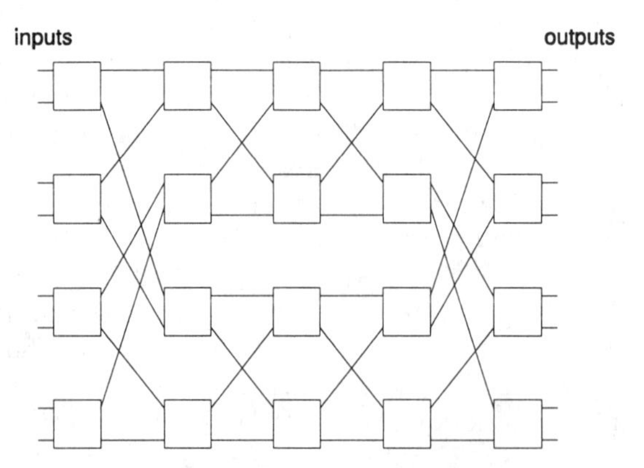

Fig. 18. An (8×8) Benes network.

is rearrangeably nonblocking as a circuit switch is Benes network [24], shown in Fig. 18. As a packet switch, it is nonblocking with the least known crosspoint complexity. However, as for the Clos network, its drawback is the time required for setting-up the paths [46], [105].

As the interest in large scale processing systems grew, many MINs were proposed and studied for the purpose of interconnecting processors and processors, and processors and memory modules in multiprocessor systems [27], [46], [72], [73], [84], [105], [159]. Among these are the networks which belong to the delta class, defined by Patel [176] as the class of MINs which possess two properties: unique path for each input-output pair, and digit-controlled routing. The latter property, known as *self-routing*, makes delta networks very attractive to high-speed packet switch designers. However, it comes at the expense of internal blocking as discussed below. Many well-known MINs belong to the delta class, such as Banyan [94], Baseline [226], Reverse Baseline [226], Omega [137], Modified Data Manipulator [226], Indirect Binary n-Cube [183], and Generalized Cube [197] networks. These networks were shown to be topologically equivalent [197], [226]. Currently, most ATM switching researchers tend to freely interchange their names; something that is usually justified by the fact that they have the same performance in under uniform traffic [73]. Some of these networks are shown in Fig. 19 for $N = 8$ (the most frequently-used names are shown). In general, a rectangular $(N \times N)$ delta-b network is constructed using unbuffered $(b \times b)$ SEs (crossbars), organized in n stages, where $N = b^n$ $(n = 1, 2, \ldots)$ and each stage has N/b SEs. In this survey, unless otherwise specified, we consider delta networks constructed with (2×2) SEs. Furthermore, we use the terms "delta" and "banyan" interchangeably when referring to any self-routing single-path $\log_2 N$ MIN.

A large number of ATM switch proposals is based on banyan networks due to their desirable features [2], [8], [46], [157], [173], [213]. Besides the self-routing property, banyan networks are modular, have the same latency for all input-output pairs, support synchronous and asynchronous modes of operation, are suitable for VLSI implementation because of their regular structure, and have a complexity of $O(N \log_2 N)$ compared to $O(N^2)$ for the crossbar switch. However, while path-uniqueness results in (the ease of) preserving cell-sequencing, it also results in banyan networks being internally blocking. This can be explained by the example shown in Fig. 20, which also illustrates the self-routing mechanism: an SE in stage i routes an incoming cell to the upper outlet if $d_i = 0$, or to the lower outlet if $d_i = 1$, where the stages of a banyan network are numbered from 1 to n $(= \log_2 N)$ starting from the input side. Internal blocking arises when two cells destined to different output ports request the same outlet of an SE; in such a situation, one cell is selected according to some policy, and the

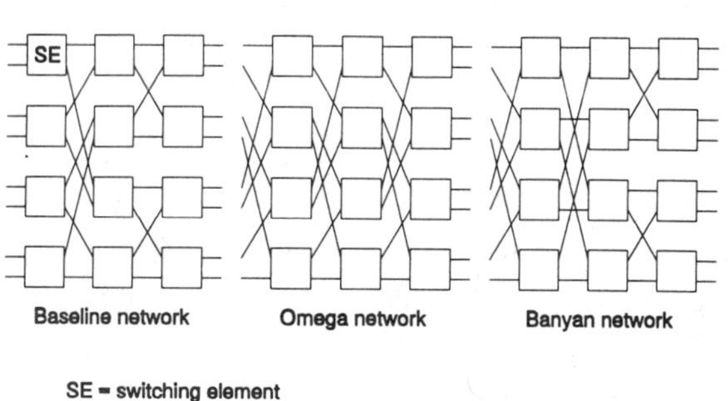

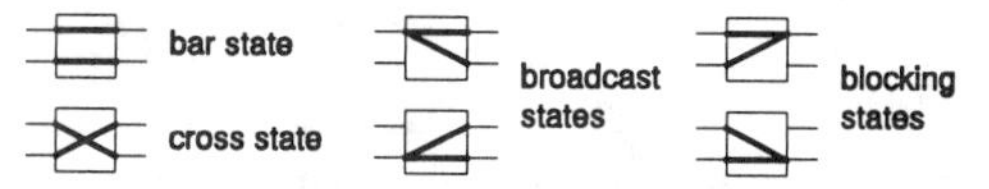

Fig. 19. Three different topologies of banyan networks.

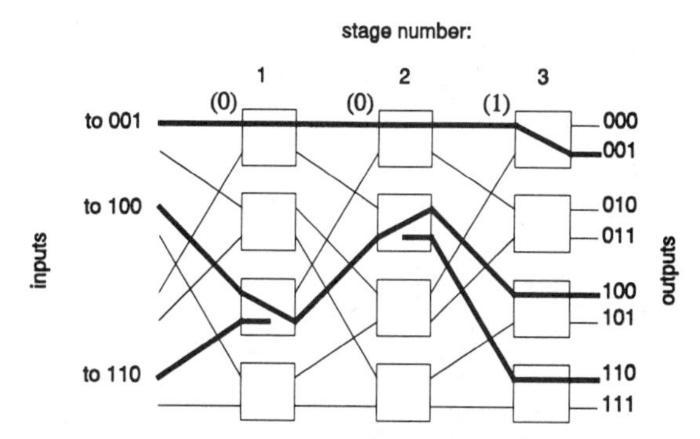

Fig. 20. Internal blocking and self-routing in banyan networks.

other cell is dropped. Furthermore, conflicts do occur among cells having the same destination.

As has been explained so far, the reduction of the crosspoint count from N^2 to $2N \log_2 N$, results in internal blocking. Patel [176] analyzed banyan networks with a random selection blocked-loss policy in the SEs, and derived the following recurrence relation $(1 \leq i \leq n)$:

$$p_i = 1 - \left(1 - \frac{p_{i-1}}{b}\right)^b, \tag{9}$$

where $p_0 = p$ (cell arrival rate), p_i is the probability that an output link of stage i carries a cell, and $(b \times b)$ is the size of each SE. The throughput is simply $TP = p_n$. Kruskal and Snir [130] derived the following approximate closed-form solution for the throughput of a banyan network:

$$TP = \frac{2b}{(b-1)n + 2\frac{b}{p}}. \tag{10}$$

Also, tight upper and lower bounds on TP were derived in [132]. TP is plotted (using Eq. (9)) as a function of $\log_2 N$ for $p = 1.0$ and $b = 2$ in Fig. 21. For the sake of comparison, the throughput of a crossbar switch with blocked-loss policy is also shown (using Eq. (8)). From the figure, it is clear that both the crossbar switch and banyan network have a performance that degrades as the switch size increases. However, the degradation is more significant for banyan networks. This is because the crossbar switch suffers only from output blocking, while banyan

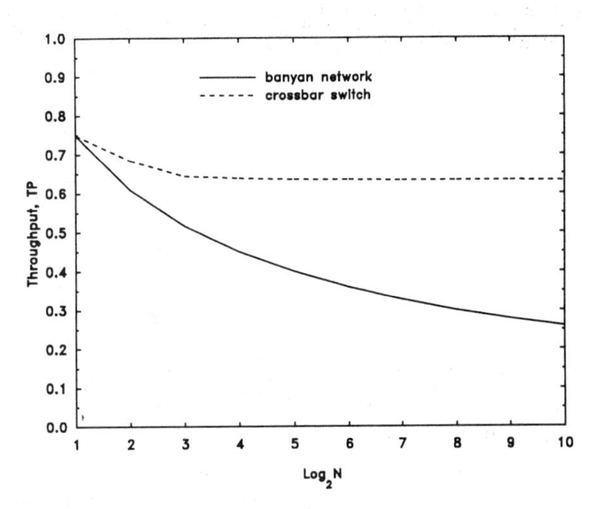

Fig. 21. Maximum throughput of banyan networks and crossbar switches as a function of the size.

networks suffer from both output and internal blocking and the latter increases with the number of stages of a banyan network. In particular, as N goes to infinity, TP_{max} (crossbar) = 0.632 and TP_{max} (banyan) = 0. From Eqs. (9) and (10), it can be seen that using SEs of sizes greater than (2×2) improves the throughput [176], [206], [207]. Increasing b reduces internal blocking by reducing the number of stages. Internal blocking is completely eliminated for $b = N$ (i.e., the crossbar switch). Switch architectures which are based on banyan networks use different methods to overcome blocking, and thus to improve the throughput performance. In the following, we classify banyan based switches according to these methods.

6.1. *Internal node buffering*

The most straightforward solution to the blocking problem in banyan networks (besides dropping blocked cells) is to place buffers within each SE, so that upon conflicts, blocked cells remain in their current buffers instead of being dropped. Typically, a buffered banyan network (BBN) is assumed to operate in a time-slotted fashion with cells advancing from a stage to the next synchronously. One of three possible modes of operation can be used: queue loss (QL), local backpressure (LB), and global backpressure (GB). With QL, a cell that arrives at a full buffer is simply dropped. With LB, a cell can advance to the next stage only if the next buffer along its path is currently not full. On the other hand, with GB a cell can advance to the next stage even if the next buffer along its path is currently full as long as it is not going to be full upon the arrival of the cell under consideration. Intuitively speaking, QL results in the worst performance and GB results in the best; the situation is reversed with regard to complexity. Notice that with either LB or GB, cell loss can happen only at the buffers of the first stage, or at the external input buffers if used.

Buffers can be placed at the inlets of each SE yielding input-BBNs. This is the simplest and most researched type of buffered banyan networks [5], [35], [45], [71], [74], [88], [89], [95], [117], [124], [131], [143], [144], [148], [189], [191], [193], [195], [205], [211], [216], [217], [218], [219], [220], [227], [232]. An (8×8) input-BBN with external input buffers is shown in Fig. 22. Alternatively, buffers can be placed at each SE outlet (output-BBNs) [77], [83], [95], [127], [130], [134], [186], [189], [205], [229]; at the crosspoints of each SE (crosspoint-BBNs) [95], [102], [131], [133], [189]; within each SE shared among all its inlets and outlets (shared-BBNs) [29], [95], [163], [218], [219]; or at both the inlets and outlets of each SE (input-output-BBNs) [164], [205]. The performance of a BBN is a function of the performance of its individual SEs: the better the performance of each SE (which depends on the strategy/size of its buffers) is, the better is the performance of

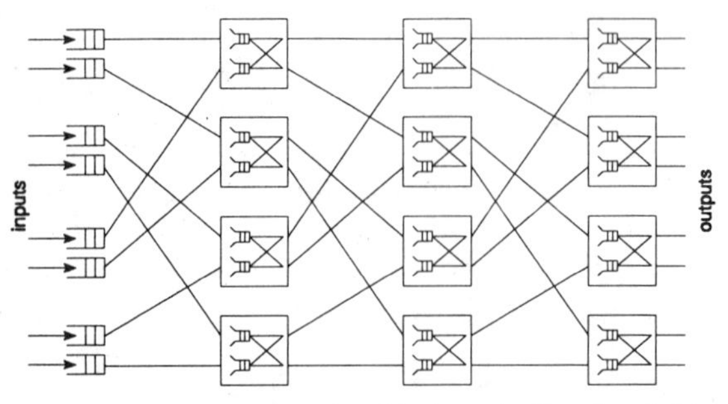

Fig. 22. A banyan network with input-buffered switching elements.

the whole network. With the assumption of infinite size buffers, output-BBNs [95], [127], [130], crosspoint-BBNs [95], and shared-BBNs [95], all achieve a 100% maximum throughput, while input-BBNs achieve only a 75% maximum throughput [71], [95], [205], for any value of N. The non-unity figure for input-BBNs is a result of HOL blocking within each SE, which in turn limits the maximum throughput of the whole structure to that of a single (2×2) SE [101].

For a 10-stage network with a single buffer at each SE inlet and GB, $TP_{max} = 0.32$ (according to simulation results [205], [211]; however, most analytical models give the optimistic figure of 0.45 [117], [191], [205], [232]). This is compared to 0.26 for an unbuffered banyan network of the same size [176]. It is clear that this improvement may not justify the added complexity. Increasing the buffer size improves the throughput performance. For example, with $N = 1024$, $TP_{max} = 0.43$ or 0.50 when 2 or 3 buffers, respectively, are used per SE inlet. Increasing the size of each SE in a BBN with single-input-buffered SEs also improves the throughput performance [35], [157], [218], [219], [232], as it is the case with unbuffered banyan networks [176]. However, with multiple buffers at each SE inlet, the throughput performance degrades as the size of the SE increases [35], [157], [218], [219], [232]. Bubenik and Turner [35] explained this by the increased negative effect of HOL blocking (which does not exist in a single-input-buffered SE) as the size of each SE increases [101]. A second reasoning [218], [219] is that the number of stages for a given N decreases as the size of the SEs increases, resulting in less buffering overall. The above suggests that it is better first to relax the FIFO queueing assumption at each SE inlet buffer (e.g., use bypass queueing [101]), and then to use SEs of sizes larger than (2×2) [35], [218], [219]. (Since the number of stages decreases with increasing the SE size, this also reduces the delay.) Adopting bypass queueing in input-BBNs built with (2×2) SEs also improves the performance [193], [195]. A cut-through mechanism [122], [216], [217] can be used to improve both the delay and throughput of a BBN [35], [157], [224]. With cut-through, a cell that arrives at an empty buffer and finds out that the desired output link is free, advances directly without first being fully received and buffered. Widjaja et al. [224] showed that the effect of cut-through switching on the delay is most significant when the load is moderate. It was also shown that the effect of this mechanism on the performance diminishes as the SE buffer size increases, since a buffer takes a longer time to become empty and hence less opportunity for cells to cut-through. The above results have been shown to be true for both input and output-BBNs. Splitting each SE inlet buffer into two parallel queues was shown to improve the throughput performance of an input-BBN [74], [220]. In [89], a change was suggested to the VLSI implementation of an n-stage input-BBN that would result in achieving the throughput performance of an $(n - 1)$-stage network. Speeding-up the internal fabric of a BBN has also been suggested [35], [95], [189], [216], [217] to improve its performance.

Szymanski and Shaikh [205] showed that BBNs built with SEs containing a combination of single input buffers and output queues outperform BBNs built with SEs containing only input queues or only output queues, given a fixed finite total amount of memory. Also, it was shown [218], [219] that while a shared-BBN offers clearly superior performance for a given amount of buffering, a parallel bypass queueing input-BBN performs impressively as well. Based on $N = 64$ and an SE storage capacity of 16 cells, Goli and Kumar [95] showed that a shared-BBN outperforms input, output, and crosspoint-BBNs, while a crosspoint-BBN provides a performance comparable to output and shared-BBNs for loads below 80%. It was argued that a crosspoint-BBN is a very good choice if both performance and complexity [133] are considered.

BBNs are known to have a performance that is very sensitive to traffic nonuniformities, regardless of the buffering strategy of the SEs. Nonuniform traffic performance evaluation was carried out for input-BBNs [5], [35], [45], [74], [88], [124], [143], [220], [227]; for output-BBNs [77], [134], [186], [229]; for crosspoint-BBNs [95], [102]; and for input-output-BBNs [164]. In general, the degradation in performance is due to the so-called *tree-saturation* effect [186], which occurs when the buffers of a last stage SE become full, and then the buffers of its two predecessors are completely filled, and so on all the way back to the input ports. Several solutions have been suggested to overcome this problem [5], [35], [45], [134], [220]. Atiquzzaman and Akhtar [5] have studied the performance of both buffered and unbuffered banyan networks under hot-spot traffic and found that the latter can have better performance.

6.2. Multiple banyans

Many ATM switch architectures are based on the use of more than one banyan network. Among these switches, three types can be identified depending on the concept behind the switch operation. In the following, we review the different types of multiple-banyan switches.

Parallel loading

Here, it is noticed that blocking in a banyan network increases with the applied load. Thus, a possible solution is to use multiple banyan networks, and to distribute the load (according to some mechanism) among them. The best example of this type of switches is the *replicated delta network* [130], [132]. It uses K banyan networks in parallel, as shown in Fig. 23 for $N=8$ and $K=2$. The ith input port of the switch is connected to the ith input of each banyan through a 1:K demultiplexer, while the ith output of each banyan is connected to one of the inputs of the ith K:1 multiplexer.

An incoming cell randomly chooses one of the K banyan networks [130], [132] (other distribution policies have also been examined in [57], [132], and [147]). One of two possible mechanisms can be used at the output ports: single acceptance (SA) or multiple acceptance (MA). With SA, only one cell out of K cells which may reach an output port is accepted [10], [11], [130], [132]. On the other hand, with MA all cells which reach an output port are accepted [11], [57], [147], [214], which necessitates the use of output buffering. (Input buffering must be used with both policies in an ATM environment for reasonable values of K.) With $N=1024$ and a target $TP_{max}=0.60$, 20 or 4 banyan networks are needed when SA [130],

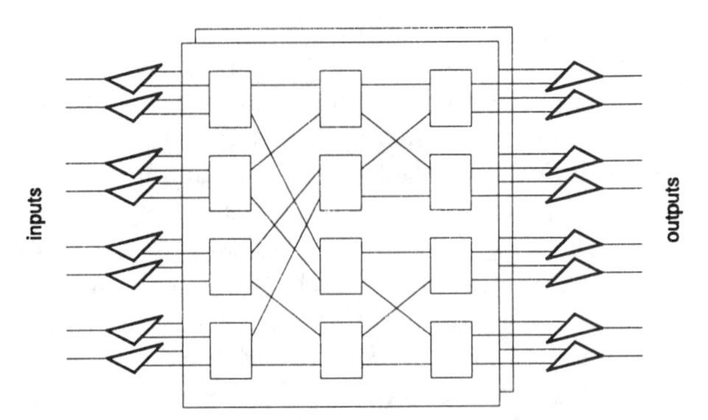

Fig. 23. An (8×8) replicated banyan network with $K=2$.

[132] or MA [213], [214], respectively, is used. Finally, cell-sequencing may be disturbed if internal buffers are used, unless cells belonging to the same connection are forced to use the same banyan network.

Sequential loading

In sequential loading multiple-banyan switches, the traffic is offered to the first banyan network, then that part of the traffic that is blocked by the first banyan is offered to the second banyan network, and so on till the last (Kth) banyan network. The traffic that remains after that is either dropped [214], [222], re-offered to one of the banyan networks along with new incoming traffic [214], or buffered at the input ports [53], [192], [225]. In all cases, output buffering is required to cope with the possibility of multiple arrivals at each output port.

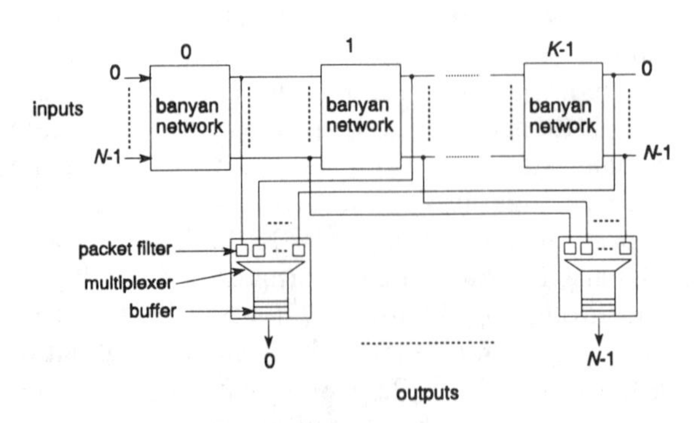

Fig. 24. The tandem-banyan switch.

The *tandem-banyan switch* [214] consists of K banyan networks in series (Fig. 24). Each output of every banyan network is connected to both the same-number input of the following banyan, and the corresponding output buffer (except for the last banyan which is connected only to the output buffers). A cell that loses contention in banyan network i $(0 \leq i \leq K-2)$ is misrouted after being marked, so as not to compete with properly routed cells in the following stages of the same banyan network. Marked cells out of network i are fed into network $i+1$ with their marks removed, while unmarked cells are extracted by the output ports. Marked cells at the outputs of the last network are lost. While the performance of delta networks is the same under uniform traffic regardless of their topology [73], when cascaded in the form of the tandem-banyan switch, their performance differs significantly [196], [214]. As an example, for $N = 1024$ and $P_{\text{loss}} = 10^{-6}$, 10 Modified Data Manipulator networks or 14 Baseline networks are required. Even better performance can be achieved if Omega networks are used [196]. Re-injecting cells that exit the last banyan network into the routing fabric for further processing, reduces the number of networks needed to achieve a given P_{loss}. In [223], it has been shown that using 2-dilated banyan networks (to be described later) reduces the number of networks needed to achieve a given P_{loss} by a factor of two or more. Finally, the tandem-banyan switch may deliver cells out-of-sequence for large values of N unless artificial delays are used.

In [53], the authors of the tandem-banyan switch [214] give an interesting discussion of the practical limitations they faced in implementing their switch. The limitation on the size of the switch comes primarily from the complexity of the function of each SE (which must, in the case of the tandem-banyan switch, examine the conflict bits in addition to the destination bits, of both cells at its inlets). Also, the arrangement of banyan networks in series renders the chip-to-chip synchronization difficult. Based on these limitations, it has been concluded that placing banyan networks in parallel rather than in series while preserving the concept of operation is a better choice. A switch architecture based on parallel banyan networks along with input and output buffering was then proposed under the name the *Memory / Space-division / Memory* (MSM) *switch*. In the MSM switch, the input controllers (one for each input port) maximizes the utilization of each banyan network by dispatching the cells to the banyan networks in sequence; thus offering the maximum load to each banyan network. This is accomplished through a two-phase operation: route set-up phase, and cell transmission phase. Cell loss can occur at both the input and output buffers. Since any individual banyan network can be accessed directly, the MSM switch can support circuit switching besides packet switching. Also, it can support multicasting.

A switch design that is similar to the MSM switch is the *Double Phase Packet Switch* (DPPS) [222]. The DPPS uses K Baseline networks in parallel with pure output buffering. Arriving cells are offered to each Baseline network sequentially, and unsuccessful cells are lost. As for the MSM switch, the DPPS is a successful pipeline application, with a two-phase pipeline that allows some requests to be issued when some cells are being transmitted; thereby increasing the average throughput of a cycle. It was shown that no more than $\log_2 N$ Baseline networks are needed to achieve a cell loss probability of 10^{-6} (under full load). In both the MSM switch and the DPPS, the state of the SEs are set prior to the actual transmission of the cells (i.e., during the first phase); thus, simplifying the design and operation of the SEs.

Another switch that is based on a parallel banyan structure is the *Pipeline Banyan* [225]. It consists of a single control plane and a number of data planes, together with input and output buffering. Cell headers are routed via the control plane to set their routing paths in the data planes. At the control plane, time is divided into slots for reservation; in each reservation slot, a data plane is selected on a round-robin basis. The SEs of the data planes are not required to perform any processing, which results in simplifying their design. However, connections are required between control plane SEs and the corresponding SEs of the data planes. Cells which cannot establish conflict-free paths, or are destined to output ports with full buffers, remain in their current input buffers. It was shown that this switch achieves a better performance compared to the tandem banyan switch [214] for the same number of banyan networks.

A switch architecture that *can* use banyan networks in parallel was proposed in [192]. It consists of parallel planes in conjunction with input and output buffering. Besides pipelining, the key feature in this switch is the use of bypass queueing in the input buffers. Bypass queueing was shown to significantly improve the throughput performance given a particular number of planes. It was shown that using a banyan network in each plane is more cost-effective than using a nonblocking architecture, given a target performance level.

Double banyan

This type is based on cascading two banyan networks, as shown in Fig. 25. (Notice the resemblance to the previously mentioned Benes network [24].) The first network is called a randomization or distribution network, depending on the particular architecture. The second network is the usual self-routing banyan network and is responsible for the delivery of cells coming out of the first network to their desired output ports. The operation of the first network affects the performance of the whole switch and is different in different switch designs. In general, the reason behind using the first network is reduce the effect of incoming traffic imbalances on the performance of the routing (i.e., second) network.

In [166], both networks are internally-unbuffered with the first one acting as a cell-basis randomization network, in which each SE sets its state randomly and not based on the destination bits of incoming cells. In [217], two buffered banyan networks were used with the first one being called a distribution network. The SEs of the distribution network ignore the destination addresses of incoming cells and route them alternately to each of their outlets; if one or both outlets are busy, the first outlet to become available is used. Unfortunately, out-of-sequence cell delivery may occur in this switch because cells belonging to the same connection may follow different paths and suffer different queueing delays. A solution to this problem could be to randomize or distribute the traffic on a connection-basis rather than on a cell-basis [66]. In this case, cells belonging to the same connection are forced to follow the same path. In [4], different connection-based path selection algorithms were suggested.

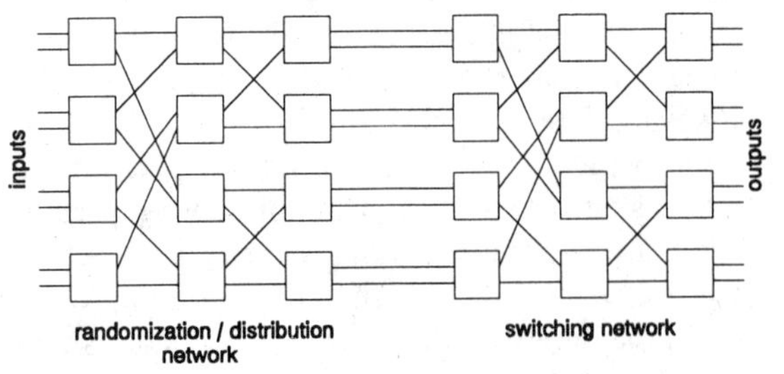

Fig. 25. A cascade of two banyan networks.

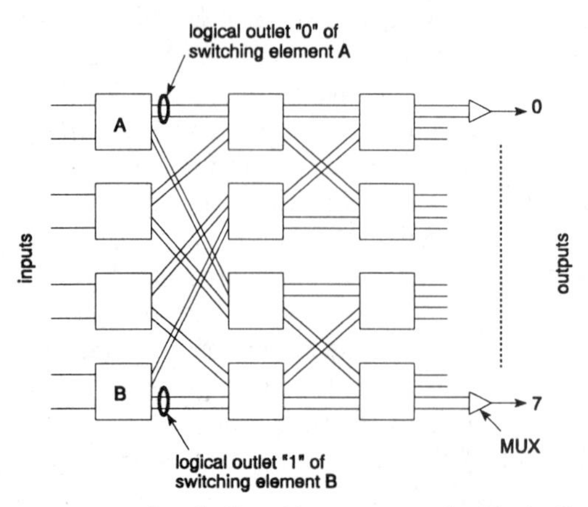

Fig. 26. An (8×8) dilated banyan network with $d=2$.

6.3. Dilation

The throughput performance of a banyan network can be improved by replacing each internal link by d links as shown in Fig. 26 for $d=2$. The resulting structure is known as the *dilated banyan network* [130], [132]. The SEs of a dilated banyan network are $(2d\times2d)$ switches with 2 output addresses. The routing algorithm here is the same as without dilation, except that internal blocking occurs only when $d+1$ or more cells all select the same logical outlet of an SE. In such a situation, d cells are randomly chosen and the others are dropped. One possible implementation of the SEs was given in [132], and is shown in Fig. 27 for $d = 4$. In this implementation, each SE consists of $2d$ 1:d demultiplexers, $2d$ d:1 multiplexers, and d $(d \times d)$ sorters. Each $(d \times d)$ sorter can be built using $d \log_2 d(\log_2 d + 1)/4$ binary comparators organized in $\log_2 d(\log_2 d + 1)/2$ stages [22].

As with the replicated banyan network, one of two acceptance policies can be used at the output side of a dilated banyan network: SA or MA. With SA, a maximum of one cell can be accepted at each output port every time slot [10], [11], [130], [132], [204], while with MA all successful cells (up to d cells) are accepted [11], [37], [91], [147], [205], [215], [223], [233]. Thus, output buffering is required only with MA. In an ATM environment, input buffering must be used with SA, while it can be ignored with MA if d is large enough to guarantee an acceptable cell loss probability within the interconnection network. It has been shown that with $d = 4$ and SA, the dilated banyan network is almost nonblocking. It becomes nonblocking when $d = 2^{\lceil n/2 \rceil}$, where $n = \log_2 N$.

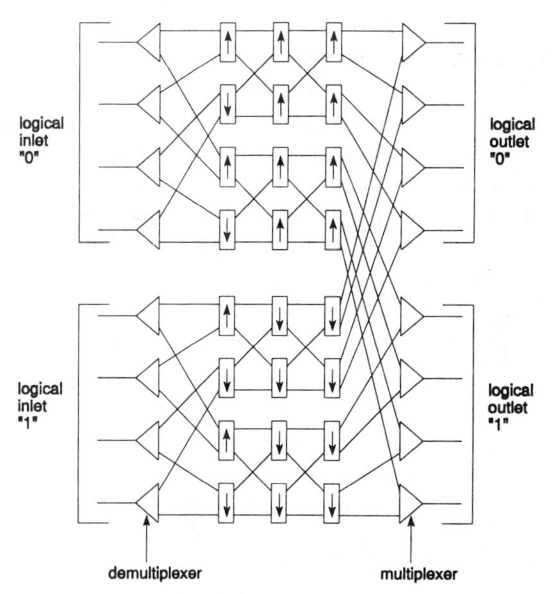

Fig. 27. A (2×2) switching element of a dilated banyan network with $d = 4$.

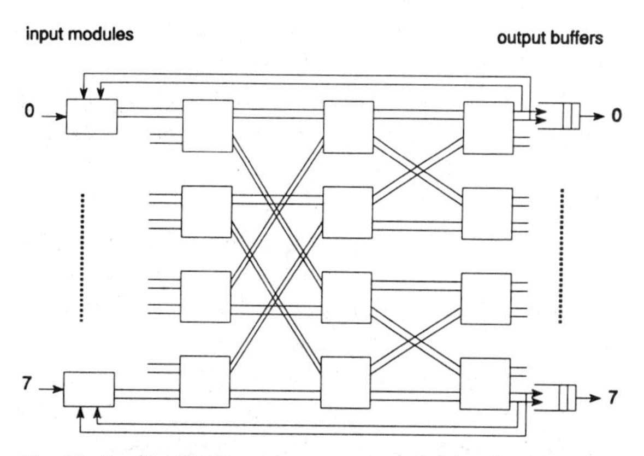

Fig. 28. An (8×8) dilated banyan network ($d = 2$) with deflection routing and recirculation.

The throughput performance of a dilated banyan network can be improved further with the MA policy. Ghosh and Daly [91] studied a dilated banyan network with pure output buffering, and showed that a TP_{max} exceeding 90% can be achieved with $d = 3$ and $B_{out} = 20$. Szymanski and Shaikh [205] showed that a dilation factor between 4 and 8 is sufficient to reduce the fraction of cells lost within the interconnection network to values between 10^{-3} and 10^{-6} under full-load conditions. Even with d as small as 3, a near-unity maximum throughput can be achieved using single output-buffered SEs [205]. However, cell-sequencing may be disturbed by this approach since it uses internal buffering in a multipath architecture. A near-unity TP_{max} can also be achieved by misrouting (i.e., deflecting) losing cells instead of dropping them, and then re-injecting them into the network through the input ports [233]. The architecture of a switch adopting this approach is shown in Fig. 28 for $N = 8$ and $d = 2$. It should be noted, however, that this approach too does not preserve cell-sequencing. Finally, assuming an internally-unbuffered network, Lee and Liew [147] addressed an important issue, namely "how does the switch complexity grow as a function of the switch size for a given loss probability requirement?" for a number of switch architectures. Using analysis, it was shown that for a given target of P_{loss}, the dilated banyan network has a lower order of complexity than the tandem-banyan switch or the replicated banyan network (with MA). It was concluded that dilation is a powerful design technique for improving performance and reducing complexity in a large switch.

6.4. Sorting

Some of banyan networks have an interesting property: a Banyan or an Omega network becomes nonblocking if incoming cells are ordered according to their output port addresses (in either an ascending or a descending order), and concentrated (i.e., compact with no gaps between active inputs), provided that output conflicts do not exist [105], [141], [165]. This is the basic idea behind all sort banyan based ATM switches. A well-known architecture that can perform the sorting function in a distributed and parallel manner is the Batcher network [22]. The operation of the Batcher network is based on the bitonic sorting principle that was introduced in the same reference. An $(N \times N)$ Batcher network can be built using $\log_2 N(\log_2 N + 1)/2$ stages, each with $N/2$ binary comparators or sorting elements. The Batcher network can sort an arbitrary set of active cells based on their output port addresses (and/or any other information contained in cell headers; this is useful when supporting prioritized traffic), and group them consecutively at the bottom or the top of its output ports. Thus, the combination of a Batcher network and a Banyan (or Omega) network solves the problem of internal blocking and renders this

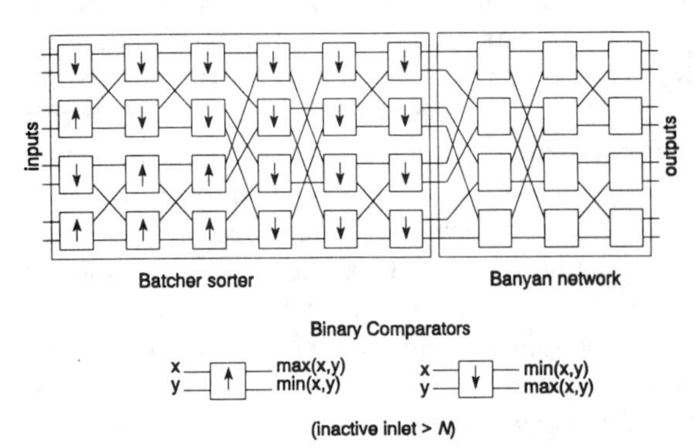

Fig. 29. The Batcher-Banyan network.

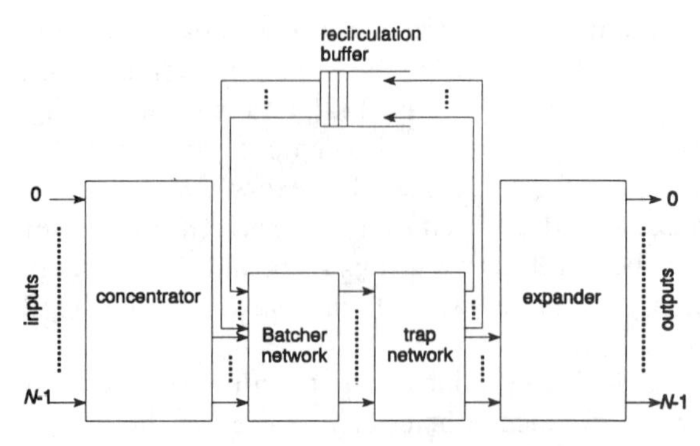

Fig. 30. The Starlite switch.

combination as effective as the crossbar switch, provided that multiple cells with the same output port address are prevented from entering the Batcher-Banyan (BB) network. The particular approach used for resolving output conflicts in a BB based switch is strongly related to the buffering strategy that the switch adopts, as will become clear below. An (8×8) BB network is shown in Fig. 29.

Although the BB network is nonblocking with less complexity than the crossbar switch, it can not be easily partitioned into integrated circuits, and maintaining synchronization across the entire structure (a total of $\log_2 N + \log_2 N(\log_2 N + 1)/2$ stages) becomes difficult as N increases [168]. The same reference noticed that most of the interest in the BB switching fabric came from the research community rather than from the industry. Some physical issues, concerning the scalability of BB networks to architectures suitable for large broadband central offices, were examined in [92].

Internal buffering

The *Starlite* switch [103] was the first sort banyan based switching fabric to be proposed in the literature. (Also, demultiplexing input traffic to cope with the very high speeds of fiber links was suggested for the first time in the same reference.) The structure of the Starlite switch is shown in Fig. 30, and it basically consists of a cascade of a concentrator, a sorting network, a trap network, and an expander. The concentrator (built from running sum adders followed by a reverse Omega network, which is the mirror image of an Omega network) is used to reduce the size of the following networks, assuming that it is likely to have a significant number of idle input ports. The outputs of the sorting (Batcher) network are fed into a trap network (built from a single stage of comparators followed by an Omega network) so that only one cell per each set of cells destined to the same output port is admitted to the expander (an Omega network), while all others are recycled back into the sorting network. Two points should be noted. The first is that cell-sequencing is not preserved by this recycling approach, but out-of-sequence cell delivery can be avoided by aging cells. Also, a buffering stage must be provided for recycled cells (i.e., internal buffering). To reduce cell loss within the recirculation buffers, a substantial fraction of the input ports must be dedicated for recirculators, which under-utilizes the switch capacity. However, sharing buffers and smoothing bursty traffic both can be achieved by this recirculation mechanism. Finally, this switch can support a special type of multicasting.

Internal-output buffering

The *Sunshine* switch [93], shown in Fig. 31, achieves high performance by utilizing both internal and output buffering. It basically combines a Batcher network with parallel Banyan networks. Incoming cells enter a Batcher network, where they are sorted according to their output port addresses and priority levels. A trap network then resolves output conflicts by selecting the K highest priority cells present for each output port address within a common time slot. Cells which exceed K for a single address must be recirculated. A concentrator (another Batcher network) separates cells to be recirculated from those to be routed, and the selector directs cells either to the recirculators for queueing, or to the Banyan networks for routing. Thus, K cells per output port address may be always switched within a given time slot, one through a different Banyan network. It was shown that this architecture can achieve the extremely low cell loss probabilities necessary for circuit emulation [93], and is robust in a bursty environment for a large enough number of Banyan networks [69]. The switch can also support output channel grouping which further improves its performance particularly under extremely bursty traffic. Finally, it is interesting to notice that when $K = 1$, the Starlite switch [103] results.

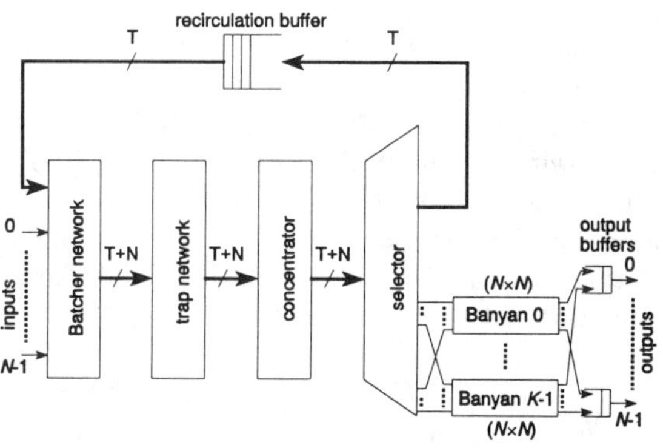

Fig. 31. The Sunshine switch.

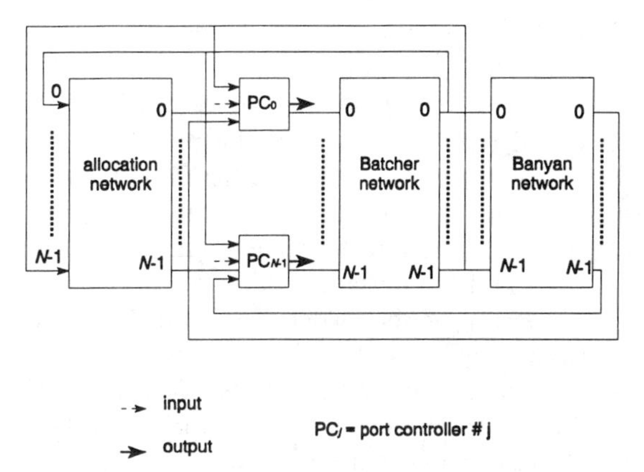

Fig. 32. The 3-phase Batcher-Banyan switch.

Input buffering

Perhaps the most straightforward approach to operate the BB network is to use FIFO input buffers and to resolve conflicts prior to switching. However, this approach has two drawbacks. First, the need for a contention resolution mechanism suitable for the BB network. Also, the maximum throughput is significantly less than unity due to HOL blocking [101], [104], [120]. The first proposal for contention resolution in an input-buffered BB network was given in [104] under the name *3-phase algorithm*, and the resulting switch was called the *BB switch*. The block diagram of the BB switch is shown in Fig. 32 [181]. Notice that same-number input and output port controllers are located in the same module PC_i ($i = 0, \ldots, N-1$) that interfaces both input port i and output port i of the switch. In phase 1 of the algorithm, request packets formed from the destination and the source addresses of each HOL cell are sorted by the Batcher network according to the destination addresses. At the outputs of the Batcher network, adjacent cells with the same destinations are purged (by the allocation network). In phase 2, winning requests are fed into the inputs of the Batcher network, and routed through the BB network using this time the source addresses, so that they acknowledge their originating input ports. Selected input ports admit their cells to the fabric in phase 3. Fig. 33 shows an example of the operation of the first two phases of this algorithm. Prioritized traffic can be easily supported by the BB switch by sorting request packets (in phase 1) according to both the destination addresses and the priority levels.

A reservation-based contention resolution mechanism for the BB network was proposed in [30]. Here, the input ports are interconnected via a ring which is used to reserve access to the output ports. Arriving cells are buffered till they can participate in the reservation process. At the beginning of each time slot, an N-bit token is circulated around the ring, where each bit corresponds to a particular output port and indicates whether that port is available or has already been reserved. When the token passes by, the HOL cell in each input buffer tries to make a reservation. When an output is successfully reserved, the corresponding HOL cell is admitted to the BB network at the beginning of the next cycle. To insure fairness, the ring is shifted one revolution plus one more position in each time slot.

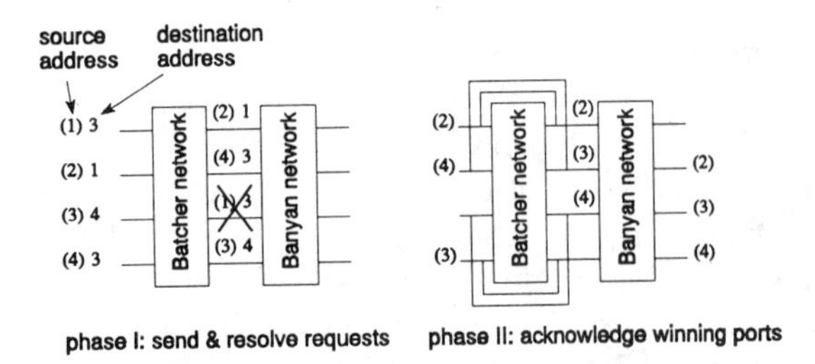

Fig. 33. An example of the first two phases of the 3-phase algorithm for the Batcher-Banyan switch.

Both the 3-phase algorithm [104] and the ring-reservation mechanism [30] suffer from drawbacks. The 3-phase algorithm requires significantly large processing overhead (e.g., 33% for $N = 1024$ and an ATM cell size) [104], and suffers from a fairness problem since either lower or higher address input ports are given priority in the allocation of the switching bandwidth when contending for the same output port [178]. On the other hand, while the ring-reservation mechanism is simpler than the 3-phase algorithm, the need to run the ring N times the speed of the external lines, to be able to poll all N input ports within each time slot, constitutes a major limitation on the scalability of the switch. Furthermore, extending the function of this mechanism to prioritized traffic does not seem to be straightforward.

To overcome the above problems, Awdeh and Mouftah [14] have recently proposed a 2-phase contention resolution algorithm that has been shown to be fair, to require less processing overhead than that required by the 3-phase algorithm [104], and to easily support prioritized traffic. The basic idea in this algorithm is to let the binary comparators of the Batcher network explicitly participate in the arbitration process. Phase 1 starts at the beginning of each time slot by forming request packets (RPs), each from a HOL cell destination address (DA) and a flag bit (FB) that is initially set to 1. RPs are then sorted by the Batcher network according to the DAs, but with the following two modifications. First, the binary sorting elements hold their settings after the RPs pass through them. Also, if a binary sorter finds out that both RPs at its inputs have the same destination address (i.e., $DA_0 = DA_1$), then it examines their FBs:

- if $FB_0 = FB_1 = 1$ (i.e., both unmarked), it *randomly* selects one of the contending RPs and marks it by resetting its FB to 0, and it treats the marked RP as the one having the largest DA;

- if $FB_0 \neq FB_1$ (i.e., one is marked and the other is unmarked), it treats the marked RP as the one having the largest DA.

In all other cases, the routing decisions are entirely based on the DAs, without altering or being affected by the FBs. This includes the case where $DA_0 \neq DA_1$. An RP wins if it remains unmarked after being processed by a last stage sorting element. ACKs are formed from the current FBs and sent back to the originating input ports through the paths that have just been set. In phase 2, input ports which receive ACKs of 1, send their full HOL cells through the BB network to self-route to their desired destination ports.

With any of the above mechanisms, the maximum throughput is 58.6% due to HOL blocking. Pattavina [177] tried to solve this problem by modifying the operation of the 3-phase BB switch [104] to support output channel grouping. In phase 1 of the modified algorithm, requests are sorted according to the desired group numbers. The channel allocation network of Fig. 32 (now consists of $\lceil \log_2 R \rceil$ stages of hardware adders, where R is the group size) computes an index for each request that identifies a specific output port in a group. Requests which cannot be accommodated in a channel group are given an incorrect index. In phase 2, an acknowledgement packet with the assigned index is sent back to the requested port, and each active port controller determines if it is a winner based on the received index. It is important to notice that an incoming cell stream (belonging to the same connection) may be divided and switched through different output ports, since the allocation of the connec-

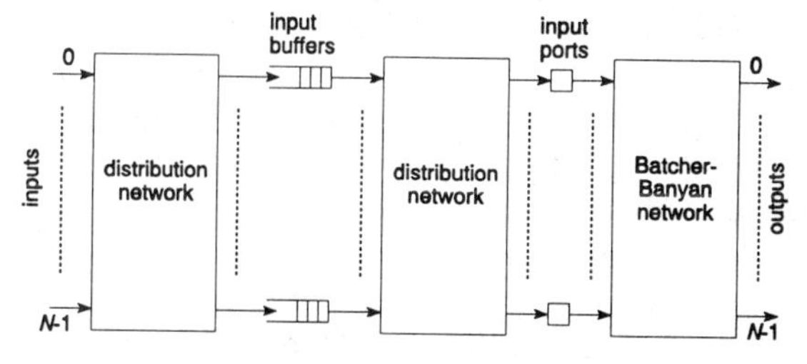

Fig. 34. A buffer subsystem for the Batcher-Banyan switch.

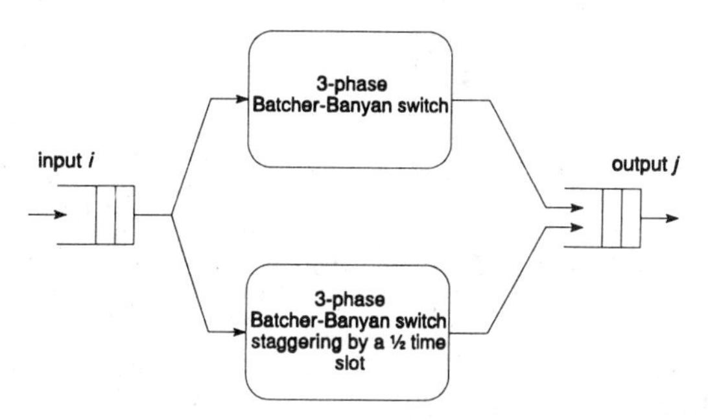

Fig. 35. The duplex switch.

tion was made to a channel group rather than a specific output port. This may affect the connection end-to-end cell sequence integrity (however, this switch still delivers cells in sequence within itself).

In [135], it was noticed that the BB switch [104] has shortcomings in its ability to share buffers, its large buffer requirement to insure a small cell loss probability, and its poor performance under imbalanced traffic. In the same reference, a buffer subsystem which attempts to solve these problems was proposed. Fig. 34 shows the block diagram of the resulting switch, which consists of a cascade of an $(N \times N)$ distribution network, N physically-separate buffers, a second $(N \times N)$ distribution network, and finally an $(N \times N)$ BB switch. The first distribution network routes incoming cells to the buffers of lower occupancy, while the second distribution network routes cells from buffers of higher occupancy to the idle input ports of the BB switch. (Clearly, this design does not preserve cell-sequencing.) Each distribution network is constructed from a Batcher sorter with feedback paths, together with a connection algorithm to perform the path set-up. Significant buffer savings were shown to be achieved by this architecture. As an example, for a switch of size 32 loaded at 90% of capacity, a total buffer size of about 30 cells is required to achieve a cell loss probability of 10^{-10}, compared to 1600 for the BB switch. Also, the adverse effect of imbalanced traffic on performance is eliminated. Despite the above, the switch still suffers from a poor throughput performance. The operation of the switch has been modified to support output channel grouping in [136], where any partition of the output ports is allowed.

Input-output buffering

The *duplex switch* [104], shown in Fig. 35, consists of two BB switch planes in parallel, with the second plane staggering by a half time slot behind the first plane. HOL cells present themselves to the first switch plane for contention resolution. At the end of phase 2, winning cells are admitted to the first plane and an updated set of contending HOL cells present themselves to the second switch plane, and so on. This process alternates between the two switch planes. Output buffering is needed to cope with the possibility of simultaneous arrivals from the two switch planes at the same output port. In general, for S BB switch planes in parallel, with adjacent planes staggered by a period of time longer than that required by phases 1 and 2 of the 3-phase algorithm [104], the capacity is equal to S times the single BB switch capacity. Furthermore, this configuration ($S > 1$) provides redundancy for enhancing system reliability.

A modular architecture for very large packet switches has been proposed by Lee [142], and its block structure is shown in Fig. 36. The inputs to the switch are partitioned into K equal subsets, with each subset being interfaced to a switch

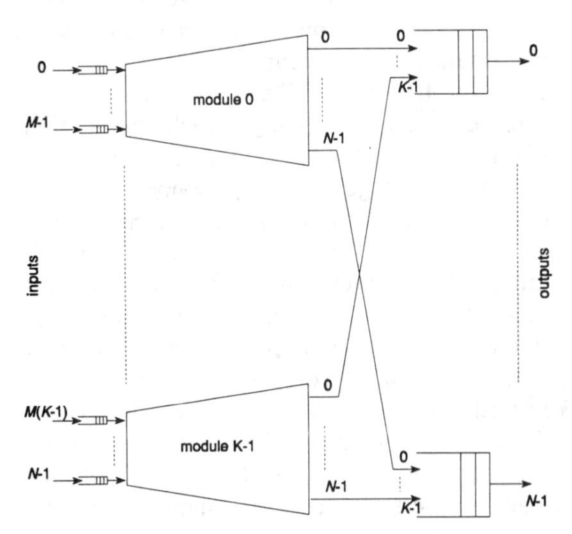

Fig. 36. Lee's modular switch architecture.

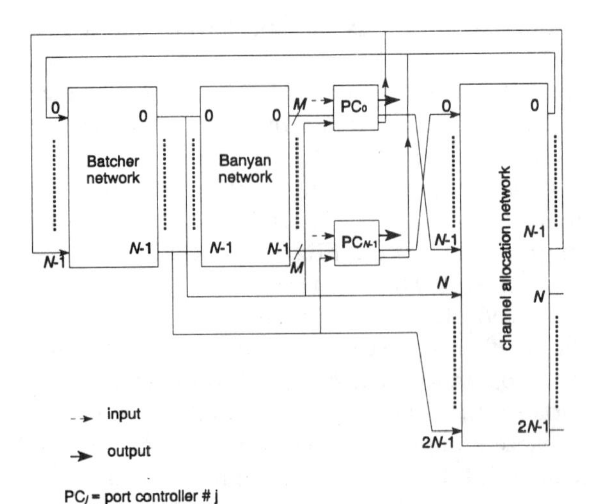

Fig. 37. Pattavina's input-output buffering switch.

module of size $(M \times N)$, where $M = N/K$. Each switch module is a cascade of an $(M \times M)$ Batcher sorter, a stack of M binary trees, and K parallel $(M \times M)$ Banyan networks. Each subset of input cells is sorted by a Batcher sorter, then partitioned into finer sub-subsets by the binary trees. In each module, the ordered cells of these sub-subsets are routed concurrently to the output buffers by K Banyan networks. Each switch module adopts a ring-reservation contention resolution mechanism [30], together with a w-windowing scheme [101]. This architecture can be considered as a result of a unification of the BB network (obtained when $K = 1$) and the Knockout switch [230] (obtained when $K = N$). Since the switch modules are interconnected only at the outputs, relaxed synchronization requirements, easy operation and maintenance, and higher reliability are achieved. It was shown that a TP_{max} greater than 90% can be achieved with $w = 2$ for $K \geq 2$. Finally, the parallel Banyan networks can be replaced with a dilated banyan network without sacrificing the nonblocking property [147]. This also applies to the Sunshine switch [93] that was described earlier.

The above described input-output buffering switches provide no means to prevent buffer overflows at the output ports. On the other hand, the switch architecture described in [179] and

[180] uses mixed input-output buffering with a backpressure mechanism. Fig. 37 shows the block structure of this switch. The switch is able to transfer up to M cells per time slot to a given output port. It is constructed using a Batcher network, a Banyan network, and a channel allocation network. The channel allocation network consists of 3 different networks: a $(2N \times 2N)$ merge network (of $n+1$ stages of (2×2) SEs), a $(2N \times 2N)$ path allocation network (of $L = \lceil \log_2 M \rceil$ stages of adders), and a $(2N \times 2N)$ concentration network (of $n+1$ stages of (2×2) SEs). The size of the Banyan network is $(N2^L \times N2^L)$ where N inputs only are used. A probe-ack contention resolution algorithm is employed to set up conflict-free paths for the cells through the interconnection network, and at the same time to insure that cell losses do not occur as a result of output buffer overflows. This makes it possible to jointly optimize the cell storage capacity at input and output buffers, and easier to define congestion control procedures. It was shown that for M as small as 3 or 4, the HOL blocking effect is significantly reduced. Also, it was shown that bursty traffic requires much more buffer capacity at the outputs to give throughputs comparable to those achieved under uniform traffic.

Output buffering

In [52], it was observed that the high complexity of the Knockout switch [230] is mainly due to the need of each bus driver to drive N loads, and also to having a separate concentrator in each of its N output bus interfaces. The same reference proposed the *Shared Concentration Output Queueing* (SCOQ) *switch*, shown in Fig. 38. The switch is composed of an $(N \times N)$ Batcher network, followed by L switching modules (SMs), each of size $(N \times K)$, where $K = N/L$. The basic idea here is that each SM concentrates and routes cells corresponding to its K outputs. Therefore, concentration is now shared among the K output ports of each SM. Also, each output of the Batcher network needs to drive only L loads. Besides the output buffers, each SM consists from a bank of address filters and L $(K \times K)$ Banyan networks. If more than L cells are destined to the same output port at the same time, at most L

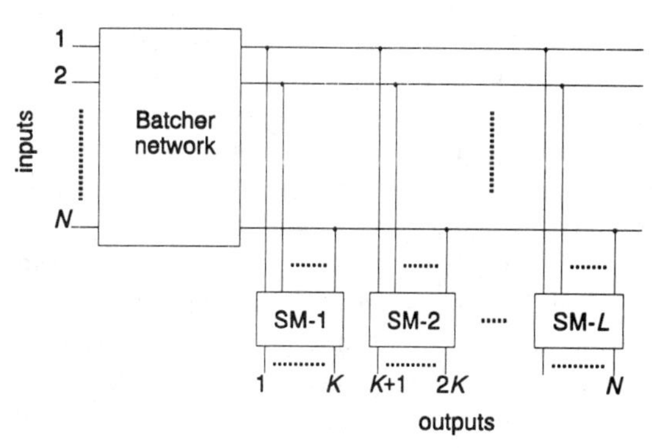

Fig. 38. The Shared Concentration Output Queueing switch.

of them will be successfully routed and the remaining cells are lost due to contentions within the Banyan networks. The SCOQ switch has been shown to compare well with the Knockout switch [230], the Sunshine switch [93], and Lee's modular switch [142], in terms of complexity.

6.5. *Deflection-routing*

Instead of dropping or internally buffering a losing cell in a multistage interconnection network, a third approach is to route that cell the *wrong* way (i.e., to misroute or to deflect it). A cell that has been deflected can get involved in the routing process again at a later stage. It should be noticed that both the tandem-banyan switch [214] and the dilated banyan network with recirculation [233] (which have been described earlier) are based on the deflection-routing principle.

Folded Shuffle switch

This is an input-output buffering switch architecture that adopts a cylindrical (i.e., the outputs of the last stage are connected to the inputs of the first) interconnection network [38]. An $(N \times N)$ switch is constructed using $n = \log_2 N$ stages, each with N unbuffered (3×3) SEs, where successive stages are interconnected through the perfect shuffle pattern [201]. In this architecture, a switch input and a switch output are associated with all SEs in each row of the network, as shown in Fig. 39 for $N = 8$. An input buffer can access any SE in the corresponding row, and an output buffer receives successful cells from any SE in the corresponding row. Thus, each input (output) buffer must be able to send (receive) up to n cells per time slot. At the beginning of each time slot, each SE computes for the cells at its internal inlets (see Fig. 39) the minimum number of stages remained to the required destinations (called cell distances). A cell with zero distance is admitted to the corresponding output buffer through the external outlet; otherwise, it is transmitted through an internal outlet according to the shortest path algorithm. Conflicts arising when more than one cell request the external outlet (leading to an output buffer) or the same internal outlet (leading to the next stage) of an SE, are resolved by deflecting a *randomly* selected cell through an internal outlet. If at least one of the internal outlets is idle, the corresponding SE allows the reception of a new cell through the external inlet (assuming that a newly received cell has no immediate access to the external outlet of the SE). Thus, cell loss is avoided between the interconnection network and the input buffers. It was shown that small input buffer sizes and considerably large output buffer sizes are needed to achieve a very small cell loss probability at reasonable loads. Finally, this design does not preserve cell-sequencing.

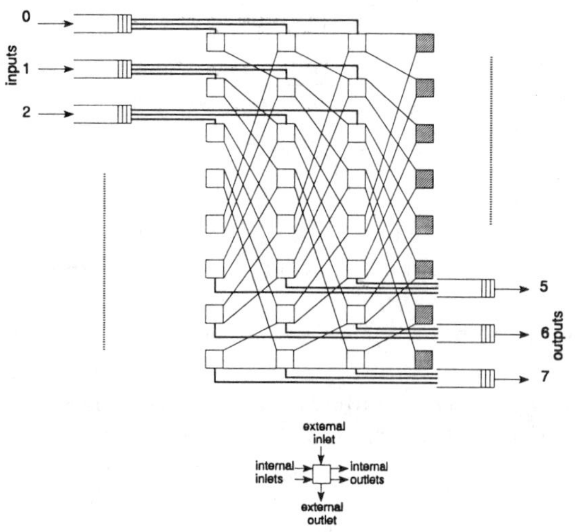

Fig. 39. An (8×8) Folded Shuffle switch.

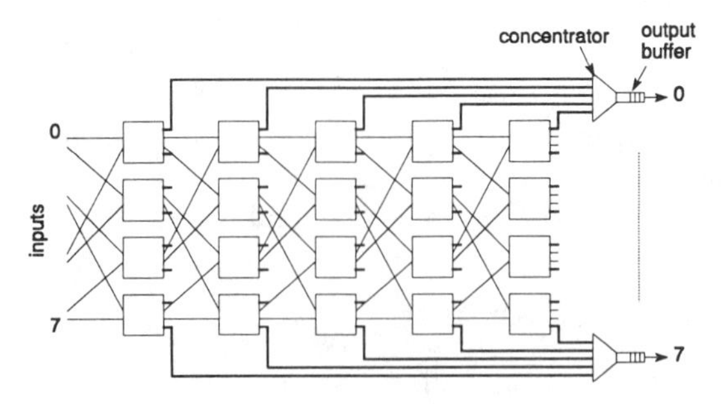

Fig. 40. An (8×8) Shuffleout switch with $K = 5$.

Shuffleout switch

Another switch architecture, in which successive stages are interconnected through the perfect shuffle pattern [201], and routing is performed according to the shortest path algorithm with deflection, has been proposed in [62] under the name the Shuffleout switch (Fig. 40). It is constructed using K $(\geq n = \log_2 N)$ stages, each with $N/2$ unbuffered (2×4) SEs, where the two outer outlets, called local outlets, are connected to the corresponding output ports. The number of buses entering an output port is equal to K; thus, concentration is required to reduce system complexity. Each SE attempts to route each received cell along its shortest path measured in remaining number of stages to destination. In case of contention, the cell *closest* to its destination wins, and the other cell is deflected through a remote outlet; if both have the same output distance, a random selection is performed. An SE computes the output distance of each incoming cell by comparing the binary address of the row in which it is located, with the $n-1$ least significant bits of the destination address of the cell. Cells which do not reach their destinations by stage K are lost. It was shown that the Shuffleout switch is robust against nonuniform traffic at the expense of few more stages compared to uniform traffic [23]. This switch architecture may deliver cells out-of-sequence, unless an additional fixed delay of $(K - s)\tau$ (where τ is the SE latency) is imposed on each cell exiting the network from stage s $(1 \leq s$

$\leq K$). Also, the Shuffleout switch is unfair since cells addressing central output ports are better served by the network than cells addressing peripheral output ports [23]. The function of the Shuffleout switch has been extended to support variable-length packets in [63].

Instead of dropping cells which do not reach their destinations after crossing the whole interconnection network, these cells are recirculated for further processing in the *closed-loop* Shuffleout switch [61]. Two different implementations have been considered: buffered and expanded. In the former, each input of the first stage is shared between an input line and a link connected to an outlet of an SE in the last stage. Input buffers are needed to cope with the possibility of two concurrent arrivals at the same input port. Priority is always given to recirculated cells. In the expanded closed-loop Shuffleout switch, recirculated cells are fed into dedicated switch inputs. Thus, an expansion of the network size is required in order to concurrently feed into the network both new and recirculated cells. A concentration stage can be used for the recirculation links before feeding them into the network inputs. In general, the expanded version requires a smaller number of stages for a given cell loss probability, compared to the buffered version at the expense of network expansion. Compared to the original switch (called the *open-loop* Shuffleout), both implementations require less number of stages for the same cell loss probability. However, they both require resequencing mechanisms

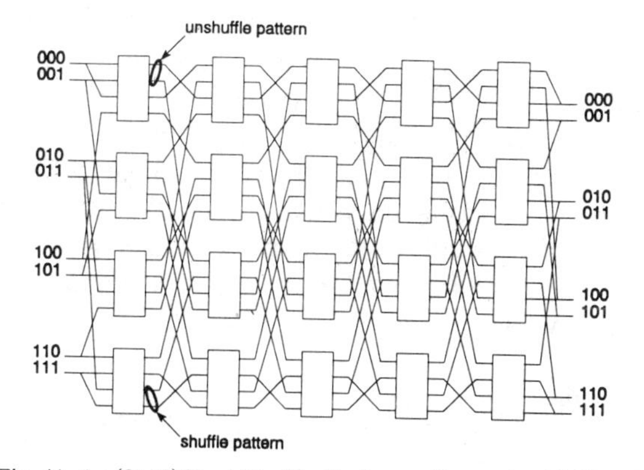

Fig. 41. An (8×8) Dual Shuffle-Exchange Network with $K = 5$.

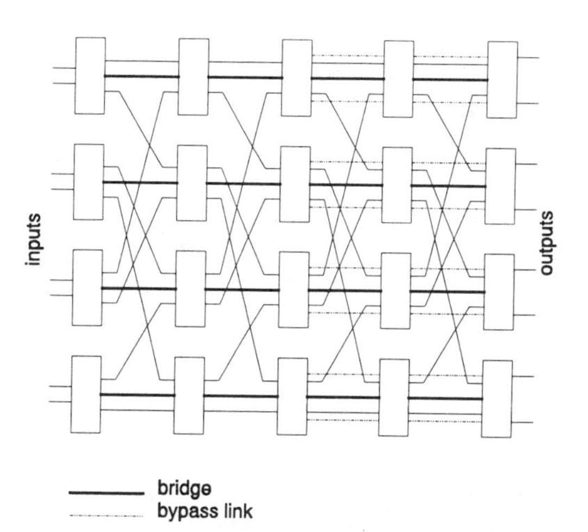

Fig. 42. An (8×8) Bridged Shuffle-Exchange Network with $K = 5$.

in the output buffers to avoid out-of-sequence cell delivery.

Dual Shuffle-Exchange Network

A Dual Shuffle-Exchange switching network (DSN) constructed by interleaving a shuffle-exchange (SN) network and an unshuffle-exchange (USN) network, has been described in [156]. (Having two complementary shuffle networks together was previously suggested in [208], although in a different context.) A DSN consists of K $(\geq n = \log_2 N)$ stages, each having $N/2$ (4×4) SEs, as shown in Fig. 41 for $N = 8$ and $K = 5$. The upper (lower) half outlets of each SE are connected to the next stage according to an unshuffling (a shuffling) pattern. Routing bits used in successive stages proceed from the least significant bit (LSB) to the most significant bit (MSB) in the USN, and from the MSB to the LSB in the SN. A routing algorithm for the DSN was described in the above reference. Basically, the SN links are used for normal routing of cells, while any error caused by deflection in the SN is corrected in the USN in a single step, so that the corresponding cell can return to its normal path in the SN again after one stage. Bypass links (not shown in the above figure) are used to collect and multiplex cells which reach their destinations. Cells which do not reach their destinations by stage K are lost. It was shown that the DSN achieves the $N\log_2 N$ lower bound on switch complexity for an arbitrary small cell loss probability. Finally, the cell-sequencing problem has not been addressed in [156].

Bridged Shuffle-Exchange Network

Zarour and Mouftah [234] noticed that the DSN [156] uses (4×4) SEs and requires a relatively complicated routing scheme. In the same reference, the Bridged Shuffle-Exchange Network (Fig. 42) was proposed and shown to achieve the $N\log_2 N$ order of complexity for an arbitrary small cell loss probability. In this architecture, a losing cell is passed without changing its status to the same-position SE in the next stage through a "bridge". Thus, the routing error is prevented and the cell can resume its routing from the following stage. However, it is possible that all arriving cells at the three inlets of an SE request the same outlet; in such a case, one cell must be deflected. Using 2 bridges from each SE to two different SEs in the following stage belonging to the same "group" (as defined in the next section) has also been considered. In both cases, cells which do not reach their destinations by the last stage are lost. A closed-loop version of the switch, analogous to the buffered closed-loop Shuffleout switch [61], was proposed in [235], and shown to require less number of stages to achieve a given cell loss probability, compared to several other deflection-routing switches. In the closed-loop design, a resequencing mechanism for the output buffers has been suggested to preserve the cell sequence integrity.

Multi Single-Stage-Shuffling switch

Up to now, we described switches which use SEs of size larger than (2×2) and require considerable amount of processing in the SEs. A deflection-routing switch architecture based on unbuffered (2×2) SEs has been proposed [7], [12], [13] under the name the Multi Single-Stage-Shuffling Switch (MS4). As with the last three described switches, the MS4 consists of K $(\geq n = \log_2 N)$ stages, each with $N/2$ SEs (Fig. 43). Successive stages are interconnected through the perfect shuffle pattern [201]. The routing algorithm here is a simple extension of that in a

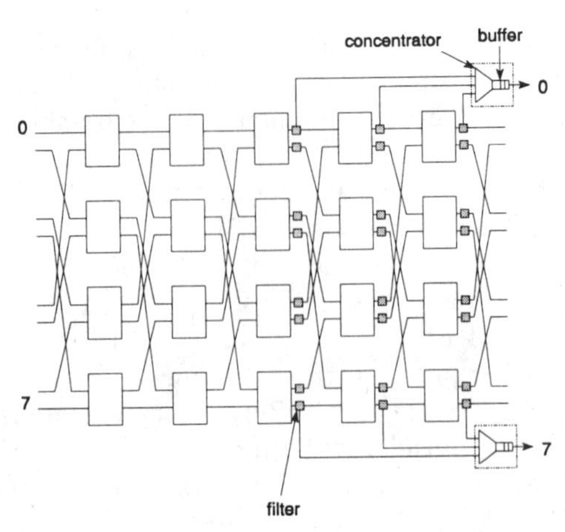

Fig. 43. An (8×8) Multi Single-Stage-Shuffling switch with $K = 5$.

banyan network. Each cell carries a counter field (C) of $\lceil \log_2 n \rceil$ bits that is initially set to n. C acts as a pointer to the particular bit (d_{n-C+1}) that an SE must examine in the destination tag of the corresponding cell. Each time a cell is correctly routed, C is decremented by one, and each time it is deflected, C is reset to n. In any conflict, the cell closest to its destination wins. A cell reaches it destination when C becomes zero. At the outputs of the last stage, cells with $C \neq 0$ are lost. It has been shown that the MS4 can achieve an arbitrary small cell loss probability with complexity of $O(N\log_2 N)$. Furthermore, the MS4 preserves cell-sequencing without using artificial delays even for strict cell loss requirements and large switch sizes. The MS4 has been shown to be robust against nonuniform traffic [7], [13]. Finally, an interesting result has been reported in [13]: for the same switch size, output concentration ratio, and output buffer size, the MS4 (with large enough K) could have better overall cell loss performance, compared to the Knockout switch [230]. This has been explained by the positive effect of presenting the traffic in an imbalanced fashion to the inputs of each output concentrator in the MS4.

Statistical Data Fork

The Statistical Data Fork (SDF) [58] is (to the best of our knowledge) the only deflection-routing switch architecture that supports output channel grouping. The switching platform of the SDF is the Omega network [137] with unbuffered (2 × 2) SEs. The output ports of the SDF are partitioned into groups and each group is assigned a codeword (a binary string). Different groups may have different priority levels in different SEs. In a similar fashion to self-routing in a banyan network, an SE in the SDF uses the codeword carried by each cell to determine its proper routing. A contention within an SE is resolved according to the priority rule associated with that SE; otherwise, random selection is used. A losing cell is deflected, and thus it *may* exit on the wrong output channel group. If a cell does not encounter too many contentions within the network, it can still exit on the correct output group. The partition of the output ports into channel groups can be done in such a way as to minimize the probability of a routing error. Filters are inserted at the outputs of the interconnection network to destroy misrouted cells. To make the SDF robust against general traffic models, a Benes network [24] can be placed at the front end of the SDF to randomize incoming traffic. Also, misrouted cells or a subset of them, can be fed back (into the randomization network) in order to increase efficiency.

6.6. Load-sharing

The concept of load-sharing in a banyan network was first introduced in [138]. It is based on

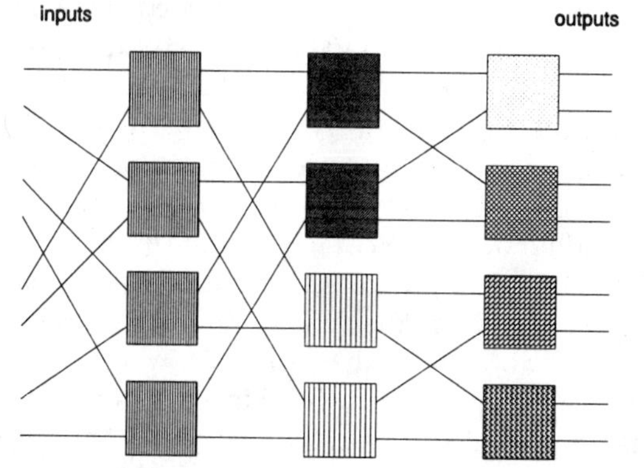

Fig. 44. Switching elements with the same type of shading belong to the same group.

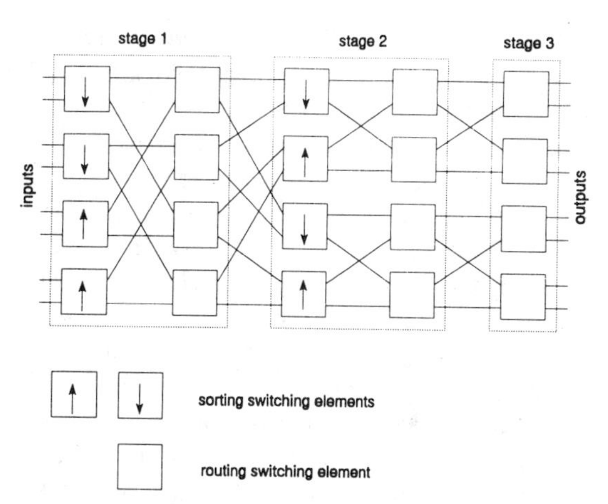

Fig. 45. A possible implementation of the load-sharing banyan network ($N = 8$).

the following observation. The SEs of any given stage of a banyan network can be organized into a number of groups in such a way that the proper routing function is not disturbed by sharing the input traffic of any two SEs in the same group. This is illustrated in Fig. 44; notice that all SEs in the first stage are in the same group, while all SEs in the last stage are in different groups. A cell can be sent to any SE in a group, and still be correctly routed to its destination. This can be useful in handling imbalanced traffic: if an SE is congested, other SEs in the same group can be used to handle incoming traffic. In the load-sharing banyan network, the idea is to allow *two* SEs in the same group to share their input traffic, which provides cells with multiple paths. SEs can be paired in such a way as to achieve the maximum possible number of alternate paths. The resulting network has higher throughput compared to banyan networks (e.g., for $N = 1024$ the load-sharing banyan network achieves at least $TP_{max} = 0.39$ compared to 0.26 achieved by a banyan network), can tolerate multiple faults, and can be diagnosed as easily as banyan networks. Finally, the concept of load-sharing is applicable to both unbuffered and buffered banyan networks. However, it might cause a cell-sequencing problem in the latter.

A self-routing implementation of the load-sharing banyan network has been reported in

[145], and is shown in Fig. 45. It is constructed by inserting sorting SEs before regular routing SEs in each stage of a banyan network, except the last stage. For an n-stage network, routing SEs in each stage, except the last one, are partitioned into 2^{n-2} groups, each of two members. Two routing SEs belonging to the same group receive cells from the same two sorting elements; thus, they share their load. Blocking occurs only when more than two cells received by any group are to be simultaneously routed to the upper or the lower output links. Both uniform and nonuniform traffic models have been considered for both unbuffered and internally-buffered routing SEs.

A self-routing ATM switch that is based on the load-sharing principle was proposed in [123], and its block diagram is shown in Fig. 46. Here, an $(N \times N)$ switch is constructed using n stages of (2×2) SEs, and $n-1$ stages of distributors. If a stage is defined to be a column of SEs and the following distributors, then stage i ($1 \le i \le n$; starting from the left-most stage) consists of $N/2$ SEs and 2^i distributors of size $(2^{n-i} \times 2^{n-i})$. The switch requires a speed-up factor of two. A distributor distributes the cells at its inputs evenly across its outputs which are connected to SEs belonging to the same group. Thus, *all* SEs of a group (except those in the first stage) share their traffic load, in contrast to the designs of [138] and [145] in which only *two* SEs in any group share their load. A distributor basically consists of a reverse Banyan network [128] with output buffers. With large enough buffers, the maximum throughput of this switch approaches 100%. Unfortunately, cell-sequencing is not preserved.

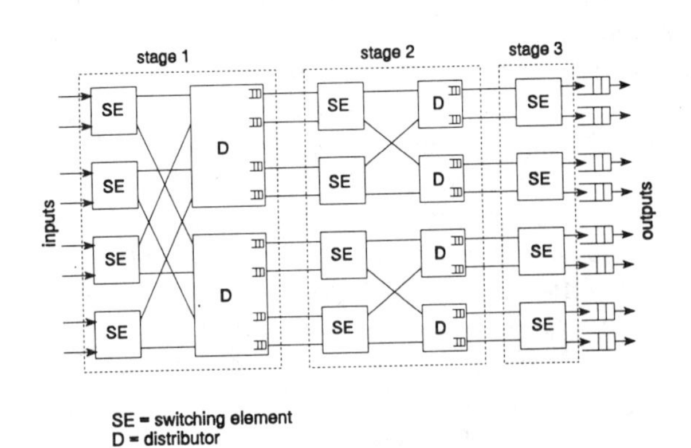

Fig. 46. Kim's self-routing multistage switch.

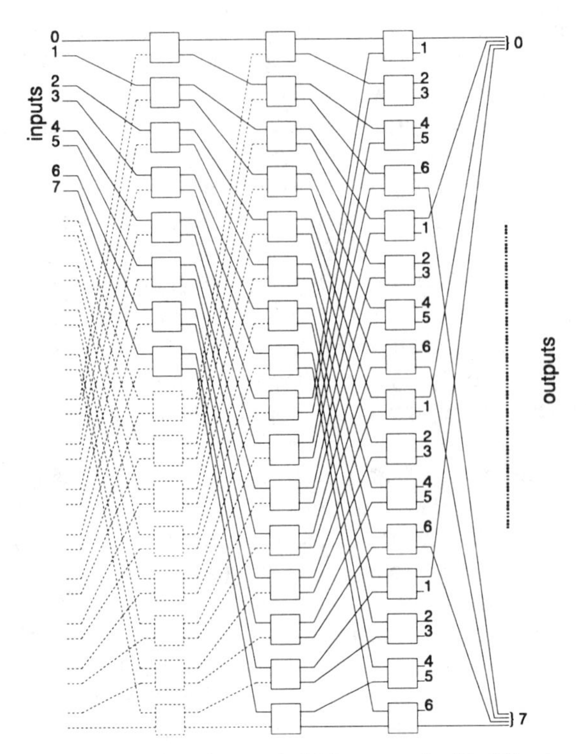

Fig. 47. An (8×8) expanded delta network with $EF = 4$.

6.7. Expansion

An approach that overcomes both internal and output blocking of a banyan network while preserving self-routing and path-uniqueness has been proposed in [9] under the name the *expanded delta network*. An $(N \times N)$ expanded delta network with expansion factor EF is constructed by interleaving EF $(N \times N)$ delta networks, as shown in Fig. 47 for $N = 8$ and $EF = 4$. The resulting network can be regarded as an $(M \times M)$ interconnection network (where $M = N \times EF$) consisting of $\log_2 N$ stages, each having $M/2$ SEs. Only N input links are used, while each output port is connected to EF of the output links. If we assume that $EF = 2^k$ where k is a positive integer, then blocking-free routing is guaranteed till stage k, inclusive. Also, $N(2^{k-1} - 1)$ SEs in the first $k-1$ stages are jobless and can be removed. Increasing EF has two joint effects:

reducing internal blocking and increasing the number of cells (up to EF cells) which may be simultaneously switched to the same output port. This results in improving the throughput performance.

At the output side of the network, either single acceptance (SA) [10], [11], or multiple acceptance (MA) [9], [11] can be used. Assuming MA, it has been shown that for a switch of size 1024 and $EF = 16$, above 90% maximum throughput can be achieved [9], compared to 26% achieved by a banyan network. Also, with either SA or MA, the expanded delta network requires the least number of crosspoints (almost always) to achieve a given target of TP_{max}, compared to the replicated and dilated delta networks [10], [11]. With MA and $1 < EF < N$, both input and output buffers are needed to reduce cell loss to an acceptable level. An ATM switch that is built around the expanded delta network has been proposed and studied in [9] under the name the *Expanded Delta Fast Packet Switch* (EDFPS). It has been shown that preventing output buffers from overflowing by a means of backpressure signals, can reduce the overall memory size needed to achieve a given P_{loss}. This can be achieved while keeping the switch operation completely distributed. The EDFPS has also been shown to be robust against bursty traffic.

An internally-buffered version of the EDFPS has been studied in [15]. This design was motivated by the fact that cell-sequencing is not disturbed by the introduction of internal buffers, since the expanded delta network is a single-path structure. It has been shown (using both analysis and simulation) that above 95% maximum throughput can be achieved for a very large switch size using single-input-buffered SEs and expansion factor as small as 4. In general, having both the expansion factor and SE buffer size/strategy as design parameters makes it easier to jointly optimize both performance and complexity.

7. Concluding remarks

This paper has presented a survey of space-division ATM switch architectures. It has given a descriptive overview of these architectures by classifying them according to the basic structure, into three categories: crossbar based, disjoint-path based, and banyan based. A considerable portion of this paper has been devoted to reviewing some necessary background, as well as discussing the various buffering strategies employed by ATM switches. While we do not claim that our survey covers *all* ATM switches proposed to-date, we hope that this paper would serve as a good starting point for those who are entering this interesting area of research. Furthermore, it should provide an ATM switching researcher with an overview of the work that has been done by other researchers in the field.

This paper has focused on electronic point-to-point switch architectures, and their performance under uniform traffic. First, despite the fact that transmission technology in broadband networks is based on fiber optics due to their enormous bandwidth, both the lack of efficient optical buffer memories and the requirement of processing cell headers, have resulted in mostly electronic solutions for switching [116]. Nevertheless, heavy research is being carried out to design fully-photonic ATM switches. Second, although some of the surveyed architectures can support multicast/broadcast connections, the extension of the function of any unicast switch to multicasting/broadcasting can be achieved by placing a copy network at the front end of the unicast switch, to replicate incoming multicast cells; each copy is then forwarded to the following unicast switch for routing. Interested readers are referred to [6], [28], [35], [36], [40], [43], [49], [50], [51], [54], [59], [80], [100], [106], [125], [140], [141], [146], [167], [188], [209], [210], [217], and [237]. Third, more research is required to evaluate and compare the different switch architectures against integrated services traffic environment, since asserting the assumption of uniformity for real-life situations may lead to optimistic evaluation of performance measures. Of course, this would require first the development of suitable traffic models [198]. Finally, it is important to extend the comparison beyond performance, and consider other factors such as implementation complexity and cost [236], fault tolerance and reliability.

References

[1] E. Aanen, J.L. van den Berg, and R.J.F. De Vries, Cell loss performance of the GAUSS ATM switch, *Proc. IEEE INFOCOM'92*, Florence, Italy, pp. 717–726, May 1992.

[2] H. Ahmadi, and W.E. Denzel, A survey of modern high-performance switching techniques, *IEEE J. Selected Areas Commun.*, Vol. 7, No. 7, pp. 1091–1103, Sep. 1989.

[3] M.K.M. Ali, and M. Youssefi, The performance of an input access scheme in a high-speed packet switch, *Proc. IEEE INFOCOM'91*, Miami, FL, pp. 454–461, Apr. 1991.

[4] G.J. Anido, and A.W. Seeto, Multipath interconnection: a technique for reducing congestion with fast packet switching fabrics, *IEEE J. Selected Areas Commun.*, Vol. 6, No. 9, pp. 1480–1488, Dec. 1988.

[5] M. Atiquzzaman, and M.S. Akhtar, Effect of nonuniform traffic on the performance of multistage interconnection networks, *IEE Proc. – Comput. Digit. Tech.*, Vol. 141, No. 3, pp. 169–176, May 1994.

[6] R.Y. Awdeh, and H.T. Mouftah, A non-typical approach for the design of multicast ATM switch architectures, *Proc. 2nd ISCA / ACM SIGCOMM Int. Conf. Computer Commun. and Networks (IC^3N)*, San Diego, CA, pp. 399–404, June 1993.

[7] R.Y. Awdeh, and H.T. Mouftah, Multi single-stage-shuffling for fast packet switching, *Proc. 36th IEEE Midwest Symp. Circuits and Systems*, Detroit, MI, pp. 1136–1139, Aug. 1993.

[8] R.Y. Awdeh, and H.T. Mouftah, Broadband packet switch architectures, *Proc. Photonics'93: 3rd IEEE Int. Workshop on Photonic Networks, Components, and Applications*, Atlanta, GA, pp. 183–188, Sep. 1993.

[9] R.Y. Awdeh, and H.T. Mouftah, The Expanded Delta Fast Packet Switch, *Proc. IEEE SUPERCOMM / ICC'94*, New Orleans, LA, pp. 397–401, May 1994.

[10] R.Y. Awdeh, and H.T. Mouftah, Approach for comparing delta-based networks with the crossbar switch, *Electronics Letters*, Vol. 30, No. 3, pp. 201–202, Feb. 1994.

[11] R.Y. Awdeh, and H.T. Mouftah, A comparative study of unbuffered interconnection networks based on crosspoint complexity, *Proc. 6th IEEE Int. Conf. Computing and Information (ICCI'94)*, Peterborough, ON, pp. 604–621, May 1994; also *CD-ROM J. Computing and Information*, to appear.

[12] R.Y. Awdeh, and H.T. Mouftah, Design and performance analysis of an output-buffering switch with complexity of $O(N \log_2 N)$, *Proc. IEEE SUPERCOMM / ICC'94*, New Orleans, LA, pp. 420–424, May 1994.

[13] R.Y. Awdeh, and H.T. Mouftah, MS4 – A high performance output buffering ATM Switch, *Computer Commun.*, to appear.

[14] R.Y. Awdeh, and H.T. Mouftah, A contention resolution algorithm for input-buffered Batcher-Banyan networks, *Int. J. Commun. Systems*, Vol. 7, No. 1, pp. 33–38, Jan.–Mar. 1994.

[15] R.Y. Awdeh, and H.T. Mouftah, Analysis and simulation of the Buffered-Expanded Delta Fast Packet Switch, *Proc. IEEE GLOBECOM'94*, San Francisco, CA, Nov. 1994.

[16] H.F. Badran, and H.T. Mouftah, Fairness for broadband integrated switch architectures under backpressure mechanisms, *Proc. IEEE ICC'91*, Denver, CO, pp. 1033–1037, June 1991.

[17] H.F. Badran, and H.T. Mouftah, Head of line arbitration in ATM switches with input-output buffering and backpressure control, *Proc. IEEE GLOBECOM'91*, Phoenix, AZ, pp. 347–351, Dec. 1991.

[18] H.F. Badran, and H.T. Mouftah, Input-output-buffered broad-band packet-switch architectures with correlated input traffic, *Canadian J. Electrical and Computer Engineering*, Vol. 18, No. 3, pp. 133–139, 1993.

[19] H.F. Badran, and H.T. Mouftah, Input-output-buffered ATM switches with delayed backpressure mechanisms, *Proc. Canadian Conf. Electrical and Computer Engineering*, pp. 771–774, 1993.

[20] J.J. Bae, and T. Suda, Survey of traffic control schemes and protocols in ATM networks, *Proc. IEEE*, Vol. 79, No. 2, pp. 170–189, Feb. 1991.

[21] P. Baran, On distributed communication networks, *IEEE Trans. Commun.*, Vol. 12, pp. 1–9, March 1964.

[22] K.E. Batcher, Sorting networks and their applications, *Proc. AFIPS Spring Joint Comp. Conf.*, Vol. 32, pp. 307–314, 1968.

[23] S. Bassi, M. Decina, and A. Pattavina, Performance analysis of the ATM Shuffleout switching architecture under nonuniform traffic patterns, *Proc. IEEE INFOCOM'92*, Florence, Italy, pp. 734–742, May 1992.

[24] V.E. Benes, Optimal rearrangeable multistage connecting networks, *Bell Syst. Tech. J.*, Vol. 43, pp. 1641–1656, July 1964.

[25] J.E. Berthold, High speed integrated electronics for communications systems, *Proc. IEEE*, Vol. 78, No. 3, pp. 486–511, Mar. 1990.

[26] D. Bertsekas, and R. Gallager, *Data Networks*, Prentice-Hall, Englewood Cliffs, NJ, 2nd ed., 1992.

[27] L.N. Bhuyan, Q. Yang, and D.P. Agrawal, Performance of multiprocessor interconnection networks, *IEEE Computer*, Vol. 22, No. 2, pp. 25–37, Feb. 1989.

[28] R.P. Bianchini Jr., and H.S. Kim, Design of a nonblocking shared memory copy network for ATM, *Proc. IEEE INFOCOM'92*, Florence, Italy, pp. 876–885, May 1992.

[29] G. Bianchi, and J.S. Turner, Improved queueing analysis of shared buffer switching networks, *IEEE / ACM Trans. Networking*, Vol. 1, No. 4, pp. 482–490, Aug. 1993.

[30] B. Bingham, and H. Bussey, Reservation-based contention resolution mechanism for Batcher-Banyan networks, *Electronics Letters*, Vol. 24, No. 13, pp. 772–773, June 1988.

[31] F. Bonomi, S. Montagna, and R. Pagkino, Busy period analysis for an ATM switching element output line, *Proc. IEEE INFOCOM'92*, Florence, Italy, pp. 544–551, May 1992.

[32] J.-Y.L. Boudec, The Asynchronous Transfer Mode: a tutorial, *Computer Networks and ISDN Systems*, Vol. 24, pp. 279–309, 1992.

[33] G. Broomell, and J.R. Heath, Classification categories and historical development of circuit switching topologies, *Computing Surveys*, Vol. 15, No. 2, pp. 95–133, June 1983.

[34] G. Bruzzi, and A. Pattavina, Performance evaluation of an input-queued switch with internal speed-up and finite output queues, *Proc. IEEE GLOBECOM'90*, San Diego, CA, pp. 1455–1459, Dec. 1990.

[35] R.S. Bubenik, and J.S. Turner, Performance of a broadcast packet switch, *IEEE Trans. Commun.*, Vol. 37, No. 1, pp. 60–69, Jan. 1989.

[36] R. Bubenik, J. Dettart, and M. Gaddis, Multipoint connection management in high speed networks, *Proc. IEEE INFOCOM'91*, Miami, FL, pp. 59–68, Apr. 1991.

[37] E.T. Bushnell, and J.S. Meditch, Dilated multistage interconnection networks for fast packet switching, *Proc. IEEE INFOCOM'91*, Miami, FL, pp. 1264–1273, Apr. 1991.

[38] P. Campoli, and A. Pattavina, An ATM switch with folded shuffle-topology and distributed access, *Proc. IEEE ICC'91*, Denver, CO, pp. 1021–1027, June 1991.

[39] H.J. Chao, A distributed modular tera-bit/sec ATM switch, *Proc. IEEE GLOBECOM'90*, San Diego, CA, pp. 1594–1601, Dec. 1990.

[40] C.-J. Chang, and C.-J. Ling, Overflow controller in copy network of broadband packet switch, *Electronics Letters*, Vol. 27, No. 11, May 1991.

[41] J.S.-C. Chen, and R. Guerin, Input queueing of internally nonblocking packet switch with two priority classes, *Proc. IEEE INFOCOM'89*, pp. 529–537, 1989.

[42] J.S.-C Chen, and T.E. Stern, Throughput analysis, optimal buffer allocation, and traffic imbalance study of a generic nonblocking packet switch, *IEEE J. Selected Areas Commun.*, Vol. 9, No. 3, pp. 439–449, Apr. 1991.

[43] X. Chen, J.F. Hayes, and M.K.M. Ali, Performance analysis of cyclic-priority input access method for a multicast switch, *Proc. IEEE INFOCOM'91*, Miami, FL, pp. 1189–1195, Apr. 1991.

[44] X. Chen, and J.F. Hayes, A shared buffer memory switch with maximum queue and minimum allocation, *Proc. Canadian Conf. Electrical and Computer Engineering*, paper 7.1, 1991.

[45] W.-S.E. Chen, Y.M. Kim, K.L. Lee, and M.T. Liu, A distributed congestion-prevention scheme for ATM switching fabrics based on buffered delta networks, *Proc. IEEE INFOCOM'91*, Miami, FL, pp. 304–313, Apr. 1991.

[46] X. Chen, A survey of multistage interconnection networks for fast packet switches, *Int. J. Digital and Analog Commun. Systems*, Vol. 4, pp. 33–59, 1991.

[47] T.H. Cheng, and D.G. Smith, Queueing analysis of a multichannel ATM switch with input buffering, *Proc. IEEE ICC'91*, Denver, CO, pp. 1028–1032, June 1991.

[48] D.X. Chen, and J.W. Mark, Performance analysis of output buffered fast packet switches with bursty traffic loading, *Proc. IEEE GLOBECOM'91*, Phoenix, AZ, pp. 455–459, Dec. 1991.

[49] X. Chen, and J.F. Hayes, Call scheduling in multicasting packet switching, *Proc. IEEE ICC'92*, pp. 895–899, 1992.

[50] X. Chen, I. Lambadris, and J.F. Hayes, A general unified model for performance analysis of multicast switching, *Proc. IEEE GLOBECOM'92*, Orlando, FL, pp. 1498–1502, Dec. 1992.

[51] W.-T. Chen, Y.-R. Chang, and C.-F. Huang, A low-cost self-routing multicast network, *Proc. IEEE ICC'93*, Geneva, Switzerland, pp. 1691–1695, May 1993.

[52] D.X. Chen, and J.W. Mark, SCOQ: a fast packet switch with shared concentration and output queueing, *IEEE/ACM Trans. Networking*, Vol. 1, No. 1, pp. 142–151, Feb. 1993.

[53] F.M. Chiussi, and F.A. Tobagi, A hybrid shared-memory/space-division architecture for large fast packet switches, *Proc. IEEE ICC'92*, pp. 904–911, 1992.

[54] J.S. Choi, C.K. Un, and B.C. Shin, Design of cascade ring multicast switch, *Electron. Lett.*, Vol. 28, No. 14, July 1992.

[55] C. Clos, A study of nonblocking switching network, *Bell Syst. Tech. J.*, Vol. 32, pp. 406–424, Mar. 1953.

[56] C.S. Cooper, High-speed networks: the emergence of technologies for multiservice support, *Computer Commun.*, Vol. 14, No. 1, pp. 27–43, Jan./Feb. 1991.

[57] G. Corazzi, and C. Raffaelli, Performance evaluation of input-buffered replicated banyan networks, *IEEE Trans. Commun.*, Vol. 41, No. 6, pp. 841–845, June 1993.

[58] R.L. Cruz, The Statistical Data Fork: a class of broadband multichannel switches, *IEEE Trans. Commun.*, Vol. 40, No. 10, pp. 1625–1634, Oct. 1992.

[59] R. Cusani, and F. Sestini, A recursive multistage structure for multicast ATM switching, *Proc. IEEE INFOCOM'91*, Miami, FL, pp. 1289–1295, Apr. 1991.

[60] G.E. Daddis Jr., and H.C. Torng, A taxonomy of broadband integrated switching architectures, *IEEE Commun. Mag.*, No. 5, pp. 32–42, May 1989.

[61] M. Decina, P. Giacomazzi, A. Pattavina, and E. Tombolini, Shuffle interconnection networks with deflection routing for ATM switching: the Closed-Loop Shuffleout, *Proc. IEEE INFOCOM'91*, Miami, FL, pp. 1256–1263, Apr. 1991.

[62] M. Decina, P. Giacomazzi, A. Pattavina, and E. Tombolini, Shuffle interconnection networks with deflection routing for ATM switching: the Open-Loop Shuffleout, *Proc. ITC-13*, pp. 27–34, 1991.

[63] M. Decina, P. Giacomazzi, and A. Pattavina, Connectionless switching by the Asynchronous Shuffleout Network, *Proc. IEEE ICC'93*, Geneva, Switzerland, pp. 701–707, May 1993.

[64] E. Del Re, and R. Fantacci, An efficient high-speed packet switching with shared input buffers, *Proc. IEEE GLOBECOM'92*, Orlando, FL, pp. 1472–1476, Dec. 1992.

[65] E. Del Re, and R. Fantacci, Performance evaluation of

input and output queueing techniques in ATM switching systems, *IEEE Trans. Commun.*, Vol. 41, No. 10, pp. 1565–1575, Oct. 1993.

[66] M. De Prycker, and M. De Somer, Performance of a service-independent switching network with distributed control, *IEEE J. Selected Areas Commun.*, Vol. 5, No. 10, pp. 1293–1301, Oct. 1987.

[67] M. De Prycker, Evolution from ISDN to BISDN: a logical step towards ATM, *Computer Commun.*, Vol. 12, No. 3, pp. 141–146, June 1989.

[68] M. De Prycker, *Asynchronous Transfer Mode: Solution For Broadband ISDN*, Ellis Horwood, Chichester, 1991.

[69] A. Descloux, Stochastic models for ATM switching networks, *IEEE J. Selected Areas Commun.*, Vol. 9, No. 3, pp. 450–457, Apr. 1991.

[70] R.J.F. De Vries, GAUSS: a single-stage ATM switch with output buffering, *Proc. IEE Int. Conf. Integrated Broadband Services and Networks*, pp. 248–252, 1990.

[71] D.M. Dias, and J.R. Jump, Analysis and simulation of buffered delta networks, *IEEE Trans. Computers*, Vol. 30, No. 4, pp. 273–282, April 1981.

[72] D.M. Dias, and J.R. Jump, Packet switching interconnection networks for modular systems, *IEEE Computer*, Vol. 14, No. 12, pp. 43–53, Dec. 1981.

[73] D.M. Dias, and M. Kumar, Packet switching in NlogN multistage networks, *Proc. IEEE GLOBECOM'84*, Atlanta, GA, pp. 114–120, Nov. 1984.

[74] J. Ding, Nonuniform traffic analysis of multistage interconnection networks with split buffers, *Proc. IEEE ICC'93*, Geneva, Switzerland, pp. 58–62, May 1993.

[75] A. Eckberg, and T.-C. Hou, Effects of output buffer sharing on buffer requirements in an ATDM packet switch, *Proc. IEEE INFOCOM'88*, New Orleans, LA, pp. 459–566, Mar. 1988.

[76] T.E. Eliazov, V. Ramaswami, W. Willinger, and G. Latouche, Performance of an ATM switch: simulation study, *Proc. IEEE INFOCOM'90*, San Fransisco, CA, pp. 644–659, June 1990.

[77] A.I. Elwalid, and I. Widjaja, Efficient analysis of buffered multistage switching networks under bursty traffic, *Proc. IEEE GLOBECOM'93*, pp. 1072–1078, 1993.

[78] N. Endo, T. Kozaki, T. Ohuchi, H. Kuwahara, and S. Gohara, Shared Buffer Memory switch for an ATM exchange, *IEEE Trans. Commun.*, Vol. 41, No. 1, pp. 237–245, Jan. 93.

[79] K.Y. Eng, M.G. Hluchyj, and Y.S. Yeh, A Knockout switch for variable length packets, *Proc. IEEE ICC'87*, Seattle, WA, pp. 794–799, June 1987.

[80] K.Y. Eng, M.G. Hluchyj, and Y.S. Yeh, Multicast and broadcast services in a Knockout packet switch, *Proc. IEEE INFOCOM'88*, New Orleans, LA, pp. 29–34, Mar. 1988.

[81] K.Y. Eng, A photonic Knockout switch for high-speed packet networks, *IEEE J. Selected Areas Commun.*, Vol. 6, No. 7, pp. 1107–1115, Aug. 1988.

[82] J. Evans, E. Duron, and Y. Wang, Analysis and implementation of a priority knockout switch, *Proc. IEEE INFOCOM'93*, pp. 1099–1106, 1993.

[83] Y. Fan, J. Wang, and C. Wang, Performance analysis of banyan network based ATM switches, *Proc. IEEE ICC'92*, pp. 1609–1613, 1992.

[84] T.-Y. Feng, A survey of interconnection networks, *IEEE Computer*, Vol. 14, No. 12, pp. 12–27, Dec. 1981.

[85] G.J. Fitzpatrick, and E.A. Munter, Input-buffered ATM switch traffic performance, *Proc. Multimedia'89*, paper 4.2, 1989.

[86] A. Fraser, Early experiments with Asynchronous Time Division networks, *IEEE Network*, Vol. 7, No. 1, pp. 12–26, Jan. 1993.

[87] V.J. Friesen, and J.W. Wong, The effect of multiplexing, switching and other factors on the performance of broadband networks, *Proc. IEEE INFOCOM'93*, pp. 1194–1203, 1993.

[88] U. Garg, and Y.-P. Huang, Decomposing banyan networks for performance evaluation, *IEEE Trans. Computers*, Vol. 37, No. 3, pp. 371–376, March 1988.

[89] U. Garg, and Y.-P. Huang, Improving the performance of banyan networks, *Computers and Electrical Engineering*, Vol. 14, No. 1/2, pp. 29–33, 1988.

[90] D.P. Gerakoulis, J. Mathew, and T.N. Saadawi, Performance analysis of a packet switch with channel assignment capabilities, *Proc. IEEE ICC'91*, Denver, CO, pp. 1527–1531, June 1991.

[91] D. Ghosh, and J.C. Daly, Delta networks with multiple links and shared output buffers: a high performance architecture for packet switching, *Proc. IEEE GLOBECOM'91*, Phoenix, AZ, pp. 949–953, Dec. 1991.

[92] J.N. Giacopelli, T.T. Lee, and W.E. Stephens, Scalability study of self-routing packet switch fabrics for very large scale broadband ISDN central offices, *Proc. IEEE GLOBECOM'90*, San Diego, CA, pp. 1609–1614, Dec. 1990.

[93] J.N. Giacopelli, J.J. Hickey, W.S. Marcus, W.D. Sincoskie, and M. Littlewood, Sunshine: a high-performance self-routing broadband packet switch architecture, *IEEE J. Selected Areas Commun.*, Vol. 9, No. 8, pp. 1289–1298, Oct. 1991.

[94] L.R. Goke, and G.J. Lipovski, Banyan networks for partitioning processor systems, *Proc. 1st Annual Symp. Computer Architecture*, pp. 21–28, Dec. 1973.

[95] P. Goli, and V. Kumar, Performance of a crosspoint buffered ATM switch fabric, *Proc. IEEE INFOCOM'92*, Florence, Italy, pp. 426–431, May 1992.

[96] A.K. Gupta, and N.D. Georganas, Analysis of a packet switch with input and output buffers and speed constraints, *Proc. IEEE INFOCOM'91*, Miami, FL, pp. 694–700, Apr. 1991.

[97] A.K. Gupta, L.O. Barbosa, and N.D. Georganas, A 16×16 limited intermediate buffer switch module for ATM networks, *Proc. IEEE GLOBECOM'91*, Phoenix, AZ, pp. 939–943, Dec. 1991.

[98] A.K. Gupta, L.O. Barbosa, and N.D. Georganas, Limited intermediate buffer switch modules and their inter-

connection networks for BISDN, *Proc. IEEE ICC'92*, pp. 1646–1650, 1992.

[99] J.G. Haro, C.C. Pastor, J.P. Aspas, and H.T. Mouftah, Evaluation study of several head-of-line selection schemes for high-performance non-blocking ATM switches, *Proc. IEEE Pacific Rim*, pp. 327–332, 1993.

[100] J.F. Hayes, R. Breault, and M.K.M. Ali, Performance analysis of a multicast switch, *IEEE Trans. Commun.*, Vol. 39, No. 4, pp. 581–587, Apr. 1991.

[101] M.G. Hluchyj, and M.J. Karol, Queueing in high-performance packet switching, *IEEE J. Selected Areas Commun.*, Vol. 6, No. 9, pp. 1587–1597, Dec. 1988.

[102] T.-C. Hou, and M. Sarraf, Internal traffic characterization of a multi-stage ATM switch, *Proc. ISCA / ACM SIGCOMM 2nd Int. Conf. Computer Commun. and Networks (IC^3N)*, San Diego, CA, pp. 391–398, June 1993.

[103] A. Huang, and S. Knauer, Starlite: a wideband digital switch, *Proc. IEEE GLOBECOM'84*, Atlanta, GA, pp. 121–125, Nov. 1984.

[104] J.Y. Hui, and E. Arthurs, A broadband packet switch for integrated transport, *IEEE J. Selected Areas Commun.*, Vol. 5, No. 8, pp. 264–273, Oct. 1987.

[105] J.Y. Hui, *Switching and Traffic Theory for Integrated Broadband Network*, Kluwer Academic Publishers, Boston, 1990.

[106] J.Y. Hui, and T. Renner, Queueing strategies for multicast packet switching, *Proc. IEEE GLOBECOM'90*, San Diego, CA, pp. 1431–1437, Dec. 1990.

[107] *IEEE Commun. Mag.*, Vol. 30, No. 8, whole issue, Aug. 1992.

[108] *IEEE Network*, Vol. 6, No. 5, whole issue, Sep. 1992.

[109] I. Iliadis, and W.E. Denzel, Performance of packet switches with input and output queueing, *Proc. IEEE ICC'90*, Atlanta, GA, pp. 747–753, Apr. 1990.

[110] I. Iliadis, Performance of a packet switch with shared buffer and input queueing, *Proc. ITC-13*, pp. 911–916, 1991.

[111] I. Iliadis, Performance of a packet switch with input and output queueing under unbalanced traffic, *Proc. IEEE INFOCOM'92*, Florence, Italy, pp. 743–752, May 1992.

[112] I. Iliadis, Synchronous versus asynchronous operation of a packet switch with combined input and output buffering, *Performance Evaluation*, Vol. 16, pp. 241–250, 1992.

[113] H. Imagawa, S. Urushidani, and K. Hagishima, A new self-routing switch driven with input-to-output address difference, *Proc. IEEE GLOBECOM'88*, Hollywood, FL, pp. 1607–1611, Nov. 1988.

[114] A. Iyengar, and M. El Zarki, Switching prioritized packets, *Proc. IEEE GLOBECOM'89*, Dallas, TX, pp. 1181–1186, Nov. 1989.

[115] A.R. Jacob, A survey of fast packet switches, *Computer Commun. Review*, Vol. 20, No. 1, pp. 54–64, Jan. 1990.

[116] A. Jajszczyk, and H.T. Mouftah, Photonic fast packet switching, *IEEE Commun. Mag.*, No. 2, pp. 58–65, Feb. 1993.

[117] Y.-C. Jeng, Performance analysis of a packet switch based on single buffered banyan networks, *IEEE J. Selected Areas Commun.*, Vol. 1, No. 6, pp. 1014–1021, Dec. 1983.

[118] Y.C. Jung, and C.K. Un, Analysis of backpressuring-type packet switches with input and output buffering, *IEE Proceedings-I*, Vol. 140, No. 4, pp. 277–284, Aug. 1993.

[119] F. Kamoun, and L. Kleinrock, Analysis of shared finite storage in a computer network node environment under general traffic conditions, *IEEE Trans. Commun.*, Vol. 28, No. 7, pp. 992–1003, July 1980.

[120] M.J. Karol, M.G. Hluchyj, and S.P. Morgan, Input versus output queueing on a space-division packet switch, *IEEE Trans. Commun.*, Vol. 35, No. 12, pp. 1347–1356, Dec. 1987.

[121] M.J. Karol, K.Y. Eng, and H. Obara, Improving the performance of input-queued ATM packet switches, *Proc. IEEE INFOCOM'92*, Florence, Italy, pp. 110–115, May 1992.

[122] P. Kermani, and L. Kleinrock, Virtual cut-through: a new computer communication switching technique, *Computer Networks*, pp. 267–286, 1979.

[123] H.S. Kim, and A. Leon-Garcia, A self-routing multistage interconnection network for broadband ISDN, *IEEE J. Selected Areas Commun.*, Vol. 8, No. 3, pp. 459–466, April 1990.

[124] H.S. Kim, and A. Leon-Garcia, Performance of buffered banyan networks under nonuniform traffic patterns, *IEEE Trans. Commun.*, Vol. 38, No. 5, pp. 648–658, May 1990.

[125] C.-K. Kim, and T. Lee, Performance of call splitting algorithms for multicast traffic, *Proc. IEEE INFOCOM'90*, San Fransisco, CA, pp. 348–356, June 1990.

[126] Y.M. Kim, and K.Y. Lee, KSMINs: Knockout switch based multistage interconnection networks for high-speed packet switching, *Proc. IEEE GLOBECOM'90*, San Diego, CA, pp. 218–223, Dec. 1990.

[127] H.S. Kim, I. Widjaja, and A. Leon-Garcia, Performance of output buffered banyan networks with arbitrary buffer sizes, *Proc. IEEE INFOCOM'91*, Miami, FL, pp. 701–710, Apr. 1991.

[128] H.S. Kim, and A. Leon-Garcia, Nonblocking property of Reverse Banyan networks, *IEEE Trans. Commun.*, Vol. 40, No. 3, pp. 472–476, Mar. 1992.

[129] H.S. Kim, Multichannel ATM switch with preserved packet sequence, *Proc. IEEE ICC'92*, pp. 1634–1638, 1992.

[130] C.P. Kruskal, and M. Snir, The performance of multistage interconnection networks for multiprocessors, *IEEE Trans. Computers*, Vol. 32, No. 12, pp. 1091–1098, Dec. 1983.

[131] M. Kumar, and J.R. Jump, Performance enhancement in buffered delta networks using crossbar switches and multiple links, *J. Parallel and Distributed Computing*, Vol. 1, pp. 81–103, 1984.

[132] M. Kumar, and J.R. Jump, Performance of unbuffered shuffle-exchange networks, *IEEE Trans. Computers*, Vol. 35, No. 6, pp. 573–577, June 1986.

[133] V.P. Kumar, J.G. Kneuer, D. Pal, and B. Brunner,

PHOENIX: a building block for fault tolerant broadband packet switches, *Proc. IEEE GLOBECOM'91*, Phoenix, AZ, pp. 228–233, Dec. 1991.

[134] T. Lang, and L. Kurisaki, Nonuniform traffic spots (NUTS) in multistage interconnection networks, *J. Parallel and Distributed Computing*, Vol. 10, pp. 55–67, 1990.

[135] P.S.Y. Lau, and A. Leon-Garcia, Design and performance analysis of a buffer subsystem for the Batcher-Banyan switch, *Proc. IEEE GLOBECOM'90*, San Diego, CA, pp. 1926–1930, Dec. 1990.

[136] P.S.Y. Lau, and A. Leon-Garcia, Design and analysis of a multilink access subsystem based on the Batcher-Banyan network architecture, *IEEE Trans. Commun.*, Vol. 40, No. 11, pp. 1757, Nov. 1992.

[137] D. Lawrie, Access and alignment of data in an array processor, *IEEE Trans. Computers*, Vol. 24, No. 12, pp. 1145–1155, Dec. 1975.

[138] C.-T.A. Lea, The load-sharing banyan network, *IEEE Trans. Computers*, Vol. 35, No. 12, pp. 1025–1034, Dec. 1986.

[139] C.-L. Lea, Design and performance evaluation of unbuffered self-routing networks for wideband packet switching, *Proc. IEEE INFOCOM'90*, San Fransisco, CA, pp. 148–156, June 1990.

[140] T.T. Lee, R. Boorstyn, and E. Arthurs, The architecture of a multicast broadband packet switch, *Proc. IEEE INFOCOM'88*, New Orleans, LA, pp. 1–8, Mar. 1988.

[141] T.T. Lee, Nonblocking copy networks for multicast packet switching, *IEEE J. Selected Areas Commun.*, Vol. 6, No. 9, pp. 1455–1467, Dec. 1988.

[142] T.T. Lee, A modular architecture for very large packet switches, *IEEE Trans. Commun.*, Vol. 38, No. 7, pp. 1097–1106, July 1990.

[143] T.-H. Lee, Analytic models for performance evaluation of single-buffered banyan networks under nonuniform traffic, *IEE Proc.-E*, Vol. 138, No. 1, pp. 41–47, Jan 1991.

[144] T.H. Lee, and S.J. Liu, Banyan network nonblocking with respect to cyclic shifts, *Electronics Letters*, Vol. 27, No. 16, pp. 1474–1476, Aug. 1991.

[145] T.-H. Lee, Design and analysis of a new self-routing network, *IEEE Trans. Commun.*, Vol. 40, No. 1, pp. 171–177, Jan. 1992.

[146] T.-H. Lee, and S.-J. Liu, A fair high-speed copy network for multicast packet switch, *Proc. IEEE INFOCOM'92*, Florence, Italy, pp. 886–894, May 1992.

[147] T.T. Lee, and S.C. Liew, Broadband packet switches based on dilated interconnection networks, *Proc. IEEE ICC'92*, pp. 255–261, 1992.

[148] K.Y. Lee, H. Yoon, and M.T. Liu, Performance evaluation of multipath packet switching interconnection networks, *J. Parallel and Distributed Computing*, Vol. 17, pp. 353–359, 1993.

[149] S.-Q. Li, and M.J. Lee, A study of traffic imbalances in a fast packet switch, *Proc. IEEE INFOCOM'89*, pp. 538–547, 1989.

[150] S.-Q. Li, Performance of a nonblocking space-division packet switch with correlated input traffic, *Proc. IEEE GLOBECOM'89*, Dallas, TX, pp. 1754–1763, Nov. 1989.

[151] S.-Q. Li, Nonuniform traffic analysis on a nonblocking space-division packet switch, *IEEE Trans. Commun.*, Vol. 38, No. 7, pp. 1085–1096, July 1990.

[152] S.-Q. Li, Performance of trunk grouping in packet switch design, *Proc. IEEE INFOCOM'91*, Miami, FL, pp. 688–693, Apr. 1991.

[153] S.C. Liew, Performance of input-buffered and output-buffered ATM switches under bursty traffic: simulation study, *Proc. IEEE GLOBECOM'90*, San Diego, CA, pp. 1919–1925, Dec. 1990.

[154] S.C. Liew, and K.W. Lu, Performance analysis of asymmetric packet switch modules with channel grouping, *Proc. IEEE INFOCOM'90*, San Fransisco, CA, pp. 668–676, June 1990.

[155] S.C. Liew, and K.W. Lu, Comparison of buffering strategies for asymmetric packet switch modules, *IEEE J. Selected Areas Commun.*, Vol. 9, No. 3, pp. 428–437, Apr. 1991.

[156] S.C. Liew, and T.T. Lee, NlogN dual shuffle-exchange network with error correcting routing, *Proc. IEEE ICC'92*, pp. 262–268, 1992.

[157] M. Listanti, and A. Roveri, Switching structures for ATM, *Computer Commun.*, Vol. 12, No. 6, pp. 349–358, Dec. 1989.

[158] H. Matsunaga, and H. Uematsu, A 1.5 Gb/s 8×8 cross-connect switch using a time-reservation algorithm, *IEEE J. Selected Areas Commun.*, Vol. 9, No. 8, pp. 1308–1317, Oct. 1991.

[159] R.J. McMillen, A survey of interconnection networks, *Proc. IEEE GLOBECOM'84*, Atlanta, GA, pp. 105–113, Nov. 1984.

[160] J.F. Meyer, S. Montagna, R. Paglino, Dimensioning of an ATM switch with shared buffer and threshold priority, *Computer Networks and ISDN Systems*, Vol. 26, pp. 95–108, 1993.

[161] S.E. Minzer, Broadband ISDN and Asynchronous Transfer Mode (ATM), *IEEE Commun. Mag.*, Vol. 27, No. 9, pp. 17–24, Sep. 1989.

[162] B. Monderer, G. Pacifici, and C. Zukowski, The cylinder switch: an architecture for a manageable VLSI giga-cell switch, *Proc. IEEE ICC'90*, Atlanta, GA, pp. 567–571, Apr. 1990.

[163] A. Monterosso, and A. Pattavina, Performance analysis of multistage interconnection network with shared-buffered switching elements for ATM switching, *Proc. IEEE INFOCOM'92*, Florence, Italy, pp. 124–131, May 1992.

[164] T.D. Morris, and H.G. Perros, Performance modelling of a multi-buffered banyan switch under bursty traffic, *Proc. IEEE INFOCOM'92*, Florence, Italy, pp. 436–445, May 1992.

[165] M. Narasimha, The Batcher-Banyan self-routing network: universality and simplification, *IEEE Trans. Commun.*, Vol. 36, No. 10, pp. 1175–1171, Oct. 1988.

[166] P. Newman, A fast packet switch for the integrated services backbone network, *IEEE J. Selected Areas Commun.*, Vol. 6, No. 9, pp. 1468–1479, Dec. 1988.

[167] P. Newman, and M. Doar, The slotted ring copy fabric for a multicast packet switch, *Proc. Int. Switching Symp.*, Stockholm, Sweden, Session C8, Paper 6, June 1990.

[168] P. Newman, ATM technology for corporate networks, *IEEE Commun. Mag.*, pp. 90–101, Apr. 1992.

[169] S. Nojima, E. Tsutsui, H. Fukuda, and M. Hashimoto, Integrated services packet network using bus-matrix switch, *IEEE J. Selected Areas Commun.*, Vol. 5, No. 10, pp. 1284–1292, Oct. 1987.

[170] H. Obara, and T. Yasushi, An efficient contention resolution algorithm for input queueing ATM cross-connect switches, *Int. J. Digital and Analog Cabled Systems*, Vol. 2, No. 4, pp. 261–267, Oct.–Dec. 1989.

[171] H. Obara, Optimum architecture for input queueing ATM switches, *Electronics Letters*, Vol. 27, No. 7, pp. 555–557, Mar. 1991.

[172] K. Ohtsuki, K. Takemura, J.F. Kurose, H. Okada, and Y. Tezuka, A high-speed packet switch architecture with a multichannel bandwidth allocation, *Proc. IEEE INFOCOM'91*, Miami, FL, pp. 155–162, Apr. 1991.

[173] Y. Oie, T. Suda, M. Murata, and H. Hiyashara, Survey of switching techniques in high-speed networks and their performance, *Int. J. Satellite Commun.*, Vol. 9, pp. 285–303, 1991.

[174] Y. Oie, M. Murata, K. Kubota, and H. Hiyashara, Performance analysis of nonblocking packet switch with input and output buffers, *IEEE Trans. Commun.*, Vol. 40, No. 8, pp. 1294–1297, Aug. 1992.

[175] M.A. Pashan, M.D. Soneru, and G.D. Martin, Technologies for broadband switching, *AT&T Tech. J.*, pp. 39–47, Nov./Dec. 1993.

[176] J.H. Patel, Performance of processor-memory interconnections for multiprocessors, *IEEE Trans. Computers*, Vol. 30, No. 10, pp. 771–780, Oct. 1981.

[177] A. Pattavina, Multichannel bandwidth allocation in a broadband packet switch, *IEEE J. Selected Areas Commun.*, Vol. 6, No. 9, pp. 1489–1499, Dec. 1988.

[178] A. Pattavina, Fairness in a broadband packet switch, *Proc. IEEE ICC'89*, Boston, MA, pp. 404–409, June 1989.

[179] A. Pattavina, A broadband packet switch with input and output queueing, *Proc. Int. Switching Symp.*, Stockholm, Sweden, pp. 11–16, June 1990.

[180] A. Pattavina, Design and performance evaluation of a packet switch for broadband central offices, *Proc. IEEE INFOCOM'90*, San Fransisco, CA, pp. 1252–1259, June 1990.

[181] A. Pattavina, Nonblocking architectures for ATM switching, *IEEE Commun. Mag.*, No. 2, pp. 38–48, Feb. 1993.

[182] A. Pattavina, and G. Bruzzi, Analysis of input and output queueing for nonblocking ATM switches, *IEEE / ACM Trans. Networking*, Vol. 1, No. 3, pp. 314–328, June 1993.

[183] M.C. Pease, The Indirect Binary n-Cube multiprocessor array, *IEEE Trans. Computers*, Vol. 26, No. 5, pp. 458–473, May 1977.

[184] S.D. Personick, and W.O. Fleckenstein, Communications switching – from operators to photonics, *Proc. IEEE*, Vol. 75, No. 10, pp. 1380–1403, Oct. 1987.

[185] J. Peterson, Throughput limitation by head-of-line blocking, *Proc. ITC-13*, pp. 659–663, 1991.

[186] G.F. Pfister, and V.A. Norton, 'Hot spot' contention and combining in multistage interconnection Networks, *IEEE Trans. Computers*, Vol. 34, No. 10, pp. 943–948, Oct. 1985.

[187] P.K. Prasanna, R. Levy, and J. Swenson, Principles and standards for Broadband ISDN, *AT&T Tech. J.*, pp. 9–14, Nov./Dec. 1993.

[188] M. Rahnema, The fast packet ring switch: a high-performance efficient architecture with multicast capability, *IEEE Trans. Commun.*, Vol. 38, No. 4, pp. 539–545, Apr. 1990.

[189] E.P. Rathgeb, T.H. Theimer, and M.N. Huber, Buffering concepts for ATM switching networks, *Proc. IEEE GLOBECOM'88*, Hollywood, FL, pp. 1277–1281, Nov. 1988.

[190] R. Rooholamini, V. Cherkassky, and M. Garver, Finding the right ATM switch for the market, *IEEE Computer*, Vol. 27, No. 4, pp. 16–28, Apr. 1994.

[191] A. Saha, and M.D. Wagh, Performance analysis of banyan networks based on buffers of various sizes, *Proc. IEEE INFOCOM'90*, San Fransisco, CA, pp. 157–164, June 1990.

[192] K.W. Sarkies, The bypass queue in fast packet switching, *IEEE Trans. Commun.*, Vol. 39, No. 5, pp. 766–774, May 1991.

[193] K. Shiomoto, M. Murata, Y. Oie, and H. Miyahara, Performance evaluation of cell bypass queueing discipline for buffered banyan ATM switches, *Proc. IEEE INFOCOM'90*, San Fransisco, CA, pp. 677–685, June 1990.

[194] H. Shi, and O. Wing, Design of a combined input/output buffered ATM switches with arbitrary switch size, buffer size, and speed-up factor, *Proc. ISCA / ACM SIGCOMM 2nd Int. Conf. Computer Commun. and Networks (IC^3N)*, San Diego, CA, pp. 377–382, June 1993.

[195] Y. Shobatake, and T. Kodama, A cell switching algorithm for the buffered banyan network, *Proc. IEEE ICC'90*, Atlanta, GA, pp. 754–760, Apr. 1990.

[196] S. Sibal, and J. Zhang, On a class of banyan networks and tandem banyan switching fabrics, *Proc. IEEE INFOCOM'93*, pp. 481–488, 1993.

[197] H.J. Siegel, *Interconnection Networks for Large-Scale Parallel Processing: Theory and Case Studies*. D.C. Health and Company, Lexington, 1985.

[198] G.D. Stamoulis, M.E. Anagnostou, and A.D. Georgantas, Traffic models for ATM networks: a survey, *Computer Commun.*, Vol 17, No. 6, pp. 428–438, June 1994.

[199] W.E. Stephens, M. DePrycker, F.A. Tobagi, and T. Yamaguchi, Guest editorial: large-scale ATM switching

systems for B-ISDN, *IEEE J. Selected Areas Commun.*, Vol. 9, No. 8, pp. 1157–1158, Oct. 1991.

[200] A. Steffora, ATM: the year of the trial, *IEEE Computer*, Vol. 27, No. 4. pp. 8–10, Apr. 1994.

[201] H.S. Stone, Parallel processing with the perfect shuffle, *IEEE Trans. Computers*, Vol. 20, No. 2, pp. 153–161, Feb. 1971.

[202] W. Strecker, Analysis of the instruction execution rate in certain computer structures, Ph.D. dissertation, Carnegie-Mellon Univ., 1970.

[203] H. Suzuki, H. Nagano, T. Suzuki, T. Takeuchi, and S. Iwasaki, Output buffer switch architecture for asynchronous transfer mode, *Proc. IEEE ICC'89*, Boston, MA, pp. 99–103, June 1989.

[204] T.H. Szymanski, and V.C. Hamacher, On the universality of multipath multistage interconnection networks, *J. Parallel and Distributed Computing*, Vol. 7, pp. 541–569, 1989.

[205] T. Szymanski, and S. Shaikh, Markov chain analysis of packet-switched banyans with arbitrary switch sizes, queue sizes, link multiplicities and speedup, *Proc. IEEE INFOCOM'89*, pp. 960–971, 1989.

[206] Q. Ta, and J.S. Meditch, A high-speed integrated services switch based on 4×4 switching elements, *Proc. IEEE INFOCOM'90*, San Fransisco, CA, pp. 1164–1171, June 1990.

[207] Q. Ta, and J.S. Meditch, A high-speed integrated services switch based on 4×4 switching elements, *Computers and Electrical Engineering*, Vol. 19, No. 5, pp. 387–397, 1993.

[208] X.-N. Tan, and K.C. Sevcik, Reduced distance routing in single-stage shuffle-exchange interconnection networks, *Performance Evaluation Review*, Vol. 15, No. 1, pp. 95–110, May 1987.

[209] C.-L. Tarng, and J.S. Meditch, A high-performance copy network for B-ISDN, *Proc. IEEE INFOCOM'91*, Miami, FL, pp. 171–181, Apr. 1991.

[210] C.-L. Tarng, J.S. Meditch, and A.K. Somani, Fairness and priority implementation in nonblocking copy networks, *Proc. IEEE ICC'91*, Denver, CO, pp. 1002–1006, June 1991.

[211] T.H. Theimer, E.P. Rathgeb, and M.N. Huber, Performance analysis of buffered banyan networks, *IEEE Trans. Commun.*, Vol. 39, No. 2, pp. 269–277, Feb. 1991.

[212] G. Thomas, and J. Man, Improved windowing rule for input buffered packet switches, *Electronics Letters*, Vol. 29, No. 4, pp. 393–395, Feb. 1993.

[213] F.A. Tobagi, Fast packet switch architectures for broadband integrated services digital networks, *Proc. IEEE*, Vol. 78, No. 1, pp. 133–166, Jan. 1990.

[214] F.A. Tobagi, T. Kwok, and F.M. Chiussi, Architecture, performance, and implementation of the tandem-banyan fast packet switch, *IEEE J. Selected Areas Commun.*, Vol. 9, No. 8, pp. 1173–1193, Oct. 1991.

[215] S. Tridandapassi, and J.S. Meditch, Priority performance of banyan based broadband ISDN switches, *Proc. IEEE INFOCOM'91*, Miami, FL, pp. 711–720, Apr. 1991.

[216] J.S. Turner, Design of an integrated services packet network, *IEEE J. Selected Areas Commun.*, Vol. 4, No. 8, pp. 1373–1380, Nov. 1986.

[217] J.S. Turner, Design of a broadcast packet switching network, *IEEE Trans. Commun.*, Vol. 36, No. 6, pp. 734–743, June 1988.

[218] J.S. Turner, Queueing analysis of buffered switching networks, *Proc. ITC-13*, pp. 35–40, 1991.

[219] J.S. Turner, Queueing analysis of buffered switching networks, *IEEE Trans. Commun.*, Vol. 41, No. 2, pp. 412–420, Feb. 1993.

[220] N.-F. Tzeng, Alleviating the impact of tree saturation on multistage interconnection network performance, *J. Parallel and Distributed Computing*, Vol. 12, pp. 107–117, 1991.

[221] W. Wang, and F.A. Tobagi, The Christmas-Tree switch: an output queuing space-division fast packet switch based on interleaving distribution and concentration functions, *Proc. IEEE INFOCOM'91*, Miami, FL, pp. 163–170, Apr. 1991.

[222] C.-M. Weng, and J.-J. Li, Solution for packet switching of broadband ISDN, *IEE Proc.-I*, Vol. 138, No. 5, pp. 384–400, Oct. 1991.

[223] I. Widjaja, Tandem-banyan switching fabric with dilation, *Electronics Letters*, Vol. 27, No. 19, pp. 1770–1772, Sep. 1991.

[224] I. Widjaja, A. Leon-Garcia, and H.T. Mouftah, The effect of cut-through switching on the performance of buffered banyan networks, *Computer Networks and ISDN Systems*, Vol. 26, pp. 139–159, 1993.

[225] P.C. Wong, and M.S. Yeung, Pipeline Banyan – a parallel fast packet switch architecture, *Proc. IEEE ICC'92*, pp. 882–887, 1992.

[226] C.-L. Wu, and T.-Y. Feng, On a class of multistage interconnection networks, *IEEE Trans. Computers*, Vol. 29, No. 8, pp. 694–702, Aug. 1980.

[227] L.T. Wu, Mixing traffic in a buffered banyan network, *Proc. ACM 9th Symp. Data Commun.*, pp. 134–139, 1985.

[228] Y. Xiong, and H. Bruneel, Approximate analytic performance study of an ATM switching element with train arrivals, *Proc. IEEE ICC'92*, pp. 1614–1620, 1992.

[229] Y. Xiong, H. Bruneel, and G. Petit, On the performance evaluation of an ATM self-routing multistage switch with bursty and uniform traffic, *Proc. IEEE ICC'93*, Geneva, Switzerland, pp. 1391–1397, May 1993.

[230] Y.S. Yeh, M.G. Hluchyj, and A.S. Acampora, The Knockout switch: a simple, modular, architecture for high-performance packet switching, *IEEE J. Selected Areas Commun.*, Vol. 5, No. 8, pp. 1274–1283, Oct. 1987.

[231] H. Yoon, M.T. Liu, and K.Y. Lee, The Knockout switch under nonuniform traffic, *Proc. IEEE GLOBECOM'88*, Hollywood, FL, pp. 1628–1634, Nov. 1988.

[232] H. Yoon, K.Y. Lee, and M.T. Liu, Performance analysis

of multibuffered packet-switching networks in multiprocessor systems, *IEEE Trans. Computers*, Vol. 39, No. 3, pp. 319–327, Mar. 1990.

[233] Y.S. Youn, and C.K. Un, Performance of dilated banyan network with recirculation, *Electronics Letters*, Vol. 29, No. 1, pp. 62–63, Jan. 1993.

[234] R. Zarour, and H.T. Mouftah, Bridged Shuffle-Exchange Network: a high performance self-routing ATM switch, *Proc. IEEE ICC'93*, Geneva, Switzerland, pp. 696–700, May 1993.

[235] R. Zarour, and H.T. Mouftah, The Closed-Loop Bridged Shuffle-Exchange Network: a high performance self-routing ATM switch, *Proc. IEEE GLOBECOM'93*, pp. 1164–1168, 1993.

[236] E. Zegura, Architectures for ATM switching systems, *IEEE Commun. Mag.*, No. 2, pp. 28–37, Feb. 1993.

[237] W. Zhong, Y. Onozato, and K. Kaniyil, A copy network with shared buffers for large scale multicast ATM switching, *IEEE / ACM Trans. Networking*, Vol. 1, No. 2, pp. 157–165, Apr. 1993.

Chapter 2

Electrooptic Photonic Switches

Switching in the optical domain has many advantages including the ability to handle signals with very large bandwidth requirements easily and efficiently. Furthermore, one of the major advantages of photonic switching is that it avoids the need for optoelectronic conversions. Such conversions not only limit the system's versatility and transparency, but can also amount to signal deterioration and to errors being introduced. With the current availability of erbium-doped fiber amplifiers (EDFAs), efficient transparent optical signal boosting has become available to compensate for the potential losses in switching arrays.

This chapter is concerned with photonic switches based on the electrooptic effect. The chapter starts with a paper by Alferness in which he reviews electrically controlled optical waveguide switch arrays realized on lithium niobate structures. The paper starts by discussing the various switch types including space division and time division, thus establishing the need for crosspoint elements. Elements such as the switched directional coupler, the Mach-Zehnder interferometer switch, and the intersecting waveguide switch are considered. The treatment highlights the trade-offs between control voltage, switching speed and sensitivity to wavelength, polarization, etc. The paper then proceeds to give a review of the titanium-diffused lithium niobate waveguide technology used. Subsequently, the treatment moves to a higher level where switch architectures are considered. This is followed by results from switch array demonstrators. The paper concludes with a discussion of the technology and its applications.

Directional-coupler-based photonic switching networks are discussed by Jajszczyk in paper 2.2. The paper proposes switching networks requiring only two switching stages containing active elements (directional couplers). Other advantages of the proposed architectures include the fact that they are nonblocking in the strict sense and that there are no crossovers provided multiple substrates are used. The structures also offer multicast capabilities. The performance of the proposed structures is assessed in terms of the insertion loss and the signal-to-noise ratio.

The performance evaluation of switching networks is continued in a paper by Chinni et al. (paper 2.3) that considers crosstalk in a lossy directional coupler. Crosstalk is shown to be inevitable due to the nonuniform losses in the waveguide and cladding regions. The work shows that crosstalk cannot be completely eliminated by adjusting the coupling length in a directional coupler fabricated on a lossy medium (material absorption). Moreover, it is shown that the coupling length for minimum crosstalk differs form that of lossless systems. The increase in crosstalk due to material absorption is shown to be as high as 20dB, thus concluding that material absorption is an important parameter that has to be considered when designing low-crosstalk devices.

A paper that addresses the integration on silicon of various devices that can be used in switching and communication concludes the chapter. The so-called silica-on-silicon technology is explored and shown to yield low-loss integrated devices that can be exploited as planar lightwave circuits (PLCs). The functionalities explored include optical circuits that provide branching, switching, and filtering. The silica-on-silicon technology is shown to possess several advantages, for instance, silicon substrates are cheaply available and in large areas, moreover, silica waveguides have the same refractive index as optical fibers and therefore good matching and coupling can be achieved. The paper starts by considering the fabrication of silica waveguides on silicon. Consideration is then given to the evolution of PLCs. Four generations are identified starting with Y-branches and directional couplers as the first-generation circuits and then moving on to the second generation composed of Mach-Zehnder interferometers and ring resonators. The third generation includes $N \times N$ star couplers and arrayed-waveguide gratings. The current fourth generation includes optical transversal and lattice filters, which can be programmed in analogy with digital electronic filters. The operation and progress made in these devices along the generations is reviewed while applications are proposed. The paper proceeds by considering hybrid optoelectronic integration on PLC platforms. An example is given of the integration of a 4×4 optical gate matrix switch with a wavelength-division multiplexing (WDM) transceiver on a PLC. The paper gives a summary and indicates future directions for this promising technology.

Waveguide Electrooptic Switch Arrays

ROD C. ALFERNESS, MEMBER, IEEE

(Invited Paper)

Abstract—**We present a review of electrooptic waveguide switch arrays with strong emphasis on those based upon titanium-diffused lithium niobate waveguides. Crosspoint and array design considerations and performance are discussed as are waveguide technology limits. Switch array demonstrations are reviewed.**

I. Introduction

LIGHTWAVE transmission via low-loss, low-dispersion single-mode fiber has become the clear choice for high-capacity point-to-point communication systems. Thousands of kilometers of fiber over terrestrial trunk lines and, more recently, transoceanic routes demonstrate that fact. Presently, fiber is beginning to penetrate into the loop, first in the feeder plant and eventually in the distribution plant, into businesses and customer premises. From the viewpoint of transmission, fiber offers the potential of enormous information bandwidth. To preserve this advantage in future lightwave networks, it is important that the total information capacity of the network not be limited by the switch that provides connectivity. Electronic switching devices have limitations on both the data rate that they will handle and the switching reconfiguration rates achievable. Optical switches will also have such limitations. However, the goal and challenge is to realize, as achieved with optical transmission, more broad-band operation.

A primary goal for optical switching is to switch very wide-band signals. The other motivation, always cited but presently not necessarily a compeling one, is that optical switches are required to avoid the need (and associated cost, power, etc.) for optical to electrical and electrical to optical conversion. This is clearly a potential advantage of optical switching for lightwave networks; especially, for example, for switching frequency-division multiplexed signals that are being routed to the same distribution point. However, presently without the availability of optical regenerators, optical to electrical conversion may be required anyway if optical switch losses are too high. Nevertheless, with advances in optical amplifiers and reduced switch losses, optical switches may, in fact, avoid the need for many, if not all, *O* to *E* or *E* to *O* conversions in local wide-band lightwave networks.

Research on optical waveguide switch arrays is motivated by both of the above goals. As we will describe, these switch arrays can switch, under electrical control, extremely wide-band signals (e.g., gigabits/second) at potential reconfiguration times on the order or less than one nanosecond. The optical signal is switched directly, no electrical conversion is required. In this paper, we will review the current status of electrically controlled optical waveguide switch arrays with strong emphasis on those based upon titanium-diffused lithium niobate waveguide technology, which is currently the most advanced electrooptic waveguide technology.

The format is as follows. In the first section, the various switch types, space division and time division, will be discussed to understand the needs of the crosspoint elements to implement these switch types. In the next section, we review the types of crosspoints elements that have been utilized in waveguide switch arrays and the particular tradeoffs between control voltage, switching speed, and sensitivity to wavelength, polarization, etc., of each type. We then give a brief review of the titanium-diffused lithium niobate waveguide technology to allow discussion of loss/voltage tradeoffs and technology limitations of device and switch array design. In the next section, we describe the array architectures that have been considered and implemented. Experimental demonstrations of switch arrays will then be described and particular parameters of the array including insertion loss, crosstalk, and switching voltage achieved by several laboratories will be discussed and compared. Results of high-speed switch arrays will also be described. We include the description of several systems experiments that have been performed using titanium-diffused lithium niobate switch arrays.

II. Switching Systems

Switching systems can be categorized based upon the switched parameter. Presently research on optical switch arrays is focused toward applications in space, and time division switching systems [1]. In a space division switch [Fig. 1(a)] each line entering the switch and exiting the switch represents a single channel. Switching is accomplished by providing a spatial path internal to the switch which connects one input line to one (or more if multicast is required) output line. In space division switching applications, the reconfiguration rate may, in fact, be relatively slow. This type of switching arrangement capitalizes on a principal advantage of optical switches which is that the information rate that can pass through the switch is much greater than in electronic systems. Note that in such space division switching, synchronization of the high-bit-rate input and output channels is not required

Manuscript received February 25, 1988; revised May 18, 1988.
The author is with AT&T Bell Laboratories, Holmdel, NJ.
IEEE Log Number 8822445.

Reprinted with permission from *IEEE Journal on Selected Areas in Communications,* Vol. 6, No. 7, pp. 1117-1130, August 1988.

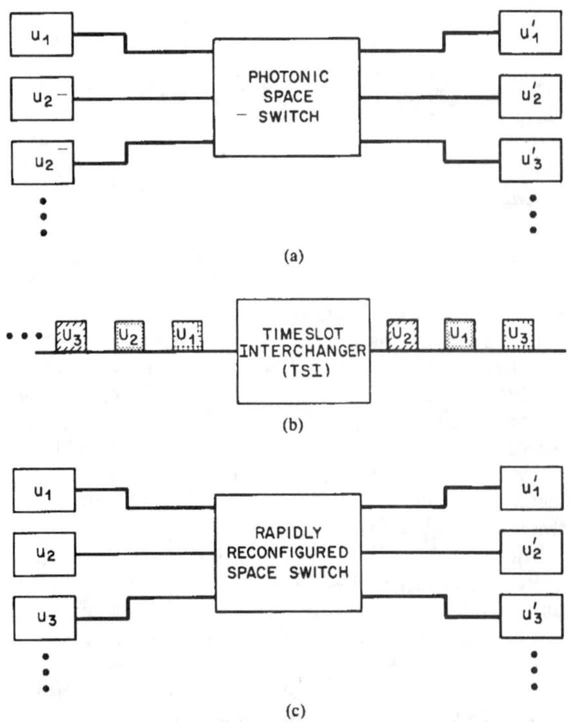

Fig. 1. (a) Space division switching system. (b) Timeslot interchange. (c) Time-multiplexed switching.

provided that glitches during switching are acceptable. Space division switching appears particularly applicable in trunk transmission applications for facility, protection, and route restoration where very broad-band signals (>2 Gbits/s) can be anticipated and where reconfiguration speed, while it must be faster than typically achievable with mechanical switches (~ 1-10 ms), need not be at rates comparable to the bit rate. Another potential application area is switching wide-band services in the loop or for local area networks. The limitation of space division switching is that each information channel requires its own input line to the switch. Therefore, the number of channels that can be connected via the switch is set by the switch dimension which may be limited.

The number of physical input and output lines to the switch can be reduced by using time division switching techniques [Fig. 1(b)]. In time division switching, each physical line contains several time multiplexed channels. A particular channel is represented by its timeslot within the periodic frame. As a result, the total number of input channels to the switch is increased by the number of time multiplexed channels per input line. To implement time division switching, a so called timeslot interchange switch [Fig. 1(b)] is required. This switch interchanges the sequence of timeslots between the input and output streams. Timeslot interchange is achieved in electronic switches simply by using dynamic access storage. Bits are stored as they enter the switch and read out in the new desired order. Presently, optical memory devices (and most other optical devices) are rather immature, making optical timeslot interchange quite difficult. Nevertheless, it has been demonstrated using optical crosspoints to selectively, in time, direct bits to the required memory element. For this application, the crosspoint reconfiguration speed is important.

For added flexibility, time division switching can be combined with time multiplexed space division switching [Fig. 1(c)]. For time multiplexed switching, the switch must be able to pass broad-band signal but must also be capable of fast reconfiguration, possibly at the bit rate, if reconfiguration is on a bit-by-bit basis. Alternately, switching can be on a block or byte basis with time guard bands between blocks to allow additional time for reconfiguration [63]. As with time division switching, the maximum bit rate that can be efficiently switched depends upon the switching speed. Very importantly, because the space switch is time multiplexed, the channels on each spatial input must be synchronized. Large electronic digital switching machines make use of timeslot interchange in combination with time multiplexed space division switching. For example, a time-space-time (TST) combination is used in telephone central office switches.

Optical switching offers the potential of switching also on the basis of wavelength. In analogy with time division switching, a device which performs wavelength conversion is required. Such devices are not yet available. However, lightwave networks based upon broadcast and wavelength dependent select architectures are being researched [2], [3]. Wavelength dependent selection can be achieved by a tunable wavelength filter, tunable laser, or by electronic filtering using heterodyne detection [2]. Wavelength division switching may ultimately be combined with time or space division switching; however, such switches are outside the scope of this review.

III. Optical Switch Crosspoint Elements

To achieve space or time division switching, arrays are built up from a basic exchange element. The basic switch function corresponding to what is referred to as the beta element is shown schematically in Fig. 2. A dual input/dual output device, the switch has two states, one the so called bar state in which an input exits from the same output port and the cross state in which the input and output lines are crossed. There are two switch types demonstrated to optically provide the beta function. The first is the generic directive switch type in which light via some structure is physically directed to one of two different outputs. It is analogous to the train yard switch. Several specific examples will be given below. An alternative approach is the gate switch (Fig. 3) in which the input signal is passively split into two parts each of which enters a device which is a simple on/off modulator or on/off gate [4]. The outputs from those gates are passively combined to provide two possible outputs. The gating device is simply an on/off modulator which may be easier to implement than a directive switch. However, the gate switch has a

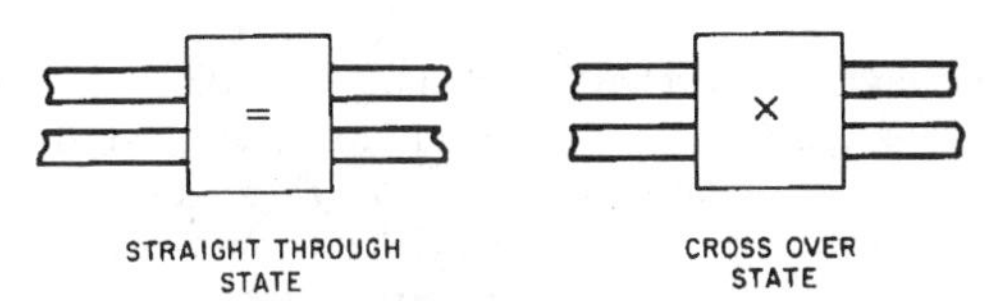

Fig. 2. Basic 2 × 2 crosspoint element with straight through ("bar") and crossover ("crossed") switch states.

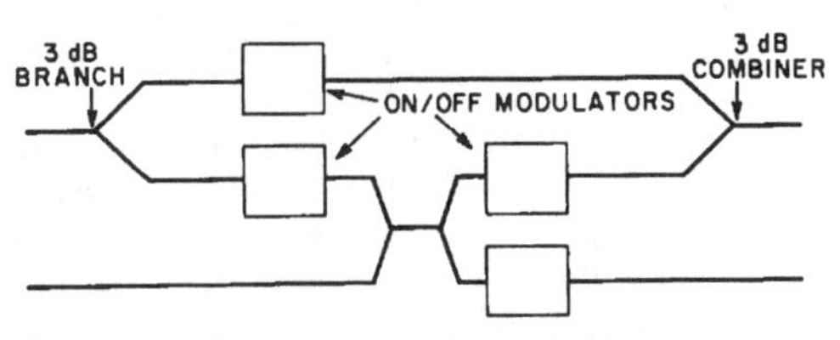

Fig. 3. Gate crosspoint switch.

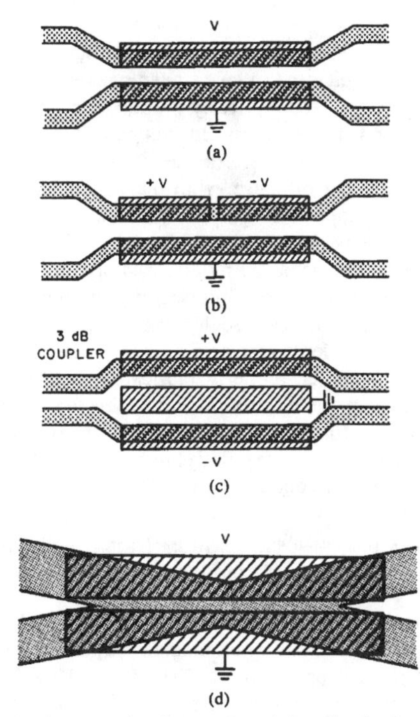

Fig. 4. Directive crosspoint elements: (a) switched directional coupler, (b) reversed $\Delta\beta$ directional coupler, (c) Mach-Zehnder interferometer, (d) intersecting waveguide or X switch.

disadvantage that there is a 3 dB split on input and a 3 dB combining loss on output. Therefore, there is inherent splitting and combining loss of 6 dB. On the other hand, the switch also makes possible 1 to many or broadcast. Potentially, the inherent loss may be overcome using optical amplifiers as the electrically controlled on/off gate [4].

IV. Directive Crosspoint Switch Types

Because of the passive splitting and combining losses of gate-type switches, directive-type switches have most frequently been used for switch arrays. The most common directive switch devices are shown schematically in Fig. 4 [5]-[7]. The directional coupler [Fig. 4(a)] consists of a pair of optical channel waveguides that are parallel and in close proximity over some finite interaction length L. Light input to one of these waveguides couples to the second waveguide via evanescent coupling. The coupling strength depends upon the interwaveguide separation and the waveguide mode size which in turn depends upon the optical wavelength and the confinement factor of the waveguide. If the two waveguides are identical, complete coupling between the two waveguides occurs over a characteristic length which depends upon the coupling strength. However, by placing electrodes over the waveguides, the difference in the propagation constants $\Delta\beta$ in the two waveguides can be made sufficiently large via the linear electrooptic effect so that no light couples between the two waveguides. Therefore, the cross state corresponds to zero applied voltage and the bar state corresponds to a finite switching voltage.

Unfortunately in order to achieve a perfect cross state the interaction length must correspond to exactly one coupling length or odd integer multiple of l [5]. Since the coupling length depends upon strip waveguide width and the waveguide-substrate index difference, all factors which depend upon fabrication, the tolerance required to reproducibly obtain exactly one coupling length is difficult to achieve. Furthermore, l depends upon wavelength.

The requirement of strict fabrication tolerances as well as the ability to achieve good switching for a relatively wide range of wavelengths can be overcome by using the so called reversed delta beta coupler shown in Fig. 4(b) [5]. In this device, the electrode is split into at least two sections. The cross state is achieved by applying equal but opposite voltages to the two-electrode task. It can be shown using the reversed delta beta coupler that over a relatively large range of coupling values one can electrooptically achieve both good cross and bar states. For example, for a two-section device, arbitrarily low crosstalk switching can be achieved for couplers with lengths between one and three coupling lengths. The switching voltage for the reversed delta beta coupler is essentially the same as for the standard directional coupler. Practical directional coupler switch arrays demonstrated to date generally use the reversed delta beta electrode structure.

The second directive crosspoint switch is the balanced bridge interferometric switch [6] shown in Fig. 4(c). The device consists of a input 3 dB coupler, two waveguides sufficiently separated so that they do not couple, electrodes to allow changing the effective path length over the two arms, and a final 3 dB coupler. Light incident in the upper waveguide is split 50/50 by the first coupler. With no voltage applied to the electrodes, the optical path lengths of the two arms are equal such that the light from the two arms enters the second coupler in phase. The second coupler acts like the continuation of the first, and all the light is crossed over to the second waveguide to provide the cross state. To achieve the bar state, voltage is applied to an electrode placed over one of the interferometer arms to electrooptically produce a π phase difference between the two arms. In this case, the two inputs from the arms of the interferometer combine at the second 3 dB coupler out of phase, with the result that light remains in the upper waveguide. Provided the splitter and combiner are perfect 3 dB couplers, both the cross and bar states can be achieved by electrical adjustment.

The third switch is the intersecting waveguide switch shown in Fig. 4(d). This device can be understood either as a modal interferometer or a zero gap directional coupler [7], [8]. When properly fabricated, both cross and bar states can be electrooptically achieved with good crosstalk performance. The required switch voltage, while difficult to write down explicitly, has a value between that of the directional coupler and interferometer [9].

V. Titanium-Diffused Lithium Niobate Waveguide Technology

The techniques for making low-loss waveguides by titanium indiffusion in lithium niobate [10] are well known. Several reviews appear in the literature [11], [12]. Here we provide a brief review to emphasize those aspects of the technology that are important in designing and optimizing switch arrays. Steps to fabricate waveguides by titanium diffusion in lithium niobate are shown in Fig. 5. The desired waveguide pattern is formed in photoresist on a polished lithium niobate crystal. Titanium metal is deposited by any of several techniques including electron beam deposition, RF sputtering, or thermal evaporation. The metal pattern is delineated either by liftoff or etching techniques. The crystal is placed into a diffusion furnace and heated to a temperature between values of 1000° and 1050°C and diffused to times that can vary from approximately 4 h to as many as 12 h. The metal thickness, diffusion time, and temperature selected depend upon the design operating wavelength and the desired waveguide confinement and mode size. Typically, to make waveguides that are highly confined, a relatively thick titanium layer, short diffusion time, and temperatures at the lower extremes are utilized. By increasing the temperature and diffusion time, the metal is diffused deeper into the crystal forming a waveguide of larger mode size assuming that sufficient metal has been deposited.

After diffusion, an insulating buffer layer may be deposited over the crystal. The most common insulator is silicon dioxide. The insulating buffer layer is required to reduce loading losses caused by an electrode placed directly on top of the waveguide particularly for the TM polarization. For some crystal orientations and electrode configurations, the buffer layer is not required. In particular, for the X cut geometry where fields required for switching are in the plane of the crystal, electrodes can be placed along the side rather than directly above the waveguide. Next the electrode pattern is photolithographically aligned and deposited. For slowly reconfigured space division switches, aluminum with a flash of chrome for adhesion is the typical electrode material. For high-speed switches and small switch arrays, used for such applications as time division multiplexing and demultiplexing, gold is the preferred material because it provides lower losses for the high-frequency, electrical control signals. Aluminum can be deposited by evaporation techniques. For very high-frequency applications, thick gold layers are formed by electroplating [13]. Thicknesses as large as 3 μm can be achieved.

Finally the crystal is cut, the crystal ends are carefully polished, and, for most applications, fiber pigtails are aligned and attached. For switch arrays, fiber arrays are required. Typically such fiber arrays utilize silicon V-groove technology to provide the accurate waveguide to waveguide spacing to be matched to that of the fiber spacing. To achieve good fiber to waveguide coupling, it is necessary to match the mode size of the waveguide to that of the fiber. Titanium-diffused lithium niobate waveguides are asymmetric; the optical mode is elliptical. The longer axis is the plane of the crystal. Typical ellipticity is about 1.7. Nevertheless, good fiber to waveguide coupling can be achieved if the geometric mean of the two-mode radii is equal to the fiber radius. This can be achieved for typical fibers by choosing the appropriate titanium metal thickness, diffusion time, and diffusion temperature. In this way, intrinsic fiber mode waveguide coupling losses for perfect lateral alignment as low as 0.3 dB have been demonstrated. Presently, fiber attachment has been most successfully achieved using UV curable epoxies. The art of fiber attachment has made significant advances and presently fibers can be permanently attached to lithium niobate crystals with attachment losses as low as a few tenths of a decibel [14]. In fact, arrays of fibers as large as 12 have been permanently attached with maximum excess attachment losses across the array of only 0.5 dB. Such arrays do show some signs of loss changes during temperature cycling but additional testing and environmental stability measurements will be required.

Relatively large mode size (~8–10 μm full width) is required for efficient fiber to waveguide coupling. However, much smaller modes are desirable to reduce switch voltages. Furthermore, a symmetric waveguide mode, most readily achieved by a buried waveguide, is appropriate for fiber coupling, while such an optical mode will not be efficiently switched with a surface electrode. The result is a tradeoff between insertion loss and voltage · length [12].

Recently, effective waveguide tapers have been demonstrated. This approach potentially allows optimal fiber coupling at the crystal ends using a buried symmetrical waveguide and a small mode near the surface in the switch interaction region. The waveguide is buried by depositing and diffusing a second layer of magnesium oxide (MgO) after the titanium-diffused waveguide is formed. Diffused MgO lowers the refractive index. Thus, by doing a shallow MgO diffusion with the deposited MgO thickness tapered to zero in the interaction region, the desired tapered

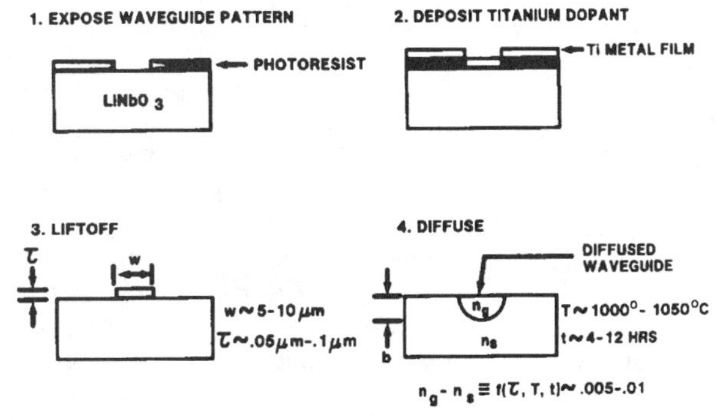

Fig. 5. Titanium-diffused lithium noibate fabrication technology.

waveguide mode has been achieved [15], [16]. Thus far, this technique that allows separate optimization of fiber coupling and device voltage has only been applied to single modulators. However, it offers even greater potential benefit for switch arrays because of the large number of devices that will share the reduced drive voltage.

Lithium niobate does demonstrate some deleterious effects. In particular, at wavelengths in the visible and near infrared it exhibits a photorefractive effect in which refractive index changes can be induced by relatively modest optical power levels within the waveguide. This effect is dramatically reduced at 1.3 μm and even more so at 1.5 μm. At 1.3 μm, measurements indicate that photorefractive effects are probably not important for power levels within typical waveguides of about 10 mW [17]. At a wavelength of 1.5 μm, power levels as high as 50 mW show no photorefractive effects in interferometric switches [18].

Drift of the effective dc bias level has also been observed in Ti:$LiNbO_3$ waveguide devices. This effect can be pronounced in devices with a poor insulating buffer layer and can, in fact, be reduced by etching the buffer layer in the electrode gap [19]. However, it has also been observed, to a much smaller degree, in devices without a buffer layer [20]. The cause, elimination, and effect of such effects is currently under study.

The desire to fabricate relatively large switch arrays has placed demands upon the titanium-diffused lithium niobate technology which, in turn, has resulted in technology advances. As discussed below, switch arrays require processing on relatively large substrates, and good uniformity of switch characteristics including switching voltages and voltage bias are required. Furthermore, low-insertion loss and acceptably low-switching voltages are necessary. If this technology is to become cost effective, processing on wafers size may be essential. Technological advances have, in fact, included the ability to do processing on 3 in wafers that has been demonstrated in several laboratories. On-going research is aimed at identifying the difficulties and stumbling blocks to achieving uniformity and reproducibility over 3 in wafers. A photograph of a processed 3 in wafer is shown in Fig. 6.

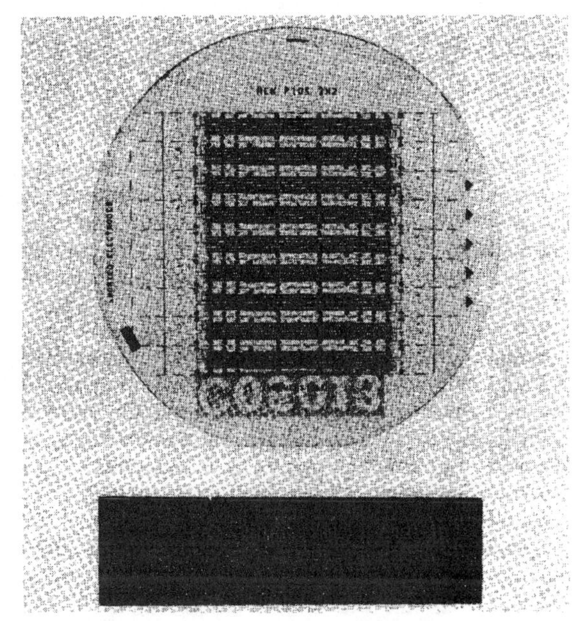

Fig. 6. Photograph of processed 3 in lithium niobate wafer. (Courtesy of R. B. Jander.)

VI. Crosspoint Considerations

Several parameters of the individual crosspoints are important in characterizing an optical switch array. The primary parameters are switch voltage, switching speed, crosstalk, and the intrinsic data rate that can be propagated through the switch. In addition, because the overall insertion loss for the switch array chip is important, excess loss within the switch element including necessary waveguide bends into and out of the crosspoint are important. For practical applications, the sensitivity of the switch to wavelength, polarization, and temperature as well as fabrication and voltage tolerances are also important. It is important to note that lithium niobate is an insulator. Therefore, negligible current flows in response to a holding voltage. Thus, for static switch applications, the switch consumes negligible power unlike electronic switches.

For each of the directive-type crosspoint switches considered above, there is a switch-type dependent figure of merit to achieve switching that can be described by the product of the electrooptically induced propagation constant change ($\Delta\beta$) times the length of the device. This so called switching condition can be written as $\Delta\beta \cdot L = p \times \pi$. p is equal to the $\sqrt{3}$ for a directional coupler and approximately that same value for a reversed $\Delta\beta$ directional coupler. It is equal to one for the interferometer and takes on a value somewhere between one and the $\sqrt{3}$ for the intersecting waveguide-type switch [11]. $\Delta\beta$, of course, depends upon the voltage, the electrooptic coefficient r, and geometric considerations including the electrode gap d, and the overlap between the electrical and optical fields Γ. It also depends both explicitly and indirectly upon wavelength λ. Thus, the switching condition for the three crosspoints can be written as

$$V \cdot L = \frac{\lambda d}{\Gamma n^3 r} p. \tag{1}$$

For fabrication parameters chosen to yield reasonable propagation and coupling loss, this voltage · length product for titanium diffusion lithium niobate directional coupler switches is found empirically to be 3.3 V · cm $\times$ $\lambda^2(\mu m)$ which corresponds to an overlap parameter of ~0.2 [12].

For time division switching-type applications, the switching speed is important. The inherent response time of the electrooptic effect in lithium niobate is subpicosecond. Switching speeds are limited practically by circuit effects. The switching speed of the above directive crosspoint-type switches depends upon the electrode type used. For low speed, space division switching applications, a lump electrode is the easiest to employ in which case the switching speed is determined by the RC time constant where the C is the capacitance of the electrode and R is the terminating resistance, typically 50 Ω. For high-speed switching applications, the traveling wave electrode has been employed. In this case, the switching speed is not limited by RC charging and discharging times but rather by the inherent mismatch in velocities of the optical and electrical signals along the electrode of the switch. The switching speed in either case can be characterized by a switching bandwidth times electrode length figure of merit. For properly designed and fabricated electrodes, that figure of merit is roughly 1 Ghz · cm for lump-type electrodes and approximately 5-6 Ghz · cm for traveling wave-type electrode devices in lithium niobate [12].

These crosspoint switches are characterized by another speed which is related to the bandwidth of information that can be propagated through the switch without distortion. The ability to propagate very broad-band information through the switch in static or space-division-type switching applications is a very important advantage of these optical switches. The ultimate bandwidth limitations are twofold. One is the actual dispersive pulse spreading of light propagating through the lithium niobate waveguides. For typical array lengths (<6 cm), these dispersive effects are not important at data rates below a terabit per second [21]. Another data rate limitation is imposed by the wavelength dependence of the crosspoint switch. If one Fourier decomposes an optical pulse into its frequency components, to avoid distortion within the crosspoint, it is important that each of those Fourier components be switched effectively by the crosspoint. Measurements on directional coupler switches, for example, indicate that crosstalk levels better than 20 dB can be achieved over a wavelength range exceeding 100 Å in the visible and probably double that in the 1.3–1.5 μm region [22]. Again this imposes a bandwidth limitation that does not become important until one approaches bandwidths of terabits per second.

In each of the three crosspoints described, measurement-limited crosstalk values better than −40 dB for both switch states have been demonstrated. However, achieving such values routinely and uniformly over a large area with uniform switching voltages remains an area of work. Because the refractive index of lithium niobate is temperature dependent, the required switching voltage can also vary with temperature so some modest temperature control may be necessary.

The electrooptic coefficient in lithium niobate depends upon the polarization of the incident light. As a result, without special designs, all of the crosspoints discussed provide low-crosstalk switch states only for a single input polarization which is chosen to utilize the largest *e/o* coefficient. For example, for switches on *Z* cut lithium niobate, light polarized perpendicular (so called TM polarized light) to the crystal plane is switched with the lowest voltage. However, devices specifically designed to achieve good crosstalk for fixed switching voltages independent of incident polarization have been demonstrated. These include a directional coupler with carefully weighted coupling [23] and, more recently, an asymmetric branching switch [24] as well as a directional coupler with extremely tight fabrication control [25]. Unfortunately, because they must be capable of switching with the weaker electrooptic coefficient, the voltage · length product in these devices is at least three (the ratio of the electrooptic coefficients for the two polarizations) and more typically five times larger than for single polarization switches. Nevertheless 1 $\times$ 16 polarization independent directional coupler switch arrays have been demonstrated with moderate (~15 dB) crosstalk performance [26] using common control voltages. More recently, 4 $\times$ 4 polarization independent switch arrays have also been reported as discussed later.

VII. Array Architecture Considerations

Switch arrays can be built up from the basic 2 $\times$ 2 crosspoints according to different cascading or interconnect architectures. In choosing an architecture, several characteristics must be considered. Of prime consideration is the issue of connectivity. If an input line requires connection to an open line through the switch and no path can be found, the connection is blocked. If this new connection can always be found without taking down and reestablishing existing connections, the switch is called strictly nonblocking. If, however, a path can always be found but only by changing a switch path of other connections, the switch architecture is called rearrangeably nonblocking. While a strictly nonblocking switch architecture is obviously desirable and in many cases essential, such switches generally require more cascaded rows or ranks of switches than do rearrangeably nonblocking architectures.

The simplest strictly nonblocking architecture is the crossbar, shown schematically for eight input and eight output lines (8×8) with directional coupler crosspoints in Fig. 7. Unfortunately, the crossbar requires a large number, N^2, of crosspoints. As importantly, it requires $2N - 1$ ranks of switches across the length of the crystal. However, it is strictly nonblocking and requires changing only two crosspoints to change a single connection. It also requires no overlapping or intersecting waveguides. The crossbar does present problems in crosstalk buildup for large arrays. For example, using relatively poor, -20 dB crosstalk crosspoints, the maximum crossbar dimension to achieve a system crosstalk of -11 dB required for signal detection is only 8×8 [27].

Although the crossbar switch is the architecture that has been used in most of the early waveguide switch array demonstrations, recently interest has focused on more efficient architectures. Several nonblocking architectures that reduce the number of crosspoints, including the Clos, Benes, and Banyan, are known from electrical switching systems. Other approaches such as the active splitter, active combiner, and binary tree architectures are beginning to be explored [27]-[32]. The primary motivation in exploring such switch configurations is twofold: to reduce the number of switch ranks required and as a result reduce loss and to avoid crosstalk buildup [27]. The need to reduce the number of switch ranks is a result of the relatively long (but narrow) space needs for individual crosspoints. From the voltage · length product quoted above, a ~ 6 mm long device is required for a 10 V crosspoint for a wavelength of 1.3 μm. An additional 1 mm is required for bends into and out of each switch and an additional ~ 2 mm on either end of the crystal to separate the input waveguides to the typical 250 μm center to center fiber separation. Thus, for a 3 in wafer, ~ 10 ranks of switches can be accommodated. Roughly double that is possible if one is willing to use 20 V control voltage which may be reasonable for slow reconfiguration applications. For a crossbar, $2N - 1$ ranks are required allowing 8×8, and with voltages of ~ 30 V, possibly a 16×16. Note that lateral space on the crystal is never an issue.

The number of ranks required for nonblocking tree-type structures is $\sim 2 \log_2 N$, significantly smaller than for the crossbar [27]. These architectures are, however, more complex than the crossbar. For example, a recently demonstrated [31], [32] 4×4 duobanyan switch is shown in Fig. 8. As shown, interconnects between switch ranks,

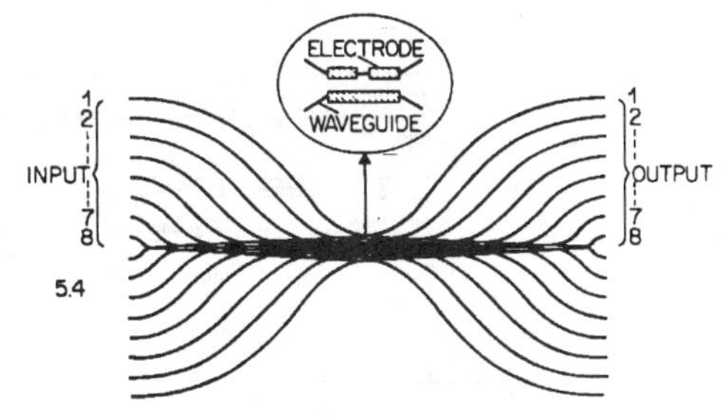

Fig. 7. Schematic of 8×8 crossbar (after [40]). (Courtesy of L. Thylen.)

even on the same crystal, may require passive waveguide intersections. Fortunately, it has been shown that such intersections can be made with relatively low crosstalk (< -35 dB) and low loss (≤ 0.2 dB) provided the intersection angle exceeds some mode confinement dependent intersection angle. For typical low-loss Ti : $LiNbO_3$ waveguides, that angle is $\sim 8°$ [33], [34]. To achieve these angles without introducing unacceptable loss requires additional real estate. Novel bend loss reduction techniques to reduce this requirement have been pursued [35]. In spite of increased complexity, significant reduction in the number of switch ranks makes these tree structures attractive for space division switching of a large number of wideband channels.

Assuming a structure which provides crosstalk avoidance [36] so that the inherent limitation to switch dimension is loss, we calculate the number of lines that could be connected without internal amplification or regeneration versus the data rate of each line. The receiver sensitivity of the latter and the transmitter power determines the allowable loss budget. We assume the receiver sensitivity of state of the art APD/FET receivers and allow a 10 dB margin. The projected number of lines versus the data rate per line is shown in Fig. 9 with the loaded loss per crosspoint as a parameter. The latter includes the total fiber-switch chip-fiber loss averaged over the number of switch ranks per chip. We, of course, allow for multiple cascaded chip switches and the external fiber interconnection loss is assumed to be part of the margin. This is perhaps overly optimistic and requires architectures that avoid crosstalk buildup. From the viewpoint of loss, Fig. 9 indicates that over 5000 lines, each carrying data at 1 Gbit/s, could be connected using switches with the loaded loss of a 1.0 dB/switch—a number which, as we indicate below, appears achievable with present technology. Also, the chip interconnection would, of course, present significant challenges as would the switch drivers. Nevertheless, optical switch arrays certainly have the potential of providing connectivity with aggregate bandwidths in the terrabit/second regime. It is worth noting that to build an $N \times N$ passive star from 2×2, 3 dB couplers requires $\log_2 N$ ranks. Assuming a perfect star with no excess loss, the loss-limited optimum star-based network corresponds to the 1.5 dB loaded loss line in Fig. 9. Therefore, from a loss viewpoint, the size of a switched network will ex-

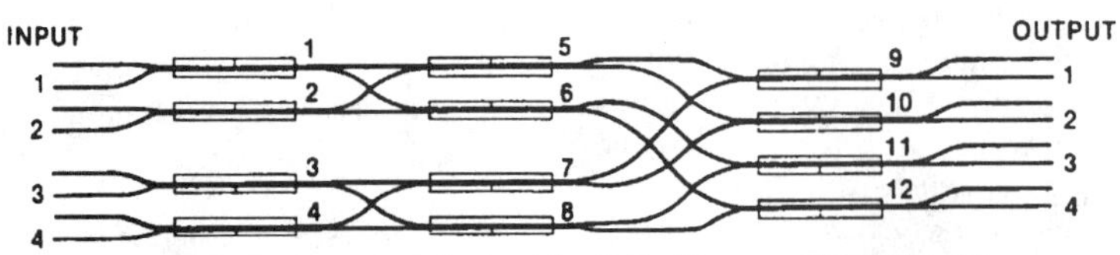

Fig. 8. Duobanyan switch array (after [31]). (Courtesy of M. Milbrodt.)

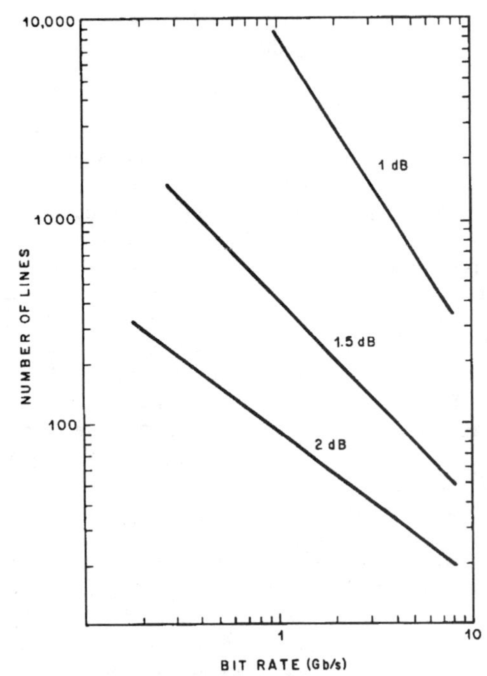

Fig. 9. Calculated maximum space division switching system dimension versus data rate per line. Assumption: APD/FET receiver with 3 dB/octave sensitivity dependence on bit rate, 10 dB margin, crosspoint loaded loss (total fiber-switch chip-fiber loss normalized by number of crosspoint ranks per chip) is given as a parameter.

ceed that of a passive star (with wavelength division selection, for example), provided the loaded loss is less than 1.5 dB/crosspoint.

To approach the above loss limit potential, architectures that avoid crosstalk buildup are absolutely essential even for the best individual crosspoints. In addition, even for modest arrays, crosstalk avoidance architectures are important to reduce the crosspoint requirements to simplify the control of the crosspoint voltages.

Even though good crosspoint crosstalk values can be achieved, until processing technology is highly developed, it is possible that the required voltages to achieve optimum crosstalk will vary somewhat across the array. In this case, array architectures that require better than −35 dB crosstalk crosspoints may require individually tuned voltages across the array. However, the voltage tolerances to achieve −20 dB crosspoints compared to −35 dB values are considerably reduced (~5 percent compared to 0.5 percent). As a result, by using architectures that avoid crosstalk buildup [30] one may be able to use common control voltages and still achieve an acceptable overall switch crosstalk. A similar argument applies to high-speed switching where typically crosstalk levels are increased somewhat compared to their dc drive values. This reasoning also applied to reversed $\Delta\beta$ directional coupler switches. The best crosspoint performance is achieved with independent control of the two sections. However, electrode-induced $\Delta\beta$ reversal with a single control voltage, with perhaps increased crosstalk levels, is preferable from a control point of view. It is important to note that efforts to make crosspoints that are drive voltage tolerant are also being pursued with promising results [24].

VIII. Switch Array Demonstration

The first chip to integrate waveguide switches was a 4 × 4 switch comprising five Ti:$LiNbO_3$ reversed $\Delta\beta$ directional couplers in a blockable configuration [36]. This early demonstration was in the visible. A low-loss version of this switch configuration for $\lambda = 1.3\ \mu m$ with pigtailed fibers was developed in 1982 [37]. The first strictly nonblocking 4 × 4 switch was a crossbar of 16 Ti:$LiNbO_3$ directional coupler switches [38]. The device did not use reversed $\Delta\beta$ electrodes; therefore, cross state crosstalk values as poor as −12 dB resulted. A path dependent loss resulted from design-related excess loss of 0.8 dB/bend. Fiber-coupled insertion loss, therefore, varied from ~5 to 10 dB but was as small as 2.7 dB for a straight waveguide on the 5.2 cm long chip. Switch voltages for the ~6 mm couplers were 11 V with a variation of ±15 percent.

Overall switch crosstalk was significantly improved by using a double-switch arrangement to avoid crosstalk buildup consisting of 32 reversed $\Delta\beta$ switches on a single chip [33]. This dual crossbar switch provided better than −30 dB (and as good as −39 dB) system crosstalk with individual crosspoints with crosstalk levels of ~ −20 dB. Using this double-switch configuration, overall switch crosstalk builds up as $N\delta^2$ rather than $N\delta$ as with the single crossbar where N is the switch dimension and δ is the crosspoint crosstalk in decibels. Because the two crossbars are arranged almost in parallel, readily available lateral space on the crystal was utilized and increased length requirements minimized. However, ~$2N$ intersections per path were required necessitating low-loss, loss-crosstalk intersections which were achieved in this case with a 7° intersection angle. Calculations indicate that a 32 × 32

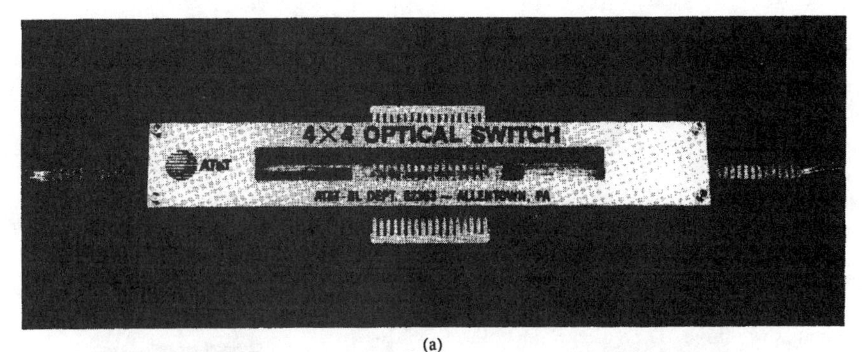

(a)

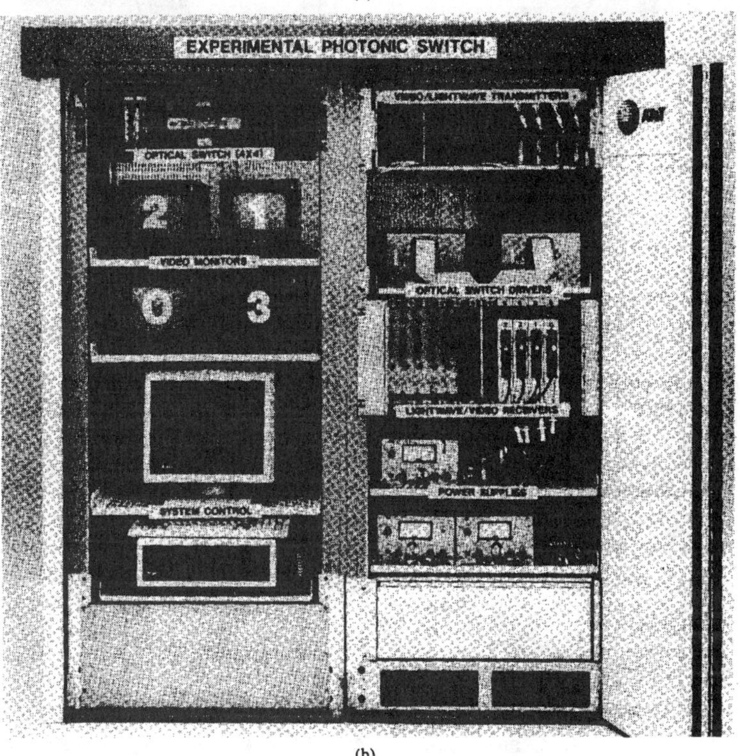

(b)

Fig. 10. (a) Photograph of pigtailed and electrically packaged 4 × 4 Ti:$LiNbO_3$ crossbar switch. (b) Photograph of photonic video switch experiment. (Courtesy of G. A. Bogert and W. A. Payne, respectively.)

switch with system crosstalk of −20 dB could be achieved with crosspoint crosstalk values of ~ −17 dB [33].

Nevertheless, with current technology, switch system crosstalk levels of −30 dB have been achieved with the simple 4 × 4 crossbar architecture and reversed $\Delta\beta$ directional couplers provided independent voltage control to the crosspoints [39]. Switch voltage between the bar and cross for the 5.5 mm long couplers (for the single polarization operation) averaged ~13 V with a variation of ~ ±2 V. Crosspoint crosstalk values of ~ −20 dB could be achieved with a common control voltage. The average fiber-coupled insertion loss was 5.2 dB. Although there was little path length or bend dependence, apparently one poor waveguide caused loss values as high as 8 dB.

Polarization preserving fiber pigtails were permanently attached to this switch. A photograph of the packaged and pigtailed switch is shown in Fig. 10(a). The switch and associated control electronics has been employed as a video space division switch in an experimental system demonstration that has been operating since 1986. A photo of the entire system is shown in Fig. 10(b). The switch switches four 90 Mbit/s video channels but could just as easily switch, for example, high-definition video [64].

A fully nonblocking 4 × 4 switch with switched intersecting waveguide crosspoints has also been demonstrated (Fig. 11) [11]. The crosspoints used a 0.9° intersection angle, with 1 mm long electrodes and yielded crosstalk levels of −22 dB with a 50 V switching voltage. A 4 cm long *Y* cut crystal was used, however, the 16 switches required a length of only 2.6 cm. Total fiber coupler insertion loss was 9–10 dB. This switch emphasizes a potential advantage of intersecting waveguide switches. They are well matched to the crossbar architecture in that only straight waveguides are required except for slow bends at the edge of the crossbar. However, except for very small intersection angles, which increase the total length, the drive voltages are large. The voltage · length product is comparable to directional couplers.

The largest single-chip switch array demonstrated to date is the 8 × 8 crossbar using reversed $\Delta\beta$ Ti:$LiNbO_3$ directional couplers [40]. To achieve the 15 ranks of directional couplers (Fig. 7) on a 6.4 cm crystal, 2 mm long crosspoints were employed. Crosspoint crosstalk values for the 64 elements ranged from −18.6 to −37.2 dB with an average value of −30.5 dB and a standard deviation of 4.1 dB. The effective voltage swing for these rather short crosspoints was 40 V with a variation of ~ ±4 percent. Path dependent losses were quite reasonable, varying from 5.3 dB for the shortest path to 6.8 dB for the longest—all fiber-to-fiber losses. This corresponds to an excess loss per crosspoint of 0.1 dB. All of these values, especially considering the factor of four increase in number of switches, represent a significant improvement in uniformity over the first 4 × 4 switches which indicates an improvement in fabrication and processing technology. This switch array has been used in an interesting demonstration of self-routing where the switch control information is contained in the optical signal and extracted using correlation techniques to set the switch control voltages [53].

Another 8 × 8 switch with similar performance is being developed for a 32 line space division switching system [41]. This system is designed to provide two-way communication via two closely spaced (1.29 and 1.32 μm) wavelengths. The fact that optical switches provided low crosstalk switching over this large a wavelength separation allows bidirectional operation through the same switch. As a result, redundant switch arrays and control electronics are not required. The 32 line bidirectional connections can be made with a folded three-stage network composed of four 8 × 8 and eight 4 × 4 switch modules. By folding, the first stage also functions as the third stage. The system loss budget is designed to provide operation at line rates up to 800 Mbits/s.

Several strictly nonblocking switches that are not of the standard crossbar architecture have recently been reported. One is shown schematically in Fig. 12 [42]. This architecture is a rectangularized switch with the same number, N^2, of crosspoints as a standard crossbar. However, it has several advantages. First, it reduces the number of ranks from $2N - 1$ to N; second, it equalizes the number of crosspoint for all paths through the switch. The first allows doubling the switch dimension for fixed cross-

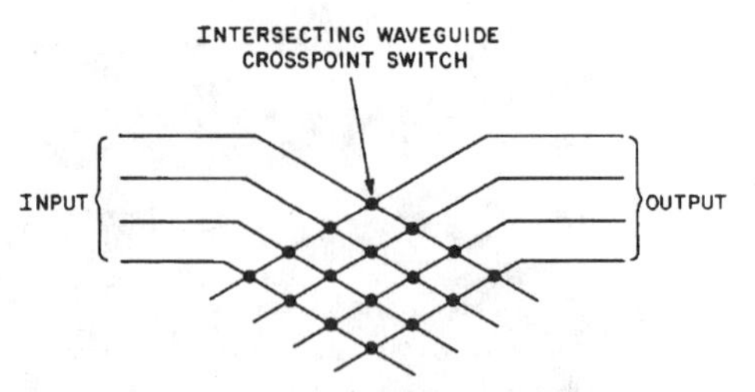

Fig. 11. Schematic of crossbar switch using intersecting waveguide switches (after [11]). (Courtesy of A. Neyer and E. Voges.)

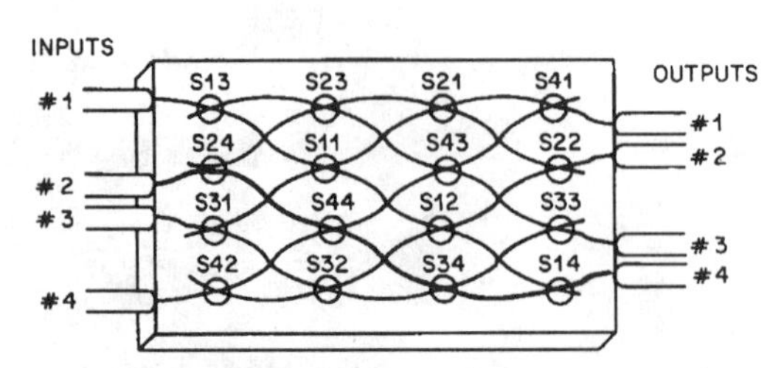

Fig. 12. Rectangular crossbar switch array (after [42]).

point length or doubling the crosspoint length for fixed dimension. The latter allows halving the switch voltage. The second advantage offers the potential of path independent loss. Both advantages were, in fact, realized. In a 4 × 4 implementation, 10 mm long directional couplers with reversed $\Delta\beta$ electrodes resulted in crosspoint switching voltage swing of less than 10 V for λ = 13 μm, the lowest achieved for a 4 × 4 array. Furthermore, the measured fiber–switch–fiber insertion loss was 4.7 dB with only a ±0.3 dB path dependent variation.

The 4 × 4 tree-structured duobayan switch architecture shown in Fig. 8 has also recently been demonstrated by several groups [31], [32]. Excellent results, demonstrating the advantages of such architectures, were achieved including average fiber-to-fiber insertion loss of 3.9 dB and overall switch crosstalk better than 35 dB. Again, because only three ranks are required, the individual crosspoints can be rather long (8.5 mm in [31]) to achieve low-switching voltages (~10 V for single-polarization operation).

In addition, a passive splitter active combiner tree-structured 4 × 4 switch that allows full broadcast has been reported [43]. The switch array has two ranks of passive splitters and two switch ranks. Total insertion loss was ~13 dB; however, 6 dB is due to passive splitting losses.

Polarization independent 4 × 4 switches that utilize tree-structured architectures have also been recently realized [44], [45]. Polarization independent switching voltages as low as 32 V using 15 mm crosspoints was achieved. Average insertion loss of ~6 dB with overall crosstalk better than −15 dB was achieved. These results were achieved by using tightly controlled diffusion parameters that equalized the waveguide mode characteristics for TE and TM modes [44]. In an alternate demon-

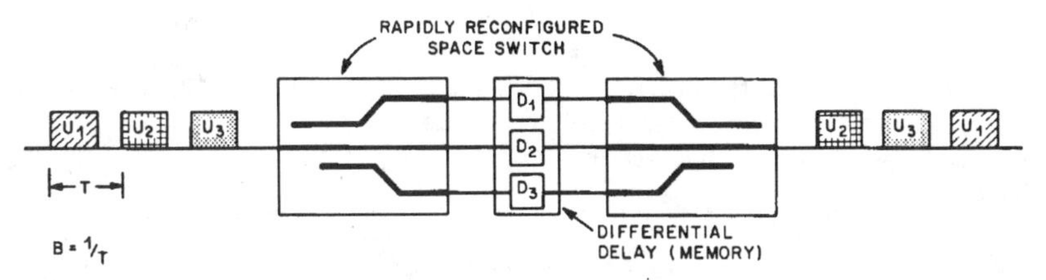

Fig. 13. Photonic timeslot interchange using high-speed 1 × 4 switch arrays and optical memory elements.

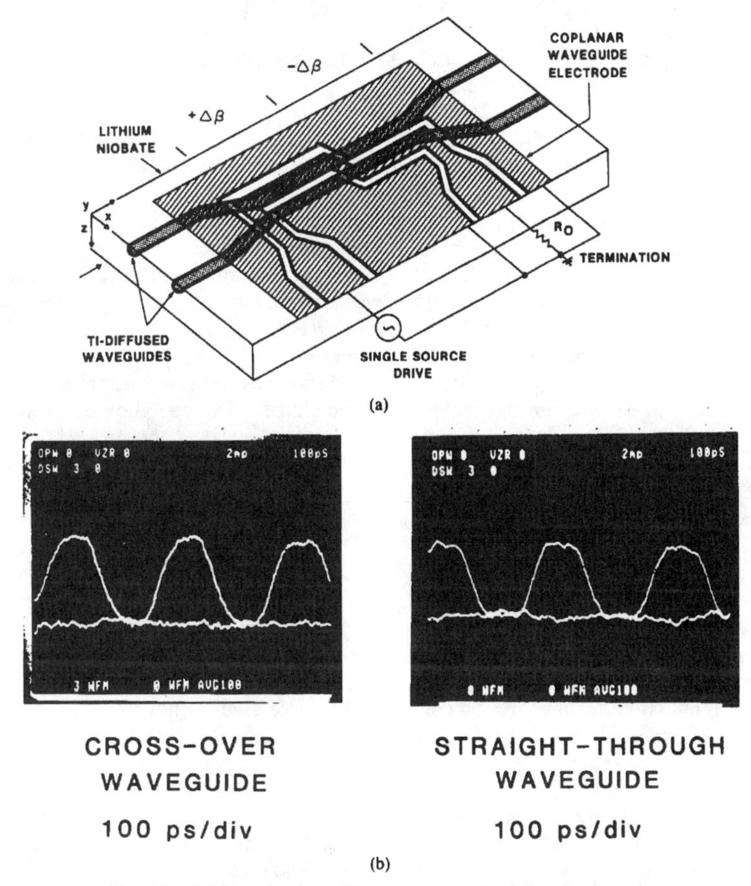

(b)

Fig. 14. (a) Schematic of traveling wave, reversed Δβ directional coupler switch. (b) Measured switching response in cross and bar states under 2 GHz sinusoidal drive.

stration, switching voltages were higher but fabrication tolerances could be relaxed [45].

Several general trends can be extracted from the space-division Ti:$LiNbO_3$ switch arrays demonstrated to date. The loss per chip can be estimated as follows. For waveguides fabricated to give good fiber waveguide coupling, the total loss per facet can be less than 0.5 dB. Propagation losses are ~0.2–0.3 dB/cm resulting in a basic chip loss for a 6 cm long crystal of 2.2–3 dB. Excess loss per switch rank is ~0.1 dB, as is the loss at appropriately designed waveguide intersections. The number of switch ranks that can be placed on a crystal depends upon the voltage one is willing to drive the crosspoints with given a voltage · length product of ~6 V · cm for $\lambda = 1.3\ \mu$m. For less than 10 V drive, four ranks have been demonstrated with a total loss of 4.5 dB or a loaded average of ~1.1 dB/crosspoint. For 40 V (2 mm couplers), the loaded average (for a crossbar with no intersections) is less than 0.5 dB/crosspoint. From a system loss point of view, these values offer the potential of a relatively large space division switch (Fig. 9) [29] by employing improved tree-structured architectures. The recent demonstration of polarization independent 4 × 4 arrays and more uniform switching voltages across the array are indicative of an improving technology.

IX. High-Speed Switch Arrays

Photonic time slot interchange has been achieved with fast 1 × N switch arrays to provide dynamic time delay (memory) selection with a generic system shown schematically in Fig. 13. The 1 × N switch reconfigures N times per frame to direct the bits of the ith time slot to one (jth) of N time delays. The reconfiguration can be on a bit or block basis. The delays can be achieved with elements with different fixed delays T, $2T$, $\cdots$ $2NT$. In this case, the required delay is determined by switching to the output port (j) connected to the appropriate delay [46], [47]. Another approach is to use reentrant delay lines to achieve dynamically determined delays [48]. In this case, the 1 × N switch sends the bits to the output ports in a fixed periodic manner. Fiber delay lines [46], [48] and bistable lasers [47] have been used as the memory elements. Because of the limited memory time of the bistable elements, bit interleaved switching was used while blocks of bits were interchanged using the fiber delay lines. In one experiment, four 64 Mbit/channels, providing a 256 Mbit/s time multiplexed data rate, were time switched using 1 × 4 and 4 × 1 Ti : $LiNbO_3$ directional coupler switches and bistable laser diodes for memory elements. Reconfiguration times of ~2 ns are required and lumped electrodes were used. Bit error rates below 10^{-10} were achieved for the system.

To time demultiplex [49] signals at bit rates above ~1 Gbit/s, traveling wave electrodes are required. For tree-type arrays, good crosstalk in both states is required. Therefore, for directional coupler devices, the reversed $\Delta\beta$ electrode may be required. Both features can be conveniently achieved using a coplanar strip electrode with electrode structure induced polarity reversal as shown in Fig. 14. Only a single control voltage is required to achieve better than −20 dB crosstalk at dc. Measurement-limited crosstalk levels of ~ −14 dB for both switch states can be achieved at 2.5 GHz switching rates with a 1 cm long electrode with a dc switching voltage of 9 V for λ = 1.3 μm. Measured results are shown in Fig. 14 [50].

A 1 × 4 Ti : $LiNbO_3$ tree array with three traveling wave directional coupler switches capable of demultiplexing a 1.6 Gbit/s data stream into four 400 Mbit/s streams has been demonstrated [51]. The 1.6 cm long couplers, which did not use reversed $\Delta\beta$, had a dc switching voltage of 5.5 V.

Using a tandem of discrete 1 × 2 Ti : $LiNbO_3$ directional couplers, a time multiplexed bit rate as high as 16 Gbits/s has been successfully demultiplexed to four, 4 Gbit/s data streams in a optical time division multiplexing transmission experiment over 8 km of fiber [52]. The short pulses were generated by mode locked semiconductor lasers, externally modulated, passively multiplexed, sent down the fiber and actively demultiplexed using a two-stage directional coupler switch tree. The first stage was a 5 mm long traveling wave reversed $\Delta\beta$ coupler operating at 8 GHz and the second were 1 cm long couplers operating at 4 GHz. Timing and phase to drive the high-speed switches were derived from one of the 4 Gbit/s detected signals. The operation of the demultiplexer on the 16 Gbit/s signal and the received eye pattern of one 4 Gbit/s received channel are shown in Fig. 15.

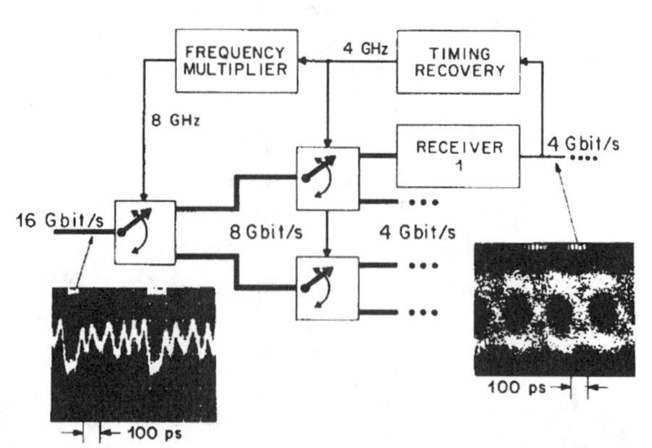

Fig. 15. Demultiplexing a 16 Gbit/s data stream with a high-speed 1 × 4 optical switch. A scope-limited oscillograph of the 16 Gbit/s data stream (left inset) and the eye pattern of one demultiplexed 4 Gbit/s channel (right inset) are shown (after [52]).

X. Discussion and Conclusions

Optical waveguide switch arrays have seen slow but sustained improvement over the last couple of years. This is a result of improvements in waveguide technology and the use of array architectures more suited to the constraints of this technology. While the largest single-chip switch, an 8 × 8, appears relatively modest in size, a 16 × 16 space switch (noncrossbar) appears possible. The largest broad-band electronic single-chip silicon modules, which are limited to signal bandwidth less than about 300 Mbit/s, are of comparable dimension [54]. The most broad-band electronic space division switch chip this author is aware of is a 4 × 4 on GaAs that can switch signals at 2 Gbits/s [55]. Unlike optical switches, this device consumes significant power, ~2 W, even when not being reconfigured. Of course, the strong and significant advantage of electronic switching is loss compensation through the switch module. To offset this disadvantage for optical switches, researchers have begun to examine cascading a very broad-band optical amplifier with optical switches [56]. This may offer important advantage as in-line amplifiers within a large switch system.

To make large switching systems, interconnection of switch modules is required for both electronic and optical versions. For spatial separations between modules exceeding several feet, optical connecting links may be essential for high bit rates (e.g., $\gtrsim$ 1 Gbit/s) even if electronic switch modules could provide the switching function [57]. In this case, optical switching modules could offer significant advantage. Alternatively, research on optoelectronic integrated circuits is being pursued in which optical receivers, electronic switch arrays, and optical transmitters are integrated on a chip of GaAs or InP. A hybrid demonstration with separate receiver, switch, and transmitter GaAs chips that provides a 4 × 4 space division switch for data rates as high as 600 Mbits/s has been reported [58].

In addition to the Ti : $LiNbO_3$ technology, electrooptic, directive-type switch crosspoints are being pursued using waveguides in GaAs and InP. No arrays have been reported to date, but significant technological advances, including low-loss waveguides, have been achieved in the last several years [59]. Compact (< 1 mm long) switches of the intersecting waveguide-type but based on total internal reflection via a very large (~ two orders of magnitude greater than the electrooptic effect) index change due to current injection have been demonstrated [60]. In addition to being compact, they are also polarization independent. Very recently, a 4 × 4 array based on this technology has been reported [61]. Another potential method to make very short, low-voltage crosspoints is to utilize the enhanced electrorefractive effect in multiple quantum well semiconductor materials [62].

While technological progress in Ti : $LiNbO_3$ and other material-based switch arrays must continue to be made, it is also important that potential system applications be explored. Because of the broad-band capability, initial application in the transmission environment—e.g., facility, protection, and route restoration—appears potentially attractive. Time division switching will require applications where synchronization is an inherent element of the system. Local switching of very broad-band lines, as for example in a computer environment, may be a potential candidate. The potential of optical switch arrays to simultaneously switch multiple wavelengths may be advantageous in multichannel coherent systems where electrical to optical conversion of each channel to allow electronic switching could be very costly. Electronic switching will, of course, continue to improve and is strongly entrenched. Therefore, switching applications that complement or enhance those that can be achieved electronically are important to identify.

Acknowledgment

The author gratefully acknowledges and thanks those who generously provided material for this review.

References

[1] See, for example, M. Sakaguchi and K. Kaede, *IEEE Communications*, vol. 25, pp. 27-32, 1987.
[2] See, for example, E. J. Bachus *et al.*, *Electron. Lett.*, vol. 22, pp. 1002-1003, 1986.
[3] See, for example, I. P. Kaminow, P. T. Iannone, J. Stone, and L. W. Stulz, *Electron. Lett.*, vol. 23, pp. 1102-1103, 1987.
[4] M. Ikeda, *CLEO '84*, Anaheim, CA, June 19-22, 1984, paper THJ4.
[5] H. Kogelnik and R. V. Schmidt, *IEEE J. Quantum. Electron.*, vol. QE-12, pp. 396-401, 1976.
[6] V. Ramaswamy, M. Divino, and R. K. Standley, *Appl. Phys. Lett.*, vol. 35, pp. 145-147, 1978.
[7] See, for example, M. Papuchon, A. Roy, and D. B. Ostrowsky, *Appl. Phys. Lett.*, vol. 31, pp. 266-267, 1977; also A. Neyes, *Electron. Lett.*, vol. 19, pp. 553-554, 1983.
[8] R. A. Forber and E. Macom, *IEEE J. Quantum Electron.*, vol. QE-22, pp. 911-919, 1986.
[9] J. Ctyroky, *J. Opt. Commun.*, vol. 1, pp. 139-143, 1986.
[10] R. V. Schmidt and I. P. Kaminow, *Appl. Phys. Lett.*, vol. 25, pp. 458-460, 1974.
[11] E. Voges and A. Neyer, *J. Lightwave Technol.*, vol. LT-5, pp. 1229-1238, 1987.
[12] S. K. Korotky and R. C. Alferness, *Integrated Optical Circuits and Components*, L. D. Hutcheson, Ed. New York: Marcel Dekker, 1987.
[13] R. C. Alferness, C. H. Joyner, L. L. Buhl, and S. K. Korotky, *IEEE J. Quantum Electron.*, vol. QE-20, pp. 301-303, 1983.
[14] E. Murphy, *IEEE J. Quantum Electron.*, to be published.
[15] K. Komatsu, M. Kondo, and Y. Ohta, *Electron. Lett.*, vol. 22, pp. 881-882, 1986.
[16] K. Komatsu, S. Yamazaki, M. Kondo, and Y. Ohta, in *Proc. IOOC'87*, Reno, NV, 1987, paper WK-5.
[17] G. T. Harvey, G. Astfak, A. Y. Feldblum, and B. Kassahun, *IEEE J. Quantum Electron.*, vol. QE-22, pp. 939-946, 1986.
[18] A. R. Beaumont, C. G. Atkins, and R. C. Booth, *Electron. Lett.*, vol. 22, pp. 1260-1261, 1986.
[19] S. Yamada and M. Minakata, *Japan. J. Appl. Phys.*, vol. 20, pp. 733-737, 1981.
[20] R. A. Becker, *Opt. Lett.*, vol. 10, pp. 417-419, 1985.
[21] A. K. Okamoto, T. Hasaka, and H. Itoh, *Opt. Lett.*, vol. 13, pp. 65-67, 1988.
[22] R. V. Schmidt and R. C. Alferness, *IEEE Trans. Circuits Syst.*, vol. CAS-26, pp. 1099-1108, 1979.
[23] R. C. Alferness, *Appl. Phys. Lett.*, vol. 35, pp. 748-750, 1979.
[24] Y. Silberberg, D. Perlmutter, and J. E. Baran, *Appl. Phys. Lett.*, vol. 51, pp. 1230-1232, 1987.
[25] M. Kondo, Y. Tanisawa, Y. Ohta, T. Aoyama, and R. Ishikawa, *Electron. Lett.*, vol. 23, pp. 1167-1168, 1987.
[26] J. E. Watson, M. Milbrodt, and T. C. Rice, *J. Lightwave Technol.*, vol. LT-4, pp. 1714-1721, 1986.
[27] R. A. Spanke, *IEEE J. Quantum Electron.*, vol. QE-22, pp. 964-967, 1986.
[28] A. M. Hill, *J. Lightwave Technol.*, vol. LT-4, pp. 785-789, 1986.
[29] M. J. Wale, J. Duthie, C. J. Groves-Kirkby, and I. Bennion, in *Proc. CLEO '87*, paper WQ5 21; E. Marom, *Opt. Comm.*, vol. 58, pp. 92-94, 1986.
[30] K. Padmanabhan and A. Netravali, in *Proc. Top. Meet. Photon. Switching*, Incline Village, NV, Mar. 18-20, 1987, paper ThB3.
[31] M. A. Milbrodt, J. J. Veselka, K. Bahadori, Y. C. Chen, G. A. Bogert, D. G. Colt, R. J. Holmes, J. R. Erickson, and W. A. Payne, in *Proc. IGWO '88*, Sante Fe, NM, Mar. 28-30, 1988, paper MF9.
[32] H. Okayama, A. Matoba, R. Shibuya, and T. Ishida, in *Proc. IGWO '88*, Sante Fe, NM, Mar. 28-30, 1988, paper MF10.
[33] M. Kondo, N. Takado, K. Komatsu, and Y. Ohta, in *Proc. IOOC/ECOC '85*, Venice, Italy, Sept. 1985, pp. 361-364.
[34] G. A. Bogert, *Electron. Lett.*, vol. 23, pp. 72-73, 1987.
[35] S. K. Korotky, E. A. J. Marcatili, J. J. Veselka, and R. W. Boswarth, *Appl. Phys. Lett.*, vol. 48, pp. 92-94, 1986.
[36] R. V. Schmidt and L. L. Buhl, *Electron. Lett.*, vol. 12, pp. 575-576, 1976.
[37] M. Kondo, Y. Ohta, M. Fujwara, and M. Sakaguchi, *IEEE J. Quantum Electron.*, vol. QE-18, pp. 1759-1765, 1982.
[38] L. McCaughan and G. A. Bogert, *Appl. Phys. Lett.*, vol. 47, pp. 348-350, 1985.
[39] G. A. Bogert, E. J. Murphy, and R. T. Ku, *J. Lightwave Technol.*, vol. LT-4, pp. 1542-1545, 1986.
[40] P. Granestrand, B. Stoltz, L. Thylen, K. Bergvall, W. Doldissen, H. Heinrich, and D. Hoffman, *Electron. Lett.*, vol. 22, pp. 816-818, 1986.
[41] S. Suzuki, M. Kondo, N. Nagashima, M. Mitsuhashi, K. Komatsu, and T. Miyakawa, in *Proc. IOOC/OFC '87*, Reno, NV, Jan. 19-22, 1987, paper WB4.
[42] I. Sawaki, T. Yamane, and H. Nakajma, in *Proc. Top. Meet. Photon. Switching*, Incline Village, NV, Mar. 18-20, 1987, paper PDP6.
[43] G. A. Bogert, in *Proc. Top. Meet. Photon. Switching*, Incline Village, NV, Mar. 18-20, 1987, paper ThD3.

[44] H. Nishimoto, S. Suzuki, and M. Kondo, in *Proc. Top. Meet. Integrated Guided-Wave Opt.*, Sante Fe, NM, Mar. 28-30, 1988, paper PD6.
[45] P. Granestrand, B. Lagerstrom, P. Svensson, L. Thylen, B. Stoltz, K. Bergvall, and H. Olofsson, in *Proc. Top. Meet. Photon. Switching*, paper PD3.
[46] M. Kondo *et al.*, in *Proc. IOOC '83*, Tokyo, Japan, June 1983.
[47] S. Suzuki, T. Terakado, K. Komatsu, K. Nagashima, A. Suzuki, and M. Kondo, *J. Lightwave Technol.*, LT-4, pp. 894-899, 1986.
[48] R. A. Thompson and P. P. Giordano, in *Proc. Top. Meet. Integrated Guided Wave Opt.*, Atlanta, GA, Feb., 1986, paper TUB4.
[49] H. Haga, M. Izutsu, and T. Sueta, *J. Lightwave Technol.*, vol. LT-3, pp. 116-120, 1985.
[50] R. C. Alferness, L. L. Buhl, S. K. Korotky, and R. S. Tucker, in *Proc. Top. Meet. Photon. Switching*, Incline Village, NV, Mar. 18-20, 1987, paper ThD6.
[51] T. Swano, K. Komatsu, M. Kondo, and Y. Onta, in *Proc. IOOC '87*, Reno, NV, Jan. 1987, paper WB3.
[52] R. S. Tucker, G. Eisentein, S. K. Korotky, L. L. Buhl, J. J. Veselka, G. Raybon, B. L. Kasper, and R. C. Alferness, *Electron. Lett.*, vol. 23, pp. 1270-1271, 1987.
[53] P. A. Perrier and P. R. Prucnal, in *Proc. CLEO '88*, Anaheim, CA, Apr. 25-29, 1988, paper TuK2.
[54] J. E. Berthold and G. A. Hayward, in *Proc. OFC '88*, New Orleans, LA, Jan. 25-28, 1988, paper THI1.
[55] Y. Nakayayama, T. Ohtsuka, H. Shimizu, S. Yokogawa, K. Kameo, and H. Nishi, *IEEE J. Solid-State Circuits*, vol. SC-21, pp. 157-161, 1986.
[56] L. Thylen, P. Granestrand, and A. Djupsjobacka, in *Proc. Top. Meet. Photon. Switching*, Incline Village, NV, Mar. 18-20, 1987, paper PDP3.
[57] J. Goodman, F. J. Leonberger, S. Y. Kung, and R. A. Athale, *Proc. IEEE*, vol. 72, pp. 850-866, 1984.
[58] T. Igama, Y. Oikawa, K. Yamaguchi, T. Horimatsu, M. Makiuchi, and H. Hamaguchi, in *Proc. IOOC '87*, Jan. 19-22, 1987, paper WG3.
[59] See, for example, P. W. McIlroy, P. M. Rogers, J. S. Singh, P. C. Spurdens, and I. D. Henning, *Electron. Lett.*, vol. 23, pp. 702-703, 1987.
[60] K. Ishida, H. Nakamura, H. Matsumura, T. Kadoi, and H. Inoue, *Appl. Phys. Lett.*, vol. 50, pp. 141-142, 1987.
[61] H. Inoue, H. Nakamura, Y. Sasaki, T. Katsuyama, and N. Chinone, in *Proc. CLEO '88*, Anaheim, CA, Mar. 25-29, 1988, paper TuR1.
[62] J. Zucker, T. Hendrickson, and C. A. Burrus, *Appl. Phys. Lett.*, to be published.
[63] J. R. Erickson, R. A. Nordin, W. A. Payne, and M. T. Ratajack, *IEEE Communications*, vol. 25, pp. 55-60, 1987.
[64] J. R. Erickson *et al.*, in *Proc. Top. Meet. Photon. Switching*, Incline Village, NV, Mar. 1987, paper ThA5.

A Class of Directional-Coupler-Based Photonic Switching Networks

ANDRZEJ JAJSZCZYK, MEMBER IEEE

Abstract—**A class of photonic switching networks composed of directional couplers is proposed. These networks require only two switching stages containing active elements. Both point-to-point and multicast architectures are presented. Various characteristics of the proposed networks are compared to those of the networks previously known, including: insertion loss, SNR, number of crossovers, and number of active elements.**

I. Introduction

WIDE-BAND optical signals can be switched under electronic control using 2 × 2 directional couplers fabricated on titanium-diffused lithium niobate ($Ti:LiNbO_3$) [1]–[3]. Several architectures for larger capacity networks composed of such couplers have been proposed, among them: crossbar [4], planar [5]–[7], tree [8]–[10], Clos and Benes [11], as well as their dilated versions [12]. Some surveys of directional-coupler-based networks have been published [1]–[3], [11].

Various optimization approaches to photonic networks were focused on one or a combination of the following parameters: system attenuation, signal-to-noise ratio, number of switch elements, number of crossovers, control complexity, device length, number of substrates, modularity, nonblocking properties, broadcast requirements, and others [11].

In this paper we propose a class of photonic networks requiring very low number of switching stages containing active elements, i.e., directional couplers. In Section II, the basic point-to-point and multicast architectures are presented. In Section III, we compare the new architectures to networks previously known. Section IV introduces an approach to simplification of the network's structure.

II. Basic Architectures

The allowed connection patterns in a single directional coupler are presented in Fig. 1(a). Application of a photonic switch which enables establishing all possible point-to-point connections, as shown in Fig. 1(b), would allow to construct new efficient optical networks. The main difference between switches of (a) and (b) is that, in the latter switch, we can block the transmission of optical signals to its output or outputs even in the case such signals are applied to an input or the inputs. We can construct a simple optical switch, shown in Fig. 2,

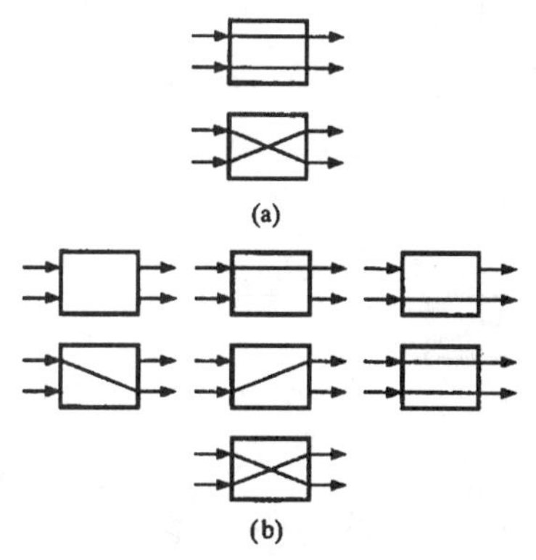

Fig. 1. Allowed connection patterns (a) in a directional coupler and (b) in an electronic switch.

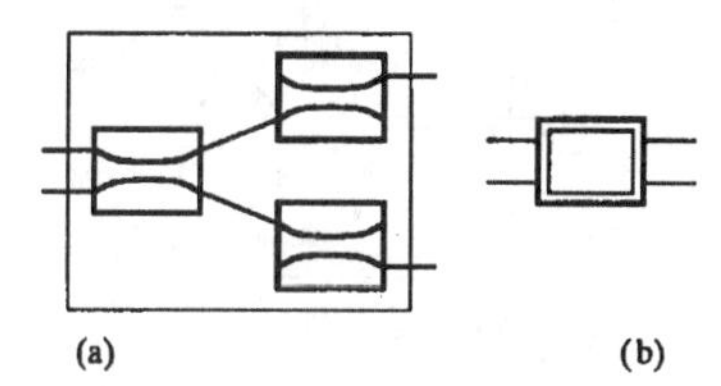

Fig. 2. A 2 × 2 seven-state optical switch; (a) structure, (b) symbol.

which can realize all states of Fig. 1(b) [13]. An alternative solution can be obtained by using a single directional coupler and two optical gates [14].

A basic structure of an $N \times N$ switching network employing elements of Fig. 2 is presented in Fig. 3. The network consists of k ($k = \lceil N/2 \rceil$, where the symbol $\lceil x \rceil$ denotes the smallest integer greater than or equal to x) active modules each containing k basic elements (of three directional couplers each), as well as $2k$ $1:k$ passive splitters and $2k$ $k:1$ passive combiners. An alternative, *nested*, structure is shown in Fig. 4. In this case, basic elements of Fig. 2 have been replaced by complete networks of Fig. 3, or other nested structures. Such an approach results in fewer crossovers between waveguides. It should be noted that nested networks also contain only two active stages, similar to their nonnested counterparts of Fig. 3. We can also note that for $N/k = 2$, Fig. 4 reduces to Fig. 3, i.e., to a nonnested network.

$1:k$ splitters and $k:1$ combiners can be implemented as tree structures composed of $1:2$ splitters and $2:1$ combiners,

Paper approved by the Editor for Optical Switching of the IEEE Communications Society. Manuscript received February 20, 1990; revised March 26, 1991.

The author is with the Franco-Polish school of New Information and Communication Technologies, ul. Mansfelda 4, 60-854 Poznań, Poland.

IEEE Log Number 92099472.

Reprinted with permission from *IEEE Transactions on Communications*, Vol. 41, No. 4, pp. 599-603, April 1993.

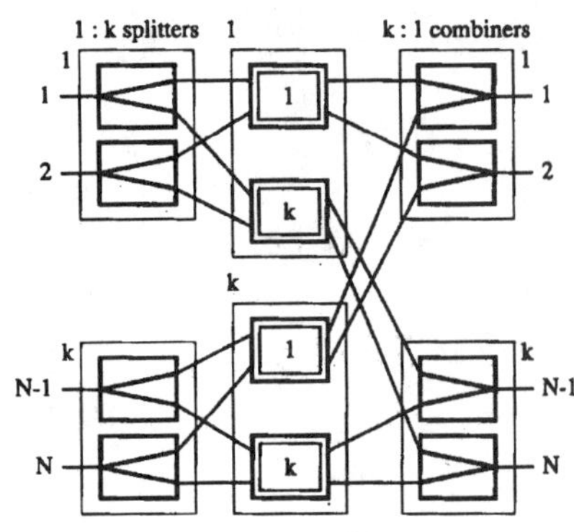

Fig. 3. An $N \times N$ switching network.

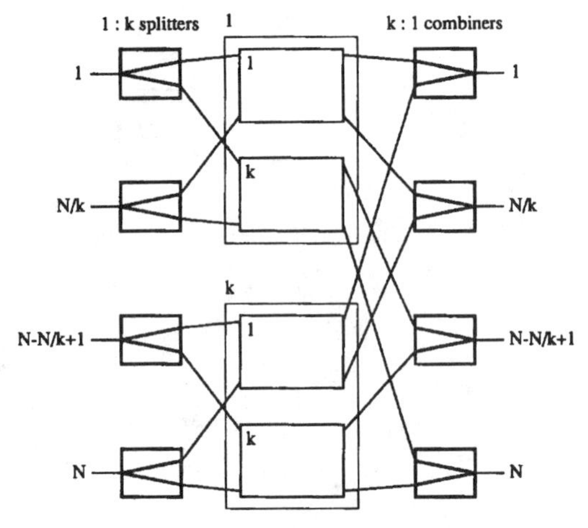

Fig. 4. An $N \times N$ nested switching network.

respectively. In such a case, the structures of Figs. 3 and 4 are isomorphic, i.e., there exists a label-preserving graph isomorphism between them. This isomorphism can be proved by arguments similar to those presented in [15]. However, the discussed photonic networks are not identical if the number of crossovers is taken into account.

The switch element of Fig. 2 makes it possible to construct point-to-point optical switching networks, i.e., networks in which one input goes to one and only one output channel. In many practical applications (e.g., video broadcasting), multicast capabilities are required. In networks having such capabilities a multiplicity of output channels can listen to the same input. A broadcast network is a limiting case in which the multicast size is equal to the number of output channels, i.e., every output can listen to any input, even if other output channels are listening to the same input.

The presented network structures can serve as multicast networks of any multicast size if we replace the switch elements of Fig. 2 by switch elements allowing broadcast states shown in Fig. 5, along with the states of Fig. 1(b). Such a switch element is shown in Fig. 6. We can note that it also contains only two active stages as the element of Fig. 2.

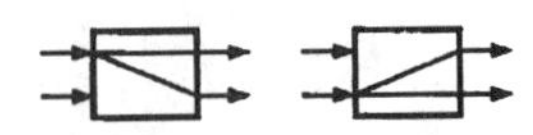

Fig. 5. Broadcast states.

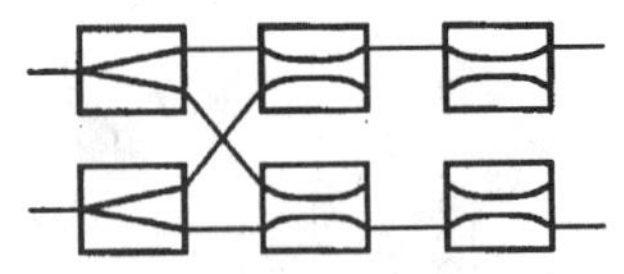

Fig. 6. A multicast 2 × 2 switch.

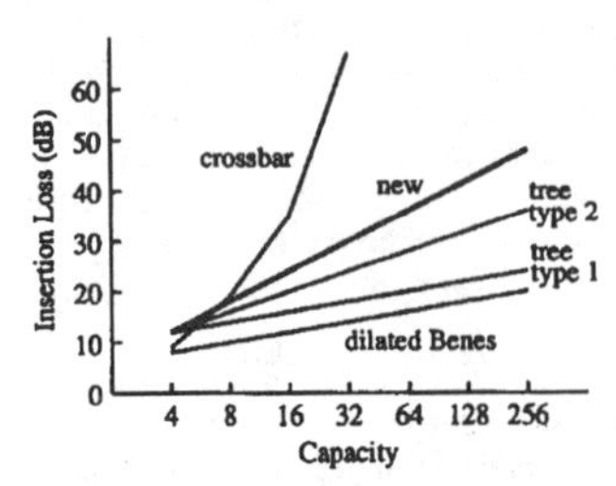

Fig. 7. The insertion loss.

III. Comparison to Other Networks

A. Attenuation Characteristics

In the proposed networks, the optical signal passes through $\log_2 N - 1$ passive splitters, the same number of passive combiners, as well as two directional couplers. The resultant insertion loss is given by

$$IL = 2(\log_2 N - 1)(3 + E) + 2L + 2W, \qquad (1)$$

where

E is the excess loss in each passive split,
L is the insertion loss in a directional coupler,
W is the waveguide-to-fiber coupling loss.

The 3 dB figure represents the 50/50 power split. Here, we assumed fiber devices for the passive components and multisubstrate architecture. If $LiNbO_3$ devices are used for the passive components, the network would have six waveguide/fiber transitions.

Fig. 7 presents the system insertion loss versus switch capacity N for various architectures. Similarly to Spanke in [11], we assumed: $L = 1$ dB, $W = 2$ dB, $E = 0$ dB. As we can see, the loss performance of the proposed network is inferior to that of dilated Benes or tree-structured networks, although it is better than the loss characteristics for the crossbar architecture. Moreover, in contrast to crossbar, the differential insertion loss is equal to zero.

B. Signal-to-Noise Ratio

We estimate the worst-case SNR, using an approach similar to that in [16]. The total power (dB) of a signal that arrives at a given output i is

$$P_{out}(i) = P_{in}(i) - IL, \qquad (2)$$

where $P_{in}(i)$ is the power entered into the input i in decibels, and IL is given by formula (1).

The noise that enters the output is the sum of the noise power that enters in the form of crosstalk. By examining Figs. 3 or 4, we can see that, in the worst case, the noise that enters the output i from input channel j can be calculated as follows:

$$P_n(i,j) = P_{in}(j) - X - 2(\log_2 N - 1)(3 + E) - 2L - 2W, \quad (3)$$

where X is the extinction ratio in decibels. The total noise in the output i is the sum of the noise power caused by $N/2$ channels (note the passive combiner at the output). Noise coming from the remaining $N/2-1$ input channels passes with the extinction ratio $2X$ and is not considered here. Therefore, taking into account that $P_n(i,j)$ are identical for all i and j, we obtain

$$P_n(total)[\text{Watts}] = (N/2) \cdot P_n(i,j)[\text{Watts}]. \quad (4)$$

Converting to decibels gives

$$P_n(total)[dB] = 10\log_{10}(N/2) + P_n(i,j)[dB]. \quad (5)$$

The worst-case SNR is

$$\text{SNR} = P_{out}(i) - P_n(total). \quad (6)$$

Substituting $P_{out}(i)$ and $P_n(total)$ by (2), (3), and (5), as well as assuming that $P_{in}(i) = P_{in}(j)$, we finally obtain

$$\text{SNR} = X - 10\log_{10}(N/2), \quad (7)$$

that is a better result than that for the crossbar architecture [11], [17].

The formula (7) gives SNR for the worst case. As in some other architectures [12], the signal-to-noise ratio highly depends on the traffic pattern inside the network.

C. Crossover Minimization

It has been shown that crossover minimization in directional-coupler-based networks has an important influence on their architecture [18]. Crossovers can cause crosstalk, signal loss, and increase the manufacturing complexity. We can minimize the number of crossovers in an optical network by shuffling its nodes, although both the starting and the resultant networks remain isomorphic in a graph-theoretical sense, i.e., they implement the same interconnection function.

If separate substrates are used for implementing splitters, combiners, and blocks of k (see Fig. 3) active switches each, no crossovers are required, which is similar to the tree-structured networks. We can note that since $k < N$, single substrate splitters and combiners can be used for networks of a greater capacity than the tree-structured optical switches in which $1:N$ and $N:1$ splitters and combiners, respectively, are used (assuming the same integration density in both cases). Now we discuss the problem of crossovers assuming only a single substrate for the whole network.

TABLE I
TOTAL NUMBER OF CROSSOVERS IN OPTIMUM NESTED NETWORKS AND NONNESTED NETWORKS

Capacity	Total Number of Crossovers	
	Nested Network	Nonnested Network
4	—	8
8	72	192
16	464	3584
32	2592	61440
64	13376	1015808
128	65664	16515072
256	311552	266338304

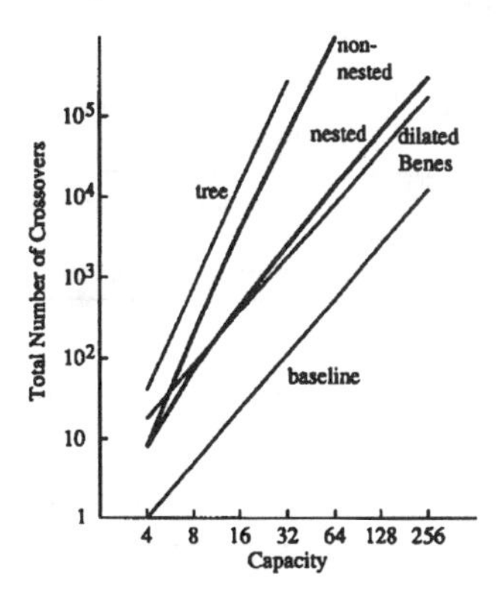

Fig. 8. The total number of crossovers.

It can be shown, as in [18], that the total number of crossovers in a point-to-point network of Fig. 4 is

$$C_{N,k} = kN(k-1)(N + N/k - 2)/4 + k^2 C_{N/k,k'}, \quad (8)$$

where k' is the parameter of the splitters and combiners in the network nested inside the network of capacity N, and $C_{2,k'} = 0$.

We can note that in the case of multicast networks, one additional crossover for one switch element should be added. Analyzing function (8), we can see that it reaches its minimum for $k = 2$, that is, the nested networks have fewer crossovers than networks of Fig. 3. Table I contains the optimum number of crossovers calculated from (8) by using a dynamic programming approach, as in [19]. For comparison, the number of crossovers in nonnested structures is also presented.

The total number of crossovers in proposed networks, as well as in some other representative photonic networks, are compared in Fig. 8. The number of crossovers both for nested ($k = 2$) and nonnested point-to-point networks are presented.

The total number of crossovers reflects the manufacturing complexity, while the worst case attenuation is related to the maximum number of crossovers between any inlet–outlet pair. It can be shown that the maximum number of crossovers between an inlet–outlet pair in a network of Fig. 4 is

$$M_{N,k} = (k-1)(N + N/k - 2) + M_{N/k,k'}, \quad (9)$$

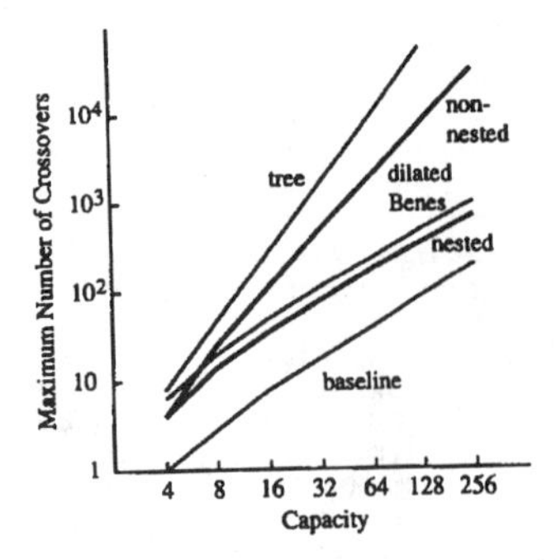

Fig. 9. The maximum number of crossovers between an inlet–outlet pair.

TABLE II
MAXIMUM NUMBER OF CROSSOVERS BETWEEN AN INLET–OUTLET PAIR

	Maximum Number of Crossovers	
Capacity	Nested Network	Nonnested Network
4	—	4
8	14	24
16	36	112
32	82	480
64	176	1984
128	366	8064
256	748	32512

where k' is the parameter of splitters and combiners in the network nested inside the network of capacity N, and $M_{2,k'} = 0$.

As in the case of the total number of crossovers, M_{N_k} is minimized for $k = 2$. Table II contains the optimum values of the maximum number of crossovers as well as such values for nonnested structures. The maximum number of crossovers for various networks is compared in Fig. 9.

D. Number of Active Switch Elements

The total number of active switch elements in a network of Fig. 3 and 4 is

$$T_N = a(N/2)^2, \tag{10}$$

where $a = 3$, for point-to-point networks, and $a = 4$, for multicast networks. Fig. 10 shows the number of active elements (directional couplers) in the proposed networks ($a = 3$), dilated Benes networks, as well as baseline networks. The values for tree-structured (type 2) networks are contained between the lines for crossbar and new networks.

E. Number of Active Stages

The most attractive feature of the networks of Figs. 3 and 4 is the fact that they contain only two active stages, i.e., signals between any inlet and any outlet cross only two active directional couplers. The number of active stages (in the worst case) in various photonic networks are compared in Fig. 11.

The proposed networks are nonblocking in the strict sense, even for multicast connections (if modules of Fig. 6 are used).

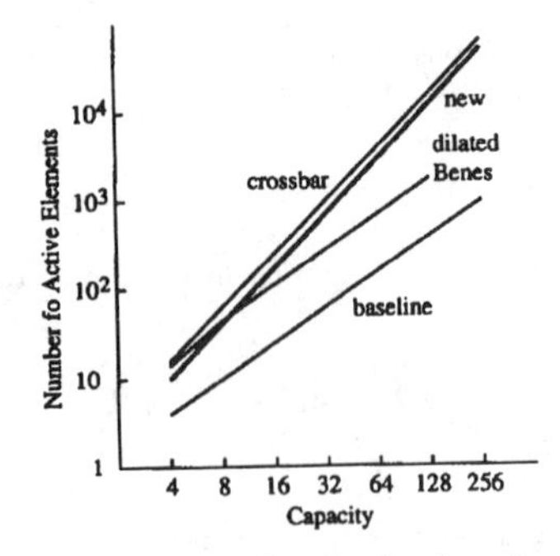

Fig. 10. The number of active elements.

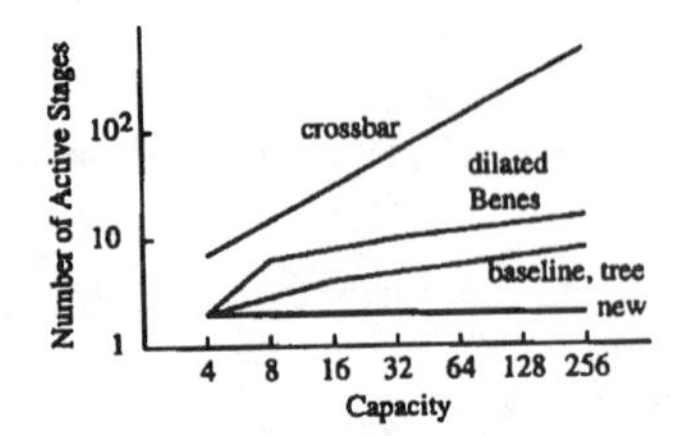

Fig. 11. The number of active stages.

The same feature has only the network proposed by Spanke (type 2) [10]. Because of only two active stages and the nonblocking properties, the networks of Figs. 3 and 4 are simple to control.

IV. SIMPLIFIED NETWORKS

In some cases, structures of the proposed networks may be simplified. The principle is illustrated by using the example of a 4 × 4 network shown in Fig. 12. Three optical elements (two directional couplers and one combiner) inside the area marked by the thick line can be replaced by a single directional coupler. If such a procedure is repeated for all elements in the last network, we obtain the network of Fig. 13. This network contains 8 directional couplers and no combiners, in contrast to 12 and 4, respectively, in the network of Fig. 12. The number of crossovers as well as the number of active stages remain unchanged. One of the differences between the basic network and its simplified counterpart is that the latter can realize fewer states. In the basic network, we can prevent propagation of signals to some or all network outputs even in the case optical signals are applied to all inputs. This is not possible in the case of the simplified network. However, we can reintroduce this feature by connecting an additional coupler to each output (or input) of a simplified network.

A more formal approach to the simplification of the networks presented in Section II is yet to be developed.

V. CONCLUSION

In this paper, a new architecture for a directional-coupler based switching network has been presented. The advantages of the proposed network can be summarized as follows:

- only two-active elements in an optical path,

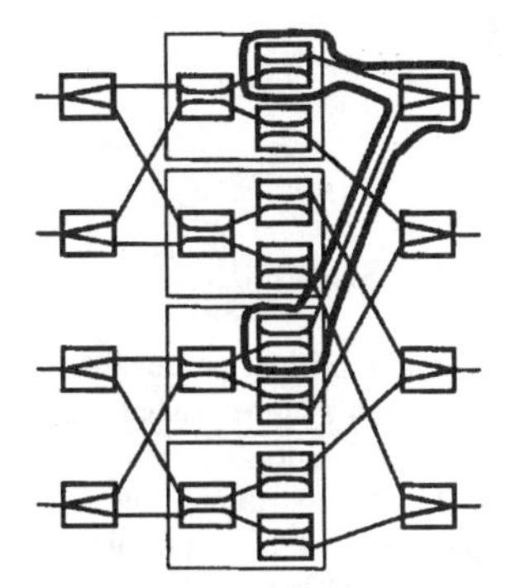

Fig. 12. Principle of simplification.

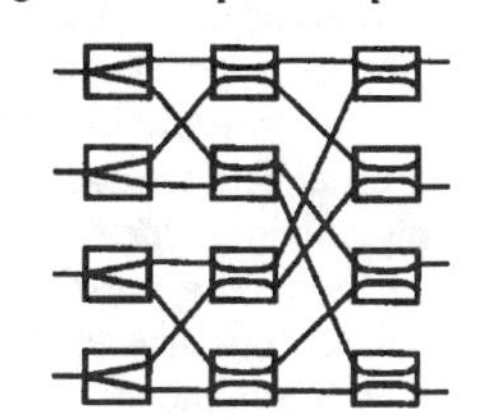

Fig. 13. A 4 × 4 simplified network.

- nonblocking in the strict sense,
- no crossovers, provided multiple substrate implementation,
- multicast capability.

The maximum capacity of a photonic network is limited mainly by the number of elements integrable on a single substrate, the maximum allowed attenuation, and the required signal-to-noise ratio. Although the required number of elements in the discussed network is relatively large, we can see that for large N the splitters/combiners become "tall" rather than "long," thus making more efficient use of the substrate [17]. As in tree-structured networks, the number of substrates, interconnection fibers, and fibers leaving a substrate edge are major limiting factors rather than the number of active and passive elements [11]. However, the number of such elements could be a problem in a single substrate implementation. If we assume the maximum attenuation allowed from network input to network output to be 30 dB, the network can grow to 32 × 32 (see Fig. 7). The most limiting factor is SNR. Since SNR characteristics of the proposed network are better than those for the crossbar architecture, we can expect maximum capacity to be at least 16 × 16 [17]. We can note, moreover, that in contrast to crossbar, the proposed network contains fewer directional couplers; and it also has lower insertion loss, and this loss is equal for all input–output pairs.

The maximum network capacity can be increased by replacing some stages of passive splitters or combiners with stages containing active elements. Therefore, we can control SNR. The limiting case is a tree-structured network with all active splitters and/or combiners. We have here some analogy to the partial dilation where a tradeoff between the amount of crosstalk and the number of switching elements is achieved [12]. The "simplification" presented in Section IV also leads to networks having more active switches and a better SNR.

References

[1] H. S. Hinton, "Photonic switching using directional couplers," *IEEE Commun. Mag.*, vol. 25, pp. 16–26, May 1987.

[2] A. Selvarajan and J. E. Midwinter, "Photonic switches and switch arrays on $LiNbO_3$, *Opt. Quantum Electron.*, vol. 21, pp. 1–15, 1989.

[3] L. Thylen, "Integrated optics in $LiNbO_3$: Recent developments in devices for telecommunications," *J. Lightwave Technol.*, vol. 6, pp. 847–861, June 1988.

[4] P. Granestrand *et al.*, "Strictly nonblocking 8 × 8 integrated optical switch matrix," *Electron. Lett.*, vol. 22, pp. 816–818, July 1986.

[5] E. Marom, "Integrated optics switch array network decomposition," *Opt. Commun.*, vol. 58, pp. 92–94, May, 1986.

[6] R. A. Spanke and V. E. Benes, "N-stage planar optical permutation network," *Appl. Opt.*, vol. 26, pp. 1226–1229, Apr. 1987.

[7] H. F. Taylor, "Optical-waveguide connecting networks," *Electron. Lett.*, vol. 10, pp. 41–43, Feb. 1974.

[8] K. Habara and K. Kikuchi, "Optical time-division space switches using tree-structured directional couplers," *Electron. Lett.*, vol. 21, pp. 631–632, July 1985.

[9] H. Okayama, A. Matoba, R. Shibuya, and T. Ishida, "Optical switch matrix with simplified $N \times N$ tree structure," *J. Lightwave Technol.*, vol. 7, pp. 1023–1028, July 1989.

[10] R. A. Spanke, "Architectures for large nonblocking optical space switches," *J. Quantum Electron.*, vol. QE-22, pp. 964–967, June 1986.

[11] R. A. Spanke, "Architectures for guided-wave optical space switching systems," *IEEE Commun. Mag.*, vol. 25, pp. 42–48, May 1987.

[12] K. Padmanabhan and A. N. Netravali, "Dilated networks for photonic switching," *IEEE Trans. Commun.*, vol. COM-35, pp. 1357–1365, Dec. 1987.

[13] A. Jajszczyk, "Application of single-stage electronic networks concepts to photonic switching," in *1990 Int. Top. Meet. Photonic Switching Tech. Dig.*, Kobe, Japan, Apr. 1990, pp. 77–79.

[14] A. Himeno, H. Terui, and M. Kobayashi, "Guided-wave optical gate matrix switch," *J. Lightwave Technol.*, vol. 6, pp. 30–35, Jan. 1988.

[15] A. Jajszczyk, "On combinatorial properties of broadband time-division switching networks," *Comput. Networks ISDN Syst.*, vol. 20, pp. 377–382, 1990.

[16] H. S. Hinton, "A nonblocking optical interconnection network using directional couplers," in *Proc. GLOBECOM '84*, Atlanta, GA, Nov. 1984, pp. 885–889.

[17] R. J. Reason, "Optical space switch architectures based upon lithium niobate crosspoints," *Brit. Telecom. Technol. J.*, vol. 7, no. 1, Jan. 1989.

[18] C.-T. Lea, "Crossover minimization in directional-coupler-based photonic switching systems," *IEEE Trans. Commun.*, vol. 36, pp. 355–363, Mar. 1988.

[19] A. Jajszczyk, "A dynamic programming approach to optimization of switching networks composed of digital switching matrices," *IEEE Trans. Commun.*, vol. COM-35, pp. 1342–1346, Dec. 1987.

Crosstalk in a Lossy Directional Coupler Switch

V.R. CHINNI, T.C. HUANG, P.-K.A. WAI,
C.R. MENYUK, SENIOR MEMBER, IEEE, AND G.J. SIMONIS

***Abstract*— Crosstalk due to material absorption in a two-waveguide, symmetric directional coupler switch is investigated. In a material with absorption, it is not possible to completely eliminate the crosstalk by adjusting the coupling length. The coupling length for minimum crosstalk differs from that of lossless systems. Theoretical limits of the lowest achievable crosstalk and the corresponding coupling lengths are calculated. The results show that the effect of absorption on crosstalk is more severe when the devices are designed for low crosstalk. The increase in crosstalk due to absorption can be as high as 20 dB. The material absorption is thus a critical parameter in designing low crosstalk devices.**

I. Introduction

Directional couplers are important guided-wave components in integrated optics. They have been used to implement switching, modulation, wavelength demultiplexing, and power splitting [1]. Crosstalk in a directional coupler is defined as the ratio of light power in the unwanted output port to the power in the desired output port. In a directional coupler, it is desirable to have all the input power coupled to the other waveguide for a cross state and no coupling at all for a bar state. However, a small amount of power always remains in the input waveguide for a cross state and couples to the other waveguide for a bar state, resulting in undesirable crosstalk. Crosstalk imposes limits on system design, especially in large switching fabrics in which the crosstalk adds. Many sources of crosstalk in directional couplers and their effects have been evaluated [2]–[10]. Crosstalk has been attributed to asymmetry of the waveguides [3], absorption loss [4], nonoptimal coupling length [1], unequal excitation of the symmetric and antisymmetric modes at the input [5], coupling of radiation modes to output, or fabrication variations [6]. Most of the analysis to date has considered lossless waveguide couplers. A notable exception is [4] in which Marcuse has shown that the losses in the surrounding medium are unsuitable to decrease the crosstalk between two dielectric waveguides as they also increase mode loss to unacceptably high levels. In this paper, we provide a detailed analysis of the effects of absorption on crosstalk in a symmetric directional coupler switch and show that absorption is a critical parameter in designing low crosstalk devices. Through a numerical example, we show that the increase in crosstalk due to absorption can be as high as 20 dB in practical directional coupler switches.

Manuscript received August 9, 1994; revised March 11, 1995. This work was supported by DOE.

V. R. Chinni was with the Department of Electrical Engineering, University of Maryland, Baltimore, MD 21228-5398 USA. He is now with AT&T Bell Laboratories, Holmdel, NJ 07733 USA.

C. R. Menyuk is with the Department of Electrical Engineering, University of Maryland, Baltimore, MD 21228-5398 USA.

P. K. A. Wai is with the Department of Electrical Engineering, University of Maryland, Baltimore, MD 21228-5398. He is also at Institute of Plasma Research, University of Maryland, College Park, MD 20742 USA.

T. C. Huang and G. J. Simonis are with the Army Research Laboratory, Adelphi, MD 20783 USA.

IEEE Log Number 9412207.

In reality, all materials exhibit finite absorption. It is generally assumed that crosstalk due to material absorption can be compensated by simply choosing the correct length for the coupling region. In this paper we show that material absorption in general leads to crosstalk in a directional coupler because the two lowest order modes, one even and one odd, have different power distributions. Ideally, these two modes beat with each other to produce a complete power transfer between the two waveguides, but, due to the different power distributions, the even and odd modes experience different amounts of absorption in a waveguide system with nonuniform absorption. Consequently, the two modes, even when excited equally, will have unequal amounts of power along the length of the coupler. So they will never be able to completely cancel each other, even when they are completely out of phase, leading to crosstalk that cannot be compensated. Consequently, switching is incomplete.

We use first order eigenmode theory to evaluate the crosstalk in a directional coupler switch. The formalism includes the effects of mode mismatch at the input and output of the coupling region. We calculate the effect of material absorption on the crosstalk and establish limits on the best achievable performance. As an example, we evaluate the crosstalk of a field-induced waveguide directional coupler switch [11]. We calculate the minimum achievable crosstalk with different waveguide spacings for different guide refractive indices. The effect of absorption on these devices is measured and the increase in the crosstalk is evaluated. Initially we will consider devices where the radiation field plays a negligible role. In Section IV, we compare the calculated results with numerical simulations using a finite difference beam propagation method [12] that models the entire field, avoiding this approximation and show that the theoretical results agree with numerical simulations. The waveguides studied in this work exhibit modal losses which not only can influence the crosstalk exhibited by the device but also can significanlty perturb the modal fields. We have used effective index method and a waveguide numerical analysis that is applicable for lossy waveguides to evaluate the mode fields and complex propagation constants [13].

II. Analysis

A typical two-waveguide symmetric directional coupler is shown in Fig. 1. The indices of refraction of the guiding and cladding regions are n_g and n_c respectively. The width of

Reprinted with permission from *IEEE Journal of Lightwave Technology*, Vol. 13, No. 7, pp. 1530-1535, July 1995.

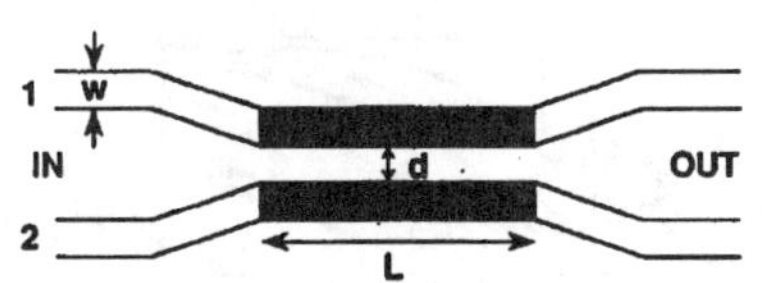

Fig. 1. Schematic diagram of a two-guide symmetric directional coupler. Width of the waveguides is w. Their edge-to-edge separation is d. Length of the coupling region is l.

the waveguide is w and their edge-to-edge separation is d. The transition regions are symmetric and adiabatic. We focus our analysis upon a directional coupler with only two guided modes in the coupling region. Accordingly, we assume that the uncoupled waveguides can only support one mode. The eigenmodes of the coupling region corresponding to the lowest order even and odd modes are ψ_e and ψ_o and their respective propagation constants are β_e and β_o. The sum of the even and odd modes gives the power distribution across the waveguides. The eigenmodes of the decoupled input waveguides are ϕ_1 and ϕ_2. The coupling of ϕ_1 to ψ_e and ψ_o depends on the geometry of the transition region. Consider launching ϕ_1 at the input waveguide. The coupling between the input excitation and the modes in the coupling region can be expressed as

$$\begin{aligned}\phi_1 &= a_e\psi_e + a_o\psi_o \\ \phi_2 &= a_e\psi_e - a_o\psi_o\end{aligned} \tag{1}$$

where a_e and a_o are the coupling coefficients. In the simple case of a straight coupler, a_e and a_o are real. The effect of the input bend region is to induce a small phase difference between the excited even and odd modes [5] so that a_e and a_o are complex. This initial phase difference between the even and odd modes can be compensated by properly adjusting the coupling length [9]. In our analysis, we assume a_e and a_o to be purely real. Note that the ratio between a_o and a_e indicates the excitation asymmetry of even and odd modes in the coupling region.

At the end of the coupling region, the modes can be projected onto the output waveguide modes with appropriate phase factors due to the output bend region. Using this approach, the output mode amplitudes can be obtained by a series of matrix multiplications from the input mode amplitudes. If we assume that the output transition region is identical to the input transition region, then by reciprocity the output coupling coefficients can also be found from (1).

After a propagation distance l, the fields at the output waveguides ϕ_1', ϕ_2' are given by [5]

$$\begin{aligned}\phi_1' &= \left[\frac{a_o^2\exp(-i\beta_e l) + a_e^2\exp(-i\beta_o l)}{4a_e^2a_o^2}\right]\phi_1 \\ &\quad + \left[\frac{a_o^2\exp(-i\beta_e l) - a_e^2\exp(-i\beta_o l)}{4a_e^2a_o^2}\right]\phi_2, \\ \phi_2' &= \left[\frac{a_o^2\exp(-i\beta_e l) - a_e^2\exp(-i\beta_l)}{4a_e^2a_o^2}\right]\phi_1 \\ &\quad + \left[\frac{a_o^2\exp(-i\beta_e l) + a_e^2\exp(-i\beta_o l)}{4a_e^2a_o^2}\right]\phi_2.\end{aligned} \tag{2}$$

From (2), the crosstalk for a propagation distance of l with $\phi_1 = 1$ and $\phi_2 = 0$ is given by

$$C(l) = \left|\frac{a_o^2\exp(-i\beta_e l) + a_e^2\exp(-i\beta_o l)}{a_o^2\exp(-i\beta_e l) - a_e^2\exp(-i\beta_o l)}\right|^2. \tag{3}$$

We will use (2) and (3) and $dC(l)/dl = 0$ to calculate the minimum possible crosstalk and the corresponding coupling length. We then calculate the best possible power transfer efficiency, and the coupling length required. In our calculations, we treat power transfer efficiency as the ratio between the output power coupled into the desired port to the power at the input.

In our analysis, we neglect the reflections at the beginning of the directional coupler and at the end of the directional coupler. We also do not consider coupling into radiation modes and the interaction between the radiation and the output modes. Reflected power is very small and also it does not affect the crosstalk calculations [5]. The radiation modes excited at the interface between the input transition region and the coupler can couple into the output. For low radiation losses (less than 1%), the influence of radiation modes on crosstalk performance is negligible [9]. The radiation can affect the results obtained for very low crosstalk devices. We have carried out numerical simulations including the effects of radiation for a field-induced waveguide directional coupler switch that we will discuss later. These simulations show that the effect of differential absorption on crosstalk is significant even when radiation is important.

In a lossy material system the propagation constant of a mode is complex, with the imaginary part representing the absorption coefficient associated with the mode. In practical systems, the material absorption is nonuniform. For example, the nonuniformity may be due to different doping concentrations or due to different material regions in the structure. As a result, the absorption associated with each mode can be different; because each mode has a different power distribution, the even and odd modes of the coupled mode structure in a nonuniform absorption system will have different absorption coefficients. In an ideal case of no asymmetry in excitation, the two modes are equally excited at the input. In reality, the excitation is unequal, establishing a floor for the minimum achievable crosstalk. Above and beyond that, due to differential absorption of the modes, each mode will have a different amount of power throughout the length of the coupler. Therefore, even when the two modes are combined in phase in a bar state, or out of phase in a cross state, the switching is not complete. Consequently, the power cannot be completely switched to any of the output waveguides, adding more crosstalk to the device, and leading, as we shall show, to a significant increase in crosstalk due to differential absorption.

III. Results and Discussion

Assuming a straight transition region, the coupling length for minimum crosstalk is obtained by differentiating (3). The coupling length for lowest crosstalk is given by solving the transcendental equation

$$\frac{a_o^4}{a_e^4} = \exp(-2\Delta\beta_{\rm im}l)\left[\frac{\Delta\beta_{\rm im} - \Delta\beta_{\rm re}\ \tan(\Delta\beta_{\rm re}l)}{\Delta\beta_{\rm im} + \Delta\beta_{\rm re}\ \tan(\Delta\beta_{\rm re}l)}\right]. \quad (4)$$

The parameters $\Delta\beta_{\rm re}$ and $\Delta\beta_{\rm im}$ are the real and imaginary parts of $\Delta\beta$, where $\Delta\beta$ is defined as the difference between the even and odd mode propagation constants, i.e., $\Delta\beta = \beta_e - \beta_o$. In the absence of material absorption, the parameter $\Delta\beta_{\rm re}$ determines the coupling length in a directional coupler. In directional couplers with absorption, $\Delta\beta_{\rm re}$ and $\Delta\beta_{\rm im}$ together determine the coupling length. The parameter $\Delta\beta_{\rm im}$ measures the difference in loss experienced by even and odd modes, indicating the effect of nonuniform material absorption. Solving (4) for the coupling length and then using (3), we obtain the lower limit of the crosstalk due to material absorption and asymmetric excitation.

As can be seen from (3) and (4), the crosstalk performance depends strongly on the asymmetry parameter a_o/a_e along with the differential modal absorption $\Delta\beta_{\rm im}$. Fig. 2 shows a_o/a_e as a function of refractive index in the guiding region. We assumed the cladding index to be 3.400 and the wavelength of operation to be 1.15 μm. The waveguide width was chosen to be 2 μm to keep the waveguide single moded in the Δn range of 0.00–0.01. Graphs corresponding to a waveguide separation d of 1–5 μm are shown in the figure. We evaluated the coupling coefficients a_o and a_e by first numerically finding eigenmodes and complex propagation constants of the coupled waveguides by using multilayer waveguide analysis [13] and then by computing the overlap integrals for the projections. As this figure indicates, the asymmetric parameter is close to unity (unity corresponding to completely symmetric case) when the waveguides are strongly guiding, or when the waveguide separation is large. The corresponding crosstalk for this device is small, but, due to strong guiding, the coupling lengths are fairly large. When the system is weakly guiding, or when the waveguide separation is small, the asymmetry parameter is smaller than unity, and the corresponding crosstalk is high, but, the coupling length is short. The even and odd modes may be considered as combinations of the individual modes of the two separate waveguides, but, due to the presence of an extra waveguide, a small perturbation of the mode profiles occurs that leads to the difference between the coupled modes and the sum or difference of the individual waveguide modes. Consequently, the input guided mode can never be equally split into the odd and even modes of the coupled system and as can be seen from the Fig. 2, the parameter a_o/a_e is never equal to unity.

Using (3) and (4), we calculate the minimum possible crosstalk and the corresponding coupling lengths as a function of $\Delta\beta_{\rm im}$. The coupling length for minimum crosstalk is shown in Fig. 3(a) and the minimum achievable crosstalk is shown in Fig. 3(b) for different values of a_o/a_e. In most of the practical devices, the loss in the guiding region is smaller than the loss in the surrounding region to keep the total modal loss as small as possible. We considered only cases in which the loss in the guiding region is smaller than the loss in the surrounding cladding. The sign of $\Delta\beta_{\rm im}$ depends on the loss experienced by even and odd modes, which is a function of field distribution of these modes and nonuniformity in material absorption. In our analysis, as shown in Fig. 3, we only show results corresponding to positive $\Delta\beta_{\rm im}$ which corresponds to the odd mode experiencing more loss than the even mode. When $\Delta\beta_{\rm im}$ is negative, results can be obtained in a similar way by solving (3) and (4).

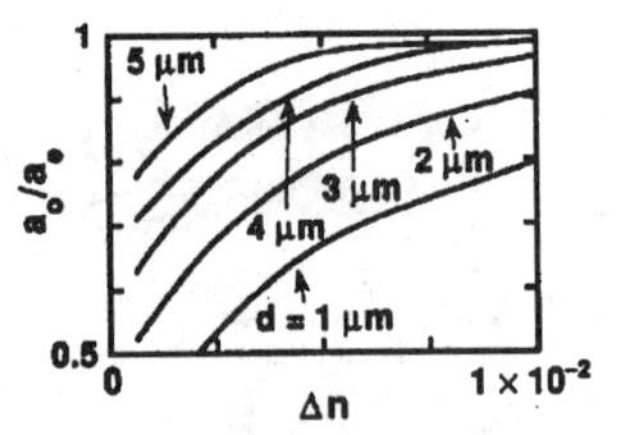

Fig. 2. Measure of asymmetry in excitation of odd and even modes as a function of difference in waveguide and cladding refractive index (Δn). Refractive index of the cladding is 3.400 and operating wavelength is 1.15 μm. Width of the waveguide w is 2.0 μm. Waveguide separation in the coupling region d is used as a parameter with values of 1–5 μm.

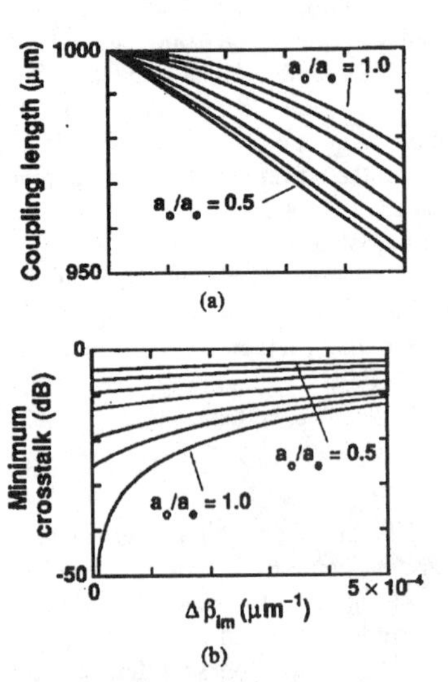

Fig. 3. (a) Coupling length corresponding to minimum crosstalk and (b) minimum crosstalk in a two-guide symmetric directional coupler switch. Differential absorption on even and odd modes cause crosstalk and also change the coupling length corresponding to lowest crosstalk. Asymmetry parameter a_o/a_e is used with values of 1.0, 0.95, 0.9, 0.8, 0.7, 0.6, and 0.5. Value of $\Delta\beta_{\rm re}$ is chosen to be $\pi/(1000\ \mu_*)$.

The minimum achievable crosstalk is a function of the ratio of the power coupled into even and odd modes at the input (a_o/a_e), the differential absorption in even and odd modes ($\Delta\beta_{\rm im}$), and the difference between the real propagation constants ($\Delta\beta_{\rm re}$). An increase in asymmetry, corresponding to lower values of a_o/a_e, increases crosstalk. An increasing differential absorption increases the crosstalk. As can be seen from the Fig. 3(b), the deterioration in crosstalk due to the differential absorption ranges from tens of dB to a few dB depending on the asymmetry of the excitation. If the directional coupler is designed to give low crosstalk, then the effect of asymmetry of excitation is minimal, and differential absorption solely determines the crosstalk. In effect, the absorption plays a crucial role in low crosstalk devices and sets a floor to minimum achievable crosstalk. As can be seen from Fig. 3(a), the coupling length in the cases with and without loss differ by about 50 μm in a coupler of length 1000 μm. The minimum crosstalk is neither at a coupling length of $l_c = \pi/|\Delta\beta|$ nor at $l_c = \pi/\beta_{re}$. Crosstalk due to absorption will be even higher if measured at the coupling length corresponding to the case without absorption; so, one has to optimize the coupling length to achieve the lowest crosstalk. From these two graphs, the designer can determine the minimum crosstalk achievable in a lossy directional coupler and find the exact coupling length to obtain that minimum crosstalk. The results of Fig. 3 present the fundamental limits on the crosstalk performance of a symmetric directional coupler with absorption.

In a device with absorption, the coupling length for maximum power transfer need not be same as the coupling length for minimum crosstalk. From (2), we solve for the coupling length corresponding to maximum power transfer to the output port. The coupling length is obtained by solving the transcendental equation

$$\frac{a_o^2}{a_e^2} = \frac{\Delta\beta_{re} a_o^2 \sin(\Delta\beta_{re} l) + \mathrm{Im}(\beta_o) a_e^2 \exp(-\Delta\beta_{im} l)}{\mathrm{Im}(\beta_e + \beta_o) a_e^2 \cos(\Delta\beta_{re} l) - \mathrm{Im}(\beta_e) a_o^2 \exp(\Delta\beta_{im} l)}, \quad (5)$$

where $\mathrm{Im}(x)$ indicates the imaginary part of x. From (4) and (5), we observe that the coupling length for maximum power in the output waveguide need not be the same as the coupling length for minimum crosstalk. When absorption is low, the difference in the coupling lengths is very small.

In Fig. 3, the graphs corresponding to $a_o/a_e = 1$ show the coupling length and minimum crosstalk when both the even and odd modes are excited equally at the input of the coupler region. The effect of absorption on crosstalk and deviation in coupling length is minimum when $a_o/a_e = 1$. In this case, the transcendental equation to solve for coupling length corresponding to minimum crosstalk reduces to

$$\Delta\beta_{re} \tan(\Delta\beta_{re} l) = -\Delta\beta_{im} \tanh(\Delta\beta_{im} l). \quad (6)$$

The coupling length can be shown to be smaller than $\pi/\Delta\beta_{re}$. When the even and odd modes are excited symmetrically, the coupling length for maximum power transfer efficiency is given by solving

$$\frac{\Delta\beta_{re} \sin(\Delta\beta_{re} l) + \mathrm{Im}(\beta_o) \exp(\Delta\beta_{im} l)}{\mathrm{Im}(\beta_e + \beta_o) \cos(\Delta\beta_{re} l) - \mathrm{Im}(\beta_e) \exp(\Delta\beta_{im} l)} = 1. \quad (7)$$

Assuming that Im β_e and Im β_o are small, and using a Taylor expansion of the arctangent, we find that the coupling length for maximum power transfer also has a value slightly smaller than $\pi/\Delta\beta_{re}$. But these coupling lengths for minimum crosstalk and maximum power transfer efficiency need not be equal as they are solutions of two different equations.

From (6), we find that if the differential absorption $\Delta\beta_{im}$ is small, the crosstalk due to absorption will be very small. For a structure with uniform loss and with a symmetric excitation, the crosstalk reduces to zero. But in reality, absorption is nonuniform, and the input excitation is never symmetric. The coupling length to obtain this idealistic zero crosstalk is not a function of the amount of absorption. The coupling length is π/β_{re}, just as in the lossless case. In this case, though the crosstalk can be zero and the corresponding coupling length is independent of absorption, the coupling length for maximum power transfer depends on the absorption and is different from that for zero crosstalk.

When losses are uniform throughout the device both longitudinally and laterally, the crosstalk as a function of asymmetry is shown as the intersection of the curves with crosstalk-axis in Fig. 3(b). The coupling length will remain unchanged for different asymmetries when the loss in the device is uniform. If the losses are uniform only laterally, but vary in the longitudinal direction, then the modes will evolve as the structure is changing in the propagation direction and a full propagation analysis is required which we did not pursue in this work. If the even and odd modes are not excited equally and if the losses are uniform or there is no loss, then the minimum possible crosstalk becomes

$$C = \left| \frac{a_e^2 - a_o^2}{a_e^2 + a_o^2} \right|^2 \quad (8)$$

as already shown in [5]. In the case where the losses are uniform throughout the device, the coupling length for maximum power transfer is given by

$$l_c = \frac{2}{\Delta\beta_{re}} \tan^{-1} \left[\frac{-\Delta\beta_{re}}{\mathrm{Im}(\beta_e + \beta_o)} \right]. \quad (9)$$

In the lossless case, the coupling length for the minimum crosstalk as well as for maximum power transfer efficiency reduces to $l_c = \pi/\Delta\beta$. Coupled mode analysis predicts zero minimum crosstalk for a lossless system with symmetric excitation, as in the case just discussed. Coupled mode analysis is applicable when the coupling between the waveguides is weak in which case the input pulse excites the even and odd modes equally. In this approximation, the normal mode and the coupled mode approaches should give identical results. However, in almost all practical situations, the odd mode spreads further outside the guiding region, and the overlap between input to the odd mode is smaller than the overlap between input to the even mode, thus contributing to the fundamental crosstalk limit. Ideally, when the waveguides are far apart and strongly guiding, the crosstalk is negligibly small.

IV. Application to a Field-Induced Directional Coupler Switch

As an example, we calculated the effects of absorption on crosstalk performance in a field-induced waveguide directional coupler switch. The coupler is designed to provide switching of the input power to the adjacent waveguide when cross voltage is applied and to remain in the bar state when no voltage is applied to the waveguide region. We investigated the crosstalk in the cross state. The field-induced waveguides are reported to have high absorption in the cladding region due to proton implantation [14]. A wavelength of 1.15 μm was used. The mode propagation constants were evaluated numerically from slab waveguide analysis [13] for the TE polarization. We assumed a straight coupler without bends. The coupling coefficients of the input to the even and odd modes

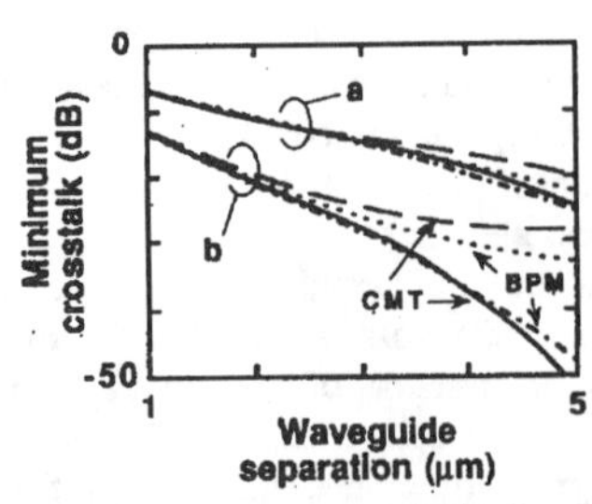

Fig. 4. The crosstalk performance of a lossy field-induced waveguide directional coupler switch. The absorption of the guiding and cladding regions are assumed to be 1 per cm and 5 per cm. The indices of refraction of the guide and cladding regions are (a) 3.405, 3.400 and (b) 3.410, 3.400 respectively. The wavelength of operation is 1.15 μm.. The waveguide width is 2 μm. The line curves and the dashed curves show the minimum crosstalk with and without absorption obtained using coupled mode theory (CMT) described in this paper. The dotted curves in (a) and (b) show the crosstalk in a system with absorption as calculated by the direct numerical simulations using the beam propagation method (BPM). Dash-dot curves in (a) and (b) indicate the crosstalk in a system without absorption as calculated by the numerical simulations.

of the coupling region are evaluated from overlap integrals. The index of the cladding region is 3.400 and the index of the guiding region is assumed to be 3.405 and 3.410 in two different cases. The absorption coefficients of the guiding and cladding regions are 1 cm^{-1} and 5 cm^{-1} respectively. We chose a waveguide width of 2 μm so that all waveguides support only the lowest order mode. The waveguide edge-to-edge separation is 1–5 μm. Shown in Fig. 4 is the minimum crosstalk achievable for each configuration and the minimum crosstalk possible when no losses are assumed. Curves in (a) indicate the results obtained with a waveguide index of $n_g =$ 3.405 and curves in (b) indicate the results obtained with waveguide index of $n_g = 3.410$. Continuous lines show the crosstalk performance without absorption, and the dashed lines show crosstalk performance with absorption. The crosstalk decreases exponentially as the waveguide separation increases. The crosstalk deterioration due to absorption becomes more severe as the crosstalk decreases as was calculated in the previous section. The crosstalk increase due to absorption is as high as 20 dB in a coupler system with a waveguide separation of 5 μm and a waveguide index of 3.410.

We simulated this directional coupler switch using a finite difference beam propagation method which gives a unified treatment of guided and radiation modes. We measured the minimum possible crosstalk for the same example used in Fig. 4, and the results are shown in Fig. 4 for comparison. Waveguide indices of (a) 3.405 and (b) 3.41 are considered. The dotted curves indicate the results obtained with absorption, and the dash-dot curves indicate the results obtained without absorption. The numerical results agree with theoretical calculations, indicating that the effect of absorption remains severe when radiation is included in the calculations. The performance deterioration is more severe for low crosstalk switches. The presence of radiation can increase the crosstalk significantly [15]. The effects of radiation depend upon the radiation present at the input and the physical dimensions of the device. The BPM calculations included the effects due to the presence of radiation, while these effects are not included in the eigenmode analysis. As the amount of radiation present is small, it can be seen that the eigenmode analysis is appropriate for evaluating the crosstalk deterioration due to the presence of absorption in the material.

V. Conclusion

In conclusion, we have shown that a two-guide symmetric directional coupler has an inevitable crosstalk due to nonuniform losses in the waveguide and cladding regions. This paper presents the fundamental limits on crosstalk performance due to intrinsic material absorption. We used normal mode analysis and evaluated the effect of absorption on crosstalk. We showed that the crosstalk due to absorption cannot be compensated by adjusting the coupling length. The coupling length for minimum crosstalk is not same as the coupling length for maximum power transfer efficiency. The difference in the coupling lengths is not significant for low absorption materials, but the increase in crosstalk due to absorption can be very large. We also calculated the minimum achievable crosstalk. The larger the nonuniformity, the larger the crosstalk. The effect of absorption on crosstalk is more severe when the devices are designed to provide low crosstalk values. The deterioration in crosstalk can be as high as 20 dB. The analysis is in good agreement with numerical simulations. This crosstalk can be compensated in asymmetric couplers, which we have not considered here, by precision tuning the geometric parameters. Even in asymmetric couplers, the elimination of crosstalk due to absorption is only possible in either the bar state or the cross state but not in both.

Acknowledgment

Computational work was carried out at SDSC and NERSC. T. C. Huang, would like to acknowledge NRC for its support.

References

[1] R. C. Alferness, "Guided-wave devices for optical communication," *IEEE J. Quantum Electron.*, vol. QE-17, pp. 946–959, 1981.
[2] D. Li, "Elimination of crosstalk in directional coupler switches," *Opt. Quant. Electron.*, vol. 25, pp. 255–260, 1993.
[3] R. V. Schmidt and H. Kogelnik, "Electro-optically switched coupler with stepped $\Delta\beta$ reversal using Ti-diffused $LiNbO_3$ waveguides," *Appl. Phys. Lett.*, vol. 28, pp. 503–506, 1976.
[4] D. Marcuse, *Light Transmission Optics.* Princeton, NJ: Van Nostrand Reinhold, 1972.
[5] K. L. Chen and S. Wang, "Crosstalk problems in optical directional couplers," *Appl. Phys. Lett.*, vol. 44, pp. 166–168, 1984.
[6] L. McCaughan and K. D. Choquette, "Crosstalk in Ti:$LiNbO_3$ directional coupler switches caused by Ti concentration fluctuations," *IEEE J. Quantum Electron.*, vol. QE-22, pp. 947–951, 1986.
[7] L. O. Wilson and F. K. Reinhard, "Coupling of nearly degenerate modes in parallel asymmetric dielectric waveguides," *Bell Syst. Tech. J.*, vol. 53, pp. 717–739, 1974.
[8] J. P. Donnelly, H. A. Haus, and L. A. Molter, "Cross power and crosstalk in waveguide couplers," *J. Lightwave Technol.*, vol. 6, pp. 257–268, 1988.
[9] J. Weber, L. Thylen, and S. Wang, "Crosstalk and switching characteristics in directional couplers," *IEEE J. Quantum Electron.*, vol. 24, pp. 537–548, 1988.
[10] H. A. Haus and N. A. Whitaker, Jr., "Elimination of cross talk in optical directional couplers," *Appl. Phys. Lett.*, vol. 46, pp. 1–3, 1985.
[11] T. C. Huang, G. J. Simonis, and L. A. Coldren, "Directional coupler optical switch constructed from field-induced waveguides," *Electron. Lett.*, vol. 28, pp. 2288–2289, 1992.

[12] Y. Chung and N. Dagli, "An assessment of finite difference beam propagation method," *IEEE J. Quantum Electron.*, vol. 26, pp. 1335–1339, 1990.
[13] K. H. Schlereth and M. Tacke, "The complex propagation constant of multilayer waveguides: An algorithm for a personal computer," *IEEE J. Quantum Electron.*, vol. 26, pp. 627–630, 1990.
[14] T. C. Huang, Y. Chung, L. A. Coldren, and N. Dagli, "Field-induced waveguides and their application to modulators," *IEEE J. Quantum Electron.*, vol. 29, pp. 1131–1143, 1993.
[15] V. R. Chinni, T. C. Huang, P. K. A. Wai, C. R. Menyuk, and G. J. Simonis, "Performance of field-induced directional coupler switches," *IEEE J. Quantum Electron.*, communicated.

Recent Progress in Silica-Based Planar Lightwave Circuits on Silicon

M. KAWACHI

Indexing terms: Fibre to the home, Hybrid optical transceivers, Integrated optics, Optical waveguides

Abstract: Low-loss integrated silica waveguides on silicon allow the efficient interaction of guided-wave optical signals and are used to form a variety of planar lightwave circuits for optical branching, switching and filtering. Silica waveguides can also be used as platforms (motherboards) for hybrid optoelectronic (OE) integration. The paper reviews recent progress on silica waveguide technologies which are currently penetrating the passive component market for optical networks and providing innovative ways of packaging OE components.

1 Introduction

Optical fibres are now used as long distance and high bit rate transmission media in telecommunication networks. The next major thrust is into the subscriber networks, which connect individual offices and homes with optical fibres [1]. Another frontier to explore is the use of multiwavelength division multiplexing systems for constructing flexible and transparent networks with wavelength-routing capabilities [2]. For these to become a reality a new family of optical components is required. These include low-cost optical transceivers, connectors, amplifiers, WDM light sources and passive optical circuits such as splitters, switches and filters [3].

Once work had started on integrated optics the use of silica waveguides on silicon substrates was proposed as a way of realisation of passive optical circuits [4–6]. The silicon substrate is low in cost and available in large areas. The silica waveguides have the same refractive index as optical fibres, so a good match with fibers can be expected.

There are a number of R & D programs concentrating on silica waveguides at, for example, LETI in Grenoble, France, BNR Europe in Essex, UK, AT & T Bell Labs in Murray Hill, US, PIRI in Columbus, US, and NTT in Ibaraki, Japan. Integrated silica waveguides with well defined core shapes provide an efficient means of controlling both the amplitude and phase of optical signals, allowing production of a variety of planar lightwave circuits (PLCs), including optical splitters, thermooptic switches and various kinds of filter [5–8].

IEE Proceedings online no. 19960493

Paper first received 27th October 1995 and in revised form 11th March 1996

The author is with NTT Opto-electronics Laboratories, Tokai, Ibaraki 319-11, Japan

Another important role for silica waveguides, especially those on silicon substrates, is to provide platforms (motherboards) for hybrid optoelectronic (OE) integration. In this hybrid integration, the silicon substrate can be used as the heat sink and optical bench on which to mount OE chips, such as laser diodes and detectors, and couple them to the silica waveguide ends.

This paper reviews recent progress on integrated silica waveguides with an emphasis on NTTs achievements with PLCs and related hybrid integration technologies.

2 Silica waveguides on silicon

There are two major approaches to forming thick silica-glass layers on planar substrates: chemical vapour deposition (CVD) and flame hydrolysis deposition (FHD). At NTT we prefer to use FHD for the fabrication of SiO_2–GeO_2 waveguides on silicon substrates.

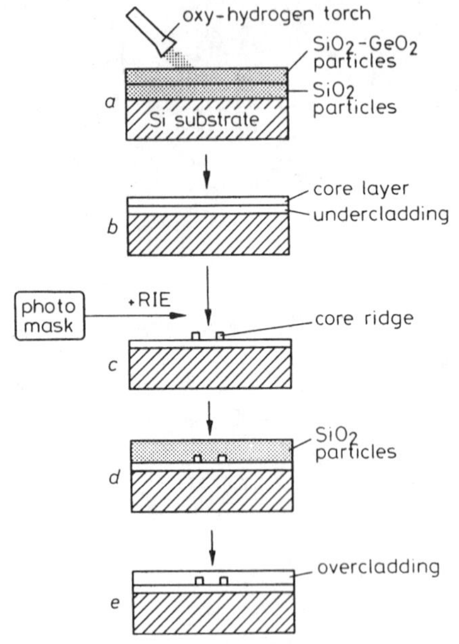

Fig. 1 *Silica waveguide fabrication process*
a FHD
b Consolidation
c Photolithography
d FHD
e Consolidation

Fig. 1 outlines the fabrication process for SiO_2–GeO_2 single-mode waveguides by the FHD method. The first step is to deposit two successive glass particle layers as

Reprinted with permission from *IEE Proceedings Optoelectronics,* M. Kawachi, "Circuits on silicon," Vol. 143, No. 5, pp. 257-262, October 1996. © 1996 by IEE.

the undercladding and core, on Si substrates by the flame hydrolysis of $SiCl_4$ and $GeCl_4$. Small amounts of PCl_3 and BCl_3 are added to lower the softening temperature of the synthesized glass particles. After deposition the Si wafers with their porous glass layers are heated to 1200~1350°C in an electric furnace for consolidation. The desired core ridges for the channel waveguides are then defined by photolithography with reactive ion etching (RIE). These ridges are finally covered with a thick overcladding layer, again with FHD. The total thickness of the glass layer on the silicon substrate is around 50μm. The FHD process, originally developed for optical fibre fabrication, has the advantage of being able to provide doped-silica films (up to ~100μm thick) at high deposition rates [5]. Moreover, glass filling is easy, even in waveguide gaps, although special attention must be paid when depositing fine glass particles.

Successful loss reduction in silica waveguides has forced us to use longer waveguides for precise loss measurements. The measured propagation loss of a 10 m-long FHD-fabricated SiO_2–GeO_2 waveguide (relative index difference $\Delta = 0.45$) with 186 bends (radius r = 15mm) on a five-inch silicon wafer was, for example, 17dB at λ = 1.3 and 1.55μm. Thus the propagation loss including bends was estimated as 0.017dB/cm [9]. The parameters of SiO_2–GeO_2 waveguides which are routinely fabricated in the NTT Labs are listed in Table 1. Low Δ waveguides are preferred for the fabrication of simple PLCs which require lower fibre-coupling losses. The higher Δ waveguides in Table 1 are used for the fabrication of larger-scale PLCs with complex waveguide bends.

Table 1: Parameters of silica waveguides on silicon

Waveguide type	low Δ	middle Δ	high Δ	super high Δ
Relative index difference Δ (%)	0.25	0.45	0.75	1.5 – 2
Core size, μm	8 × 8	7 × 7	6 × 6	4.5 × 4.5 – 3 × 3
Loss, dB/cm	< 0.1	< 0.1	<0.1	0.1
Fibre coupling loss*, dB/point	< 0.1	0.1	0.5	2
Minimum bending radius**, mm	25	15	5	2

* Coupled to conventional single-mode fibre (with index-matching oil)

** Bending loss in 90° arc waveguide is less than 0.1 dB at λ = 1.55 μm

The introduction of rare-earth ions like Er into integrated silica waveguides is attractive with regard to the fabrication of waveguide lasers, amplifiers and functional circuits. P_2O_5 co-doping in the core has been used to achieve a high Er concentration (~0.5wt%). A total gain of 24dB was thus attained with a pigtailed 35cm-long bent waveguide [10]. Further efforts are required to enable the local doping of Er ions along otherwise passive waveguides. It may also be interesting to study photoinduced grating formation in Er-doped or passive silica waveguides with a view to extending possible PLC functions [11]. The electro-optic effect in poled silica waveguides may also be attractive with a view to exploring novel PLC functions [12].

3 Evolution of silica-based PLCs

A variety of planar lightwave circuits have already been developed based on low-loss silica waveguides on silicon substrates [7, 8]. Well defined core geometry enables us to design complicated circuits by using various numerical simulation techniques. Device performance can be predicted precisely and the desired PLCs can be fabricated with high precision and reproducibility. These PLCs can be divided into four generations as shown in Fig. 2. The first generation includes simple Y-branches and directional couplers, which can be used as optical splitters and taps. The second generation includes Mach–Zehnder (MZ) interferometers and ring resonators, which have historically been designed by analogy with microwave or millimetre-wave filters and are finding many applications such as optical switches and filters. The third generation includes $N \times N$ star couplers and arrayed-waveguide grating (AWG) multiplexers, which are associated with radiative slab waveguides and are analogous with phased-array antennas. The fourth generation includes optical transversal filters and lattice filters, which can be programmed to perform different lightwave functions in an analogy with electronic digital filters.

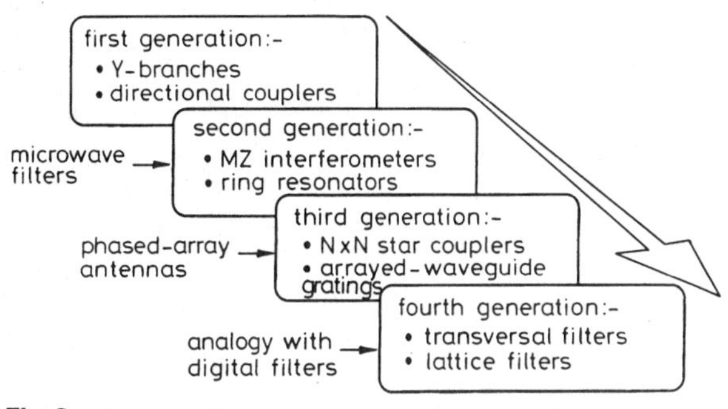

Fig.2 *Evolution of silica-based planar lightwave circuits*

In the following, recent progress on these PLCs is briefly described with particular emphasis on $1 \times N$ splitters with Y-branching units, MZ interferometers, AWG multiplexers, and programmable lattice filters, representing each of the four generations.

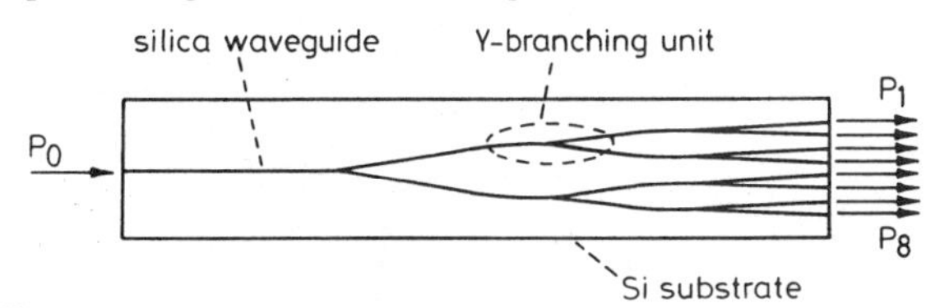

Fig.3 *Configuration of 1 × 8 optical splitter*
PLC chip as 1 × 8 splitter (top view)

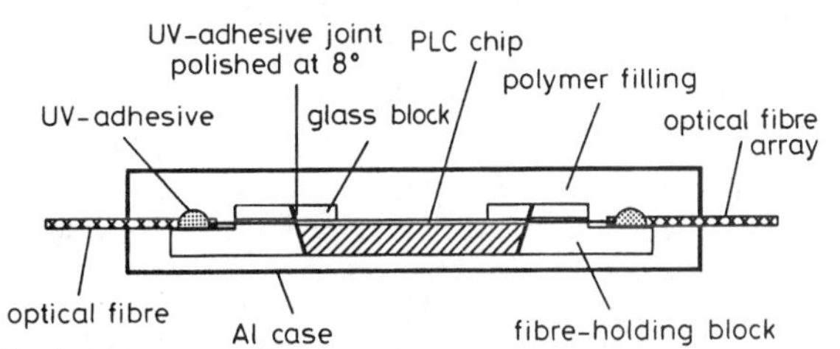

Fig.4 *Configuration of 1 × 8 optical splitter*
PLC splitter module (cross sectional view)

3.1 1 × N splitters with Y-branching units

Multiple Y-branching units can be integrated together to form $1 \times N$ optical splitters (N = 4, 8, 16, 32).

Figs. 3 and 4 show the configuration of a 1 × 8 optical splitter which consists of seven Y-branching units. The PLC chip is about 4 × 30mm in size with low-Δ silica waveguides. The chip is connected to input and output fibre arrays (fibre pitch = 250μm) using UV-curable adhesive [13]. The module size after packaging is 7 × 10 × 80mm. The fibre-to-fibre excess loss for these 1 × 8 splitter modules is typically 1dB. Low reflectivity (< –50dB) is achieved by polishing both the waveguide and fibre end faces at an angle (typically 8°). This kind of integrated 1 × N splitters, which meets the Bellcore specification for long-term durability, is now commercially available for practical use. The major market for PLC splitters is in the construction of passive double star (PDS) systems for optical access networks.

3.2 Mach–Zehnder interferometers

An MZ intefferometer which consists of two directional couplers and two waveguide arms integrated on a Si substrate can have various lightwave functions depending on the path difference ΔL between the two directional couplers. These functions include optical switching, wavelength-insensitive coupling, WDM and optical frequency-division multiplexing (FDM) as illustrated in Fig. 5 [3]. Thermo-optic (TO) phase shifters (thin-film heaters) can be deposited on the waveguide arms so as to provide passive silica waveguides with optical tuning or switching functions by utilising the TO coefficient of silica glass (1 × 10^{-5}/°C) [5]. In the optical FDM region, the MZ waveguide arm is loaded with a stress-applying film (typically an a-Si film) to balance the waveguide birefringence between the longer and shorter arms. This stress-applying film can be trimmed by YAG-or Ar-laser irradiation so that the phase difference between the TE and TM modes are adjusted to zero or $2N\pi$ where N is an integer [14]. A number of MZ interferometer switch or FDM units have been integrated together on a single substrate to form, for example, 128-channel optical frequency selective filters [15] and 8 × 8 optical matrix switches [16, 17].

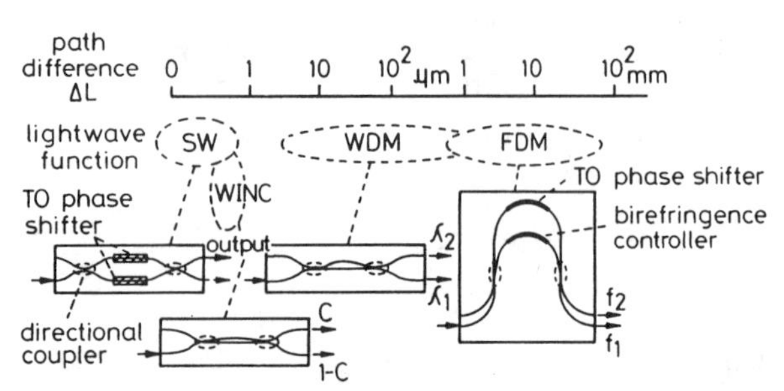

Fig.5 *Mach–Zehnder optical interferometer family*
SW: optical switching; WINC: wavelength-insensitive coupling; WDM: wavelength-division multiplexing; FDM: optical freequency-division multiplexing

An MZ interferometer with a very small and critical path difference of ΔL = 0.6μm can operate as a wavelength-insensitive coupler (WINC) used, for example, in tapping optical signals irrespective of wavelength [18]. The basic idea of this device is to introduce an out-of-phase factor between the two directional couplers to prevent the coupling ratio from increasing monotonically with wavelength. A coupling ratio of 20 ± 2% over a wide wavelength range of 1.25 – 1.65mm is, for example, achieved with a fibre-to-fibre excess loss of less than 0.5dB. 20%-WINC arrays with low-Δ waveguides are now being introduced in the field by NTT as an optical fibre transmission line testing system.

A 50%-WINC can be applied as the top 2 × 2 coupler in the 2 × N-type splitter configurations. A wavelength-insensitive switch (WINS) has also been proposed in which simple 50% directional couplers are replaced with 50%-WINCs in the MZ switch geometry [19]. This WINS can operate in the 1.3 and 1.55μm bands simultaneously.

3.3 Arrayed-waveguide grating multiplexers

Fig. 6 shows an AWG multiplexer configuration for dense WDM or optical FDM applications with a channel spacing of a few nm or less [20, 21]. This multiplexer consists of input/output waveguides, two focusing fan-shaped slab waveguide regions and a phased-array region of multiple-channel waveguides with an optical path difference ΔL between any two adjacent waveguides. The light beam from one of the input waveguides radiates in the fan-shaped slab waveguide and is then coupled into the arrayed channel waveguides. After travelling through the arrayed waveguides, the light converges on a focal line where the ends of the output waveguides are located. Multibeam lightwave interaction in the circuit can provide a dense WDM function similar to that of a classical bulk-optic grating, and hence the light converges on different positions on the focal line, depending on the wavelengths of the optical signals. Polarisation-insensitive AWG operation can be achieved by inserting a thin-film polyimide half-waveplate as a TE/TM mode converter in the middle of the arrayed-waveguide region [22].

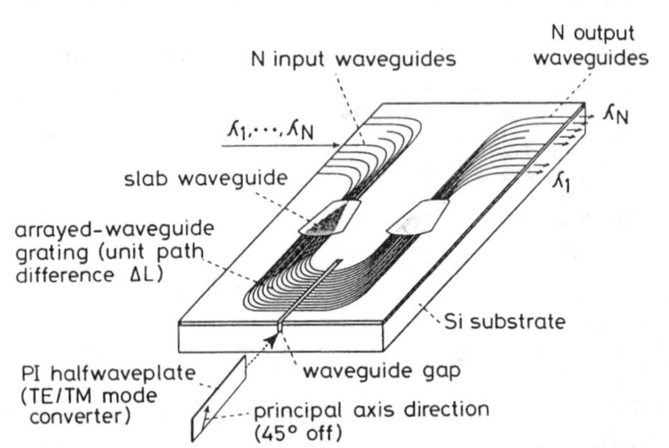

Fig.6 *Configuration of arrayed-waveguide grating multiplexer*

A series of AWG multiplexers, ranging from a 15-nm spaced 8-channel AWG to a 0.4-nm (50-GHz) spaced 64-channel AWG, have been successfully fabricated with fibre-to-fibre insertion losses of about 3~5dB and an interchannel crosstalk of about –30dB [23]. The number of the arrayed waveguides is 30 to 160 depending on the channel number N of the multiplexer. The path difference ΔL is 12.8μm for the 15-nm spaced 8-channel AWG and 63μm for the 0.4-nm spaced 64-channel AWG. The temperature sensitivity of these multiplexers is 0.015nm/°C, reflecting the TO coefficient of silica glass.

The 16-channel optical add/drop multiplexer (ADM) operation has already been demonstrated by integrating three 16 × 16 AWGs and 16 MZ switches on a single Si substrate [24].

The present AWG multiplexers and related circuits, when combined with tunable DFB lasers, are expected to be the key components in constructing wavelength-addressed networks or switch systems [25, 26]. Some AWG multiplexers are now commercially available from NEL (NTT Electronics Technology Corp.) in Ibaraki, Japan.

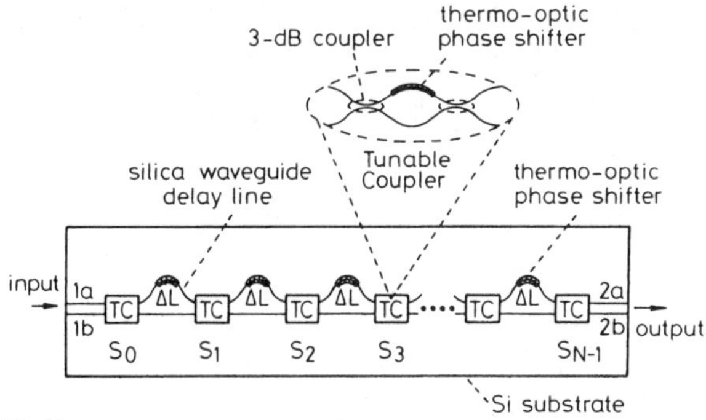

Fig. 7 *Logical arrangement of lattice filter*

3.4 Programmable lattice filters

The logical arrangement of an $(N-1)$-stage optical lattice filter as a programmable lightwave filter is shown in Fig. 7. With this arrangement, arbitrary frequency characteristics of

$$H(\omega) = \sum_{k=0}^{N-1} a_k \exp(-jk\omega\tau) \qquad (1)$$

can be produced by tuning the coefficient a_k, where τ is the unit delay time caused by an optical delay line with a path difference ΔL. Arbitrary coefficients a_k $(k = 0, \ldots N-1)$, including complex numbers, can be expressed by controlling the electric-field amplitude and carrier phase of optical signals using TO tunable couplers and TO phase shifters [27].

Eqn. 1 is related to the well known transfer function of finite impulse response (FIR) digital filters

$$H(z) = \sum_{k=0}^{N-1} a_k z^{-k} \qquad (2)$$

where z^{-1} $(= \exp - j\omega\tau)$ corresponds to the unit time delay in discrete digital systems. This relation can be used to synthesise useful PLC functions by analogy with conventional electronic digital signal processing, as recently reported by K. Jinguji [28]. Programmable PLC functions include matching the patterns and equalizing the dispersion of optical signals.

A six-stage $(N = 7)$ lattice filter, for example, has been fabricated by using high-Δ waveguides and applied to compensate for the chromatic dispersion of a 50km-long 1.3μm zero-dispersion fibre at 1.55μm [29]. This is the first stage in the development of programmable PLC filters as variable-dispersion synthesisers.

4 PLC platforms for hybrid OE integration

Early prototypes of hybrid integration have been already proposed. These include a 4 × 4 optical gate matrix switch and a 1.3/1.55μm WDM transceiver with silica-based PLCs [30, 31]. In these hybrid circuits, however, the silicon substrates could not be used as the heat sink for OE chips because the silicon surface was covered with a thick silica waveguide layer. To overcome this problem, a PLC platform structure with a terraced silicon substrate has recently been proposed, as illustrated in Fig. 8 [32]. The silica waveguide is formed only on the ground plane of the silicon substrate, and the terraced surface in Fig. 8 acts as an optical bench on which to mount OE chips such as laser diodes and gates and couple them to the waveguide ends. If required, high-speed electrical wiring can be formed on the platform for the multigigabit operation of OE chips [33].

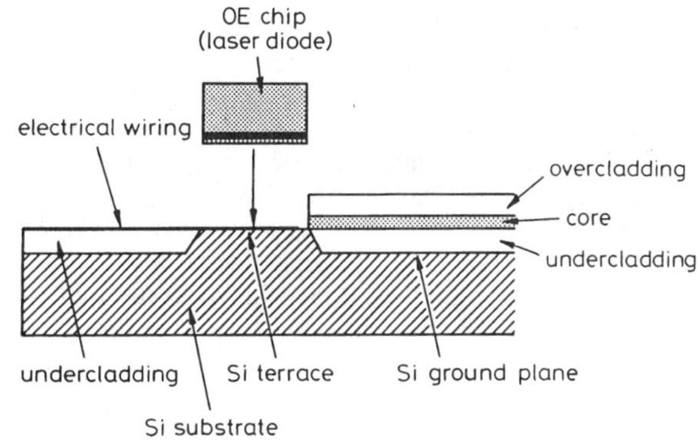

Fig. 8 *PLC platform for hybrid integration (cross sectional view)*

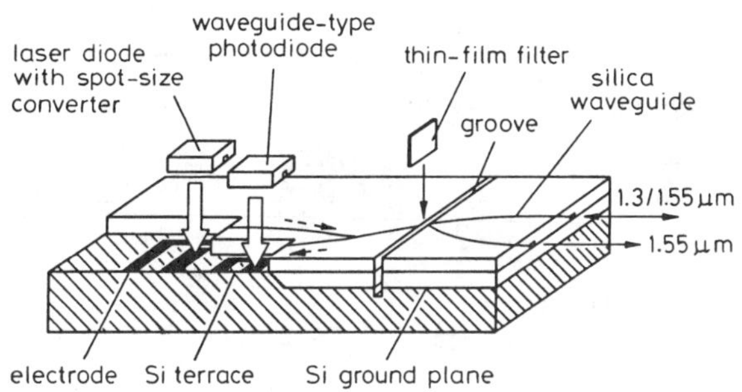

Fig. 9 *Hybrid integrated WDM transceiver on PLC platform*

OE chips which provide a good match with silica waveguides are eagerly awaited as they will enable coupling efficiency to be increased and positioning tolerances to be relaxed. Fig. 9 shows a hybrid integrated optical WDM transceiver incorporating a 1.3μm laser diode with an integrated spot-size converter and a waveguide photodiode which are flip-chip bonded on the silicon terrace for bidirectional TDM operation at 30-50 Mbit/s [34, 35]. The thin-film filter, which is inserted in the waveguide gap in Fig. 9, is to separate 1.55μm optical signals for video distribution. Two optical fibres, one for the common port and the other for the 1.55-μm output port, are connected to the waveguide end. A low coupling loss of 3dB between the expanded-mode laser and waveguide has been attained without the use of any additional lenses.

The present types of PLC platforms, perhaps combined with nonhermetic packaging, will allow practical innovations to be made in the massproduction of optoelectronic modules, including low-cost WDM transceivers for fibre-to-the-home (FTTH) [36].

5 Summary

Recent progress in silica waveguide technologies has been reviewed with particular emphasis on NTT's achievements with silica-based PLCs. There are two major roles for silica-based PLCs, as summarised in

Fig. 10. The first is to offer passive optical paths for networks with optical branching, switching and filtering functions. The time is now ripe for actual introduction of simple PLC splitters and couplers into optical access networks. PLC switches and filters will then open the door to the construction of more advanced optical networks including dense WDMs. The second role of PLCs is to offer platforms for hybrid integration, where laser diodes, photodetectors, and OEICs are integrated together with passive silica optical waveguides on silicon. Hybrid optical transceivers with the minimum number of components on the PLC platform will result in a drastic cost reduction that is required if FTTH is to become a reality. Thus with continuous improvements in optical circuit design, waveguide processing and device packaging, the present PLC and related hybrid integration technologies will penetrate the low-cost commodity markets for FTTH and progress to the higher level integration of optics and electronics for the next generation of telecommunication systems.

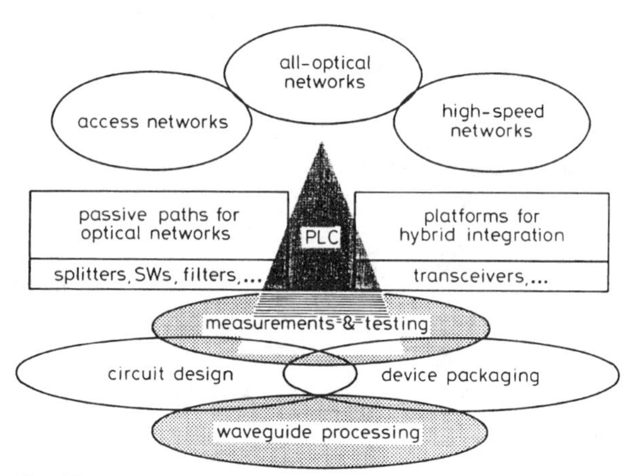

Fig. 10 *Future roles of silica-based PLCs*

6 Acknowledgments

A large number of people, including most of the members of the Photonic Component Laboratory at NTT Optoelectronics Laboratories, have been engaged in R & D activities on NTT PLC technologies. The author thanks all of them although it is impossible to cite individual names in this limited space. He also expresses gratitude to PLC users in the related NTT System Laboratories who bravely employed early PLCs in their system experiments and provided useful clues for further device improvements. The author is also grateful to the staffs of PIRI and NEL who are deeply involved with the commercial side of PLC manufacture.

7 References

1 MIKI, T.: 'Fiber-optic subscriber networks and system development', *IEICE Trans.*, 1991, **E74**, pp. 93–100

2 See, for example, the special issue on broad-band optical networks in *J. Lightwave Technol.*, May/June 1993, **11**

3 IKEGAMI, T., and KAWACHI, M.: 'Passive optical paths for networks', *Phys. World*, September 1991, pp. 50–54

4 See, for example, the review paper of HICKERNELL, F.S.: 'Optical waveguides on silicon', *Solid State Technol.*, Nov. 1988, pp. 83–87

5 KAWACHI, M.: 'Silica waveguides on silicon and their application to integrated-optic components', *Opt. Quant. Electron.*, 1990, **22**, pp. 391–416

6 GRANT, M.F.: 'Glass integrated optical devices on silicon for optical communication', *Critical Reviews of Science and Technology*, 1994, **CR53**, pp. 55–80

7 KAWACHI, M.: 'Special issue on planar lightwave circuit technologies', *NTT R & D*, 1994, **43**, pp. 1273–1318 (in Japanese)

8 KAWACHI, M.: 'Recent progress in planar lightwave circuits'. Paper ThB1–1, Presented at IOOC'95, 1995

9 HIDA, Y., HIBINO, Y., OKAZAKI, H., and OHMORI, Y.: '10 m long silica-based waveguide with a loss of 1.7 dB/m'. Paper IThC6, Presented at IPR'95, 1995

10 KITAGAWA, T., HATTORI, K., SHUTO, K., OGUMA, M., TEMMYO, J., SUZUKI, S., and HORIGUCHI, M.: 'Erbium-doped silica-based planar amplifier module pumped by laser diodes'. Paper ThC12.11, Presented at ECOC'93, 1993

11 KITAGAWA, T., BILODEAU, F., MALO, B., THERIAULT, S., ALBERT, J., JOHNSON, D.C., HILL, K.O., HATTORI, K., and HIBINO, Y.: 'Single-frequency Er-doped silica-based planar waveguide laser with integrated photo-induced Bragg reflectors', *Electron. Lett.*, 1994, **30**, pp. 1311–1312

12 ABE, M., KITAGAWA, T., HATTORI, K., HIMENO, A., and OHMORI, Y.: 'Electro-optic switch with a poled silica-based PLC on a Si substrate. To be presented at IPR'96, 1996

13 HIBINO, Y., HANAWA, F., NAKAGOME, H., ISHII, M., and TAKATO, N.: 'High reliability optical splitters composed of silica-based planar lightwave circuits', *J. Lightwave Technol.*, 1995, **13**, pp. 1728–1735

14 SUGITA, A., JINGUJI, K., TAKATO, N., and KAWACHI, M.: 'Laser-trimming adjustment of waveguide birefringence in optical FDM components', *IEEE J. Sel. Area Commun.*, 1990, **8**, pp. 1128–1131

15 TAKATO, N., SUGITA, A., ONOSE, K., OKAZAKI, H., OKUNO, M., KAWACHI, M., and ODA, K.: '128-channel polarisation-insensitive frequency-selection-switch using high-silica waveguides on Si', *IEEE Photon. Technol. Lett.*, 1990, **2**, pp. 441–443

16 KAWACHI, M., OKUNO, M., KATO, K., KATOH, K., OHMORI, Y., and HIMENO, A.: 'Silica-based optical matrix switch with intersecting Mach-Zehnder waveguides for larger fabrication tolrances'. Presented at OFC'93, Paper TuH4, 1993,

17 NAGASE, R., HIMENO, A., OKUNO, M., KATO, K., YUKIMATSU, Y., and KAWACHI ,M.: 'Silica-based 8 × 8 optical matrix switch module with hybrid integration driving circuits and its system application', *J. Lightwave Technol.*, 1994, **12**, pp. 1631–1639

18 JINGUJI, K., TAKATO, N., SUGITA, A., and KAWACHI, M.: 'Mach-Zehnder interferometer type optical waveguide coupler with wavelength-flattened coupling ratio', *Electron. Lett.*, 1990, **26**, pp. 1326–1327

19 KITOH, T., TAKATO, N., JINGUJI, K., YASU, M., and KAWACHI, M.: 'Novel broad-band optical switch using silica-based planar lightwave circuit', *IEEE Photon. Technol. Lett.*, 1992, **4**, pp. 735–737

20 TAKAHASHI, H., SUZUKI, S., KATO, K., and NISHI, I.: 'Arrayed-waveguide grating for wavelength division multi/demultiplexer with nanometer resolution', *Electron. Lett.*, 1990, **26**, pp. 87–88

21 DRAGONE, C., EDWARDS, C.A., and KISTLER, R.C.: 'Integrated N × N multiplexer on silicon', *IEEE Photon. Technol. Lett.*, **3**, pp. 896–899

22 INOUE, Y., OHMORI, Y., KAWACHI, M., ANDO, S., SAWADA, T., and TAKAHASHI, H.: 'Polarization mode converter with polyimode half waveplate in silica-based planar lightwave circuits', *IEEE Photon. Technol. Lett.*, 1994, **6**, pp. 626–628

23 OKAMOTO, K., MORIWAKI, K., and SUZUKI, S.: 'Fabrication of 64 × 64 arrayed-waveguide grating multiplexer on silicon', *Electron. Lett.*, 1995, **31**, pp. 184–185

24 OKAMOTO, K., TAKIGUCHI, K., and OHMORI, Y.: '16-channel optical add/drop multiplexer using silica-based arrayed-waveguide gratings', *Electron. Lett.*, 1995, **31**, pp. 723–724

25 TACHIKAWA, Y., INOUE, Y., KAWACHI, M., TAKAHASHI, H., and INOUE, K.: 'Arrayed-waveguide grating add-drop multiplexer with loop-back optical paths', *Electron. Lett.*, 1993, **29**, pp. 2133–2134

26 OKAMOTO, K.: 'Silica-based planar lightwave circuits for WDM systems'. Paper ThB1, Presented at OFC'95, 1995

27 KAWACHI, M., and JINGUJI, K.: ' Planar lightwave circuits for optical signal processing'. Paper FB7, Presented at OFC'94, 1994

28 JINGUJI, K., and KAWACHI, M.: 'Synthesis of coherent two-port lattice-form optical delay-line circuit', *J. Lightwave Technol.*, 1995, **13**, pp. 73–82

29 TAKIGUCHI, K., JINGUJI, K., and OHMORI, Y.: 'Variable group-delay dispersion equaliser based on a lattice-form programmable optical filter', *Electron. Lett.*, 1995, **31**, pp. 1240–1241

30 YAMADA, Y., TERUI, H., OHMORI, Y., YAMADA, M., HIMENO, A., and KOBAYASHI, M.: 'Hybrid-integrated 4 × 4 optical gate matrix switch using silica-based optical waveguides and LD array chips', *J. Lightwave Technol.*, 1992, **10**, pp. 383–389

31 TERUI, H., KOMINATO, T., YOSHINO, K., ICHIKAWA, F., HATA, S., SEKINE, S., KOBAYASHI, M., YOSHIDA, J., and OKADA, K.: 'Optical module with a silica-based planar lightwave circuit for fibre-optic subscriber systems', *IEEE Photon. Technol. Lett.*, 1992, **4**, pp. 660–662

32 YAMADA, Y., TAKAGI, A., OGAWA, I., KAWACHI, M., and KOBAYASHI, M.: 'Silica-based optical waveguide on terraced silicon substrate as hybrid integration platform', *Electron. Lett.*, 1993, **29**, pp. 444–445
33 MINO, S., YOSHINO, K., YAMADA, Y., YASU, M., and MORIWAKI, K.: 'Optoelectronic hybrid integrated laser diode module using planar lightwave circuit platform', *Electron. Lett.*, 1994, **30**, pp. 1888–1890
34 YAMADA, Y., SUZUKI, S., MORIWAKI, K., HIBINO, Y., TOHMORI, Y., AKATSU, Y., NAKASUGA, Y., HASHIMOTO, T., TERUI, H., YANAGISAWA, M., INOUE, Y., AKAHORI, Y., and NAGASE, R.: 'A hybrid integrated optical WDM transmitter/receiver for optical subscriber utilising planar lightwave circuit platform'. Paper PD12, Presented at OFC'95, 1995
35 SAKAI, Y., KUROSAKI, T., SATO, R., TOHMORI, Y., SUZUKI, S., MITOMI, O., and YOSHIDA, J.: 'Coupling characteristics of 1.3 μm laser diode integrated with spot-size converter to silica planar lightwave circuit'. Presented at CLEO/Pacific-Rim'95, Paper FU3, 1995
36 UCHIDA, N., HIBINO, Y., SUZUKI, Y., KUROSAKI, T., ISHIHARA, N., NAKAMURA, M., HASHIMOTO, T., AKAHORI, Y., INOUE, Y., MORIWAKI, K., YAMADA, Y., KATO, K., TOHMORI, Y., WADA, M., and SUGIE, T.: 'Passively aligned hybrid WDM module integrated with a sopt-size cinverted laser diode and waveguide photodiode on a PLC platform for fiber-to-the-home'. Paper PD15, Presented at OFC'96, 1996

Chapter 3

Optical-Amplifier-Based Switches

Optical amplifiers are widely used to compensate for losses either in transmission or in photonic switching arrangements that typically exhibit considerable losses. However, photonic switches can be directly constructed using optical amplifiers. For example, semiconductor optical amplifiers (SOA) and passive couplers can be directly used to realize switching circuits. Moreover, erbium-doped fibers have been used in a Mach-Zehnder arrangement to realize switching functions.

Lithium niobate switching structures are well established. Furthermore, directional couplers based on this technology are natural switches as outlined in Chapter 2. There are, however, several limitations to such switches. They are characterized in many cases by very high insertion losses and the crosstalk can be considerable between the switching ports. Additionally, their versatility is limited since the natural states for a directional-coupler-based switch are the bar and the crossed states. Therefore, disconnecting the inputs or having a broadcast arrangement are options that are not available in the simple directional coupler switch.

Optical amplifier-based switches offer several advantages compared to their lithium niobate counterparts. For example, structures can be realized with unity gain (i.e., zero losses). Furthermore, crosstalk figures as low as -60 dB can be achieved. These switches are, however, not ideal and can have the same problems associated with SOA, such as fiber coupling losses and temperature stability.

This chapter is concerned with photonic switches based on optical amplifiers. It starts with a paper by Evankow and Thompson (paper 3.1) that discusses switching modules based on laser diode amplifiers. Different switching modules are considered and attention is given to their blocking properties, number of amplifiers used, broadcast ability, fault tolerance, and total noise power. A nonblocking architecture is given for a 2×2 switching element comprising four SOAs and four directional couplers. The device and system aspects are analyzed. Subsequently attention is given to multistage 4×4 switching elements and a 2×3 element. The paper concludes with a summary that outlines the advantages and limitations of each of the nine structures considered.

A versatile device, semiconductor laser amplifier in a loop mirror (SLALOM), is introduced and evaluated by Eiselt et al. in paper 3.2. Various operating modes are reported for this device including single-pulse switching, and two-pulse switching together with different functionalities such as pulse shaping, decoding, retiming, and time division multiplexing. The paper outlines the device principle of operation, the basic switching configurations, and applications in photonic systems.

The treatment continues with a contribution by Kalman et al. (paper 3.3) in which they evaluate space division switches based on semiconductor optical amplifiers. In particular, they focus their attention on establishing the size limitations on SOA-based switching fabrics. They take into account limiting factors such as signal-to-noise ratio and saturation constraints. Their results indicate that the SOA saturation power limits the number of switches that can be cascaded. The implication is that an output saturation power of 100 mW, for example, limits the size of switching fabrics to 100 64×64 or 200 8×8 switches. The paper starts by considering the behavior of the SOA switch and analyzes factors such as the amplifier spontaneous emission, saturation, and signal-to-noise ratio constraints, then moves on to consider switch architectures. Subsequently the size limitations are established and summarized in the conclusions.

The last paper in this chapter focuses on the analysis of nonlinear optical switching in erbium-doped fiber. A model is given for the nonlinear phase shift and the associated signal loss attributed to absorption resonance in doped fiber. The model relates both the phase shift and the signal loss to the important system parameters such as dopant spectroscopic parameters, the fiber parameters, pump power, and the pump and signal wavelengths. The change in the doped fiber refractive index as a result of laser pumping can be used to produce a phase shift. Phase shifts as large as π were observed in doped fiber length of about 1 m with pump powers of just few milliwatts. The exploitation of such a phase shift in interferometric-based switches is explored and examples are given of switches based on the Mach-Zehnder configuration, and a further example is given of an experimental switch based on a two-mode fiber (TMF) interferometer. The latter is shown to be less susceptible to temperature and environmental changes than the Mach-Zehnder configuration.

Photonic Switching Modules Designed with Laser Diode Amplifiers

JOSEPH D. EVANKOW, JR. AND RICHARD A. THOMPSON, SENIOR MEMBER, IEEE

Abstract—**Photonic switching elements are designed from semiconductor optical amplifiers and passive couplers with fiber-to-fiber unity gain and low crosstalk. Designs for a 2 × 2 and an asymmetric 2 × 3 element, and several designs for 4 × 4 elements, are presented. While most amplifier analyses have stressed the importance of ultra-low facet reflectivities for high-gain operation, with protection against external reflections with optical isolators, modest facet reflectivities are satisfactory for these elements. It is also shown that substantial amounts of external reflection can be tolerated. The various architectures are compared according to amplifier count, blocking characteristic, broadcast potential, noise power (amplified spontaneous emission), and fault tolerance.**

I. Introduction

WHILE semiconductor optical amplifiers (SOA's) are effective linear amplifiers [1], [2], they can also function as switches [3], [4]. Although much of photonic switching has relied on lithium niobate technology, with optical amplifiers compensating for losses [5], optical switching elements can be made directly from SOA's and passive couplers [6], [7].

Described here is a series of 2 × 2, 2 × 3, and 4 × 4 unity-gain photonic switches made with fiber passive couplers and SOA's. These networks, establishing connectivity between any pair of input and output ports and not switching "by-the-bit," makes it possible to manipulate data at rates that exceed the modulation bandwidth of an individual amplifier. In the next section, device and systems considerations are used to compare the various elements. In the following sections, designs for one-stage 2 × 2 and 4 × 4 elements, two designs for multiple-stage 4 × 4 elements built from the 2 × 2 element, a 2 × 3 element, another multiple-stage 4 × 4 designed using 2 × 2 and 2 × 3 elements, and optimizations of these with some applications are presented.

A. Device Considerations

When SOA's are operated at high levels of gain (>20 dB), in the presence of even minute amounts of reflection from either the facets or any other dielectric discontinuity, periodic Fabry–Perot variations in the gain curve cause temperature instability. While most amplifier reflectivity analyses [8], [9] have stressed the importance of ultra-low facet reflectivities for high-gain operation, together with optical isolators to protect against system reflections, it is shown here that moderately high-facet reflectivities are satisfactory for the operation of each module. Although system reflections, regenerative feedback, adversely affect all SOA's, these cascaded modules distribute the gain over several SOA's, allowing low-gain operation for each amplifier and thus robustness in reflectivity for the overall module. Since the degree of Fabry–Perot variation, nonresonant gain ratio, at a given level of gain, depends on the product of the internal single-pass gain and the geometric mean of all associated reflectivities, distributed low-gain operation offers an advantage to complex cascaded structures. Since recent efforts in packaging [10]–[12] have shown that SOA's can function away from optical tables, cascaded configurations in field transmission systems is envisioned.

Manuscript received October 21, 1987; revised April 14, 1988.
The authors are with AT&T Bell Laboratories, Murray Hill, NJ 07974.
IEEE Log Number 8822557.

B. System Considerations

Photonic switching networks in lithium niobate have been built and described in the literature for sizes 4 × 4 [13]–[15] and even 8 × 8 [16]. Architectures capable of larger sizes have been proposed and analyzed [17], [18]. Two major drawbacks to lithium niobate implementations are high-insertion loss and high crosstalk (compared to even the worst electronic switches). There has been considerable research, worldwide, to produce better devices and to understand the limits of the technology. There has been a corresponding effort to invent architectural solutions, since network architectures have been designed for years around near-perfect electromechanical or solid-state devices. Clever architectural concepts have been designed around photonic device limitations yielding an overall network performance that exceeds the quality of the individual devices. An example is the proposed use of *dilation* to improve overall crosstalk in a network built from 2 × 2 switches [19]. Although the lithium niobate directional coupler is a natural 2 × 2 photonic switching element, it only operates in one of two states: *bar* and *crossed*. Broadcasting, in which one input is connected to BOTH outputs, is desirable but probably impractical. It would also be useful to disconnect inputs, when those inputs are temporarily connected to noise or other unwanted signals. Unless ports are deliberately unused, systems built from such elements tend to "conserve" noise because every input is steered to a destination.

Lithium niobate based technology is certainly well es-

Reprinted with permission from *IEEE Journal on Selected Areas in Communications,* Vol. 6, No. 7, pp. 1087-1095, August 1988.

tablished; however, the switching elements proposed here can easily achieve the following characteristics:

- zero net insertion loss (unity gain),
- crosstalk ≤ -60 dB, typical (dependent on cavity length).

The purpose of this paper is to convince the reader that these characteristics are not only possible, but they may border on being trivial. While temperature stabilization for the amplifiers is an unfavorable constraint, especially at higher nonresonant gain ratios, the control electronics and device drivers are similar to lithium niobate technology. During the discussion of each of the switching elements, the following system's issues will be considered:

- the number of amplifiers in an element,
- its *blocking* properties,
- its ability to broadcast or simply communicate point-to-point,
- its fault tolerance,
- and total noise power.

II. The 2 × 2 Element

The first photonic switching element, shown in Fig. 1, is a nonblocking 2 × 2. It is made from four discrete packaged SOA's and four fiber directional couplers splitting and combining the signal equally (50–50 percent, −3 dB). It is an implementation of a traditional *beta element* [20], enabling connectivity in both the *bar* and *crossed* states, similar to a directional coupler fabricated in lithium niobate. The upper input, for example, is connected to the lower output by enabling the top SOA in the figure.

A. Device Considerations

Shown in Fig. 1 are the major losses that each SOA must overcome for unity-gain operation of the module. A major question, from a device standpoint, is whether the gain needed to overcome all the associated losses (i.e., coupling into and out of the SOA, and splitter/combiner losses) requires unrealistically stringent tolerances for the fabrication of the AR coatings [21], [22]. For example, if a TE polarized reflectivity of $R \leq 10^{-3}$ were required for a 1.30 μm laser cavity, with typical dimensions, then the thickness (h) and the refractive index (n) of the film would have the following tolerances [21]: $\Delta h \simeq \pm 50$ Å, and $\Delta n/n \simeq \pm 3$ percent. While these values are obtainable, large scale production becomes difficult. The lower the reflectivity values, the tighter the tolerances on the film's parameters. Although film tolerances for the fabrication of the AR coatings is an important consideration, it will not be analyzed here. Instead, the focus will be on the required facet reflectivity and the susceptibility to external reflection of each architectural design. Since these architectural designs do not require ultra-low reflectivities, the film tolerances will, of course, be correspondingly more relaxed.

The equations relating the gain and reflectivity are

$$G(f) = \frac{KG_o(1 - R_1)(1 - R_2)}{(1 - G_oR)^2 + 4G_oR\sin^2(\pi f/FSR)} \tag{2.1}$$

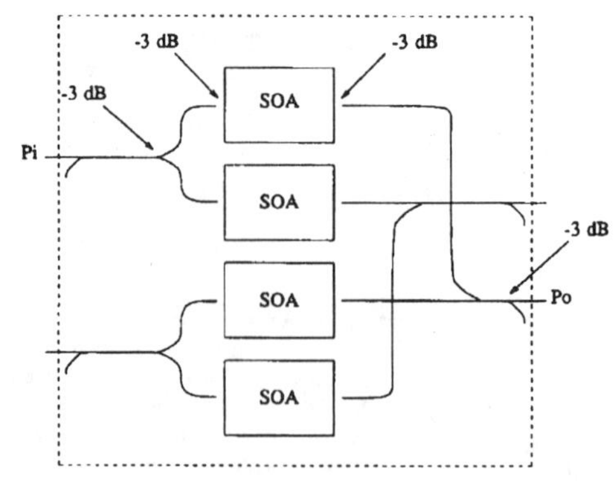

Fig. 1. 2 × 2 one-stage element with four SOA's.

where G_o is the single-pass gain ($\equiv e^{\Gamma(g-\alpha)L}$), $R_{1,2}$ are the facet reflectivities, R is the geometric mean of the facet reflectivities ($R^2 = R_1R_2$), FSR is the free-spectral range ($c/2nL$), L is the cavity length, K is the coupling coefficient, c is the speed of light in a vacuum, and Γ is the optical confinement factor. The ratio of the minimum to the maximum gain is defined as the nonresonant gain ratio [9] $G_{nr} \equiv G_{min}/G_{max}$. As this value approaches zero, the gain curve is characterized by the familiar high-finesse Fabry–Perot interference fringes, resulting from a high single-pass gain-reflectivity product, while a ratio approaching one is characteristic of an ideal traveling-wave amplifier (low finesse) arising from a "reflectionless" interaction inside the gain medium,

$$G_{nr-FPA} = 0 \qquad G_{nr-TWA} = 1. \tag{2.2}$$

The nonresonant gain ratio can be represented in terms of the single-pass gain and the geometric mean of the facet reflectivity as

$$G_{nr} = \frac{(1 - G_oR)^2}{(1 + G_oR)^2}. \tag{2.3}$$

The net peak value of the gain is

$$G_{p(\max)} = \frac{K(1 - G_{nr})(1 - R_1)(1 - R_2)}{4RG_{nr}}. \tag{2.4}$$

Rearranging (2.4) and assuming $R = R_1 = R_2$ yields

$$\frac{(1 - R)^2}{R} = \frac{4G_{p(\max)}G_{nr}}{K(1 - G_{nr})}. \tag{2.5}$$

With this last expression, it is possible to obtain the required facet reflectivities at each level of gain needed for unity-gain operation in an optical element.

Although many questions related to system requirements have not been answered, a $G_{nr} = -3.0$ dB is assumed. Since unity-gain for the circuit shown in Fig. 1 requires $G_{p(\max)} = +6$ dB and assuming a coupling coefficient of $K = -6.0$ dB, facet reflectivities of $R = 1.52$ percent cause a 3.0 dB variation in gain (G_{nr}) over the wavelength region near the center of the gain curve. While this resonant variation seems severe, the temperature can easily be stabilized over $\Delta T = \pm 0.5°C$. Earlier studies have shown [23] that net gains of $G_p = 6$ dB, with a resonant variation of slightly greater than $G_{nr} = -3$ dB, result in only a $\Delta G = 0.25$ dB over $\Delta T = \pm 0.75°C$.

This reflectivity value ($R = 1.52$ percent) is much greater than the reflectivities required for higher levels of gain. As the required gain increases, so does the need for low reflectivities. If, for example, the gain was $G_{p(\max)} =$ 30 dB, then the facet reflectivity would have to be $R =$ 0.025 percent, for a resonant variation of $G_{nr} = -3$ dB and perfect coupling ($K = 1$). For the same resonant gain variation ($G_{nr} = -3$ dB) with a realistic coupling coefficient ($K = -6$ dB), a net gain of $G_{p(\max)} = 30$ dB yields a required reflectivity of $R = 0.006$ percent. The previously calculated value of $R = 1.52$ percent enables higher yields in the fabrication of the AR coatings. In addition, for high levels of gain, external reflections from other amplifier facets, or even scattering in long lengths of fiber can cause the net reflectivity to quickly exceed the previously specified limits for the facets and thus significantly alter the nonresonant gain ratio. Optical isolators alleviate external reflection concerns but add to the overall cost and complexity. It will be shown that these cascaded architectural configurations, with amplifiers having nonzero facet reflectivities, alter but do not exceed previously specified system parameters. It is envisioned that optical isolators will need to be used only on the input/output ports to protect against reflections from long lengths of fiber and the lasers generating the signals.

As previously stated, an important concern is the effect of external reflection on the resonant behavior of this element. In other words, how much external reflection can each element withstand before the nonresonant gain ratio moves outside some prescribed limit?

Fig. 2 is a graph of the percentage change away from the original nonresonant gain ratio, for the optical element shown in Fig. 1, in the presence of external reflection at the input/output of the module. Since drastic changes in this ratio will cause adverse affects with the stable operation, small changes (< 10 percent) were monitored. If, for example, a nonresonant gain ratio of $G_{nr} = -5$ dB with an allowed change of $\Delta G_{nr} = 10$ percent are the system constraints, then an external reflectivity of $R_{ext} = 16.5$ percent can be tolerated.

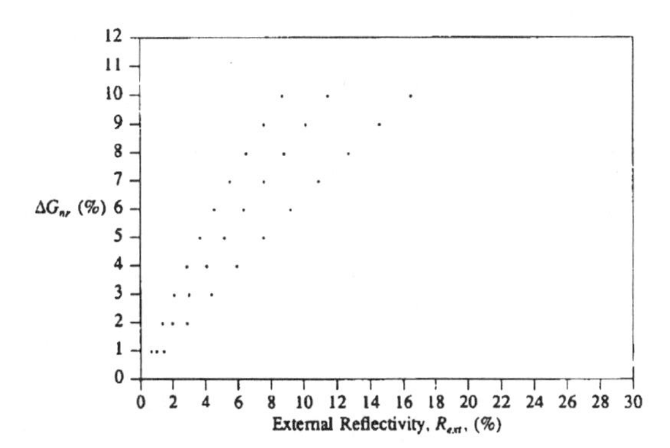

Fig. 2. Percentage change in nonresonant gain ratio for the 2 × 2 element versus external reflectivity for $G_{nr} = -1$ dB, $G_{nr} = -3$ dB, and $G_{nr} = -5$ dB, for each curve left to right, respectively.

B. System Considerations

Referring to Fig. 1, if one input is connected to either of the outputs, the remaining input can be connected to the remaining output. Thus, the architecture is completely nonblocking as is the corresponding implementation in lithium niobate.

If the top two SOA's in the figure are enabled, then data on the upper input fiber appear on both output fibers, at zero net loss. The architecture is seen to support broadcast, or one-to-two communication, as well as dual point-to-point communication. Equivalent operation by a single directional coupler in lithium niobate is difficult and probably impractical. An equivalent architecture could be implemented in lithium niobate by replacing each block in Fig. 1 labeled SOA by a simple on-off switch, but the corresponding loss and crosstalk characteristics can probably never be matched.

If the upper input in Fig. 1 must be connected to the lower output, the only path is via the top SOA. If it *fails*, then the connection is impossible. Thus, there is no alternate routing and the element is sensitive to faults. With no knowledge of the failure statistics of SOA's, it is difficult to conjecture the importance of this drawback, or to compare its reliability to other implementations. If the issue is critical, potential solutions include improving the reliability of an individual SOA, replicating each SOA, $n + k$ sparing the SOA's in an element, or embedding the element in a larger architecture that is tolerant of faults in its building blocks.

Since each amplifier produces spontaneous emission, successive amplification in cascaded architectures is of concern. The analysis [24], [25] for the total amplified spontaneous emission has assumed uniform coupling losses and that each successive SOA amplifies all of the previous amplifier's spontaneous emission (i.e., an ideal match between all gain curves). The noise power of each amplifier is

$$N = \frac{Bh\nu n_{sp}\chi}{\eta}(G_o - 1), \tag{2.6}$$

where

$$n_{sp} = \frac{N_e}{N_e - N_o}, \tag{2.7}$$

N_e is the carrier concentration, N_o is the carrier concentration for optical transparency, B is the filter bandwidth, $h\nu$ is the photon energy,

$$\eta = \frac{\Gamma g - \alpha}{\Gamma g}, \tag{2.8}$$

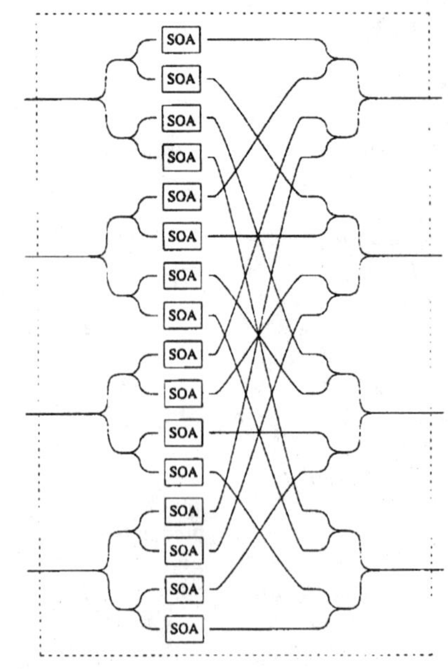

Fig. 3. 4 × 4 one-stage element with 16 SOA's.

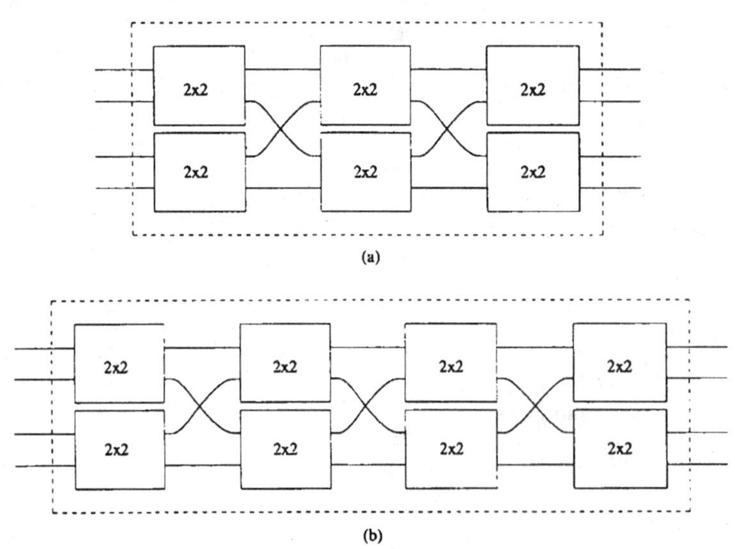

Fig. 4. 4 × 4 multistage elements with (a) 24 SOA's, and (b) 32 SOA's.

and

$$\chi = \frac{(1 + RG_o)(1 - R)}{(1 - RG_o)^2}. \tag{2.9}$$

The simplified results for each switching element are placed in Table I (see Section VIII)—multiple values indicate a route dependent intensity.

III. A One-Stage 4 × 4 Element

A 4 × 4 nonblocking optical element, shown in Fig. 3, is a generalization of the 2 × 2 element of Fig. 1. It uses a minimum number of active devices, namely 16. A similar element was described by Kobayashi *et al.* [6].

All fiber couplers are −3 dB splitters/combiners, and while the device count is minimized, it comes at the cost of a required decrease in facet reflectivity. Since there are two levels of couplers at the input and output, there is a total coupler loss of −12 dB. If the coupling coefficient is again assumed as $K = -6$ dB, then the overall chip gain must be $G = +18$ dB. This requires facet reflectivities of $R = 0.38$ percent ($G_{nr} = -3$ dB), a realistic value, but one that makes fabrication of the AR coatings much harder to obtain. This element is about as sensitive to external reflection, for $G_{nr} < -3$ dB and $\Delta G_{nr} < 5$ percent, as the 2×2 element shown in Fig. 1 because of the second tier of fiber couplers (see Section II-A). For $G_{nr} > -3$ dB and $\Delta G_{nr} > 5$ percent, this 4×4 is more tolerant of external reflections (i.e., +2 percent at $G_{nr} = -5$ dB, $\Delta G_{nr} = 10$ percent).

As in the 2×2 case, this 4×4 element is completely nonblocking, it supports both broadcast and point-to-point communication, and is not fault tolerant. Except for its low count of SOA's, this element does not compare well to the 4×4 elements to be described below. It has more stringent requirements on the reflectivity and the lack of alternate routes for fault tolerance. However, the reflectivity requirements are not severe and alternate routing may be obtained by using the element in a fault-tolerant super-architecture.

IV. Multistage 4 × 4 Elements

The 2×2 element in Fig. 1 can be used as a basic unit from which other elements can be built. Elements using six and eight such 2×2 units are shown in Fig. 4(a) and (b).

Since these two elements are built from the element in Fig. 1, the required reflectivity is similar, however, each stage affects the next, since it introduces reflections in the cascaded form that were not taken into account with the earlier calculation. While a calculated change will occur, the nonresonant gain ratio will change $\Delta G_{nr} = 4.0$ percent from its original value of $G_{nr} = -3$ dB.

The element in Fig. 4(a) will be shown to be rearrangeably nonblocking in Section VII-B. Adding a fourth stage to the element of Fig. 4(a) makes a new element, shown in Fig. 4(b) that is completely nonblocking. Since each component 2×2 unit supports broadcast, the entire element is seen to support broadcast, as well as point-to-point, communication.

To be rearrangeably nonblocking and fault tolerant, an architecture must provide multiple paths between any pair of endpoints. The failure of any single SOA does not make either element inoperable, not even for specific pairs of endpoints. For example, let the second SOA down in the upper left 2×2 unit in Fig. 4(a) fail. Then the only constraint on the 4×4 element is that the top input fiber cannot use a path through the network requiring the top fiber between the upper left unit and the upper center unit. The overall network can always be rearranged around this constraint. If, for example, the top two SOA's in the upper left unit both failed, then the upper input fiber is isolated from the rest of the network. Thus, both elements are seen to tolerate single faults, although not double faults, even though both are made from 2×2 units that are not fault tolerant.

V. Optimized Multistage 4 × 4 Elements

The next pair of elements are shown in Figs. 5 and 6. By using all output ports of the passive splitters, an optimum 4×4 can be made.

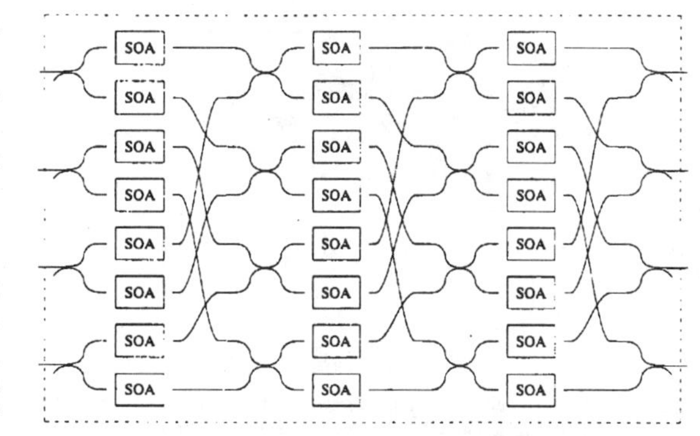

Fig. 5. Optimized 4 × 4 three-stage rearrangeably nonblocking element.

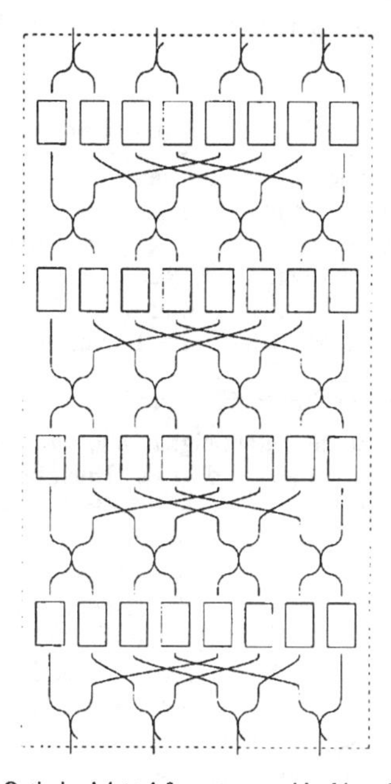

Fig. 6. Optimized 4 × 4 four-stage nonblocking element.

While the SOA count is the same as in the corresponding elements in Fig. 4, the total number of couplers is reduced. Starting with the three-stage configuration in Fig. 4(a), the output couplers on the left-hand units are effectively combined with the input couplers on the center units and the output couplers on the center units with the input couplers on the right-hand units. The result is the element in Fig. 5. The use of all four ports of these combined passive fiber couplers enables a more efficient use of the amplified signal, thereby reducing the constraints on the facet reflectivity.

Fig. 7 compares the robustness against changes in external reflectivity for the optimized switching elements in Figs. 5 and 6.

The blocking characteristic, support of broadcast, and fault tolerance for the elements in Figs. 5 and 6 are the

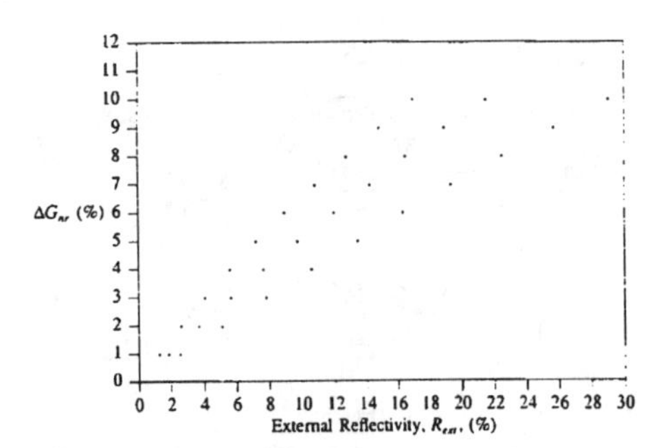

Fig. 7. Percentage change in the nonresonant gain ratio of the three- and four-stage optimized 4 × 4 switching elements versus external reflectivity for $G_{nr} = -1$ dB, $G_{nr} = -3$ dB, and $G_{nr} = -5$ dB, for each curve left to right, respectively.

same as for the corresponding elements in Fig. 4(a) and (b).

VI. A 2 × 3 Element

Another optical switching element, to be used as a basic building block for more complex elements, is shown in Fig. 8. The two stages of fiber couplers first split the input signal 1/3-2/3 with the larger signal proceeding to the second tier. If the signal is then split 50-50 percent, each of the three SOA's will receive 1/3 of the original signal.

A. Device and System Considerations

To combine the output signal onto one fiber, the output coupler adds another −3 dB loss. Thus, the total loss contributed by the couplers is −7.77 dB. If it is assumed that the coupling loss is $K = -6$ dB, the total loss that the SOA must overcome is about −14 dB. This translates into a facet reflectivity of $R = 1.02$ percent, if the nonresonant gain ratio is $G_{nr} = -3$ dB, as before. While this required facet reflectivity is lower than the 2 × 2 element shown in Fig. 1, it is still a reasonable value. Fig. 9 is a graph showing the susceptibility of a 2 × 3 element to changes in external reflection.

As with all the elements proposed in this paper, this element supports both broadcast and point-to-point communication. As with the other one-stage elements proposed in this paper, shown in Figs. 1 and 2, this element is completely nonblocking but is intolerant of even single faults (as a standalone element).

B. Nonblocking Clos Networks

Since such an asymmetric architecture is unlikely with directional couplers in lithium niobate, little architectural work has been done around such a unit. A three-stage 4 × 4 switching element, built with a stage of 2 × 2 elements (Fig. 1) in the center and stages of 2 × 3 elements (Fig. 8) on the edges is shown in Fig. 10. Like the other multistage elements, since the component units support broadcast, the entire element supports both broadcast and point-to-point communication. Like the other multistage elements, this element also tolerates single faults. However, it also tolerates failures in any two SOA's, although its blocking characteristic degrades to rearrangeably nonblocking. Unlike any other element proposed here, no double fault isolates any input or output fiber.

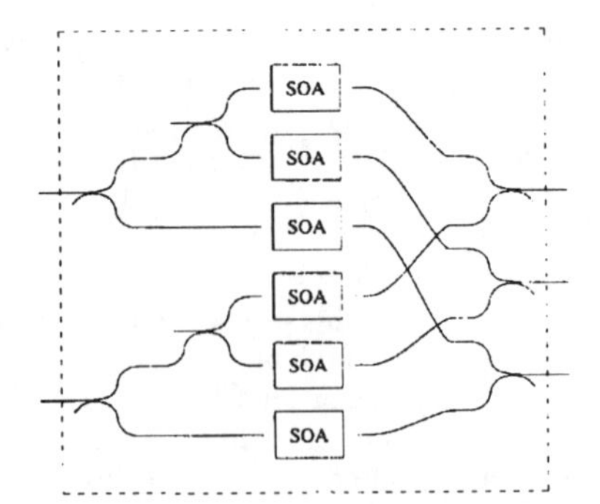

Fig. 8. 2 × 3 one-stage element with six SOA's.

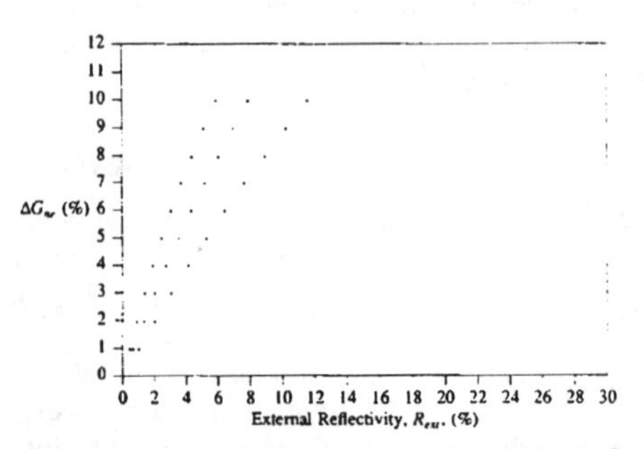

Fig. 9. Percentage change in the nonresonant gain ratio for the 2 × 3 element versus external reflectivity for $G_{nr} = -1$ dB, $G_{nr} = -3$ dB, and $G_{nr} = -5$ dB, for each curve left to right, respectively.

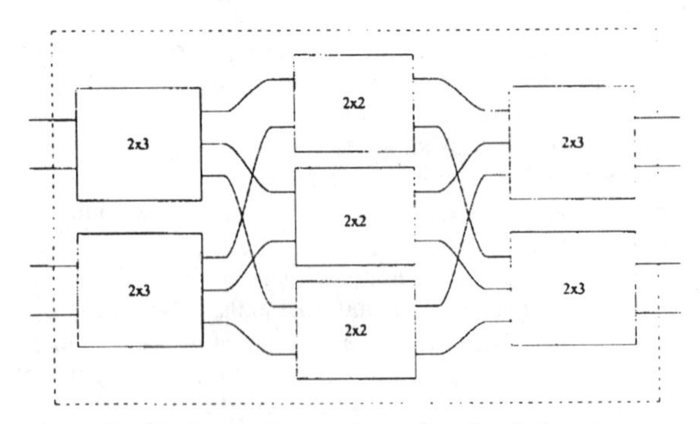

Fig. 10. 4 × 4 multistage element from 2 × 3 elements.

Clos described general symmetric three-stage networks [26]. Let the left and right stages be identical, with r distinct $n \times m$ switches and let the center stage have m distinct $r \times r$ switches, all appropriately interconnected. Thus, the overall network has rn inputs and rn outputs and each input-output pair has m different interconnecting paths, one through each distinct switch in the center stage. Two sufficient, but not necessary, conditions proven about Clos networks are

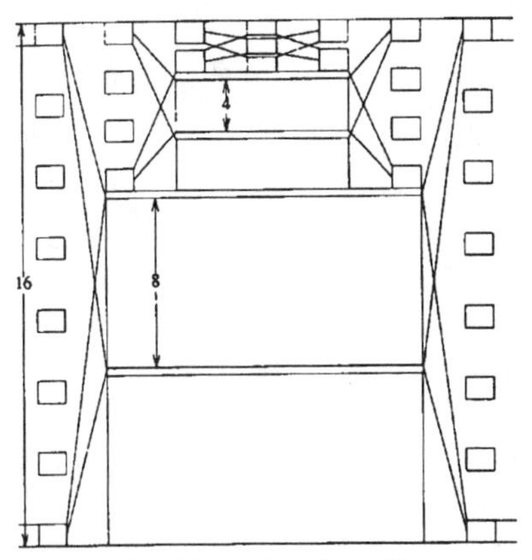

Fig. 11. $2^n \times 2^n$ recursive Clos architecture.

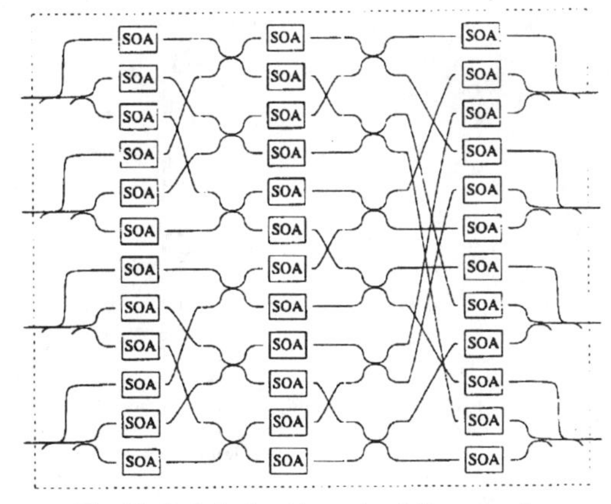

Fig. 12. Optimized multistage 4 × 4 Clos network.

- if $m \geq n$, then the network is rearrangeably nonblocking;
- if $m \geq 2n - 1$, then the network is completely nonblocking.

Referring to the three-stage element in Fig. 4(a) or 5, $n = m = 2$ satisfies the first condition and the element is rearrangeably nonblocking. Referring to the three-stage element in Fig. 10, $n = 2$ and $m = 3$ satisfies the second condition and the element is completely nonblocking. The element of Fig. 10 uses 36 SOA's, compared to 32 used in the only other element that is both completely nonblocking and fault tolerant (although only for single faults)—the four-stage element of Fig. 4(b) or 6.

Benes showed a recursive construction for $2^n \times 2^n$ rearrangeably nonblocking Clos networks [27]. The *beta element*, shown in Fig. 1, is the basis of the recursion, for $n = 1$. One constructs a three-stage $2^n \times 2^n$ network by using two switches with size $2^{n-1} \times 2^{n-1}$ as the center stage and by using a layer of $n/2$ *beta elements* as the left and right stages. Each *beta element* connects to a pair of ports on one side and to each switch in the center stage on the other side. Fig. 4(a) shows a 4 × 4 Benes network, the first step of the recursion, for $n = 2$. For every step of the recursion, $n = m = 2$, satisfying the condition for rearrangeability. If the 2 × 2 units support broadcast, then the overall Benes network does. All the Benes networks are single fault tolerant, possibly becoming blocking networks in the process, but never isolating any inputs or outputs.

We present a similar recursive construction for completely nonblocking Clos networks, built from 2 × 3 elements. One constructs a three-stage $2^n \times 2^n$ network by using three switches with size $2^{n-1} \times 2^{n-1}$ as the center stage and by using a layer of $n/2$ 2 × 3 elements as the left and right stages. Each 2 × 3 element connects to a pair of ports on one side and to each switch in the center stage on the other side. Fig. 10 shows a 4 × 4 recursive Clos network, the first step of the recursion, for $n = 2$. For every step of the recursion, $n = 2$ and $m = 3$, satisfying the condition for completely nonblocking networks.

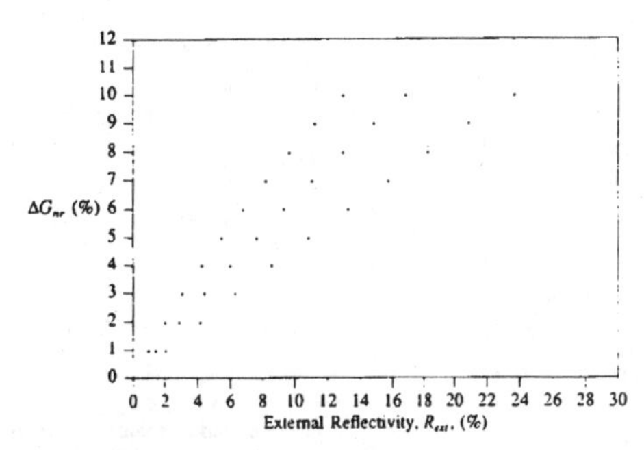

Fig. 13. Percentage change in the nonresonant gain ratio for the optimized multistage Clos network versus external reflectivity for $G_{nr} = -1$ dB, $G_{nr} = -3$ dB, and $G_{nr} = -5$ dB, for curves left to right, respectively.

If the 2 × 3 and 2 × 2 units support broadcast, then the overall recursive Clos network does. All these networks are double fault tolerant, possibly becoming blocking networks in the process, but never isolating any inputs or outputs.

Unlike the Benes construction, this recursion for Clos networks becomes inefficient for large networks (large n). For example a 16 × 16 network would have nine 4 × 4 elements, like that of Fig. 10, embedded in the center. A network with similar properties made from 4 × 7 elements would only require seven of them. Other variations on the 4 × 7 element include a 4 × 8, that would be fault-tolerant and still completely nonblocking, and a 4 × 6, that might be *virtually nonblocking* (depending on the traffic statistics). This area of architectural research is open.

VII. Optimized Multistage 4 × 4 Clos Network

The elimination of passive couplers in between the 2 × 3 and 2 × 2 elements reduces the transfer loss from one stage to the next. The networks in Figs. 10 and 11 simplify to the circuit shown in Fig. 12.

Fig. 13 shows the effect of external reflection on this network.

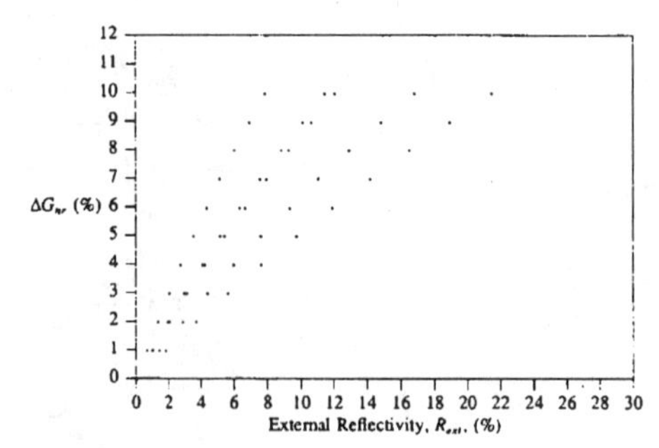

Fig. 14. Comparison of various switching element's susceptibility to changes in external reflectivity for an initial $G_{nr} = -3$ dB. The curves, from left to right, are the 2 × 3, 2 × 2, one-stage 4 × 4, 4 × 4 optimized multistage Clos, and optimized Benes multistage 4 × 4 network.

TABLE I
COMPARISON OF ELEMENTS AND NETWORKS

Switch Type	Device Count	*Robustness*† R_{ext}	*Noise*‡ *Power*	$R_{facet,max}$ @G_{nr}=−3dB	Fault-Tolerance	Blocking Characteristic
2x3	6	7.9%	1.3N, 2N, 2.7N	1.02%	0 - Fault	non-blocking
2x2	4	11.6%	N	1.52%	0 - Fault	non-blocking
4x4	16	12.2%	2.1N	0.38%	0 - Fault	non-blocking
4x4, 3-stage	24	11.6%	3N	1.42%	1 - Fault	rearrangeably n-b
4x4, 4-stage	32	11.6%	4N	1.42%	1 - Fault	non-blocking
4x4, Clos	36	7.9%	4N, 8N	0.90%	2 - Fault	non-blocking
4x4, Clos opt.	36	16.9%	3.4N, 6.9N	2.00%, 1.52%	2 - Fault	non-blocking
4x4, 3-stage opt.	24	21.7%	3N	2.94%, 1.42%	1 - Fault	rearrangeably n-b
4x4, 4-stage opt.	32	21.7%	4N	2.94%, 1.42%	1 - Fault	non-blocking

†—$G_{nr} = -3$ dB, $\Delta G_{nr} = -10$ percent.
‡—normalized to the output intensity of the 2 × 2 (multiple values indicate route dependent intensity).

VIII. Summary

Fig. 14 shows a comparison of the various elements and networks with regard to susceptibility to changes in external reflectivity. Clearly, the optimized multistage Benes 4 × 4 network is most robust; however, there are other considerations that are also important. From the standpoint of fault tolerance, the optimized multistage Clos 4 × 4 is the best choice. If total amplified spontaneous emission is a major concern, then the single-stage 4 × 4 is the best choice among all the 4 × 4's.

Table I compares the advantages and disadvantages of each switching element and network. Photonic switching elements are designed from semiconductor optical amplifiers and passive couplers with fiber-to-fiber unity-gain and low crosstalk. Designs for a 2 × 2 and an asymmetric 2 × 3 element, and several designs for 4 × 4 elements, are presented. While most amplifier analyses have stressed the importance of ultra-low facet reflectivities for high-gain operation, with protection against external reflections with optical isolators, modest facet reflectivities are satisfactory for these elements. It is also shown that substantial amounts of external reflection can be tolerated. The various architectures are compared according to amplifier count, blocking characteristic, broadcast potential, noise power, and fault tolerance.

Acknowledgment

Special thanks to V. E. Kelly for his assistance with the computational analysis.

References

[1] Y. Yamamoto, "Characteristics of AlGaAs Fabry–Perot cavity type laser amplifiers," *IEEE J. Quantum Electron.*, vol. QE-16, pp. 1047-1052, Oct. 1980.

[2] T. Mukai and Y. Yamamoto, "S/N and error rate performance in AlGaAs semiconductor laser preamplifier and linear repeater systems," *IEEE Trans. Microwave Theory Tech.*, vol. MTT-30, pp. 1548-1556, Oct. 1982.

[3] M. Ikeda, "Switching characteristics of laser diode switch," *IEEE J. Quantum Electron.*, vol. QE-19, pp. 157-164, Feb. 1983.

[4] J. Hegarty and K. A. Jackson, "High-speed modulation and switching with gain in a GaAlAs traveling-wave optical amplifier," *Appl. Phys. Lett.*, vol. 45, pp. 1314-1316, Dec. 15, 1984.

[5] L. ThyLén, P. Granestrand, and A. Djupsjöbacka, "Optical amplification in switching networks," presented at Top. Meet. Photon. Switching, Incline Village, NV, Mar. 1987.

[6] M. Kobayashi, A. Himeno, and H. Terui, "Guided-wave optical matrix switch," in *Proc. IOOC-ECOC '85*, Venice, Italy, 1985, pp. 73-76.

[7] J. D. Evankow, Jr. and R. A. Thompson, "Photonic switching with laser diode amplifiers," unpublished.

[8] R. M. Jopson and G. Eisenstein, "Optical amplifiers for photonic switches," in *Proc. Top. Meet. Photon. Switching*, Incline Village, NV, March 18-20, 1987, FC1-1-FC1-3, pp. 116-118.

[9] G. Eisenstein and R. M. Jopson, "Measurements of the gain spectrum of near-traveling wave and Fabry-Perot semiconductor optical amplifiers at 1.5 μm," *Int. J. Electron.*, vol. 60, pp. 113-121, 1986.

[10] R. T. Ku *et al.* "A packaged 1.5 μm InGaAsP laser amplifier," to be published.
[11] I. W. Marshall, M. J. O'Mahony, and P. D. Constantine, "Optical system with two packaged 1.5 μm semiconductor laser amplifier repeaters," *Electron. Lett.*, vol. 22, pp. 253-255, Feb. 27, 1986.
[12] H. Kataoka and M. Ikeda, "Laser-diode optical switch module," *Electron. Lett.*, vol. 20, pp. 438-439, May 24, 1984.
[13] H. S. Hinton, "A nonblocking optical interconnection network using directional couplers," in *Proc. GLOBECOM'84*, 1984, pp. 26.5.1-26.5.5.
[14] T. Shimoe, K. Hajikano, and K. Murakami, "Path-independent insertion-loss optical space switch," in *Proc. OFC/IOOC '87*, 1987, paper WB2.
[15] G. A. Bogert, "4 × 4 Ti:LiNbO3 switch array with full broadcast capability," presented at OSA Top. Meet. Photon. Switching, 1987, paper ThD3.
[16] P. Granestrand *et al.*, "Strictly nonblocking 8 × 8 integrated optic switch matrix in Ti:LiNbO3," presented at IGWO '86, 1986, paper WAA3.
[17] S. Suzuki, M. Kondo, K. Nagashima, K. Komatsu, and T. Mikakawa, "Thirty-two line optical space-division switching system," presented at OFC/IOOC '87, 1987, paper WB4.
[18] R. A. Spanke, "Architectures for guided-wave optical space switching systems," *IEEE Communications*, vol. 25, May 1987.
[19] K. Padmanabhan and A. Netravali, "Dilated networks for photonic switching," presented at OSA Top. Meet. Photon. Switching, 1987.
[20] A. E. Joel, "On permutation switching networks," *Bell. Syst. Tech. J.*, vol. 47, pp. 813-822, May 1968.
[21] G. Eisenstein, "Theoretical design of single-layer antireflection coatings on laser facets," *AT&T Tech. J.*, vol. 63, pp. 357-364, Feb. 1984.
[22] T. Saitoh, T. Mukai, and O. Mikami, "Theoretical analysis and fabrication of antireflection coatings on laser-diode facets," *J. Lightwave Technol.*, vol. LT-3, pp. 288-293, Apr. 1985.
[23] J. D. Evankow, Jr., N. A. Olsson, and R. T. Ku, "Performance of packaged near-traveling-wave semiconductor laser amplifier with multi-longitudinal mode input," *J. Lightwave Technol.*, to be published.
[24] C. H. Henry, "Theory of spontaneous emission noise in open resonators and its application to lasers and optical amplifiers," *J. Lightwave Technol.*, vol. LT-4, pp. 288-297, Mar. 1986.
[25] P. S. Henry, "Lightwave primer," *IEEE J. Quantum Electron.*, vol. QE-21, pp. 1862-1879, Dec. 1985.
[26] C. Clos, "A study of nonblocking switching networks," *Bell Syst. Tech. J.*, vol. 32, pp. 406-424, 1953.
[27] V. E. Benes, *Mathematical Theory of Connecting Networks and Telephone Traffic*. New York: Academic, 1965.

SLALOM: Semiconductor Laser Amplifier in a Loop Mirror

M. EISELT, W. PIEPER, AND H.G. WEBER

Abstract—**The processing of optical signals in the optical domain is an important issue resulting from the desire to take advantage of the full bandwidth of the optical fiber. In this paper, we present detailed investigations on a device, which utilizes a semiconductor laser amplifier in a loop mirror configuration (SLALOM). Different modes of operation are reported like nonlinear single pulse switching and two-pulse switching at different operation speeds (1–100 Gb/s). Furthermore, a number of applications of the SLALOM in photonic systems, like pulse shaping, decoding, retiming and time-division demultiplexing, are presented. In addition, the SLALOM can be used for an estimate of the linewidth enhancement factor α and the carrier lifetime τ_e in an SLA.**

I. INTRODUCTION

FIBER optic transmission is a well established technique today. However, routing, switching and other signal-processing operations are still performed electrically after optoelectronic conversion. An important question is, which improvements of today's communication systems are possible, when no optoelectronic conversion is involved and when signal-processing is performed on the signal while it is still in the form of light. One expects a more efficient use of the transmission bandwidth of the optical fiber and perhaps also simpler and less expensive devices for signal-processing. Especially, devices in which light signals are controlled by light are very promising because data signals and control signals may be transmitted in the same optical fiber. One such device is the nonlinear Sagnac interferometer or nonlinear optical loop mirror (NOLM). The NOLM has two important properties, the inherent stability of its interferometer arrangement and the intrinsically very fast response time, based on the fiber nonlinearity, which enables ultrafast signal-processing [1]–[3]. However, the NOLM needs long fiber lengths and large optical powers. For instance pulses of about 1 W peak power in 1 km of optical fiber length are required to make the NOLM operating as optically controlled optical switch.

Recently, we reported on a new device [4], which is based on the same Sagnac interferometer principle as the NOLM. However, its operation does not depend on the optical nonlinearity of the fiber but on the optical nonlinearity of a semiconductor laser amplifier (SLA) in the fiber loop. The SLALOM (SLA in a Loop Mirror) is a device which may be used as nonlinear gain element providing shaping and contrast enhancement of optical pulses [5], as optical correlator providing retiming of optical data signals [6] and header decoding in optical packet switching experiments [7], and as optical demultiplexer for data rates up to 100 Gb/s [8]. The demultiplexer application of the SLALOM is based on a work by Sokoloff *et al.* [9]. As compared to the NOLM, the SLALOM has two advantages. Firstly, the device may be very compact so that integration on a chip is possible. Secondly, the required optical power is of the order of 1 mW. A disadvantage is that the operation speed is generally lower. It is of the order of a few GHz for most applications of the SLALOM except the applications as demultiplexer.

In this contribution, we explain the principle of operation of the SLALOM (Section II). After a brief overview of the gain dynamics in a semiconductor laser amplifier (Section III), which is based on a simplified model, we discuss the different modes of operation of the SLALOM (Section IV). Here, theoretical results, which are derived from the gain dynamics, are compared with experimental results of the basic SLALOM arrangements. In Section V, applications of the SLALOM in systems for optical signal processing are reported. Finally, in Section VI, a short consideration of the operation speed is given.

II. PRINCIPLE OF OPERATION

Fig. 1 depicts the basic elements of the SLALOM. It consists of a directional coupler (ideally a 50:50 coupler) with two branches connected forming a loop. The loop contains a polarization control (PC) and an SLA. A signal with electrical field $E_{in}(t)$ at the angular frequency ω is split in the coupler and travels in clockwise (cw) and counter-clockwise (ccw) direction through the loop. Assuming a spatially concentrated SLA, the clockwise travelling signal will be amplified by the complex gain

$$\underline{g}_{cw}(t) = g_{cw}(t) \cdot \exp\left[-j\varphi_{cw}(t)\right]$$

with

$$g_{cw}(t) = g\left(t + \frac{t_d}{2} - \frac{T}{2}\right)$$

and

$$\varphi_{cw}(t) = \varphi\left(t + \frac{t_d}{2} - \frac{T}{2}\right). \quad (1)$$

Here, t_d is the pulse round trip time in the loop, as indicated in Fig. 1.

Manuscript received March 20, 1995. This work was supported by the Land Berlin and the Bundesminister für Bildung, Wissenschaft, Forschung and Technologie of the Fed. Rep. of Germany under contract 01 BS 202/4.

The authors are with Heirich-Hertz Institute, 10587 Berlin, Germany.

IEEE Log Number 9414402.

Reprinted with permission from *IEEE Journal of Lightwave Technology,* Vol. 13, No. 10, pp. 2099-2112, October 1995.

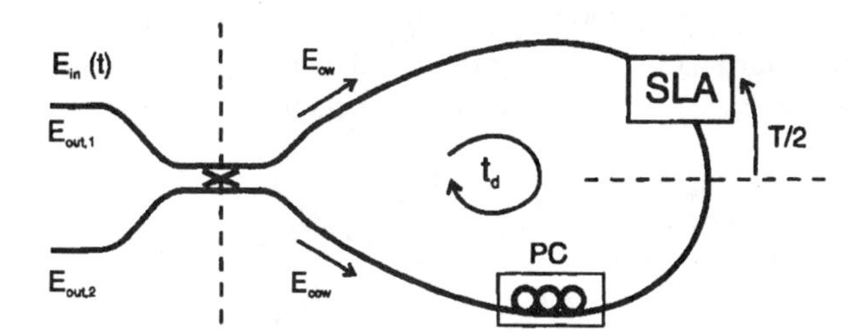

Fig. 1. Basic configuration of the SLALOM. PC: polarization control, SLA: semiconductor laser amplifier.

The counter-clockwise signal that arrives by the time T later at the SLA experiences the gain

$$\underline{g}_{ccw}(t) = g_{ccw}(t) \cdot \exp\left[-j \cdot \varphi_{ccw}(t)\right]$$

with

$$g_{ccw}(t) = g\left(t + \frac{t_d}{2} + \frac{T}{2}\right)$$

and

$$\varphi_{ccw}(t) = \varphi\left(t + \frac{t_d}{2} + \frac{T}{2}\right). \qquad (2)$$

The complex gain depends on the saturation of the amplifier by the preceding signals and the gain recovery. In Section III we will develop a coarse model to determine the amplifiers complex gain.

Having traversed the loop, the clockwise and counter-clockwise travelling signals superimpose at the coupler, if the polarization control is properly matched. With the coupling ratios k and d for the cross and through coupling, respectively, the field at the upper branch in Fig. 1 results to

$$\begin{aligned}\underline{E}_{out,1}(t) &= jdk\underline{E}_{in}(t - t_d) \cdot e^{-j\omega t_d} \\ &\cdot [\underline{g}_{cw}(t - t_d) + \underline{g}_{ccw}(t - t_d)]\end{aligned} \qquad (3)$$

and the output field at the lower branch is

$$\begin{aligned}\underline{E}_{out,2}(t) &= \underline{E}_{in}(t - t_d) \cdot e^{-j\omega t_d} \\ &\cdot [d^2 \cdot \underline{g}_{cw}(t - t_d) - k^2 \cdot \underline{g}_{ccw}(t - t_d)].\end{aligned} \qquad (4)$$

For an ideal 50:50 coupler, $d^2 = k^2 = 1/2$. In this ideal case, and if $\underline{g}_{cw} = \underline{g}_{ccw}$, the injected signal will be totally reflected to the upper output while there is no signal at the output 2 of the coupler. For the further investigations, we will mainly be interested in the output 2 signal so that we define the output power P_{out} only for output 2

$$\begin{aligned}P_{out}(t) &= \tfrac{1}{2}\,\underline{E}_{out,2}(t) \cdot \underline{E}^{*}_{out,2}(t) \\ &= P_{in}(t - t_d) \cdot \{d^4 \cdot g^2_{cw}(t - t_d) + k^4 \\ &\cdot g^2_{ccw}(t - t_d) - 2d^2k^2 \cdot g_{cw}(t - t_d)g_{ccw}(t - t_d) \\ &\cdot \cos\left[\varphi_{cw}(t - t_d) - \varphi_{ccw}(t - t_d)\right]\} \\ P_{out}(t) &= g^2_{cw}(t - t_d) \cdot P_{in}(t - t_d) \\ &\cdot \left\{d^4 + k^4 \cdot \frac{g^2_{ccw}}{g^2_{cw}} - 2d^2k^2\right. \\ &\left.\cdot \frac{g_{ccw}}{g_{cw}} \cdot \cos\left[\varphi_{cw} - \varphi_{ccw}\right]\right\}.\end{aligned} \qquad (5)$$

It can be seen that the output power depends on the gain ratio $\Delta g = (g_{ccw}/g_{cw})$ and the phase difference $\Delta\varphi = \varphi_{cw} - \varphi_{ccw}$.

There are mainly three ways to obtain this phase and gain difference:

1) Injection of a single pulse and saturation of the SLA by the clockwise travelling pulse. If T is smaller than the gain recovery time of the SLA, the counter-clockwise travelling pulse sees a saturated SLA.
2) Injection of two pulses, separated by time $\tau \approx T$, and saturation of the SLA by the counter-clockwise travelling first pulse, such that the clockwise travelling second pulse sees the saturated SLA, or vice versa.
3) Injection of a single (signal) pulse at the coupler input branch and of a second (control) pulse at an additional coupler in the loop. The control pulse saturates the SLA between the transition of clockwise and counter-clockwise travelling signal pulses.

In Section IV we will investigate realizations of the three switching principles and compare them with theoretical results, which are based on a simplified model of the SLA's saturation characteristics. This model is sketched in the following section.

III. Gain Dynamics in a Semiconductor Laser Amplifier

Generally, the gain of an SLA is a function of the inverted carrier distribution in the active region. The carrier density in the conduction band, N, is a measure for the gain per cm of the device. The carrier density at a specific longitudinal coordinate z in the amplifier at time t is governed by the rate equation

$$\begin{aligned}\frac{\partial N(z, t)}{\partial t} &= \frac{I}{eV} - \frac{N(z, t)}{\tau_e} \\ &- \frac{P(z, t) \cdot \Gamma \cdot g \cdot [N(z, t) - N_T]}{h\nu A}.\end{aligned} \qquad (6)$$

With I: effective injection current; e: electron charge; V: active volume of SLA; τ_e: carrier lifetime; Γ: mode confinement factor; g: gain coefficient; N_T: carrier density for transparency; $h\nu$: photon energy; A: cross-section of active region. The first term on the right hand side of (6) describes the pumping by the injection current, the second term accounts for the decrease of carrier density by radiative and nonradiative recombination, and the third term denotes the decrease of carrier density by stimulated emission and thus by amplification of the optical power in the amplifier. The optical power $P(z, t)$ propagating in positive z-direction changes with

$$\frac{\partial P(z, t)}{\partial z} = P(z, t) \cdot \Gamma \cdot g \cdot [N(z, t) - N_T] - P(z, t) \cdot \alpha_D \quad (7)$$

where α_D is the waveguide loss coefficient.

For small input signals to the SLA, the carrier density $N(z, t)$ is independent of z and $P(z, t)$. It has the steady-state value $N(z, t) \equiv N_s$. From (7), the small signal gain of an SLA with length L is obtained

$$G_0 = \frac{P_{out}}{P_{in}} = \exp\left[\Gamma g(N_S - N_T)L - \alpha_D \cdot L\right]. \qquad (8)$$

In the following, the SLA is regarded as a spatially concentrated device. For this reason, the total number of carriers per cross-section, $N_{tot}(t)$, which are available for the amplification process, is calculated by integration over the length L of

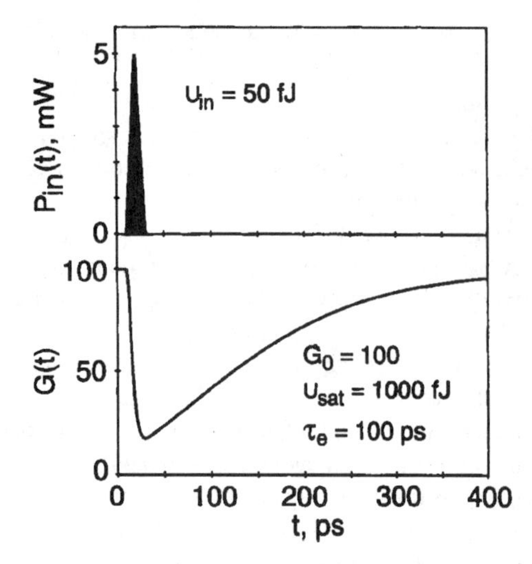

Fig. 2. Input power to an SLA and calculated change of gain.

the SLA

$$N_{tot}(t) = \int_{z=0}^{z=L} [N(z, t) - N_T]\, dz. \tag{9}$$

Now, (7) is integrated over the length L of the amplifier. Using (9), we obtain

$$G(t) = \frac{P(L, t)}{P(0, t)} = \exp[\Gamma g N_{tot}(t) - \alpha_D \cdot L]. \tag{10}$$

In the two succeeding subsections, the calculation of $N_{tot}(t)$ is performed separately for the saturation of the amplifier by a short optical pulse and for the period of gain recovery, when no optical signal is injected.

A. Saturation by a Short Optical Pulse

For the amplifier saturation period we follow an approach similar to that published by Siegmann [10]. For the short term of saturation by an optical pulse, we neglect the first and second term of (6). A change of carrier density is assumed to be based only on the amplification process.

This simplified (6) is integrated over the length L of the amplifier. As N_T is invariant with time, we obtain

$$\frac{d}{dt}\int_0^L N(z, t)\, dt = \frac{d}{dt} N_{tot}(t) = -\frac{1}{h\nu A}\int_0^L P(z, t)\Gamma g \cdot [N(z, t) - N_T]\, dz. \tag{11}$$

The right hand side of (11) is substituted by (7). With the approximation

$$\int_{z=0}^{L} P(z, t)\, dz \approx \frac{L}{\ln G_0}[P_{out}(t) - P_{in}(t)] \tag{12}$$

we obtain the temporal change of N_{tot} as

$$\frac{dN_{tot}(t)}{dt} = -\frac{1 + \dfrac{\alpha_D L}{\ln G_0}}{h\nu A}[P_{out}(t) - P_{in}(t)]. \tag{13}$$

Substituting $[P_{out}(t)]/[P_{in}(t)]$ by the gain $G(t)$ from (10) and integrating (13) over the time, a relation between the input power $P_{in}(t)$ and the gain $G(t)$ is obtained

$$\ln \frac{1 - \dfrac{1}{G_0}}{1 - \dfrac{1}{G(t)}} = \frac{\Gamma g\left(1 + \dfrac{\alpha_D L}{\ln G_0}\right)}{h\nu A}\int_0^t P_{in}(t')\, dt. \tag{14}$$

After defining the accumulated injected pulse energy as

$$U_{in}(t) = \int_0^t P_{in}(t')\, dt' \tag{15}$$

and the saturation energy of the SLA as

$$U_{sat} = \frac{h\nu A}{\Gamma g\left(1 + \dfrac{\alpha_D L}{\ln G_0}\right)} \tag{16}$$

we obtain the momentary gain of the SLA during the pulse transition

$$G(t) = \frac{1}{1 - \left(1 - \dfrac{1}{G_0}\right)\exp\left(-\dfrac{U_{in}(t)}{U_{sat}}\right)}. \tag{17}$$

With use of (10), $N_{tot}(t)$ can be obtained.

Equation (17) shows that the gain of the SLA decreases as long as the (short) pulse is injected into the SLA. Thus, the saturation time can be as short as some picoseconds.

B. Gain Recovery

After the short optical pulse has saturated the SLA according to (17), the gain recovers due to injection of carriers by the injection current. For this gain recovery period, we can neglect the stimulated recombination term (third term) of (6). The resulting equation is integrated over the length of the SLA and yields with (9)

$$\frac{dN_{tot}(t)}{dt} = \left(\frac{I}{eV} - \frac{N_T}{\tau_e}\right)\cdot L - \frac{N_{tot}(t)}{\tau_e}. \tag{18}$$

The differential equation (18) is solved resulting in $N_{tot}(t)$. This is combined with (10) and the momentary gain during the recovery period is obtained

$$G(t) = G_0 \cdot \left[\frac{G(t_s)}{G_0}\right]^{\exp[-(t-t_s)/(\tau_e)]}, \quad t \geq t_s. \tag{19}$$

Here, $G(t_s)$ is the gain after the saturation by the optical pulse. It can be seen that the gain recovers with the time constant τ_e, which is some hundred picoseconds for an SLA.

As an example, Fig. 2 shows the calculated gain change in an SLA using the above equations and a short input pulse as shown in the figure.

C. Phase Change

The amplitude and phase change of the optical field after traversing the SLA can be described as

$$\begin{aligned}\frac{\underline{E}_{out}}{\underline{E}_{in}} &= \exp\left(-j\frac{2\pi \underline{n} L}{\lambda}\right)\\ &= \exp\left(-j\frac{2\pi n' L}{\lambda}\right)\cdot\exp\left(-\frac{2\pi n'' L}{\lambda}\right)\\ &= e^{-j\varphi_s}\cdot\sqrt{G}\end{aligned} \tag{20}$$

using the complex refractive index

$$\underline{n} = n' - jn''. \tag{21}$$

The real part of the refractive index is a measure for the phase delay φ_s, whereas the imaginary part describes the attenuation or gain of the device.

Both quantities are functions of the carrier density. Usually, the following relation is assumed

$$\frac{dn'}{dN} = \alpha\cdot\frac{dn''}{dN}. \tag{22}$$

Here, α is the linewidth enhancement factor. From (20) and (22) it follows that the phase differences and gain ratios for two different states of saturation of the SLA are coupled by

$$\varphi_1 - \varphi_2 = -\frac{\alpha}{2}\ln\left(\frac{G_1}{G_2}\right). \tag{23}$$

Therefore, using (23) $P_{out}(t)$ from (5) can be written as

$$\begin{aligned}P_{out}(t) = P_{in}(t-t_d)\cdot G_{cw}(t-t_d)\cdot\Bigg\{d^4 + k^2\cdot\frac{G_{ccw}}{G_{cw}}\\ -2d^2k^2\cdot\sqrt{\frac{G_{ccw}}{G_{cw}}}\cdot\cos\left(\frac{\alpha}{2}\ln\frac{G_{ccw}}{G_{cw}}\right)\Bigg\}.\end{aligned} \tag{24}$$

In this equation, the power gain G is related to the field gain g of (5) as

$$G = g^2.$$

Equation (24) is the basic equation for the theoretical investigation of the SLALOM device. For the symmetrical case ($G_{cw} = G_{ccw}$), this equation yields

$$\begin{aligned}P_{out,symm}(t) &= P_{in}(t-t_d)\cdot G_{cw}(t-t_d)\\ &\quad\cdot\{d^4 + k^4 - 2d^2k^2\}\\ &= P_{in}(t-t_d)\cdot G_{cw}(t-t_d)\cdot\{d^2 - k^2\}^2\end{aligned}$$

For an optimum coupling ratio of 50:50 ($d^2 = k^2 = 0.5$) the output power will be zero. In the following Section, we will investigate how an asymmetric case ($G_{cw} \neq G_{ccw}$) can be obtained.

IV. Basic Switching Configurations

In this Section, we report on three basic configurations to obtain a fiber loop, which is asymmetric for optical pulses, and thus can switch pulses to the SLALOM output.

TABLE I
Parameters of MQW-SLA at Different Values of Injection Current I: G_0 and τ_e Taken from [11], U_{sat} from Saturation Power Measurements

I, mA	G0, dB	τ_e, ps	Usat, fJ
120	17.5	270	1215
140	21.6	210	1450
170	25.0	150	1601
200	26.9	115	1777
250	28.7	100	2039
300	29.6	95	2084

A. Data of Used SLA's

The experimental verifications of the basic configurations were implemented using two different SLA's. One of them was manufactured by NTT and had an MQW structure with a linewidth enhancement factor $\alpha = 4$ [11]. The most important parameters are listed in Table I.

The second amplifier we used had a bulk structure and was manufactured by the integrated optics division of HHI. It had a fiber-fiber-gain of about 5 dB and $\alpha \cong 6$. The carrier lifetime was measured to about $\tau_e \cong 300$ ps. The residual facet reflectivity is an important parameter for the SLA application in a SLALOM setup. Firstly, the facet reflectivity influences the phase shift in the amplifier, secondly, reflected pulses will appear at the SLALOM output with a higher amplitude than ideally extinguished pulses. These pulses may be misinterpreted in a practical setup. The MQW-SLA in our experiments had a buried window structure and AR coated facets. The remaining reflectivity was $r \approx 2\cdot 10^{-4}$ per facet. This value was good enough for the reported applications. The bulk structured SLA had a facet reflectivity of about $5\cdot 10^{-3}$. This resulted in additional pulses at the SLALOM output, which had to be accounted for in the measurement.

B. Single Pulse Switching

In the single pulse mode of operation, the position of the SLA in the loop is only slightly noncentric. That is, the eccentricity T is smaller than the carrier lifetime τ_e in the SLA. In this case, the clockwise (cw) travelling pulse is amplified by the gain G_{cw} and saturates the SLA while traversing it. A time of T later, the counter-clockwise (ccw) travelling pulse reaches the SLA, which is still saturated, and the pulse is amplified by the gain $G_{ccw} < G_{cw}$.

From (17) it can be seen that the SLA gain saturation is a function of the pulse energy and so will be the gain ratio G_{ccw}/G_{cw}, which is the essential parameter determining the output power, according to (24). Another important value for the output power is the gain of the amplifier for the clockwise travelling pulse, which is reduced for higher pulse powers. Fig. 3 shows the SLALOM gain $G_{SLALOM} = [P_{out}(t+t_d)/P_{in}(t)]$ for the pulse peak powers as a function of the input pulse peak powers. The theoretical results, depicted in solid lines, were obtained using the above described saturation characteristics and a loop eccentricity of $T = 40$ ps. The assumed carrier lifetime depends on the SLA's single pass gain for different values of the injection current and is $\tau_e = 100$ ps, 150 ps, and 270 ps for a gain of 28.7 dB, 25.0 dB,

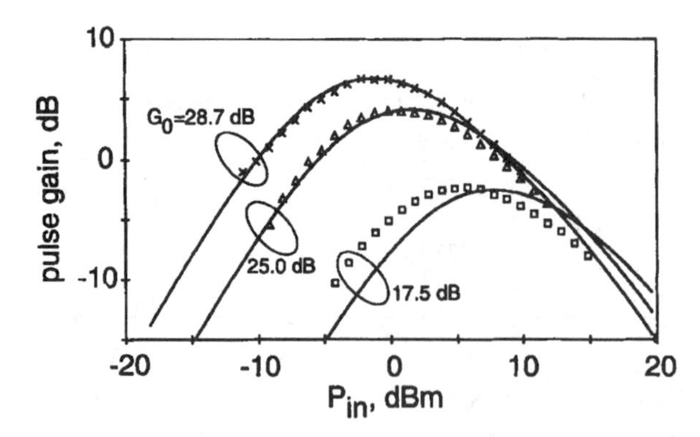

Fig. 3. Pulse gain of the SLALOM in ist nonlinear mode of operation. Theoretical (lines) and experimental (dots) results for different gain values of the SLA.

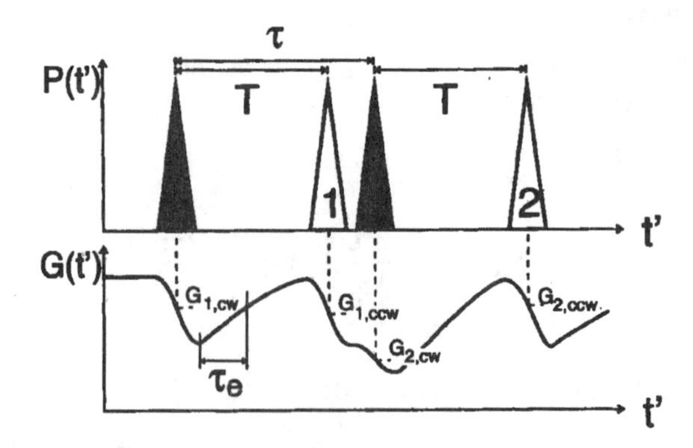

Fig. 4. Schematic of the power at the SLA in the loop and resulting SLA gain for eccentricity $T <$ pulse separation τ, grey: clockwise (cw), white: counter-clockwise (ccw) travelling pulses.

and 17.5 dB, respectively. The values were taken from [11] (see Table I) to enable a comparison between theoretical and experimental results.

To verify the theoretical results, a SLALOM was assembled with a semiconductor laser amplifier, described above, positioned in the loop with an eccentricity of $T = 40$ ps. Single pulses with a width of 45 ps and variable pulse peak power were injected into the coupler, and the peak powers of the output pulses were measured. By varying the injection current to the SLA, appropriate values of the SLA's gain were selected to compare the measured results with the theoretical predictions. The measured values for the SLALOM pulse gain are shown as dots in Fig. 3. They are in good agreement with the theoretical values shown as solid lines.

C. Two-Pulse Correlation

If the eccentricity T of the SLA in the loop is made somewhat larger than the carrier lifetime τ_e in the SLA, the carrier density in the amplifier has already recovered when the counter-clockwise travelling pulse arrives. In this case, no gain and phase differences will be obtained and no pulse is seen at the output, if a single pulse is injected.

However, when two pulses, separated by τ, are injected into the SLALOM, the situation changes. Fig. 4 shows, in principle, the optical input power $P(t')$ to the SLA in the loop and the resulting gain $G(t')$ as a function of time. The clockwise component of pulse 1, injected at time t, arrives at the SLA at $t' = t + (t_d/2) - (T/2)$. It is amplified by $G_{1,cw}$ and saturates the SLA. If the pulse separation τ is larger than the eccentricity T, the next pulse at the SLA is the counter-clockwise pulse 1 arriving by the time T later. As $T > \tau_e$, the SLA gain has recovered between the cw and ccw pulse, so that cw and ccw travelling pulses 1 see the same carrier density in the SLA and ccw pulse 1 sees gain $G_{1,ccw}$. The next pulse at the SLA is the cw travelling pulse 2, which arrives by the time τ later than the cw pulse 1, that is a time of $\tau - T$ after the ccw pulse 1. If $\tau - T < \tau_e$, this pulse still sees the saturated carrier density and experiences the gain $G_{2,cw}$. Finally, after time T, when the ccw pulse 2 arrives, the gain has recovered and the pulse experiences the gain $G_{2,ccw}$. When, after circulation in the loop, the pulses meet again at the coupler, pulse 1 is reflected

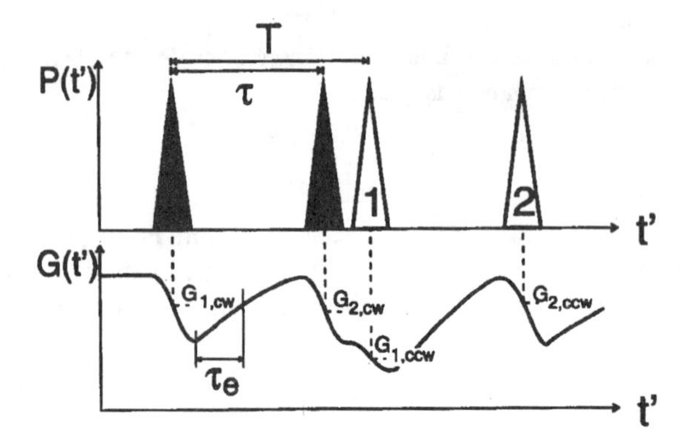

Fig. 5. as Fig. 4, but $T > \tau$.

to the input as $G_{1,cw} = G_{1,ccw}$. But pulse 2 can be seen at the output as $G_{2,cw} < G_{2,ccw}$. The gain ratio $(G_{2,cw}/G_{2,ccw})$ is a function of the difference of arrival times, $\tau - T$.

If the initial pulse separation τ is smaller than the eccentricity T (see Fig. 5), the counter-clockwise pulse 1 will see the saturated gain as a function of $T - \tau$. In this case, the first pulse will be seen at the SLALOM output, whereas pulse 2 will be reflected to the input.

To verify this mode of operation, two pulses ($P_{peak} = 12.7$ dBm, $FWHM \approx 35$ ps) were injected into the SLALOM, separated by a variable time between 11 and 13 ns. In the SLALOM loop, the eccentricity of the amplifier was $T = 11.78$ ns. At the SLALOM output, the first and second pulse were still separated by the pulse separation τ. Their peak powers were measured and are shown in Fig. 6 as a function of the pulse separation τ. It can be seen, that for $\tau < T$ the first pulse can be seen at the output, whereas the second pulse is mainly suppressed, and vice versa for $\tau > T$. For $\tau = T$ there is a sharp change between the first and the second pulse to be seen at the output.

The slopes of the output powers (1st pulse for $\tau < T$ and 2nd pulse for $\tau > T$) can be explained by using the gain recovery of (19) in combination with (24). As one of the two cases we investigate the case $\tau < T$ (see Fig. 5), and therefore the output power of the first pulse as a function of the initial pulse separation. The clockwise travelling first pulse sees the unsaturated gain of the SLA, thus $G_{1,cw} = G_0$. After

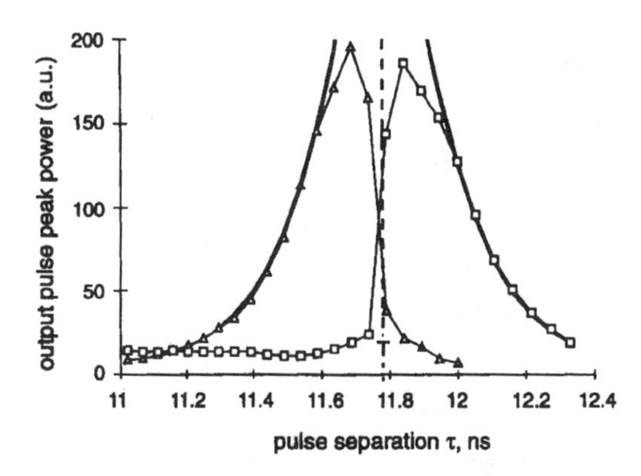

Fig. 6. Correlator output power of first (triangles) and second (squares) pulses as function of input pulse separation. Eccentricity T of the loop was 11.78 ns. Thick solid lines: theoretical values (see text).

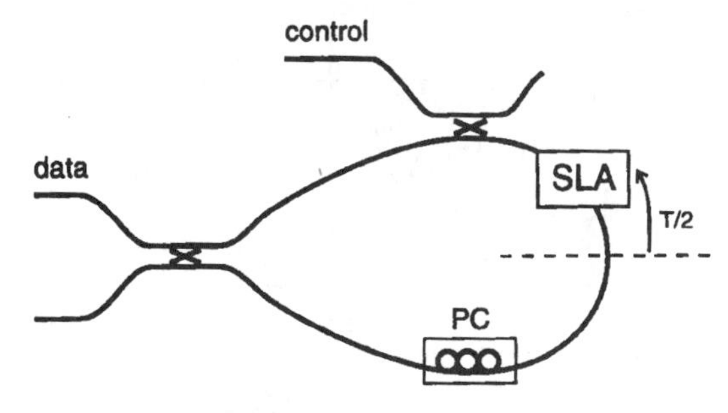

Fig. 7. Configuration of the SLALOM for switching with control pulse.

the transition of the clockwise second pulse, the gain of the SLA is saturated to the value $G(t_s) = G_s$. The gain recovers according to (19). Therefore the ccw pulse 1 experiences the gain

$$G_{1,ccw} = G(t_s + T - \tau) = G_0 \cdot \left[\frac{G_s}{G_0}\right]^{\exp[-(T-\tau)/(\tau_e)]} \quad (25)$$

If we neglect the effect of gain saturation by the ccw pulse 1, we obtain for the gain ratio

$$\frac{G_{ccw}}{G_{cw}} = \left[\frac{G_s}{G_0}\right]^{\exp[-(\tau-T)/(\tau_e)]} \quad (26)$$

When (26) is inserted into the SLALOM equation (24), the SLALOM output power as a function of the pulse separation τ is obtained. The solid lines in Fig. 6 show the results which were matched to the experimental data using an initial saturation of the amplifier

$$\frac{G_s}{G_0} = 0.54.$$

The calculation of the SLA saturation does not cover the overlap of two counter-propagating pulses in the amplifier. Thus, the region of $|T - \tau| < 100$ ps is not described correctly by the theory.

D. Pulse Switching with Control Pulse

For the control-pulse switching mode, a somewhat extended SLALOM set-up can be used. In the fiber loop an additional coupler is introduced by which a control pulse is injected to saturate the SLA (Fig. 7). If the wavelength of the control pulse is different from the data pulse, wavelength selective couplers in the loop can be used to couple the control pulse to and from the loop, so that the data pulse is not influenced and the control pulse is not seen at the SLALOM output [12]. Another way to suppress the control pulse is the use of a filter at the SLALOM output.

Depending on the position of the SLA in the SLALOM, we can distinguish between two operation speeds. With an eccentricity T of the SLA in the loop, which is larger than the gain recovery time τ_e, an optical switch for medium switching periods (about 100 ps) can be realized. If, on the other hand, the eccentricity T is small compared to the gain recovery time, very short switching periods (some ps) can be obtained.

E. Medium Speed Switching

If a short control pulse is injected into the loop such that it saturates the SLA at time t_s, the gain $G(t)$ of the SLA will first be reduced very fast according to (17). For $t > t_s$, the gain will recover with the time constant τ_e according to (19). That means, all data signals whose clockwise (cw) components pass the SLA before t_s and whose counter-clockwise (ccw) components traverse it shortly after t_s , will experience a gain ratio $(G_{ccw}/G_{cw}) < 1$ and can be seen at the SLALOM output. Similarly, if the cw component of a data signal passes the SLA shortly after t_s, the ccw component will see the recovered amplifier gain, as $T > \tau_e$. This results in $(G_{ccw}/G_{cw}) > 1$ so that the data signal is switched to the SLALOM output. Therefore, a single control pulse opens two windows in the time domain, during which a data signal can pass from the SLALOM input to the output.

If we restrict our inspection to the second time window, where the cw data signal passes the amplifier at $t > t_s$, we can define the gain ratio as

$$\frac{G_{ccw}}{G_{cw}} = \frac{G(t+T)}{G(t)} = \frac{G_0 \cdot \left[\frac{G(t_s)}{G_0}\right]^{\exp[-(t+T-t_s)/\tau_e]}}{G_0 \cdot \left[\frac{G(t_s)}{G_0}\right]^{\exp[-(t-t_s)/\tau_e]}} = \left[\frac{G(t_s)}{G_0}\right]^{\{\exp[-(t+T-t_s)/\tau_e] - \exp[-(t-t_s)/\tau_e]\}} \quad t > t_s. \quad (27)$$

For $T \gg \tau_e$, the term $\exp[-(t + T - t_s)/\tau_e]$ tends to zero, that means $G(t + T) = G_0$: the ccw signal experiences the unsaturated gain. Using the result of (27) in (24), we obtain the temporal shape of the switching window. Fig. 8 shows the calculated SLALOM gain P_{out}/P_{in} for different values U_{in}/U_{sat}. An unsaturated gain $G_0 = 500$, carrier lifetime $\tau_e = 115$ ps, $\alpha = 4$, and $T = 10$ ns were assumed for the calculations. It can be found that the maximum SLALOM gain doesn't change with the saturating pulse energy, but the width of the switching window as well as the temporal position of the maximum gain change. Furthermore, for a high pulse energy,

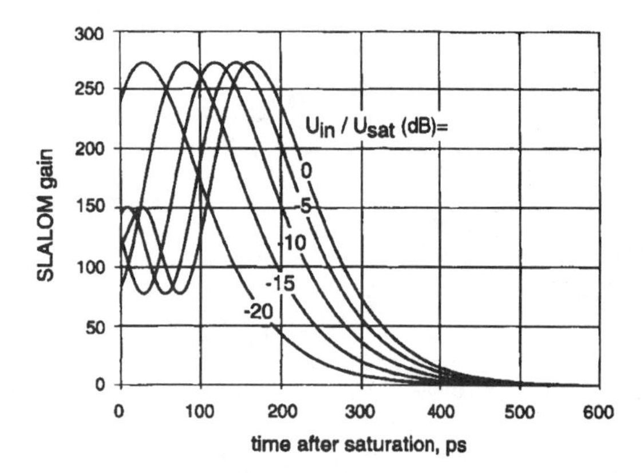

Fig. 8. SLALOM gain (theoretical) as function of time after saturation. Parameter: saturating pulse energy. $G_0 = 500$, $\tau_e = 115$ ps, $\alpha = 4$, $T = 10$ ns.

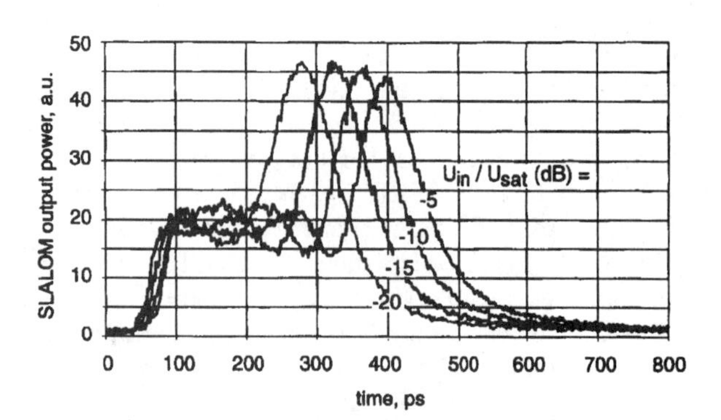

Fig. 9. SLALOM output power (measured) as function of time in control pulse switching operation.

oscillations in the SLALOM gain can be seen. These are due to alternating constructive and destructive interference of cw and ccw signals at the SLALOM coupler, as the phase difference $\varphi_{cw} - \varphi_{ccw}$ in (5) passes through several π.

To verify the theoretical results, an experiment was performed. A continuous wave signal at $\lambda = 1540$ nm with 240 μW power was fed into the SLALOM. The eccentricity T was 10 ns. A pulse at $\lambda = 1550$ nm with FWHM ≈ 25 ps and variable energy was coupled to the loop by an additional 3-dB coupler in the loop. The 1540 nm SLALOM output signal was filtered and displayed on an oscilloscope. Fig. 9 shows the result. It is very similar to the theoretically obtained results. Especially, the oscillations in the output power or SLALOM gain can be observed.

These oscillations allow to calculate the SLA linewidth enhancement factor and the carrier lifetime from the measured curves. In the appendix, a detailed analysis can be found. Here, we report the approximated results. The first (latest) maximum and minimum output power values $P_{out,\max}$ and $P_{out,\min}$, respectively, are connected to the linewidth enhancement factor α as

$$\frac{P_{out,\min}}{P_{out,\max}} = (1 - e^{-\pi/\alpha})^2. \tag{28}$$

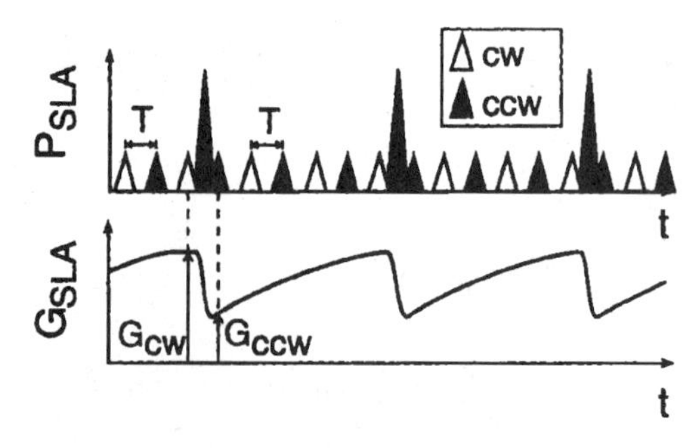

Fig. 10. Schematic of power at the SLA in the loop and resulting SLA gain in fast pulse switching operation, white: clockwise (cw), grey: counter-clockwise (ccw) travelling pulses.

Solving this equation for α, we obtain

$$\alpha = \frac{-\pi}{\ln\left(1 - \sqrt{\frac{P_{\min}}{P_{\max}}}\right)}. \tag{29}$$

The time interval between the first maximum and minimum can be used as a measure of the carrier lifetime. As an approximation, the lifetime reveals to

$$\tau_e = \frac{t(P_{out,\max}) - t(P_{out,\min})}{\ln 2}. \tag{30}$$

Taking the values from Fig. 9, we measure $(P_{out,\max} / P_{out,\min}) = 0.31$ and $t(P_{out,\max}) - t(P_{out,\min}) = 80$ ps.

Using (29) and (30), we obtain $\alpha = 3.9$ and $\tau_e = 115$ ps. These are quite good results compared to the values reported in [11] and listed in Table I.

On the other hand it should be mentioned that the obtained results are only an approximation, as the dependence of α and τ_e on the carrier density N is not considered due to averaging over a large range of N.

E. Fast Pulse Switching

The fast switching function is based on the effect that the amplifier can be saturated very quickly and the gain recovers with a much longer time constant (see Section III). The principle was first shown by Salin *et al.* [13]. The eccentricity of the SLA in the loop is small ($T < \tau_e$).

The control pulse is injected via the wavelength selective coupler. Fig. 10 shows the input power to the SLA, namely cw and ccw travelling data pulses (low power) as well as a high power control pulse travelling only in one direction. The control pulse saturates the SLA. If a data pulse is injected into the SLALOM such that its cw component arrives at the amplifier shortly before the control pulse and its ccw component traverses the SLA after the control pulse, the gain for cw and ccw pulses will be different. The gain ratio (G_{ccw}/G_{cw}) is defined by the control pulse energy.

If a data pulse is injected into the SLALOM at a later time, cw and ccw travelling components both will traverse the SLA after the saturating control pulse. They experience nearly the same gain, as the gain recovery time is large compared to the time difference T. Therefore, only a short time slot, which is mainly determined by T, is switched to the SLALOM output.

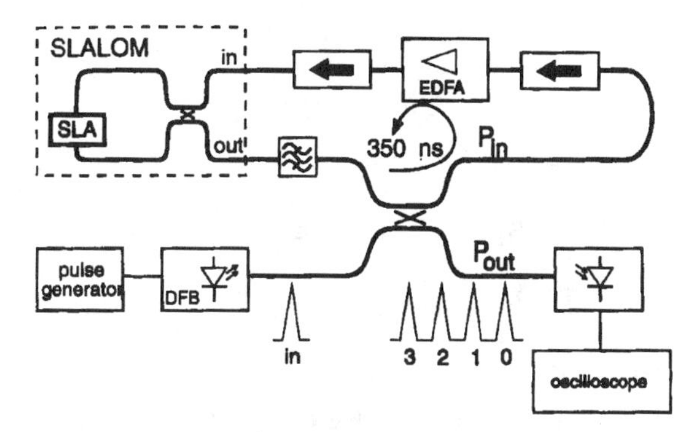

Fig. 11. Set-up for investigation of nonlinear pulse reshaping in a fiber ring.

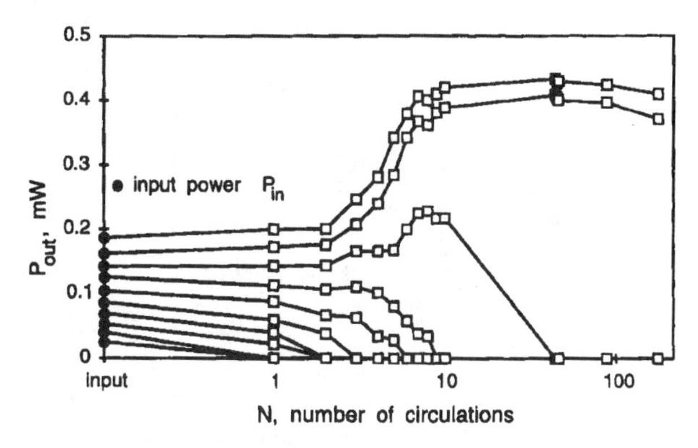

Fig. 12. Fiber ring output power after N circulations for different values of input power.

If the control pulses are injected with a fixed repetition rate, it is not necessary that the SLA gain has completely recovered when the next control pulse arrives. If a quasisteady state has been reached, the amount of saturation by each control pulse is constant. An optimum switching function is obtained, if this saturation results in a phase difference of $\Delta\varphi = \pi$ for the switched pulses [14].

Detailed investigations on the shape of the switching window, taking into account the length of the amplifier and the width of the control pulses, can be found in [15].

V. Applications in Photonic Systems

In this section we report on applications of the SLALOM using the different modes of operation. The nonlinear single-pulse operation was used for pulse regeneration in a fiber loop buffer [5]. The correlation mode of operation was used to decode time-interval coded signals [7]. Finally, the pulse switching with control pulse was used as a retiming gate in an optical regenerator [6] and also for demultiplexing high-speed optical signals [8].

A. Nonlinear Pulse Regeneration

For optical switching applications, fiber loop buffers [16] are a possible method to store high bit rate data signals in the optical form for a limited amount of time (some μs). A packet of optical data is fed to a fiber loop, where it circulates until it is recalled by switching it to the loop output. While circulating in the loop, the signal has to be amplified to overcome the waveguide and coupler losses. Noise during this repetitive amplification processes and other interferences restrict the storage time to about 100 circulations [16], [17].

Here, the nonlinear characteristic of the SLALOM can be used to reshape the pulse after each circulation and to remove the noise from the signal. In the experiment depicted in Fig. 11 we investigated how the nonlinear input-output characteristic of the SLALOM affects pulse propagation in a fiber ring (length 70 m).

Pulses with a low repetition rate, generated by a gain switched DFB laser, are injected into the fiber ring, via a 3-dB coupler. The fiber ring contains an erbium doped fiber amplifier (EDFA) with two isolators and a SLALOM in the setup for nonlinear operation. An optical filter is needed to prevent gain saturation of the EDFA due to amplified spontaneous emission of the SLA. After each circulation in the loop, a part of the pulse power is coupled to the output where it is detected and observed on an oscilloscope. The remainder of the pulse power continues circulating in the loop until it is cancelled by switching off the SLA in the SLALOM. Thereafter, the next pulse is injected into the fiber ring. Thereby, a series of output pulses can be seen on the oscilloscope, from which one pulse with a defined number of circulations can be selected for measurement.

Fig. 12 depicts measurements of P_{out} (see Fig. 11) versus the number N of circulations for different values of the input power P_{in} with $P_{out} = P_{in}$ for $N = 0$. These results indicate that stable circulations are only possible above a certain power level P_{in}. Above this threshold, the effective fiber ring gain (including the nonlinear gain of the SLALOM according to Fig. 3) is larger than one.

Above threshold the number of circulations seems to be unlimited. Experiments were performed for up to $N = 10^6$ circulations, without significant change of the pulse shape (see Fig. 13). Due to the nonlinear input-output characteristic there was also no significant increase of the noise level with the number of circulations. The measurements in Fig. 13 were performed with a fast photodiode and a sampling oscilloscope having a bandwidth of about 20 GHz. For each sample a new start of the measurement procedure had to be undertaken and 512 samples had to be taken for recording one pulse with for instance 1 000 000 circulations. The spikes associated with the pulses in Fig. 13 were not reproducible and we attribute them to the measurement technique. Part of the noisy structure of the pulses is certainly caused by signal-spontaneous beat noise. This noise seems to be present also in the tail of the pulse which is about as long as two times the gain recovery time of the SLA.

Another problem became apparent, when a pair of pulses was circulating in the fiber ring. We observed that after N circulations the position of the second pulse relative to the first one was not fixed, but statistically distributed with a variance σ^2. The measured σ^2 was proportional to N and a minimum jitter of $\sigma^2 = 1, 2$ ps^2 per circulation was obtained. Presumably, this jitter is related to a pulse shaping process during the amplification in the SLA. When the pulse is amplified, the SLA is saturated and the pulse peak is shifted

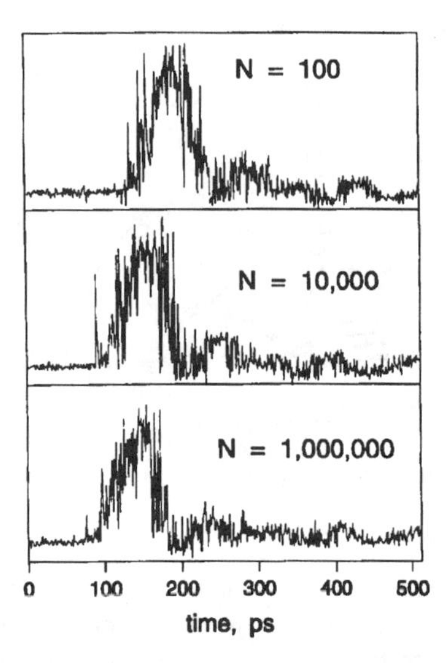

Fig. 13. Fiber ring output pulse shapes after $N = 100$, 10^4, and 10^6 circulations.

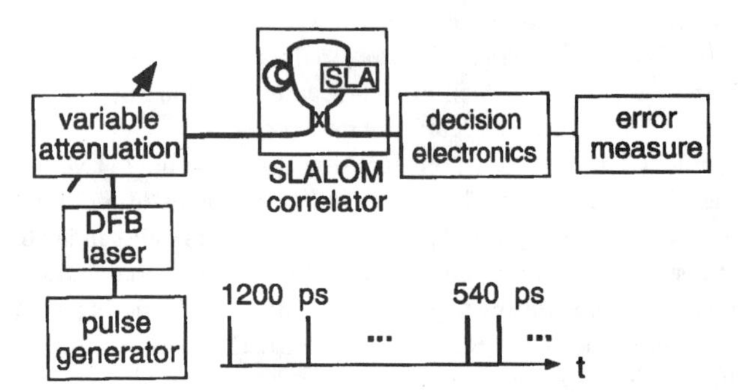

Fig. 14. Experimental setup for pulse-interval decoding. Alternating, pulse pairs with 1200 ps and 540 ps were generated.

slightly. As the saturation depends on the preceding processes (saturation by ASE, earlier pulses) and on the pulse energy, this pulse shift is random and causes the pulse jitter. The jitter can be minimized if the circulating pulse power and the SLA gain is chosen as small as possible. It can be further reduced using narrower pulses.

B. Pulse-Interval Decoding

The pulse correlation operation of the SLALOM was used for decoding a pulse-interval coded binary signal in an optical ATM-switching node. In an OATM switching system, data packets are routed through 2×2 switches according to the routing information (routing tag) which precedes the data packets. The routing tag is decoded in each switching node and the corresponding switch is set to its cross or bar state [18].

As the speed of the switching operation is partly determined by the decoding time, optical processing of the routing tag is required to increase the speed. The SLALOM was used to decode a routing tag, in which the information is contained in the separation of two optical pulses (pulse-interval coding). The experimental setup to measure the decoding error can be found in Fig. 14. The decoder consists of a SLALOM with an internal eccentricity of $T = 480$ ps, followed by a photodiode and an electronic decision gate to detect the SLALOM output pulse. Alternating, pulse pairs with a separation of 1200 ps and 540 ps were generated and sent to the SLALOM. The probability of a detection error (false detected 1200 ps separation or not detected 540 ps separation) was measured as a function of the pulse energy (Fig. 15). For pulse energies of about 30 fJ a misdetection probability of 10^{-9} was obtained. A further increase of the pulse energy to 280 fJ didn't deteriorate the performance, indicating a dynamic range of the decoder of at least 10 dB.

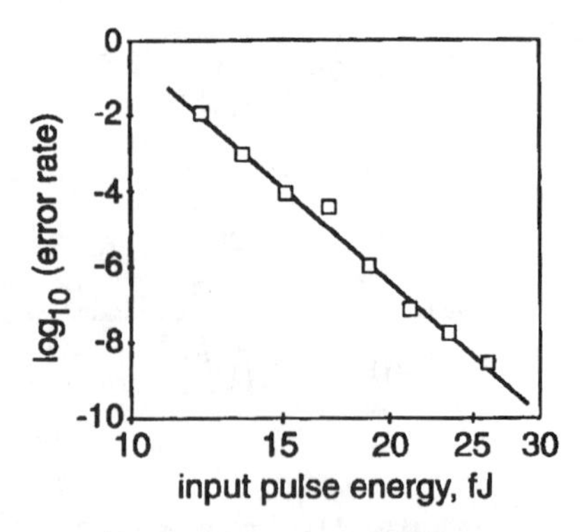

Fig. 15. Detection error probability as function of pulse energy.

C. SLALOM as Optical Retiming Device

All-optical regenerators are essential components for photonic switching systems. In general these regenerators operate as follows. The optical data signal is combined with an optical clock signal (synchronized with the data signal) and both are coupled into a decision gate. Demonstrations of an optical clock recovery using a mode-locked fiber ring laser [19] or a bistable DFB laser [20] have been reported. Here, we consider only the operation of the decision gate and not the synchronization of the clock pulses. Decision gates based on a SEED (data rate 5 kb/s [21]), on a bistable three electrode DFB laser diode (data rate 200 Mb/s [22]), on a bistable Fabry–Perot amplifier (data rate 10 Gb/s [23]) and on a nonlinear Sagnac interferometer switch (data rate 5 Gb/s [3]) have been demonstrated.

Here, we report on the application of the SLALOM in its medium speed switching mode of operation as an optical decision gate. The optical data pulses, which have experienced a timing and amplitude jitter during the fiber transmission, are combined with the locally generated clock pulses at a different wavelength. Both are fed to a SLALOM with an internal eccentricity of T. Upon combination, data and clock pulses have an average separation of somewhat more than T, so that the clock pulse is switched to the SLALOM output, if the data pulse is present within a broad time slot, given by the correlation width of the SLALOM. Figuratively speaking, the data pulse opens a gate, through which the corresponding clock pulse passes to the output. At the SLALOM output, a filter selects only the clock pulse wavelength, so that unwanted components of the data pulses are suppressed.

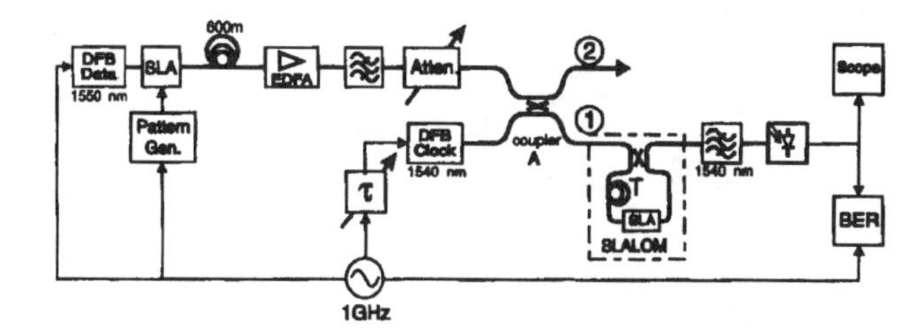

Fig. 16. Experimental setup for data retiming.

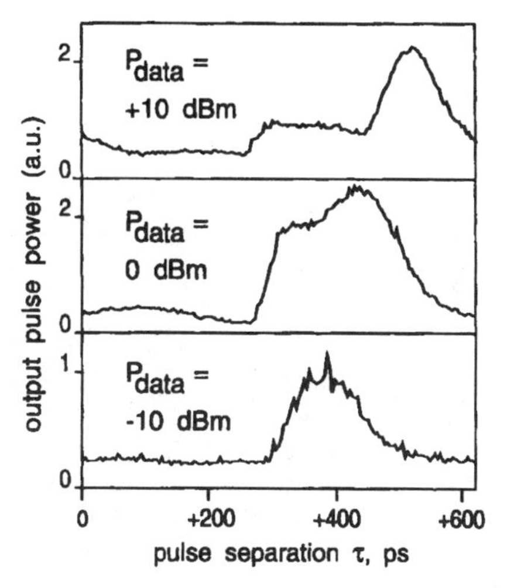

Fig. 17. SLALOM output power of clock pulse as function of separation of data and clock pulses. clock pulse input power: -6.5 dBm; data pulse input power: $+10$, 0, -10 dBm.

In the experiment we report here, the data rate is 1 Gb/s. The internal eccentricity T in the SLALOM is chosen to 310 ps. Data and clock pulses are combined so that the data pulse precedes the clock pulse by an average time of 410 ps. This delay time τ is variable to evaluate the locking range of the retiming device. The experimental setup can be found in Fig. 16.

At the coupler A, data pulses ($\lambda d = 1550$ nm, peak power between -5 dBm and $+10$ dBm) and clock pulses ($\lambda c = 1540$ nm, peak power -6.5 dBm) are combined. Both pulse sources (gain switched DFB lasers) are driven by the same master oscillator (1 GHz). The data pulses are modulated by an SLA gate, which is driven by a pattern generator (PRBS $2^{23}-1$), and pass an Erbium doped fiber amplifier (EDFA, 26-dB gain), an optical filter and a variable attenuator to obtain the desired power at the coupler A. A tunable electrical delay line (time delay τ) between the master oscillator and the clock pulse source is used to vary the time delay between clock and data pulses for testing the jitter tolerance of the optical regenerator. A relative jitter may also be provided by frequency modulation of the master oscillator in combination with the different fiber lengths in the clock and data paths. The filter at the output of the SLALOM is tuned to the clock pulses ($\lambda c = 1540$ nm).

To investigate the locking range of the optical regenerator we performed the following experiment. The time separation τ between data pulse and clock pulse was changed continuously and the output pulse power behind the filter ($\lambda c = 1540$ nm) at the output of the SLALOM in Fig. 16 was measured versus

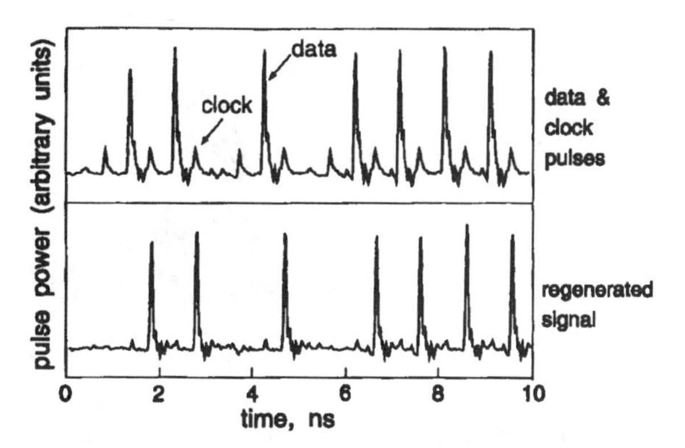

Fig. 18. upper trace: input signal to SLALOM retiming device; lower trace: output signal after filtering.

τ for the fixed power -6.5 dBm of the clock pulse and for three powers P_{data} of the data pulse. For $\tau > 0$ the data pulse arrives prior to the clock pulse at the SLALOM. Fig. 17 shows the results. The pulse separation $\tau_{\max}$ associated with the maximum output pulse power increases with increasing power of the data pulse. Moreover, the locking range of the regenerator (temporal width of the switching window) seems to have a maximum near $P_{data} = 0$ dBm. Consequently we expect an optimum for the jitter tolerance of the SLALOM based regenerator for data pulse powers near 0 dBm.

Fig. 18 depicts, in the upper part, data and clock pulses at output 2 of the coupler A and, in the lower part, the regenerated signal, where the information of the data pulses has been transferred to the clock pulses. The regenerated signal is independent of the polarization of the data signal for $\lambda c \neq \lambda d$. In general the only requirement for λc and λd is, that both fit into the bandwidth (60 nm) of the SLA. Fig. 19 depicts the eye diagrams of the data signals before and after regeneration, where the data pulses were jittered intentionally.

To test the jitter tolerance and the dynamic range of the SLALOM based regenerator, we performed BER measurements of the regenerated signal by electrically varying the relative delay between the input data pulses and the clock pulses as well as the data pulse peak power. Fig. 20 shows curves for equal bit-error rates at the BER values 10^{-5}, 10^{-7}, and 10^{-9}. At BER $= 10^{-9}$ (circles in Fig. 20), the measurements revealed an optimum jitter tolerance of 150 ps for a data pulse power of about 0 dBm and also an increase of τ with increasing data pulse power, both in agreement with the results in Fig. 17. For a pulse separation τ of about 450 ps, the dynamic range is 15 dB. Both values (large jitter tolerance and large dynamic range) cannot be obtained simultaneously. As a compromise, the shaded area in Fig. 20 defines a region for BER $< 10^{-9}$, where a dynamic range of 7 dB is obtained together with a jitter tolerance of 100 ps.

D. SLALOM as Demultiplexer

The application of the SLALOM as a fast demultiplexer was first demonstrated by Sokoloff *et al.* [9]. He called his extended SLALOM setup "TOAD" (terahertz optical asymmetric demultiplexer). The setup has now widely been used to demultiplex high data rates of up to 160 Gb/s down to 10 Gb/s,

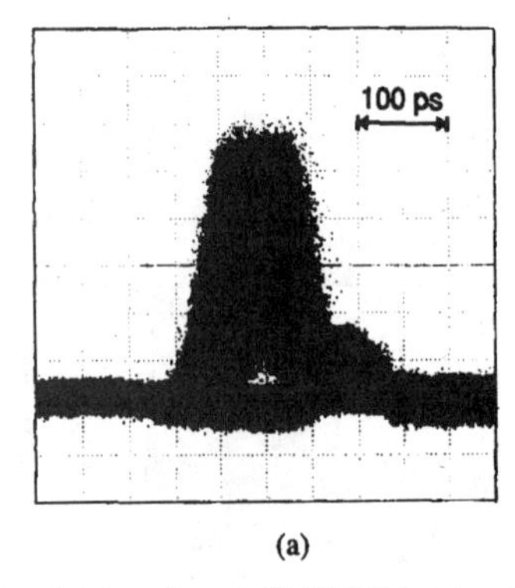

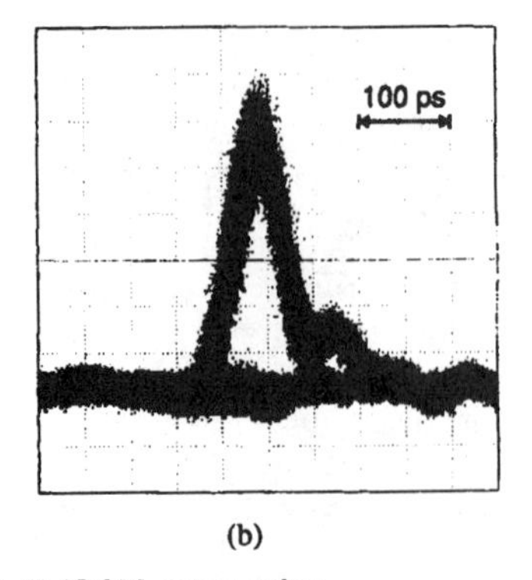

Fig. 19. Eye diagrams of (a) jittered data pulses at SLALOM input and (b) SLALOM output pulses.

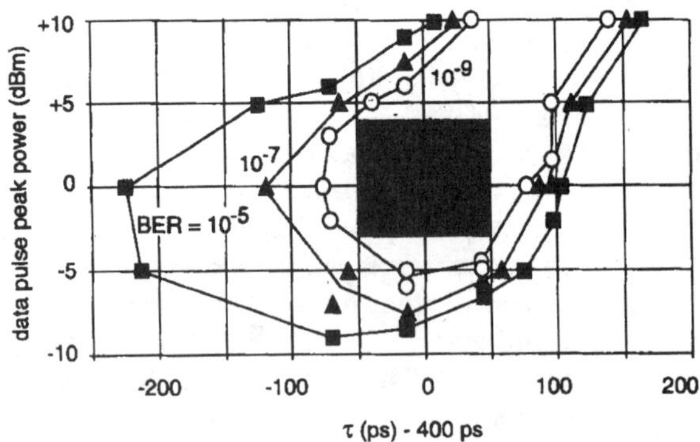

Fig. 20. Error rate of retimed data signals as function of data and clock pulse separation τ and data input power. Shown are curves for equal bit-error rate (BER) 10^{-5} (□), 10^{-7} (Δ) and 10^{-9}(○). The shaded rectangle demonstrates jitter tolerance of 100 ps and dynamic range of 7 dB.

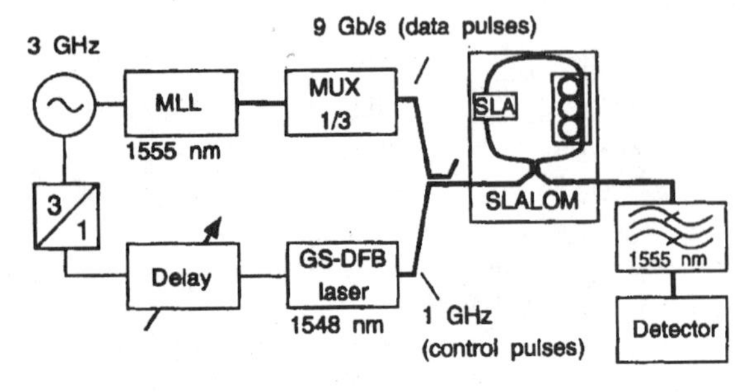

Fig. 21. Experimental setup for time-division demultiplexing.

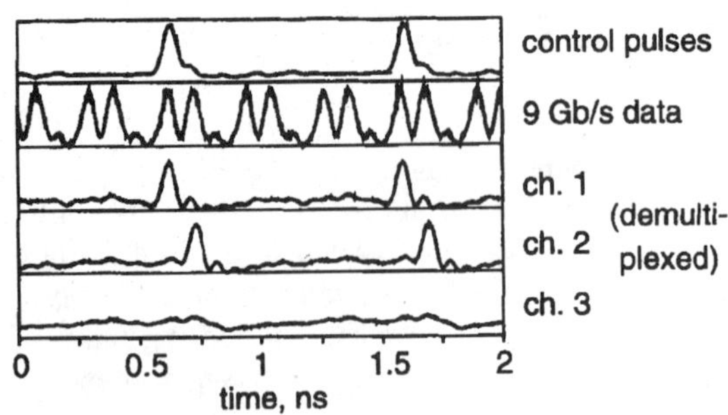

Fig. 22. Results of demultiplexing with SLALOM.

which can be processed electronically [12], [24], [25]. In this paper, we report on a rather basic experiment demonstrating the principle of the SLALOM-demultiplexer, where control and data pulses both were injected into the SLALOM by the same input port [8].

Fig. 21 depicts the experimental setup. The data signal was a 9 Gb/s bit stream with the periodic RZ bit pattern 110. It was generated by an actively modelocked (3 GHz) external cavity laser (MLL) operating at 1555 nm and a passive 1:3 multiplexer consisting of two fiber couplers and an appropriate delay line. Control pulses with 1 GHz repetition rate were obtained from a gain switched DFB-laser (GS-DFB) operating at 1548 nm. The 1 GHz modulation signal was derived from the generator driving the MLL. A variable delay enabled a shift of the control pulses relative to the data pulses. Both pulse streams were combined by a 3-dB coupler and were fed into the SLALOM. At the input of the SLALOM the peak power was 75 μW for the data pulses and 1.2 mW for the control pulses, and the pulse width was 20 ps for the data pulses and 35 ps for the control pulses. There is no need to adjust the polarizations of the data and control pulses relative to each other. The SLA in the SLALOM was arranged such that the propagation times of the pulses from the coupler to the SLA via the two arms differed by a time $T = 65$ ps. The SLA was the bulk structured laser described in Section IV. At the output of the SLALOM only the data pulses passed an optical filter (1555 nm) and were detected.

Fig. 22 shows the traces of the injected control and data pulses as well as the filtered output pulses for three different positions of the control pulses with respect to the data pulses. It can be seen that a switching of one time slot can be obtained, while neighboring slots are suppressed. In principle a slot width of as small as 4 ps can be achieved [25].

VI. Limitations of High Speed Operation

All of the reported applications are restricted in the data rate by the gain recovery time of the SLA. The maximum data rate is some Gb/s. For the demultiplexer operation, the demultiplexed data rate is restricted by the gain recovery time, whereas the line data rate is mainly limited by the geometrical length of the SLA [15].

To reduce the carrier lifetime, mainly two approaches have been made. One of them is to implant recombination centers into the active region of the SLA by proton- or nitrogen-ion bombardment [26]. The other approach is to saturate the SLA by a strong continuous wave signal, thus increasing the induced recombination and reducing the effective carrier lifetime. Using this second method a reduction of the carrier lifetime from some hundreds of ps to about 30 ps has been obtained [14]. Unfortunately, a reduction of SLA gain comes along with the reduction of τ_e in both methods. Therefore, a compromise has to be found. Using an SLA with reduced τ_e in a SLALOM configuration, an operation on data rates of more than 10 Gb/s can be expected. For the demultiplexer even 80 Gb/s for the lower data rate can be anticipated [14]. That means, the SLALOM configuration is a promising candidate for optical signal processing in the very high bit rate regime.

VII. Conclusion

We reported on the SLALOM (semiconductor laser amplifier in a loop mirror) and its applications for optical signal processing. Four modes of operation were described:

- single pulse processing based on the nonlinear gain characteristics,
- two-pulse correlation,
- pulse switching for medium switching times (times larger than the gain recovery time),
- fast pulse switching (times smaller than the gain recovery time).

Applications of the SLALOM in data transmission and processing systems have been shown for each of the four modes of operation: The nonlinear characteristics of the SLALOM were used to reshape pulses circulating in a fiber ring buffer. One million circulations of a single pulse were obtained. A decoder for pulse-interval coded signals was implemented using the correlation mode of operation of the SLALOM. With 30 fJ optical pulse energy, a misdetection probability of 10^{-9} was obtained. In the medium time switching mode, the SLALOM was used as a retiming gate for an optical regenerator. A retiming bit error rate of 10^{-9} was obtained with a jitter tolerance of 100 ps and a dynamic range of 7 dB. Finally, the fast pulse switching was demonstrated by demultiplexing a 1 Gb/s bit stream from a 9-Gb/s data stream.

The operation speed of the SLALOM is mainly determined by the SLA's gain recovery. By reducing the gain recovery time, an operation on data rates of more than 10 Gb/s can be expected. For the demultiplexer operation, a lower data rate of 80 Gb/s and a slot width of 4 ps can be achieved.

The variety of applications together with the in principle polarization insensitive operation and feasibility of device integration make the interferometer with an SLA a promising device for all-optical signal processing.

Appendix
Estimate of Linewidth Enhancement Factor and Carrier Lifetime

In the medium speed switching experiment, described in Section IV, the measured output power is a function of time, according to (5) and (19). As a slight modification from (5), the SLALOM output power is

$$P_{out} = P_{in} \cdot G_{ccw} \left\{ d^4 \frac{G_{cw}}{G_{ccw}} + k^4 - 2d^2k^2 \sqrt{\frac{G_{cw}}{G_{ccw}}} \cdot \cos(\varphi_{cw} - \varphi_{ccw}) \right\} \quad \text{(A1)}$$

d and k are the coupling coefficients of the coupler in the SLALOM.

With $G_{ccw} = g_{ccw}^2 = G_{SLA}(t+T)$, $G_{cw} = G_{SLA}(t)$, and $G_{SLA}(t) = G(t)$ according to (19), (A1) describes the temporal change of the output power after the saturation at $t = t_s$.

According to (23), the relation between $\Delta\varphi = \varphi_{cw} - \varphi_{ccw}$ and (G_{cw}/G_{ccw}) can be used

$$\frac{G_{cw}}{G_{ccw}} = \exp\left(-\frac{2\Delta\varphi}{\alpha}\right). \quad \text{(A2)}$$

Therefore, P_{out} can be written as a function of the (constant) gain G_{ccw} and the time dependent $\Delta\varphi$

$$P_{out} = P_{in} \cdot G_{ccw} \cdot \left\{ d^4 \cdot \exp\left(-\frac{2\Delta\varphi}{\alpha}\right) + k^4 - 2d^2k^2 \cdot \exp\left(-\frac{\Delta\varphi}{\alpha}\right) \cos\Delta\varphi \right\}. \quad \text{(A3)}$$

The maximum and minimum of P_{out} with respect to $\Delta\varphi$ is obtained solving $(dP_{out}/d\Delta\varphi) = 0$.

As the first solution will be found at $\Delta\varphi \approx \pi$, the following approximations are made

$$\exp\left(-\frac{\Delta\varphi_{\max}}{\alpha}\right) \approx \exp\left(\frac{-\pi}{\alpha}\right);$$
$$\cos\Delta\varphi_{\max} \approx -1;$$
$$\sin\Delta\varphi_{\max} \approx \pi - \Delta\varphi_{\max}.$$

Thereby, we obtain

$$\Delta\varphi_{\max} \approx \pi - \frac{d^2 \exp\left(-\frac{\pi}{\alpha}\right) + k^2}{\alpha \cdot d^2}. \quad \text{(A4)}$$

Similarly, the first minimum of the output power can be found at $\Delta\varphi_{\min} \approx 2\pi$.

Using the approximations

$$\exp\left(-\frac{\Delta\varphi_{\min}}{\alpha}\right) \approx \exp(-\frac{2\pi}{\alpha}),$$
$$\cos\Delta\varphi_{\min} \approx 1,$$

and

$$\sin\Delta\varphi_{\min} \approx \Delta\varphi_{\min} - 2\pi$$

we obtain

$$\Delta\varphi_{\min} \approx 2\pi + \frac{d^2 \exp\left(-\frac{2\pi}{\alpha}\right) - k^2}{\alpha \cdot d^2}. \quad \text{(A5)}$$

For practical relevant values of $0.4 \leq d^2 \leq 0.6$ and $\alpha \geq 2$, the approximations reveal results for $\Delta\varphi$ which differ by less than 0.1 from the numerically obtained zeros of $(dP_{out}/d\Delta\varphi)$.

Using $\Delta\varphi_{\max}$ and $\Delta\varphi_{\min}$ from (A4) and (A5), respectively, (A3) reveals $P_{out,\max}$ and $P_{out\ \min}$. The ratio

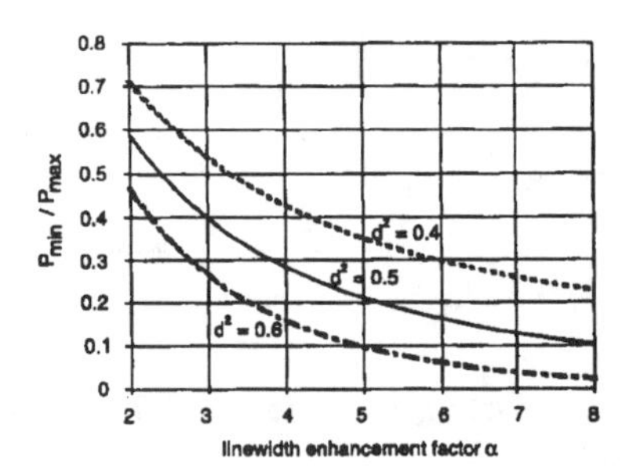

Fig. A1. Ratio of minimum and maximum SLALOM output power as function of linewidth enhancement factor α d^2 defines the coupling ratio of the SLALOM coupler.

$(P_{out,\min}/P_{out,\max})$ is a function of d^2, k^2, and α. The function $(P_{out,\min}/P_{out,\max}) = f(\alpha)$ is plotted in Fig. A1 for the three coupling ratios $d^2 = 0.4$, $d^2 = 0.5$ (optimum coupler), and $d^2 = 0.6$.

If the calculation of $\Delta\varphi_{\min}$ and $\Delta\varphi_{\max}$ is simplified further, we reveal $\Delta\varphi_{\min} \approx 2\pi$ and $\Delta\varphi_{\max} \approx \pi$. Using these results in (A3), we obtain

$$\frac{P_{out,\min}}{P_{out,\max}} \approx \frac{(d^2 e^{-2\pi/\alpha} - k^2)^2}{(d^2 e^{-\pi/\alpha} + k^2)^2}$$

For $d^2 = k^2 = 1/2$, this results to

$$\frac{P_{out,\min}}{P_{out,\max}} \approx (1 - e^{-\pi/\alpha})^2. \quad \text{(A6)}$$

This approximation was used in Section IV.

The carrier lifetime τ_e in the SLA can be calculated using the time interval between the minimum and maximum output power of Fig. 10. As described before, the ratio of the gain for cw and ccw signals determines the output power. The gain for the ccw signal was assumed to be the unsaturated SLA gain, therefore the gain ratio is determined as

$$\frac{G_{cw}}{G_{ccw}} = \frac{G_{cw}}{G_0}.$$

Using (19), the temporal change of G_{cw} is determined.

With use of (A2), the times $t_{\min}$ and $t_{\max}$ for minimum and maximum output power can be determined as a function of $\Delta\varphi_{\min}$ and $\Delta\varphi_{\max}$, respectively:

$$t_{\min} - t_s = -\tau_e \cdot \ln \frac{-2\frac{\Delta\varphi_{\min}}{\alpha}}{\ln \frac{G(t_s)}{G_0}} \quad \text{(A7)}$$

$$t_{\max} - t_s = -\tau_e \cdot \ln \frac{-2\frac{\Delta\varphi_{\max}}{\alpha}}{\ln \frac{G(t_s)}{G_0}}. \quad \text{(A8)}$$

Using (A7) and (A8), the time separation $t_{\max} - t_{\min}$ reveals

$$t_{\max} - t_{\min} = \tau_e \cdot \ln \frac{\Delta\varphi_{\min}}{\Delta\varphi_{\max}}. \quad \text{(A9)}$$

As an approximation for $d^2 = k^2 = 1/2$, we can again use $\Delta\varphi_{\min} \approx 2\pi$ and $\Delta\varphi_{\max} \approx \pi$ and we obtain $t_{\max} - t_{\min} \approx \tau_e$ ln 2, which has been used in (30).

Acknowledgment

The authors wish to thank their colleagues R. Ludwig, R. Langenhorst, R. Schnabel, L. Küller, and E. Dietrich for technical support and discussions. They also thank K. Magari from NTT and M. Möhrle from HHI for providing the SLA's used in the experiments.

References

[1] D. M. Patrick, A. D. Ellis, and D. M. Spirit, "Bit-rate flexible all-optical demultiplexing using a nonlinear optical loop mirror," *Electron. Lett.*, vol. 29, pp. 702–703, 1993.

[2] P. A. Andrekson, N. A. Olsson, J. R. Simpson, D. J. Digiovanni, P. A. Morton, T. Tanbun-Ek, R. A. Logan, and K. W. Wecht, "64 Gbit/s all-optical demultiplexing with the nonlinear optical loop mirror," *IEEE Photon. Technol. Lett.*, vol. 4, pp. 644–647, 1992.

[3] M. Jinno and M. Abe, "All-optical regenerator based on nonlinear fiber Sagnac interferometer," *Electron. Lett.*, vol. 28, pp. 1350–1352, 1992.

[4] M. Eiselt, "Optical loop mirror with semiconductor laser amplifier," *Electron. Lett.*, vol. 28, pp. 1505–1507, 1992.

[5] M. Eiselt, W. Pieper, G. Großkopf, R. Ludwig, and H. G. Weber, "One million pulse circulations in a fiber ring using a SLALOM for pulse shaping and noise reduction," *IEEE Photon. Technol. Lett.*, vol. 4, pp. 422–424, 1993.

[6] M. Eiselt, W. Pieper, and H. G. Weber, "A decision gate for all-optical data retiming using a semiconductor laser amplifier in a loop mirror configuration," *Electron. Lett.*, vol. 29, pp. 107–109, 1993.

[7] ——, "New optical correlator technique for header recognition in self-routing networks," in *OFC/IOOC'93, Tech. Dig.*, pp. 199–200.

[8] ——, "SLALOM as high speed multiplexer/demultiplexer," *Electron. Lett.*, vol. 29, pp. 1167–1168, 1993.

[9] J. P. Sokoloff, P. R. Prucnal, I. Glesk, and M. Kane, "A terahertz optical asymmetric demultiplexer (TOAD)," in *Photonics in Switching 1993*, Palm Springs, postdeadline paper PD4.

[10] A. E. Siegman, *Lasers*. Mill Valley, CA: Univ. Sci. Books, 1986, p. 362 ff.

[11] R. Ludwig, K. Magari, and Y. Suzuki, "Properties of 1.55 μm high gain polarization insensitive MQW semiconductor laser amplifiers," in *Opt. Amplifiers and Their Applicat., Tech. Dig. 1992*, OSA, Washington, DC, 1992, vol. 17, pp. 148–151.

[12] A. D. Ellis and D. M. Spirit, "Compact 40 Gbit/s optical demultiplexer using a GaInAsP optical amplifier," *Electron. Lett.*, vol. 29, pp. 2115–2116, 1993.

[13] F. Salin, J. Squier, and K. Clay, "Ultrafast gating using a nonlinear Sagnac interferometer," *Optics Commun.*, vol. 88, pp. 151–156, 1992.

[14] R. J. Manning and G. Sherlock, "Recovery of a π phase shift in ~12.5 ps in a semiconductor laser amplifier," *Electron. Lett.*, vol. 31, pp. 307–308, 1995.

[15] J. P. Sokoloff, P. R. Prucnal, I. Glesk, and M. Kane, "Performance of a 50 Gbit/s optical time domain multiplexed system using a terahertz optical asymmetric demultiplexer," *IEEE Photon. Technol. Lett.*, vol. 6, pp. 98–100, 1994.

[16] R. Langenhorst, M. Eiselt, W. Pieper, G. Großkopf, R. Ludwig, L. Küller, E. Dietrich, and H. G. Weber, "Fiber loop optical buffer," *J. Lightwave Technol.*, submitted for publication.

[17] W. Pieper, M. Eiselt, G. Großkopf, R. Langenhorst, A. Ehrhardt, and H. G. Weber, "Investigation of crosstalk interference in a fiber loop optical buffer," *Electron. Lett.*, vol. 30, pp. 435–436, 1994.

[18] A. Ehrhardt, M. Eiselt, G. Großkopf, L. Küller, R. Ludwig, W. Pieper, R. Schnabel, and H. G. Weber, "Semiconductor laser amplifier as optical switching gate," *J. Lightwave Technol.*, vol. 11, pp. 1287–1295, 1993.

[19] K. Smith and J. K. Lucek, "All-optical clock recovery using a mode-locked laser," *Electron. Lett.*, vol. 28, pp. 1814–1816, 1992.

[20] M. Möhrle, U. Feiste, J. Hörer, R. Molt, and B. Sartorius, "Gigahertz self-pulsation in 1.5 μm wavelength multisection DFB lasers," *IEEE Photon. Technol. Lett.*, vol. 4, pp. 976–978, 1992.

[21] C. R. Giles, T. Li, T. H. Wood, and C. A. Burrus, "All-optical regenerator," *Electron. Lett.*, vol. 24, pp. 848–850, 1988.

[22] M. Jinno and T. Matsumoto, "Optical retiming regenerator using 1.5 μm wavelength multielectrode DFB LD's," *Electron. Lett.*, vol. 25, pp. 1332–1333, 1989.

[23] K. Weich, E. Patzak, and J. Hörer, "Fast all-optical switching using two-section injection locked semiconductor lasers," *Electron. Lett.*, vol. 30, pp. 493–494, 1994.

[24] K. Suzuki, K. Iwatsuki, S. Nishi, and M. Saruwatari, "160 Gbit/s single polarization subpicosecond transform limited pulse signal demultiplex-

ing using ultrafast optical loop mirror including MQW travelling wave semiconductor laser amplifier," *Electron. Lett.*, vol. 30, pp. 660–661, 1994.

[25] I. Glesk, J. P. Sokoloff, and P. R. Prucnal, "Demonstration of all-optical demultiplexing of TDM data at 250 Gbit/s," *Electron. Lett.*, vol. 30, pp. 339–341, 1994.

[26] D. J. Derickson, R. J. Helkey, A. Mar, J. R. Karin, J. G. Wasserbauer, and J. E. Bowers, "Short Pulse Generation Using Multisegment Mode-Locked Semiconductor Lasers," *IEEE J. Quantum Electron.*, vol. 28, pp. 2186–2201, 1992.

Space Division Switches Based on Semiconductor Optical Amplifiers

R.F. KALMAN, L.G. KAZOVSKY, AND J.W. GOODMAN

Abstract—**Multiple space-division switches based on semiconductor optical amplifiers (SOA's) can be cascaded to obtain larger switching fabrics. We present a general analysis of optical switching fabrics using SOA's, considering noise and saturation effects associated with amplified spontaneous emission. We find that the SOA saturation output power limits the number of switches which can be cascaded. For example, a saturation output power of 100 mW limits the size of switching fabrics to 100 64 × 64 switches or 200 8 × 8 switches if distributed gain matrix-vector multiplier (MVM) switches or Benes switches are employed. The corresponding limit for lumped gain MVM fabrics is 10 64 × 64 or 100 8 × 8. The Benes switch may be more suitable for large switch sizes ($N > 16$) because it requires fewer SOA's.**

I. Introduction

All optical space-division switches can be constructed using optical splitters, combiners, and semiconductor optical amplifiers (SOA's) [1] which provide gain and binary on/off switching in 1–10 ns. Multiple switches can be interconnected to form larger switching fabrics.

Switches built from SOA's have been examined by a number of authors [1], [2], and at least one paper has considered the size limitations of a specific SOA-based switch [3]. The objective of this letter is to evaluate the limitations on the size of switch fabrics constructed using SOA-based switches. The results we derive are based on signal-to-noise ratio and saturation constraints. In Section II, we analyze the behavior of systems based on SOA's. In Section III, we evaluate the size limitations of three specific switch implementations. Discussion and conclusions are presented in Sections IV and V, respectively.

II. SOA and Switch Behavior

A. The Evolution of Spontaneous Emission

The output optical power from a SOA, P_{out}, is related to the input power, P_{in}, by

$$P_{\text{out}} = L_{\text{in}} L_{\text{out}} G P_{\text{in}} + L_{\text{out}} p n_{\text{sp}} (G-1) h\upsilon_c \Delta\upsilon \quad (1)$$

where G is the internal SOA gain, and L_{in} and L_{out} are the input and output coupling losses to the SOA, respectively, υ_c is the center frequency of the amplifier bandpass, and $\Delta\upsilon$ is the effective amplifier optical bandwidth.

Manuscript received January 22, 1992; revised May 18, 1992. This work was supported by DARPA under NOSC Contract N66001-86-C-0382 and ONR Contract N00014-91-J-1857.

The authors are with Optivision, Inc., Palo Alto, CA 94304 and the Department of Electrical Engineering, Stanford University, Stanford, CA 94305.

IEEE Log Number 9202665.

n_{sp} is the excess spontaneous emission factor [4] and p is a factor which ranges from 1 for a device which amplifies only one polarization to 2 for a polarization-insensitive device.

Consider a cascade of M identical stages, each stage consisting of a system loss, L_s, and a SOA with its associated coupling losses. Using (1), we find that the power output from the Mth stage, P_M, is given by

$$P_M = G_{\text{sig}} P_{\text{in}} + G_{\text{sp}} n_{\text{sp}} p_{\text{eff}} h\upsilon_c \Delta\upsilon_{\text{eff}} \quad (2)$$

where $\Delta\upsilon_{\text{eff}}$ is the effective overall gain bandwidth due to the cascade of SOA's, and p_{eff} is the effective polarization-dependent factor which ranges from 1 to 2 in a manner similar to p. The net signal gain through the system, G_{sig}, is given by

$$G_{\text{sig}} = (L_s L_{\text{in}} L_{\text{out}} G)^M. \quad (3)$$

By analogy to the signal gain, we can define a spontaneous emission gain, G_{sp},

$$G_{\text{sp}} = \left(\frac{G_{\text{sig}}^{1/M}}{L_s L_{\text{in}}} - L_{\text{out}} \right) \frac{1 - G_{\text{sig}}}{1 - G_{\text{sig}}^{1/M}} \cong M \left(\frac{G_{\text{sig}}^{1/M}}{L_s L_{\text{in}}} - L_{\text{out}} \right) \frac{G_{\text{sig}} - 1}{\ln G_{\text{sig}}} \quad (4)$$

The second equality in (4) holds for large M.

B. Saturation and Signal-to-Noise Ratio Constraints

Due to the high spontaneous emission level emerging from a cascade of SOA-based stages, the post-detection noise in a receiver placed at the output of the last SOA is typically dominated by signal-spontaneous and spontaneous-spontaneous beat noise. This leads to a post-detection signal-to-noise ratio (SNR), γ, of approximately [5]

$$\gamma = \frac{(G_{\text{sig}} P_{\text{in}})^2}{4 G_{\text{sp}} n_{\text{sp}} h\upsilon_c B_e (G_{\text{sig}} P_{\text{in}} + G_{\text{sp}} n_{\text{sp}} h\upsilon_c B_o)} \quad (5)$$

where B_e is the electrical noise bandwidth of the receiver and $B_o (B_o \leq \Delta\upsilon_{\text{eff}})$ is the bandwidth of an optical filter placed in front of the receiver. To achieve a 10^{-9} bit error ratio (BER), we require $\gamma > 144$.

SOA's exhibit nonlinear distortion characterized by a saturation output power, P_{sat}, at which the gain has dropped to $1/e$ of its unsaturated value [6]. Saturation leads to a number of undesirable effects: a decrease in gain, intersymbol interference (ISI) and, in frequency-division multiplexed systems, crosstalk. We consider the simple saturation constraint:

Reprinted with permission from *IEEE Photonics Technology Letters,* Vol. 4, No. 9, pp. 1048-1051, September 1992.

$$(G_{sig} P_{in} + G_{sp} p_{eff} n_{sp} h\nu_c \Delta\nu_{eff}) \frac{1}{L_{out}} \le P_{sat}. \quad (6)$$

Equation (6) indicates that the total power emerging from the endface of the last SOA in a cascaded amplifier system (i.e., before its output coupling loss) must be less than P_{sat}.

C. Specific Switch Architectures

In this letter, we examine switches based on two versions of the matrix-vector multiplier (MVM) crossbar switch [7], [8] (Fig. 1) and the Benes switch [1] (Fig. 2). The structure of the three switches allows the direct application of the analysis of Section II-A. For each of the switches, Table I gives the number of stages traversed by a signal, M_s, the system loss per stage, L_s, and the number of SOA's required to implement a switch with N inputs/outputs.

The MVM switches allow completely general nonblocking interconnections between the inputs and outputs [7]. The distributed gain MVM (DGMVM) switch differs from the lumped gain MVM (LGMVM) switch in that it incorporates additional SOA's after each 1×2 splitter and 2×1 combiner. The Benes switch provides a rearrangeably nonblocking interconnection [1]. From Table I, note that the number of SOA's in the MVM switches grows as N^2 versus only $N \log_2 N$ for the Benes switch.

Multiple switches can be interconnected to form larger switching fabrics. For practical reasons, these larger switch fabrics will tend to be dilute and thus highly blocking. A signal traveling though a switch fabric containing M_c cascaded switches encounters a total of M_t stages where

$$M_t = M_s M_c. \quad (7)$$

III. System Size Limitations

We consider system size limitations based on two phenomena: required signal-to-noise ratio (SNR) at the system output, and saturation of the SOA's. For a given value of the signal gain G_{sig}, we can solve (5) and (6) simultaneously to find the maximum allowable spontaneous gain, $G_{sp,max}$, and the optimum input signal level. Using G_{sig} and $G_{sp,max}$, and solving (4) for the maximum permissible total number of stages, M_{max}, we find

$$M_{max} = G_{sp\,max} \left(\frac{G_{sig}^{1/M}}{L_s L_{in}} - L_{out} \right)^{-1} \frac{\ln G_{sig}}{G_{sig} - 1}. \quad (8)$$

Utilizing the relationships between M_s, L_s, and N for the three switch types (Table I) and using (7) and (8), we can find the maximum number of switches of a given size which can be cascaded. Fig. 3 shows the maximum number of 8×8 switches which can be cascaded for the three switch types as a function of P_{sat}. Similarly, Fig. 4 shows the maximum number of 64×64 switches which can be cascaded.

IV. Discussion

From Figs. 3 and 4, the number of switches which can be cascaded can be seen to be proportional to the saturation output power of the SOA's. Current SOA's exhibit saturation power levels in the range of 1–100 mW (the latter being achieved in quantum well devices).

Because their system loss per stage is small, DGMVM and Benes switches typically utilize low gain SOA's which have relatively small on/off extinction ratios, leading to crosstalk. At the expense of increasing n_{sp}, the off-state absorption of a SOA can be made arbitrarily large by increasing its length. Figs. 3 and 4 assume a n_{sp} of 5, which corresponds to an extinction ratio of 50 dB for a device with an internal gain of 10. This high absorption prevents crosstalk from impacting system performance. Because they typically utilize high gain SOA's, LGMVM switches tend not to exhibit crosstalk problems.

Due to the large numbers of required SOA's, achieving practical switches requires integration of SOA's and optical waveguides. Electrical pin-out constraints of packages suggest that monolithic integration of drive electronics with the optical devices may also be necessary to achieve $N > 8$. As noted in Table I, for large switch sizes ($N > 16$) the Benes switch requires considerably fewer SOA's than the MVM structures and may thus be the preferred architecture for large switches.

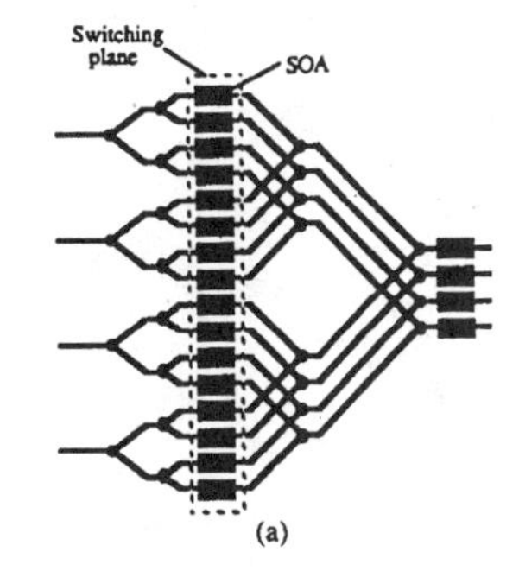

(a)

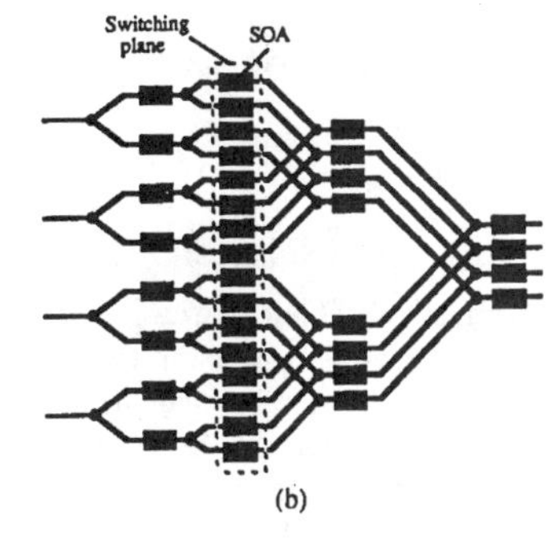

(b)

Fig. 1. MVM crossbar switches: (a) lumped-gain; (b) distributed gain. As an example 4 × 4 switches are shown.

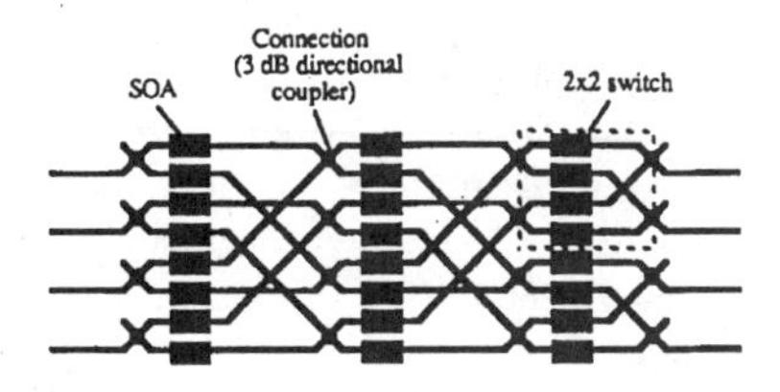

Fig. 2. Benes switch. As an example, a 4 × 4 switch is shown.

TABLE I
NUMBER OF STAGES (M), SYSTEM LOSS PER STAGE (L_s), AND NUMBER OF SOA'S, FOR LGMVM, DGMVM, AND BENES SWITCHES OF SIZE N

	LGMVM	DGMVM	Benes
Number of stages (M_s)	2	$2 \log_2 N$	$2 \log_2 N - 1$
System Loss per Stage (L_s)	$1/N$	0.5	0.5
Number of SOA's	$N(N+1)$	$3N(N-1)$	$2N(\log_2 N - 1)$

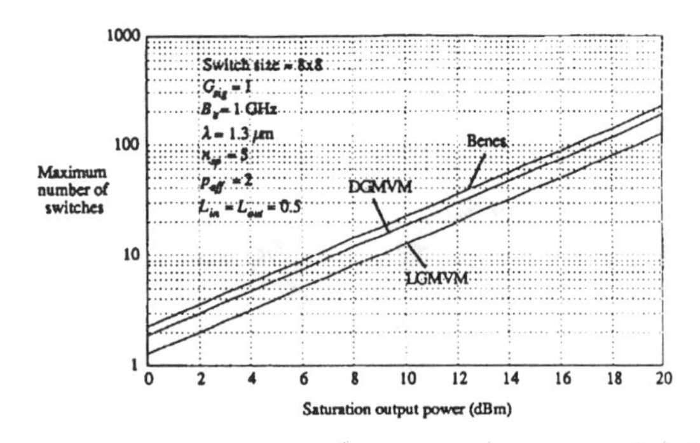

Fig. 3. Maximum number of cascaded 8 × 8 switches versus P_{sat} for the three switch types.

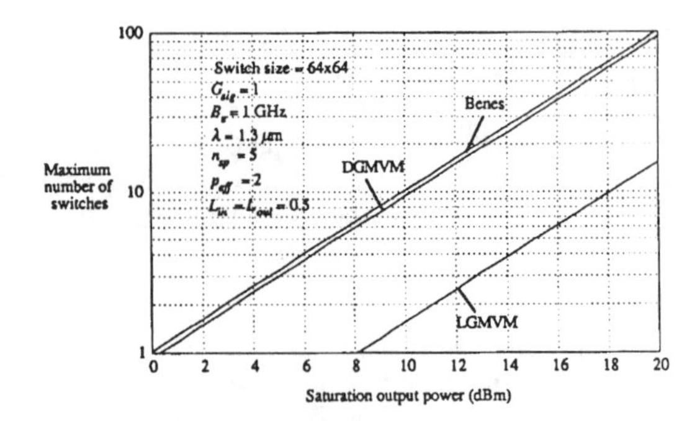

Fig. 4. Maximum number of cascaded 64 × 64 switches versus P_{sat} for the three switch types.

V. Conclusions

The size of optical switching fabrics based on SOA's are ultimately limited by signal-to-noise ratio (SNR) and saturation considerations, both of which are associated with spontaneous emission from the SOA's. The number of switches which can be cascaded is proportional to the saturation output power of the SOA's.

Three specific space division switch architectures were considered. Based on consideration of saturation and signal-to-noise ratio, it was found that, for the distributed gain matrix-vector multiplier (DGMVM) switch and the Benes switch, cascades of 100 64 × 64 switches or nearly 200 8 × 8 switches can be achieved for SOA's with saturation output powers of 100 mW. For lumped gain matrix-vector multiplier (LGMVM), approximately 10 64 × 64 switches or 100 8 × 8 switches can be cascaded for SOA's with saturation output powers of 100 mW. For large switch sizes, the DGMVM and Benes switches, both of which distribute gain throughout the switch, are notably better than the LGMVM switch, which lumps gain in two "planes." The Benes switch may be more suitable for large switch sizes ($N > 16$) because it requires fewer SOA's.

Acknowledgment

We would like to thank the reviewers for their helpful suggestions.

References

[1] J. D. Evankow and R. A. Thompson, "Photonic switching modules designed with laser diode amplifiers," *IEEE J. Select. Areas Commun.*, vol. 6, pp. 1087–1094, 1988.

[2] M. Ikeda, "Laser diode switch," *Electron. Lett.*, vol. 17, no. 22, pp. 899–900, 1981.

[3] M. Gustavson and L. Thylen, "Switch matrix with semiconductor laser amplifier gate switches: A performance analysis," *OSA Proc. Photon. Switching*, vol. 3, pp. 77–79, 1989.

[4] C. H. Henry, "Theory of spontaneous emission noise in open resonators and its application to laser and optical amplifiers," *J. Lightwave Technol.*, vol. 4, no. 3, pp. 288–297, 1986.

[5] N. A. Olsson, "Lightwave systems with optical amplifiers," *J. Lightwave Technol.*, vol. 7, no. 7, pp. 1071–1082, 1989.

[6] A. A. M. Saleh and I. M. I. Habbab, "Effects of semiconductor-optical-amplifier nonlinearity on the performance of high-speed intensity-modulation lightwave systems," *IEEE Trans. Commun.*, vol. 38, no. 6, pp. 839–846, 1990.

[7] A. R. Dias, R. F. Kalman, J. W. Goodman and A. A. Sawchuk, "Fiber-optic crossbar switch with broadcast capability," *Opt. Eng.*, vol. 27, no. 11, pp. 955–960, 1988.

[8] A. Himeno and M. Kobayashi, "4 × 4 optical gate matrix switch," *J. Lightwave Technol.*, vol. 3, no. 2, pp. 230–235, 1985.

Analysis of Nonlinear Optical Switching in an Erbium-Doped Fiber

R.H. PANTELL, M.J.F. DIGONNET, R.W. SADOWSKI, AND H.J. SHAW

Abstract— **We present a mathematical model of the strong, resonantly enhanced nonlinear phase shift recently reported in Er-doped fibers, which relates the phase shift and signal loss to the fiber parameters and the pump and signal wavelengths. Predictions are in fair agreement with the phase shift and loss measured in an experimental Er-doped fiber switch based on this effect. A strong, nearly wavelength independent contribution to the nonlinear phase shift is observed in the switch. The model suggests that this effect is due to the same nonlinear effect arising from one or more VUV transitions in Er^{3+}.**

I. Introduction

THE presence of a strong nonlinear phase shift in optically pumped Er^{3+}-doped fibers has recently been the object of increasing interest [1]–[5]. This phase shift arises from the imaginary part of the electronic susceptibility associated with one of the many absorption transitions of Er^{3+}. When the fiber is optically pumped, the electronic distribution of the Er^{3+} ions is modified and therefore the magnitude of this phase shift changes. This effect was first modeled by Desurvire, who used a density-matrix operator approach to model the phase shift in an Er-doped fiber amplifier (EDFA) [1]. More recently, this effect was used to produce self phase modulation and switching in a two-core Er-doped fiber pumped at 514.5 nm [2], [3]. It was also observed experimentally in an EDFA pumped at 980 nm [4]. Our own work demonstrated a potentially compact fiber switch based on the same principle. With one particular two-mode Er-doped fiber pumped with a 1.48-μm diode laser, full switching of a 906-nm signal was obtained with an absorbed pump power of only 8 mW, for a signal absorption loss of only 0.25 dB [5].

This paper presents a theoretical model of the nonlinear phase shift and the associated signal absorption loss due to an absorption resonance in a doped fiber. Previous reports investigating switches based on this principle were limited to phenomenological models which did not connect the phase shift to the dopant spectroscopy [2], [3]. The object of this model is to provide an understanding of the physical process involved, and to relate both the phase shift and the signal loss to the dopant spectroscopic parameters, the fiber parameters, pump power, and the pump and signal wavelengths. The main purpose of this model was to understand and analyze the phase shift observed in our Er-doped fiber switch. It shows, as confirmed by experiments, that by selecting a signal wavelength far enough from the peak of absorption, the signal absorption loss can be negligible while the phase shift remains reasonably large. Experimentally, the phase shift is found to arise from both a resonant term and a strong, almost wavelength-independent term. The model provides support for the hypothesis that this offset originates from the same nonlinear process involving a high-energy (VUV) absorption transition of Er^{3+}. This contribution is very important, as it enabled us to make a fiber switch with a low loss, a low drive power, and a weaker dependence on signal wavelength.

II. Basic Principles

It is well known that optical nonlinearities can be considerably enhanced near an absorption resonance [6]. An electronic transition A– >B is associated with a linear complex susceptibility χ, whose imaginary part χ'' is responsible for photon absorption and real part χ' contributes to the refractive index of the material. The magnitude of the absorption and index contributions are proportional to the difference in population density ΔN between levels A and B. Consequently the refractive index can be modified by changing ΔN, which may be accomplished by optical pumping. As an example, consider a fiber made of a material exhibiting an isolated absorption transition centered at a wavelength λ_0. When a pump beam with a wavelength λ_p near λ_0 is launched into the fiber, pump photons are absorbed and excite electrons into the excited state. This change in the electronic distributions of the ground and excited states causes a change in the refractive index of the material for any signal with a wavelength λ_s in the vicinity of λ_0. This index change is small, but in a long enough fiber and with sufficient pump power it can result in a sizeable phase shift. Rare earth doped fibers, which possess many absorption transitions in the visible and infrared spectrum, are good test beds to demonstrate this effect. Er-doped fibers, for example, have been recently shown to produce a phase shift of π over lengths of the order of a meter and with pump powers of just a few mW [2], [3], [5]. Our own work [5] suggests an improvement in the launched power-length product of about 6000 over what was previously observed in undoped fibers, which relied on the weak Kerr effect of silica [7].

This principle can produce a number of interesting devices based on the modulation of the phase of a signal, such as interferometric optical switches. Two examples of interferometric fiber switches are shown in Fig. 1. The first configuration (Fig.

Manuscript received August 4, 1992; revised February 6, 1993. This work was supported by Litton Systems, Inc. and the Department of Energy under Contract \# DE-FG03-92ER12126. R. W. Sadowski acknowledges the Fannie and John Hertz Foundation for financial support.

The authors are with the Edward L. Ginzton Laboratory, Stanford University, Stanford, CA 94305.

IEEE Log Number 9210148.

Reprinted with permission from *IEEE Journal of Lightwave Technology*, Vol. 11, No. 9, pp. 1416-1424, September 1993.

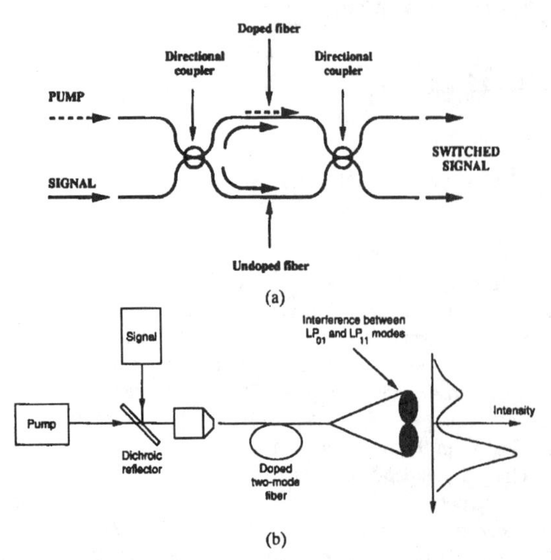

Fig. 1. Two types of optically-pumped interferometric fiber switches, based on (a) a Mach–Zehnder configuration, and (b) a two-mode fiber configuration.

1(a)) is an all-fiber Mach–Zehnder (MZ) interferometer, in which one or both arms are made of doped fiber but the pump is only coupled into one of the arms. If in the absence of pump all of the signal power comes out at output port 2, when the pump is turned on with enough power to induce a phase of π, the signal fully switches from output port 2 to output port 3.

The second configuration, which is used in our experimental switch, is the two-mode fiber (TMF) switch (Fig. 1(b)). It utilizes a fiber which carries two signal modes but only the LP_{01} mode at the pump wavelength. The signal is launched into the fiber so as to excite equal power in each mode (which is done in practice by offsetting the input laser beam with respect to the fiber core). At the fiber output the power spatial distribution resulting from interference between these two modes produces the typical pattern shown in Fig. 1(b). When the differential phase shift $\Delta\phi$ between the two modes is equal to $2m\pi (m = \text{integer})$, the power adds constructively in the vicinity of one of the two lobes of the LP_{11} mode and destructively in the vicinity of the other lobe. This distribution is reversed when $\Delta\phi = (2m+1)\pi$. When a pump beam is launched into the fiber, the spatial overlap of the pump mode intensity with the LP_{01} signal mode intensity is greater than with the LP_{11} signal mode intensity. Thus, the phase shift is larger for the LP_{01} than for the LP_{11} mode, and at the fiber output the spatial distribution changes. With a differential phase shift of π the signal power is switched from one lobe to the other.

The advantage of this approach over the MZ interferometer is that the two modes utilize the same guiding region and therefore react similarly to environmental changes. Consequently, a TMF interferometer is far less susceptible to temperature changes and other environmental parameters than a MZ interferometer, which generally requires stabilization. The penalty for this increased stability is that in a TMF interferometer both modes experience a nonlinear phase shift, so that for a given pump power the net phase shift is smaller than in a MZ switch. Thus, the pump power requirement of a TMF switch is generally larger than for an equivalent MZ switch, by a typical factor of 2–4. However, the nonlinear effect discussed here requires small pump powers, and this difference is not too severe. Furthermore, as discussed in Section III-C, the power requirement of a TMF switch can be made as low as that of a MZ switch by proper design of the dopant profile.

III. Model

The parameters to be determined for an all-optical fiber switch are the pump power requirement, the fiber length, the signal loss, and the response time. In this section, we show that these parameters (with the exception of the time response, which will be the object of a separate publication) may be obtained from a simultaneous solution of the rate equations with expressions for the refractive index and absorption in the fiber.

A. Basic Spectroscopy Assumptions

We consider a fiber with a core dopant exhibiting one or more absorption transitions. the case of a single absorption transition, illustrated in Fig. 2(a), is relevant to dopants with an isolated, two-level system. The pump frequency ν_p (wavelength λ_p) and the signal frequency ν_s (wavelength λ_s) then both lie in the vicinity of the same transition center frequency ν_0 (wavelength λ_0). The other situation of interest is a multiple-level configuration (Fig. 2(b)), which simulates our experimental Er-doped fiber switch. The signal and the pump lie in the vicinity of two different ground state absorption transitions, which terminate on levels well above the metastable level. This situation is interesting because 1) it benefits from the advantages of pumping a three-level system, 2) it is often desirable in practice to use different signal and pump wavelengths, and 3) it is relevant to most rare earths, which generally possess multiple absorption transitions.

Independently of the number of levels involved, the contribution of a single $i->j$ transition to the refractive index n_{ij} at frequency ν_s is given by: [6]

$$n_{ij} = \frac{e^2 g_i f_{ij} K}{m} \frac{\lambda_{ij}}{8 n_0 c \epsilon_0} \Delta N_{ij} g'_{ij}(v_s) \tag{1}$$

where e and m are the charge and the mass of the electron, respectively, ϵ_0 is the permittivity of vacuum, n_0 is the index of refraction of the material, c is the speed of light in vacuum, and f_{ij} is the oscillator strength of the transition. K is the Lorentz correction factor, given by $K = (n_0^2+2)^2/9$. $\Delta N_{ij} = N_i/g_i - N_j/g_j$ is the difference in population density between levels i and j, of degeneracy g_i and g_j, respectively.

The term $g'_{ij}(\nu_s)$ in (1) is the real part of the normalized lineshape of the transition. It is a Lorentzian for a purely homogeneously broadened transition, a Gaussian for a purely inhomogeneously broadened transition, and a convolution of these two functions for mixed broadening [6]. For example, for a Lorentzian the real and imaginary parts are:

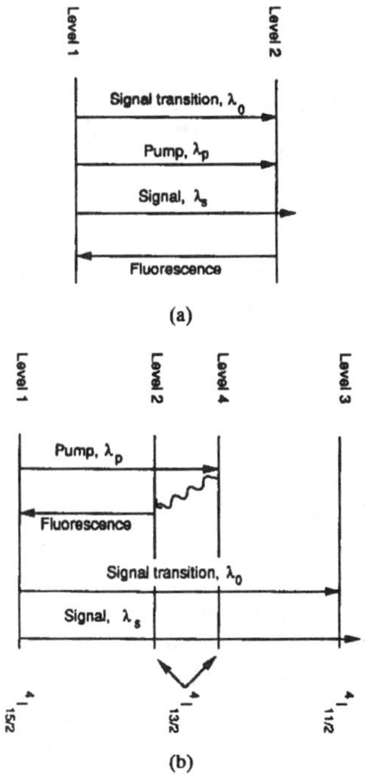

Fig. 2. Energy diagrams and transitions wavelengths of interest (a) for a two-level system, and (b) for a multiple-level system, showing the energy levels of Er^{3+} relevant to the experimental device.

$$g'_{ij}(v) = \frac{1}{\pi}\frac{\Delta v_{ij}}{\Delta v_{ij}^2 + \left(\frac{\delta v_{ij}}{2}\right)^2} \tag{2a}$$

$$g''_{ij}(v) = \frac{1}{2\pi}\frac{\delta v_{ij}}{\Delta v_{ij}^2 + \left(\frac{\delta v_{ij}}{2}\right)^2} \tag{2b}$$

where $\Delta\nu_{ij} = \nu - \nu_{0,ij}$ is the detuning from linecenter $\nu_{0,ij}$ and $\delta\nu_{ij}$ is the full width at half maximum (FWHM) of the transition. These functions are normalized such that the integral of $g''_{ij}(\nu)$ over the frequency space is unity. Fig. 3 shows the dependence of these two functions on the normalized detuning from linecenter $\Delta\nu_{ij}/\delta\nu_{ij}$, both for a Lorentzian and a Gaussian lineshape [6]. Since n_{ij} is proportional to $g'_{ij}(\nu_s)$ (see (1)), the index contribution is zero when the signal is at linecenter and maximum for a detuning $\Delta\nu_s = \pm\delta\nu_{ij}/2$.

As stated earlier, the index contribution n_{ij} is proportional to the population difference ΔN_{ij} (see (1)). It implies that when a pump is applied, the population difference ΔN_{ij} decreases and therefore n_{ij} decreases. In the following, we use laser rate equations to derive expressions for ΔN_{ij}, the phase shift and the absorption loss for these two configurations. The same formalism can be easily applied to treat the case of more complex spectroscopic configurations.

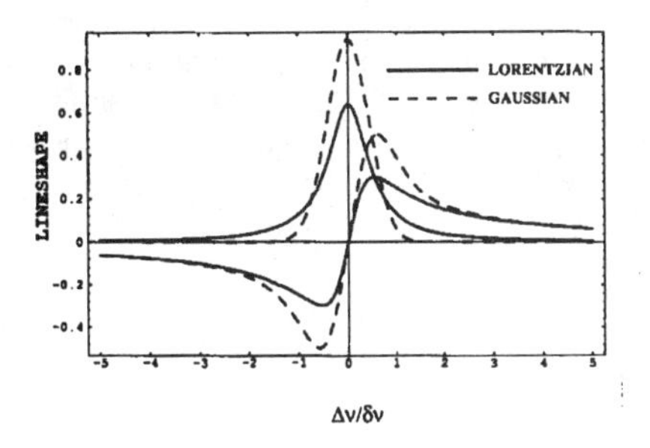

Fig. 3. Dependence of $g'(\nu)$ and $g(\nu)$ on the relative detuning for a Lorentzian and a Gaussian lineshape.

B. Two-Level Configuration

1) *Index Change:* We refer here to the case of Fig. 2(a), where the signal and the pump lie in the vicinity of the same transition. The relevant population density difference in (1) is then $\Delta N_{12} = N_1/g_1 - N_2/g_2$. To write the index contribution in a more tractable form, it is useful to express the fluorescence lifetime of level 2 as a function of the oscillator strength f_{12} of the transition: [6]

$$\tau_{f,2} = \frac{g_2}{g_1}\frac{m}{e^2 f_{12} K}\frac{\epsilon_0 c \lambda_0^2}{2\pi n_0} \tag{3}$$

Inserting this expression in (1) gives:

$$n_{12} = g_2 \frac{\lambda_0^3}{16\pi n_0^2 \tau_{f,2}} \Delta N_{12} g'_{12}(v_s) \tag{4}$$

2) *Population Change:* To calculate the nonlinear index change as a function of the pump power we must relate the population density change ΔN_{12} to the pump power $P_p(0)$ launched into the fiber. Referring to Fig. 2(a), absorption takes place from the ground state (level 1) to the excited state (level 2) with a rate W_{12}. The pump photons also stimulate electrons back down to level 1 at a rate W_{21}. We assume that all degenerate levels within each level have the same population, i.e., that they essentially all lie within less than $\Delta E = kT$, where k is the Boltzman constant and T is the absolute temperature. Under these assumptions, the rates W_{12} and W_{21} satisfy [6]

$$\frac{W_{21}}{W_{12}} = \frac{\sigma_{21}(v)}{\sigma_{12}(v)} = \frac{g_1}{g_2} \tag{5}$$

where $\sigma_{12}(\nu)$ and $\sigma_{21}(\nu)$ are the absorption and stimulated emission cross sections of the transition, respectively, and g_1 and g_2 degeneracy of levels 1 and 2.

For a two-level system, it can be shown using the laser rate equations that the steady-state populations are [8]

$$N_1 = \frac{1 + W_{21}\tau_2}{1 + (W_{21} + W_{12})\tau_2} N_0 \tag{6a}$$

$$N_2 = \frac{W_{12}\tau_2}{1 + (W_{21} + W_{12})\tau_2} N_0 \tag{6b}$$

where $N_0 = N_1 + N_2$ is the total dopant concentration. τ_2 is defined as the total lifetime of level 2 in the absence of a pump or signal ($W_{12} = W_{21} = 0$). It is related to the fluorescence lifetime $\tau_{f,2}$ and the nonradiative lifetime $\tau_{nr,2}$ (due to phonon coupling to either level 1 or intermediate levels not shown in Fig. 2(a) by:

$$\frac{1}{\tau_2} = \frac{1}{\tau_{f,2}} + \frac{1}{\tau_{nr,2}}. \tag{7}$$

The pump absorption cross section $\sigma_p = \sigma_{12}(\nu_p)$ is related to the pumping rate W_{12} by:

$$W_{12} = \frac{\sigma_p}{h\nu_p}\frac{P_p}{A} \tag{8}$$

where P_p is the pump power and A is the area illuminated by the pump beam. The difference $\Delta N_{12} = N_2/g_2 - N_1/g_1$ is therefore:

$$\frac{\Delta N_{12}}{N_0} = -\frac{1}{g_1}\frac{1}{1+\dfrac{P_p}{AI_{p,\text{sat}}}} \tag{9}$$

where $I_{p,\text{sat}} = (h\nu_p)/(\gamma\sigma_p\tau_2)$ is the saturation intensity of the transition, with $\gamma = 1 + g_1/g_2$.

In an end-pumped fiber, the pump power launched at $z = 0$ is gradually absorbed along the fiber, which means that $P_p, \Delta N_{12}$ and n_{ij} depend on the position z along the fiber. To account for this z-dependence, we write the evolution of the pump power along the fiber as:

$$\begin{aligned}\frac{dP_p}{dz} = &- \sigma_{12}(v_p)N_1(z)P_p(z)\\ &+ \sigma_{21}(v_p)N_2(z)P_p(z)\end{aligned} \tag{10}$$

where the first term arises from absorption from the ground state, and the second term from stimulated emission from the excited state. With the help of (2) and (9), (10) can be written as:

$$\frac{dP_p}{dz} = -\frac{\alpha_p}{1+\dfrac{P_p(z)}{AI_{p,\text{sat}}}}P_p(z) \tag{11}$$

where $\alpha_p = \sigma_p N_0$ is the small-signal absorption coefficient at the pump frequency. The solution of (11) is:

$$\ln\left(\frac{P_p(z)}{P_p(0)}\right) + \frac{1}{AI_{p,\text{sat}}}[P_p(z) - P_p(0)] = -\alpha_p z \tag{12}$$

which is the well-known equation for the transmission of an optical signal through a two-level saturable medium. An improvement parameter in this model is the pump power absorbed by the fiber, of length l, which is given by:

$$P_{\text{abs}} = P_p(0) - P_p(l) \tag{13}$$

obtained by solving (12) for the unabsorbed pump power $P_p(l)$.

3) *Nonlinear Phase Shift:* The phase shift $\Delta\phi$ accumulated along a length l of fiber is obtained by integrating the incremental phase shift induced by each length dz of fiber:

$$\Delta\phi = \int_0^l \frac{2\pi}{\lambda_s}\Delta n(z)dz \tag{14}$$

where $\Delta n(z)$ is the index change when the pump is on, i.e., $\Delta n(z) = n_{12}(P_p(z)) - n_{12}(P_p(0) = 0)$. Using (4) for $n_{12}(z)$, in which $\Delta N_{12}(z)$ is replaced by (9), we write the nonlinear index change as:

$$\Delta n(z) = -\frac{g_2}{g_1}\frac{\lambda_0^3}{16\pi n_0^2\tau_{f,2}}N_0 g_{12}'(v_s)\frac{\dfrac{P_p(z)}{AI_{p,\text{sat}}}}{1+\dfrac{P_p(z)}{AI_{p,\text{sat}}}}. \tag{15}$$

From (11), it appears that the last ratio in this expression is proportional to dP_p/dz, so that (14) is easily integrated in z. Using the expression of the saturation intensity, we find:

$$\Delta\phi = -\left(1+\frac{g_2}{g_1}\right)\frac{1}{8n_0^2hc}\frac{\lambda_p\lambda_0^3}{\lambda_s}\frac{P_{\text{abs}}}{A}\frac{\tau_2}{\tau_{f,2}}g_{12}'(v_s) \tag{16}$$

This final expression shows that the nonlinear phase shift is proportional to the pump power absorbed by the fiber and to the reciprocal of the core area. Since the fiber must essentially carry a single mode at the wavelength of operation ($\lambda_s \approx \lambda_p \approx \lambda_0$), the core area A is approximately proportional to λ_0^2, which means that the phase shift scales approximately linearly with λ_0. Everything else being equal, the phase shift is larger in the infrared portion of the spectrum.

Using (16), the power P_{abs}^{π} required to induce the π phase shift needed in an interferometric switch is:

$$P_{\text{abs}}^{\pi} = \frac{8\pi n_0^2 hc}{1+\dfrac{g_2}{g_1}}\frac{\lambda_s}{\lambda_p\lambda_0^3}\frac{\tau_{f,2}A}{\tau_2|g_{12}'(v_s)|}. \tag{17}$$

An interesting observation is that P_{abs}^{π} is independent of fiber length.

Since, in general, not all the pump power is absorbed by the fiber, the launched pump power $P_p^{\pi}(0)$ required for a π phase shift is also a useful measure of the power requirement. This quantity is obtained from (12)

$$P_p^{\pi}(0) = \frac{P_{\text{abs}}^{\pi}}{1 - \exp\left(-\alpha_p l + \dfrac{P_{\text{abs}}^{\pi}}{AI_{p,\text{sat}}}\right)} \tag{18}$$

in which P_{abs}^{π} is given by (17). Since $P_p^{\pi}(0)$ must be positive, (18) only applies when $\alpha_p l \geq P_{\text{abs}}^{\pi}/(AI_{p,\text{sat}})$. It means that a π phase shift can be obtained only if the fiber length exceeds a minimum value l_{min}, which is found from (17) to be:

$$l_{\text{min}} = \frac{g_1}{g_2}\frac{8\pi n_0^2\tau_{f,2}}{N_0}\frac{\lambda_s}{\lambda_0^3}\frac{1}{|g_{12}'(v_s)|}. \tag{19}$$

The physical reason behind the existence of a minimum length is that for a given concentration N_0 and fiber length, the phase shift is maximum when the populations of the two levels have been equalized. The maximum phase shift is then limited by the product $N_0 l$. If this product is too small, there just are not enough ions for the phase shift to reach π. As the fiber length approaches $l_{\min}$ the launched pump power requires goes to infinity.

4) *Signal Loss:* The loss experienced by the signal is a critical switch parameter. Because the signal frequency is close to an absorption resonance, in the absence of a pump the signal suffers a finite "cold' loss due to absorption from the ground state. Also, when the pump is turned on the ground state population decreases and the signal loss decreases. In practice it is desirable to minimize both the cold loss and the pump-induced change in loss. Since the latter is necessarily smaller than the former, in the following we only investigate the cold loss.

The signal loss can be characterized in terms of the signal absorption coefficient α_s:

$$\alpha_s = \sigma_{12}(v_s)N_0. \tag{20}$$

The small-signal absorption cross section $\sigma_{12}(\nu_s)$ is related to the spectroscopic parameters of the two-level system by [6]

$$\sigma_{12}(v_s) = \frac{g_2}{g_1}\frac{\lambda_0^3}{8\pi n_0^2 \tau_{f,2}\lambda_s} g''_{12}(v_s) \tag{21}$$

where $g''_{12}(\nu_s)$ is the imaginary part of the transition lineshape, given by (2b). Inserting (21) into (20) yields the following expression for the signal loss $L_{\text{dB}} = 4.343\alpha_s l$, in decibels

$$L_{\text{dB}} = 4.343\frac{g_2}{g_1}\frac{\lambda_0^3}{8\pi n_0^2 \tau_{f,2}\lambda_s} N_0 g''_{12}(v_s)l. \tag{22}$$

Since the phase shift is zero at linecenter (see (16) and Fig. 3), the signal needs to be detuned to experience a finite phase shift. Referring to Fig. 3 and examining the case of a Lorentzian lineshape, the phase shift is maximum for a detuning of $\Delta\nu_s = \pm\delta\nu_{12}/2$. Operating at this frequency is beneficial because it minimizes the pump power requirement. However, for strongly resonant systems it also results in a high signal absorption loss. For larger detunings ($\Delta\nu_s \gg \delta\nu_{12}/2$) the signal loss decreases but the power requirement increases, which means that a trade-off must be made between pump power and loss. Far from resonance, the loss decreases as $\delta\nu^{-2}$, while the nonlinear phase shift decreases only as $\delta\nu^{-1}$, i.e., the loss decreases faster than the phase shift (see Fig. 3). This conclusion also applies to a Gaussian lineshape (see Fig. 3). It implies that by properly selecting the signal wavelength, both the loss and the pump power can be kept at reasonably low levels.

C. Multiple-Level System

The derivation of the phase shift and loss for the more general multiple-level system of Fig. 2(b), relevant to our experimental work, is treated in much the same way as a two-level system except that it presents two major differences. The first one is in the behavior (and mathematical expression) of the population density difference entering in (1): in a two-level system a strong pump equalizes the populations, whereas in a three-level system it can create a significant population inversion. The second difference is that the signal transition now lies far from the transition from the ground state to the metastable level (as in our Er-doped fiber switch, probed in the 900–960 nm region). In this case, we can safely assume that most of the nonlinear index change arises from the 1− > 3 transition (see Fig. 2(b)), i.e., from $\Delta N_{13} = N_1/g_1 - N_3/g_3$. Assuming that level 3 has a lifetime negligible compared to level 2, it does not build up a significant population; hence $N_3/g_3 \ll N_1/g_1$. From (1), the index change is now given by:

$$n_{13} = g_3 \frac{\lambda_{13}^3}{16\pi n_0^2 \tau_{f,3}} \frac{N_1}{g_1} g'_{13}(v_s) \tag{23}$$

where $\tau_{f,3}$ is the fluorescence lifetime of level 3 and λ_{13} is the center wavelength of the 1− > 3 transition. N_1/g_1 is easily obtained from the rate equations.

Using these assumptions and carrying out the same calculation as outlined in Section III-B, we obtain the following expressions for the nonlinear phase shift, the absorbed pump power for a π phase shift, and the signal cold loss:

$$\Delta\phi = -\frac{g_3}{g_1}\frac{1}{8n_0^2 hc}\frac{\lambda_p\lambda_{13}^3}{\lambda_s}\frac{P_{\text{abs}}}{A}\frac{\tau_2}{\tau_{f,3}} g'_{13}(v_s)\xi \tag{24}$$

$$P_{\text{abs}}^{\pi} = \frac{g_1}{g_3} 8\pi n_0^2 hc \frac{\lambda_s}{\lambda_p \lambda_{13}^3}\frac{\tau_{f,3}A}{\tau_2 |g'_{13}(v_s)|}\frac{1}{\xi} \tag{25}$$

$$L_{\text{dB}} = 4.343\frac{g_3}{g_1}\frac{\lambda_{13}^3}{8\pi n_0^2 \tau_{f,3}\lambda_s} N_0 g''_{13}(v_s)l. \tag{26}$$

In these expressions, the pump power absorbed by the fiber is still obtained by solving (12) for $P_p(l)$ and inserting the latter in (13). The launched pump power required for a π phase shift, $p_p^{\pi}(0)$, is still given by (18) with P_{abs}^{π} replaced by (25).

An overlap factor ξ has been introduced into (24) and (25) to account for the difference in phase shift between the two arms of the interferometer, which depends on the type of interferometer used. For a plane wave interferometer, $\xi = 1$. For a fiber MZ interferometer (Fig. 1(a)), assuming the dopant is only in the core and uniformly distributed, only the core index of the fiber is changed, and the phase shift experienced by the guided signal is reduced from the plane wave case. It can be shown that ξ is then equal to $\xi_{\text{MZ}} = \eta_{01}$, where η_{01} is the fractional power of the LP_{01} signal mode contained in the core. For a TMF interferometer (Fig. 1(b)), the phase shift experienced by the LP_{01} and the LP_{11} mode is proportional to η_{01} and η_{11}, respectively, where η_{11} is the fractional power of the LP_{11} signal mode in the core. Hence $\xi = \xi_{\text{TMF}} = \eta_{01} - \eta_{11}$.

From these considerations, it appears that the power requirement for a TMF switch is generally larger than for a MZ switch, by the factor $\xi_{\text{MZ}}/\xi_{\text{TMF}} = (\eta_{01} - \eta_{11})/\eta_{01}$. This ratio is typically in the range of 2–4. However, it can be reduced

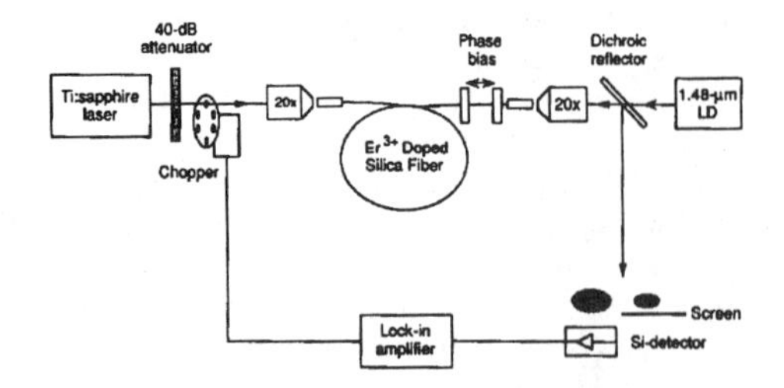

Fig. 4. Diagram of the experimental nonlinear two-mode fiber switch.

TABLE I
PARAMETERS OF THE ERBIUM-DOPED FIBER

Core diameter	$a = 3.35\ \mu m$
Numerical aperture	NA=0.13
Er^{3+} concentration	$N_0 = 1.3\ 10^{19}\ cm^{-3}$
Ground state degeneracy ($^4I_{15/2}$)	$g_1 = 16$
Pump state degeneracy ($^4I_{13/2}$)	$g_2 = 14$
Upper state degeneracy ($^4I_{11/2}$)	$g_3 = 12$
Fluorescence lifetime ($^4I_{13/2}$)	$\tau_{f,2} = 7.95$ msec
Non-radiative lifetime ($^4I_{13/2}$)	$\tau_{nr,2} = \infty$

to very close to unity by concentrating the dopant near the center of the core, as can be accomplished in EDFA's to improve the pumping efficiency. The spatial overlap between the dopant distribution and the LP_{11} mode, whose electric field is zero at the center of the core, is then vanishingly small. The overlap factor ξ_{TMF} then approaches ξ_{MZ}, and the pump power requirements of a TMF switch becomes as low as that of a MZ switch. In addition, the spatial overlap between the more strongly confined dopant distribution and the LP_{01} mode, whose electric field is maximum at the center of the core, is larger and $\xi_{TMF} = \xi_{MZ}$ approach unity.

IV. Application to an Er-Doped Fiber Switch

In this section we apply this model to analyze the performance of an experimental Er-doped two-mode fiber switch reported elsewhere [5]. The diagram of the experimental setup is shown in Fig. 4. The parameters of the fiber, fabricated by GTE Laboratories, Inc., are summarized in Table I. The signal, tuned over the 900–960 nm region to explore the $^4I_{15/2} -^4 I_{11/2}$ absorption transition of Er^{3+} (see Fig. 2(b)), was provided by a Ti-sapphire laser. It was launched to inject approximately equal powers in the LP_{01} and LP_{11} modes. The 1.48-μm pump power, near the $^4I_{15/2} -^4 I_{13/2}$ absorption of Er^{3+} (see Fig. 2(b)), was provided by an InGaAsP laser diode and coupled into the fiber in the backward direction. The fiber was single mode at this wavelength. Between 65 and 98% of the launched pump power was absorbed, depending on pump power and saturation. The signal power was kept well below the absorption saturation level to avoid wavelength-dependent, self-phase modulation of the signal. The weak signal was mechanically chopped around 1.7 kHz and analyzed with a lock-in amplifier.

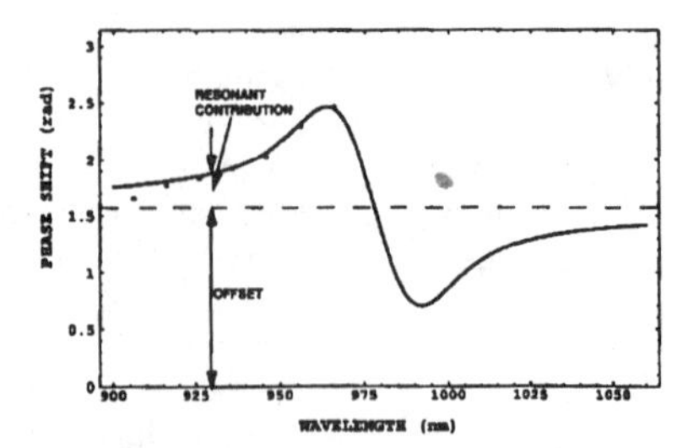

Fig. 5. Measured phase shift as a function of signal wavelength for $P_{abs} = 10$ mW. The solid curve is a fit to a Gaussian lineshape centered at 978 nm, with a linewidth of 24 nm and a constant offset (dashed line) of 1.57 radians.

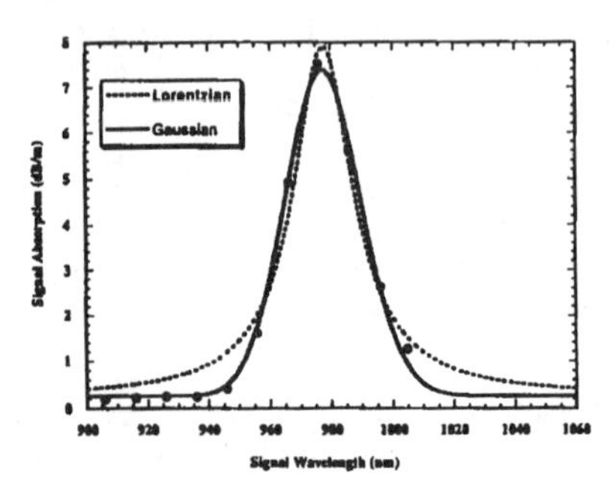

Fig. 6. Absorption spectrum for the switch of Fig. 6, with $P_{abs} = 10$ mW. The solid line is a fit to a Gaussian lineshape with a FWHM = 29 nm, a centerline of 978 nm and an offset of 0.26 dB/m. The dashed line is a fit to a Lorentzian lineshape (FWHM = 24 nm, same centerline and offset).

A. Experimental Results

With a fiber length of 3.4 m, a phase shift of π was found to require an absorbed pump power of 15.5 mW [5]. The dependence of the phase shift on the signal wavelength, measured for a fixed absorbed pump power $P_{abs} = 10$ mW, is shown in Fig. 5. Data was not obtained close to the absorption peak, where the signal was too strongly absorbed, and for long wavelengths, where the LP_{11} mode was too close to cut-off. Fitting the experimental data points of Fig. 5 to either a Gaussian or a Lorentzian lineshape $g'(\nu)$ did not give satisfactory fits. However, fitting these data points to either a Gaussian or a Lorentzian lineshape with a constant offset gives excellent results. This is illustrated in Fig. 5 for the case of a Gaussian. The fit is the solid line. The values of the fitting parameters are a linecenter $\lambda_0 = 978$ nm, a FWHM of 24 nm, and an offset phase shift of 1.57 rd. This result suggests that the observed phase shift has two contributions, namely the expected resonant term associated with the 978-nm transition (represented by the Gaussian part of the fit), and a term that is essentially independent of wavelength over the measurement range (represented by the horizontal dashed line in Fig. 5). The resonant contribution at the peak ($\lambda_s = 966$ nm) is $\Delta\phi_R = 0.89$ rd for 10 mW of absorbed pump power, or $\Delta\phi_R/P_{abs} = 0.089$ rd/mW. The contribution from the offset is $\Delta\phi_{NR} = 1.57$ rd, or $\Delta\phi_{NR}/P_{abs} \approx 0.16$ rd/mW.

The small-signal absorption spectrum for the LP_{01} mode, measured over the same wavelength range as Fig. 5, is shown

TABLE II
COMPUTED SWITCH PARAMETERS

Lineshape	$\tau_{f,3}$	$\Delta\phi_R/P_{abs}$ (@ 966 nm)
Lorentzian	7.6 ms	0.13 rd/mW
Gaussian	9.6 ms	0.18 rd/mW

in Fig. 6. The measured spectrum was again fitted to either a Gaussian (solid line) or a Lorentzian (dashed line), in both case with a small offset. The Gaussian provides the best fit (which is the reason for using the Gaussian fit in Fig. 5). For the Gaussian fit, the values of the fitting parameters are a linecenter $\lambda_0 = 978$ nm, a FWHM of 29 nm, and a small offset loss of 0.26 dB/m. The values of λ_0 and the FWHM are in good agreement with the values inferred from the fit to the phase shift data of Fig. 5.

Figs. 5 and 6 illustrate several important characteristics of this switch. If the signal wavelength is selected at the maximum of the resonant phase shift (966 nm), the resonant phase shift is large (≈ 0.87 rd) but also the cold loss quite high, about 17.4 dB (see Fig. 6). However, away from resonance, e.g., at 906 nm, the resonant phase shift is still sizeable (≈ 0.2 rd) while the resonant cold loss is considerably smaller, about 0.5 dB. Another important point is that away from resonance the resonant contribution is much smaller than the contribution of the offset, as illustrated in Fig. 5 at 906 nm. The offset term is therefore quite important in practice, since it contributes to most of the useful phase shift while introducing negligible absorption loss. The possible physical origin of this offset is discussed in Section IV-C.

B. Interpretation of the Resonant Term

To verify the validity of the model, the peak value of $\Delta\phi_R/P_{\text{abs}}$ was calculated using (24), which requires the lifetime $\tau_{f,3}$ of the $^4I_{11/2}$ level. This value was calculated as follows. The peak absorption for the LP_{01} mode at 978 nm is 7.5 dB/m (Fig. 6). At this wavelength the relative power in the core η_{01} is calculated [9] to be 0.873, so that the plane-wave peak absorption coefficient is $\alpha_0 = 7.5/0.873/4.343 = 1.98$ m^{-1}. From this value and the fiber dopant concentration N_0 (see Table I) the peak absorption cross section is $\sigma_0 = \alpha_0/N_0 = 1.52\,10^{-21}$ cm ^{2}. The value of $\tau_{f,3}$ can then be inferred using (21). For this last step we need to look individually at each kind of lineshape, since $g''(\nu)$ enters in (21).

Table II lists the calculated values of $\tau_{f,3}$ (21) and $\Delta\phi_R/P_{\text{abs}}$ (24), assuming either a Lorentzian or a Gaussian lineshape. We used the calculated values of $\eta_{01} = 0.873$ and $\eta_{11} = 0.561$, i.e., $\xi_{\text{TMF}} = 0.312$. These inferred values of $\tau_{f,3}$ are in agreement with published values (7.7–10 ms) [10]. The computed values of $\Delta\phi_R/P_{\text{abs}}$ range from 0.13 rd/mW for a Lorentzian to 0.18 rd/mW for a Gaussian lineshape. They are within a factor of 2 of the measured value of 0.089 rd/mW. This discrepancy has not been resolved. We believe it is due in part to the fact that amplified spontaneous emission (ASE) was not included in the model. ASE will tend to reduce the steady-state population change and therefore reduce the phase shift predicted by this model. The phase shift may also depend on the relative polarization of the pump and the signal, much like it does for the Kerr effect [11]. These various issues are now being investigated. Some of the error may also come from the values given to the parameters $\tau_{f,3}$ and ξ, which were inferred rather than measured. However, considering the complexity of the physical problem at hand, the present theory is in fair agreement with experimental observations. A greater accuracy can be expected from straightforward refinements.

C. Interpretation of the Wavelength-Independent Offset

The likely origins for the observed offset are a thermal effect or a contribution of the same nonlinear origin discussed in previous sections, but arising from other transitions. A thermal effect may be expected since non-radiative relaxation from the pump band to the metastable level generates heat in the fiber core and a slight rise in fiber temperature. In the Appendix, we present a calculation of the thermal phase shift in a TMF and a MZ switch. Under our experimental conditions, the steady-state gradients of temperature across the entire fiber (center of core to edge of cladding) are predicted to be negligible. The differential thermal phase shift between the LP_{01} and LP_{11} modes is also calculated to be much smaller than 10^{-2} rd under the conditions of Fig. 5, i.e., negligible. This is largely due to the fact that in a TMF switch the two modes share the same guiding region and therefore experience essentially the same thermal phase shift. In contrast, in a MZ interferometer the thermal contribution is predicted to be quite large, about 7.3 rd (see Appendix), which illustrates one of the benefits of a TMF configuration. These considerations strongly suggest that the observed offset does not have a thermal origin, in agreement with observations reported in [3] for a twin-core fiber.

Instead, we believe the offset originates from nonlinearities due to one or more other absorption transitions of Er^{3+}. Comparison between Er-doped fibers with different core sizes and numerical apertures shows that the offset increases with increasing confinement of the optical waves [5], which further supports an electronic origin. The transition or combination of transitions responsible for this effect have not yet been identified. The model presented here could easily predict the cumulative effect on the refractive index of all the visible and IR transitions of Er^{3+}, including ground state absorption (GSA) and excited state absorption (ESA) transitions since the excited state is significantly populated when the pump is on. However, the cross sections of many of these transitions have never been measured. Since contributions to the phase shift from GSA and ESA transitions have opposite signs (when the pump is turned on, the relevant ΔN decreases for GSA but increases for ESA), a small error in cross sections would result in a large error in predictions.

An interesting possibility is that the offset may be due to a transition to a VUV level (or several such transitions). In Er^{3+} and other rare earths, the lower lying energy levels arise mostly from f-shell electrons, while levels in the UV and VUV originate from d-shell electrons. Transitions between levels from the same shell, which have the same parity, are not allowed. This is the reason why in rare earth ions the visible and IR transitions exhibit very low oscillator strengths, typically in the range of 10^{-5}–10^{-6}, [12] with the result that most low-energy levels have very long radiative lifetimes. However, UV-VUV transitions occur between the ground state and a d-shell level, which have opposite parities. They are therefore allowed and should exhibit a high oscillator strength. Since n_{ij} is proportional to the oscillator strength (see (1)), such transitions are expected to have a large contribution. Also, because their linecenter is at short wavelengths, in the visible and IR n_{ij} is essentially independent of signal wavelength, and the phase shift is expected to vary weakly with wavelength (as $1/\lambda_s$). Under this hypothesis the offset would have the same physical origin as the resonant term, except that it is a far from resonance (or nonresonant) contribution from a much higher energy transition.

To assess this concept, consider a hypothetical VUV level 100 nm above the $^4I_{15/2}$ ground state. Let f_{14} and f_{24} be the respective oscillator strengths of the two allowed transitions, from the ground state and the excited state, respectively, to this level. Using (1), it can be shown that the index change arising from these two transitions is:

$$\Delta n(z) = \frac{e^2}{m}\frac{KN_2(z)}{8n_0 c\epsilon_0}\left[\frac{f_{14}\lambda_{14}^2}{g_1} - \frac{f_{24}\lambda_{24}^2}{g_2}\right]\xi_{\mathrm{TMF}} \quad (27)$$

where we assumed Lorentzian lineshapes and a signal detuning much larger than the linewidths of these two transitions, i.e., $g'_{14}(\nu_s) \approx -\lambda_{14}/(\pi c)$ and $g'_{24}(\nu_s) \approx -\lambda_{24}/(\pi c)$. Following the procedure outlined earlier, namely using the expression of N_2 (6b) and of the pump power (11) and integrating from $z = 0$ to $z = 1$, provides the following expression for the phase shift due to the VUV level:

$$\Delta\phi_{\mathrm{NR}} = \frac{e^2}{mc^2\epsilon_0}\frac{K\tau_2}{4\lambda_s n_0 h\nu_p}\left[\frac{f_{14}\lambda_{14}^2}{g_1} - \frac{f_{24}\lambda_{24}^2}{g_2}\right]\xi_{\mathrm{TMF}}\frac{P_{\mathrm{abs}}}{A}. \quad (28)$$

In the following we take $\lambda_{14} = 100$ nm, $\lambda_{24} = 107$ nm (since the difference in energy between the $1- > 4$ and $2- > 4$ transitions is the energy of a $1.55\,\mu$m photon), and assume $f_{14} = f_{24}$, which is reasonable since levels 1 and 2 have nearly the same energy compared to level 4. Using our fiber parameters and $P_{\mathrm{abs}} = 10$ mW, (28) predicts that an oscillator strength $f_{24} = f_{24} = 0.12$ would produce a nonresonant phase shift $\Delta\phi_{\mathrm{NR}}$ of 1.57 rd, i.e., equal to the offset observed experimentally (Fig. 5). The sign is also right: (28) and (24) predict that for wavelengths shorter than the 980-nm resonance, the resonant and non-resonant term have the same sign, as is indeed the case (see Fig. 5). The value of 0.12 is reasonable, as it corresponds to a strongly allowed transition and satisfies the oscillator strength conservation rule $f_{14} + f_{24} \leq 1$.

This calculation lends support to the hypothesis that the strong, weakly wavelength dependent nonlinearity observed in our test fiber originates from one or more allowed high-energy transitions. Independent measurements are needed to confirm this proposed origin. However, from a practical standpoint this term readily appears to be very important. In the switch described here it produces a sizable, nearly wavelength independent phase shift over a wide range of wavelengths, while introducing a negligible loss.

V. Conclusions

In rare earth doped fibers the pump-induced refractive index change associated with an electronic transition can be sizeable, both near and far from a transition. This effect was utilized to demonstrate a low-loss, low-power fiber switch made of a two-mode Er-doped fiber pumped with a 1.48-μm diode laser. Two components, a resonant term and a weakly wavelength-dependent offset, were observed in the experimental phase shift. The second component is very important because it dominates at signal wavelengths far from an absorption resonance, and contributes to a sizeable, wavelength-independent index change at the cost of negligible loss. It enabled us to achieve a π phase shift in a 3.4-m fiber with an absorbed pump power of only 15.5 mW and a loss of only 0.5 dB. In another fiber, the same goal was achieved with a length of 0.95 m, a power of 8 mW and a loss of 0.25 dB. These results indicate a third-order nonlinear susceptibility $\chi^{(3)}$ about three orders of magnitude larger than in pure silica.

A simple analytical model was developed to relate the magnitude of the nonlinear phase change to the quantum electronic parameters of Er^{3+}. It provides simple mathematical expressions for the main parameters of the fiber switch, in particular its pump power requirement and signal loss as a function of the fiber parameters (length, core size, NA) and signal and pump wavelengths. This model shows, as confirmed experimentally, that a sizeable resonant phase shift and a negligible signal absorption loss can be obtained if the signal wavelength is properly selected. This model should prove helpful in designing switches with low losses and low pump power requirements based on fibers doped with rare earths or other impurities.

Several major conclusions were drawn from the interpretation of experimental data. First, the model predicts a resonant term which is within a factor of 2 of the observed resonant phase shift. Despite this discrepancy, it gives credence to the physical mechanism involved in the nonlinear process. Second, it is shown analytically that pump-induced heating of the fiber is not responsible for the observed offset. Third, we propose that the offset term arises from nonresonant contributions from one or more high-energy absorption transitions in Er^{3+}. In particular, a transition in the VUV with a high oscillator strength (0.12) would produce the observed nonresonant phase shift. Further measurements are needed to confirm the origin of this important contribution.

Appendix
Thermal Contribution to the Phase Shift

Heating of the fiber core is expected to occur due to the generation of phonons by the pump in the fiber core. Assuming a uniform distribution of Er $^{3+}$ in the core and an undoped cladding, the steady-state temperature profile across the fiber cross section can be obtained by solving the heat flow equation subject to the condition that the temperature be continuous at the core-cladding boundary. The solution for the steady-state temperature difference between the center and the edge of the core, ΔT_c, is:

$$\Delta T_c = \frac{qP_{\text{abs}}}{4\pi kl} \tag{1A}$$

where P_{abs} is the pump power absorbed by the core, q is the fraction of P_{abs} converted into photons, l is the fiber length, and k the thermal conductivity of the fiber.

The relevant values for Fig. 5 are $l = 3.4$ m, $k = 1.38$ W/m/°K for silica at room temperature, and $P_{\text{abs}} = 10$ mW. For every pump photon absorbed by the fiber the phonon energy released in the fiber is equal to the difference in energy of the pump level $^4I_{11/2}$ and the metastable level $^4I_{13/2}$. This energy is $6.06\,10^{-21}$ J for a pump photon energy of $1.34\,10^{-19}$ J, or $q = 4.5\%$, whereupon (1A) gives $\Delta T_c = 7.6\,10^{-6}$°C.

Similarly, the steady-state temperature difference between the core-cladding boundary and the outer edge of the cladding can be shown to be:

$$\Delta T_{cl} = \frac{qP_{\text{abs}}}{2\pi kl} \ln\left(\frac{b}{a}\right) \tag{2A}$$

where a and b are the core and cladding radius, respectively. Again, using our fiber parameters, $b = 40\,\mu$m and $a = 3.35\,\mu$m, we find a negligible temperature difference, $\Delta T_{cl} = 3.8\,10^{-5}$°C.

The temperature across the entire fiber is therefore almost uniform. The steady-state temperature change ΔT_f of the fiber with respect to its surrounding is equal to:

$$\Delta T_f = \frac{qP_{\text{abs}}}{4\pi kl}\left[1 + \frac{2k}{hd} + 2\ln\left(\frac{d}{a}\right)\right] \tag{3A}$$

where d is the radius of the jacket and h the heat transfer coefficient of the fiber material. Here it has been assumed that the jacket has the same thermal conductivity as the fiber. However, the dominant term is by far heat exchange at the surface of fiber (the middle term in (3A), so that ΔT_f is essentially independent of k, and this assumption has no impact on the final answer. Using $h \approx 10$ W/m^2/°K for silica at room temperature, [13] and $d = 59\,\mu$m, the predicted temperature rise is $\Delta T_f = 3.6\,10^{-2}$°C.

In a TMF switch the temperature profile in the region where the modes are confined can be approximated by a constant temperature $T_0 + \Delta T_c + \Delta T_{cl}$ in the core and a constant temperature T_0 in the cladding, where T_0 is the room temperature. The temperature step $\Delta T = \Delta T_c + \Delta T_{cl}$ is obviously much larger than the temperature step actually seen by the modes, so that this assumption provides a worst case figure for the thermal phase shift between the two modes. The latter is given by:

$$\Delta\phi_T = \frac{2\pi l}{\lambda}\frac{\partial n}{\partial T}\Delta T\xi \tag{4A}$$

where $\partial n/\partial T = 1\,10^{-5}$°C^{-1} for silica, $\Delta T = \Delta T_c + \Delta T_{cl} = 4.5\,10^{-5}$°C and $\xi = \xi_{\text{TMF}} = 0.312$. Thus $\Delta\phi_T = 3.3\,10^{-3}$ rd, which is negligible compared to the measured value of the wavelength-independent offset of $\approx \pi/2$.

For comparison, it is instructive to look at a MZ switch, in which only one arm is pumped and the thermal contribution is significantly larger. The relevant parameter is then the temperature difference between the pumped and the unpumped fiber arm, given by ΔT_f (3A). In a MZ switch with the same parameters as the TMF switch reported herein, and $P_{\text{abs}} = 10$ mW, the thermal phase shift between the two arms would be given by (4A) where $\Delta T = \Delta T_f = 3.6\,10^{-2}$°C and $\xi = \xi_{\text{MZ}}$, i.e., $\Delta\phi_T = 7.3$ rd. This is at least three orders of magnitude larger than in a TMF or a twin-core fiber interferometer.

Acknowledgment

The authors would like to thank W. Miniscalco, with GTE Laboratories, Incorporated, for providing the Er-doped fiber used in this work, as well as for helpful discussions.

References

[1] E. Desurvire, "Study of the complex atomic susceptibility of erbium-doped fiber amplifiers," *J. Lightwave Technol.*, vol. 8, no. 10, pp. 1517–1527, Oct. 1990.

[2] R. A. Betts *et al.*, "Nonlinear refractive index in erbium doped optical fiber: Theory and experiment," *IEEE J. Quantum Electron.*, vol. 27, no. 4, pp. 908–913, Apr. 1991.

[3] P. L. Chu and B. Wu, "Optical switching in twin-core erbium-doped fibers," *Opt. Lett.*, vol. 17, no. 4, pp. 255–257, Feb. 1992.

[4] S. C. Fleming and T. J. Whitley, "Measurement of pump induced refractive index change in erbium doped fibre amplifier," *Electron. Lett.*, vol. 27, no. 21, pp. 1959–1961, Oct. 1991.

[5] R. H. Pantell, R. W. Sadowski, M. J. F. Digonnet, and H. J. Shaw, "Laser-diode-pumped nonlinear switch in erbium-doped fiber," *Opt. Lett.*, vol. 17, no. 14, pp. 1026–1028, July 1992.

[6] R. H. Pantell and H. E. Puthoff, *Fundamentals of Quantum Electronics*. New York: Wiley, ch. 2 and 5, 1969.

[7] H. G. Park, C. C. Pohalski, and B. Y. Kim, "Optical Kerr switch using elliptical core two-mode fiber," *Opt. Lett.*, vol. 13, no. 9, pp. 776–778, Sept. 1988.

[8] R. Loudon, *The Quantum Theory of Light*. Oxford: Oxford Univ. Press, 2nd ed., Oxford, 1983.

[9] D. Gloge, "Weakly guiding fibers," *Appl. Opt.*, vol. 10, pp. 2252–2258, Oct. 1971.

[10] E. Desurvire, "Gain saturation effects in high-speed, multichannel erbium-doped fiber amplifiers at $\lambda = 1.53\,\mu$m," *J. Lightwave Technol.*, vol. 7, no. 12, pp. 2095–2104, Dec. 1989.

[11] P. D. Maker and R. W. Terhune, "Study of optical effects due to an induced polarization third order in the electric field strength," *Phys. Rev. A*, vol. 137, p. 801, 1978.

[12] W. Miniscalco, in *Electronic and Optical Properties of Rare-Earth Doped Fibers, Rare Earth Doped Fiber Lasers and Amplifiers*, M. Digonnet, Ed. New York: Marcel Dekker, 1993, ch. 2.

[13] F. Kreith, in *Principles of Heat Transfer*. New York: Harper and Row, 1973, p. 14, 3rd ed.

Chapter 4

Optical Memories for Switching

Devices that can route and switch signals optically can be broadly divided into two categories: passive devices and devices that interact with the optical signal and sense the individual bits offering more versatility. Passive devices carry out mappings from inputs to outputs based on some control signals. This category is completely transparent and because the devices are passive and do not extract information from the optical signals, they are completely transparent and have huge bandwidths. The devices outlined in Chapters 2 and 3 come under this broad category. Another category of optical switches is the one in which an optical signal interacts with the device in such a way that it controls the state of the switching. These devices sense individual bits and can perform various Boolean logic operations. However, due to this strong interaction, the effective speed of this class of switching devices is limited. This can be attributed, for example, to the fact that once bits are sensed, several set-reset operations have to be performed before the next data bit arrives, thus limiting the maximum data rate that can be handled. An optical device belonging to this category is the self-electrooptic effect device (SEED).

Grindle and Midwinter (paper 4.1) argue for the classification given above and go on to consider SEED devices in optical switching systems. They indicate that up to recently, several problems have hindered the practical exploitation of SEED-based switches. They list factors such as high power, critical biasing, and accurate temperature control as contributing factors. However, recent progress has alleviated many of these problems and in particular devices such as the symmetric SEED (S-SEED) have overcome many of these problems. The paper gives a detailed account of the device theory and principle of operation, illustrating that there are two main modes of operation. These are the "logic" or "bit-sensitive" mode and the "relational" or "bit-insensitive" mode. The paper continues to consider applications of this device and outlines functionalities such as magnitude comparison between two binary signals and functionalities resembling those of the Batcher-Banyan or "Starlite" switching fabric. Experimental results are then given and packet routing capabilities are outlined. The paper also indicates future prospects for this technology.

McCormick et al. (paper 4.2) further investigate S-SEED devices in this chapter. This paper demonstrates the parallel operation of a 32×16 symmetric self-electrooptic effect device array. The two-dimensional array consists of 512 devices and various modes of operation are demonstrated including bistable operation, static operation, and operation as a random access memory (RAM). Each S-SEED device consisted of two multiple-quantum-well (MQW) modulators connected in series, each of the modulators driven by a 4-mW laser diode. The device operation is demonstrated and bistability curves are given for 64 of the devices. The paper also gives a discussion of several of the issues that arise in the operation of large optical switching device arrays.

Zhou et al. study the bistable characteristics of MQW DFB lasers in paper 4.3. They report bistable characteristics and all-optical set-reset operation of 1.5-μm MQW DFB lasers. The lasers are characterized by a high extinction ratio of 20 dB, high lasing output power of 4 mW, fast switch-off time of about 2.5 ns, and a low peak power of set-reset pulses (500 μW). A switch-on time in the subnanosecond range was demonstrated. The authors conclude that the good results obtained in terms of a high extinction ratio with a high optical output power and a fast switching response with a lower switching power requirement are all very attractive features in all-optical switching and computing applications that demand cascade combinations.

Paper 4.4, which describes all-optical devices obtained using nonlinear fiber couplers, concludes the chapter. The devices considered include a bandpass filter, an AND/OR logic gate, and a power-dependent switch utilizing saturation. The paper starts by giving the model used for the coupler, then moves on to consider applications in bandpass power filters under two conditions: equal and unequal core Kerr coefficients. Logic gates are subsequently dealt with and the power-dependent switch is developed. The paper concludes by indicating that the power flow (or mode) portrait (PFP) model developed can be utilized to realize a host of other all-optical devices based on similar principles.

Greatly Enhanced Logical Functionality of the S-SEED for use in Optical Switching Systems

R.J. GRINDLE AND J.E. MIDWINTER

Indexing terms: S-SEED, Self routing switching, Optical switching systems

Abstract: Sophisticated optical functionality can be obtained from the S-SEED by the use of simple optical control techniques. Two modes of operation are described in which the S-SEED combines the advantages of both relational and logical optical switching elements. Theoretical and experimental results of the S-SEED being used for possible applications in self routing switching fabrics are presented and discussed.

1 Introduction

Optics offers many attractive features for use in telecommunications, computing and in the high capacity interconnection of electronics [1]. Optical fibre has now, owing to its low cost and huge capacity, almost entirely displaced coaxial cable in long distance terrestrial communications links [2]. Recently, the advent of the optical fibre amplifier [3], has enabled optical signals to be amplified without the signal first having to be converted into an electrical one. As transmission becomes less of an issue in telecommunications networks, more attention is being directed towards switching, and how it can benefit overall network performance. With the all optical transmission of telecommunication signals over larger and larger distances, it becomes more and more desirable to be able to perform switching and logical processing operations on optical signals without first having to have recourse to the speed limiting, electronic domain.

Asynchronous transfer mode (ATM) is a packet based transmission format in which data is grouped into 48 byte bundles contained within a packet or cell of 53 bytes. The cell also consists of a 5 byte header in which the cell's destination address is contained. Individual cells are 'posted' over the network to their respective destinations, with each ATM cell being treated exactly the same on its routing through the network, regardless of its service origin. Higher bit rate services such as TV simply send out more ATM cells to satisfy their increased bandwidth requirements. ATM allows different bit rate services to make efficient use of available network bandwidth by the statistical multiplexing of many cells on high capacity broadband channels [4]. Also, priorities can be assigned to delay sensitive or loss sensitive services. However, ATM switching requires very much more computing and processing than circuit switched ISDN telephony channels at 64 kbits/s which have relatively long holding times. An evolutionary approach from the proposed synchronous network (SDH or SONET [5]) to full scale ATM would include the asynchronous multiplexing of many ATM cells (going towards the same destination) onto the synchronous payload envelope (SPE) of a much larger synchronous packet or timeframe [5]. This would require switching at a far lower rate than necessary for individual ATM cells. Given the dramatic increase in processing overhead by switching individual ATM cells rather than by using current circuit switched techniques, it may be desirable for individual ATM cells to be able to self route through a switch to their network destination. One such switch type is the Batcher–Banyan routing network of which the 'Starlite' switch is an example [6]. At each stage in this switch, a simple logical operation is performed on the cell header which determines whether the cell is exchanged or bypassed. This is a nonblocking switch architecture whose size scales very favourably with switch size. It would be very convenient to be able to perform this logical operation and self routing, optically and 'on the fly', thereby distributing control of the switch and the routing of cells over the entire switch fabric. To implement in electronics, this normally requires a small processor with the order of 100 gates per exchange bypass stage.

A number of technologies exist which can switch or route optical signals. These can be divided into two main categories based on their device characteristics in photonic switching architectures [7]. First, 'relational' or 'passive' devices are optical switches which perform a mapping function between given inputs and outputs depending on the state of some control signal. These devices are transparent to the service bit rate passing through them, and therefore their bandwidth can be very large. However, they have no ability to extract information from the bits throughput. Optical switches of this kind would include mechanical fibre switches [8] and

Paper 9192J (E13), first received 4th June and in revised form 22nd September 1992

The authors are with the Department of Electronic & Electrical Engineering, University College London, Torrington Place, London WC1E 7JE, United Kingdom

The authors would like to acknowledge BT Laboratories and the Science and Engineering Council (SERC) for the funding of this work.

Thanks to I. Burnett and Dr. N. Whitehead (BT Laboratories) for their most helpful thoughts and discussions and Dr. M. Whitehead (University College London) for the use of the MQW devices used in this experiment. Thanks to Dr. J.S. Roberts (University of Sheffield) for growing the material used in these devices and to A. Rivers (University College London) who was responsible for device fabrication and device mounting.

planar waveguide switches, for example, in $LiNbO_3$ [9]. The second category of optical switches are 'logic' devices in which an optical signal interacts with the device in such a way that it controls the state of the switching device (i.e. ON or OFF). These devices sense individual bits and can perform various Boolean logic operations. However, because of their strong interaction with individual bits, the maximum transmitted bit rate that may be conveyed is limited because of the need for some of the devices in a system to be able to change states or switch as fast or faster than the signal bit rate. An optical logic device of this type is the bistable self-electro-optic effect device (SEED) [10, 11], which uses the quantum confined Stark effect [12] in semiconductor multiple quantum wells to alter its absorption of light.

Because of the relative characteristics of the various logic devices available (i.e. relational devices: high bandwidth and data transparency, but no logical operation and logical devices: optical processing functionality but much lower bandwidth, perhaps the advantages of both can be used in some 'hybrid' network switching fabric. Up until quite recently, however, it has been difficult to consider the usage of current optical logic devices in real systems applications. High power, critical biasing and accurate temperature control device requirements have meant the slow evolution of these devices from the laboratory into practical systems. However, the SEED and, in particular, the symmetric or S-SEED, overcome many of these difficulties [13]. Its dual rail differential logic operation, low switching energies and wide bistable hysteresis width lend it to some very interesting systems applications. The S-SEED has already prove useful in optical computing applications such as optical clock extraction [14] and wavelength modulation/conversion [15]. It can also be used as a self linearised modulator [16] or an optical 'tap' [17]. A whole range of SEED devices can now be fabricated in large arrays [18], and with varying operating characteristics, such as very high contrast [19] and low voltage [20].

Current S-SEED computing architectures, however, are operated in a rigidly clocked synchronous mode. Logical functionality is obtained in return for the penalty of increased system complexity arising from a number of optical beams being incident on individual devices and from the requirements for strict clocking techniques [21]. It will be shown in this paper how the S-SEED can be used in a more flexible manner, combining the properties of both relational and logic devices, enabling quite sophisticated functionality and a degree of transparency to be achieved for a number of very simple control algorithms, in the context of a self routing, optical packet switch. Single switch stage functionality is demonstrated for this nonregenerative, 'intelligent' optical switch.

2 Theory

An S-SEED consists of two PIN, multiple quantum well (MQW) modulators reverse biased in series [13], with centre voltage, V_{centre} between the two S-SEED diodes (Fig. 1*a*. The MQW modulators are operated at a wavelength corresponding to the unbiased el-hhl exciton peak, and exhibit decreasing absorption and photocurrent for increasing applied voltage. They therefore have negative differential resistance (i.e. increasing device absorption/photocurrent for decreasing applied bias and vice versa), the condition required for positive feedback and bistability [10]. If the optical power incident on one of the diodes is sufficiently greater than the optical power incident on the other diode, positive feedback causes the entire supply voltage to be seen across the diode with the least light incident on it; the other diode experiencing a very low voltage. One diode is operating in a low voltage,

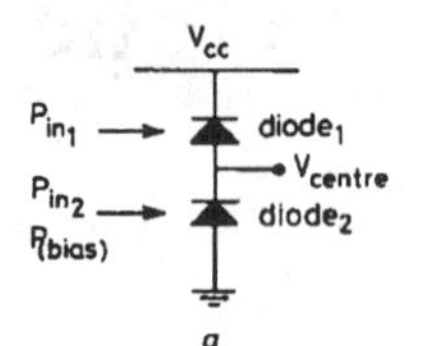

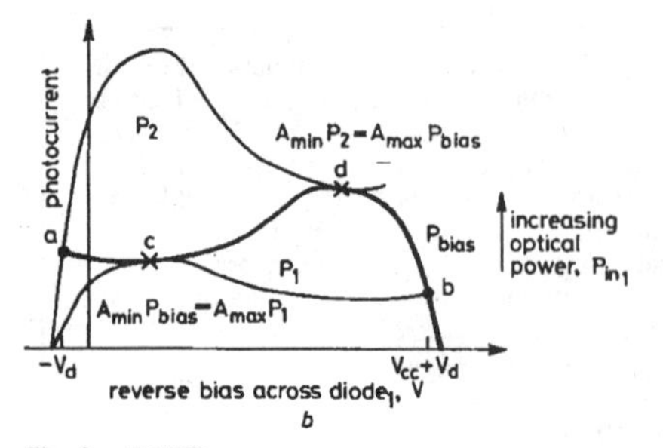

Fig. 1 *S-SEED*
a schematic diagram
b graphical loadline analysis

high absorbing state with maximum absorption A_{max} [%], and the other is experiencing an high electric field, and operates in a low minimum absorbing state, A_{min}. The S-SEED is bistable in the ratio of its two optical input signals.

S-SEED operation can be easily understood by the use of current loadlines (Fig. 1*b*). The two light solid curves are the photocurrent versus voltage characteristics of diode$_1$ of the S-SEED, for two optical input powers, P_1 and P_2. The heavy solid curve is the photocurrent of diode$_2$ (diode$_1$'s load) for a fixed incident optical bias power, $P_{in_2} = P_{bias}$. Optical power incident on diode$_2$ is held constant at P_{bias}, and the optical power, P_{in_1} incident on diode$_1$ is ramped up and down. If P_{in_1}, the optical power incident on diode$_1$, is sufficiently greater than P_{bias} (i.e. $P_{in_1} > P_2$), the only S-SEED loadline intersection point is at **a**, with all the supply voltage across diode$_2$. If P_{in_1} is decreased to a value sufficiently less than P_{bias} (i.e. $P_{in_1} < P_1$), the only possible operating solution is reversed to that at **b**, the entire supply voltage now being seen across diode$_1$. For optical powers P_{in_1}, with $P_1 < P_{in_1} < P_2$, the S-SEED is bistable, with a total of three possible operating points, two of which are stable [10]. For optical powers within these limits, the S-SEED remains in either of the states **a** or **b** because the difference between the two input powers is not sufficient to cause the S-SEED to alter its state. For reflective SEEDs there is a corresponding change in device reflectivity as the S-SEED switches states. In either stable state, one of the MQW diodes is more reflecting than the other, owing to the change in device absorption with voltage (quantum confined Stark effect [12]).

It can be seen from Fig. 1*b* that, when P_{in_1} is increased to a power P_2, such that the two curves intersect at points **d** and **a**, a further increase in P_{in_1} will cause the device to switch it operating point from **d** to **a**. Just at the point **d**, the *minimum* photocurrent generated by diode$_1$

for an incident optical power P_2 is equal to the *maximum* photocurrent generated by diode$_2$ for the incident power P_{bias}, i.e. $A_{min}(e\lambda/hc)P_2 = A_{max}(e\lambda/hc)P_{bias}$. If P_{in_1} is increased beyond P_2, the photocurrent from diode$_1$ acts to charge up diode$_2$ and the S-SEED changes its operating point to that at **a**. Conversely, at operating point **c**, the minimum photocurrent generated by diode$_2$ for P_{bias} is equal to the maximum photocurrent through diode$_1$ for an optical power P_1 (i.e. $A_{min} P_{bias} = A_{max} P_1$). If P_{in_1} is made smaller than P_1, diode$_1$ charges up, diode$_2$ discharges and the device switches its stable operating point to **b**, with all the supply voltage across diode$_1$. The voltage between the two S-SEED diodes (V_{centre}) can be monitored, giving an indication of the S-SEED's present state.

It is thus seen that $P_2 = (A_{max}/A_{min})P_{bias}$ and $P_1(A_{min}/A_{max})P_{bias}$. The required input contrast ratio of the two input signals is therefore $1 : C$, where $C = (A_{max}/A_{min})$, to cause S-SEED switching. The width of the bistable S-SEED region, $\Delta P = (P_2 - P_1)$ for a given bias beam P_{bias} is $\Delta P = P_{bias}(A^2_{max} - A^2_{min}/A_{max}A_{min})$. A_{max} and A_{min} can be varied by device design; therefore S-SEEDs may be designed with specific hysteresis widths to suit the intended systems application.

A schematic S-SEED bistable, hysteresis plot is shown in Fig. 2. Optical power incident on diode$_2$ of the

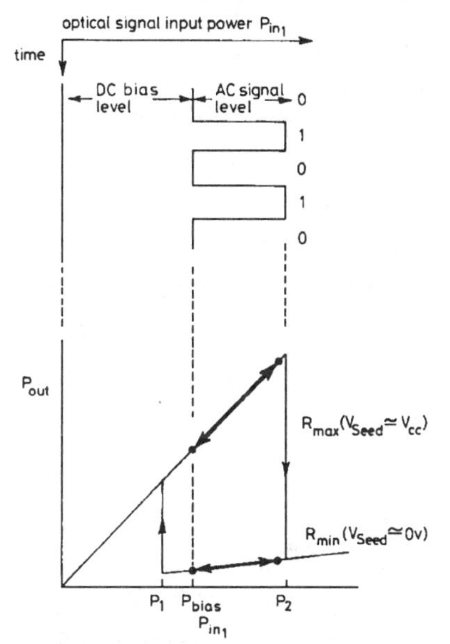

Fig. 2 *Schematic S-SEED hysteresis plot*

S-SEED is held constant at P_{bias} and power on the other input, P_{in_1} is ramped up and down. P_{out} reflected from diode$_1$ is shown on the *y* axis. By driving the S-SEED with input powers $P_{in_1} = P_2$ and $P_{in_2} = P_{bias}$, a further increase in P_{in_1} causes the photocurrent from diode$_1$ (charging diode$_2$) to exceed the photocurrent discharging diode$_2$. Upon switching with $P_{in_1} > P_2$, device reflectivity changes from a maximum, R_{max} (corresponding to low absorption, A_{min} at operating point **d**), to a minimum reflectivity R_{min}, corresponding to maximum absorption, A_{max} for low electric field (operating point **a**). The greater the difference in the input powers and photocurrents, the faster the charge transfer and the more quickly switching occurs. For slowly varying input beams, switching time from one state to the other is $\Delta t \simeq C_{tot} V_{cc}/S(P_{in_2} - P_{in_1})$, where S is the average MQW device responsivity (A/W) [13]. The optical switching energy E_{opt} required to switch the S-SEED is constant, and can be approximated by $E_{opt} \simeq \Delta t \, \Delta P'$, where $\Delta P'$ is the difference in the optical input powers ($P_{in_2} - P_{in_1}$). For rapidly varying input signals, and when P_{in_1} is only infinitesimally greater than P_2, or infinitesimally smaller than P_1, the S-SEED will take a long time to switch. This is the characteristic of critical slowing down [22]. Therefore, to quickly change the state of the S-SEED, P_{in_1} should be much greater than P_2 or much less than P_1, for a given bias power $P_{in_2} = P_{bias}$.

It is important to note that the constant biasing power $P_{in_2} = P_{bias}$ on diode$_2$ does not correspond to the power at the exact centre of the hysteresis curve. This is because the absolute change in device photocurrent with voltage depends on the absolute power incident on the device. For $(A_{max}/A_{min}) = 2$ and $P_{bias} = 100\,\mu$W, $P_2 = 200\,\mu$W and $P_1 = 50\,\mu$W. A different real change in power on one of the S-SEED inputs is required depending on the direction of switching (i.e. from V_{centre} going from high to low, or vice versa). The relative position of P_{bias} in the hysteresis plot is given by:

$$P_{bias(rel)} = \frac{P_{bias} - P_1}{P_2 - P_1} = \left(\frac{A_{max}A_{min} - A^2_{min}}{A^2_{max} - A^2_{min}}\right)$$

where $P_{bias(rel)} = 0.5$ when P_{bias} coincides with the exact centre of the hysteresis plot, for the nonbistable case when $A_{max} = A_{min}$. For all values of $A_{min} < A_{max}$, $P_{bias(rel)} < 0.5$, the position of P_{bias} moves to the left of the hysteresis plot.

For 'AC' optical input signals in which a logic 0 corresponds to zero power incident at an S-SEED input, signal combinations 10 (i.e. a logic 1 incident on diode$_1$, logic 0 on diode$_2$) and 01 will always be sufficient to cause the S-SEED to switch, because, when $P_{in_2} = P_{bias} = 0$, P_{in_1} = logic 1 is always greater than P_2, since $P_2 = (A_{max}/A_{min})P_{bias}$. The greater the power level corresponding to logic level 1, the faster the S-SEED changes states since the difference between P_{in_1} and P_{in_2} is larger. The S-SEED switching time Δt, should be sufficiently smaller than the incident signal bit rate t_b, so that the S-SEED can fully latch on each successive input bit pairs ($\Delta t \ll t_b$). When both inputs are high, however, (i.e. 11), the S-SEED remains latched in its previous state because the difference between its two input signals should not be sufficient to cause the S-SEED to change its state; the S-SEED is operating in its bistable region. If a constant optical bias control signal is added to the two AC digital signal inputs, the control signal power effectively degrades the input contrast ratio of the two input signals. If a large enough control bias is simultaneously applied on both S-SEED inputs, the input contrast ratio falls below that required to induce switching and the S-SEED remains latched in the same state as it was in prior to the application of the control bias. The S-SEED operates within the confines of its own bistable hysteresis loop, remaining latched in either of its two stable states, regardless of the AC component of the two input data (streams (Fig. 2). This is a similar manifestation of the effect used in conventional SEED systems. High power optical beams are used to read out the state of the S-SEED which operates within its bistable hysteresis loop. This is known as time sequential gain. However, for device operation as described here, the bias beams are not used to read out the state of the device but to give to the device additional logical functionality.

The condition for nonswitching is that the maximum power difference between the two input channels does not exceed the S-SEED input contrast ratio, $C = (A_{max}/A_{min})$. This implies that, when a 0 and 1 are present at the S-SEED inputs, the AC + control bias sinal power (corresponding to logic 1) is not sufficiently greater than the control signal power on its own (corresponding to logic 0), for the S-SEED to be caused to change states, i.e.

$$\text{AC + control must be less than} \left(\frac{A_{max}}{A_{min}}\right) \text{control}$$

$$\Rightarrow \quad \text{AC} < \left(\frac{A_{max}}{A_{min}} + 1\right) \text{control}$$

(S-SEED nonswitching signal format)

Conversely, the condition for the S-SEED to change states (i.e. S-SEED switching), is that the ratio of the two input intensities is greater than (A_{max}/A_{min}).

$$\Rightarrow \quad \text{AC} > \left(\frac{A_{max}}{A_{min}} - 1\right) \text{control}$$

(S-SEED switching signal format)

Another variation on this way of using S-SEEDs is due to the fact that, when no optical power is incident on the S-SEED, once the S-SEED has stabilised through its leakage currents, each diode of the S-SEED experiences the same voltage across it, $V_{centre} \simeq V_{cc}/2$. If two equal input power beams are applied, the device operates in this unstable state. A small difference in optical powers will push the S-SEED in one direction or the other, to one of its stable states. In this case, not all the switching energy is derived from the signal beams. Some of it is obtained from the bias beams so that a small difference in input powers is enough to cause the S-SEED to latch. This is known as the 'optical sense amplifier' mode of operation [11]. If an optical signal incident on the S-SEED is in nonswitching format, and the S-SEED is operating at $V_{centre} \simeq V_{cc}/2$, if the first bit of each of the input signals are different (i.e. 01 or 10), this difference is enough to cause the S-SEED to latch to whichever state. Once the S-SEED has latched on the first bit difference of the two inputs, it remains latched for the remainder of the optical signal input, since the signal inputs are in S-SEED nonswitching format. For this technique to be successful, the power difference between 1 and 0 signals (i.e. the amplitude of the AC information signal) must be sufficiently greater than the control bias level for the S-SEED to latch within one bit interval, for a given signal power (i.e. $\Delta t \ll t_b$). If the first input bits are 00 or 11, the two input signal powers should be sufficiently similar for the S-SEED not to latch in one bit interval. The S-SEED should remain unlatched until the first bit *difference* of the two data signals.

3 Application

There are therefore two possible modes of operation of the S-SEED. 'The first of these is the logic' or 'bit sensitive' mode, in which when the signal inputs to the S-SEED are AC, the S-SEED changes its state depending on whether its inputs are 01 or 10. The S-SEED is acting as an optical logic device whose present state depends on its current inputs. The second mode of operation is the 'relational' or 'bit insensitive' mode, in which, when control optical bias beams are applied to the two S-SEED inputs, the S-SEED remains latched in its state prior to the application of the control bias. The S-SEED is now acting as a passive or relational device which suppresses one of its input signals with respect to the other owing (to the difference in the two diode reflectivities), depending on which of the two stable states the S-SEED has latched. The S-SEED now passes or suppresses one of each of its inputs (to varying degrees, depending on S-SEED component device parameters of absorption, A_{max} and A_{min} and reflection, R_{max} and R_{min}) and passively relates the inputs to their outputs in a bit transparent mode of operation. It is shown schematically in Fig. 3, that, in 'logic' mode, the S-SEED changes its state with the inputs 10 and 01, remaining in that state for the inputs 00 and 11. When the optical control bias is applied, the S-SEED remains latched regardless of the combination of bits at its inputs (relational mode) with one diode highly absorbing and the other reflecting. Operating the S-SEED in this simple manner (i.e. the adding or removing of a control optical bias signal which controls whether the device can switch its state or not), combines the advantages of both optical logic and relational devices. Using S-SEEDs in these two modes of operation can, with a few very simple control algorithms, provide for relatively complex optical processing functionality in photonic switching fabrics.

If the signals arriving at the inputs to an S-SEED are in the form of two, bit synchronised optical data streams in least significant bit (LSB) first format, the S-SEED changes its state as the inputs vary from 01 to 10. If these data signals are followed by a control bias signal in nonswitching format, the S-SEED remains latched in a state corresponding to the most significant bit (MSB) difference of the two data streams. If the serial binary number at input_1 is greater than that at input_2, the S-SEED latches in the state $V_{centre} \simeq V_{cc}$; if the reverse is true, the S-SEED remains latched with $V_{centre} \simeq 0$. A magnitude comparison between two binary optical signals has there-

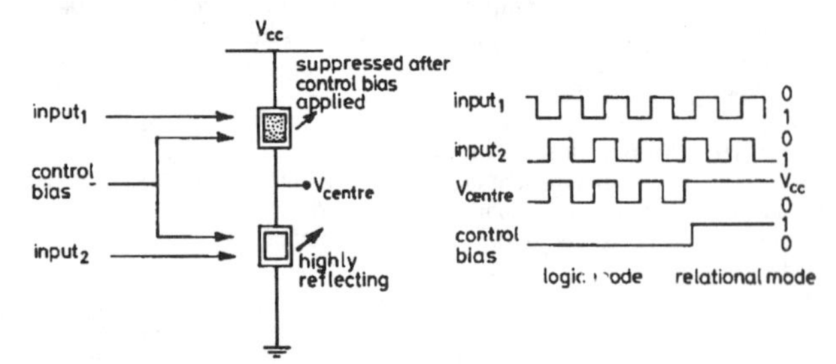

Fig. 3 *Two modes of S-SEED operation*

fore taken place. This represents a very simple task for an S-SEED operating in this fashion but it would be a significant feat for conventional electronics, requiring of the order of 100 logic gates. This type of functionality, combined with switching on the first bit difference of two data streams as described above, can lend itself to some interesting system applications. This type of functionality is similar to that required in a Batcher–Banyan or 'Starlite' type switching fabric, where optical packets or cells need to be routed according to the magnitude of their header destination addresses. Two cells are input into a switch and are exchanged or bypassed depending on which header address if the greater, or whether a particular bit in a header is either 1 or 0. Once the S-SEED has latched on an incident header, the ensuing data, in S-SEED nonswitching format, is passively reflected and related independently of its bit rate. The advantages of both logic and relational devices (i.e. bit sensitivity and transparency), are thus utilised by this way of using S-SEEDs.

4 Experiment

The SEED devices used in this experiment were of an asymmetric Fabry–Perot, multiple quantum well (MQW), type structure as reported in [23]. The MQW device structure was grown by MOVPE and is a normally off (bias-reflecting) asymmetric Fabry–Perot modulator [24] consisting of a 15 × (150 Å GaAs well/60 Å $Al_{0.3}Ga_{0.7}As$ barrier) intrinsic region sandwiched between *p*-type and *n*-type layers (Fig. 4). The unbiased

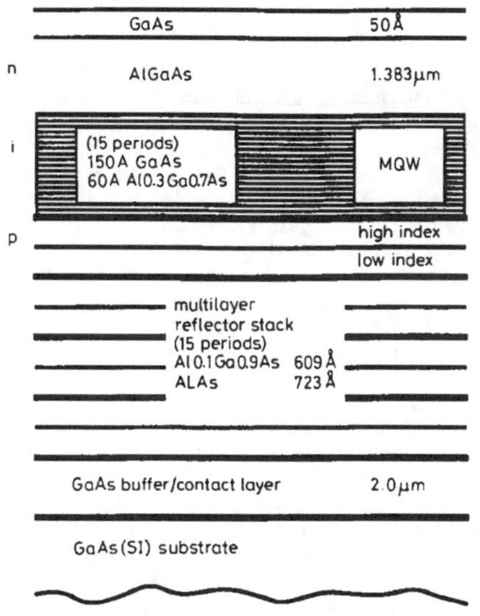

Fig. 4 *SEED structure used*

el–hhl esciton peak wavelength was aligned with an asymmetric Fabry–Perot cavity resonance, and cavity reflection at zero field is made low by virtue of matching the top and effective bottom mirror reflectivities. Upon increasing the applied electric field across the device intrinsic region, by way of the quantum confined Stark effect (QCSE) [12], the el–hhl exciton redshifts and loses oscillator strength. As cavity absorption falls, the cavity becomes asymmetric and device reflection increases. The front and back mirrors are formed by the air/GaAs interface ($R = 0.3$) and an integrated *p*-type semiconductor multiple quarter wave stack of $R \geqslant 0.95$, respectively. Maximum absorption for these MQW devices is $A_{max} \simeq 0.80$ for a forward bias voltage of about 1 volt (to counteract the diode's built-in field) and minimum device absorption, $A_{min} \simeq 0.44$ for a reverse bias voltage of 2.5 volts. A corresponding device reflection change from 10% to approximately 50% occurs for the same voltage swing, i.e. a 5 : 1 reflection contrast ratio. [For the devices used (A_{max}/A_{min}) was measured from photocurrent spectra to be $\simeq 1.8$]. The devices used in this experiment were large (540 μm × 540 μm) with optical window size, 400 μm × 400 μm. These devices therefore exhibited relatively large capacitances (approximately 50 pF) which would limit the maximum speed of systems made with these devices owing to their high switching energies.

The experimental setup was as shown in Fig. 5. An argon–ion laser pumped tunable Ti : sapphire laser was used as a source of continuous light at 860 nm, from which four separate beams were obtained. Two of these beams were 'AC' modulated by simultaneously passing them through an optical chopper. The chopper was arranged so as to modulate the two signals with each of the possible combinations of 1 and 0 in a repeating series (i.e. 00, 01, 11, 10, etc.). (The effective bit rate of these AC header and data signals could be caried by slowing down or speeding up the optical chopper.) Each of the AC signal beams was directed onto the corresponding input diode of a discrete device S-SEED which was configured so that it could easily be changed into two single SEEDs reversed bias in series with a resistive load (R-SEEDs). Two constant power or control beams were also made incident on each S-SEED input. The magnitudes of the AC and control optical signals incident on the two S-SEED inputs were measured by reverse biasing the two SEED devices, separately, in series with a resistive load (R-SEED configuration [10]). Photocurrent flowing through the SEED devices was measured as a voltage drop across the R-SEEDs, 33 kΩ resistive loads. A direct comparison of the relative magnitudes of the incident AC and control powers incident on the two inputs could therefore be made on the oscilloscope screen. The magnitudes of the signal and control optical signals incident on the two S-SEED inputs were controlled using variable attenuators and neutral density filters. Because discrete devices were used to make up the S-SEED, devices manufactured from slightly different parts of the wafer exhibited slightly different behaviour, with the consequence that, as with electronic devices, it was difficult to match up the devices exactly. By examining the magnitude of the photocurrents flowing through the separate devices configured as R-SEEDs, rather than the optical power incident, it was easier to allow for any differences in their maximum and minimum absorptions, ensuring that their respective values of (A_{max}/A_{min}) were similar. Measuring photocurrents rather than incident optical power also allowed accurate matching of the photocurrents by misaligning the light incident of the individual devices.

Two photodetectors D_1 and D_2 were used to monitor the two optical input signals (P_{in_1} and P_{in_2}) incident on the S-SEED. The two optical input channels and the voltage between the two devices, V_{centre} (showing the present state of the S-SEED), were observed on an ocilloscope screen. The first order diffracted beam of an acoustooptic modulator located at (A), was used to generate the control signal beams incident on the S-SEED. The control signal beams could therefore be easily turned on and off at a variable frequency and duty cycle, controlled by a variable function generator via the blanking input to the acoustooptic modulator supply. The blanking input to the acoustooptic modulator was used to trigger the oscilloscope so that the effect of adding or removing the optical control bias to the information signal, on the S-SEED, could be observed. By removing or allowing the optical control bias, the difference between header and data signals was simulated. This was used to demonstrate the S-SEED latching on the MSB difference of two input packet headers. The acoustooptic modulator was subsequently moved to position (B), so that all light incident on the S-SEED could be either totally transmitted or suppressed. The transmission of light through the modulator on to the S-SEED was used to simulate the arrival of an optical packet signal after a period of switch inactivity. This was used to demonstrate S-SEED latching on the first bit difference of two input signals in nonswitching format.

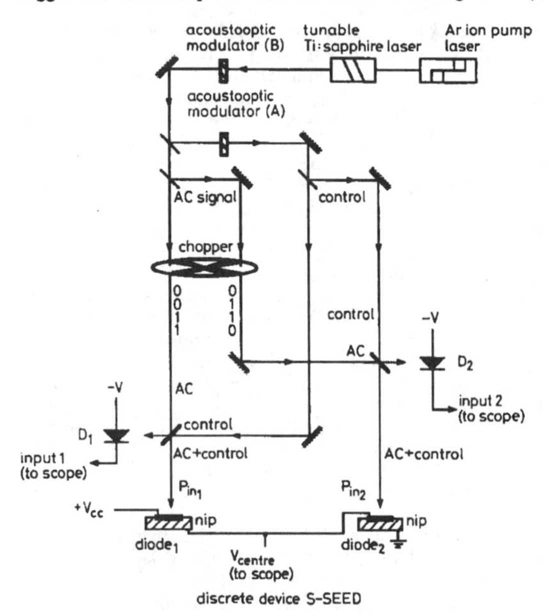

Fig. 5 *Experimental setup used*

5 Results

The S-SEED was supplied with a reverse bias of $V_{cc} = 2.0$ volts. Upon application of two constant power input beams, the S-SEED remained in either of two states, $V_{centre} = 2.5$ volts or -0.8 volts. The S-SEED exhibited the expected bistable behaviour, and it was observed that, if either of the two beams were momentarily removed or broken, the S-SEED would switch its point of operation to the corresponding stable state, and remain in that state when both input beams were reinstated. The S-SEED was bistable for a wide range of supply voltage greater than 0 volts. However, the value of V_{cc} was chosen such that, upon switching, the MQW diodes making up the S-SEED would experience the voltage swing yielding the largest possible change in their respective absorption of incident light. V_{centre} can be greater than V_{cc}, because, in the latched state, the PIN diodes' built-in voltages, V_d, act to reinforce V_{cc}. Upon switching to either of the stable states, one of the diodes experiences a reverse bias of approximately $V_{cc} + V_d$, whilst the other experiences a small forward bias, $-V_d$. (Intersection of the S-SEED operating curves occurs at a voltage greater than V_{cc} and less then zero because the photocurrent/internal quantum efficiency of the individual MQW diodes only falls off after the very significant built-in field, for these relatively thin devices, is counteracted: operating points a and b, Fig. 1*b*.) The magnitude of the constant power inputs was varied, and it was verified that the required input contrast ratio C, was the same as expected from the previous theory (i.e. it was noted that $C \simeq A_{max}/A_{min} \simeq 1.8$, $P_2 = CP_{bias}$ and $P_1 = C^{-1}P_{bias}$). The difference in optical powers required to switch the S-SEED was measured directly by observing the difference in magnitudes of photocurrent flowing through the two R-SEEDs, converted from the S-SEED, just after switching had taken place.

The two AC digital signal beams were used as inputs to the S-SEED (Fig. 5). The AC signals were designed to input to the S-SEED a repeating sequence of all the possible combinations of 1 and 0 (i.e. 00, 01, 11 and 10 etc). For the input combination diode$_1$ = 1 and diode$_2$ = 0 (i.e. 10), the S-SEED switched to stable state $V_{centre} \simeq 2.5$ volts (point **a**, Fig. 1*b*). For input combination 01, the S-SEED switched to the stable state, $V_{centre} = -0.8$ volts (point **d**, Fig. 1*b*). For the 11 input combinations, the S-SEED remained latched in its previous state at a voltage slightly less than 2.5 volts for the high state and slightly greater than -0.8 volts in the low state. This was because the S-SEED is now operating in its bistable region, and the operating points a and b are quite high on the operating curves (Fig. 1*b*). For optimum performance, the voltage difference between points a and b should be the same as the desired voltage swing for the maximum change in device absorption upon switching, thereby maximising the extent to which one input is suppressed/reflected with respect to the other. Thus, for the repeating sequence of 01 and 10 input combinations, the S-SEED switched its operating point correspondingly, remaining latched for the intervals 00 and 11. It was noticed that, for AC input combination 00 when no light is incident on either input of the S-SEED, the S-SEED diodes slowly discharge because of their respective leakage currents. The time taken for this discharging, however, was large compared to the input signal data rate, and so had little effect on the expected device behaviour.

As expected, upon application of the control bias signals, the S-SEED remained latched in the state prior to the application of the control bias powers. This was observed by taking a photograph of the oscilloscope screen showing the two input signal channels and V_{centre} (indicating the current state of the S-SEED). The oscilloscope was single shot triggered by the blanking input to the acoustooptic modulator. When the control signals are applied to the S-SEED, the oscilloscope triggered once, exposing a photographic film which allows the interdependence of the inputs and control signals on device behaviour to be assessed. Fig. 6 shows the two AC

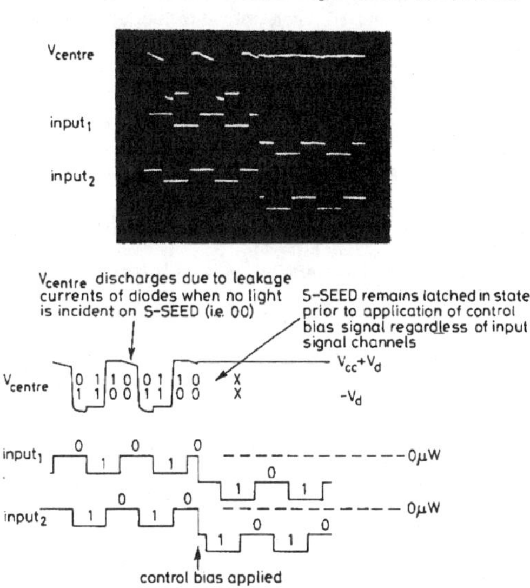

Fig. 6 *S-SEED switching on most significant bit difference of two data streams*

modulated input signals, $input_1$ and $input_2$, detected by D_1 and D_2, and their effect on V_{centre} (which shows the state of the S-SEED at a given time). It was seen that, for input combination 01, the S-SEED switches to the stable state $V_{centre} = -0.8$ volts, and for the input 10, $V_{centre} =$ 2.5 volts. For the state 00, the centre voltage decays slightly owing to the dark currents of the two diodes. However, because this decay is slow, it has little effect on the S-SEED's performance. Just prior to the application of the control bias signals to the S-SEED inputs (Fig. 6), the input state is 00, the S-SEED thus remains in its previous state for 10, $V_{centre} \simeq 2.5$ volts. It can be seen that, when the control bias offset is applied to the two inputs, the S-SEED remains latched in its present state $V_{centre} \simeq$ 2.5 volts, regardless of the combination of the AC signals present, XX (X = don't care). The S-SEED then remains latched for the remainder of the following data signal, in S-SEED nonswitching format, passively relating to varying degrees its inputs to its outputs. This effectively demonstrates S-SEED latching on the MSB difference of two optical packet headers.

With no light incident on the S-SEED, V_{centre} was found to be equal to zero volts, presumably because of unequal leakage currents in the two devices. Two 470 kΩ resistors in series were placed in parallel with the S-SEED, with the centre voltage between the two series resistances tied to the centre voltage, V_{centre}, of the S-SEED. The resistive values were chosen to be of low impedance compared to the reverse bias impedance of the S-SEED under zero illumination and of high impedance relative to the impedance of the illuminated S-SEED. This anchored V_{centre} to $V_{cc}/2$ when no light was incident on the S-SEED. Under illuminated conditions, the resistors in parallel had no discernible effect on device behaviour. When the acoustooptic modulator transmitted, optical signals in S-SEED nonswitching format were made incident on the S-SEED. Depending on the time at which the signals are applied to the S-SEED, relative to the input data signal (i.e. whether the first header bits input are 01 or 10, the S-SEED latches to either of its stable states and remains latched in that state. When the incident light is removed (i.e. once the optical packet has passed), the S-SEED returns to its 'middle' state, $V_{centre} \simeq V_{cc}/2$. This can be observed in Fig. 7. Initially the

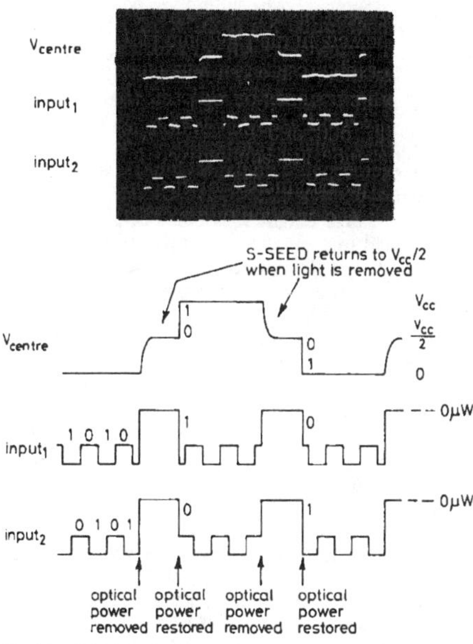

Fig. 7 *S-SEED switching on first significant bit difference of two data streams*

S-SEED is latched in the state $V_{centre} \simeq 0$ and continues to do so regardless of the combination of 1s and 0s at its inputs. When the optical power is removed, the centre voltage of the S-SEED returns to $V_{cc}/2$, as the respective leakage currents of the two diodes act to equalise the voltages across themselves, aided by the series resistances in paralle. When the optical signals in S-SEED nonswitching format are restored, the first signal inputs can be seen to be 10, therefore the S-SEED patches to $V_{centre} \simeq V_{cc}$. When the optical power is subsequently removed and restored, the S-SEED can be seen to latch to $V_{centre} \simeq 0$, owing to the fact that the first bit difference input was 0 and 1, as can be observed in the photograph. This demonstrates the principle of routing the optical cell according to the first bit difference of two optical cell headers, as required by the Banyan routing algorithm. It should be noted again that, when the first incident bits are different, 10 or 01, the S-SEED must latch more quickly than one bit interval. However, when the first incident bits are either 00 or 11, the incident signals must be sufficiently similar so as not to cause latching *until* their first bit difference. The tolerance in the degree of similarity required between the 00 and 11 signals is relaxed at high data speeds. However, for the relatively low data rates (~1 kbit/s) used in this experiment, any difference in optical powers was very significant, since the S-SEED latching time was small compared to the bit interval of the signal data.

Using the S-SEED in this way (i.e. latching on the most significant bit or the first significant bit difference of two input signals), the behaviour of the S-SEED is both stable and repeatable. By adjusting the frequency of the acoustooptic modulator's blanking input, the phase of the control bias signal to the S-SEED inputs could be matched so that it was possible to select regular and repeating latching of the S-SEED in either direction. For example, it was possible to simulate optical packet switching for cells latching the S-SEED in either direction either continuously, or pseudorandomly. The system was also immune to noise in the optical beams used owing to the wide hysteresis width of the S-SEED for a given optical power intensity.

6 Device and switch optimisation

MQW device optimisations applicable to this way of using S-SEEDs are chiefly to allow a good degree of tolerance in the relative magnitudes of the AC and control signals used, and to keep switching times to a minimum. Maximising (A_{max}/A_{min}) implies that a relatively small control bias is required to convert the signal into 'S-SEED nonswitching' format. Also, maximising (A_{max}/A_{min}) means that the separation between diode photocurrents upon switching is greater. Thus, after taking into consideration critical slowing down, switching times will be slightly faster. Making A_{max} as large as possible increases the device maximum responsivity for a given optical power, and also increases the SEED switching contrast ratio. [If device absorption in the low bias case is very high, reflection is very low ($R_{min} \simeq 0$) and the ratio of the reflection coefficients of the two stable states (i.e. the contrast ratio) is high.] Having low values of A_{min} is important in reducing the insertion loss the S-SEED since the greater the maximum reflectivity, R_{max}, the less lossy is a system of cascaded S-SEEDs. For the Banyan routing algorithm, and fastest possible latching on the first bit difference, it is required that the difference between the two signals 10 or 01 is as large as possible but that it still remains in S-SEED nonswitching format. Ideally, AC + control signal is just less than (A_{max}/A_{min}) control, with (A_{max}/A_{min}) being as large as possible so that, with a greater difference in optical powers and photocurrent levels, the faster the charge transfer between the two MQW diodes, and the faster the S-SEED switches. Maximising (A_{max}/A_{min}) also allows better definition in logic levels in the S-SEED nonswitching signal format. Significantly, the maximisation of (A_{max}/A_{min}) for this method of using S-SEEDs is a similar optimisation to that arrived at in [25], using S-SEEDs in cascaded, synchronous systems.

7 Conclusions

We have shown that the advantages of both optical relational and logical switching devices can be combined by using the symmetric-SEED in the way described in this paper. Quite complex optical functionality can be achieved for very simple control algorithms, and a degree of system intelligence and transparency can be instilled in an S-SEED system simply by use of a suitable signal format.

Functionality very similar to that required in a single element of a self routing packet switched network has been successfully demonstrated. However, this device does not readily lend itself to use in cascaded processing systems owing to it being nonregenerative. Also the signal beam tolerances for switching on the first significant bit of two incident headers might prove restrictive. However, the single element functionality and the additional 'intelligence' that can be achieved by the use of simple optical control techniques give us an indication of what can be achieved when optical technology is applied resulting in perhaps much more simple and yet functional devices than were previously possible.

8 References

1 MIDWINTER, J.E.: '"Light" electronics, myth or reality', *IEE Proc. J*, 1985, **132**, (6), pp. 371–393
2 COCHRANE, P.: 'Future directions in long haul fibre optic systems', *Br. Telecom Technol. J.*, 1990, **8**, (2), pp. 5–16
3 NAKAGAWA, K., and SHIMADA, S.: 'Optical amplifiers in future optical communication systems', *IEEE Lightwave Commun. Syst.*, 1990, **1**, (4), pp. 57–62
4 MINZER, S.E.: 'Broadband ISDN and asynchronous transfer mode (ATM)', *IEEE Commun. Mag.*, 1989, **27**, (9), pp. 17–24
5 HEALEY, E.M.: 'SONET: synchronous optical network standards', *Int. J. High Speed Electr.*, 1990, **1**, (2), pp. 169–181
6 HUANG, A., and KNAUER, S.: 'Starlite: a wideband digital switch'. Proceedings of the IEEE *Global Telecommunications* Conference, Atlanta, Georgia, USA, 1984, pp. 121–125
7 HINTON, H.S.: 'Photonic switching fabrics', *IEEE Commun. Mag.*, 1990, **28**, (4), pp. 71–89
8 YOUNG, W.C., and CURTIS, L.: 'Single-mode fiber switch with simultaneous loop-back feature'. IEEE/OSA Top. Meeting on *Photonic Switching*, Incline Village, NV, 1987
9 HA, W.L., FORTENBURY, R.M., and TUCKER, R.S.: 'Demonstration of photonic fast packet switching at 700 Mbit/s data rate', *Electron. Lett.*, 1991, **27**, (10), pp. 789–790
10 MILLER, D.A.B., CHEMLA, D.S., DAMEN, T.C., WOOD, T.H., BURRUS, C.A., GOSSARD, A.C., and WIEGMANN, W.: 'The quantum well self-electrooptic effect device: optoelectronic bistability and oscillation, and self-linearized modulation', *IEEE J.*, 1985, **QE-21**, (9), pp. 1462–1475
11 CHIROVSKY, L.M.F., LENTINE, A.L., and MILLER, D.A.B.: 'Symmetric self electro-optic effect device as an optical sense amplifier', *in* 'Technical digest, conference on lasers and electro-optics' (CLEO) (Washington DC: Optical Society of America, 1989) Vol. 11, paper MJ2
12 MILLER, D.A.B., CHEMLA, D.S., DAMEN, T.C., GOSSARD, A.C., WIEGMANN, W., WOOD, T.H., and BURRUS, C.A.: 'Electric field dependence of optical absorption near the band-gap of quantum well structures', *Phys. Rev. B.*, 1985, **32**, pp. 1043–1060
13 LENTINE, A.L., HINTON, H.S., MILLER, D.A.B., HENRY, J.E., CUNNINGHAM, J.E., and CHIROVSKY, L.M.F.: 'Symmetric self-electro-optic effect device: optical set-reset latch, differential logic gate and differential modulator/detector', *IEEE J.*, 1989, **QE-25**, (8), pp. 1928–1936
14 GILES, C.R., LI, T., WOOD, T.H., BURRUS, C.A., and MILLER, D.A.B.: 'All-optical regenerator', *IEE Electron. Lett.*, 1988, **24**, pp. 848–850
15 BAR-JOSEPH, I., SUCHA, G., MILLER, D.A.B., CHEMLA, D.S., MILLER, B.I., and KOREN, K.: 'Self-electro-optic effect device and

modulation convertor with InGaAs/InP multiple quantum wells', *Appl. Phys. Lett.*, 1988, **52** (1), pp. 51–53
16 MILLER, D.A.B., CHEMLA, D.S., DAMEN, T.C., WOOD, T.H., BURRUS, C.A., GOSSARD, A.C., and WIEGMANN, W.: 'Optical-level shifter and self-linearized optical modulator using a quantum well SEED', *Optics Lett.*, 1984, **9**, (12), pp. 567–569
17 GRINDLE, R.J., and MIDWINTER, J.E.: 'A self-configuring optical fibre-tap/photodetector–modulator with very high photodetection efficiency and high extinction ratio', *Electron. Lett.*, 1991, **27**, pp. 2170–2172
18 LENTINE, A.L., McCORMICK, F.B., NOVOTNY, R.A., CHIROVSKY, L.M.F., D'ASARO, L.A., KOPF, R.F., KUO, J.M., and BOYD, G.D.: 'A 2 kbit array of S-SEEDs', *IEEE Phot. Tech. Lett.*, 1990, **2**, (1), pp. 51–53
19 GRINDLE, R.J., MIDWINTER, J.E., and ROBERTS, J.S.: 'An high contrast, low-voltage, symmetric-self-electro-optic effect device (s-seed)', *Electron. Lett.*, 1991, **27**, pp. 2327–2329
20 WEINER, J.S., GOSSARD, A.C., ENGLISH, J.H., MILLER, D.A.B., CHEMLA, D.S., and BURRUS, C.A.: 'Low-voltage modulator and self-biased self-electro-optic-effect device', *IEE Electron. Lett.*, 1987, **23**, pp. 75–77
21 McCORMICK, F.B., TOOLEY, F.A.P., CLOONAN, T.J., BRUBAKER, J.L., LENTINE, A.L., MORRISON, R.L., WALKER, S.L., HINTERLONG, S.J., and HERRON, M.J.: 'S-SEED-based photonic switching network demonstration', *in* HINTON, H.S., and GOODMAN, J.W. (Eds.): 'OSA proceedings on photonic switching' (Optical Society of America, Washington DC, 1991), vol. 8, pp. 44–47
22 GARMIRE, E., MARBURGER, J.H., ALLEN, S.D., and WINFUL, H.G.: 'Transient response of hybrid bistable devices', *Appl. Phys. Lett.*, 1979, **34**, pp. 374–378
23 WHITEHEAD, M., RIVERS, A., and PARRY, G.: 'Low-voltage multiple quantum well reflection modulator with >100 : 1 on : off ratio', *IEE Electron. Lett.*, 1989, **25**, pp. 984–985
24 WHITEHEAD, M., RIVERS, A., PARRY, G., and ROBERTS, J.S.: 'A very low voltage, normally-off asymmetric Fabry–Perot reflection modulator', *Electron. Lett.*, 1990, **26**, pp. 1588–1590
25 LENTINE, A.L., MILLER, D.A.B., CHIRIVSKY, L.M.F., and D'ASARO, L.A.: 'Optimization of absorption in symmetric self-electrooptic effect devices: a systems perspective', *IEEE J.*, 1991, **QE-27**, (1), pp. 2431–2439

Parallel Operation of a 32 × 16 Symmetric Self-Electrooptic Effect Device Array

F. B. McCormick, A. L. Lentine, R. L. Morrison, S. L. Walker, L. M. F. Chirovsky, and L. A. D'Asaro

Abstract—**We demonstrate the parallel operation of a two-dimensional array of 512 symmetric self electrooptic effect devices. Both bistable operation and "static" operation as a random access memory (RAM) are shown. Some optical issues in the operation of large optical switching device arrays are discussed.**

FREE-SPACE optics for digital optical computing or for electrooptic interconnections promises the potential for large numbers of high bandwidth input-output connections ("pinouts") [1]. Reasonable device switching energies have been achieved [2], and system demonstrations have shown increasing parallelism [3]–[8]. This letter presents results of an experiment in which 512 symmetric self electrooptic effect devices (S-SEED's) [9] were simultaneously operated. In this experiment, simultaneous continuous bistable operation was shown for the 32 × 16 array of S-SEED's as well as simultaneous optical latching of optical data in a random access manner onto the array. Each S-SEED received two inputs and reflected two outputs, thus the total number of pinouts demonstrated was 2048. Each S-SEED consists of two multiple quantum well modulators (the "*S*" and "*R*" modulators) connected in series (Fig. 1). Two AlGaAs laser diodes with output powers of 4 mW were used. One laser was used to drive all of the *S* modulators, and the other laser, the *R* modulators. After collimation, the outputs of the lasers were combined on a knife-edge mirror using the optical isolator arrangement shown in Fig. 2 [3]. This pair of beams was replicated 512 times (32 × 16) by a two-dimensional binary phase grating (BPG) [10], forming an array of "beamlets."

To eliminate vignetting and ensure telecentric imaging [11] onto (and off) the device array, the BPG, which acts as an aperture stop in the system, is imaged through the PBS into the front focal plane of the objective lens. This was accomplished by using an afocal image relay, as in Fig. 2. The choice of relay lens type was determined by the need for diffraction limited performance over +/− 3°, moderate speed ($\approx f/8$), an external aperture stop, and ease of assembly from inexpensive, off-the-shelf lenses. One lens form satisfying these requirements is a symmetric, or Plossl eyepiece [12]. This type of lens is easily constructed of two identical achromatic doublets, and the use of these eyepieces rather than simple doublets in this case decreases the wavefront aberration at the full image field by over an order of magnitude. Another useful characteristic of this lens form is that the focal length can be adjusted by varying the spacing between the two doublets. This provides a simple mechanism for correcting the focal length errors in the doublets or objective lens (typically +/− 1%). This correction is required since a 1% focal length error will result in a 1% change in the spacing of the spots generated by the BPG. Over the 439 μm half image field of this experiment, a 1% error in the spot spacing will move the spots at the edge of the image almost completely off the 5 μm S-SEED windows! The ease of construction, low cost, and performance of this lens form make it extremely useful in many free space optical computing/switching applications.

The array of beamlets pass through the relay stage, and are reflected by the PBS and the dichroic mirror to be imaged by the objective lens onto the S-SEED array. Low loss is achieved by using an optical isolator arrangement. The reflected outputs are imaged onto a camera and a movable detector.

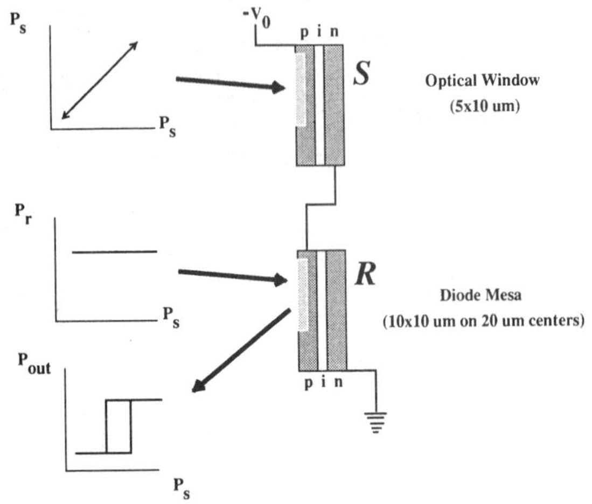

Fig. 1. Schematic diagram of S-SEED showing input and output beams.

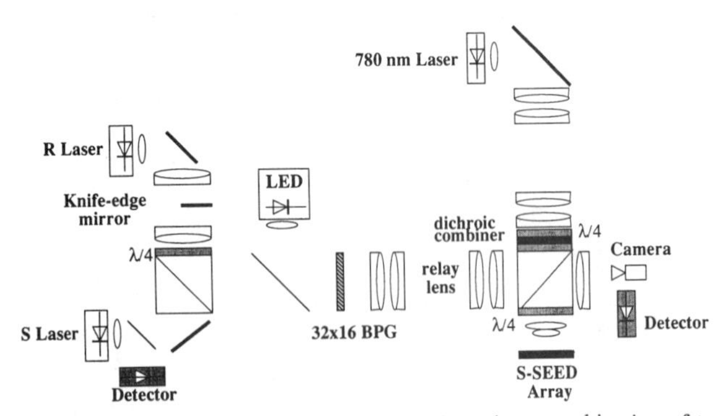

Fig. 2. Experimental setup showing knife-edge mirror combination of two lasers, symmetric eyepiece relay lenses, and dichroic beam combiner. Output is directed to camera and movable detector.

Manuscript received December 7, 1990.
F. B. McCormick, A. L. Lentine, R. L. Morrison, and S. L. Walker are with AT&T Bell Laboratories, Naperville, IL 60566.

L. M. F. Chirovsky and L. A. D'Asaro are with AT&T Bell Laboratories, Murray Hill, NJ 07974.
IEEE Log Number 9143139.

Reprinted with permission from *IEEE Photonics Technology Letters,* Vol. 3, No. 3, pp. 232-234, March 1991.

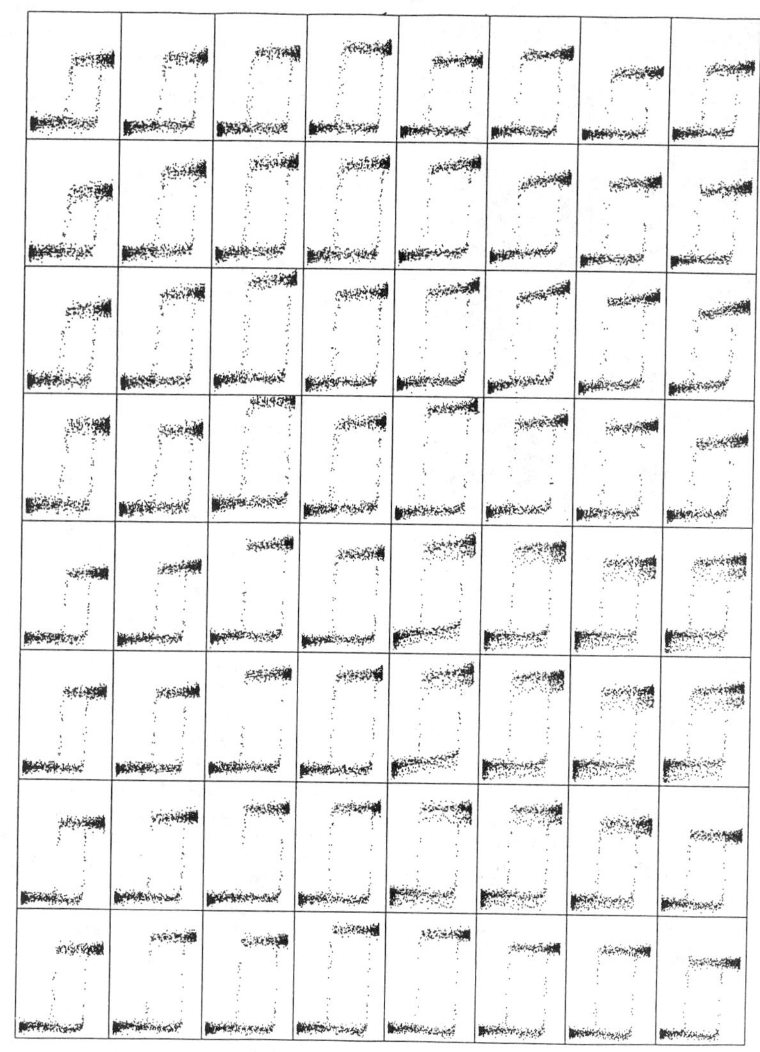

Fig. 3. Bistability curves ($P_{R\text{-out}}$ versus $P_{S\text{-in}}$) of 64 of the 512 S-SEED's. Every other device in every fourth row of the 32 × 16 array is shown.

We observed simultaneous bistable operation of all of the 512 devices by keeping the power delivered by the *S* laser constant while ramping the power from the *R* laser. The exciton absorption peak of these devices had been previously measured to be 849 nm. The nominal wavelengths of the *S* and *R* lasers were 849 and 846 nm, respectively. Bistable operation of most of the array was observed at bias voltages as low as 6 V; however, photocurrent nonuniformities introduced by spot intensity variations, misalignments, etc., prevented bistable operation of some devices at this low voltage. The effects of these nonuniformities can be seen from the plots of Fig. 3, where the reflected *R* output powers, $P_{R\text{-out}}$, are plotted against the *S* input powers, $P_{S\text{-in}}$, (bistability curves) for 64 of the 512 devices. The bistability curves of every other S-SEED in every fourth row of the 32 × 16 S-SEED array are shown. The data in Fig. 3 were collected using a video image acquisition system [13], which was used to collect operational data on all of 512 devices. An average contrast ratio of 3:1 was achieved at 20 V of bias across the S-SEED array at the stated wavelengths. The optical power throughput (laser to S-SEED array) was measured to be 25%, which matches the predicted throughput calculated from measured individual component losses.

By varying the *R* laser power slowly (0.5 Hz) and monitoring the devices' reflected outputs with an infrared camera and television, we can obtain a graphic visual indication of photocurrent uniformity across the S-SEED array. The photocurrents are determined by the incident spot intensity, the alignment of the spot onto the device input window, and the device responsivity. During the cyclic ramping up and down of the *R* laser power, the time at which a particular device switches state is determined by its bistable characteristic and its photocurrent. Since the devices in this array all have essentially identical characteristics, devices with lower photocurrents will switch at a later time during the ramping up part of the cycle (and earlier during the down-ramping part). Very slight misalignments and nonuniformities can thus be detected. Absolute determination of the magnitude of the misalignments and nonuniformities requires separation of the various causes and is currently being investigated. For example, a slight magnification error can cause a mismatch between the spot spacing and the spacing of the devices. This error accumulates radially, so that devices at the edge of the array receive less optical power than those at the center. The edge devices thus switch from absorbing to reflective after the center devices. The image pattern created by this effect is a radially expanding "wave." In our experiment, since the two lasers had slightly different wavelengths, the spot spacing of the two images was slightly different. That is the outer spot positions varied by 3/849 or 0.3%, which introduces about

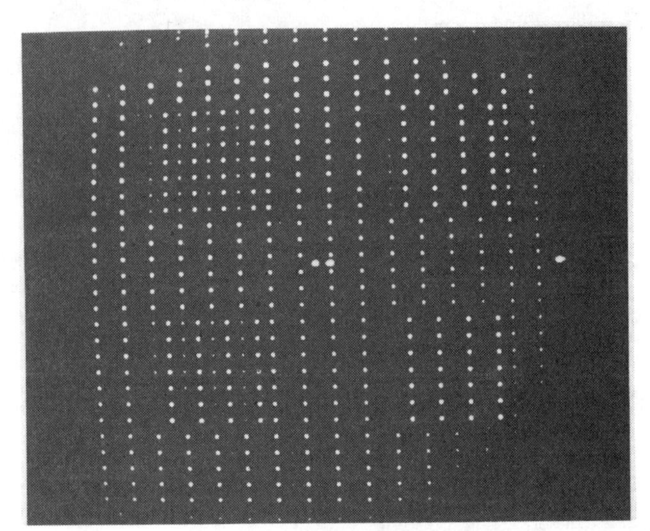

Fig. 4. Demonstration of S-SEED RAM: four 4 × 7 device "blocks" within the 32 × 16 array have been written to the complementary state.

a 1.5 μm shift. This misalignment within the 5 μm S-SEED windows was enough to cause the "wave" pattern to be observed. Analogous patterns can be seen for rotation/translation errors and optical and polarization aberrations errors [14].

By adjusting the two lasers for equal powers, and introducing a third laser beam to individually set and reset the S-SEED's, we operated the devices as S/R latches. This effectively implemented a 512-bit all-optical static RAM, as shown in Fig. 4. Data were input by a 780 nm laser providing 3 μW of power that was scanned across the array. Only 300 nW total optical power per device (1.2 mW total input power) was required to hold the state of the array.

We operated 32 × 16 array symmetric self electrooptic effect devices in parallel. Both bistable operation and "static" operation as a random access memory (RAM) have been demonstrated. The system was constructed using a simple but well corrected telecentric imaging system. A novel technique was developed to observe and correct the effects of spot array misalignment, optical and polarization aberration, and laser wavelength shifts.

References

[1] See e.g., H. S. Hinton, "Architectural considerations for photonic switching networks," *J. Selec. Areas Commun.*, vol. 6, p. 1209, 1988.

[2] A. L. Lentine, L. M. F. Chirovsky, L. A. D'Asaro, C. W. Tu, and D. A. B. Miller, "Energy scaling and subnanosecond switching of symmetric self electrooptic effect devices," *IEEE Photon. Technol. Lett.*, vol. 1, p. 129, 1989.

[3] F. B. McCormick, A. L. Lentine, L. M. F. Chirovsky, and L. A. D'Asaro, "An all-optical shift register using symmetric self electro-optic effect devices," OSA, *Photon. Switch.*, paper ThC5-1, J. E. Midwinter and H. S. Hinton, Eds., 1989.

[4] M. E. Prise, N. C. Craft, R. E. LaMarche, M. M. Downs, S. J. Walker, L. A. D'Asaro, and L. M. F. Chirovsky, "A module for optical logic circuits using symmetric self electrooptic effect devices," *Appl. Opt.*, vol. 29, May 1990.

[5] F. B. McCormick, A. L. Lentine, R. L. Morrison, S. L. Walker, L. M. F. Chirovsky, and L. A. D'Asaro, "Simultaneous parallel operation of an array of symmetric self-electrooptic effect devices," *OSA Ann. Meet. Tech. Dig. Series*, vol. 18, OSA, Washington, DC, 1989, paper MII4.

[6] B. S. Wherrett, R. G. A. Craig, J. F. Snowdon, G. S. Buller, F. A. P. Tooley, S. Bowman, G. S. Pauley, I. R. Redmond, D. McKnight, M. R. Tagizadeh, A. C. Walker, and S. D. Smith, "Construction and tolerancing of an optical CLIP," *Dig. Opt. Comput. II*, R. Arrathoon, Ed., *Proc. SPIE*, p. 1215, 1990.

[7] T. J. Cloonan, M. J. Herron, F. A. P. Tooley, G. W. Richards, F. B. McCormick, E. Kerbis, J. L. Brubaker, and A. L. Lentine, "A 3-D crossover switching network based on S-SEED arrays," *IEEE Photon. Technol. Lett.*, vol. 2, p. 6, 1990.

[8] F. B. McCormick, F. A. P. Tooley, T. J. Cloonan, J. L. Brubaker, A. L. Lentine, S. J. Hinterlong, and M. J. Herron, "A digital free space photonic switching network demonstration using S-SEEDs," *Conf. Lasers and Electro-Optics, 1990 Tech. Dig. Series*, Vol. 7, OSA, Washington, DC 1990, postdeadline paper CPDP1.

[9] A. L. Lentine, F. B. McCormick, R. A. Novotny, L. M. F. Chirovsky, L. A. D'Asaro, R. F. Kopf, J. M. Kuo, and G. D. Boyd, "A 2 kbit array of symmetric self electrooptic effect devices," *IEEE Photon. Technol. Lett.*, vol. 2, p. 51, 1990.

[10] R. L. Morrison and S. L. Walker, "Binary phase gratings generating even numbered spots arrays," *App. Opt.*, 1990.

[11] M. Born and E. Wolf, *Principles of Optics*, 6th ed. New York: Pergamon, 1980, p. 187.

[12] W. J. Smith, *Modern Optical Engineering*. New York: McGraw-Hill, 1966.

[13] R. L. Morrison, "Image analysis for diagnostics in photonic switching," *Proc. SPIE Opt. Eng. Midwest*, Sept. 27, 1990.

[14] R. A. Chipman, "Polarization analysis of optical systems," *Optical Engineering 28*, No. 2, pp. 90–99, 1989.

Bistable Characteristics and All-Optical Set-Reset Operations of 1.55-μm Two-Segment Strained Multiquantum-Well DFB Lasers

J. ZHOU, M. CADA, G.P. LI, AND T. MAKINO, MEMBER, IEEE

Abstract— **Bistable characteristics and all-optical set-reset operations in 1.55-μm two-segment InGaAsP–InP strained multiquantum-well (MQW) DFB lasers were studied. An extinction ratio as high as 20 dB with a lasing output power of 4 mW was obtained, partially due to the dispersion effect of strained MQW DFB structures. The detailed transient dynamics of the optical set-reset operations were observed for the first time, with optical injection from a single-mode laser, implying a potential for high-speed applications. A switch-on time in subnanosecond region and a switch-off time of 2.5 ns were measured using input pulses with a peak power of 500 μW.**

OPTICAL bistability in two-segment diode lasers is promising for use in optical communications and information processing. Of the variety of possible functions of an optical bistable device, the optical set-reset operation is one of the most fundamental ones [1]–[8]. In conventional set-reset configurations [2], the bistable on/off state is set by an optical pulse, while the reset is achieved by an electrical pulse. Recently, the reset by a positive optical pulse has attracted much attention for all-optical signal processing [4]–[8]. The optical set-reset operations, using an optical input from a single-wavelength light, were also demonstrated in conventional Fabry–Perot type bistable lasers [7], [8]; however, the switching time was not mentioned. More recently, the set-reset dynamics were reported with a side-light injection MQW bistable laser and the switch-off time of 2 ns was obtained [4]. However, two injecting lasers and a peak power of 200 mW for reset pulses were required.

In this letter, we report bistable characteristics and all-optical set-reset operations of 1.55-μm two-segment strained multiquantum-well (MQW) DFB lasers with a higher extinction ratio (20 dB), a higher lasing output power (4 mW), a fast switch-off time (2.5 ns) and a lower peak power of set-reset pulses (500 μW). The detailed transient dynamics were studied, for the first time, with optical injection from a single-mode laser. The results indicate a potential for fast high-level optical signal processing applications.

Manuscript received May 2, 1995; revised June 15, 1995. This work was supported by Bell-Northern Research Ltd. (BNR) and Natural Sciences and Engineering Research Council (NSERC).

J. Zhou and M. Cada are with the Department of Electrical Engineering, Technical University of Nova Scotia, P.O. Box 1000, Halifax, NS B3J 2X4 Canada.

G. P. Li and T. Makino are with Bell-Northern Research, P.O. Box 3511, Station C, Ottawa, ON K1Y 4H7 Canada.

IEEE Log Number 9414279.

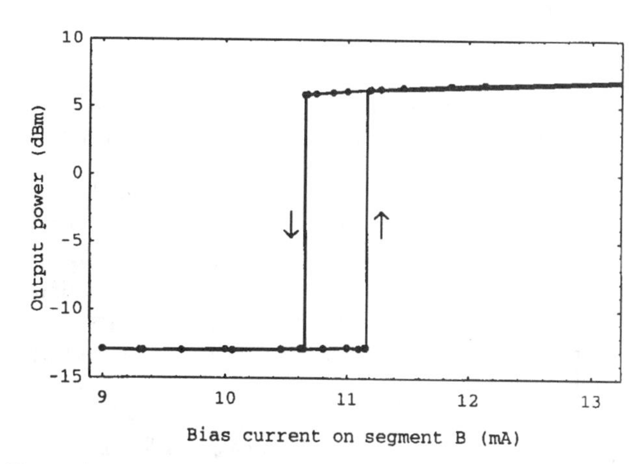

Fig. 1. Optical output power versus bias current on the segment B with a bias current of 52 mA on the segment A of a two-segment strained MQW DFB bistable laser.

The bistable lasers used were 1.55-μm two-segment ridge waveguide strained MQW DFB structures whose details are given in [9]. The lengths of segments A and B were 240 and 120 μm, respectively. The cleaved facet of the segment A, used as the front facet in the measurement, was coated with a 5% antireflection film. Bistable characteristics were observed at a higher bias on the segment A and a lower bias on the segment B. Fig. 1 shows the optical output versus the current on the segment B with a 52-mA bias on the segment A, with an extinction ratio as high as 20 dB and a lasing output power of 4 mW. It was found that the switch-on current (on the segment B) and the extinction ratio of the bistable loop increase with an increase of the bias current on the segment A; the dispersion effect of strained MQW DFB structures appears to enhance the extinction ratio [10].

The test bed for optical set-reset measurements is shown in Fig. 2. The bistable laser and the injecting laser as a trigger were thermally stabilized within 0.1 °C. The injecting laser, a gain-coupled DFB single-mode laser, was driven by a pulse generator, thus producing optical pulses with a repetition frequency of 500 kHz, a pulse width of about 10 ns and rise/fall times of less than 5 ns. The output of the injecting laser was collimated by a lens followed by a dual-stage optical isolator with an isolation over 75 dB, after which the TE-polarized output remained unchanged. This output beam was split by a beam splitter; one portion, as a trigger, was injected into the bistable laser using a 60 × lens, and the other portion was used

Reprinted with permission from *IEEE Photonics Technology Letters,* Vol. 7, No. 10, pp. 1125-1127, October 1995.

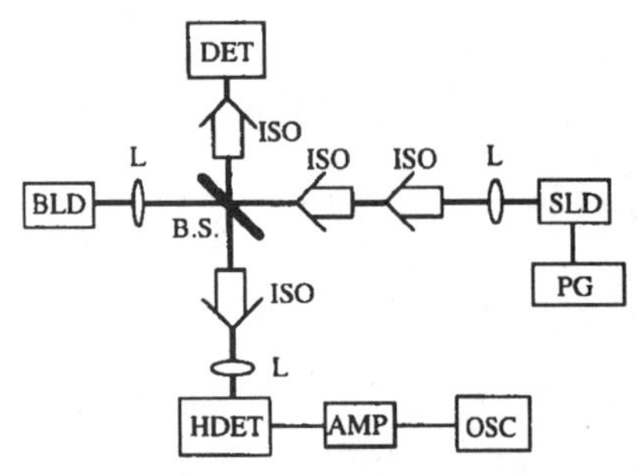

Fig. 2. Experimental test bed for all-optical set-reset measurements of bistable lasers. BLD: Bistable laser; SLD: Single-mode injecting laser as a trigger; ISO: Isolator; L: Lens; HDET: High-speed detector; AMP: Amplifier; B.S.: Beam splitter; DET: Power detector; PG: Pulse generator; OSC: Oscilloscope.

to monitor the optical power. The optical output of the bistable laser was extracted through the same beam splitter, an isolator with an isolation of about 40 dB, and a lens. A high-speed detector (45 GHz) was used for detection. The output signal from the detector was amplified by a wide-band amplifier, followed by an oscilloscope. The total response time of the test system was about 800 ps limited by the electric amplifier.

To obtain optical set-reset operations, the bistable laser was biased at currents of 52.4 mA (segment A) and 10.73 mA (segment B) which made the laser operate within its bistable loop (10.65–11.20 mA). It was single-mode ($\lambda_{out} = 1.5595$ μm) in the lasing state. The injecting laser ($\lambda_{in} = 1.556$ μm) was adjusted to one of the residual Fabry–Perot side modes outside the stopband of the DFB bistable laser to obtain a lower switch-off threshold. Figs. 3 and 4 show the pulsed optical input and the set-reset optical output of the bistable laser, respectively. The insert of Fig. 3 shows the pulse shape. The set and reset pulses were the same in this experiment. The peak power of the optical pulses injected into the active waveguide of the bistable laser was estimated to be ~500 μW. As shown in Fig. 4, the on/off state remained unchanged between two adjacent input pulses with an interval of 2 μs. The slopes in the on/off states were due to the low-frequency cut off of the wideband amplifier (30 KHz). The extinction ratio was 20 dB with the output power of 4 mW in the "on" state, the same as under the dc operation (Fig. 1). We believe that the set function was mostly due to absorption saturation in the absorption region by the set optical pulse, while the reset function was mainly due to gain quenching caused by the optical depletion of the carrier density in the gain region by the reset optical pulse, though the dispersion effect may have been contributing to the switching mechanisms [8]. It should be noted that, unlike those in [7] and [8], the input wavelength was (3.5 nm) shorter than that of the bistable laser, and the optical energy for the reset operation is the same as for the set operation. The detailed physical mechanism is under study.

The inserts of Fig. 4 show the switch-on/off transients. A subnanosecond switch-on time was observed (left insert); it was less than 200 ps [2], [4]. The evolution in the set transient was due to the carrier-density change in the bistable laser cavity. On the other hand, a switch-off time of 2.5 ns, whose transient includes an initial slow decrease plus an accelerated drop, was measured (right insert). This value is

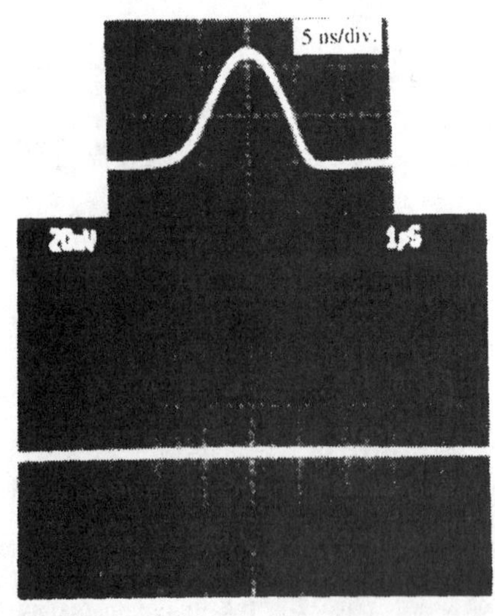

Fig. 3. Pulsed optical input of the bistable laser as a trigger. The insert shows the pulse shape.

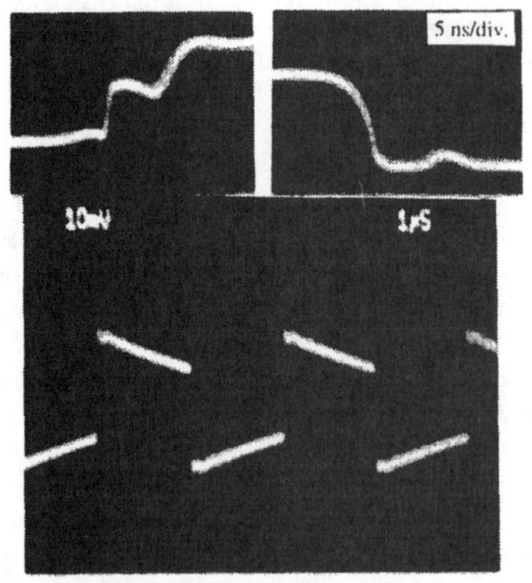

Fig. 4. Set-reset optical output of the bistable laser, with bias currents of 52.4 mA (segment A) and 10.73 mA (segment B, within its bistable loop). The slopes in the on/off states are due to the low-frequency cut off of the wideband amplifier (30 KHz). The inserts (left and right) show the switch-on and switch-off transients, respectively. The trigger signal is shown in Fig. 3.

possibly limited by the rise time of the reset pulse (5 ns); a subnanosecond switch-off time could be expected using a higher intensity pulse with a fast rise time [4], [6]. A small recovery envelope following the switch-off was also seen, which will eventually limit the maximum repetition frequency achievable. The maximum repetition frequency may be ultimately limited by the carrier recovery time [3]; Gbit/s applications could be possible, with some improvements in device parameters.

To study physical mechanism behind the set-reset dynamics, the pulsed optical input measurements were also carried out at a lower bias of 42.2 mA on the segment A, with which a smaller bistable loop (10.89–10.93 mA) is obtained. Fig. 5 shows the pulsed optical output of the bistable laser due to absorption saturation at a bias of 10.48 mA on the segment B (below its electric switch-off current), with the same pulsed op-

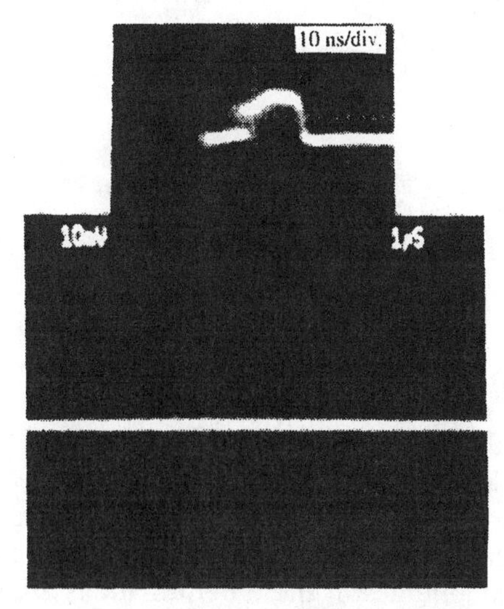

Fig. 5. Pulsed optical output of the bistable laser with bias currents of 42.2 mA (segment A) and 10.48 mA (segment B, below its electrical switch-off current). The insert shows the pulse shape. The trigger signal is shown in Fig. 3.

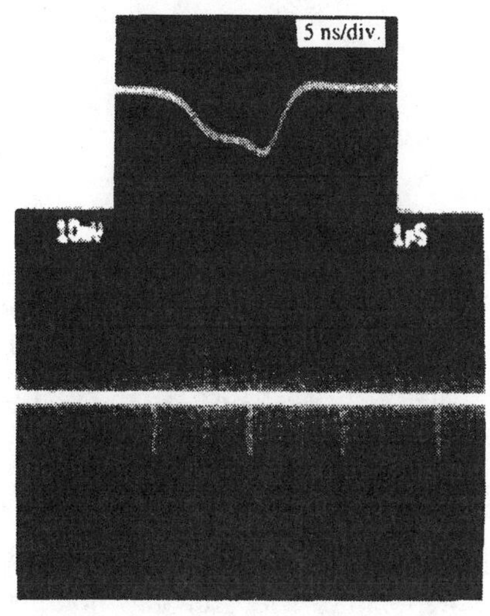

Fig. 6. Inverted pulsed optical output of the bistable laser with bias currents of 42.2 mA (segment A) and 11.24 mA (segment B, above its electrical switch-on current). The insert shows the pulse shape. The trigger signal is shown in Fig. 3.

tical input as in Fig. 4. The insert of Fig. 5 is the pulse shape, which shows subnanosecond rise/fall times, though some turn-on jitter is apparent. Fig. 6 shows the inverted pulsed optical output of the bistable laser due to gain quenching under the same conditions as in Fig. 5, except for a bias of 11.24 mA on the segment B (above its electric switch-on current). The insert of Fig. 6 shows the pulse shape whose fall/rise times appear to be related to the rise/fall times of the input optical pulse. These inherently different switching dynamics are attributed to the different mechanisms (absorption saturation and gain quenching).

In summary, we demonstrated all-optical set-reset operations of 1.55-μm two-segment strained MQW DFB bistable lasers with an extinction ratio as high as 20 dB and a lasing output power of 4 mW. Dynamic behaviors of the optical set-reset operations were investigated, for the first time, using a single-mode injecting DFB laser. The measured switch-off time of 2.5 ns was limited by the rise time (5 ns) of reset pulses with a peak power of 500 μW. The results imply, in addition to the measurements of the pulsed optical output with a pulsed optical input, that the mechanisms of absorption saturation and gain quenching are responsible for the inherent difference between the set-reset dynamics. A high-extinction ratio with a high optical output power and a fast-switching response with a lower switching power requirement obtained are very attractive for all-optical switching and computing applications that require cascade combinations.

Acknowledgment

The authors wish to thank Dr. R. Hui of BNR for his helpful discussions.

References

[1] H. Kawaguchi, "Progress in optical functional devices using two-section laser diode/amplifiers," in *IEEE Proc.*, vol. 140, no. 1, pp. 3–15, 1993.

[2] P. Blixt and U. Ohlander, "19 ps switching of a bistable laser diode with 30 fJ optical pulse," *IEEE Photon. Technol. Lett.*, vol. 2, no. 3, pp. 175–177, 1990.

[3] H. F. Liu, Y. Hashimoto, and T. Kamiya, "Switching characteristics and maximum repetitive frequency of InGaAsP/InP bistable injection lasers," *IEEE J. Quantum Electron.*, vol. 24, no. 1, pp. 43–51, 1988.

[4] H. Uenohara, Y. Kawamura, H. Iwamura, K. Nonaka, H. Tsuda, and T. Kurokawa, "Set and reset operation dependence on input light intensity of a side-light-injection MQW bistable laser," *Electron. Lett.*, vol. 29, no. 18, pp. 1609–1611, 1993.

[5] K. Inoue and K. Oe, "Optically triggered off-switching in a bistable laser diode using a two-electrode DFB-LD," *Electron. Lett.*, vol. 24, no. 9, pp. 512–513, 1988.

[6] J. H. White, J. E. Carroll, and R. G. Plumb, "Room-temperature optically triggered bistability in twin-stripe lasers," *Electron. Lett.*, vol. 19, no. 14, pp. 558–560, 1983.

[7] T. Odagawa and S. Yamakoshi, "Optical set-reset operations of bistable laser diode with single-wavelength light," *Electron. Lett.*, vol. 25, no. 21, pp. 1428–1429, 1989.

[8] M. Okada, H. Kikuchi, K. Takizawa, and H. Fujikake, " The effect of a detuned optical input on bistable laser diodes with inhomogeneous current injection," *IEEE J. Quantum Electron.*, vol. 29, no. 1, pp. 109–120, 1993.

[9] G. P. Li and T. Makino, "Longitudinal-mode switching in two-segment multi-quantum-well distributed-feedback laser," *J. Appl. Phys.*, vol. 77, no. 9, pp. 1–3, 1995.

[10] J. Zhou and M. Cada, unpublished.

All-Optical Devices Using Nonlinear Fiber Couplers

David R. Rowland

***Abstract*—The coupler parameters for various new all-optical devices using weakly coupled nonlinear fibers are anticipated from a graphical method of representing extremums of core power called the "Power Flow Portrait." These devices are then analyzed in detail both analytically and numerically. Devices considered include a bandpass power filter, various schemes for logic gates, and a novel power-dependent switch utilizing saturation.**

I. Introduction

The twin-core coupler with Kerr law nonlinearity has been extensively studied for both its general properties and its possible uses in various all-optical devices such as switches and amplifiers [1]–[8]. Starting with Jensen [1] and Maier [2] who studied the symmetric coupler, various authors have extended this work by considering various effects caused by having nonidentical cores, both with regards to the linear propagation constants and the Kerr constants of each core [4], [6], [8], [9]. The effects of non-Kerr-law materials [10] and saturation [11], [12] on device operation have also been considered.

Building on the work of Snyder and Mitchell [9], [13], it has proved possible to tackle the general problem of the nonlinear directional coupler (NLDC) in a more systematic, directly physical way than has been done previously by way of the power flow (or mode) portrait (PFP) [14]. The PFP has been used in this paper to anticipate the coupler parameters necessary for various all optical devices utilizing an NLDC. Devices considered include a bandpass power filter, an AND/OR logic gate and a novel alternative design, using saturation, of a power dependent switch.

We first describe the model we are using as well as the relevant approximations and notation. Each device type mentioned above is then discussed in turn. Various analytical results are derived in the appendices.

II. Model and Notation

The starting point of our analyses is a twin-core coupler consisting of two axially uniform, weakly coupled cores which are single moded in isolation and matched in the linear {i.e., zero power} limit. With these assumptions, the total coupler field ψ can be written as [15], [16]

$$\psi(x, y, z) = a_1(z)\psi_1(x, y) + a_2(z)\psi_2(x, y) \qquad (1)$$

where $\psi_i(x, y)$ is the modal field of core i in isolation normalized so that $P_i = |a_i|^2$ is the power in core i and $i = 1$ or 2.

From this structure we consider in general both axially uniform linear perturbations, leading to a linear detuning of the cores, and axially varying perturbations induced by an intensity dependent refractive index. We also assume that the amplitudes of these perturbations are sufficiently small and slowly varying that scalar theory is valid and that forward to backward mode coupling and radiation can be neglected [16].

Power coupling is determined by the mismatch parameter M, given by [16]

$$M = (\beta_1 - \beta_2)/2C \qquad (2)$$

where β_1 and β_2 are the local mode propagation constants of each core considered in isolation and C is the coupling coefficient for the unperturbed matched cores. Note that although we restrict our analyses to cases where $\beta_1 \cong \beta_2$ so that the approximation $C_{12} \cong C_{21} \cong C$ is always valid (see Appendix B and [16], [22]), the mismatch parameter M can nevertheless be arbitrarily large.

Details of how to construct and read the PFP are given in [14]. A brief summary of pertinent ideas and results are as follows. The physical quantities of interest are the powers P_1 and P_2 in cores 1 and 2 as functions of distance 'z' along the coupler, the characteristic length scale of which is given by the linear coupling length $L_c = \pi/2C$. Since P_1 and P_2 vary periodically (except for singular cases) with 'z', the extrema of P_1 and P_2 are then of primary interest. These extrema occur when the phase difference ϕ between the fields in each core are zero or π. These extrema are illustrated graphically on the lower and upper sheets of the PFP respectively where the fraction of power in core 1, P_1/P, is plotted against total power $P = P_1 + P_2$ (see Fig. 7). Regions for which P_1 is a maximum are separated from regions where it is a minimum by curves indicating where no power is transferred between the cores—i.e., *the (nonlinear) modes*. Stable modes are indicated by solid lines and the unstable modes are indicated by dotted lines. (The modes are calculated using (C4).) For here it suffices to know that the arrows on the PFP indicate the direction of initial power flow and that:

1) the arrows point inwards at the edges of the PFP and reverse direction at each mode.

Manuscript received January 9, 1991; revised April 25, 1991. This work was supported in part by the Australian Telecommunications and Electronics Research Board.

The author is with the Optical Sciences Centre, Australian National University, Canberra, Australia.

IEEE Log Number 9101283.

Reprinted with permission from *IEEE Journal of Lightwave Technology*, Vol. 9, No. 9, pp. 1074-1082, September 1991.

2) If a given input is associated with an upward arrow on the PFP, then P_1 is at a minimum and the allowed value for the subsequent maximum (half a period later) must have a larger value of P_1 and it must be associated with a downward arrow on the PFP (total power must also be conserved). The reverse result holds for inputs associated with downward arrows.

3) Points on the PFP near a stable mode remain near the stable mode whereas points near an unstable mode can travel far from the unstable mode.

4) It is shown in Appendix D that a necessary condition to achieve 100% power transfer from core 1 to core 2 at some power P_{100}, is that a mode be located at $P_1/P = 1/2$ on the PFP at this power.

(For more details on reading the PFP, see Section III of [14]).

It should be stressed that the nonlinear modes referred to above do not have the same properties as linear modes such as superposition and beating. Their usefulness lies in the fact that they demarcate regions on the PFP of maximum and minimum values of power in core 1 [9], [13], [14].

The structure of PFP is determined by the linear mismatch parameter, $\overline{M} = (\overline{\beta}_1 - \overline{\beta}_2)/2C$ and the ratio of the Kerr coefficients for each core $\kappa_r = \kappa_2/\kappa_1$ {[9] and Section V of [14]}. The Kerr coefficient κ_i is defined in Appendix C and is proportional to the nonlinear constant of core i. $\overline{\beta}_i$ is the propagation constant in the linear limit of the mode of core i when core i is taken in isolation. Power is conveniently scaled relative to the "critical power" [1] $P_c = 4C/\kappa_1$. Exact results can be obtained by numerically solving the power flow equations written conveniently in terms of the Stokes' parameters (see Appendix A and [3]).

III. Bandpass Power Filters

A bandpass power filter transmits a large signal for input powers within a certain range, and little of the signal for input powers above or below that range. Such a device might find application in a novel multistate logic system or as a tuned receiver on a broadcast ring network. A nonlinear directional coupler can be used to construct such a device by launching all the power into core 1 and then monitoring the output from core 2. We assume for this section ideal Kerr law nonlinearities.

A. Cores with Unequal Kerr Coefficients {i.e., $\kappa_r \neq 1$}

The choice of parameters leading to the "best" bandpass power filter, is most easily done from the perspective of the PFP. An "ideal" filter, which selectively passed P_{100} only, would thus have a PFP like that shown by the dotted lines in Fig. 1(a). That is, since the input corresponds to $P_1/P = 1$, to achieve zero power transfer between the cores for $P < P_{100}$ and for $P > P_{100}$, thus requires a mode to exist at $P_1/P = 1$ at these powers. Also, to achieve 100% power transfer at P_{100}, requires a mode to exist at $P_1/P = 1/2$ at this power, thus giving the indicated PFP. From Appendix C, it can be seen that couplers with $\overline{M} < 0$, $|\overline{M}| >> 1$, and $\kappa_r < 0$, have PFP's most closely approximating the desired one. In fact, from (C11) we have that the $\phi = 0$ (i.e., in phase excitation) PFP's which have zero gradient at $P_1/P = 1/2$, have parameters which satisfy $\kappa_r = (1 + \overline{M})/(1 - \overline{M})$. Now, the vertically rising parts of the "ideal" PFP can only be obtained in the limit as $\overline{M} \rightarrow -\infty$ which thus implies that $\kappa_r \rightarrow -1_+$ in this limit (note from (D3) that P_{100}/P_c also becomes infinite in this limit). Representative PFP's and device characteristics are shown in Fig. 1 for finite $\overline{M}$. The length of the coupler was chosen to be the half period of power transfer for an input power of P_{100}. The plots and coupler length were calculated numerically.

Note that for $-1 \leq \overline{M}|1 + \kappa_r|/2(1 - \kappa_r) \leq -1/2$, an unstable mode appears in the $\phi = 0$ PFP (see Figs. 6 and 7) but that by (D4), 100% power transfer is still possible. However, since the periods of trajectories in phase space which pass near the unstable mode vary rapidly [3], [7], this can lead to unwanted oscillations in the device characteristics (see Fig. 2), thus complicating design.

B. Cores with Equal Kerr Coefficients {i.e., $\kappa_r = 1$}

In case A above, a bandpass power filter was achieved by choosing the device parameters $\Delta\overline{\beta}$ and κ_r so that little power would be transferred, *regardless of the device length*, except around P_{100}. An alternative approach for realizing a bandpass power filter relies on the fact that the period of power transfer also varies with the input power. An example of this alternative strategy is as follows.

When the Kerr constants are identical, the conditions for 100% power transfer (from (D1) and (D2)) become: $\Delta\overline{\beta} = 0$ and $P < P_c$. These conditions effectively make the device of length L_c a low- or a high-pass power filter, depending on which core the output is taken from. However, due to the fact that the period of power transfer is a function of power [1], this means that powers just below P_c can be selected by using a device of length $2L_c$ (see Fig. 3). Longer devices of length $2nL_c$, where 'n' is an integer, will also have the desired low and high-power characteristics, but will also have 'n' bands just below P_c instead of just the one.

It should be noted that for the two cases discussed above, the passband power can be set arbitrarily since P_c, and hence P_{100}, can be chosen, given κ, by varying the coupling coefficient C, by for example, altering the core separation. There is however, a tradeoff between low-power operation and the length of the coupler [17].

IV. And/Or Logic Gates

The idea of a logic gate is that square pulses whose logic state is determined by their peak power, are manipulated according to the rules of logic algebra by the nonlinear coupler devices. The scheme suggested here allows for easier concatenation of such devices than a comparable previously suggested scheme [18].

In this scheme, logical "0" corresponds to a signal of zero power and logical "1" corresponds to a signal of

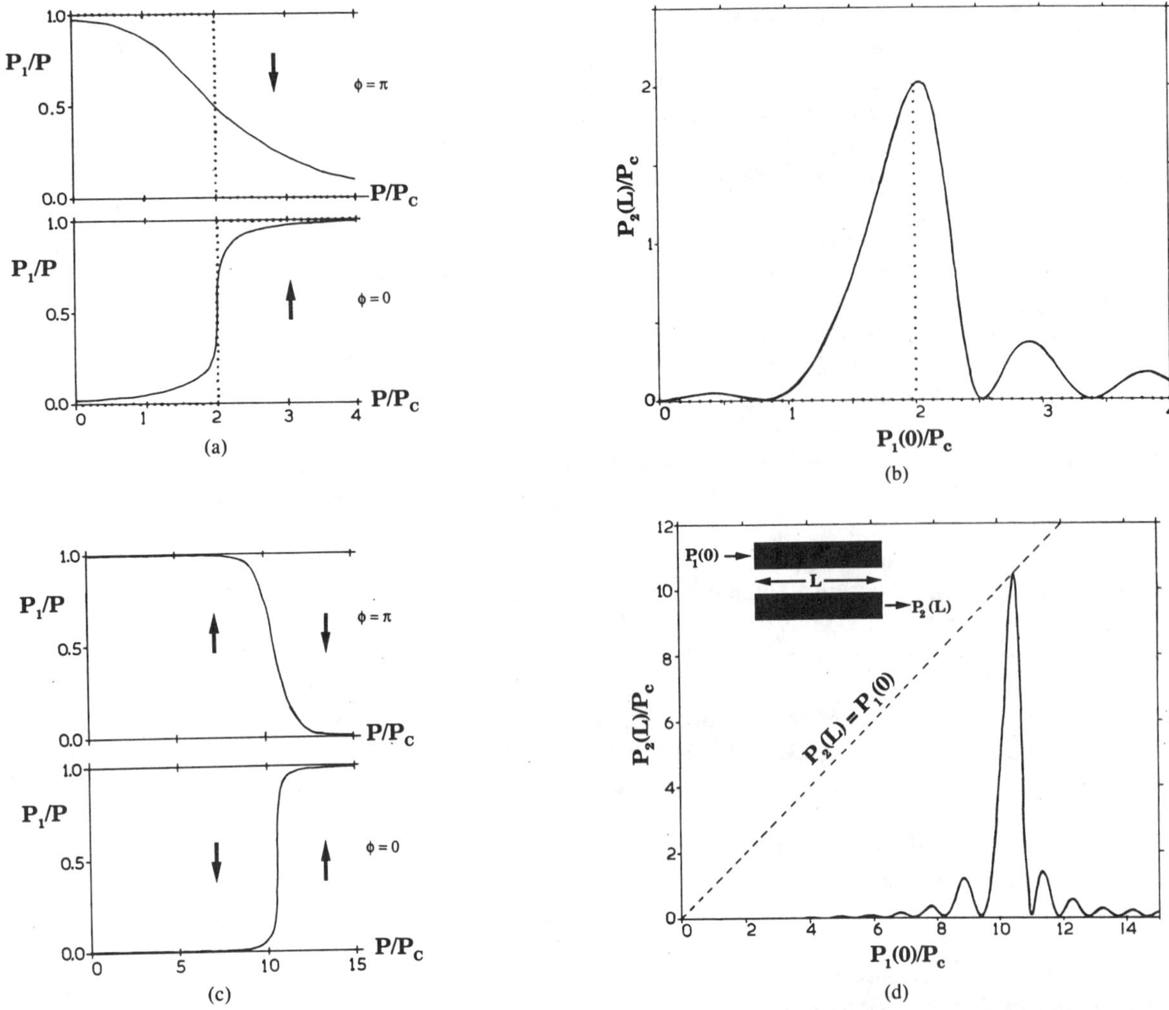

Fig. 1. Power flow portraits are shown in (a) and (c) and the corresponding device output characteristics are shown in (b) and (d). The PFP's are shown drawn with the $\phi^0 = 0$ (in phase excitation) modes on the bottom and the $\phi^0 = \pi$ (out of phase excitation) modes on the top. The arrows on the PFP indicate the direction of initial power flow in each region. The dotted curves in (a) and (b) represent those for an ideal power selective filter. In both cases, the input is into core 1 and the output is taken from core 2. Note that trajectories in phase-space are localized about the $\phi^0 = \pi$ mode for $P < P_{100}$ and about the $\phi^0 = 0$ mode for $P > P_{100}$. In (a) and (b), $\overline{M} = -3$, $\kappa_r = -0.5$, and $L/L_c = 1.0725$. In (c) and (d), $\overline{M} = -20$, $\kappa_r = -0.904762$, and $L/L_c = 1.073$.

power P_0, where P_0 is determined by the device parameters. The signals to be operated upon logically are added in phase and injected into core 1 with the output taken from either core 1 or core 2 with one core leading to AND and the other core leading to OR operation. Thus core 1 can be considered to have two inputs P_{11} and P_{12} such that $(P_{11}^{\text{in}}, P_{12}^{\text{in}}) \in ((0, 0), (1, 0), (0, 1), (1, 1))$. Choosing the corresponding outputs to be $(P_1^{\text{out}}, P_2^{\text{out}}) \in ((0, 0), (1, 0), (1, 0), (1, 1))$ leads to logical AND operation from core 2 and logical OR operation from core 1. Conversely, choosing the corresponding outputs to be $(P_1^{\text{out}}, P_2^{\text{out}}) \in ((0, 0), (0, 1), (0, 1), (1, 1))$ leads to logical OR operation from core 2 and logical AND operation from core 1. (A scheme previously published in the literature [18], used the equivalent system of a birefringent fiber with x- and y-polarizations corresponding to cores 1 and 2. Their device however, was an "algebraic" AND gate, i.e., "1" AND "1" actually gave "2". This would cause complications when concatenating such devices but this problem is avoided by our scheme.)

Both versions of operations are possible provided 100% power transfer is possible at $P = P_0$ and 50% power transfer is possible at $P = 2P_0$ and the device length is chosen correctly. By allowing for the possibility of both types of operation, the chances of there being a short device length satisfying the desired requirements are increased. Of course, one could choose to have greater than 50% power transfer at $P = 2P_0$ and then choose the device length so that only 50% of the power is transferred in that length. The advantage of choosing to have exactly 50% power transferrable at $P = 2P_0$, is that a small error in the device length would then only lead to a second-order variation in the fraction of power in each core whereas the alternative approach would lead to a first-order variation in the fraction of power in each core [19]. From (D3) and (E1), 100% power transfer is possible at $P = P_0$ and 50% power

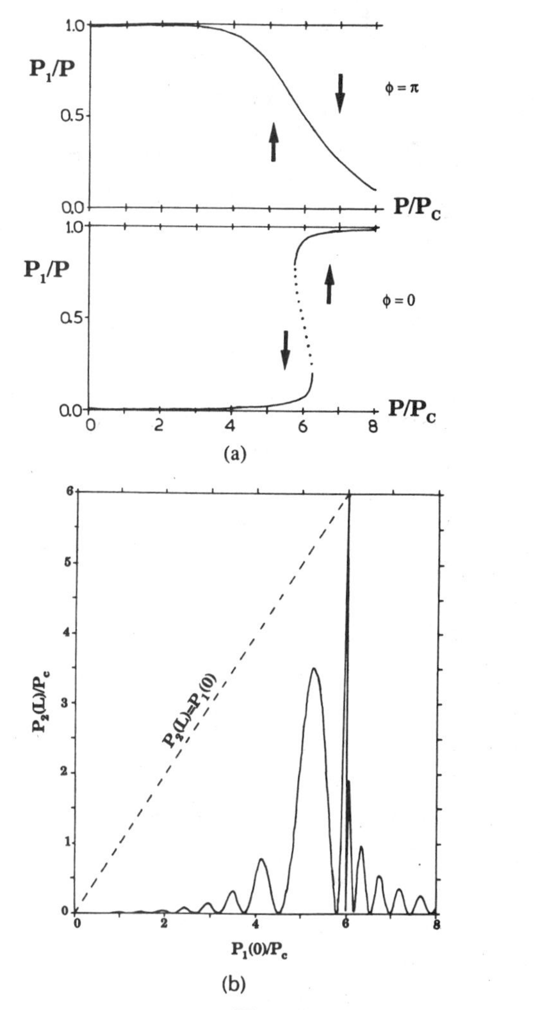

Fig. 2. As per Fig. 1, but now with $\overline{M} = -10$, $\kappa_r = -0.67$ and $L/L_c = 2.08$. The dotted line in the mode portrait represents an unstable mode.

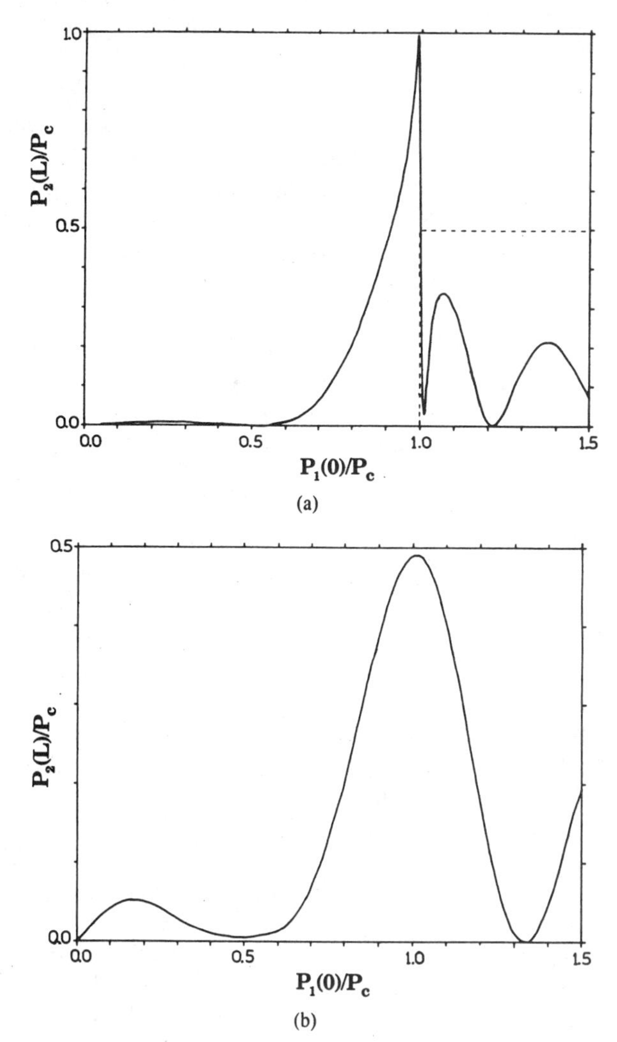

Fig. 4. Output versus input curves for couplers acting as AND gates with $P_0/P_c = 0.5$. The device parameters are for (a) $\overline{M} = 0$, $\kappa_r = 1.0$, $L/L_c = 2.15$, and for (b) $\overline{M} = -1$, $\kappa_r = -1$ and $L/L_c = 2.06$.

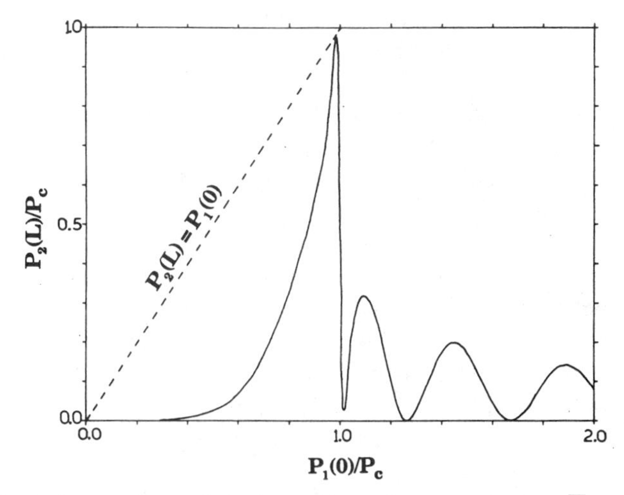

Fig. 3. Output characteristics for the identical core coupler, i.e., $\overline{M} = 0$, $\kappa_r = 1.0$, which is $2L_c$ long.

transfer is possible at $P = 2P_0$ provided the device parameters satisfy

$$\overline{M} = (\kappa_r - 1)/2 \tag{3}$$

and the logical "1" signal satisfies

$$\frac{P_0}{P_c} = \frac{1 - \overline{M}}{3 - \kappa_r}. \tag{4}$$

The length of the coupler required for the desired output characteristics is most easily determined numerically.

The device characteristics for two simple choices of $\overline{M}$ and κ_r are shown in Fig. 4. For the chosen parameters $P_0 = 0.5P_c$. Note that in Fig. 4(a), the sharp edge near $P_1(0) = 2P_0 = P_c$ makes this choice of device parameters extremely intolerant to small errors in P_0, thus making concatenation of such devices difficult. In this regard, the choice of parameters shown in Fig. 4(b) are much better. However, due to differences in the periods of power flow for $P_1^{in} = P_0$ as compared to when $P_1^{in} = 2P_0$, perfect device operation is difficult to obtain for this choice of device parameters, as is evidenced by the fact that for the length chosen in Fig. 4(b), neither "0" nor "1" is exactly obtained.

V. Alternative Design for a Power Dependent Switch

The "ideal" power dependent switch is in reality equivalent to an ideal high- (low-) pass power filter, i.e., 100% of the input power is transmitted for high (low) input powers and no power is transmitted for low (high) input powers. Such a device based on an NLDC, was first suggested by Jensen [1]. The purpose of this section is to develop an alternative equivalent device.

If a given NLDC configuration acts as one type of filter, it will also act as the other type merely by taking the output from the other core. Thus we shall, without loss of generality, consider the design of a high-pass power filter only. The device suggested in [1] consists of identical cores, is L_c long, and has the output taken from core 1. The alternative design considered here will be for a high-pass power filter with output taken from core 2 rather than core 1. Such a device would then have 0% power transfer at low powers and 100% power transfer at high powers.

Taking into consideration the fact that the period of power transfer is generally power dependent, this means that to obtain the required device characteristics, the cores must be highly mismatched at low powers and must satisfy the conditions for 100% power transfer ((D1) and (D2)), for all powers above some switching power. Consequently, the desired PFP would have modes near $P_1/P = 1$ below the switching power and near $P_1/P = 1/2$ above the switching power and a rapid transition between the two at the switching power (see Fig. 5(a)). The method presented in Appendix C for generating a PFP given $\beta_i(P_i)$ $\{i = 1, 2\}$, generalizes readily to arbitrary nonlinearity. This procedure can be reversed so that by making some simplifying assumptions, the qualitative nature of $\beta_i(P_i)$ can be anticipated given the required PFP. The required PFP for this problem leads to a consideration of classes of possible $\beta_i(P_i)$'s of which one solution is considered on a more mathematical footing below.

Now, a necessary condition for 100% power transfer to occur over some length is that the average mismatch $\{\langle\Delta\beta\rangle\}$ over this length be zero. From Appendix F, we find that assuming an arbitrary power law expansion for the propagation constants leads to the conclusion that 100% power transfer is not possible for a range of powers except for the special case of identical cores. However, if we choose the parameters such that $\langle\Delta\beta\rangle \to 0$ as $P \to \infty$ $\{$see Appendix D$\}$, then a reasonable high-pass power filter can still be achieved. We next show how this can be done.

Now, for realistic material models, the series expansions for β_i must converge at all powers (ideal Kerr law behavior is the low-power limit of these series expansions). Using this assumption with (F2) for the mismatch and (F4) for the average mismatch, it is easily shown that this requires that both β_1 and β_2 approach the same high-power "saturation" limit β^{sat}. Following Stegeman *et al.* [11], we shall assume exponentially saturating media and for simplicity assume that $|\overline{\beta}_1 - \beta^{\text{sat}}| = |\overline{\beta}_2 - \beta^{\text{sat}}| = |\Delta\overline{\beta}|/2$. Thus

$$\beta_i = \overline{\beta}_i \pm \frac{|\Delta\overline{\beta}|}{2}(1 - \exp(-2P_i/P_s)) \tag{5}$$

where the plus sign is for $i = 1$, the minus sign is for $i = 2$, and $P_s = |\Delta\overline{\beta}|/\kappa$ is a scaling "saturation power." (Note that this gives ideal Kerr law behavior in the low-power limit, with Kerr constants of $\pm\kappa$.) It is shown at the end of Appendix F that the length required for 100% power transfer when all the power is initially launched into core 1, approaches L_c in the limit as $P/P_s \to \infty$.

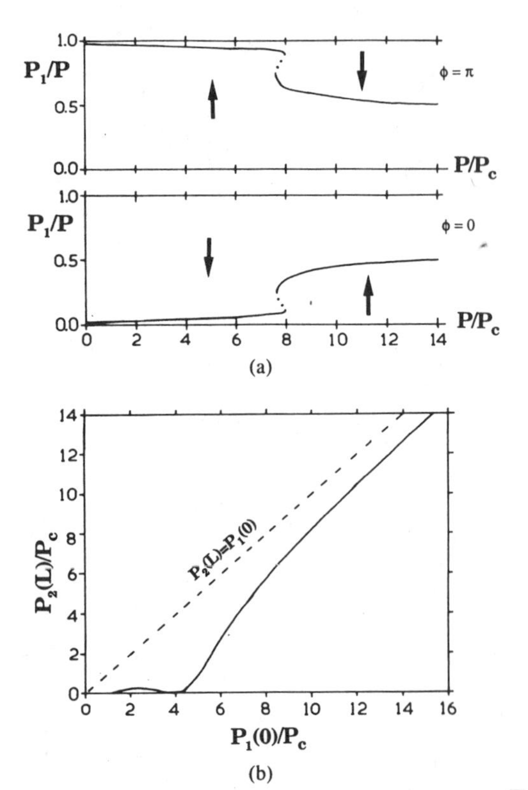

Fig. 5. (a) PFP and (b) output characteristics of a coupler with $\overline{M} = -5$ and $L = L_c$ and cores which saturate exponentially to the same high-power propagation constant.

A representative mode portrait and output characteristic is illustrated in Fig. 5. It was discovered that as $\overline{M} \to 0_-$, the unstable modes in the mode portrait disappeared without qualitatively changing the output characteristics. Note the absence of oscillations in the output characteristics curve at high power in contrast to the case for the Kerr law switch [1]. Although $P_2(L_c) \to P_1(0)$ very slowly with increasing power, thus resulting in significant power output from core 1 at high powers, reasonable power-dependent switching is still evidenced.

VI. Conclusions

We have illustrated the power of the method of working from the perspective of the power flow portrait for the NDLC in three cases, i.e., for bandpass power filters, for an AND/OR logic gate, and for a novel power-dependent switch which is a high-pass power filter. By anticipating the mode portrait required to obtain the desired output characteristics, the required device and material parameters needed to achieve these portraits was determined, in the case of the bandpass power filter, from a catalogue of NLDC mode portrait types, and by "extrapolation" in the case of the power-dependent switch. Undoubtedly, further interesting devices may be anticipated from the above-mentioned catalogue.

Appendix A
Power Flow Equations and Invariants for Nonlinear Couplers

We assume that we have two axially uniform, but not necessarily identical, weakly coupled cores which are single moded in isolation. With these assumptions, the total coupler field can be written as [15], ch. 29 of [16]

$$\psi(x, y, z) = a_1(z)\psi_1(x, y) + a_2(z)\psi_2(x, y) \quad \text{(A1)}$$

where $\psi_i(x, y)$ is the modal field of core i in isolation normalized so that $P_i = |a_i|^2$ is the power in core i with $i = 1$ or 2. It proves convenient to work with the Stokes' parameters defined as [3], [20], [21]

$$S_0 = |a_1|^2 + |a_2|^2 = P_1 + P_2$$

$$S_1 = |a_1|^2 - |a_2|^2 = P_1 - P_2$$

$$S_2 = a_1^* a_2 + a_1 a_2^* = 2\sqrt{P_1 P_2} \cos(\phi)$$

$$S_3 = i(a_1^* a_2 - a_1 a_2^*) = 2\sqrt{P_1 P_2} \sin(\phi) \quad \text{(A2)}$$

where, within the weak coupling approximation and assuming a lossless medium, S_0 is the conserved total power, i.e., $S_0^2 = S_1^2 + S_2^2 + S_3^2 = P^2$ and $\phi = \phi_1 - \phi_2$ is the difference in the phases of the amplitudes of the fields in cores 1 and 2. The remaining (A2), together with the coupled-mode equations [15], p. 570 of [16], lead to the following evolution equations [3], [13]:

$$\frac{dS}{dz} = \Omega \times S \quad \text{(A3)}$$

where $S = [S_1, S_2, S_3]$, $\Omega = [\Delta\beta, 2C, 0]$, $\Delta\beta = \beta_1 - \beta_2$ where β_i is the local mode propagation constant of core i in isolation and C is the coupling coefficient. (We assume in this work that $C_{12} \cong C_{21} \cong C$ (see Appendix B).)

If the local mode propagation constant in each core depends only on the power in each core, then (A3) has a second invariant given by [13]

$$\Gamma = \gamma(S_1) + S_2 \quad \text{(A4)}$$

where

$$\gamma(S_1) = \frac{1}{2C}\int_0^{S_1} \{\beta_1(S_1) - \beta_2(S_1)\}\, dS_1. \quad \text{(A5)}$$

Consequently, solutions to (A3) can be represented in phase space as intersections of the cylinder given by (A4) and a sphere of radius P [3], [19].

Appendix B
Validity of $C_{12} \cong C_{21} \cong C$

From [22] we have that

$$C_{12} - C_{21} = N_{12}(\beta_1 - \beta_2) \quad \text{(B1)}$$

where $N_{12} = \int_{A\infty} \psi_1 \psi_2\, dA$ is the interaction coefficient defined by the overlap integral of the modal fields of cores 1 and 2. Note that N_{12} decreases approximately exponentially with increasing core separation and is usually considered to be negligible (see [21] for some plots of N_{12}). Consequently, the approximation $C_{12} \cong C_{21} \cong C$ is valid provided that

$$\frac{|C_{12} - C_{21}|}{C_{12} + C_{21}} = \frac{N_{12}|\beta_1 - \beta_2|}{C_{12} + C_{21}} \ll 1. \quad \text{(B2)}$$

This condition can always be satisfied for an arbitrarily large mismatch parameter $M = \Delta\beta/2C$, provided N_{12} is small enough, i.e., provided that cores are sufficiently well separated. Note that increasing the core separation also causes C_{12} and C_{21} to decrease exponentially. Therefore, to keep M fixed, $|\beta_1 - \beta_2|$ must also be decreased by a corresponding amount.

Note that this restriction carries with it the penalty that an increase in $|\overline{M}|$ will require a corresponding increase in device length through the fact that $L_c = \pi/2C$ increases with increasing core separation (see for e.g., [17]).

Appendix C
Catalogue of Power Flow Portraits for Kerr Law Couplers

Reproduced here for convenience and in our notation, is a catalogue of PFP's and a summary of pertinent background theory from [14]. All symbols are defined as in the text.

Assuming an arbitrary Kerr nonlinearity in each core {i.e., the refractive index depends linearly on the intensity I as $n = n_0 + n_{2i}I$ with $i = 1$ or 2}, standard perturbation theory for the local mode propagation constant of each core [16] leads to

$$\beta_i = \overline{\beta}_i + \kappa_i P_i \quad \text{(C1)}$$

when the nonlinearity induced change in the refractive index can be considered to be a small perturbation, which is always the case in this paper. Consequently, it follows that the Kerr coefficient is given by

$$\kappa_i = (2\pi n_{2i}/\lambda) \int_{A\infty} \psi_i^4\, dA \Big/ \left(\int_{A\infty} \psi_i^2\, dA\right)^2 \quad \text{(C2)}$$

and without loss of generality, it is assumed that $\kappa_1 > 0$. From (2), this determines the mismatch parameter to be

$$M = \overline{M} + (1 - \kappa_r)P/P_c + (1 + \kappa_r)S_1/P_c. \quad \text{(C3)}$$

From [9], the equation defining the {nonlinear} modes of this structure is given by

$$\frac{S_1}{\pm\sqrt{P^2 - S_1^2}} = M \tag{C4}$$

which upon rearrangement and substitution of (C3) gives

$$p = \frac{1}{[(1-\kappa_r) + (1+\kappa_r)s]}\left\{\frac{s}{\pm\sqrt{1-s^2}} - \overline{M}\right\} \tag{C5}$$

where $p = P/P_c$, $s = S_1/P$, and P_1/P which is plotted in the PFP's is obtained from "s" via

$$P_1/P = (1+s)/2. \tag{C6}$$

The plus and minus signs in (C4) and (C5) correspond to modes with an initial phase difference ϕ^0 between the amplitudes of the modes in each core of zero and π respectively.

A. Linear Limit

As $p \to 0$, (C5) gives that

$$s \to \frac{\pm\overline{M}}{\sqrt{1+\overline{M}^2}} \tag{C7}$$

which corresponds to the modes of the linear structure. (Again, the plus (minus) sign refers to the $\phi^0 = 0$ (π) mode.)

B. Infinite Power Limit

1) $\phi^0 = 0$ modes: As $p \to \infty$, there are three modes if κ_r is positive and only one mode when κ_r is negative. These modes are as follows:

$$s \to \left\{\begin{array}{ll} +1, \quad -1, \quad \dfrac{\kappa_r - 1}{\kappa_r + 1}, & \text{when } \kappa_r > 0 \\ +1, & \text{when } \kappa_r \le 0 \end{array}\right\}. \tag{C8}$$

2) $\phi^0 = \pi$ modes: As $p \to \infty$

$$s \to \left\{\begin{array}{ll} -1, & \text{iff } \kappa_r < 0 \\ \dfrac{\kappa_r - 1}{\kappa_r + 1}, & \text{iff } \kappa_r > 0 \end{array}\right\}. \tag{C9}$$

From (C8) and (C9), it can be seen that the line $\kappa_r = 0$ is the boundary between mode portraits of qualitatively distinct characteristics (see Fig. 7, (e) versus (f) and (c) versus (d)).

C. Degenerate Modes

From (C5), it can be shown that the fractional power split required for the $\phi^0 = 0$ linear mode, gives a mode at all powers provided

$$\overline{M} = \frac{\kappa_r - 1}{2\sqrt{\kappa_r}} \tag{C10}$$

which requires $\kappa_r > 0$. Thus the line $\overline{M} = (\kappa_r - 1)/2\sqrt{\kappa_r}$, (see Fig. 6), divides regions where the degenerate mode portrait has its symmetry broken in different ways (see Fig. 7(b) versus (c) versus (e)).

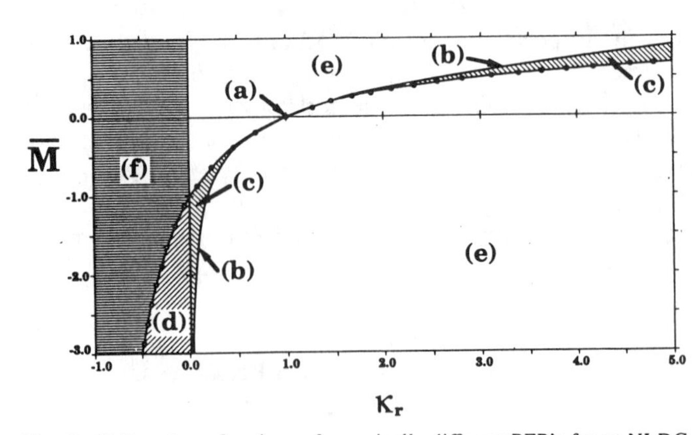

Fig. 6. Delineation of regions of generically different PFP's for an NLDC with parameters such that the linear (zero power) mismatch is given by $\overline{M}$ and the ratio of the Kerr constants is given by κ_r. The solid curve (b) is given by $\overline{M} = (\kappa_r - 1)/2\sqrt{\kappa_r}$ and the broken curve is given by $\overline{M} = (\kappa_r - 1)/(\kappa_r + 1)$.

D. Unstable Modes in the $\phi^0 = 0$ Linear Mode Branch

The boundary between parameter regions for which the branch of the $\phi^0 = 0$ sheet of the PFP, which is continuous with the linear $\phi^0 = 0$ mode, has and doesn't have unstable modes, is determined by the condition that $dp/ds = d^2p/ds^2 = 0$ for some value of "s." This condition leads to the following equation for the boundary (see Fig. 6):

$$\overline{M} = \frac{\kappa_r - 1}{\kappa_r + 1}. \tag{C11}$$

The boundaries discussed above and a collection of representative mode portraits for the various delineated parameter regions are shown in Figs. 6 and 7.

Appendix D
Conditions for 100% Power Transfer

100% power transfer between the cores corresponds geometrically to the conditions that the trajectory given by (A4) passes through the points $[P, 0, 0]$ and $[-P, 0, 0]$ in phase space, but does not pass outside the sphere of radius P in doing so [19]. Using (A4) to state these conditions algebraically gives

$$\gamma(P) = \gamma(-P) = \Gamma \tag{D1}$$

and

$$R^2 = S_1^2 + [\gamma(P) - \gamma(S_1)]^2 < P^2 \tag{D2}$$

when $-P < S_1 < P$. Physically, (A5) and (C1) can be used to interpret (D1) as meaning that the average mismatch is zero over the length required for 100% power transfer {i.e., $\langle\Delta\beta\rangle = 0$, where $\Delta\beta = \beta_1 - \beta_2$}. This seems intuitively reasonable from the results of linear coupler theory [16], [23].

The results for Kerr law nonlinearities are as follows. By substituting the expression for the mismatch, (C3), into (A5), and then using the condition for 100% power transfer, (D1), the power for which 100% power transfer over

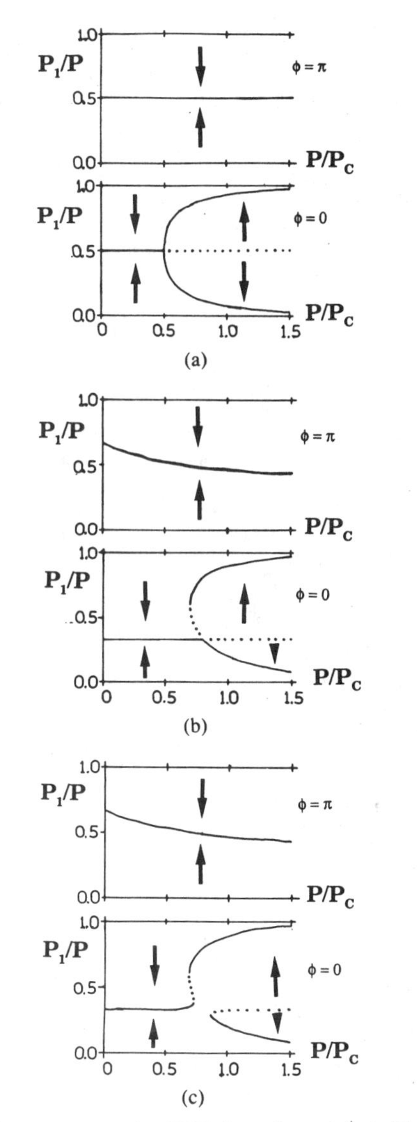

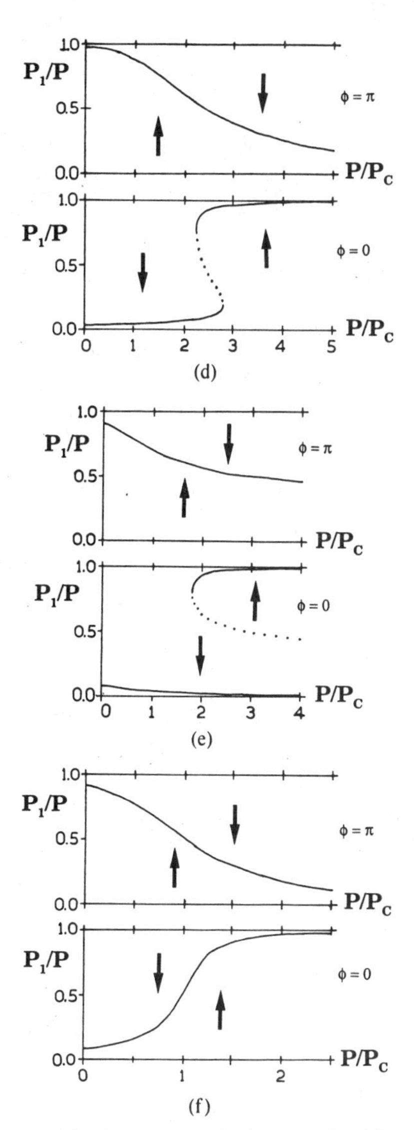

Fig. 7. (a)–(f) are representative PFP's from the regions in Fig. 6 with the corresponding letters. (a) is of course a particular example of (b) and represents the symmetric coupler. The respective parameters are as follows: (a) $\overline{M} = 0$, $\kappa_r = 1.0$, (b) $\overline{M} = -1/2\sqrt{2}$, $\kappa_r = 0.5$, (c) $\overline{M} = -0.35$, $\kappa_r = 0.5$, (d) $\overline{M} = -3.0$, $\kappa_r = -0.25$, (e) $\overline{M} = -1.5$, $\kappa_r = 0.5$, (f) $\overline{M} = -1.5$, $\kappa_r = -0.5$.

some length is possible, P_{100}, is given by:

$$\frac{P_{100}}{P_c} = \frac{-\overline{M}}{1 - \kappa_r} \tag{D3}$$

provided (from (D2))

$$\frac{\overline{M}|1 + \kappa_r|}{2(1 - \kappa_r)} \geq -1. \tag{D4}$$

Note also from (C5) and (C6) that P_{100} is that power for which a mode occurs when $S_1 = 0$, i.e., when $P_1/P = 1/2$, where from (C3), the mismatch is also zero.

Appendix E
Conditions For 50% Power Transfer

The condition required for 50% power transfer is that the trajectory determined by (A4) passes through the points $[P, 0, 0]$ and $[0, P, 0]$ in phase space without passing outside the sphere of radius P in doing so. Using (A4) and (C2) to state these conditions algebraically gives

$$\frac{P_{50}}{P_c} = \frac{2(1 - \overline{M})}{3 - \kappa_r} \tag{E1}$$

provided (D2) again holds.

Appendix F
Equations For General Nonlinearities

We shall assume that for a general nonlinearity that the propagation constants in each core can be expanded as a power series in P_i, i.e.,

$$\beta_1 = \overline{\beta}_1 + \sum_{n=1}^{\infty} a_n P^n \tag{F1a}$$

$$\beta_2 = \bar{\beta}_2 + \sum_{n=1}^{\infty} b_n P^n \quad \text{(F1b)}$$

so that the mismatch is given by

$$\Delta\beta = \Delta\bar{\beta} + \sum_{n=1}^{\infty} \sum_{k=0}^{n} \left(\frac{(a_n - (-1)^k b_n)}{2^n} \, {}^nC_k P^{n-k} S_1^k \right) \quad \text{(F2)}$$

where ${}^nC_k = n!/(n-k)!k!$.

We next consider the conditions necessary to obtain 100% power transfer. A necessary condition for 100% power transfer is that (see Appendix D)

$$\langle \Delta\beta \rangle = \frac{1}{2P} \int_{-P}^{P} \Delta\beta \, dS_1 \equiv 0. \quad \text{(F3)}$$

Equations (F2) in (F3) give that the average mismatch is

$$\langle \Delta\beta \rangle = \Delta\bar{\beta} + \sum_{n=1}^{\infty} \frac{(a_n - b_n)}{n+1} P^n. \quad \text{(F4)}$$

Evidently, for a given set of parameters, this expression cannot be identically zero for a range of powers. (The one exception is the special case of identical fibers for which $a_n = b_n$ and $\Delta\bar{\beta} = 0$.)

As an example, we use (5) to obtain the mismatch and then substituting into (A4) and (A5), leads to the following expression for S_2 corresponding to having all the power launched into core 1:

$$S_2 = \frac{|\Delta\bar{\beta}|}{2} \exp(-P/P_s) \{\sinh(S_1/P_s) - \sinh(P/P_s)\}. \quad \text{(6)}$$

Thus $S_2 \to 0$ as $P/P_s \to \infty$ and hence the length required for 100% power transfer approaches L_c in this limit [19].

Acknowledgment

The author thanks A. Ankiewicz and L. Poladian for helpful comments regarding the manuscript and D. J. Mitchell and A. W. Snyder for illuminating discussions on the conceptual approach via the PFP.

References

[1] S. M. Jensen, "The nonlinear coherent coupler," *IEEE J. Quantum Electron.*, vol. QE-18, no. 10, pp. 1580–1583, 1982.

[2] A. A. Maier, "Optical transistors and bistable devices utilizing nonlinear transmission of light in systems with unidirectional coupled waves," *Soviet J. Quantum Electron.*, vol. 12, no. 11, pp. 1490–1494, 1982.

[3] B. Daino, G. Gregori, and S. Wabnitz, "Stability analysis of nonlinear coherent coupling," *J. Appl. Phys.*, vol. 58, no. 12, pp. 4512–4514, 1985.

[4] S. Trillo and S. Wabnitz, "Nonlinear nonreciprocity in a coherent mismatched directional coupler," *Appl. Phys. Lett.*, vol. 49, no. 13, pp. 752–754, 1986.

[5] S. Wabnitz, E. M. Wright, C. T. Seaton, and G. I. Stegeman, "Instabilities and all-optical phase-controlled switching in a nonlinear directional coupler," *Appl. Phys. Lett.*, vol. 49, no. 14, pp. 838–840, 1986.

[6] A. Ankiewicz, "Interaction coefficient in detuned nonlinear couplers," *Opt. Commun.*, vol. 66, no. 5, 6, pp. 311–314, 1988.

[7] A. Ankiewicz, "Design features for an optical fibre transistor," *IEE Proc.*, vol. 136, pt J, no. 2, pp. 111–117, 1989.

[8] G. Peng, A. Ankiewicz, and A. W. Snyder, "Optical fibre nonlinear couplers with differing core nonlinearities," *Int. J. Optoelectron.*, vol. 4, no. 5, pp. 389–396, 1989.

[9] D. J. Mitchell and A. W. Snyder, "Modes of nonlinear couplers—building blocks for physical insight," *Opt. Lett.*, vol. 14, no. 20, pp. 1143–1145, 1989.

[10] D. J. Mitchell, Y. Chen, and A. W. Snyder, "Directional couplers composed of non-Kerr-law material," *Opt. Lett.*, vol. 15, no. 10, pp. 535–537, 1990.

[11] G. I. Stegeman, C. T. Seaton, C. N. Ironside, T. Cullen, and A. C. Walker, "Effects of saturation and loss on nonlinear directional couplers," *Appl. Phys. Lett.*, vol. 50, no. 16, pp. 1035–1037, 1987.

[12] E. Cagliotti, S. Trillo, S. Wabnitz, B. Daino, and G. I. Stegeman, "Power-dependent switching in a coherent nonlinear directional coupler in the presence of saturation," *Appl. Phys. Lett.*, vol. 51, no. 16, pp. 293–295, 1987.

[13] A. W. Snyder and D. J. Mitchell, "Description of nonlinear couplers by power conservation," *Opt. Lett.*, vol. 14, no. 20, pp. 1146–1148, 1989.

[14] A. W. Snyder, D. J. Mitchell, L. Poladian, D. R. Rowland, and Y. Chen, "Physics of nonlinear fiber couplers," *J. Opt. Soc. Amer. B*, to be published.

[15] A. W. Snyder, "Coupled-mode theory for optical fibers," *J. Opt. Soc. Amer.*, vol. 62, pp. 1267–1277, 1972.

[16] A. W. Snyder and J. D. Love, *Optical Waveguide Theory.* London: Chapman and Hall, 1983.

[17] A. W. Snyder and Y. Chen, "Nonlinear fiber couplers: Switches and polarization beam splitters," *Opt. Lett.*, vol. 14, no. 10, pp. 517–519, 1989.

[18] K. Kitayama, Y. Kimura, and S. Seikei, "Fiber-optic gate," *Appl. Phys. Lett.*, vol. 46, no. 4, pp. 317–319, 1985.

[19] D. R. Rowland, "Phase-space analysis of alternating Δβ couplers," *Optical and Quantum Electron.*, vol. 22, pp. 369–373, 1990.

[20] E. Hecht and A. Zajac, *Optics.* Reading, MA: Addison-Wesley, 1974.

[21] A. Ankiewicz, "Novel effects in nonlinear coupling," *Optical and Quantum Electron.*, vol. 20, pp. 329–337, 1988.

[22] A. W. Snyder and A. Ankiewicz, "Optical fiber couplers—Optimum solution for unequal cores," *J. Lightwave Technol.*, vol. 6, no. 3, pp. 463–474, 1988.

[23] A. W. Snyder, Y. Chen, D. R. Rowland, and D. J. Mitchell, "Mismatched directional couplers," *Opt. Lett.*, vol. 15, no. 7, pp. 357–359, 1990.

Chapter 5

Free-Space Optical Switching

High-speed digital systems seem to be reaching the performance limits of existing interconnection technologies. The continuing increase in clock speed and electronic complexity of today's digital systems have led to widespread investigation of new interconnect and packaging solutions. Indeed, optical interconnects may be one such solution. Free-space optical interconnects offer an evolutionary means of extending the performance of electronic technology by alleviating the communication bottlenecks in systems needing high-speed, high-density interconnections. Light beams can cross in free space without interfering with one another, and they have a high-frequency bandwidth. Free-space optical interconnects use the dimension perpendicular to the planes containing the electronics, transmitters, and receivers, rather than concentrating the optical energy via waveguides.

The main advantages of using free-space optical interconnects relative to electronic interconnects include:

- Higher interconnection densities with lower power dissipation
- Lower signal distortion and dispersion
- Reduced crosstalk and sensitivity to electromagnetic interference
- Decreased skew of system clock and data signals
- Higher reliability, lower costs, and higher utilization of chip areas for processing rather than communication

However, the primary limitations for implementing free-space interconnects are the alignment and packaging of optical transmitters, optical channel elements, and optical receivers; the optoelectronic transmitter and receiver cost, reliability, and fabrication; and the loss of optical power along the interconnection path. The delay of the signals, although constant, is unlikely to be less than in electronic interconnections. These limitations pose a lot of challenges to the researchers working on the implementation of free-space optical systems. This chapter presents some of the important results obtained by researchers trying to overcome these limitations.

Paper 5.1 describes five free-space switching fabric demonstrators constructed in a research laboratory. The paper gives an overview of the self-electrooptic effect device (SEED) technology, which was the device platform used by the demonstrators, then a discussion of the architecture, optics, and optomechanics developed for each of the five demonstrators. The major attributes of the five system demonstrators may be summarized as follows. The optics and optomechanics developed for these demonstrators have allowed (1) the number of supported information channels to be increased from 2 to 1024; (2) the system bit rate has gone from 10 to 50 kb/s; (3) the flat field angle of the supported spatial bandwidth has gone from 1° to 6.7°; (4) the maximum number of effective pin-outs per chip has gone from 12 to 10,240; (5) both the optics and mechanics have gone from off-the-shelf to custom components; (6) the system area at the same time has been reduced from ~32 to ~1 ft^2; (7) the assembly and alignment time has been reduced from days to hours; and (8) the measured lifetime of the system has increased from hours to weeks. Perhaps the most important progress that has been made is the development of the new custom optics and mechanics that will be required to make this technology realizable.

In paper 5.2, Kawai presents new optical interconnections suitable for three-dimensional combining of lens arrays, along with a new multistage interconnection network with a self-routing function. He demonstrates the feasibility of these interconnections through the use of planar microlens arrays. Two separate lens arrays are fabricated to construct an 8×8 network of this design. When a planar microlens array is integrated with a surface-emitting laser diode array and a photodetector array, electronic devices may be densely interconnected through optical signals handling pathways, and massively parallel processing systems using such optical interconnections may be achieved.

J. W. Parker, in paper 5.3, presents a review of the European Stategic Progam for Research in Information Technology II Optical Interconnections for VLSI and Electronic Systems (ESPRIT II OLIVES) program under which components for practical demonstrations of optical interconnections for high-performance processors are under development. The project is focused on the construction of four demonstrators of optical interconnections at the module, backplane, multichip module, and chip levels, and the development of the required optical pathway, hybridization, and optoelectronic component technologies to realize these demonstrators. Significantly, it also includes a task under which the potential performance of the demonstrators is compared with the requirements in systems of commercial interest to the industrial partners. These demonstrator-led activities are supplemented by the theoretical investigation of free-space interconnects between adjacent parallel boards and optical backplane buses, and by efforts to develop the technology for monolithic integration of multiple-quantum-well (MQW) modulators on complementary metal oxide semiconductor (CMOS) circuitry.

Electrooptical components such as the laser driver and optical receiver used by an optical interconnect system usually have to operate in a noisy environment. In this environment, switching noise and crosstalk generated by the nearby digital circuits or adjacent optical interconnects might be coupled into the electrical path of an optical interconnect, either through power distribution or through parasitic capacity and inductive coupling among neighboring interconnects. Indeed, in paper 5.4 Li and Stone reported their research results confirming that the switching noise is a significant interference source in a transmitter array and a receiver array. A differential driver array can provide less waveform distortion and smaller switching noise compared to those that can be achieved by the single-ended configuration. In this paper, Li and Stone present and analyze a differential optical interconnect architecture. This interconnect technique provides a symmetric channel for transmitting optical signals. It relaxes the system and circuit constraints posed by the traditional interconnect architecture, namely, the dependency of the absolute magnitude of decision threshold and the vulnerability to the power supply and the ground noise. Though this architecture is less vulnerable to many types of noise, it is sensitive to channel mismatch. The mismatch between channels causes the cross-coupling among the common-mode and differential-mode signals of the previous stage and thus changes the optimum decision threshold.

At the end of this chapter, Leight and Willner describe in paper 5.5 a reduced switching delay in wavelength division multiplexed two-dimensional multiple-plane optical interconnections using multiple-wavelength vertical cavity surface emitting lasers (VCSEL) arrays. The wavelength-division multiplexing (WDM) system outperforms single-wavelength systems in terms of internodal switching delay, even when the number of wavelengths is small relative to the total number of two-dimensional planes. This WDM optical interconnection provides simultaneous and reconfigurable communication among a network of nodes while most efficiently utilizing substrate area. Leight and Willner have analyzed the bus, dual-bus, and ring architectures, all of which can be supported by this interconnection. Their analysis includes results for when each node can access an entire plane of pixels, a row (or column) of pixels, or an individual pixel. They claim that the integration of WDM into this multiple-plane system provides vastly improved performance over electronic interconnection techniques and current optical interconnection solutions.

Although there has been tremendous progress in the development of this connection-intensive free-space technology, it needs to be understood that in order to become a reality in the marketplace, a packaging technology will be required that is stable, reliable, inexpensive, and manufacturable. Although several organizations have made significant advances with their free-space system demonstrators, most of these objectives have not been fully realized. More focus is needed on the design and testing issues associated with building stable, reliable optical hardware. As with the existing high-speed electronic technology, the development of hardware design rules, testing tools, and verification procedures will be required for the eventual deployment of this free-space digital optical technology.

Free-Space Digital Optical Systems

H. SCOTT HINTON, SENIOR MEMBER IEEE, TOM J. CLOONAN, SENIOR MEMBER, IEEE, FREDERICK B. MCCORMICK, JR., ANTHONY L. LENTINE, MEMBER, IEEE, AND FRANK A.P. TOOLEY

Invited Paper

Within the past 15 years there has been significant progress in the development of two-dimensional arrays of optical and optoelectronic devices. This progress has, in turn, led to the construction of several free-space digital optical system demonstrators. The first was an optical master–slave flip-flop using Hughes liquid-crystal light valves as optical logic gates and computer-generated holograms as the gate-to-gate interconnects. This was demonstrated at USC in 1984. Since then there have been numerous demonstrations of free-space digital optical systems including a simple optical computing system (1990) and five switching fabrics designated **System**$_1$ *(1988),* **System**$_2$ *(1989).* **System**$_3$ *(1990),* **System**$_4$ *(1991) and* **System**$_5$ *(1993). The main focus of this paper will be to describe the five switching fabric demonstrators constructed be AT&T in Naperville, IL. The paper will begin with an overview of the SEED technology which was the device platform used by the demonstrators. This will be followed by a discussion of the architecture, optics, and optomechanics developed for each of the five demonstrators.*

I. Introduction

As the technology-fed, information processing markets of telecommunications, computing, and entertainment continue to grow and mature, they demand larger, faster, and more intelligent hardware platforms to support their software-controlled applications. At the current time, these hardware platforms are based on digital integrated circuits (IC) and their supporting interconnection technologies. Future applications for these markets also indicate that there will be continued pressure to improve the performance of digital hardware technology. As an example, the future of the telecommunications carriers and equipment suppliers will be heavily influenced by the evolution of new broadband services to both business and the residential customers. These future switching fabrics will also need to support both conventional circuit switching capabilities and packet services such as ATM cells embedded in Sonet data streams at costs per port similar to existing plain old telephone service (POTS) [1]. These new services will force the evolution of a new broadband switching office that can support in excess of 10 000 users with broadband channel rates exceeding 100 Mb/s, implying switching fabrics with aggregate throughputs greater than 1 terabit per second (Tb/s). These systems requirements will then force printed circuit boards (PCB), multichip modules (MCM), and even IC's to be capable of supporting the same or even greater aggregate throughputs.

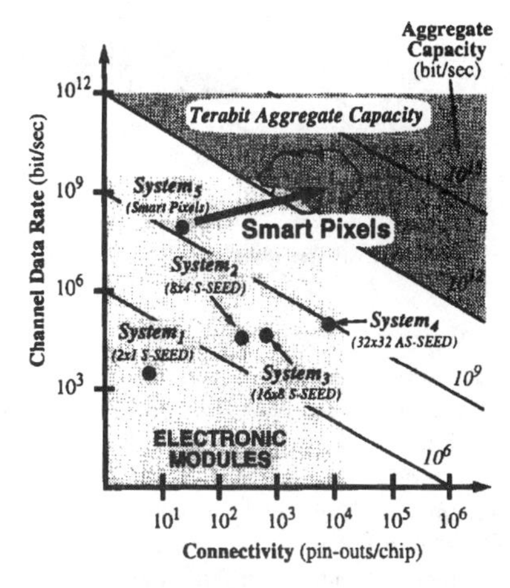

Fig. 1. Device aggregate capacity.

As the performance requirements of these large digital systems has increased, the need for bringing photonics into the digital interconnection hierarchy has been realized [2]. At the current time, single-channel optical data links (ODL) found in existing products provide both frame-to-frame and shelf-to-shelf interconnection. One dimensional optical data links (1D-ODL), linear arrays of ODL's, are moving from the research laboratories to next-generation

Manuscript received January 4, 1994; revised July 21, 1994.

H. S. Hinton is with the University of Colorado, Department of Electrical and Computer Engineering, Boulder, CO 80309-0425 USA.

T. J. Cloonan and A. L. Lentine are with AT&T Bell Laboratories, Naperville, IL 60566 USA.

F. B. McCormick is with Call/Recall Corporation, San Diego, CA 92121 USA.

F. A. P. Tooley is with Heriot-Watt University, Physics Department, Riccarton, Edinburgh, EH14 4AS, United Kingdom.

IEEE Log Number 9405338.

Reprinted with permission from *Proceedings of the IEEE,* Vol. 82, No. 11, pp. 1632-1649, November 1994.

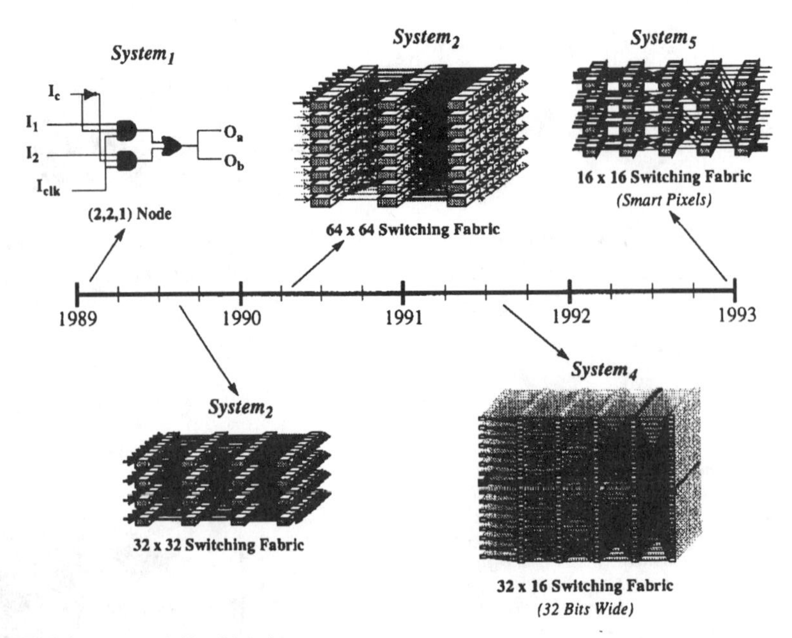

Fig. 2. Evolution of AT&T free-space switching fabric demonstrators.

products [3] for both frame-to-frame and shelf-to-shelf interconnection.

Applications requiring increased connectivity deeper within the interconnection hierarchy, such as MCM-to-MCM and chip-to-chip, are constrained by the existing metal interconnects. The metal interconnection technology is limited by several defects such as the skin effect, which results in a greater attenuation in transmission lines at high frequencies, and parasitic inductance and capacitance, which reduces the usable bandwidth of the interconnections. In addition to these temporal bandwidth limitations, there is also a growing problem of interconnect density. Finally, the energy required per pin-out for electrical interconnects limits the number of pin-outs, which in turn limits the number of digital gates that can be integrated onto a single IC, MCM, or PCB [3], [4]. This, in turn, leads to increased latency and cost.

One of the approaches to overcome these limitations is to take advantage of both the temporal bandwidth (channel data rate) and spatial bandwidth (connectivity) available with optical and optoelectronic structures. Figure 1 illustrates the aggregate capacity as a function of connectivity and per channel data rate [5], [6]. The capabilities of electronic modules (PCB's, MCM's, and IC's) are located in the lower left corner of the figure. The upper right is the high-performance region supporting greater than a terabit aggregate capacity that can be accessed through smart pixels and optical interconnection.

In addition to supporting and enhancing the current electronic infrastructure, this new interconnection capability should foster both new architectures and fabrics that will allow the required broadband systems for both telecommunications and computing to be economically realized.

This increased interest in connectivity and the potential advantages of free-space digital optics has led to the implementation of several system demonstrators. The first was an optical master–slave flip-flop based on Hughes liquid-crystal light valves as optical logic gates with holographic interconnects between the gates. This was demonstrated at USC in 1984 [7]. Since then there have been numerous demonstrations of free-space digital optical systems including an simple optical computing system (1990) [8] and several switching fabrics *System*$_1$ (1988) [9], *System*$_2$ (1989) [10], *System*$_3$ (1990) [11], *System*$_4$ (1991) [12], and *System*$_5$ (1993) [13] demonstrated by AT&T. The evolution of the AT&T switching fabric demonstrators is shown in Fig. 2.

The main focus of this paper will be to outline and describe the five system demonstrators constructed by AT&T in Naperville, IL. The paper will begin with a discussion of the SEED technology which became the device platform for the demonstrators. This will be followed by a discussion of the architecture, optics, and optomechanics developed for each of the demonstrators.

II. SEED Technology

The self-electrooptic effect device (SEED) technology [14] is based on multiple quantum-well (MQW) modulators. Multiple quantum wells consist of thin, alternating layers of narrow- and wide-bandgap materials such as GaAs and AlGaAs. Because of confinement of carriers in the quantum wells, the absorption spectrum shows distinct peaks which are termed exciton peaks. When an electric field is applied perpendicular to the plane of the quantum wells, the position of the peaks shift. This electro-absorption

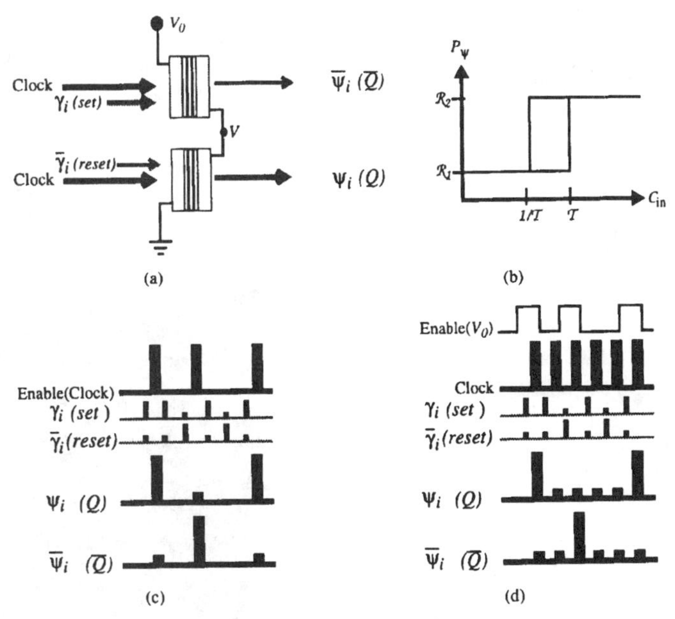

Fig. 3. Symmetric-self electrooptic effective-device (S-SEED). (a) S-SEED with inputs and output, (b) Power transfer characteristics. (c) Optically enabled S-SEED. (d) Electrically enabled S-SEED.

mechanism is called the quantum-confined Stark effect (QCSE) [15] [16], and it is strong enough that a 1-μm-thick multiple quantum-well stack can have changes in absorption coefficient of a factor of two or so for a 5-V change across the stack. By placing the multiple quantum-well material in the intrinsic region of a reverse-biased p-i-n diode, the resulting device can modulate light in response to a change in voltage. The same device can also detect light.

The evolution of the SEED technology includes the Resistor-SEED [17]–[19], the symmetric-SEED (S-SEED) [20]–[22], the Logic SEED (L-SEED) [23], [24], and currently the integration of FET's with MQW modulators (FET-SEED) [25]–[27] to create a viable smart pixel [29] technology. This section will give a brief description of the basic operation of S-SEED, Transmission Gate SEED's (TG-SEED) (a subset of L-SEED's), and FET-SEED technologies and how they can be used as switching nodes.

A. The Symmetric-SEED

The S-SEED, which behaves like an optical inverting S-R latch, is composed of two electrically connected Multiple Quantum Well (MQW) p-i-n diodes as illustrated in Fig. 3(a). In this figure, the device inputs include the entering signals, γ_i (set) and $\overline{\gamma}_i$ (reset), and the clock signals. To operate the S-SEED the γ_i and $\overline{\gamma}_i$ inputs are also separated in time from the clock inputs as shown in Fig. 3(c) and (d). The γ_i and $\overline{\gamma}_i$ inputs, which represent the incoming data and their complement, are used to set the state of the device. When $\gamma_i > \overline{\gamma}_i$, the S-SEED will enter a state where the lower MQW p-i-n diode will be transmissive forcing the upper diode to be absorptive. When $\gamma_i < \overline{\gamma}_i$, the opposite condition will occur. Low switching intensities are able to change the device's state when the clock signals are not present. After the device has been put into its proper state, the clock beams are applied to both inputs. The ratio of the power between the two clock beams should be approximately one, which will prevent the device from changing states. These higher energy clock pulses will transmit the state of the device to the next stage of the system. Since the γ_i and $\overline{\gamma}_i$ inputs are low-intensity pulses and the clock signals are high-intensity pulses, a large differential gain may be achieved. This type of gain is referred to as time-sequential gain.

The operation of an S-SEED is determined by the power transfer characteristic shown in Fig. 3(b) [11]. The optical reflected power (ψ_i), when the clock signal is applied, is plotted against the ratio of the total optical signal power impinging on the set and reset windows (when the clock signal is not applied). Assuming the clock power is incident on both signal windows, the output power is proportional to the reflectivity $\mathcal{R}_i$. The ratio of the input signal powers is defined as the input contrast ratio $C_{\rm in} = P_\gamma / P_{\overline{\gamma}}$. As $C_{\rm in}$ is increased from zero, the reflectivity of diode 1 switches from a low value $\mathcal{R}_1$ to a high value $\mathcal{R}_2$ at a $C_{\rm in}$ value approximately equal to the ratio of the absorbances of the two optical windows: $\mathcal{T} = (1 - \mathcal{R}_1)/(1 - \mathcal{R}_2)$. Simultaneously, the reflectivity of diode 2 switches from $\mathcal{R}_2$ to $\mathcal{R}_1$. The return transition point (ideally) occurs when $C_{\rm in} = (1 - \mathcal{R}_2)/(1 - \mathcal{R}_1) = 1/\mathcal{T}$. The ratio of the two reflectivities, $\mathcal{R}_2/\mathcal{R}_1$, is the output contrast $C_{\rm out}$. Typical measured values of the preceding parameters include $C_{\rm out} = 3.2$, $\mathcal{T} = 1.4$, $\mathcal{R}_2 = 50\%$, and $\mathcal{R}_1 = 15\%$ [11].

1) S-SEED-Based 2-Module: The operation of an S-SEED as a 2-module (switching node) can be accomplished

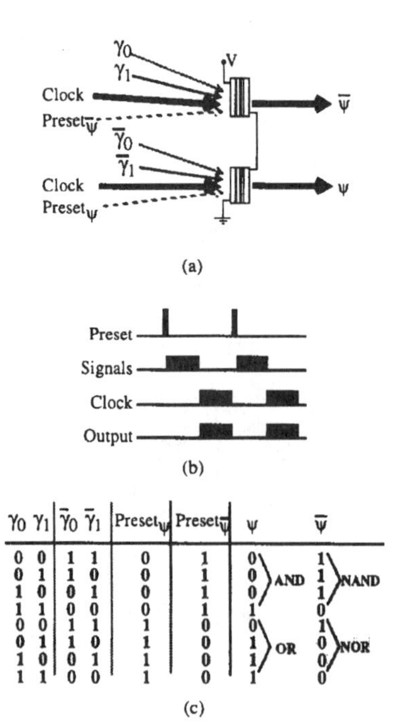

γ_0 γ_1	$\overline{\gamma}_0$ $\overline{\gamma}_1$	$Preset_\psi$	$Preset_{\overline{\psi}}$	ψ		$\overline{\psi}$	
0 0	1 1	0	1	0	AND	1	NAND
0 1	1 0	0	1	0		1	
1 0	0 1	0	1	0		1	
1 1	0 0	0	1	1		0	
0 0	1 1	1	0	0	OR	1	NOR
0 1	1 0	1	0	1		0	
1 0	0 1	1	0	1		0	
1 1	0 0	1	0	1		0	

Fig. 4. Logic using S-SEED devices.

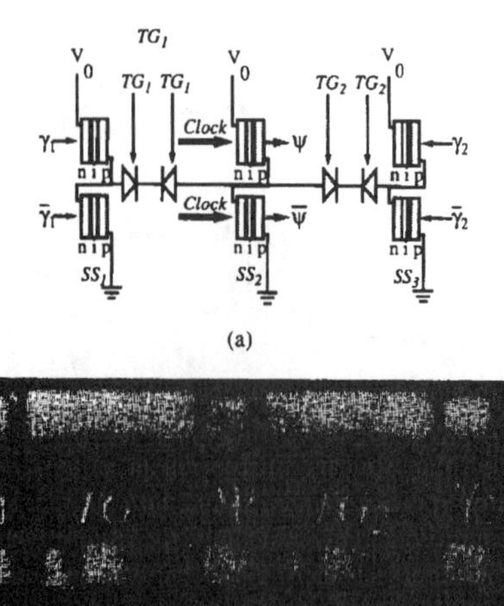

Fig. 5. Schematic diagram and picture of integrated (2, 2, 1) node.

by either optically or electrically enabling the individual S-SEED's. To optically enable an S-SEED array, a spatial light modulator can be used to select which S-SEED's receive clock pulses. If an S-SEED receives a clock pulse, the information previously latched into the device will be transferred to the next stage of the network. If no clock is received the information cannot be transferred. This is illustrated in of Fig. 3(c). On the other hand, the S-SEED's can also be electrically enabled by controlling the voltage applied to the devices as shown in Fig. 3(d). If the appropriate voltage is present, the S-SEED's behave as previously described and the information will be transferred. If no voltage is present both MQW p-i-n diodes will become absorptive preventing the stored information from transferring to the next stage.

2) S-SEED-Based Optical Logic Gates: The S-SEED is also capable of performing optical logic functions such as NOR, OR, NAND, and AND [21]. This allows S-SEED's to be used to implement more complex switching building blocks such as 2×1 and 2×2 switching nodes. The inputs will also be differential, thus still avoiding any critical biasing of the device. A method of achieving logic gate operation is shown in Fig. 4. The logic level of the inputs will be defined by the ratio of the optical power on the two optical windows. When the power of the signal incident on the γ_i input is greater than the power of the signal on the $\overline{\gamma}_i$ input, a logic "1" will be present on the input. On the other hand, when the power of the signal incident on the γ_i input is less than the power of the signal on the $\overline{\gamma}_i$ input, a logic "0" will be incident on the input.

For the noninverting gates, OR and AND, we can represent the output logic level by the power of the signal coming from the ψ output relative to the power of the signal coming from the $\overline{\psi}$ output. As before, when the power of the signal leaving the ψ output is greater than the power of the signal leaving the $\overline{\psi}$ output, a logic "1" will be represented on the output. To achieve AND operation, the device is initially set to its "off" or logic "0" state (i.e., ψ low and $\overline{\psi}$ high) with preset pulse $Preset_{\overline{\psi}}$ incident on only one p-i-n diode as shown in Fig. 4. If both input signals have logic levels of "1" (i.e., set = 1, reset = 0), then the S-SEED AND gate is set to its "on" state. For any other input combination, there is no change of state, resulting in AND operation. After the signal beams determine the state of the device, the clock beams are then set high to read out the state of the AND gate. For NAND operation, the logic level is represented by the power of the $\overline{\psi}$ output signal relative to the power of the ψ output signal. That is, when the power of the signal leaving the $\overline{\psi}$ output is greater than the power of the signal leaving the ψ output, a logic "1" is present on the output. The operation of the OR and NOR gates is identical to the AND and NAND gates, except that preset pulse $Preset_\psi$ is used instead of the preset pulse $Preset_{\overline{\psi}}$. Thus a single array of devices can perform any or all of the four logic functions and memory functions with the proper optical interconnections and preset pulse routing.

3) (2, 1, 1) Nodes Based on Transmission-Gate SEED's: Another functional operation that can be performed with the SEED technology is the "transmission gate." These devices consists of back-to-back quantum-well photodiodes that can transfer the voltage from the center tap of one S-SEED to another. An example of the application of this

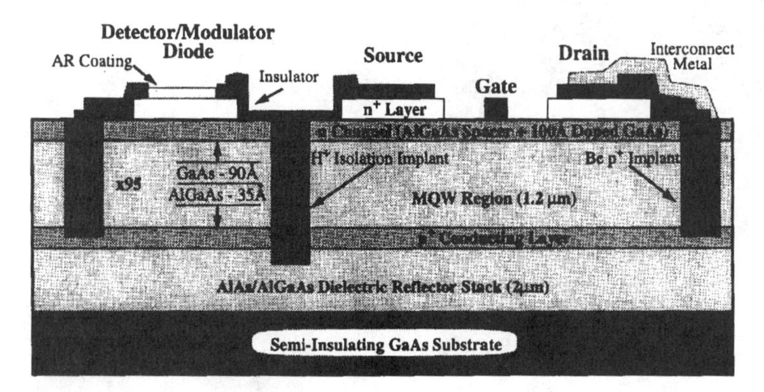

Fig. 6. FET-SEED layer structure.

device is an S-SEED based (2, 2, 1) node "smart pixel" as shown in Fig. 5. The triplet notation represents: (number of inputs $[n_i]$, number of outputs $[m_i]$, capacity of the node $[c_i]$) [31]. This third parameter, c_i, indicates the number of channels that can be actively passed through the node at a given time. Neighboring S-SEED's (SS_1, SS_2,, and SS_3) are connected by optoelectronic transmission gates, TG_1 and TG_2, consisting of a pair of back-to-back quantum-well photodiodes. These photodiodes transfer the voltage from one S-SEED to another. Input signals, γ_i and $\overline{\gamma}_i$, set the states of S-SEED's SS_1 and SS_3. Transfer of the information from these S-SEED's to the output S-SEED SS_2 is accomplished by applying an optical enable signal to the appropriate transmission gate. For example, to transfer the information from S-SEED SS_1 to SS_2 requires the application of enable TG_1. A 10 × 16 array of switching these integrated switching nodes has been fabricated and demonstrated. A shift register using these devices has also been demonstrated.

B. FET-SEED's

To take further advantage of the spatial bandwidth available in the optical domain, integrated electronic circuits need to be integrated with optical detectors (inputs) and modulators (outputs) to create smart pixel arrays [27]. This mixture of the processing capabilities of electronics and the communications capabilities of optics will allow the implementation of connection-intensive architectures with more complex nodes than simple switches. In addition, the gain provided by the electronic devices should allow higher speed operation of the nodes.

The field-effect-transistor self-electrooptic effect device (FET-SEED) technology consists of doped-channel field-effect transistors, MQW modulators, and p-i-n MQW detectors integrated on a single common layer substrate as shown in Fig. 6 [27]. This provides a device platform that supports high-performance buffered-FET logic electronic circuits capable of optical inputs and outputs. The performance of the integrated FET's (g_m = 80 mS/mm) and SEED's (contrast ratio >3) is equivalent to the separately processed devices.

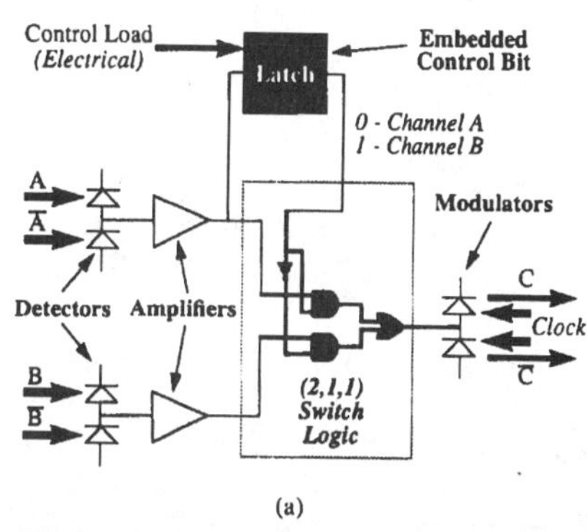

(a)

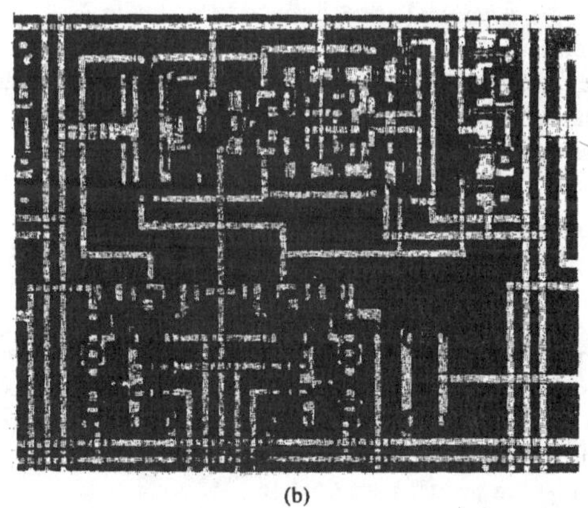

(b)

Fig. 7. *Systems* FET-SEED-based embedded control node. (a) Schematic. (b) Picture.

1) FET-SEED Based (2, 1, 1) Embedded Control Node: Using the FET-SEED technology a 4 × 4 array of (2, 1, 1) switching nodes has been fabricated and tested [32]. A schematic diagram and a picture of the node are shown in Fig. 7. This node is composed of an electronic (2, 1, 1) switching node that can switch the information from either input A or B to output C. The active input is determined by the bit located in the latch. If a "0" is in the latch the information entering input A will be directed to the output, if a "1" is present then the information entering input B

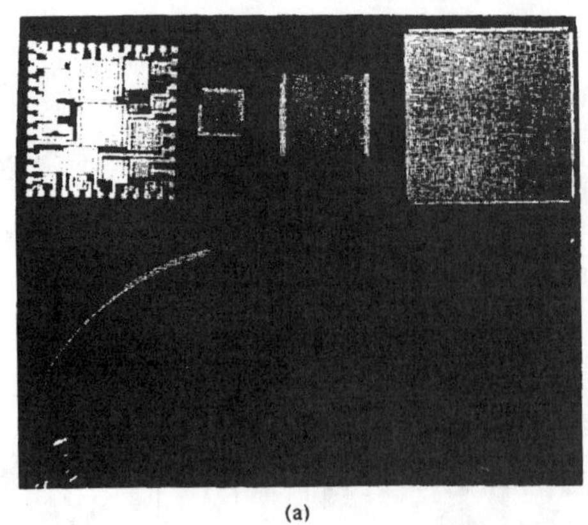

(a)

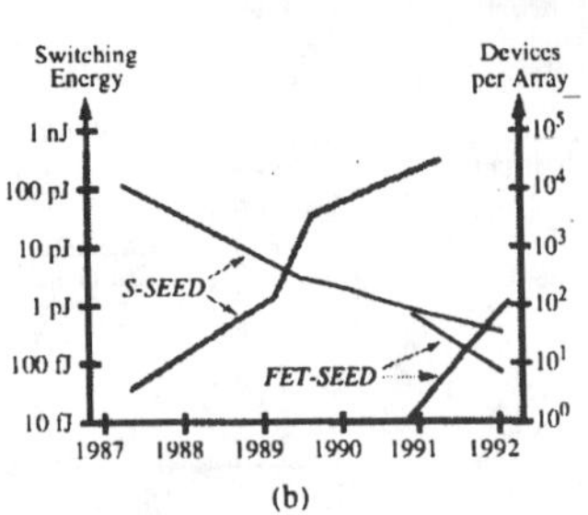

(b)

Fig. 8. Advances in SEED Technology. (a) Generation of S-SEED arrays including 1) 16 × 8 , 2) 64 × 32 , 3) 64 × 128 . and 4) 128 × 256. (b) Plot of progess in switching energy and devices per array.

will be connected to the output. The bit located in the latch is loaded from input A when the electrical control load signal is activated. This allows control information embedded in the data stream to control the switching node (Section III-E).

Each node is composed of 25 GaAs depletion-mode FET's and 17 p-i-n diodes. Thus the entire 4 × 4 array integrated 400 FET's and 272 diodes on the same substrate. The nodes are on a 210-μm pitch with 7 × 7 μm^2 detector and modulator windows (The first stage array had 11 × 11 μm^2 detector window separated by 70 μm to be compatible with the $System_5$ input fiber bundle). Individually, the nodes operated up to 222 Mb/s, requiring only 75 fJ of optical energy for the optical inputs. The nominal contrast ratio at 5.5 V was greater than 3.5:1 for operation at 850 nm.

C. SEED Technology Evolution

Over the past five years there has been considerable progress at AT&T in developing the SEED technology. There have been four generations of fabricated S-SEED arrays including: a 16 × 8 array (1988), a 64 × 32 array (1989), a 128 × 64 array (1990), and finally a 256 × 128 array (1991). These four generations are illustrated in Fig. 8 [33]. The switching energies have decreased from hundreds of picojoules with the first S-SEED's to less than 100 fJ in the most recent FET-SEED's in the same period of time.

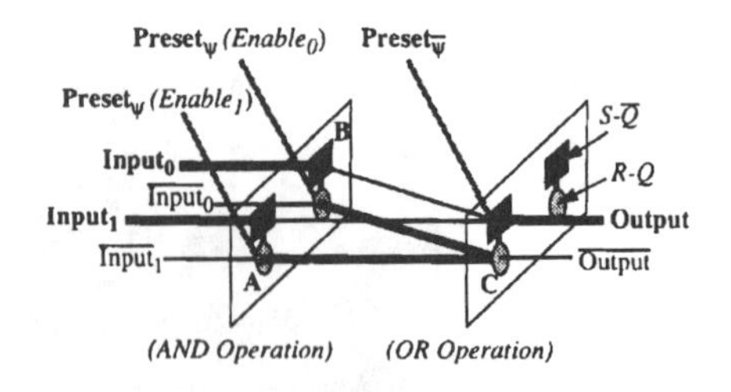

(a)

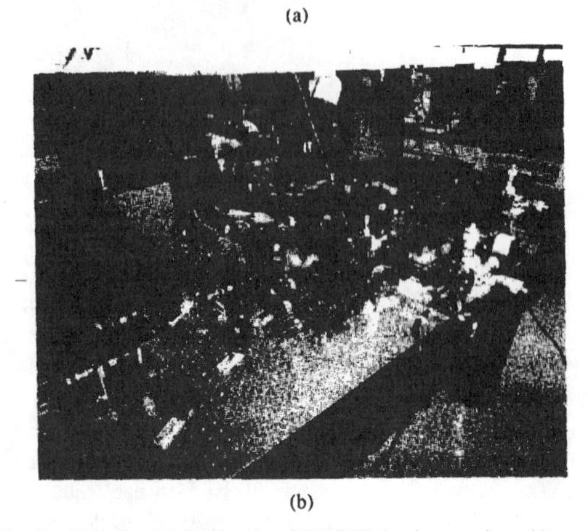

(b)

Fig. 9. $System_1$. (a) Schematic of S-SEED implementation. (b) Picture of demonstrator.

III. System Demonstrators

This section will review the five system demonstrators developed by AT&T. The discussion will include a review of the architecture, optics, optomechanics, and performance for each demonstrator.

A. $System_1$

The first attempt at an S-SEED-based switching demonstrator was to interconnect two S-SEED arrays and create the (2, 1, 1) node shown in Fig. 9 [9]. To realize this node the S-SEED arrays were to be used as optical logic gates as described in Section II-A2. The S-SEED's in the first array need to provide the functionality of digital AND gates while the second S-SEED array must function as an OR gate as shown in Fig. 9(a) of the figure. A picture of the demonstrator is shown in Fig. 9(b).

1) Optical and Optomechanical Hardware: For this first demonstrator all the components were off-the-shelf catalog components. Figure 10 shows the layout of the components required for this initial demonstrator.

Prior to receiving any input signals the preset signals for both S-SEED arrays ($Enable_0$, $Enable_1$, $Preset_{\overline{\Psi}}$) are triggered to set the S-SEED's into their proper state prior to the reception of any valid input signals. These preset signals are encoded on a 780-nm light source, passed through the dichroic beam splitters ($DCBS\#1$, $DCBS\#2$), through the knife edge mirrors ($KE\#1$, $KE\#3$), through the polarization beamsplitters ($PBS\#1$, $PBS\#3$), and then imaged (infinite conjugate imaging) onto the appropriate opti-

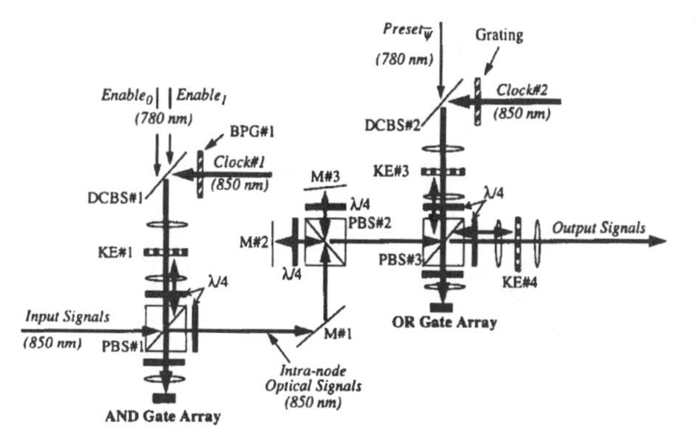

Fig. 10. *System$_1$* demonstrator component layout.

cal windows of the S-SEED arrays. The $\lambda/4$ plates in this system are aligned with the fast axis at 45° with respect to the PBS plane of incidence.

The input signals (*Input$_1$*, *Input$_2$*) then enter the system through $PBS\#1$ where they are reflected up, passed through a quarter-wave plate and imaged onto $KE\#1$ made of evaporated palladium on 2-mm glass substrates. These signals are then reflected, pass a second time through the $\lambda/4$ plate to complete the rotation of the polarization of the input signals to allow them to pass through $PBS\#1$ and be imaged on the first S-SEED array.

After the information from the input signals has been loaded into the "AND" S-SEED arrays, the clock signal (*Clock*#1) is enabled. This signal, operating at 850 nm, passes through the binary phase grating ($BPG\#1$) which redistributes the incident optical power into four equal intensity light beams (see Section II-A2). These clock signals are reflected down by $DCBS\#1$, pass through $KE\#1$ and $PBS\#1$, and then imaged onto the optical windows of the S-SEED array.

The reflected signals leaving the S-SEED arrays (modulated clock signals) are directed to $PBS\#2$ via $PBS\#1$ and $M\#1$ (the light entering $PBS\#2$ is circularly polarized). This second beamsplitter provides the optical interconnect between the two S-SEED arrays. In this case, one of the polarization components of each entering signal is directed to $M\#2$ to provide the straight connection between the arrays, while the other polarization component is directed to the tilted mirror $M\#3$ which provides the shifted interconnect (see Section III-B3b for more detail on crossover interconnects). The signals, of both linear polarizations, leaving the optical interconnect PBS enter the second beam combination polarization beamsplitter ($PBS\#2$). These signals are directed to either $KE\#3$ or $KE\#4$, depending on their polarization, where they are reflected, pass again through $PBS\#3$ and then imaged onto the second S-SEED array (see the discussion on beam combination of Section III-B4c). The output is then derived from the "OR gate" S-SEED device modulating the *Clock*#2 signals and then directing these output signals through $KE\#4$ where they can be monitored by either a detector or a camera.

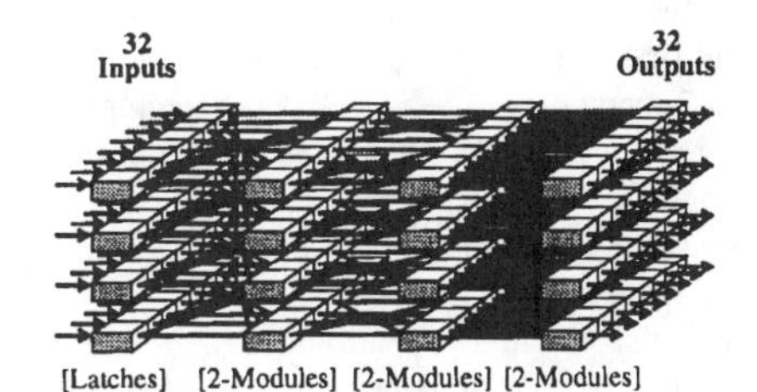

(a)

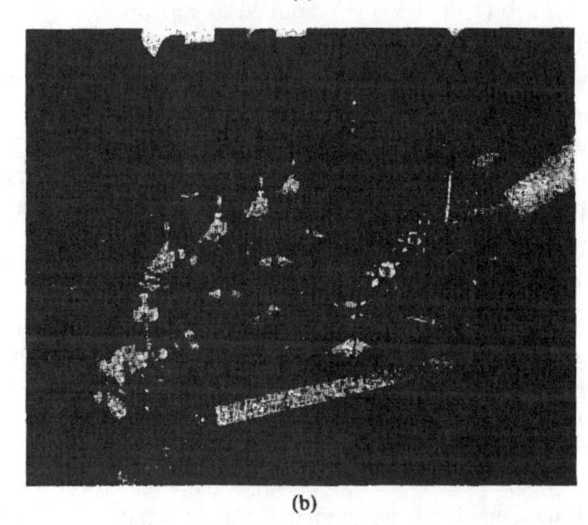

(b)

Fig. 11. *System$_2$*. (a) Network topology. (b) Picture of demonstrator.

2) Performance: The S-SEED's for this demonstration had rectangular windows of $10\times$ 20 μm spaced 80 μm apart, a contrast ratio of 3:1, and a switching energy of ~ 12 pJ. The clock signals reached the S-SEED windows with ~50 μW of optical power, while the power in the optical signals reaching the "OR gate" windows was only 0.9 μW. This led to a maximum measured system bit rate of 5 kb/s.

B. System$_2$

The second system demonstrator had the focus of integrating the optical and optomechanical hardware into a single Optical Hardware Module (OHM) [10]. This was the first attempt towards reducing the large number of degrees

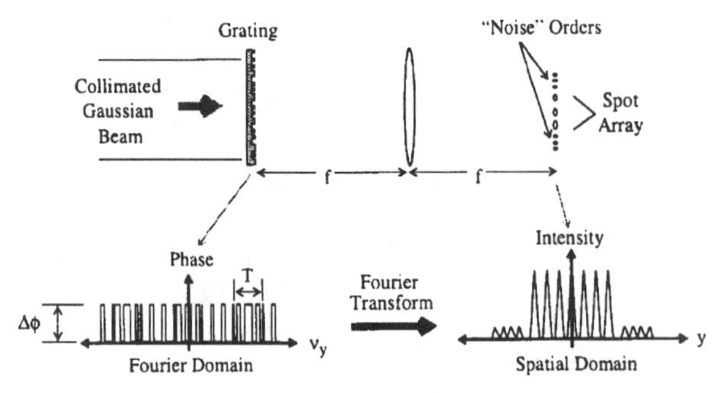

Fig. 12. Fourier-plane spot array generation.

of freedom that were required for systems based on off-the-shelf catalog components. The demonstrator was a four-stage multistage interconnection network [30]–[34] (MIN) based on 2-modules (see Section II-A1). The topology for the fabric and a picture of the completed demonstrator are shown in Fig. 11.

1) Optical and Optomechanical Hardware: With the objective of reducing the size, complexity, and cost of these systems. This system was partitioned into four optical hardware modules (OHM), where each OHM contained all the components required for a given stage in the switching fabric. Each OHM was then required to support four major functions: 1) switching nodes (8 × 4 S-SEED's), 2) spot array generation, 3) stage-to-stage optical interconnection, and 4) beam combination. Each of these is discussed in more detail below.

a) Spot array generation: The two-dimensional (2D) arrays of SEED technology based switching nodes that have been previously discussed require an optical power supply to clock the devices [37]. The generation of 2D arrays of uniform intensity spots requires two basic components. The first is a high-power, single-frequency, diffraction-limited laser that can provide the appropriate power per pixel required to meet the system speed requirements. The second component in a spot array generator requires some mechanism to equally and uniformly divide the power from the laser and distribute it to the optical windows of the S-SEED's. There have been several different approaches to the distribution of optical power to both S-SEED's and FET-SEED smart pixels [37]. The approach that has been pursued in the AT&T system demonstrators is Fourier-plane spot array generation using binary phase gratings (BPG) to uniformly distributing the optical power to the S-SEED's [38]. These phase gratings are made by etching glass with a repetitive multilevel pattern. For the case of a binary-phase grating, there are two thicknesses of glass. This grating is illuminated by a plane wave from a laser source as illustrated in Fig. 12. The light transmitted through the grating is Fourier transformed at the back focal plane of a lens which is the output plane of the spot array generator and the location of a smart pixel array.

For $System_2$ only the BPG and Fourier lens required to create the needed 8 × 8 array were located on the OHM. The laser source and collimating optics were constructed with catalog optics and optomechanics and mounted on the optical bench adjacent to the OHMs.

b) Optical interconnect: To create the MIN shown in Fig. 11 requires a crossover interconnect [39] between each of the stages. Figure 13(a) illustrates the topology of the crossover interconnect. The output of each node is split into two components: a straight-through and a crossed interconnect. To convert to a 3D network, the 2D network can be fan-folded on the fold lines shown in Fig. 13(a) to create the 3D network of Fig. 13(b) [10]. Note that the interconnect between each stage of the 3D network is still a 2D interconnect. Figure 13(c) shows the operation of the interconnect system. An input image enters the system from the left. The light associated with each "pixel" or "spot" is circularly polarized. The perpendicular component of the input light will be imaged onto the mirror. The reflected light from the mirror will have its polarization rotated, allowing it to pass through the polarizing beamsplitter where it can be imaged onto the output image plane. The parallel component of the input light will pass through the beamsplitter and be imaged onto the retro-reflector grating (ruled aluminum prismatic mirror array). The light incident on one mirror of each retro-reflector grating will be reflected to the opposing mirror where it will be redirected back into the beamsplitter. This retro-reflection process shifts the position of each spot imaged onto the grating. This shifting implements the crossover patterns of the crossover interconnection topology. The light leaving the retro-reflector grating will have its polarization rotated allowing it to be redirected and focused onto the output image plane. Thus both the straight through and crossed interconnects have been achieved. The amount of shift in space is related to the period of the retro-reflector grating.

c) Beam combination: Free-space photonic switching fabrics based on S-SEED arrays require that each device must be able to receive two input signals plus a clock [39], [40]. In addition, the reflected output signal must be directed from the device to the next stage of the switching fabric. The major constraints of this problem are that the spots must be small (< 5 μm), often requiring the entering signals to use the full aperture of an imaging lens, and that the signals must not interfere at the device's optical window.

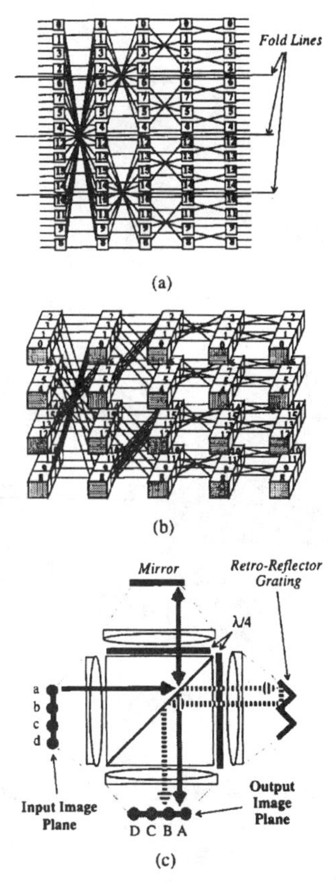

Fig. 13. Crossover interconnect. (a) 2D topology. (b) 3D topology. (c) Optical hardware.

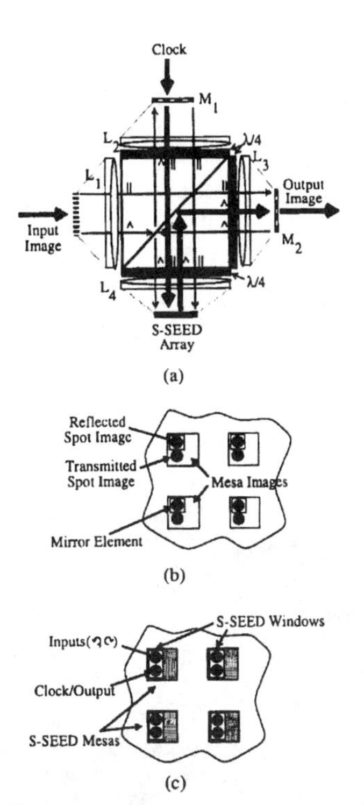

Fig. 14. Image-division (space-multiplexed) beam combination. (a) Beam combiner. (b) Mask. (c) S-SEED array.

An example of a beam combination system using image-division (space-multiplexing) is illustrated in Fig. 14. This beam combiner is composed of a polarizing beamsplitter that is surrounded on three sides by $\lambda/4$ plates. The lenses, L_{1-4}, are used in the infinite conjugate mode. At the bottom of the beam combiner is the S-SEED array. The location of the beams incident on the S-SEED array are shown in Fig. 14(c). The input image is composed of an array with each of the two input signals being associated with one of the two linear polarizations. As an input image enters the beam combination system on the left, the perpendicular component is directed upward while the parallel component passes through the beamsplitter. Both components travel through $\lambda/4$ waveplates where the linear polarization is converted to circular. The light traveling upward is imaged onto patterned mirrors of mask M_1 as shown in Fig. 14(b). This mask consists of an array of small mirrors that are located in the image plane of L_2 such that light reflected from them will be imaged onto the upper half of the rectangular window of the S-SEED array. Therefore, the image traveling upward reflects off M_1, passes again through a $\lambda/4$ waveplate which changes the polarization from circular to parallel allowing the light to be passed through the polarization beamsplitter onto the upper half of the S-SEED array.

The parallel component of the input image passes though the polarization beamsplitter, is imaged onto the small patterned mirrors of mask M_2, reflected by the polarization beamsplitter, and then imaged onto the upper half of the S-SEED array. Since these two inputs will have orthogonal circular polarizations they will not interfere.

The optical clock will enter the beam combination unit by being imaged onto the transparent region below the small mirrors of mask M_1. This array of high-power spots will be imaged onto the bottom half of the S-SEED array. The reflected clock signal, which then becomes the output signal of the S-SEED array, will have its polarization rotated allowing it to be reflected by the polarization beamsplitter and then imaged onto the transparent region below the small mirrors of mask M_2. This output image can then be collected and used by an optical interconnect or another beam combination unit.

d) System$_2$ optical hardware module: The *System$_2$* OHM was the first attempt at integrating the S-SEED switching nodes, crossover optical interconnect, and the image-division beam combination onto a single optomechanical structure. It included several optomechanical mounts mounted onto a stainless-steel baseplate in addition to gluing the $\lambda/4$ substrates directly to the polarization beamsplitters for both the optical interconnection and beam combination units. A schematic diagram and a picture of this OHM is shown in Fig. 15. For this structure the optical interconnect is composed of PBS_1, L_2, L_4, M_1, and the prism grating, the beam combiner includes

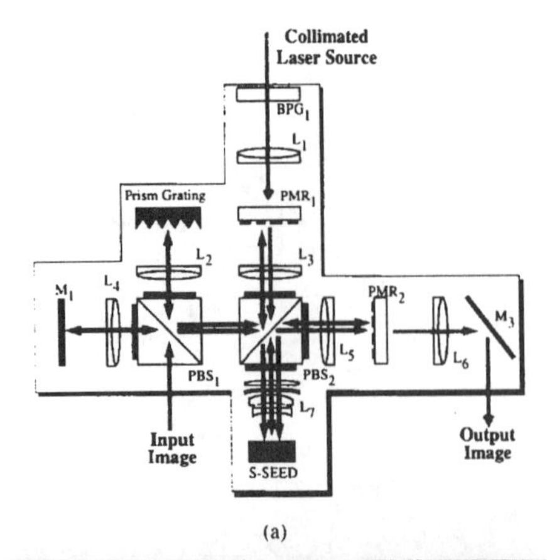

(a)

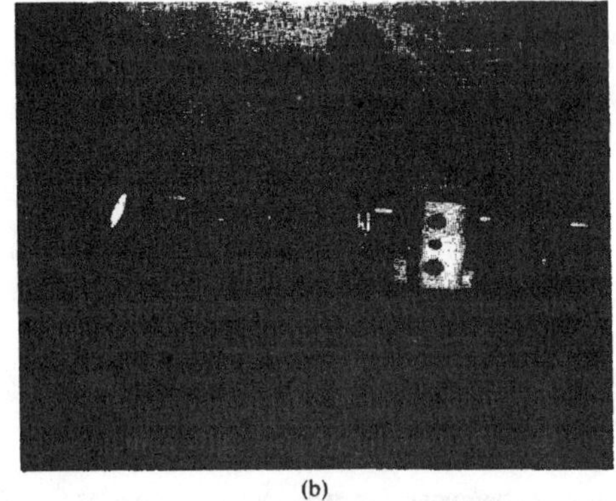

(b)

Fig. 15. *System*$_2$ optical hardware module. (a) Schematic. (b) Picture.

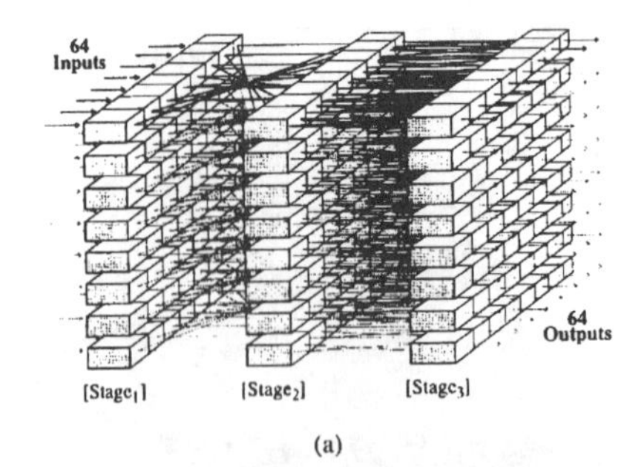

(a)

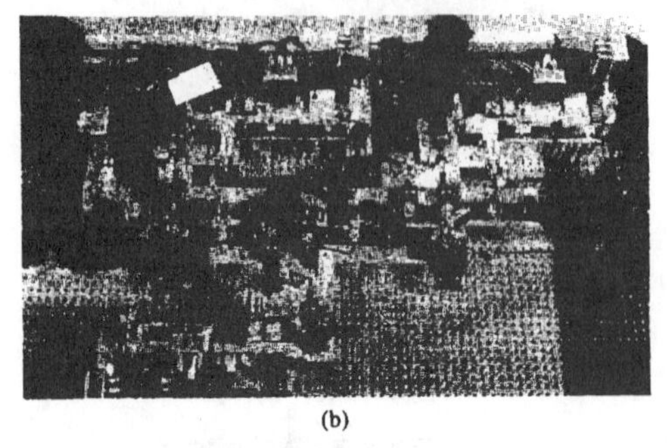

(b)

Fig. 16. *System*$_3$. (a) Network topology. (b) Picture of demonstrator.

L_7, PBS_2, L_3, L_5, and PMR_{1-2}, and the portion of the spot array generator included is L_1 and BPG_1.

On both the input and output stage there was an 8 × 8 fiber bundle carrying 32 dual-rail signals. Each bundle contained 64 100-μm core multimode fibers that were bundled into a square array with 400-μm spacings.

6) Performance: The S-SEED's used in this demonstration had rectangular windows of 5 × 10 μm spaced 20 μm apart with a contrast ratio of 3:1 and a switching energy of 3.8 pJ. Power measurements made at various points within the system indicated that ~0.6% of the optical power from the clock laser (10 mW) would arrive at the S-SEED array associated with the next node stage. Thus each S-SEED optical window would receive ~ 1.8 μW. This system achieved a maximum measure bit-rate of 55 kb/s which was limited by the optical power available from the laser sources (10 mW/laser), and the losses resulting from optical aberrations, vignettings, alignment, Fresnel reflections, and nonuniformities in the gratings.

C. *System*$_3$

The third system demonstrator continued to push the optics and optomechanics necessary to increase the spatial bandwidth or chip-to-chip connectivity of free-space digital optical systems [11]. This was accomplished by constructing a three-stage crossover network with 128 (8 × 16 array) S-SEED's per stage. The network topology and a picture of the demonstrator are shown in Fig. 16. This system was composed of three OHM's. The first OHM (*Stage*$_1$) provided the input interface function as well as storing the control information for the optical logic gate based (2, 1, 1) nodes created with the following two stages. The input to this OHM was an 8 × 8 (100-μm core) fiber bundle (same as *System*$_2$). The last two OHM's were mirror images of each other with the exception of the prism gratings required to provide the different crossover interconnects. The outputs from the third OHM were either imaged onto a camera or into an output 8 × 8 (100-μm core) fiber bundle.

1) Optical and Optomechanical Hardware: This third system demonstrator focused on reducing the total number of optomechanical variables with the objective of increasing the stability of the system. This was accomplished by building a system based on commercially available optics and a custom-designed optomechanical mounting system. This mounting system was based on V grooves milled into jig-plate aluminum. By using these V grooves with fold mirrors and precentered lenses, the optical and mechanical axes within each stage were collinear to a precision of less than 1 arcmin. This V-groove mounting technique was an important step in the evolution of free-space optomechanical technology because it focused attention on minimizing costs, increasing the system stability, and reducing the size per OHM.

A schematic diagram and picture of the OHM for *System*$_3$ is shown in Fig. 17. In this figure M_1, L_4, the polarizing beam splitter PBS_1, L_2, and the retro-reflector grating are the basic parts of the optical interconnect (same as

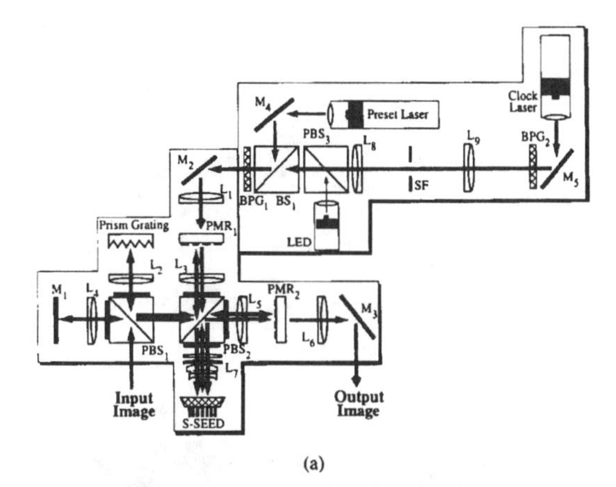

(a)

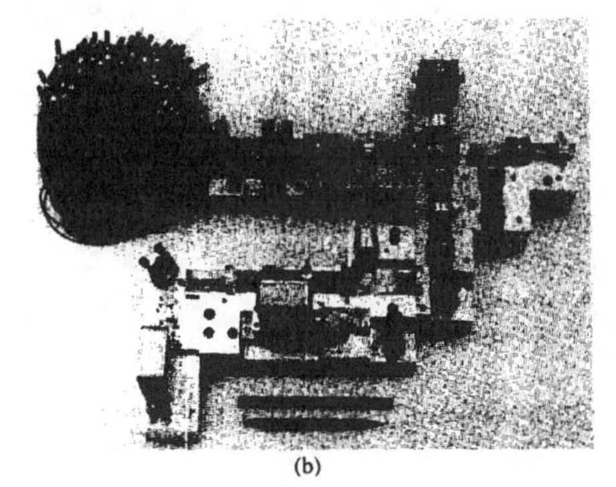
(b)

Fig. 17. $System_3$ optical hardware module.

$System_2$). The beam combination unit (same as $System_2$) is composed of L_7, PBS_2, L_5, L_3, the patterned mirror reflectors PMR_1, and PMR_2, L_1, L_6, M_2, and M_3. This system was the first attempt to integrate the entire spot array generator (SAG) into the OHM. The SAG included the preset signals, the clock, and LED illuminator to provide general illumination for system alignment and testing. The preset signals required for optical logic operation of the S-SEED's are generated by the preset laser (3 mW at 850 nm) and then directed to the beam combination unit by M_2, the beamsplitter BS_1, and the binary phase grating $BPG_1(8 \times 16)$ grating. The clock signal is generated by the clock laser (3 mW at 850 nm) and then distributed to the beam combiner by M_4, $BPG_2(1 \times 2$ grating), L_9, SF (spatial filter), L_8, PBS_3, BS_1, and BPG_1.

2) System Experiments: There were three major systems level experiments that were based on the $System_3$ optical and optomechanical hardware that had been developed, each dependant on the functionality assigned to the different S-SEEDs arrays. These three experiments (see Table 1) include the demonstration of 1) a three-stage 2-module-based crossover network (128 nodes per stage), 2) 64 parallel optical-logic-gate-based (2, 1, 1) nodes, and 3) a single stage of 64 parallel transmission-gate-based (2, 1, 1) node.

Table 1 $System_3$ Experiments

Experimental Network	$Stage_1$	$Stage_2$	$Stage_3$
2-Module Crossover Network	Latch (S-SEED)	2-Module (S-SEED)	2-Module (S-SEED)
64 Parallel (2, 1, 1) Nodes (S-SEED-based optical logic gates)	Latch (S-SEED)	AND (S-SEED)	OR (S-SEED)
64 Parallel (2, 1, 1) Nodes (Transmission gate-based smart pixels)	Latch (S-SEED)	(2,1,1) Node Smart Pixels (TG-SEED)	—

The first system experiment, the 2-Module-Based Crossover Network, required the S-SEED's to act as latches, storing the information that is incident upon them (Section I-A1). Information was transferred between stages by blocking one of the arms of the optical interconnect. This guarantees that only one signal will be present on the inputs of the 2-modules in the following stage (requirement for 2-module). Thus information was successfully transferred through all three stages of the network. To make this system a fully operational 2-module-based 3-stage MIN would require either individually electronically disabled S-SEED's or a spatial light modulator located in the same position as PMR_1. This spatial light modulator would have to block the Clock signal going to the S-SEED 2-modules that are supposed to have no optical outputs. Neither of these control schemes was implemented in $System_3$.

The second systems experiment, (2, 1, 1) Nodes Based on S-SEED Optical Logic Gates, implemented 64 parallel (2, 1, 1) nodes, the S-SEED arrays acting as optical logic devices (Section II-A2). For this experiment, $Stage_1$ was used as a latch to store the information entering the system from the fiber bundles and prepare for entry into the (2, 1, 1) nodes. The (2, 1, 1) nodes were created by forcing both the S-SEED's in $Stage_2$ to function as optical AND and the S-SEED's in $Stage_3$ to function as optical OR gates. Thus with the crossover interconnection between $Stage_2$ and $Stage_3$ 64 parallel (2, 1, 1) nodes are created. Although this demonstration was set up to use embedded control [45] no individual node control was implemented.

The final experiment, (2, 1, 1) Nodes Based on Transmission Gate Based Smart Pixel, used $Stage_1$ as a latch to store the information entering the system from the fiber bundles and prepare for entry into the (2, 1, 1) node smart pixels (Section II-A3) located in $Stage_2$. This simple experiment demonstrated the successful operation of the TG-SEED smart-pixel arrays.

3) Performance: The S-SEED's used in this demonstration had rectangular windows of 5×10 μm space 20 μm apart with a contrast ratio of 3.2:1 and a switching energy of 3.8 pJ. Power measurements made at various points within the system indicated that ~1.5% of the optical power from the clock laser (10 mW) of $Stage_i$ would arrive at the $Stage_{i+1}$ S-SEED array. Thus each S-SEED optical window would receive on average ~ 600 nW (900 nW for the high state and 300 nW for the low state). This includes an ~ 3.5-dB loss from the output of an S-SEED in $Stage_i$ to the input of an S-SEED in $Stage_{i+1}$. The losses in the system were the result of the optical signal passing through 132 surfaces, 11 lens passes, and 6 image planes (clock laser ⇒ S-SEED $Stage_i$ ⇒ S-SEED $Stage_{i+1}$). This system achieved a maximum measure bit rate of 33 kb/s which was primarily limited by the optical power available from the laser sources (10 mW/laser), the losses in the input fiber bundle, and alignment. The alignment tolerances for the system included a ± 10-μm defocus limit and a ± 1.5-μm lateral shift limit. These limits were achieved by establishing the following system parameters guidelines: $\Delta x, \Delta y < \pm 0.5$ μm; $\Delta z < \pm 5$ μm; tilt $(x, y) < \pm$ 40 arcmin; roll $(z) < \pm 8$ arcmin; $\Delta\lambda < \pm 1$ nm.

D. $System_4$

The fourth system demonstrator continued the evolutionary trend in pushing the optics and optomechanics to increase the available system spatial bandwidth [12]. This demonstrator was composed of a six stage 16 × 32 Banyan network [41] (32-bit-wide data path) with 1024 (32 × 32 array) Application-Specific SEED's (AS-SEED) per stage. The AS-SEED's were specially designed and fabricated S-SEED's that included small metal mirrors located at predetermined locations on the optical windows to block unwanted signals. The network topology and a picture of the demonstrator are shown in Fig. 18. This system was composed of six OHM's mounted onto a single baseplate. The first OHM ($Stage_1$) provided the input interface function. The AS-SEED array of this first stage was used as a spatial light modulator (SLM) to enter information into the remaining five stages. For this SLM all the AS-SEED's were electrically ganged together so that each of the 16 rows of information could be individually modulated and entered into the remaining fives stages of the system at any time (all the bits were the same—either logical "1" or "0"). The last five OHM's were identical with the exception of the optical interconnection BPG's which provide the different link-stage interconnects. The outputs from the sixth OHM were imaged onto either a camera or a detector.

1) Optical and Optomechanical Hardware: The goals of the $System_4$ optical and optomechanical hardware included: 1) ensure mechanical and thermal stability, 2) minimize the number and complexity of system components to reduce fabrication and assembly tolerances and costs, 3) provide for simple repair and replacement procedures, and 4) provide a compact system package. To achieve these objectives the system optomechanics attempted to minimize the number of adjustment mechanisms (knobs), use modular kinematic or pseudokinematic mounting, and avoid springs, flexures, and cantilevers. This new mounting hardware included slots (15 mm wide and 6 mm deep) milled into a metal baseplate (22.86 cm × 31.75 cm × 2.5 cm) made of low-carbon rolled steel with electroless nickel plating (to

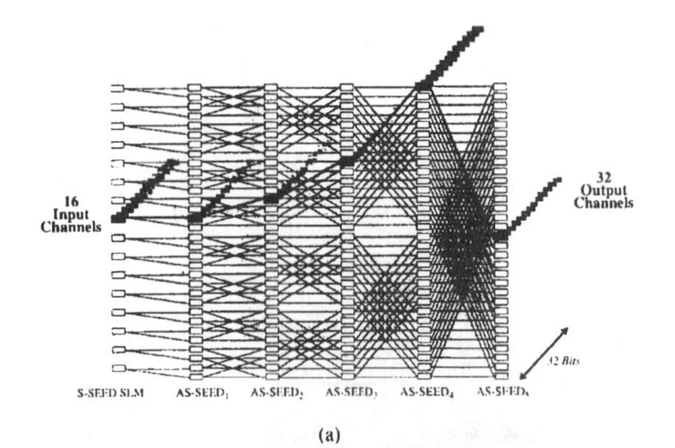

(a)

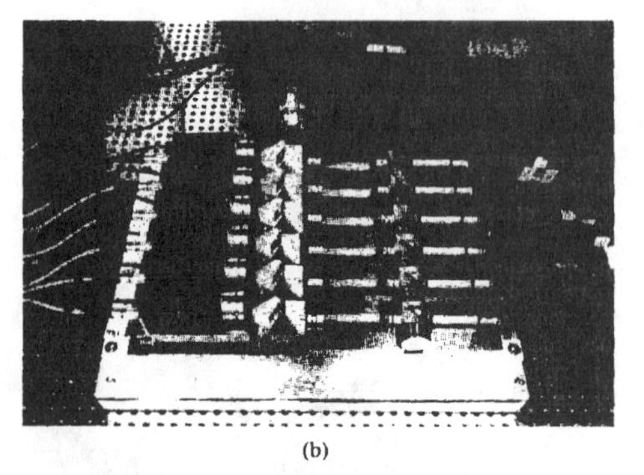

(b)

Fig. 18. $System_4$. (a) Network topology. (b) Picture of demonstrator.

prevent corrosion and reduce the stick-slip friction between the baseplate and the mounted components) and precentered optical components (mounted in 25-mm cylindrical cells) held in the slots by flat ceramic magnets located at the bottom of the slots. Slots were used instead of V grooves to hold all the components with the exception of the laser pens (laser, collimator, and Brewster telescope) and PBS's. The finished plate was flat to within 2.5 μm with slot parallelism and orthogonality better than 0.5 arcmin.

The spot array generator hardware used by AT&T's $System_4$ is shown in Fig. 19. In this subsystem, the laser-drive electronics and the thermoelectric cooler are used to control the hand-picked 60-mW semiconductor 850-nm laser. The output light is initially collimated and then circularized by the Brewster telescope. The analyzer and $\lambda/4$-wave plate are used as an isolator to reduced the antistabilizing effects of back reflections on the laser. The collimated light then passes through the Risley prisms which are used to register the spot arrays on the SEED device photosensitive windows. Finally, the light passes through a 64 × 32 BPG, which in conjunction with the objective lens of the system (see Fig. 22) redistributes the input optical power into 2048 equal power spots of light.

The $System_4$ optical interconnect is used to create a 3D network composed of a 2D $N \times M$ network replicated X times to create X parallel $N \times M$ networks. This is equivalent to an $N \times M$ network that is X bits deep. This type of network can be implemented with a simple

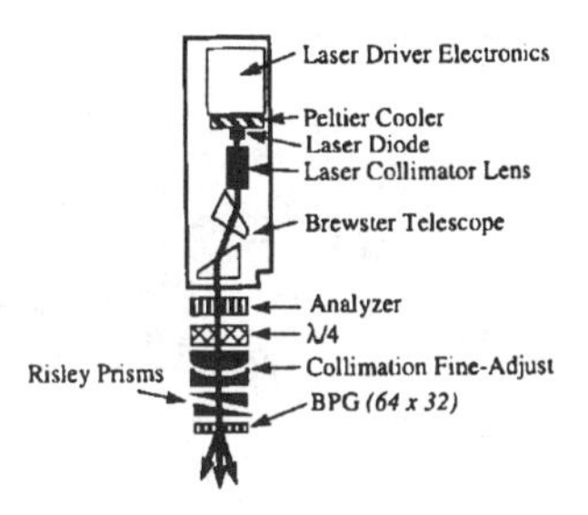

Fig. 19. *$System_4$* spot array generation.

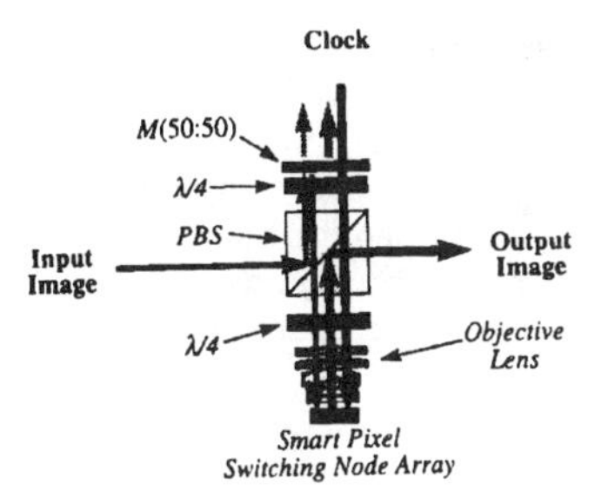

Fig. 21. Lossy beam combination.

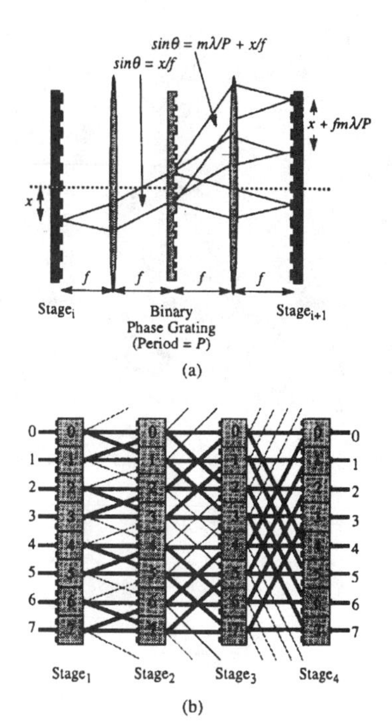

Fig. 20. Pupil-plane interconnection. (a) 1 × 3 BPG interconnect. (b) Banyan network using 1× 3 interconnects.

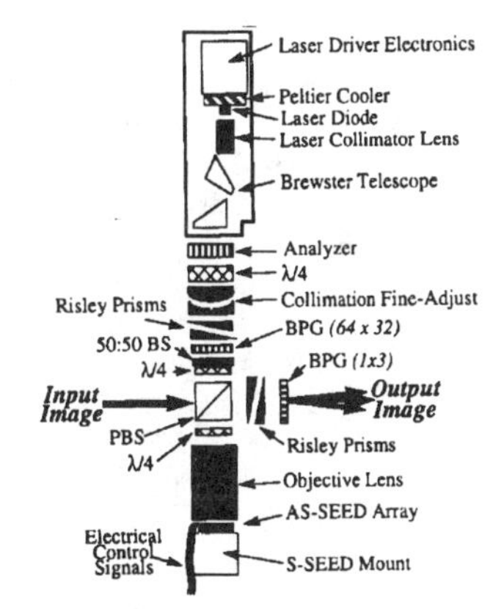

Fig. 22. *$System_4$* optical hardware module.

2D interconnect. An example of the interconnect is shown in Fig. 20, where the output of each node is directed to the pupil-plane where a binary phase grating splits the signal into three equal parts. These copies of the original signal are then directed to the inputs of the next stage of the network. Figure 20(b) illustrates how a Banyan network (thick lines) can be created using these interconnects. The thick lines represent active connections between nodes in $Stage_i$ and nodes in $Stage_{i+1}$; the thin lines represent connections between stages in the network that are blocked by placing metal masks in front of the optical windows of specific nodes in $Stage_{i+1}$. If all three signals created by the interconnect are used, instead of just two as described above, a TIADM network can be implemented [46]. Note that the splitting angle of the BPG is different for each stage in the network. The Banyan network shown in Fig. 18 is an example of a 16 × 32 network in which each channel is 32 bits deep.

$System_4$ used a simple but lossy beam combination method to combine the 2 input signals, the clock, and the output signals (although the losses were similar to those of *$System_3$*). The beam combination hardware included the polarization beamsplitter, PBS, and a partially reflecting mirror, $M(50:50)$, shown in Fig. 21 [47]. The input image enters the beam combination unit s-polarized, thus it is reflected up through a $\lambda/4$ plate to the partially reflecting mirror M. Half of the power is reflected, passing through the $\lambda/4$ plate again, which will rotate the image to p-polarized light. This reflected image will then pass through the polarization beamsplitter, PBS, through another $\lambda/4$ plate, the objective lens L, and onto the array of switching nodes. For the case S-SEED nodes, the information present on the input image is latched into the S-SEED's. The circularly polarized *Clock* can then pass through all the elements (half the *Clock* power is reflected (lost) at $M(50:50)$), be modulated by the information stored in the switching node array, and then be reflected out of the beam combination unit to provide the output image to the next part of the system.

Integrating the new SAG hardware, the Banyan optical interconnect hardware, and the new lossy beam combination hardware yields the *$System_4$* OHM shown in Fig. 22.

2) System Experiment: The six-stage Banyan network previously described was demonstrated. The control for the system was accomplished by electrically disabling rows of the AS-SEED arrays. When the AS-SEED's were disabled they absorbed the *Clocks* on both inputs/outputs preventing a usable signal from being passed onto the next stage of the system. When the voltage for the AS-SEED's was on, they behaved as 2-modules, capturing and storing the bits of information to pass onto the next stage when clocked. The entire system was under interactive computer control allowing any paths to be set up through the network in real time.

3) Performance: The S-SEED's used in this demonstration had rectangular windows of 6 × 21 μm spaced 40 μm apart with a contrast ratio of 3:1 and a switching energy of ~ 4 pJ at 10 V. Power measurements made at various points within the system indicated that ~1% of the optical power from the clock laser (100 mW) of $Stage_i$ would arrive at the $Stage_{i+1}$ S-SEED array. This includes a > 8 dB loss from the output of an S-SEED in $Stage_i$ to the input of an S-SEED in $Stage_{i+1}$. This system achieved a maximum measure bit rate of 100 kb/s which was limited by the optical power available from the laser sources (100 mW/laser) and alignment variations. The alignment tolerances for the system included a ± 10-μm defocus limit and a ± 2.0-μm lateral shift limit. These limits were achieved by establishing the following system parameters guidelines: $\Delta x. \Delta y < \pm$ 0.5 μm; $\Delta z < \pm$ 3 μm; tilt $(x, y) < \pm$ 20 arcmin; roll $(z) < \pm$ 2 arcmin; $\Delta\lambda < +$ 0.5 nm; and $\Delta f < \pm$ 15 μm (difference in focal lengths of objective lenses).

E. $System_5$

The fifth system demonstrator had the objective of increasing both the functionality and the temporal bandwidth of a switching fabric by implementing a packet-based network that uses FET-SEED smart pixels as the switching nodes [13]. The general architecture and a picture of the system are shown in Fig. 23.

$System_5$ is a packet-switching fabric designed to switch ATM-like cells of data. The input cells initially enter and are stored by the electronic input buffers (double buffers). While in the input buffers, the destination address for all the cells are sent to the electronic control where a fast path hunt algorithm developed for EGS networks [42]–[44] calculates the paths that need to be set up throughout the Banyan multistage network [46]. After the path hunt is completed, a routing address is prepended onto each cell. This routing address includes $s - 1$ "0" bits followed by s control bits. The transfer of information to and from the electronic control as well as the complete MIN path hunt must be completed in one cell time (for ATM cells: 53 octets × 8 bits/octet/155 Mb/s ~ 2.7 μs). Prior to sending the cells through the MIN the paths need to be set up, this is accomplished by loading the s control bit into the control latches of the (2, 1, 1) smart pixel switching nodes (Section II-B1). This is accomplished by initially loading the "0" bits, one stage at a time, into the control latches. This puts all the (2, 1, 1) node smart pixels into the state where the A input channels are available to both the smart-pixel outputs and control latches. The s embedded control bits are then serially shifted into the smart pixels of the s stages. When the first control bit reaches the sth stage of the MIN the electronic control sends a signal loading all the s control bits into the control latches of the smart pixels. With this completed, all the paths are set up throughout the MIN allowing cells to be directed to their desired output buffers. This processes is repeated for each cell entering the MIN.

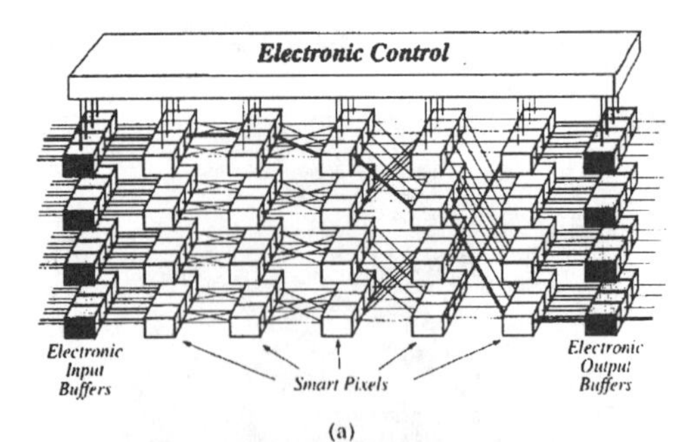

(a)

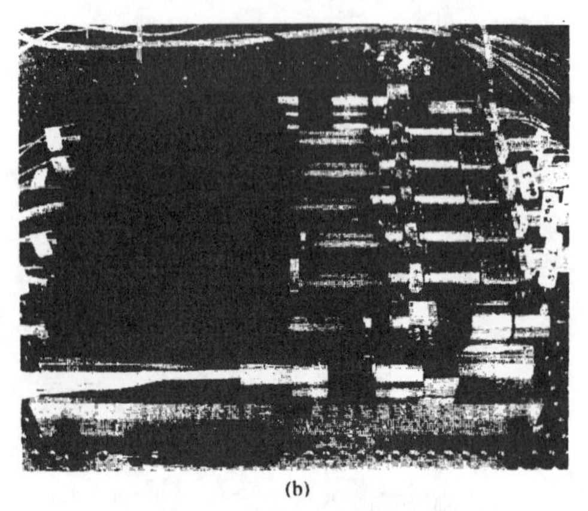

(b)

Fig. 23. $System_5$. (a) Network topology. (b) Picture of demonstrator.

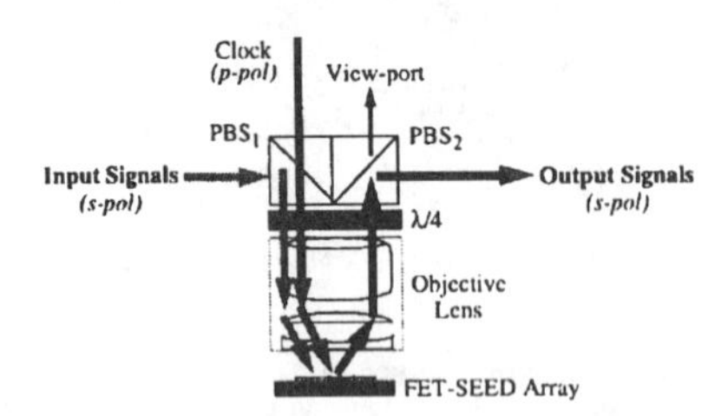

Fig. 24. $System_5$ pupil-division beam combination.

1) Optical and Optomechanical Hardware: The optomechanical hardware (baseplate, optical interconnect) used by $System_5$ was improved but essentially the same as that used by $System_4$ with the exception of the beam combination hardware. This system used pupil-division beam combination rather than the image-division or amplitude-division approaches that had been utilized in the previous demonstrators. Figure 24 illustrates the basic components associated with this new approach.

The input signals (s-polarization) are directed down through the objective lens to the FET-SEED array by

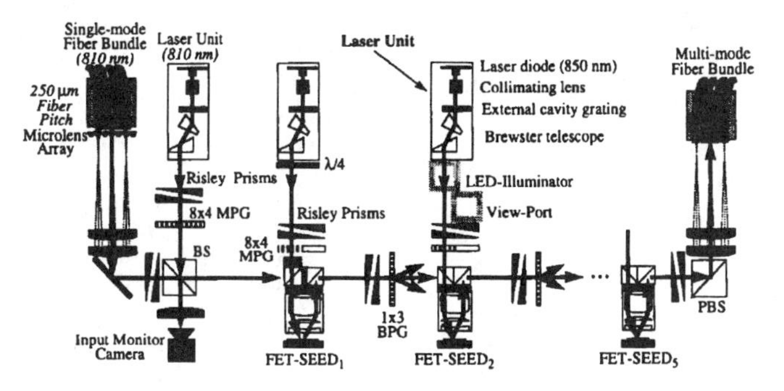

Fig. 25. *System*$_5$ optical hardware layout.

Table 2 AT&T Free-Space Photonic Switching System Specifications

	System$_1$ (1988)	*System*$_2$ (1989)	*System*$_3$ (1990)	*System*$_4$ (1991)	*System*$_5$ (1993)
Supported Channels	2	32	64	1024	32
System Bit Rate	5 kb/s	55 kb/s	33 kb/s	100 kb/s	50 Mb/s
Flat Field Angle	1°	1.6°	3.2°	6.7°	4.4°
Chip/System Pin-outs	12/20	256/1024	1024/3072	10 240/61 440	192/960
Optics	catalog	catalog	catalog	custom	custom
Mechanics	catalog	catalog and custom	custom	custom	custom
System Area	32 ft^2	16 ft^2	6ft^2	0.78 ft^2	1.16 ft^2
Control	optical	optical	optical	electrical	optical
Assembly and Alignment Time	days	days	8 h	3 h	3 h
Lifetime	< 8 h	< 24 h	< 6 days	> 5 weeks	> 8 weeks

PBS_1. The *Clock* signal (p-polarization) passes through PBS_1 and the objective lens and onto the FET-SEED array where it is reflection-modulated and directed up to PBS_2 where it is directed to the right as an output signal. The advantages of this approach over the *System*$_4$ hardware include: 1) four times more optical power can be imaged onto the FET-SEED arrays than the amplitude-division approach, 2) because the beam diameter is 1/2 of *System*$_4$ the accumulation of aberrations are decreased, and 3) a view-port is available at every stage to ease test and alignment. The main disadvantage is that the spot size on the FET-SEED arrays has to increase by a factor of 1.5.

The optical hardware used in this system is outlined in Fig. 25. The first stage in this system, as in the previous demonstrators, is used to latch and switch the incoming data that have been entered into a single-mode fiber bundle. The last four stages are identical, with the exception of the optical interconnect (Banyan). Finally, the output of the final smart-pixel array is imaged onto a multimode fiber bundle to be delivered to the electronic output buffers.

2) Performance: The FET-SEED smart pixels used in this demonstration had a contrast ratio of 3.5:1 and a switching energy of ~ 75 fJ at 222 Mb/s. Power measurements made at various points within the system indicated that ~2% of the optical power from the clock laser (60 mW) of *Stage*$_i$ would arrive at the *Stage*$_{i+1}$ S-SEED array. This includes a > 5-dB loss from the output of an S-SEED in *Stage*$_i$ to the input of an S-SEED in *Stage*$_{i+1}$. This system (15 channels were operational) achieved a maximum measure bit rate of 50 Mb/s which was limited by the optical power available from the laser sources (60 mW/laser) and alignment nonuniformities. The alignment tolerances for the system included a ± 10-μm defocus limit and a ± 2.0-μm lateral shift limit. These limits were achieved by establishing the following system parameters guidelines: $\Delta x, \Delta y < \pm$ 1.0 μm; $\Delta z < \pm$ 5 μm; tilt $(x, y) < \pm$ 20 arcmin; roll $(z) < \pm$ 3 arcmin; $\Delta\lambda < \pm$ 0.5 nm; and $\Delta f < \pm$ 15 μm (difference in focal lengths of objective lenses).

IV. Discussion

Some of the major attributes of the five system demonstrators are listed in Table 2. From this table it can be seen that the optics and optomechanics developed for these demonstrators has allowed 1) the number of supported information channels to be increased from 2 to 1024, 2) the system bit rate has gone from 10 kb/s to 50 Mb/s, 3) the flat field angle of the supported spatial bandwidth has gone from 1° to 6.7°, 4) the maximum number of effective pin-outs per chip has gone from 12 to 10 240, 5) both the optics and mechanics have gone from off-the-shelf to custom components, 6) the system area at the same time has reduced from $\sim$ 32 ft^2 to $\sim$ 1 ft^2, 7) the assembly and alignment time has been reduced from days to hours, and 8) the measured lifetime of the system has increased from hours to weeks. Perhaps the most important progress that has been made is the development of the new custom optical and mechanics that will be required to make this technology realizable.

Although there has been tremendous progress in the development of this connection-intensive free-space technology, it needs to be understood that in order to become a reality in the marketplace, a packaging technology will be required that is stable, reliable, inexpensive, and manufacturable. Although AT&T and others have made significant advances with their free-space system demonstrators, most of these objectives have not been fully realized. More focus is needed on the design and testing issues associated with building stable, reliable optical hardware. As with the existing high-speed electronic technology, the development of hardware design rules, testing tools, and verification procedures will be required for the eventual deployment of this free-space digital optical technology.

ACKNOWLEDGMENT

The authors would like to gratefully acknowledge M. G. Beckman, D. B. Buchholz, J. L. Brubaker, S. J. Hinterlong, E. Kerbis, R. L. Morrison, R. A. Novotny, G. W. Richards, J. Sasian, S. L. Walker and M. J. Wojcik for their contributions in developing the free-space photonic switching demonstrators.

REFERENCES

[1] W. E. Stephens and K. C. Young, "Terabit-per-second through-put switches for broadband central offices: an overview," *IEEE LC S*, vol. 1, no. 4, pp. 20–27, Nov. 1990.

[2] R. R. Tummala and E. J. Rymaszewski, *Microelectronics Packaging Handbook*. New York: Van Nostrand Reinhold, 1989.

[3] R. A. Nordin *et al.*, "A systems perspective on digital interconnection technology," *J. Lightwave Technol.*, vol. 10, no. 6, pp. 811–827, June 1992.

[4] D. A. B. Miller, "Optics for low–energy communication inside digital processors: quantum detectors, sources, and modulators as efficient impedance converters," *Opt. Lett.*, vol. 14, no. 2, pp. 146–148, Jan. 15, 1989.

[5] H. S. Hinton and D. A. B. Miller, "Free-space photonics in switching," *AT&T Tech. J.* pp. 84–92, Jan./Feb. 1992.

[6] H. S. Hinton *et al.*, "Digital switching systems based on the SEED technology and free-space optical interconnects," *Photonic Switching and Interconnects* A. Marrakchi, Ed. New York: Marcel Dekker. pp. 213–247.

[7] A. A. Sawchuk and T. C. Strand, "Digital optical computing," *Proc. IEEE*, vol. 72, no. 7, pp. 758–779, July 1984.

[8] M. E. Prise, N. C. Craft, M. M. Downs, R. E. LaMarche, L. A. D'Asaro, L. M. F. Chirovsky, and M. J. Murdocca, "Optical digital processor using arrays of symmetric self-electro-optic effect devices," *Appl. Opt.*, vol. 30, no. 17, pp. 2287–2296, June 10, 1991.

[9] E. Kerbis, T. J. Cloonan, and F. B. McCormick, "An all–optical realization of a 2 × 1 free-space switching node," *IEEE Photon. Technol. Lett.*, vol. 2, no. 8, pp. 600–602, Aug. 1990.

[10] T. J. Cloonan, M. J. Herron, F. A. P. Tooley, G. W. Richards, F. B. McCormick, E. Kerbis, J. L. Brubaker, and A. L. Lentine, "An all-optical implementation of a 3D crossover switching network," *IEEE Photon. Technol. Lett.*, vol. 2, pp. 438–440, 1990.

[11] F. B. McCormick, F. A. P. Tooley, T. J. Cloonan, J. L. Brubaker, A. L. Lentine, R. L. Morrison, S. J. Hinterlong, M. J. Herron, S. L. Walker, and J. M. Sasian, "Experimental investigation of a free-space optical switching network by using symmetric self-electro-optic-effect devices," *Appl. Opt.*, vol. 31, no. 26, pp. 5431–5446, Sept. 10, 1992.

[12] F. B. McCormick, T. J. Cloonan, F. A. P. Tooley, A. L. Lentine, J. M. Sasian, J. L. Brubaker, R. L. Morrison, S. L. Walker, R. J. Crisci, R. A. Novotny, S. J. Hinterlong, H. S. Hinton, and E. Kerbis, "A six-stage digital free-space optical switching network using *S*-SEEDs," *Appl. Opt.*, vol. 32, no. 26, Sept. 10, 1993, pp. 5153–5171.

[13] F. B. McCormick, T. J. Cloonan, A. L. Lentine, J. M. Sasian, R. L. Morrison, M. G. Beckman, S. L. Walker, M. J. Wojcik, S. J. Hinterlong, R. J. Crisci, R. A. Novotny, and H. S. Hinton, "A 5-stage free-space optical switching network with field-effect transistor self-electro-optic-effect device smart-pixel arrays," *Appl. Opt.*, vol. 33, no. 8, pp. 1601–1618, Mar. 10, 1994.

[14] H. S. Hinton, "Multiple quantum-well technology takes SEED," *Circuits and Devices*, vol. 9, no. 2, pp. 12–18, Mar. 1993.

[15] D. A. B. Miller *et al.*, "Band edge electro-absorption in quantum well structures: The quantum confined stark effect," *Phys. Rev. Lett.*, vol. 53, pp. 2173–2177, 1984.

[16] D. A. B. Miller *et al.*, "Electric field dependence of optical absorption near the bandgap of quantum well structures," *Phys. Rev. Lett.*, vol. B32, p. 1043, 1985.

[17] D. A. B. Miller *et al.*, "Novel hybrid optically bistable switch: The quantum well self-electro-optic effect device," *Appl. Phys. Lett.*, vol. 45, pp. 13–15, 1984.

[18] D. A. B. Miller *et al.*, "The quantum well self-electrooptic effect device: optoelectronic bistability and oscillation, and self-linearized modulation," *IEEE J. Quantum Electron.*, vol. QE-21, no. 9, pp. 1462–1476, Sept. 1985.

[19] D. A. B. Miller, "Quantum well self electro-optic effect devices," *Opt. Quantum Electron.*, vol. 22, pp. 561–598, 1990.

[20] A. L. Lentine *et al.*, "Symmetric self-electro-optic effect device: Optical set-reset latch," *Appl. Phys. Lett.*, vol. 52, no. 17, pp. 1419–1421, Apr. 25, 1988.

[21] A. L. Lentine, H. S. Hinton, D. A. B. Miller, J. E. Henry, J. E. Cunningham, and L. M. F. Chirovsky, "Symmetric self-electro-optic effect device: optical set-reset latch, differential logic gate, and differential modulator/detector," *IEEE J. Quantum Electron.*, vol. 25, no. 8, pp. 1928–1936, Aug. 1989.

[22] L. M. F. Chirovsky *et al.*, "Large arrays of symmetric self-electro-optic effect devices," *OSA Proc. on Photonic Switching*, H. S. Hinton and J. W. Goodman, Eds. (Optical Society of America, Washington, DC, 1991), vol. 8, pp. 56–59.

[23] A. L. Lentine *et al.*, "Logic self electro-optic effect devices: Quantum well optoelectronic multiport logic gates, multiplexers, demultiplexors and shift registers," *IEEE J. Quantum Electron.*, vol. 28, no. 6, pp. 1539–1553, June 1992.

[24] A. L. Lentine, T. J. Cloonan, and F. B. McCormick, "Photonic switching nodes based on self electro-optic effect devices," *Opt. Quantum Electron.*, vol. 24, pp. S443–S464, 1992.

[25] D. A. B. Miller, M. D. Feuer, T. Y. Chang, S. C. Chunk, J. E. Henry, D. J. Burrows, and D. S. Chemla, "Field-effect transistor self-electrooptic effect device: integrated photodiode, quantum well modulator and transistor," *IEEE Photonics Technol. Lett.*, vol. 1, no. 3, pp. 62–64, Mar. 1989.

[26] P. Wheatley, P. J. Bradley, M. Whitehead, G. Parry, J. E. Midwinter, P. Mistry, M. A. Pate, and J. S. Roberts, "Novel nonresonant optoelectronic logic device," *Electron. Lett.*, vol. 23, p. 92, 1987.

[27] L. A. D'Asaro *et al.*, "Batch fabrication and operation of GaAs-$Al_xGa_{1-x}As$ field-effect transistor-self-electrooptic effect device (FET-SEED) smart pixel arrays, *IEEE J. Quantum Electron.*, vol. 29, no. 2, pp. 670–677, Feb. 1993.

[28] H. S. Hinton, *An Introduction to Photonic Switching Fabrics*. New York: Plenum, 1993.

[29] ——, "Architectural considerations for photonic switching net-

works," *J. Selected Areas Commun.*, vol. 6, pp. 1209–1226, Aug. 1988.
[30] H. S. Hinton *et al.*, "Space division switching," in *Photonics in Switching II*, J. E. Midwinter, Ed. New York: Academic Press, 1993, pp. 119–167.
[31] G. M. Masson, G. C. Gingher, and S. Nakamura, "A sampler of circuit switching networks," *Computer*, pp. 32–48, June 1979.
[32] A. L. Lentine *et al.*, "4× 4 arrays of FET-SEED embedded control 2 × 1 switching nodes," in *Topical Meet. on Smart Pixels* (Santa Barbara, CA, Aug. 1992), postdeadline paper.
[33] P. J. Anthony, "Review of SEED technology," in *OSA Annu. Meet. of the Optical Society of America* (Albuquerque, NM, 1992).
[34] T.-Y. Feng, "A Survey of interconnection networks," *Computer*, pp. 12–27, Dec. 1981.
[35] H. S. Hinton, *An Introduction to Photonic Switching Fabrics*. New York: Plenum, 1993.
[36] H. S. Hinton *et al.*, "Space division switching," *Photonics in Switching II*, J. E. Midwinter, Ed. New York: Academic Press, 1993, pp. 119–167.
[37] N. Streibl, "Beam shaping with optical array generators," *J. Modern Opt.*, vol. 36, no. 12, pp. 1559–1573, 1989.
[38] R. L. Morrison, "Symmetries that simplify the design of spot array phase gratings," *J. Opt. Soc. America A*, 1992.
[39] J. Jahns and M. J. Murdocca, "Crossover networks and their optical implementation," *Appl. Opt.*, vol. 27, no. 15, pp. 3155–3160, Aug. 1, 1988.
[40] M. E. Prise, M. M. Downs, F. B. McCormick, S. J. Walker, and N. Streibl, "Design of an optical digital computer," in *Optical Bistability IV*, W. Firth, N. Peyhambarian, and A. Tallet, Eds. Paris, France: Les Editions de Physique, 1988, pp. C2-15-C2-18.
[41] F. B. McCormick and M. E. Prise, "Optical circuitry for free space interconnections," *Appl. Opt.*, vol. 29, no. 14, pp. 2013–2018, 1990.
[42] L. R. Goke and G. J. Lipovski, "Banyan networks for partitioning multiprocessor systems," in *Proc. 1st Annu. Symp. on Computer Architecture*, 1973, pp. 21–28.
[43] G. W. Richards, U.S Patents 4 993 016 and 4 991 168.
[44] T. J. Cloonan *et al.*, "A complexity analysis of smart pixel switching nodes for photonic extended generalized shuffle switching networks," *IEEE J. Quantum Electron.*, vol. 29, no. 2, pp. 619–634. Feb. 1993.
[45] T. J. Cloonan, G. W. Richards, F. B. McCormick, and A. L. Lentine, "Extended generalized shuffle network architectures for free-space photonic switching," in *OSA Proc. on Photonic Switching*, H. S. Hinton and J. W. Goodman, Eds. (Optical Society of America, Washington, DC, 1991), vol. 8, pp. 43–47.
[46] T. J. Cloonan *et al.*, "Architectural issues related to the optical implementation of an EGS network based on embedded control," *Opt. Quantum Electron.*, vol. 24, pp. S415–S442, 1992.
[47] T. J. Cloonan and M. J. Herron, "Optical implementation and performance of one dimensional and two-dimensional trimmed augmented data manipulator networks for multiprocessor computer systems," *Opt. Eng.*, vol. 28, no. 4, pp. 305–314, 1989.
[48] F. B. McCormick, F. A. P. Tooley, J. L. Brubaker, J. M. Sasian, T. J. Cloonan, A. L. Lentine, R. L. Morrison, R. J. Crisci, S. L. Walker, S. J. Hinterlong, and M. J. Herron, "Design and tolerancing comparisons for S-SEED-based free-space switching fabrics," *Opt. Eng.*, vol. 31, no. 12, pp. 2697–2711, 1992.

Free-Space Multistage Optical Interconnection Networks Using Micro Lens Arrays

Shigeru Kawai

(Invited Paper)

Abstract—**Presented here are new optical interconnections suitable for three-dimensional combining of lens arrays. Also described is a new multistage interconnection network with a self-routing function. The number of light paths which this network is estimated to be capable of handling is roughly ten or more times that of previously reported self-routing networks. Two separate lens arrays are fabricated to construct an 8 × 8 network of this design, and the feasibility of the proposed new interconnections is successfully demonstrated.**

I. Introduction

TECHNICAL advances in fabrication have permitted electrical devices to be packed ever more densely, producing a consequent increase in the number of signal paths. This, combined with ever-increasing demand for signal speed, has resulted in problems of signal skew and bandwidth limitation. Such problems might be overcome by using light beams in place of electrical signals. Light beams can cross in free-space without interfering with one another, and they have a high frequency bandwidth. This makes optical systems particularly promising for further advances in high-density and high-speed communications.

Optical interconnections can be classified into two types, free-space and waveguide. Free-space interconnections have no physical pathways, while waveguide interconnections have physical pathways. Free-space techniques are suitable for interconnections between processors on different planes, e.g., board-to-board interconnections, because light beams can cross in free-space with no mutual interference. Previous proposals have been made for free-space interconnections [1]-[5], and among them, those using lens arrays appear to be the most appropriate [2], [3]. Their use of imaging produces higher light-power efficiency, and their simpler architectures permit chip-like integration.

In this study, I present some new optical interconnections suitable for use with lens arrays and describe a new multistage interconnection network with a self-routing function. I also demonstrate the feasibility of these interconnections through the use of planar micro lens (PML) arrays [6], and, further, I discuss optical interconnection systems for massively parallel processing using free-space and waveguide techniques.

Manuscript received April 17, 1991.

The author is with C&C Information Technology Research Laboratories, NEC Corporation 1-1, Miyazaki 4-chome, Miyamae-ku, Kawasaki 216, Japan.

IEEE Log Number 9102560.

II. Optical Interconnections Using Lens Arrays

A. Principles

Lenses may be used to focus an output beam from a light source array on to a photodetector array, as shown in Fig. 1. By changing the position and the size of lenses, it is possible to control the direction of outgoing beams and to make desired interconnections. The proper positions for light sources, photodetectors, and lenses can be calculated geometrically as shown in Fig. 2. When an output beam from a light source at point (x_s, y_s, z_s) is incident to a lens at point (x_l, y_l, z_l), it may be focused on the point

$$x_d = (1 + m)x_l - mx_s \quad (1)$$

$$y_d = (1 + m)y_l - my_s \quad (2)$$

$$z_d = (1 + m)z_l - mz_s \quad (3)$$

where m is the lateral magnification of the lens.

When two connections are carried out at the same time using one lens, the position for the lens may be calculated in the following way.

Assume that light sources and photodetectors are arranged at pitch p. The relationship between a light source at point (x_s, y_s, z_s) and another light source at point (x_s', y_s', z_s), as well as that between a photodetector at point (x_d, y_d, z_d) and another photodetector at point (x_d', y_d', z_d) may be described as

$$|x_i' - x_i| = k_i p \quad (4)$$

$$|y_i' - y_i| = l_i p \quad (5)$$

where $i = d$ or s, and where k_i and l_i are integers. Further, when an output light beam from point (x_s', y_s', z_s) is incident to a lens at point (x_l, y_l, z_l) and focused on point (x_d', y_d', z_d), the following relations are obtained:

$$k_d = k_s m \quad (6)$$

$$l_d = l_s m. \quad (7)$$

When lenses are put at positions which satisfy these conditions, two connections are achieved at the same time

Reprinted with permission from *IEEE Journal of Lightwave Technology*, Vol. 9, No. 12, pp. 1774-1779, December 1991.

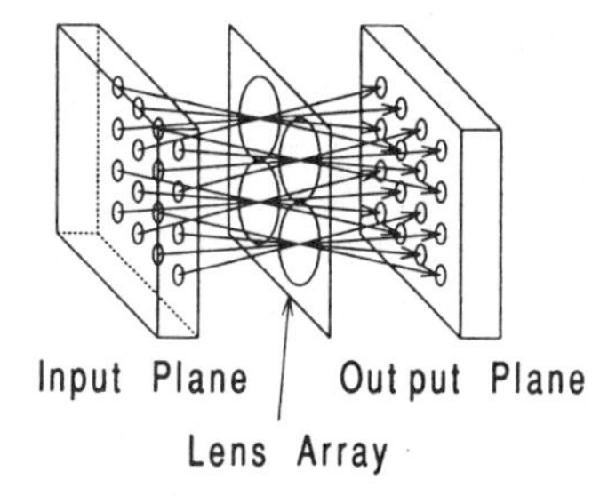

Fig. 1. Optical interconnections using lens arrays.

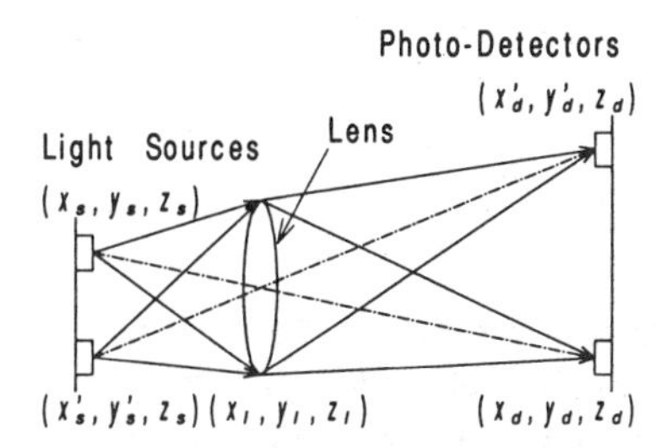

Fig. 2. Positions for optical devices in interconnections.

with a single lens. Lens positions which will produce more than two connections may be similarly obtained.

B. New Interconnections

If multistage network interconnections, most notably of the shuffle- and butterfly-permutation type [7], could be arranged in three dimensions, the number of interconnected processors and computer memories could be increased tremedously.

Free-space optical interconnections, which have no fixed pathways, are most suitable for three-dimensional (3-D) interconnections. Shuffle-permutations have already been carried out using a free-space interconnection technique [2], and omega [8] and baseline [9] networks may be produced by combining the shuffle-permutations. Fig. 3 shows two conventional designs used to produce a 4-shuffle, which shuffles eight elements into two groups of four elements each. Fig. 3(a) shows an ordinary electrical wiring arrangement, while Fig. 3(b) shows the light paths in an ordinary optical system arrangement whose outputs are the inverse arrangement of the inputs. Fig. 4 illustrates the interconnection pattern proposed here. Note that its odd and even inputs are in a reverse arrangement to that of the inputs in Fig. 3(a), while the outputs return to their original odd/even order of 0, 1; 2, 3; etc. We will refer here to these reverse arrangement techniques as "cross interconnections" and to ordinary electrical wiring techniques as "normal interconnections." One set of interconnections may be connected to another set by 2 × 2, 4 × 4, etc. switches which are controlled to reverse once again the odd/even positioning so that the output is again set to the 1, 0; 3, 2; etc. order. A comparison of Fig. 3(b) and 4 shows that the proposed pattern provides less aberration and higher light-power efficiency.

This design is significant in that such techniques can be applied not only to shuffle-permutations but to many other kinds of interconnections, including the butterfly-permutations used to construct banyan networks [10] (see Fig. 5) and to the basic interconnections used to construct full-crossover networks [11] (see Fig. 6). We should remember, however, that such networks are by their nature more suited to use with electrical pathways or other optical interconnection techniques than with free-space optical interconnections using lens arrays.

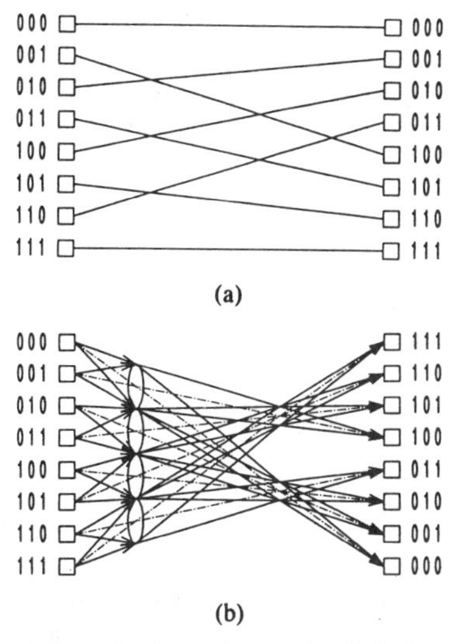

Fig. 3. Ordinary interconnection pathways for 4-shuffle permutation. (a) Electric interconnections. (b) Optical interconnections.

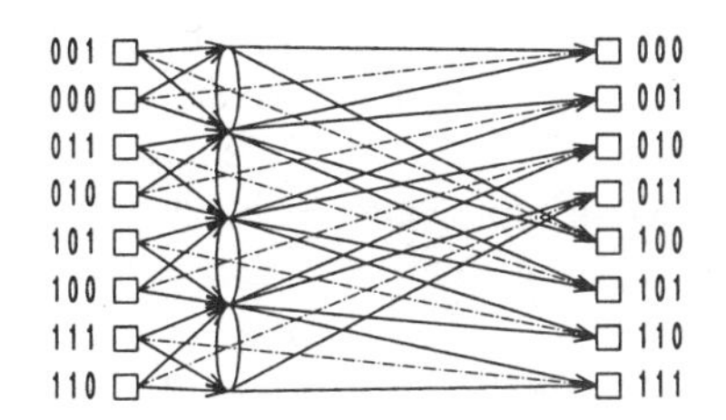

Fig. 4. Proposed new optical interconnections.

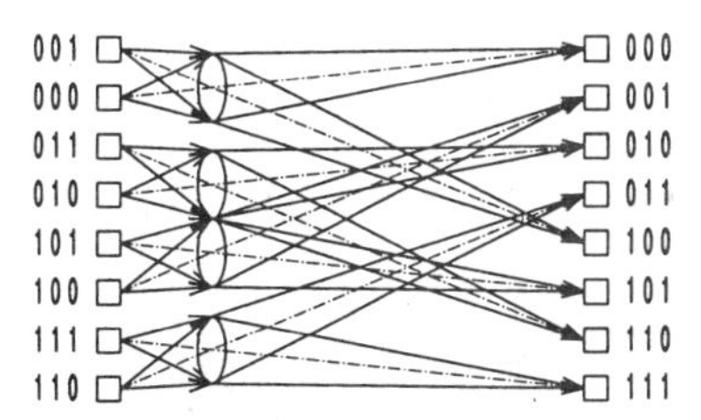

Fig. 5. Butterfly-permutation using the proposed interconnection.

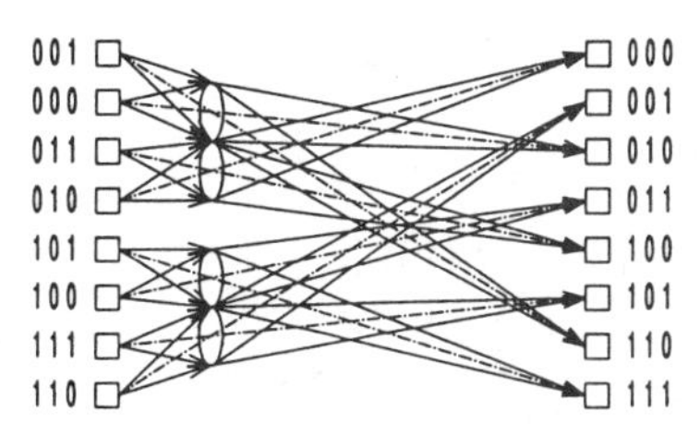

Fig. 6. Crossover interconnection using the proposed interconnection.

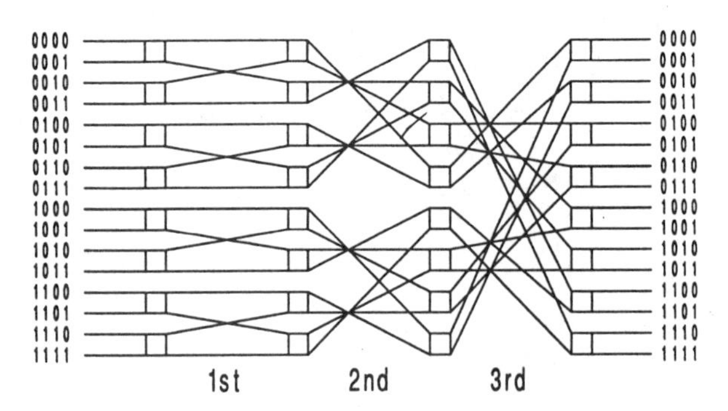

Fig. 7. New multistage interconnection network with self-routing function.

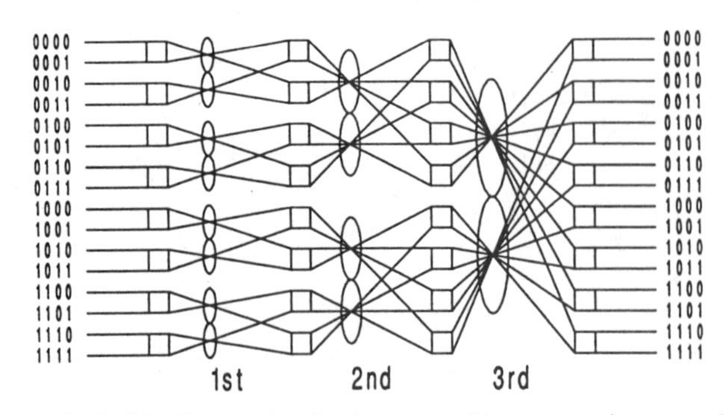

Fig. 8. Optical implementation for the proposed interconnection network.

III. New Self-Routing Network

Fig. 7 shows a new self-routing network design which is more suitable for use with optical pathways. Squares between interconnections in Fig. 7 show 4 × 4 switches. This network is suitable for use with lens arrays because it provides low aberration with high light-power efficiency. It consists of 2^k groups of size 2-cross interconnections where k is a stage number. Fig. 8 shows its optical implementation. When a total stage number is even, e.g., the network in Fig. 7, its odd stages consist of cross interconnections and its even stages consist of normal interconnections. When the number is odd, its odd stages consist of normal interconnections and its odd stages consist of cross interconnections. This network has the same functions as do such conventional self-routing networks as banyan, baseline, and omega. Furthermore, it is better suited than other networks for use with lens arrays.

Lens positions (x_i, y_i, z) are obtained as described in the previous section.

$$x_i = \begin{cases} \left\{\left[\dfrac{i + \text{step}\,(i)}{2}\right] a + \text{sgn}\,(i) \cdot \dfrac{2a+1}{6}\right\} p & (i\text{: even}) \\ \left\{\left[\dfrac{i + \text{step}\,(i)}{2}\right] a - \text{sgn}\,(i) \cdot \dfrac{2a+1}{6}\right\} p & (i\text{: odd}), \end{cases} \tag{8}$$

$$y_j = \begin{cases} \left\{\left[\dfrac{j + \text{step}\,(j)}{2}\right] a + \text{sgn}\,(j) \cdot \dfrac{2a+1}{6}\right\} p & (j\text{: even}) \\ \left\{\left[\dfrac{j + \text{step}\,(j)}{2}\right] a - \text{sgn}\,(j) \cdot \dfrac{2a+1}{6}\right\} p & (j\text{: odd}), \end{cases} \tag{9}$$

$$z = \frac{3}{2} f \tag{10}$$

where $i, j = \pm 1, \pm 2, \cdots, a = 2^k$, [ ] is Gauss' notation and

$$\text{sgn}\,(x) = \begin{cases} 1 & (x > 0) \\ -1 & (x < 0) \end{cases} \tag{11}$$

$$\text{step}\,(x) = \begin{cases} 1 & (x \geq 0) \\ 0 & (x < 0). \end{cases} \tag{12}$$

IV. Optical Performance for Self-Routing Networks

A. Aberrations

Aberrations are commonly proportional to a lens' incident beam axis heights. Ideally, optical systems should have low beam axis heights. Fig. 9 shows the ratios of maximum beam axis heights to lens radii for the networks described above. The ratio for my proposed network is smaller than those for the omega and crossover networks, and it has lower aberration.

B. Light-Power Efficiency

Image size is proportional to the square of the lateral magnification. When an image is too large, part of the output beam which produces it may be incident to neighboring detectors, while such spreading of the image may also produce the further undesirable result of insufficient light signal power incident to the desired detector. This spreading of the image may be solved by lessening lens magnification size. Fig. 10 compares light-power efficiency for various networks. The lateral magnification for both the proposed network and the omega network is always sufficiently small to prevent light beams from being incident to neighboring detectors. Light-power efficiency does not depend on the number of interconnection paths for these two networks, while for the other networks, S/N ratios decrease to less than 1 at more than 1000 (32 × 32) interconnections.

C. Imaging Performance

When the lenses used are of the same numerical aperture size, the total size of the optical system will be roughly proportional to their lateral magnification size. Fig. 11 gives the optical system sizes and lens radii for the various networks. For 1000 interconnections, the distance between the input and the output plane in the proposed network and in the omega networks is about 1 mm, while in the other networks, it is more than 100 mm.

D. Total Performance

Both the proposed and the omega networks offer the possibility of high light-power efficiency and high density

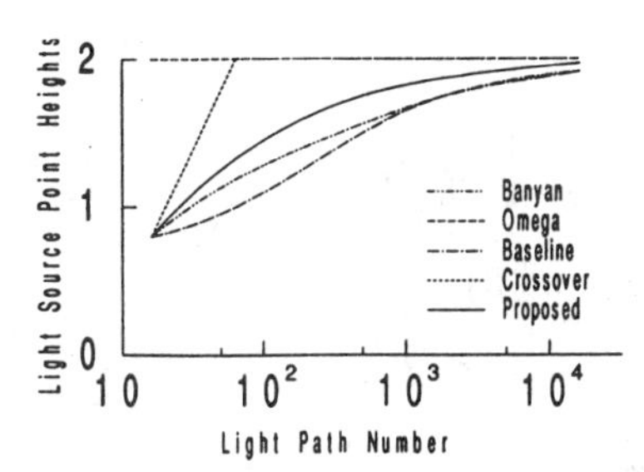

Fig. 9. Ratio of maximum beam axis heights to lens radii for the various networks.

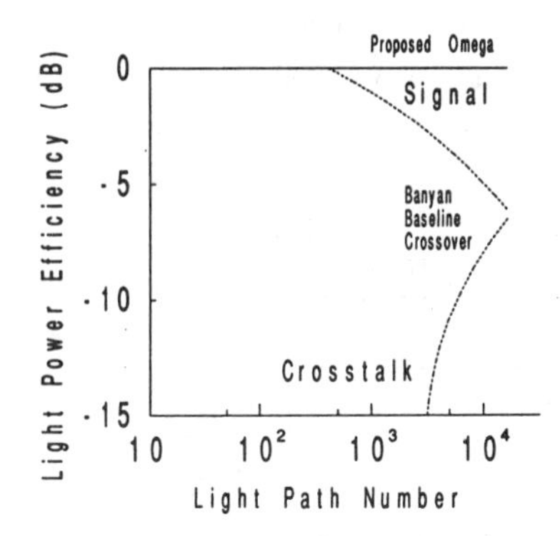

Fig. 10. Light-power efficiency for the various networks.

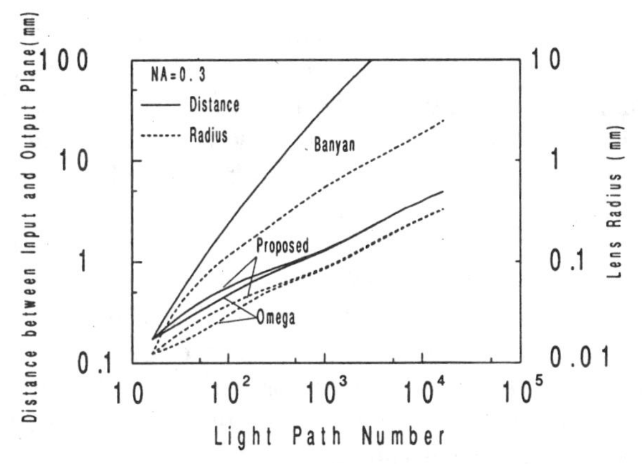

Fig. 11. Optical system sizes and lens radii for the various networks.

of interconnections. Since the omega network has a higher aberration level and one additional interconnection stage, the proposed network appears to be the more appropriate approach.

V. Experimental Results

A. Ray-Trace Estimations

In order to estimate the light path number limit for this network, ray tracings were carried out on lenses which were, roughly speaking, biconvex spherical lenses. Under the conditions of an S/N ratio greater than 10 dB and a system size smaller than 10 mm, more than 16 000 (128 $\times$ 128) processors or computer memories, ten times the number in previous equivalent networks (banyan, base-

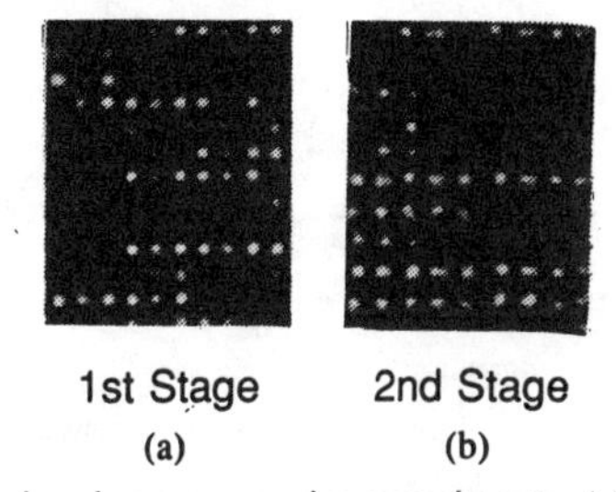

Fig. 12. Planar micro lens arrays using experiments. (a) 1st stage. (b) 2nd stage.

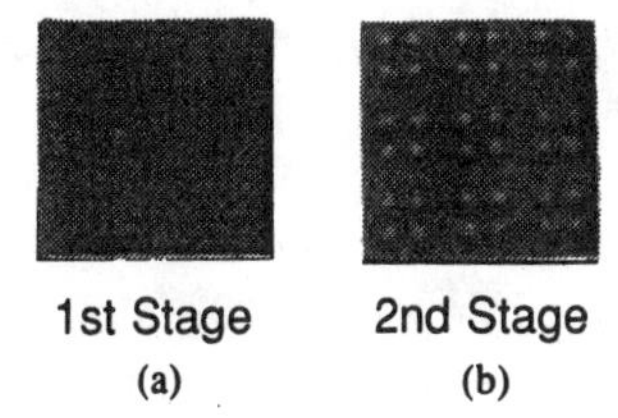

Fig. 13. Imaging experiment results. (a) 1st stage. (b) 2nd stage.

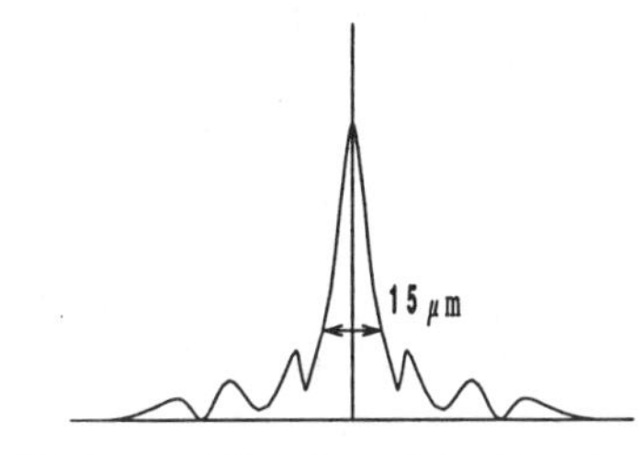

Fig. 14. Focused light beam intensity using a PML.

line, etc.) were successfully connected to one another. If aspherical lenses were to be used, aberrations might be further lessened and the number of light paths might be further increased.

B. Optical Experiments

For further confirmation of feasibility, I conducted imaging experiments using PML arrays. PMLs have a refractive-index distribution and can converge laser beams to near their diffraction limits. Fig. 12 shows intensity distribution for a focused laser beam using a PML. It has almost ideal convergence characteristics and is suitable for interconnection purposes. In my experiments, in which 31.25-μm pitch light sources and photodetectors were assumed, two kinds of PML arrays were constructed to produce an 8 $\times$ 8 network. Fig. 13 shows 1st stage PML arrays, whose average pitch is 62.5 μm, and 2nd stage PML arrays, whose average pitch is 125 μm. An input mask with 10-μm apertures arranged at a 31.25-μm pitch were illuminated by incoherent diffused light beams, and it is imaged using produced lenses as shown in Fig. 14. The desired optical light handling pathways were successfully produced, however, ghost spots and beam intensity variability occurred. Because, light beams can pass through the part out of lens in the array, and the light sources used in these experiments can not be considered

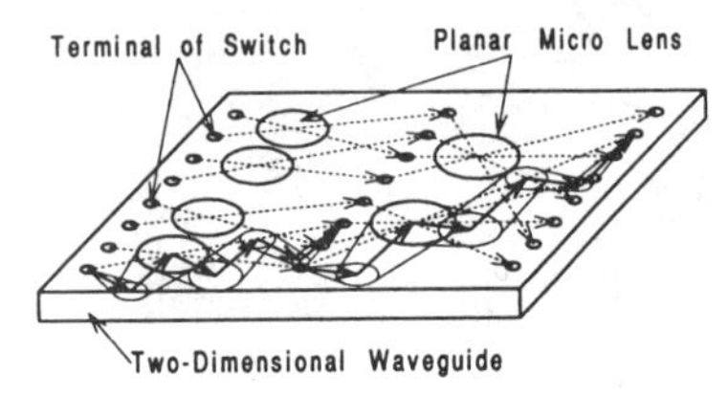

Fig. 15. New optical interconnection using waveguides.

as point sources, undesired light beams were present. Optical interconnections with low aberration and high light-power efficiency may be performed with a surface-emitting laser diode array.

VI. Implementations of Waveguide Techniques

Free-space techniques are very suitable for board-to-board interconnections because signal paths from processors can be obtained from the center of a board, and high density and small signal skew interconnections may be achieved. Two-dimensional waveguides may be required, however, for self-routing networks to perform in-board interconnections, and some approaches have already been proposed for achieving free-space optics using waveguide techniques [12].

Fig. 15 shows a new waveguide interconnection system for the self-routing networks proposed here. The PML arrays are also used in this system. They are also used for lateral focusing [13]. Beams from light sources in switching terminals are made incident to a waveguide through the use of grating couplers or prisms. The light beams whose incident angles are greater than the waveguide's critical angle will move through it. The light beams are reflected along within incident to lenses fabricated into it, and are focused onto photodetectors in subsequent stage switching terminals. The relationship between light sources, lenses and photodetectors will be obtained in the same way as described in the previous section. The self-routing networks described the previous section may be also produced by connecting light sources, lenses and photodetectors in series with 2 × 2 switches. Such systems are functionally equivalent to systems using the free-space techniques. Si or GaAs plates may be used for the 2-D waveguides, because infrared light beams can propagate in these materials [14]. If such a waveguide were to be realized, both waveguides and electric circuits would be made by the same materials. It may be easier to integrate with them.

VII. Optical Interconnections for Parallel Processing

Fig. 16 shows a processing module using such optical interconnections in massively parallel processing systems. Optical interconnections in these systems include various techniques. Table I shows hierarchy levels for optical interconnections in parallel processing systems. Interconnections between racks and between modules are performed almost entirely with optical fiber, while inter-

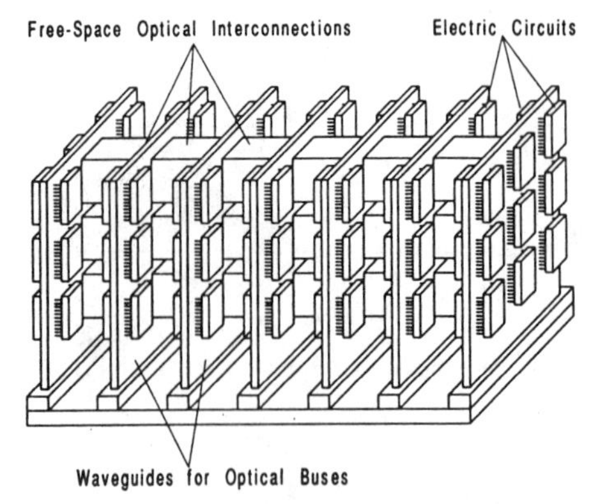

Fig. 16. Massively parallel processing system using the various optical interconnections.

TABLE I
Interconnections for Massively Parallel Processing

Interconnection Level	Techniques
Racks	Optical Fiber
Modules	Optical Fiber/Free-Space
Boards	Free-Space
Chips	2-D Waveguide

connections between boards and between chips may be performed with waveguide techniques [14] and with the free-space techniques proposed here. Signal paths are obtained from the center of the board in free-space techniques, while all the signals are gathered in connectors at the edge of the board, and signal skew may be brought between processors and connectors in waveguide techniques. Furthermore, free-space techniques provide many signal paths, while the number of signal paths are limited by the size of connectors resembling electrical wiring in waveguide techniques. Then, free-space techniques are suitable for interconnections between boards, while waveguide techniques may be suitable for interconnections between chips on boards.

Optical devices using free-space techniques are difficult to align. This problem may be solved by integrating a PML array with a surface-emitting laser diode array and a photodetector array, as shown in Fig. 17. In such optical chips, these devices are required to be arranged with accuracy in Table II. These tolerance limits may be easily obtained using latest technology.

VIII. Summary

New optical interconnections suitable for use with lens arrays have been presented, and a new multistage network with a self-routing function has been described. Their feasibility has been successfully demonstrated through the use of PML arrays. When a PML array is integrated chip-like with a surface-emitting laser diode array and a photodetector array, electronic devices may be densely interconnected through optical light handling pathways, and massively parallel processing systems using such optical interconnections may be achieved.

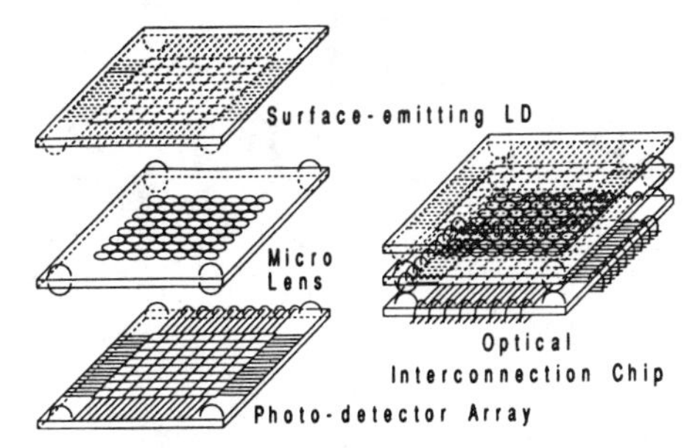

Fig. 17. Example of free-space optical interconnection chip.

TABLE II
TOLERANCES FOR ARRANGEMENT
(In the case of light power loss = 3 dB, light source size = 10 $\mu m\phi$, light source pitch = 125 μm, detector size = 100 $\mu m\phi$ and PML focal length = 500 μm.)

Factors	Tolerance Limits
Lateral Distance	50 μm
Longitudinal Distance	310 μm
Angle	33 mrad

ACKNOWLEDGMENT

The author gratefully thanks M. Oikawa of Nippon Sheet Glass Co., Ltd. for supplying the planar micro lens arrays used in these experiments. He also thanks K. Iinuma, T. Temma, J. Tajima, and Mr. M. Mizoguchi, for their suggestions and encouragement.

REFERENCES

[1] J. W. Goodman, F. I. Leonberger, S. Y. Kung, and R. A. Athale, "Optical interconnections for VLSI systems," *Proc. IEEE*, vol. 72, no. 7, pp. 850-866, July 1984.
[2] A. A. Sawchuk and I. Glaser, "Geometries for optical implementations of the perfect shuffle," *Proc. SPIE*, vol. 963, pp. 270-282, 1989.
[3] S. Kawai, "Free-space multistage optical interconnection networks using micro lens arrays," in *Photonic Switching II*, K. Tada and H. S. Hinton, Eds. Berlin: Springer-Verlag, 1990, pp. 216-129.
[4] J. Jahns and M. J. Murdocca, "Crossover networks and their optical implementation," *Appl. Opt.*, vol. 27, no. 15, pp. 3155-3160, Aug. 1988.
[5] A. Guha, J. Bristow, C. Sullivan, and A. Husain, "Optical implementation for massively parallel architectures," *Appl. Opt.*, vol. 29, no. 8, pp. 1077-1093, Mar. 1990.
[6] M. Oikawa, E. Okuda, K. Hamanaka, and H. Nemoto, "Integrated planar microlens and its applications," *Proc. SPIE*, vol. 898, pp. 3-11, 1988.
[7] R. W. Hockney and C. R. Jesshope, Parallel computers. Adam Hilger Ltd, Bristol, 1981, pp. 158-178.
[8] D. H. Lawrie, "Access and alignment of data in an array processor systems," *IEEE Trans. Comput.*, vol. C-24, no. 12, pp. 1145-1155, Dec. 1975.
[9] C. Wu and T. Feng, "On a class of multistage interconnection networks," *IEEE Trans. Comput.*, vol. C-29, no. 8, pp. 694-702, Aug. 1981.
[10] D. S. Wise, "Compact layout of Banyan/FFT networks," in *VLSI Systems and Computations*, H. T. Kung, B. Sproull, and G. Steele, Eds. Rockville: Computer Science, 1981, pp. 186-195.
[11] L. R. Goke and G. J. Lipovski, in *Proc. 1st Annu. Symp. Comput. Arch.*, p. 21, 1973.
[12] J. Jahns and S. J. Walker, "Imaging with planar optical systems," *Opt. Comm.*, vol. 76, no. 5, 6, pp. 313-317, May 1990.
[13] D. Intani, A. Akiba, T. Baba, and K. Iga, "Planar microlens relay optics utilizing lateral focusing," in *Tech. Dig. Gradient-Index Optical Systems Topical Meet.* (Monterey, CA), Apr. 1991, pp. 82-85.
[14] S. Kawai and M. Mizoguchi, "Two-dimensional optical buses for parallel processing," in *Tech. Dig. Optical Computing* (Salt Lake City, UT), Mar. 1991, pp. 136-139.

*

Optical Interconnection for Advanced Processor Systems: A Review of the ESPRIT II OLIVES Program

James W. Parker

(Invited Paper)

Abstract—**OLIVES is a three year collaborative project under the European Community's ESPRIT program under which components for and practical demonstrations of optical interconnections for high-performance processors are under development. A review of the major aims of the project and the progress to date is given.**

I. Introduction

THE performance of advanced electronic systems is increasingly limited by constraints imposed by interconnects and this limitation will inevitably become more serious as processor power increases. Optics have already proved their worth over distances of hundreds of meters or more both in telecommunications and more recently in interconnecting computers and computer peripherals. Emerging standards such as FDDI will further increase this penetration and promise vast improvements in intercabinet connectivity in the years to come. These systems are, however, in many ways very similar to telecommunication links in that the terminal equipment (i.e., the optical transceivers) occupy a volume comparable with or greater than a packaged silicon IC and the interconnect topology is, at the physical level, point to point. Future systems will increasingly incorporate optical interconnects within a single cabinet and these interconnects will require new component technologies and optical pathway concepts more appropriate to the shorter distances. These components must permit a much higher density of integration than is required for traditional applications, will in many cases include fanout and must be cost competitive. They must be developed in conjunction with a clear view of the system requirements since the most important applications of optical interconnects over these short distances will not arise from direct substitution of the existing electrical technology with optics, but rather from architectural changes made possible by the different characteristics of the optical technology. This is because the architecture and detailed design of electronic systems have been carefully honed to take advantage of the strong points of electrical interconnects and, where possible, avoid the weak points.

In this paper I shall review some aspects of the progress to date in this area, which has been made within collaborative project optical interconnections for VLSI and electronic systems (OLIVES) which is filling the gap between optical technology and electronic systems. This is a three year project under the European Strategic Programme for Research in Information Technology (ESPRIT). It commenced in January 1989 and combines the complementary skills of four major electronics companies (STC (now part of Northern Telecom), Siemens, Plessey (now GEC-Marconi Materials Technology), and Thomson-CSF), a chemical company (Akzo) and five academic institutions (University College London (UCL), Foundation for Research and Technology, Hellas/RCC, Centro Nacional de Microelectronica (CNM), Interuniversitair Microelectronic Centrum (IMEC), and Eldgenössische Technische Hochschule, Zurich (ETH)). The project is focused on the construction of four demonstrators of optical interconnections at the module, backplane, multichip module and chip levels, and the development of the required optical pathway, hybridization, and optoelectronic component technologies to realize these demonstrators. Significantly, it also includes a task under which the potential performance of the demonstrators are compared with the requirements in systems of commercial interest to the industrial partners. These demonstrator-led activities are supplemented by the theoretical investigation of free-space interconnects between adjacent parallel boards and optical backplane buses, and by efforts to develop the technology for monolithic integration of MQW modulators on CMOS circuitry.

II. Demonstrators

Electrical systems are constructed in a hierarchical fashion and the properties required of the interconnects are different at the different levels. A whole range of electrical interconnect technologies have been developed to deal with this and with the differing requirements of sys-

Manuscript received April 12, 1991; revised May 28, 1991. This work was partially supported by the Commission for the European Communities under project 2289.

The author is with BNR Europe Limited, London Road, Harlow, Essex, England, CM17 9NA.

IEEE Log Number 9102561.

Reprinted with permission from *IEEE Journal of Lightwave Technology,* Vol. 9, No. 12, pp. 1764-1773, December 1991.

tems. Likewise optical interconnections must be tailored to the application, although commercial considerations will dictate that common components will be used wherever possible. Many proposals have been made for interconnects at the backplane, board, and chip levels. Within the OLIVES program, four subsystem demonstrators are under construction as follows:

TABLE I
SUBSYSTEM DEMONSTRATORS

Interconnection	Demonstrator
Module to Module	Multifiber Bus
Backplane	Mastercard
Multichip Module	Waveguide Array
Chip	Chip level clock distribution

These are supplemented by major technology demonstrators of low-power high-density optical interfaces, described in the following section, and of GaAs/Si technology. The rationale for and progress to date in construction of the subsystem demonstrators is now outlined.

A. Multifiber Bus

Fig. 1 shows an optical realization of a conventional electrical time division multiplexed bus. A number of nodes (eight in this case) by 0.5–5 m apart, are connected by ribbon fiber through an array of passive star couplers. At each of the nodes, there are array transmitters and receiver modules in which an array of lasers and corresponding drive circuitry (or receivers) transmits into (receives from) the ribbon fiber. These are shown in the inset to Fig. 1. In a full scale implementation of such a system, 12-fiber ribbon would be used; with one of the fibers allocated for transmission of the common clock signal and a further fiber for parity and/or control information. With a data rate of 3.2 Gb/s per fiber, the aggregate data rate would be 32 Gb/s. If necessary, several such basic building blocks would be used in parallel to achieve the overall data bandwidth required.

One particular example of the application of such a bus is in high-performance mainframe computers, where a relatively small number of powerful processors must be interconnected with a shared memory [1]. By appropriate communication between the processors, the whole can be made to appear to operate as a single machine. A bus interconnection is very attractive in such a case for several reasons. These include the reuse of well understood protocols for dealing with such issues as bus contention and cache coherence, and the greater modularity of design which buses allow which leads to a reduction in inventory to the benefit of both supplier and customer. However, the overall throughput of any shared memory machine is a strong function of the bandwidth of the processor/memory connection. The most ambitious conventional electrical bus is the proposed Futurebus+, which is considered to represent the ultimate performance achievable in electrical technology. Initially it will be 32 bits wide at 25 Mb/s per line, rising to 256 bits wide at 125 Mb/s per line. This final figure corresponds to a total throughput of 32 Gb/s, equal to just one multifiber bus building block. Furthermore, the maximum end to end separation of Futurebus+ is limited to about 0.5 m. These figures are inadequate in the machines of highest performance and only an optical solution is possible.

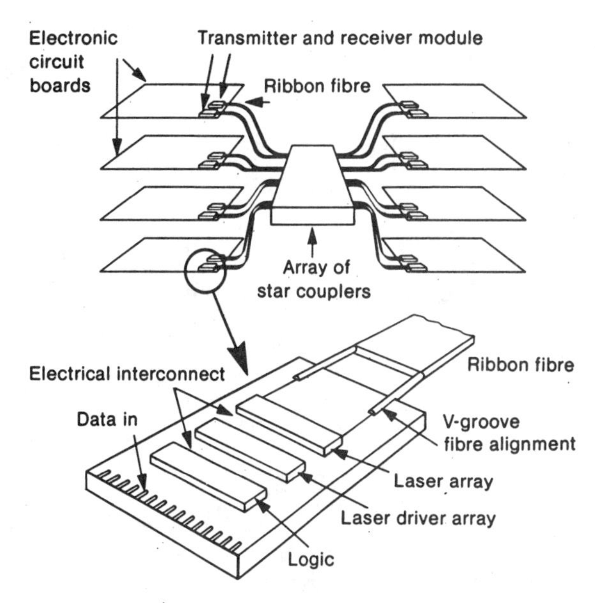

Fig. 1. Schematic of multifiber bus.

A demonstration of the multifiber bus is under construction [2]. The ribbon fiber transmitters and receivers are based on silicon opto-hybrid technology. This uses silicon *v* grooves to align the fibers and provide reflective structures, solder bump self alignment of the laser and receiver arrays, and a high density interconnect on the silicon substrate to make electrical connections to the hybridized driver chips and passive components. The mask set for the array transmitter module, which will have six operational channels at 1 Gb/s per channel, has been designed and fabricated. This incorporates provision for the mounting of six 1 Gb/s laser driver chips, associated decoupling and load resistors, an 8-element laser array and an 8 way ribbon fiber together with provision for monitoring of the back faces of the individual elements of the laser array. Fig. 2 shows a plot of the mask layout for this submount. Fig. 3 shows the laser array with segmented electrodes for solder bumping, and the backface detector photodiode array. The receiver will be based on the device described in Section III below.

Measurements of the modal noise characteristics of such a system have been made. The choice between multimode and single-mode fiber is an important issue, since the relaxed alignment tolerances of multimode fiber improves the power budget and reduces cost. Extensive studies of this subject have been made for long haul applications, but little work has been done for such short distance interconnects. Two components of modal noise can, in general, be distinguished, a slow component originating from

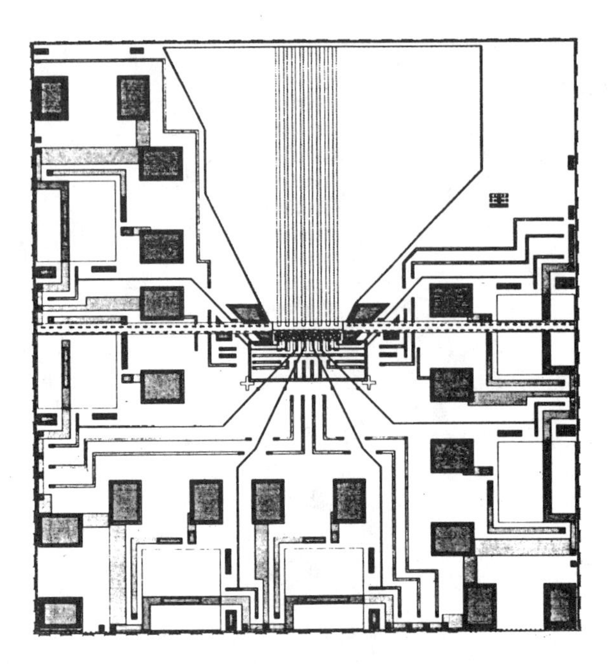

Fig. 2. Multifiber bus transmitter submount.

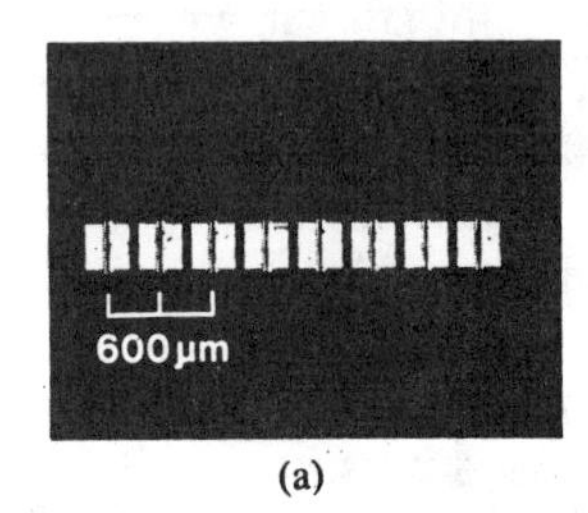

(a)

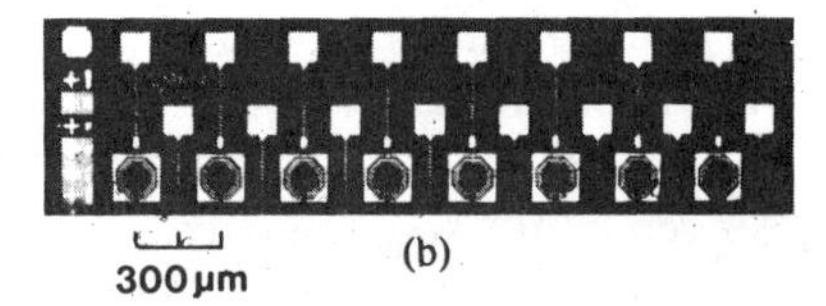

(b)

Fig. 3. (a) Eight element laser array. (b) Backface detector photodiode array.

environmental influences and a fast component due to laser mode hopping. With a suitable DFB laser the latter is entirely absent. Detailed measurements of the environmentally induced drift in received power have been made with DFB lasers and both 50/125 and 62.5/125 fiber. With a commercial 8 × 8 star coupler as the mode-selective loss element (corresponding to an estimated mode-selective loss of 1.7 dB), a penalty due to the "slow" component of modal noise of 0.8 dB was measured. This figure excludes any penalty which would result from the necessary gain equalization at the receiver; however, this component of signal level drift is relatively small compared with other sources. This experiment indicates that multimode fiber may be considered for the application.

B. The Mastercard Demonstrator

Fig. 4 shows the concept of the mastercard demonstrator [3] for backplane interconnects. The 'mastercard' uses total internal reflection in a glass slab to guide collimated

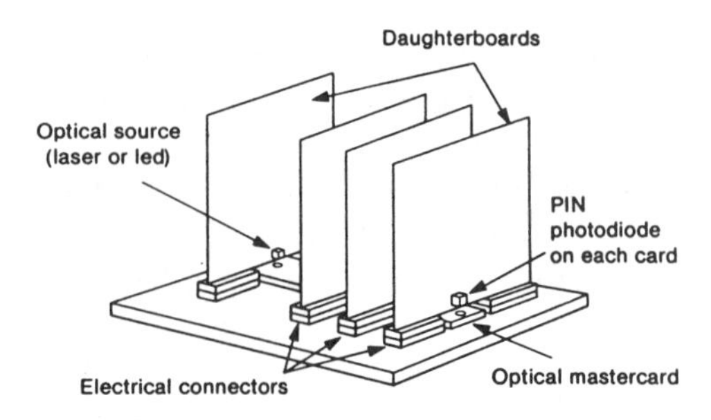

Fig. 4. Schematic of mastercard demonstrator.

beams, from a source located on one daughterboard, to several receivers located on the other daughterboards. The mastercard slab is provided with holograms which perform beam directing splitting and possibly focusing functions.

In real time processors, particularly in avionic systems, a key issue is the volume occupied by the backplane interconnects. For example, a typical system of this nature would operate with a clock frequency of 100 MHz or more, at which rate coaxial cables would be used to distribute the common system clock to all boards. Assuming that the coaxial cables have a minimum bend radius of 30 mm, and a diameter of 2 mm, a volume of approximately 120 cm^3 [3] is required for the clock distribution tree in an 8 board rack. The corresponding mastercard occupies only 5 cm^3. Similarly, where operation at a high data rate is required, an increased interconnect density may be achievable.

The first approach to the construction of the mastercard is illustrated in Fig. 5. This is a one-to-one demonstration in which analogue holograms provide both beam deflection and focusing. This assembly was constructed and operated at up to 350 MHz, with a measured bit error rate at 90 Mb/s of 10^{-11}.

This simple arrangement, in which the collimation is performed by the hologram, suffers from a number of problems, however. An alignment tolerance of ± 10 μm between the daughterboard and mastercard was required. While this could be repeatedly achieved it is not compatible with standard tolerances for electrical boards. Furthermore, the loss for this point to point interconnect was 18 dB, which would be unacceptably high if compounded by fanout. Finally, the holographic elements were realized in photoresist which exhibited some instability.

These problems have been overcome in the second realization, which is shown in Fig. 6. Here, a microlens placed in front of the source (CD Laser THY ML1346) is used to collimate the beam and a similar lens is used in front of each receiver. The laser microlens has a diameter of 3.9 mm and a focal length of 2.2 mm. The microlenses in front of the receiver p-i-n have a diameter of 3.9 mm and focal length of 2 mm. Computer generated holograms etched directly into the borosilicate glass mastercard are used. These perform only the deflection and beam splitting functions.

Table II shows the calculated and measured diffraction efficiencies of the emitting and receiving holograms.

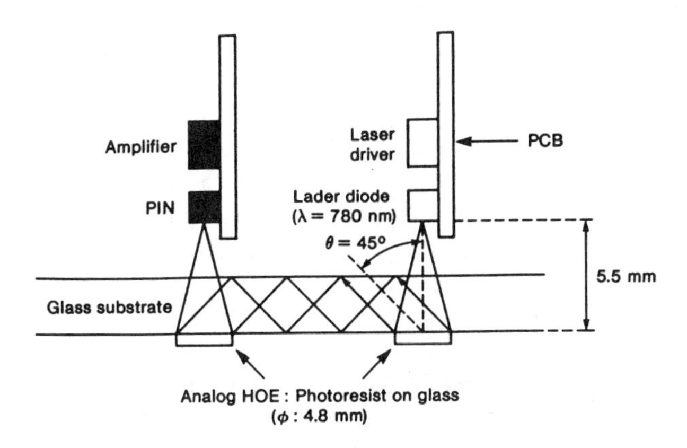

Fig. 5. First approach to mastercard.

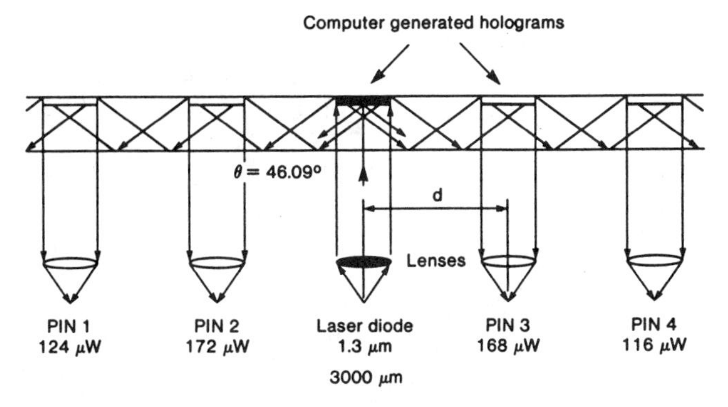

Fig. 6. Second approach to mastercard.

TABLE II
CALCULATED AND MEASURED DIFFRACTION EFFICIENCIES

	Emitting CGH		Receiving CGH	
Order	Calculated	Measured	Calculated	Measured
−1	49%	40%	—	—
0	2%	14%	79%	71%
1	49%	40%	18%	21%
2	—	—	2.5%	7%

The measured power budget of this assembly is shown in the Fig. 6. The assembly has a fanout of 4. The excess loss is 7 dB and the nonuniformity between the 'receiving' elements is 0.9 dB. The demonstration is intended for clock distribution; the measured optical clock skew was 100 ps. The positioning tolerances for the laser diode and p-i-n diode were ±200 and ±150 μm, respectively.

In a further development it is planned to implement multiple board to board connections to provide full interconnectivity between the boards in the rack.

C. The Waveguide Array Demonstrator

As clock frequencies of computer systems increase, electrical crosstalk between parallel tracks will become increasingly problematic, even at the multichip module level, and the reduction of this crosstalk is only possible with some sacrifice of line density. Optical waveguides offer the potential to overcome this limitation. Waveguides separated by only 10 μm or so show negligible levels of crosstalk even if run in parallel for many tens of centimeters. Fig. 7 shows a schematic of a silicon multichip module with an overlay providing parallel optical data interconnects for the longer distance tracks (shorter interconnections will still be handled electrically), to overcome these crosstalk limitations.

A demonstration of this 'waveguide array' interconnect is under construction [4]. This will comprise an 8-channel parallel link using a single-mode waveguide array, laser diode arrays and photo-receiver arrays; a cross section of the demonstrator is shown in the lower part of Fig. 7. The pitch of the waveguides, which are fabricated by the flame hydrolysis technique, is 125 μm in this first realization. Laser diode arrays operating at 850 nm will be used. These and the corresponding receiver arrays will be mounted on etched silicon submounts; a diagram of the photodiode submount appears in Fig. 8.

In the application for this demonstrator the data interconnects are latency-critical, that is to say that it is vital to minimize the path delay. This is made up, in the optical case, of the gate delays in the laser driver and receiver IC's, the laser turn-on time which is a function of the driving characteristics and the propagation time. Fig. 9 show a comparison of the electrical and optical latency. These are plotted for signals at 100 Mb/s ('Low bit rate') and 1 Gb/s ('High bit rate'), for a variety of driver technologies viz: ECL, GaAs, and silicon bipolar. Certain assumptions have been made which are described in detail in [4]. For the optical interconnections, these assumptions are that the laser diodes have a low threshold compared to the drive current, so that the turn on time is small (around 20 ps), that the delay through the optical receiver is 0.3 ns. For the electrical interconnects, additional buffer stages are inserted at intervals to restore the signal which is attenuated due to the small track cross section. This accounts for the steps in the curves for high bit rate. The figure predicts that the latency of the guided-wave optical interconnect drops below that of the electrical alternative at distances greater than 2–5 cm for 1 Gb/s signals, confirming that the multichannel waveguide optical interconnect is attractive. It should be noted that this calculation assumes that there is no fanout in the interconnection; with fanout, the advantages of the guided wave optics will become evident at smaller distances.

D. The Chip Level Clock Distribution Demonstrater

Within a single chip, the delay (typically 0.5 ns or more with state of the art technology) caused by the conversion of electrical signals to optical signals and back again makes the use of optical interconnection for data unattractive in most instances. Clock signals, however, are distinguished by a requirement to minimise *differential* delay. The superior fanout capability of optics allows electrical buffer stages to be eliminated and path length differences minimized, the main sources of chip-level skew.

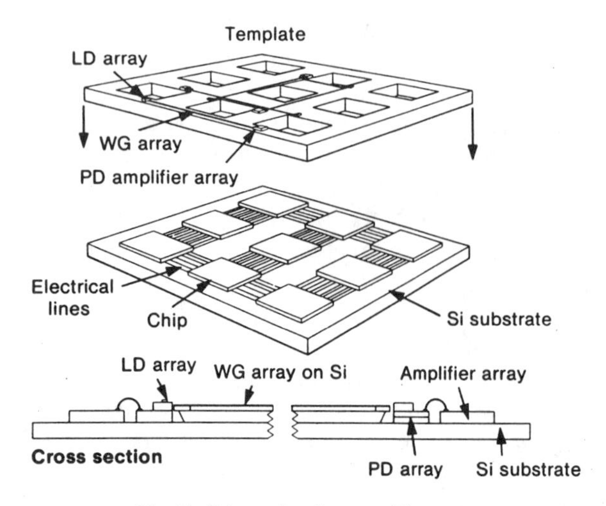

Fig. 7. Schematic of waveguide array.

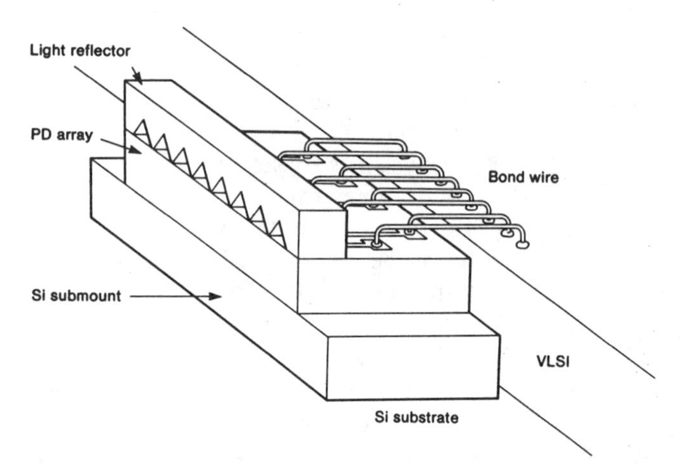

Fig. 8. Schematic of photodiode submount.

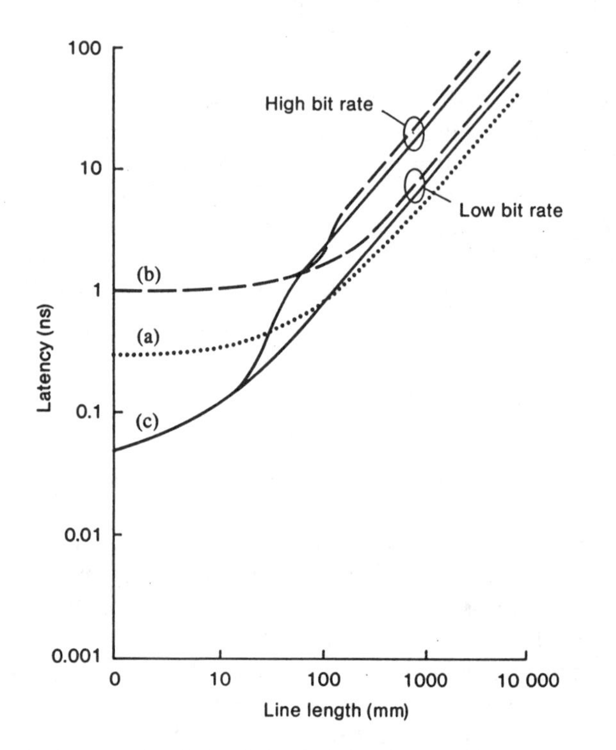

Fig. 9. Latency of optical and electrical signals: (a) guided wave optical, (b) electrical with ECL drivers, (c) electrical with high-speed bipolar devices.

The fanout achievable optically is bounded at the top by power consumption in the receivers [5] and at the lower end by the minimum required to give sufficient reduction in skew compared with the electrical alternative. Fig. 10 shows the optical power incident on each photodetector (assumed to be 100 μm diameter), and power consumption calculated for an optical distribution network based on a holographic element as the power splitter, with an assumed diffraction efficiency of 40%. In this calculation the clock frequency is 1 GHz and ECL is assumed. The power is dominated by the consumption of the receiver amplifier, which is assumed to be 10 mW per stage with a gain of 4 per stage. The figure shows that, for very large fanouts, the power consumption in the receiver chips becomes unacceptable. A practical limit might be 5 W at which power level the maximum fanout is 100.

Calculations of the improvement in clock skew in a typical chip give an estimated reduction from 2 ns to 350 ps with a fanout of as low as 17 which is easily achievable, corresponding to an increase in maximum speed from 50 MHz to nearly 300 MHz.

A schematic of the first experimental setup of this chip level clock distribution demonstrator is shown in Fig. 11. A laser diode adjacent to the chip is reflected onto a computer generated holographic element which is on the bottom side of the cover plate. The dimensions of the (dummy) VLSI chip are 10 mm × 10 mm; four photodiodes are mounted on this in a symmetrical pattern (this is not a layout restriction as appropriate design of the hologram can allow the PD's to be located anywhere).

Two types of computer generated holographic element have been considered for this demonstrator, a multiplexed element (Fig. 12(a)) and a faceted element (Fig. 12(b)). Early work was based on the multiplexed design in which every part of the HOE contains the interference pattern from all waves (i.e., the incoming LD beam and four fanout beams). Later work has been based on the faceted design in which the element is divided into four sections, each of which corresponds to only a single LD to PD connection. This has two principal advantages, namely, that the amount of data storage required during writing of the pattern is very much reduced (10 Kbyte compared with about 1 Gbyte) and the feature size for a given routing capability is larger. The penalty is that the beams incident on the photodetectors are less well focused, the calculated beam size for the multiplexed design is 4 μm × 8 μm as opposed to 8 μm × 16 μm for the faceted design.

A four-facet CGH as illustrated in Fig. 12(b) has been realised as a binary relief grating etched in silicon, patterned using a 5× reticle. An SF_6 plasma was used for the etching and the resulting relief structure metallized with Ti/Au to improve the reflectivity. The measured diffraction efficiency of 39% is close to the theoretical maximum of 40.5%, indicating near-perfect fabrication of the HOE. Although this implies that 60% of the light is diffracted into the zeroth order, this light is spread over a wide area on the chip so that the resulting light intensity is about 40 dB below that in the spots themselves, giving

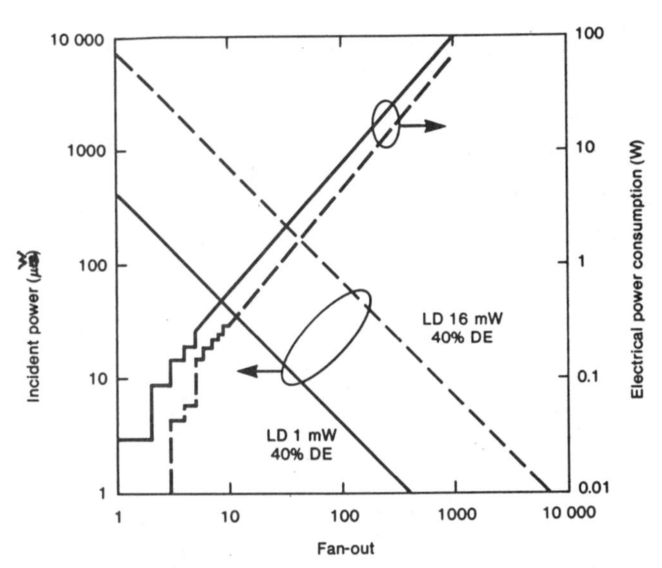

Fig. 10. Electrical power consumption (right ordinate) and power incident on photodiode (left ordinate) for holographic clock distribution as a function of fanout.

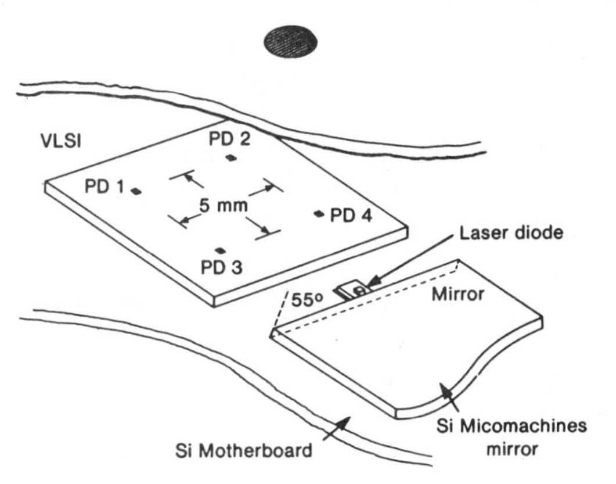

Fig. 11. Schematic of chip level clock distribution experimental arrangement.

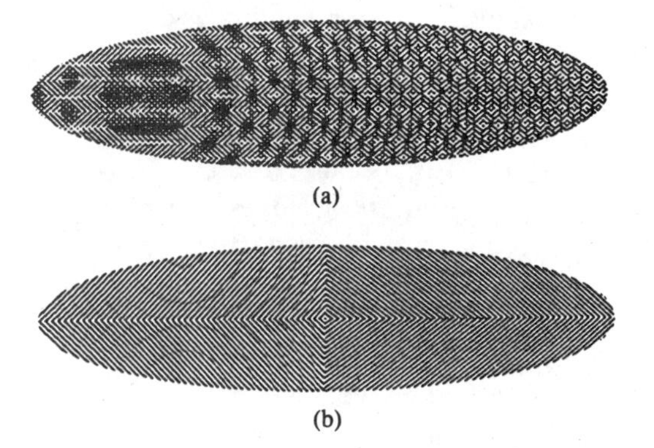

(a)

(b)

Fig. 12. Holographic elements for chip-level clock distribution (a) multiplexed, (b) facetted.

adequate contrast. Assembly of the complete demonstrator has commenced.

III. Technology Development

The requirement for new components and assembly techniques optimised for the needs of optical interconnects was alluded to above. These include optical pathways (waveguides and holographic elements), optoelectronic interfaces and component mounting and hybridization.

Some of the specific optical pathway elements which have been developed have been described above. In this Section I consider the optoelectronic interfaces and the techniques for component mounting and hybridization.

A. Optoelectronic Interfaces

Optoelectronic interfaces are key to the success of optical interconnects. Low cost arrays of devices occupying a small area and with low power consumption are necessary. The following optoelectronic interfaces have been designed and fabricated within the program:

8-element Laser diode arrays at 0.85 μm and 1.5 μm,
8- and 64-element MQW modulator arrays,
8- and 64-element photodiode arrays,
8-channel receiver arrays.

Most of the above components have been designed for flip chip mounting.

Modulator arrays are discussed elsewhere in this issue [6] and laser arrays have been considered in outline above. Here I concentrate on a custom receiver array which has been designed and fabricated within the OLIVES project and will be used in the multifiber bus demonstrator, in the waveguide array demonstrator and in demonstrations of low power chip-to-chip interconnects based on modulator arrays. This receiver chip is based on the CD1014 gate array [7] fabricated in a 1-μm silicon bipolar process. Two versions of the chip have been fabricated, one optimized for solder bump mounting of a photodiode array and the second for wire bonding.

Fig. 13 shows the building blocks of the chip. Each of the eight receiver channels has four stages, a transimpedance preamplifier, a differential amplifier, a comparator and an output buffer. One additional preamplifier common to all eight stages is used as a temperature compensated reference for the differential amplifiers. The output is fully ECL compatible and the necessary reference voltages are provided by on-chip bias generators. A single -4.5 V source is required. A full description of the optimization of this chip is given elsewhere [8].

A plot of the mask layout for the solder bump version of the chip is given in Fig. 14. Samples of the chip have been tested. The first results have been obtained by injecting light from an 800-nm laser directly into the front transistor of the preamplifier. Table III summarizes the results obtained during the measurements and simulations: All the figures are for the entire array of eight receivers including the output buffers. No measurable interchannel crosstalk has been detected. The area of a single channel is equivalent to that of 4 ECL gates, the whole device is about 2.3-mm square.

The simulated response to a 1 ns pulse is shown in Fig. 15 for various input optical power levels; the measured

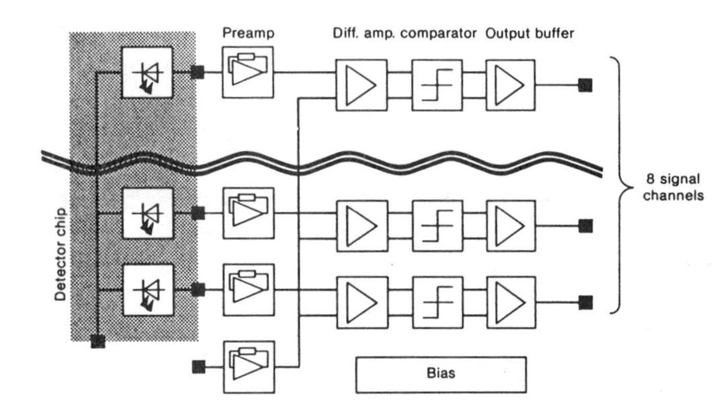

Fig. 13. Functional schematic of receiver array.

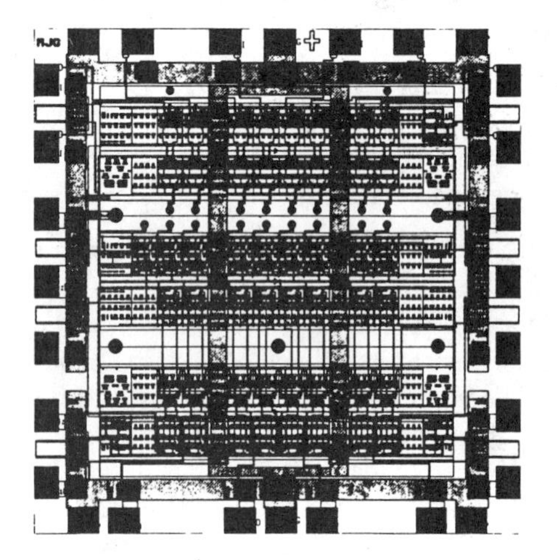

Fig. 14. Receiver array mask plot.

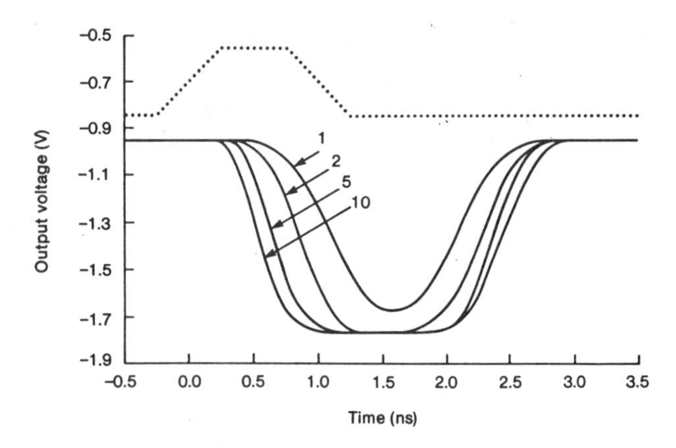

Fig. 15. Simulated response of receiver to 1 ns pulse for various degrees of input light level overdrive.

TABLE III
MEASURED AND SIMULATED RECEIVER PERFORMANCE

Parameter	Measured Value	Simulation
Supply current	62 mA	82 mA
Power consumption	280 mW	368 mW
Minimum input level	7 μA	2.5 μA
Delay for minimum input	1.4 ns	1.1 ns
Delay change for 5 fold overdrive	0.4 ns	0.4 ns

results are in good agreement with the simulations [8]. The pulse broadening visible when the drive level is greater than the minimum value is attributable to the dc coupling used throughout the chip, which causes the receiver to flip into the 'on' state earlier in a low to high transition for larger input levels. This translates into a limit on the dynamic range (of about 3 dB) which can be achieved at the design bit rate of 1 Gb/s or alternatively a limitation in bit rate if a higher dynamic range is required.

B. Component Hybridization

The mounting and positioning of optoelectronic components is frequently done using active techniques, i.e., the chip is powered up or illuminated and the fiber, lens or hologram is positioned while monitoring the output power in an appropriate plane. For optical interconnects at the cabinet level or below, this will not be acceptable on cost grounds and dead reckoning must be used. However the tolerances required are frequently of the order of one micron or even less, and this therefore presents very severe problems. Two techniques have been extensively developed and enhanced for use in the demonstrators described above which offer the potential to overcome many of the difficulties. These are silicon *v*-groove micro-machining and flip-chip self-aligned solder bump mounting.

Anisotropic etching of silicon is a well established technique for forming *v*-grooves, etched pits and other structures. It has found application in the production of a wide range of devices, including pressure sensors and accelerometers. More recently, it has been used in integrated optics for the precision alignment of fiber waveguides to other optical devices.

In order to etch silicon *v*-grooves, an SiO_2 or Si_3N_4 layer is grown on the surface and then patterned, using photolithography, to form an etch mask. An etchant such as KOH, diluted in IPA and water, is then used to attack the ⟨100⟩ surfaces preferentially. The widths and depths of *v*-grooves formed in this way can be controlled to less than a micron. Also, because the etch mask is defined photolithographically, the grooves can be positioned to submicron accuracies with respect to other features; for example, the bond pads of optoelectronic components. Fig. 16 shows an example of the application of this technique to the positioning of an array of fibers for the multifiber bus described above. The technique is used extensively also in the waveguide array and edge-chip to area demonstrators.

Flip-chip (or solder bump) bonding was first developed in the late 1960's to overcome the limitations of conventional wire bonding techniques [9]. Traditionally, the inputs and outputs (I/O's) from a chip have been placed around its perimeter, and aluminium or gold bond wires used to connect the chip to interconnects on the underlying substrate. As device complexity has increased, the available perimeter for bonding has proved insufficient for the number of I/O's required. Flip-chip bonding has sev-

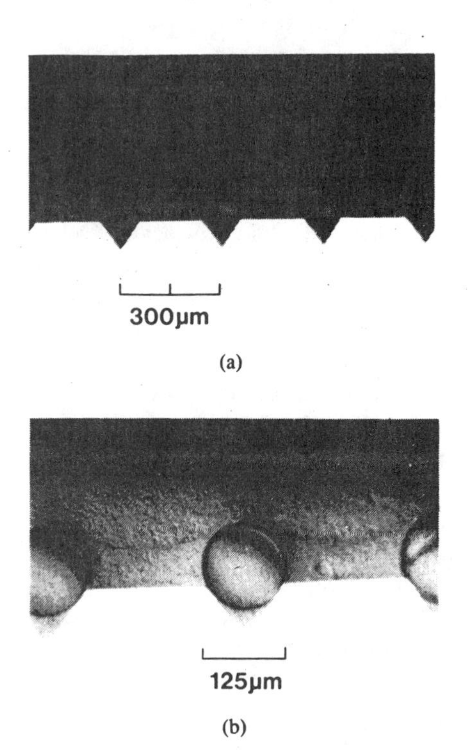

Fig. 16. Fiber array in silicon v-grooves (a) array, (b) close-up of single fiber.

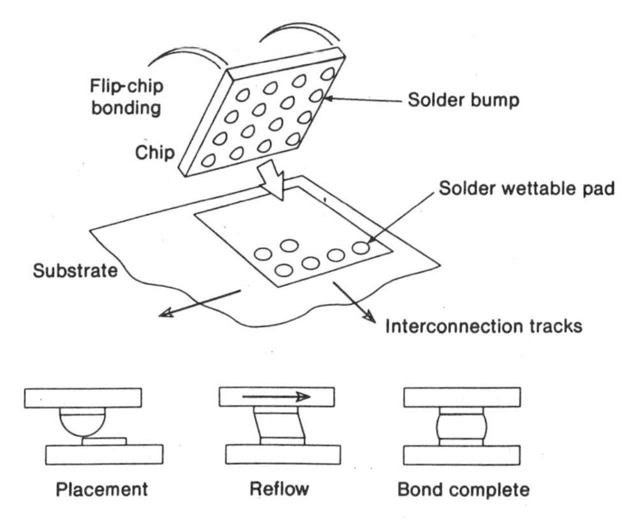

Fig. 17. Flip-chip solder bonding—basic concept.

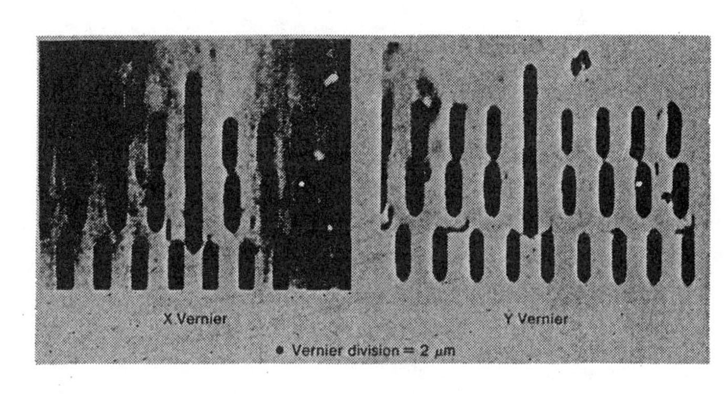

Fig. 18. Solder verniers.

eral advantages over conventional techniques. Firstly, a large number of bonds can be formed simultaneously with very high yield: the simultaneous formation of 12 000 bonds has been demonstrated. Secondly, area-to-area bonds are possible, and so the restriction that perimeter bonding places on device geometry and number of I/O's is lifted. Thirdly, the resistance, capacitance and inductance of flip-chip bonds are lower than obtained by conventional wire bond techniques. Lead capacitances of less than 0.001 pf and resistances of 1.2 mΩ per bond (compared with 0.025 pF and 122 mΩ obtained wire bonds) have been reported. This fact makes flip-chip bonding an attractive technique for interconnecting very high frequency components.

These electrical properties are also very attractive for optical interconnect applications, but more important still is the self aligning property. The application of solder bump bonding to optoelectronic components was first reported in 1989 [10] and the technique has since been extensively developed.

To understand the mechanism behind self alignment, the solder bonding process must be considered. The bonding process is illustrated in Fig. 17. Matching solder wettable pads (of, for example, Cr/Cu/Au or Ti/Pt/Au) are defined on both chip and substrate. A controlled volume of solder is then deposited on one set of pads using vacuum evaporation or electro-plating, and reflowed to form bumps. The chip and substrate are brought into approximate alignment, and then heated. When the solder melts, wetting occurs and a metallurgical bond is formed between chip and substrate. While still molten, the high surface tension of the solder acts to pull the chip into accurate alignment with the substrate. Typically the solder is 95%Pb–5%Sn which melts at 310°C, 60%Pb–40%Sn, which melts at 180°C, or 20%Au/80%Sn which melts at 280°C.

To assess the accuracy of this technique, solder verniers have been deposited. Examples of these are shown in Fig. 18. These have 2 μm spacing and alignment in both in-plane directions to better than 2 μm (which represents the limit of direct measurement) is routinely achieved. Other measurements based on assessments of the loss at fiber/waveguide interfaces allow the interference to be made that submicron tolerances have been achieved. Direct measurement of the accuracy of the height of bonds has confirmed that, with careful control of the deposition and reflow parameters, submicron vertical positioning accuracies can be reproducibly achieved.

Within the program, this technique is used to position the components of the multifiber bus demonstrator, and to assemble optoelectronic interface subassemblies. Three layer assemblies have also been fabricated, comprising a modulator array, a silicon mount together with a (simulated) diffused lens array; these are described elsewhere in this issue [6]. A fluxless technique for the mounting of lasers and laser arrays based on AuSn eutectic solder has also been developed [2].

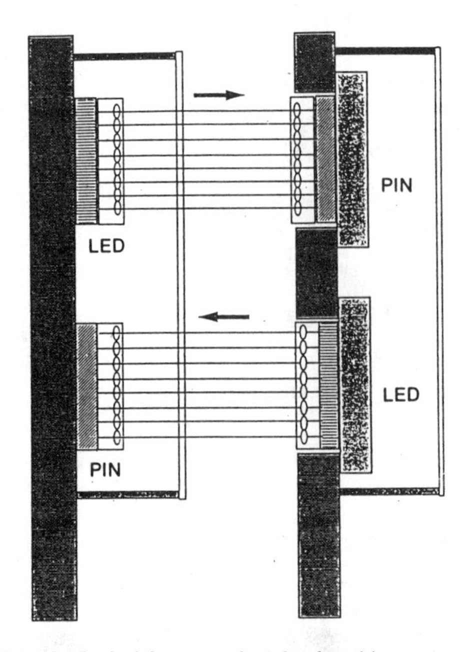

Fig. 19. Optical free space board to board interconnect.

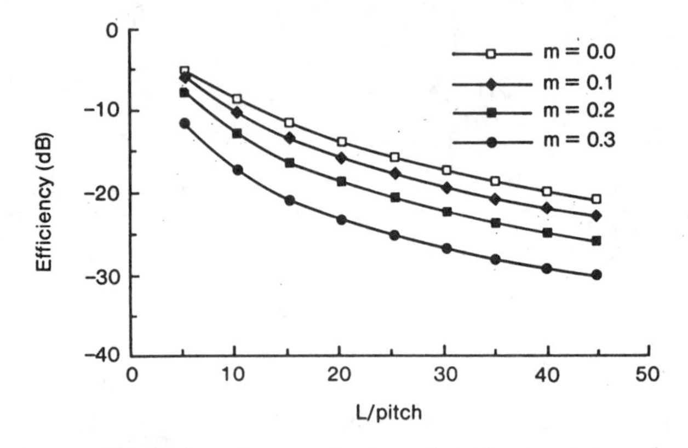

Fig. 20. OFSI efficiency for several values of misalignment (m) as a function of separation between source and detector microlenses (all values normalized to pitch).

IV. Other Activities

The activities described above are all directed toward the fabrication of the demonstrators and assessing their utility in practical systems. In addition to these, a number of other tasks are being undertaken directed toward future developments. One of particular interest is a theoretical investigation of high-density free space board-to-board interconnects, based initially on a technologically relatively simple scheme using LED arrays and microlenses. This optical free-space interconnect (OFSI) is illustrated schematically in Fig. 19. Here an array of LED's on the "transmitting" board illuminates a lens array which collimates the beams. The collimated beams illuminate another lens array which focuses the beams onto a detector array. In many ways this can be thought of as an optical zero insertion-force connector. In parallel processing systems, where an area interconnect is a natural requirement, this would allow the close coupling of processors in three dimensions as opposed to two with electrical interconnects.

Several aspects of the performance and design of this scheme have been analyzed [11]. Fig. 20 shows a calculation of the power efficiency as a function of longitudinal separation of the lenses, L, for various values of lateral misalignment, m, between the boards. The diagram is normalized to the LED pitch. Because an LED source emits with a lambertian profile, the efficiency is fairly low, however this is not a serious limitation in the design since no fanout is involved. Taking a reasonable example pitch of 250 μm and board spacing of 6.25 mm ($L = 25$), a misalignment between the boards of 25 μm ($m = 0.1$) would give an efficiency of about -16 dB. Additional calculations have confirmed the effect of angular misalignment and, in particular, have shown that misalignments of up to about 1.5° can be tolerated.

Other activities within the program, which have not been described above, include the investigation of the requirements for backplane buses, a small task to determine the suitability of nonlinear polymers to optical interconnect applications and an ambitious task to demonstrate the technology for monolithic integration of MQW modulators with CMOS circuitry. A key achievement of this activity is the demonstration of both growth and pregrowth substrate preparation at a temperature of less than 400°C. Further details of this activity can be found elsewhere [12], [13].

V. Conclusions

In conclusion, the major aims of the OLIVES collaborative project have been described. Many of the key components required to realise practical optical interconnects have been developed and subsystem demonstrators utilizing these components are under construction. The prospects for the use of optics in real systems have been significantly enhanced by close cooperation between the systems designers and the optical interconnect technologists and it is to be expected that these technologies will find commercial application within a very few years.

Acknowledgment

The author would like to acknowledge the contributions of all members of the OLIVES team at all the participating organizations.

References

[1] W. A. Crossland *et al.*, "Some applications of optical networks in the architecture of electronic computers," *IEE Colloquium Optical Connection and Switching Networks for Commun. Computing* (London), May 1990.
[2] R. G. Peall *et al.*, "Developments in silicon opto-hybrid technology for high performance opto-electronic modules," *IEE Colloquium on Advances in Interconnection Technol.* (London), Apr. 1991.
[3] C. Sebillotte, "Holographic optical backplane for boards interconnection," *International Symposium on Advances in Interconnects and Packaging* (Boston), Nov. 1990.
[4] H. Karstensen *et al.*, "Optical multichannel chip to chip data distribution," *Proc. SPIE 1281 (Proc. Optical Interconnections and Networks* (The Hague)), Mar. 1990.
[5] H. Zarschisky *et al.*, "Holographic optical elements for free space

clock distribution," in *Proc. SPIE 1281 (Proc. Optical Interconnections and Networks* (The Hague)), Mar. 1990.
[6] M. Goodwin *et al.*, "Optoelectronic component arrays for optical interconnection of circuits and subsystems," *J. Lighwave Technol.*, vol. 9, no. 12, Dec. 1991.
[7] S. Duncan *et al.*, "A 1 μm trench isolated high speed bipolar process," in *Proc. IEEE VLSI Symp.* (San Diego), 1888.
[8] J. Wieland and H. Melchior, "Optical receivers in ECL for 1 GHz parallel links," *SPIE/IEEE Intern. Symp. Advances in Interconnects and Packaging* (Boston), Nov. 1990.
[9] L. F. Miller, "Controlled collapse reflow chip joining," *IBM J. Res. Develop.*, vol. 13, no. 3, pp. 238–250, 1969.
[10] M. J. Wale *et al.*, "A new self-aligned technique for assembly of integrated optical devices with optical fibre and electrical interfaces," *European Conf. Optical Commun.* (Sweden), 1989.
[11] B. Dhoedt *et al.*, "Micro-optic imaging systems for optical board to board interconnect: a comparison of lens arrays with a single lens system," *Graded Index Opt. Syst. Topical Meet.*, Apr. 1991.
[12] P. Panayotatos *et al.*, "MBE GaAs on Si: Material and Devices for Optical Interconnects," *Intern. Conf. Physical Concepts of Materials for Novel Optoelectron. Dev. Applications* (Eurogress Aachen, FRG), Oct./Nov. 1990.
[13] Y. González *et al.*, "Low temperature process of GaAs/AlAs heterostructures on Si by atomic layer molecular beam epitaxy," *EURO MBE, 1991* (Finland), Apr. 1991.

*

Differential Board/Backplane Optical Interconnects for High-Speed Digital Systems Part I: Theory

Chung-Sheng Li, *Member, IEEE*, and Harold S. Stone, *Fellow, IEEE*

Abstract— **Switching noise is one of the major performance bottlenecks in a dense optical interconnect system. In this paper, we show that differential configuration and balanced operation at both the transmitter and the receiver sides are essential to achieve low switching noise ($\leq$ 5%) with large array size ($\geq$ 16). A fully differential configuration is proposed in this paper to minimize the possible switching noise. Several candidate structures for differential optical interconnects are investigated. Based on these structures, the signal-to-noise ratio (SNR) and the system penalty due to channel mismatch are analyzed and the results are compared to those of a single-ended interconnect with similar driver and receiver structures. From this analysis, we show that (1) The SNR of differential optical interconnects is similar to that of single-ended structures, (2) System penalty due to mismatch is negligible if there exists a slight channel mismatch ($\leq$ 1 dB). However, a power penalty ($\simeq$ 1.52 dB) exists when the mismatch between differential channels is significant ($\geq$ 2 dB). A prototype based on this fully differential interconnect concept has been designed and fabricated and its performance is reported in [54].**

I. Introduction

OPTICAL interconnects at the backplane and the board levels have the potential to provide much larger bandwidth, higher density, superior immunity to ground-loop noise and reflection noise than metal interconnects. However, electro-optical components such as the laser driver and optical receiver used by an optical interconnect system usually have to operate in a noisy environment. In this environment, switching noise and crosstalk generated by the nearby digital circuits or adjacent optical interconnects might be coupled into the electrical path of an optical interconnect, either through power distribution or through parasitic capacitive and inductive coupling among neighboring interconnects.

A simulation approach has been used in this paper to determine the amount of crosstalk that arises from a high-density optical transmitter and receiver array in a hybrid or monolithic integration environment. This paper presents evaluations of single-ended and differential structures for transmitter arrays and receiver arrays. We show that switching noise can be one of the dominant sources of crosstalk in both transmitter and receiver array with single-ended structures. A large power supply decoupling capacitor ($\gg$100 nF) is usually necessary to suppress the switching noise. A driver array consisting of the same number of differentially configured balanced drivers, each with a dummy load that imitates the laser loading conditions, can significantly reduce the switching noise to $\leq$ 5 % at 1 Gb/s. We also show that it is possible to drive a large size ($\geq$ 32) receiver array synchronously at 1 Gbps. The system considered uses a differentially configured receiver structure to reduce noise coupling through the power supply. In contrast, a single-ended receiver array introduces much larger switching noise ($\geq$ 25% of the total voltage swing, or -12 dB) and thus hampers the operation of a large array of receivers. A large decoupling capacitance ($\sim$ 100 nF) is required for suppressing the switching noise in a receiver array with single-ended structure while a much smaller decoupling capacitance ($\sim$ 1 nF) is necessary for a differentially configured receiver array.

A fully differential structure in which both data and its complement are simultaneously transmitted minimizes the interference between adjacent channels and maximizes the immunity to external power and/or ground disturbances at the transmitters and the receivers. This structure can be embodied within an optical interconnect system. Some potential advantages of such an interconnect structure include:

- The optimal threshold at the output of the receiver does not depend upon the absolute value of the signal levels.
- Complicated AGC circuits at the receiver side might be simplified or eliminated.
- This structure allows photodetectors to be DC-coupled into a receiver with a differential front-end and therefore does not require scrambling or encoding of the transmitted data stream.
- Due to its balanced operation throughout the interconnect, this architecture is less vulnerable to common-mode noise generated from the power supply, ground, switching, and parasitic coupling than single-ended interconnect structures.

All these advantages can be achieved without having to add substantial complexity to the existing electrical circuits associated with the transmitters and the receivers.

The fully differential optical interconnect concept was first applied in free space optical interconnect using S-SEED's (symmetric self-electrooptic devices) [1] due to the poor distinction ratio[1] of the SEED device [2–5]. The generalization of this concept to optical interconnects using other electrooptic devices and a comprehensive system analysis of this architecture has yet to be developed.

Manuscript received August 7, 1992; revised January 4, 1993.

The authors are with IBM Research T. J. Watson Research Center, Yorktown Heights, NY 10598.

IEEE Log Number 9210149D.

[1] Distinction ratio is defined as P_0/P_1 throughout this paper where P_0 is the power for logical ZERO while P_1 is the power for logical ONE.

Reprinted with permission from *IEEE Journal of Lightwave Technology,* Vol. 11, No. 7, pp. 1234-1249, July 1993.

In this paper, several candidate structures for differential optical interconnects are investigated. The most straightforward implementation of a differential optical interconnect uses two laser diodes, two spatially-separate interconnects (waveguides or fibers) and two photodetectors for each channel. The signal-to-noise ratio (SNR) and the system penalty due to channel mismatch of this architecture are analyzed and compared to those from a single-ended interconnect with similar driver and receiver structures. The analysis shows:

- The SNR of differential optical interconnects is similar to that of single-ended structures.
- System penalty due to mismatch is negligible for small mismatches in the channel ($\leq$1 dB). However, a power penalty ($\geq$1.52 dB) exists when the mismatch between differential channels is significant ($\geq$2 dB).

A chip set consisting of a 6-preamplifier array, a 12-preamplifier array, a 6-full-receiver array, and a laser driver based on the concept described in this paper has been designed and fabricated. The receiver array uses a Rockwell GaAs E/D MESFET process while the driver uses an IBM advanced bipolar process [6]. Based on the receiver design reported in [6], the performance of the differential optical interconnect structure is evaluated in further detail for various operation conditions in Part II [7]. In more sophisticated implementations, differential channels can be combined in the frequency domain (FSK), wavelength domain (WDM), time domain (Manchester encoded), and polarization domain, resulting in interconnect architectures that have the same number of waveguides (or fibers) as in a single-ended interconnect structure.

The organization of this paper is as follows: The switching noise of transmitter and receiver arrays is investigated in Section II and III, respectively, for single-ended and differential configurations. Section IV compares and summarizes the relative merits of different optical interconnect structures that can be used in a dense optical interconnect environment. Section V examines feasible implementation techniques of differential optical interconnects. System performance degradation due to channel mismatch is studied in Section VI. This paper is summarized in Section VII.

II. ELECTRICAL INTERACTIONS IN A TRANSMITTER ARRAY

Crosstalk among laser array elements has been a subject of continuous interest. Fabrication and characterization of one-dimensional individually addressable laser or LED arrays have been reported in [8]–[11]. Recently, two-dimensional vertical cavity surface-emitting laser diode arrays or surface-emitting LED arrays have received much attention and have emerged as a very promising light source for two-dimensional optical interconnects. The performance of these LED's and laser diodes is reported in [12]–[16] but the electrical crosstalk data has yet to be established. Most of the laser or LED driver circuits in use today were published in the late 70's and early 80's [17]–[23]. Both monolithic integration [24] and hybrid integration [25], [26] of the driver array with the laser array have been exploited. Crosstalk in these designs is usually determined through experiments or simulations for a specific data point, and a systematic study of the crosstalk due to switching noise is yet to be addressed.

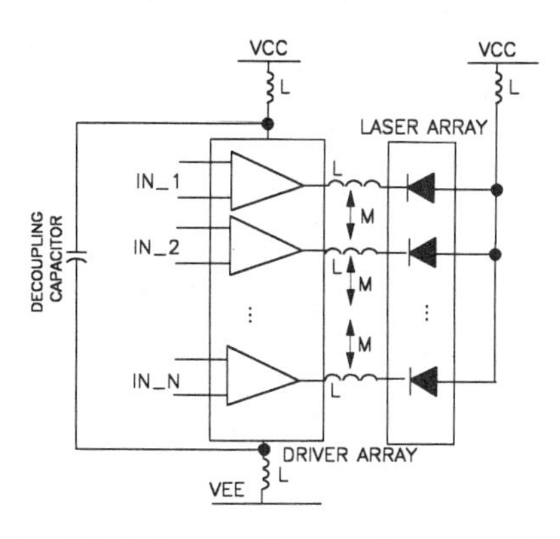

Fig. 1. Block diagram of a transmitter array of size N.

Possible interactions in a dense transmitter array are

- Electrical crosstalk between laser diodes and drivers due to the sharing of a common substrate,
- Electrical crosstalk due to parasitic capacitance and mutual inductance between adjacent channels,
- Switching noise due to the sharing of a common power supply and ground.

These interactions increase with channel density, modulation speed, and modulation current of the transmitter.

In this section, both analysis and simulation are used to determine the system penalty due to switching noise.

A. Transmitter Array Modeling

A transmitter array is shown in Fig. 1. It consists of a driver array and a laser diode array hybrid integrated on the same substrate. The equivalent circuit of this transmitter array also includes parasitic coupling between bonding wires, as well as the power supply inductance, ground inductance, and the decoupling capacitance. The substrate of the laser array is die bonded to the package, which usually includes a thermoelectric cooler to stabilize the temperature of the laser diodes.

1) Laser Modeling Each laser is modeled by its series resistance R_s, an ideal diode, and a capacitor C_d in parallel with the resistor and the diode as shown in Fig. 2 [27]. Good electrical isolation between laser elements can be obtained by providing a deep trench from the surface to the semi-insulating layer [11]. The laser diode array is assumed to share a common resistive substrate. We neglect the substrate coupling between adjacent laser diodes by assuming the substrate resistance between adjacent laser diodes is much larger than the resistance between active layers.

2) Single-Ended Driver Circuit The single-ended transmitter consisting of a single-ended driver and a laser diode is shown in Fig. 3. The driver circuit includes a bias control transistor Q1 and a modulation transistor Q2.

3) Differential Driver Circuit The equivalent circuit of a differential driver that drives a single laser is shown in Fig. 4. In the driver circuit shown in this figure, transistors Q1 and Q2 provide the input buffering, Q3 and Q4 are emitter coupled and provide the current drive to the laser through the collector of Q4. Transistors Q5 and Q6 compose a current

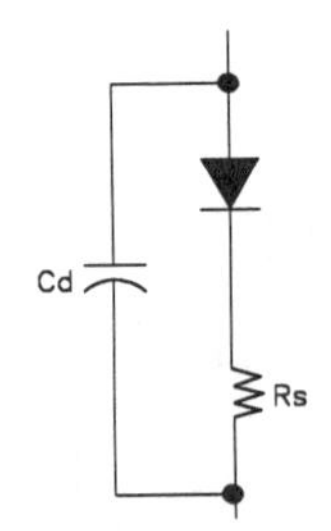

Fig. 2. Circuit model of a laser diode.

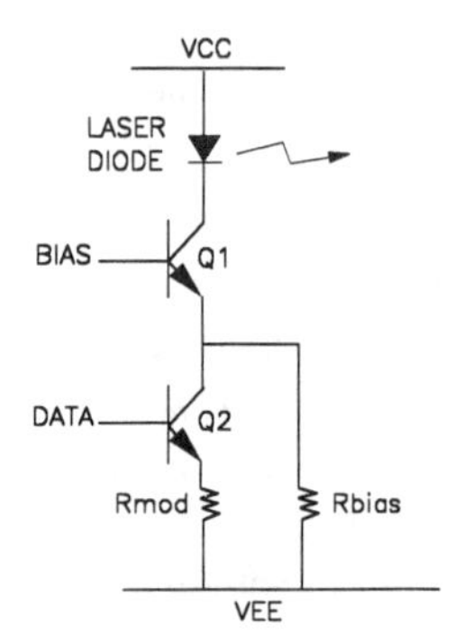

Fig. 3. Circuit diagram of a single-ended laser driver.

mirror which provides the modulation current control, while Q8 and Q9 compose another current mirror which provides the bias current control for the laser diodes.

The bipolar transistors used in the simulations in this paper have been assumed to have an f_T larger than 20 GHz, which is quite typical for an advanced self-aligned silicon bipolar process.

4) Differential Driver Circuit with a Dummy Load A fully differential circuit for driving a sink type laser array is shown in Fig. 5. This design contains separate external control circuits for adjusting the biasing current and the modulation current. The adjustment mechanism is connected to every driver in an array.

The major difference between this laser driver and the one described in (3) above is, in addition to the dummy load, the additional biasing circuitry necessary to provide biasing current for the dummy load. These two biasing circuits share the same current mirror and cannot be individually adjusted.

B. Switching Noise Analysis of Single-Ended Driver

The driver circuit shown in Fig. 3 is very vulnerable to power supply and ground disturbances when used in an array configuration (Fig. 1), as demonstrated from the following analysis.

The current demand from the power supply V_{CC} and the current injection into the power supply V_{EE} changes from I_{bias} to $I_{bias} + I_{mod}$ over a period of time t_r when the data input switches from logical ZERO to logical ONE. For a current transient of di/dt, V_{CC} decreases by Ldi/dt while V_{EE} increases by Ldi/dt where L is the inductance of the bonding wire of the power supply. Assuming the voltage reference to the bias control V_{bias} remains constant, the total

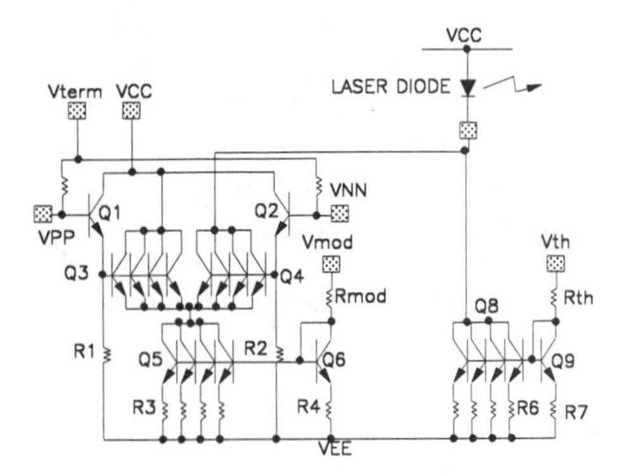

Fig. 4. Circuit diagram of a differential laser driver.

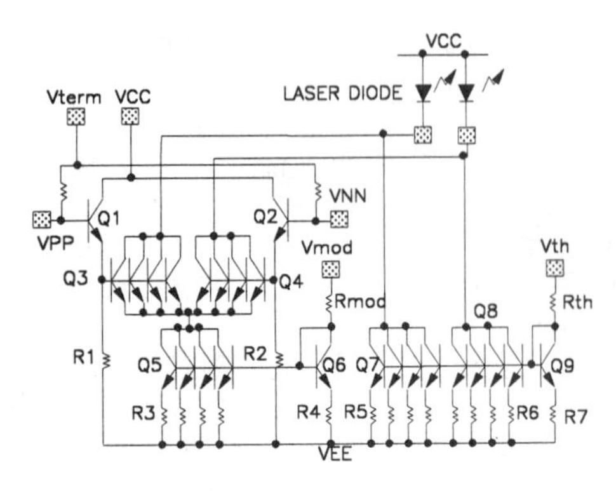

Fig. 5. Circuit diagram of a fully differential driver. This circuit is for a sink-type laser array in which the p-type substrate serves as the common ground.

bias current is reduced since

$$I_{bias} = \frac{V_{bias} - V_{BE} - V_{EE}}{R_{bias}} \quad (1)$$

and the modulation current is also reduced since

$$I_{mod} = \frac{V_{data} - V_{BE} - V_{EE}}{R_{mod}} \quad (2)$$

Assuming channel 1 in the array has a constant input of logical ONE while all the other channels switch from logical ZERO ($I_{mod} = 0$) to logical ONE, the total current flowing through the laser diode of channel 1 is reduced due to the decrease of both modulation and bias current. The fractional *reduction* of the current flowing through the laser diode of the quiescent channel equals

$$\frac{\Delta I}{I_{mod}} = NL\frac{1}{t_r}\left(\frac{1}{R_{bias}} + \frac{1}{R_{mod}}\right) \quad (3)$$

if there are a total of $N+1$ drivers. In this expression, t_r is the rise time of the signal and L is the inductance of the bonding wire. Assuming N is 16, L is 0.5 nH, t_r is 200 ps, R_{bias} and R_{mod} are both equal to 100Ω, the fractional switching noise is 80%!

In this first-order approximation,[2] the switching noise is linearly proportional to the number of channels as well as the bonding wire inductance, and inversely proportional to the rise time of the signal. There is no power supply noise cancellation

[2] We have ignored such effects as the Early effect which might further reduce the modulation and bias current.

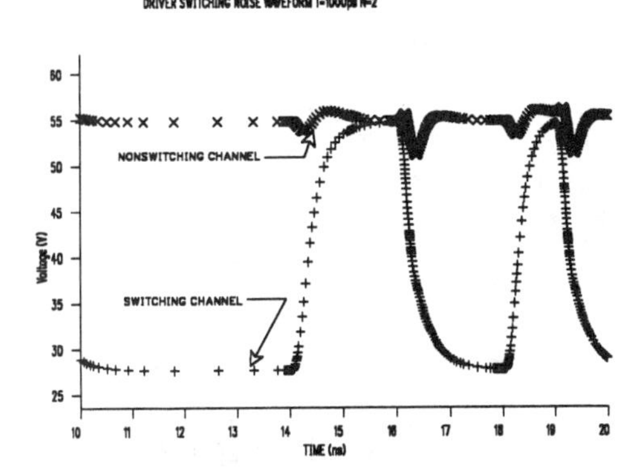

Fig. 6. Switching noise waveform of the laser driver. $N = 2, T = 1000$ ps, and $t_r = 200$ ps. Output waveform from a switching driver is also displayed for comparison.

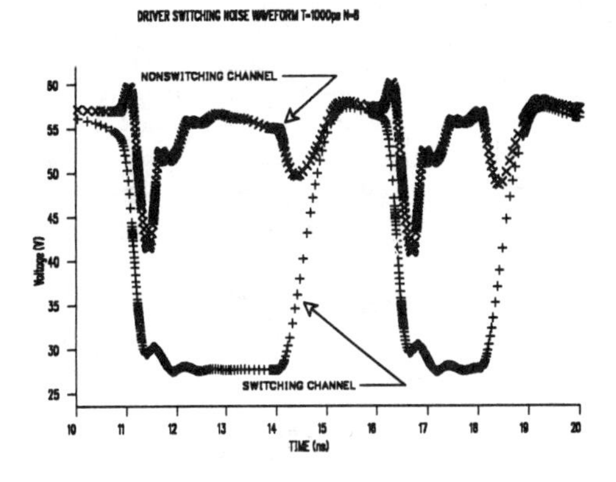

Fig. 7. Switching noise waveform of the laser driver. $N = 8, T = 1000$ ps, and $t_r = 200$ ps. Output waveform from a switching driver is also displayed for comparison.

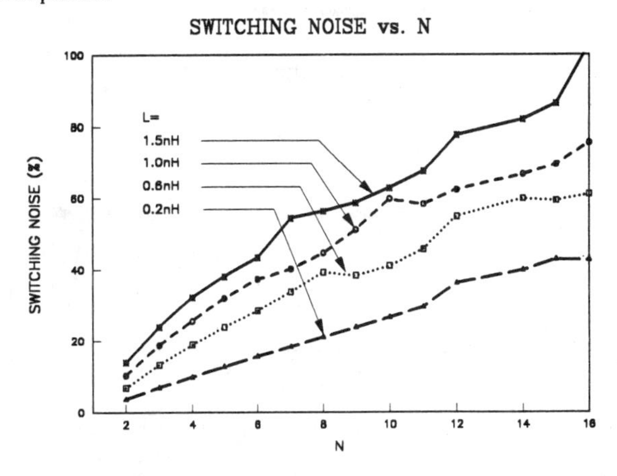

Fig. 8. Switching noise as a function of the array size. Bit rate = 1 Gb/s.

mechanism, so all the disturbances on the power supply will be coupled into the output laser driver current.

C. Simulation of Single-EndedInterconnect Using Differential Driver

1) Simulation Methodology A worst-case simulation methodology has been used to obtain the maximum switching noise under various packaging conditions. In this model, the worst-case scenario is obtained by transmitting a pseudo-random sequence synchronously into receivers $1, 2, \ldots, N, N + 2, \ldots, 2N + 1$, so that a total of $2N$ out of $2N + 1$ drivers switch simultaneously in the same direction while the input to driver N remains silent. This scenario incurs the maximum current excursion from the power supply as well as introduces a maximum adjacent channel coupling. The output optical power of channel N is then observed and the maximum power spike is compared with the full output swing of an active channel.

Sending identical data patterns to all channels as well as switching synchronously are both essential in obtaining the worst case scenario since

- the switching noise generated from Ldi/dt,
- the adjacent channel crosstalk generated from Cdv/dt and Mdi/dt

all maximize when there is an exact temporal alignment of the transition period of all the active channels with all of the transitions going in the same direction. The switching noise is defined as

$$N_{sw} = \frac{\Delta S_{\max}}{S_{nominal}} \tag{4}$$

where ΔS_{max} is the maximum relative current or voltage excursion of a quiescent channel from its quiescent value while $S_{nominal}$ is the nominal current or voltage swing of a channel.

2) Simulation Results The simulated waveform shown in Figure 6 assumes a driver array in which each driver has the same circuit as shown in Figure 4 with $N = 2$ and $L_{VCC} = L_{VEE}$ 1.5 nH. As the laser drivers of the active channels switch between ON's ($\simeq$ 55 mA) and OFF's ($\simeq$ 28 mA), the current in the quiescent channel also fluctuates around its ON state ($\simeq$ 55 mA). The switching noise spikes invariably coincide with the transitions of the waveforms of the active channels. The switching noise increases with size of the array, as clearly indicated in Fig. 7 for $N = 8$.

Fig. 8 shows the switching noise as a function of the number of channels for various values of bonding wire inductance. As indicated in this figure, the switching noise increases linearly with the number of channels and the bonding wire inductance. Simulations also show that balancing the driver load with a resistor that matches the resistance of the laser diode suppresses only a small fraction of the switching noise.

D. Simulation of a Differential Transmitter Array with Dummy Loads

The hybrid integrated transmitter array system assumed for the simulation is shown in Fig. 9 in which a bipolar driver array is wire-bonded to a laser diode array. The bonding wires between the driver array and laser diode array introduce crosstalk and waveform distortion for the signals and have been accounted for in the simulations. The same worst-case methodology as described in the previous section is used here to obtain the maximum switching noise.

The simulated switching noise waveform for an array of size 16 consisting of identical design is shown in Fig. 10. Switching noise vs. array size is shown in Fig. 11. As demonstrated by both figures, the maximum switching noise is less than 5% even for an array size of 16. This is significantly less than

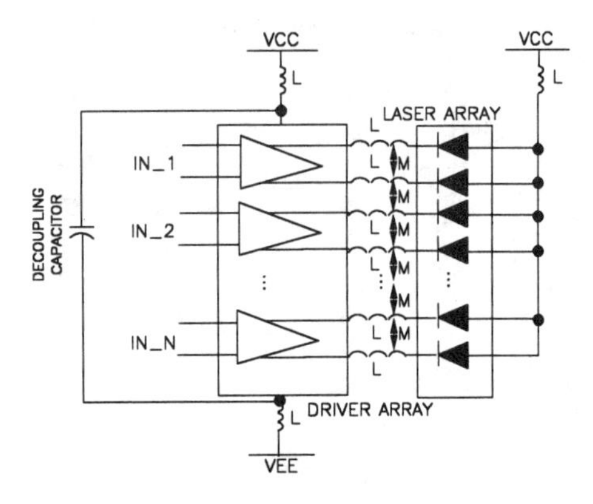

Fig. 9. Block diagram of a differential transmitter array of size N.

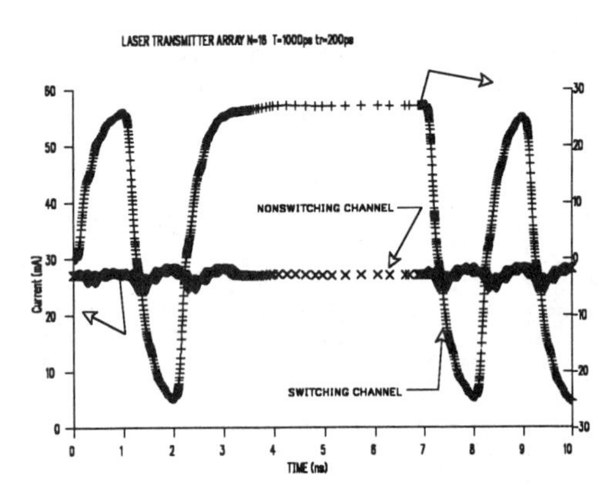

Fig. 10. Simulation of switching noise for a driver array at 1 Gb/s. $N = 16, t_r = 200$ ps, $L = 1.5$ nH for the signal bonding wire, and 0.5 nH for the power supply and ground bonding wire. Output waveform from an active driver is also displayed for comparison.

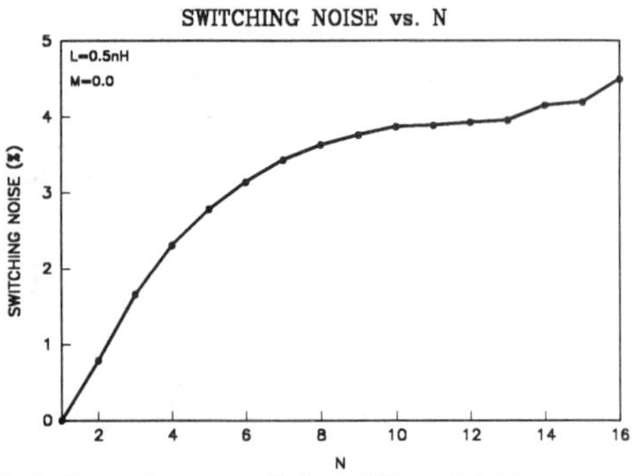

Fig. 11. Switching noise versus N for a differential driver array. A common p-substrate laser diode array is assumed.

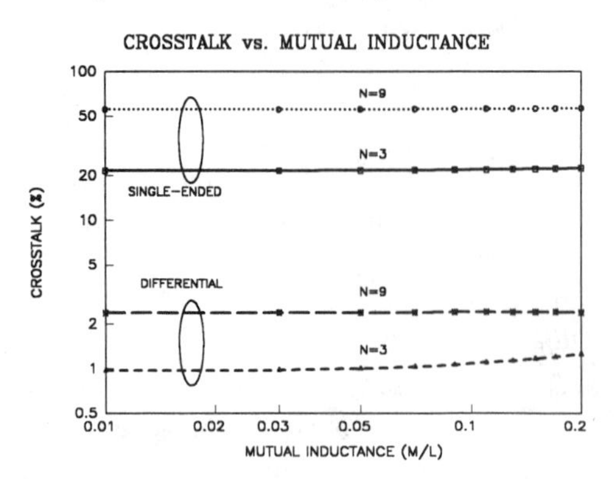

Fig. 12. Crosstalk versus mutual inductance for various driver array configurations. Mutual inductance is measured with respect to the bonding wire inductance, i.e., $M/L.L$ is assumed to be 0.5 nH while power decoupling capacitance is assumed to be 10 nF in all cases.

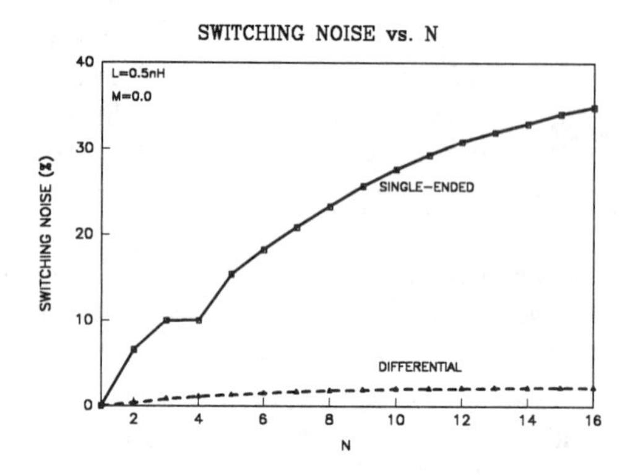

Fig. 13. Comparison of switching noise versus N between single-ended and differential driver array. A common n-substrate laser diode array is assumed.

those reported in Fig. 8 in which a similar driver circuit is used to drive a single-ended configuration. The crosstalk of a driver array is plotted as a function of mutual inductance in Fig. 12 for both single-ended and differential configuration. The ratio between mutual inductance and the bonding wire inductance in this figure varies from 0.01 to 0.20. Apparently, the dominant effect of crosstalk comes from switching noise (as illustrated between $N = 3$ and $N = 9$) rather than mutual inductance. The maximum crosstalk rises less than 1% as the ratio between the mutual inductance and the lead inductance varies from 0.01 to 0.2. Switching noise vs. the size of the array for both single-ended and differentially configured driver array using BiCMOS technology to drive a common n-substrate laser diode array is compared in Fig. 13. The crosstalk is significantly less in the differential case as compared to the single-ended case. This result is consistent with the case in which a common p-substrate laser diode array is driven by a bipolar driver array even though the basic circuit technologies for these two laser driver array are different.

III. Electrical Interactions in a Receiver Array

At the receiver side, maximizing the receiver sensitivity is usually the primary objective in conventional receiver design for long-haul optical communication systems. For such systems, minimizing the receiver thermal noise, or equivalently maximizing the receiver sensitivity, is sufficient to optimize such performance parameters as the repeater spacing or the diameter of a network. On the other hand, coupling noise generated by the interactions among elements in a receiver array, together with thermal noise, limits the system performance of a dense optical interconnect system.

The literature describes a number of receiver array designs using either hybrid integration [25], [28] or monolithic integration [26], [29]–[33] technology. Up to 6 and 8 channels/chip have been demonstrated thus far with monolithic [26] and hybrid [28] technology, respectively. Electrical crosstalk between photodetectors in a p-i-n array has been previously examined in [34], [35]. It was concluded in [34] that the common substrate of a p-i-n array introduces negligible DC crosstalk. A majority of the crosstalk came from the parasitic coupling

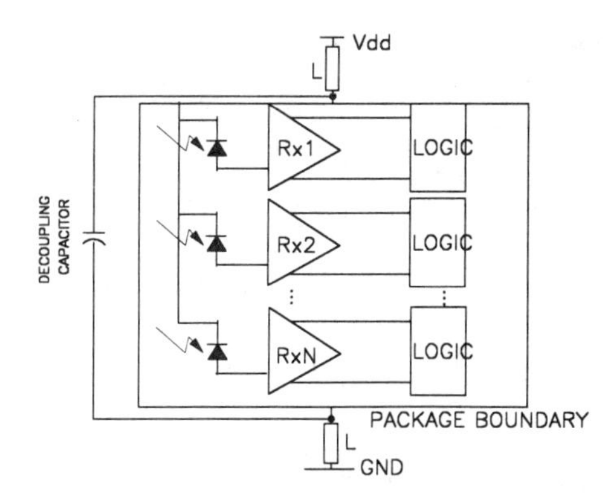

Fig. 14. Basic configuration of a receiver array with common power supply and ground.

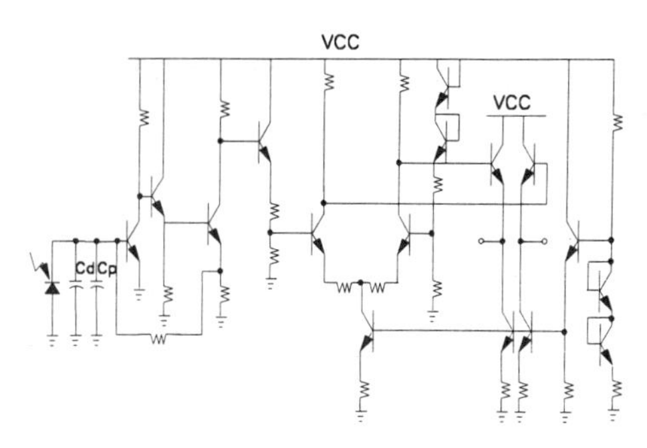

Fig. 15. Circuit configuration of a single-ended receiver. This receiver circuit ([35]) is used in Fig. 14 for single-ended array simulation.

between the bonding wires connecting between photodetectors and receivers. This type of crosstalk, however, is not present in a monolithic integration environment. For this reason, the monolithic integration environment has been assumed in this paper, and the dominant source for coupling noise is due to switching noise.

Possible interactions among elements in a receiver array include:

- Electrical crosstalk among photodetectors due to the sharing of a common substrate,
- Electrical crosstalk due to parasitic capacitance and inductance between adjacent bonding wires,
- Switching noise due to finite bonding wire inductance of the shared power supply and ground distribution, and
- Power supply noise due to external disturbances on the power supply from other digital circuits.

All these interactions increase with the component density, the modulation speed, and the input signal level of a receiver.

A. Receiver Array Model

In this section, we develop the equivalent circuits for both single-ended and differential receiver arrays. These models include the bonding wire inductance of the power supply in order to predict the switching noise introduced by the sharing of a common power supply.

The basic configuration for a receiver array is shown in Fig. 14. All of the receivers in this array have identical structure and device parameters. They share the same power supply and ground through an on-chip power and ground distribution network.

The receiver model used for the single-ended configured receiver array is shown in Fig. 15 [36]. Each receiver in this array consists of a shunt-series feedback preamplifier stage, a single-ended to balanced conversion stage, and an output driver stage. The photodetector in this receiver is DC-coupled to the input of the preamplifier. The circuit model used for the differential receiver is shown in Fig. 21 [37]. The MSM-photodetector in this case is AC-coupled into the front-end of the preamplifier.[3] The biasing networks of the photodetectors are not shown in Fig. 15 and Fig. 21. In both cases, each photodetector is modeled by an ideal current source in parallel with a capacitor representing the capacitance of the photodetector.

For the simulations performed in this section, the power supply inductance varied from 0.05 to 2 nH. The photodetector capacitance of the receiver varied from 100 fF to 2 pF. The bit rate varied from 100 Mb/s to 1 Gb/s, while the rise time and fall time of the signal varied from 6/100 to 8/10 of a bit interval. A silicon bipolar technology is assumed for the single-ended receiver while the GaAs E/D MESFET (enhancement/depletion metal semiconductor field effect transistor) technology is assumed for the differential receiver. Each receiver was loaded with an RC filter to emulate the post-amplifier and the decision circuitry.

B. Simulation Methodology

The simulation strategy used for simulating the electrical interference in a dense transmitter array was also used for the receiver simulation in that $N-1$ receivers of an N-receiver array switched from ONE to ZERO or from ZERO to ONE simultaneously, and simultaneously observed the receiver that did not switch. The results obtained give an upper bound on the switching noise. In the simulations, N varied from 2 to 32.

C. Simulation of Single-Ended Receiver Array

For a single-ended configured transimpedance receiver array with each receiver using the same configuration shown in Fig. 15, a significant switching noise exists at a bit rate greater than 100 Mbps if a decoupling capacitor between the power-supply pad and the ground pad is not provided. Fig. 16 shows the simulated output waveform from the channel with quiescent input when the other 15 channels switch synchronously. This waveform is overlaid with the simulated waveform from the channel with an active input, showing a maximum of 10% and 20% of switching noise is reached when the corresponding effective bonding-wire inductance of the power supply is 0.2 and 0.5 nH, respectively.

[3] Since an MSM-photodetector usually requires a large bias voltage across its terminals ($\geq$1 V) for reasonable responsivity, it is not possible to tie the terminals of an MSM-photodetector directly into the inputs of a differential receiver which is very sensitive to input offset voltage. Therefore, AC-coupling is usually necessary for an MSM photodetector to be used in conjunction with a receiver with a differential front end. A recent design reported in [38], however, successfully uses a pair of mismatched front-end transistors to allow DC-coupling of a PIN photodetector to the front-end of the receiver while maintaining the DC biasing of the photodetector.

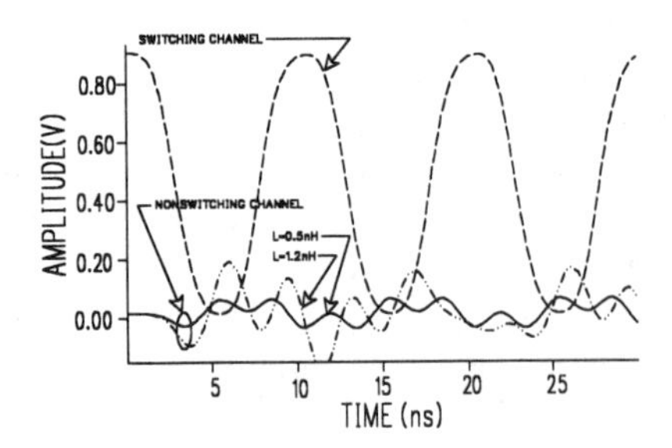

Fig. 16. Switching noise waveform. This waveform is the output from a single-ended configuration receiver with quiescent input and simultaneous switching of the other 15 receivers for an effective lead inductance of 0.5 and 1.2 nH. The output of a switching receiver is also shown for comparison. Bit rate and rise time of the signal is 100 Mb/s and 2 ns, respectively. The photodetector capacitance, C_d, and the package capacitance, C_p, are both equal to 0.5 pF. No decoupling capacitance is provided in this case.

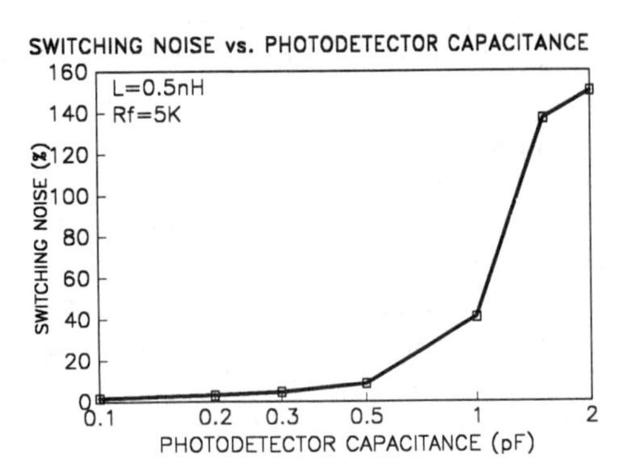

Fig. 17. Switching noise versus photodetector capacitance. Lead inductance is assumed to be 0.5 nH, while feedback resistance is 5 KΩ.

As the photodetector capacitance increases, the coupling between the power supply and the input of the preamplifier increases and the amount of switching noise increases accordingly, as shown in Fig. 17. Fig. 18 shows a small decoupling capacitor (0.5 nF ~ 2 nF) helps to reduce the switching noise, but a large decoupling capacitor ($\gg$ 2 nF) is necessary to eliminate the switching noise completely. However, a large decoupling capacitor is very difficult to incorporate into an integrated circuit due to its excessive area requirement,[4] indicating that more power supply and ground pads are required to reduce the effective power supply and ground lead inductance. Furthermore, we can observe from Figure 18 that the value of the lead inductance where a maximum switching noise occurs decreases with increasing decoupling capacitance. By fixing the value of the lead inductance (e.g., 1 nH), the switching noise increases at first as the decoupling capacitance increases from zero to 1 nF before the switching noise drops as the decoupling capacitance increases above 2 nF. This suggests a small decoupling capacitance ($\leq$2 nF) for a single-ended receiver does not suppress the switching noise. The shift of the lead inductance value where a maximum switching noise occurs is due to the shift of the LC-resonant condition formed between the decoupling capacitance and the lead inductance.

Figs. 19 and 20 show switching noise as a function of the number of receivers for various values of decoupling capacitance. The switching noise has a nonlinear dependency

[4] On a GaAs chip, each picofarad of capacitance requires an area of 92μm $\times$92μm.

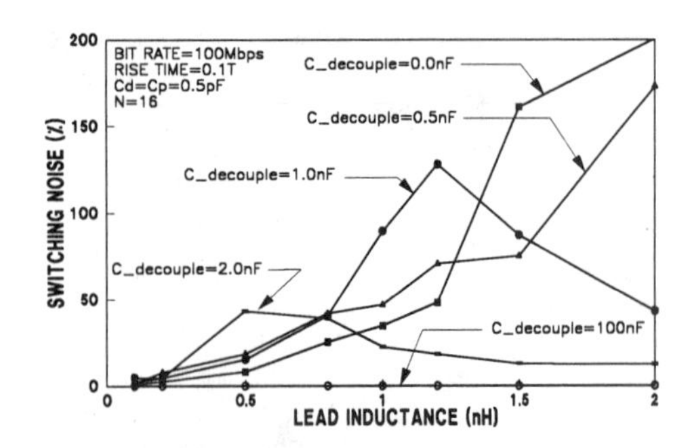

Fig. 18. Switching noise versus lead inductance for a single-ended configured receiver array with $N = 16$. Bit rate and rise time of the optical signal are 100 Mb/s and 1 ns, respectively. Other operating conditions are similar to those of Fig. 16.

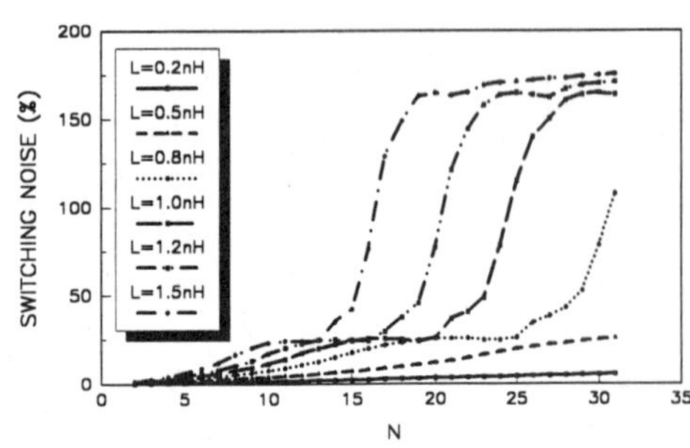

Fig. 19. Switching noise generated by a single-ended configured receiver as a function of array size. Bit rate is 100 Mb/s, rise time is 1 ns, and no decoupling capacitor is provided.

on the number of receivers when no decoupling capacitance is provided. Assuming the lead inductance equals 1.5 nH, the amount of maximum switching noise first increases linearly with N until $N = 10$, after which the switching noise remains essentially constant at 25% between $N = 10$ and $N = 13$. The switching noise rises sharply from 25% to 150% as N increases from 13 to 18. The amount of switching noise levels off to 175% at $N = 18$ since the preamplifier saturates at this point. As the lead inductance decreases in Figure 19, the switching noise levels off at a larger value of N. Also, the slope of the increase in switching noise as a function of N decreases as the lead inductance decreases. From Fig. 19, we conclude that a maximum receiver array size of 20 is allowed for a lead inductance $\leq$1 nH, assuming the maximum allowable switching noise is 25%(-12 dB).

As the decoupling capacitance increases from zero, the dependence of switching noise on N becomes linear, as shown in Figure 20. Note that switching noise *increases* for $N \leq 18$ and decreases for $N \geq 20$ as a decoupling capacitance of 1 nF is introduced between the power and the ground. The nonlinear dependency between the switching noise and the size of an array as well as the increase in switching noise for small size arrays in this receiver configuration are possibly due to the forming of an LC-resonance condition between the decoupling capacitor and the lead inductance as explained earlier.

D. Simulation of a Differential Receiver Array

Fig. 22 shows the switching noise waveform generated from a receiver array of size 4 with each receiver using a

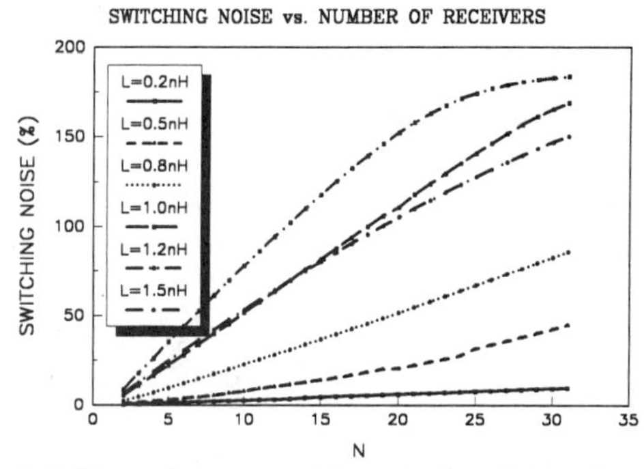

Fig. 20. Switching noise generated by a single-ended configured receiver as a function of array size. All the simulation condition is similar to Fig. 19 except a decoupling capacitor of 1 nF is provided between power and ground.

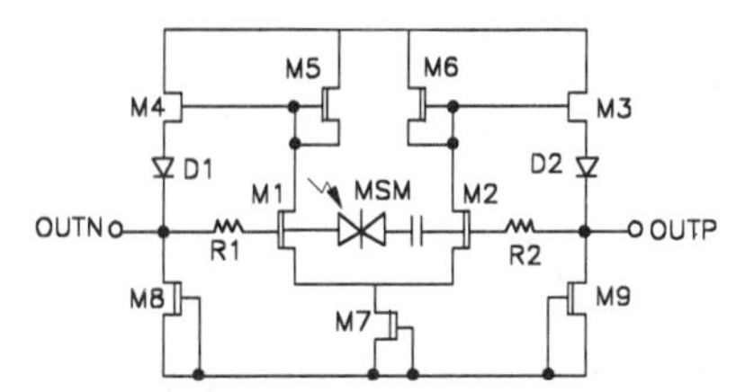

Fig. 21. Circuit configuration of a differential receiver. This receiver circuit ([36]) is used in Fig. 14 for differential receiver array simulation.

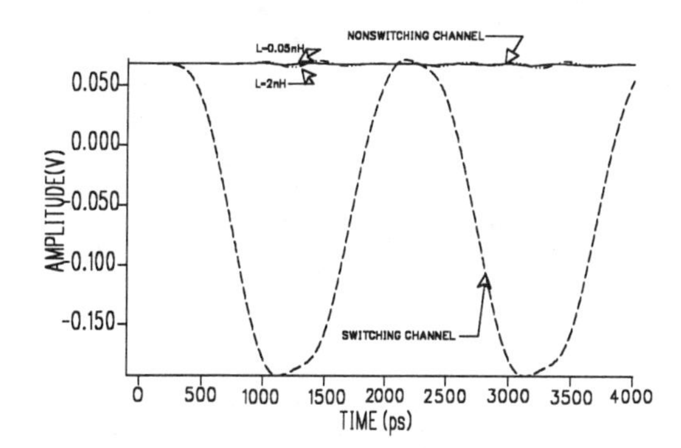

Fig. 22. Switching noise generated by a differentially configured receiver array. The array size is 4, and the lead inductance varies from 0.05 to 2 nH. Each receiver is operating at 1 Gb/s with a rise time 100 ps. No decoupling capacitor is provided in this sumulation. The output of a switching receiver is also shown for comparison.

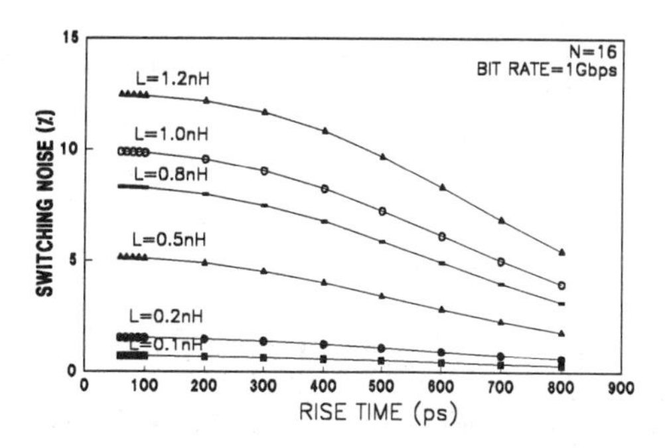

Fig. 23. Switching noise versus signal rise time for a differential receiver array with $N = 16$. Bit rate is 1 Gb/s, no decoupling capacitor is provided in this case.

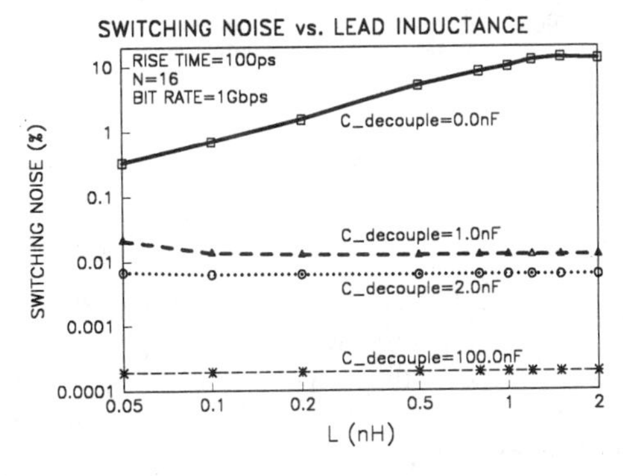

Fig. 24. Switching noise versus lead inductance for a differentially configured preamp. Rise time of the optical signal is 100 ps, size of the receiver array is 16, and bit rate is 1 Gb/s.

differential configuration shown in Fig. 21. The switching noise is suppressed by the common-mode rejection of the differential input-stage at the sacrifice of higher thermal noise ($\leq$ 3 dB) because of the differential front-end. The amount of switching noise only slightly increases as the rise time decreases, as shown in Fig. 23. The switching noise as a function of lead inductance for various decoupling capacitance is shown in Fig. 24. The maximum switching noise for 2 nH lead inductance is less than 12%(−18.4 dB) at 1 Gb/s for a receiver array of size 16 even without a decoupling capacitor. In contrast to the case of the single-ended receiver, even a small amount of decoupling capacitance is effective in eliminating the switching noise. For a decoupling capacitance equal to 1.0 nF, the maximum switching noise is less than 0.05% for almost any practical range of lead inductance value.

IV. Optical Interconnect Architectures

In Sections II and III, we have shown that balanced operation at either the transmitter side or the receiver side in a dense array environment can significantly reduce the switching noise and enhance the power supply noise immunity. A fully balanced interconnect system benefits from these advantages. In this section, we first examine possible optical interconnect structures using combinations of the transmitter and receiver structures discussed in Sections II and III. The merits and drawbacks of each structure are then evaluated qualitatively and quantitatively.

A. Single-Ended Interconnect

There are six possible structures for a single-ended interconnect, as summarized in Fig. 25. Possible driver structures for a single-ended interconnect can be either single-ended or differential. When a differential current switch is used to drive a laser, the unused branch of the current switch is usually tied to the power supply.

Similarly, both single-ended and differential front-end structures can be used for the receiver. When a differential front-end is used, the photodetector is usually AC-coupled with the receiver because the photodetector and the front-end of the receiver have different biasing requirements. A loss of low-frequency spectral components from the data stream is unavoidable for an AC-coupled front-end. This dictates that the data has to be encoded or scrambled in order to eliminate the DC and low frequency spectral components.

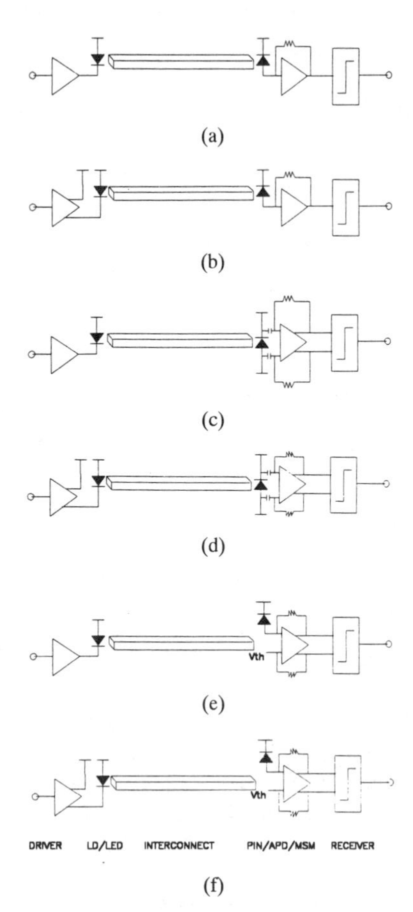

Fig. 25. Possible structures for a single-ended optical interconnect. (a) Single-ended driver, single-ended receiver. (b) Differential driver, single-ended receiver. (c) Single-ended driver, differential receiver. (d) Differential driver, differential receiver. (e) Single-ended driver, differential receiver. (f) Differential driver, differential receiver.

Assuming equal probabilities for transmitting ONE's and ZERO's, it can be shown [39] that the bit-error rate (BER) of architecture (a) is $0.5erfc(P_{av}RR_f/\sqrt{2N_{tot}})$ where $erfc(.)$ is the complementary error function, P_{av} is the average optical power received by a photodetector, R_f is the transimpedance of the amplifier, R is the responsivity in A/W, and N_{tot} is the total receiver noise observed at the output. Using the architecture in Fig. 25(a) as a reference, we compare the performance of other interconnect architectures:

- The same optical power is required by the architecture in (b) to achieve the same BER as in (a), except twice as much electrical power is consumed at the driver side (and thus half E/O conversion efficiency) due to the use of a constant current source for supplying the laser modulation current.
- Both the signal voltage swing and the total electrical noise power are doubled in (c), regardless whether an AC-coupling capacitor is used. Since the signal power is the square of the signal voltage, the signal-to-noise ratio in this scheme is doubled as compared to (a). The BER is $0.5erfc(P_{av}RR_f/\sqrt{N_{tot}})$ and thus 1.5 dB less optical power is required in architecture (c) to achieve the same BER as compared to architecture (a) while the same electrical power is consumed by the driver. Note that the detector current is directly proportional to the optical power, and thus 3 dB electrical power difference translates into 1.5 dB optical power difference.
- Architecture (d) is similar to architecture (c), and thus 1.5 dB less optical power is required in architecture (d) to achieve the same BER as compared to architecture (a) though twice as much electrical power is consumed (and thus half of the E/O conversion efficiency).
- For architecture (e), the signal strength is the same as in architecture (a) while the noise power is doubled, and the BER thus equals $0.5erfc(P_{av}RR_f/2\sqrt{N_{tot}})$, and 1.5 dB more in optical power is required to maintain the same BER as in architecture (a). The E/O conversion efficiency is the same as in architecture (a).
- Architecture (f) is similar to architecture (e), and 1.5 dB more optical power is required to achieve the same BER as compared to architecture (a). The E/O conversion efficiency at the driver is one half of architecture (a).

The trade-offs of using these structures in a dense array are compared as follows.

Disadvantages The structure in (a) is not suitable for dense optical interconnect applications since it incurs a large switching noise at both the transmitter and the receiver ends. The combinations that use a differential driver with a single LD/LED device, as shown in Fig. 25(b), (d), and (f), have the following disadvantages:

- Matching the load of the driver arms with appropriate passive and/or active devices for these configurations is difficult due to the low resistance ($\sim 5\Omega$) and high capacitance ($\sim$ 5 pF) of an LD/LED device. If the LD/LED's are not monolithically integrated with the drivers, the lead inductance of the bonding wire between the LD/LED's and the drivers introduces additional imbalance and waveform distortion for high-speed operations.
- A slight imbalance at the load of a differential laser driver induces nonnegligible switching noise.

It is also possible to have the laser driver drive two laser diodes with the light from only one of the laser diode coupled into the interconnect. The purpose of having a second laser diode is to serve as a dummy load for the laser driver to reduce the switching noise of the driver array.

Now let us consider the combinations consisting a single PIN/APD with a differential receiver, as shown in Fig. 25(c), (d), (e), and (f). These combinations have the following disadvantages:

- An AC-coupled design in architecture (c) and (d) requires a large chip area.[5] In principle, two AC coupling capacitors can be used to couple the terminals of a photodetector to the front-end of the receiver. However, this configuration usually requires a large area in order to achieve a low cut-off frequency, which is determined by the impedance of the detector bias network and the coupling capacitance. Alternatively, it is possible to AC-couple one terminal of the photodetector and DC-

[5]The required area of an AC-coupling capacitor depends on the cutoff frequency of the data spectrum. Lower cutoff frequency requires larger AC-coupling capacitor. Line coding of the data stream usually can help to increase the data cutoff frequency and reduce the area required by the AC-coupling capacitor.

couple the other one. But this configuration requires level restoration circuitry to cancel the possible DC level shifting at the other terminal due to different light input levels.

- An AGC circuit or a limiting-receiver design is required [40]. The detection of the received optical signals is sensitive to the absolute value of the threshold since the received optical power can vary over a wide dynamic range because of the possible large variation in propagation loss, coupling loss, and the need to provide fan-out. Additional tolerance has to be provided to allow for device variation and aging. Therefore either an automatic gain control mechanism or a limiting amplifier stage is necessary to maintain a constant logic swing. However, using AGC circuitry requires a DC-balanced data stream while using a limiting amplifier results in an unequal noise power for logical ONE's and ZERO's.

Both the requirements for AC-coupling (for easier receiver design) and AGC (to enhance the dynamic range) necessitate line coding techniques such as Manchester [41] or 8B/10B code [42] to maintain a DC-balanced data stream. However, encoding/decoding implies a reduction in the effective bandwidth available to the data and an increase in latency. The usefulness of this type of optical interconnect is restricted to applications that can tolerate the latency and bandwidth loss due to encoding and decoding to maintain a DC-balanced signal.

Advantages The single-ended configuration requires the least number of optical and optoelectronical components at the expense of additional complexity in the electronic components. Moreover, mismatch between device components does not introduce significant system penalty in these configurations as in the differential cases. If the interconnect is not in a critical path of a digital system and can afford encoding/decoding delay, single-ended interconnect has the potential of offering the highest interconnect density for asynchronous channels, only limited by the switching noise incurred at the photodetectors.

B. Differential Interconnect

There are two possible structures for a differentially configured optical interconnect, as summarized in Fig. 26. The driver for this type of interconnect has to be differentially configured, while the receiver can be either single-ended or differentially configured. The photodetectors can be either AC- or DC-coupled to the receiver.

The performance of the architectures listed in Fig. 26 are summarized below:

- In Fig. 26(a), the signal voltage swing is doubled as compared to Fig. 25(a) while the noise remains essentially the same, the BER thus equals 0.5 $erfc\ (\sqrt{2}P_{av}RR_f/\sqrt{N})$ and 3 dB less in optical power is required to support the same BER as compared to Fig. 25(a). The E/O conversion efficiency is the same.
- In Fig. 26(b), the signal voltage swing is doubled as compared to Fig. 25(a) while the noise power is also doubled, the BER thus equals is 0.5 $erfc(P_{av}RR_f/\sqrt{N})$ and 1.5 dB less in optical is required to support the same BER as compared to that in Fig. 25(a). The E/O conversion efficiency is the same.

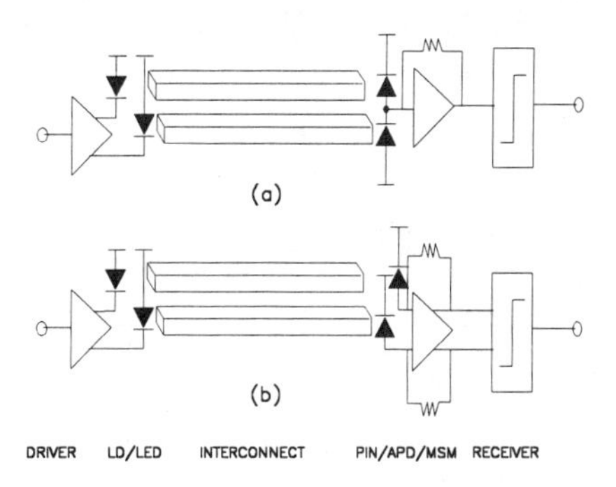

Fig. 26. Possible structures for a differentially configured opitcal interconnect. (a) Single-ended receiver. (b) Differential receiver.

The trade-offs of using these structures are compared as follows.

Disadvantages The number of optical and opto-electronical components such as the waveguides, photodetectors, laser diodes, and, depending on the modulator structure, some types of modulators are doubled in this configuration. This has an adverse impact on interconnect density and system reliability. Furthermore, mismatch of the parameters along a differential link such as the threshold current and differential quantum efficiency of the laser diodes, the attenuation of the interconnects, and the quantum efficiency of the photodetectors can introduce additional system penalty.

Advantages In contrast to the single-ended interconnect configurations, a fully differential optical interconnect architecture enjoys the following advantages:

- Self-thresholding. Since the noninverted and inverted signals are both present in a differential optical interconnect, the determination of a logical ONE vs. a logical ZERO can be achieved by simply subtracting the signals coming out of the inverting channel from the signals coming out of the noninverting channel, and using differential zero as the threshold to distinguish between ONE's and ZERO's.[6]
- Less vulnerable to common-mode noise.
- Allows easy DC-coupling to the front-end of a differential receiver.[7]
- Allows DC-unbalanced data streams, and thus eliminates the requirements for encoding/decoding.

V. Implementation of Differential Optical Interconnects

The differential optical interconnect structure discussed in the previous section assumes a straightforward implementation

[6] The comparison between differential optical interconnects and single-ended optical interconnects is similar to that between frequency shift keying (FSK) and on-off keying (OOK) in classical communications. It has been well-known in communication theory [41] that given the same symbol distance and hence noise immunity, the performance of FSK is exactly the same as OOK in terms of the required energy per bit to achieve the same bit-error-rate. In practice, however, a constant decision threshold can be maintained in an FSK environment while the optimum threshold depends on the absolute amplitude of the light intensity in an OOK environment.

[7] However, long stream of ONE's might have a heating effect on a laser diode, increases its threshold current and reduces its optical power output if the bias current of the laser driver is not adjusted accordingly.

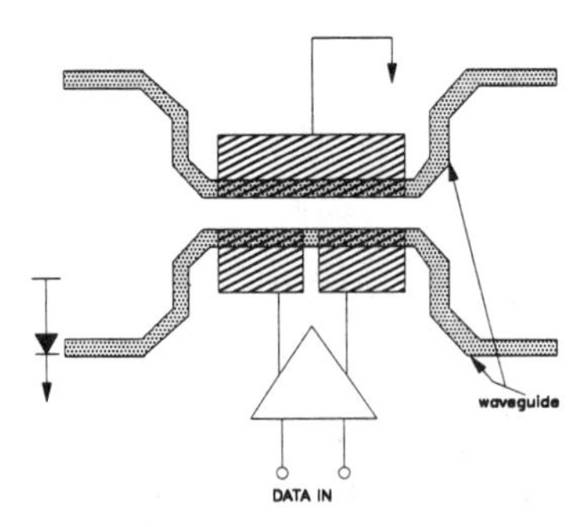

Fig. 27. Structure of a directional coupler modulator.

in which two laser diodes, two waveguides, and two photodetectors for each channel are used. In this section, other techniques that can be used to implement a differential optical interconnect system are discussed and evaluated in terms of their expected BER. These techniques have the potential of reducing the number of laser diodes or/and the number of waveguides.

A. External Modulators

Instead of using two lasers for each channel, it is possible to use just one laser and one external modulator (or two external modulators if the quantum well modulator is used) to generate the differential optical signals. The number of light sources can be further reduced by having several channels sharing the same *optical power supply* . The number of waveguides or photodetectors are not affected in this approach.

External modulators such as a Mach–Zehnder type interferometer, directional coupler, or total internal reflection (TIR) modulator [43]–[45] on $LiNbo_3, GAAs, INP$, or active polymer substrates can be used for generating differential light output from a single light source. A representative structure illustrating the generation of complementary light signals using a directional coupler and a single laser diode is shown in Fig. 27. Most of these devices are based on the electro-optic effect, in which the refractive index of the material ($LiNbO_3$ or $GaAs$) and therefore the *effective* propagation path length changes when a voltage is applied. The refractive index change can also be induced by carrier injection or depletion in semiconductor materials ($Si, GaAs$ or InP) [46]. Alternatively, the differential optical signals can be generated by using quantum well modulators [2]–[5]. Using external modulators can usually achieve more balanced operation as compared to laser diodes due to the elimination of the sensitivity of device performance to variation in laser threshold current and differential quantum efficiency. Detailed performance analysis of using external modulators for differential optical interconnects will be given in [7].

Since the receiver structure is not modified by this approach, The BER derived in Section IV-D still holds.

B. Frequency Division Multiplexing Technique

The laser's emission frequency (wavelength) is modified at different levels of current injection due to *adiabatic chirp*. An optical interconnect system using FSK modulation, as shown in Fig. 28, can be implemented by biasing the laser well above

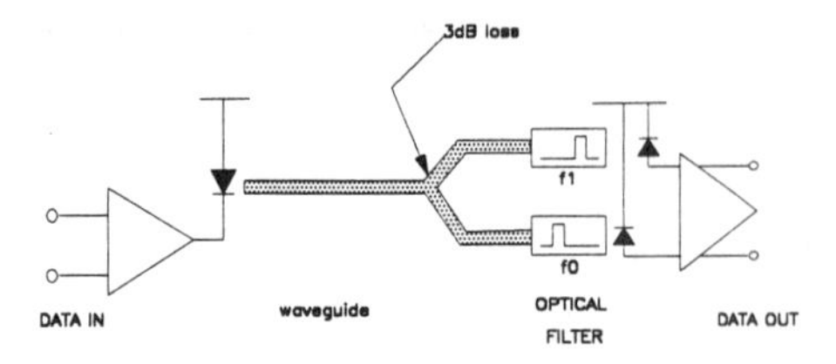

Fig. 28. Structure of differential optical interconnect using FSK.

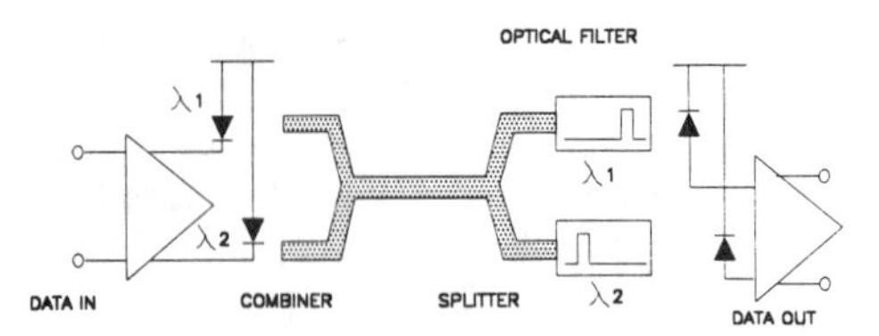

Fig. 29. Structure of a differential optical interconnect using WDM.

lasing threshold and modulating the laser with a small current. At the receiver side, the light is split into two filters via a splitter (or a grating demultiplexer) with one filter positioning at the center frequency of ONE while the other one positioning at the center frequency of ZERO. The output light signals from the filter is sent to a differential receiver. This structure uses the same total number of waveguides and lasers as that of a single-ended optical interconnect at the expense of requiring a more complicated receiver structure (requiring two filters and single-mode lasers).

If a grating demultiplexer is used, this structure requires the same optical power to achieve the same BER as those analyzed in Section IV-B. On the other hand, a 3-dB power penalty is suffered if a power divider is used to route the optical signals to the receiver.

B. Wavelength Division Multiplexing

The number of interconnecting fibers or waveguides can also be reduced by using wavelength-division multiplexing (WDM). Differential signals are sent at different wavelengths through the same waveguide or fiber, as shown in Fig. 29. For single-wavelength LD/LED's and PIN/ APD's, fiber or waveguide combiners and splitters can be used to multiplex and demultiplex optical signals. A moderate wavelength selectivity is required since only two wavelengths need to coexist within the same interconnect. There has been a successful demonstration [47] of a monolithically integrated detector chip which includes a Bragg grating wavelength demultiplexer and PIN photodetectors. It is also possible to fabricate dual-wavelength devices so that two wavelengths can be coupled in and out of the fiber and waveguide without having to use waveguide combiners/splitters. There already exist dual-wavelength LED's, LD's [48]–[51], as well as dual-wavelength PIN's [52], [53] that can serve this purpose.

If the wavelengths used by the differential signals are too close, it might be difficult to fabricate fused PIN's that can distinguish two wavelengths. A small overlap in the responsivity versus wavelength of a fused dual-wavelength PIN is not necessarily harmful since common-mode response can be suppressed by the differential receiver.

As long as a grating demultiplexer is used, wavelength division multiplexing can achieve the same BER as those structures analyzed in Section IV-B. Otherwise a 3-dB power penalty is suffered as a result of the power splitting.

D. Polarization Division Multiplexing Technique

The differential optical signals can also be multiplexed in the same waveguide/fiber by using different polarizations (TE and TM). A system using this technique is shown in Fig. 30. A polarization modulator [43] is used at the source to modulate the incoming single-polarization light (TE or TM) so that the polarization state of the light is toggled (TE→TM/TM→TE) when the modulating voltage is in the ON state. At the receiving end of the interconnect, there is a polarization splitter so that TE signals come out from the bottom waveguide while TM signals come out from the top waveguide. The outputs from the polarization splitter are then applied to a differential receiver. This architecture requires a polarization maintaining waveguide or fiber and usually needs large modulation voltage.

Since no power splitting is involved in a polarization splitting, a polarization division multiplexed interconnect can achieve the same BER as those structures analyzed in the previous section.

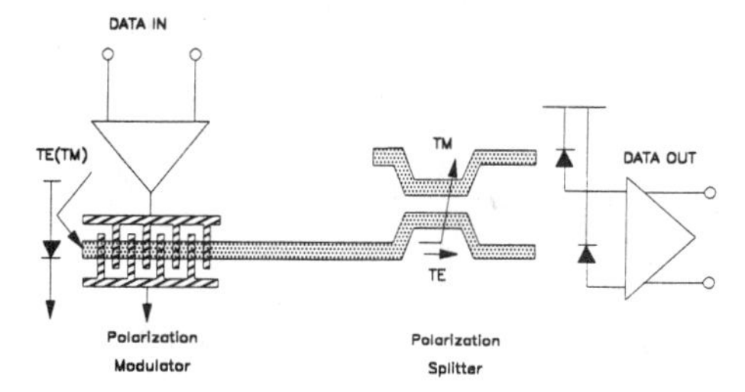

Fig. 30. Structure of a differential optical interconnect using polarization division multiplexing.

VI. System Degradation Due to Channel Mismatch

In practice, the differential gain of LD/LED's, the attenuation through the fibers/waveguides, the responsivity of the photodetector can be different between differential channels. Furthermore, process-related offset voltages in the circuitry of the laser driver, preamplifier, and postamplifier introduce additional imbalance between differential channels. In the worst case, the imbalance between differential channels can drive the preamplifier or postamplifier into clamping and the channel will fail. Even if the outright failure does not occur, the waveform at the output of the receiver might be severely distorted as clamping is approached and thus degrade the system performance. However, it is likely that the offset voltage problem can be reduced significantly in the future as technologies such as silicon bipolar or GaAs on silicon become more mature. In this section, we analyze the effect of parameter mismatch on system performance.[8]

A. Threshold Offset

The input signals to the positive and the negative terminals of the receiver, respectively, are

$$i_{in,+} = R_+ A_+ p_+(t) \qquad i_{in,-} = R_- A_- p_-(t) \tag{5}$$

where R is the responsivity of the PIN/APD while A is the attenuation either due to propagation or due to fan-out[9] of an LD/LED. $p_+(t)$ and $p_-(t)$ are the output power of the laser driver at time t. Appendix A shows that the DC offset of the decision threshold is

$$\frac{TH_{offset}}{\Delta V_{out,diff}} = \Delta M \frac{P_{av}}{2(P_1 - P_0)} \tag{6}$$

where $\Delta V_{out,diff}$ is the voltage swing of the signal while ΔM is the total channel mismatch, defined as

$$\Delta M = \frac{\Delta R}{R} + \frac{\Delta A}{A}. \tag{7}$$

We have assume the channel mismatch is small and is dominated by the mismatch of the waveguide attenuation as photodetector responsivity. If the laser has a zero extinction ratio (i.e., $P_0/P_1 = 0$), then $P_0 = 0$ and $P_{av} = \frac{1}{2}P_1$. The fractional threshold change, in this case, is equal to one quarter of the total channel mismatch, ΔM. As the extinction ratio of the laser increases, the fractional threshold change increases for the same amount of total channel mismatch. The fractional threshold change equals the total channel mismatch when the extinction ratio equals 0.6.

From this analysis, we can argue that a low-extinction-ratio laser is desirable in a fully differential optical interconnect to minimize the threshold offset due to channel mismatch. Furthermore, at least one quarter of the total channel mismatch will be reflected in the threshold offset even if lasers with zero threshold current are used.

If it is not feasible to adjust the thresholds of the individual circuits, this offset either has to be compensated by circuit techniques or be absorbed by the noise tolerance of the logic.

B. System Penalty

Assuming the probability of transmitting ONE's and ZERO's are equally likely, the error probability of a

[8] In addition to the DC mismatches discussed here, there are several AC mismatch sources which could occur in a differential optical interconnect and degrade the channel performance. These mismatch sources include mechanical vibration, thermal gradient of the laser, and the laser coherence time. The first one is not likely to be a serious problem for a differential optical interconnect since this type of noise can be cancelled at the receiver due to its large common-mode component. A small thermal gradient exists between adjacent lasers when the transmitted data is not DC-balanced. However, the thermal gradient of a laser array is much reduced when it is operated in a differential mode because of the spatially balanced nature of a differential optical interconnect array. Modal noise due to the mismatch of laser coherence in a differential optical interconnect will be twice as much as in a single-ended interconnect when lasers with large coherence time are used with multimode waveguides. However, this problem can be overcome by using self-pulsating lasers or by premodulating the lasers at a frequency comparable to their relaxation oscillation frequency.

[9] Unfortunately, the symbols used here conflict with the symbols used earlier in this paper. But we keep these symbols in order to be consistent with the conventions used in the literature. The meaning of these symbols will be clarified if they are not clear from the context.

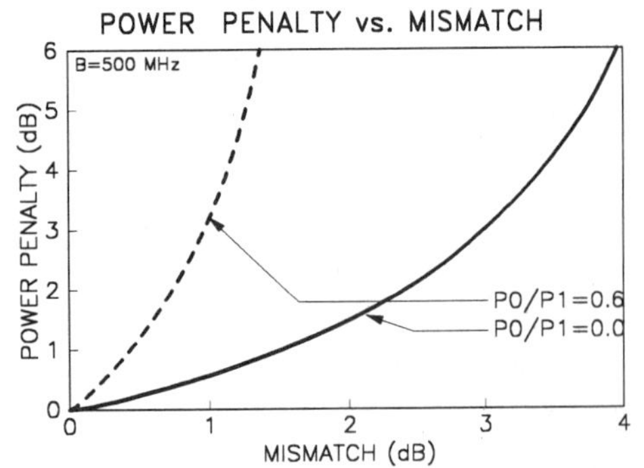

Fig. 31. Power penaly as a function of mismatch. Extinction ratio (P_0/P_1) is set to 0.6 adn 0. The receiver bandwidth B is 500 MHz.

differential optical interconnect system is

$$P_e = \frac{1}{4}\,\mathrm{erfc}(\frac{V_1 - V_{th}}{\sqrt{2}\sigma_{N,1}}) + \frac{1}{4}\,\mathrm{erfc}(\frac{V_{th} - V_0}{\sqrt{2}\sigma_{N,0}}) \tag{8}$$

where V_{th} is the threshold voltage and V_1, V_0, $\sigma_{N,1}$, and $\sigma_{N,0}$ have been derived in the previous section. The optimal threshold $V_{th,opt}$ is the solution to $\partial P_e/\partial V_{th} = 0$.

When the actual threshold is not the optimal threshold, more optical power is required to maintain the same bit-error-rate and thus an nonoptimal threshold incurs a power penalty. The power penalty as a function of the channel mismatch is plotted in Fig. 31. From this figure, we can see that a mismatch of 2 dB results in 1.52 dB power penalty when the laser is operated with zero extinction ratio (i.e., $P_0/P_1 = 0$). The same amount of power penalty is produced by 0.6 dB of mismatch as the extinction ratio of the laser increases to 0.6. Note that this analysis assumes the interconnect is operated in its linear region. These results do not hold if any of the components are driven into saturation or clamping and become nonlinear. A more complete simulation model is provided in [7].

VII. Summary

In summary, we found that the switching noise is a significant interference source in a transmitter array and a receiver array. A differential driver array can provide less waveform distortion and smaller switching noise compared to those that can be achieved by the single-ended configuration. Furthermore, we show that the switching noise in a driver array with either single-ended or differential configuration is linearly proportional to the size of the array and the lead inductance of the power supply. The switching noise of a large single-ended driver array (N≥8) is very significant (≥ 25%), indicating that a very large power decoupling capacitor (≫100 nF) is necessary. In contrast, the maximum switching noise of a differential driver array with dummy load at 1 Gb/s is less than 5% (or −26 dB). Similarly, it is possible to drive a large number (~32) of differentially configured receivers at 1 Gb/s in a monolithic integration environment using the criterion of −12 dB maximum allowable crosstalk with a single power supply. It is more difficult to obtain similar performance from a single-ended configuration without the help of a large decoupling capacitor (≥100 nF). Therefore, a fully differential interconnect architecture is more appropriate for a dense optical interconnect environment where switching noise is the dominant noise source.

In this paper, we presented and analyzed a differential optical interconnect architecture. This interconnect technique provides a symmetric channel for transmitting optical signals. It relaxes the system and circuit constraints posed by the traditional interconnect architecture, namely, the dependency of the absolute magnitude of decision threshold and the vulnerability to the power supply and the ground noise.

There are various ways of implementing a differential optical interconnect without increasing the interconnect density. We have investigated combining differential channels in the time division, in the frequency division, in the wavelength division and in the polarization division. All of these techniques are realizable using today's technology. It also has interesting implications for possible future device structures suitable for this application.

We have shown in this paper that the signal-to-noise performance of a differential optical interconnect is almost the same as a single-ended optical interconnect using differentially configured laser drivers and receivers. Differential optical interconnect is slightly better due to the omission of the biasing resistors which reduces the thermal noise of the front-end.[10]

Though this architecture is less vulnerable to many types of noise, it is sensitive to channel mismatch. The mismatch between channels causes the cross-coupling among the common-mode and differential-mode signals of the previous stage and thus changes the optimum decision threshold. We showed that a total channel mismatch of 2 dB will cause a 10% of threshold offset, and a 1.52 dB power penalty, assuming the extinction ratio of the laser is zero. This is usually tolerable in most of the applications. When high sensitivity is required in some applications, it becomes necessary to use circuit techniques to reduce the system penalty created by this channel mismatch.

Appendix
Derivation of Threshold Offset as a Function of Mismatch

Defining the detector responsivity and channel attenuation are, respectively,

$$R = \frac{R_+ + R_-}{2}$$
$$\Delta R = R_+ - R_- \tag{A1}$$
$$A = \frac{A_+ + A_-}{2}$$
$$\Delta A = A_+ - A_- \tag{A2}$$

Therefore, the detector output currents are

$$i_{in,+}(t) = (R + \frac{\Delta R}{2})(A + \frac{\Delta A}{2})p_+(t)$$
$$= RA(1 + \frac{\Delta R}{2R})(1 + \frac{\Delta A}{2A})p_+(t)$$
$$i_{in,-}(t) = (R - \frac{\Delta R}{2})(A - \frac{\Delta A}{2})p_-(t) \tag{A3}$$

[10] In practice, thermal noise performance is not an essential criterion for deciding between single-ended or differential optical interconnect structure since these short distance interconnects are usually not operated in their sensitivity region.

$$= RA(1 - \frac{\Delta R}{2R})(1 - \frac{\Delta A}{2A})p_-(t).$$

Assuming the mismatches are small, the above equation can be approximated by

$$i_{in,+}(t) = RA(1 + \frac{\Delta R}{2R} + \frac{\Delta A}{2A})p_+(t)$$
$$i_{in,-}(t) = RA(1 - \frac{\Delta R}{2R} - \frac{\Delta A}{2A})p_-(t) \quad \text{(A4)}$$

Defining a parameter, ΔM, *total channel mismatch*

$$\Delta M = \frac{\Delta R}{R} + \frac{\Delta A}{A} \quad \text{(A5)}$$

as well as the differential mode and the common-mode of the LD/LED output optical power

$$p_c(t) = \frac{p_+(t) + p_-(t)}{2}$$
$$p_d(t) = p_+(t) - p_-(t). \quad \text{(A6)}$$

The differential-mode of the input current to the receiver amplifier then equals

$$i_{diff}(t) = i_{in,+}(t) - i_{in,-}(t)$$
$$= RA(p_d(t) + \Delta M p_c(t)) \quad \text{(A7)}$$

and the common-mode of the input current to the receiver amplifier is

$$i_{common}(t) = \frac{i_{in,+}(t) + i_{in,-}(t)}{2}$$
$$= RA(p_c(t) + \frac{1}{4}\Delta M p_d(t)). \quad \text{(A8)}$$

If there is no mismatch within the receiver, the differential output is therefore equal to

$$v_{out,diff}(t) = R_{dm}RA(p_d(t) + \Delta M p_c(t)) \quad \text{(A9)}$$

where R_{dm} is the differential transimpedance of the receiver.

If the optical output powers from the channels of a differential transmitter are denoted respectively as $P_{i,+}$ and $P_{i,-}$ where $i = 1$ represents the logical ONE while $i = 0$ represents the logical ZERO, the common mode and differential mode of the optical signals then equal:

$$P_i = \frac{P_{i,+} + P_{i,-}}{2}$$
$$\Delta P_i = P_{i,+} - P_{i,-} \quad \text{(A10)}$$

When a logical ONE is sent, the corresponding differential mode and common mode optical power are

$$P_{d,1} = (P_1 + \frac{\Delta P_1}{2}) - (P_0 - \frac{\Delta P_0}{2})$$
$$P_{c,1} = \frac{1}{2}[(P_1 + \frac{\Delta P_1}{2}) + (P_0 - \frac{\Delta P_0}{2})] \quad \text{(A11a)}$$

On the other hand, the corresponding differential mode and common mode optical power when a logical ZERO is sent are

$$P_{d,0} = (P_0 + \frac{\Delta P_0}{2}) - (P_1 - \frac{\Delta P_1}{2})$$
$$P_{c,0} = \frac{1}{2}[(P_0 + \frac{\Delta P_0}{2}) + (P_1 - \frac{\Delta P_1}{2})] \quad \text{(A11b)}$$

Substituting (A11a), (A11b), into (A9), we can obtain the differential voltage for logical ONE and ZERO:

$$v_{out,diff,one} = R_{dm}RA[(P_1 + \frac{\Delta P_1}{2}) - (P_0 - \frac{\Delta P_0}{2})]$$
$$+ \frac{1}{2}R_{dm}RA\Delta M[(P_1 + \frac{\Delta P_1}{2})$$
$$+ (P_0 - \frac{\Delta P_0}{2})]$$
$$v_{out,diff,zero} = R_{dm}RA[(P_0 + \frac{\Delta P_0}{2}) - (P_1 - \frac{\Delta_1}{2})]$$
$$+ \frac{1}{2}R_{dm}RA\Delta M[(P_0 + \frac{\Delta P_0}{2})$$
$$+ (P_1 - \frac{\Delta P_1}{2})] \quad \text{(A12)}$$

The ideal decision threshold should thus be set at half way between the voltages representing ONE and ZERO:

$$TH_{offset} = \frac{1}{2}R_{dm}RA[(\Delta P_1 + \Delta P_0) + \Delta M(P_1 + P_0)] \quad \text{(A13)}$$

The total voltage swing of the received signal is

$$\Delta V_{out,diff} = R_{dm}RA[2(P_1 - P_0) + \frac{1}{2}\Delta M(\Delta P_1 - \Delta P_0)] \quad \text{(A14)}$$

If the mismatch is dominated by the waveguide attenuation as well as the photodetector responsivity, i.e., $\Delta P_1 \ll P_1$ and $\Delta P_0 \ll P_0$, (A13) can be simplified to

$$TH_{offset} = R_{dm}RA\Delta M P_{av} \quad \text{(A.15)}$$

where $P_{av} = \frac{1}{2}(P_1 + P_0)$, and (A14) can be simplified to

$$\Delta V_{out,diff} = 2R_{dm}RA(P_1 - P_0) \quad \text{(A16)}$$

Equations (A15) and (A16) can be combined, yielding

$$\frac{TH_{offset}}{\Delta V_{out,diff}} = \Delta M \frac{P_{av}}{2(P_1 - P_0)} \quad \text{(A17)}$$

ACKNOWLEDGMENT

The authors are grateful for the receiver models provided by K. Y. Toh and D. Rogers of IBM T. J. Watson Research Center and the careful review of the manuscript by Profs. D. G. Messerschmitt and J. M. Kahn of University of California, Berkeley, and Dr. Y. H. Kwark of IBM T. J. Watson Research Center.

REFERENCES

[1] A. Dickinson and M. E. Prise, "An integrated free space optical bus," *Proc. ICCD*, pp. 62–65, 1989.

[2] D. A. B. Miller *et al.*, "Novel hybrid optically bistable switch: The quantum well self-electro-optic effect devices," *Appl. Phys. Lett.*, July 1984.

[3] D. A. B. Miller *et al.*, "The quantum well self-electrooptic effect device: Optoelectronic bistability and oscillation and self-linearized modulation," *IEEE J. Quantum Electron.*, Sept. 1985.

[4] D. A. B. Miller, J. E. Henry, A. C. Gossard, and J. H. English, "Integrated quantum well self-electro-optic device: 2×2 array of optically bistable switches," *Appl. Phys. Lett.*, Sept. 1986.

[5] A. L. Lentine *et al.*, "Symmetric self-electro-optic device: Optical set-reset latch," *Appl. Phys. Lett.*, Apr. 1988.

[6] C.-S. Li, P. Lim, Y. Kwark, C. M. Olsen, and H. S. Stone, "A chip set for fully differential board/backplane optical interconnections," *Proc. Optical Fiber Commun.*, p. ThC4, Feb. 1993.
[7] C. M. Olsen and C.-S. Li, "Differential board/backplane optical interconnects for high-speed digital systems Part II: Simulation results," *J. Lightwave Technol.*, vol. 11, no. 7, July 1993.
[8] J. P. Van Der Ziel, R. A. Logan, and R. M. Mikulyak, "A Closely Spaced (50μm) Array of 16 individually Addressable Buried Heterostructure GaAs Lasers" *Appl. Phys. Lett.*, no. 41, p. 9, 1982.
[9] P. P. Deimal *et al.*, "Electrical and optical integration of a monolithic 1×12 array of InGaAsP/InP ($\lambda = 1.3\mu$m) light emitting diodes," *J. Lightwave Technol.*, vol. 3, 1985.
[10] D. Botez, J. C. Connolly, D. B. Gilbert, M. G. Harvey, and M. Ettenberg, "High-power individually addressable monolithic array of constricted double heterojunction large-optical-cavity lasers, " *Appl. Phys. Lett.*, no. 41, 1982.
[11] L. A. Koszi, B. P. Segner, H. Temkin, W. C. Dautremont-Smith, and D. T. C. Huo, "1.5 μm InP/GaInAsP linear laser array with twelve individually addressable elements," *Electron. Lett.*, vol. 24, no. 4, pp. 217–219, 1988.
[12] L. A. Koszi, H. Temkin, B. H. Chin, S. G. Napholtz, and B. P. Segner, "Fabrication and performance of an InP/InGaAsP monolithic 12×12 element matrixed LED array," *Electron. Lett.*, no. 23, pp. 284–286, 1987.
[13] D. L. McDaniel, Jr., J. G. McInerney, M. Y. A. Raja, C. F. Schaus, and S. R. J. Brueck, "Vertical cavity surface-emitting semiconductor laser with CW injection laser pumping," *IEEE Photon. Technol. Lett.*, vol. 2, no. 3, Mar. 1990.
[14] R. S. Geels, S. W. Corzine, J. W. Scott, D. B. Young, and L. A. Coldren, "Low threshold planarized vertical-cavity surface emitting lasers," *IEEE Photon. Technol. Lett.*, vol. 2, no. 4, Apr. 1990.
[15] Y. H. Lee *et al.*, "Characteristics of top-surface-emitting GaAs quantum-well lasers," *IEEE Photon. Technol. Lett.*, vol. 2, no. 9, Sept. 1990.
[16] J. L. Jewell *et al.*, "Surface-emitting microlasers for photonic switching and interchip connections," *Opt. Eng.*, vol. 29, no. 3, Mar. 1990.
[17] M. Uhle, "The influence of source impedance on the electrooptical switching behavior of LED's," *IEEE Transs Electron Dev.*, vol. 23, pp. 438–441, 1976.
[18] R. Olshnasky and D. Fye, "Reduction of dynamic linewidth in single-frequency semiconductor lasers," *Electron. Lett.*, no. 20, pp. 928–929, 1984.
[19] L. Bickers and L. D. Westbrook,"Reduction of laser chirp in 1.5 μm DFB lasers by modulation pulse shaping," *Electron. Lett.*, no. 21, pp. 103–104, 1985.
[20] M. A. Karr, F. S. Chen, and P. W. Shumate,"Output power stability of GaA as laser transmitter using an optical tap for feedback control," *Appl. Opt.*, no. 18, pp. 1262–1265, 1979.
[21] P. W. Shumate, F. S. Chen, and P. W. Dorman, "GaAIAs laser transmitter for lighwave transmission systems," *Bell Syst. Tech. J.*, vol. 57, pp. 1823–1836, 1978.
[22] D. W. Smith and M. R. Matthews, "Laser transmitter design for optical fiber systems," *IEEE J. Select. Area Commun.*, pp. 515–523, 1983.
[23] R. G. Schwatz and B.A. Wooley, "Stabilized biasing of semiconductor lasers," *Bell Syst. Tech. J.*, vol. 62, pp. 1923–1936, 1983.
[24] O. Wada *et al.*, "Optoelectronic integrated four channel transmitter array incorporating AlGaAs/GaAs quantum well lasers," *J. Lightwave Technol.*, vol. 7, no. 1, pp. 186–197, Jan. 1989.
[25] K. Kaede *et al.*, "12-Channel Parallel Optical-Fiber Transmission Using a Low-Drive Current 1.3-μm LED Array and a p-i-n PD Array," *J. Lightwave Technol.*, pp. 883–, June 1990.
[26] N. Yamanaka, M. Sasaki, S. Kikuchi, T. Takada, and M. Idda, "A Gigabit rate five-highway GaAs OE-LSI chipset for high-speed optical interconnections between modules or VLSI's," *J. Lightwave Technol.*, vol. 9, no. 5, June 1991.
[27] K. P. Jackson, C. Harder, P. Buchmann, and K. Datwyler, "High-speed characterization of a monolithically integrated GaAs-AlGaAs quantum-well laser-detector," *IEEE Photon. Technol. Lett.*, vol. 2, no. 11, Nov. 1990.
[28] J. Wieland and H. Melchior, "Optical receivers in ECL for IGHz parallel links," *SPIE Proc. Internat. Symp. Advances Interconnect. Packaging*, vol. 1389, pp. 659–664, 1990.
[29] W.-P. Hong *et al.*, "High-functionality waveguide/MSM/HEMT integrated receiver prepared by one-step OMCVD grown on patterned InP substrates," *Proc. OFC'91*, p. 5, Feb. 1991.
[30] J. D. Crow, "Optical interconnects for high-performance data processing systems," *Proc. IOOC'89*, 1989.
[31] J. D. Crow *et al.*, "A GaAs MESFET IC for optical multi-processor network" *IEEE Trans. Electron Dev.*, vol. 36, p. 263, 1989.
[32] N. Yamanaka and T. Takada, "A 1.5 Gb/s GaAs four-channel selector LSI with monolithically integrated newly structured GaAs ohmic contact MSM photodetector and laser driver," *IEEE Photon. Technol. Lett.*, vol, no. 10, pp. 310–312, 1989.
[33] M. Makiuchi *et al.*, "A monolithic four-channel photoreceiver integrated on GaAs substrate using metal-semiconductor-metal photodiodes and FET's," *IEEE Electron. Dev. Lett.*, vol. 6, no. 12, pp. 634–635, 1985.
[34] D. R. Kaplan and S. R. Forrest, "Electrical crosstalk in p-i-n arrays, Part I: Theory," *J. Lightwave Technol.*, vol. 4, no. 10, Oct. 1986.
[35] M. G. Brown *et al.*, "Monolithically integrated 1×12 array of planar InGaAs/InP photodiodes," *J. Lightwave Technol.*, vol. 4, no. 3, pp. 283–287, Mar. 1986.
[36] R. G. Meyer and R. A. Blauschild,"A wide-band low-noise monolithic transimpedance amplifier," *IEEE J. Solid-State Circuits*, vol. 21, no. 4, pp. 530–533, Aug. 1986.
[37] D. Rogers, "Monolithic integration of a 3-GHz detector/preamplifier using a refractory-gate, ion-implanted MESFET process," *IEEE Electron Dev. Lett.*, vol. 7, no. 11, pp. 600–602, Nov. 1986.
[38] G. K. Chang *et al.*, "Novel differential transimpedance receiver for high-bit-rate SONET systems," in *Proc. OFC'92*, 1992, pp. 235–236.
[39] J. Gowar, *Chapter 14: The Receiver Amplifier, Optical Communication System* Englewood Cliffs, NJ: 1984.
[40] R. Reimann and H.-M. Rein, "Bipolar high-gain limiting amplifier IC for optical fiber receivers operating up to 4 Gbit/s," *IEEE J. Solid-State Circuits*, vol. 22, no. 4, pp. 504–511, Aug. 1987.
[41] E. A. Lee and D. G. Messerschmitt, *Digital Communication*. Boston: Kluwar Academic Publishers 1988.
[42] A. Widmer and P. Franaszek, "A dc-balanced partitioned-block, 8B/10B transmission code," *IBM J. Res. Develop.*, Sept. 1983.
[43] R. C. Alferness, "Waveguide electro-optic modulators," *IEEE Trans. Microwave Theory and Tech.*, Aug. 1982.
[44] E. Voges and A. Neyer, "Integrated-optic devices on for optical communication," *J. Lightwave Technol.*, Sept. 1987.
[45] H. Kogelnik and R. V. Schmidt, "Switched directional coupler with alternating," *IEEE J. Quantum Electron.*, July 1976.
[46] K. Ishida, H. Nakamura, H. Matsumura, T. Kadoi, and H. Inoue, "InGaAsP/InP optical switches using carrier induced refractive index change," *Appl. Phys. Lett.*, vol. 50, no. 3, pp. 141–142, 1987.
[47] G. Winzer *et al.*, "Monolithically integrated detector chip for a two-channel unidirectional WDM link at 1.5 μm," *J. Select. Area Commun.*, vol. 8, no. 6, Aug. 1990.
[48] T. P. Lee, C. A. Burrus, and A. G. Dentai, "Dual wavelength surface emitting InGaAsP LEDs," *Electron. Lett.*, vol. 16, p. 845, 1980.
[49] J. C. Campbell, A. G. Dentai, T. P. Lee, and C. A. Burrus, "Improved two-wavelength demultiplexing InGaAsP photodetector," *J. Quantum Electron.*, vol. 16, p. 601, 1980.
[50] K. Ogawa, T. P. Lee, C. A. Burrus, J. C. Campbell, and A. G. Dental, "Wavelength division multiplexing experiment employing dual-wavelength LEDs and photodetectors," *Electron. Lett.*, vol. 17, p. 857, 1981.
[51] A. K. Chin, B. H. Chin, I. Camlibel, C. L. Zipfel, and G. Minneci, "Practical dual-wavelength light-emitting double diode," *J. Appl. Phys.*, vol. 57, p. 5519, 1981.
[52] J. C. Campbell, C. A. Burrus, J. A. Copeland, and A. G. Dentai, "Wavelength-discriminating photodetector for lightwave system," *Electron. Lett.*, vol. 19, pp. 672–674, 1983.
[53] T. H. Wood *et al.*, "Wavelength-selective voltage-tunable photodetector made from multiple quantum wells" *Appl. Phys. Lett.*, vol. 47, pp. 190–192, 1985.
[54] C.-S. Li, H. S. Stone, Y. Kwark, and C. M. Olsen, "Fully differential optical interconnects for high-speed digital systems,"*IEEE Trans. VLSI Syst.*, vol. 1, no. 2, pp. 151–163, 1993.

Reduced Switching Delay in Wavelength Division Multiplexed Two-Dimensional Multiple-Plane Optical Interconnections Using Multiple-Wavelength VCSEL Arrays

JAMES E. LEIGHT AND ALAN E. WILLNER, SENIOR MEMBER, IEEE

***Abstract*— We calculate the expected number of internodal hops for a network established with a wavelength division multiplexed (WDM) two-dimensional (2-D) multiple-plane optical interconnection. This WDM optical interconnection incorporates WDM pixels consisting of multiple-wavelength vertical-cavity surface-emitting laser (VCSEL) arrays and wavelength-selective detectors. The WDM interconnection can support simultaneous and reconfigurable communication among a network of nodes. Using the expected number of hops as a measure of internodal switching delay, we show that the integration of WDM into the interconnection results in a significantly reduced delay as compared to single-wavelength systems. Substantial delay reduction results even when the number of wavelengths is small relative to the number of 2-D planes. We analyze the bus, dual-bus, and ring architectures since they define the means of communication between pixels. For each architecture, we analyze three configurations which provide each node access to i) an entire plane of pixels, ii) a row (or column) of pixels, or iii) an individual pixel. When each network node has access to an entire plane of pixels, the proposed WDM interconnection incurs substantially shorter delay than single-wavelength optical interconnections. By allowing a node to access an entire row or column of pixels, the interconnection benefits from the incorporation of spatial division multiplexing (SDM) and the number of nodes connected can grow substantially with negligible added delay. Finally, when a node can access only a single pixel, a large number of independent processors can be interconnected exhibiting far less switching delay than other electronic or optical interconnections of comparable size.**

I. Introduction

HIGH-CAPACITY, low-switching-delay interconnections are necessary for high-speed data transfer among a multitude of network users. Applications of such systems include the interconnection of high-speed data processors, multiple two-dimensional (2-D) smart-pixel arrays, or stacked layers in a printed-circuit board. The existence of high-speed traffic bottlenecks and large power consumption in electrical interconnections [1], [2] has resulted in an increased interest in the use of optics as an interconnect technology [3], [4]. In this paper, we compare the network internodal switching delay of a proposed optical interconnection using wavelength division multiplexing (WDM) [5], [6] with the switching delay of optical interconnections using a single wavelength.

Manuscript received February 7, 1995; revised January 26, 1996. This work was supported by the Advanced Research Projects Agency (ARPA) under the "Ultra-Dense Optical Interconnects" initiative.

The authors are with the Department of Electrical Engineering—Systems, University of Southern California, Los Angeles, CA 90089 USA.

Publisher Item Identifier S 0733-8724(96)04602-6.

Multiple-plane optical interconnections can support the switching and routing of data among user-accessed 2-D arrays of optoelectronic pixels. A fixed-size 2-D array of pixels is arranged on each of a multitude of parallel substrate planes. An optical link can be established between two pixels if they occupy the same spatial position on different planes (we describe these pixels as being colocated). To date, there are two proposed single-wavelength optical solutions to the multiple-plane interconnection problem.

In the first method (denoted as "plane-to-plane") a data packet is detected at each plane [see Fig. 1(a)]. If the data is not intended for that plane, it is retransmitted to the next plane. This system is characterized by an inherently long average switching delay, an optoelectronic bottleneck, and large power consumption. The plane-to-plane method demonstrates an inefficient use of optoelectronic hardware because many of the pixels are occupied retransmitting data packets and are unable to transmit additional data. A second method, illustrated in Fig. 1(b), uses large through-wafer holes to eliminate the redundant processors and provide a direct optical path between pixels [7], [8]. This method eliminates unnecessary optoelectronic hardware, saves power, and substantially reduces the switching delay. However, this "via-window" method supports only a static interconnection and therefore does not allow reconfiguration of the interconnection. Furthermore, this method fails to efficiently utilize costly substrate area.

In a previous paper [9], we proposed the incorporation of (WDM) into the 2-D multiple-plane interconnection and derived formulae for the average capacity (i.e., average number of simultaneous links) of such an interconnection. The WDM system was realized by incorporating several multiple-wavelength vertical-cavity surface-emitting lasers (VCSEL) [10] identically into each transmitting pixel and incorporating wavelength-selectivity into each subsequent detecting plane so that each plane absorbed only one wavelength and was transparent to the others. A basic WDM multiple-plane interconnection, in which the first plane consists of transmitting pixels and the remaining planes consist of detecting pixels, is illustrated in Fig. 2.

Reprinted with permission from *IEEE Journal of Lightwave Technology*, Vol. 14, No. 6, pp. 1467-1479, June 1996.

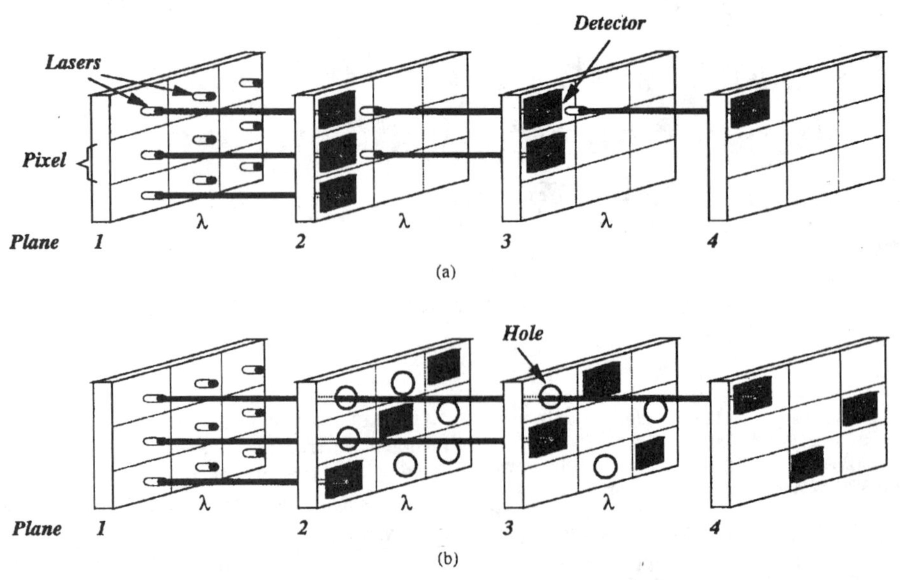

Fig. 1. Single-wavelength 2-D multiple-plane optical interconnections: (a) plane-to-plane, (b) via-window.

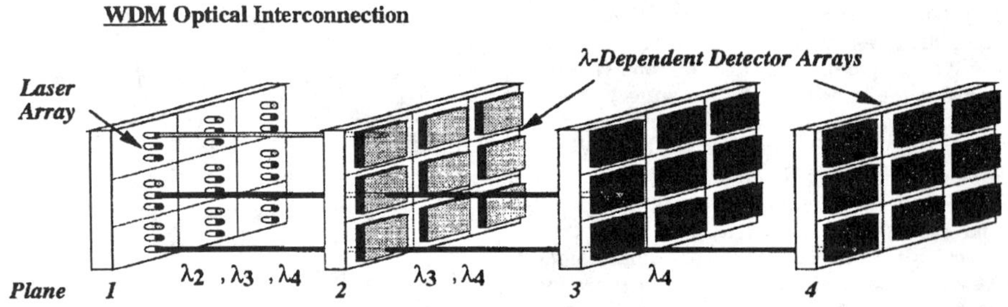

WDM Optical Interconnection

Laser Array

λ-Dependent Detector Arrays

Plane 1 $\lambda_2, \lambda_3, \lambda_4$ 2 λ_3, λ_4 3 λ_4 4

Fig. 2. WDM multiple-plane optical interconnection.

Beginning at the first plane, every pixel is designed with a detector spectral response that is slightly offset from the response of the corresponding pixel on the next plane, with all detectors on the same plane being characterized by the same spectral response. The detectors are designed so that the cutoff wavelength (determined by the material composition [11], [12]) increases for each subsequent plane as illustrated in Fig. 3. Each detecting plane absorbs all the wavelengths shorter than the detector cutoff and is transparent to signals at wavelengths longer than the detector cutoff. As an example, if the first of four planes is transmitting signals at the wavelengths $\{\lambda_2 < \lambda_3 < \lambda_4\}$, then detectors on Plane 2 absorb λ_2 only and are transparent to λ_3 and λ_4; detectors on Plane 3 absorb λ_3 and are transparent to λ_4; and detectors on Plane 4 absorb λ_4. The spectral spacing between the multiple-wavelengths $(\Delta\lambda)$ is determined by both the shape of the detector responsivity curve and minimum power-penalty requirements [9]. An experimental demonstration has been performed to establish simultaneous communication from one pixel transmitting in the 900–1000 nm wavelength range to three one-pixel detecting planes at 155 Mb/s with a wavelength separation of 43 nm [13]. The next generation of devices are expected to perform closer to Gb/s data rates with a wavelength separation near 10 nm.

The VCSEL arrays consist of at most S lasers emitting at distinct wavelengths spaced equally apart by $\Delta\lambda$. If multiple lasers per pixel are allowed to operate simultaneously, a user can either broadcast the same data or transmit independent data to many users. However, these types of functionality require additional driver electronics that can greatly increase the complexity and size of each pixel. Fortunately, even when only one **individual** laser per pixel is allowed to operate at any time, the system functionality is enhanced since spatially colocated signals can be established at different wavelengths and will not interfere with each other. We assume that only one **individual** laser per pixel is allowed to operate at a time. The reconfigurable switching from one plane to many detecting planes is accomplished by turning "ON" the appropriate laser in the VCSEL array. The optical beams propagate in paths orthogonal to the 2-D planes. Consequently, an individual pixel can optically link only to those pixels on the other planes in its "line-of-sight".

The WDM multiple-plane optical interconnection consists of M parallel planes, each plane hosting an $N \times N$ pixel

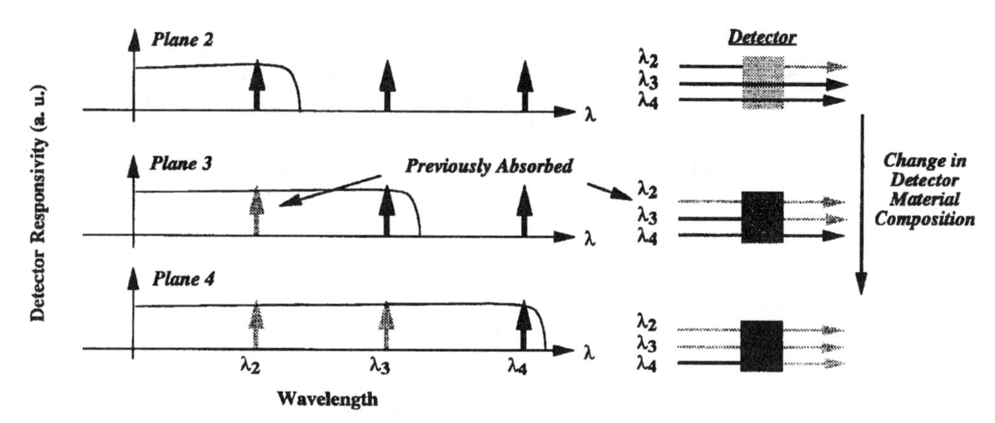

Fig. 3. Wavelength-selectivity. Wavelength-selective detector planes detect only the shortest remaining wavelength in the signal.

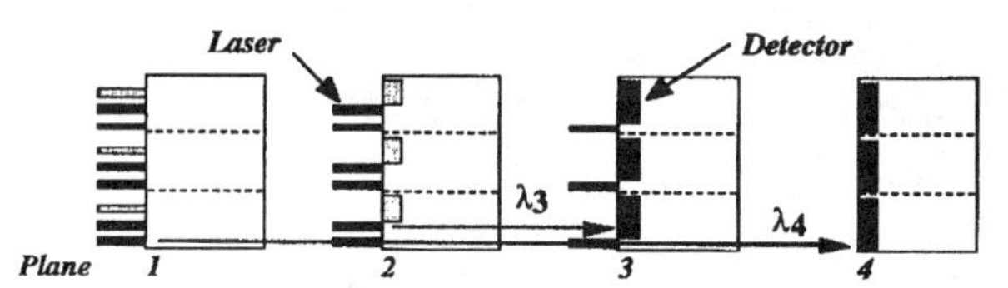

Fig. 4. WDM interconnection in which intermediate pixels can transmit and receive data.

array. The interconnection provides the switching fabric for a network of processing nodes. A single node accesses a pixel or group of pixels on one of the parallel planes. The WDM system we analyze is a slight modification of Fig. 2. Every pixel consists of a laser array **and** a detector providing each network node the capability of sending and receiving data (see Fig. 4).

In this paper, we assess the advantage of low switching delay resulting from the incorporation of WDM into the multiple-plane optical interconnection relative to the single-wavelength interconnection. The switching-delay reduction is a consequence of the unique absorption/transmission characteristics of the parallel wavelength-selective detecting planes. Many direct internodal links between nonadjacent planes can be established, reducing the number of hops required to send data between nodes. This paper addresses only the dependence of the switching delay on the physical structure of the interconnection; network control and contention resolution issues affecting the total transmission delay are not examined in this work.

A WDM system increases the network capacity by providing each pixel with multiple optical channels. The proposed WDM optical interconnection is a tunable-transmitter/fixed-receiver system in which the user transmits data on one of a multitude of available optical wavelengths to establish a communication link. The transmitting wavelength is chosen so that the signal is detected at the desired receiving pixel. One advantage of WDM is that multiple links can be established on separate wavelengths simultaneously; optical signals that are colocated spatially but at different wavelengths do not interfere with each other. For example, as illustrated in Fig. 4, the WDM system allows one communication link to be established between pixels on two outer planes (Plane 1 and Plane 4) and an **additional** communication link to be established at a different wavelength between pixels on two inner planes (Plane 2 and Plane 3). Furthermore, the WDM interconnection: i) can easily be reconfigured, ii) allows simultaneous communication among many planes, and iii) delivers a large aggregate capacity to many users.

If enough multiple-wavelength lasers are on each pixel, the system can provide a one-hop connection between colocated pixels on any two planes. We define an optical "hop" as the detection or retransmission of data. However, because of technological limitations, there may be fewer lasers on a pixel than the number required for all one-hop connections. For such interconnections, we must resort to "multihop" techniques and connections between nodes consist of multiple detections and retransmissions of the same data. The multihop system is a compromise between the plane-to-plane interconnection and the one-hop fully multiplexed WDM system.

As a measure of the network switching delay, we calculate the expected (i.e., average) number of internodal hops for a network supported by the WDM optical interconnection. We find substantial delay reduction even when the number of wavelengths is small relative to the number of 2-D planes. For all structures analyzed in this work, the expected number of hops is nearly inversely proportional to the number of wavelengths per pixel. Consequently, the addition of even one wavelength to a single-wavelength system nearly halves the expected switching delay. By allowing a network node to access an entire row (or column) of pixels, we incorporate space-division-multiplexing (SDM) [14] into the interconnection. The resultant system is a hybrid WDM/SDM interconnection that can support an increasing number of nodes with only marginal additional delay relative to the plane configuration. Finally, by allowing the node to access only an individual pixel, we show that a large number of processors can be interconnected in a system exhibiting far less switching delay than electronic or optical interconnections of comparable size.

II. System Model

A. Network Configurations

There are three possible configurations that define the number of pixels per plane each node can use to access the

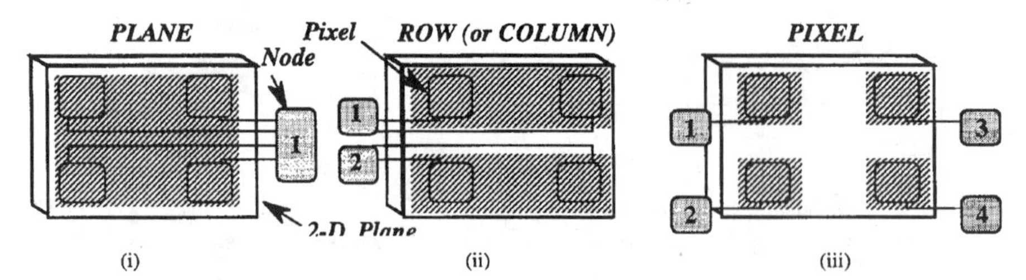

Fig. 5. Three nodal access configurations: (i) plane; (ii) row/column; (iii) pixel.

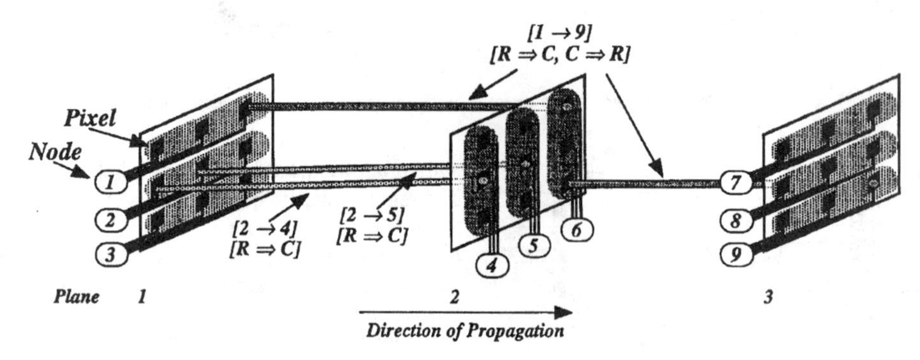

Fig. 6. WDM multiple-plane interconnection configured so that each node has access to a row [R] or column [C] of pixels.

network and consequently, how many nodes can be supported by an interconnection with M 2-D planes. The configurations, illustrated in Fig. 5, are defined as follows:

i) **Plane:** A node has access to an entire plane of pixels.

   For this configuration, each network node has access to all N^2 pixels per plane. As a result, each node has many choices for the spatial channel as well as the wavelength channel on which to send data. This configuration is useful for networks with large throughput requirements per node. Furthermore, data contention can be significantly reduced, because of the added flexibility to route traffic on unused spatial channels. A limitation is that the resulting interconnection can support only M nodes.

ii) **Row or Column:** A node has access to a row or column of pixels.

   In this configuration, each node has access to either a row or a column of N pixels. The interconnection can now support $N * M$ nodes. For a fully connected network, the orientation of the node configuration alternates between row and column from one plane to the next as illustrated in Fig. 6. This alternating orientation provides the added benefit that each node can now access in one hop **all** of the N nodes on the next plane. In the illustration of Fig. 6, node 2 can send data to nodes 4, 5, or 6. This resulting system is a hybrid WDM/SDM interconnection. A user now chooses a destination node by selecting the laser at the proper transmitting **wavelength** and the pixel at the proper **spatial location** so that the transmitting pixel is in the same line as a pixel accessed by the destination node.

iii) **Pixel:** A node has access to an individual pixel.

   In this configuration, each node has access to one "smart" pixel, and $N^2 * M$ nodes can be interconnected. Although for large M, this configuration can support a large number of nodes, it is not practical when traffic on the network is heavy because of the increased likelihood of contentions. A challenge with this configuration is that only a relatively small number of node-to-node links can be made optically since each node can link only to those nodes in the same line on the other planes. In addition to the optical interconnection, there must be an electrical interconnection between pixels on the same plane in order to establish a fully connected network. We will discuss the topology of such an electrical interconnection, and its integration into the system, later in the paper.

B. Network Architectures

For each of the three configurations described above, there are three network architectures that can be implemented with the 2-D multiple plane system. Each architecture influences the functionality, complexity, and cost of the interconnection. The architecture is closely linked to the physical structure of the interconnection since it determines the direction of communication as well as which pairs of **pixels** can establish communication links. The architectures are described below:

1) **Unidirectional Bus** (Fig. 7): The bus consists of M parallel wavelength-dependent planes. An optical link is established between two pixels when a laser at the appropriate wavelength on a pixel is turned on so that the light is detected at a specific destination pixel.

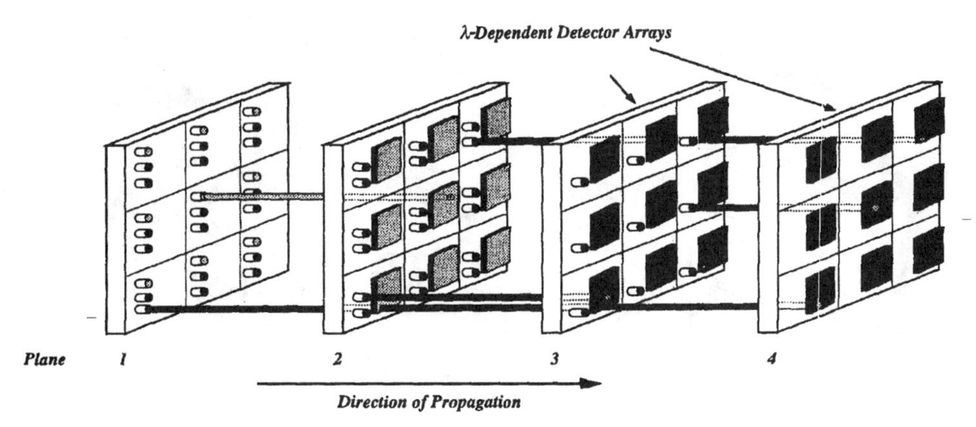

Fig. 7. Unidirectional bus architecture.

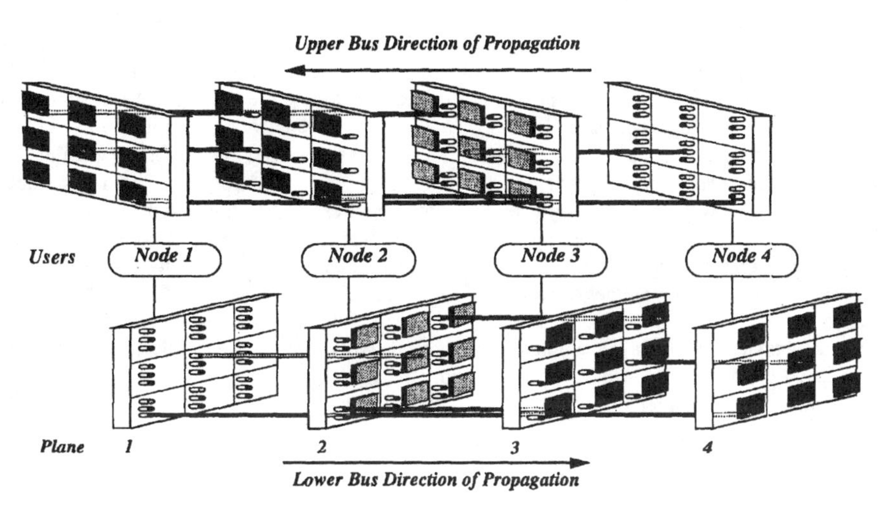

Fig. 8. Dual-bus architecture.

The functionality of this architecture is limited, since it only provides unidirectional communication from lower numbered planes to higher numbered planes, but an analysis of the bus is useful as it is the building block for other architectures.

2) **Dual Bus** (Fig. 8): The dual bus consists of two unidirectional buses, with light on the lower bus propagating from left-to-right and light on the upper bus propagating from right-to-left. If a node has access to a set of pixels on each bus, it can communicate (i) with a higher numbered node through the lower bus and (ii) with a lower numbered node through the upper bus. Although the amount of optical hardware is doubled, this architecture supports a network in which every node can establish a link with all other nodes.

3) **Unidirectional Ring** (Fig. 9): The unidirectional ring consists of a bus in conjunction with beam-deflecting elements, such as angled mirrors, to route light from the final plane back to the first plane. This architecture provides for a fully connected network with the use of only one multiple-plane system, but its physical realization (i.e., adding the angled mirrors) might be difficult to accomplish.

We derive formulae for the expected number of hops for each of the configurations and architectures described above. We analyze multihop systems where the number of wavelengths per pixel (S) varies from 1 to $M-1$, where M is the total number of planes.

At this juncture, the relevance of the physical structure of these WDM systems as outlined should be made clear. In the system model, we have not focused on the network logical topology (i.e., the mathematical representation for the connections between nodes) to be associated with this interconnection so as to make more clear the advantage of a WDM system over its single-wavelength counterpart; we are interested in the actual connections rather than the characteristics of a network utilizing this interconnection. The logical topology for the network supported by this interconnection is only marginally dependent on the architectures and configurations that we have defined. For our analysis, we have chosen the node configurations such that the network topology matches the physical architecture. This is done to avoid obscuring the relationship

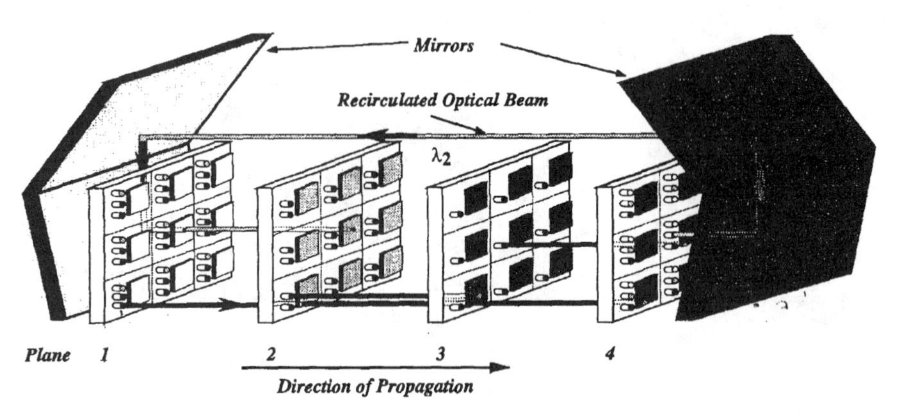

Fig. 9. Ring architecture.

between the physical structure of the system and the resultant performance. Bus and ring topologies belong to a family of interconnection patterns characterized by less efficient **linear** performance. Consequently, we would expect the switching delay to likewise exhibit a linear relationship to the number of planes. Of course, there are other topologies such as 2-D Benes, Banyan, and Shuffle Networks that can be used to establish network connections and are characterized by much better performance. We expect that the use of WDM will offer additional advantages for each of these topologies as well. For example, the possibility to skip stages in a Shuffle Network could significantly reduce the switching delay encountered by such a network. We leave the realizations of these topologies supported by the 2-D multiple-plane interconnection as the subject of future work.

III. Mathematical Model

A. Background

For the WDM multihop system, the number of hops encountered by an observed data stream is a random variable with value depending on i) which source/destination pair an observed data stream is moving between and ii) the number of intermediate nodes utilized to relay the data. The expected number of hops $\bar{H}$ is a statistical average over all possible node pairs of the number of hops required to send any given data stream. The expected number of hops is frequently used as a measure of the average switching delay incurred by data packets in a multihop network.

The total transmission delay for an observed data stream is comprised of numerous components. Data packets are delayed by optical-to-electrical/electrical-to-optical (OE/EO) conversion, optical and electrical propagation, and buffering or deflection routing schemes used to alleviate contention. To provide a robustness to our analysis, we have disregarded delay components not associated with switching data between nodes. We disregard propagation delays because they depend on the specific physical structure of the system; they are determined by factors such as the 2-D plane spacing and electrical connection layout. We disregard delay associated with buffering or deflection routing because these delays depend on the specific contention resolution scheme as well as the topology and traffic load experienced by the network. We instead determine the average delay associated with switching data through the network, assuming a fixed OE/EO conversion and node processing delay. The switching delay of a data stream is directly proportional to the number of hops the data requires to be routed from source to destination. By determining the expected number of hops we can acquire an accurate measure of the relative performance between two systems. Therefore, we use the number of hops as a figure of merit. With the additional knowledge of a specific physical and network design, we would gain a more accurate picture of the total transmission delay experienced by a data stream on the network.

B. Basic Equations

We first present a general formulation for the expected number of hops, $\bar{H}$, between nodes interconnected with the multiple-plane system. The derivation is for the minimum expected number of hops, assuming that the shortest path between two nodes is used. The expected number of hops can be expressed in one of two equivalent forms:

$$\bar{H} = \sum_{i=1}^{K} \sum_{\substack{j=1 \\ j \neq i}}^{K} H(i,j) P(i,j) \tag{1}$$

$$\bar{H} = \sum_{i=1}^{K} \bar{H} \mid (i,^*) P(i,^*) \tag{2}$$

where K is the total number of nodes interconnected, $H(i,j)$ is the minimum number of hops from source node i to destination node j and $P(i,j)$ is the probability that an observed data stream is being sent from source node i to destination node j. The * in (2) indicates than the destination node can be any node in the system other than the source node. Therefore, $\bar{H} \mid (i,^*)$ is the expected number of hops for links where the source is node i, and $P(i,^*)$ is the probability of the union of events where data is being sent between node i and any arbitrary destination node. Equation (1) is simply the definition of the expected value of a random variable and (2)

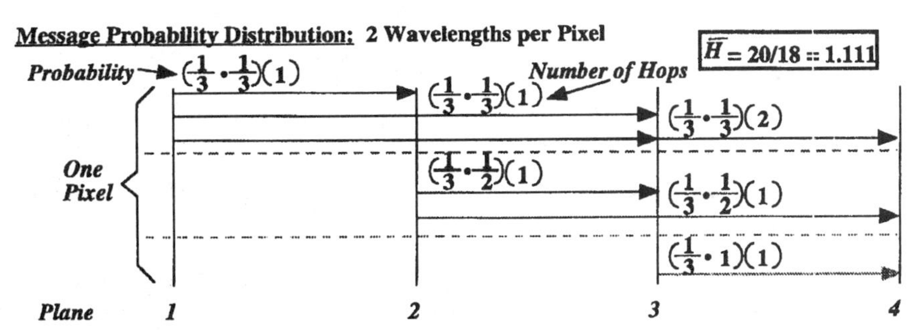

Fig. 10. Calculation of the expected number of hops for a two wavelength per pixel bus system. The message probabilities are given assuming each node is given fair access to the interconnection.

is the result of using conditional expectation and Bayes' Law of Total Probability.

In evaluating the expected number of hops for a randomly observed data stream, we must determine the minimum possible **number of hops** in each internodal connection as well as the **probability** of each connection. The number of hops between nodes is determined by the architecture, the configuration, and the number of wavelengths per pixel. The probability of the connection depends on the network control algorithm. Although the focus of this work is not on the details of the algorithm, we do assume that network control is established to allow each node to send the same number of messages per unit time. If a message being switched through the interconnection is chosen at random, the probability that it originated from Node 1 must be the same as the probability that it originated from Node 2, etc. For the dual-bus and ring architectures, since every possible pair of nodes can communicate, the probability model imposed on the data messages translates to every connection being equally likely. However, for the bus architecture, with every node sending the same number of messages in a unit time interval, the connections are not equally likely. The probability and the number of hops is shown in Fig. 10 for every possible connection in a four plane system with two wavelengths per pixel. For example, if we observe a message at random, the probability that it is being sent from Node 1 to Node 3 is $\frac{1}{3} \cdot \frac{1}{3} = \frac{1}{9}$. One third of the time, the message originated from Node 1 and one third of that time, the message is being sent to Node 3. When the network access rules are different, the resulting probabilities will change accordingly.

Most mathematical models allow for a node to establish a link with itself, since closed-form solutions usually exist when this allowance is made. This type of link, possibly used for self-diagnostics or feedback oriented signal processing, would be established infrequently and we therefore assume it occurs with zero probability; the deviation in the results is minimal. The formulation of $\bar{H}$ is determined by using the simple assumption that with S wavelengths per pixel ($1 \leq S \leq M-1$), each pixel can link optically to those pixels in the same spatial position on the following S planes in **one** hop.

The expected number of hops will be a function of the number of available wavelengths S the number of rows or columns N and the number of planes M. We first derive $\bar{H}$ for the plane configuration with the bus architecture since this is a basic building block for the other configurations and architectures. We then extend this result to determine the formulations for other system structures.

IV. Results

A. Plane Configuration

We first give the formulation for the plane configuration in the bus architecture. We use (2) to calculate the expected number of hops after determining the values of $P(i,*)$ and $\bar{H}(i,*)$ for each value of i. We assume that an observed data stream originates from any of the transmitting nodes with equal probability $1/(M-1)$ where, for the plane configuration, the number of nodes is equal to the number of planes and only the first $M-1$ nodes can transmit data. The final node sends no data in the unidirectional bus architecture because no destination planes follow it. The value of $\bar{H} \mid (i,*)$ is different for each i, because each progressive node sees fewer destination nodes. The expected number of hops from source node i in an M plane system is the same as the expected number of hops from source node 1 in a $(M-i+1)$ plane system. As an illustrative example consider a four-plane system. Plane 1 can send data to three planes: Plane 2, 3, or 4. Plane 2 can send data to two planes: Plane 3 or 4. From the point of view of Plane 2, it is as if it was accessing a pixel on Plane 1 in a $(M-i+1=3)$ plane system. For each transmitting plane i, we calculate the expected number of hops to the following receiving planes j. We do this by determining, for each value of i, the number of hops to the receiving plane j [15]. An expression for the number of hops is derived using the fact that for each transmission, the data can at most move S planes ahead. The expected number of hops for the bus architecture with the plane configuration is

$$\bar{H} = \frac{1}{M-1} \sum_{i=1}^{M-1} \left(\frac{1}{M-i} \sum_{j=i+1}^{M} \left\lceil \frac{j-1}{S} \right\rceil \right) \tag{3}$$

where i represents the transmitting plane, j represents the receiving plane, and $\lceil x \rceil$ is the nearest integer greater than or equal to x, since the number of hops must always be an integer. A simple explanation of the routing rule used to determine $\bar{H}$ is that at each intermediate node, the laser with the longest wavelength is used to move the maximum number (S) of

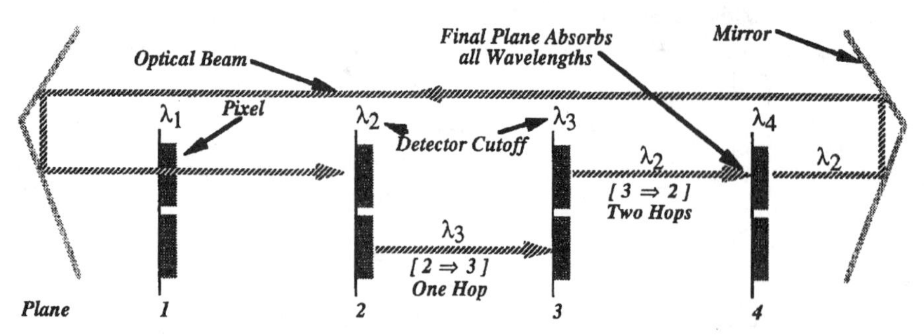

Fig. 11. Illustration of extra hops resulting from light absorption at the final plane.

planes ahead at each hop until the destination is no more than S planes away in which case, the appropriate laser is turned on to reach the destination in one final hop.

We use the same idea to derive an expression for the expected number of hops for the plane configuration in the dual-bus architecture. In this architecture, all of the M nodes are capable of sending data, and the probability that node i is the source, $P(i, *)$ is $1/M$. When using the dual-bus architecture, communication from left to right is accomplished on the lower bus and communication from right to left is accomplished on the upper bus. Again, the number of hops between planes is the difference of the transmitting and receiving plane indices divided by the number of wavelengths. We now account for the possibility $i > j$ by using the absolute value of the difference. The resulting expression for the expected number of hops is

$$\bar{H} = \frac{1}{M}\sum_{i=1}^{M}\left(\frac{1}{M-1}\sum_{j=1}^{M}\left\lceil\frac{|j-i|}{S}\right\rceil\right) \tag{4}$$

where $|x|$ denotes the absolute value of x.

For the plane configuration in the ring architecture, the formulation is slightly more complex. Since all of the M nodes in the system can send data, the probability that node i is the source, is $1/M$. In determining the number of hops for links in the bus and dual-bus architectures no light passed through the final plane. However, in the ring architecture, there are some node to node connections for which data is routed through the final plane. Since detectors on the final plane have the longest cutoff wavelength, they absorb **all** of the light in the system regardless of the wavelength. Therefore, an optical link can never bypass the final plane. As an example (see Fig. 11), if the number of wavelengths per pixel is 3 and a node on Plane 3 sends data to Plane 2 on wavelength λ_2, the link will require not one, but **two** hops: a hop from Plane 3 to the last plane where λ_2 is absorbed and a hop from the last plane to the second plane. Even if the number of lasers per pixel is the maximum allowable $(M-1)$, nodes can only establish one-hop optical links with those nodes in the same line on planes in the **forward** direction. To account for this, we note that for links (i, j) where $i > j$ the total number of hops is the sum of the hops for the link (i, M) and the hops for the link (M, j). The expression for the expected number of hops from transmitting plane i now consists of two summations; one summation with two terms for links to destination planes previous to the transmitting plane, and a second summation with a single term for links to destination planes following the transmitting plane:

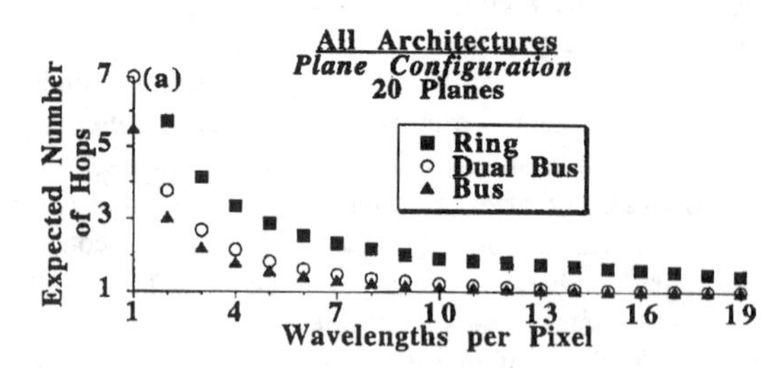

Fig. 12. Expected number of hops as a function of the number of available wavelengths for the bus, dual-bus, and ring architecture.

$$\bar{H} = \frac{1}{M}\sum_{i=1}^{M}\frac{1}{M-1} \times \left(\sum_{j=1}^{i-1}\left(\left\lceil\frac{M-i}{S}\right\rceil + \left\lceil\frac{j}{S}\right\rceil\right) + \sum_{j=i+1}^{M}\left\lceil\frac{j-1}{S}\right\rceil\right). \tag{5}$$

The expected number of hops given by (3), (4), and (5) as a function of the maximum number of wavelengths per pixel (S), is plotted in Fig. 12. Notice the dramatic decrease in hops with increased number of wavelengths for all architectures. As the number of available wavelengths continues to increase, there is little change in the expected number of hops. For as few as five WDM wavelength channels, the average number of hops is less than 3. The results for the bus and dual-bus follow each other closely. The WDM system with only a few wavelengths per pixel can provide full connectivity among the nodes with little switching delay.

The average number of hops is nearly inversely proportional to the number of wavelengths per pixel. To clearly illustrate this, we plot in Fig. 13 the expected number of hops as a function of the number of planes (M) for various values of the number of wavelengths per pixel. The average number of hops varies linearly with the number of planes; the slope of the curves decrease with S. This improved performance is possible

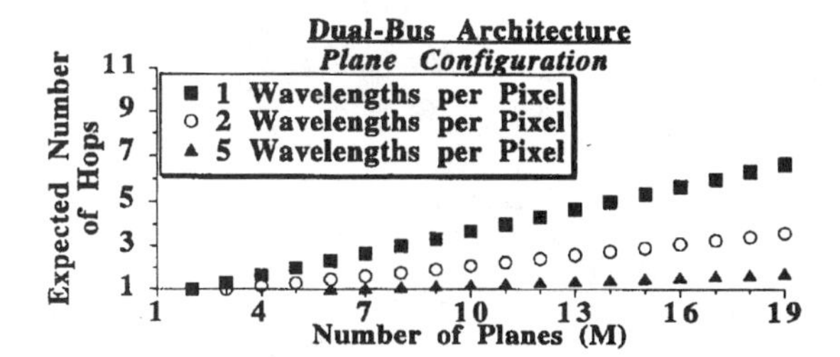

Fig. 13. Expected number of hops as a function of the number of planes for 1, 2, and 5 wavelengths per pixel.

because more nodes can be reached in a single hop with additional wavelength channels. This result originates from the novel use of wavelength-selectivity in the interconnection as planes can be bypassed to establish communication links. As a result, more node pairs can be connected in fewer hops. For example, if there are two wavelengths for each pixel, then for most of the source nodes, two nodes can be reached in one hop, two nodes can be reached in two hops, etc. If there are five wavelengths per pixel, then five nodes can be reached in one hop, five nodes in two hops, etc. Hence, the resulting inverse trend. However, even when there are five wavelengths, there are some source nodes that are unable to reach five nodes in one hop, two hops, etc. The expected number of hops is therefore not exactly inversely proportional to the number of wavelengths, although the curves follow the inverse relation closely.

B. Row/Column Configuration

The solution to the row/column configuration is a more general derivation of (3), (4), and (5) for the plane configuration. Using (2), we must determine the probability that a node i can establish a link and the expected number of hops from source node i. An important mathematical property of the row/column configuration that simplifies the calculation of the expected number of hops is that the number of hops from a given node is the same as the number of hops from any other node on the same plane, i.e., each node on the same plane can link to the same number of nodes in 1 hop, 2 hops, etc. As a result, we need only calculate $\bar{H} \mid (i,^*)$ for a single node on each plane and then average the results over the probability distribution of the source planes. For each architecture, $P(i,^*)$ is the same as for the plane configuration because now we are concerned with the probability that the source node is located on Plane i.

There are a few subtelties in calculating the number of hops between a pair of nodes for the row/column configuration. The first issue is that the relative orientation between the nodes on the source plane and the destination plane affects the number of hops. If the orientation of the nodes on the receiving plane is orthogonal to the source node, then the number of hops for all N nodes are the same as calculated in the plane configuration. If the orientation of the nodes on the receiving plane is parallel to the source node, then **one** node is the usual number of hops away and the remaining $N-1$ nodes require an additional hop. This feature is a consequence of the space-multiplexing (i.e., spatial parallelism) inherent in the row/column configuration. When the relative orientations of the source and destination nodes are parallel, the data needs to be detected at an intermediate plane with orthogonaly oriented nodes, so that the light beam can be spatially shifted to a position in line with the receiving node. When the relative orientations of the source and destination nodes are orthogonal, the transmitting node can pick a pixel on which to transmit the data in line with the receiving node. A second issue concerns the number of wavelengths. Assume that the source and destination nodes are parallel and separated by more than S planes. If the number of wavelengths is even, as the data is passed using the longest wavelength available from one plane to the next, the data never stops on a plane with an orthogonal orientation and an extra hop is required at some point to shift the light beam to the right position. If in the same situation, the number of wavelengths is odd, then after the first hop, the beam can be shifted to the correct position. However, if the destination plane is less than S planes away (i.e., $j-i<S$), an additional hop is needed for those nodes at a spatial position different from the transmitting node. Thus the relative orientation and the number of wavelengths affect the routing and the expected number of hops is different for even or odd values of S. Before deriving an expression for the expected number of hops in the row/column configuration, it is useful to define the following quantities:

$$\alpha_{ij} = \left\lceil \frac{j-1}{S} \right\rceil \tag{6}$$

$$\beta_{ij} = \left\lceil \frac{|j-i|}{S} \right\rceil \tag{7}$$

$$\gamma_{ij} = \left\lceil \frac{M-i}{S} \right\rceil + \left\lceil \frac{j}{S} \right\rceil. \tag{8}$$

These terms appeared in the expressions for the plane configuration and will also appear in the expressions for the row/column configuration. α_{ij} is the number of hops between two planes where the destination plane follows the transmitting plane, β_{ij} allows for $i>j$ and γ_{ij} is the sum of hops to the final plane and hops to the destination plane in the ring architecture when $i>j$.

For the row/column configuration in the bus architecture, nodes on all but the last plane are able to send data and consequently the probability that a source node on Plane i is able to establish a link, $P(i,^*)$ is $1/(M-1)$. In deriving an expression for the expected number of hops, we use the fact that from each source node the number of possible communication links is the product of the number of planes that can be chosen to receive data $(M-i)$ and the number of row/column nodes per plane (N). The resulting expression for even S is

$$\bar{H} = \frac{1}{M-1} \sum_{i=1}^{M-1} \frac{1}{N(M-i)} \times \left(\sum_{\substack{j=i+1 \\ j-i\,\text{even}}}^{M} \alpha_{ij} + (\alpha_{ij}+1)(N-1) + \sum_{\substack{j=i+1 \\ j-i\,\text{odd}}}^{M} \alpha_{ij} N \right) \tag{9}$$

and for odd S

$$\bar{H} = \frac{1}{M-1}\sum_{i=1}^{M-1}\frac{1}{N(M-i)} \times \left(\sum_{\substack{j=i+1\\ j-i\,\text{even}}}^{M} (i-u(j-i-S))(1+2(N-1)) \right.$$
$$\left. + u(j-i-S)\alpha_{ij}N + \sum_{\substack{j=i+1\\ j-i\,\text{odd}}}^{M}\alpha_{ij}N \right) \quad (10)$$

where $u(x)$ is the unit step function: $u(x) = 0, x \leq 0$ and $u(x) = 1, x \geq 0$. This function is used to add in the first term if $j - i < S$ and the second term if $j - i \geq S$.

For the row/column configuration in the dual-bus architecture the same types of links are found. We recognize that nodes on the same plane (when $j - i = 0$) are separated by two hops, one on the lower bus and one on the upper bus. The expression for the expected number of hops for even S is

$$\bar{H} = \frac{1}{M}\sum_{i=1}^{M}\frac{1}{NM-1} \times \left((N-1) + \sum_{\substack{j=1\\ j-i\,\text{even}}}^{M} \beta_{ij} + (\beta_{ij}+1)(N-1) \right.$$
$$\left. + \sum_{\substack{j=1\\ j-i\,\text{odd}}}^{M} \beta_{ij}N \right) \quad (11)$$

and for odd S

$$\bar{H} = \frac{1}{M}\sum_{i=1}^{M}\frac{1}{NM-1} \times \left(2(N-1) + \sum_{\substack{j=1\\ j-i\,\text{even}}}^{M} (1-u(|j-i|-S)) \right.$$
$$\times (1+2(N-1)) + u(|j-i|-S)\beta_{ij}N$$
$$\left. + \sum_{\substack{j=1\\ j-i\,\text{odd}}}^{M} \beta_{ij}N \right). \quad (12)$$

For the row/column configuration in the ring architecture, we calculate the expected number of hops accounting for all the types of links relevant to this configuration and architecture. The resulting expression for even S is

$$\bar{H} = \frac{1}{M}\sum_{i=1}^{M}\frac{1}{NM-1} \times \left(\sum_{\substack{j=1\\ \{i\,\text{odd}\}\cup\{j\,\text{odd}\}}}^{i-1} \gamma_{ij}\; N + \sum_{\substack{j=1\\ \{i\,\text{even}\}\cap\{j\,\text{even}\}}}^{i-1} \gamma_{ij} \right.$$
$$+ (\gamma_{ij}+1)(N-1) + \sum_{\substack{j=i+1\\ j-i\,\text{even}}}^{M} \alpha_{ij}$$
$$+ (\alpha_{ij}+1)(N-1) + \sum_{\substack{j=i+1\\ j-i\,\text{odd}}}^{M} \alpha_{ij}N$$
$$\left. + \begin{cases} \gamma_{ij}(N-1) & \text{odd } i \\ (\gamma_{ij}+1)(N-1) & \text{even } i \end{cases} \right) \quad (13)$$

and for odd S

$$\bar{H} = \frac{1}{M}\sum_{i=1}^{M}\frac{1}{NM-1} \times \left(\sum_{\substack{j=1\\ \{i\,\text{odd}\}\cup\{j\,\text{odd}\}}}^{i-1} \gamma_{ij}N \right.$$
$$+ \sum_{\substack{j=1\\ \{i\,\text{even}\}\cap\{j\,\text{even}\}}}^{i-1} (1-u(M-i-S))$$
$$\times \left[\left(1+\left\lceil \frac{j}{S} \right\rceil\right) + \left(2+\left\lceil \frac{j}{S} \right\rceil (N-1)\right) \right]$$
$$+ u(M-i-S)\gamma_{ij}N$$
$$+ \sum_{\substack{j=i+1\\ j-i\,\text{even}}}^{M} (i-u(j-i-S))(1+2(N-1))$$
$$+ u(j-i-S)\alpha_{ij}N + \sum_{\substack{j=i+1\\ j-i\,\text{odd}}}^{M} \alpha_{ij}N + (N-1)$$
$$\left. \times \begin{cases} \gamma_{ij} & \text{odd } i \\ (1-u(M-i-S))(2+\lceil \frac{i}{S} \rceil) \\ \quad + u(M-i-S)\gamma_{ii} & \text{even } i \end{cases} \right). \quad (14)$$

Although the expressions appear daunting, the terms are easily recognizable from the expressions for the plane configuration. This derivation is only valid for an even number of planes because if there is an odd number of planes, nodes on the last plane will have an orientation **parallel** to nodes on the first plane. One of the assumptions that has been made is that nodes on adjacent planes are separated by **one** hop. In the ring architecture, the first plane is adjacent to the final plane and this condition is violated when the number of planes is odd. To avoid this problem, we assume an even number of planes. This is an acceptable constraint that should not affect the realization of the ring architecture.

The expected number of hops for the dual-bus architecture is plotted in Fig. 14. The plots are given for three values of N where $N = 1$ is equivalent to the plane configuration. Additionally, an upper limit is given to illustrate the maximum possible deviation from the plane-to-plane system. The results for the bus and ring architectures follow the results for the

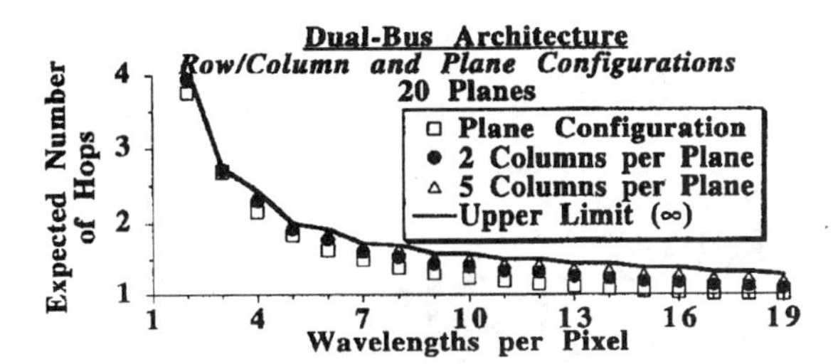

Fig. 14. Expected number of hops as a function of the number of wavelengths per pixel for three different numbers of rows or columns (N) and an upper limit, all for the dual-bus architecture.

dual-bus closely, so we only provide curves for the dual-bus architecture. The most significant feature of these plots is the small dependence on the number of rows or columns in the configuration. This feature illustrates the benefit gained from the integration of SDM into the system through the alternating row/column orientation. Regardless of the value of N, the fraction of the total number of nodes reachable in one hop, two hops, etc., is nearly the same. In other words, as N increases, the number of nodes increases, but at the same time the number of nodes which are one hop away on the next plane also increases. Using the row/column configuration, we can therefore increase the total number of nodes supported by the system with little increase in delay.

C. Pixel Configuration

The multiple-plane interconnection is designed so that the optical beams propagate orthogonal to the 2-D plane axes. As a result, each individual pixel may only establish an optical link with a pixel in the same spatial position on one of the succeeding planes. For the configuration in which each node has access only to an individual pixel, it is impossible to fully connect the network optically. To achieve full connectivity, data now must be routed to the correct destination plane **optically** and to the correct pixel on that destination plane **electrically**. The number of message hops is now the sum of the hops between planes (optical hops) and hops between pixels on a single plane (electrical hops). For each architecture, the expected number of optical hops between planes is simply the solution as derived for the plane configuration. The expected number of electrical hops between pixels on the same plane must be calculated for the specific electrical interconnection topology used to connect those pixels. We first discuss the performance requirements for this electrical interconnection and then illustrate one solution.

The electrical interconnection between nodes on the same plane must satisfy certain design criteria. The number of hops between nodes on the same plane should be as low as possible since electronic bottlenecks between these nodes will limit the performance of the entire system. The interconnection should also require few electronic components per pixel and as few overlapping connection lines as possible. Electronics comprise the majority of substrate area, and additional connections can result in extra fabrication processing steps as well as electromagnetic interference and crosstalk.

The simplest electrical interconnection is the unidirectional mesh (UniMesh) shown in a Fig. 15(a) [16]. This is a two-in, two-out interconnection with constraints on the boundary nodes. Each node is linked to one or two nearest neighbors, and the directed links are set up so as to form a set of coupled four-node rings. The alternating orientation of the ring connections allows for all possible source-destination pairs with none of the connections overlapping; internodal connections will overlap with connections to the off-board bonding pads.

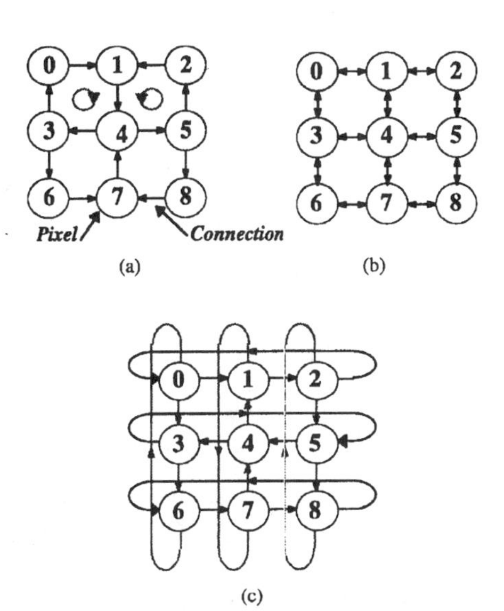

Fig. 15. Topological layout for three possible electronic interconnects: (a) unidirectional mesh, (b) bidirectional mesh, and (c) Manhattan Street Network.

By increasing the pixel complexity, we can alternatively interconnect the pixels in a bidirectional mesh (BiMesh) [Fig. 15(b)] [16]. This is simply an interconnection between nearest neighbors with boundary constraints similar to the unidirectional mesh. This interconnection will provide improved performance but at the expense of greatly increased pixel size and complexity; each pixel must now contain more electronic driver and receiver circuitry resulting in much larger pixels and consequently fewer pixels per plane.

A final possibility for an electrical interconnection topology, is shown in Fig. 15(c); it is a Manhattan Street Network (MSN) [17], [18] in which alternating vertical and horizontal torodial connections are made between pixels. We do not limit our interconnection to having an even number of rows and columns as in a standard MSN. This type of interconnection is a two-in, two-out connection with no boundary conditions. Consequently, all pixels have identical designs resulting in easier fabrication. There are other topologies that provide further performance improvements but at the cost of increased pixel complexity. One example is the Triangularly Arranged Network [19], which is simply an extension of the MSN with more connections between pixels to decrease the number of hops between nodes. We will not consider these other topologies because their complexity is prohibitive in providing simple connections for a large number of nodes.

To determine the expected number of hops for each of these topologies, we use an "All-pairs, Shortest Path" algorithm [20]. For each topology, we set up an $N^2 \times N^2$ adjacency matrix in which a "1" in position (i, j) indicates a directed connection

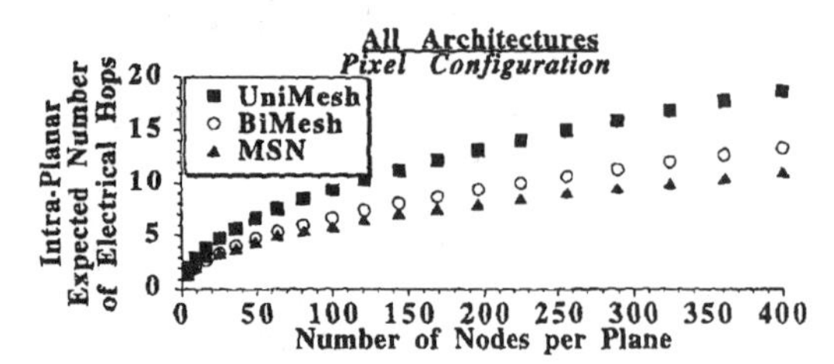

Fig. 16. Expected number of hops for the unidirectional mesh, bidirectional mesh, and Manhattan Street Network as a function of the number of nodes per plane (N^2).

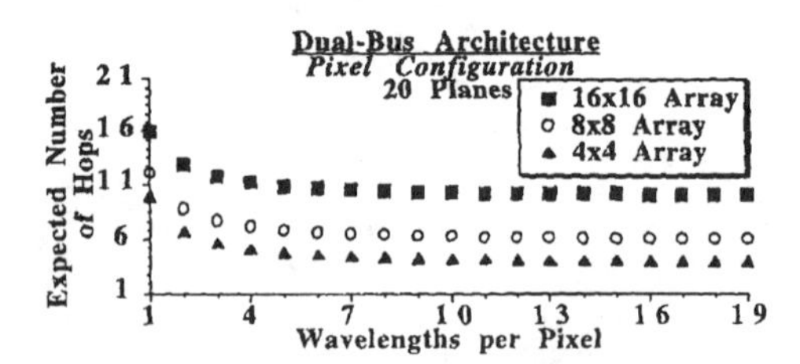

Fig. 17. Expected number of hops as a function of the number of wavelengths per pixel for three different $N \times N$ pixel arrays. Each curve is for a dual-bus architecture with the modified Manhattan Street Network electrical interconnection.

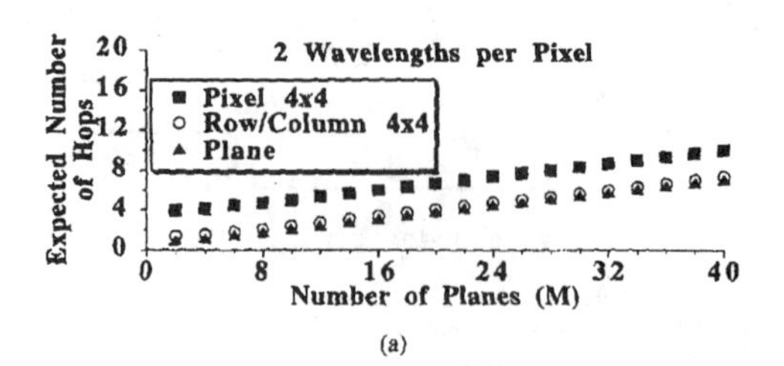

(a)

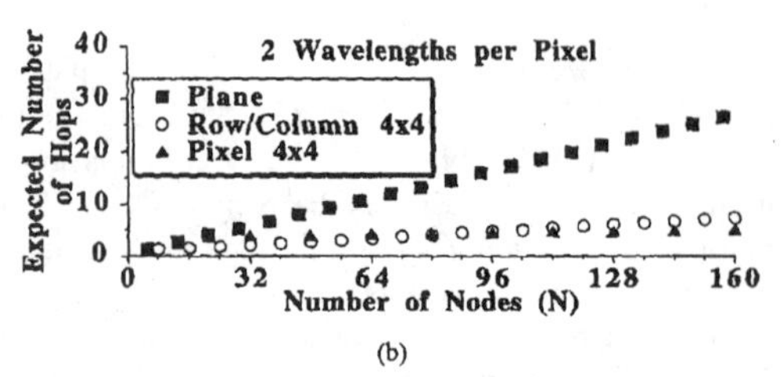

(b)

Fig. 18. Expected number of hops as a function of (a) number of planes and (b) number of nodes; for the plane, row/column, and pixel configurations with the same 4×4 pixel array and two wavelengths per pixel.

from node i to node j. This adjacency matrix is input to the shortest-path algorithm and the expected number of hops is computed from an $N^2 \times N^2$ output matrix with internodal distances as elements. We set the diagonal elements equal to zero since we are not concerned with connections from a node to itself. The expected number of hops as a function of the number of nodes on the plane (N^2) is plotted in Fig. 16. Of the three topologies, the MSN shows the best performance. We plot in Fig. 17 the expected number of hops as a function of the number of available wavelengths (S) for a dual-bus architecture with a MSN electrical topology. For even a few wavelengths, the number of hops quickly becomes limited by the number of electrical hops on the individual plane. The electronic bottleneck in this interconnection dominates the switching delay when the number of nodes per plane is large.

D. Overall Comparison

Thus far, we have shown the average number of hops as a function of the number of lasers per pixel and the number of planes. However, it may also be useful to know how the number of hops will vary with the total number of nodes interconnected. If we are given an architecture and per node traffic load, we need some guidelines in determining which configuration is optimal. In Fig. 18(a) and (b) we plot the expected number of hops for the dual-bus architecture with the plane, row/column and pixel configurations.

Fig. 18(a) shows the expected number of hops as a function of the total number of planes, whereas Fig. 18(b) shows the expected number of hops as a function of the total number of nodes. There is little difference in the curves of Fig. 18(a), while the curves of Fig. 18(b) show a marked difference. For 160 nodes spread out over several planes, the pixel configuration results in less than one fifth the average number of hops as compared to those hops for the plane configuration. On average, the number of hops increases with the number of planes and **not** with the number of nodes.

From the above analysis, we can conclude that for a fixed number of nodes which must be interconnected, i) the row/column configuration requires N times more planes as the pixel configuration and ii) the plane configuration requires N^2 times more planes as the pixel configuration. We can qualitatively state that the optimal interconnection is one that uses as few planes as possible while meeting the minimum traffic load requirement. The trade-off is between incorporating more nodes in the network but at the cost of decreased connectivity resulting in more message contentions and added delay. For each source node, when the number of accessible pixels decreases, there is less flexibility in choosing a transmission channel that is unused. Each node is allowed access to a smaller fraction of the interconnection as the number of nodes per plane increases from 1 to N to N^2. For low network traffic levels, the pixel configuration is optimal in connecting a large number of nodes. As seen in Fig. 18(b), there is little difference in the use of the plane, row/column, or pixel configurations for less than 20 total nodes and the plane or row/column configurations can be used to provide more throughput to each node with no delay penalty. Given the above, we formulate a qualitative system design rule as follows.

a) Choose the **pixel** configuration to connect extremely large numbers of nodes with low or bursty traffic loads.
b) Choose the **row/column** configuration to connect large numbers of nodes with moderate to heavy traffic levels.
c) Choose the **plane** configuration to connect few nodes with heavy traffic loads.

V. SUMMARY

We have shown that a wavelength division multiplexed 2-D multiple-plane optical interconnection delivers decreased internodal switching delay (as measured by the expected number of hops between node pairs) compared to single-wavelength optical interconnections. The WDM system outperforms single-wavelength systems even when the number of wavelengths is small relative to the total number of 2-D planes. This WDM optical interconnection provides simultaneous and reconfigurable communication among a network of nodes while most efficiently utilizing substrate area. We have analyzed the bus, dual-bus and ring architectures, all of which can be supported by this interconnection. Our analysis includes results for when each node can access an entire plane of pixels, a row (or column) of pixels, or an individual pixel.

The WDM interconnection always incurs substantially fewer hops between network nodes than single-wavelength interconnections. We have described a unique hybrid WDM/SDM system (the row/column configuration) and have shown the performance advantages of such an approach. We have discussed design criteria for the appropriate choice of nodal access configuration, which is dependent on the number of network nodes connected and predicted traffic load requirements. The plane configuration is best chosen to connect few nodes with heavy traffic levels. The row/column configuration can be used to connect a large number of nodes with moderate traffic levels. The pixel configuration delivers the best performance when connecting an extremely large number of nodes with low average traffic levels. Overall, the integration of WDM into this multiple-plane system provides vastly improved performance over electronic interconnection techniques and current optical interconnection solutions.

ACKNOWLEDGMENT

The authors acknowledge the assistance of D. M. Alexander-Zapalla in providing the source code for the "All Pairs, Shortest Path" algorithm.

REFERENCES

[1] H. S. Hinton, "Photonics in switching," *IEEE LTS*, pp. 26–35, Aug. 1992.
[2] M. R. Feldman, S. C. Esener, C. C. Guest, and S. H. Lee, "Comparison between optical and electrical interconnects based on power and speed considerations," *Appl. Opt.*, vol. 27, pp. 1742–1751, May 1, 1988.
[3] A. Yariv, "The beginning of integrated optoelectronic circuits," *IEEE Trans. Electron Devices*, vol. ED–31, pp. 1656–1661, Nov. 1984.
[4] J. W. Goodman, F. J. Leonberger, S. Y. Kung, and R. A. Athale, "Optical interconnections for VLSI systems," in *Proc. IEEE*, vol. 72, pp. 850–866, July 1984.
[5] C. A. Brackett, "Dense wavelength division multiplexing networks: Principles and applications," *IEEE J. Select. Areas Commun.*, vol. 8, pp. 948–964, Aug. 1990.
[6] I. P. Kaminow, "FSK with direct detection in optical multiple-access FDM networks," *IEEE J. Select. Areas Commun.*, pp. 1005–1014, Aug. 1990.
[7] A. Dickinson and M. E. Prise, "Free-space optical interconnection scheme," *Appl. Opt.*, vol. 29, no. 14, pp. 2001–2005, May 10, 1990.
[8] A. D. Norte, A. E. Willner, W. Shieh, and A. R. Tanguay Jr., "Multiple-layer optical interconnections using through-wafer hollow-dielectric-waveguide vias," *IEEE Photon. Technol. Lett.*, vol. 6, pp. 851–854, July 1994.
[9] A. E. Willner, C. J. Chang-Hasnain, and J. E. Leight, "2-D WDM optical interconnections using multiple-wavelength VCSEL's for simultaneous and reconfigurable communication among many planes," *IEEE Photon. Technol. Lett.*, vol. 5, no. 7, pp. 838–841, July 1993.
[10] C. J. Chang-Hasnain, J. P. Harbison, C. E. Zah, M. W. Maeda, L. T. Florez, N. G. Stoffel, and T. P. Lee, "Multiple wavelength tunable surface-emitting laser arrays," *IEEE J. Quant. Electron.*, vol. 27, pp. 1368–1376, June 1991.
[11] T. P. Lee and T. Li, "Photodetectors," in *Optical Fiber Telecommunications*, S. E. Miller and A. G. Chynoweth, Eds. Orlando, FL: Academic, 1979, pp. 593–626.
[12] T. P. Lee, J. C. Campbell, K. Ogawa, A. R. McCormick, A. G. Dentai, and C. A. Burrus, "Dual-channel 1.5 Mb/s lightwave receiver employing an InGaAsP wavelength-demultiplexing detector," *Electron. Lett.*, vol. 15, pp. 388–389, 1979.
[13] J. E. Leight, S. Homan, A. E. Willner, G. Giaretta, M. Li, and C. J. Chang-Hasnain, "Experimental demonstration of reconfigurable and simultaneous wavelength-division-multiplexed multiple-plane optical interconnections," *IEEE Photon. Technol. Lett.*, vol. 8, no. 2, Feb. 1996.
[14] H. S. Hinton, J. R. Erickson, T. J. Cloonan, and G. W. Richards, "Space-division switching," in *Photonics in Switching*, J. E. Midwinter, Ed. San Diego, CA: Academic, 1993, vol. II, pp. 119–167.
[15] A. S. Acampora, M. J. Karol, and M. G. Hluchyj, "Terabit lightwave networks: The multihop approach," *AT&T Tech. J.*, vol. 66, no. 6, pp. 21–34, Nov./Dec. 1987.
[16] K. Hwang, *Advanced Computer Architecture*. New York: McGraw-Hill, p. 93, 1993.
[17] N. F. Maxemchuk, "The Manhattan street network," in *Proc. GLOBECOM'85*, New Orleans, LA, Dec. 1986, pp. 255–261.
[18] N. F. Maxemchuk, "Regular mesh topologies in local and metropolitan area networks," *AT&T Tech. J.*, vol. 64, no. 7, pp. 1659–1685, Sept. 1985.
[19] G. E. Myers and M. El Zarki, "Routing in TAC—A triangularly-arranged network," in *Proc. INFOCOM'90*, Silver Spring, MD, 1990, pp. 481–486.
[20] T. H. Cormen, C. F. Leiserson, and R. L. Rivest, *Introduction to Algorithms*. Cambridge, MA: M.I.T. Press, 1992, pp. 550–578.

Chapter 6

Photonic Switch Architectures

Despite the advanced state of electronic switching technology, photonic switching technology can offer several advantages. In photonic switches, the information signal remains in optical form during switching. Cost, complexity, and failure rate of the switch are reduced since no optoelectronic conversion is required. Photonic switches have been shown to have a very large transmission bandwidth (comparable to that of the optical channel) and faster reconfiguration time than their electronic counterparts. Though photonic switches are expected to provide high interconnectivity, large photonic switches have not yet been demonstrated.

Three classes of switches have been investigated in the optical domain: time-division switches, wavelength-division switches, and space-division switches. Most reported optical time-division switches are based on architectures similar to their electronic counterparts. Optical implementations of $N \times N$ time-division interchangers have used integrated-optic $1 \times N$ and $N \times 1$ switch matrices as optical write and read gates, and fiber-optic delay lines or bistable laser diodes as optical memories. These approaches seem to have, however, several shortcomings, as reported in the papers of this chapter. Wavelength-division switches are attractive because of their potentially high throughput. They are either based on the broadcast-and-select configuration or the active wavelength routing configuration. They too are subject to some limitations unless combined with other techniques, and in particular with space-division multiplexing, as reported in several papers in this book. Most optical space-division switches use basic building blocks composed of four-port switching elements (crosspoints), such as electrooptic crystal in integrated-optic form, and are either of the directional coupler or X type. Lithium niobate ($LiNbO_3$) remains the dominant material in the integrated-optic technology because of its excellent piezoelectric, electrooptic, and waveguiding properties, and because of the relative maturity of its processing technology compared to other materials. Photonic switch arrays that provide a disjoint physical path for each signal from an input to an output port have been implemented by interconnecting several of these $LiNbO_3$ optical crosspoints. The collection of papers in this chapter deals with these issues.

H. S. Hinton reviews in paper 6.1 photonic switching fabrics based on both guided wave structures and free-space structures. The discussion on guided wave structures extends to fabrics based on space channels, time channels, and wavelength channels and that on free-space channels includes an evaluation of free-space relational switching fabrics and digital free-space switching fabrics. The paper highlights the prevailing emphasis to find effective methods of using the temporal bandwidth of relational devices or the spatial bandwidth available in free-space digital optical systems.

Blumenthal et al. in paper 6.2 review the progress in photonic switching toward the realization of photonic packet switches that have the versatility to build complete switching networks. In particular, attention is given to all-optical photonic packet switches in which the data portion of the packet remains all-optical throughout the switching and the routing process, from source to destination. The control portion, however, may or may not be regenerated optoelectronically. Also in this paper, an overview of the design issues for all-optical photonic packet switching is given and contrasted with electronic packet switch implementations. Low-level functions that have been experimentally implemented include routing, contention resolution, synchronization, and header regeneration.

The theme of packet switching architectures continues in paper 6.3 by Jajszczyk and Mouftah who consider photonic fast packet switching architectures. They present architectures that exploit wavelength, time, code, and space to achieve the required functions of fast packet switching. They conclude that among various solutions, those based on a wavelength-division transport network and an electronic controller are the most mature. Further progress in this area is mostly related to the development of more advanced optoelectronic devices, such as tunable lasers and filters.

Granestrand et al. report in paper 6.4 on a tree-structure based architecture for an 8×8 polarization-independent switch matrix comprising 112 digital optical switches. The matrix, which has been pigtailed and packaged in an open package for convenient system demonstrator use, exhibits a worst-case insertion loss of less than 15 dB. This work was carried out within the Research and Development in Advanced Communications in Europe (RACE) project on Multiwavelength Transport Network (MWTN): RACE-MWTN project (RACE II-R2028).

The chapter is concluded by Perrier and Prucnal (paper 6.5) who introduce a switch architecture that has the potential to achieve high dimensionality. The architecture resembles a collapsed network and does not use optical crosspoints; rather a dedicated path is provided for all input/output port connections on a common high-bandwidth transmission medium. This eliminates the limitations associated with classic space-division switch architectures employing 2×2 switching elements. The demonstration of a fully connected 120×120 space-division time-multiplexed photonic switch is reported.

Photonic Switching Fabrics

H. SCOTT HINTON

WITHIN RECENT YEARS THERE HAS BEEN A significant amount of interest in applying the new and developing photonics technology in telecommunications switching systems [1]. This has been viewed as being increasingly more important as the telecommunications industry is anticipating broadband capabilities such as Broadband Integrated Services Digital Network (BISDN). There are several devices that have emerged within the past few years that have the potential of meeting this goal. These devices can be arranged into two major classes according to the function they perform [2]. The first of these classes, called "relational" devices (also referred to as analog or passive devices), perform the function of establishing a large bandwidth "relation," or a mapping between the inputs and the outputs. This relation is a function of the control signals to the device and is independent of the signal or data inputs. As an example, if the control signal is not enabled, the relation between the inputs and the outputs of a 2 × 2 device might be upper input → upper output and lower input → lower output. When the control is enabled, the relationship might be upper input → lower output and lower input → upper output. This change in the relation between the inputs and outputs corresponds to a change in the state of the device. Another property of this device is that the information entering and flowing through the devices cannot change or influence the current relation between the inputs and outputs. This type of device is used in fabrics that are exploiting the bandwidth transparency in broad bandwidth devices. Examples of some of the fabrics based on these type of devices include the directional coupler, optical amplifier, and star coupler based fabrics, which include both time- and wavelength-division utilization of the available bandwidth. Thus, the strength of relational devices is that they cannot sense the presence of individual bits passing through them, they can only pass them. Each channel through a device can be viewed as a broad-bandwidth analog transmission channel to the passing bits. This bandwidth transparency allows relational devices to support either high-bit-rate channels, multiple time-multiplexed channels of differing bit rates, or a large collection of dense wavelength-division channels. The weakness of relational devices is that they cannot sense the presence of individual bits that are passing through them; they can only pass them. This inability to sense the passing bits prevents these devices from reading and responding to packet headers or other line-rate control information.

The second class of devices will be referred to as "logic" devices. In these devices, the data or information-bearing signal that is incident on the device controls the state of the device in such a way that some Boolean function or combination of Boolean functions is performed on the inputs. These are the type of devices required for digital switching fabrics. Each device must be able to change states, or switch, as fast as or faster than the signal bit rate in addition to regenerating or restoring the incident signal level. This high speed requirement for logic devices will limit the bit rates of signals that can eventually flow through their fabrics to less than those that can pass through relational fabrics. Examples of fabrics based on these devices are the free-space networks based on either optical logic gates or smart pixels. Thus, the strength of logic devices is the added flexibility that results from their ability to sense the bits that are passing through them; while their weakness is that they sense the bits that pass through them, which limits the maximum bit rate that they can handle.

This article will begin by reviewing the strengths and limitations of the photonic technology. There will then be a discussion on photonic switching fabrics based on guided-wave devices. This will include a review of switching fabrics based on space channels, time channels, and wavelength channels. Finally, photonic switching fabrics based on free-space devices will be described. This will include a section on free-space relational switching fabrics, a section on the basic hardware required for digital free-space optical fabrics, and a section on digital free-space switching fabrics.

The Strengths and Limitations of the Photonic Technology

Prior to discussing either photonic devices or their applications, it is important to understand both their potential and limitations. This section has the purpose of discussing the strengths and weaknesses of the photonics switching technology. It will begin by discussing the temporal bandwidth limitations of photonic devices. Then it will focus on spatial bandwidth, commonly referred as the parallelism of optics, and how it can be used in photonic fabrics.

Temporal Bandwidth

The term "photonic switching" to most telecommunications engineers brings to mind switching fabrics that can control channels with bandwidths capable of supporting signal rates in excess of 100 Gb/s. For fabrics based on relational devices these channels could be transparent to virtually any bit rate. Since the devices themselves do not have to change states, the fabric bandwidth will be the transmission bandwidth of the devices. The main issue with relational devices is how to effectively use the available bandwidth. Typically, there are physi-

Reprinted with permission from *IEEE Communications Magazine,* Vol. 28, No. 4, pp. 71-89, April 1990.

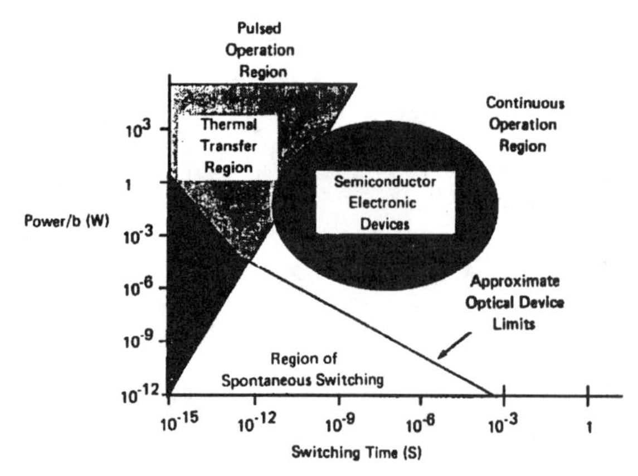

Fig. 1. Fundamental switching limits at 850 nm.

cal space channels used to connect users at point x to users at point y with an available channel bandwidth B_c. At each entrance point to these physical space channels there are users requiring a bandwidth B_u between these two locations. When the available bandwidth of the channel equals the desired bandwidth of the user ($B_c = B_u$), one space channel should be assigned between each pair of users. On the other hand, when the available channel bandwidth is much greater than the user bandwidth ($B_c \gg B_u$), it is desirable to share the available channel bandwidth between several users by allowing multiple access to the same space channel. This multiple access can be accomplished by multiplexing several users in either the temporal or spectral domain. To multiplex several users onto the same channel in the time domain, the pulse widths of the information passing through the channel are shortened until they fill the available bandwidth. Some of the methods currently used to share the available temporal bandwidth of a single space channel include Time-Division Multiplexing (TDM), Time-Division Multiple Access (TDMA), Code-Division Multiple Access (CDMA), and packet switching.

The second multiplexing method that can be used to fully utilize the available channel bandwidth is to operate in the spectral domain rather than the time domain. This can be accomplished through such techniques as Wavelength-Division Multiplexing (WDM), Wavelength-Division Multiple Access (WDMA), or Frequency-Division Multiple Access (FDMA). For the case of switching using WDM, each user is assigned a fixed transmitting (receiving) wavelength but has the capability to receive (transmit) the wavelengths of all the other users. As an example, for the case of a fixed transmitting wavelength per user, the information to be transported from one user to the other is modulated on to its assigned wavelength λ_1. The receiving user can then lock its tunable receiver onto the wavelength λ_1 and receive the information [3]. FDMA, on the other hand, electronically multiplexes several different frequencies together, and then uses this composite signal to modulate an optical carrier. This is also referred to as subcarrier multiplexing [4].

The maximum bit rate of optical logic devices and the reconfiguration rate of relational devices is limited by how fast the devices can change states, or switch. This bandwidth will be referred to as the switching bandwidth. Switching, in this case, refers to the changing of the present state of a device to an alternate state, as opposed to the "switching" that is analogous to an interconnection network reconfiguration. In the normal operating regions of most devices, a fixed amount of energy, the switching energy, is required to make them change states. This switching energy can be used to establish a relationship between both the switching speed and the power required to change the state of the device. Since the power required to switch the device is equal to the switching energy divided by the switching time, then a shorter switching time will require more power. As an example, for a photonic device with an area of 100 μm^2 and a switching energy of 1 fJ/μm^2 to change states in 1 ps requires 100 mW of power instead of the 100 μW that would be required if the device were to switch at 1 ns. Thus, for high power signals the device will change states rapidly, while low power signals yield a slow switching response.

Some approximate limits on the possible switching times of a given device, whether optical or electrical, are illustrated in Figure 1 [5]. In this figure the time required to switch the state of a device is on the abscissa while the power per bit required to switch the state of a device is on the ordinate. The region of spontaneous switching is the result of the background thermal energy that is present in a device. If the switching energy for the device is too low, the background thermal energy will cause the device to change states spontaneously. To prevent these random transitions in the state of a device, the switching energy required by the device must be much larger than the background thermal energy. To be able to differentiate statistically between two states, this figure assumes that each bit should be composed of at least 1,000 photons [6]. Thus, the total energy of 1,000 photons sets the approximate boundary for this region of spontaneous switching. For a wavelength of 850 nm, this implies a minimum switching energy on the order of 0.2 fJ.

For the thermal transfer region Smith assumed that for continuous operation the thermal energy present in the device cannot be removed any faster than 100 W/cm^2 (1 μW/μm^2). There has been some work done to indicate that this value could be as large as 1,000 W/cm^2 [7]. This region also assumes that there will be no more than an increase of 20°C in the temperature of the device [5]. Devices can be operated in this region using a pulsed rather than continuous mode of operation. Thus, high-energy pulses can be used if sufficient time is allowed between pulses to allow the absorbed energy to be removed from the devices.

The cloud represents the performance capabilities of current electronic devices. This figure illustrates that optical devices will not be able to switch states orders of magnitude faster than electronic devices when the system is in the continuous, rather than the pulsed, mode of operation. There are, however, other considerations in the use of optical computing or photonic switching devices than how fast a single device can change states. Assume that several physically small devices need to be interconnected so that the state information of one device can be used to control the state of another device. To communicate this information, there needs to be some type of interconnection with a large bandwidth that will allow short pulses to travel between the separated devices. Fortunately, the optical do-

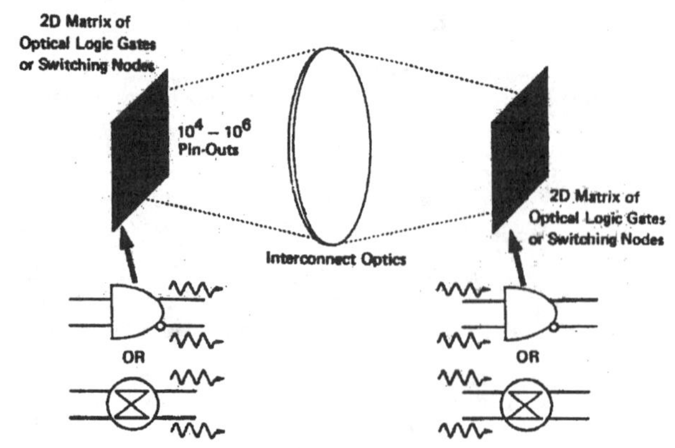

Fig. 2. Optical parallel interconnections.

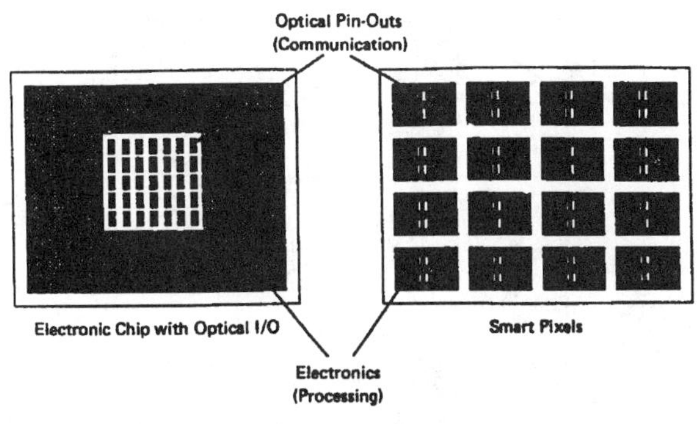

Fig. 3. Optical pin-outs and smart pixels.

main can support the bandwidth necessary to allow bit rates in excess of 100 Gb/s, which will allow high-speed communication between these individual switching devices. In the electrical domain the communications bandwidth between two or more devices is limited by the resistance, capacitance, and inductance of the path between the different devices. Therefore, even though photonic devices cannot switch orders of magnitude faster than their electronic counterparts, the communications capability or transmission bandwidth present in the optical domain should allow higher-speed systems than are possible in the electrical domain [8].

Therefore, networks composed of relational devices will have their signal bit rates limited by the transmission bandwidth and their reconfiguration rates limited by the switching time of the devices, while switching networks based on optical logic will have both their signal bit rates and reconfiguration rates limited by the switching time of their devices.

Spatial Bandwidth

Another method of increasing the capacity of a system, in addition to operating at higher speeds, is to operate on information in parallel instead of in serial. In pursuing this parallelism, attention has recently been placed on free-space optics. These types of systems are normally composed of multiple Two-Dimensional (2D) arrays of optical devices that are interconnected through bulk optics. Figure 2 shows the optical interconnection between two 2D arrays of optical elements. The interconnection in this case is a simple lens system. The optical elements, which will be referred to as pixels, could be optical NOR gates, optical light valves, etc. The number of pixels that can be interconnected in this manner is limited by the resolution of the optical interconnection system. In principle, even relatively inexpensive optical imaging systems exhibit resolutions on the order of 10 μm over a 1-mm field. This provides access to 100 × 100 or 10^4 pixels. If each pixel can be equated to a pin-out then for a 2D array there can be greater than 10^4 pin-outs. Since the optical pin-outs can be much smaller that there electronic counterparts, there is the potential for pin-outs/chip that exceed 10^4. As an example, a typical bonding pad requires an area of $(75\ \mu m)^2$. Using optical pin-outs with an area of $(5\ \mu m)^2$ and 10 μm center-to-center spacings implies 64 optical pin-outs in the same area required by one electrical bonding pad, and this does not include the area required for the driver circuitry required for the electrical bonding pad. The maximum number of pixels or pin-outs that can be supported by a lens or any optical system is referred to as its Space-Bandwidth Product (SBWP) [9], or the degrees of freedom of the system. Satellite imaging systems have been made that have a SBWP of 10^8 pixels.

The optical elements of the 2D arrays could be optical NOR gates, optical light valves, or even a mixture of electronic and optical devices (smart pixels or electronic chips with optical Input/Output, or I/O) as shown in Figure 3. This mixture of electronic and optical devices is designed to take advantage of the strengths in both the electrical and optical domains. The optical devices include detectors to convert the signals from the previous 2D array to electronic form and modulators (surface emitting lasers or Light-Emitting Diodes, LEDs) to enable the results of the electronically processed information to be transferred to the next stage of 2D arrays. The electronics does the intelligent processing on the data. Since the electronics is localized with short interconnection lengths, the speed of these smart pixels should be very fast.

In switching fabrics, the need for a large number of connections is more evident. An example is the multistage interconnection network based on perfect shuffle interconnects shown in Figure 4. The number of stages of these networks determines its blocking probability. If there are $\log_2 N$ stages, the network will be fully connected but blocking, whereas if there are $3\log_2 N - 4$ stages, the network will be rearrangeably nonblocking and capable of realizing all possible permutations [10]. As N becomes large, so does the number of connections required per stage. One solution to this problem is to create 2D arrays of switching nodes, or exchange/bypass or sorting nodes, and then interconnect them with free-space optical interconnects. This should eventually reduce the hardware cost of such a fabric in addition to providing the capability of implementing a large space-division switching fabric.

Thus, the objective of free-space digital optics is to integrate intelligent nodes onto a single two-dimensional array, thus

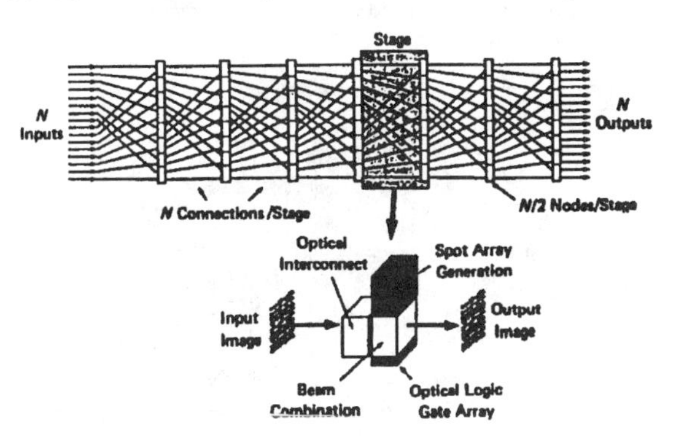

Fig. 4. Shuffle interconnection network.

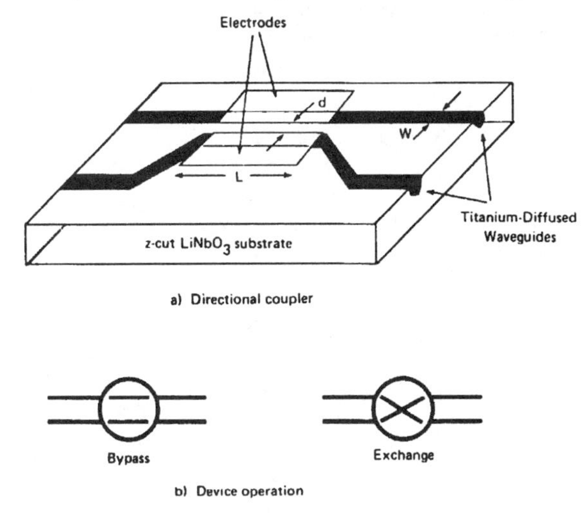

Fig. 5. Directional coupler.

gaining the advantages of Very Large Scale Integration (VLSI), in addition to using bulk optics to interconnect the 2D arrays of nodes. These principles will be further discussed in upcoming sections.

Switching Fabrics Based on Guided-Wave Devices

This section will discuss some of the proposed photonic switching fabrics that are based on guided-wave devices. It will begin with a discussion of the switching fabrics based on space channels. The devices used to implement these fabrics include directional couplers and optical amplifiers. It will then discuss switching fabrics based on time channels. The discussion of these time-based fabrics will include active reconfigurable fabrics based on TDM, time-slot interchangers, and universal time slots, in addition to passive shared media fabrics. The section will then outline some of the switching fabrics that have been proposed using wavelength channels. Finally, there will be a brief review of multi-dimensional fabrics.

Switching Fabrics Using Space Channels

A space channel is a physical channel that has been established between two users. Such a channel can be viewed as a transparent channel whose full bandwidth is available to both users. The three devices that are the basis of the discussed space channel fabrics are directional couplers, optical amplifiers, and spatial light modulators. Examples, of each type of fabric will be discussed in the following section.

Fabrics Based on Directional Couplers

A directional coupler is a device that has two optical inputs, two optical outputs, and one control input, as shown in Figure 5. The control input is electrical and is capable of putting the device in the "bar" state, in which the upper (lower) optical inputs are directed to the upper (lower) optical outputs, or the cross state, in which the upper (lower) optical inputs are directed to the lower (upper) optical outputs [11]. The most advanced implementations of these devices have occurred using Ti:$LiNbO_3$ technology [12]. The strength of directional couplers is their ability to control extremely high bit-rate information. They are limited by several factors: the electronics required to control them limits their maximum reconfiguration rate; the long length of each directional coupler prevents large scale integration; and the losses and crosstalk associated with each device limit the maximum size of a possible network un-

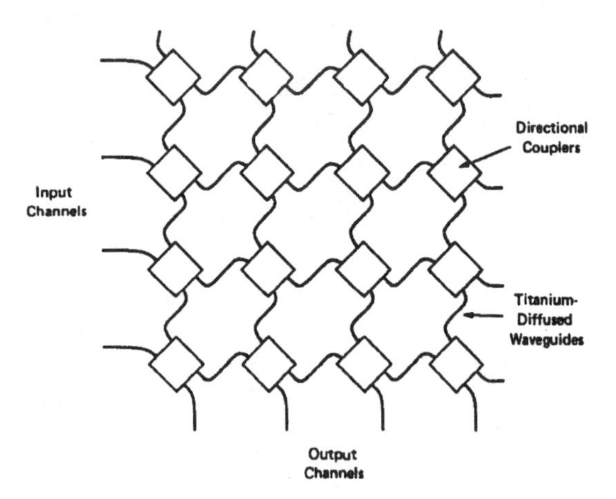

Fig. 6. 4 × 4 crossbar interconnection network.

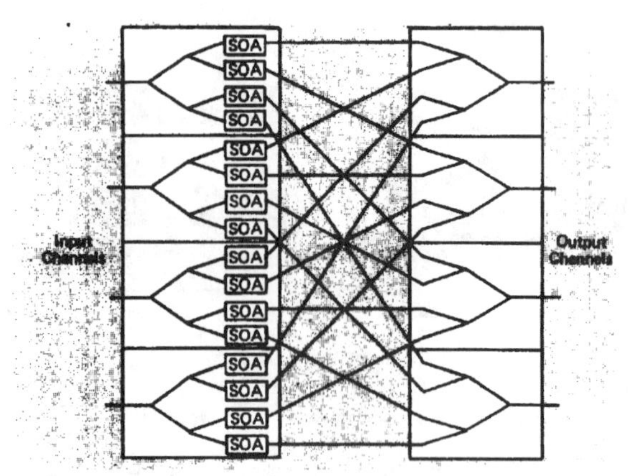

Fig. 7. Switching fabric.

less some type of signal regeneration is included at critical points within the fabric [13]. A modest number of these devices have been integrated onto a single substrate to create larger photonic interconnection networks, such as an 8 × 8 crossbar interconnection network [14]. As another example, in Figure 6 a 4 × 4 crossbar interconnection network composed of 16 integrated directional couplers is illustrated. All the integrated directional couplers have crosstalk less than −35 dB with an average fiber-to-fiber insertion loss of less than 5.2 dB [15].

The implementation of a large space switch requires the interconnection of many smaller photonic switches that are used as building blocks. These building blocks will most likely have dimensions less than 16 × 16 because of the long length of directional couplers and the large bending radii required in the integrated waveguides. Two examples of topologies for these building blocks are the crossbar interconnection network [13] and the broadcast network proposed by Spanke [16]. For point-to-point networks, the interconnection of these building blocks to construct a larger switching fabric can be done with Clos, Benes, banyan, omega, or shuffle networks. If video information is to be a main component of the fabric traffic, then a broadcast environment becomes important. A good topology for a broadcast network is a Richards network [17].

A good application of directional-coupler-based fabrics is a protection switch. In this environment, the only time the switch will need to be reconfigured is when a failure occurs in an existing path. Thus, high bit rates can be passed through the switch with moderate reconfiguration rate requirements. This application matches the capabilities of the directional coupler—it requires long hold times with moderate reconfiguration rates. Once a path has been set up, high-speed data, multiplexed speech, or video can be transferred through the relational fabric.

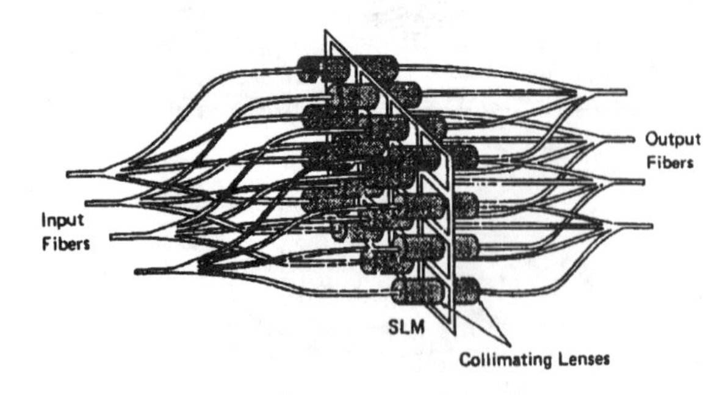

Fig. 8. Fiber optic crossbar interconnection network based on an SLM.

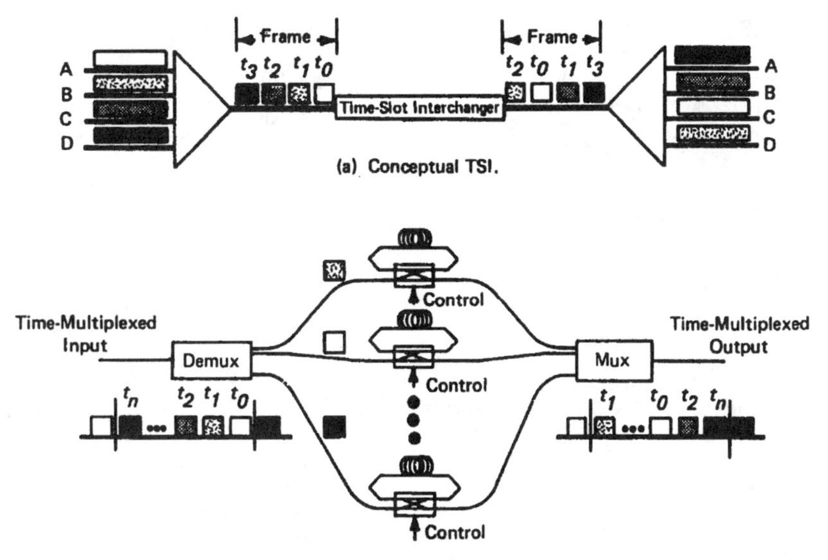

Fig. 9. Time-division switching.

Fabrics Based on Optical Amplifiers

A Semiconductor Optical Amplifier (SOA), when appropriately biased, can provide optical gain through stimulated emission to any entering signal. On the other hand, an SOA will act as an absorber to the incoming light if the bias is removed. This on/off switch can be combined with a fan-out interconnection network topology to create a switching fabric. An example is shown in Figure 7. On the left side of the figure, each of the N input channels is split or divided into N separate channels. On the right side of the figure, each output channel will combine the light received from the N fan-out modules. In the middle of the switching fabric is a column of SOAs that are individually controlled. By turning "on" an SOA the entering signal is amplified, creating a path from an input to an output. When an SOA is turned "off" the entering signal is absorbed, blocking that path through the fabric. Thus, by controlling the paths available through the network by the SOAs, a strictly non-blocking network has been established. These fabrics also offer the signal transparency of relational devices. Detailed calculations have shown that 32 × 32 switching fabrics with an output SNR of 20 dB at 1 GHz could be fabricated [18].

Fabrics Based on Spatial Light Modulators

A Spatial Light Modulator (SLM) is a 2D array of optical modulators [19]. Each of these modulators is independent of the others and has the capability of modulating the incident light. For the applications described in this section, the modulators will be assumed to be digital, in that they possess two states: transparent to the incoming light (on) and opaque to the incoming light (off). An SLM that is currently available in the marketplace is based on the magneto-optic effect [20]. These arrays are electrically controlled such that an electrically enabled pixel will be transparent while a disabled pixel will block the incident light. Some other SLMs include the Liquid Crystal Light Valves (LCLV) [21], Lead Lanthanum Zirconate Titanate (PLZT) modulators [22], deformable mirrors [23], and GaAs Multiple-Quantum Well (MQW) modulators [24].

An implementation of a crossbar interconnection network using an SLM is illustrated in Figure 8 [25]. In this figure, each input fiber channel is split into four fiber channels. Prior to passing through the SLM windows, the light in each fiber is collimated. The light passing through each SLM window is then collected by the collimating lenses and directed into another fiber. The light from each row of the SLM is combined in the fiber and directed to an output channel. As with all relational structures, high signal bit rates pass through the switch with the speed limitation being the fabric reconfiguration time.

Switching Fabrics Using Time-Channels

As a result of the large signal bandwidth available in relational devices, the signal bit rate passing through the device can be much larger than the bit rate of any single user. In this situation, the information from the users can be compressed (in time) and share the relational devices with many other users. There will be two types of time-division switching fabrics discussed in this section. The first type will consist of active fabrics made of relational devices that can be reconfigured for each time slot. An example is a Time-Slot Interchanger (TSI). The second type capitalizes on the ability to have multiple access to parallel time channels. Typically, all users connect to a passive shared media such as a bus or star coupler, and access each other through multiple access techniques. Examples include TDMA and CDMA fabrics.

Active Reconfigurable Fabrics

An active reconfigurable fabric is based on the ability to reconfigure the time slots present in a TDM stream of information. A conventional TDM signal is normally composed of either a bit-multiplexed or block-multiplexed stream of information. A bit-multiplexed data stream is created by interleaving the compressed or sampled bit-synchronized bits from each of the users. This type of multiplexing is the method of choice for most transmission systems since it only requires the storage of one bit of information for each user at any time. Unfortunately, most of the bit-multiplexed transmission systems are further complicated by adding pulse-stuffing and other special control bits to the data stream.

Block-multiplexing, on the other hand, stores a frames worth of information from each of the user's and then orders the bits entering the channel such that each users data is contiguous. When used in a switching environment, this multiple access method requires the switching fabric to reconfigure only at

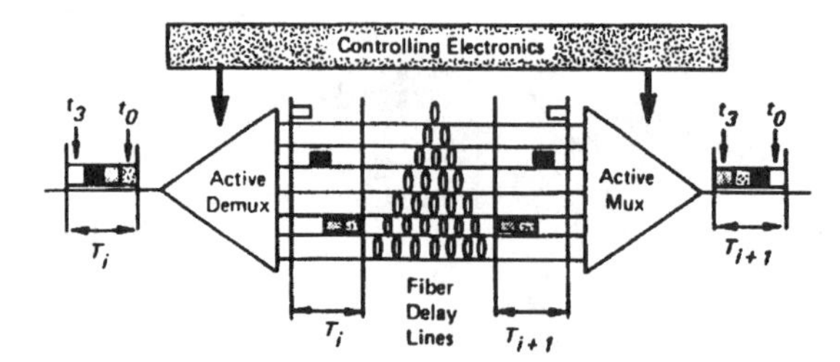

Fig. 10. TSI using variable-length fiber-delay lines.

block boundaries. By allowing a small amount of dead time between the blocked-multiplexed information, the requirements on the reconfiguration time of the fabric can be relaxed. This can be attractive for switching fabrics, such as directional coupler based fabrics, that have slow reconfiguration times [26].

• *Time Slot Interchangers*

Switching can be achieved by interchanging the position, in time, of the time slots in a frame of time-multiplexed information stream. This is illustrated in the TSI shown in Figure 9. In 9(a), the four input signals are time-multiplexed onto a single space channel. User A is put on the bus first, with user D being last. The TSI provides the function of interchanging these time slots of information in time. For this example, user A's time slot has been moved into the third time slot. Since the TDM demultiplexer will direct the first time slot to user A, a connection has been made between users A and C. Also, notice the connections between users B and D, users C and B, and finally users D and A. Figure 9(b) illustrates a proposed photonic implementation of the TSI [27]. The input time slots of the TSI are directed to fiber-delay lines where they can recirculate until needed at the output. The fiber-delay lines must create a time delay equal to the duration of a time slot. As an example, the input time slot t_0 will need to pass through the fiber-delay line $n + 2$ times, while input time slot t_n will pass through the fiber loop only once.

A second example of a TSI is illustrated in Figure 10 [28]. On the left side of the figure, a frame of a time-multiplexed information stream enters the active demultiplexer. The frame in this case is composed of four time slots and has a duration T_i. The objective of the TSI is to rearrange the time slots in the frame and deliver it to the output channel delayed by T_i. By comparing the input frame to the output frame, it can be seen that the following interchanging of time slots has to take place: $t_0 \rightarrow t_2$, $t_1 \rightarrow t_3$, $t_2 \rightarrow t_1$, and $t_3 \rightarrow t_0$. To accomplish this task, the controlling electronics directs the input time slot to the fiber delay channel that has the number of equivalent time-slot delays equal to the number of time slots in the frame (four in this case) plus the difference in time slots between the input and desired output time slots. As an example, the controlling electronics direct the first time slot, through the active demultiplexer, to the channel composed of six equivalent time-slot delays (4TS + 2TS = 6TS). The last time slot needs to be transferred to the first time slot, thus requiring (4TS − 3TS = 1TS) on the fiber-delay loop. After these two time slots pass through the fiber delay lines, they are directed by the active multiplexer, under the control of the controlling electronics, to the output channel, where they will be placed in the desired time slots. Note that this structure can be used with either bit- or block-multiplexed information, although the reconfiguration rate of the active mux and demux units would be significantly reduced with block-multiplexed information.

• *Universal Time Slots*

A good application of the bandwidth transparency of optical fiber is through the use of universal time slots [29]. A universal time slot is a partitioned section of time that can contain information transmitted at any bit rate. This is illustrated in Figure 11 where a standard frame is decomposed into 256 time slots. Each time slot can contain information of any bandwidth. For example, a time slot of voice would require approximately 100 kb/s, while an adjacent time slot could contain video information at a bit rate in excess of 1 Gb/s. Through the use of TSIs, as previously described, and a space switch composed of directional couplers, a time-space-time scheme could be implemented that could handle information ranging from voice to video.

Passive Shared Media Fabrics

As opposed to the previous fabrics, which interchange a deterministic and periodic time-multiplexed stream of time slots under a centralized control structure, these fabrics are based on either deterministic or statistical multiple access to a common interconnect such as a bus or star coupler. The control structure can be either centralized or distributed.

Ring networks are examples of switching fabrics based on a passive shared medium. The passive shared medium is typically an optical fiber that is accessed in time with either passive taps such as fiber couplers or directional couplers operating as active taps. For a synchronous ring structure, each user is assigned a unique piece of time (time slot) to read the information from the ring. Other users can send information to a user by entering information into the destination user's time slot. The multiple access to the time slots is arbitrated by the centralized control. There are also many other schemes for using ring structures in switching applications, both with centralized control and distributed asynchronous control schemes based on packet structures [30].

Instead of using a single fiber as the shared passive media, a star coupler can be used. A star coupler is a device with N inputs and N outputs that combines all the input channels and redistributes them equally to all the outputs [31]. A TDMA fabric would then consist of time encoders on each input, the star coupler to combine and redistribute all the input signals, and, finally, time decoders to select which input should be received. A fabric is referred to as a Fixed-Transmitter Assignment (FTA) network if the encoders or transmitters are fixed, and the decoders or receivers can be adjusted to select any input. Conversely, a Fixed-Receiver Assignment (FRA) network has fixed receivers and tunable transmitters [32].

An example of an FTA network is illustrated in Figure 12 [32]. For this fabric, the address associated with each output channel is the position, in time, of the sampled input signal. Thus, the effective address for the upper output channel is one unit of delay, while the address of the lower output channel is N units of delay. In this figure all synchronous inputs are sampled and directed to a tunable TDMA encoder. The TDMA encoder sets the appropriate delay for the sampled input to match the delay required by the desired output channel. The outputs from all the TDMA encoders are then combined and distribut-

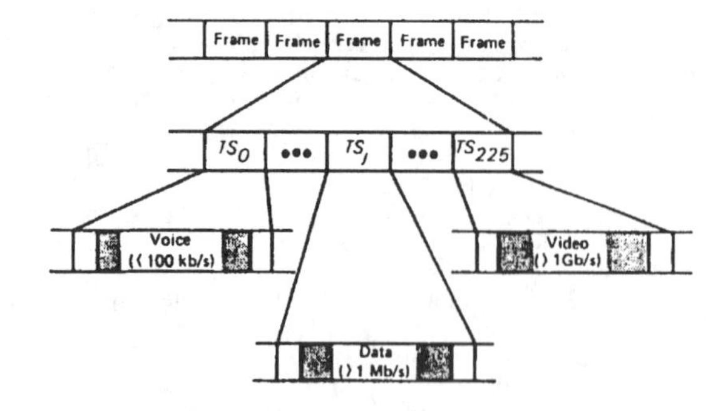

Fig. 11. Universal time slots.

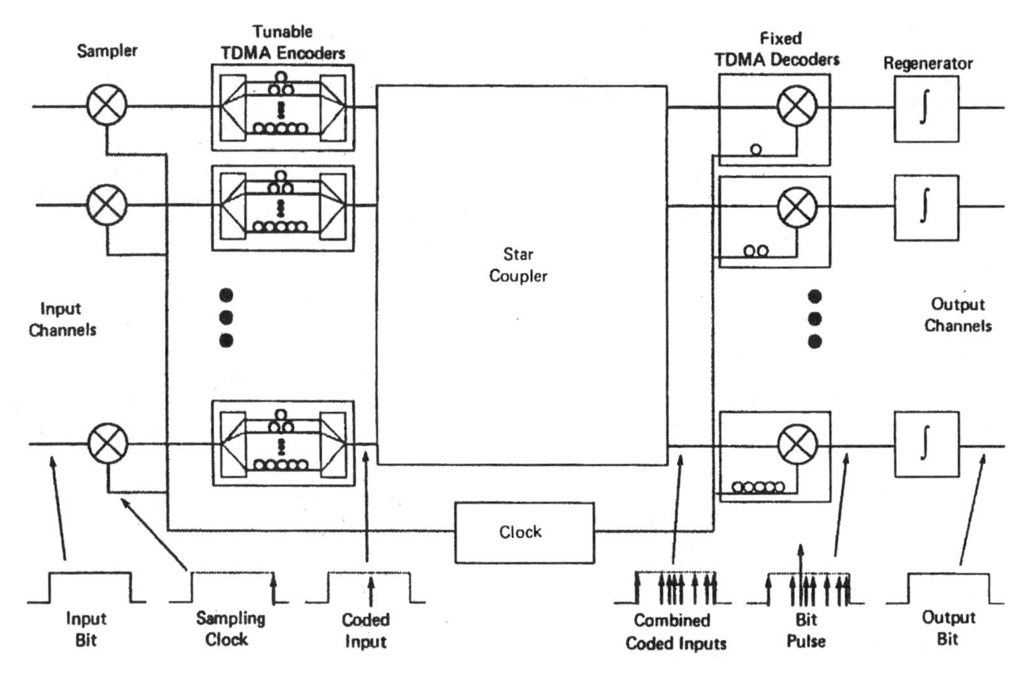

Fig. 12. TDMA switching.

ed to all the decoders. Each decoder delays the clock signal the appropriate amount and then incoherently combines it with whatever light is present. If a sample is present, the combination of the delayed clock and the sample will combine to trigger a thresholding device which will indicate that a bit is present. The sample will then be converted to a bit of the proper duration.

Another method of multiple access through time is code-division multiplexing. This type of multiplexing is accomplished through the use of either orthogonal or pseudo-orthogonal codes to represent both the bits and the users [33–35]. In Figure 13(a), different code sequences, one associated with each user, are used to represent the bits. Each bit, then, is represented by the unique code of the user. When no bit is present, there will be no information present on the user's input channel. In 13(b), a conceptual implementation of a switching fabric using CDMA is illustrated. Assuming bit-synchronized inputs, the CDMA scheme begins by the generation of a short pulse for every bit entering the fabric. In the figure, this operation is labeled as the pulse generator. This pulse is then split among k fiber-delay lines. The code sequence representing a given input channel is then composed of a unique collection of pulses (chips) of different delays. The code sequences from all the encoders are then combined and distributed to the output channels by the star coupler. Each star coupler output contains the superposition of all the different code sequences generated by the encoders. The CDMA decoders, like the encoders, begin by splitting the optical energy among a group of fiber-delay lines. The decoder for user j has to undo the code sequence generated by its corresponding encoder. This implies that each decoder must contain the inverse of all delay loops present in the encoders. The tunable fiber delay lines in the decoder are set at the appropriate lengths to combine all the individual pulses of a desired code sequence into a single pulse at the end of a normal bit duration. Since the bit codes are either orthogonal or pseudo-orthogonal, a simple thresholding decision determines whether a bit is present or not. Finally, the output of the decoder, assuming a bit is present, has to be integrated or stretched to the appropriate bit duration to communicate with the outside world.

The strength of this CDMA switching fabric is that the high-speed portion of the fabric control is both distributed and photonic. This distributed control is the result of the code sequence being an effective address read by the designated decoder. The role of the controlling electronics is to determine which fiber-delay lines are to be included for a given decoder. The weakness of these CDMA switching fabrics is that $k \ll N$, which limits them to smaller fabrics with low-occupancy environments, such as local area networks.

Switching Fabrics Using Wavelength Channels

Like the time-division fabrics discussed, fabrics based on wavelength channels can either rearrange or reconfigure the information present on the different wavelengths, or share those wavelength channels through multiple access techniques. This section will begin with a discussion of a proposed Wavelength Interchanger (WI). It will then be followed by a review of the work on switching fabrics based on sharing wavelength channels through multiple access.

Wavelength-Interchanger

Just as in the case of a TSI, where a switching function can be performed by interchanging the time slots in a time-multiplexed information stream, a WI can provide a switching function for a wavelength-multiplexed channel. This is illustrated in Figure 14, in which a wavelength-multiplexed signal enters the λ-switch [36]. Since each user is associated with a unique wavelength, a connection can be made between two users by converting the transmitter's wavelength (λ_t) to the receiver's wavelength (λ_r). The WDM signal enters the λ-switch, where the power is equally divided among n channels. Each of these channels will go through a coherent detection process, where the information on the desired wavelength can be detected. This information is then used to modulate a fixed wavelength laser. The outputs of the fixed lasers, all of different wavelengths, will be combined onto a single fiber. As a specific example, assume the information modulated on λ_n needs to be moved to the carrier λ_j. The fabric control will adjust the tunable laser associated with the fixed laser generating the λ_j carrier. This tunable laser will select the information on λ_n. This information will then modulate the fixed output laser of wavelength λ_j. Thus, the information on λ_n has been transferred to λ_j.

Figure 15 illustrates how a collection of WIs can be connected to create a larger-dimensional fabric. The WIs are connected into a three-stage fabric through the λ-multiplexers and -demultiplexers. Recognizing that the combination of the λ-multiplexer, the λ-switch, and the λ-demultiplexer is equivalent to an $n \times n$ switch allows the use of interconnection network topologies, such as shuffle or Benes networks, as the interconnection patterns between WIs. If the λ-mux and λ-demux are such that an $n \times m$ switch is implemented, ther other network topologies, such as Clos networks, provide the fixed interconnections between WIs.

Passive Shared Media Fabrics

Another type of relational architecture that has received a considerable amount of attention is wavelength-division multiple access. This is schematically shown in Figure 16. In this figure, the entering information is used to modulate a light source that has a unique wavelength for each input. All the optical energy is combined and redistributed by a star coupler to all the output channels. The tunable filter on each output is adjusted such that it only allows the wavelength associated with the desired input channel to pass to the detector. Thus, by varying the tunable filter, an output has access to any or all of the input channels. Several approaches to tunable filters have been pursued. The first is to use movable gratings [37]. A second type of tunable filter could be a tunable Fabry-Perot etalon [38]. Finally, coherent detection could be used to select the desired wavelength [39].

Multi-Dimensional Fabrics

In the early days of telecommunications switching, the switching fabrics used were space-division. With the advent of digitized voice, it became apparent that electronic hardware in the fabric itself could be reduced by adding the dimension of time to the space-division fabric. As an example, if a 1,024 × 1,024 space-division switch were able to switch 128 time slots/frame (1 frame = 125 μs), then a switching fabric with a dimensionality of approximately 128,000 × 128,000 could be made (e.g., 4ESS™).

™4ESS is a trademark of AT&T.

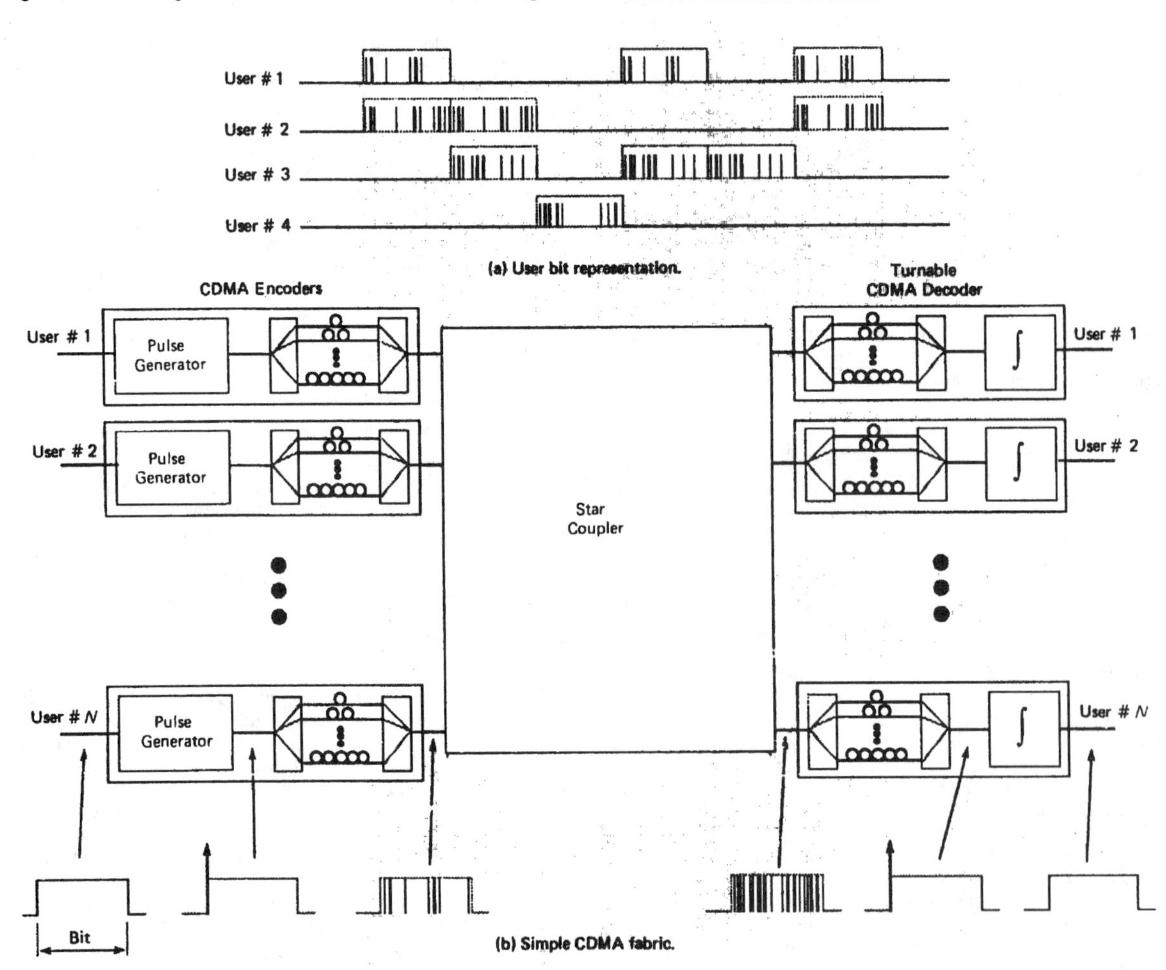

Fig. 13. CDMA switching fabric.

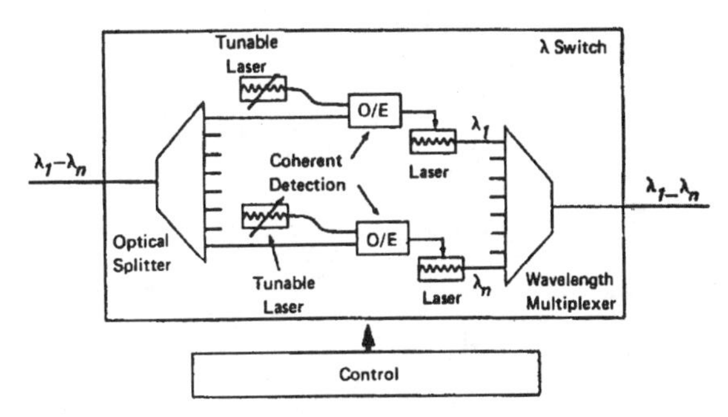

Fig. 14. Wavelength interchanger.

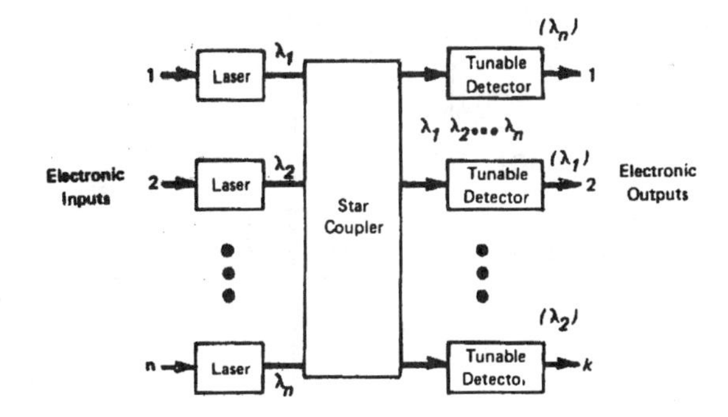

Fig. 16. WDMA using a passive shared medium.

An example of a 512 × 512 Time-Space-Time (TST) switch is shown in Figure 17. In this figure, the input lines are partitioned into sections of 32 lines, which are time-multiplexed onto a single space channel. Thus, each channel consists of 32 time slots. If the bit rate of the input signals is 150 Mb/s, then the time-multiplexed information stream will require a bit rate > 4.8 Gb/s (200 ps/b). This time-multiplexed signal then enters the TSI where the 32 time slots can be interchanged. From there, the information enters the time-multiplexed space-division switch (the advantage of multi-dimensional switching is that the size of the space switch can be small). The output of the space switch is directed to the output TSI, which is then demultiplexed to the output space channels. The difficulty with TST configurations is the timing requirements imposed upon the centralized control. As an example, to avoid any phase discontinuities on the output channels from the space switch, there needs to be bit alignment of the time-multiplexed information stream entering the 16 × 16 switch. Assuming a 5-Gb/s bit rate implies that each bit has a pulse duration of 200 ps. Thus, to prevent these phase discontinuities on the output channels, all the input bits should be bit-aligned to within 10 ps of each other. This timing burden will be placed on the initial time-division multiplexer, or else an elastic store will have to be placed on the input to the space switch (this assumes that the controlling electronics can recognize variations of ≈ 10 ps). To illustrate the critical packaging problem, if the length of fiber from two TSIs differs by 1 cm (assuming an index of refraction of 1.5 in the fiber), there will be a 50-ps difference in the bit arrival times at the space switch. In addition to the bit and frame alignment required by the space switch, each TSI will require the alignment of bit and frame boundaries to prevent phase discontinuities on its output channel.

The strength of the multi-dimensional switching structures, such as the TST switch previously shown, is the minimal amount of hardware required to build them. The TST switch of Figure 17 requires ≈ 2,000 fibers, 32 multiplexers (demultiplexers), 1 16 × 16 space-division switch, and ≈ 1,000 directional couplers for the TSIs. Even less hardware is required for a ring or linear bus network (512 × 512), which would require 512 couplers and 512 fibers. The disadvantage of the ring structure is the ≈ 80-Gb/s bit rate on the single time-division channel. Thus, the advantage of minimized hardware comes at the cost of increased timing complexity.

Another example of a multi-dimensional fabric is a packet switch. Such a switch is basically a space-division fabric that can reconfigure itself rapidly, allowing the sharing of space channels in time. HYPASS [40], a high-performance packet switch, is an example of a packet-switching fabric that has been proposed using WDMA. It is illustrated in Figure 18. In this fabric, the packetized information enters the fabric from the left, where it is initially stored in a First In First Out (FIFO). The objective is to modulate the tunable laser, tuned to the fixed wavelength of the designated output port; pass the information through the transport star coupler; and then receive the information at the desired output port. Prior to accessing the star coupler, it is necessary to check to see if the desired output port is busy. This is accomplished through the specialized control hardware. If an output port is available, the protocol processor associated with the fixed wavelength receivers will turn on the laser associated with the particular output port, allowing light to enter the control star coupler. The tunable receivers attached to the control star coupler can tune to the wavelength of any of the output channels. If the signal is present, it will signal the input channel decoder to tune the laser to the appropriate wavelength and then command the FIFO to send the current packet to the desired output channel. Note that in this fabric the packet address is converted to the specific wavelength of

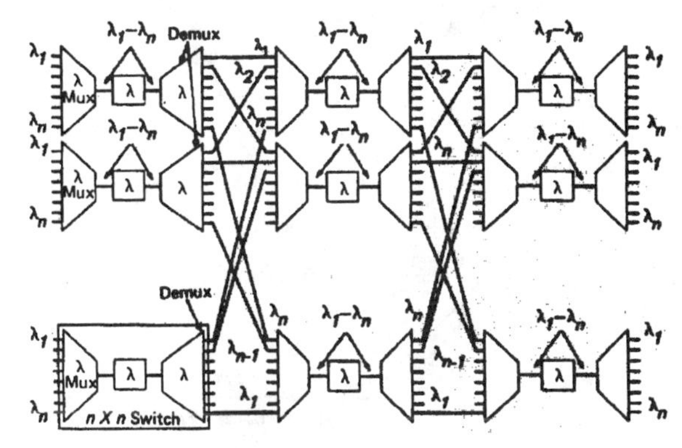

Fig. 15. Wavelength-space network.

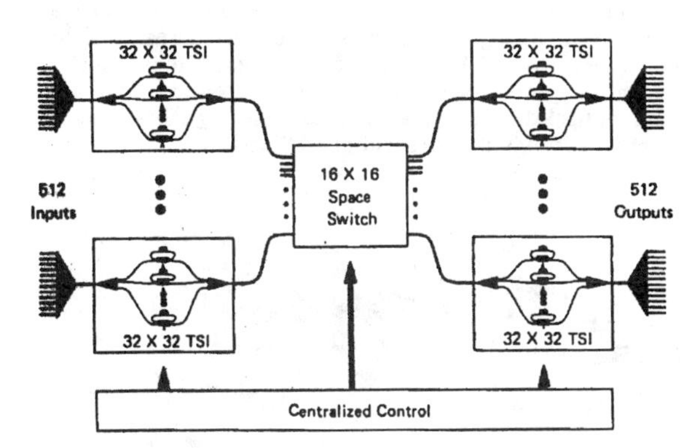

Fig. 17. 512 × 512 TSI switch.

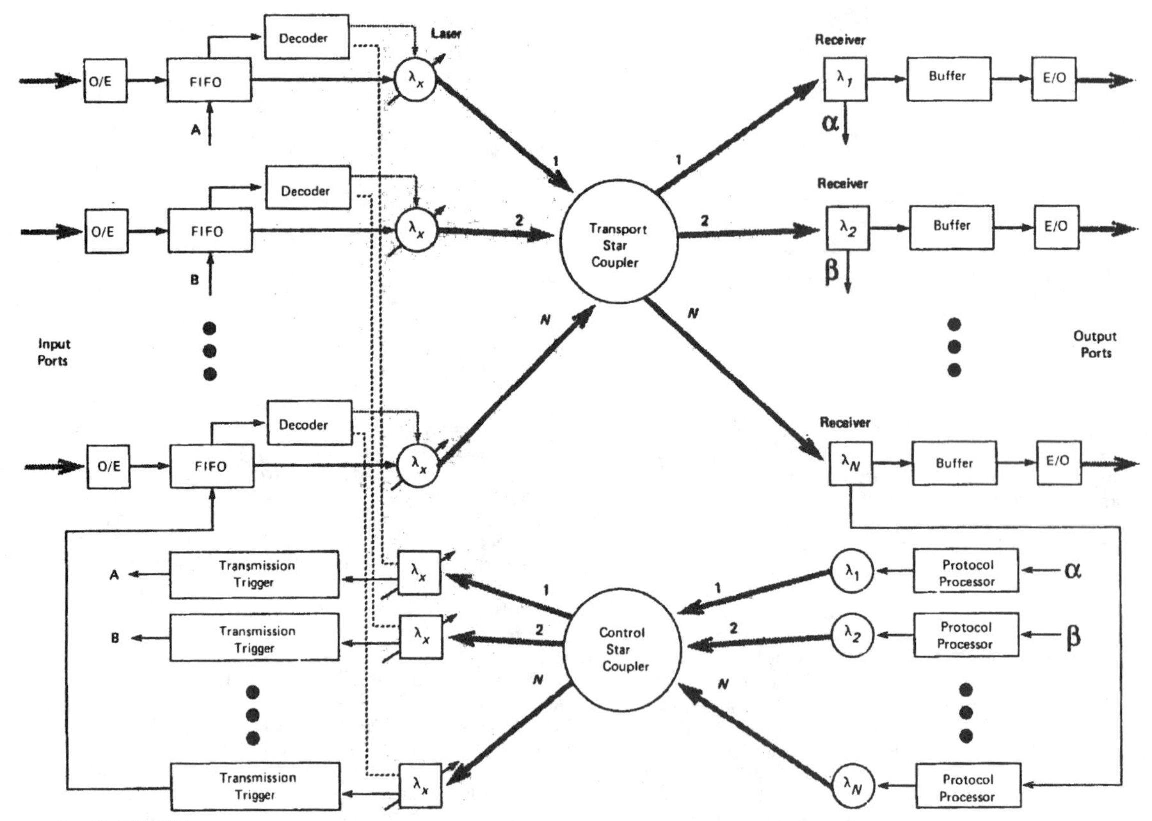

Fig. 18. HYPASS.

the output channel. Thus, the address in the fabric is the wavelength of light entering the transport star coupler.

Switching Fabrics Based on Free-Space Devices

The second, and least understood, strength of the optical domain is the spatial bandwidth provided by imaging systems. This spatial bandwidth can be viewed as connections or optical pin-outs [41]. The purpose of this section is to outline some of the switching fabrics that have been proposed that are based on free-space devices. The section will begin with a discussion of a relational or analog switching fabric based on spatial light modulators. This will be followed by a discussion of the optical hardware required for switching fabrics based on free-space digital optical devices. Finally, some switching fabrics based on free-space digital optics will be outlined.

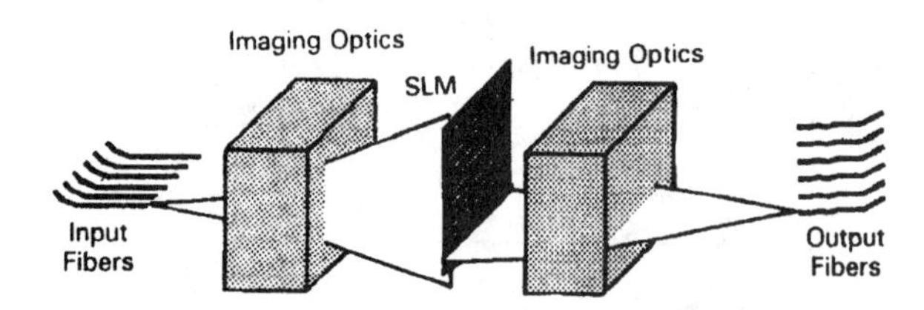

Fig. 19. Optical crossbar interconnection network.

Free-Space Relational Fabrics

An example of a free-space analog (relational) fabric based on a spatial light modulator is illustrated in Figure 19 [25] [42] [43]. In this figure, the fiber inputs are horizontally aligned as a row of inputs. The inputs are aligned to associate each fiber with a unique column of the SLM. A lens system is used to

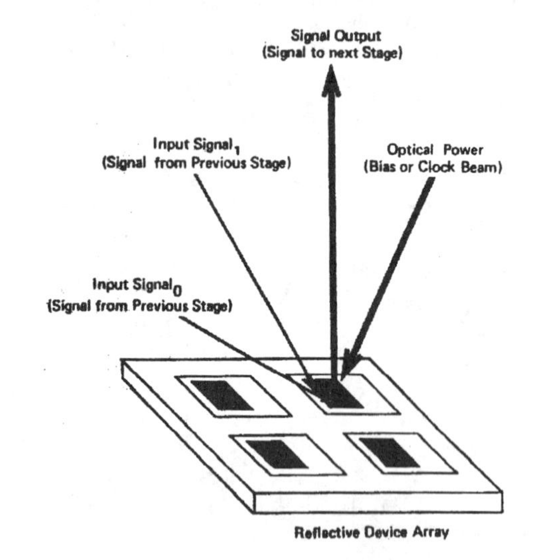

Fig. 20. Beam combination.

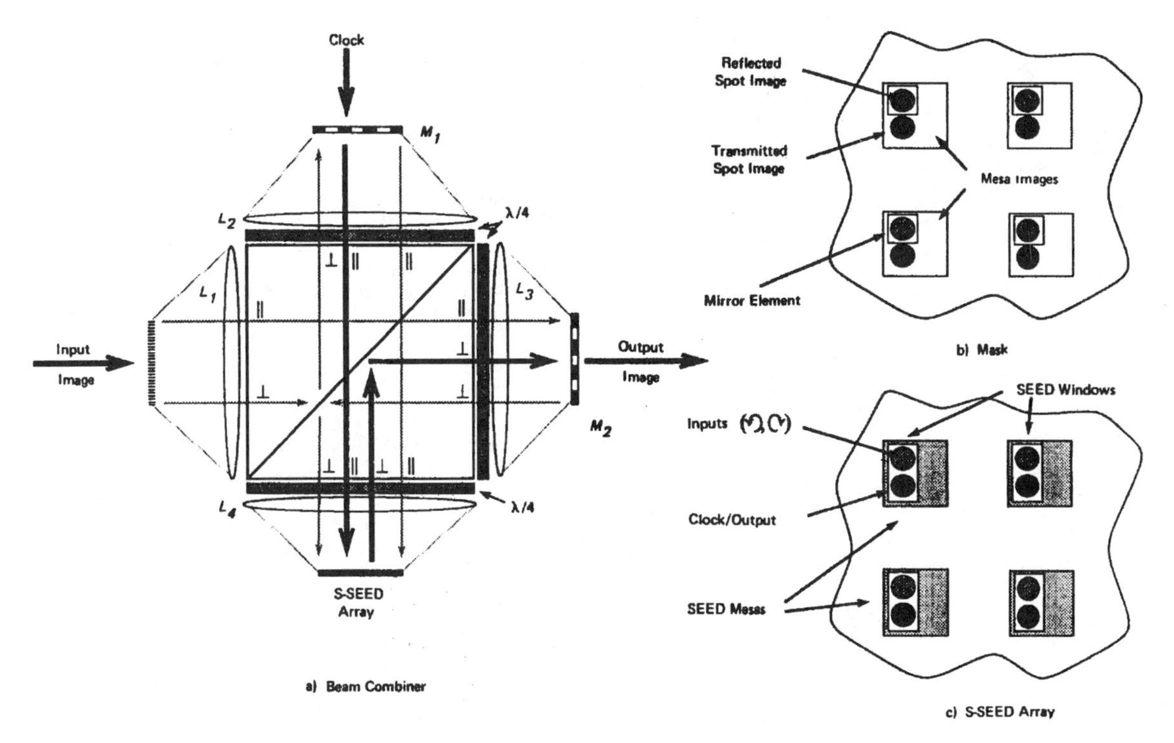

Fig. 21. Space-multiplexed beam combination.

spread these inputs vertically so that the light emitted from each input is spread over all the elements of the SLM's associated column. The appropriate pixels of the SLM are enabled before the data passes through the fabric. The one or more enabled pixel(s) in each column allow the incident light to be transmitted through the device while the remaining pixels block the incident light. The output column of fibers accepts the light that is passed through the SLM. An important restriction for this type of structure is that only one pixel on each row can be enabled at any time. The relational nature of this structure is evident in that each row of the SLM acts like an $N \times 1$ switch where N is the number of pixels per row. The total structure is topologically equivalent to a nonblocking crossbar interconnection network.

Free-Space Digital Optics Hardware

This section will discuss the basic hardware components that are required to implement a switching fabric based on free-space digital optics. These basic components are illustrated in the bottom part of Figure 4. They include logic devices (assuming that the devices are limited to a fan-in and fan-out of two and that the two input signals to any device are different polarizations), free-space optical interconnects, a beam combiner, and a spot array generator. Each of these components will be discussed in more detail in the following sections.

Devices

There are several devices that have been proposed for use in switching fabrics based on free-space digital optics. The leading contenders at the current time are Non-Linear Fabry-Perot (NLFP) devices [44–46], the VSTEP devices [47], and devices based on or derived from the Self Electro-optic Effect Device (SEED) [48–50]. From a systems point of view, one of the most attractive devices is the Symmetric-SEED (S-SEED) [51] [52]. These devices have the following attributes:

- They have differential outputs and inputs, which reduces the required system contrast ratio.

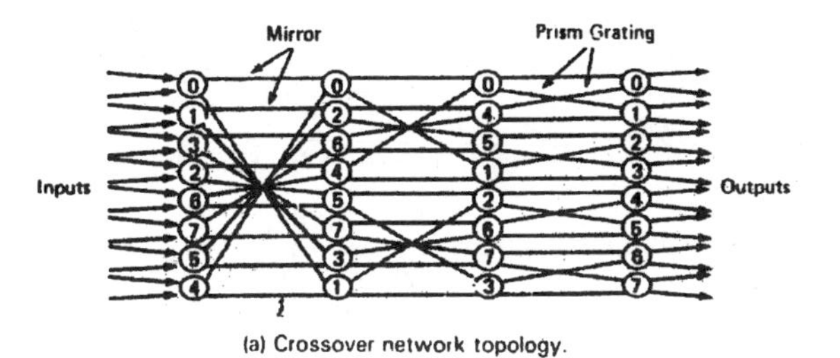

(a) Crossover network topology.

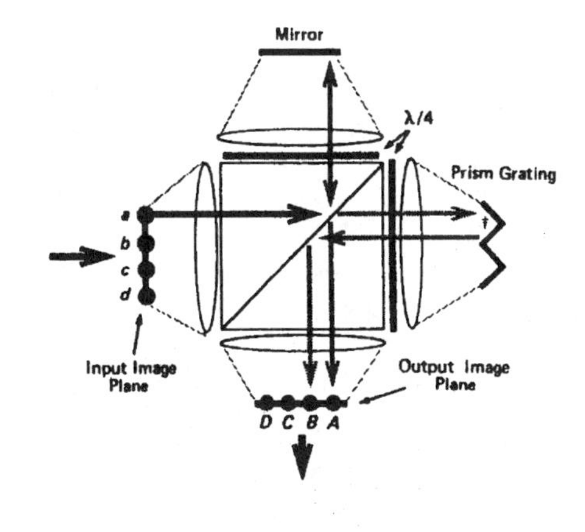

(b) Crossover network implementation.

Fig. 22. Crossover interconnection network.

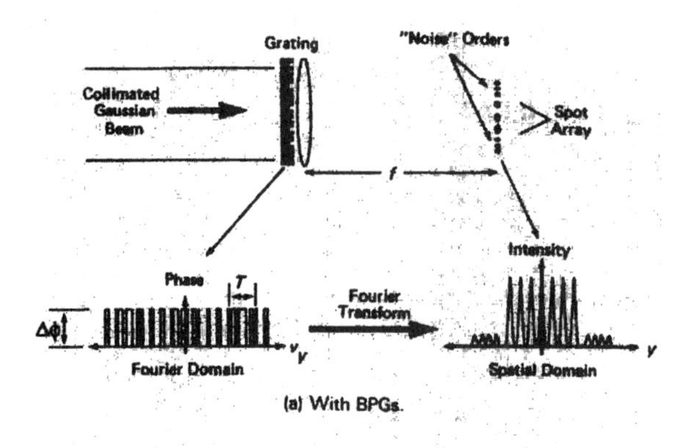

(a) With BPGs.

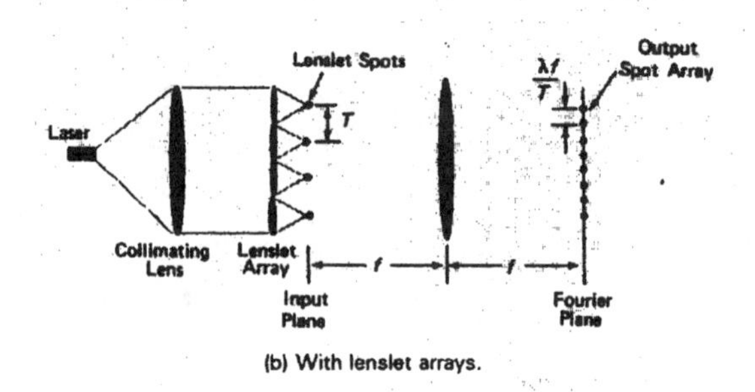

(b) With lenslet arrays.

Fig. 23. Spot array generation.

- They avoid the critical biasing problems by switching or changing states as a ratio of the two inputs instead of thresholding or triggering to a specific intensity value.
- They provide high gain through the use of time-sequential gain.
- They provide the functional operation of an S-R latch or optical logic gates (AND, OR, NAND, NOR, etc.) [41] [52].

Another type of device that has more functionality than the basic optical logic gates is the smart pixels [41], which are composed of a mixture of electronics to perform the processing and optical pin-outs to provide the connection between different devices. The electronics in these 2D-OEICs can provide many operational features, such as the exchange/bypass function for shuffle exchange nodes or the additional functionality of sorting nodes for Batcher-Banyan fabrics.

Beam Combination

The beam combination problem is illustrated in Figure 20 [53] [54]. Here each device must be able to accept two input signals plus a clock or bias beam. In addition, the output signal must be directed from the device to the output port. The major constraints of this problem are that the spots must be small ($< 5\ \mu m$), forcing the entering signals to use the full aperture of an imaging lens, and that the signals must not coherently interfere at the device's optical window.

An example of a beam combination system using space-multiplexing is illustrated in Figure 21. This beam combiner is composed of a Polarizing Beam Splitter (PBS) that is surrounded on three sides by $\lambda/4$ plates. The lenses, $L_1 - L_4$, are used in the infinite conjugate mode. At the bottom of the beam combiner is the S-SEED array. The location of the beams incident on the S-SEED array are shown in Figure 21(c). As an input image enters the beam combination system on the left (this image is composed of an array with each of the two input signals being associated with one of the two linear polarizations),

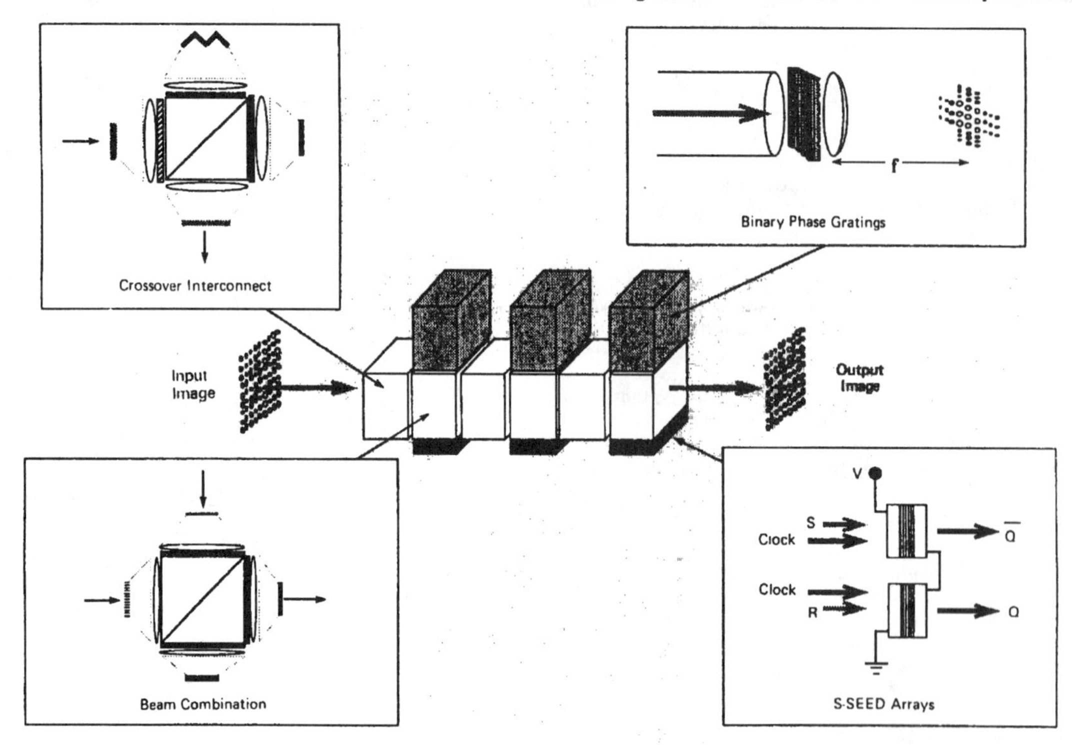

Fig. 24. Switching fabrics based on free-space digital optics.

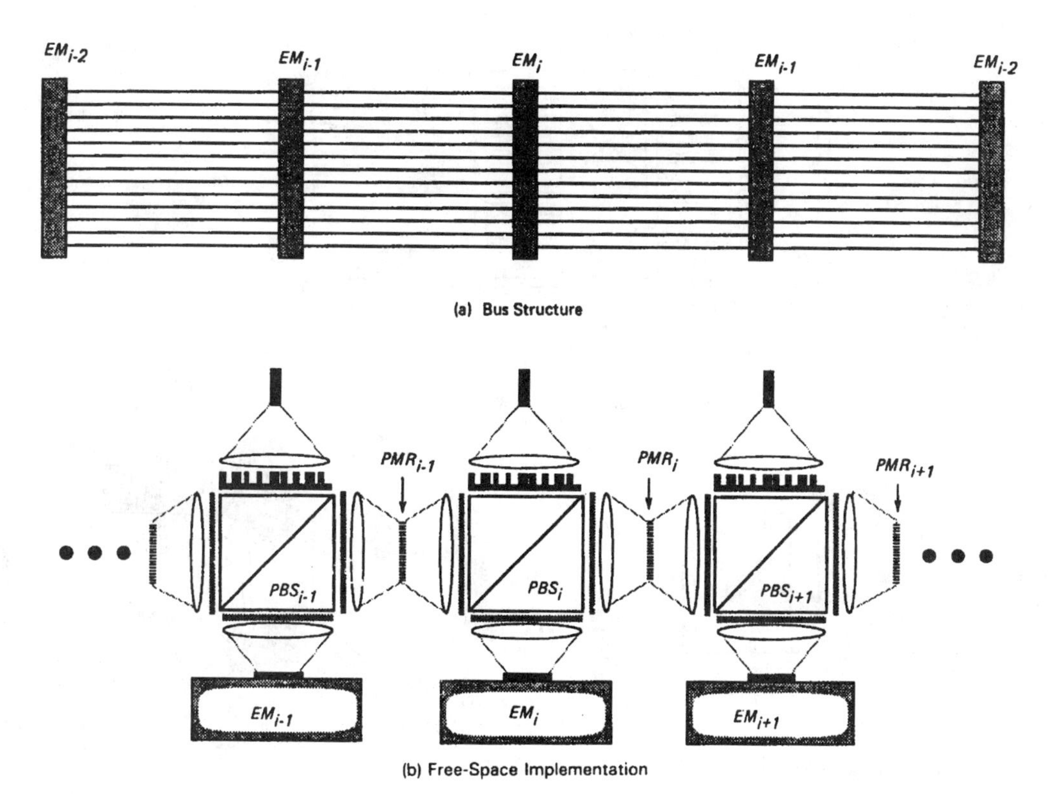

Fig. 25. Free-space bus/backplanes.

the perpendicular component is directed upward while the parallel component passes through the PBS. Both components travel through $\lambda/4$ waveplates, where the linear polarization is converted to circular. The light traveling upward is imaged onto patterned mirrors of mask M_1, as shown in Figure 21(b). This mask consists of an array of small mirrors that are located in the image field of L_2 such that light reflected from them will be imaged onto the upper half of the rectangular window of the S-SEED array. Therefore, the image traveling upward reflects off M_1 and passes again through a $\lambda/4$ waveplate, which changes the polarization from circular to parallel, allowing the light to be passed through the PBS onto the upper half of the S-SEED array.

The parallel component of the input image passes though the PBS and is imaged onto the small patterned mirrors of mask M_2, reflected by the PBS, and then imaged onto the upper half of the S-SEED array. Since these two inputs will have different orientations of circular polarizations, they will not interfere.

The optical clock will enter the beam combination unit by being imaged in the transparent region below the small mirrors of mask M_1. This collection of high-power spots will be imaged onto the bottom half of the S-SEED array. The reflected output signal of the S-SEED array, derived from the clock, will have its polarization rotated, allowing it to be reflected by the PBS and then imaged onto the transparent region below the small mirrors of mask M_2. This output image can then be collected and used by an optical interconnect or another beam combination unit. This same beam combination unit can also work for smart pixels or any other device that requires two signal inputs, a clock input, and a reflected output.

Optical Interconnects

The optical interconnect has the role of providing the interconnection between the devices on different arrays. Space-variant interconnects have been implemented using both bulk optics and holograms [55] [56]. The optical interconnect that has received the most attention has been the perfect shuffle [57–59]. This interconnect has become the basis for many multistage interconnection networks. An alternative to a perfect shuffle are interconnects that are topologically equivalent. An example of such an interconnect is the crossover network [60], which is illustrated in Figure 22. Figure 22(a) illustrates the topology of the interconnect. The output of each node is split into two components: a straight-through and a crossed interconnect. Figure 22(b) contains a conceptual demonstration of the operation of the interconnect system. An input image enters the system from the left. Each pixel or spot is circularly polarized. As an example, circularly polarized light enters at input location *a*. The perpendicular component of the light will be imaged onto the mirror. The reflected light from the mirror will have its polarization rotated, allowing it to pass through the PBS where it can be imaged onto the output image plane. The parallel component of the input light will pass through the beam splitter and be imaged onto the prism grating. The light incident on the prism grating will be reflected to the opposing mirror where it will be redirected back into the beamsplitter. This reflection process shifts the position, in space, of the light being imaged onto the grating corresponding to the crossover patterns of the crossover interconnection topology. The light reflected off the prism grating will have its polarization rotated by the $\lambda/4$ plate, allowing it to be redirected and focused onto the output image plane. Thus, both the straight-through and crossed interconnects have been achieved. The amount of shift in space is related to the period of the prism grating.

Spot Array Generation

The 2D arrays of optical logic gates that have been previously discussed will require an optical power supply to either clock

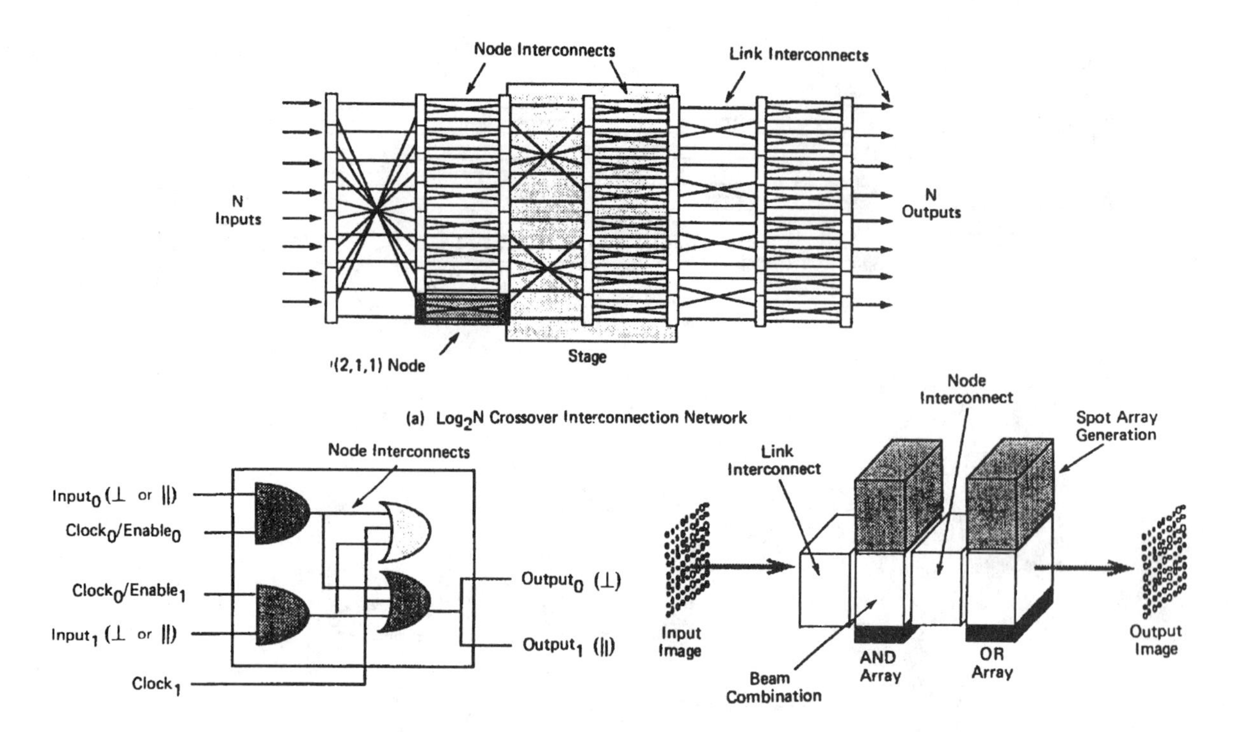

Fig. 26. Free-space digital network.

the devices, as in the case of S-SEED structures, or provide the optical bias power necessary to place the devices in the nonlinear region. These optical power supplies must be 2D arrays of uniform spots. There have been several approaches to this problem. One example is through the use of a repetitive array of diffractive elements referred to as phase gratings. These gratings are made by etching glass with a repetitive pattern of different thicknesses. For the case of a Binary Phase Grating (BPG), there would be two thicknesses of glass. This grating is illuminated by a plane wave from a laser source, as illustrated in Figure 23(a). The light transmitted through the grating is transformed at the back focal plane of a lens, which is the output plane of the spot array generator. One approach to designing BPGs assumes an even number of transition points and a phase difference of π between the two thickness [61] [62]. A second approach uses an odd number of transition points and does not constrain the phase difference to π [63]. Multilevel phase gratings have also been demonstrated [64].

Another approach to spot array generation is shown in Figure 23(b). For this approach, a collimated source illuminates a lenslet array. The output plane of the lenslet array will be transformed to the output plane. Through the Fourier transformation process, closely spaced, uniform-intensity spots are created.

Switching Fabrics Based on Free-Space Digital Optics

Using the hardware building blocks that have been discussed, photonic switching fabrics can be implemented, as illustrated in Figure 24. This section will briefly describe free-space bus/backplanes, free-space shuffle networks, and finally, free-space sorting networks.

Free-Space Bus/Backplane

Buses and backplanes have typically been one of the main bottlenecks in switching fabrics. This section will begin by describing a bus/backplane structure that can support up to 10^4 connections. Figure 25 is an illustration of an interconnect fabric that could be used for either a bus or a backplane [65]. It is a synchronized bus that regenerates the signal at each module. It also has the capability of bidirectional operation.

Each of the electronic modules, EM_i, will have a collection of optical pin-outs, which might consist of pin detectors for inputs and modulators for outputs. The electronic signal to be transferred from EM_i to either EM_{i-1} or EM_{i+1} is used to modulate one of the modulators outputs on EM_i. The carrier to be modulated will be derived from a laser source driving a spot array generator. This array of spots will be incident upon the EM's modulators. The output of these reflective modulators will be reflected up and directed to the right by the PBSs. The image from the optical pin-outs of EM_i will be imaged onto PMR_i (PMR stands for Patterned Mirror Reflector). Signals passing through PMR_i will pass through PBS_{i+1} and eventually be reflected by PMR_{i+1}. These redirected signals will then be reflected by PBS_{i+1} and then imaged onto EM_{i+1}. On the other hand, the signals reflected off PMR_i will pass through PBS_i and PMR_{i-1}, and then be reflected downward by PBS_{i-1}, where they will be imaged onto the detectors of EM_{i-1}. The interconnections to be established between the different EMs is determined by the location of the small space-variant mirrors of the PMRs.

Shuffle Networks

Switching networks that are based on a perfect shuffle interconnect or its topological equivalent can be implemented using the building blocks that have been previously described. Examples include blocking networks, such as the baseline networks [66], and rearrangeably nonblocking permutation networks [67]. The nodes required for such a network can be implemented with either smart pixels or implementing the logical function of the node with multiple stages of S-SEEDs or other optical logic devices. Such a fabric would follow the pattern outlined in Figure 4. The input image, which consists of all the separate inputs put in the parallel format of a 2D image, passes initially through the crossover interconnect (perfect shuffle). The image then passes through the beam combination unit prior to being imaged onto the node inputs (smart pixel inputs). To read the information stored in the node (smart pixel array), a clock signal is generated through the spot array generation process and then directed to the array of devices. The impinging array of spots is modulated by the information stored in the devices and reflected through the beam combination unit to the next interconnect stage. This process is repeated for the desired number of stages ($\log_2 N$ for blocking networks). Finally, an output interface would be reached that would convert the parallel image into the serial output channels destined for the outside world.

Another example of a space-division fabric using free-space digital optics is a network based on a (2,1,1) node made with optical logic gates incorporated in any network based on perfect shuffle interconnects [69]. The (2,1,1) triplet represents the following: (Number of inputs, Number of outputs, Capacity of the node). The first two parameters of the triplet represent the number of inputs and outputs; thus a (2,1,1) node has two inputs and one output. The third parameter indicates the number of channels that can be passed through the node at a given time. A logic diagram of a (2,1,1) node is shown in part (b) of Figure 26.

A $\log_2 N$ crossover interconnection network is a specific example of a network that can be based on (2,1,1) nodes. In this example, there are eight inputs and eight outputs. The network is composed of three stages, where each stage is composed of two arrays of optical logic gates. The first logic gate array in each node will logically select which of the two input channels will be allowed to pass its traffic to the node output. The second logic gate array provides an OR function. The output of the node will have a fan-out of two, half of its energy in each of the two linear polarizations. A two-dimensional array of signals will enter each stage through a link interconnect. This interconnect will have a different period for each stage and provide the node-to-node connectivity required by the network topology. The crossover interconnect, topologically equivalent to the perfect shuffle as previously discussed, will implement this function, providing the straight-through connection to one of the polarizations while providing the crossover function for the other [70]. This permuted 2-D array of signals will then pass through the beam combination unit onto the AND logic gates. The enable signal will allow only one of the two possible signals to pass through the node interconnect to the OR gate. Note that the node interconnect is a fine period crossover network. The output of the OR gate will then pass through the beam combination unit to the next stage of the fabric. A means of implementing the AND function is to use an S-SEED as a latch to capture and store the information for both inputs. By placing an SLM in one of the image planes of the spot array generator, the S-SEEDs receiving a clock pulse can be selectively controlled. Using this clock control method, only the input AND gate containing the information to be moved to the output would be clocked. Thus, the paths through the network are controlled through the state of the SLM.

Sorting Networks

A sorting network is a switching fabric that can monitor the addresses of input packets entering the fabric and order them in either ascending or descending order at the output of the fabric. An example of such a network is the Batcher sorting network [68]. For this type of structure, the optical interconnect has to be a 2D shuffle (requires a 3D volume) [71] instead of the 1D interconnects (planar structures) [72]. In these structures, the input consists of a 2D array of inputs rather than a single row of entering channels. The advantage of this approach is that the size of the shuffle-based sorting fabric is equal to the number of smart pixels that can be fabricated onto a 2D-OEIC.

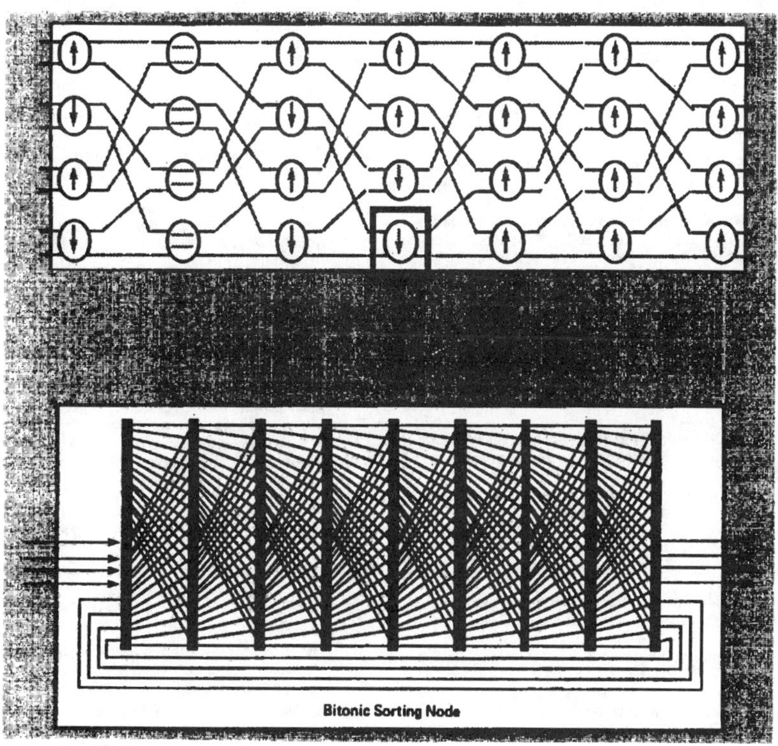

Fig. 27. Sorting network using optical PLAs.

An all-optical approach to a sorting network can be implemented through the use of the optical Programmable Logic Arrays (PLAs) that have been proposed by Murdocca [73] [74]. The basic concept of this type of fabric is illustrated in Figure 27. A sorting network based on perfect shuffle interconnects is shown in the upper half of the figure. The hardware required to implement a single sorting node has been expanded and shown in the lower half of the figure.

Discussion

This paper has reviewed a good portion of the photonic switching fabrics that have been proposed in the research community. The prevailing emphasis has been to find effective methods of using the temporal bandwidth of relational devices or the spatial bandwidth available in free-space digital optical systems. At the current time, no one approach has emerged as the unanimous choice of all researchers. It should also be pointed out that none of these proposed fabrics are available in the marketplace today.

There is another issue surrounding the future of photonic switching fabrics that remains hidden when fabrics by themselves are reviewed. The switching fabric is a small part of an entire switching system. A complete switching system includes transmission equipment for the line interfaces (this cost dominates the total hardware cost) and software (which dominates the total system cost). As new transmission formats are introduced, it is important to remember that at each line interface of a switching system will be a piece of hardware required to convert the transmission protocol to a protocol that can be handled by switching systems. Because of this lack of coordination between switching and transmission, there will always be a dominant cost at each input to the switching system.

References

[1] P. W. Smith, "On the Role of Photonic Switching in Future Communications Systems," *IEEE Circuits and Devices Mag.*, pp. 9–14, May 1987.

[2] H. S. Hinton, "Photonic Switching Technology Applications," *AT&T Tech. J.*, vol. 66, issue 3, pp. 41–53, May/June 1987.

[3] B. S. Glance *et al.*, "WDM Coherent Optical Star Network," *J. of Lightwave Tech.*, vol. 6, no. 1, pp. 67–72, Jan. 1988.

[4] T. E. Darcie, "Subcarrier Multiplexing for Multiple-Access Lightwave Networks," *J. of Lightwave Tech.*, vol. LT-5, no. 8, pp. 1,103–1,110, Aug. 1987.

[5] P. W. Smith, "On the Physical Limits of Digital Optical Switching and Logic Elements," *Bell Sys. Tech. J.*, vol. 61, no. 8, pp. 1,975–1,993, Oct. 1982.

[6] S. L. McCall and H. M. Gibbs, "Conditions and Limitations in Intrinsic Optical Bistability," C. M. Bowden, M. Ciftan, and H. R. Robl (eds.), *Optical Bistability*, pp. 1–7, New York: Plenum Press, 1981.

[7] D. B. Tuckerman and R. F. W. Pease, "High-Performance Heat Sinking for VLSI," *IEEE Electron. Device Lett.*, vol. EDL-2, no. 5, pp. 126–129, May 1981.

[8] D. A. B. Miller, "Optics for Low-Energy Communication Inside Digital Processors: Quantum Detectors, Sources, and Modulators as Efficient Impedence Converters," *Opt. Lett.*, vol. 14, no. 2, pp. 146–148, Jan. 15, 1989.

[9] J. W. Goodman, *Introduction to Fourier Optics*, New York: McGraw-Hill, Inc., 1968.

[10] A. Varma and C. S. Raghavendra "Rearrangeability of Multistage Shuffle/Exchange Networks," *IEEE Trans. on Commun.*, vol. 36, pp. 1,138–1,147, Oct. 1988.

[11] H. S. Hinton, "Photonic Switching Using Directional Couplers," *IEEE Commun. Mag.*, vol. 25, no. 5, pp. 16–26, May 1985.

[12] R. V. Schmidt and R. C. Alferness, "Directional Coupler Switches, Modulators, and Filters Using Alternating Db Techniques," *IEEE Trans. on Circuits and Sys.*, vol. CAS-26, no. 12, pp. 1,099–1,108, Dec. 1979.

[13] H. S. Hinton, "A Non-Blocking Optical Interconnection Network Using Directional Couplers," *Proc. of the IEEE Global Telecom. Conf.*, vol. 2, pp. 885–889, Nov. 1984.

[14] P. Granestrand *et al.*, "Strictly Nonblocking 8 × 8 Integrated Optical Switch Matrix," *Elect. Lett.*, vol. 22, no. 15, July 17, 1986.

[15] G. A. Bogert, "A Low Crosstalk 4x4 Ti:LiNbO3 Optical Switch with Permanently Attached Polarization-Maintaining Fiber Arrays," Topical Meeting on Integrated and Guided-Wave Optics, Atlanta, GA, pp. PDP 3.1–3, Feb. 1986.

[16] R. A. Spanke, "Architectures for Large Nonblocking Optical Space Switches," *IEEE J. of Quantum Electron.*, vol. QE- 22, no. 6, pp. 964–967, June 1986.

[17] G. W. Richards and F. K. Hwang, "A Two-Stage Rearrangeable Broadcast Switching Network," *IEEE Trans. on Commun.*, vol. COM-33, no. 10, pp. 1,025–1,035, Oct. 1985.

[18] M. Gustavsson and L. Thylen, "Switch Matrix with Semiconductor Laser Amplifier Gate Switches: A Performance Analysis," *OSA Proc. on Photonic Switching*, J. E. Midwinter and H. S. Hinton, eds., Optical Soc. of America, Washington, D.C., 1989, vol. 3, pp. 77–79.

[19] A. D. Fisher, "A Review of Spatial Light Modulators," Topical Meeting on Opt. Comp., Incline Village, Nevada, Mar. 18–20, 1985.

[20] W. E. Ross, D. Psaltis, and R. H. Anderson, "2D Magneto Optic Spatial Light Modulator for Signal Processing," SPIE Conf., Crystal City-Arlington, VA, May 3–7 1982.

[21] A. R. Tanguay, "Materials Requirements for Optical Processing and Computer Devices," *Opt. Eng.*, pp. 2–18, Jan./Feb. 1985.

[22] A. Himeno and M. Kobayashi, "4 × 4 Optical-Gate Matrix Switch," *J. of Lightwave Tech.*, vol. LT-3, no. 2, pp. 230–235, Apr. 1985.

[23] D. R. Pape and L. J. Hornbeck, "Characteristics of the Deformable Mirror Device for Optical Information Processing," *Opt. Eng.*, vol. 22, pp. 675–681, 1983.

[24] G. Livescu *et al.*, "Spatial Light Modulator and Optical Dynamic Memory Using Integrated Self Electo-Optic Effect Devices," *Proc. of the Conf. on Lasers and Electro-Optics* (postdeadline paper), pp. 283–284, Apr. 26–May 1, 1987.

[25] A. R. Dias, R. F. Kalman, J. W. Goodman, and A. A. Sawchuk, "Fiber-Optic Crossbar Switch with Broadcast Capability," *Opt. Eng.*, vol. 27, no. 11, pp. 955–960, Nov. 1988.

[26] K. Oshima *et al.*, "Fiber-Optic Local Area Passive Network Using Burst TDMA Scheme," *J. of Lightwave Tech.*, vol. LT-3, no. 3, pp. 502–510, June 1985.

[27] R. A. Thompson and P. P. Giordano, "An Experimental Photonic Time-Slot Interchanger Using Optical Fibers as Reentrant Delay-Line Memories," *J. of Lightwave Tech.*, vol. LT-5, no. 1, pp. 154–162, Jan. 1987.

[28] H. Goto, K. Nagashima, and S. Suzuki, "Photonic Time-Division Switching Technology," *Photonic Switching: Proceedings of the First Topical Meeting*, T. K. Gustafson and P. W. Smith (eds.), pp. 151–157, New York: Springer-Verlag, 1987.

[29] R. A. Thompson, R. V. Anderson, J. V. Camlet, and P. P. Giordano, "Experimental Modular Switching System with a Time-Multiplexed Photonic Center Stage," *OSA Proc. on Photonic Switching*, J. E. Midwinter and H. S. Hinton, eds., vol. 3, pp. 212–218, Washington, DC: Optical Society of America, 1989.

[30] M. Skov, "Implementation of Physical and Media Access Protocols for High-Speed Networks," *IEEE Commun. Mag.*, pp. 45–53, June 1989.

[31] A. A. M. Saleh and H. Kogelnik, "Reflective Single-Mode Fiber Optic Passive Star Couplers," *J. of Lightwave Tech.*, vol. 6, no. 3, pp. 392–398, Mar. 1988.

[32] P. R. Prucnal and P. A. Perrier, "A New Direction in Photonic Switching: A Collapsed-Network Space-Division Switching Architecture," *OSA Proc. on Photonic Switching*, J. E. Midwinter and H. S. Hinton (eds.), vol. 3, pp. 212–218, Washington, DC: Optical Society of America, 1989.

[33] P. R. Prucnal, M. A. Santoro, and T. R. Fan, "Spread Spectrum Fiber Optic Local Area Network Using Optical Processing," *J. of Lightwave Tech.*, vol. LT-4, no. 5, pp. 547–554, May 1986.

[34] P. R. Prucnal, M. A. Santoro, and S. K. Sehgal, "Ultrafast All-Optical Synchronous Multiple Access Fiber Networks," *IEEE J. of Sel. Areas in Commun.*, vol. SAC-4, no. 9, pp. 1,484–1,493, Dec. 1986.

[35] G. J. Foschini and G. Vannucci, "Using Spread-Spectrum in a High-Capacity Fiber-Optic Local Network," *J. of Lightwave Tech.*, vol. 6, no. 3, pp. 370–379, Mar. 1988.

[36] M. Fijiwara *et al.*, "A Coherent Photonic Wavelength-Division Switching System for Broadband Networks," *Proc. of the 14th Euro. Conf. on Opt. Commun. (ECOC'88)*, Brighton, England, pp. 139–142, Sept. 11–15, 1988.

[37] H. Kobrinski *et al.*, "Demonstration of High Capacity in the LAMBDANET Architecture: A Multiwavelength Optical Network," *Elect. Lett.*, vol. 23, pp. 824–826, 1987.

[38] I. P. Kaminow, P. P. Iannone, J. Stone, and L. W. Stulz, "FDM-FSK Star Network with a Tunable Optical Filter Demultiplexer," *Elect. Lett.*, vol. 23, pp. 1,102–1,103, 1987.

[39] B. Glance *et al.*, "Densely Spaced FDM Coherent Star Network with Optical Signals Confined to Equally Spaced Frequencies," *J. of Lightwave Tech.*, vol. 6, no. 11, pp. 1,770–1,781, Nov. 1988.

[40] M. S. Goodman *et al*, "Demonstration of fast wavelength tuning for a high performance packet switch," Fourteenth European Conference on Optical Communications, Brighton, England, September 11–15, pp. 255–258, 1988.

[41] H. S. Hinton, "Architectural Considerations for Photonic Switching Networks, *J. of Sel. Areas of Commun.*, vol. 6, pp. 1,209–1,226, August 1988.

[42] J. W. Goodman, A. R. Dias, and L. M. Woody, "Fully Parallel, High-Speed Incoherent Optical Method for Performing Discrete Fourier Transforms," *Opt. Let.*, vol. 2, no. 1, pp.1–3, January 1978.

[43] A. A. Sawchuk, B. K. Jenkins, C. S. Raghavendra, and A. Varma, "Optical Crossbar Networks," *IEEE Computer*, vol. 20, no. 6, pp. 50–60, June 1987.

[44] S. D. Smith, "Optical Bistability, Photonic Logic and Optical Computation," *Appl. Opt.*, vol. 25, pp. 1,550–1,564, May 15, 1986.

[45] J. L. Jewell, M. C. Rushford, and H. M. Gibbs, "Use of a Single Nonlinear Fabry-Perot Etalon as Optical Logic Gates," *Appl. Phys. Let.*, vol. 44, pp. 172–174, Jan. 15, 1984.

[46] J. L. Jewell *et al*, "GaAs-AlAs Monolithic Microresonator Arrays," *Appl. Phys. Let.*, vol. 51, pp. 94–99, July 13, 1987.

[47] K. Hara, K. Kojima, K. Mitsunaga, and K. Kyuma, "Differential Optical Switching at Subnanowatt Input Power," *IEEE Photonics Techn. Let.*, vol. 1, no. 11, Nov. 1989, pp. 370–372.

[48] D. A. B. Miller *et al*, "Novel Hybrid Optically Bistable Switch: The Quantum Well Self-Electro-Optic Effect Device," *Appl. Phys. Let.*, vol. 45, pp. 13–15, 1984.

[49] D. A. B. Miller *et al*, "The Quantum Well Self-Electro-Optic Effect Device: Optoelectric Bistability and Oscillation, and Self-Linearized Modulation," *IEEE J. of Quantum Electron.*,vol. QE-21, pp. 1,462–1,476, Sept. 1985.

[50] D. A. B. Miller *et al*, "Integrated Quantum Well Self-Electro-Optic Device: 2 x 2 Array of Optically Bistable Switches," *Appl. Phys. Let.*, vol. 49, pp. 821–823, 1986.

[51] A. L. Lentine *et al*, "Symmetric Self-Electro-Optic Effect Device: Optical Set-Reset Latch," *Appl. Phys. Let.*, vol. 52, pp.1,419–1,421, Apr. 25, 1988.

[52] A. L. Lentine *et al*, "Symmetric Self-Electrooptic Effect Device: Optical Set-Reset Latch, Differential Logic Gate, and Differential Modulator/Detector," *IEEE J. of Quantum Electron.*, vol. 25, no. 8, pp. 1,928–1,936, Aug. 1989.

[53] M. E. Prise *et al*, "Design of an Optical Computer," *J. de Physique*, Colloque C2, 49, juin 1988.

[54] M. E. Prise, N. Streibl, and M. M. Downs, "Optical Considerations in the Design of Digital Optical Computers," *Opt. nd Quantum Electronics*, vol. 30, pp. 49-77, 1988.

[55] A. A. Sawchuk and T. C. Strand, "Digital Optical Computing," *Proc. IEEE*, vol. 72, pp. 758–779, July 1984.

[56] B. K. Jenkins *et al*, "Architectural Implications of a Digital Optical Processor," *Appl. Opt.*, vol. 23, pp. 3,65–3,74, Oct. 1, 1984.

[57] A. W. Lohmann, W. Stork, and G. Stucke, "Optical Perfect Shuffle," *Appl. Opt.*, vol. 25, pp. 1,530–1,531, May 15, 1986.

[58] G. Eichmann and Y. Li, "Compact Optical Generalized Perfect Shuffle," *Appl. Opt.*, vol. 26, pp. 1,167–1,169, Apr. 1, 1987.

[59] K.-H. Brenner and A. Huang, "Optical Implementation of the Perfect Shuffle Interconnection," *Appl. Opt.*, vol. 27, no. 1, pp. 135–137, Jan. 1, 1988.

[60] J. Jahns and M. J. Murdocca, "Crossover Networks and Their Optical Implementation," *Appl. Opt.*, vol. 27, pp. 3,155–3,160, Aug. 1, 1988.

[61] H. Dammann and K. Gortler, "High-Efficiency In-Line Multiple Imaging by Means of Multiple Phase Holograms," *Opt. Commun.*, vol. 3, no. 5, pp. 312–315, July 1971.

[62] H. Dammann and E. Klotz, "Coherent Optical Generation and Inspection of Two-Dimensional Periodic Structures," *Optica Acta*, vol. 24, no. 4, pp. 505–515, 1977.

[63] U. Killat, G. Rabe, and W. Rave, "Binary Phase Gratings for Star Couplers with High Splitting Ratio," *Fiber and Integrated Opt.*, vol. 4, no. 2, pp. 159–167, 1982.

[64] J. Jahns, N. Streibl, and S. J. Walker, "Multilevel Phase Structures for Arrays Generation," OE/LASE'89 Los Angeles, CA, 1989.

[65] A. Dickinson and M. E. Prise, "Free-Space Optical Interconnection Scheme," Top. Mtg. on Photonic Switch., Salt Lake City, UT, Feb. 27–Mar. 1, 1989.

[66] C. Wu and T. Feng, "On a Class of Multistage Interconnection Networks," *IEEE Trans. Computers*, vol. 29, no. 8, pp. 694–702, Aug. 1980.

[67] C. Wu and T. Feng, "Universality of the Shuffle-Exchange Network," *IEEE Trans. Computers*, vol. C-30, no. 5, pp. 324–332, May 1981.

[68] K. E. Batcher, "Sorting Networks and Their Applications," 1968 Spring Joint Computer Conf., *AFIPS Proc.*, vol. 32, pp. 307–314, 1968.

[69] H. S. Stone, "Parallel Processing with the Perfect Shuffle," *IEEE Trans. on Computers*, vol. C-20, pp. 153–161, Feb. 1971.

[70] J. E. Midwinter, "'Light' Electronics, Myth or Reality?" *IEEE Proc.*, vol. 132, pp. 371–383, Dec. 1985.

[71] C. W. Stirk, R. A. Athale, and C. B. Friedlander, "Folded Perfect Shuffle Optical Processor," *Appl. Opt.*, vol. 27, pp. 202–203, Jan. 15, 1988.

[72] M. Taylor and J. E. Midwinter, "A Novel Two Dimensional Perfect Shuffle Network," Topical Meeting on Photonic Switching, Postdeadline paper, Salt Lake City, UT, 1989.

[73] M. Murdocca, "Optical Design of a Digital Switch," *Appl. Opt.*, vol. 28, no. 13, July 1, 1989, pp. 2,505–2,517.

[74] M. Murdocca *et al*, "Optical Design of Programmable Logic Arrays," *Appl. Opt.*, vol. 27, pp. 1,651–1,660, May 1, 1988.

Photonic Packet Switches: Architectures and Experimental Implementations

DANIEL J. BLUMENTHAL, MEMBER, IEEE, PAUL R. PRUCNAL, FELLOW, IEEE
AND JON R. SAUER

Invited Paper

Photonic packet switches offer high speed, data rate and format transparency, and flexibility required by future computer communications and cell-based telecommunications networks. In this paper, we review experimental progress in state-of-the-art photonic packet switches with an emphasis on all-optical guided-wave systems. The term all-optical implies that the data portion of a packet remains in optical format from the source to the destination. While the data remain all-optical, both optical and optoelectronic techniques have been used to process packet routing functions based on extremely simple routing protocols. An overview of the design issues for all-optical photonic packet switching is given and contrasted with electronic packet switch implementations. Low-level functions that have been experimentally implemented include routing, contention resolution, synchronization, and header regeneration. System level demonstrations, including centralized photonic switching and distributed all-optical multihop networks, will be reviewed.

I. Introduction

In future fiber-optic packet-switched communication networks, the high transmission link data rates as well as the large number of packets transmitted per second will place severe demands on the aggregate network bandwidth. For packet-based applications (e.g., computer communications, ATM-based telecommunications), packets or cells are individually routed and the switch reconfiguration speed is of prime importance. Although electronic technology can achieve high switching speeds, it is not well matched to the transmission bandwidths of fiber-optic links. Photonic switches provide both the high switching speeds and a transmission bandwidth compatible with the fiber-link bandwidth. Perhaps more importantly, photonic packet switches open the possibility for new network architectures that are transparent to the packet data rate and format, extending the success of point-to-point links to the switched network level.

Manuscript received December 21, 1993; revised July 7, 1994.

D. J. Blumenthal is with the School of Electrical and Computer Engineering, Microelectronics Research Center, Georgia Institute of Technology, Atlanta, GA 30332 USA.

P. R. Prucnal is with the Department of Electrical Engineering, Center for Photonics and Optoelectronics Materials, Princeton University, Princeton, NJ 08544-5263 USA.

J. R. Sauer is with the Department of Electrical and Computer Engineering, Optoelectronic Computing Systems Research Center, University of Colorado at Boulder, Boulder, CO 80309 USA.

IEEE Log Number 9405125.

In this paper we review progress towards the realization of photonic packet switches that have the functionality required to build complete switching networks. Design and implementation issues are very different from their electronic counterparts and new approaches in switch architecture and protocol design are required. Current progress towards experimental implementation of "all-optical" photonic packet switches is examined. The term all-optical implies that the data portion of a packet is maintained in optical format from the source to the destination. However, the control portion of the packet may or may not be optoelectronically regenerated at each switch depending on the control technique used. Optical and optoelectronic approaches to routing control are discussed.

The basic required functionality for a packet switch can be summarized by the five low-level functions described below in (a–e):

a) *Routing:* Routing of packets from switch input to switch output or from source to destination in a distributed network. Packet headers are carried with individual packets, separated from the data at each switch, and processed to set the correct switch state.

b) *Flow Control and Contention Resolution:* Traffic in the switch and network must be regulated to prevent packets from running into each other and congesting resources. Contention resolution is needed to mediate flow of packets through internal switch links (internal block) and switch output ports (output port contention). Buffering, blocking, dropping, and deflection are examples of techniques used to control traffic flow and resolve contention.

c) *Synchronization:* Time alignment of packets at multiple switch input ports in order to correlate the packet positions with actual switching events.

Reprinted with permission from *Proceedings of the IEEE,* Vol. 82, No. 11, pp. 1650-1667, November 1994.

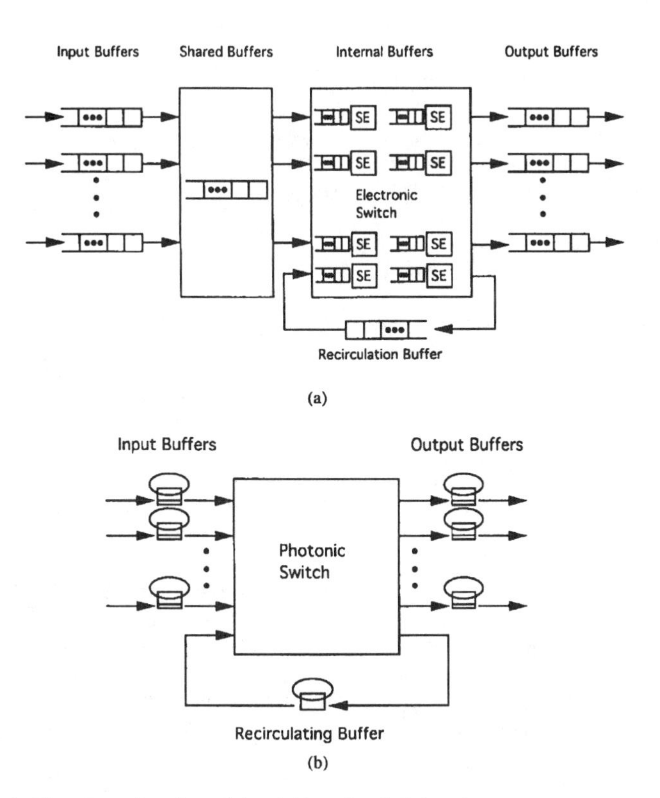

Fig. 1. Possible buffer locations in (a) an electronic packet switch and (b) an all-optical photonic packet switch.

d) *Header Regeneration/Reinsertion:* The calculation of a new packet header and its reinsertion with payloads at the appropriate switch output. For multihop switching header regeneration/reinsertion should be independent of the packet lifetime in the network.

e) *Cascadability:* The ability to route packets through multiple switches in terms of routing, contention, and timing and level restoration.

After exploring the differences between electronic and photonic packet switches in Section I-A, we discuss characteristics of all-optical switches in Section II. Optical coding techniques that have been experimentally demonstrated are reviewed in Section III. Routing control techniques and experimental demonstrations are presented in Sections IV and V, respectively. Experimental demonstrations of contention resolution using deflection routing in recirculating fiber loop input buffers are discussed in Section VI. Header regeneration and a multihop switching experiment are described in Sections VII and VIII. Packet synchronization is discussed in Section IX.

A. A Comparison of Photonic and Electronic Packet Switches

There are fundamental differences between photonic and electronic switch technologies. Photonic switching offers distinct advantages in terms of combined bandwidth and switch speed. However, the difficulty of routing data optically has provided a need to develop novel packet switching architectures and techniques. One distinction between switching signals electronically and optically is that electronic data can be stored statically, while optical data cannot be stored statically and must be processed and switched on the fly. Other factors that influence a packet-switch architecture come from the relative cost of switch facilities versus link bandwidth. Electronic packet switching has been influenced by the drastic reduction in cost of switch crosspoints and memory following the invention of integrated electronic circuits. Low-cost switch elements have made the electronic link bandwidth relatively expensive. Therefore, routing techniques were developed that heavily utilized memory for synchronization, flow control, and contention resolution, and placed an emphasis on efficiently utilizing the link bandwidth. High-performance electronic packet switching architectures and techniques have been extensively studied [1]. Aggregate throughput requirements for state-of-the-art electronic ATM packet switches will be on the order of 10 Gb/s for a 128 × 128 switch with line rates of 150 Mb/s [2]. The situation is reversed for photonic switching systems. The link bandwidth of optical fibers is relatively inexpensive, whereas photonic switch elements are more expensive than their electronic counterparts, and optical memories are both expensive and currently impractical for system level use.

In a typical electronic switch, buffers are used for synchronization, contention resolution, and flow control, and can be located at the switch inputs, switch outputs, internally, or shared as illustrated in Fig. 1(a). The term "bufferless" or "memoryless" in electronic switching usually refers to an absence of buffers within the switch fabric itself; therefore, we only refer to buffered electronic switches. The buffer size or depth is chosen to handle the mismatch between the input flow of packets and the estimated output flow. Sequential electronic buffers are characterized by clocked operation, where packets are synchronously transferred from stage to stage of the buffer, and removed first-in-first-out (FIFO) or last-in-first-out (LIFO). Sequential input buffers suffer an undesirable characteristic called head-of-the-queue blocking, where a blocked packet at the head stops unblocked packets in the queue from being routed even though they can be routed without contention. Shared, recirculating, and output buffers alleviate the head-of-the-queue blocking by allowing packets to be removed from any stage of the queue. Random-access memories (RAM) are used to implement shared buffers, but do not scale well in terms of access time for increasing number of switch inputs. Studies have shown that higher performance is reached when using output buffers [3] for FIFO buffering. Buffering is also used to pipeline packets through the switch, providing an increase in throughput with a tradeoff in latency.

From a broadband communications perspective (e.g., simultaneous support of multiple traffic types, simultaneous support of digital and analog signals), electronic buffers can be a limiting factor. The fastest bit rate supported is based on the bsic buffer clock rate. Therefore, all traffic must conform to the buffer increments while entering the switch. Additionally, a future increase in transmission line bit rates will require the electronic buffers to be upgraded.

B. Data-Rate and Format Transparency

Photonic switch research over the past ten years has focused on how to best utilize the wide bandwidth to increase performance or fill requirements difficult to perform with electronic switching. One example is the capability of photonic switches in conjunction with optical fibers to maintain data in all-optical format from the source to final destination.This characteristic allows the simultaneous transport of multiple data rates (data-rate transparency) and multiple formats (format transparency). Photonic packet switches are characterized as elastic-buffered or passive-buffered. Here, the term passive-buffered implies that only passive delay lines are used, typically fiber loops. Elastic-buffered all-optical switches are characterized by the available optical memory technologies useful for packet storage. These buffers are different from their electronic counterparts. The primary type of optical buffer in use recirculates packets instead of holding them statically or stationary in memory. Optical buffers can be used at the input, output, and as recirculatory buffers as illustrated in Fig. 1(b). Examples of fiber-based memories are shown in Fig. 2: Fig. 2(a)—a simple fiber loop delay; Fig. 2(b)—a programmable fiber-optic delay line [4]; Fig. 2(c)—a photonic switch time slot interchanger [5]–[7], and Fig. 2(d)—an active switched recirculating delay line [8]. The passive fiber-loop delay line can compensate for fixed, known delays. The programmable delay line shown uses a $1 \times N$ switch that selects a certain delay, and other configurations have been investigated [8]–[11]. This structure has high loss and does not scale well for a large number of required delays, where the longest delay must accommodate the maximum number of buffered packet slots. The time slot interchanger efficiently buffers packets by shifting a packet back in time relative to the current packet stream, with one stage required for each level in the packet queue. The active switched delay line is compact and can achieve a variable numbers of delays if optical amplification is used in the feedback loop. While optical amplifier noise limits the number of circulations, nonlinear gain compression can be used to realize up to 10^6 pulse recirculations [12]. To date, single packet buffering has been achieved [13], [14]. The loop delay line and recirculating delay line buffers can be used for holding or reordering packets but not for synchronization. Optical synchronization can be performed using time slot interchangers as discussed at the end of this paper.

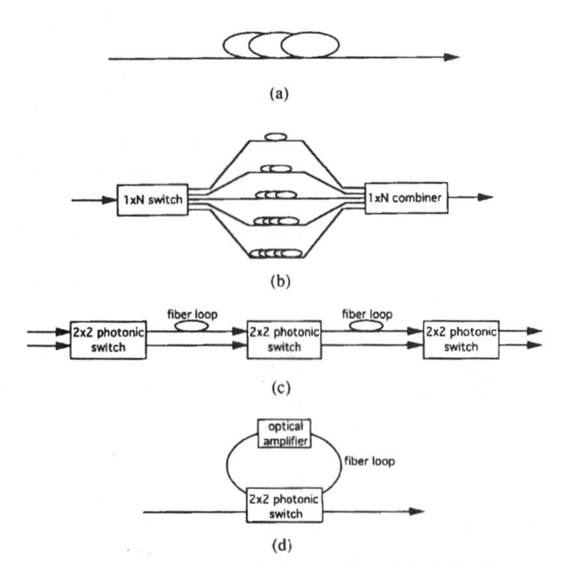

Fig. 2. Optical buffer implementations. (a) Fiber delay line. (b) Programmable fiber delay line. (c) Feed-forward time slot interchanger. (d) Active switched recirculating delay line.

Another important difference between electronic and photonic switching is the speed mismatch between the time required for the switch to change its state and the processing time required to determine the appropriate state of the switch. Core-and-edge logical network structures [15] for packet switches distribute the processing burden of low-level functions, such as routing, within the core of the network, and perform high-level functions, such as session setup, at the periphery of the network. The high-level functions require a large amount of slow processing, and can be performed easily with electronics. The low-level functions, on the other hand, require relatively simple processing, but must be performed at high speed and completed in a time interval t' less than the packet length T_p

$$t' \leq T_p \tag{1}$$

so that the switch is ready to route the next packet as soon as it arrives. This is particularly important when packets are routed on a packet-by-packet basis as with process synchronization and memory fetches in a computer interconnection network.

One proposed method of performing the low-level functions in a photonic switch is to use an electronic overlay network [16] where the header of each packet is converted to an electronic signal and sent to a self-routing electronic switch that is topologically identical to the photonic switch. In the electronic overlay network, the routing information contained in the packet header determines the state of each electronic switching element, which in turn determines the state of the optical switching element. Once the electronic switch is completely set, the optical packet (including the header and the data packet) is sent through the photonic switch. To avoid a data flow bottleneck at the switch input, the time t' required to electronically process the routing information and set the switching elements, must be less than the packet period T_p.

The processing speed of the low-level functions can be increased by using parallel processing or pipelining [17]. A K-fold increase in speed can be obtained by connecting a system of K processors in parallel, and sequentially allocating the input data to the individual processors. However, the hardware required to replicate the processing units K times and sequentially allocate the data to the K individual processors may be complex. A K-fold increase in speed can also be obtained by forming a pipeline which partitions the low-level processing functions into a sequence of K discrete processing stages, of duration t'_i, for $i = 1, \cdots, K$. In this case, processing of the low-level functions must be completed at each stage of the pipeline in a time interval less than the packet length, that is

$$t'_i \leq T_p, \quad \forall i = 1, \cdots, K. \tag{2}$$

As bit rates increase, it will become increasingly expensive for electronic processing to satisfy (2), even if parallel processing or pipelining is used. This suggests turning to optical processing of low-level functions to take advantage of its speed, and in some cases, its parallelism.

II. Photonic Switch Architectures

Photonic switch networks can be categorized into two classes: centralized or distributed. In a centralized switch, sources and destinations communicate through switch elements at the switch periphery (see Fig. 3(a)). Centralized switches are appropriate for geographically localized communications such as single multiprocessor computing systems or inter-office communications. Distributed switches allow sources and destinations to communicate directly through most or all switch elements (see Fig. 3(b)) and are appropriate for architectures where nodes are geographically separated such as a distributed computing environment or a broadband digital services network (BDSN). Combined centralized and distributed switching is often found in real networks, where each distributed switch is an $N \times N$ switch.

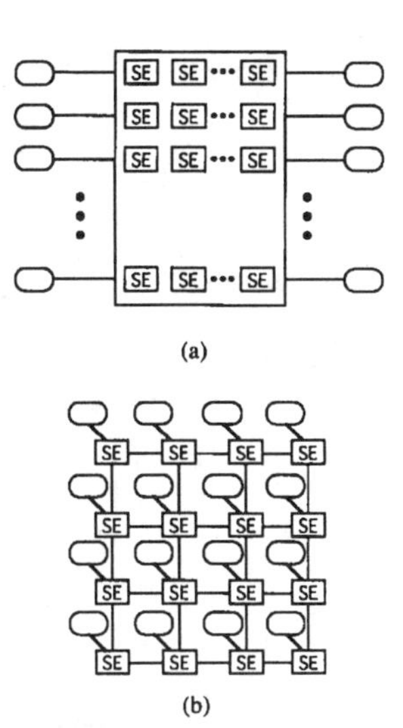

Fig. 3 Relationship between switches, sources, and destinations in (a) centralized and (b) distributed switches.

Photonic packet switches can be modeled in terms of the constituent subsystems: the switch fabric, the routing control processor (RCP), and the input and output interface units as illustrated in Fig. 4. The switch fabric forms optical connections between inputs and outputs; this connectivity may be one-to-one (permutation), broadcasting of one-to-many, or concentration of many-to-one. The RCP sets the connection state of the switch, mediating internal blocking states and output-port contention. Symmetric switches, where the number of inputs equals the number of outputs, are considered in this paper. For distributed switch architectures, the RCP also mediates the connections between incoming network links, a local host, and outgoing network links.

A. Design Principles

The following set of principles are important in designing an all-optical photonic switch:

- preservation of the optical bandwidth and end-to-end optical transparency throughout the routing process;
- minimization of the number of optical buffers or memories;
- use of extremely simple routing protocols;
- reduction of optical loss and crosstalk;
- reduction of the number of photonic switch elements;
- use of synchronization techniques amenable to switching without static buffers;
- meeting telephony synchronization standards.

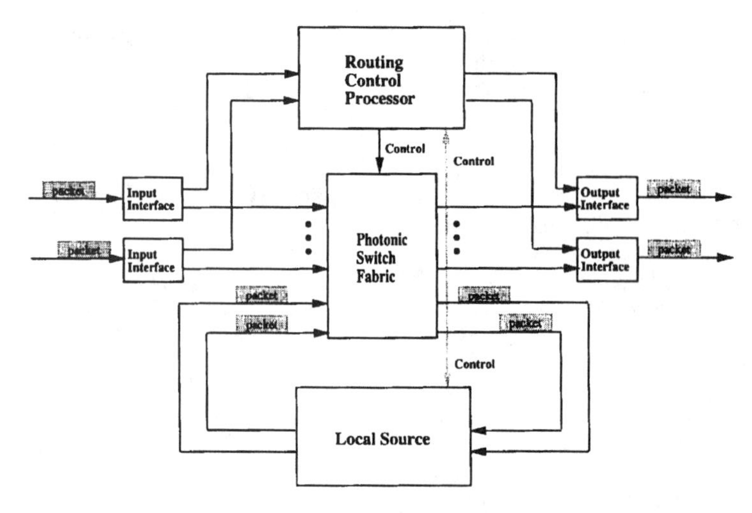

Fig. 4 General photonic switch architecture illustrating switch fabric, I/O interfaces, routing control processor, and local information source/destination.

III. Optical Packet Coding Techniques

Packetized communication is the transportation of data as a discrete unit or cell accompanied by some form of routing information. A packet can be functionally described using a layered model. At the physical network level there are effectively two layers, the payload and the header. The payload contains information processed only by the sources and destinations; the header contains information processed only by the switches. Information that might be included in the payload include data, packet number, and source address. Examples of information in the header include destination address, priority, packet empty–full bit, and packet length.

Several optical packet coding techniques have been experimentally demonstrated, and are summarized in Fig. 5. Each technique exploits the data rate and format transparency of photonic switches, and are classified according to the way the information is transported in the fiber. Three basic categories are bit-serial, out-of-band-signaling, and bit-parallel. Bit-serial coding demonstrations have been performed at the bit level, where each bit carries routing information, and at the packet level, where a string of bits are transported in the payload portion of a packet following a routing header. Bit-level coding has been demonstrated using optical code division multiplexing (OCDM) [18] where a series of pulses, called chips, represents an orthogonal destination address for that bit. Each bit is correlated with a fiber-optic tapped-delay-line transversal filter to determine if the bit should be switched to the local destination or passed through to the next switch. OCDM coded bits have been generated at chip rates up to 12.5 Gchips/s with a bit width T_b of 10 ns [19].

Optical pulse interval and mixed-rate time-domain techniques have been employed to generate packets with multiple information fields in a bit-serial format. Optical pulse interval coding is structured so that the header can be processed optically, allowing the routing control processor to operate at the same bandwidth as the photonic switch and optical fiber [20]. The packet is of duration T_p, and consists of a header followed by a data payload. The duration of a data bit is t. The header is comprised of a framing pulse of duration t followed by several identification fields. The framing pulse is used to extract routing information in a self-timed manner. Field #1 contains the packet destination address and field #2 contains the packet length. Packets with 10-ns-long header and 80 ps per slot have been experimentally generated [21].

Mixed-rate time-domain coding is structured for electronic processing of the header and photonic routing of the payload. The header information is transmitted at a slower rate than the payload so that electronics can easily perform high-level routing functions, yet the high-bandwidth information is routed through the photonic switch. Experimental generation of mixed-rate bit-serial packets has been reported at 100 Mb/s for the header and 700 Mb/s for the payload [22].

Out-of-band signaling involves transmission of a signaling or control channel on a frequency-band separate from the data channel. In photonic packet switching, two types of out-of-band signaling techniques have been experimentally demonstrated. In both approaches, the payload and header are transmitted in parallel, on separate channels, within the same fiber link. We refer to this technique as out-of-band/in-fiber signaling. The first approach is to use subcarrier multiplexing (SCM) to encode the payload and header as radio frequency (RF) sidebands on the optical carrier, each at a distinct sideband frequency. The sideband separation is dictated by the payload and header data rates. Packets have been transmitted on an optical carrier at 1.3 μm, with a baseband payload rate of 2.5 Gb/s and a 40-Mb/s header coded on a 3-GHz subcarrier [23]. The second approach is to code the payload and header at two separate

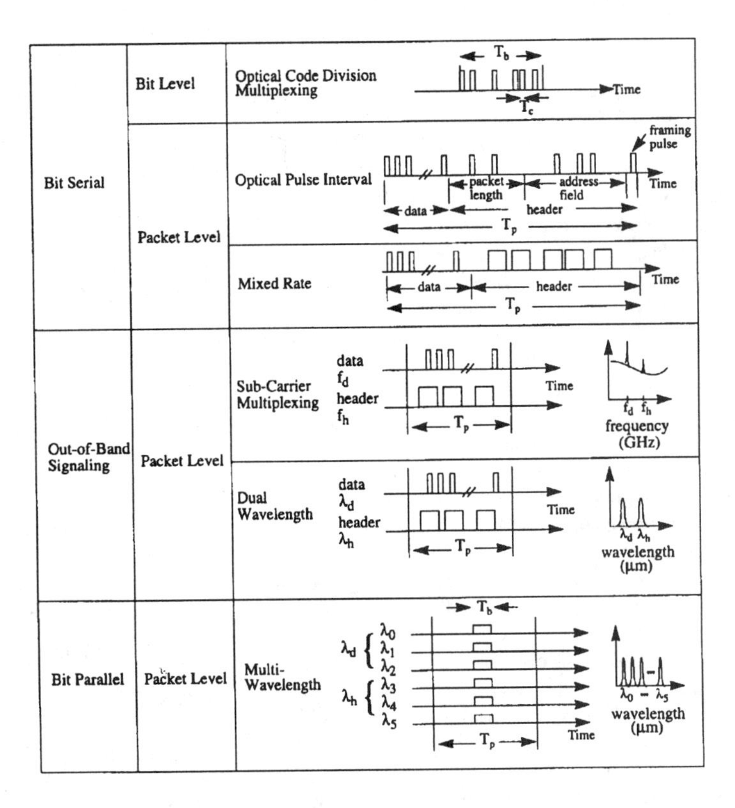

Fig. 5. Summary of experimentally demonstrated optical packet coding techniques.

optical wavelengths λ_p and λ_h. In [24], a packet switching demonstration transported the payload at 1.3 μm at a rate of 933 Mb/s and the header at 1.55 μm at a rate of 155 Mb/s.

Optical packets can also be transmitted bit-parallel with each bit in the packet at a separate frequency in a single-mode fiber [25]-[27]. At transmission, the bit-parallel packet occupies a time frame slightly larger than a single bit duration. Fiber dispersion will spread the bits leading to bit skew. 6-bit packets have been experimentally transmitted with a channel spacing of 2 nm [28]. The packet consists of a 4-bit payload at 1.3 μm and a 2-bit header at 830 nm. The individual bit rate per wavelength was 50 Mb/s, yielding a bit-parallel payload rate of 200 Mb/s. The generation of bit-parallel packets can also be accomplished using a mixture of multiwavelength and subcarrier multiplexing [29].

A. Comments on Coding Techniques

Bit-level coding using OCDM utilizes the high fiber and switch bandwidths to allocate header overhead to each payload bit. Generally, streams of bits are logically assembled into packets, and the abundant bandwidth may be better applied to support other low-level functions in addition to the destination address. Optical pulse interval coded packets allow higher level functions to be implemented, and the use of a framing pulse provides a synchronization mechanism on a packet-by-packet basis. The current state-of-the-art all-optical processing techniques make it difficult to implement the full set of low-level functions optically. While this may change in the future, direct electronic processing of high-bandwidth header information is more difficult than switching high-bandwidth payloads with a photonic switch. Therefore, either the high-speed header must be slowed down, or low-speed headers transmitted with high-speed payloads. Mixed-rate time-domain coding allows the electronics to process a relatively low-speed header. The consequence for performing mixed-rate coding bit-serially is added latency and decreased throughput due to the fact that the header can occupy a significant portion of the packet.

Transmitting the payload and header in parallel increases throughput since the header can occupy the same duration as the payload and be processed in parallel. A primary consideration for out-of-band-signaling is the need to keep the payload and header logically connected throughout the routing process. For known inter-switch distances, delay compensation can be used between two wavelengths for realignment between the header and payload. An issue common to both out-of-band-signaling and bit-parallel techniques is crosstalk and signal degradation due to fiber-optic nonlinearities [30]. While an SCM header will minimize separation of the payload from the header, a disadvantage of subcarrier multiplexing of the header is the added complexity of active microwave mixing components in the transmitter and at each switch, and that the bit rate is constrained to be lower than the subcarrier frequency. A power penalty is also paid by requiring the transmitter to supply signal power for both the payload and header simultaneously [22]. Nonlinear crosstalk effects in the fiber may limit this approach due to the close channel spacing of the SCM channels. Additionally, the SCM passes through the photonic switch with the payload, an undesirable quality in most architectures. In a dual-wavelength system, passive optical filtering can be used to extract out-of-band headers, leading to a lower complexity for header extraction. Two tuned optical sources are required at each transmitter, and the payload and header power requirements are decoupled. Source and optical filter stability, in addition to fiber dispersion, are critical issues in this approach.

Finally, bit-parallel multiwavelength packets offer the advantages of complete wavelength division multiplexing in that the electronics at both the network edges and core need only run at the packet rate instead of the bit rate. For applications where information is initially bit-parallel (e.g., computer communications), this technique avoids bottlenecks associated with high-speed parallel-to-serial and serial-to-parallel conversion. As with any bit-parallel transmission technique, bit skew caused by channel delay variation (e.g., dispersion) must be compensated for in order to be increased over a bit-serial approach. In this case, dispersion compensation can be used. The complexity and availability of multiwavelength sources are also an important issue as is the need for inter-node wavelength synchronization.

IV. Routing Strategies

Routing is the method used to choose a preferred path to send a packet from its source to destination, whether it be input and output ports for a switch or endpoints in a multiswitch network. Routing control may be centralized or distributed. Centralized control involves a single processor monitoring the network and setting up the switch states according to the routing requests. As networks become more widely distributed or incorporate more switches, centralized control degrades latency and throughput, and increases processing complexity. With distributed routing, packets carry destination information for processing at each node. This form of processing reduces the burden on a centralized processor and increases the switch throughput. Additionally, distributed routing decisions are based on local information, whereas with centralized routing decisions in a wide-area network are made on global and perhaps obsolete information. Mixtures of centralized and distributed routing may provide optimal performance since global information is often useful in computing routing paths.

Routing strategies based on simple algorithms can reduce information bottlenecks at the switch. A routing protocol must be able to handle both switch-level routing and contention resolution. The appropriate routing strategy or algorithm strongly depends on the switch or network topology. Centralized and distributed switches can be classified

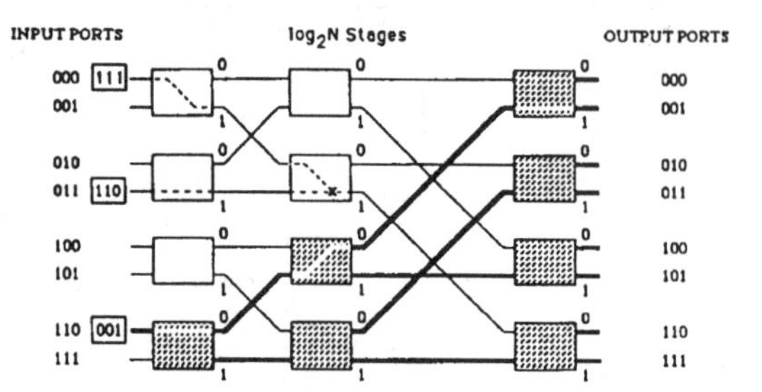

Fig. 6. Destination tagged self-routing through a Banyan switch [32].

as randomly (or semi-randomly) connected or regularly connected. In real-world applications, distributed-switch networks evolve in a semi-random manner, where a mathematical relationship between the addresses of neighboring nodes is not guaranteed, and the packet destination address must be carried throughout the routing process. In this case, the packet destination address is mapped at each switch to a local switch output port. This type of routing strategy is sometimes termed "fixed-directory" routing and requires a routing table at each switch that contains an outgoing link for each destination address [31]. Generally, routing tables must be maintained as the network evolves. Electronically implemented routing tables grow in complexity and access delay with increasing network size and can pose a switch bottleneck; therefore, parallel access optical [32] and opto-electronic [33] routing tables are desirable to maintain high switch performance.

Centralized switches and locally connected distributed switches can be organized into a regular topology where a mathematical relationship exists between the addresses of connected switches. In this case, the position of each switch element (e.g., a 2 × 2 switch) within the network can be used to determine the appropriate path to the destination. Simplifying the routing rule in this way allows the complexity of the routing control processor to be reduced. One example of a self-routing centralized switch that has been experimentally investigated is the Banyan-type network [34]. A typical construction of multistage architectures starts with four-port switching elements arranged in a binary tree (shown as the cross-hatched boxes and heavy lines in Fig. 6). In principle, an arbitrarily large $1 \times N$ binary tree could be constructed with only $N - 1$ 2×2 switches (crosspoints) arranged in $\log_2 N$ stages (if N is a power of 2) interconnected by fiber amplifiers. However, optical crosstalk and loss may prove to be a practical limitation to eventual switch sizes.

The distributed routing procedure in the binary tree is particularly simple. If the destination address of a packet is represented in binary form, then each bit of the address determines whether a stage should direct the packet to its upper (bit 0) or lower (bit 1) output, where the most significant bit controls the first stage, and each successive bit controls successive stages. Each stage need only examine one bit of the address. For example, in Fig. 6 the address 001 indicates that the first stage should route the packet

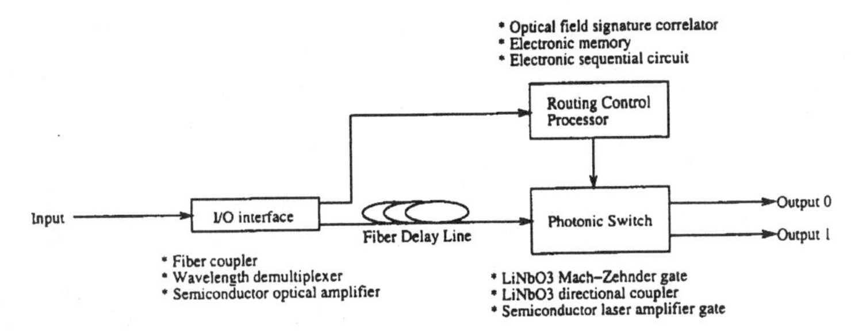

Fig. 7. Experimental block diagram for 1×2 self-routing switch demonstrations.

to the upper output, the second stage to the upper output, and the third stage to the lower output. It is easily verified that using this procedure, the packet self-routes to output port 001. An $N \times N$ multistage switch constructed from distinct $1 \times N$ binary trees would consist of N binary trees for demultiplexing and N for multiplexing, requiring a total of about twice the number of crosspoints as a crossbar switch.

The required number of crosspoints in a multistage switch can be reduced to $(N/2)\log_2 N$ if the inputs share the binary trees. For example, the 8×8 Banyan switch shown in Fig. 6 requires only 12 crosspoints. The reduction in the number of crosspoints achieved by the Banyan architecture is accompanied by a major limitation: internal blocking can occur if two packets are destined for the same output of the same crosspoint. Such conflicts result in blocking one of the packets and correctly routing the other. Thus, not all input–output permutations are possible with the Banyan switch, and the throughput of this architecture is severely limited. This is seen in Fig. 6, where two packets with destination addresses 111 and 110 need to be simultaneously routed to the lower output port of the second crosspoint in the second stage, resulting in an internal conflict. Resolution of internal and output port contention states is addressed later in this paper.

V. Experimental Self-Routing $1 \times N$ Switches

Self-routing was the first type of photonic packet switch experiment performed since it is the lowest required level of functionality, and can be demonstrated using a simple photonic gate (pass or block) or a 1×2 photonic switch. Extension of self-routing to a 2×2 switch and larger sizes requires processing of contention resolution as discussed in the next section. In order to show self-routing, it is first necessary to generate an optical packet that contains a payload and header (packet generation techniques were discussed in Section III). The first experiments coded headers that contained the destination address. Self-routing is then demonstrated by injecting an optical packet into the switch, stripping the header from the payload, processing the header by mapping the destination address to a switch state, setting the switch state, and passing the payload through the switch.

The general form of a self-routing photonic packet switch experiment is illustrated in Fig. 7. At the switch input, packet headers are directed to the RCP through the input interface. Three different methods have been used: tapping of optical power using a fiber coupler [18], [20]–[23], separation of wavelength multiplexed headers from the payload [24], [27], [28], and detection of gain modulation in a semiconductor laser amplifier [35]. Power tapping removes a portion of the complete optical packet with loss that can be compensated for with an optical amplifier; wavelength multiplexing removes only the header, allowing the payload to pass through the switch with only a small decrease in optical power due to excess losses; detection of gain modulation in a laser amplifier allows monitoring of the header while simultaneously amplifying the complete packet. Several types of photonic switch elements have been used in self-routing 1×2 switch demonstrations and are illustrated in Fig. 8: Fig. 8(a)—a Mach–Zehnder gate [36]; Fig. 8(b)—a semiconductor laser amplifier gate [37]; and Fig. 8(c)—a passive optical splitter with semiconductor laser amplifier gates [38]. The 1×1 switches (modulators) were used to simulate routing between two outputs using the on and off states.

In 1987, the first reported photonic packet switch experiment involved bit-level destination address coding using optical code-division multiplexing at 100 Mchip/s with 32-chip packet headers in a $LiNbO_3$ electrooptic Mach–Zehnder integrated-optic modulator [18]. In a subsequent demonstration, this experiment was scaled up to a $LiNbO_3$ 8×8 crossbar photonic switch with 12.5 Gchips/s with 125-bit packet headers and a switch reconfiguration rate of 1.33 GHz [19]. This experiment, shown in Fig. 9, was used to investigate optical and electronic crosstalk issues involved with self-routing at high speeds in a complex photonic switch structure as well as high-speed reconfigurability and data-rate transparency.

Numerous experiments using optical correlation followed, and a comprehensive review is given in [39]. These experiments addressed more complex issues for low-level routing functions in optically controlled switches such as:

- Optical time-division routing [4], [39]–[41] to improve coding bandwidth efficiency.
- Optical self-clocked pulse-interval routing to allow timing acquisition on a packet-by-packet basis [42].
- Optoelectronic correlation technique using a photoconductive "AND" gate [34], [43] to increase the speed and throughput of the header recognition process.

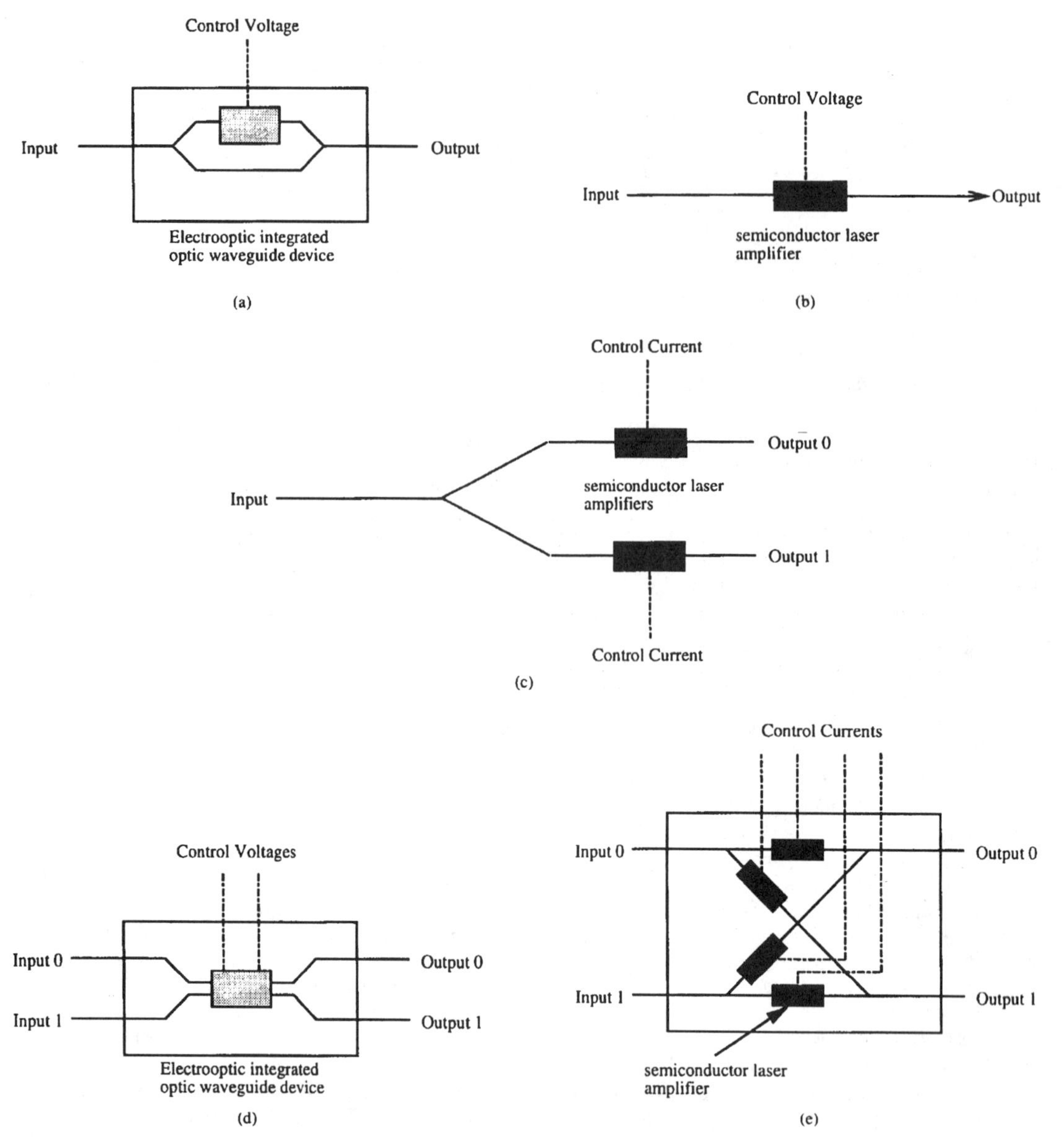

Fig. 8. Photonic switch fabrics used in experimental demonstrations. (a) Mach–Zehnder interferometer gate. (b) Semiconductor laser amplifier (SLA) gate. (c) Y-branch with SLA. (d) Directional coupler. (e) SLA gate array tree switch.

Beginning in 1991, a series of electronically controlled packet switch experiments were reported. A self-routing experiment that demonstrated header extraction by monitoring a semiconductor laser amplifier bias current [35] is shown in Fig. 10. The laser amplifier bias current is processed electronically to self-route the packet through a 1×2 $LiNbO_3$ switch. The optical amplifier also provides gain to the payload, providing an overall gain of 2 dB. The packets were coded as mixed-rate bit-serial, with a payload data rate of 700 Mb/s and header data rate 100 Mb/s.

A self-routing photonic packet switch that transported header and payload in parallel using SCM [23] is shown in Fig. 11. This technique simplifies clock and signal recovery over the bit-serial mixed-rate technique. The switch routed packets with a 128-byte payload at 2.56 Gb/s non-return-to-zero (NRZ) by electronically decoding a 2-byte header at 40 Mb/s NRZ. An important issue is the integrity with which the header information can be transmitted and decoded. Errors in decoding a packet header lead to unintentional misrouting of a packet. The modulation

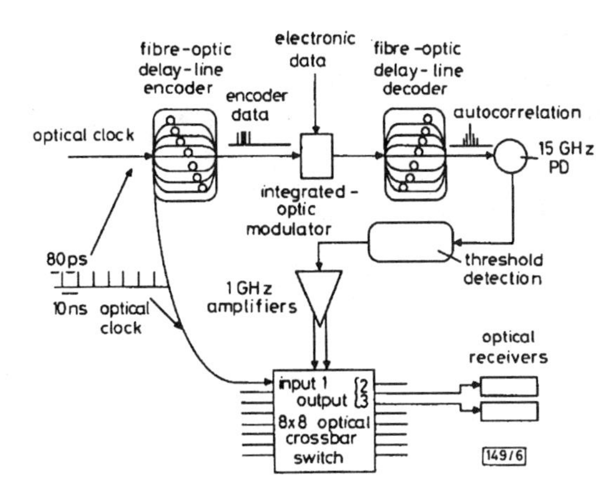

Fig. 9. Experimental demonstration of self-routing in an 8×8 $LiNbO_3$ photonic switch [19].

depths of the header and baseband payload were adjusted to obtain a measured header bit error rate (BER) of 10^{-9}. For computer communications, where a BER of 10^{-15} or better is required, error correction may be necessary for this technique.

Out-of-band-signaling using the dual-wavelength technique was first demonstrated in [27] for a bit-parallel deflection routed switch, and later in [24] in a dual-wavelength demonstration for ATM cell switching. In the later experiment, the header was transmitted at 1550 nm at a rate of 155 Mb/s and the payload transmitted at 1300 nm at a maximum data rate of 933 Mb/s.

A. Self-Routing Through Cascaded 1×2 Switches

An important step in experimental photonic packet switching was to demonstrate that multiple 1 × 2 self-routed switches could be cascaded. Distributed control of a 1 × 4 $LiNbO_3$ tree switch using pulse interval coding [34] is shown in Fig. 12. This experiment demonstrated destination-tag routing where each bit in the header is used to control a specific stage in the switch. Note that the control structure is distributed by stage instead of by switch crosspoint, leading to an efficient utilization of the optical control resources. A subsequent demonstration of electronic self-routing in two cascaded 1 × 2 semiconductor amplifier gate array switches using mixed-rate, bit-serial packets was reported in [38], [56]. This experiment demonstrated cascaded operation of amplified photonic switch elements. BER measurements were made with a received signal power of −24 dBm and the combined RCP and switching rise time were on the order of 8 ns.

VI. Contention Resolution in Photonic Switches

The next important development in photonic packet switching was to extend the previous self-routing work in 1 × N switches to $N \times N$ switches. This required demonstration of contention resolution of optical payloads. For the most part, contention resolution methods have previously been classified according to the manner in which they resolve packet collisions: Buffering, blocking, dropping, or deflecting [44]. Experimental demonstrations of contention resolution have focused on deflection routing and buffering with a small number of buffers as these techniques are well matched to currently available photonic techniques. $N \times N$ photonic switch fabrics used in this type of experiment are illustrated in Fig. 8(d) and (e): $LiNbO_3$ directional couplers [36] and semiconductor laser optical amplifier gate arrays [45].

A. Deflection Routing

Store-and-forward techniques, where packets are stored in static memories, are well-suited for electronic switches in which flow-control algorithms and buffers can be implemented inexpensively with electronic logic. If storage buffers are expensive, as is presently the case with optical technology, then deflection or "hot-potato" routing [46] can be used, provided the number of input links to a crosspoint is the same as the number of output links [47]. The deflection routing protocol is ideally suited to nodes in which storage is difficult [26]. In deflection routing, internal conflicts between two packets are resolved by correctly routing one packet and deflecting the other to an available outgoing link. In this way, all packets are forwarded, without elastic buffering and without packet loss.

Routing decisions are based on destination addresses and packet priorities. For example, when packets simultaneously entering both inputs of a 2 × 2 switch are destined for the same output port, the packet with higher priority is directed to the desired output port and the other packet deflected to the remaining output port. The case of equal priorities is handled with a fair resolution measure described below. Essentially, the links between switches are used as passive delay-line buffers for deflected packets. Deflection routing techniques require that the network topology be multipath or recirculatory (e.g., shuffle exchange networks [26], Manhattan Street networks [47]), so that deflected packets can be routed to the destination over an alternate path. The priority of deflected packets is increased to reduce end-to-end latency and to avoid deflecting a packet indefinitely [26]. Several different measures of priority have been used, including age and distance to final destination, the choice of which impacts the overall network performance [48].

A great deal of analysis has been performed to determine the throughput and delay penalties associated with deflecting packets in various network architectures. In the late 1970's, the commercially produced HEP supercomputer utilized electronic deflection routing [49]. Maxemchuck has shown that, compared to a Manhattan Street network with an infinite number of buffers, 55–70% of the throughput can be obtained using deflection routing [47]. The throughput of hot-potato routing has also been compared to store-and-forward routing for networks with symmetric connectivity diagrams, such as the recirculating perfect shuffle or rectangular grid [50]. It was shown that the performance of hot-potato routing decreases monotonically

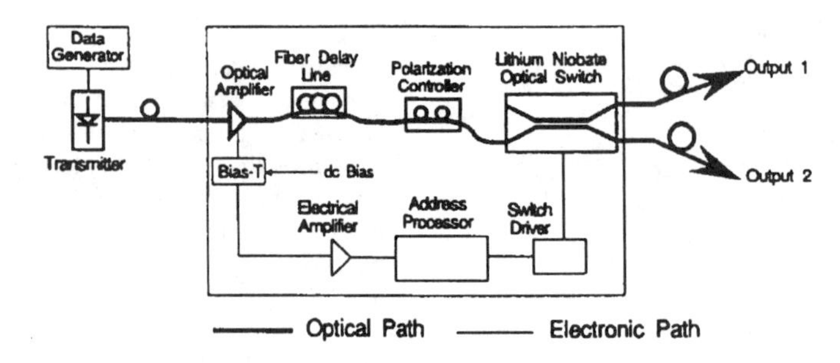

Fig. 10. Extraction of packet header by sensing current in semiconductor optical amplifier. Figure reproduced from [35].

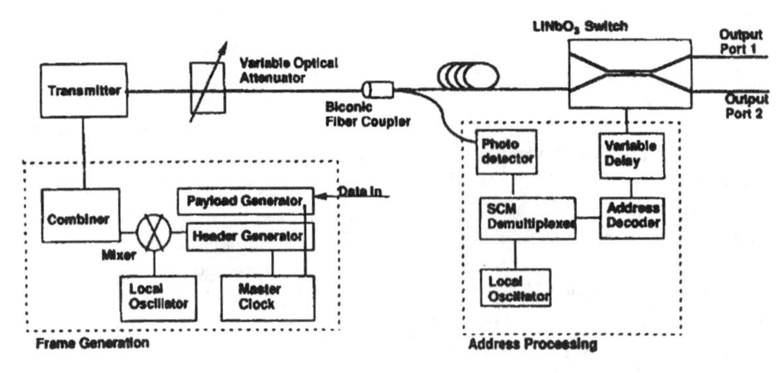

Fig. 11. Self-routing using subcarrier multiplexing (SCM) to code header in parallel with payload. Figure reproduced from [23].

as the number of nodes increases; with several hundred nodes, the throughput of hot-potato routing is 30% of store-and-forward routing. This performance degradation can be compensated either by increasing the link speed or reducing the probability of deflection as discussed in a later section on contention avoidance.

The first experimental demonstration of deflection routing in a photonic packet switch was reported in [27]. This was also the first demonstration of a full 2 × 2 switch where packets were injected into both input ports, producing real contention conditions for the same output port. Contention conditions were created and resolved in a 2 × 2 $LiNbO_3$ switch using the experimental setup shown in Fig 13(a). Independent packets were injected into both ports by optical power splitting of a serially generated packet stream, delaying one arm by a single packet delay, thereby aligning sequential packets in parallel at the switch inputs. Contention could then be forced by programming two sequential packets to go to the same output port, with equal or unequal priority. This switch routed bit-parallel multiwavelength packets, employing out-of-band signaling to transport the payload and header at separate wavebands (1300 and 830 nm).

A photograph of the experiment is shown in Fig. 13(b). This switch utilized a routing table for randomly connected networks, and the final destination addresses were carried with the packet throughout the routing process. The header field consists of 2 bits that correspond to four final destination addresses and one priority bit. The routing table maps the four addresses into one of the two output ports. Contention resolution is processed by storing one of the two switch control states (bar or cross) in memory locations corresponding to the 26 possible input request states. Minterms that represent contention with equal priorities are not stored and are handled with finite-state logic that precedes the routing table. In the case of contention with equal priority the switch was maintained in its prior state. This method promotes fairness for statistically independent packets. In order to fully support deflection routing by updating the priority of deflected packets, this switch reinserts new headers as discussed in Section VII.

B. Shared Buffer Deflection Routing

The performance of deflection routing in terms of packet latency variance can be improved by reducing the probability of deflection through the use of a small number of buffers [51], [52]. The use of feedforward buffers (e.g., time slot interchangers) in photonic switches is desirable from the perspective that they are simple to control, preserve the fiber bandwidth, and do not suffer the buildup of optical amplifier noise as is the case with amplified recirculatory buffers. A recently demonstrated experimental

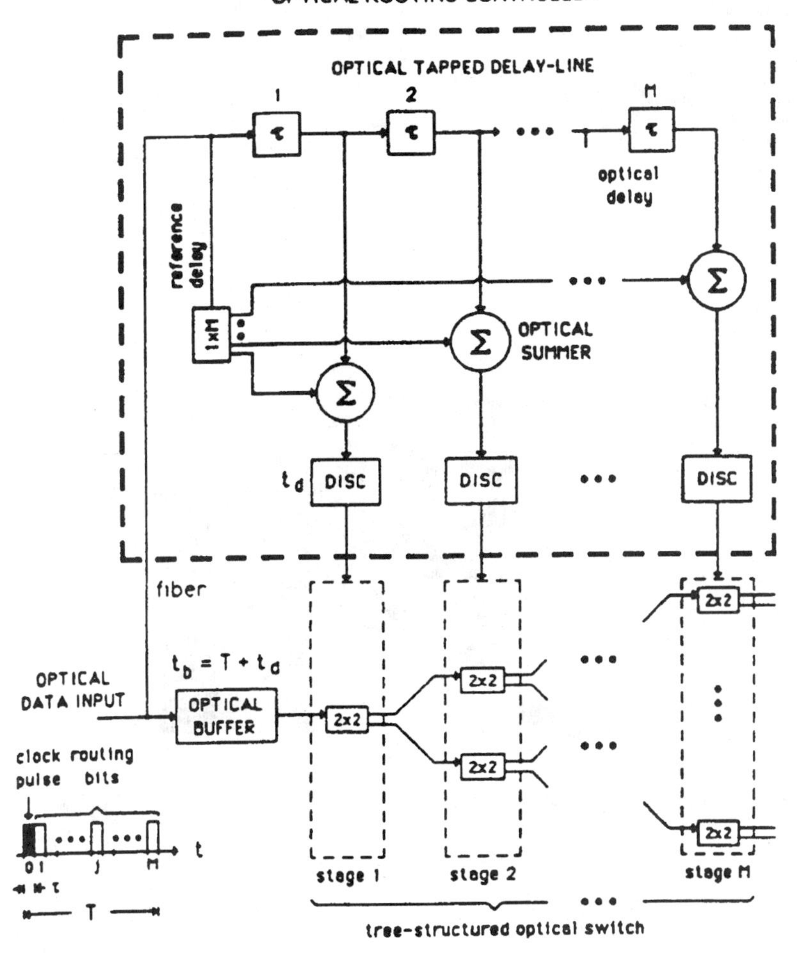

Fig. 12. Optical routing processor used to control tree-structured photonic packet switch using pulse interval packet coding. Figure reproduced from [34].

2×2 photonic packet switch operating at 1.25 Gb/s with drop/add capabilities [53] is illustrated in Fig. 14. The single packet feedforward delay line memory is shared by both inputs, and provides one level of resolution for output port contention. In the case that the buffer cannot resolve a contention case, a packet is deflected. Packets entering the node at $i1$ or $i2$ or contained in M are perceived by the controller in one of five possible ways: empty (E), for the node (FN), caring to exit on output $o1$ (C1), caring to exit on output $o2$ (C2), or don't care (DC) (e.g., both outputs provide equivalent shortest paths to their destination). Deflections occur when packets at the input of $SW3$ vie for the same output. When $i1$ and $i2$ are FN, one is missed. The objective of the controller is to maximize the node's throughput by minimizing the number of deflections. Switch $SW2$ is just for absorption/injection, and routing switch $SW3$ is controlled with a simple nonpriority hot-potato routing of its input packets [53].

C. Input Buffer Routing

Experimental demonstration of active buffering to resolve output-port contention [14] is shown in Fig. 15. Recirculating fiber-optic input buffers alleviates the problem of head-of-the-queue blocking associated with electronic sequential input buffering. In order to compensate for losses due to multiple recirculations, optical amplifiers are inserted into the switched loops. Amplified spontaneous emission (ASE) can accumulate in multiple passes, resulting in amplifier gain saturation and intersymbol interference in the bit stream. To reduce instabilities in the loop, the optical amplifier must be gated off each time a packet passes it, making it difficult to store multiple packets in a single memory. In order to have an acceptable reduction in contention probability, multiple recirculating buffers are required at each input port adding to the switch complexity and losses. This technique could also be combined with

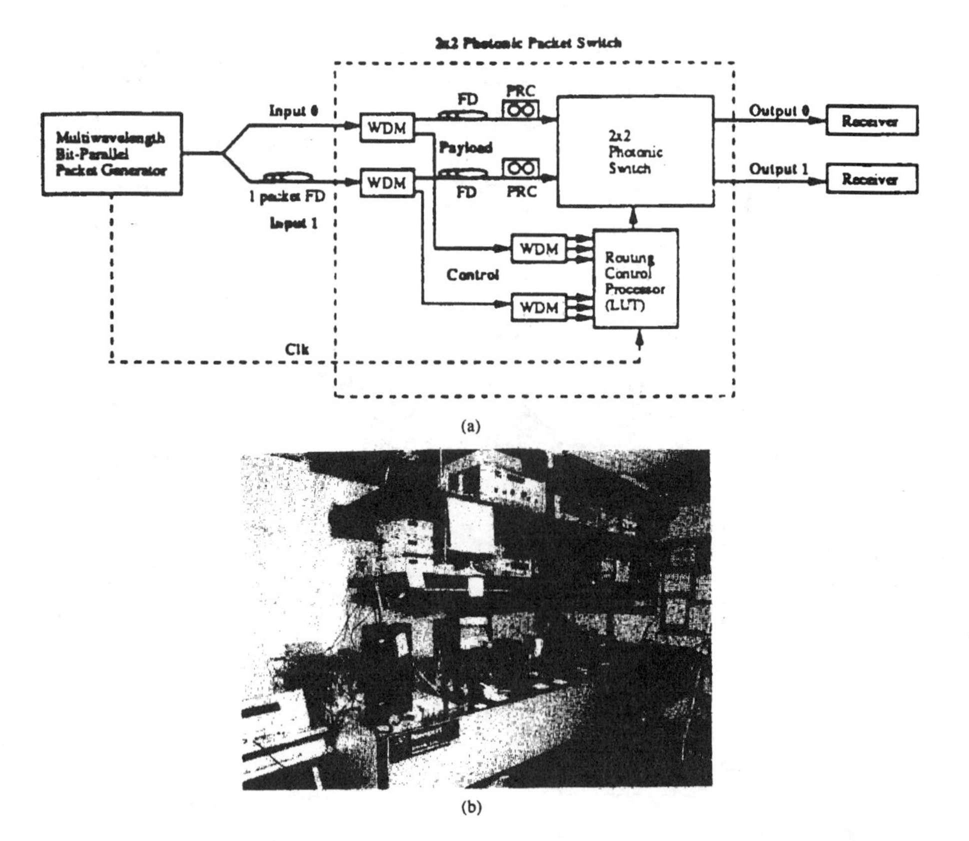

(a)

(b)

Fig. 13. Experimental demonstration of deflection routing in a 2 × 2 photonic packet switch. (a) Block diagram. (b) Photograph of experimental setup.

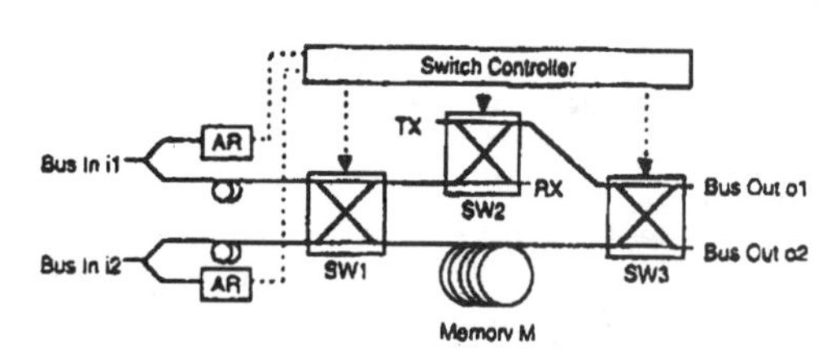

Fig. 14. Experimental demonstration of single-input buffer deflection routing using three 2 × 2 photonic switches. Figure reproduced from [53].

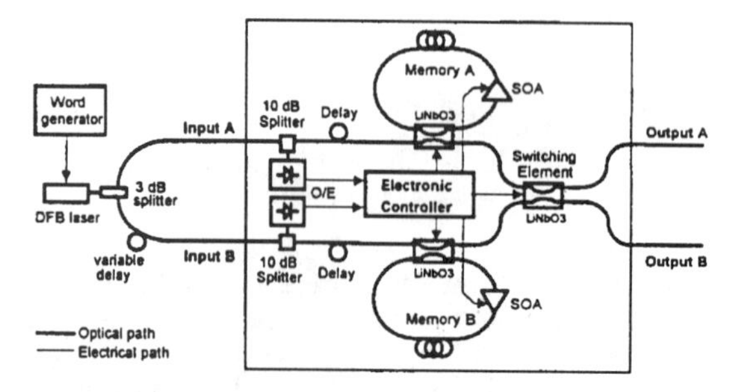

Fig. 15. Optical input buffering to resolve contentions using active recirculating switched fiber delay lines. Figure reproduced from [14].

deflection routing, however, optical bandwidth is better preserved and the switch is less complex if feedforward delay lines are used as shared memory or output port buffering as discussed above.

VII. Header Regeneration

Header regeneration is the process of computing, generating, and reinserting a header with the associated payload at the appropriate switch output port. There are several circumstances where this functionality is required: i) in all-optical photonic switches where the header is completely removed from the payload for processing (e.g., multiwavelength out-of-band-signaling) or ii) where routing strategies require a modification of the packet header (e.g., priority updating in deflection routed switches, cell routing in ATM switches). Generally, it is important that a header regeneration technique be employed that can operate for cascaded switches, and is independent of the number of switches a packet traverses. Reinsertion of a new optical header with a through-going packet can be acheived by impressing header information on a continuous-wave (CW) period of light embedded within an optical packet [55]. The

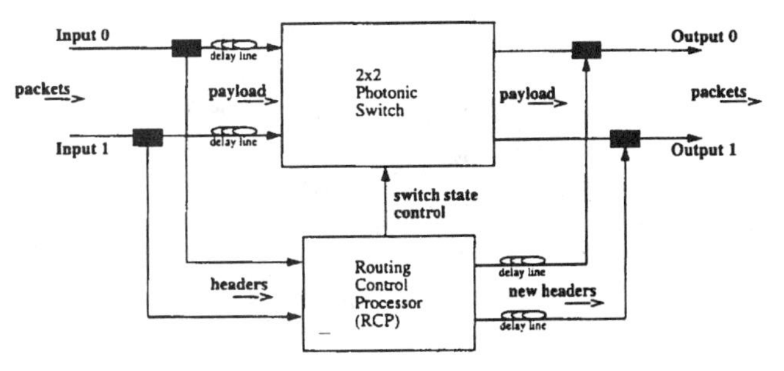

Fig. 16. Block diagram of an all-optical 2×2 switch for multihop networks [28].

same semiconductor amplifier laser gate that is used to route the packet is also used to modulate the CW packet section. The critical disadvantage of this technique for multihop networks is that it only works for a one-switch pass.

A more general approach has been demonstrated that allows regeneration of new headers with each switch pass [28]. The overall switch architecture is shown in Fig. 16 and is similar to the deflection routing 2×2 switch described in Section VI with the addition of an electronic routing control processor that generates new optical header information by directly modulating lasers at the header waveband. This information is merged with the outgoing payloads using wavelength-division multiplexers at the output interfaces. The parallelism of out-of-band-signaling greatly simplifies reassembly of the packets as timing is less critical than with a bit-serial approach. Fiber delay lines are required at the output stages to match alignment of headers with the payload.

VIII. All-Optical Multihop Routing

All-optical multihop routing is the complete routing of packets through multiple general-purpose photonic switch stages without optoelectronic conversion of the payload except at the endpoints. At each stage, the full functionality of routing, contention resolution, header regeneration, and synchronization is required since switch nodes have multiple inputs and outputs. A necessary condition for implementation of deflection routing is to provide a means for deflected packets to reach the destination by an alternate path or through retransmission. The first demonstration of all-optical multihop packet routing that performed routing, contention resolution, and header regeneration for each packet at each stage was reported in [28]. This switch routed packets three all-optical payload hops without optical amplification. This experiment is illustrated in Fig. 17. Packets are routed through three switches from input port 0 of the first switch to output port 0 of the third switch.

The experimental block diagram in Fig. 18 illustrates how this degree of functionality for multiple all-optical switch hops is demonstrated. Packets are originally generated by a multiwavelength bit-parallel transmitter and inserted at input port 0 of the switch. The receiver, located at output port 0, demultiplexes the individual payload bits for packet verification. Output port 1 is directly connected to input port 1 through a fiber-optic delay-line in order to present the switch with independent packets and to demonstrate multihop routing. Packets enter input port 0 at a rate equal to the time it takes for a packet to traverse the feedback loop. For demonstration purposes, new optical header information is inserted in real time with the outgoing payload at one of the switch output ports. The new header consists of a new port address and updated priority bit.

Individual packet headers are extracted from the payload at each of the switch input ports. The control bits for each input port are further wavelength-demultiplexed and processed by the RCP. The RCP contains a combinatoric circuit that performs deflection routing and an arbitration circuit that accommodates the livelock condition. The arbitration circuit follows the state of the priority bits as long as they are different. If priorities are the same, the arbitration circuit maintains its previous state. Therefore, the combinatoric circuit makes decisions on states other than priority saturation. The delays are required to properly align the clock for pipelining. The output switch state is latched for a complete packet cycle while the new address and priority bits are latched for only a single bit period. The output of the RCP controls the 2×2 switch state of cross or bar, and generates two new control bits for the packet exiting output port 1. Whichever packet is selected to exit output port 1 has its current control bits binary complemented for reinsertion as the new control bits. This technique is arbitrary and is used to facilitate verification of the switch operation.

The switch routes multiple optical wavelengths simultaneously, as demonstrated by the switching of bit parallel payloads. The electronics were only required to run at the packet rate due to WDM transmission of bit parallel packets. Packets were shown to traverse the switch up to three times and circulate up to two times around the fiber feedback loop before excess losses limited detection. The excess losses per round trip switch pass were measured at -13 dB. In a more mature system, optical amplifiers and optimized WDM components would be used to compensate for losses due to switching, coupling, and attenuation. The routing decision time, including demultiplexing delay, detection delay, and RCP delay was approximately 100

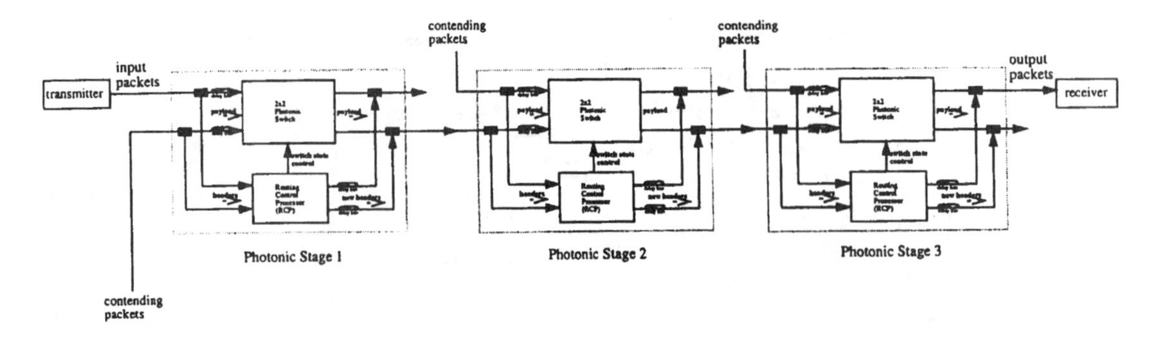

Fig. 17. Experimental demonstration of 3 all-optical hops with self-routing, contention resolution, and header regeneration/reinsertion performed at each stage [28].

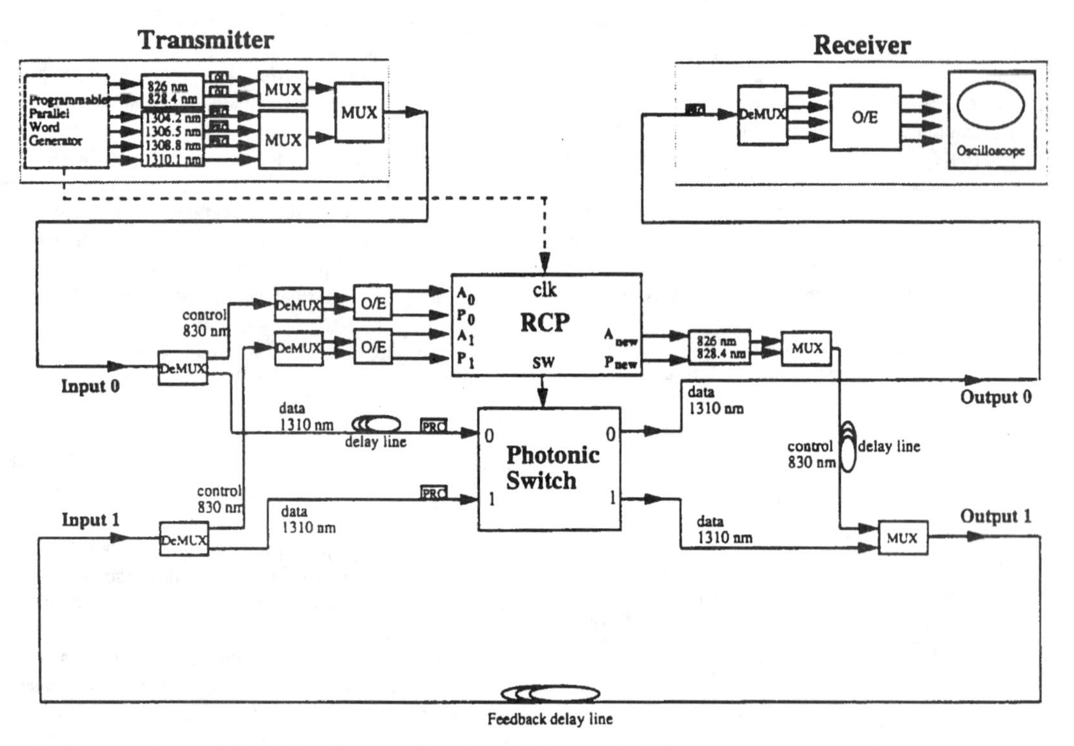

Fig. 18. Experimental setup of all-optical multihop photonic packet switch demonstration [28].

ns. Although packets were injected at a rate determined by the feedback path length, the maximum packet rate is determined by the packet duration of 50 ns. Assuming a guard band of 5 ns per packet, to account for finite switching times and other uncertainties, the throughput per input port is 18.2 Megapackets per second (Mp/s) corresponding to an aggregate switch throughput of 109 Mb/s for 3-bit packets. The laser drivers and receivers used are capable of 500-ps rise-time operation. Implementation of the simple RCP in ECL would allow a conservative 5-ns packet width plus 5-ns guardband yielding a potential switch throughput of 200 Mp/s or 600 Mb/s for a 3-bit packet. Addition of more lasers for more parallel bits will increase this throughput proportionally.

IX. Synchronization

Synchronization is a fundamental issue that must be addressed when combining individual photonic switches into a centralized switch or a distributed switch network. In order for the routing process to function properly, alignment of packets within the switch fabric and headers within the RCP must be considered. For simplicity, consider a 2×2 switch with local access. Two independent packets or messages arrive at the incoming network links from different sources and request access to any of the two outgoing links or the local destination. Depending on the architecture, the input interface units perform one or more of the following functions: buffering, synchronization, separation of header

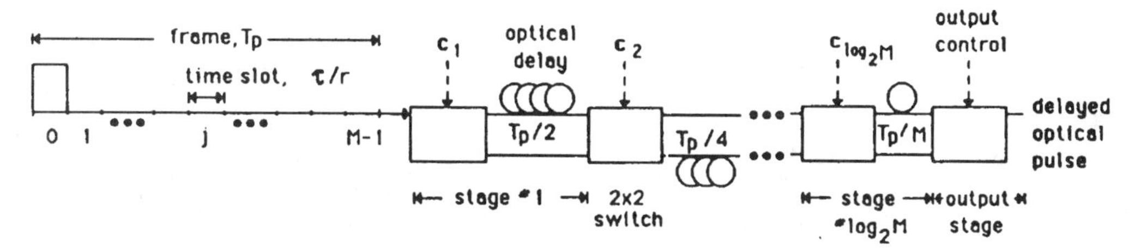

Fig. 19. An experimental synchronization processor using 2 × 2 photonic switches and feedforward fiber-optic delay lines [32].

from payload, and optical or optoelectronic conversion. Additionally, packets or messages can be generated or received at a local source/destination.

In a photonic packet switch, the output interface units may perform any combination of the following: buffering, synchronization, rejoining of header with payload, and optical or optoelectronic conversion. Synchronous switching requires realignment of data at both switch inputs to coincide with a change in the switch state. For optoelectronic RCP's headers can be resynchronized at the interface electronics, however, optical routing controllers require some form of header synchronization. Synchronization requires clock transmission or extraction between nodes, and must be acquired on a packet-by-packet basis unless an out-of-band channel is used to distribute the clock. Alternatively, the inputs can be asynchronously processed requiring a policy such as first come first serve. The RCP must keep track of units in the switch so information is not prematurely switched out. This issue is the most complicated since unbalanced loading of the switch may result, e.g., one input dominating access to the desired output.

An experimentally demonstrated synchronization processor is shown in Fig. 19. It consists of $\log_2 r(L+1)$ stages where the resolution of the framing pulse recognition is determined by $1/r$, which corresponds to a fraction of a time slot τ, and $L+1$ is the total number of time slots in a maximum-length packet. The input packet to the first stage arrives with an arbitrary phase. Since no *a priori* information is available about the packet length, it is assumed that the framing pulses are spaced no closer than the duration of the maximum length packet $(L+1)\tau$. The packet is directed to both the 1×2 photonic switch and the framing pulse recognizer. A passive delay is introduced before the 1×2 switch to match the propagation delay through the framing pulse recognizer and the delay of the gating pulse generator, so that the gating pulse and the packet arrive simultaneously at the switch.

X. Summary

The main challenges in using photonic switches for packet-switched networks relate to providing high-speed, high-throughput control functions such as routing, contention resolution, synchronization, cascadability, header reinsertion, and multihop operation, as well as minimizing the use of elastic buffers, which are currently not practical with optical technology. Progress has been made in several areas required for photonic packet switching. These include the development of:

- a deeper understanding of the beneficial tradeoffs involved in combining optical and electronic technologies in packet switched networks;
- the development of "lightweight" optical protocols suitable for packet-switched network architectures with all-optical interiors;
- demonstration of basic switch functions for multihop photonic packet switching.

A review of experimental demonstrations of these photonic packet-switch control functions has been provided. High-speed optical header encoding and decoding has been demonstrated bit-serially up to 12.5 Gb/s, bit-parallel with up to six discrete wavelengths, and hybrid serial/parallel using optical out-of-band signaling techniques. Routing controllers based on optical and optoelectronic processing techniques were reviewed. Simplified routing strategies, such as 1 bit per Banyan stage, are the most easily implemented with all-optical processing, whereas more complex functions can be handled by optoelectronically detecting and regenerating the headers at each switch stage. Experimental demonstrations of high-speed self-routing through photonic switches were reviewed, in which bit rates up to 12.5 Gb/s and out-of-band wavelength signaling and header reinsertion have been acheived. Extensive analysis of deflection routing with limited (or no) elastic buffering has been carried out and experimentally demonstrated. Experimental demonstrations of multihop photonic packet switching and packet synchronization were discussed and are important steps towards demonstrating the functionality required to construct large-scale all-optical packet-switched networks. Photonic packet switching is at the stage where optoelectronic and photonic integration will make this a potentially viable commercial technology as well as reduce the cost and improve reliability of these systems. For the immediate future, integration is essential to scale network demonstrators to the level where higher level network functions can be developed and interaction between multiple switches in a realistic network environment can be studied.

References

[1] H. Ahmadi and W. E. Denzel, "A survey of modern high-performance switching techniques," *IEEE J. Selected Areas Commun.*, vol. 7, no. 2, pp. 1091–1103, Sept. 1989.

[2] E. W. Zegura, "Architectures for ATM switching systems," *IEEE Commun. Mag.*, vol. 31, no. 2, pp. 28–37, Feb. 1993.

[3] M. J. Karol, M. G. Hluchyi, and S. P. Morgan, "Input versus output queuing on a space-division packet switch," *IEEE Trans. Commun.*, vol. COM-35, pp. 1347–1356, Dec. 1987.

[4] P. R. Prucnal, M. A. Santoro, and S. K. Sehgal, "Ultrafast all-optical synchronous multiple access fiber networks," *IEEE JSAC*, vol. SAC-4, no. 9, pp. 1484–1493, 1986.

[5] R. A. Thompson and P. P. Giordano, "An experimental photonic time slot interchanger using optical fibers as reentrant delay line memories," *J.Lightwave Technol.*, vol. LT-5, no. 1, pp. 154–162, Jan. 1987.

[6] P. R. Prucnal, P. A. Perrier, and M. W. Chbat, "Demonstration of a self-clocked time-division interchanger," in *Proc. OSA Topical Meet. on Photonic Switching* (Salt Lake City, UT, Mar. 1-3, 1989), pp. 219–225.

[7] D. K. Hunter, I. Andonovic, B. Culshow, and P.E. Barnsley, "Experimental test-bed for optical time-domain switching fabrics," in *OFC '93* (San Jose, CA, Feb. 1993) paper TuO2, 21-26, pp. 71–72.

[8] H. Avramopoulos and N. A. Whitaker, Jr., "Addressable fiber-loop memory," *Opt. Lett.*, vol. 18, no. 1, pp. 22–24, Jan. 1, 1993.

[9] P. R. Prucnal, M. F. Karol, and J. L. Stacy, "Optical TDMA encoder," *IEEE Photonics Technol. Lett.*, vol. 3, no. 2, pp. 170–172, 1991.

[10] H. Goto *et al.*, "An experiment on optical time-division digital switching using bistable laser diodes and optical switches," in *Proc. IEEE Global Telecommunications Conf.* (Atlanta, GA, Nov. 26–29, 1984), paper 26.4, pp. 880–884.

[11] S. Suzuki *et al.*, "An experiment on high-speed optical time-division switching," *IEEE J. Lightwave Technol.*, vol. LT-4, no. 7, pp. 894–899, July 1986.

[12] M. Eiselt *et al.*, "One million pulse circulations in a fiber ring using a Slalom for pulse shaping and noise reduction," in *Proc. 18th European Conf. on Optical Communication* (Berlin, Germany, 1992), p. 875.

[13] M. Calzavara *et al.*, "Optical fibre loop memory for multi-wavelength packet buffering in ATM switching applications," in *OFC/IOOC '93*, paper TUE3, 1993.

[14] J. Spring and R. S. Tucker, "Photonic 2 x 2 packet switch with input buffers," in *Topical Meeting on Photonics in Switching* (Palm Springs, CA, Mar. 15–17, 1993), pp. 11–14.

[15] P. Cinato and A. de Bosio, "Optical technology applications to fast packet switching," in *Photonic Switching (OSA Proceedings)*, J. E.. Midwinter and H. S.. Hinton, Eds., vol. 3, pp. 233–236, 1989.

[16] A. deBosio, C. DeBernardi, and F. Melindo, "Deterministic and statistic circuit assignment architectures for optical switching systems," *Topical Meeting on Photonic Switching* (Tech. Dig. Ser.), vol. 13, paper ThB2, pp. 35–37, 1987.

[17] L.T. Wu, S. H.. Lee, and T. T.. Lee, "Dynamic TDM—A packet approach to broadband networking," in *Proc. IEEE Int. Conf. on Communications*, vol 3, paper 46, pp. 1585–1592, 1987.

[18] P. R. Prucnal, D. J. Blumenthal, and P. A. Perrier, "Self-routing optical switch with optical processing," in *Topical Meeting on Photonic Switching* (Tech. Dig. Ser.), vol. 13, paper ThB2, pp. 193–195, 1987.

[19] D. J. Blumenthal, P. R.. Prucnal, L. Thylen, and P. Granestrand, "Performance of an 8 x 8 $LiNbO_3$ switch matrix as a gigahertz self-routing switching node," *Electro. Lett.*, vol. 23, no. 25, pp. 1359–1360, 1987.

[20] P. A. Perrier and P. R. Prucnal, "High-dimensionality shared-medium photonic switch," *IEEE Trans. Commun.*, vol. 41, no. 1, pp. 224–236, Jan. 1993.

[21] ——, "Self-clocked optical control of a self-routed photonic switch," *J. Lightwave Technol.*, vol. 7, no. 6, pp. 983–989, June 1989.

[22] W. L. Ha, R. M. Fortenberry, and R. S. Tucker, "Demonstration of photonic fast packet switching at 700 Mbit/s data rate," *Electron. Lett.*, vol. 27, no. 10, pp. 789–790, May 9, 1991.

[23] A. Budman, E. Eichen, J. Schalafer, R. Olshansky, and F. McAleavey, "Multigigabit optical packet switch for self-routing network with subcarrier addressing," in *OFC '92* (San Jose, Ca, 1992), paper TuO4, pp. 90–91.

[24] C. J. Moss *et al.*, "Experimental results for fast, high-capacity optical switching architectures," in *Topical Meet. on Photonics in Switching* (Palm Springs, CA, 1993), paper PWB4–1, pp. 204–207.

[25] M. L. Loeb and J. G. R. Stilwell, "High-speed data transmission on an optical fiber using a byte-wide WDM system," *J. Lightwave Technol.*, vol. 6, pp. 1306–1311, Aug. 1988.

[26] J. R.. Sauer, "A multi-Gb/s optical interconnect," in *OE/Lase 1990 (Digital Optical Computing)*, paper #22, Conf 1215, vol. 2.

[27] D. J. Blumenthal, K. Y. Chen, J. Ma, R. J. Feurerstein, and J. R. Sauer: "Demonstration of a 2 × 2 photonic switch for computer interconnects," *IEEE Photonics Technol. Lett.*, vol. 4, no. 2, Feb. 1992.

[28] D. J. Blumenthal, R. J. Feuerstein, and J. R. Sauer, "First demonstration of all-optical multihop photonic packet switching," *IEEE Photonics Technol. Lett.*, vol. 6, no. 3, pp. 457–460, Mar. 1994.

[29] C. S. Ih, R. S. Tian, H. X. Zhou, and X. Q. Xia, "Dense all-optical WDM-SCM technology for high speed computer interconnects," in *Proc. Optoelectronic Interconnects , SPIE OE-LASE '93*, (Los Angeles, CA, 1993), vol. 1849-37,

[30] A. R. Chraplevy, "Limitation on lightwave communications imposed by optical-fiber nonlinearities," *J. Lightwave Technol.*, vol. 10 no. 8, pp. 1548–1557, 1990.

[31] P. R. Bell and K. Jabbour, "Review of point-to-point network routing algorithms," *IEEE Commun. Mag.*, vol. 24, no. 1, pp. 34–38, 1986.

[32] P. R. Prucnal, "Optically processed self-routing, synchronization, and contention resolution for 1D and 2D photonic switching architectures," *IEEE J. Quantum Electron.*, vol. 29, pp. 600–612, 1993.

[33] D. J. Blumenthal, J. R. Sauer, H. Lee, and B. Van Zeghbroeck, "A real-time packet-rate reconfigurable 2 × 2 photonic switch," presented at the IEEE/LEOS Topical Meeting on Optical Multi-Access Networks, Santa Barbara, CA, Aug. 1992.

[34] K. K. Goel *et al.*, "Demonstration of packet switching through an integrated-optic tree switch using photo-conductive logic gates," *Electron. Lett.* vol. 26, no. 5, pp. 287–288, 1990.

[35] R. M. Fortenberry, W. L. Ha, and R. S. Tucker, "Photonic fast packet switch with gain," in *Topical Meet. on Photonic Switching* (Salt Lake City, UT, Mar. 6–8, 1991), pp. 128–131.

[36] L. Thylen "Integrated optics in $LiNbO_3$: Recent developments in devices for telecommunications," *J. Lightwave Technol.*, vol. 6, no. 6, pp. 847–861, June 1988.

[37] J. C. Simon "Semiconductor laser amplifier for single mode optical fiber communications," *J. Opt. Commun.*, vol. 4, pp. 51–62, 1983.

[38] M. Eiselt, W. Pieper, and H. G. Weber, "Photonic packet switching using cascaded semiconductor optical amplifier gates," in *OFC '92* (San Jose, CA, 1992), p. 130.

[39] P. R. Prucnal and P. A. Perrier, "Self-routing photonic switching with optically-processed control," *Opt. Eng.*, vol. 29, no. 3, 170–182, 1990.

[40] P. R. Prucnal, M. A. Santoro, S. K. Sehgal, and I. P. Kaminow, "TDMA fiber optic network with optical processing," *Electron. Lett.*, vol. 22, no. 23, pp. 1218–1219, 1986.

[41] P. R. Prucnal, D. J. Blumenthal, and M. A. Santoro, "A 12.5 Gbps fiber-optic network using all-optical processing," *Electron. Lett.*, vol. 23, no. 12, pp. 629–630, 1987.

[42] P. A. Perrier and P. R. Prucnal, "Optical self-routing of 12.5 Gbit/s time-division multiplexed data," in *Proc. Conf. on Lasers and Electro-Optics* (Tech. Dig. Ser.) (Anaheim, CA, Apr. 25–29, 1988), vol. 7, paper TUK2, pp. 74–75.

[43] E. Desurvire, B. Tell, I. P. Kaminow, G. J. Qua, K. F. Brown-Goebeler, B. I. Miller, and U. Koren, "High-contrast in In-GaAs:Fe photoconductive optical "AND" gate for time division demultiplexing," *Electron. Lett.*, vol. 24, no. 7, pp. 396–397, 1988.

[44] W. Dally, *VLSI and Parallel Computation*. Los Angeles, CA: Morgan Kauffman, 1990, ch. 3.

[45] M. Gustavson *et al.*, "Monolithically integrated 4 × 4 In-GaAsP/InP laser amplifier gate switch arrays," in *Topical Meet. on Optical Amplifiers and their Applications* (Santa Fe, NM, June 24–26, 1992), pp. 39–42.

[46] P. Baran, "On distributed communications networks," *IEEE Trans. Commun. Syst.*, pp. 1-9, Mar. 1964.

[47] N. F. Maxemchuck, "Regular and mesh topologies in local and metropolitan area network, " *AT&T Tech. J.*, vol. 64, no. 7, pp. 1659–1686, 1985.

[48] Z. Zhang and A. S. Acampora, "Performance analysis of multihop lightwave networks with hot potato routing and

distance-age-priorities," in *IEEE INFOCOM '91*, Apr. 1991, pp. 1012–1021.

[49] B. J. Smith, "A pipelined, shared resource MIMD computer," in *Proc. 1978 Int. Conf. on Parallel Processing* (Bellaire, MI, 1978), pp. 6–8.

[50] A. S. Acampora, Center for Telecommun. Res., Annu. Tech. Rep., pp. 1–4, Nov. 1989.

[51] A. V. Ramanan, "Space-time switching networks for multiprocessors," Ph.D. dissertation, Univ. of Colorado at Boulder, 1993.

[52] F. Forghieri, A. Bononi, and P. R. Prucnal, "Analysis and comparison of hot-potato and single buffer deflection routing in very high bit rate optical mesh networks," to be published in *IEEE Trans. Commun.*

[53] R. K. Boncek, P. R. Prucnal, A. Bononi, J. P. Sokoloff, J. L. Stacy, and H. F. Bare, "1.24416 Gbit/s demonstration of a transparent optical ATM packet switch node," *Electron. Letters*, vol. 30, no. 7, pp. 579–580, Mar. 1993.

[54] A. Bononi, F. Forghieri, and P. R. Prucnal, "Design and channel constraint analysis of ultrafast multihop all-optical networks with deflection routing employing solitons," *J. Lightwave Technol.*, vol. 11, no. 12, pp. 2166–2176, Dec. 1993.

[55] R. Fortenberry, A. J. Lowery, W. L. Ha, and R. S. Tucker, "Photonic packet switch using semiconductor optical amplifier gates," *Electron. Lett.*, vol 27, no. 14, pp. 1305–1306, July 4, 1991.

[56] I. Glesk, J. P. Solokoff, and P. R. Prucnal, "All-optical address recognition and self-routing in a 250Gbit/s packet-switched network," *Electron. Lett.*, vol. 30, no. 16, pp. 1322–1323, Aug. 1994.

Photonic Fast Packet Switching

Research is in its infancy, but wavelength-, time-, code-, and space-division approaches are all promising.

Andrzej Jajszczyk and H. T. Mouftah

Broadband integrated services digital networks (B-ISDN) are designed to offer a variety of services with bit rates ranging from several kb/s (e.g., teleactions) to hundreds of Mb/s (e.g., HDTV), and in some cases approaching Gb/s (e.g., in interconnections of high-speed LANs). A multiplicity of rates and the burstiness of traffic sources lead naturally to systems based on the fast packet switching (or Asynchronous Transfer Mode) concept. The requirements of data buffering and processing of packet headers have resulted in electronic solutions for most of the existing fast packet switching nodes. On the other hand, transmission technology in broadband networks is based on optics. The optical medium offers the enormous bandwidth approaching 30 THz. The full potential of this bandwidth, however, cannot be used in systems having electronic nodes in the transmission path. Therefore, it is not surprising that a significant effort is put on merging two approaches: photonics and fast packet switching.

This paper's purpose is to present various current approaches to photonic fast packet switching. Previous surveys concentrated on wavelength-division packet networks [1-3], or optically processed routing [4]. We will present here wavelength-, time-, code-, and space-division approaches, including free-space photonic fast packet switching.

Wavelength-Division Systems

The most mature of today's photonic fast packet experiments use a multiwavelength approach that takes advantage of the vast bandwidth of fiber optics. This bandwidth may provide a large number of optical channels in a single medium and makes it possible to construct switches based on shared media architectures, both multi- and single-hop. The need for wavelength reuse, caused by technology limitations, leads to switches employing active fabrics. Table 1 contains a comparison of some representative multiwavelength fast packet switches.

Single-Hop Shared Media Switches

In a single-hop approach a passive star coupler is used as broadcast medium to relay information on all wavelengths to every node (terminal) in the network. Thus, each node directly sends packets to any other node without going through any intermediate nodes. The passive star architecture (Fig. 1) is now a dominant configuration in photonic fast packet experiments. A passive star coupler is the heart of the fabric. Various arrangements of transmitters and receivers are possible: either the transmitters contain tunable lasers and the receivers are of fixed wavelength or fixed-wavelength transmitters and tunable receivers are used. The latter solution enables multicasting, but the fabric requires side signaling to inform the receiver of the appropriate wavelength to which it must be tuned. In principle, both transmitters and receivers may be tunable, which can lead to a reduction of the number of the required wavelengths at the expense of nonblocking properties.

In passive-star switching fabrics with tunable transmitters, the input laser can be tuned to one of the N wavelengths, depending on the desired output channel to be connected to the tuned input. The incoming data packets modulate light of the input lasers. All signals are then combined in the passive star coupler and broadcast to the fixed-wavelength receivers of the output channels. In the described arrangement, only point-to-point connections are possible.

In switching fabrics with tunable receivers, the incoming information modulates light sources having a unique fixed-wavelength each. All the optical energy is combined and then split in a passive star coupler. A tunable receiver is associated with each output. It is tuned such that it only allows the wavelength of the desired input to pass. A multicast connection is established by letting a set of receiving nodes tune to the same wavelength.

Optical-passive-star-based switching fabrics usually are controlled by electronic devices, but sometimes supplemented by some optical components. For example, the FOX and the HYPASS [2, 6] systems contain wavelength-division, passive-star

ANDRZEJ JAJSZCZYK is a visiting scientist at the Department of Electrical Engineering, Queen's University, Kingston, Ontario, Canada.

H. T. MOUFTAH is a professor in the Department of Electrical Engineering, Queen's University at Kingston, Ontario, Canada.

Reprinted with permission from *IEEE Communications Magazine,* Vol. 31, No. 2, pp. 58-65, February 1993.

Switch	Switching fabric	Control & contention resolution	Buffers	Ref.
LAMBDANET	Passive star coupler. Each node contains a fixed-wavelength transmitter and a set of fixed-wavelength receivers.	Distributed electronic control associated with each node.	Electronic buffers at outputs.	[2]
SYMFONET	Passive star coupler. Each node contains a fixed-wavelength transmitter and a set of fixed-wavelength receivers.	Distributed electronic control associated with each node. All nodes are synchronized.	Memories of parallel processors interconnected by the switch.	[5]
FOX	Passive star coupler. Tunable transmitters and fixed-wavelength receivers.	Wavelength-division control network with tunable transmitters and fixed-wavelength receivers. Contention resolution by response indicators and retransmissions.	Memories of parallel processors interconnected by the switch.	[6]
HYPASS	Passive star coupler. Tunable transmitters and fixed-wavelength receivers.	Wavelength-division control network with fixed-wavelength transmitters and tunable receivers. Input-buffered/output-controlled arbitration procedure.	Electronic buffers at inputs.	[2]
BHYPASS	Passive star coupler. Tunable transmitters and fixed-wavelength receivers.	Electronic Batcher-banyan network.	Electronic buffers at inputs.	[2]
Photonic Knockout Switch	Asymmetrical passive star configuration Fixed-wavelength transmitters and tunable receivers.	Electronic knockout contention controller.	Electronic buffers at outputs.	[7]
Star-Track	Passive star coupler. Fixed-wavelength transmitters and tunable receivers.	Electronic token-ring-type control network.	Electronic buffers at inputs.	[2]
Distributed photonic switch	Two passive star couplers: information coupler and reference carrier alignment coupler. Fixed-wavelength transmitters and tunable receivers in each network interface unit.	Electronic control. Scheduling algorithm based on the priority and packet time delay constraints.	Electronic buffers.	[8]
AOTF-based switch	Passive star coupler. A pair of fixed-wavelength transmitters and a pair of tunable receivers for each node.	Electronic control. Flow of data packets controlled by a reservation scheme. Control packets use a slotted ALOHA protocol.	Electronic buffers.	[9]
Switch using multiwavelength matching	Passive star coupler. Multiwavelength pattern generators at transmitters and correlators at receivers.	Photonic control based on multiwavelength matching.	—	[10]
Large ATM switch	Passive optical star couplers and electronic shared memory switches.	Electronic contention resolution device.	Electronic shared buffers at inputs and outputs.	[11]
TeraNet	Multihop network with passive optics.	Electronic control using the Asynchronous Time Sharing approach.	Electronic buffers at outputs of Network Interface Units.	[12]
ATM opto-electronic node	Active routing architecture. Space-frequency switching fabric.	Electronic controller.	Optical buffers at inputs.	[13]
PSRM-based switch	Multistage switching fabric composed of photonic self-routing modules (PSRMs).	Distributed.	Multiple-input/single output optical memories in each PSRM.	[14]

■ **Table 1.** *Comparison of multiwavelength fast packet switch proposals.*

control networks to resolve output contention. In some switches, contention resolution schemes used in electronic fast packet switching nodes are adopted. The Photonic Knockout Switch [7] uses a knockout controller, while BHYPASS [2] employs a Batcher-banyan network. An electronic token-ring-type control network has been applied in the Star-Track switch [2].

A self-routing method in passive-star-based switches, using multiwavelength matching, has been proposed [10]. In such a method routing addresses are identified by combination patterns representing several wavelengths. Data is concurrently transferred using multiple wavelengths corresponding to the address pattern. At the receiver, the incoming signal is correlated with the receiver's address pattern.

To avoid packet loss in case of receiver input contention, passive-star-based systems usually contain electronic buffers at the transmitters or the receivers.

Summarizing, the passive star, single-hop archi-

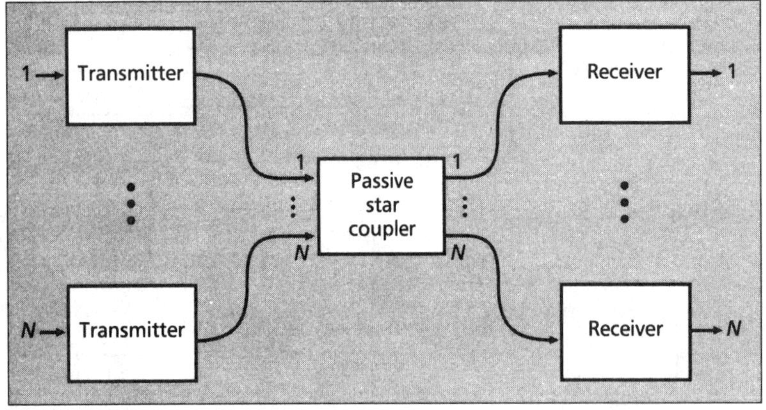

■ **Figure 1.** *Passive star architecture.*

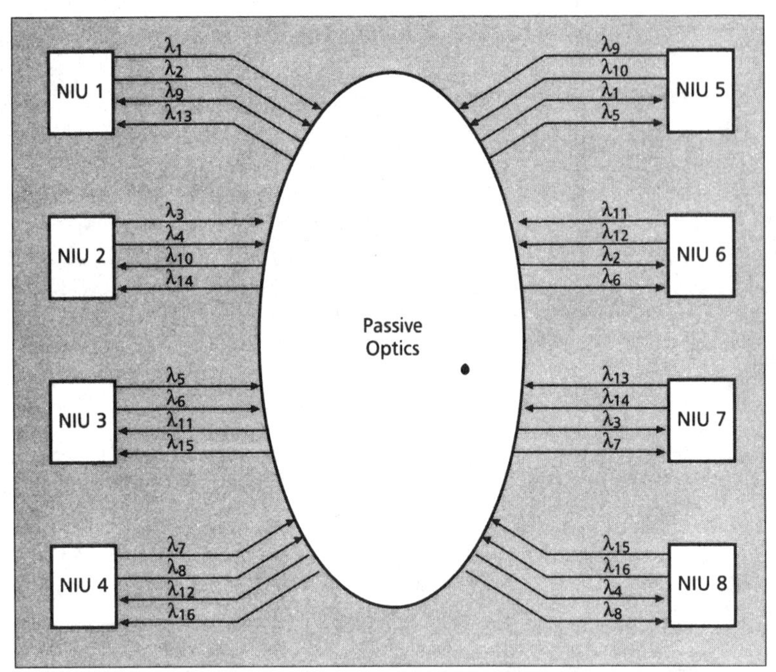

■ **Figure 2.** *Eight-node multihop network; network interface unit (NIU).*

tecture has the following advantages:

- Simple, passive hardware for the switching fabric.
- Simple control protocols.
- Growability.

The size of the switch, however, is strongly constrained by the limits of: tunability, power budget, and signaling capacity.

Multihop Networks

Multiwavelength networks based on the passive star configuration require rapidly tunable transmitters or receivers. By using a multihop approach, however, we can exploit optical bandwidth having only fixed-wavelength transmitters and receivers [1]. An example of an eight-node multihop network is shown in Fig. 2. The network is composed of network interface units (NIU) and a passive optical medium. Each NIU is connected to the transmission medium by two input and two output ports. An additional two-way port is used for connection of a user. Although 16 channels of different wavelengths are transmitted through the passive optics, each NIU has only two fixed-wavelength transmitters and two fixed-wavelength receivers (all wavelengths are different). If the source NIU and the destination NIU have no channel in common, packets are routed through intermediate nodes. For example, if NIU 2 sends a packet to NIU 7, a common channel with λ_3 can be selected and the communication is achieved in one hop. However, if NIU 2 wants to send a packet to NIU 3, an intermediate node NIU 8 should be selected to transmit the packet in two hops, using λ_4 for the transmission from NIU 2 to NIU 8 and λ_{15} for the transmission from NIU 8 to NIU 3. The interconnection pattern of Fig. 2 represents the recirculation perfect shuffle, but other logical topologies can be obtained by appropriately assigning wavelengths to the network interface units. The contention is resolved in the NIUs. The multihop concept has been implemented in an experimental TeraNet system [12].

Systems with Active Fabrics

Along with the passive star configurations, active routing architectures have been studied. In switching fabrics with active routing, the wavelength determines the path which is taken through the fabric by a signal. De Bosio *et al* studied a fast packet switch with an active fabric that contains N sets of tunable filters and modulators for each of its N inputs [13]. Each set connects the input to a different output. The fabric is centrally controlled according to addresses extracted from the packet headers at the input interfaces. The switch employs optical input buffers implemented as reentrant fiber loops.

S. Kuroyanagi, *et al* have proposed a photonic self-routing module (PSRM) which can be used as a building block for larger-capacity networks [14]. The input interface of the switch is a wavelength converter that assigns (to each packet) the wavelength related to the destination address. The packet selector, consisting of a passive splitter and fixed-wavelength filters, routes the packet to one of the optical buffers. The buffer contains a wavelength-division switching network and an optical shift register composed of 2 x 1 switches and fiber delay lines. The switching network directs the packet to the appropriate point of the optical shift register.

Time- and Code-Division Switches

The number of experiments with photonic time-division fast packet switches has been very limited since the presence of buffer memories is an inherent feature of such architectures and efficient optical buffer memories are not available now. However, the time-division concept has been used in conjunction with other concepts (e.g., wavelength division).

The ULtrafast PHotonic ATM (ULPHA) switch, based on a passive-star topology, employs this approach [15]. The switch uses time-division multiplexing with time compression. Incoming packets in an electrical form enter input interfaces where they are converted to optical signals and their lengths are compressed. The destination address and data bits modulate light of two different wavelengths. The passive star coupler multiplexes the packets in time by using different fixed delays for each input line. The multiplexed packets are then broadcast to all output interfaces where they are selected (according to their destination addresses), buffered, and converted to electrical signals of the appropriate bit rate.

Code-division multiplexing can also be used for switching purposes. Such kind of multiplexing is based on the assignment of orthogonal or pseudo-orthogonal codes to the address of each user. In code division multiple access (CDMA), in the electric domain also referred to as spread spectrum technique, each bit is encoded with its destination address. The topology of a CDMA switch can resemble that of Fig. 1. Each bit incoming to the switch is encoded with an optical code sequence, representing the destination address of that bit. The code sequences from all the encoders are combined in a passive star coupler and then distributed to all decoders. The decoders extract by autocorrelation only bits destined to them.

A combination of the wavelength-division and the code-division concepts also seems to be a viable solution. Schemes based on the wavelength division can exploit enormous throughput potential of optical domain by using sophisticated opto-electronic technology, while the schemes based on the code-division multiplexing are tolerant to a variety of system inadequacies [16]. However, they are poor in terms of spectral efficiency. By combining both concepts we can obtain systems of relatively high throughput, but requiring a moderate level of technology.

Space-Division Photonic Switches

Waveguide-Based Switches

Electronic Header Processing — Optical header processing poses numerous technical problems. Therefore, in many photonic switching experiments a data signal is transmitted in optical form through a switching fabric, while control functions are performed in the electrical domain.

A general structure of a space-division photonic switch is shown in Fig. 3. The optical energy of an incoming line is split at an input interface. A small fraction of this energy is converted to an electrical form and sent to the electronic controller. The output interfaces feed the outgoing lines and can be used for the header translation as well as for queueing outgoing packets.

In general, the optical switching fabric may have one of the topologies developed for waveguide switches. Practical demonstrations, however, have been limited to simple structures (e.g., for 1 x 2 switches based on a single lithium niobate directional coupler or semiconductor optical amplifier gates. A 4 x 4 InP-based nonblocking matrix has been used in another experiment. The major limiting factor in these fast packet switching demonstrations is control circuitry rather than the switching fabric technology.

The major challenge for the designers of photonic fast packet switching systems involves the contention resolution. Although the internal contention can be avoided by using nonblocking fabrics, the problem of the output contention still remains. The major obstacle here is the lack of the optical equivalent of electronic buffer memories. One way to bypass this problem is to convert an optical payload into electrical form. Such an approach has the following advantages:

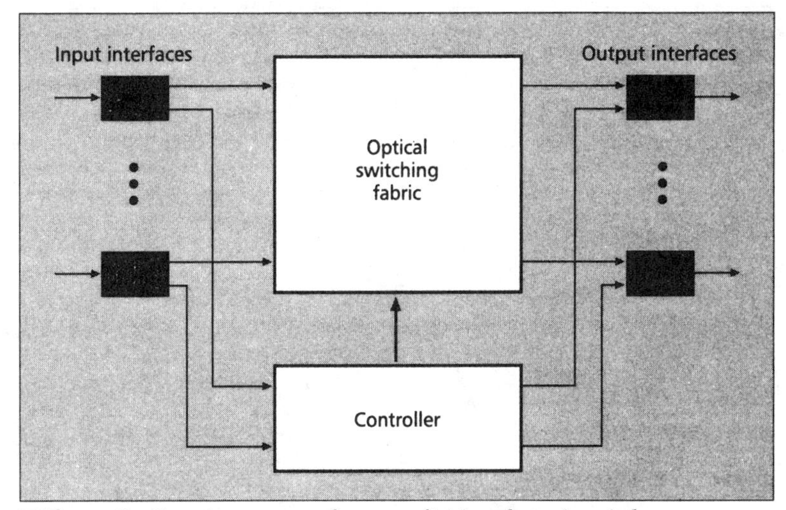

Figure 3. *Generic structure of a space-division photonic switch.*

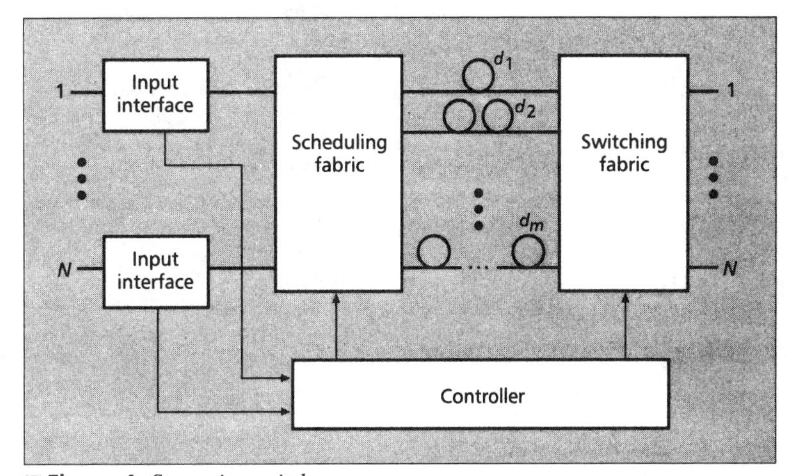

Figure 4. *Staggering switch.*

- Electronic buffer memories can be applied
- Electronic processing of the payload signal is possible

The major disadvantages include:

- Need for optical/electrical and electrical/optical conversions
- Bandwidth limitations by electronic devices

In the extreme case of a photonic fast packet switch, optical technology is applied exclusively to the high-speed switching fabric, while all remaining parts of the switch are electronic devices.

When the optical form of the payload has to be preserved in all parts of the switch (this does not include electronic processing of the header) some means of optical buffering have to be implemented. Optical fiber delay lines are used as memory elements in the staggering switch proposed by Haas [17] and shown in Fig. 4. The switch contains two fabrics: the scheduling and the switching fabric interconnected by a set of optical delay lines, where the d_i delay line delays by i packet time units. The scheduling fabric distributes packets to the delay lines in such a way that, at any time slot, no two packets arriving at the switching stage are destined to the same output, thus resolving the output contention. Other arrangements of the delay lines (e.g., re-entrant fiber loops) also can be used in this architecture.

Self-routing photonic switching can be implemented by processing packet headers in an electronic overlay control fabric having the same topology as the photonic transport fabric.

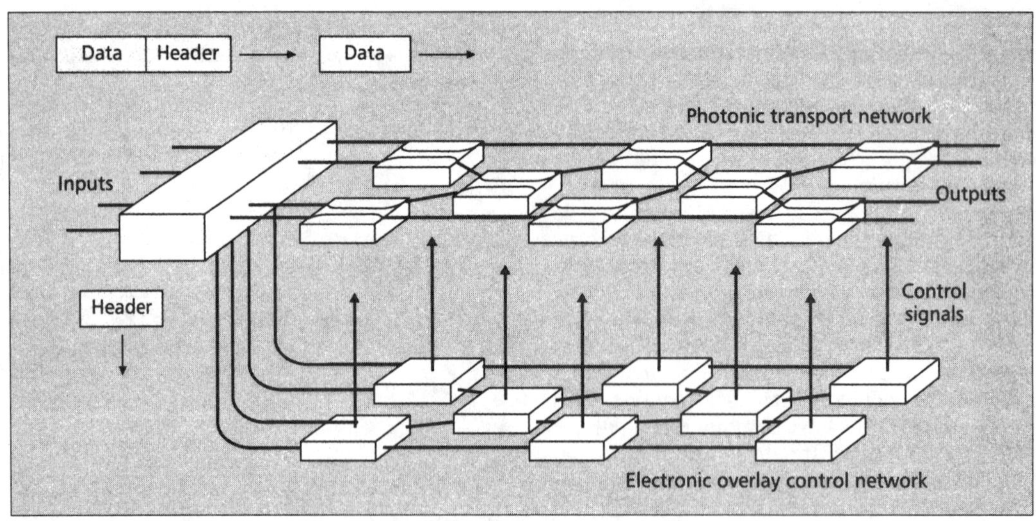

■ **Figure 5.** *4 x 4 self-routing switch using planar fabrics.*

An alternative way to resolve the output contention can be based on deflection routing. In such type of routing the need for buffering is eliminated by routing packets to an alternate path instead of storing it at the node as in store-and-forward networks. For the deflection routing, a bit per wavelength encoding technique can be used. In this case a packet is transmitted in parallel wavelength channels in one time slot. To simplify header extraction, control and payload can occupy different wavelength bands.

A subcarrier addressing also can be used in space-division photonic fast packet switches with electronic header processing. In such a case, a low bandwidth header is placed on an electrical subcarrier above the baseband frequencies occupied by the packet payload. The header and the payload occupy the same time slot and are transmitted simultaneously. A low bandwidth header makes it possible to use low speed electronics to process the address. Moreover, only a small portion of the optical energy needs to be tapped to detect the header, in contrast to systems with headers operating at the data transmission rate.

In the presented considerations, we indirectly assumed that all routing decisions are taken by a centralized controller. Self-routing photonic switching can be implemented by processing packet headers in an electronic overlay control fabric having the same topology as the photonic transport fabric. Routing decisions taken by switches of the electronic network are used for a direct control of corresponding switches of the photonic network. Processing time of short packet headers by the electronic fabric matches transmission time of significantly longer packet information fields in the high-speed optical transport fabric. One of the problems facing designers of such systems is the selection of a topology which is suitable both for photonic and self-routing electronic switching. It has been shown that a planar or a hypercube topology can be used in such applications. An example of a 4 x 4 self-routing photonic switch employing a planar architecture is shown in Fig. 5 [18].

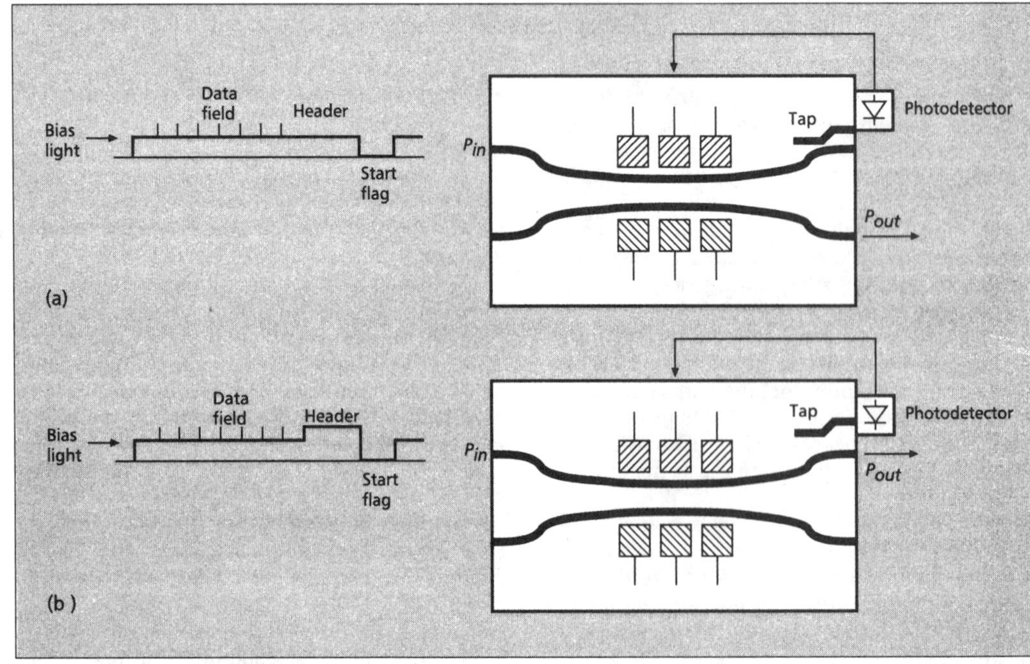

■ **Figure 6.** *Bistable directional coupler: (a) cross state; (b) bar state,* P_{in} *– input power,* P_{out} *– output power.*

Optical Header Processing

In the preceding section, we described fast packet switching systems requiring electronic processing of packet headers. Such an approach involves opto-electronic conversions resulting in higher hardware complexity and failure rate of the node as well as limiting the transmission bandwidth. These problems can be resolved by using fully optical switches. Various schemes for fast packet switches with optical header processing have been proposed. We divide all these schemes into the following two broad categories:

- Switches with control circuitry inseparable from the switching element structure.
- Switches with separate controllers for switching elements.

The first category includes novel designs of switching elements based on various principles, while the second uses known waveguide switching elements (e.g., conventional directional couplers), but employing optical processing for the control. We will present some representative examples of the switches belonging to both groups.

One of the switching elements directly controlled by an optical signal is shown in Fig. 6. The bistability is achieved by a feedback between an output and a driver of multisection electrodes. The device operates with a constant bias light. The packet is composed of a start flag, a header and a data field. In the start flag no bias light is transmitted to reset the coupler, i.e., to set up the cross state. If only the bias light enters the switch, the cross state is preserved (Fig. 6a). A header containing both bias and address light changes the state from cross to parallel (Fig. 6b). The state is maintained due to the feedback. To avoid state changes caused by interference between bias light and data pulses separate wavelengths are used. A lattice-type ATM network with fully photonic self-routing using the described switches has been proposed [19].

Laser diode technology was applied in another fully optical self-routing switch. The switching fabric has a Benes structure and is composed of 2 x 2 switches. The topology of a single 2 x 2 switch is shown in Fig. 7a and the header structure is presented in Fig. 7b [20]. The header starts with a specific bit pattern (frame bits). For each network stage, one routing bit is necessary. Some surplus bits are inserted between routing bits to compensate delays for header decoding. Let us suppose that an optical packet arrives at input I_1 of the switch. The packet is branched to optical gain waveguides W_{11} and W_{12} and induces terminal voltage changes in electrodes J_1 and J_2. Such terminal voltage changes are converted into electronic pulses which are decoded by the electronic control circuits. If the packet is to be sent to O_1, the forward current is injected in electrode J_1 and waveguide W_{11} changes from an absorption to a gain guide. The optical signals are amplified and emitted from O_1. If the injection current is applied to electrode J_2 the light signal is emitted from O_2.

Now we give a brief survey of photonic fast packet switches with separate controllers associated with switching elements. They usually are based on code-division and time-division encoding techniques applied to packet headers [4]. A generalized architecture of a self-routing switching node

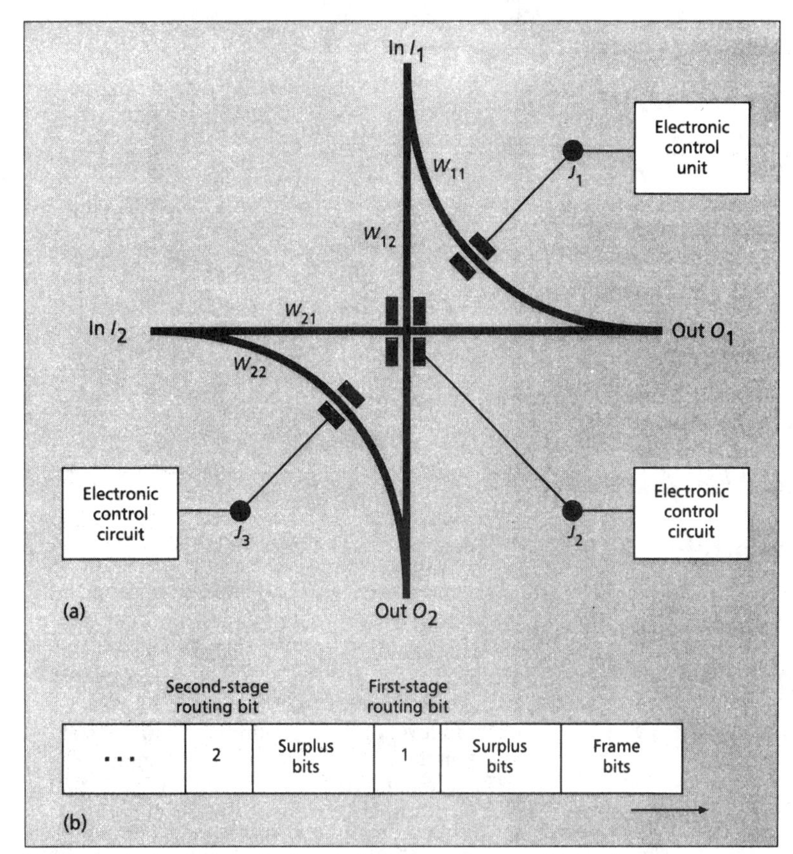

■ **Figure 7.** *2 x 2 laser diode switch: (a) architecture; (b) header structure.x*

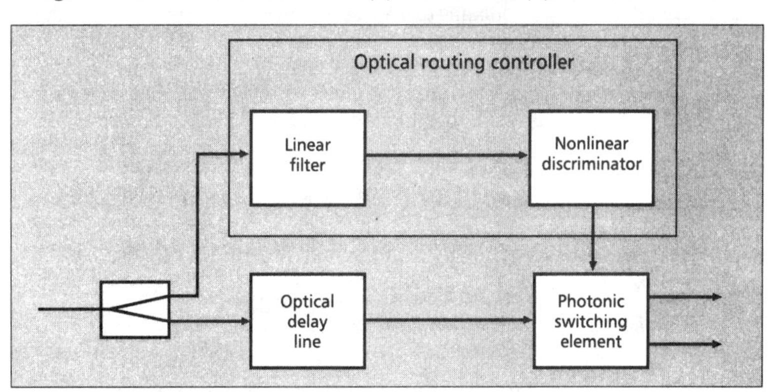

■ **Figure 8.** *Self-routing with optical processing architecture.*

using one of these techniques is shown in Fig. 8. For simplicity of the description we assume that the node contains only one input and two outputs. The node is composed of a photonic switching element (e.g., a directional coupler), an optical delay line, and an optical routing controller. The optical routing controller recognizes the address field in a packet, determines the outgoing link, and generates a control signal to set up the appropriate path inside the photonic switching element. The controller contains two separate stages: linear and nonlinear.

The linear stage (linear filter) can use a correlator consisting of a set of delay lines. The correlator recognizes the destination address of each packet.

The nonlinear stage contains a discriminator which detects the threshold at the output of the correlator. The optical fiber-based fixed delay line is used for the synchronization between the control signal changing the state of the photonic switching element and the information signal.

The inherent parallelism of free-space optics is an attractive feature which can be employed in photonic fast packet switching systems.

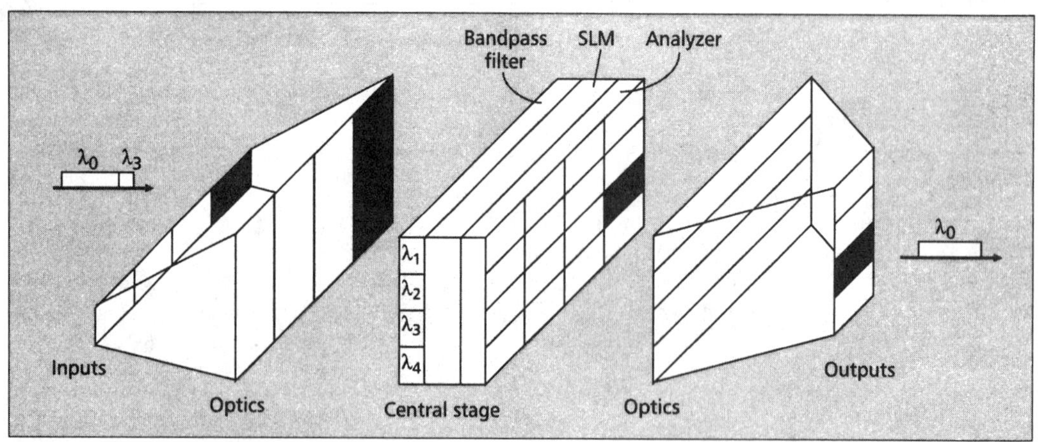

■ **Figure 9.** *Free-space self-routing crossbar switch.*

A packet header, which is to be switched, generates an auto-correlation peak at the linear filter output. This peak is detected by the discriminator which produces a gating pulse of a length equal to the packet's length. The pulse changes the state of the photonic switching element. Packet headers which are not to be switched generate a cross-correlation function with a peak below the threshold level of the discriminator and the state of the switching element remains unchanged.

The recognition process of an optical destination address requires an appropriate method of address coding. In the code-division technique, each packet header is encoded with a waveform corresponding to a code sequence of N^2 chips representing the destination address, where N is the number of different addresses. The time-division encoding technique requires only an address sequence of length N. The destination address of a packet is represented by the position of the single pulse in the packet header. The discussed time-division approach requires an optical clock signal which must be synchronously distributed to all switching nodes in the switching network. To overcome the problem of synchronization, a pulse-interval time-division encoding technique has been developed. In such a technique, the destination address is encoded in the position of an address pulse relative to a reference (clock) pulse within a time frame. No external synchronization is required. The pulse-interval time-division encoding technique was demonstrated experimentally. The application of this technique was proposed for a self-routing tree-structured switching node, as well as for fabrics containing two active stages. The discussed technique suffers from problems associated with a sensitivity to uneven attenuation of optical signals coming from different sources, which may lead to improper address pulse detection.

It should be noted that in all examples presented in this section, there were no means to prevent contention. Such a prevention without conversions from optical to electrical signals seems to be very difficult at the current level of technology unless cumbersome optical delay line structures are used.

Free-Space Switches

Free-space optics exploits the spatial bandwidth offered by the optical domain. The inherent parallelism of free-space optics is an attractive feature which can be employed in photonic fast packet switching systems. The free-space technology can be used both in active switching devices and as optical interconnects between optoelectronic switching elements.

The concept of an active free-space fast packet switching element will be illustrated by using an example of a crossbar switch shown in Fig. 9 [21]. The switch contains a set of inputs, a set of outputs, a crossbar mask, and two sets of lenses (optics). The crossbar mask is an array of independent optical modulators. Usually, the modulators can be in one of two disjoint states: transparent to the incoming light, or opaque to the light. Each input is associated with a unique column of the crossbar mask and each output is associated with a unique row of the mask. By appropriate enabling of mask elements we can obtain a desired connection between an input and an output.

Several methods to enable crossbar mask pixels by packet headers have been studied. The packet address was encoded either by a wavelength, time position, or polarization state. In the first case, the incoming packets contain header pulses of wavelength λ_i, where i is the destination address, and an information field of a common wavelength λ_0 (see Fig. 9 [21]). The crossbar mask is composed of three layers: bandpass filters (wavelength λ_i for row i), optically addressable spatial light modulators and analyzers. The incoming packet is distributed over the entire column associated with the input. Only the horizontal bandpass filter with the center wavelength identical to the wavelength of the packet header allows the packet to pass to the spatial light modulator, which modulates the polarization state of the information field. The analyzer passes only the modulated light.

The time position encoding was used in an experiment with vertical-to-surface transmission electro-photonic device (VSTEP) arrays [22]. Headers of the data packets contained only single pulses located in one of N time slots, where N was the number of outputs. Electrical bias signals were applied to each row of the array. Each bias voltage was raised to a high level in a different time slot. If the header bit and the high bias signal coincided the VSTEP element was turned to the ON state and enabled the transmission of the data packet. A 2x2 free-space crossbar switch also can be controlled by recognizing different polarization states. Such elementary switches can be used in more complex switching structures.

Collisions of packets at the outputs of the switching fabric can be avoided by using a priority scheme based, for example, on different optical power levels (depending on priority) of the packet headers. In case of a conflict, only the highest priority packets could be transmitted through the crossbar mask. By half reflection coating on the reverse side of the crossbar mask the accepted packets are transmitted not only to outputs, but also to respective input ports. This enables the buffer memory element at inputs to retransmit the rejected packets during the next transmission cycle.

One of the most promising directions in the photonic fast packet switching research involves a combination of optoelectronic switching devices and free-space interconnects. This approach takes advantage of the maturity of today's VLSI technology and the inherent high bandwidth and parallelism of free-space optics. The advantages of free-space interconnects over metallic-based interconnections can be listed as follows [23]:

- No need for the source electronics to drive large capacitive loads
- No need for impedance matching terminating transistors that increase the power requirements of the source electronics
- Higher density of pin-outs between chips
- Lower level of crosstalk

The active switching stages can be implemented with GaAs or Si digital circuits. On top of these circuits GaAs modulators and detectors may be grown. As an alternative, the whole stage can be implemented with self-electrooptic effect devices (SEEDs). Such an approach has been used in an optical version of the Starlite switch [24]. The free-space interconnects for this switch employ crossover interconnections that are relatively complex and also tend to decrease the amount of optical power arriving at the detectors. These drawbacks can be reduced by applying switch architectures that require only one-to-one imaging hardware used in one-chip-to-one-chip interconnections [23].

Since optics here is used only for interconnections between electronic stages, the methods of contention resolution developed for electronic fast packet switches can be used (e.g., the recirculation of blocked packets).

Conclusion

The approaches to photonic fast packet switching systems discussed herein show that the research in this area, although very promising, is still in its infancy. Among various solutions, those based on a wavelength-division transport network and an electronic controller are most mature. Further progress in this area is mostly related to the development of more advanced optoelectronic devices, such as tunable lasers and filters.

Lack of the optical equivalent of electronic buffer memories results in the mixed electro-optical solutions of photonic fast packet switches. Although fully optical switches have been demonstrated, their practical applications do not seem to be viable in near futùre. One very attractive and promising direction of research in the area of photonic switching involves the application of optoelectronic technology to switching stages and free-space optics to interconnections. This mixture of technologies combines advantages of high-integration density of electronic devices and high parallelism, large number of pin-outs, as well as resistance against electromagnetic interference offered by free-space optical systems.

The approaches to photonic fast packet switching systems discussed here show that the research in this area, although very promising, is still in its infancy.

References

[1] A. S. Acampora and M. J. Karol, "An Overview of Lightwave Packet Networks," *IEEE Network*, vol. 3, pp. 29-41, Jan. 1989.

[2] M. S. Goodman, "Multiwavelength Networks and New Approaches to Packet Switching," *IEEE Commun. Mag.*, vol. 27, pp. 27-35, Oct. 1989.

[3] B. Mukherjee, "WDM-based Local Lightwave Networks. Part I: Single-hop Systems," *IEEE Network*, vol. 6, pp. 12-27, May 1992.

[4] P. R. Prucnal and P. A. Perrier, "Optically Processed Routing for Fast Packet Switching," *IEEE LCS Mag.*, vol. 1, pp. 54-67, May 1990.

[5] R. J. Westmore, "SYMFONET: Interconnect Technology for Multi-node Computing," *Electron. Lett.*, vol. 27, pp. 697-98, April 1991.

[6] E. Arthurs, *et al.*, "Multiwavelength Optical Crossconnect for Parallel Processing Computers," *Electron. Lett.*, vol. 24, pp. 119-20, Jan. 1988.

[7] K. Y. Eng, "A Photonic Knockout Switch for High-speed Packet Networks," *IEEE J. Sel. Areas Commun.*, vol. 6, pp. 1107-16, Aug. 1988.

[8] A. Fioretti and S. R. Treves, "A Novel Distributed Photonic Switch," *Proc. XIII Int. Switching Symp.*, Stockholm, Sweden, vol. I, pp. 147-52, May-June 1990.

[9] K. W. Cheung, *et al.*, "Wavelength-selective Circuit and Packet Switching Using Acousto-optic Filters," *Proc. IEEE GLOBECOM '90*, San Diego, Calif., pp. 1541-47, Dec. 1990.

[10] K. Hagishima and Y. Doi, "An Optical Self-routing Switch Using Multiwavelength Matching," *Proc. IEEE Int. Conf. Commun. '89*, Boston, Mass., pp. 745-48, June 1989.

[11] A. Cisneros and C. A. Brackett, "A Large ATM Switch Based on Memory Switches and Optical Star Coupler," *Proc. IEEE Int. Conf. Commun. '91*, Denver, Colo., pp. 721-28, June 1991.

[12] R. Gidron and A. Temple, "TeraNet: A Multihop Multichannel ATM Lightwave Network," *Proc. IEEE Int. Conf. Commun. '91*, Denver, Colo., pp. 602-08, June 1991.

[13] A. de Bosio, *et al.*, "ATM Photonic Switching Node Architecture Based on Frequency Switching Techniques," in *Photonic Switching II*, K. Tada and H. S. Hinton, eds., pp. 300-03 (Springer-Verlag, 1990).

[14] S. Kuroyanagi, T. Shimoe, and K. Murakami, "Photonic ATM Switching Network," in *Photonic Switching II*, K. Tada and H. S. Hinton, eds., pp. 296-99 (Springer-Verlag, 1990).

[15] Y. Shimazu, M. Tsukada, and S. Kikuchi, "Ultrafast Photonic Packet Switch with Optical Output Buffer," in *Photonic Switching II*, K. Tada and H.S. Hinton, eds., pp. 292-95 (Springer-Verlag, 1990).

[16] G. Vannucci, "Combining Frequency-division and Code-division Multiplexing in a High-capacity Optical Network," *IEEE Network*, vol.3, pp. 21-30, March 1989.

[17] Z. Haas, "The Staggering Switch: An Almost-all Optical Packet Switch," *Conf. on Optical Fiber Commun. '92 Rec.*, San Jose, Calif., p. 133, Feb. 1992.

[18] H. Obara, *et al.*, "Self-routing Planar Network for Guided-wave Optical Switching Systems," *Electron. Lett.*, vol. 26, pp. 520-21, April 1990.

[19] N. Ogino and M. Fujioka, "Photonic Lattice-type Self-routing Switching Network," in *Photonic Switching II*, K. Tada and H. S. Hinton, eds., pp. 312-15 (Springer-Verlag, 1990).

[20] R. Kishimoto and M. Ikeda, "Optical Self-routing Switch Using Integrated Laser Diode Optical Switch," *IEEE J. Select. Areas Commun.*, vol. 6, pp. 1248-54, Aug. 1988.

[21] M. Hashimoto, M. Fukui, and K. Kitayama, "All-optical Self-routing Crossbar Switch," in *Photonic Switching II*, K. Tada and H. S. Hinton, eds., pp. 309-11 (Springer-Verlag, 1990).

[22] S. Suzuki and K. Kasahara, "Photonic Packet Switch Based on VSTEP Two-dimensional Array," in *Photonic Switching Topical Meeting*, Salt Lake City, Utah, pp. 124-27, March 1991.

[23] T. J. Cloonan and A. L. Lentine, "Self-routing Crossbar Packet Switch Employing Free-space Optics for Chip-to-chip Interconnections," *Appl. Optics*, vol. 30, pp. 3721-33, Sept. 1991.

[24] M. Murdocca and T. J. Cloonan, "Optical Design of a Digital Switch," *Appl. Optics*, vol. 28, pp. 2505-17, July 1989

Pigtailed Tree-Structured 8 x 8 $LiNbO_3$ Switch Matrix with 112 Digital Optical Switches

P. GRANESTRAND, B. LAGERSTRÖM, P. SVENSSON, H. OLOFSSON,
J.-E. GALK AND B. STOLTZ

***Abstract*—We report on integrated optics Ti:$LiNbO_3$-based 8 × 8 switch matrices. The matrix design uses a tree structure with 112 digital optical switches as switch elements. The matrix, which has been pigtailed and packaged in an open package for convenient system demonstrator use, exhibits a worst case insertion loss of less than 15 dB's.**

I. Introduction

In the future, optical switch matrices are expected to be important components in telecommunication systems, as they are capable of performing code- and frequency-transparent switching of optical signals without conversion to electrical form.

A number of matrices have been reported (e.g. [1]-[5]). Up to now, however, matrices using a full tree structure [6], [7] have been limited to sizes up to 4×4. In this paper we report on a polarization-independent 8 × 8 switch matrix using the tree structure, and with digital optical switches as switch elements.

II. Matrix Design

The tree structure layout shown in Fig. 1 is used in the matrix [8]. This structure is strictly non-blocking, has good crosstalk performance and is convenient for broadcasting applications. The switch element used, the digital optical switch, has a number of important features, such as: polarization independence, large wavelength operation range, uncritical voltage supply (which implies good stability performance), and simple polarization-independent power-splitting operation (at zero voltage). In the full 8×8 tree structure used in the matrix (Fig. 1) 112 1 × 2 switches are required. The switch element, shown in Fig. 2, comprises a Y-branch shaped in accordance with Ref. 9, adapted to waveguide parameters corresponding to the ordinary polarization, but with a limitation on the maximum branching angle in the first part of the branch (near the apex).

This limitation will reduce the excess losses in the branches and increase tolerances for waveguide parameter variations (important in this case, where the two polarizations have substantially different waveguide parameters). The total length of the switch element (including the input taper, measured to the end of the shaped bend) is 4.342 mm. To be more specific, the switch is shaped with the following (design) parameters (notation according to Ref. 9): $\gamma_{33} = 0.37\ \mu m^{-1}$, $X_0 = 0.15$, $\gamma = 0.0889$, $K_0 = 7.51\ mm^{-1}$ (coupling coefficient at zero waveguide gap). γ_{33} is the transverse decay-parameter of the field for the isolated waveguide; X_0 is $\Delta\beta/2K_0$ where $\Delta\beta$ is the difference between the propagation constants for the two output channels (at the "design" voltage) and γ is a crosstalk parameter defined in [9]. Note that these parameters have been adapted to the TE polarization, which requires the highest voltages. The shaped bend is terminated when the gap between the waveguides is 11.4 μm. This design is not expected to give any excess loss compared to a straight switch.

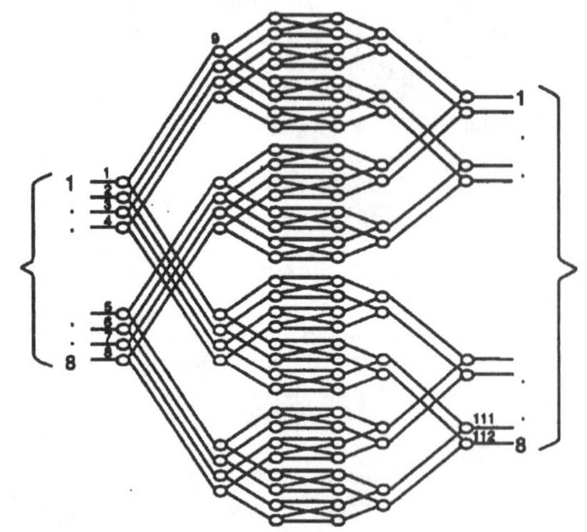

Fig. 1. The tree structure (Refs. 6, 7) used in the 8 × 8 matrix (arranged in analogy with Ref. 8). The number of switch elements is 112 ($2n(n-1)$ for a $n \times n$ matrix).

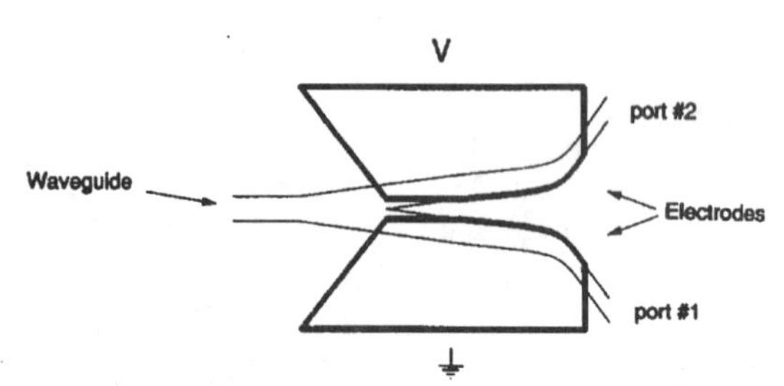

Fig. 2. 1×2 Digital Optical Switch used in the matrix. The branch curvature is shaped in accordance with Ref. 9, but with an angle limitation to 4 mrad in the first part of the branch (near the apex).

Manuscript received September 8, 1993; revised November 1, 1993.
P. Granestrand, B. Lagerström and P. Svensson are with the Fiber Optics Research Center, Ericsson Telecom AB S-126 25 Stockholm, Sweden.
H. Olofsson, J.-E. Falk and B. Stoltz are with Ericsson Components, S-164 81 Stockholm, Sweden.
IEEE Log Number 9214963

Reprinted with permission from *IEEE Photonics Technology Letters,* Vol. 6, No. 1, pp. 71-73, January 1994.

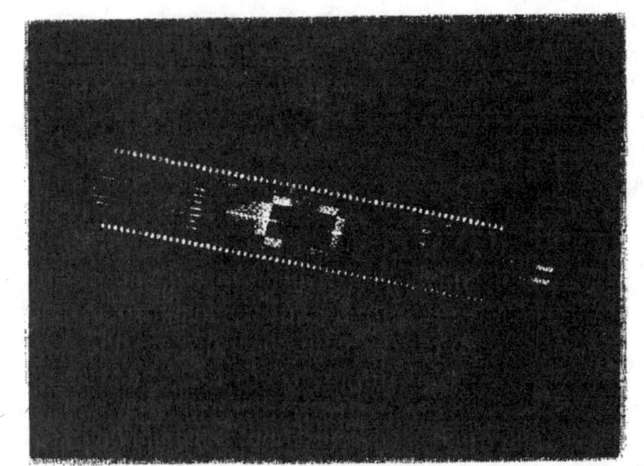

Fig. 3. Photograph of the matrix chip. Dimensions are $\sim$80 $\times$ 15 mm^2.

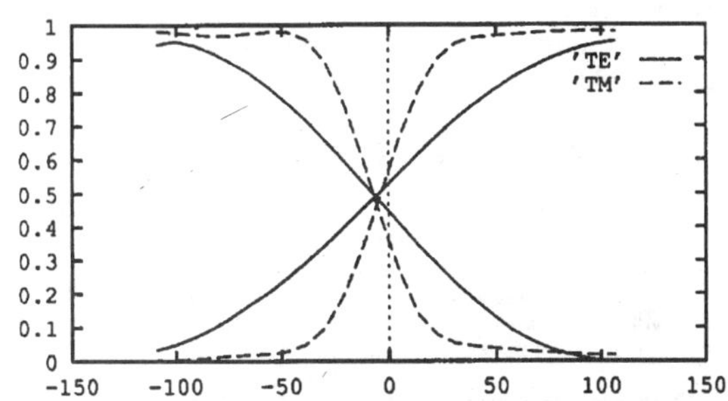

Fig. 4. Transfer functions for one switch element for the TM (extraordinary) and TE (ordinary) polarizations, shown with dashed and solid lines respectively.

The bend radius in the waveguide bends is 35 mm (constant radius type, with straight lines inserted in the middle if necessary), the Ti-strip width prior to diffusion is 6 μm, and the minimum crossover angle in the waveguide crossovers is 8.2 degrees.

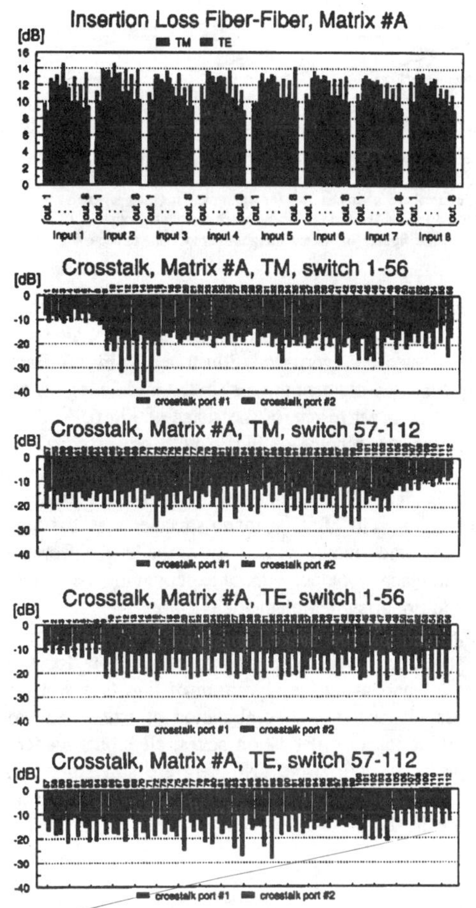

Fig. 5. Experimental results for one pigtailed switch matrix. Drive voltages of $\pm$105 volts were used in the measurements.

III. Fabrication and Experimental Results

Chips in the Z-cut, Y-propagating orientation comprising matrices according to section 2, were fabricated using process parameters adapted to give well-confined, single-mode waveguides at 1.55 microns wavelength (850 Å thick Ti e-beam evaporated, diffusion performed at 1050 centigrades for 9 hours). The chip dimensions were $\sim$15 $\times$ 80 mm^2 (Fig. 3). Quartz buffer layers and gold electrodes, both evaporated, were utilized. After Anti-Reflection (AR) coating and initial measurements the chips were pigtailed with conventional single-mode fibres (by gluing on the fibres one by one) and packaged in a sandwich-type package made up of layers of alumina and the two Printed Circuit Boards (PCB) for electrical connections (not a hermetically sealed package). On the chip the ground electrodes for each "rank" of switches where connected together, reducing the number of bond-connections to 124. This package was then mounted on a 'mother' PCB for convenient use in systems experiments and demonstrators.

The measurements were performed at 1.55 microns wavelength. Figure 4 shows transfer functions for one of the switch elements of one of the matrices for both polarizations (measured before packaging). As can be seen in the figure, the transfer functions exhibit a deviation from the ideal 50/50 power splitting at zero voltage, typical for the switch elements of the matrix.

Fig. 5 shows experimental results for one of the fabricated matrix modules ($\#A$), measured after pigtailing. The worst-case loss (worst-case path and worst-case polarization) was +14.6 dB. Note that the crosstalk values given in the figure are for the individual switch element; the total crosstalk of the tree-structured matrix is often better than that of the single switch element.

In some applications the loss variation exhibited by the matrices is prohibitive. In these cases it is possible to use feedback control to adjust the losses in the matrix to the maximum value (this of course requires that the signal be measured). When the matrix is set in a full point-to-point state, 48 switch elements will have their states defined by the paths (8 paths $\times$ 6 levels). The remaining switches are usually set to a 3 dB state. However, when the above-mentioned loss equalization is to be applied, better crosstalk performance can

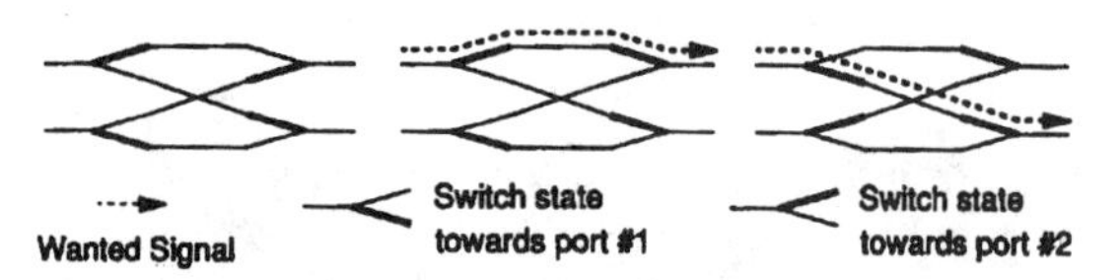

Fig. 6. Example of improved matrix setting algorithm used to maintain good crosstalk properties even when the losses are equalized by active feedback control, of switches 1–8 for example. In cases where switch states are undefined in the 16 middle "four-switch-element groups", configurations according to the figure (or mirror images thereof) are chosen.

be achieved by using a somewhat more elaborate algorithm for setting the state of the undefined switch elements. One simple example of such an algorithm is illustrated in Fig. 6. Each of the 8 'four-switch-element groups' in the middle region of the switch matrix is set to one of the states of Fig 6.

Combining the scheme of Fig. 6 with power equalization on the 8 input switches (switches 1–8 in Fig. 1) ensures that no crosstalk signal from a 'detuned' input switch element (this crosstalk signal will often be much stronger than the wanted signal in the other output branch, due to the large difference in losses) will reach an output without experiencing the extinction in two additional switch elements.

By scanning all possible connection cases and polarization combinations in a computer with measured data for switch module $\#A$ and using the algorithm of Fig. 6, a worst-case total crosstalk (worst-case combination of paths and worst-case combination of polarization states, all inputs loaded) of -15.5 was obtained. (Note that there are still some switch elements that have undefined switch states; these are set to the 3-dB state in the simulation.)

IV. Discussion

The design employed could be improved in a number of ways. For example, by using waveguide tapers at the input and output of the switch matrix, better mode matching between the TM polarized mode and the field of the conventional single-mode fibre can be obtained. This would give lower loss difference between the two polarizations as well as lower maximum insertion loss. To improve the loss uniformity in the matrix, a number of extra waveguide crossovers could be included in the middle region, e.g. between the third and fourth layers (in depth) in the matrix, to increase the losses for the paths with lowest loss in the present design. The differential losses are mainly due to excess losses in the waveguide crossovers. By using improved waveguide crossovers, as described in Ref. 10 for example, these differential losses can be reduced (along with the maximum insertion loss). Also the switch element used can be improved with respect to drive voltage. If the shape of the switch element is carefully examined one finds that it is very moderately shaped. It does not deviate much from a straight switch. A more radically shaped design would probably have given somewhat lower drive voltages, but it would also have been more sensitive to fabrication tolerances.

V. Summary

We have reported an 8 × 8 switch matrix in a full tree structure comprising 112 digital optical switches. The matrix was packaged and pigtailed with standard single-mode fibres. Measurements indicate adequate performance for use in system demonstrators.

This work was carried out within the RACE-MWTN project (RACE II-R2028).

References

[1] P. Granestrand, B. Stoltz, L. Thylén, K. Bergvall, W. Döldissen, H. Heinrich and D. Hoffman, "Strictly non-blocking 8 × 8 integrated optics switch matrix," *Electron. Lett.*, vol. 22, (1986), pp. 816–818.

[2] P. Granestrand, B. Lagerström, P. Svensson, L. Thylén, B. Stoltz, K. Bergvall, J.-E. Falk and H. Olofsson, "Integrated optics 4 × 4 switch matrix with digital optical switches," *Electron. Letters*, vol. 26 (1990), pp. 4–5.

[3] H. Nishimoto, M. Iwasaki, S. Suzuki and M. Kondo, "Polarization independent $LiNbO_3$ 8 × 8 matrix switch," *IEEE Photonics Techn. Lett.*, vol. 2 (1990), pp. 634–636.

[4] P. J. Duthie and M. J. Wale, "16 × 16 Single chip optical switch array in lithium niobate," *Electron. Lett.*, vol. 27 (1991), pp. 1265–1266.

[5] T. O. Murphy et al., "A 16 × 16 Ti:$LiNbO_3$ dilated Benes photonic switch module," *Photonic Switching, Technical Digest, Optical Society of America*, 1991, pp. PDP7–9.

[6] K. Habara and K. Kikuchi, "Optical time-division space switches using tree-structured directional couplers," *Electron. Lett.*, vol. 21 (1985), pp. 631–632.

[7] R. A. Spanke, "Architectures for large nonblocking optical space switches," *IEEE Journ. of Quantum Electron.*, vol. QE-22 (1986), pp. 964–967.

[8] G. A. Bogart, "4 × 4 Ti:$LiNbO_3$ switch array with full broadcast capability," *First Topical Meeting on Photonic Switching* (1987), paper ThD3.

[9] W. K. Burns, "An analytic solution for mode coupling in optical waveguide branches," *IEEE Journ. of Quantum Electron.*, vol. QE-16 (1980), pp. 446–454.

[10] T. O. Murphy *et al.*, "Reduced waveguide intersection losses for large tree-structured Ti:$LiNbO_3$ switch arrays," *OSA Proceedings on Photonic Switching*, vol. 3 (1989), pp. 117–120.

[11] P. Granestrand and B. Lagerström, P. Svensson, H. Olofsson, J.-E. Falk and B. Stoltz, "Tree-structured 8 × 8 $LiNbO_3$ switch matrix with digital optical switches," *Proceedings European Conference on Integrated Optics 1993 (ECIO'93), Neuchâtel, Switzerland*, pp. 10-12–10-13.

High-Dimensionality Shared-Medium Photonic Switch

PHILIPPE A. PERRIER AND PAUL R. PRUCNAL, FELLOW, IEEE

Abstract— **A space-division photonic switch which has the potential to achieve high dimensionality is presented. The proposed switch, which resembles a collapsed network, does not use optical crosspoints; rather a dedicated path is provided for all input/output port connections on a common high-bandwidth transmission medium. This eliminates the restrictions imposed by 2 × 2 switching elements in classical space-division switching fabrics. The demonstration of a fully connected 120 × 120 space-division time-multiplexed photonic switch is reported. The dimensionality and blocking performance of a shared-medium photonic switch that uses time-multiplexing is analyzed.**

I. Introduction

THE use of optical fiber transmission in communications systems permits broad-band signals to be transmitted reliably and cheaply. In present fiber-optic communications networks, however, routing, switching and other signal-processing operations are still carried out electronically, so that optical signals must be converted to electrical signals at each end of a transmission link. For future communications systems, there is a growing interest in postponing the interface with electronics past the transmission link and therefore in performing some signal processing operations, previously carried out electronically, in the optical domain [1].

Progress in semiconductor technology suggests that electronic switches could be used to switch optical signals. The optical signal at each input port would first be converted into an electrical signal by a photodetector before being switched electronically, and then converted back into optical form at the output port for further transmission. Strictly nonblocking electronic switches with 16 input/output ports [2], with 2 Gbit/s transmission bandwidth and switching speed on the order of a few megahertz [3], have been demonstrated.

Despite the advanced state of electronic switching technology, photonic switching technology can offer several advantages. There is, first, an interest in completing the optical path through the switch, avoiding the optical-to-electrical conversions at the input and output ports of a switch [4], [5]. Opto-electronic interfaces increase the switch hardware complexity and thus the potential failure rate of the switch, as well as its cost. Second, it is anticipated that the demand for system bandwidth in future communications systems will be very large, and the required high bit-rates will place severe demands on the electronic parts. The low speed of the signal processing electronics, relative to the information rate on the transmission link, would result in a bottleneck at the switch input.

In photonic switches, the information signal remains in optical form during switching. Cost, complexity, and failure rate of the switch are reduced since no opto-electronic conversion is required. Photonic switches have been shown to have a very large transmission bandwidth (comparable to that of the optical channel) and faster reconfiguration time than their electronic counterparts (in the sub-picosecond region) [6]. Though photonic switches are expected to provide high-interconnectivity, large photonic switches have not yet been demonstrated.

Three classes of switches have been investigated in the optical domain: time-division (TDS), wavelength-division (WDS), and space-division (SDS) switches.

In previously reported optical TDS's (also referred to as time-slot interchangers (TSI's)), switching is carried out with architectures similar to their electronic counterparts: the time-multiplexed optical input data stream is first sequentially demultiplexed; the data in each time slot is written in an optical memory cell; finally, the contents of each storage element are read out in the desired order and multiplexed onto an optical output highway, thus accomplishing the desired time-slot permutation [7]. Optical implementations of $N \times N$ TSI's have used integrated-optic $1 \times N$ and $N \times 1$ switch matrices as optical write and read gates, respectively [8], [9], and fiber-optic delay lines [9] or bistable laser diodes (BLD's) [8] as optical memories. These approaches seem to have, however, several shortcomings. Implementing optical memories with fiber-optic delay lines requires $2N$ delays if all possible time-slot permutations are to be accommodated. This number can be reduced to N if reentrant delay lines are used, at the expense of requiring optical amplification and added control complexity in the delay selection for each time slot [9]. Implementing optical memories with BLD's can also be difficult since BLD's can store only one bit of information at a time and require precise bias current and temperature control. To store B bits per time slot, as necessary in a circuit-switched environment, would therefore require $B \times N$ BLD's.

WDS's are attractive because of their potentially high throughput [10]. Their development is still at an early stage, however, due to the lack of wavelength converters. As a

Paper approved by the Editor for Communications Switching of the IEEE Communications Society. Manuscript received January 12, 1989; revised October 15, 1990. This paper was presented at the IEEE ICC'88, 1988 Conference Record 3 (IEEE Press, NJ), pp. 1485–1489, and Proceedings 9th International Conference on Computer Communications, 1989 (North-Holland, NY), pp. 149–154.

P. A. Perrier is with Alcaltel CIT, France.

P. R. Prucnal is with the Department of Electrical Engineering, Princeton University, Princeton, NJ 08544.

IEEE Log Number 9201635.

Reprinted with permission from *IEEE Transactions on Communications*, Vol. 41, No. 1, pp. 224-236, January 1993.

consequence, most of the research on photonic switching has centered around SDS's.

The basic building block of SDS's is a four-port switching element (crosspoint). Several types of optical switching elements have been developed [11]. Optical switching elements that have received the most attention use an electro-optic crystal in integrated-optic form, and are either of the directional coupler or X type [12]. Lithium niobate ($LiNbO_3$) remains the dominant material in the integrated-optic technology because of its excellent piezo-electric, electro-optic and waveguiding properties, and because of the relative maturity of its processing technology compared to other materials [13]. Photonic switch arrays that provide a disjoint physical path for each signal from an input to an output port have been implemented by interconnecting several of these $LiNbO_3$ optical crosspoints [12]. Many of the conventional electronic switching fabrics developed for telecommunication applications or processors/memories interconnections [14], have been considered for planar optical SDS's.

The development of integrated-optics in semiconductor materials (silicon, and III–V compound materials) is progressing rapidly. In particular, improvements in material growth technology and waveguide design for III–V materials have allowed the demonstration of low-voltage, submillimeter switches [15]. Low-loss arrays of these switches remain to be demonstrated [16]. This technology is particularly attractive because of the potential for monolithic integration of optical and electronic devices.

The fabrication of large optical switches using integrated optics is limited by several factors. First, the minimum cross-sectional dimension required to confine light to a waveguide is approximately equal to the wavelength of the light [17]. Furthermore, in the case of directional coupler switches, the minimum interaction length required to achieve switching between optical waveguides is determined, in part, by the electro-optic strength of the material [18]. Together, these factors restrict the minimum size of an optical switching element. The maximum substrate size, determined by the wafer processing technology (100 × 75 mm at the present time), in turn, limits the maximum number of switching elements that can be fabricated on a single substrate and the maximum number of switch ports. Second, the density of components is further limited by the minimum bend radius with acceptable loss of the waveguides connecting two switching elements. Third, the cumulative effects of the waveguide propagation loss and the crosstalk contributed by each switching element in a large switch would be unacceptably high. As a consequence, and though an integrated-optic 8 × 8 crossbar switch matrix has been experimentally demonstrated [19], optical switch fabrics larger than 16 × 16 appear difficult to construct on a single substrate, if $LiNbO_3$ switching elements are used [12].

Several approaches can then be considered to achieve a larger integrated-optic switch fabric. One approach is to develop efficient optical signal regenerators to compensate for the loss in individual switching elements. If a semiconductor material is used, the integration of regenerators on the same substrate as the switch matrix is possible. On the other hand, if a dielectric material is used, and it is an active gain medium as well, then amplification could potentially be achieved by pumping. A second approach is to reduce the interaction length required for switching by increasing the switching voltage. However, higher voltages limit the switching speed. As pointed out earlier, a large portion of the optical chip real estate is devoted to the interconnection of switching elements and fiber/chip interfaces. A third alternative would be to reduce the losses for small radius bends so that the length necessary for the interconnection of switching elements is decreased. Several techniques such as offset of the bends, enlargement of the bend effective index, or reduction of the index at the bend boundaries have been investigated [20]. New fiber/chip interconnection techniques have also been investigated. For instance, the use of passive optical "wiring" chips for fiber/chip coupling would free some space on the active switch matrix substrate thus allowing a larger number of switching elements [20]. A fourth alternative is to reduce the number of switching elements that are to be integrated for a given switch dimension. Such new switching fabrics have already been developed [21]–[24]. However, for a given switch dimension, a decrease in the number of crosspoints is usually accompanied by the introduction of blocking and by an increase in the switch control complexity. A fifth approach is to partition the optical switch into several smaller optical switch arrays and to butt-couple the multiple substrates [24] or interconnect them with optical fibers. Then, the dimension of the largest manufacturable optical switch would be determined by the number of optical fibers (or waveguides) that could be coupled to the edge of the substrate and by the complexity of the interconnection field [25], [26]. With the appropriate choice of architecture [27], it is expected that switches with size 100 × 100 will be feasible.

As in electronic switching systems, larger photonic switches could be implemented by cascading TDS's and SDS's . These hybrid switches would simultaneously reduce the required size of the optical SDS's and the required size of the memory bank of the optical TDS's. However, this scheme would introduce new problems, such as the need to reconfigure the SDS at the slot rate.

It is apparent from the above discussion that, regardless of which integrated-optic technology is used, the planar switching architectures developed for electronic switches, which provide a dedicated physical path from an input to an output port, are not necessarily optimal for optical switches. Nonplanar two-dimensional [28], [29] and three-dimensional [30], [31] optical switches exploiting the highly-parallel and noninterfering nature of light have been demonstrated. Their integration in practical switching networks is, however, only at the proposal stage.

A nonblocking optical switching architecture that would not use crosspoints (disjoint physical paths) would represent a significant step toward a switching architecture more suited to the optical domain.

In this paper, an optical SD switch, designated by "shared-medium switch", is presented which has the capability to achieve high dimensionality. The shared-medium switch circumvents the problems associated with switching fabrics constructed from 2 × 2 optical switching elements by permitting

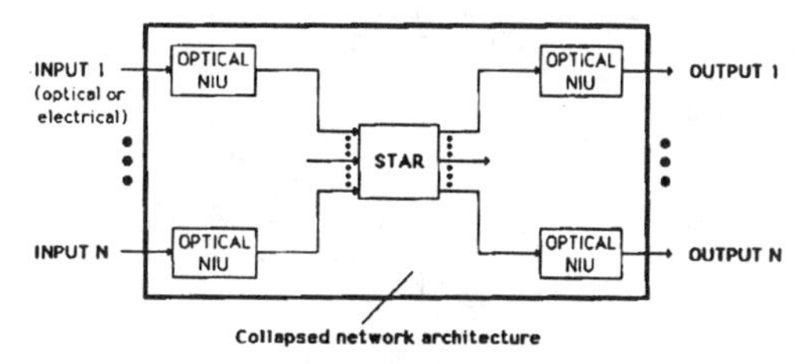

Fig. 1. Block diagram of the shared-medium space-division photonic switch architecture.

all input/output port connections to share a common high-bandwidth transmission path.

The configuration of the shared-medium switch is described and its features are detailed in Section II. The experimental demonstration of a 120 × 120 switch is reported in Section III. The switch dimensionality is analyzed in Section IV. Section V presents a scheme that takes advantage of the bandwidth of optical fibers to reduce the collision blocking probability and discusses possible extensions of the shared-medium switch.

II. Shared-Medium Switch Configuration

A. Description of the Switch

A block diagram of the shared-medium switch with N input and output ports is shown in Fig. 1 [32]. This switch exploits the large bandwidth of optical fiber by first multiplexing the signals from all input ports onto a common optical channel. The multiplexed signals are then broadcast to all output ports. The desired input port signal is demultiplexed at the appropriate output port, thus achieving the switching function. The multiplexing operation at each input port is performed using a network interface unit (NIU). Its functions are to sample the input signal and to encode it in such a way that the correct input signal can be retrieved at the desired output port. The demultiplexing operation is performed at an output port NIU whose function is to select the desired encoded input signal from the combined signals.

The shared-medium switch configuration is that of a collapsed network. As in the case of a network that can support several simultaneous users, the necessary network aggregate information capacity is many times larger than the data rate of each user. Similarly, the network-like switch aggregate information capacity must be many times larger than the individual input port data rate. If a large number of ports is to be supported, the multiplexing and demultiplexing functions must be carried out using optical signal processing. Indeed, optical processing permits a higher aggregate information capacity than could be achieved with electronic multiplexing. Several well-known, optically-processed, multiple-access techniques can be used to multiplex all input port signals onto the same optical channel: code-division multiple access (CDMA) [33]–[35], time-division multiple access (TDMA) [35]–[37], wavelength-division multiple access (WDMA) [38]–[40], and frequency-division multiple access (FDMA) [41].

In fixed-receiver assignment (FRA) TDMA (CDMA, WDMA, FDMA), each time slot (sequence, wavelength, frequency) is associated with an output port. On the other hand, in fixed-transmitter assignment (FTA) TDMA (CDMA, WDMA, FDMA), each time slot (sequence, wavelength, frequency) is associated with an input port. For the switch to be strictly nonblocking, N (the number of time slots, sequences, wavelengths, or frequencies) must be greater than or equal to the number of input/output ports.

1) Shared-Medium Switch Using TDMA: In the time-multiplexed switch configuration, each data bit of duration T is divided into N time slots of width $\tau(N = T/\tau)$. A central clock source produces a low duty-cycle optical train of pulses of width τ and repetition rate $1/T$. The optical clock is synchronously distributed to each input port and output port NIU of the switch. At each input port NIU, the input signal is sampled by the optical clock, encoded with the port address (using either FTA or FRA), and injected into the passive optical star coupler. Each output of the star coupler carries time-multiplexed encoded signals from all input ports. At each output port NIU, the desired input signal is selected by demultiplexing the combined encoded input signals.

2) Shared-Medium Switch Using CDMA: In the code-multiplexed switch configuration, the bit interval T is divided into N^2 chips of width $\tau'(N^2 = T/\tau'$, in the case prime code sequences [33] are used) from which N orthogonal sequences can be generated. The input signal at each input port NIU is sampled and optically encoded with the sequence corresponding to the port address (using either FTA or FRA). All sequence-encoded input signals are multiplexed in the star coupler. At each output port NIU, the desired input signal is demultiplexed by optically correlating the appropriate output port sequence with the combined code-multiplexed signals from the star coupler. Note that in this case no clock is necessary and the switch can be operated asynchronously. However, asynchronous operation not only requires that the sequences, used as addresses, be orthogonal, but also that any combinations of time-shifted sequences be orthogonal as well. This condition limits the number of sequences that can be simultaneously used in the switch.

3) Shared-Medium Switch Using WDMA or FDMA: In the wavelength- or frequency-multiplexed switch configuration, the optical spectrum is partitioned into N distinct wavelengths or frequencies. The input signal at each input port NIU is encoded with the wavelength or frequency corresponding to the port address (using either FTA or FRA). All wavelength- or frequency-encoded input signals are multiplexed in the star coupler. At each output port NIU, the desired input signal is recovered by demultiplexing the combined signals from the star coupler. Again in this case, the switch operates asynchronously.

B. Features of the Shared-Medium Switch

In a conventional $N \times N$ planar integrated-optic SDS, N concurrent, physically disjoint paths are maintained through the switch. Connections are established in a way that attempts to avoid conflict with already established paths.

The shared-medium switch exhibits several advantages over conventional switching fabrics that use distinct physical paths:

a) The collapsed-network structure of the shared-medium photonic switch overcomes the problems associated with classical SDS's by eliminating crosspoints altogether. Simultaneous routing is achieved by exploiting the large bandwidth of the optical transmission channel. After multiplexing the encoded input signals in the passive star coupler, all signals share a common physical path.

b) The $N \times N$ shared-medium switch is strictly nonblocking, i.e., an idle input port can always be connected to an idle output port.

c) The complexity of the shared-medium switch grows linearly with the number of ports, N. (Planar integrated-optic switching fabrics do not usually grow linearly as additional ports are added.) Furthermore, the shared-medium switch allows a modular growth: to add a new port to the switch, an input and output NIU are attached to the star coupler. The existing input port NIU's (or output port NIU's, in the case of FTA) must be able to accommodate an additional port address, that is, an additional time slot, sequence, wavelength, or frequency must be made available.

d) By eliminating crosspoints, the crosstalk associated with each switching element in an integrated-optic SDS is also eliminated, resulting therefore in lower total crosstalk. The absence of crosspoints does not, however, preclude the possibility of crosstalk due to adjacent time slots, wavelengths, frequencies, or close sequences. A discussion of the crosstalk resulting from synchronization errors, timing jitter, dispersive broadening and temperature drift in the case of a shared-medium switch using TDMA will be carried out in Section IV-B.

e) In the FTA mode, the shared-medium switch can be used for broadcasting.

f) The control of the shared-medium switch is simple. To establish a desired I/O port connection, the routing control consists of selecting the appropriate port address. Due to the present state of the optical processing technology, the processing required in establishing an I/O port connection cannot be carried out optically. Thus, the availability of routing information in the electrical domain is presupposed for all multiple-access schemes considered here. The speed at which the switch can be reconfigured is dictated by the time it takes to select the appropriate code (in the case of CDMA) or delay (for TDMA) or to tune a laser or filter to the desired wavelength (for both WDMA or FDMA).

g) Since an isolated failure affects only one port (unless the star coupler fails, in which case the whole switch fails; the possibility of a passive component such as a star coupler failing, however, is remote), the shared-medium switch is reliable. This is not the case in integrated-optic switches in which the failure of a single switching element might affect several output ports.

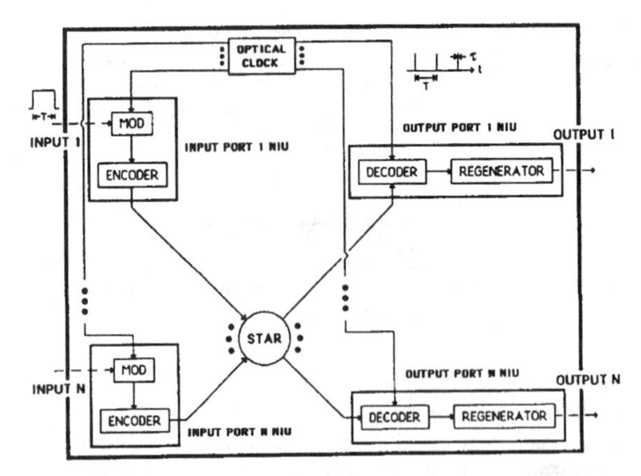

Fig. 2. Block diagram of the shared-medium space-division photonic switch architecture using time-division encoding.

C. Shared-Medium Photonic Switch Using TDMA

In this section, a time-multiplexed shared-medium switch is considered in more detail. A block diagram of this switch configuration is shown in Fig. 2. The details of a connection between an input and output port are shown in Fig. 3. The optical clock is distributed to each input port and output port NIU by single-mode fibers. At each input port NIU, the optical clock pulses are gated by the data sequence on the input port. If the input signals are electrical, an electro-optic modulator can be used; in the case the input signals are optical, an opto-optic modulator would be used. In either case, the required bandwidth of the modulator corresponds to the input port data rate. In this way, the original NRZ (electrical or optical) sequence is converted into a low-duty cycle optical signal at the same rate, without resorting to high-speed circuitry. The modulator output is then fed into an optical time-division (TD) encoder. Assuming that FRA/TDMA is used, the encoder positions the optical pulse from the modulator output in the time slot of the desired output port by selecting the delay corresponding to that time slot. An attractive design of the TD encoder will be discussed in Section IV-C. After passively combining the optical signals from all N encoder outputs in the star coupler, the composite pulse stream is broadcast to all output port NIU's. There, the desired input signal is demultiplexed by incoherent optical processing. Optical sampling of the correct time slot is performed by adding the TDMA frame from the star coupler to the output port NIU clock, delayed to the time slot corresponding to that output port. This raises the amplitude of the sampled signal above that of the surrounding time slots. The optical decoder output is then threshold-detected and a pulse of duration T is regenerated at the intended output port, thus accomplishing the desired switching function. The rise time of the regenerator must be less than τ and the fall time less than T. In principle, threshold-detection and data regeneration could be performed optically [42]. (If the output of the switch is required to be electrical, the regenerator output would be photodetected.) In practice, however, the regenerator would consist of a fast photodetector followed by a Schmitt trigger circuit.

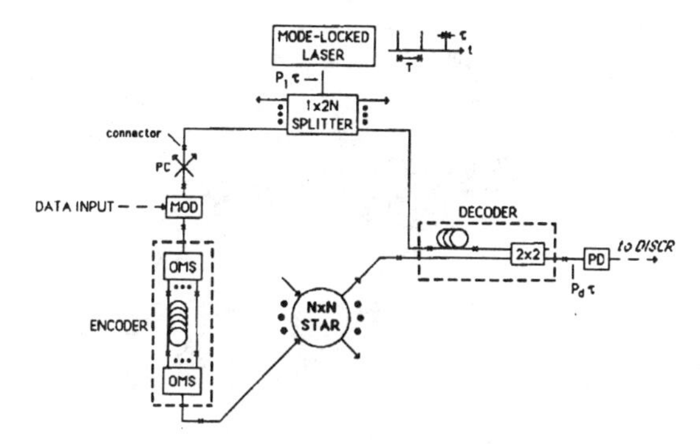

Fig. 3. Details of an input to output port connection.

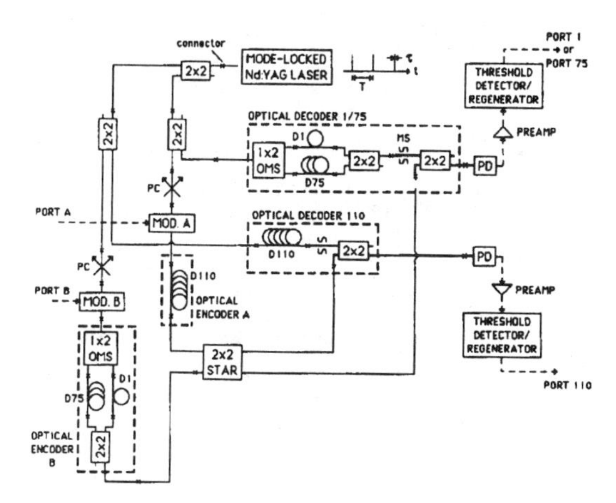

Fig. 4. Experimental set-up of FRA/TDMA shared-medium space division photonic switch.

III. Experimental Demonstration of a Time-Multiplexed Shared-Medium Photonic Switch

A. Experimental Setup

An experimental demonstration of the shared-medium photonic switch was carried out using FRA/TDMA, though FTA/TDMA, optical CDMA, WDMA, or FDMA is also possible. The experimental setup is shown in Fig. 4. As an optical clock, a central mode-locked Nd:YAG laser generates optical pulses of width $\tau = 80$ ps at 100-MHz repetition rate ($T = 10$ ns, and corresponds to a bit interval). This TDMA frame thus consists of $N = 125$ time slots, which can accommodate as many as 125 input/output ports. The emission wavelength is 1.319 μm and the pulse peak power at the laser output is 150 W. The coupling efficiency to a single-mode fiber was 37.5%.

Though 125 input/output ports could be accommodated with this switch, two active input ports and three active output ports were configured for the demonstration. As seen in Fig. 5, signals arriving at input port A are directed to output port 110, whereas data arriving at input port B can be routed either be output port 1 or output port 75. Owing to the fact that opto-optic modulators and optical regenerators are not readily available, both the input and output signals were taken to be

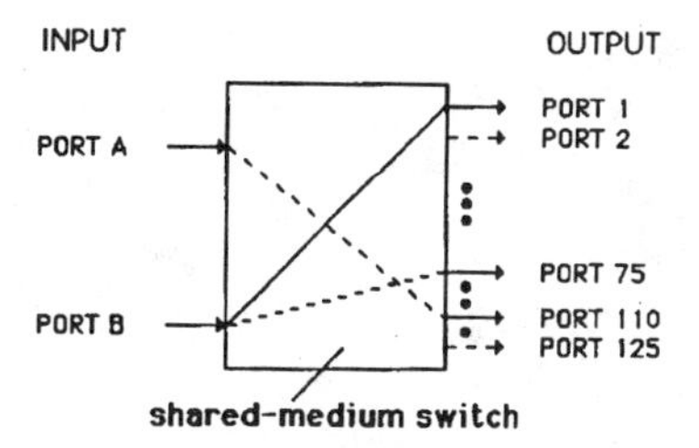

Fig. 5. Experimental switching configuration.

in electrical form. At each of the two input ports, A and B, a 3-GHz $LiNbO_3$ integrated-optic waveguide modulator is used to convert a 100 Mbit/s electrical pseudo-random sequence into a 0.8% duty cycle optical signal. The output of modulator A is fed into optical encoder A, consisting of fiber-delay D_{110}, which delays the sampled data to the 110th time slot. On the other hand, the data appearing on input port B can be routed to either output port 1 or output port 75. The proper slot, time slot #1 or #75, is selected in optical encoder B by choosing between fiber-delay D_1 or D_{75} with a 1×2 opto-mechanical switch (OMS).

The two time-encoded input signals are multiplexed in the star coupler. Since $T/\tau = 125$, a maximum aggregate rate of 12.5 Gbits/s can be supported. The multiplexed pulse stream is broadcast to the output ports 1, 75, and 110. Each output port decoder samples the appropriate time slot by incoherently adding its delayed optical clock to the frame received from the star coupler. The addition is performed with a multimode 2×2 combiner (thick lines in Fig. 4 represent multimode fibers), with a mode scrambler at one of its inputs to average out coherence effects between pulses. The addition raises the sampled signal above the surrounding multiplexed data which can then be threshold-detected and regenerated. The output of each decoder is detected by a 5.3 GHz Ge pn diode photodetector, amplified by a 1 GHz broad-band amplifier, thresholded, and regenerated to 10 ns long electrical pulses by a Schmitt trigger circuit.

B. Experimental Results

The experimental results, displayed on a 1-GHz oscilloscope, are shown in Fig. 6. Modulator A is driven by a 100 MHz, periodic 1010$\cdots$ data sequence, generated by a pattern generator (trace A). Modulator B is driven by a 100 MHz 10000000 repetitive data sequence, generated by a bit-error-rate test set (trace B). For each bit "1" at input port A, an 80 ps optical pulse is positioned in slot 110 of the 10 ns frame. For every bit "1" at input port B, optical encoder B is set, in the first case, to select slot 1. The star coupler output is shown in Trace C. In the second case, optical encoder B is set to select slot 75 for every bit "1" at input port B. The output of the star coupler for this case is shown in trace D.

At output port decoder 110, the optical clock is delayed to the 110th time slot. The sum of the star coupler output and the delayed clock is shown in trace E. Each double-amplitude peak in trace E triggers the generation of a 10 ns long electrical pulse. The resulting sequence at output port 110

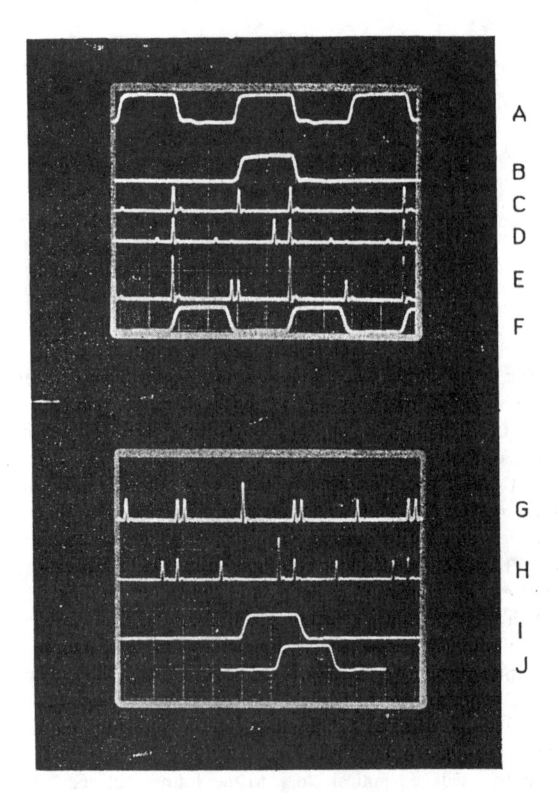

Fig. 6. Switching demonstration through the shared-medium space-division photonic switch. Trace A: Input port A data sequence. Trace B: Input port B data sequence. Trace C: Star coupler output. Each bit "1" in port A (B) input sequence is sub-encoded with an 80 ps pulse in time slot 110 (1). Trace D: Star coupler output in the case each bit "1" in port B input sequence is subencoded with an 80 ps optical pulse in time slot 75. Each bit "1" in port A input sequence is still sub-encoded with a pulse in time slot 110. Trace E: Port 110 NIU ouput. Port 110 NIU optical clock is added to the star coupler output, raising data pulses in every slot 110 above surrounding multiplexed data pulses. Trace F: Port 110 output. Each double amplitude peak at port 110 NIU output is threshold-detected and a 10 ns electrical pulse is generated. The original port A input sequence has been switched to output port 110. Trace G: Port 1 NIU output. Port 1 NIU optical clock is added to the star coupler output, raising data pulses in every slot 1 above surrounding multiplexed data pulses. Trace H: Port 75 NIU output. Trace I: Data sequence at input port B is switched to output port 1. Trace J: Data sequence at input port B is switched to output port 75.

is shown in trace F and corresponds to the original sequence at input port A (Trace A). This demonstrates switching of information from input port A to output port 110.

At output port 1, the frame from the star is added to the optical clock delayed to slot 1. In the first case, when the output of the star coupler is shown in trace C, the resulting addition is shown in trace G. At output port 75, the frame from the star coupler is added to the optical clock delayed to slot 75. In the second case, when the output of the star coupler is shown in trace D, the resulting addition is shown in trace H. The regenerated data, switched from port B to output port 1, is shown in trace I. The regenerated data, switched from port B to output port 75, is shown in trace J.

The simultaneous switching of pseudo-random sequences between input port A and output port 110 and input port B and output port 1 or 75 was achieved with a bit-error-ratio (BER) less than 10^{-9}.

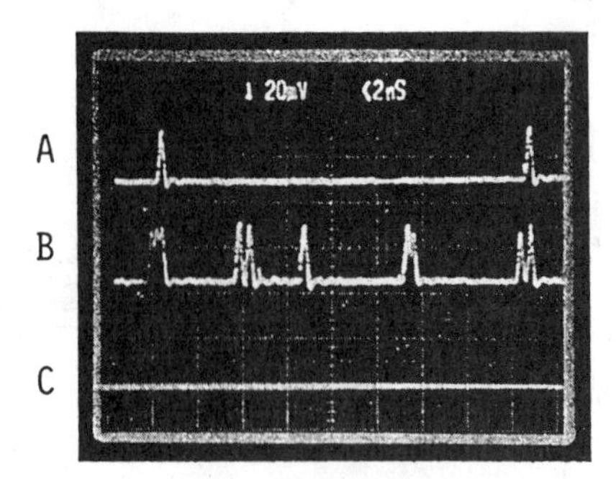

Fig. 7. Crosstalk measurement.

The interference level between two adjacent slots is also measured. In Fig. 7, the delayed clock at the output port 45 (trace A) is added to the TDMA frame from the star coupler with data pulses present in the 43rd and 44th slots (trace B). The interchannel interference from channel 44 to channel 45 was measured to be less than 10^{-9} (trace C).

IV. Switch Dimension Analysis

The maximum number of ports a shared-medium photonic switch can accommodate is limited, to a first approximation, by either the duty cycle of the mode-locked laser or the optical power budget. An upper bound on the switch size is first determined in Section IV-A for the time-multiplexed shared-medium switch discussed in Section III. Factors that could further restrict the switch size and the reconfiguration speed of the switch are investigated in the following two sections. Throughout this analysis, it is assumed that both the input and output signals are in electrical form.

A. Power Budget Analysis

Though both the input and output signals are taken to be electrical, a similar power budget analysis would hold if any one or both signals were in optical form. The detailed system considered for the power budget analysis is illustrated in Fig. 3.

The total switch insertion loss is given by

$$\begin{aligned} L_T = & 10\log 2N + L_{1\times 2N} + L_{\rm MOD} + L_{\rm ENC} \\ & + 10\log N + L_{N\times N} + 10\log(1/\alpha) \\ & + L_{2\times 2} + 6L_c \quad \text{(dB)} \end{aligned} \tag{1}$$

where $10\log 2N$ and $10\log N$ represent the splitting losses of the $1\times 2N$ splitter and $N\times N$ star coupler, respectively, and $L_{1\times 2N}$ and $L_{N\times N}$ are their associated excess losses; $L_{\rm MOD}$ and $L_{\rm ENC}$ are the modulator and encoder insertion losses, respectively, and L_c is the connector insertion loss. $10\log(1/\alpha)$ represents the dividing loss of the variable 2×1 combiner of the receiver (α denotes the splitting ratio of the signal power from the star coupler to the combiner output), and $L_{2\times 2}$ is its excess loss. To optimize the discrimination process at each output port NIU, the power of the delayed clock signal and that of the signal from the star coupler at the combiner output should be made equal [35]. This can be achieved by adjusting α to equalize the losses in the delayed clock and signal paths.

The TD optical encoder at each input port NIU is assumed here, as in the experimental demonstration, to consist of a set of N optical fiber delay lines (one for each time slot) between two opto-mechanical switches (OMS's), though more attractive implementations are possible (a practical design will be presented in Section IV-C). Encoding of each optical pulse with its TD address is achieved by choosing the proper fiber-optic delay line with the OMS's. For that particular implementation, the encoder insertion loss is given by

$$L_{\text{ENC}} = 2L_{\text{OMS}} + 2L_c \quad \text{(dB)}.$$

The total insertion loss can be rewritten as

$$L_T = 20 \log \sqrt{2} N + L_{ex} \quad \text{(dB)} \tag{2}$$

where L_{ex} is given by

$$L_{ex} = L_{1\times 2N} + L_{\text{MOD}} + 2L_{\text{OMS}} + L_{N\times N} + 10 \log (1/\alpha) + L_{2\times 2} + 8L_c \quad \text{(dB)} \tag{3}$$

Fiber attenuation has been neglected and the power is assumed equally divided among the $2N$ and N output fibers of the $1 \times 2N$ splitter and $N \times N$ star coupler, respectively.

To estimate the excess loss of the $1 \times 2N$ splitter and $N \times N$ star coupler, it is further assumed that both were realized by cascading stages of 2×2 couplers (with one splice between two stages). All 2×2 couplers are taken to have the same excess loss, $L_{2\times 2}$. With this assumption, the excess loss of the $1 \times 2N$ splitter is given by

$$L_{1\times 2N} = L_{2\times 2}(\lceil \log_2 2N \rceil) + L_c(\lceil \log_2 2N \rceil - 1) \quad \text{(dB)}. \tag{4}$$

where $\lceil \ \rceil$ denotes "the smallest integer greater than" operator. Similarly, the $N \times N$ star coupler excess loss is given by

$$L_{N\times N} = L_{2\times 2}(\lceil \log_2 N \rceil) + L_c(\lceil \log_2 N \rceil - 1) \quad \text{(dB)}. \tag{5}$$

L_{ex} can then be rewritten as

$$L_{ex} = (L_{2\times 2} + L_c)(\lceil \log_2 2N \rceil + \lceil \log_2 N \rceil + 1) + 5L_c + L_{\text{MOD}} + 2L_{\text{OMS}} + 10 \log (1/\alpha) \quad \text{(dB)}.$$

In particular, L_{ex} can be estimated from the following specifications of commercially available components: $\alpha = 0.5$, $L_{2\times 2} = 0.05\,\text{dB}, L_{\text{MOD}} = 5\,\text{dB}, L_{\text{OMS}} = 1\,\text{dB}$, and $L_c = 0.1\,\text{dB}$. These values yield

$$L_{ex} = 0.15(\lceil \log_2 2N \rceil + \lceil \log_2 N \rceil + 1) + 10.51 \quad \text{(dB)}. \tag{6}$$

The minimum required laser pulse energy, P_{1T}, to yield a signal level at the input to the photodetector, P_{dT}, that will ensure triggering of the discriminator with a given bit-error rate is given by

$$P_{dT} = P_{1T} - L_T \quad \text{(dB)} \Rightarrow P_{1T} = P_{dT} 10^{(L_T)/10} \quad (J) \tag{7}$$

where L_T is specified by (2).

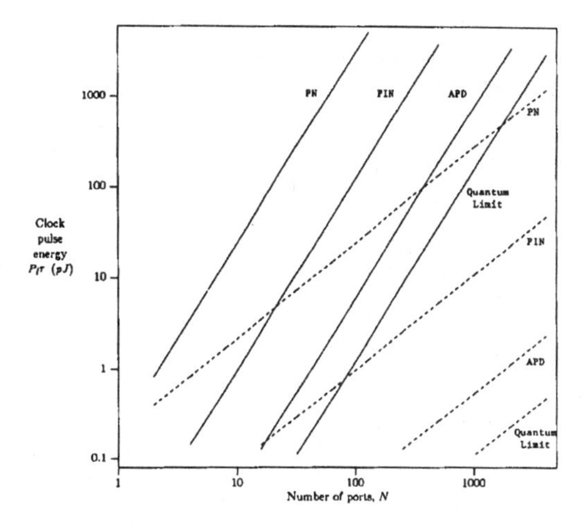

Fig. 8. Switch dimension determined from the optical clock pulse energy as a function of three detector sensitivities and the quantum limit for 10^{-9} bit error rate.

Three photodetector sensitivities and the quantum limit for 10^{-9} bit-error-rate are considered: 8 fJ for the pn diode used in the experimental demonstration, 0.32 fJ for a p-i-n diode, 15.8 aJ for an avalanche photodiode (APD), and 3.2 aJ for the quantum limit (at 1.3 μm).

From (7), the minimum laser pulse energy P_{1T} versus the number of ports N is plotted in Fig. 8 for the three detector sensitivities and the quantum limit (solid lines).

In the present demonstration, the laser pulse energy coupled in the fiber was 4.5 nJ (37.5% coupling efficiency), providing sufficient energy for 120 input/output ports with the pn diode detector used. Since 125 time slots could be supported, the present system was power limited. If an APD were used instead of the pn diode, then about 2270 ports could, in principle, be accommodated. In such a large switch, depending on the input data rate, other pulse-compression techniques may have to be utilized to arrive at the necessary pulse width.

The optical clock can also be generated in a distributed fashion using synchronized mode-locked semiconductor lasers. Accurate synchronization of the sources would be facilitated due to their close physical proximity. The achievable switch dimension for the case a mode-locked semiconductor laser is associated with each input/output port pair is considered next. In this configuration, the $1 \times 2N$ splitter is replaced by a 1×2 coupler at each port pair; the remainder of the system is unchanged. The total system insertion loss becomes

$$L_T = 10 \log N + L_{ex} \quad \text{(dB)}. \tag{8}$$

Again, 10 log N is the dividing loss of the star coupler and L_{ex} is given by

$$\begin{aligned} L_{ex} &= (10 \log 2 + L_{2\times 2}) + L_{\text{MOD}} + 2L_{\text{OMS}} \\ &\quad + L_{N\times N} + 10 \log (1/\alpha) + L_{2\times 2} + 8L_c \quad \text{(dB)} \\ &= (L_{2\times 2} + L_c)(\lceil \log_2 N \rceil + 2) + 5L_c \\ &\quad + 10 \log 2 + 10 \log (1/\alpha) + L_{\text{MOD}} + 2L_{\text{OMS}} \quad \text{(dB)} \end{aligned} \tag{9}$$

where (5) has been substituted for the star coupler excess loss.

The minimum clock pulse energy P_{1dT} versus the number of ports N as a function of the same three detector sensitivities and the quantum limit, is shown by the set of dashed lines in Fig. 8. The same component specifications as before were used to estimate L_{ex}. From Fig. 8, a switch with 1000 input and output ports could be accommodated if a mode-locked semiconductor laser with 0.5 pJ pulse energy is used in conjunction with an APD at each input/output pair.

In the above analysis, it was assumed that all mode-locked semiconductor lasers had the same output power. In practice, however, the output intensities and pulse widths will differ from laser to laser. Both of these imperfections will reduce the theoretical dimension of the switch: the former, because of the unequal power distribution across the output ports, and the latter, because the minimum slot width will have to be equal to the largest pulse width.

B. Feasibility Analysis of a Time-Multiplexed Shared-Medium Photonic Switch

There are many other factors to consider, beyond a simple power budget analysis, to determine the scalability of the shared-medium switch. The speed of the photodetector, the uniformity of the power splitting and the crosstalk are the most difficult obstacles in realizing a large dimension shared-medium switch using time-multiplexing. These factors are discussed below.

1) Detection Process: Given the individual port rate, the detection speed will place a fundamental restriction on the number of ports the shared-medium switch can accommodate. Two techniques have been considered to demultiplex the desired time slot. In the first technique, the delayed clock is incoherently added to the frame from the star coupler and the double-amplitude pulse, if any, is threshold detected, as previously explained. This is the technique that has been considered throughout the paper. The fastest photodetectors commercially available in the 1300 nm region have a 3 dB frequency response of 25 GHz [43], allowing pulses 40 ps long to be observed and have a required minimum pulse power of 10 μW. The above technique could be significantly improved if the two signals for the same minimum pulse power were added coherently before detection. Pulses of width as narrow as 5 ps could be observed. However, insuring coherence between the two signals would increase the complexity of the shared-medium switch. Instead of using incoherent pulse summation followed by threshold detection, a second approach would be to demultiplex the desired input signal using pulse-coincidence through a photoconductive 'AND' gate [44]. The use of the photoconductive "AND" gate for optical TDM demultiplexing has already been demonstrated [45]. In this device, the two optical inputs are semiconductor gaps connected by a microstrip line. An electrical bias voltage is applied at one end of the microstip line; the electronic output is taken at the other end. Only when the delayed optical clock and the optical pulse in the time slot to be demultiplexed are simultaneously exciting the two gaps will a high-speed electronic pulse result at the gate output. This approach has the advantage of performing the time-slot selection and the threshold detection in one step. The speed and minimum pulse power of demonstrated photoconductive switches are, however, lower than for photodetectors (65 ps and 1 mW, respectively).

2) Non-Uniform Power Splitting in the Couplers: Equal power distribution at the output of the $N \times N$ star coupler and the $1 \times 2N$ splitter is critical since the demultiplexing operation, in the time-multiplexed configuration of the shared-medium switch, relies on selecting pulses out of a background with only a 3 dB margin (for the optimal choice of the power splitting ratio α). If the power distribution is unequal, pulses in different time slots will have unequal amplitude, and false threshold detection could result, decreasing the overall system performance. It should be noted, however, that to reduce the occurrence of false threshold detection, the clock signal at each output port NIU can be increased by adjusting the power splitting ratio of the combiner. Even though star couplers with low excess loss and small port-to-port deviation in output power distribution have been reported [46], large single-mode components ($N \geq 100$) with uniformity better than 3 dB appear difficult to construct with today's technology.

Several alternatives that would alleviate this constraint have been studied [47]. Another approach is to implement splitters/star couplers of large dimension by cascading smaller size units. One way that would provide acceptable uniformity across all output ports would be to interconnect units of size $\sqrt{N} \times \sqrt{N}$ arranged in two columns of $\sqrt{N}$ units each.

3) Crosstalk: The different mechanisms responsible for crosstalk in the TD configuration of the shared-medium switch are investigated below.

a) Dispersive Broadening: Spreading of pulses into an adjacent time slot due to dispersion could result in crosstalk. In standard single-mode fibers, total dispersion is dominated by the material dispersion of fused silica [48]. A maximum dispersion of 3.5 ps/(nm linewidth $\times$ km of fiber) has been specified in the 1285–1330 nm wavelength region [49]. Material dispersion together with the use of an optical clock source with nonzero spectral width could therefore contribute significantly to pulse broadening and crosstalk. Within the switch, however, dispersion of short pulses would be minimal due to the short length of the fiber links between input and output ports. Assuming that the maximum length between any pair of input/output ports is 100 m and that the laser has a linewidth of 1 nm, then pulses would overlap into adjacent time slots by about 0.18 ps. Assuming time slots 40 ps in duration (minimum pulse width detectable), this broadening represents less than 1% of the time slot interval. The resulting crosstalk would be minimal.

b) Synchronization: Crosstalk could also result from the lack of synchronization between the delayed optical clock and the frame from the star coupler. The close proximity of all NIUs greatly facilitates the synchronous distribution of the optical clock to all input port and output port NIUs. For the optical clock to arrive synchronously at all input and output port NIU's, the distance between each NIU and the central optical clock source must be an integral multiple of the TDMA frame length, L. It is also required that the frames arrive at the output port NIU's in synchronism with the clock, so that, at a given output port, the clock signal superimposes exactly with

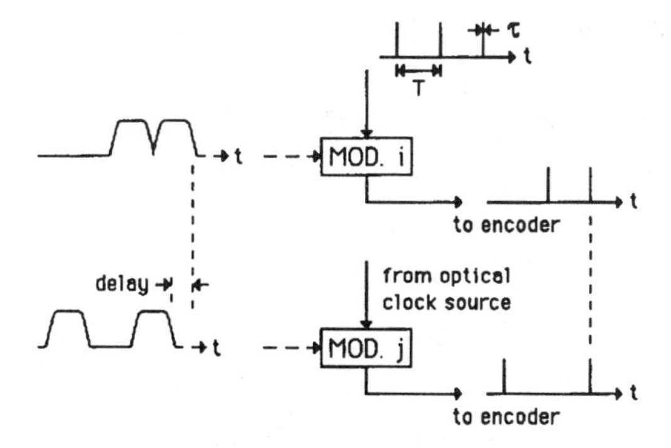

Fig. 9. Input signals synchronization by external modulation.

its associated time slot. Therefore, the length of fiber between each input port NIU and each output port NIU must also be an integral multiple of L.

Input signals need not arrive time-aligned at the switch input ports. Indeed, even though incoming bits on different input ports might be out of synchronization, they gate a synchronized clock. Synchronization between input ports is achieved by the external modulation, provided the optical clock pulses are not gated during the transition times of the electrical input signals (Fig. 9).

c) Timing Jitter: As experimentally observed in Section III, there is no interference between adjacent channels provided each of the concurrent signals occupies a distinct time slot. It follows that to avoid interference between adjacent channels in the multiplexing process, the length of the required delay lines must be precisely controlled. The accuracy of the fiber lengths is given by

$$\Delta L = |(c/n)\tau.\Delta\tau| \quad (m) \tag{10}$$

where $\Delta\tau$ is the allowed percentage of slot overlap (tolerance) and τ is the time-slot width. For 10 ps pulses and 5% tolerance, fiber length would have to be kept within 100 μm. Such precision in fiber lengths would not be practical; it could however be achieved by using integrated waveguides with adjustable optical path length (integrated delays), controlled by the electro-optic effect. Since electro-optically induced refractive index changes might not be large enough to guarantee tuning over the entire frame, a combination of both technologies would be needed.

d) Thermal Drift: The treatment of both synchronization and timing jitter did not take into account the change in propagation delay Δt caused by a temperature variation ΔT, and given by

$$\Delta t = \zeta.L.\Delta T \quad (m) \tag{11}$$

where ζ is the thermal expansion coefficient and L is the length of fiber. A typical value for ζ of unjacketed fiber is approximately 40 ps/km °C[50]. If left uncorrected, this variation of the optical path length could lead to synchronization errors and timing jitter and therefore cause crosstalk. The former would arise from a difference in length between the delayed clock and the signal path whereas the latter would result from a change in length of the encoder fiber-optic delay lines. The amount of thermal stability necessary to prevent crosstalk due to thermal drift is, however, not excessive. Various technique have been devised that can easily regulate an environment to within 1°C. If the switch is placed in such an environment, so that all path lengths will be affected in a similar fashion, the change in delay associated with temperature fluctuations will not cause synchronization errors or timing jitter. In summary, keeping the size of the shared-medium switch compact would allow precise control of the drift mechanisms and ensure that the crosstalk is smaller than in integrated space division switches.

C. Speed of Reconfiguration

The maximum port speed (bps/port) of the shared-medium switch is determined by the rate at which the clock can be modulated by the digital information. This rate is presently of the order of a few tens of GHz for electro-optic modulators [51], [52]. The input port speed will determine the duration of the TDMA frame and, therefore, the number of ports given the width of the optical clock pulses.

The reconfiguration speed of the time-multiplexed shared-medium switch is ultimately limited by the time it takes to select the appropriate delay in the TD encoder. As already seen, one possible implementation of the encoder consists of N different delay lines between two commutators. The commutator could consist of a $1 \times N$ (or $N \times 1$) binary tree built from elemental 2×2 lithium niobate directional couplers with one unused input (or output). Each binary tree requires $N-1$ switches and an optical pulse traverses $2\log_2 N$ switches from the input to the output of the encoder. In this design, the number of switches that can be integrated on a substrate would limit the number of delays and therefore the number of switch ports that can be accommodated [53]. On the other hand, $1 \times N$ opto-mechanical switches, with N large, have been demonstrated; their switching speeds is, however, of the order of ms. Both these implementations are undesirable due to either the small number of delays that can be accommodated or the slow switching speed.

A variable-integer-delay line implementation for the encoder, shown in Fig. 10, allows a large number of delays with fast reconfiguration. The feed-forward structure [54], [55] consists of $\log_2 N$ delay stages and an output stage with a single output. Each delay stage consists of a single 2×2 optical switching element, a reference delay at one output, and a fixed optical delay in excess of the reference delay at the other output. The value of the fixed excess delay for the jth stage is $T/2^j$. Only one input is used in the first stage. The output stage consists of a 2×2 optical switching element with only one used output. Each optical switching element can be set in either the bar or cross state. The state of a switching element is set by the electrical control input, where a binary 0 at the control sets the 2×2 switching element in the bar state, whereas a binary 1 sets the 2×2 switching element in the cross-state.

Say, for the sake of argument, that an optical pulse at the modulator output is to be positioned in the jth time slot. The control of the TD encoder is straight forward. First, time slot j, as well as all other time slots, is represented by a $\log_2 N$-bit long binary sequence $(b_1, b_2, \cdots, b_{\log_2 N})$ where b_1

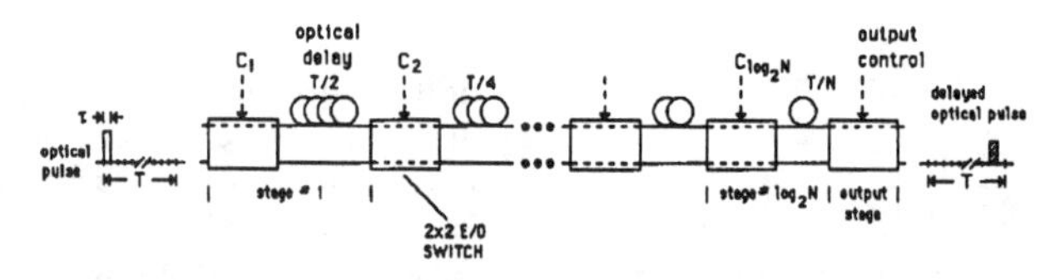

Fig. 10. Variable-integer-delay line implementation of the optical time-division encoder.

is the most significant bit. At the encoder, a control sequence $(c_1c_2, \cdots, c_{\log_2 N})$ is generated from the binary representation of time slot j. The first bit in the control sequence c_1 is equal to b_1. The second control bit c_2 equals 0 if $b_2 = b_1$; otherwise it is equal to 1. The third control bit c_3 equals 0 if $b_3 = b_2$; otherwise it is equal to 1. As a general rule, control bit c_i, with $i = 2, \cdots, \log_2 N$, equals 0 if $b_i = b_{i-1}$; otherwise it is equal to 1. Each control bit is used to set the corresponding stage of the TD encoder. Finally, the control of the output stage is set equal to 0 if the parity of the control sequence is even, or to 1 if the parity of the control sequence is odd. The output stage serves only to ensure that the delayed pulse always exits at the chosen output of the last 2 × 2 optical switching element.

The output port binary sequence may be sent, for example, in front of the information signal entering an input port of the time-multiplexed shared-medium switch. The control sequence is then used to set the state of the TD encoder. After the control sequence has set the encoder as described above, the modulated optical clock pulses that follow will be delayed by an amount $j\tau$ in excess of the reference delays, accomplishing the desired time-division encoding operation of the output port address.

V. Discussion

A. Reduction of the Collision Probability

Even though the shared-medium switch is strictly nonblocking, in fixed receiver assignment multiple access, a collision will occur if two or more signals at different input ports are to be routed to the same output port. This collision will result in at least one I/O link that cannot be established.

An $N \times N$ shared-medium switch is considered and an homogeneous model is assumed: a signal arrives at each input port independently with probability $\rho = 1$ (worst case analysis), and each incoming signal is equally likely destined to any output port.

The probability that no collision will occur is given by

$$P_{nc} = \frac{N!}{N^N} \tag{12}$$

where $N!$ is the number of incoming signals arrangements in which no two incoming signals are to be routed to the same output port, and N^N is the total number of incoming signal arrangements. The probability that a collision will occur at any output port, is given by

$$P_c = 1 - \left(\frac{N!}{N^N}\right). \tag{13}$$

As N grows large, the probability of collision also increases.

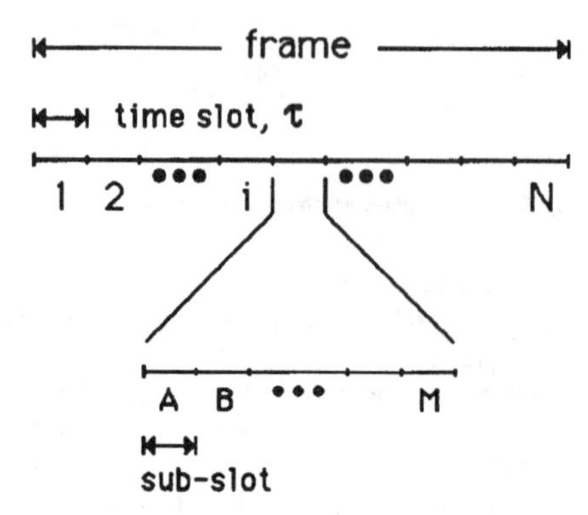

Fig. 11. Encoding of each time slot with M subslots to reduce link blocking probability.

One way to resolve this contention problem is to divide each time slot into M subslots as shown in Fig. 11, and to allocate signals for the same output port to different sub-slots within that time slot. At each output port, buffers can be used to store the different signals.

If $M = N$, then no collisions will ever occur at the expense of a large increase in required bandwidth. As will be shown below, the probability of collision is small even when $M << N$. An estimate of the probability of collision $P\{\text{collision}/N, M\}$ is now calculated for various values of N and M. Though an analytical expression for $P\{\text{collision}/N, M\}$ with $M = 2, \cdots, N-1$ is difficult to obtain, an estimate can be obtained through Monte Carlo simulations as follows. A sequence of equiprobable integers n_j is generated where j represents the input port number $(1 \leq j \leq N)$ and n represents the output port number $(1 \leq n \leq N)$. Such a sequence is generated K times.

For a given value of N, an estimate of the probability of collision is calculated as a function of M as follows. For the case $M = 1$, if $n_i = n_j$ for any i and j, then a collision occurs. For the case $M = 2$, if $n_i = n_j = n_k$ for any i, j, and k, then a collision occurs; and so on. After K iterations of generating a sequence, an estimate of the probability of collision is given by

$$P\{\text{collision}/N, M\} = (\text{number of times a collision has occured})/K \tag{14}$$

for each $M, 1 \leq M \leq N$.

The probability of collision as a function of M is plotted in Fig. 12 for $N =$ 10, 100, and 1000. The number of iterations used was $K = 10^5$. It is seen from Fig. 12 that $M =$7, 9,

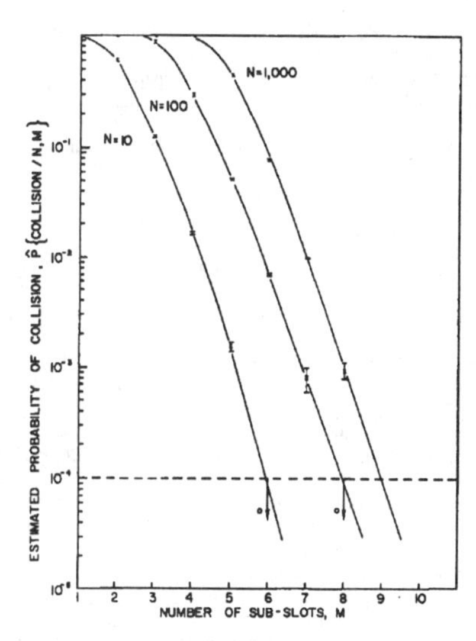

Fig. 12. Estimated probability of collision as a function of the number of sub-slots for N = 10, 100, and 1000.

and 10 subslots are sufficient to ensure a 10^{-4} probability of collision for N = 10, 100, and 1000, respectively.

This procedure was carried out several times. The range of values obtained for each probability of collision is represented on the figure by error bars.

This scheme to reduce the collision probability is not restricted to FRA/TDMA. It can easily be adapted to CDMA and WDMA by associating several code sequences and wavelengths to each output port, respectively.

If a contention protocol is used, some (at least M) I/O connections will be successful even in the presence of collisions. This means that the probability of losing data is lower than the probability of collisions. The analysis to determine this probability was carried out for the packet-switching architecture referred to as the Knockout Switch by Yeh *et al.* [56]. Though the shared-medium switch topology is quite different from that of the Knockout Switch, the way blocking is handled in both topologies can be seen as analogous, and therefore the same analysis holds. In the Knockout Switch, as in the shared-medium switch, all input signals are broadcast to all output ports. For the Knockout Switch, however, this broadcast is achieved via a fully interconnected mesh of buses. Again, congestion will occur if multiple packets simultaneously arriving at different input ports are destined for the same output port. At each output port, a bus interface filters out the packets not intended for that output. The packets which are passed then contend for a set of L buffers through a knockout tournament, referred to as the knockout contention scheme. These L buffers can be regarded as the equivalent of the M subslots in the shared-medium switch. If more than L packets simultaneously arrive at a particular output, the remaining packets are simply discarded. They have shown that the probability of losing packets in this manner is extremely small: for a load $\rho = 1$ and $N \to \infty$, L (or M in the case of the shared-medium switch) = 8 guarantees a probability of lost packets of 10^{-6}.

B. Extensions and Applications of the Shared-Medium Photonic Switch

Though in the present experiment each time slot represents the output port address of a bit, time slots can also be assigned in a variety of other ways. For example, in fixed-transmitter assignment TDMA, each time slot represents an input port address. FTA/TDMA is suitable for broadcasting. Here, each input port NIU has a unique fixed delay which positions the pulse in an assigned time slot. Each output port NIU includes a slot selector. To switch to a desired input port, the output port NIU selects the appropriate delay. Several or all output port NIU's can select the same input port simultaneously, establishing either multicast or broadcast connections.

As mentioned in Section II-A, the multiple-access technique used to encode the port address in not restricted to TDMA. Other encoding schemes, such as code-division (CDMA), wavelength-division (WDMA) or frequency-division multiple access (FDMA) can also be used. Here again, either FTA or FRA can be selected.

Though we limited our discussion to an $N \times N$ switch, an architecture with dimension $N \times M$, with $M > N$, can also be achieved.

Also, a variety of topologies (e.g., ring, bus, star) can be chosen for the shared-medium switch architecture. For example, a ring topology could be chosen to link transmitting and receiving sections of the switch. Or, a star topology could be used to connect all input ports to output ports while the optical clock is distributed to all port transmitters and receivers through a bus topology. The passive optical star configuration was chosen because it gives the largest switch dimension for a given power budget.

The input and output signals need not be restricted to the electrical domain. The shared-medium switch architecture also applies if the transmitted data are in optical format. Two modifications are required: opto-optic modulators are required in the input port NIU's and optical discriminators are required in the output port NIU's. Optical discriminators have recently been demonstrated [57]; optically controlled modulators are currently under investigation.

As mentioned in Section II-Bf), the reconfiguration speed of the shared-medium switch is limited by the time it takes to tune the encoder. This restriction could limit the utility of the switch in some applications. The shared-medium switch would be particularly well suited as the $N \times N$ space division switch in a Digital Access Cross-connect System (DACS) [58]. Indeed, with the deployment of high-speed optical transmission in long-haul networks, we have seen that the speed and size requirements of the switches are becoming more and more severe. As mentioned in the introduction, though high-speed electronics switches could fulfill the current requirements, the DACS application is one case where completing the optical path through the switch might be beneficial.

Each of the k input lines to the DACS carries, say, m time-multiplexed channels (or time-slots). The data in each slot in successive frames belong to the same session or message. In the DACS, these m channels are first demultiplexed, then switched to the proper output port of the $N \times N$ space-division switch (where $N = mk$), before being multiplexed onto k output lines. Once a session between an input and output line is established, it usually remains connected for a long period of time, so that the rate at which the $N \times N$ space-division switch must be reconfigured is quite slow. If the shared-medium switch is used as an $N \times N$ space-division switch, when a session is terminated, a new connection can be established by tuning the encoder (in time, code-sequence, wavelength, or frequency).

In time-multiplexed circuit-switched applications, the interconnection pattern of the space-division switch needs to be changed on a slot-by-slot basis. For high slot rates, this reconfiguration speed would be difficult to achieve with most space-division switch technologies, including the shared-medium switch, where the encoder would need to be tuned at the slot rate.

In packet-switched applications, the space-division switch must be reconfigured on a packet-by-packet basis. Since the length of a packet is usually much longer than the duration of a time slot, the reconfiguration speed of the space-division switch would not need to be high.

VI. Conclusion

A photonic space-division switch, which resembles a collapsed network, is presented. Since this "shared-medium switch" does not rely on the use of distinct physical paths between input and output ports, the limitations found in integrated-optic arrays are avoided and very large configurations are possible. The demonstration of a 120×120 time-multiplexed shared-medium switch, with an aggregate throughput of 12 Gbit/s, is reported. The feasibility of scaling this architecture to larger dimensions is discussed, including the power-budget, high-speed time-division demultiplexing, crosstalk, synchronization, timing jitter and thermal drift. The design of a high speed time-division encoder is presented, and is demonstrated elsewhere [55]. A method for reducing the probability of collision, using subslot assignment to the input ports, is described and analyzed.

Many issues remain to be solved before the shared-medium switch architecture can be implemented in practical systems. Some of these issues are the choice of a control algorithm to tune the encoder, interfacing this control algorithm with the switching fabric, synchronization of the incoming signals, and implementation of a contention resolution protocol, such as the subslot assignment described above. When these issues have been addressed, we believe that a 1000×1000 switch, with an aggregate throughput of 1 Tbit/s, will be possible.

References

[1] P. W. Smith, "On the role of photonic switching the future communications systems," in *Proc. IEEE Int. Conf. Commun.*, Seattle, WA, June 7–10, 1987, vol. 3, paper 45.4, pp. 1570–1574.

[2] G. A. Hayward, A. M. Gottlieb, D. G. Boyer, and J. E. Berthold, "High-speed 16 × 16 CMOS crosspoint switch," *Electron. Lett.*, vol. 21, no. 20, pp. 923–925, 1985.

[3] L. Ronn and P. Deichmann, "16 × 8 space switch for 2.24 Gbit/s transmission," *Electron. Lett.*, vol. 23, no. 5, pp. 236–237, 1987.

[4] J. E. Midwinter, "Photonic switching components: Current status and future possibilities," *Topical Meet. Photon. Switching*, Incline Village, NV, Mar. 18–20, 1987, *Tech. Digest*, vol. 13, paper WB1, pp. 8–10.

[5] H. S. Hinton, "Applications of the photonic switching technology for telecommunications switching," in *Proc. IEEE Int. Conf. Commun.*, Seattle, WA, June 7–10, 1987, vol. 13, paper 45.2, pp. 1559–1564.

[6] B. Saleh and M. Teich, *Fundamentals of Photonics.* New York: Wiley, ch. 21, 1991.

[7] M. Schwartz, *Telecommunication Networks: Protocols, Modeling and Analysis.* Reading, MA: Addison-Wesley, 1987.

[8] S. Suzuki, T. Terakado, K. Komatsu, K. Nagashima, A. Suzuki, and M. Kondo, "An experiment on high-speed optical time-division switching," *J. Lightwave Technol.*, vol. LT-4, pp. 894–899, July 1986.

[9] R. A. Thompson and P. P. Giordano, "An experimental photonic time-slot interchanger using optical fibers as reentrant delay-line memories," *J. Lightwave Technol.*, vol. LT-5, pp. 154–162, Jan. 1987.

[10] S. Suzuki and K. Nagashima, "Optical broadband communications network architecture utilizing wavelength-division switching technologies," *Topical Meeting Photon. Switching*, Incline Village, NV, Mar. 18–20, 1987, *Tech. Dig.*, vol. 13, paper ThA2, pp. 21–23.

[11] S. F. Su, L. Jou, and J. Lenart, "A review on classification of optical switching systems," *IEEE Commun. Mag.*, vol. 24, pp. 50–55, May 1986.

[12] A. Selvarajan and J. E. Midwinter, "Photonic switches and switch arrays on $LiNbO_3$," *Opt. Quantum Electron.*, vol. 21, no. 1, pp. 1–15, 1989.

[13] F. J. Leonberger, "Progress in Ti:$LiNbO_3$ and InP waveguide devices for signal-processing applications," *4th Int. Conf. Integrat. Opt. Optical Fiber Commun.*, Tokyo, Japan, June 27–30, 1983, *Tech. Dig.*, paper 30B2-1, pp. 240–241.

[14] T.-Y. Feng, "A survey of interconnection networks," *Comput.*, vol. 14, no. 12, pp. 12–27, 1981.

[15] R. C. Alferness, "Recent advances in integrated optics," *Conf. Lasers Electro-Optics*, Baltimore, MA, Apr. 26,–May 1, 1987, paper WQ1, pp. 198–200.

[16] H. Inoue, H. Nakamura, Y. Sasaki, T. Katsuyama, N. Chinone, and K. Morosawa, "Eight-millimeter long nonblocking 4 × 4 optical switch array," *Conf. Lasers Electro-Optics*, Anaheim, CA, Apr. 25–29, 1988, paper TuK1, pp. 72–74.

[17] P. W. Smith, "On the physical limits of digital optical switching and logic elements," *Bell Syst. Tech. J.*, vol. 61, no. 8, pp. 1975–1993, Oct. 1982.

[18] L. D. Hutcheson, "Integrated optics: Evolution and prospects," *Optics News*, vol. 14, no. 2, pp. 7–9, 1988.

[19] P. Granestrand and L. Thylen, "Strictly nonblocking 8 × 8 integrated-optic switch matrix in Ti:$LiNbO_3$," *Topical Meeting Integrat. Guided-Wave Opt.*, Atlanta, GA, Feb. 26–28, 1986, *Tech. Dig.*, Paper WAA3, pp. 4–6.

[20] H. Heidrich and D. Hoffmann, "Review on integrated-optics switch matrices in $LiNbO_3$," *Trans. Inst. Electron., Inform. Commun. Eng.*, vol. E 73, no. 1, pp. 94–98, 1990.

[21] K. Habara and K. Kikuchi, "Optical time-division space switches using tree-structured directional couplers," *Electron. Lett.*, vol. 21, no. 14, pp. 631–632, 1985.

[22] P. J. Duthie and M. J. Wale, "Rearrangeably nonblocking 8 × 8 guided wave optical switch," *Electron. Lett.*, vol. 24, no. 10, pp. 594–596, 1988.

[23] M. A. Milbrodt, J. J. Veselka, K. Bahadori, Y. C. Chen, G. A. Bogert, D. G. Coult, R. J. Holmes, J. R. Erickson, and W. A. Payne, "A tree-structured 4 × 4 switch array in lithium niobate with attached fibers and proton-exchange polarizers," *Topical Meeting Integrat. Guided-Wave Opt.*, Santa Fe, NM, Mar. 28–30, 1988, *Tech. Dig.*, paper MF9.

[24] J. E. Watson, M. A. Milbrodt, K. Bahadori, M. F. Dautartas, C. T. Kemmerer, D. T. Moser, A. W. Schelling, T. O. Murphy, J. J. Veselka, and D. A. Herr, "A low-voltage 8 × 8 Ti:$LiNbO_3$ switch with a dilated-Benes architecture," *J. Lightwave Technol.*, vol. 8, pp. 794–801, May 1990.

[25] R. A. Spanke, "Architecture for large nonblocking optical space switches," *IEEE J. Quantum Electron.*, vol. QE-22, pp. 964–967, June 1986.

[26] R. A. Spanke and V. E. Benes, "N-stage planar optical permutation network," *Appl. Opt.*, vol. 26, no. 7, pp. 1226–1229, 1987.

[27] K. Padmanabhan and A. Netravali, "Dilated networks for photonic switching," *IEEE Trans. Commun.*, vol. COM-35, pp. 1357–1365, Dec. 1987.

[28] A. D. Fisher and J. N. Lee, "The current status of two-dimensional spatial light modulator technology," in *SPIE Vol. 634: Optic. Hybrid Comput.*, Leesburg, VA, Mar. 24–27, 1986, pp. 352–371.

[29] A. A. Sawchuck, B. K. Jenkins, C. S. Raghavendra, and A. Varma, "Optical crossbar networks," *Comput.*, vol. 20, no. 10, pp. 50–60, 1987.

[30] A. Sawchuck and B. K. Jenkins, "Dynamic optical interconnections for parallel processors," in *Proc. SPIE Vol. 625: Optic. Comput.*, Los

Angeles, CA, Jan. 23–24, 1986, pp. 143–153.

[31] H. Laor, "Optical switching using piezoelectric-driven beams for fiber to fiber coupling," *6th Topical Meeting Opt. Fiber Commun.*,, New Orleans, LA, Feb. 28–Mar. 2, 1983, *Dig. Tech. Papers*, paper TuG8, pp. 52–54.

[32] P. R. Prucnal and P. A. Perrier,"Passive shared-media systems," in *Proc. IEEE Int. Conf. Commun.*, Atlanta, GA, Apr. 16–19, 1990, paper 328.3, pp. 1135–1139; "A new direction in photonic switching: A collapsed network space-division switching architecture," in *OSA Proc. Photon. Switching*, (Optical Society of America, Washington, DC, 1989), vol. 3, pp. 200–207; also, "A new direction in photonic switching," in *Proc. 9th Int. Conf. Comput. Commun.*, Tel Aviv, Israel, Oct. 30–Nov. 4, 1988, pp. 149–154; "Demonstration of a high-dimensionality local area network photonic switch," in *14th European Conf. Opt. Commun.*, Brighton, England, Sept. 11–15, 1988; also in *Cof. Publ.*, part 1, pp. 264–267.

[33] P. R. Prucnal, M. A. Santoro, and T. R. Fan, "Spread spectrum fiber-optic local area network using optical processing," *IEEE J. Lightwave Technol.*, vol. LT-4, pp. 547–554, May 1986.

[34] J. A. Salehi, "Emerging optical code-division multiple access communications systems," *IEEE Network*, vol. 3, pp. 31–39, Mar. 1989.

[35] P. R. Prucnal, M. A. Santoro, and S. K. Seghal, "Ultrafast all-optical synchronous multiple access fiber networks," *IEEE J. Select. Areas Commun.*, vol. SAC-4, pp. 1484–1493, Dec. 1986.

[36] P. R. Prucnal, M. A. Santoro, S. K. Sehgal, and I. P. Kaminow, "TDMA fibre-optic network with optical processing," *Electron. Lett.*, vol. 22, no. 23, pp. 1218–1219, 1986.

[37] P. R. Prucnal, D. J. Blumenthal, and M. A. Santoro, "12.5 Gbit/s fibre-optic network using all-optical processing," *Electron. Lett.*, vol. 23, no. 12, pp. 629–630, 1987.

[38] E. Arthurs, M. S. Goodman, H. Kobrinski, and M. P. Vecchi, "A broad-band optoelectronic packet switching system," *IEEE Trans. Commun.*, vol. 37, pp. 645–648, June 1989.

[39] M. S. Goodman, "Multiwavelength networks and new approaches to packet switching," *IEEE Commun. Mag.*, vol. 27, pp. 27–35, 1989.

[40] I. P. Kaminow, "Non-coherent photonic frequency-multiplexed access networks," *IEEE Network Mag.*, vol. 3, pp. 4–12, 1989.

[41] R. A. Linke, "Frequency division multiplexed optical network using heterodyne detection," *IEEE Network, Mag.*, vol. 3, pp. 13–20, 1989.

[42] W. Sharfin and M. Dagenais, "The role of nonlinear diode laser amplifiers in optical processors and interconnects," *Optic. Quantum Electron.*, vol. 19, pp. 547–566, 1987.

[43] BTD Catalog, BT&D Technologies, Wilmington, DE.

[44] E. Desurvire, B. Tell, I. P. Kaminow, G. J. Qua, K. F. Brown-Goebeler, B. I. Miller, and U. Koren, "High contrast GaInAs:Fe photoconductive optical AND gate for time-division demultiplexing," *Electron. Lett.*, vol 24, no. 7, pp. 396–397, 1988.

[45] K. K. Goel, P. R. Prucnal, Y. Shimazu, M. Milbrodt, E. Desurvire, and B. Tell, "Demonstration of packet switching through an integrated-optic tree switch using photoconductive logic gates," *Electron. Lett.*, vol. 26, no. 5, pp. 287–288, 1990.

[46] C. Dragone, C. H. Henry, I. P. Kaminow, and R. C. Kistler, "Efficient multichannel integrated optics star coupler on silicon," *IEEE Photon. Technol. Lett.*, vol. 1, pp. 241–243, Aug. 1989.

[47] P. A. Perrier, "Optical processing for self-clocking and self-routing in photonic switching networks," Ph.D. dissertation, Columbia Univ., New York, 1989.

[48] S. E. Miller and I. P. Kaminow, *Optical Fiber Telecommunications II*. New York: Academic, 1988.

[49] CCITT Recommendation G.652

[50] L. Cohen and J. Fleming, "Effects of temperature on transmission in lightguides," *Bell Syst. Tech. J.*, vol. 58, no. 4, pp. 945–951, 1979.

[51] S. K. Korotky, G. Eisenstein, R. S. Tucker, J. J. Veselka, and G. Raybon, "Optical intensity modulation to 40 GHz using a waveguide electro-optic switch," *Appl. Phys. Letter*, vol. 50, no. 23, pp. 1631–1633, 1987.

[52] D. W. Dolfi, M. Nazarathy, and R. L. Jungerman, "40 GHz electro-optic modulator with 7.5 V drive voltage," *Electron. Lett.*, vol. 24, no. 9, pp. 528–529, 1988.

[53] J. E. Watson, M. A. Milbrodt, and T. C. Rice, "A polarization-independent 1 × 16 guided-wave optical switch integrated on lithium niobate," *J. Lightwave Technol.*, vol. LT-4, pp. 1717–1721, Nov. 1986.

[54] R. A. Thompson, "Optimizing photonic variable-integer-delay circuits," in *Topical Meeting Photon. Switch.*, Incline Village, NV, Mar. 18–20, 1987, *Tech. Digest*, vol. 13, paper FD4, pp. 241–243.

[55] P. R. Prucnal, M. F. Krol, and J. L. Stacy, "Demonstration of a rapidly tunable optical time-division multiple-access coder," *IEEE Photon. Technol. Lett.*, vol. 3, pp. 170–172.

[56] Y.-S. Yeh, M. G. Hluchyi, and A. S. Acampora, "The knockout switch: A simple, modular architecture for high-performance packet switching," *IEEE J. Select. Areas Commun.*, vol. SAC-5, pp. 1274–1283, Oct. 1987.

[57] C. R. Giles, T. Li, T. H. Wood, C. A. Burrus, and D. A. B. Miller, "All-optical regenerator," *Electron. Lett.*, vol. 24, no. 14, pp. 848–850, 1988.

[58] J. A. Gerrish and D. A. Morrison, "The DACS door to the subrate world," *Bell Lab. Rec.*, pp. 12–19, Jan. 1986.

Chapter 7

TDM Optical Switch Architecture

Most photonic ATM switches proposed thus far have been wavelength-division asynchronous transfer mode (ATM) switches whose wavelengths are used for cell routing. However, these wavelength systems require sensitive wavelength tuners and complicated electrical controllers to avoid cell contention. In order to solve this problem a number of researchers have concentrated their efforts on the use of time-division multiplexing (TDM) in the design of photonic switch architectures. The results of some of these efforts are reported in this chapter. High-speed optical memories and write/read gates are necessary for the construction of the optical time-division switch. An optical fiber delay line can be used as the optical memory element. Furthermore, several kinds of optical bistable memory devices have been developed. An electrooptic switch and laser diode switch can be used as high-speed optical write/read gates. The first three papers (papers 7.1, 7.2, and 7.3) are concerned with ATM optical switch architectures. The rest of the papers discuss various other time domain switch architectures.

The paper by Masetti et al. (paper 7.1) describes the work carried out within the Research and Development in Advanced Communications in Europe (RACE) project R2039 ATMOS (asynchronous transfer mode optical switching). Four different switch architectures have been proposed, investigated, and developed, all based on a high-speed optical routing matrix electrically controlled at lower speed. The basic optical key components and subsystems (wavelength converters, space switches, and optical buffers) are described in detail. In particular, system demonstration of wavelength conversion at 10 and 20 Gb/s has been realized and its performance has been reported.

The discussion on ATM photonic switch architecture is extended by Tsukada and Nakano (paper 7.2) who describe an ultrafast photonic ATM (ULPHA) switch. The switch is an output buffer-type ATM switch based on a time-division broadcast-and-select network. The hardware description focuses on the cell coder for generating ultrafast optical cells, the cell buffer with cell selection function, and the cell decoder for detecting ultrafast optical cells. Experimental results are reported at 25 Gb/s for a video distribution switching experiment.

The implementation of optical packet-switched networks requires that the problems of resource contention, signaling, and local and global synchronization be resolved. A technique for resolving potential resource contention in such networks is reported by Chlamtac et al. (paper 7.3). The contention resolution by delay lines (CORD) methodology exploits optical delay lines and switching matrices. Signaling is dealt with using subcarrier multiplexing of packet headers. Synchronization takes advantage of clock tone multiplexing techniques, and of digital processing for global packet-slot alignment. This work was carried out by the Advanced Research Project Agency (ARPA)-sponsored CORD Consortium which has been formed to explore the practical and reliable feasibility and effectiveness of these key optical networking techniques.

The discussion on switching in the time domain is continued by Qiao et al. (paper 7.4) who propose a time domain approach for avoiding crosstalk in optical multistage interconnection networks. Such an approach can be regarded as "dilating" a network in time, instead of space. Here, the connections that need to use the same switch are established during different time slots, so that path conflicts are automatically avoided. The time domain dilation is useful for overcoming the limits on the network size while utilizing the high bandwidth of optical interconnects. While the space domain approach trades hardware complexity for crosstalk-free communications, the time domain approach trades off time complexity. A comparison of the proposed time domain to the space domain approach is carried out in this paper by analyzing the trade-offs involved in both approaches.

Tree-type optical switch architectures play an important role in the design of photonic switching networks. Jajszczyk and Mouftah discuss in paper 7.5 all known tree-type architectures in a unified framework, and propose a number of new solutions. The discussed networks can be implemented with guided-wave-based switching elements, or laser diodes and passive splitters and combiners. Techniques for improving signal-to-noise ratio as well as waveguide crossover minimization are presented. Implementation issues on single and multiple substrates are discussed and the various characteristics of tree-type networks are compared.

High Speed, High Capacity ATM Optical Switches for Future Telecommunication Transport Networks

F. Masetti, *Member, IEEE,* J. Benoit, *Member, IEEE,* F. Brillouet, J. M. Gabriagues, A. Jourdan, M. Renaud, D. Böttle, G. Eilenberger, K. Wünstel, M. Schilling, *Member, IEEE,* D. Chiaroni, P. Gavignet, J. B. Jacob, G. Bendelli, *Member, IEEE,* P. Cinato, P. Gambini, *Member, IEEE,* M. Puleo, *Member, IEEE,* T. Martinson, P. Vogel, T. Durhuus, C. Joergensen, K. Stubkjaer, *Member, IEEE,* R. Baets, *Member, IEEE,* P. Van Daele, *Member, IEEE,* J. C. Bouley, R. Lefèvre, M. Bachmann, W. Hunziker, *Member, IEEE,* H. Melchior, *Fellow, IEEE,* A. McGuire, F. Ratovelomanana, and N. Vodjdani, *Member, IEEE*

(Invited Paper)

Abstract—**This paper describes the work carried out in the RACE Project R2039 ATMOS (asynchronous transfer mode optical switching). The project is briefly illustrated, together with its main goal: to develop and assess concepts and technology suitable for optical fast packet switching. The project's technical approach consisted in the exploitation of the space and wavelength domains for fast routing and buffering: The major achievements are then reported. Four different switch architecture concepts have been proposed, investigated and developed, all based on a high speed optical routing matrix electrically controlled at lower speed. The basic optical key components and subsystems (wavelength converters, space switches and optical buffers) are described in detail, with the outstanding results obtained and the corresponding projected performance. In particular, system demonstration of wavelength conversion at 10 and 20 Gb/s has been realized, to show the usefulness of the ATMOS technology both to implement optimized high performance optical packet-switching fabrics as well as transparent optical circuit-routing nodes. Four rack-mounted, reduced size demonstrators of basic switching matrices have been designed and implemented scalable to real system sizes. The obtained good results in terms of bit error rate and hardware integration are reported, showing that ATM switches are feasible with state-of-the-art optical technology.**

I. Introduction

THE GROWTH of existing and new services will create a large increase of traffic flow in telecommunication networks in the coming years [1]. The range of future services will be very diverse in terms of required channel capacity (i.e., bit rate), channel occupancy (continuous or bursty), connection duration, connection set-up time and frequency. In metropolitan and local area networks, the growth of services such as high-speed computer file transfers, interactive computer data-exchange, multimedia connections, requires high capacity networks and nodes capable of dynamically sharing the bandwidth between different types of users. In this heterogeneous environment of current telecommunication networks, routing and multiplexing are performed electrically, optics being confined to transmission, since the transmission bandwidth offered by optical fibers is virtually unlimited. However, future broadband transport networks should reach beyond the possibilities of today's electronic switch implementations, and photonic technologies will play an essential role also in routing and switching.

The research and development in advanced communications in Europe (RACE) program has its major goal in realizing the implementation of integrated broadband communications (IBC) networks all over Europe. Fed by input links with bit rates ranging from 155 Mb/s to 2.5 Gb/s [2], future IBC nodes will require switching capacities of several Tb/s. Since the switch bandwidth and speed could become a network bottleneck, the introduction of photonics appears to be an attractive way to ensure high performance by increasing the node throughput, speed and flexibility. As far as flexibility is concerned, it is likely that the network operation will rely on the asynchronous transfer mode (ATM), a key multiplexing and routing technique for future IBC networks supporting various services at different bit rates and adopting standard format fixed-length cells [3].

The feasibility of optical fast packet switching and routing has been demonstrated by the RACE Project R2039 ATMOS, a consortium started in January 1992 and composed by 11 partners. The main objective of ATMOS was to design, fabricate and assess under quasi real traffic conditions, new fast packet switching and routing nodes fulfilling the mentioned speed ($\geq$2.5Gb/s) and capacity (Tb/s) requirements for future ATM-based IBC transport networks.

Manuscript received April 6, 1995; revised August 26, 1995. This work was supported by European Commission and the RACE Program.

F. Masetti, J. Benoit, F. Brillouet, J. M. Gabriagues, A. Jourdan, M. Renaud, and M. Bachmann are with Alcatel Alsthom Recherche, Département Fibers et Systemes Photoniques, 91460 Marcoussis, France.

D. Böttle, G. Eilenberger, K. Wünstel, and M. Schilling are with Alcatel SEL, Stuttgart, Germany.

D. Chiaroni, P. Gavignet, and J. B. Jacob are with Alcatel CIT, Villarceaux, France.

G. Bendelli, P. Cinato, P. Gambini, and M. Puleo are with Centro Studi E Laboratori Telecommunicazioni, Turin, Italy.

T. Martinson and P. Vogel are with ASCOM, Bern, Switzerland.

T. Durhuus, C. Joergensen, and K. Stubkjaer are with the Technical University of Denmark, Lingby, Denmark.

R. Baets and P. Van Daele are with IMEC—University of Gent, Belgium.

J. C. Bouley and R. Lefèvre are with Centre National d'Etudes des Telecommunications, Bagneux, France.

W. Hunziker and H. Melchoir are with ETH, Zürich, Switzerland.

A. McGuire is with BT Laboratories, Ipswich, U.K.

F. Ratovelomanana and N. Vodjdani are with Thomson CSF, Orsay, France.

Publisher Item Identifier S 0733-8716(96)03684-0.

Reprinted with permission from *IEEE Journal on Selected Areas in Communications,* Vol. 14, No. 5, pp. 979-998, June 1996.

To reach this objective, ATMOS has developed the required advanced optical technology, with a very important effort on the design and manufacturing of new key components adapted to low loss, polarization insensitive switching matrices. In particular, the component reconfiguration time has to be short enough to be compatible with the severe constraints imposed in a high speed ATM environment.

The originality and the strength of the ATMOS consortium has been twofold.

Very close contacts and relationships between system and component activities have been established, to bring and improve the technology from basic concepts to high performance components according to the aimed system specifications: a very good example of this approach is the all-optical wavelength converter, a key device described later.

The developed components and functional sub-blocks, combined differently to show their flexibility, have then been implemented within different optical switching demonstrators, all exploiting space and/or wavelength switching for routing and multiplexing, time and/or wavelength switching for contention resolution. Four different system concepts have been proposed in ATMOS for the optical switching matrix internal architecture [4]–[7], three targeting public IBC networks and the fourth developed for private networks. Electronics has been used for complex functionalities, especially for control, where a lower speed is not a limiting factor. The demonstrators, scalable to real system dimensions, were shown in public exhibitions in 1993 and 1994.

The main objectives have almost been achieved in the first three years, and the project continued a reduced effort activity in 1995 to improve the technology, to provide a stronger basis for the longer term objective of optimized switching systems. However, the outstanding results obtained show that optical switching is already feasible with state-of-the-art technology and could lead to practical applications in the very near future.

This paper is organized as follows. Section II reports the different system concepts proposed and developed, describing in detail their functionalities. In Section III, the ATMOS reference models for the matrix and node are illustrated. Section IV discusses the key components and technology developments, and in Section V, the switching demonstrators are reported, with their characteristics and the obtained performance. Conclusions are drawn in Section VI.

II. Node System Concepts

A broadband switching node consists of a set of link interfaces that perform signal regeneration and clock synchronization, and the switching fabric itself, which is responsible for the routing functions. Photonic implementation of switching matrices, to be used as building blocks for large switching fabrics, is now being investigated widely in the world.

The ATMOS system concepts proposed for the switching matrix were based on realistic analysis of the photonic technology potential capabilities and performance. In particular, on one hand it appears feasible to implement high speed matrices with state-of-the-art optical technology, to increase performance simultaneously removing critical limits of large electronic systems, such as problems caused by dense interconnections, electro-magnetic interference and power consumption. On the other hand, purely high speed photonic switching is today limited by the difficulty of synchronizing, decoding and buffering cells directly in the photonic domain. A limited cell header processing and decoding can be provided, but optical memories capable of large data storage for ATM switching, have not yet been realized. Yet, even considering a small switching fabric (e.g., 16 inlets/outlets) fed by uniform input traffic, a storage capacity of at least 10–15 cells per queue is required, in order to ensure an acceptable cell loss probability (e.g., 10^{-9}) in heavy load operating conditions (e.g., from 0.5 Erlang for output queuing up to 0.9 for completely shared buffering [8]).

Four different approaches for ATM buffering (cell contention resolution) have been proposed, investigated, and developed in ATMOS, each framed into a different system concept for the switching matrix. The common point of these matrices, briefly described hereafter, is that they are all composed of an all-optical high-speed routing network electrically controlled. The matrix functional sub-blocks have been developed assuming that the incoming cells arrive in phase at the N inlets, with their associated n-bit long tags already added. This can be accomplished by equipping the switching fabric with electrical interfaces in a first step, all-optical interfaces being a longer term goal. The matrix main design parameters are the number of inputs n and the buffer capacity per output m.

A. Fiber Delay Line Switch Concept

The fiber delay line switch concept [4] relies on dynamic wavelength encoding and multiplexing for fast switching of ATM cells: signals are routed toward their destination on the basis of their assigned wavelength. In addition, contention between cells with the same destination is solved by means of calibrated optical fiber delay lines, adjusted to multiples of the cell duration. The n inputs, n outputs switching matrix, shown in Fig. 1, consists of four functional blocks: the cell-encoder block, incorporating n fast tunable optical wavelength converters that assign to each cell the wavelength corresponding to its target output; the time-switching and buffering block, consisting of $n \times m$ fast optical gates, providing to each wavelength converter access to all the p wavelength division multiplexed (WDM)-operated fiber delay lines: these fiber delay lines have increasing lengths, from $0 \times T$ up to $(m-1) \times T$, where T is the cell period; the wavelength demultiplexer block, including n bandpass filters, each tuned to a different wavelength, thus, defining the output address; the electronic control block, implementing the switching algorithm: on the basis of the internal tags attached to the incoming cells, this block drives the wavelength converters and manages with a first-in first-out (FIFO) discipline the n queues of the buffer by activating appropriately the optical gates providing access to the fiber delay lines, so as to realize an output queuing buffer. From the traffic performance viewpoint, an $n = 16$ matrix needs $m = 16$ to achieve a throughput of 0.4 Erlang under uniform traffic patterns satisfying the ATM requirements.

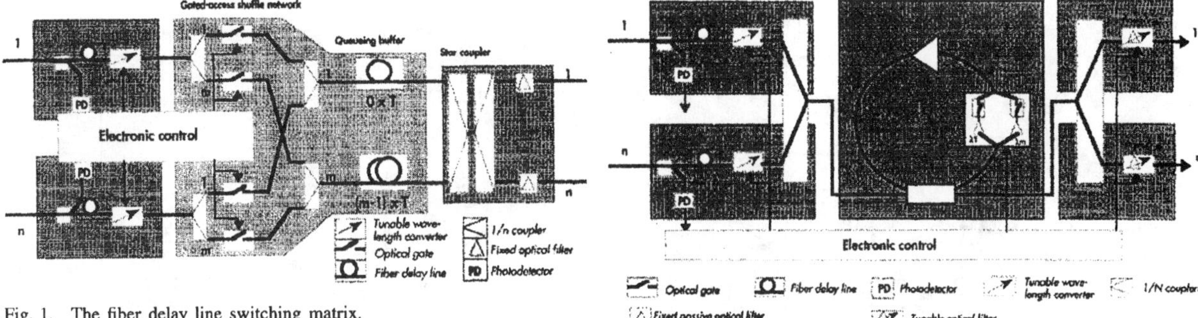

Fig. 1. The fiber delay line switching matrix.

Fig. 2. The fiber loop memory switching matrix.

B. Fiber Loop Memory Switch Concept

The fiber loop memory switch concept [5] is based on a fiber loop memory operating in a WDM regime. The multistage switching network is based on a basic matrix, shown in Fig. 2, characterized by n input and n output ports. The fiber loop length is equal to one cell period, while the capacity of the memory corresponds to a set of m optical wavelengths (memory positions) to which the input cells are converted before being set into the loop. The operation is synchronous on a cell period basis; in each matrix, in case of contention, the input cells are converted to one available wavelength in the loop and are kept circulating (stored) in the WDM loop memory by activating the corresponding passive fixed filter (i.e., by tuning on the related semiconductor optical amplifier gate). At the input of the memory loop, half of the power enters the loop, and half goes toward the outputs through the passive coupler, so that when the contention is solved, the cell is routed to the destination link simply by properly tuning the corresponding output tunable filter. At the same time, the passive filter in the loop is turned off to erase the cell in the memory. The electronic control, on the basis of the internal tags attached to the incoming cells, manages the m buffer positions so as to realize a completely shared buffer. This buffering system allows the simultaneous reading of several cells from the loop, thus, overcoming the problem of head-of-line blocking and of multiple accesses of the equivalent electronic buffering solutions. From the traffic performance viewpoint, an $n = 16$ matrix, with partially shared buffer architecture (four inputs per buffer) needs $m = 14$ in each loop to achieve a throughput of 0.4 Erlang under uniform traffic patterns satisfying the ATM requirements.

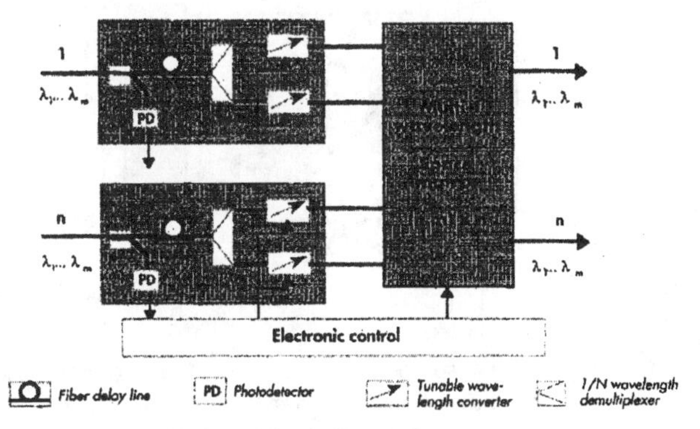

Fig. 3. The multidimensional switching matrix

C. Multidimensional Switch Concept

The multidimensional switch concept [6] aims at exploiting the wavelength domain for contention resolution to minimize the amount of optical buffers required in a multistage switch network arrangement. The basic matrix, shown in Fig. 3, switches the incoming cells by exploiting the wavelength (with wavelength converters) and space (with space switches) domains: on the basis of the internal tags attached to the incoming cells, the electronic control drives the wavelength converters and the multiwavelength space matrix. In case of contention, the cells competing at the same time for the same output of a switching element are converted to m free additional wavelengths and transmitted at the same time slot instead of being delayed in a buffer and then transmitted consecutively: this is achieved by activating appropriately the additional optical wavelength converters on each inlet. Consequently, the number of internal wavelengths of a switching element must be extended, being the price to pay for avoiding the use of the time domain. From the traffic performance viewpoint, an $n = 16$ matrix needs $m = 5$ additional wavelengths per input (with routing to a group of four outputs) to achieve a throughput of 0.8 Erlang under uniform traffic patterns satisfying the ATM requirements, contention being shifted to downstream stages. In this concept, buffers can be eliminated except at the last stage of the switch, where optical or optoelectronic buffers have to be added to perform also the concentration function on each output.

D. Corporate Optical ATM Network Concept

The corporate optical ATM network (COATN) concept [7], shown in Fig. 4, aims at realizing a high speed optical ATM backbone with straightforward interfaces to the long distance public network. In the basic architecture, suitable for both access and private networks, ATM electronic switches are interconnected with an optical double fiber ring through dedicated optical access nodes. Optical access nodes are based on elementary optical space switches that perform the following functions: transfer of the ATM stream between the fiber bus and a specific output; coarse and fine synchronization by means of opto-electronic correlation; stream demultiplexing

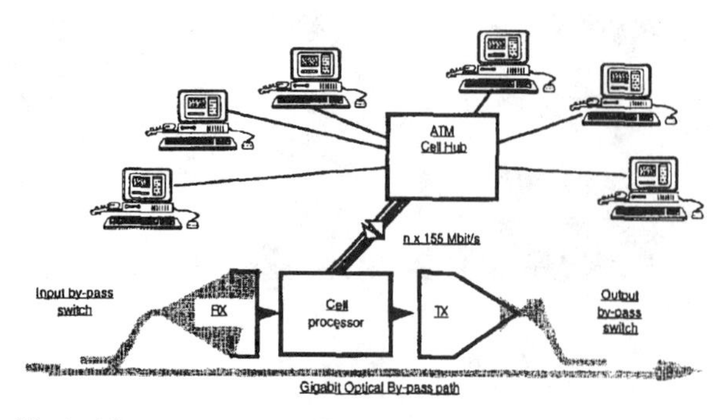

Fig. 4. The corporate optical ATM network and its access node.

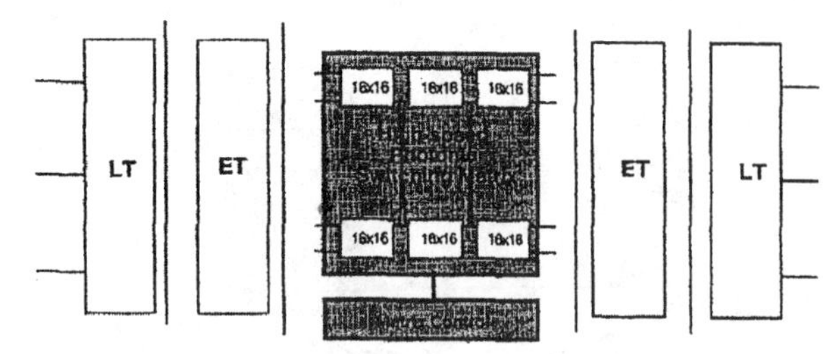

Fig. 5. Schematic representation of a high capacity ATM photonic switching node.

to separate high speed optical time multiplexed channels; backbone reconfiguration on a bypass path, to provide fast protection. ATM cells are inserted and dropped at the bit level using multifunctional high speed switches with dedicated hybrid drivers. Since only the space domain is exploited, ATM cells are electrically buffered at lower speed before being inserted into the backbone.

III. NODE REFERENCE MODEL

A switching system with 256 inputs and outputs operating at 2.5 Gb/s was defined for the node as a reference configuration for theoretical calculations and numerical simulations. The feasibility of such an optical switching fabric has been analyzed simultaneously from a system point of view and from a technological point of view. Adherence to ATM transport requirements (cell loss probability $P_L < 10^{-9}$ with 0.8 Erlang load under uniform traffic) has also been evaluated. Fig. 5 shows schematically the internal structure of this ATM optical switching node: incoming signals cross the line termination (LT) and the exchange termination (ET) before being routed by the multistage photonic switching matrix at high speed.

Large, monolithic optical switching matrices can be hardly realized, since physical limitations of current optical and opto-electronic devices introduce implementation constraints, thus, limiting the practical size of a single-stage matrix. Therefore, multistage matrices have to be realized by interconnecting elementary switching modules of limited size. The maximum size of the module must be determined on the basis of power budget considerations, to avoid degradation of optical signals leading to sensitivity penalties at the detection side. Basically, these impairments result from the combined effect of nonideal behavior of optical key components—wavelength converters, space switches and filters—(leading to degradation of the

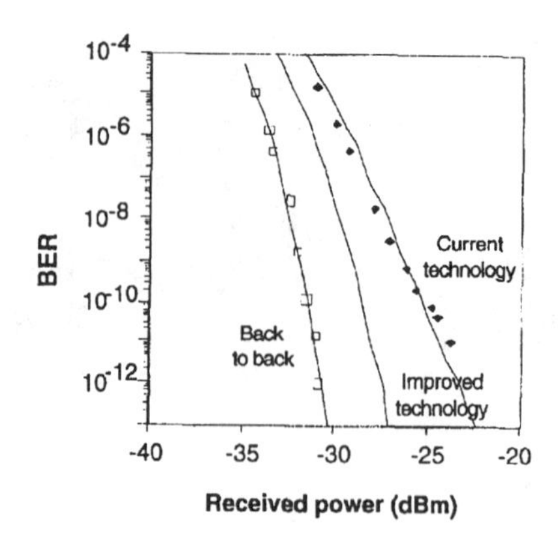

Fig. 6. Projected switching performances of a 16 × 16 fiber delay line switching matrix evaluated from: a) 4 × 4 experimental results at 2.5 Gb/s (standard technology), and b) optimized wavelength converters and optical gates developed in ATMOS.

signal extinction ratio); the accumulated noise (coming from amplified spontaneous emission of the amplifiers); the nonideal synchronization of the signals on output.

It has been demonstrated in the project that 16 × 16 modules is the optimal choice to find a compromise between traffic performance, optical signal transfer properties and implementation effort for the total switching fabric (in terms of number of components and control complexity). In fact, the feasibility of 16 × 16 switching matrices has been demonstrated and cascadeability can be anticipated, provided the use of optimized devices capable of certain conditions (e.g., extinction ratio regeneration and/or signal amplification at each stage), as shown by Fig. 6 in the case of the fiber delay line switching matrix [9]. So multistage (three, five, ...) optical Clos-like networks are physically feasible to implement the switching fabric [10].

Traffic performance of the different switching matrix system concepts and the related multistage Clos networks have been evaluated in depth, considering also the issues of limited buffering capacities and of traffic concentration at the output of the switching fabric. Since different options are available for the fabric topology, the routing policy, the buffering technique and the interstage flow control, it has been demonstrated in the project that it is always possible to find an architectural configuration for each system concept capable of reaching an acceptable throughput while matching the ATM requirements. This has been studied by means of simulation, as shown in Fig. 7 in the case of the fiber delay line matrix, which adopts queuing of cells on output [11].

Obviously, different internal dimensionings (especially of the central stage) can be obtained, depending if the concentration at the output of the fabric is realized electrically (with electrical buffers) or optically (with optical buffers). In the case of optical buffers, it has been demonstrated that in a Clos network made of matrices with limited buffers, the critical stages for the cell loss are the first and the last [11].

Then, different trade-off options between performance and complexity have been investigated, both for the elementary switching matrix (Fig. 8, in the case of the multidimensional

TABLE I

RESULTS OF THE DIMENSIONING STUDY OF THE FIBER LOOP MEMORY SWITCHING FABRIC, WITH THE EVALUATION OF THE TOTAL AMOUNT OF KEY OPTICAL DEVICES (LAST 3 COLUMNS) NEEDED TO REALIZE THE REFERENCE STRUCTURE. DIFFERENT SIZES AND BUFFERING SCHEMES ARE EVALUATED FOR THE BASIC MATRIX AND THE 256 × 256 MULTISTAGE NETWORK IS ASSUMED TO BE A SINGLE PLANE FOR A 0.8 ERLANG TRAFFIC THROUGHPUT, AND A DOUBLE PLANE (D256 IN THE TABLE), FOR A 0.4 ERLANG TRAFFIC THROUGHPUT PER PLANE

	N	ρ	M	C	S	n. circ.	matr. size	FC per mod	PF=A per mod	TF per mod	FC (Total)	PF=A (Total)	TF (Total)
SHARED BUFFER ARCHITECTURE	4	.8	50	4	7	40	256	4	50	4	1792	22400	1792
	4	.4	16	4	8	12	512	4	16	4	4096	16384	4096
	4	.4	16	4	7	12	D 256	4	16	4	3584	14336	3584
	8	.8	»50	8	5	46	256	8	...	8	1280	...	1280
	8	.4	28	8	6	13	512	8	28	8	3072	10752	3072
	8	.4	28	8	5	13	D 256	8	28	8	2560	8960	2560
	16	.8	»100	16	3	48	256	16	...	16	768	...	768
	16	.4	46	16	4	14	512	16	46	16	2048	5888	2048
	16	.4	46	16	3	14	D 256	16	46	16	1536	4416	1536
PARTIALLY SHARED BUFFER ARCHITECTURE	4	.8	30	2	7	40	256	4	60	8	1792	26880	3584
	8	.8	25	2	5	46	256	8	100	32	1280	16000	5120
	8	.8	43	4	5	46	256	8	86	16	1280	13760	2560
	16	.8	23	2	3	48	256	16	184	128	768	8832	6144
	16	.8	37	4	3	48	256	16	148	64	768	7104	3072
	16	.4	14	4	4	14	512	16	56	64	2048	7168	8192
	16	.4	14	4	3	14	D 256	16	56	64	1536	5376	6144
INPUT BUFFER ARCHITECTURE	4	.8	18	1	7	40	256	4	72	16	1792	32256	7168
	8	.8	17	1	5	46	256	8	136	64	1280	21760	10240
	16	.8	15	1	3	48	256	16	240	256	768	11520	12288

LEGENDA:
N = number of I/O
M = n. of memory positions in the buffer
C = sharing ratio = n. inputs per buffer
S = n. of stages for a 256x256, 512x512 or double plane 256x256 network (D 256)
p = traffic load at the input of the module

FC = n. Frequency Converter
PF = n. Passive Filters = A = n. Amplifiers (gates)
TF = n. Tunable Filters
FC tot = total number of FC in the multistage network
PF tot = total number of PF in the multistage network
TF tot = total number of TF in the multistage network

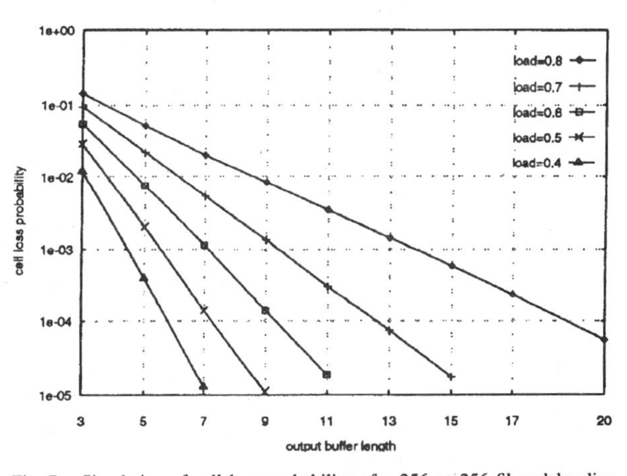

Fig. 7. Simulation of cell loss probability of a 256 × 256 fiber delay line multistage switching fabric (realized with 16 × 16 matrices) as a function of the matrix buffer length.

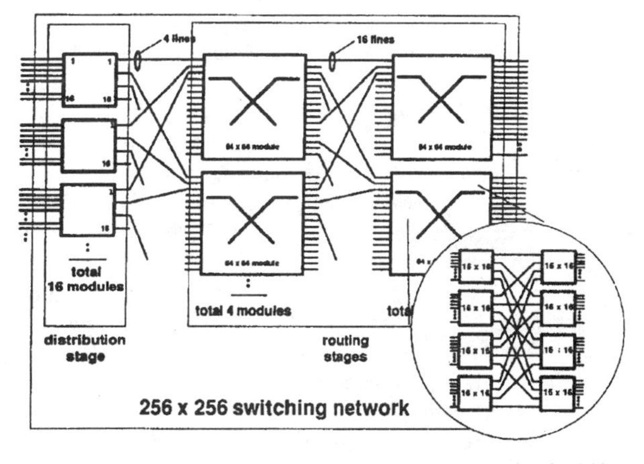

Fig. 8. Internal structure of the distribution and routing matrix of a 256 × 256 multidimensional switching fabric (based on 16 × 16 matrices).

matrix, and Fig. 9, in the case of the fiber loop memory matrix) and for the switching fabric (Table I, in the case of the fiber loop memory matrix) organization [12]. It is worth noting that the hardware dimensioning of Table I did not take into account the complexity of LT's and ET's. As expected, in a large size multistage switch, some thousands of optical and optoelectronic devices are required.

Finally, the opportunities for the introduction of ATM optical switches in future IBC networks have also been evaluated, developing layered models to identify novel optical solutions that complement existing ATM network proposals and which minimize the networks processing requirements.

Implicitly, we have assumed the switch internal operating mode to be synchronous and time-slotted, with the time slot duration, at most, equal to the external link cell transmission time. Thus, at the link interface, a cell processor is required to handle the incoming cells performing the synchronization and link level protocol functions (cell header processing). Then an n-bit long tag is generated for internal routing purposes and added to the cell, to be used by the matrix electronic control logic to select the requested outgoing link.

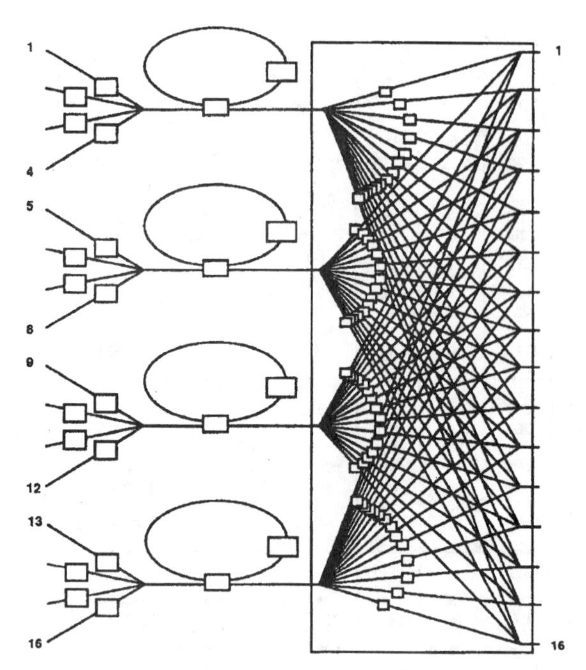

Fig. 9. Schematic of the 16 × 16 fiber loop memory switching matrix with partially shared buffer architecture.

Optical interfaces can be realized with state-of-the-art optical technology [13], requiring or not a new cell format. In the ATMOS project it has been demonstrated that the current electronic-oriented ATM cell format requires necessarily electronic interfaces, especially for ATM cells transported within SDH frames [14]. However, optical clock recovery and coarse and fine optical synchronization at the input and at the output of the switching fabric have been demonstrated at Gb/s bit rates in ATMOS by using correlation techniques [15] (in the case of the corporate ATM optical network access node) or by designing a cell/packet synchronizer [16] (in the case of the fiber loop memory switching fabric). These two sub-blocks, both based on key components developed by ATMOS (optical space switches, fiber delay lines and wavelength converters), can be used if a new cell/packet format is adopted, suitable for all-optical processing, thereby allowing optical end-to-end transport and switching in the network.

The network layered model and the study of optical switching nodes equipped with different types of network-node interfaces also provided a framework for a number of evolutionary scenarios [17]. In particular, four layers have been identified.

- *ATM Layer:* This is based on existing standards where ATM cells can be transported within SDH frames or as a continuous cell stream. Switches have standard electronic interfaces. The core of the ATM switch may be electronic or optical.
- *Optical ATM Layer:* Conventional ATM cells are encapsulated within an optical ATM cell that allows a reduction in the amount of processing at the switch interfaces that can become all-optical. This layer is divided into partitions in which routing occurs by means of optically reading the virtual path identifier (VPI) field in an ATM cell header but in which VPI translation is not mandatory, and each partition is known as a VPI island. In transferring between islands, VPI translation is required, by means of, e.g., standard electronic interfaces.
- *Supercontainer Layer:* The amount of processing required can be reduced further by using super containers which consist of a number of standard ATM or optical ATM cells and an optical header supporting routing, synchronization and guardbands. This introduces in the network an additional multiplexing level, which can be regarded as a "virtual trunk" level. Super containers are switched as single units and can use similar switch fabrics to those proposed in ATMOS.
- *Transparent Optical Layer:* Routing occurs at the wavelength channel level without the need to process the contents of this channel. This coarse grain routing and switching requires optical circuit-based cross-connects. This type of routing was out of the scope of ATMOS, although it has been demonstrated that also circuit-based cross-connects can be implemented with components and technology developed in ATMOS.

Although the concept of layered networks is normally associated with public networks, it can also be applied to private networks such as the corporate optical ATM network proposed in this project, thus, developing ideas for a layered network model where optical ATM cells or packets are used to bridge the granularity gap between access networks and the core backbone of optical transport networks.

IV. Key Technology and Sub-Blocks

The ATMOS technology must be based on switching matrices with a short reconfiguration time (fast routing, carried out on a cell basis) and capability to resolve the contention of simultaneously incoming optical cells destined to the same switch output (time switching or time slot interchange). Therefore, the ATMOS technology approach for optical routing and time switching of ATM cells at high speed was, first, to select the relevant optical techniques compatible with high quality transmission performances, traffic requirements, and scalability to real system dimensions. This imposed severe functional constraints (high speed, buffer capacity) and technological requirements (polarization insensitivity, low loss, low noise, power consumption ...) which has led to a careful selection of the technologies.

Fast switching matrices have been studied, based on wavelength routing with fast tunable wavelength converters and filters through a passive optical network, or on fast active waveguide space-switches. These matrices have been developed scalable up to 16 × 16 dimensions, exhibiting good transmission performances, that can be improved in the very near future by the use of recently developed optimized components. However, due to the finite turn on/off time of the different switching devices, the reconfiguration time of the switching matrices is today limited to a few ns: this implies the use of an internal cell format including a short guardband between consecutive cells in order to reconfigure the switching matrix.

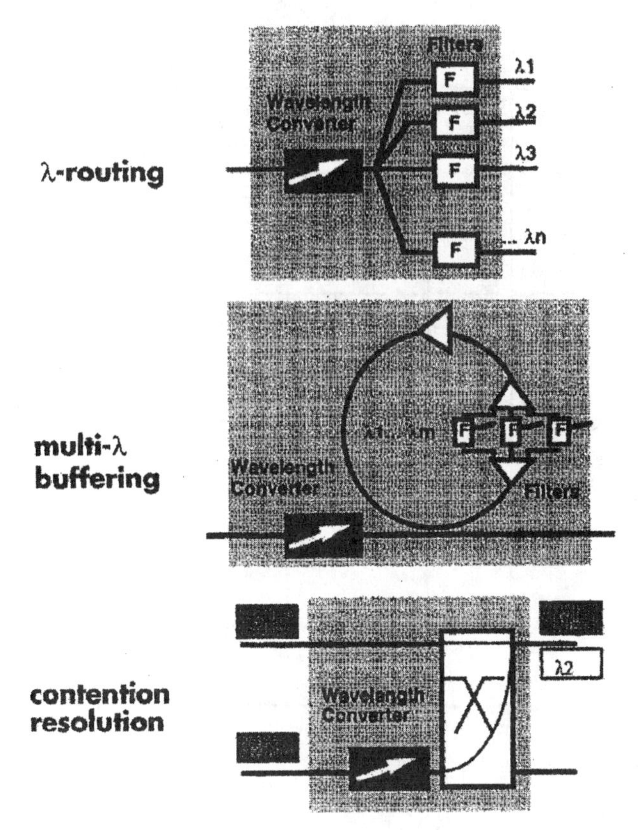

Fig. 10. Applications of tunable wavelength converters in ATMOS.

Resolution of contention is achieved by cell wavelength conversion and buffering in multiwavelength optical fiber memories, controlled by fast switching devices (gates or tunable filters), that can store or delay during the appropriate time the cells destined to the same outlet in order to avoid collisions.

The basic functionalities of the ATMOS switches have been implemented in the 1.55 μm wavelength region using for optical and optoelectronic active devices, mostly the InP-based technology. The device performance required for high speed and high quality optical transmission have requested specific studies and developments of packaged modules integrating high speed electronics and low-loss optical fiber interconnections. As results, a number of fast switching devices fulfilling the system requirements has been studied, produced and integrated in the system demonstrators (and reported in about 300 technical papers published by the consortium).

A. Wavelength Converters

The most significant contribution of ATMOS in the component area is the definition, feasibility demonstration, optimization and implementation in various system demonstrators of a very important new device for all-optical systems and networks: the all-optical wavelength converter (AOWC), providing means to perform optically the translation of the wavelength of a very high bit rate stream (beyond 20 Gb/s [18]) over more than 20 nm without penalty and with some signal regeneration.

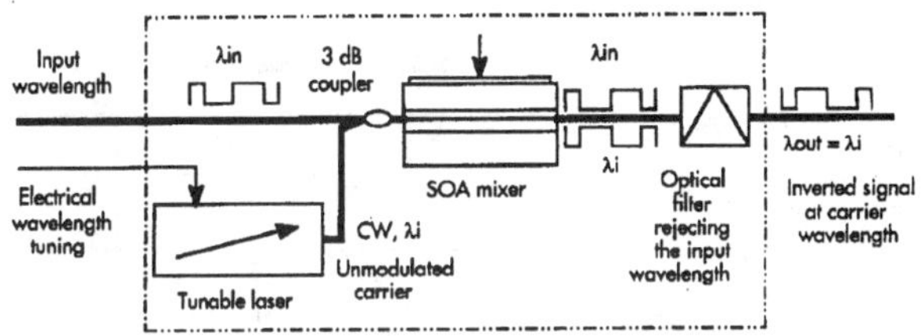

Fig. 11. Principle of operation of an SOA-AOWC, obtained using an SOA in the gain saturation regime as a mixer and a tunable laser as a carrier generator.

This device has been integrated and successfully assessed in various subsystem functional blocks, used for different applications, as shown in Fig. 10.

ATMOS has pioneered several technical approaches [18]–[27] to realize all-optical wavelength conversion, most of them being based on carrier depletion in semiconductor lasers or in semiconductor optical amplifiers (SOA's). These approaches have been intensively studied and compared with special attention to system requirements (maximum operating speed, wavelength operation range, wavelength tunability, extinction ratio enhancement, polarization sensitivity, ...) for all-optical switching and routing in future networks [19].

Table II gives an overview of the respective performances of four structures studied by ATMOS, two laser- and two SOA-based wavelength converters. The performances reported in this table are experimental values obtained on nonoptimized devices; therefore, they can be improved in many cases.

Lasers structures include their tuning element. The DBR–AOWC [20], whose wavelength and polarization sensitivity has been reduced using both anti-reflective coating and a square shaped active waveguide, has a tuning range limited to a few nm. The Y-interferometric laser [21] is characterized by a very broad tuning range (more than 20 nm) of the converted wavelength.

Although it needs an additional tunable CW sources, the AOWC based on gain saturation in polarization insensitive SOA has been found to be a very efficient and practical approach to implement high speed wavelength converters in system demonstrations [22]; this approach is shown in detail in Fig. 11. This was done at increasing bit rates at the three last ECOC exhibitions on the RACE booth from 1993 to 1995. In 1993, four SOA-AOWC's were used at 2.5 Gb/s at the input of a 4 $\times$ 4 optical ATM switching matrix (see Section V-A). In 1994, in cooperation with the R2018 GAIN project, ATMOS demonstrated wavelength routing by converting at 10 Gb/s, the wavelength of one among eight WDM channels in an 8 $\times$ 10 Gb/s transmission experiment (Figs. 12 and 14 [23]). Finally, in 1995, 20 Gb/s was demonstrated (Figs. 13 and 15) where the wavelength converted stream was used to carry high quality broadband services (HDTV). A practical technology limit of 40 Gb/s has been determined theoretically and recently assessed experimentally. This needs to use a long ($\geq$1 mm) SOA operated at high bias current with a large confinement factor ($\Gamma \approx 0.6$) in order to increase the photon

TABLE II
COMPARATIVE PERFORMANCE OF ALL-OPTICAL WAVELENGTH CONVERTERS DEVELOPED IN ATMOS

Devices	Maximum obtained bit-rate	Input wavelength range (λ sensitivity)	Output tunability	Switching speed	Polarisation sensitivity	ER enhancement capability
Igain Iphase IBragg, λi, λo, Filter, DBR	10 Gbit/s	30 nm I λi-λo I>2nm	6nm (16 modes) λi≠λo	2-7 ns	<0.5 dB	yes
I1 I2 I3 I4, λi, λo, Y-laser	5 Gbit/s	30 nm (high)	20nm (one current) λi≠λo	2-5ns	high (reducible)	yes
Ibias, λi, λo, CW, λo, SOA	20 Gbit/s	30 nm (no)	CW source (better for λo<λ peak)	CW source	<0.5 dB	no
λi, SOA, SOA, λo, CW, λo, I-AOWC	10 Gbit/s	30 nm (no)	CW source	CW source	<0.5 dB	yes

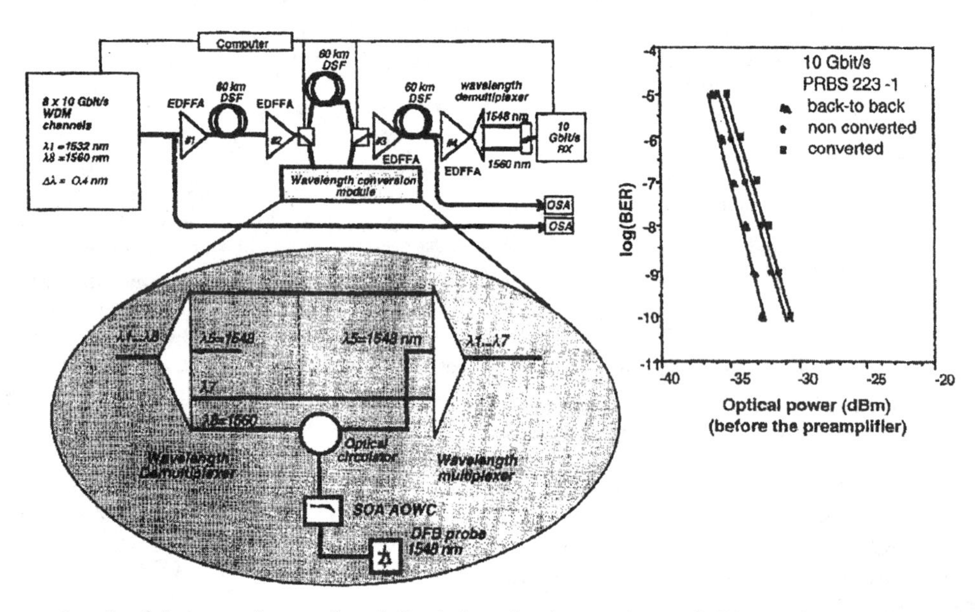

Fig. 12. Scheme and results of the system demonstration of all-optical wavelength conversion at 10 Gb/s in a 80 Gb/s, eight-channel WDM transmission experiment.

density and, hence, to reduce the carrier lifetime. The main drawback of this device is a degradation of the output signal extinction ratio (compared to the input one) which precludes the cascadability of several devices where imposed by system or network requirements, and the chirp induced by the phase modulation associated to the gain modulation.

This phase modulation can be exploited if SOA-AOWC's are integrated in an interferometric structure, such as a Mach–Zehnder (MZI) or a Michelson (MI) interferometer, to provide efficient high speed wavelength conversion (Fig. 16). This structure, proposed in 1993 by ATMOS, presents decisive advantages.

- Thanks to the nonlinear transfer characteristics of the interferometer, extinction ratio enhancement of the output signal is possible.
- FM to AM conversion allows chirp-free, or chirp-compressed output signals (thanks to the possibility of using the device operated with a negative α factor).

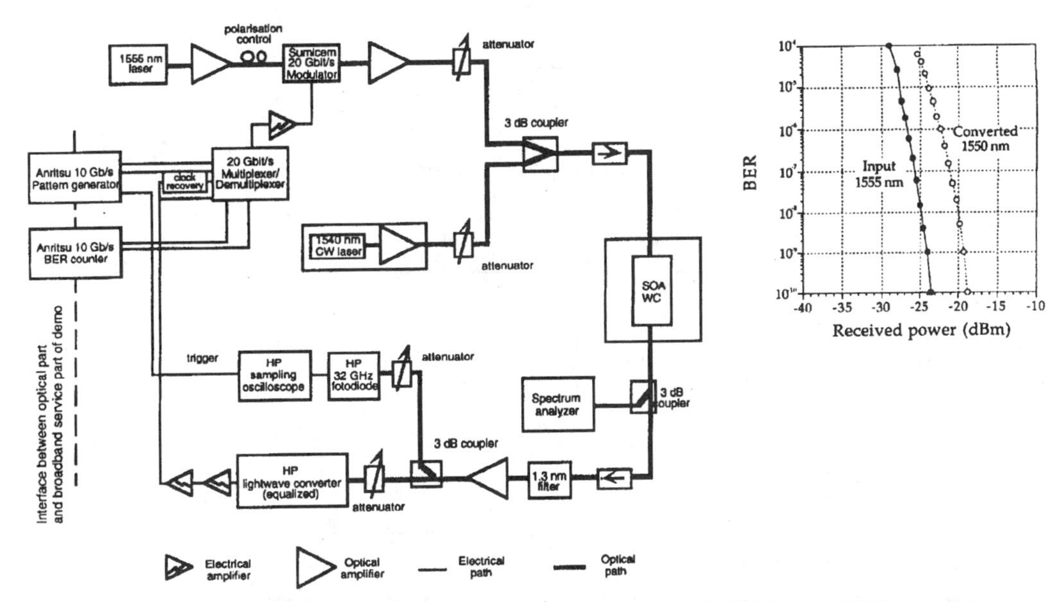

Fig. 13. Scheme and results of the system demonstration of all-optical wavelength conversion at 20 Gb/s featuring HDTV transmission.

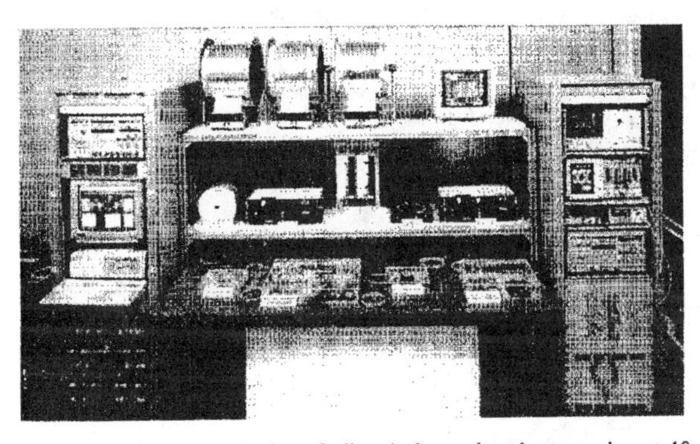

Fig. 14. System demonstration of all-optical wavelength conversion at 10 Gb/s in a 80 Gb/s, eight-channel WDM transmission experiment.

Fig. 15. System demonstration of all-optical wavelength conversion at 20 Gb/s featuring HDTV transmission.

This structure, first demonstrated in a hybrid version in 1993 [24], has been monolithically integrated, exploiting the technologies of the polarization insensitive SOA's on one hand and of the Y laser on the other hand, leading to the realization of two technologies: one called all-active, based on a compact MQW structure, the other active–passive, based on the integration of a square shaped bulk SOA in a low-loss passive interferometric structure. Both have been tested in transmission systems at 2.5 Gb/s (Fig. 17 [25]) and 10 Gb/s (Fig. 18 [26]), respectively. These results show that this device of moderate complexity (important point when considering manufacturability) allows to meet the most important system requirements, in particular, regarding signal reshaping and chirp compression, as demonstrated by Fig. 17, where a penalty-free transmission over 190 km of standard fiber at 1.55 μm has been achieved, using the converted signal of an all-active wavelength converter.

B. Space Switches

High performance InP space switching devices have also been developed in ATMOS, integrated and successfully assessed in various functional blocks for different applications as illustrated in Fig. 19.

ATMOS has studied three different approaches based on InP waveguide switches:

- association of polarization insensitive SOA's used as gates [22], [27] within passive shuffle networks;
- Mach–Zehnder Interferometric switches exploiting the polarization independent Franz Keldish effect [28]–[29]; and
- carrier depletion directional couplers [30].

Table III gives an overview of the performances of these devices that have been successfully tested and assembled into systems. It is worth mentioning that the complexity of integration of these fast switches is typically limited to 4 $\times$ 4 matrices, due to their non-negligible electrical power consumption, especially for reverse biased switches exhibiting a capacitive behavior, and also due the non-negligible optical losses in non-amplifying switches.

A very illustrative example of the progresses achieved by ATMOS in the field of high speed, high performance (low loss,

TABLE III
CHARACTERISTICS OF THE OPTICAL SPACE SWITCHES DEVELOPED IN ATMOS

	Semiconductor Optical Amplifier (SOA) Gates	Mach-Zender Interferometric (MZI) Switches	Carrier Depletion Directional Coupler (CDDC) Switches
Optical transmission	discrete: +20dB between lensed fibers; arrays: +14 dB, 4 channels fiber-to-fiber	for 2 switches in series - 8 dB fiber-to-fiber	for 2 switches in series - 16 dB fiber-to-fiber - 8 dB fiber-to-detectorr
Polarization independence	yes	yes	no
Optical bandwidth extinction ratio	40-50 nm 50 dB	> 60 nm 20 dB	>60 nm 20 dB
Driving voltage / current	75 mA	5 V	10 V
Maximum speed	200 ps	< 200 ps	100 ps
Switching time	< 260 ps with InP HBT current drivers	< 260 ps with InP HBT voltage drivers	1 ns with hybrid drivers, < 300 ps with InP HBT current drivers
Complexity demonstrated	4 SOA Gates with tilted faces	5 MZI space switches with high speed electrodes and integrated tapers	3 CDDC space switches with integrated waveguides, photodetectors and 2 passive splitters

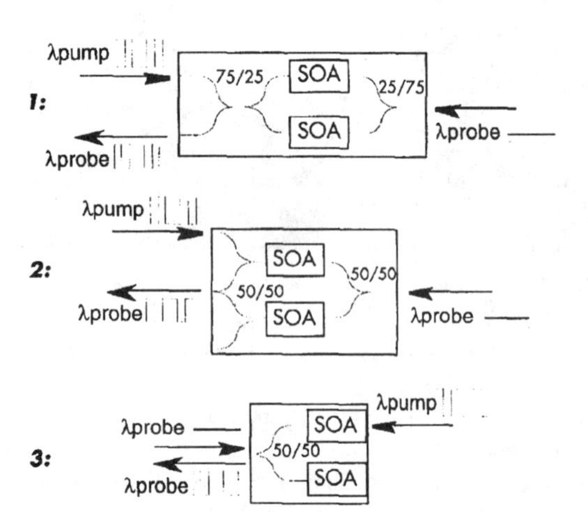

Fig. 16. Schematics of possible structures for interferometric AOWC: 1: asymmetric MZI, 2: symmetric three-port MZI, and 3: symmetric MI.

polarization insensitivity, low driving electrical power, ...) InP space switches, is the hybrid integration of a four-SOA gate module [31], including low-loss multiple fiber interconnection [32]–[33], high speed InP HBT drivers [34] and a monolithic array of four polarization insensitive SOA's; the module is shown in Fig. 20. In order to achieve simultaneously high performance and a high fabrication yield, a special design of the individual devices (e.g., SOA with tilted facets, ...) and of the hybridization process (flip-chip mounting on V-grooved silicon motherboards, ...) was successfully performed. The very interesting performances [35], [36] of this device can

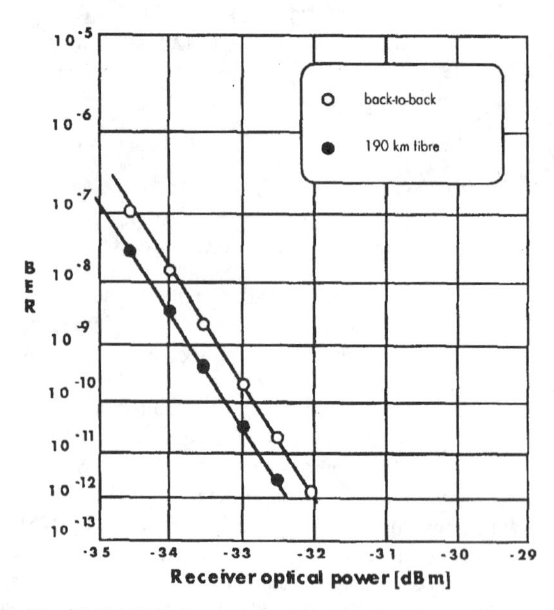

Fig. 17. 2.5 Gb/s BER characteristic of all-active I-AOWC capable of chirp compression (converted, noninverted data, after 190 km).

be significantly improved by integrating clamped gain SOA gates [37].

C. Packaged Modules

Packaged switching devices/modules integrating in hybrid assemblies, fast InP switching devices (space-switches, tunable

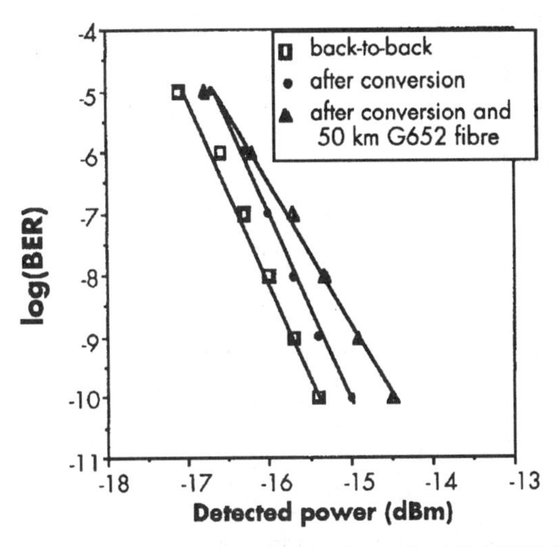

Fig. 18. 10 Gb/s BER characteristic of active-passive MI–AOWC (converted, noninverted data, after 50 km).

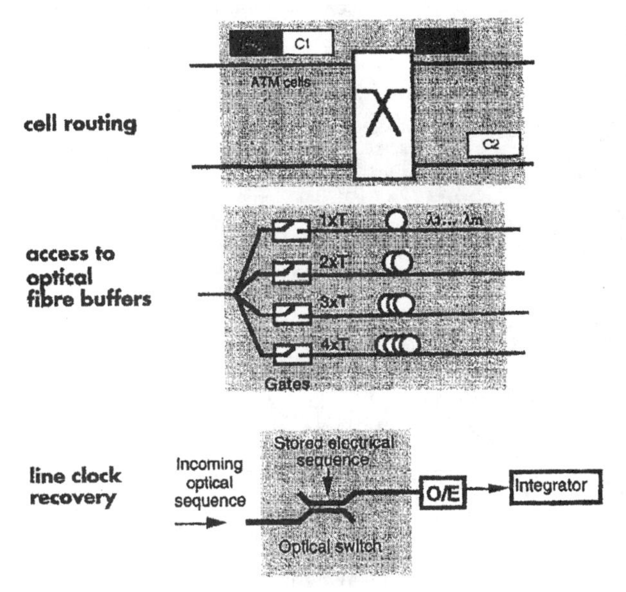

Fig. 19. Applications of optical space switches in ATMOS.

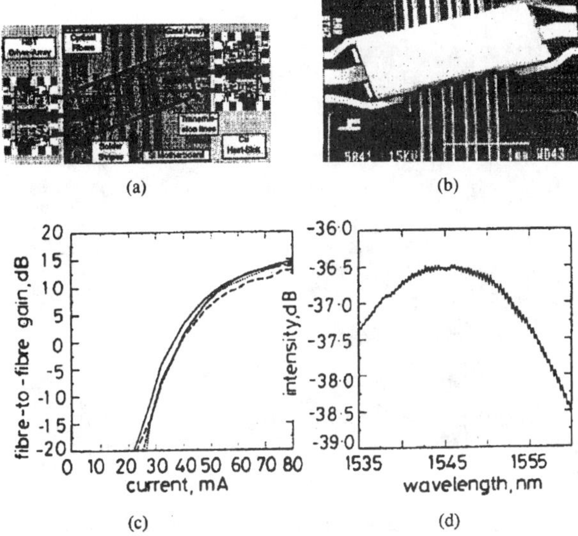

Fig. 20. Gate array module for the fiber delay line matrix: (a) module central part including high speed InP HBT drivers and multiple fiber interconnection to the monolithic array of four polarization insensitive SOA's with tilted waveguides, (b) flip-chip mounted gate array on Si motherboard, (c) module fiber-to-fiber gain at 1553 nm, and (d) spontaneous emission spectrum at 80 mA amplifier current.

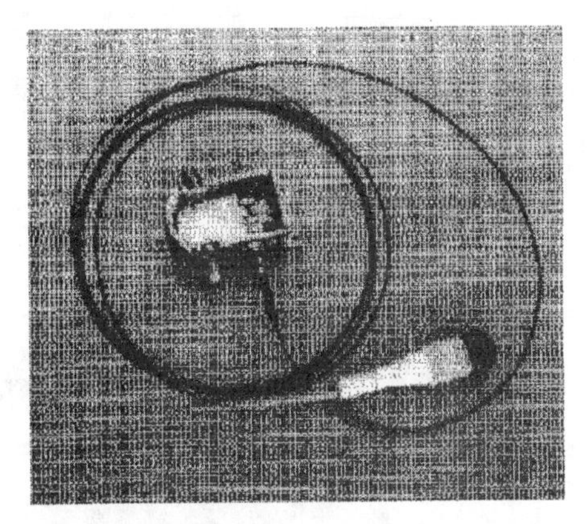

Fig. 21. High speed wavelength encoding module integrating a fast tunable DBR, laser or wavelength converter, with fast GaAs P-HEMT driving IC's.

filters, lasers, AOWC's, ...) efficiently coupled with interconnection fibers and high speed III–V electronics [38], [39] (GaAs FET and HEMT integrated circuits from commercial foundries, or InP HBT integrated circuits developed within the project) have been designed, fabricated and successfully tested in ATMOS. A further example relevant to a high speed wavelength encoding module integrating a fast tunable DBR, laser or wavelength converter, with fast GaAs P-HEMT driving integrated circuits is represented in Fig. 21 [39]. The main characteristics of this module are a tuning range of 2 nm (tuning over four DBR modes) for a driving voltage 0.8 V and a driving current of 70 mA within 5 ns. The speed of this module is today limited by the carrier lifetime in the Bragg section of the DBR, but new quickly tunable structures based on electrooptic effects have been investigated [40], [41] in order to overcome this limitation.

D. Optical Buffers

Optical buffers based on optical fibers used as delay lines have been developed in ATMOS [42], [43]. Passive fiber-based buffers offer several advantages, such as immediate availability, suitability for delaying data, absence of electrical power consumption and negligible insertion loss. Indeed, their capacity can be simply enhanced by efficiently exploiting the fiber spectral domain, via dense WDM: whenever the chromatic dispersion effects can be neglected, a number of cells at different wavelengths can be buffered simultaneously with the same optical hardware. Its main drawbacks are the significant physical size of the buffer and an inherent lack of flexibility (the buffer can operate only at a fixed data rate). In practice, one meter of fiber causes the light to be delayed

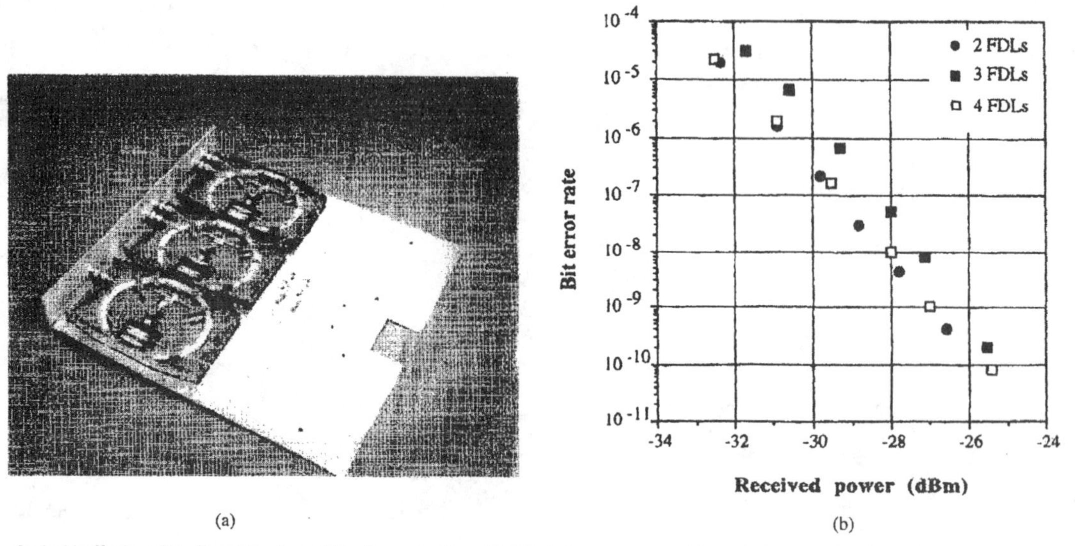

(a) (b)

Fig. 22. Optical buffer based on fiber delay lines: (a) buffer assembly and (b) BER for one output of the switching matrix using two, three, and four delay lines.

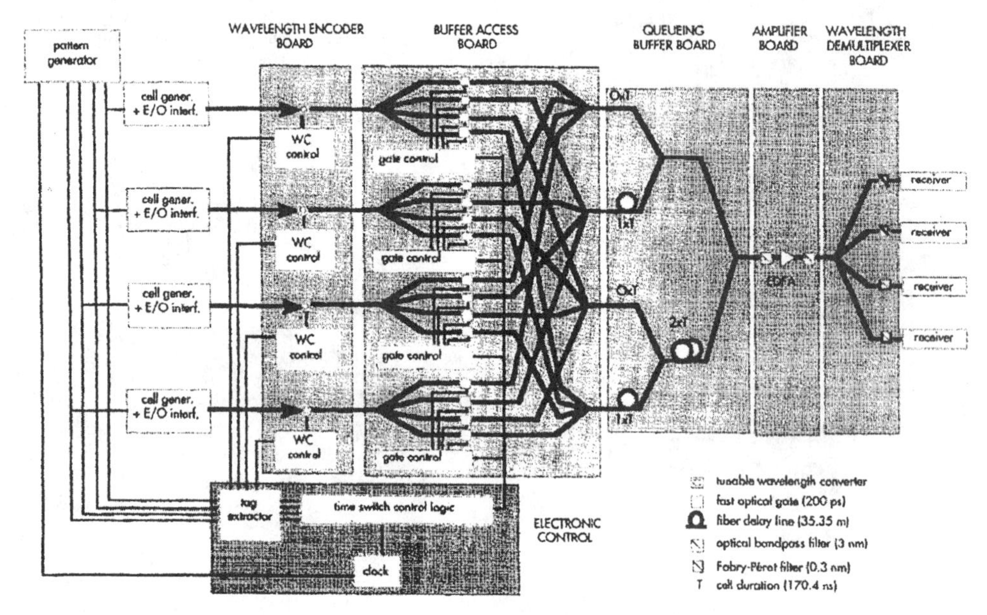

Fig. 23. Scheme of the fiber delay line matrix demonstrator.

by about 5 ns, in the near infrared region, thus, a standard ATM cell at 2.5 Gb/s can be stored for a time slot in about 35.4 meters.

The operation of such a buffer has been successfully tested placing the $m = 4$ binary tree-shape assembly, accessed via 16 SOA gates, within the 4×4 rack-mounted demonstrator of the fiber delay line switching matrix, fed by ATM cells at 2.488 Gb/s [44]. In order to evaluate the contribution of the FDL's to the penalty of the total system (ATM matrix equipped with the buffer), different traffic configurations have been used, with patterns presenting contentions requiring the use of two, three and four delay lines, respectively. Fig. 22 shows the assembled optical buffer and the obtained BER for one output of the matrix as a function of the output power on the receiver.

V. Switching Demonstrators

Four optical switching demonstrators, one for each system concept, have been realized and rack-mounted in ATMOS. They are briefly described in the following.

A. The Fiber Delay Line Switch Demonstrator

A fully-equipped demonstrator of a 4×4 fiber delay line switching matrix with four buffer positions, whose scheme is shown in Fig. 23, has been implemented and mounted in a rack. The different components, in particular, four quickly

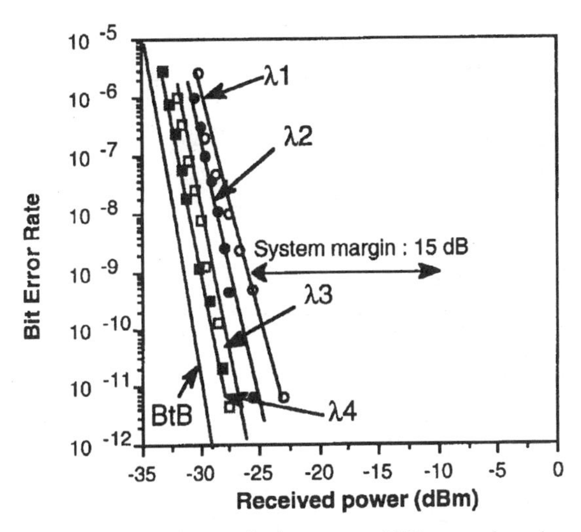

Fig. 24. Fiber delay line matrix demonstrator BER curves in a 4 × 4 cell-switching experiment at 2.488 Gb/s (using all-optical converters).

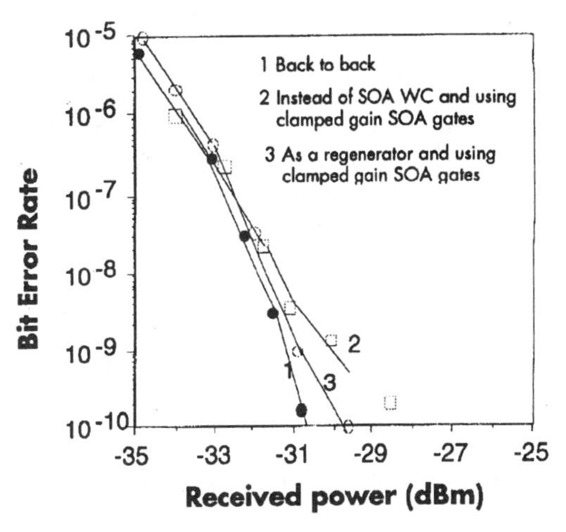

Fig. 25. Fiber delay line matrix demonstrator experimental BER curves in a 16 × 16 configuration at 2.488 Gb/s (using I-AOWC as a regenerator or as a wavelength converter).

tunable AOWC's made from polarization insensitive SOA combined with tunable DBR, 16 fast polarization insensitive SOA gates, one four-position fiber delay line buffer, one Erbium doped fiber amplifier and four Fabry–Perot adjustable filters separated by 0.6 nm to select the appropriate output, were integrated on separate boards, as indicated in Fig. 23. Special care was paid to balance the optical paths of the matrix so that all signals were synchronized to within ±40 ps. The key functionalities of routing and time-switching of 2.5 Gb/s cells were demonstrated with excellent overall system performance while exploiting the full 4 × 4 matrix and running error-free (bit-error rates $< 10^{-13}$, as shown in Fig. 24). Less than 3 dB sensitivity penalty with respect to the back-to-back configuration was obtained. The demonstrator was presented for the first time at the ECOC'93 exhibition in Montreux, Switzerland (Fig. 32).

Using the first generation components from this 4 × 4 demo, modeling has shown that the optical signal-to-noise ratio (SNR) in a 16 × 16 configuration could be close to unacceptable limits, due mainly to the accumulation of spontaneous emission noise from the gates. With SOA gates capable of operating at a higher output level, such as clamped gain SOA gates, it is possible to overcome this limitation, especially when, in addition, interferometric AOWC's capable of signal extinction ratio regeneration, such as a MI–AOWC, are used instead of a simple SOA–AOWC. Early experimental simulations of a 16 × 16 fiber delay line switching matrix using improved technology, have been already shown in Fig. 6. The use of this optimized technology, i.e., clamped gain SOA gates and MI-AOWC, has experimentally demonstrated negligible system penalty (Fig. 25), allowing to assert that 256 × 256 fiber delay line switching fabrics with excellent performances can be built by cascading 16 × 16 elementary matrices in a three-stage Clos network.

B. Fiber Loop Memory Switch Demonstrator

An experimental demonstration of a 2 × 2 fiber loop memory matrix has been realized, whose scheme is shown in Fig. 26, with four wavelength channels in the loop, corresponding to four memory positions. This demonstrator includes two fast tunable DBR–AOWC modules at the input and two wavelength-matched fast tunable DFB filters at the output routing stages. Both modules include their high speed driving GaAs electronics made from P–HEMT integrated circuits. The fiber loop memory is composed by an array of gateable optical Fabry–Perot filters (four for the demonstrator), a compact Erbium doped fiber amplifier followed by a large passband optical filter, and a looped fiber length, such that the total loop delay equals one cell duration, with a total loop gain ≈ 1. A close spacing of the wavelength channels (≈0.4 nm) is required to implement a large buffer capacity and to reduce the necessary tuning range of sources and filters.

Resolution of cell contention has been experimentally demonstrated and BER measurements, carried out at the switch output, are reported in Fig. 27, which shows the back-to-back configuration, a straight-through cell and a stored cell. Influence of wavelength crosstalk, number of circulations in the loop and wavelength allocation is still being studied.

The different components, electronics, optics and drivers have been integrated and assembled on board and then mounted into standard rack units (Fig. 33), mounting all CW current drivers and temperature control into one subrack unit. The demonstrator was presented at the CNR workshop in June 1994, Turin, Italy.

C. The Multidimensional Switch Demonstrator

A system demonstrator of the switching matrix of the Multidimensional switch concept has been realized, in a subequipped 4 × 4 arrangement with four wavelengths, and is schematically illustrated in Fig. 28. The demonstrator showed the functionalities and basic characteristics of the described concept, the switching of 2.5 Gb/s ATM cells in wavelength and space domain and the contention resolution of cells directly in the wavelength domain. This matrix has been improved both in terms of performance and integration using recently developed components and modules, such as

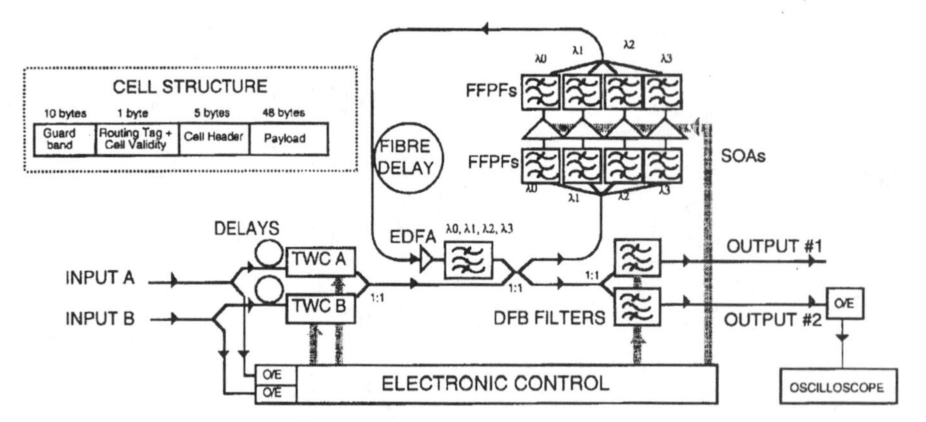

Fig. 26. Scheme of the fiber loop memory matrix demonstrator.

all-active MZI–AOWC's and multifunctional InP fast space switch hybrid modules, with broadcasting capability and low polarization sensitivity (≈1.5 dB). These improvements lead to penalty-free wavelength conversion at 2.5 Gb/s as well as space switching, with set-up times lower than 500 ps, allowing also broadcasting with a channel unfairness of less than 1.5 dB.

System experiments have been carried out on a complete end-to-end path through the 4 × 4 demonstrator configuration to evaluate the overall performance. Referring to Fig. 28, the end-to-end path starts with one asymmetrical Y-laser as tunable source (externally modulated with a 2.5 Gb/s ATM cell stream), followed by an MZI–AOWC and the 4 × 4 space switch represented by two cascaded switch modules (1 × 4 and 4 × 1). The BER measurements, reported in Fig. 29, showed less than 0.8 dB overall penalty. In presence of crosstalk from an adjacent input of the matrix, the overall penalty reached 1.8 dB, which was caused by the limited crosstalk attenuation characteristics of the sample switch modules.

The rack-mounted demonstrator (Fig. 34) was presented at the ECOC '94 exhibition in Florence, Italy.

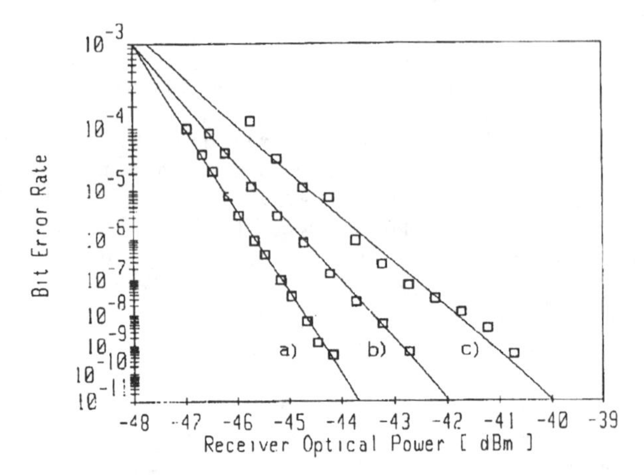

Fig. 27. BER measurement for different cells at the fiber loop memory matrix demonstrator output: a) back-to-back, b) straight-through cell, and c) stored cell.

D. The COATN Demonstrator

A demonstrator investigated the key features of the COATN concept. In a first part, the optical ATM layer functionalities have been tested incorporating dedicated high-speed space switching opto-electronic circuits to perform optoelectronic correlation for synchronization at the ATM cell and bit levels, as shown in Fig. 30. The results would allow clock recovery at and beyond 10 Gb/s with the same technology.

In a second part the optical add-drop-transfer functionalities have been tested at 1.25 Gb/s with a complete ATM layer realization. The complete scheme of the demonstrator is given in Fig. 31. A rack mounted version of the nodes (Fig. 35) has been assembled in 1995 and will be implemented into a pilot system at the University of Cambridge (RACE HIPERNET project).

E. Electronic Control

Finally, within each demonstrator, one single electronic control board includes all the control logic: the several-bit long tags added to incoming cells, with different internal proprietary format adapted to each demonstrator, address at lower speed (typically 622 Mb/s) the control parameter tables that deliver the required current and voltage values to be applied to configure the wavelength converters, space switches and filters, in order to route the cells towards their destination outlets. During the processing of the tags, the optical cells are delayed in a suitable fiber length with a transit delay duration of at most one time slot ($T = 170$ ns), to allow enough processing time to the control logic. The routing tags are also used to resolve contentions and manage the queues. For instance, in the case of the fiber delay line matrix, they are transferred to processors located on the "queue management" daughterboards that act essentially as pointer registers, one associated to each wavelength (i.e., one outlet), then, the associated pointer is incremented/decremented according to the traffic and the register value is fed back to the "tag extract and gate selection" daughterboard corresponding to the input that has emitted the cell: this information is used to activate/deactivate accordingly the gates giving access to the correct fiber delay line of the buffer.

VI. Conclusion

Continuous progress in photonic technology and components will allow the development of new architectures and

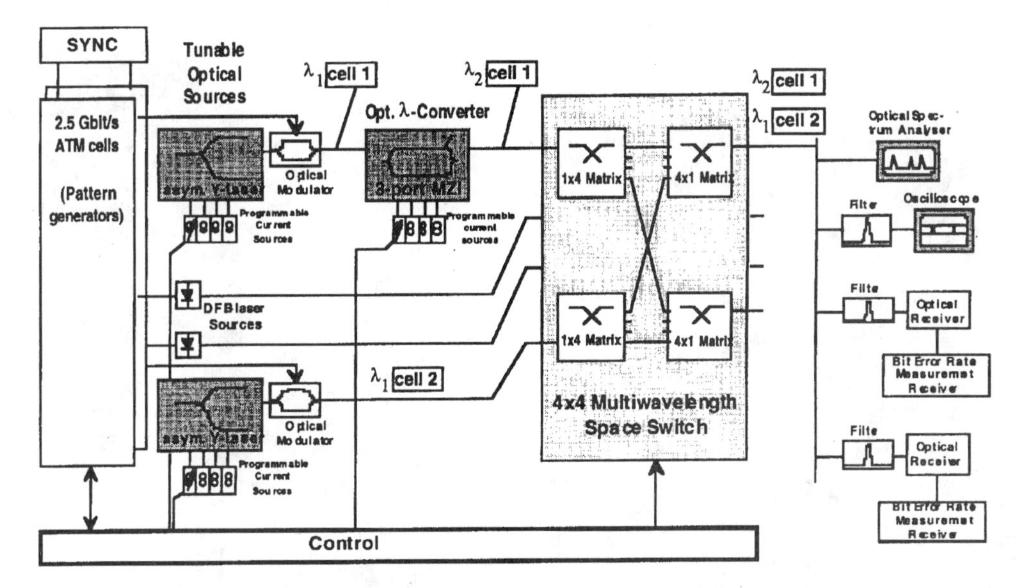

Fig. 28. Scheme of the multidimensional matrix demonstrator.

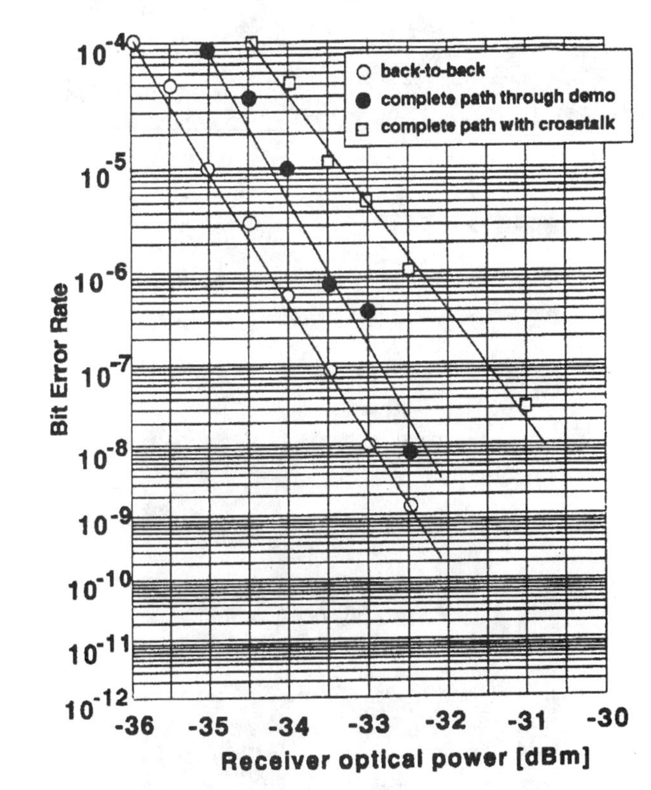

Fig. 29. Overall performance of the 4 × 4 multidimensional matrix demonstrator.

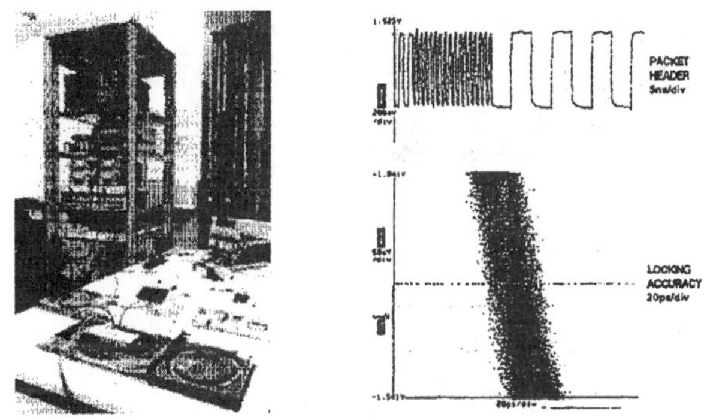

Fig. 30. Multifunctional testbed of the corporate optical ATM network and synchronization results obtained by means of optoelectronic correlation.

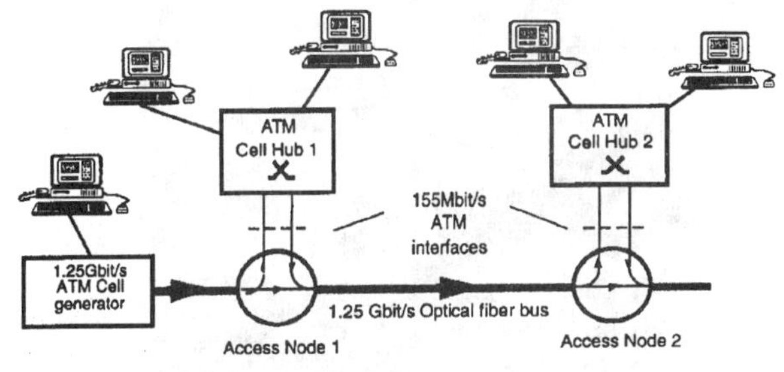

Fig. 31. Scheme of the corporate ATM optical network demonstrator.

systems meeting the need of processing and switching in the terabit range, by removing critical limits of large electronic systems, such as, volume of the central switching fabric, and problems caused by dense interconnections, electro-magnetic interference and power consumption. The ATMOS project has developed system concepts and the basic technology to implement high performance all-optical nodes, supporting high speed and applying optical fast packet switching techniques starting from the ATM principles as a reference.

Three system concepts for large nodes in future public networks and one for small nodes in private networks have been established. Feasibility of the proposed system concepts and the required advanced technology has been assessed with calculations and simulations, and also in four system high-

Fig. 32. The fiber delay line matrix demonstrator (ECOC'93).

Fig. 34. The multidimensional matrix demonstrator (ECOC'94).

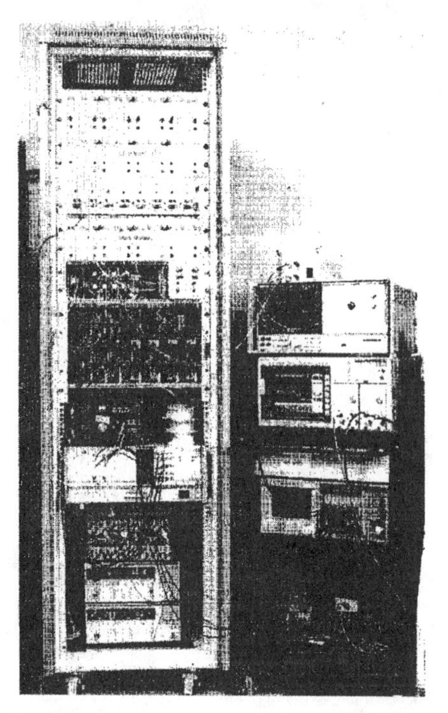

Fig. 33. The fiber loop memory matrix demonstrator (CNR WS'94).

Fig. 35. The corporate optical ATM network demonstrator (1995).

speed demonstrators, which showed very promising results for the potential applications in future telecommunication transport networks.

The severe requirements imposed to technology allowed to develop key devices (all-optical wavelength converters, semiconductor optical amplifiers, optical gates, space switches, ...) with outstanding results capable of reaching well beyond the capabilities of the rack-mounted demonstrators. In particular, system demonstrations of wavelength conversion at 10 and 20 Gb/s have been made, to show the usefulness of the ATMOS technology also to build routing devices for future multigigabit WDM transport networks.

In conclusion, the ATMOS project has provided the system and technological basis to develop and assess the role of fast packet switching and routing techniques in the future high speed, all-optical telecommunication networks, showing that there may not be pure photonic switching by the new millennium, but there certainly will be photonics in switching.

ACKNOWLEDGMENT

The authors wish to thank all the partners of the ATMOS Consortium, in particular, all of the people involved in the project.

REFERENCES

[1] *Performance of the Telecommunications Sector up to 2010 under Different Regulatory and Market Options.* Cambridge, U.K.: Analysis Publications, Feb. 1992.

[2] CCITT, "Synchronous digital hierarchy," Geneva, Switzerland, Recommendations G.707, G.708, G.709, 1988.

[3] M. De Prycker, *Asynchronous Transfer Mode: Solution for B-ISDN.* London, UK: Ellis Horwood, 1990.

[4] J. M. Gabriagues, D. Chiaroni, D. De Bouard, P. Gavignet, C. Chauzat, J. B. Jacob, F. Masetti, P. A. Perrier, and M. Sotom, "Design and implementation of a gigabit ATM photonic switching matrix," *J. High Speed Networks*, vol. 4, no. 4, Oct. 1995.

[5] P. Cinato, G. Bendelli, M. Burzio, and P. Gambini, "Architectural analysis, feasibility study and first experimental results for a photonic ATM switching module to be employed in large size switching networks," in *Proc. ISS '95*, Berlin, Germany, Apr. 1995, paper C7.3.

[6] G. Eilenberger, D. Boettle, and K. Wuenstel, "Concepts for an optical transport network and experimental results," in *Proc. ISS '95*, Berlin, Germany, Apr. 1995, paper C7.2.

[7] T. M. Martinson, "Corporate ATM optical transport network," in *Proc. EFOC&N'93*, The Hague, The Netherlands, June–July 1993, pp. 166–169.

[8] M. J. Karol, M. G. Hluchyj, and S. P. Morgan, "Input versus output queuing on a space-division packet switch," *IEEE Trans. Comm.*, vol. COMM-5, no. 12, Dec. 1987, pp. 1347–1356.

[9] D. Chiaroni, P. Gavignet-Morin, J. B. Jacob, J. M. Gabriagues, J. Jacquet, D. De Bouard, G. Da Loura, and C. Chauzat, "Feasibility demonstration of a 2.5 Gbit/s 16 × 16 ATM photonic switching matrix," in *Proc. OFC'93*, San Jose, CA, Feb. 1993, paper WD2.

[10] D. Chiaroni, P. Gavignet-Morin, and A. Jourdan, "Theoretical feasibility analysis of a 256 × 256 ATM optical switch for broadband applications," in *Proc. ECOC'93*, Montreux, Switzerland, Sept. 1993, paper We19.

[11] M. Casoni, G. Corazza, C. Raffelli, F. Masetti, and J. B. Jacob, "Clos architectures for the design of large photonic ATM switches," in *Proc. EFOC&N'94*, Heidelberg, Germany, June 1994, pp. 106–110.

[12] G. Bendelli and P. Cinato, "An ATM photonic switching module: Dimensional and architectural analysis," in *Proc. ECOC'94*, Florence, Italy, Sept. 1994, pp. 821–824.

[13] F. Masetti and J. M. Gabriagues, "Multigigabit ATM optical cell processors: Design guidelines and implementation issues," in *Proc. EFOC&N'93*, The Hague, The Netherlands, June–July 1993, paper 81.

[14] J. B. Jacob and J. M. Gabriagues, "Photonic technology for an access node in ATM-ISDN," in *Proc. 6th World Telecommun. Forum, Tech. Symp.*, Geneva, Switzerland, Oct. 1991, vol. I, pp. 111–118.

[15] T. Martinson, M. H. Fitzpatrick, D. Jasnoch, M. Renaud, J. P. Hebert, and I. Privat, "Experimental verification of correlation based synchronization in high-speed packet networks," in *Proc. ECOC'94*, Florence, Italy, Sept. 1994, pp. 601–604.

[16] M. Burzio, P. Cinato, R. Finotti, P. Gambini, M. Puleo, E. Vezzoni, and L. Zucchelli, "Optical cell synchronization in an ATM optical switch," in *Proc. ECOC'94*, Florence, Italy, Sept. 1994, pp. 821–824.

[17] D. Boettle, M. Burzio, P. Cinato, G. Eilenberger, J. B. Jacob, T. Martinson, F. Masetti, A. McGuire, M. Sotom, P. Voge, and J. Benoit, "ATMOS (ATM optical switching)—System perspective," in *Proc. Telecom '95 Tech. Forum*, Geneva, Switzerland, Oct. 1995.

[18] B. Mikkelsen, M. Vaa, R. J. Pedersen, T. Durhuus, C. Joergensen, C. Braagaard, N. Storkfelt, K. E. Stubjkäer, P. Doussiére, P. Garabédian, C. Graver, E. Derouin, T. Fillion, and M. Klenk, "20 Gb/s polarization insensitive wavelength conversion in semiconductor optical amplifiers," in *Proc. ECOC'93*, Montreux, Switzerland, Sept. 1993.

[19] K. Stubkjäer, B. Mikkelsen, T. Durhuus, C. Joergensen, S. L. Danielsen, M. Vaa, N. Vodjdani, F. Ratovelomanana, A. Enard, G. Glastre, P. Pagnod, P. Doussiére, P. Garabèdian, C. Graver, A. Jourdan, J. Jacquet, D. Leclerc, M. Erman, M. Schilling, W. Idler, D. Baums, E. Lach, G. Laube, and K. Wünstel, "Optical wavelength converters and their applications," in *IOOC'95*, Singapore, June 1995, paper ThB3–1.

[20] S. Gurib, A. Jourdan, J. G. Provost, J. Jacquet, "All-optical wavelength conversion at 10 Gbit/s by injection locking in a DFB laser," in *CLEO'94*, Anaheim, CA, May 1994, paper CTuT1.

[21] O. Hildebrand, M. Schilling, D. Baums, W. Idler, K. Dütting, G. Laube, and K. Wünstel, "The Y-laser: A multifunctional device for optical communication systems and switching networks," *IEEE J. Lightwave Technol.*, vol. 11, p. 2066, 1993.

[22] K. Stubkjäer, B. Mikkelsen, T. Durhuus, C. Joergensen, C. G. Joergensen, T. Nielsen, B. Fernier, D. Leclerc, and J. Benoit, "Semiconductor optical amplifiers as linear amplifiers, gates and wavelength converters," *ECOC'93*, Montreux, Switzerland, invited paper, Sept. 1993.

[23] B. Clesca, A. Jourdan, S. Artigaud, G. da Loura, L. Hamon, G. Soulage, J. C. Jacquinot, P. Doussiére, P. de Vivie de Régie, D. Bayart, J. L. Beylat, M. Sotom, C. Joergensen, T. Durhuus, K. Stubkjaer, M. Semenkoff, and M. Guibert, "A joint RACE experiment: 80 Gbit/s WDM transmission combining wideband optical fiber amplifiers with an in-line 10 Gbit/s all-optical wavelength converter," in *Proc. ECOC'94*, Florence, Italy, Sept. 1994, postdead, paper, pp. 27–30.

[24] B. Mikkelsen, T. Durhuus, C. Joergensen, R. J. S. Pedersen, C. Braagaard, and K. E. Stubkjäer, "Polarization insensitive wavelength conversion of 10 Gbit/s signals with SOA's in a Michelson interferometer," *Electron. Lett.*, vol. 30, pp. 260–261, Feb. 1994.

[25] W. Idler, D. Baums, K. Daub, E. Lach, G. Laube, U. Kroemer, M. Schilling, and K. Wünstel, "Chirp-free 2.5 Gbit/s signal regeneration by monolithic Mach–Zehnder interferometer wavelength converter and transmission over 188 Km of standard fiber," in *Proc. OFC'95*, San Jose, CA, Feb.–Mar. 1995, paper TuO5.

[26] N. Vodjdani, F. Ratovelomanana, A. Enard, G. Glastre, D. Rondi, R. Blondeau, C. Joergensen, S. L. Danielsen, T. Durhuus, B. Mikkelsen, and K. Stubkjäer, "Integrated optics all optical wavelength converters," in *Proc. ECIO'95*, Delft, The Netherlands, Apr. 1995.

[27] P. Doussiére, P. Garabedian, C. Graver, D. Bonnevie, T. Fillion, E. Derouin, M. Monnot, J. G. Provost, D. Leclerc, and M. Klenk, "1.55 μm polarization independent semiconductor optical amplifier with 25 dB fiber to fiber gain," *IEEE Photonics Tech. Lett.*, vol. 6, no. 2, Feb. 1994.

[28] M. Bachmann, E. Gini, and H. Melchior, "Polarization insensitive waveguide modulator using InGaAsP/InP Mach–Zenhder interferometer," in *Proc. ECOC'92*, Berlin, Germany, Sept. 1992.

[29] R. Krähenbühl, M. Bachmann, W. Vogt, T. Brenner, H. Duran, R. Baucknecht, W. Hunziker, R. Kybruz, C. Holtmann, E. Gini, and H. Melchior, "High-speed low loss InP space switch matrix for optical communication systems, fully packaged with electronic drivers and single mode fibers," in *Proc. ECOC'94*, Florence, Italy, Sept. 1994, pp. 511–514.

[30] M. Renaud, I. Privat, J. P. Hébert, J. Le Bris, J. L. Peyre, A. Pinquier, J. F. Vinchant, and T. M. Martinson, "High Speed multifunctional InP module for optical signal processing," in *Proc. ECOC'94*, Florence, Italy, Sept. 1994, pp. 515–519.

[31] D. Leclerc, P. Brosson, F. Pommereau, R. Ngo, P. Doussière, F. Mallecot, T. Fillion, G. Gelly, C. Artigue, E. Grard, G. L. Gentner, A. Pinquier, J. G. Provost, J. F. Vinchant, I. Wamsler, G. Laube, W. Hunziker, W. Vogt, and H. Melchior, "High performance semiconductor amplifier array for self-aligned packaging using Si V-groove flip-chip technique," in *Proc. ECOC'94*, Florence, Italy, Sept. 1994, postdead paper, pp. 117–120.

[32] T. Brenner and H. Melchior, "Highly efficient fiber-waveguide coupling achieved by InGaAsP/InP integrated optical mode shape adapters," in *Proc. ECOC'93*, Montreux, Switzerland, Sept. 1993.

[33] W. Hunziker, W. Vogt, and H. Melchior, "OEIC packaging with self-aligned fiber to waveguide array coupling by Si-V-groove flip-chip technique," invited paper at *Integrated Photonics Research Topical Meeting*, San Francisco, CA, Feb. 1994.

[34] R. Baucknecht, H. Duran, M. Schmatz, and H. Melchior, "InGaAs/InP double heterostructure bipolar transistors for high speed and high voltage driver circuit applications," in *Proc. IPRM'93*, Paris, Apr. 1993.

[35] P. Gavignet, M. Sotom, J. C. Jacquinot, D. Leclerc, P. Brosson, F. Pommereau, R. Ngo, and P. Doussière, "Penalty free 2.5 Gbit/s photonic switching using a semiconductor 4-gate array module," *Electron. Lett.*, vol. 31, no. 6, pp. 487–488, 1995.

[36] W. Hunziker, W. Vogt, H. Melchior, P. Gavignet, M. Sotom, J. C. Jacquinot, P. Brosson, D. Leclerc, F. Mallecot, T. Fillion, I. Wamsler, and G. Laube, "Self-aligned flip-chip packaging of tilted semiconductor

optical amplifier arrays on Si motherboard," *Electron. Lett.*, vol. 31, no. 6, 1995, pp. 488–489.
[37] P. Doussière, A. Jourdan, G. Soulage, P. Garabédian, C. Graver, T. Fillion, E. Derouin, and D. Leclerc, "Clamped gain travelling wave semiconductor optical amplifier for WDM applications," in *Proc. 14th IEEE Int. Semiconductor Laser Conf.*, Hawaii, Sept. 1994.
[38] T. Morf, C. Brys, I. Pollentier, P. De Dobbelaere, P. Van Daele, P. Demeester, M. Renaud, J. L. Peyre, and T. Martinson, "Monolithic optical receiver for high speed access nodes using epitaxial lift-off," in *Proc. ECOC'93,* Montreux, Switzerland, Sept. 1993.
[39] R. Lefèvre, J. Hourany, M. Billard, E. Grard, M. Monnot, A. Coquelin, T. Fillion, A. Jourdan, and J. Jacquet, "High-speed module for wavelength encoding in ATM optical switching," in *Proc. OFC'94,* San Jose, CA, Feb. 1994, paper ThQ6.
[40] F. Delorme, S. Slempkes, B. Rose, A. Ramdane, and H. Nakajima, "Ultra fast optical switching operation of DBR lasers using an electro-optical tuning section," in *Proc. ECOC'94,* Florence, Italy, Sept. 1994, pp. 1019–1022.
[41] H. Nakajima, J. Charil, S. Slempkes, D. Robein, A Gloukhian, J. Landreau, B. Pierre, and J.-C. Bouley, "Absorption-controlled tunable DFB amplifiers-filters," to appear in *Electronics Letters.*
[42] F. Masetti and P. Gavignet-Morin, "Optical fiber buffer for high-performance broadband switching," *European Trans. Telecom. & Rel. Technol.*, vol. 4, no. 6, Nov.–Dec. 1993.
[43] M. Calzavara, P. Gambini, M. Puleo, M. Burzio, P. Cinato, E. Vezzoni, F. Delorme, and H. Nakajima, "Resolution of ATM cell contention by multiwavelength fiber loop memory," in *Proc. ECOC'94,* Florence, Italy, Sept. 1994, pp. 567–570.
[44] P. Gavignet, F. Masetti, D. Chiaroni, and J. B. Jacob, "Design and implementation of a fiber delay line optical buffer for multi-gigabit photonic switch fabrics," *Eur. Trans. Telecom. Rel. Technol.*, vol. 7, no. 1, Jan. 1996.

Ultrafast Photonic ATM (ULPHA) Switch and a Video Distribution Experiment

Masato Tsukada and Hidetoshi Nakano

Abstract—**We describe the hardware implementation and experimental results for each function of our ultrafast photonic ATM (ULPHA) switch, which is an output buffer type ATM switch based on a time-division broadcast-and-select network. The hardware description focuses on the cell coder for generating ultrafast optical cells, the cell buffer with cell selection function, and the cell decoder for detecting ultrafast optical cells. Experiments demonstrate the generation and detection of 23 Gb/s, 4 b optical cells, including bit-error-rate measurements, and the operation of the cell buffer with the cell selector function. In a video distribution switching experiment, the switching of two time-division-multiplexed channels (25 Gb/s, 4 b optical cells) was achieved.**

I. Introduction

THE asynchronous transfer mode (ATM) is very attractive because it supports various bit-rates of services in future communication networks. While electronic-based cell switches for the ATM have been extensively researched, their throughput is limited by the speed of their electronic circuits. The major speed-limiting factors in high-speed data transfer circuits are the waveform distortion due to electromagnetic induction and the R-C time constants of the electronic circuits. In addition, high power consumption and heat production are problems in high-speed electronic circuits because more power is required for faster operation of logic gate transistors. One terabit per second is thus considered the practical limit of throughput that can be achieved by purely electronic technologies [1]. Optical circuits, on the other hand, are free from these problems because optical signals can be transmitted without distortion, even at rates higher than 1 Tb/s. Moreover, the low-loss characteristics of optical fiber allow the high bandwidths in the transmission system to be fully utilized. Photonic technologies are becoming increasingly attractive for use in switching systems because the high bandwidth of photonic technology is a key to breaking through the switching throughput limitation of purely electronic circuits [1] and [2]. Most of the photonic cell switches proposed thus far have been wavelength-division cell switches [3] and [4], whose wavelengths are used for cell routing. However, these wavelength systems require sensitive wavelength tuners and complicated electrical controllers to avoid cell contention. In order to solve this problem, we proposed an ultrafast photonic ATM (ULPHA) switch based on a time-division "broadcast and select" network [5] and [6]. The ULPHA switch has a high throughput and excellent traffic characteristics, since it uses ultrafast optical pulses for cell signals and avoids cell contention by using novel optical output buffers.

Manuscript received July 10, 1995; revised July 1, 1996.

The authors are with the NTT Network Service Systems Laboratories, Tokyo 180, Japan.

Publisher Item Identifier S 0733-8724(96)07661-X.

This paper describes recent studies on the hardware implementation of an ULPHA switch and also presents a design example for 102 Gb/s throughput. Experiments demonstrated the generation and detection of optical cells and the operation of the optical cell buffer, including bit-error-rate measurements. A buffer with a cell selection function was also demonstrated. In a video distribution switching experiment, the switching of two time-division-multiplexed channels (25 Gb/s, 4 b optical cells) was demonstrated. Finally, channel crosstalk is also discussed.

II. Configuration of ULPHA Switch

The ULPHA switch is an n-input, n-output cell switch based on an optical time-division "broadcast and select" network (Fig. 1). The bit-rate of electrical input and output is V and that of the optical time-division (TD) highway is $n \times V$. All the cells generated by the cell coder are multiplexed onto the optical TD highway. Certain cells addressed to each output are selected from the highway with their throughput controlled by the buffer. All these functions are processed optically. The operations of the ULPHA switch are as follows.

First, the cell coder generates ultrafast optical data cells and slow optical address cells. Then, the star coupler time-multiplexes these cells onto the optical TD highway. A timing chart of optical pulses in the cell coder output and the star coupler output is also shown in Fig. 1. Electrical cells arriving at n-inputs are made from 2^k bit of data and several bits of address, and have a bit rate of V. Here, ultrafast data cells are generated by a cascade compressor in the cell coder using ultra-short optical pulses. At the same time, slow address cells are generated by direct modulation of the laser diode in the cell coder. They are wavelength multiplexed onto the same time position. The star coupler time-multiplexes all cells from each cell coder onto the optical line by giving them proper delays. The multiplexed cells are separated by a guard-time needed for switching by tree structure switches at output buffers.

Next, the cell selector, the cell buffer, and the cell decoder select cells from the optical TD highway and adjust their throughputs and bit-rates to V. The cell selector analyzes the address cells and passes the corresponding data cells to the buffer only when it detects matching addresses. The selected cells are buffered in the buffer so that they leave with proper cell intervals. This is necessary to get an output of the bit-rate

Reprinted with permission from *IEEE Journal of Lightwave Technology*, Vol. 14, No. 10, pp. 2154-2161, October 1996.

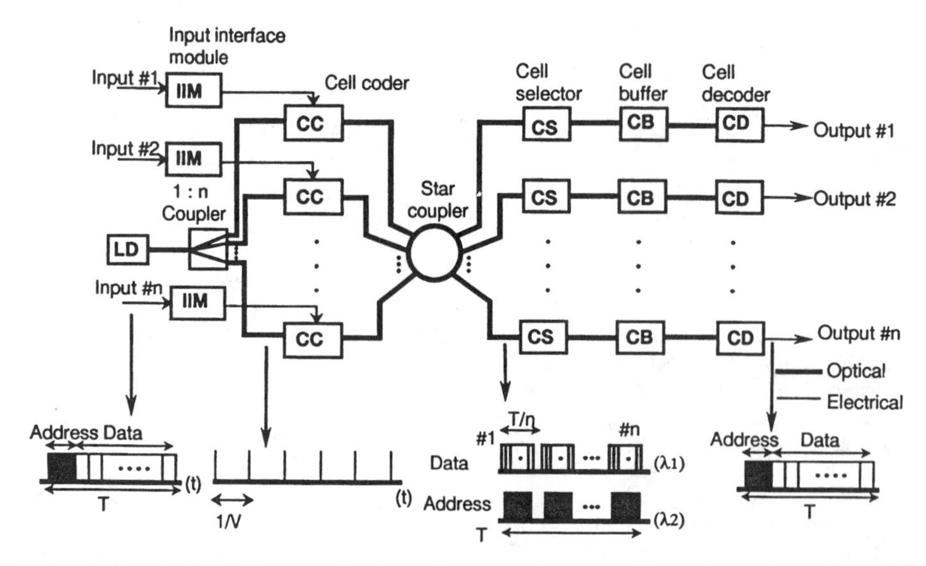

Fig. 1. Block diagram of the ULPHA switch. The bit rate of the electrical inputs and outputs is V and that of the optical TD highways is $n \times V$. In each input interface module, the cells are regenerated and reclocked by the system clock at rate V, and individual cells are inserted into synchronous time slots. The cell coder and star coupler multiplex cells made up of ultrashort optical pulses onto the TD highways. The cell selector, cell buffer, and cell decoder select certain cells from the highways and adjust their throughputs and bit rates to V. All these functions are optically processed on a cell-by-cell basis.

V. The cell decoder converts the cell bit-rate back to V and outputs it as an electrical signal.

The ULPHA switch has a high throughput because it uses ultra-short optical pulses for the optical data cells. It can avoid contention without using complicated controllers and algorithms. Moreover, address detection and analysis are easy because simple WDM without complicated controllers is used.

III. Hardware Implementation

A. Cell Coder and Star Coupler

The cell coder and star coupler configuration is shown in Fig. 2(a). An electrical cell with a length of $T(= 2^k/V)$ is stored in a register, and its data and address are read separately. The data cell is converted into modulated optical pulses (wavelength: λ_1) with a bit-rate of V. The modulated optical pulses travel through a cascade compressor consisting of a number of "COMs" (COM#1,$\cdots$, COM#k) and an optical gate switch. A "COM" consists of a 1×2 optical coupler, a pair of fiber delay lines, and a 2×1 optical coupler. The relationship between optical pulse trains and the control signal for the optical gate switch is shown in Fig. 2(b). In the cascade compressor, a pair of fiber delay lines in COM#j results in a relative delay of $2^{j-1}(1/V - \Delta t)$ between two optical pulse trains divided by the optical coupler. At the compressor's output, the optical pulse groups that consist of 2^k optical pulses with a separation of Δt are obtained. This type of pulse group is called an "optical cell." Finally, only the optical cell that has the correct sequence is selected by the optical gate switch. As the cascade compressor is composed of passive optical elements, the system does not limit the bandwidth.

The address cell is converted into an optical signal (wavelength: λ_2) by direct LD modulation. Finally, the ultrafast data cells (λ_1) and relatively slow address cells (λ_2) are wavelength multiplexed and sent to the star coupler.

B. Cell Buffer with Cell Selection Function

The circuit described here has the same structure as proposed in [7]. The functions of cell selector and cell buffer are achieved at the same time by the circuit shown in Fig. 3(a). Its timing chart is shown in Fig. 3(b). This circuit consists of an optical single-output FIFO buffer memory and a cell interval regulating circuit. The buffer memory can accept as many as n data cells in a duration of T seconds, but only one of them will be output from it during that time. The regulating circuit can regulate the interval between any two consecutive cells so that the interval is always kept to T seconds.

C. Cell Decoder

The proposed configuration of the cell decoder and its conversion timing chart for the case where the input is an N-bit optical cell are shown in Fig. 4(a) and (b), respectively [8]. At the decoder, optical cell bursts with bit-rate $Vc(= 1/\Delta t)$ and N bit $(2^{k-1} < N \leq 2^k)$ enter pulse-interval-expander (PIE) #1, which consists of a 1×2 switch, a delay line, and a 2×1 coupler. In PIE#1, the 1×2 switch first sends the optical pulses alternately to its dual ports according to the RF driving signal of frequency $Vc/2$. The delay line then gives one of these pulses a relative delay of $(1/V - 1/Vc)$, which corresponds to the length of one electrical bit. The 2×1 coupler recombines the two pulse streams. Subsequently, in PIE#k, the 1×2 switch, which is driven by an RF signal of frequency $Vc/2^k$, sends the pairs of pulses alternately to its dual ports, and then the delay line gives one pulse stream a relative delay of $2^{k-1}(1/V - 1/Vc)$. In this way, the decoder acts as a binary demultiplexer, whereby optical cells of bit-rate Vc are finally converted into cells of bit-rate $Vc/(1 + VcT)$ with the correct sequence of pulses. The RF driving signals applied to the 1×2 switches can be generated by a series of 1/2 frequency dividers and distributed to the 1×2 switches in each PIE. In this distribution, it is necessary

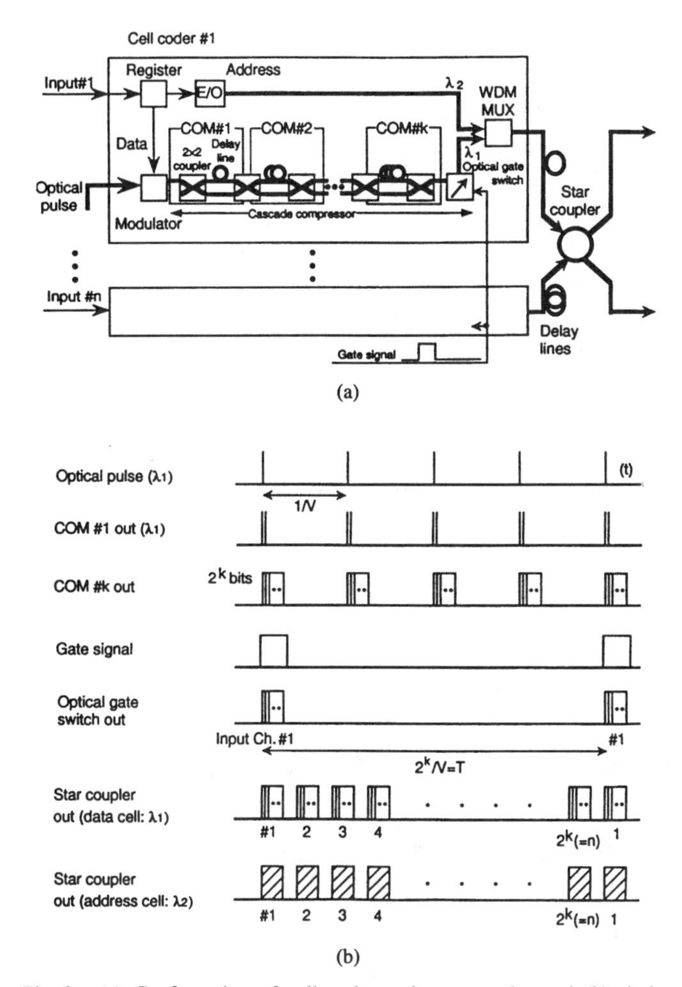

Fig. 2. (a) Configuration of cell coder and star coupler and (b) timing diagram for the cell coder. The cell coder generates 2^k-bit ultrafast optical data cells of wavelength λ_1. At the same time, the cell coder generates slow address cells of wavelength λ_2. The data and address cells are wavelength-multiplexed at the cell coder output.

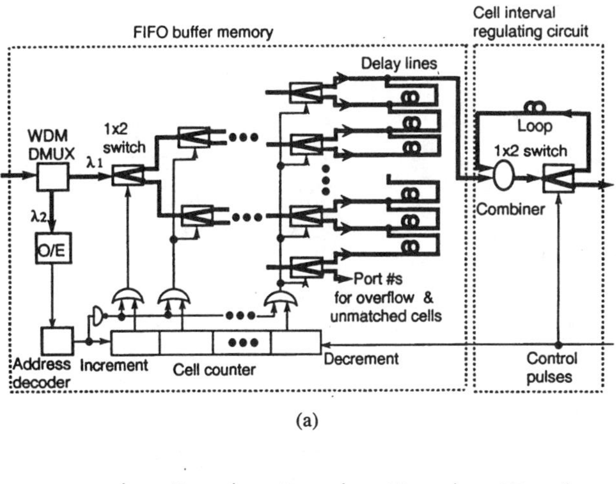

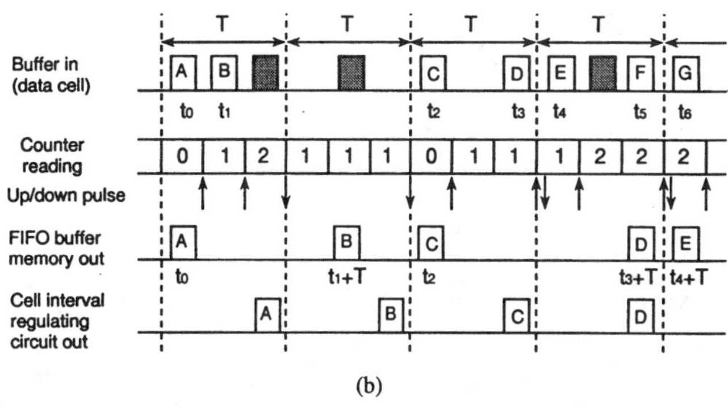

Fig. 3. (a) Configuration of optical buffer having a cell selector function. The optical FIFO buffer memory gives each optical cell an appropriate delay so that no more than one cell exists in a duration of T seconds at the output. The cell interval regulating circuit can regulate the interval between any two consecutive cells such that the interval is always kept to T s and (b) timing chart for the buffer. Cells A-G are allocated to each duration of T s. Hatched cells are destined for another output and discarded from port #s of the switch.

to control the phase of each RF signal by electrical delay lines so that the 1×2 switches are driven at the correct timing. This configuration requires less hardware, such as optical switches, fiber delay lines, and couplers, than a tree-structure switch [5]. Fig. 5 shows the number of 1×2 switches versus the number of bits in one cell for both methods. It is assumed in this case that the $1 \times n$ tree-type switch is composed of $(n-1)$ 1×2 switches. The proposed decoder requires far fewer switches than the other one when there are a large number of bits in one cell. For instance, when the number of bits is 512, the new decoder requires only nine switches whereas the other requires 511. This advantage applies to the fiber delay lines and the optical couplers.

Another advantage of this converter is that broad-band switching technology is not required. The key technology is high-speed and narrow-band switching. Recently, a binary multiplexer/demultiplexer based on an electrooptic Ti:LiNbO$_3$ switch has been demonstrated at a speed of up to 72 Gb/s [9], and electrooptic on/off gates using two cascaded Ti:LiNbO$_3$ Mach–Zehnder intensity modulators have been demonstrated at a speed of up to 49.6 Gb/s [10]. Applying these techniques to the 1×2 switches in the PIE's should make it possible to detect optical signals with speeds of over 100 Gb/s.

IV. Switch Design

This section describes the relationship between the number of channels and the number of bits in one cell in the ULPHA switch. The number of address bits required to express n different outputs is

$$\log_2 n. \tag{1}$$

The length of a data cell with 2^k bit is

$$2^k \cdot \Delta t. \tag{2}$$

Now, if the data cell length including guard-time $\Delta\tau$ is equal to the address cell length, they are related by

$$2^k \cdot \Delta t + \Delta\tau \geq \frac{\log_2 n}{V}. \tag{3}$$

Furthermore, we can get the number of channels limited by the relationship between the input/output bit rate and the optical pulse width as

$$(2^k \cdot \Delta t + \Delta\tau)n \leq \frac{2^k}{V} + \frac{\log_2 n}{V}. \tag{4}$$

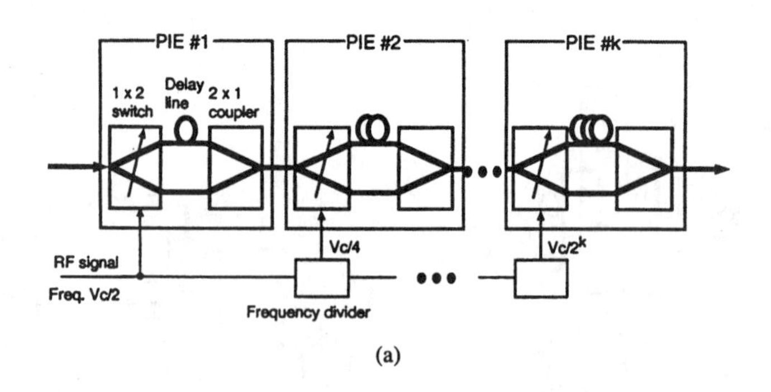

(a)

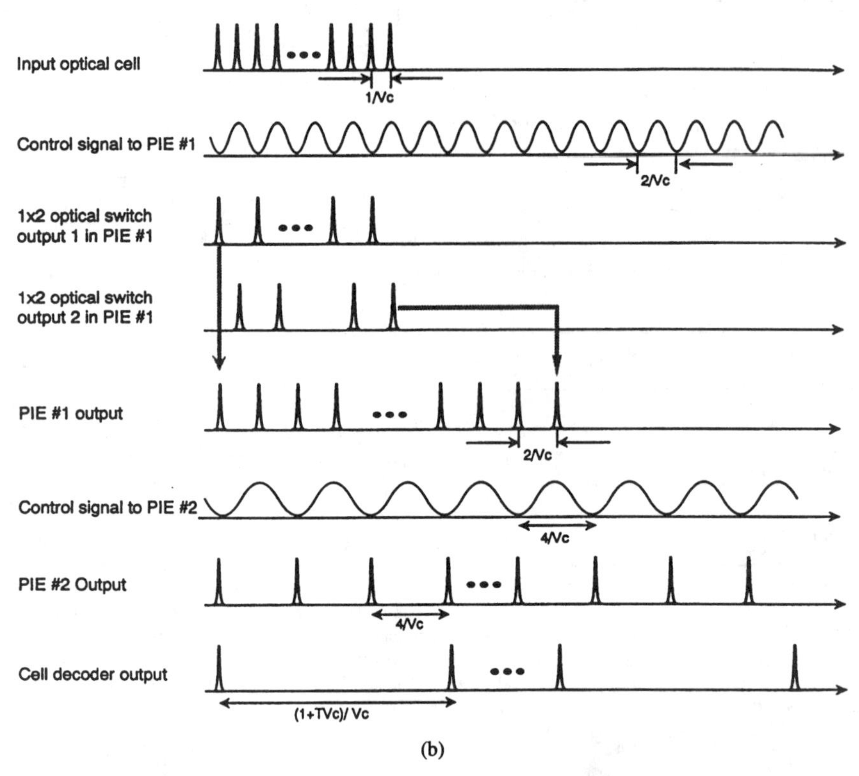

(b)

Fig. 4. (a) Configuration of cell decoder and (b) timing chart of the cell decoder. The cell decoder acts as a binary demultiplexer, whereby optical cells of bit rate Vc are finally converted into cells of bit rate $Vc/(1 + VcT)$ with the correct sequence pulses.

Considering conditions (3) and (4) and the following assumptions:

- input bit rate (V): 1.6 Gb/s,
- number of bits in one cell (2^k): 512,
- guard-time ($\Delta\tau$): 1 ns, and
- pulse interval of optical cell (Δt): 6 ps,

the following system can be designed:

- number of input channels (n): 64,
- optical TD highway bit rate ($n \times V$): 102 Gb/s.

Furthermore, if such switch modules are interconnected to construct a multi-stage network [11], a much larger capacity, e.g., 4096×4096 with 1.6 Gb/s input, will be possible in the future. Then, the total throughput would be 6.5 Tb/s.

V. Experiment

A. Generation and Detection of Optical Cells

We performed an experiment to detect ultrafast 4 b optical cells using the proposed coder and decoder. Fig. 6 shows the experimental set-up. A mode-locked DFB-LD (1.31 μm) was used as a light source; it generated 20 ps optical pulses with a 1.6 GHz repetition rate. A Ti:$LiNbO_3$ intensity modulator was used as a gate switch to obtain a pulse train with a 100 MHz repetition rate. Next, these pulses were modulated by an LD intensity modulator based on a 100 Mb/s electrical NRZ signal with a "1011" bit pattern. The modulated optical pulses entered the cascade compressor. After passing through the compressor, the bit rate of the 4 b cells became 23 Gb/s. These cells then entered the cell decoder, which has two PIE's. Switching signals with frequencies of 11.6 and 5.8 GHz were applied to the two 2×2 Ti:$LiNbO_3$ switches in PIE#1 and PIE#2, respectively. The optical 3-dB bandwidths of these switches were 20 GHz [12] and 8 GHz, respectively. The delay lines give the pulses relative delays of 9.96 and 19.92 ns, respectively. Finally, a 100-Mb/s electrical signal with the correct sequence ("1011") was obtained. The bit-error-rate was measured after the LD intensity modulator and the cell decoder. The results are shown in Fig. 7. The receiver sensitivity was -33.5 dBm at a BER of 10^{-9}. The bit rate

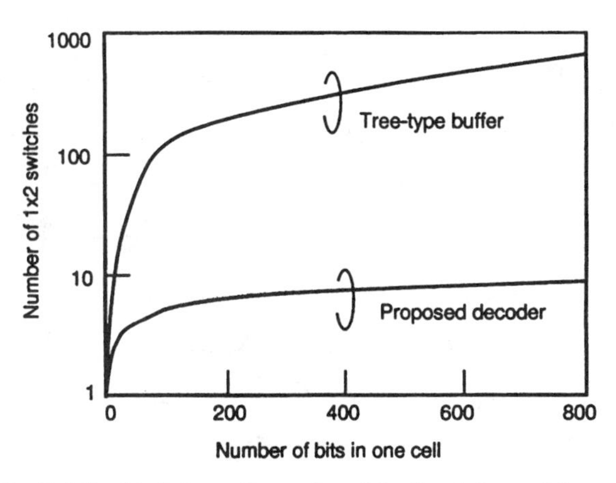

Fig. 5. Relationship between the number of 1 × 2 switches and the number of bits in one cell.

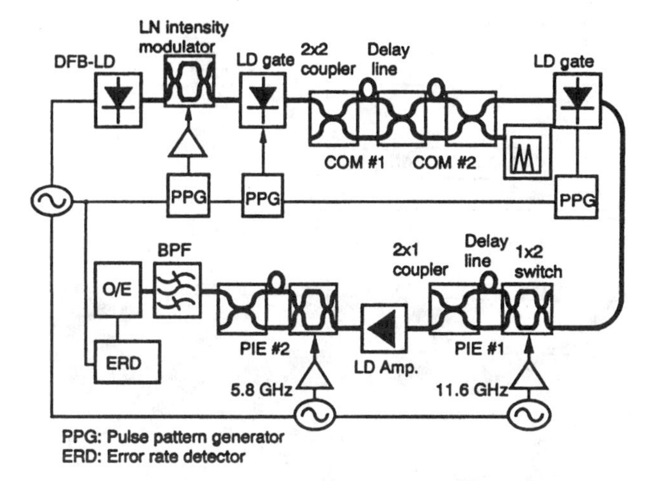

Fig. 6. Experimental set-up for optical cell generation and detection.

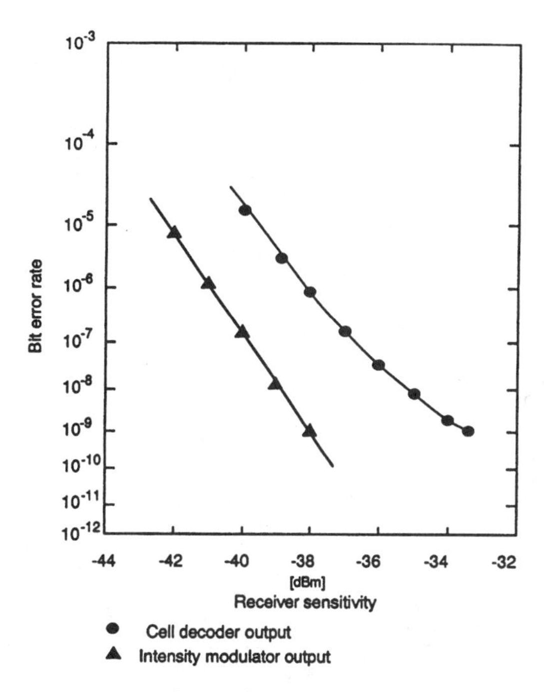

Fig. 7. Bit error rate against receiver sensitivity.

conversion resulted in a power penalty of less than 4.5 dB at a BER of 10^{-9}. This could be attributed to crosstalk, caused by switching at the PIE's. Since the product of the pulse width and the bit-rate was not so small in this experiment, the remaining optical power corresponding to the pedestals of the pulses influenced the adjacent bits as crosstalk. We discuss this problem in Section VII.

B. Cell Buffer with Cell Selection Function

An experiment on the FIFO buffer memory with the cell selection function in the buffer was performed using a 1 × 4 optical switch and an electrical control circuit. Fig. 8 shows an experimental set-up for the FIFO buffer memory. For data cells, 1.6 Gb/s, 8 b (including 2 b guard-time for switching) optical cells are generated by direct modulation of a 1.32 μm DFB-LD. For address cells, 1.6 Gb/s, 8 b electrical cells are generated by a pulse pattern generator. The electrical controller consists of an address decoder, an up-down counter, and several electrical gate circuits. In the address decoder consisting of an 8 b demultiplexer and a comparator, the electrical address cell is judged by comparing it to a preset 8-bit address. The 1 × 4 Ti:$LiNbO_3$ [13] optical switch was driven by signals from the electrical controller. Output ports #1, #2, and #3 of the 1 × 4 optical switch were connected with 0, T (40 ns), and $2T$ (80 ns) fiber delay lines, respectively, while output port #4 was prepared for discarded or overflowed cells. In this experiment, only optical data cells with the address pattern of "11 010 111" were selected by this buffer. The selected cells were switched to ports #1, #2, and #3 according to the control signal for the first and second stages of the 1 × 4 optical switch [(Q1, Q2): (0, 0), (0, 1), and (1, 0)], respectively. The optical data cells with other address patterns were switched to port #4 by the control signal (1, 1). Fig. 9 shows the experimental results. Traces (a)–(f) show input optical data cells, electrical address cells, electrical control signals for the first stage (Q1), electrical control signals for the second stage (Q2), output optical data cells, and discarded cells, respectively. According to traces (b)–(d), the first optical data cell was guided to port #1, the third and sixth cells were guided to port #2, and the eighth cell was guided to port #3. The other cells were guided to port #4 for discarding. Thus, we obtained good buffering and selection, in which there were no more than two selected cells in a time interval of 40 ns.

VI. 2 × 1 Video Distribution Experiment

A. System Description

To demonstrate the fundamental operation of the ULPHA switch, we performed a video distribution experiment [14]. The block diagram and the timing chart of the experimental system are shown in Figs. 10 and 11, respectively. Gain-switched DFB-LD's operating at 1.303 μm were used as sources of ultrashort optical pulses for data signals. Pulse duration was compressed to 14 ps (FWHM) by linear chirp compensation using dispersion shifted optical fiber. The repetition rate of the laser pulse was 100 MHz. Laser diode gate switches driven by 100 Mb/s signals from video coders were used as intensity modulators. At the cell coder, the modulated

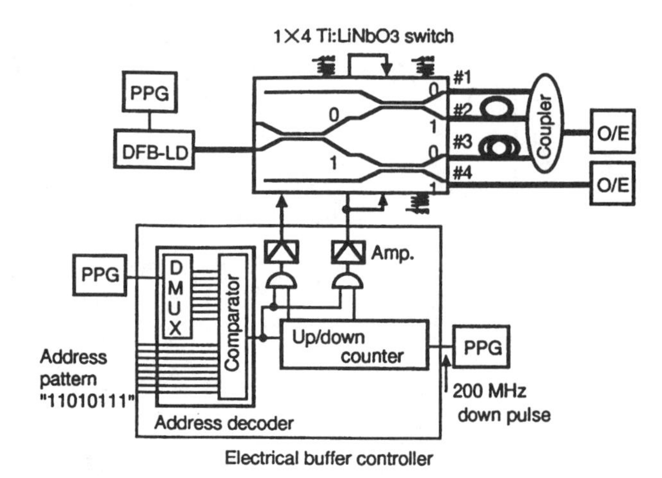

Fig. 8. Experimental set-up for the FIFO buffer memory.

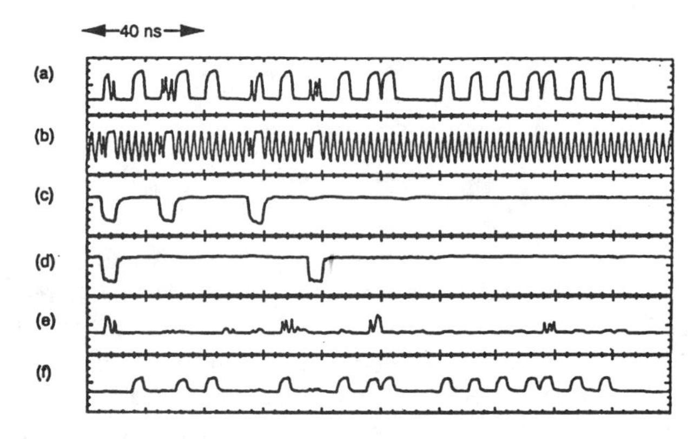

Fig. 9. Experimental results for the FIFO buffer memory.

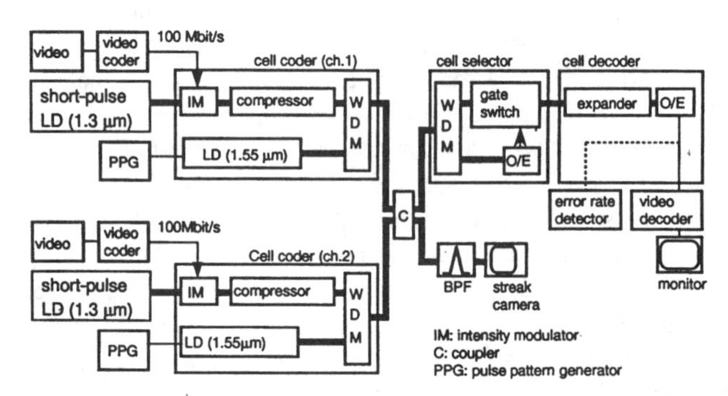

Fig. 10. Block diagram of the 2 × 1 video distribution experiment.

optical pulse train was converted into 4 b optical cells with a bit-rate of 25 Gb/s. The address signal was generated by direct modulation of a DFB–LD operating at 1.552 μm and was attached to the corresponding data. An optical coupler combined the cells from two channels to generate a TDM-highway. At each cell selector, according to the 1.552 μm address signal, the LD gate switch was driven to select the address-matched data with a wavelength of 1.303 μm from the TDM highway.

The selected optical cells, which had a bit-rate of 25 Gb/s, were converted into 100 Mb/s cells by the cell decoder. Bit-rate down-converted cells were detected by a photodetecter and changed into electrical ones. In the cell decoder, $LiNbO_3$

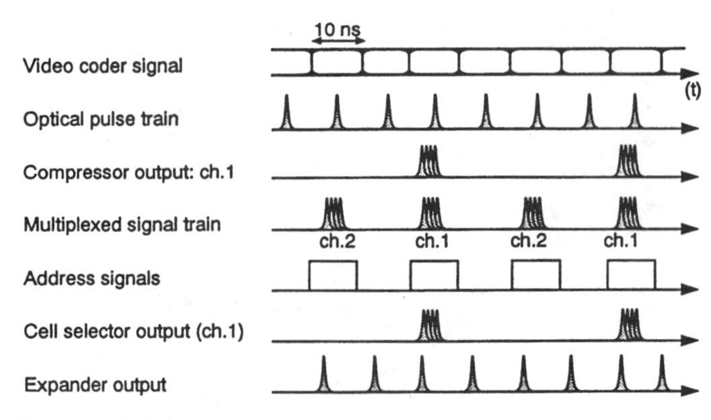

Fig. 11. Timing chart of the 2 × 1 video distribution experiment.

TABLE I
SPECIFICATIONS OF THE EXPERIMENTAL 2 × 1 VIDEO DISTRIBUTION SYSTEM

Video coder	100 Mbit/s mB1C bit sequence
Optical pulse	100 MHz repetition rate 14 ps pulse-width
Wavelength	Data (optical pulse): 1.3 μm Address: 1.55 μm
Intensity modulator (IM)	LD gate switch
Compressor	100 Mbit/s→25 Gbit/s
Expander	25 Gbit/s→100 Mbit/s

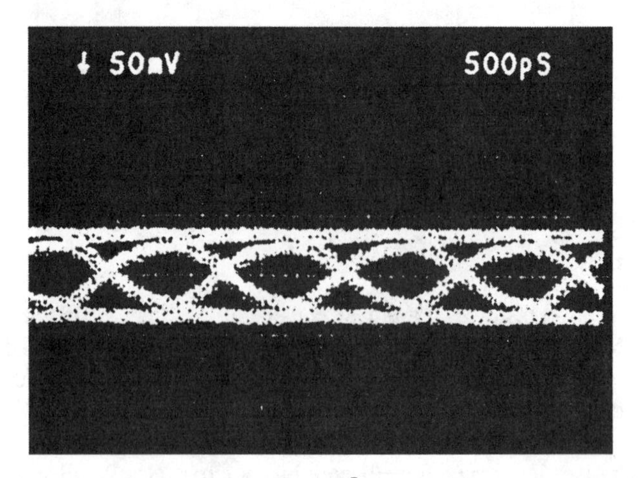

Fig. 12. Eye-diagram at 1 Gb/s PRBS (2^7-1) modulation in one of LD gate switch.

switches with an optical 3-dB bandwidth of 20 GHz [12] were used as high-speed optical switches. Table I shows the specifications of this experimental system.

B. Switching Speed and Extinction Ratio of LD Gate Switch and Wavelength Crosstalk

In this experiment, the LD gate switches were used at the intensity modulators, the cell coders, and the cell selector. Fig. 12 shows the eye-diagram at 1 Gb/s PRBS (2^7-1) modulation in one of the LD gate switches. The switching speed was about 2 ns. Furthermore, a static extinction ratio was more than 30 dB. The others switches had similar switching speeds and extinction ratios. In addition, wavelength crosstalk at the optical receiver was less than −35 dB. These influences on system performance can be considered to be very small.

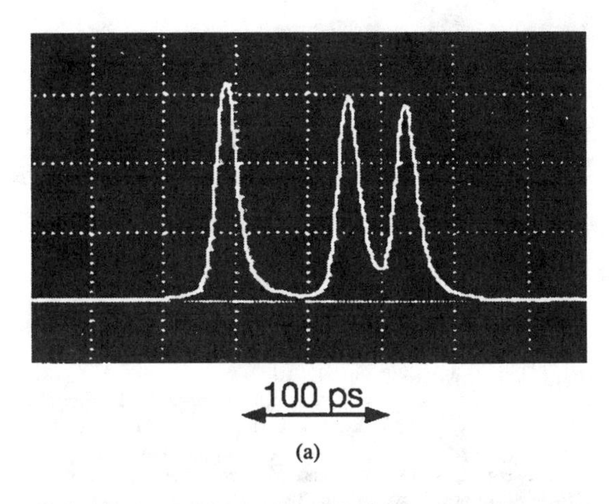

(a)

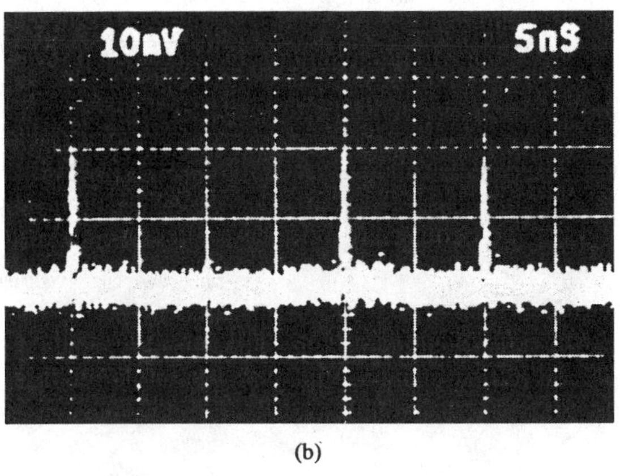

(b)

Fig. 13. (a) Example of a 25 Gb/s, 4 b optical cell waveform monitored by a synchronous streak camera. This waveform corresponds to the bit pattern "1011" and (b) example waveforms of the 100 Mb/s, 4 b ("1011") signal at the cell decoder output.

C. Switching Characteristics

Fig. 13 shows pulse waveforms at the output of the cell coder and the cell decoder when optical pulses were modulated by a ("1011") fixed pattern. The former was monitored by a streak camera and the latter was detected by a pin-diode. A bit error rate of 10^{-6} at -32 dBm was measured. We could not measure a bit error rate of more than 10^{-6} because the maximum power at the output of the cell decoder was -32 dBm in these experiments. Thus, we could get two kinds of good video image by switching the bit pattern from the pulse pattern generators for address signals. The results show the feasibility of switching video sources.

VII. Discussion

In this section, we discuss channel crosstalk which is generated in the cell decoder. Fig. 14 shows the state of optical pulses in one bit period ($1/V$). The channel crosstalk is defined by the ratio of the output power of a signal pulse to the sum of the output powers of other bits' pulses in one bit period. When the number of bits in one cell is 2^k and excess loss of 1×2 switches is neglected, the output power of the signal pulse is written as

$$\frac{P_{\text{in}}}{2^k} \prod_{j=1}^{k} (1 - R_j) \tag{5}$$

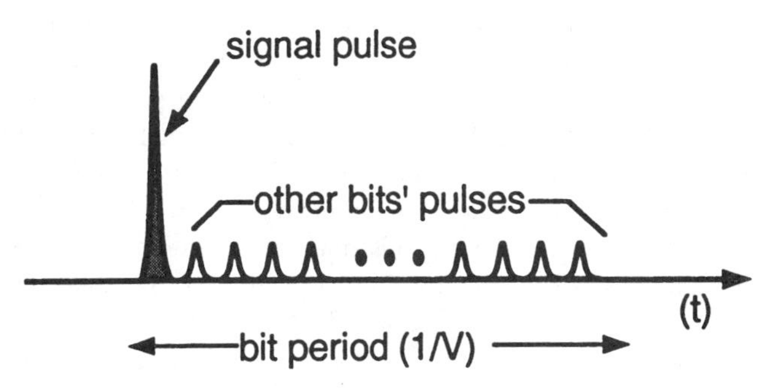

Fig. 14. State of optical pulses in one bit period.

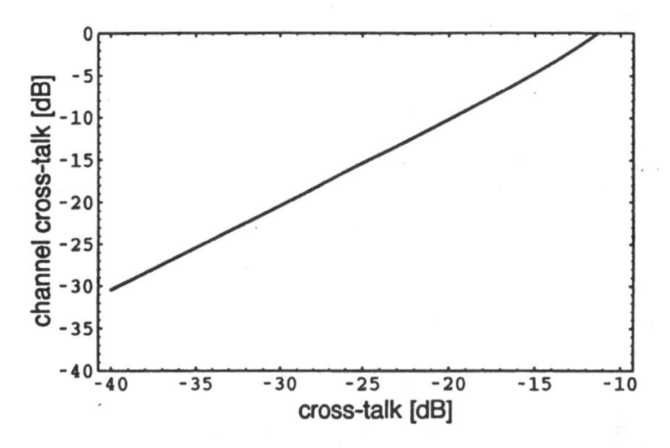

Fig. 15. Relationship between the crosstalk and the channel crosstalk.

where R_j is the crosstalk generated by switching in PIE#j and P_{in} is the peak power of input pulses.

The sum of the output powers of the other bits' pulses is written as

$$\begin{cases} \dfrac{P_{\text{in}} R_1}{2^k} & k = 1 \\ \dfrac{P_{\text{in}} R_1}{2^k} \left\{ 1 + \displaystyle\sum_{c=2}^{k} \left[\prod_{j=2}^{c} (1 - R_j) \right] \right\} & k \geq 2. \end{cases} \tag{6}$$

For simplicity of discussion, we assume that each crosstalk R_j is equal to R_1, which has the worst value in R_j. Fig. 15 shows the relationship between the crosstalk and the channel crosstalk. To keep the channel crosstalk below -15 dB, the crosstalk must be less than -25 dB.

R is determined by the product of a modulation waveform and an optical pulse profile. When a Gaussian profile is assumed as the optical pulse and a Ti:$LiNbO_3$ Mach–Zehnder switch is used in each PIE, a crosstalk of less than -25 dB is obtained when static crosstalk of the 1×2 switch is -30 dB and the normalized channel spacing $(1/V)/\Delta t$, which is defined by the ratio of the bit period to the pulse interval, is 6 [10]. Thus, when Δt is 6 ps, a pulse width of 1 ps is required, so the channel crosstalk is -15 dB.

VIII. Conclusion

This paper described recent studies on the hardware implementation of an ULPHA switch, and also presented a design example of 102 Gb/s throughput. 23 Gb/s, 4 b optical cells were experimentally generated and detected, and an optical cell buffer was demonstrated. The bit-error-rate was measured at the intensity modulator output and the cell decoder output. The receiver sensitivity was -33.5 dBm at a BER of 10^{-9}, and the bit rate conversion resulted in a power penalty of less than 4.5-dB at a BER of 10^{-9}. The optical cell buffer with cell selection function achieved good buffering and selection, in which there were no more than two selected cells in a time interval of 40 ns. A video distribution switching experiment demonstrated the switching of two (25 Gb/s, 4-bit optical cells) time-division-multiplexed channels. In addition, we discussed channel crosstalk in the cell decoder. The relationship between the channel crosstalk and the crosstalk generated by switching in the PIE was obtained. These results show the great potential of the ULPHA switch.

Acknowledgment

The authors would like to thank Dr. Miyazawa for providing the devices. They would also like to thank Dr. K. Yukimatsu, Dr. Shimazu, and Dr. Wen for their encouragement and discussions.

References

[1] K. Yukimatsu and T. Aoki, "Advanced switching technologies toward tera-bit communications networks," in *Singapore ICCS' 90 Conf. Proc.*, Nov. 1990, vol. 1 of 2, no. 11.1.1, pp. 397–401.

[2] T. Aoki, "Trends in electronic communication switching technologies," in *Tech. Dig. Photon. Switch. Top. Meeting,* 1990, vol. 12C-3, pp. 1–5.

[3] H. Kobrinski, "An optoelectronic packet switch utilizing wavelength tuning," in *GLOBECOM'88 Conf. Rec.*, Nov. 1988, pp. 948–950.

[4] S. Kuroyanagi, T. Shimoe, and K. Murakami, "Photonic ATM switching network," in *Top. Meeting Photon. Switching '90,* 1990, vol. 14B-2, pp. 223–225.

[5] Y. Shimazu and M. Tsukada, "Ultrafast photonic ATM switch with optical output buffers," *J. Lightwave Technol.*, vol. 10, pp. 265–272, Feb. 1992.

[6] M. Tsukada and Y. Shimazu, "Ultrafast photonic ATM switch based on time-division broadcast-and-select network," in *GLOBECOM'91 Conf. Rec.*, Nov. 1991, vol. 34.5.

[7] W. D. Zhong, M. Tsukada, K. Yukimatsu, and Y. Shimazu, "Terahipas: A modular and expandable terabit/second hierarchically multiplexing photonic ATM switch architecture," *J. Lightwave Technol.*, vol. 12, pp. 1307–1315, July 1994.

[8] M. Tsukada, Y. Shimazu, and H. Nakano, "Detection of 23-Gbit/s 4-bit optical cells using a new bit-rate down converter," *Electron. Lett.*, vol. 29, no. 13, p. 1203, 1993.

[9] S. K. Korotky and J. J. Veselka, "Efficient switching in a 72-Gb/s $Ti:LiNbO_3$ binary multiplexer/demultiplexer," in *OFC'90, Conf. Rec.*, 1990, vol. TUH2.

[10] M. Jinno, "Ultrafast time-division demultiplexer based on electrooptic on/off gate," *J. Lightwave Technol.*, vol. 10, pp. 1458–1465, Oct. 1992.

[11] G. J. Fitzpatrick and E. A. Munter, "Input-buffered ATM switch traffic performance," in *Proc. Multimedia'89*, Apr. 1989, no. 4.2.

[12] H. Suzuki, H. Nagano, T. Suzuki, T. Takeuchi, and S. Iwasaki, "Output buffer switch architecture for asynchronous transfer mode," in *ICC'89 Conf. Rec.*, June 1989, pp. 99–103.

[13] T. Nozawa, M. Yanagibashi, and H. Miyazawa, "A broad-band and low-driving voltage 2×2 $Ti:LiNbO_3$ optical switch," in *OSA Proc. Photonic Switching*, Apr. 1990, vol. 14D-7(PD), pp. 34–36.

[14] H. Miyazawa, T. Nozawa, and M. Yanagibashi, "A high-speed 1×4 $Ti:LiNbO_3$ optical switch," in *OEC'90, Conf. Rec.*, July 1990, vol. PD-2, pp. 8–9.

[15] H. Nakano, M. Tsukada, and K. Yukimatsu, "Photonic ATM switching system," in *Proc. APPC'93*, 1993, vol. 1C.1, pp. 52–55.

CORD: Contention Resolution by Delay Lines

IMRICH CHLAMTAC, FELLOW, IEEE, ANDREA FUMAGALLI L.G. KAZOVSKY, FELLOW, IEEE,
PAUL MELMAN, WILLIAM H. NELSON, SENIOR MEMBER, IEEE, PIERLUIGI POGGIOLINI,
MAURO CERISOLA, A.N.M. MASUM CHOUDHURY, SENIOR MEMBER, IEEE,
THOMAS K. FONG, R. THEODORE HOFMEISTER, MEMBER, IEEE,
CHUNG-LI LU, ADISAK MEKKITTIKUL, DELFIN JAY M. SABIDO IX,
CHANG-JIN SUH, AND ERIC W.M. WONG

Abstract—**The implementation of optical packet-switched networks requires that the problems of resource contention, signaling and local and global synchronization be resolved. A possible optical solution to resource contention is based on the use of switching matrices suitably connected with optical delay lines. Signaling could be dealt with using subcarrier multiplexing of packet headers. Synchronization could take advantage of clock tone multiplexing techniques, of digital processing for ultra-fast clock recovery, and of new distributed techniques for global packet-slot alignment.**

To explore the practical feasibility and effectiveness of these key techniques, a consortium was formed among the University of Massachusetts, Stanford University, and GTE Laboratories. The Consortium, funded by ARPA, has three main goals: investigating networking issues involved in optical contention resolution (University of Massachusetts), constructing an experimental contention-resolution optical (CRO) device (GTE Laboratories), and building a packet-switched optical network prototype employing a CRO and novel signaling/synchronization techniques (Stanford University).

This paper describes the details of the project and provides an overview of the main results obtained so far.

I. INTRODUCTION

BOTH circuit and packet switching have been proposed to provide the user with the required flexibility for an efficient access to optical networks. Recently, a trend has developed that places the emphasis on packet switching solutions to provide the added flexibility for supporting bulk data transfer, multimedia, and other advanced applications [1]–[3]. These solutions appear to be preferable in the field of supercomputing and local-area and metropolitan-area network (LAN/MAN) backboning, and are potentially more flexible and efficient in allocating bandwidth to multiple users with different transmission needs, especially at high transmission speeds. In addition, the packet switching approach is fostered by the worldwide emergence of the ATM standard.

Manuscript received January 15, 1995; revised October 31, 1995. This work was sponsored by ARPA under Contracts MDA972-93-1-0028, MDA972-93-1-0031, and MDA972-93-C-0057.

I. Chlamtac, A. Fumagalli, C.-J. Suh, and E. W. M. Wong are with the Department of Electrical and Computer Engineering, University of Massachusetts, Amherst, MA 10003 USA.

L. G. Kazovsky, P. Poggiolini, M. Cerisola, T. K. Fong, R. T. Hofmeister, C.-L. Lu, A. Mekkittikul, and D. J. M. Sabido IX, are with the Optical Communicatins Research Laboratory, Stanford University, Stanford, CA 94305-4055 USA.

P. Melman, W. H. Nelson, and A. N. M. Masum Choudhury are with GTE Laboratories, Waltham, MA 02254 USA.

Publisher Item Identifier S 0733-8716(96)03687-6.

In this context, *all-optical*[1] solutions are attracting a considerable interest, whereby some or all of the switching and routing operations are done without conversion of packets between the optical and electronic domain within the network.

Even in all-optical packet networks, the fundamental problem of *resource contentions* still arises. When multiple packets, arriving on different fibers or wavelengths, need to simultaneously access one or more of the optical network resources (for instance receivers, switches or wavelengths), the resulting contention must be resolved. In an all-optical network, a solution maintaining the packets in the optical domain is desirable.

One possible way to tackle the contention problem in packet switched systems with WDM-TDM access techniques is the use of switched optical delay lines (SDL). An SDL unit is an optical fabric based on a combination of optical switching matrices and fiber delay lines. It allows to redistribute packets contending for the same resource over time and space, while keeping them in optical form. This way, rescheduling of resource access by the contending packets can be obtained, and contention can be reduced.

SDL's for optical packet switched networks were originally proposed in [4] to deal with receiver contentions in a WDM star network. The technique, termed Quadro, was later proposed for rings [5], [6], and multihop optical Manhattan Street Networks (MSN's) [7]. Since then several research groups have explored the concept, in various contexts, to deal with different contention problems [8]–[12]. An approach for simultaneously dealing with resource contentions as well as time-misaligned frames was proposed in [13].

An experimental prototype of an SDL fabric was presented in [14]. In Europe, the ATMOS project employed delay lines [15], [16]. In Japan [17] and Australia [18], recirculating delay lines have been demonstrated in photonic switching applications. A multihop packet switching node making use of SDL's was reported on in [19].

The actual practicality of the SDL approach, however, still needs to be proved. For instance, the traditional $LiNbO_3$ interferometric switching devices, on which virtually all prototypes have been based so far, are very difficult to control and are polarization sensitive. A key step that needs to be taken in

[1] In this paper we will term "all-optical" any approach in which the optical packet is not converted to the electronic domain prior to reaching the final destination node.

Reprinted with permission from *IEEE Journal on Selected Areas in Communications*, Vol. 14, No. 5, pp. 1014-1029, June 1996.

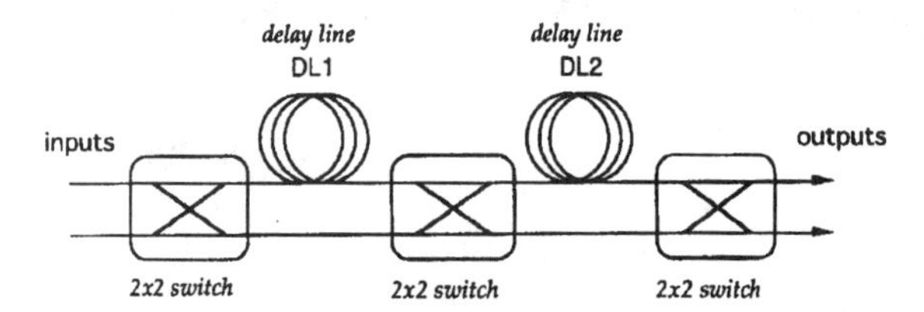

Fig. 1. Block diagram of a two-stage SDL architecture. This schematic also describes the CRO that is being built by GTE and will be used in the network prototype assembled at Stanford University.

view of actual applications is the fabrication of a polarization-insensitive device, easy to control and adjust, that features *integrated* switches and possibly integrated amplifiers.

Another major problem in all-optical packet-switched networks is that of signaling. A routing or receiving node must be able to retrieve header information from simultaneously incoming WDM-multiplexed packets, on the fly. To accomplish this task, subcarrier multiplexing of packet headers has recently been proposed [20]–[26]. This concept still needs to be tested in conjunction with fully operational transmission and synchronization subsystems. Several implementation issues, such as crosstalk between baseband payload and headers, need to be further investigated.

Finally, synchronization, both at the packet time-slot level and clock level, is a very critical aspect of all-optical packet networks. Slot synchronization is often required so that packets arrive at the network routing or receiving nodes with their boundaries aligned in time. This calls for some form of global slot alignment. Clock recovery must be done on a packet-by-packet basis, since in adjacent time-slots a node may be looking at packets coming from different transmitters. Ultra-fast digital techniques, potentially capable of recovering the clock in a few bits, have been recently proposed. A more traditional technique, consisting in sending the clock tone along with the packet data has also been proposed [20]. Both the implementation and the integration of these techniques within a complete network testbed are still open issues.

In this paper we describe our current effort to build a prototype of a WDM all-optical packet switched star network where receiver contentions are substantially alleviated by means of an SDL receiver front-end, called contention resolution optics (CRO). Subcarrier multiplexing of packet headers is used to encode and decode header information. A novel global packet-slot synchronization technique is used. A fully digital clock recovery techniques is used to decode header bits, whereas for packet data bits the clock is directly extracted from a multiplexed tone.

This work is carried out by an ARPA sponsored Consortium consisting of the University of Massachusetts at Amherst, Stanford University and GTE Laboratories.

II. The CORD Consortium

The objectives of the ARPA sponsored CORD Consortium (COntention Resolution by Delay Lines) are to demonstrate the reliable feasibility and effectiveness of the CRO principle and of its enabling devices, as well as to develop strategic technologies for the implementation of packet-switched all-optical networks, such as subcarrier-based packet header encoding-decoding, global packet-slot synchronization and ultra-fast clock recovery. A network prototype will combine the protocol, system and device technologies developed by the Consortium in order to assess their viability and effectiveness. By developing and demonstrating these key technologies, the CORD project hopes to pave the way toward WDM all-optical packet switched network solutions which may take full advantage of the potential capacity of the optical medium.

The main project tasks are the following:

- Investigate networking issues and develop optimal CRO control strategies (University of Massachusetts).
- Develop the key devices and construct an experimental CRO fabric based on integrated semiconductor technologies (GTE).
- Build an optical packet-switched network prototype, utilizing the CRO, subcarrier-based packet header encoding-decoding, global packet-slot synchronization and ultra-fast clock recovery (Stanford University).

More specifically, the tasks of the University of Massachusetts are the following. To conduct a general investigation of the applicability, effectiveness and scalability of the concept of optical contention resolution in the context of optical networking. To design and evaluate an optimal CRO structure and control strategies, optical channel access management, network resource allocation and traffic management. In particular, the objectives of the CRO control strategy are the minimization of residual contentions and of the time spent by a packet in the CRO. Lastly, to evaluate and optimize global network performance in communication systems using CRO based nodes.

The task of GTE is the development of key devices for a rugged and reliable implementation of a CRO. The functional diagram of the CRO that GTE will construct is shown in Fig. 1. GTE has specifically developed a fast semiconductor optical switch and will achieve its integration with an optical amplifier on a InP substrate. Dielectric waveguide interconnection technology based on the concept of mode transformation in uniform waveguides is used to provide the necessary low loss fiber-to-switch and switch-to fiber pigtailing. The optical amplifiers are included in the CRO module to offset the device insertion losses and ease the constraints on the switch and interconnection tolerances, which may impair its practicality.

Stanford University is building a two-node network prototype which includes the CRO fabric provided by GTE. Among the specific tasks of Stanford University there are system analysis and design and implementation of: high speed optical digital transmitters and receivers, subcarrier-based packet header encoders and decoders, high speed node and CRO control electronics, global time-slotting circuitry, ultra-fast clock recovery for header and payload bits, ATM-size packet generation and reception logic at 2.488 Gb/s and error checking and network performance measurement circuitry. Network performance evaluation tests will be conducted to assess the effectiveness of the CRO approach.

The principles and technologies developed by the Consortium are flexible and are expected to find future application, together or separately, in a variety of topologies and network architectures. The CRO concept is directly applicable to other key problems in optical networks, such as all-optical time-slot synchronization, all-optical bridging, and packet/circuit integration in optical networks. The multiple subcarrier header encoding-decoding and the slot and clock synchronization techniques that will be mastered within this project are expected to yield a substantial contribution toward the solution of the critical problems of signaling and ultra-fast synchronization in all-optical packet-switched networks.

In the following sections we describe in detail the background and the current work being conducted by each Consortium group.

III. UMass—Optical Contention Resolution: Principles, Control Strategies and Network Performance

Contention resolution using delay lines is the key concept used by this Consortium to optically deal with fundamental resource contentions in optical (WDM or TDM) packet-switched networks without resorting to electronic conversion of the transmitted data and thus avoiding electronic bottlenecks. To show this problem and the principle of the solution, consider packets arriving at a node on multiple channels (space separated fibers, and/or WDM channels). The packets are statistically distributed in time, and can be either destined to the node or need to be switched to outgoing channels. The multiple simultaneously arriving packets give rise to contentions occurring when two or more packets are competing for the same resource. This resource can be a receiver (when packets are destined to the node), it can be an outgoing channel (when packets destined to another network node need to be switched through), or packets may require the same switching element or temporary storage regardless of their final destination. It is important to observe that in some topologies only certain types of contentions will exist. For instance, in a WDM star topology in which the number of receivers is smaller than the number of wavelengths, only receiver contentions need to be resolved at the destination node. On the other hand, in a multihop topology, in addition to possible receiver contentions between packets destined to the node, packets passing through the node will contend for common outgoing links. The principle of optical contention resolution considered by the Consortium is based on the use of SDL's which shift contending packets in space and in time, to resolve the packet overlap, i.e., contention, at all relevant optical resources. Thus the same solution, using a single hardware mechanism, can be applied not only to different types of contentions, but also to resolve a combination of different contention types occurring simultaneously at a given node. The presented approach utilizes optical 2×2 switches reconfigurable on a packet-by-packet basis, and delay lines for optically storing the contending packets in the switching process. The net result is a switched delay line strategy, which interleaves the arriving packets onto one or more contention-resolved, and highly reduced packet loss, streams. The interleaving operation is controlled by local node intelligence, driven by inband or outband signaling.

A. Switched Delay Line (SDL) Operation

Since the SDL approach is applicable to different network systems, topologies, types of contention, as well as single or multiple wavelength channels, we introduce a generic hardware design, presented here as an isolated SDL module, which can be incorporated in different switching node designs. For the sake of demonstration simplicity Fig. 1 shows a diagram of a two-stage SDL. In principle, the state of the optical switches is set so that if two contending packets arrive at the node in the same time slot, one is delayed in optical delay line DL_1 by one slot. If contention persists, (e.g., new contending packets arrive at the node) this operation is repeated in DL_2. The following example (Fig. 2) demonstrates this principle on an actual sequence of arriving packets. Consider a sequence of six packets arriving in pairs from two different input channels in three consecutive slots at a node performing both packet reception and packet switching operations (e.g., a node in a mesh topology). The first two white packets (w) denote packets passing through the node that need to be switched on outgoing channel l_1, the following two pairs of black packets (b) are destined to the node's receiver. It is trivial to see that without SDL half of the packets, one white and two black, would be lost, unless they are stored electronically and/or devices (receivers) are replicated. Fig. 2 shows the operation of a 2-DL SDL node, starting with packets w_1 and w_2 in the two delay lines. In the current slot (slot no. 1) arriving black packet b_1 is propagated to the single receiver (reachable through channel l_2), and white packet w_1 is switched out of the delay line on the correct channel l_1. Packet w_2 is switched to DL_2, and packet b_2 is placed on DL_1. In the following slot (slot no. 2) two more black packets b_3, b_4 arrive at the node. In this slot packet b_2 is switched to the receiver (channel l_2), packet w_2 is propagated out of the node on channel l_1, while arriving packets b_3, b_4 are placed in delay lines DL_2 and DL_1 respectively for reception in subsequent time slots. Thus, with two delay lines the SDL node resolves both the receiver and channel contentions shown in this example.

This example demonstrates the following: 1) the possibility to resolve two types of contentions (for receiver and channel) using the same mechanism and a small number of DL's, 2) the fact that packets remain in the optical domain while contention is resolved, and 3) only information available from packets present at the node (e.g., their destination) is considered in the contention resolution process.

B. SDL Structure

The original two-input (wavelength) SDL design [7] is based on the following two optical components: d delay lines (in Fig. 1 $d = 2$, $\mathrm{DL}_1, \mathrm{DL}_2$) and $(d+1)$ wavelength *insensitive* 2×2 optical switches. The length of each DL_i is equal to m_i slots, with $m = \sum_i m_i$ the *total storage capacity* of the SDL. DL_i can "optically store" up to m_i packets,

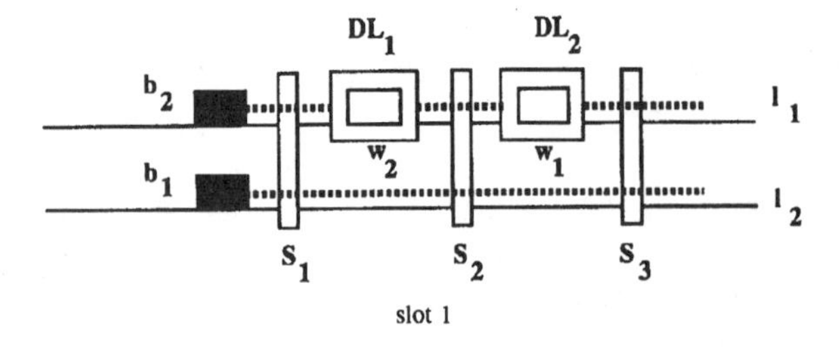

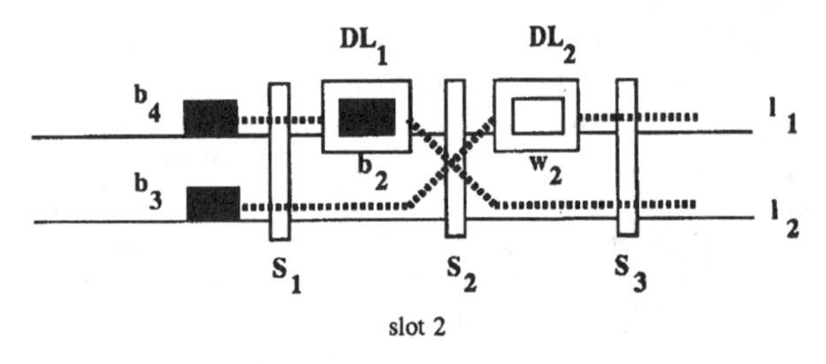

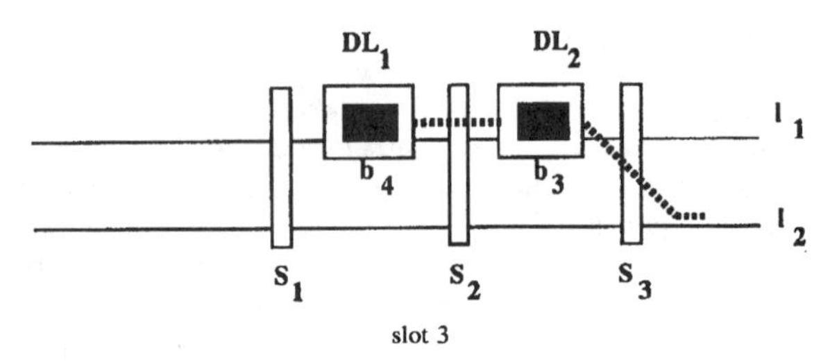

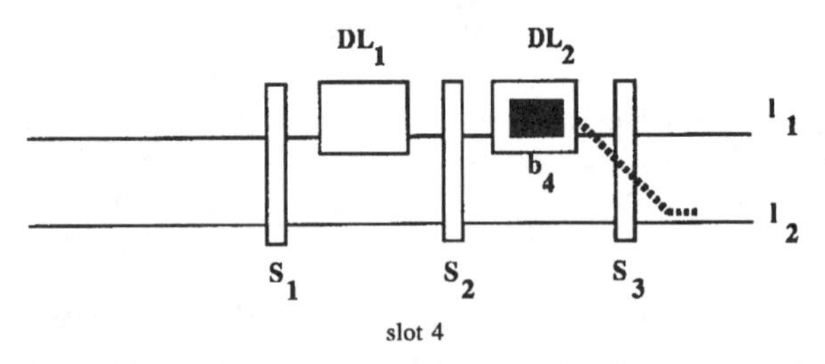

Fig. 2. Example of contention resolution in a node of a mesh network. Contention occurs between two arriving white packets (w_1, w_2) which need to be routed to output link l_1, and among four black packets (b_1, b_2, b_3, b_4) destined to the node. A two-stage SDL architecture is used to resolve contentions. (S_i: 2 × 2 switch, DL_i: optical delay line.)

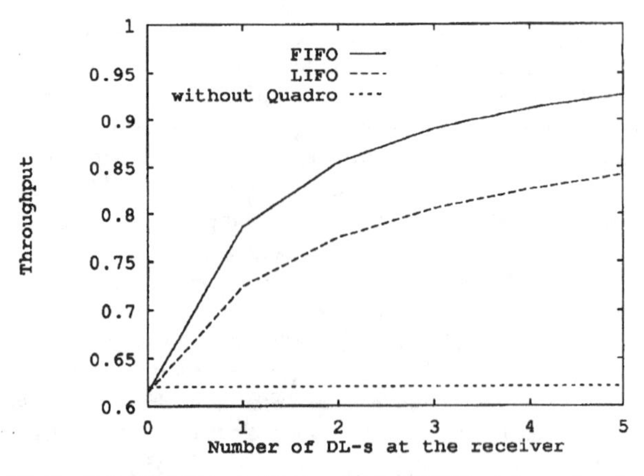

Fig. 3. Normalized throughput versus number of DL's at the receiver for a ten-node, ten-wavelength star network. Bottom line shows throughput when SDL (Quadro) is not used. The other two curves show throughput improvement when FIFO and LIFO reception strategies are used to control the SDL [28].

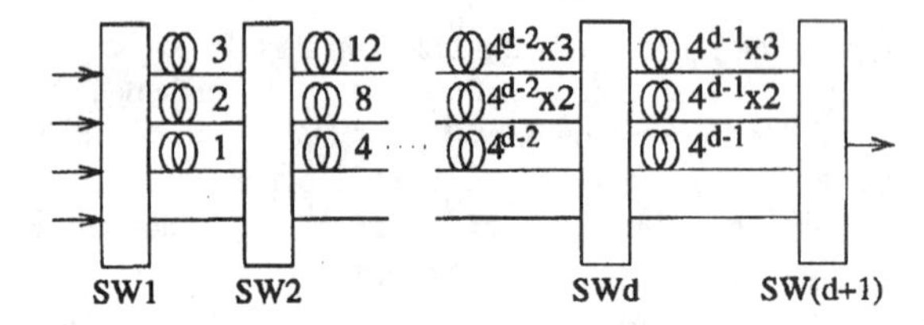

Fig. 4. 4×1 nonblocking SDL.

with each packet being stored for m_i time slots. *Single-buffer* DL based SDL is such that $m_i = 1\ \forall i = 1, \cdots, d$. *Multibuffer* DL-based SDL is such that al least one $m_i > 1$. Each switch can be set up electronically either in the *bar* state or in the *cross* state to route packets into and out of DL's. By optically propagating each packet through delay lines and optical switches comprising the SDL, packets can be switched without O/E and E/O conversions of the packet payload and fast electronic buffering of in-transit information is avoided. A two-stage SDL, like the one shown in Fig. 2 with $m_1 = m_2 = 1$, is the device that will be installed in the final CORD testbed.

Note that the proposed SDL design can be inherently fault tolerant. A fault-tolerant route-through passive switch is obtained when an electronic failure occurs by using 2×2 switches that stay in the default bar state when the electronic control signal is lost.

One possible solution to extend the above two-input SDL to a N-input (wavelength) d-stage SDL is shown in Fig. 4 for $N = 4$. It consists of $d + 1$ $N \times N$ photonic switches and $(N - 1)d$ DL-s. The length of each DL_{ij} is m_{ij} slots, with i the stage, counting from the left, and j the position of the DL within the stage, counting from the bottom. Single-buffer DL-based SDL is such that $m_{ij} = 1\ \forall i = 1, \cdots, d$ and $\forall j = 1, \cdots, N - 1$. The multibuffer DL approach has at least one $m_{ij} > 1$. Fig. 4 shows the multibuffer SDL with $m_{ij} = N^{i-1}j$.

C. SDL Control Strategies

The goal of the SDL control strategy is to manage the flow of packets in SDL so as to resolve contentions. An SDL strategy selects the packet to be transmitted on each output link, and chooses the DL's in which to store packets not able to leave the switch in the current slot. In addition, the design of SDL strategies is subject to unique optical hardware constraints. As opposed to electronic switches, in SDL a packet can be held in the node for a limited time determined by the propagation delay of the optical signal through the DL's visited by the packet. It is relatively easy to see that many different strategies are possible, and finding the optimal strategy may not be trivial.

Several strategies have been identified and evaluated depending on the SDL architecture in use. Single-buffer SDL's can be controlled using a FIFO strategy that selects for reception in the current slot one of the packets in the rightmost position of the SDL [28]. Multibuffer SDL's can be controlled using either a strategy that guarantees FIFO reception [34] or the *sneak* strategy that offers slightly better performance (lower packet loss probability) but does not guarantee the FIFO order among the received packets [33].

Combined with the SDL hardware design, the SDL strategy presents a number of fundamental advantages. First, its operation is localized in time, executed so that no memory of earlier states needs to be maintained, and it is localized in space, requiring only information derived from packets found in the switch. The principle of localized contention resolution simplifies the strategy, avoids the need for coordination with other (remote) nodes and booking of other nodes' activities. This makes the SDL approach applicable to different single or multihop topologies. Secondly, in terms of performance, SDL remains robust under non homogeneous [4] and bursty [33] traffic conditions.

D. Control Complexity and Packet Signaling

SDL control electronics executes the control strategy, handling the flow of packets in the SDL. The operation is slot (rather than bit) time oriented, where the slot time can be assumed to be a fraction of μs. In single-buffer SDL, algorithms with complexity that grows only linearly with the number of DL's in the switch are easily obtainable, e.g., by selecting the packet to be propagated out of the switch through a linear search. In multibuffer SDL, algorithms whose complexity is proportional to the number of DL's rather than m can be found [33], thus making possible to increase the SDL total storage capacity without increasing the complexity of its control electronics by simply lengthening the DL's.

The SDL control strategy utilizes information about arriving packets. To provide this information the CORD Consortium has chosen the header subcarrier multiplexing technique. The details of this technique are found in Section IV-B.

E. Network Topologies and Performance

In the CORD project two widely used network topologies are being examined for the use of SDL to obtain and evaluate the potential of an optical network, the star and the ring. Due to space limitation only the star topology is discussed in this paper. Description and detailed performance analysis of WDM ring networks made by nodes equipped with SDL can be found in [5] and [35].

We consider a common system-wide clock in which time is divided into fixed slots, whose duration equals the packet transmission time plus receiver's tuning time. Each node can synchronize its own transmission by receiving its transmitted reference signal (after a round trip through the network) and comparing it to the other nodes' transmission as explained in Section IV-C. Also, each node has to know the destinations of the arriving packets prior their reception. The signaling techniques discussed in Section IV-B can provide the node with this information.

WDM stars constitute a good solution for interconnecting very high speed nodes requiring a dedicated wavelength channel for each transmission. Of course, this solution is possible only when the number of nodes to be connected does not exceed the number of wavelength channels provided by the existing technology. Note that in the ring networks presented in [5] and [35] the number of nodes is not bounded by the number of available wavelengths.

In a star topology, each node is connected to the hub by two fibers, carrying optical signals from the node (input fiber of the hub), and to the node (output fiber of the hub). By assumption each node has one (fixed) transmitter and one (tunable) receiver for transmission of data packets. Each node is assigned a wavelength for transmission so that none of the wavelengths is shared by two or more nodes. Collisions associated with multiple transmissions occurring on the same wavelength thus do not occur in this system. However, control of transmissions must deal with (receiver) contentions. These contentions occur when several packets are arriving at a receiver at the same time slot on different wavelengths and can be dealt with using SDL. Under contention occurrence, using a deterministic system-wide process the receiver will accordingly select one packet for reception and, executing the same selective process, the transmitter can learn if its packet has been the one received [27]. Without SDL, this approach can guarantee the reception of one of the packets sent to it, and looses the remaining conflicting packets. This loss and penalty of subsequent retransmissions significantly reduce performance, limited to about 62% of the channel capacity.

We observe a 10-node (10 wavelength) system under homogeneous and nonbursty traffic conditions, i.e., source and destination nodes for each new packet are uniformly chosen. Fig. 3 shows the maximum throughput achievable (packet per slot per receiver) versus SDL *effective storage capacity* c, defined as the maximum number of packets that can be stored in the SDL without originating packet loss. c is a function of m and of the SDL architecture. Naturally, $c \leq m$. The maximum throughput of a WDM star remains constant at the value of 0.62 when SDL is not used (this case also corresponds to the system discussed in [27]). On the other hand, with storage capacity $c < 6$, a SDL node based star yields throughput already surpassing 0.9. A detailed analysis of the general case with w wavelengths in the star can be found in [28].

The effective storage capacity in single-buffer SDL's is equal to the number of stages, thus $c = d$. For the multibuffer SDL architecture shown in Fig. 4, the effective storage capacity is given by $c = N^d - 1$ and it is plotted in Fig. 5 as a function of the number of 2×2 switches[2] for several values of N. To achieve targeted values of c in systems with a large number of nodes (wavelengths) the required number of 2×2 switches (proportional to cost) may be quite considerable in both single and multibuffer architectures. However, even for large capacity systems the SDL *depth* defined as the maximum number of 2×2 switches crossed by a packet prior reception (proportional to crosstalk and power loss) remains limited, expecially in multibuffer SDL. Fig. 6 shows c versus depth

[2]Under the assumption of using $N \times N$ Benes switches.

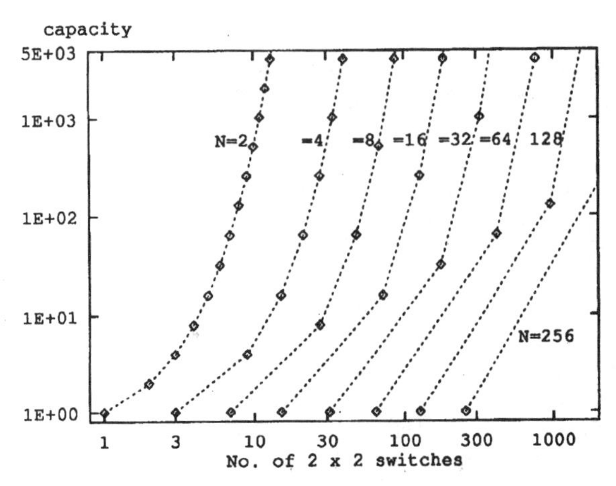

Fig. 5. Effective storage capacity c versus number of 2×2 switches in the $N \times 1$ SDL architecture shown in Fig. 4, for several values of N.

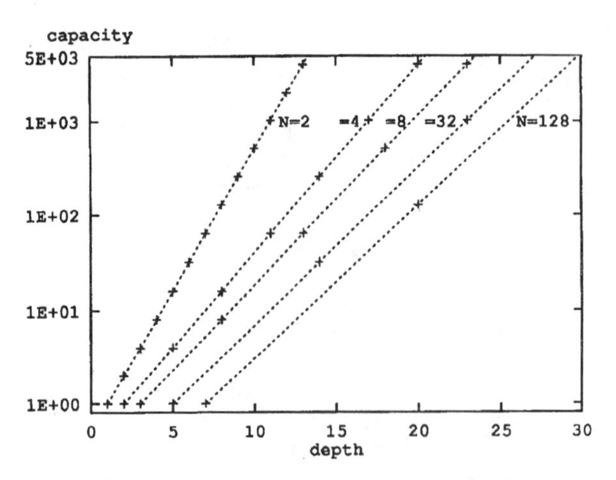

Fig. 6. Effective storage capacity c versus depth in the $N \times 1$ SDL architecture shown in Fig. 4, for several values of N.

of the multibuffer SDL shown in Fig. 4, for several values of N. In a 128 wavelength system, an effective capacity of about 100 packets is achieved with depth equal to 20.

IV. Stanford University—Experimental CORD

To test the SDL-based CRO concept, the CORD Consortium has focused on a WDM packet-switched star network testbed, that is being built at the Optical Communications Research Laboratory of Stanford University. The CRO fabric itself will be provided by GTE. To construct the testbed network, numerous key technologies for optical packet networks, such as subcarrier header encoding, global time-slot synchronization and ultra-fast clock recovery, have been developed and implemented.

In the following, the testbed will be described in detail along with the main experimental results obtained so far. At the time this paper is written the testbed implementation is nearing completion.

A. General Description

The testbed (Fig. 7) consists of two nodes. Each node has one fixed-wavelength transmitter, operating at $\lambda_1 = 1310$ nm and $\lambda_2 = 1320$ nm for node 1 and 2, respectively,[3] and one receiver. Each node transmits ATM-sized packets at 2.488 Gb/s, addressed both to the other node and to itself, thus originating receiver contentions. At the receiver, a WDM demultiplexer separates the two wavelengths. One of the nodes is equipped with an optical switch, following the WDM demultiplexer, that allows it to select either one of the two wavelengths for reception. If two packets addressed to that node arrive at the same time on both wavelengths, only one is received, and the other dropped. The other testbed node (blown up in Fig. 8) is equipped with a CRO whose inputs are fed by the outputs of the WDM demultiplexer. As explained in the previous sections, the CRO helps relieve receiver contentions and therefore packet loss probability will be lower for the node equipped with the CRO.

For the network to operate as described, packets need to arrive at the nodes aligned in time. The nodes also have to learn about the presence of incoming packets and read their addresses in order to set up the CRO or receiver switch. Finally, clock recovery for both the address information and the ATM payload must be performed on a packet-by-packet basis.

The specific techniques to achieve these goals and the actual testbed implementation are described in detail in the following.

B. Testbed Implementation

The packet destination address, or "header," must be read before the packet enters the CRO, so that the CRO can be properly configured. This must be done simultaneously on all wavelengths and without actually decoding the high-speed ATM cell data. This is accomplished using a technique based on subcarrier-multiplexing of the headers [20]–[26]. Each node directly modulates the laser with a signal that encodes the high-speed ATM cell data at baseband, while the header is simultaneously encoded on a subcarrier. The subcarrier frequency must be higher than the tails of the data spectrum and each node must use a different, fixed subcarrier frequency (Fig. 11). The subcarrier-encoded header information need not be the ATM cell header, since it is only used internally by the optical network and is discarded upon reception. The header and ATM cell signals are combined by means of a microwave coupler to form the laser modulating current. In the testbed the subcarriers are generated using commercial voltage controlled oscillators (VCO's) operating at $f_1 = 3$ GHz and $f_2 = 3.5$ GHz for node 1 and 2, respectively. The VCO's are directly modulated to FSK encode the header bits. The frequency deviation is ±45 MHz.

The node receivers are equipped with a "header detector" that consists of a 10/90 splitter that takes out of the fiber 10% of the light arriving at the node and feeds it to a commercial 5 GHz photodiode (Fig. 8), followed by commercial 50 Ω

[3] The choice of the 1300-nm window rather than the 1550-nm window was due to GTE's semiconductor technology for CRO component fabrication, which was more mature and well-established in the second window.

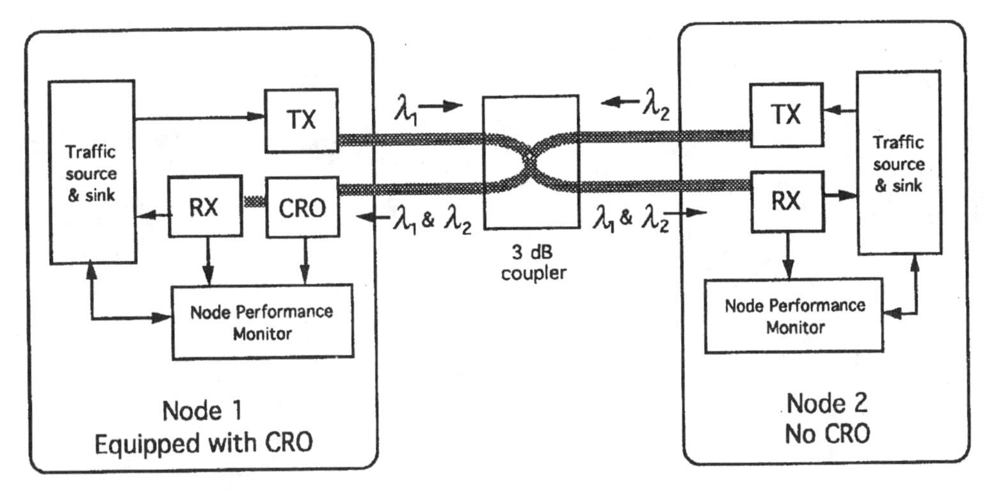

Fig. 7. Block diagram of the optical network testbed being implemented at Stanford University.

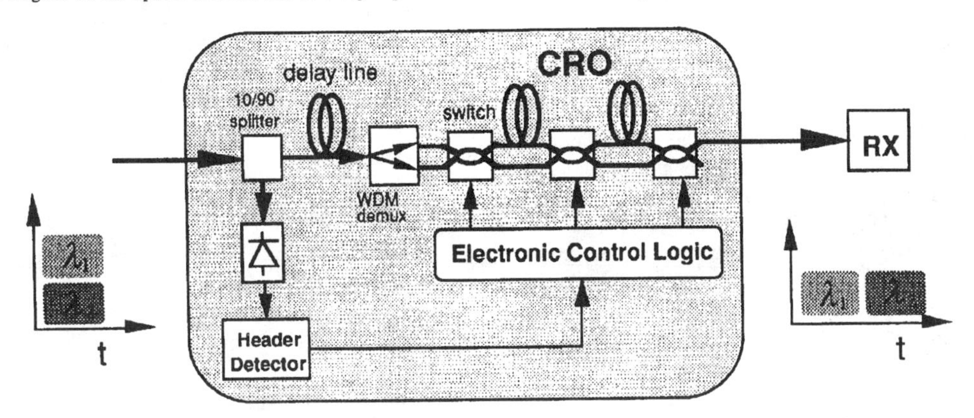

Fig. 8. Block diagram of the optical network testbed node comprising the CRO device. Thanks to the CRO, two simultaneously incoming packets destined to the node are time-resolved all-optically and reception of both can take place.

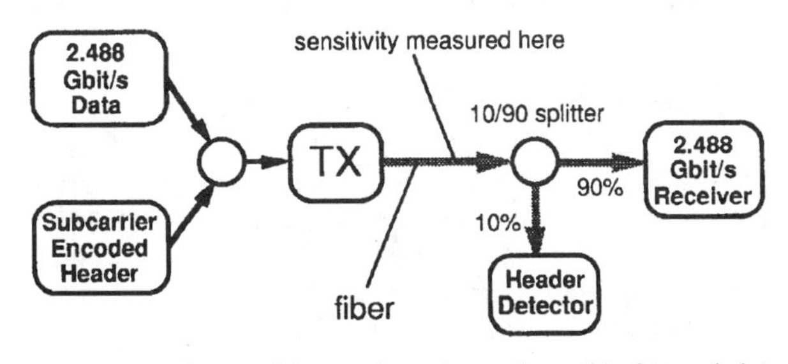

Fig. 9. Block diagram of the experimental setup for combined transmission of subcarrier encoded headers at 80 Mb/s and baseband 2.488 Gb/s data.

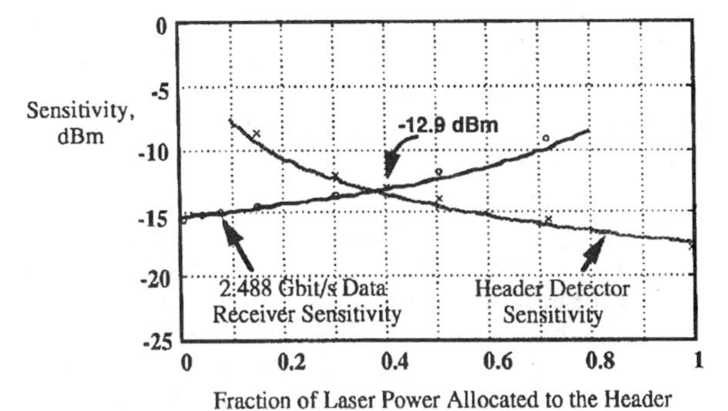

Fig. 10. Sensitivity of the 2.488 Gb/s receiver and of the header detector, versus the fraction of the total laser power devoted to header transmission. The fraction of power allocated to 2.488 Gb/s data is 1 minus the abscissa.

amplifiers and a pair of band-pass filters, centered at 3 and 3.5 GHz. The filters are followed by delay-and-multiply FSK demodulators that recover header data from both subcarriers simultaneously. No wavelength sensitive or tunable devices are needed. If two packets arrive simultaneously on the two wavelengths, the ATM cell data jams at baseband (in the header detector photodiode current) and is undecodable. The headers instead do not interfere with each other because they arrive on two distinct subcarriers (Fig. 12), and can be decoded in parallel. This technique is scalable by assigning to each node a unique subcarrier frequency.

Headers have been chosen to be 20-b long, and are transmitted over a full packet time-slot. The duration of the slot is set at 250 ns, to comfortably fit an ATM cell (whose duration is 170 ns at 2.488 Gb/s) and allow time for guard bands and clock recovery (see Section IV-E). The time slot format is shown in Fig. 13. The resulting header bit rate is 80 Mb/s.

The header bits tell the receiving node whether an incoming packet is addressed to it. The subcarrier frequency on which

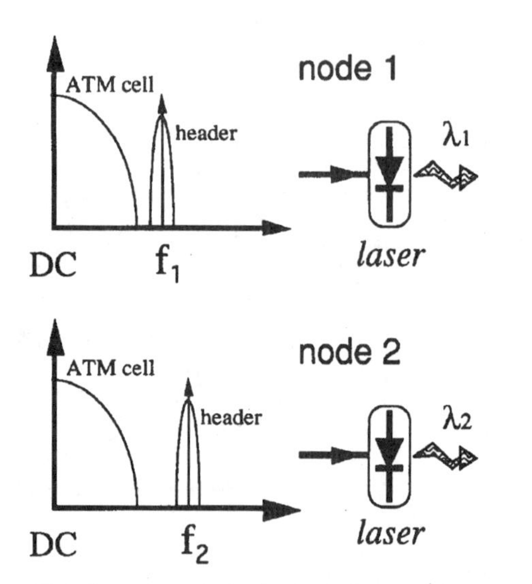

Fig. 11. ATM cells are transmitted at baseband. Headers are multiplexed on a subcarrier.

the header is encoded tells the node which node sent the packet and as a result also identifies the wavelength on which the packet is traveling. Using this information, the node can set up its receiver and (if it has it) the CRO. An optical delay line is inserted between the header detector and the receiver switch (or CRO) to allow time for processing of the headers and receiver switch (or CRO) setup (Fig. 8).

The 2.488 Gb/s ATM cell receivers were constructed using commercial photodiodes and 50 Ω amplifiers. The data reception low-pass filter eliminates the header subcarrier from the detected signal.

The power spectrum of the ATM cell data has the typical $\sin^2(\pi fT)/(\pi f)^2$ shape. The first sidelobe of this spectrum extends between 2.488 and 4.976 GHz, thus interfering with the subcarriers. As simulation showed and experiments have proved, to avoid serious performance degradation this interference must be reduced. To this purpose, a data shaping filter was inserted in the transmitter which suitably attenuates the data first sidelobe.

The results of combined transmission of both headers at 80 Mb/s and a continuous data stream at 2.488 Gb/s, tested in the experimental setup of Fig. 9, are shown in Fig. 10. They have been preliminarily reported in [36], [37]. The abscissa indicates how much of the laser power is allocated to the subcarrier-encoded header. All the remaining power is allocated to high-speed data. The lasers used in the testbed are LASERTRON DFB's. Sensitivity is measured before the 10/90 coupler and is affected by its presence. Header reception sensitivity, when all the optical power is allocated to headers, is −17 dBm, i.e., approximately −28 dBm at the header detector photodiode. High-speed data reception sensitivity, when all the optical power is given to data, is −15.5 dBm (−16.5 dBm at photodiode). The best combined transmission operating point is obtained when 40% of the power is given to the subcarrier encoded header and 60% to high-speed data. The combined sensitivity is −12.9 dBm. This relatively low value is due to the use of PIN photodetectors and 50 Ω amplifiers.

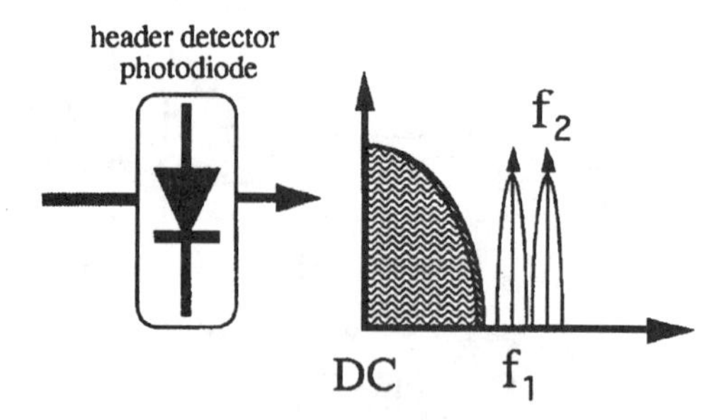

Fig. 12. When two packets arrive simultaneously at the header detector photodiode on the two wavelengths, ATM data jams but the headers do not interfere because they arrive on two distinct subcarrier frequencies.

A combined sensitivity of −20.9 dBm has been obtained by replacing the high-speed receiver PIN photodiode with an APD [38].

The experiments have shown the header subcarrier-encoding technique to be robust and reliable. Subcarrier frequencies well within the first data side lobe can be used and the power allocation ratio has proved not very critical. In the testbed the subcarrier channel also conveys slot synchronization signals, as explained later.

C. Time Slotting and Slot Synchronization

Packet timeslots must necessarily be aligned in time on all wavelengths for proper CRO operation. To ensure this, a global slot synchronization method has been devised. The method is scalable to an arbitrary number of nodes over a star network. It operates as follows.

One of the nodes takes on the role of the "master" node. It generates a 3-b long (37.5 ns) slot marker called a "ping" which is transmitted on the 3-GHz subcarrier that the master node also uses for header transmission. The ping is sent at the beginning of each time slot (Fig. 13). The other nodes, called "slave" nodes, also send their own pings and headers every timeslot, over their own unique subcarrier frequencies (in CORD there is just one slave node using $f_2 = 3.5$ GHz). Each slave node then listens to both the master ping and its own returning ping, and tries to make them arrive at its receiver simultaneously by properly altering the frequency and phase of its own ping transmission. This guarantees timeslot synchronization at both the slave and master node receiver inputs. This technique is scalable to any number of WDM nodes in an optical network as long as it is possible to assign a narrowband dedicated subcarrier to each one of them. All slave nodes operate independently, since they only listen to the master ping and their own. Also, any slave node can take on the role of the master node. As a result, this technique is fully distributed and fault-tolerant.

A very critical implementation aspect is the locking circuitry between the master ping and the returning slave ping. In principle it could be done with a PLL deriving its error signal from a comparison of the arrival times of the two pings. This solution is inadequate because the delay across the star between slave ping transmission and reception may cause the PLL to oscillate and never properly lock. Instead we sought a fully digital implementation, which is conceptually similar to

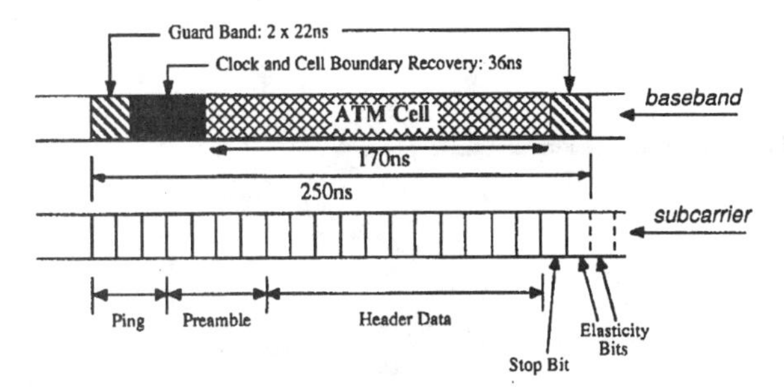

Fig. 13. Format of the timeslot used in CORD.

the bit stuffing and dropping techniques used for instance in SONET/SDH. The arrival time of the master and slave pings is digitally compared and when it exceeds the equivalent of a bit (12.5 ns) the subsequent time-slot is made either one bit longer or one bit shorter, as needed, by transmitting both or none of the two elasticity bits shown in Fig. 13. Ideally, this ensures slot synchronization with respect to the master ping to within a bit, i.e., ±12.5 ns. The channel delay problem is easily coped with by disabling the feedback signal after a correction has been made, for the amount of time needed for the time-shifted slots to come back to the slave node.

In CORD, the arrival time comparison as well as bit stuffing or dropping has been implemented using commercial fast FPGA's clocked at 80 MHz. We conducted an experiment where the slave node was connected to the star through a 6 km (12 km round-trip) fiber. The subcarrier power impinging on the header receiver photodiode was set at −28 dBm, which corresponds to the power level needed to receive the header bits (that are encoded on the same subcarrier; see the previous section). The results are shown in Fig. 14. The bottom trace is the master ping received by the slave, whereas the top trace is the returning slave ping. The oscilloscope was triggered by the master node internal slot synch signal.

The traces show that the maximum slave ping fluctuation with respect to the master ping was 18 ns. This value is lower than expected (±12.5 ns) because one of the two clocks was consistently faster than the other, therefore triggering only corrections of one type (namely, dropping one bit). By swapping the transmitter quartzs a similar but inverted situation was seen. The total variation envelope over the two cases was approximately 30 ns. This value is slightly higher than the ideal ± 1-b (25 ns) because of the presence of noise on the channel. The plot does not show the header bits because it is derived from a specific ping detecting circuit that is triggered by the presence of three successive "ones," a sequence which is allowed only for pings.

This technique works with any slot-length and is independent of the packet payload data rate. It could be used in a ring configuration, with minor modifications. Its flexibility and reliability make it very attractive for a wide range of optical networking applications.

D. Header Clock Recovery

Headers are transmitted immediately after the ping, on the node unique subcarrier frequency. They are sent *every* time slot: if no ATM cell is transmitted, a specific header address indicating an empty slot is sent.

As a result, clock synchronization could have been performed by using conventional PLL techniques, because a continuous stream of data is received on each subcarrier. However, this would not have solved the problem of synchronizing the receiving node internal 80 MHz clock with the recovered 80 MHz clock. Higher internal clock rates or special asynchronous circuitry would have been needed to perform this further task. In addition, PLL's are too slow for slot-by-slot clock recovery, and therefore this approach would have failed to address the more general problem of fast clock synchronization in all-optical packet networks, in which packets from different nodes may arrive at a receiver in adjacent time-slots with no clock frequency or phase consistency among one another.

For the CORD project, a novel and specific ultra-fast technique was developed which enables header clock recovery on a slot-by-slot basis. The technique has been called digital phase alignment (DPA) and its block diagram is shown in Fig. 15.

The concept assumes that the remote node transmitter clock and the receiving node clock do not drift substantially with respect to each other over the time-length of a single slot, a requirement that is satisfied even by the most inexpensive quartz oscillators. Therefore, only the relative phase of the transmission and reception clocks needs to be aligned.

Immediately after the ping, a 4-b preamble sequence (0101) is inserted (Fig. 13). The DPA circuit detects the bit transitions (i.e., zero-crossings) in the preamble. The best sampling time for header reception is half a bit after the bit transition time and the DPA circuit time-shifts the header for best reception.

The header time-shifting is accomplished by means of multitap delay lines, which are commercially available as inexpensive dual-in-line components. The DPA circuit simply chooses the delay-line tap corresponding to the best alignment with the sampling clock.

A critical aspect of this technique is the algorithm which is used during the preamble to estimate where in time the bit transitions occur. In theory, a 1-b preamble, i.e., observing a single transition, would be sufficient. However, due to noise, transitions can be substantially jittery so that making the estimate on only one transition may lead to substantial errors. The current implementation allows the DPA circuit to use a one, two or four-bit preamble and therefore it can look at up to four consecutive transitions to estimate the average zero-crossing time. Averaging effectively reduces the jitter. Provisions were also made to reject possible glitches and inconsistencies in the received preamble pattern. Another critical aspect is how finely spaced the delay-line taps are, which sets the resolution for the transition time estimate. Currently, the spacing is 2 ns, i.e., six taps per bit.

An experimental test of the DPA technique over an optical channel is shown in Fig. 16. The ideal curve was obtained with a BER tester. The penalty due to a four-bit preamble DPA is about one dB (optical). Almost another dB would be lost if only a two-bit preamble was used. No BER floors were observed.

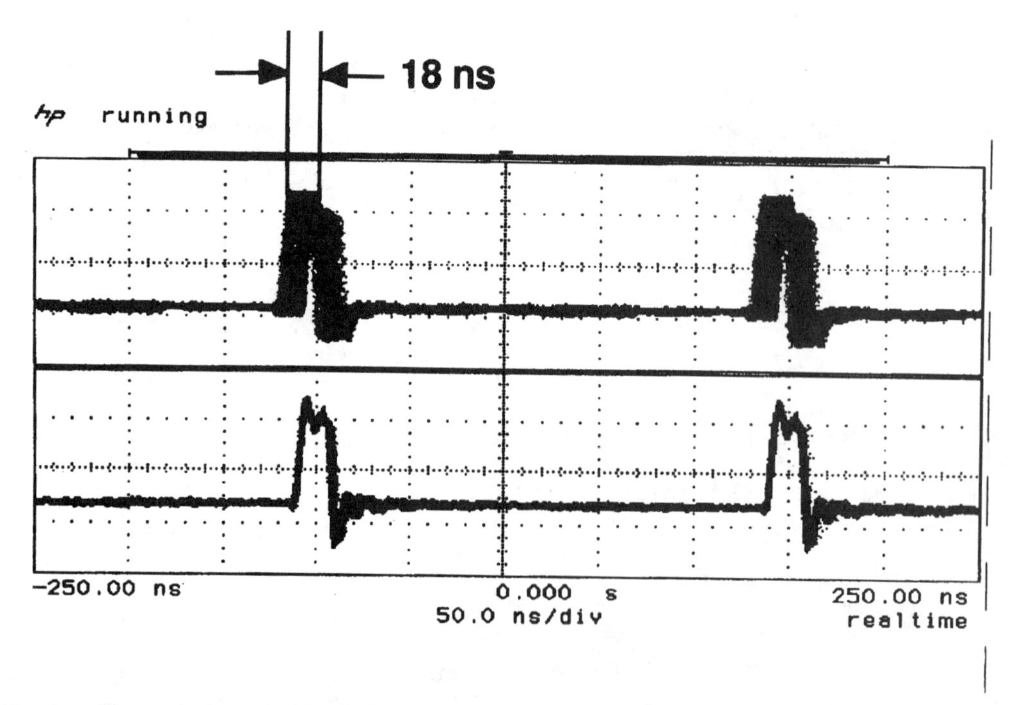

Fig. 14. Bottom trace: The master ping received by the slave. Top trace: the returning slave ping. The oscilloscope was triggered by the master node internal slot synch signal.

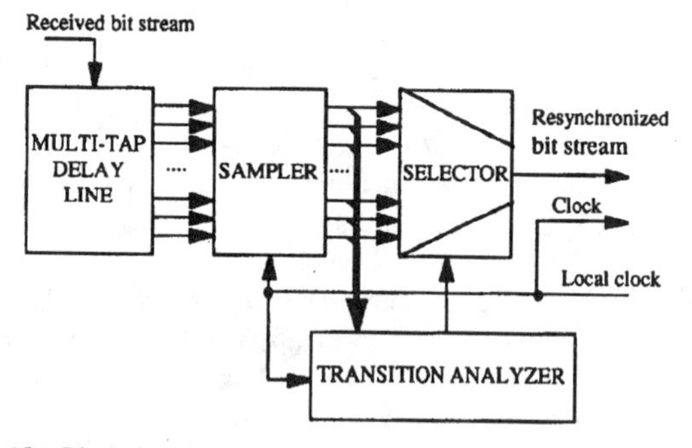

Fig. 15. Block diagram of the DPA circuit.

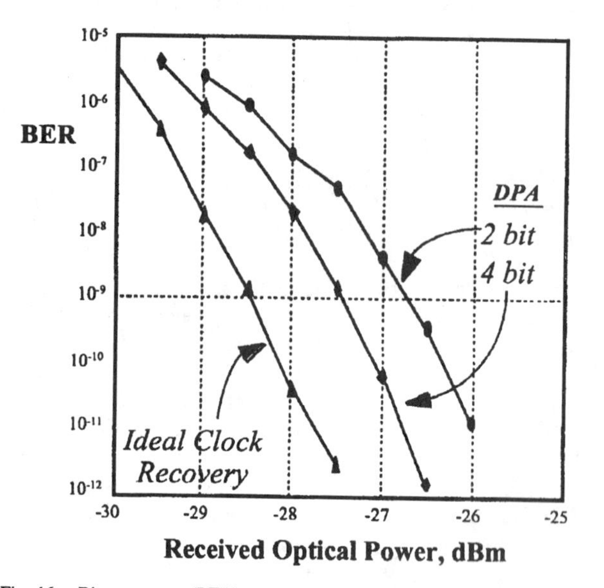

Fig. 16. Bit-error rate (BER) measurement of subcarrier-encoded header reception using the BER tester (ideal clock recovery curve) and the DPA circuit with two and four preamble bits.

In summary, DPA has proved effective and reliable. It allows ultra-fast clock alignment on a slot-by-slot basis, with a small sensitivity penalty. These features make it attractive for use in various all-optical network architectures. The speed of the DPA circuitry is currently limited to 80 Mb/s, due to its FPGA-based implementation. An independent group at Stanford University is now attempting to integrate the DPA in a custom CMOS chip, with the objective of upgrading its speed to 2.5 Gb/s.

E. Data Clock Recovery

The actual CORD slot payload is an ATM cell (Fig. 13). For payload reception two circumstances make slot-by-slot clock recovery necessary: 1) there is no synchronization among transmission clocks of different nodes; 2) packets are detected after possibly going through the CRO, which alters the phase relationship even of cells coming from the same transmitter. PLL-based techniques are then inadequate because too slow. Digital oversampling techniques, similar to the DPA technique developed for headers, are promising but to date none seems to be available at this speed.

In the CORD testbed, payload clock recovery is performed by extracting a clock tone at 2.488 GHz which is transmitted along with the payload. The clock tone falls exactly at a notch of the payload transmission spectrum and therefore relatively

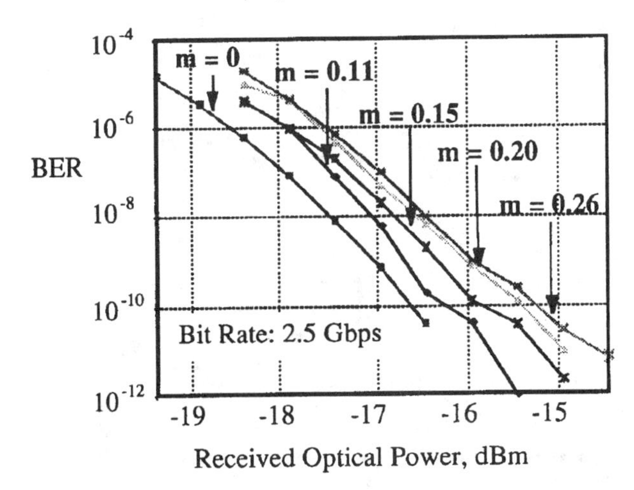

Fig. 17. BER measurement of 2.488 Gb/s payload data reception using the BER tester and the clock recovered from the clock tone, for different values of the fraction m of the total laser power devoted to clock tone transmission. The $m = 0$ curve refers to the ideal case of back-to-back connection of the clock signal to the BER tester input.

low interference from the payload signal is incurred if a narrow band-pass filter is used to extract it.

A number of trade-offs need be carefully balanced. The narrower the filter, the longer it takes for the clock tone to be satisfactorily recovered. On the contrary, a relatively wide clock filter would allow for fast clock recovery. However, it would also increase the amount of payload signal intereference and noise affecting the recovered clock. This, in turn, would cause extra clock jitter and therefore a higher BER. To overcome the extra jitter, more optical power would have to be devoted to clock tone transmission, at the expense of some system penalty.

The best way to go about optimizing these trade-offs is by first setting the desired clock recovery time. This also sets the clock filter bandwidth. Then, the optical power splitting ratio between payload signal and clock tone can be adjusted for best BER performance.

In Fig. 17, the experimental results are shown. The band-pass filter was 60 MHz-wide. Its output stabilizes in the equivalent of 40 bits at 2.488 Gb/s (16 ns). Before decision, the payload signal was passed through a band-reject filter to suppress the clock tone. The recovered clock was used to drive the BER tester. The plot shows that when 89% of the power is devoted to data transmission and 11% to the clock tone, only a 0.4 dB optical penalty is incurred with respect to the ideal case when all the optical power is assigned to the payload and the clock signal is connected back-to-back to the BER tester input. No floors were found in the BER curve.

In summary, this technique is fairly simple and robust, and is capable of recovering the clock in a few tens of bits with an SNR penalty of only 2 dB.

F. CRO Preprototype

As explained in Section V, an integrated implementation of the CRO is being pursued by GTE. In the meantime, A "CRO preprototype" (henceforth just "CRO") made of discrete components has been assembled. A layout of the current version is shown in Fig. 18. It makes use of $LiNbO_3$ switches from GEC Marconi and semiconductor optical amplifiers (10 to 12 dB gain) made by HP. The WDM demux was custom made by JDS-FITEL. Most of the fiber is polarization maintaining, including the delay-lines. The semiconductor amplifiers are polarization insensitive and their pigtails consist of conventional single-mode fiber.

This CRO turned out to be difficult to adjust and operate because of the presence of polarization-sensitive switches. Despite their good nominal extinction-ratio (better than 16 dB), even slight polarization misalignments in certain parts of the CRO would reduce it to consistently less than 12 dB. The problem has been completely solved by carefully re-aligning the polarization of all the switch pigtail connectors for maximum extinction ratio and by adding several spans of single-polarization[4] (SP) fiber at critical locations where random polarization fluctuations could be generated.

The CRO can now operate for hours without readjustments. Residual polarization fluctuations at the input or inside the CRO only translate into slow and generally small *power* fluctuations at the output. The switch extinction ratios are 17 dB and 19 dB and never degrade thanks to the SP fibers.

A theoretical upper-bound calculation of both coherent and incoherent crosstalk impact was carried out, showing that for a one-stage CRO, as in Fig. 18, a minimum extinction ratio of 16 dB is needed to guarantee less than 0.5 dB penalty. Our switches have better than 16 dB crosstalk, so we conducted an experiment to check if the above prediction was correct. All possible CRO switch configurations were explored. No observable change in BER performance was seen when the interfering channel was suppressed to remove incoherent crosstalk and/or when the fiber in the CRO branch that would carry coherent crosstalk was disconnected. According to our calculations, our current switch performance would ensure that a two-stage (three-switch) CRO would have less than 2.5 dB penalty.

From preliminary experimental results and theoretical calculations the expected sensitivity, when the 2.488 Gb/s ATM cells are sent along with the subcarrier header and clock tone, is -10 dBm at the node input. This relatively low sensitivity is due to the use of 50-Ω microwave preamplifiers in the receivers and of semiconductor optical amplifiers in the CRO. Better overall performance and scalability could be obtained with APD's [38] or by converting to 1.55-μm technology with EDFA's. The GTE integrated CRO is expected to significantly improve this result as well.

G. Status and Future Evolution

As of now, all of the critical subsystems that are needed for the star network testbed have been built and tested. Circuitry to perform 2.488 Gb/s serial-to-parallel-64 conversion has been

[4] SP fibers bleed off completely one polarization and effectively behave as fiber polarizers. From our experience with the CRO, we developed the convinction that bad switch extinction ratios are often due to poor polarization alignment. It would probably be recommendable that polarization sensitive devices be pigtailed with SP fiber rather than PM fiber, because this greatly reduces the negative impact of polarization misalignments in the setup.

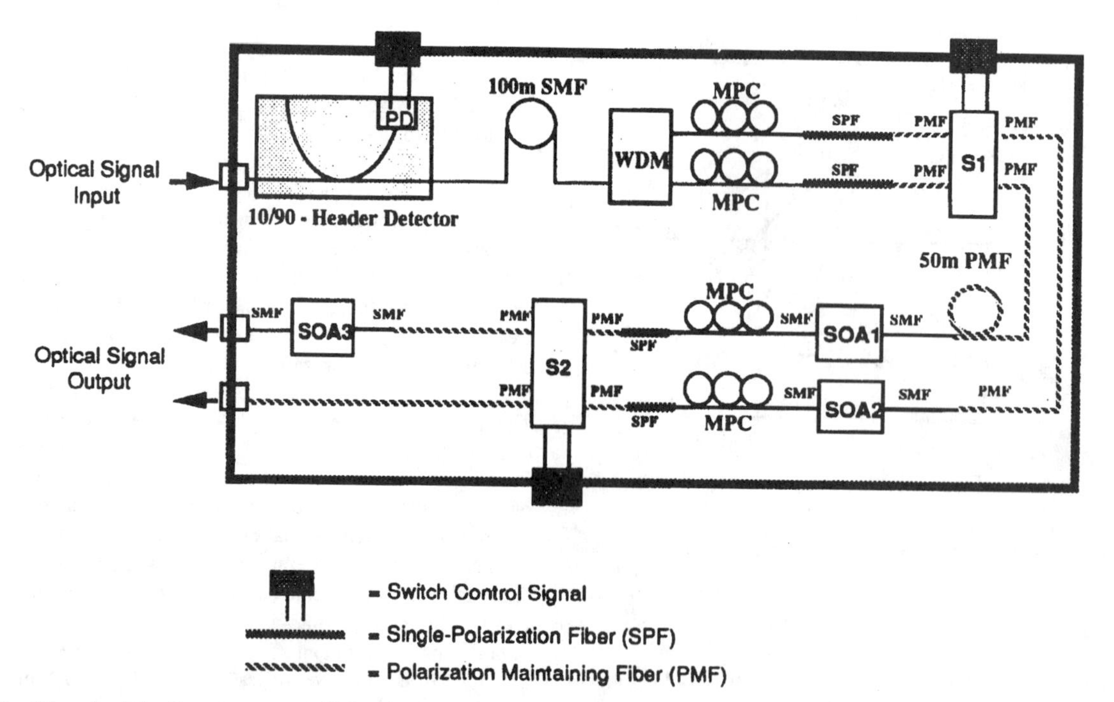

Fig. 18. Schematic of the discrete-component CRO preprototype. S_1 and S_2 are $LiNbO_3$ switches. SOA: semiconductor optical amplifier. MPC: manual polarization controller. WDM: WDM demultiplexer.

built for the ATM payload receivers, and a 64-parallel-to-serial board is being implemented to be used at the transmitters. Error checking and packet traffic generation logic is nearing completion. The testbed is expected to be fully operative in the first half of 1996.

When available from GTE the integrated semiconductor CRO version will replace the discrete-component preprototype.

V. GTE—Optical and Optoelectronic Components

The preprototype has shown that implementing a CRO device using discrete optical components results in substantial difficulties in achieving good performance. The inclusion of optical amplifiers is also unavoidable and as a consequence the overall size of the final device is also considerable, making its practicality questionable.

These observations alone are sufficient motivation to seek device implementation on III–V semiconductor substrates where integration of active and passive components has been already demonstrated.

One of the challenges in this work is to fabricate the CRO module with low enough signal losses and low back reflections. Coupling of fibers to waveguides and waveguiding devices is accompanied by a mode mismatch which gives rise to both losses and reflections and demonstrates clearly the essential role of integration in fabrication of opto-electronic modules.

A tentative schematic of an integrated CRO module is shown in Fig. 19. The two key devices in the CRO module are optical switches and amplifiers. Both of these components can be integrated to provide higher functionality modules.

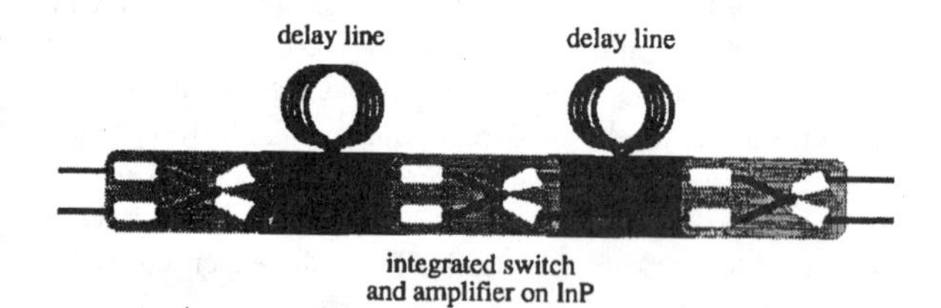

Fig. 19. Rough schematic of the integrated version of the CRO. The gray substrate is InP. The black lines are optical waveguides within the semiconductor material. White patches are electrodes and are used to control the X-shaped digital optical switches and to bias the optical amplifiers.

This approach is part of the project effort and includes the development of semiconductor amplifiers and digital optical switches on a InP substrate.

A. Digital Optical Switches

The ideal optical switch would be independent of polarization, wavelength, environmental effects, and insensitive to electrical drive variations. The modal-evolution switch comes closest to meeting these requirements. It operates by selecting the propagation path of the optical signal through two intersecting waveguides by inducing a small index asymmetry between the waveguides This asymmetry is achieved by injecting current through an electrode into the waveguide, or by reverse biasing the electrode to deplete charge carriers in the waveguide. For example, with no injected current, the output waveguides are symmetric and the incoming light is equally divided between them. When current is injected in one output waveguide, the symmetry is broken, and the light propagates down the waveguide with the larger index. The physical mech-

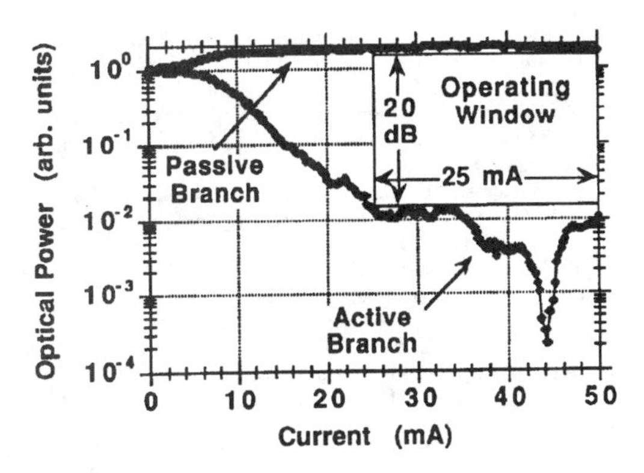

Fig. 20. Optical output power of a Y-branching switch versus applied current showing the wide range of control current (50% of maximum) which gives an extinction ratio of over 20 dB.

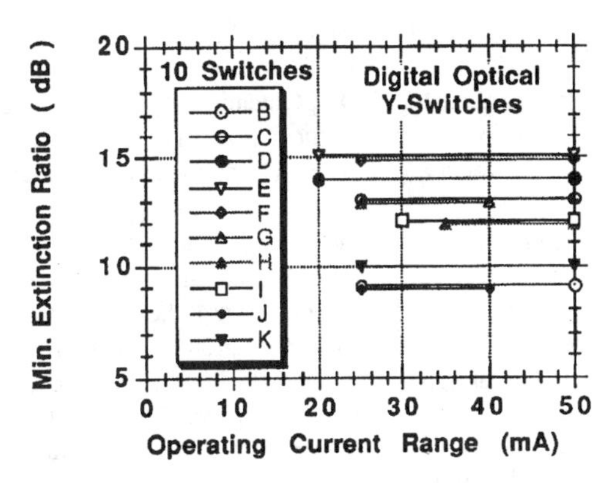

Fig. 21. Minimum extinction ratio achieved from both output arms of the 10 best switches on one bar of 30 1/2° Y-branching switches versus range of applied current which achieves that minimum extinction ratio.

anisms by which the current reduces the index in the injected waveguide are plasma dispersion and band filling. Since these processes are independent of wavelength and polarization, the switching is wavelength and polarization independent. These switches are extremely fast (≈ nanoseconds) and have an advantage over gain switches in power requirements and reduced noise. Compared to the interferometric switches, they are more fabrication tolerant, and are polarization and wavelength insensitive. But the key attraction of these switches is their digital nature: once sufficient index asymmetry is induced, they switch completely, and the control signal does not have to be held at a precise value. Due to this digital nature they are sometimes referred to as digital optical switches.

Digital optical switches in a 1 × 2 Y-configuration had been demonstrated in lithium niobate [39] and in a semiconductor (InP) at 1.55 μm [40], but a robust packaged 2×2 semiconductor X-switch operating at 1.3 μm needed to be developed for the CORD project. GTE is well on its way to achieving this goal. It was the first to extend the InP digital optical Y-switch to 1.3-μm wavelength operation in 1992 [41], [42]. For unpackaged devices, GTE demonstrated polarization independent 1×2 Y-switches and polarization independent blocking 2×2 X-switches, both with extinction ratios exceeding 20 dB, in 1993 [43]–[46]. And it demonstrated polarization and wavelength independent operation, with an extinction ratio exceeding 20 dB at both 1.3 μm and 1.55 μm, in the same Y-switch in 1994 [48]–[50] (in particular, see [49, Figs. 3–5]). Finally, it demonstrated the first nonblocking 2×2 digital X-switch, with a preliminary extinction ratio of 10 dB, in 1995 [51]–[52]. In addition, during this period, GTE developed a substantially more compact practical switch by increasing the waveguide's opening angle from 1 to 15 milliradians [43]–[52], while permitting a wide tolerance of switching current as shown in Figs. 20 and 21.

The packet switched network built in the CORD project is designed to operate with a packet slot size of 250 ns. Therefore, the switching time needed is a few nanoseconds. The digital optical switch in the injection mode responds on the scale of the carrier lifetime, which is indeed on the order of a few nanoseconds. Small-signal operation at 600 MHz, capacitance limited, has been demonstrated previously [53] and optimization of the contact geometry is expected to improve this figure further.

But to be practical, a switch needs to be robustly packaged with low fiber-to-fiber loss and no internal reflections. To this end, GTE has been developing a discrete packaged version of its 2×2 X-switch. This device integrates the previously demonstrated 0.5° opening angle X-switch with three other features on one chip. First, selective differential loss higher-order-mode stripping ridge-waveguides on the inputs, for improved extinction ratio. Second, new low-loss curved input and output waveguides for compactness. And third, angled 7° fiber-to-chip interfaces, to eliminate reflections [54]. However, this packaged switch, which is expected to be available at the beginning of 1996, will still have a chip-to-fiber coupling loss.

B. Integrated Semiconductor Optical Amplifiers

To make up for the inevitable chip-to-fiber coupling loss, optical amplifiers must be included in the completed module. The best way to include amplifiers is monolithic integration with the semiconductor switches on the same semiconductor substrate. This goal is being pursued by GTE and in 1996, utilizing the dual active/passive weak binding switch waveguides developed this past year, 2×2 X-switches will be integrated with optical amplifiers to give a first zero loss fiber-to-fiber digital optical switch.

Semiconductor optical amplifiers present several challenges. One of them is achieving flat spectral gain. This requires reducing the facet reflectivity from the usual 32% to about 0.01%. This can be done by tilting the amplifier facet by about 7° with respect to the guiding ridge. In other words, the facet is no longer perpendicular to the active strip, so that reflections do not feed back into the active region. This technique is by far the most practical way of making optical amplifiers, and has produced state-of-the-art amplifiers with 31 dB internal gain and 21 dB fiber-to-fiber gain [54]. As mentioned before, it is also being used for the stand-alone switch modules.

VI. Summary

This paper has presented an ongoing Consortium effort aimed at demonstrating the feasibility of an all-optical solution to deal with the problems of resource contention and of signaling/synchronization in optical packet networks. The Consortium, named CORD (COntention Resolution by Delay Lines), was formed among the University of Massachusetts, Stanford University, and GTE Laboratories, and is funded by ARPA.

The University of Massachusetts has investigated networking issues involved in optical contention resolution. GTE Laboratories has been working toward the fabrication of a switched-delay-line contention-resolution optical fabric (CRO) based on digital optical switches integrated with optical amplifiers on an InP semiconductor substrate. Stanford University has successfully developed and implemented techniques to achieve global time-slot synchronization, ultra-fast clock recovery, and in-band signaling by means of subcarrier header multiplexing. Stanford University is currently asssembling a WDM network testbed designed to bring together the above signaling/synchronization techniques and test the CRO concept. A discrete-component CRO is currently used, to be replaced by the GTE integrated version when available.

All these principles and technologies are flexible and are expected to find an application, together or separately, in a variety of topologies or network architectures. The innovative concepts in components, subsystems and networking theory developed within this project are expected to yield a substantial contribution toward the practical exploitation of ultra-fast all-optical packet-switched networks.

ACKNOWLEDGMENT

The authors would like to acknowledge the contributions of E. Meland in wafer processing and R. Bryant in mask fabrication.

REFERENCES

[1] S. B. Alexander *et al.*, "A precompetitive consortium on wide-band all-optical networks," *J. Lightwave Technol.*, vol. 11, nos. 5/6, pp. 714–735, May/June 1993.

[2] C. B. Brackett *et al.*, "A scalable multiwavelength multihop optical network: A proposal for research on all-optical networks," *J. Lightwave Technol.*, vol. 11, nos. 5/6, pp. 736–753, May/June 1993.

[3] P. E. Green, Jr. *et al.*, "All-optical packet-switched metropolitan-area network proposal," *J. Lightwave Technol.*, vol. 11, nos. 5/6, pp. 754–763, May/June 1993.

[4] I. Chlamtac and A. Fumagalli, "QUADRO-star: High performance optical WDM star networks," in *Proc. IEEE GLOBECOM'91*, Phoenix, AZ, Dec. 1991.

[5] ——, "All-optical double ring networks," in *3rd Workshop on Very High Speed Networks*, Greenbelt, Maryland, Mar. 1992; also *J. High Speed Networks*, vol. 2, no. 4, pp. 355–371, 1993.

[6] ——, "CROWNs: All-optical WDM multi-ring topologies," in *Proc. EUROPTO Int. Symp. Fiber Optic Networks Video Commun. (SPIE)*, Berlin, Germany, Apr. 1993.

[7] ——, "An optical switch architecture for Manhattan networks," *IEEE J. Select. Areas Commun.*, vol. 11, no. 4, pp. 550–559, May 1993.

[8] A. V. Ramanan, H. F. Jordan, and J. R. Sauer, "Space-time switching in fiber optic packet-switched networks," in *Proc. Conf. Multigigabit Fiber Commun. Syst., Int. Symp. Opt. Appl. Sci. Eng. (SPIE)*, San Diego, CA, July 1993.

[9] A. Bononi, F. Forghieri, and P. R. Prucnal, "Throughput limitations in ultra-fast all-optical soliton mesh networks using deflection routing," in *Proc. OFC'93*, San José, CA, Feb. 1993.

[10] Z. Haas, "The 'Staggering Switch': An electronically controlled optical packet switch," *J. Lightwave Technol.*, vol. 11, nos. 5/6, pp. 925–936, May/June 1993.

[11] M. J. Karol, "Shared-memory optical packet (ATM) switch," in *Proc. Conf. Multigigabit Fiber Commun. Syst., Int. Symp. Opt. App. Sci. Eng. (SPIE)*, San Diego, CA, July 1993.

[12] D. K. Hunter and I. Andronovic, "Optical contention resolution and buffering modulae for ATM networks," *IEE Elett. Lett.*, vol. 29, no. 3, pp. 280–281, Feb. 1993.

[13] D. K. Hunter and D. G. Smith, "An architecture for frame integrity optical TDM switching," *J. Lightwave Technol.*, vol. 11, nos. 5/6, pp. 914–924, May/June 1993.

[14] D. K. Hunter, I. Andronovic, B. Culshaw, and P. E. Barnsley, "Experimental testbed for optical time-domain switching fabric," in *Proc. OFC'93*, paper Tu02, San José, Feb. 1993, pp. 71–72.

[15] M. Calvazara, P. Gambini, M. Puleo, B. Bostica, P. Cinato, and E. Vezzoni, "Optical fiber loop memory for multiwavelength packet buffering in ATM switching applications," in *Proc. OFC'93*, San José, Feb. 1993.

[16] P. Gavignet *et al.*, "Multiwavelength optical buffer based on fiber delay lines for gigabit packet switching," in *Proc. OFC'93*, San José, Feb. 1993.

[17] K. Habara, K. Sasayama, A. Himeno, and S. Suzuki, "Photonic ATM switch using a frequency-routing-type time-division interconnection network (FRONTIERNET) with FDM output buffers," in *Proc. OFC'94*, San José, CA, Feb. 1994.

[18] J. Spring and R. S. Tucker, "Photonic 2×2 packet switching with input buffers," *Electron. Lett.*, no. 29, pp. 284–285, 1993.

[19] D. J. Bluementhal, R. J. Feurstein, and J. R. Sauer, "First demonstration of multihop all-optical packet switching," *IEEE Photon. Technol. Lett.*, vol. 6, no. 3, pp. 457–460, Mar. 1994.

[20] A. Budman, E. Eichen, J. Schlafer, R. Olshansky, and F. McAleavey, "Multigigabit optical packet switch for self-routing networks with subcarrier addressing," in *Proc. OFC'92*, paper Tu04, Feb. 1992.

[21] W. I. Way *et al.*, "Self-routing WDM high-capacity SONET ring network," in *Proc. OFC'92*, paper Tu02, Feb. 1992.

[22] M. W. Maeda *et al.*, "Wavelength-division multiple-access network based on centralized common-wavelength control," *IEEE Photon. Technol. Lett.*, vol. 5, pp. 83–85, Jan. 1993.

[23] S. Fong Su and R. Olshansky, "Performance of WDMA networks with baseband data packets and subcarrier multiplexed control channels," *IEEE Photon. Technol. Lett.*, vol. 5, pp. 236–239, Feb. 1993.

[24] I. Chlamtac, A. Fumagalli, L. G. Kazovsky, and P. T. Poggiolini, "A multi-Gbit/s WDM optical packet network with physical ring topology and multi-subcarrier header encoding," in *ECOC'93*, Montreux, Switzerland, Sept. 1993.

[25] S. F. Su, A. R. Bugos, V. Lanzisera, and R. Olshansky, "Demonstration of a multiple-access WDM network with subcarrier-multiplexed control channels," *IEEE Photon. Technol. Lett.*, vol. 6, pp. 461–463, Mar. 1994.

[26] P. Poggiolini and S. Benedetto, "Theory of subcarrier encoding of packet headers in quasiall-optical broadband WDM networks," *J. Lightwave Technol.*, pp. 1869–1881, Oct. 1994.

[27] M.-S. Chen, N. R. Dono, and R. Ramaswami, "A media-access protocol for packet-switched wavelength division multiaccess metropolitan area networks," *IEEE J. Select. Areas Commun.*, vol. 8, no. 6, pp. 1048–1057, Aug. 1990.

[28] I. Chlamtac and A. Fumagalli, "Performance of Reservation Based (Quadro) WDM Star Networks," in *Proc. IEEE INFOCOM'92*, Florence, Italy, May 1992; *IEEE Trans. Commun.*, Aug. 1994.

[29] W. L. Ha, R. M. Fortenberry, G. M. Wluka, and R. S. Tucker, "Demonstration of photonic fast packet switching at 700 Mb/s data rate," *Electron. Lett.*, vol. 27, no. 10, pp. 789–790, May 1991.

[30] Z. Haas and R. Gitlin, "Optical distribution channel: An 'almost-all' optical LAN based on field-coding technique," in *Proc. IEEE INFOCOM'92*, Florence, Italy, May 1992.

[31] I. M. I. Habbab, M. Kavehrad, and C.-E. W. Sundberg, "Protocols for very high-speed optical fiber local area networks using a passive star topology," *J. Lightwave Technol.*, vol. 5, no. 12, pp. 1782–1794, Dec. 1987.

[32] M. J. Karol, M. G. Hluchyj, and S. P. Morgan, "Input versus output queueing on space-division packet switch," *IEEE Trans. Commun.*, vol. COM-35, no. 12, pp. 1347–1356, Dec. 1987.

[33] I. Chlamtac, A. Fumagalli, and C. J. Suh, "A delay line receiver architecture for all-optical networks," in *Proc. IEEE INFOCOM'96*, San Francisco, CA, Mar. 1996.

[34] ——, "N-input first-in first-out delay line receiver architecture for all-optical networks," Univ. of Massachusetts, Amherst, Tech. Rep. \#TR-95-CSE-07, Sept. 1995.

[35] E. W. M. Wong, I. Chlamtac, and A. Fumagalli, "Performance evaluation of CROWNs: WDM multi-ring topologies," in *Proc. ICC'95*, Seattle, WA, June 1995.

[36] M. Cerisola, T. K. Fong, R. T. Hofmeister, L. G. Kazovsky, C. L. Lu, P. Poggiolini, and D. J. M. Sabido IX, "Performance of the control channel for an optical network implemented with subcarrier multiplexing and a novel ultra-fast clock recovery technique," in *Proc. OFC'95*, San Diego, CA, Feb. 26, 1995.
[37] C.-Li Lu, D. J. M. Sabido IX, P. Poggiolini, R. T. Hofmeister, and L. G. Kazovsky, "CORD—A WDMA optical network: Subcarrier-based signaling and control scheme," *IEEE Photon. Technol. Lett.*, vol. 7, no. 5, pp. 555–557, May 1995.
[38] ——, "Power budget optimization of a WDM network with multi-channel subcarrier multiplexing control," *IEE Electron. Lett.*, accepted for publication.
[39] M. Erman, "Semiconductor based switches within the RACE photonic switching program," in *OSA Proc. Third Top. Switching Meeting*, 1991.
[40] J. A. Cavailles *et al.*, "First digital optical switch based on InP/GaInAsP double heterostructure waveguides," *IEE Electron. Lett.*, vol. 27, no. 9, 1991.
[41] W. H. Nelson, A. N. M. Choudhury, M. Abdalla, R. Bryant, W. Niland, E. Vaughan, and W. Powazinik, "Performance and design of high extinction = ratio X- and Y-junction InP/InGaAsP digital optical switches," in *Proc. IEEE LEOS Annu. Mtg.*, paper EOS/OTA3.6, Boston, MA, Nov. 16–19, 1992.
[42] A. N. M. Masum Choudhury, W. H. Nelson, M. Abdalla, W. Niland, R. Bryant, E. Vaughan, and W. Powazinik, "Fabrication of high-performance digital optical switches in intersecting InP/InGaAsP ridge waveguides," in *Proc. IEEE LEOS Annu. Mtg.*, paper EOS/OTA3.7, Boston, MA, Nov. 16–19, 1992.
[43] W. H. Nelson, A. N. M. Musum Choudhury, M. Abdalla, R. Bryant, W. Niland, E. Vaughn, and W. Powazinik, "Large angle 1.3 μm InP/InGaAsP digital optical switches with extinction ratio exceeding 20 dB," in *Proc. OSA 4th Top. Photon. Switch. Mtg.*, paper PDP-6, Palm Springs, CA, Mar. 15–17, 1993.
[44] A. N. M. Musum Choudhury, W. H. Nelson, M. Abdalla, R. Bryant, W. Niland, M. Rothman, and W. Powazinik, "1.3 μm InP/InGaAsP digital optical switches with extinction ratio of 30 dB," in *Proc. LEOS'93*, San José, CA, 1993, pp. 494–495.
[45] W. H. Nelson, A. N. M. Musum Choudhury, M. Abdalla, R. Bryant, E. Meland, W. Niland, and W. Powazinik, "Issues in the performance of InP/InGaAsP digital optical switches: Theoritical predictions vs. experimental results," in *Proc. IPR'94*, San Francisco, CA, 1994, pp. 32–34.
[46] ——, "Large angle 1.3 μm InP/InGaAsP digital optical switches with extinction ratio exceeding 20 dB," in *Proc. OFC'95*, San José, Feb. 1994, pp. 53–54.
[47] W. H. Nelson, A. N. M. Masum Choudhury, and J. S. LaCourse, "Optical switching expands communications-network capacity," *Laser Focus World*, June 1994
[48] W. H. Nelson, A. N. M. Masum Choudhury, M. Abdalla, R. Bryant, E. Meland, and W. Niland, "Wavelength independent InP/InGaAsP digital optical switches with extinction ratio exceeding 20 dB at both 1.3 μm and 1.5 μm," in *Proc. ECOC'94*, Florence, Italy, Sept. 1994, pp. 523–526.
[49] ——, "Wavelength and polarization independent large angle InP/InGaAsP digital optical switches with extinction ratio exceeding 20 dB," *IEEE Photon. Technol. Lett.*, vol. 6, pp. 1332–1334, 1994.
[50] ——, "Digital optical switches for wavelength division multiplexing with extinction ratio exceeding 20 dB at both 1.3 μm and 1.5 μm," in *Proc. LEOS'94*, Boston, MA, 1994, pp. 259–260.
[51] W. H. Nelson, "Progress on digital optical switches and switch arrays," in *Proc. OFC'95*, San Diego, CA, Feb. 1995, pp. 280–281.
[52] W. H. Nelson, "InP optical switches," in *Proc. CLEO/PR'95*, Chiba, Japan, 1995, pp. 81–82.
[53] M. Renaud *et al.*, "Monolithically integrated InP photonic circuits for optical processing in very high speed optical loops," in *OSA Proc. Photon. Switch. Conf.*, Salt Lake City, Utah, Mar. 1991.
[54] R. Boudreau, R. Morrison, R. Sargent, R. Holmstrom, W. Powazinik, E. Meland, E. Wilmot, and J. LaCourse, "High gain (21 dB) packaged semiconductor optical amplifiers," *Electron. Lett.*, vol. 27, no. 1845, 1991.

A Time Domain Approach for Avoiding Crosstalk in Optical Blocking Multistage Interconnection Networks

CHUNMING QIAO, MEMBER, IEEE, RAMI GEORGES MELHEM,
DONALD M. CHIARULLI, AND STEVEN P. LEVITAN

***Abstract*— Crosstalk in Multistage Interconnection Networks can be avoided by ensuring that a switch is not used by two connections simultaneously. In order to support crosstalk-free communications among N inputs and N outputs, a *space domain* approach dilates an $N \times N$ network into one that is essentially equivalent to a $2N \times 2N$ network. Path conflicts, however may still exist in dilated networks.**

This paper proposes a *time domain* approach for avoiding crosstalk. Such an approach can be regarded as "dilating" a network in time, instead of space. More specifically, the connections that need to use the same switch are established during different time slots. This way, path conflicts are automatically avoided. The time domain dilation is useful for overcoming the limits on the network size while utilizing the high bandwidth of optical interconnects.

We study the set of permutations whose crosstalk-free connections can be established in just two time slots using the time domain approach. While the space domain approach trades hardware complexity for crosstalk-free communications, the time domain approach trades time complexity. We compare the proposed time domain to the space domain approach by analyzing the tradeoffs involved in these two approaches.

I. Introduction

Broadband switching networks can be built from 2×2 electro-optical switches such as lithium-niobate switches [2], [7], [8], [21]. Each switch has two active inputs and two active outputs. Optical signals carried on either input can be coupled to either output by applying an appropriate voltage to the switch. Fig. 1 shows the two logic states of such a switch, namely, "straight" and "cross," and a multistage interconnection network (MIN) with the *generalized cube* (GC) topology built from these switches.

One of the problems associated with these electro-optical switches is *crosstalk*, which is caused by undesired coupling between signals carried in two waveguides. For example, when a switch is in state "straight," a certain amount of signal power from input "A" may be coupled to "D". Similarly, a certain amount of signal power from input "B" may also be coupled to output "C". Crosstalk also occurs when the switch is in state "cross". When both inputs of the switch carry signals, output signals are affected by first order crosstalk. This is a significant factor which affects the signal-to-noise ratio (SNR).

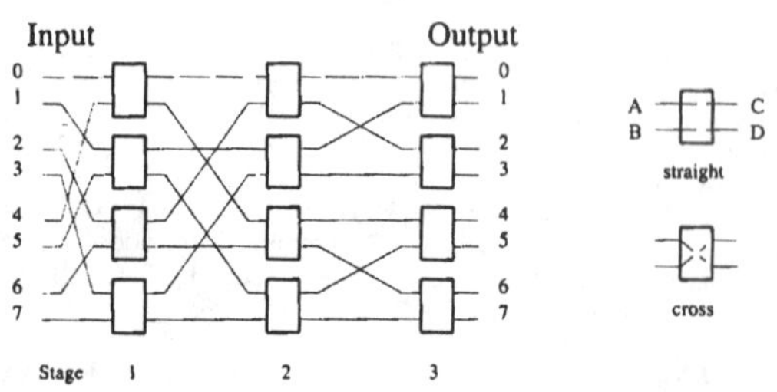

Fig. 1. An 8×8 generalized cube (GC) MIN built from 2×2 switches.

First order crosstalk at switches (hereafter abbreviated as crosstalk) can be avoided by ensuring that only one input of every switch is active at any given time. In other words, no two connections should use the same switch simultaneously. For MIN's, this can be accomplished with a space domain approach, called *network dilation* [9], [10], [24], [25]. Using this approach, an $N \times N$ network is dilated into a network that is essentially equivalent to a $2N \times 2N$ network, in which half of the input and output ports are used for N inputs and outputs, respectively. In the resulting dilated network, the connections of certain permutations can be established simultaneously without crosstalk. This concept has been generalized into *wavelength dilation* which is used to suppress crosstalks between closely separated wavelengths in wavelength-routing networks [19]. The space domain approach is reviewed in more details in Section II-A.

If the size of the network is limited to $N \times N$, crosstalk-free communications between $N/2$ inputs and outputs can be supported using the space domain approach. One of the objectives of this research is to support crosstalk-free communications among up to N inputs and outputs using the $N \times N$ network. For this, time-division multiplexing (TDM) can be employed to utilize the high bandwidth offered by the optical interconnects [1], [12]. An alternative would be to construct a $2N \times 2N$ network, which, depending on the value of N, may or may not be economically and/or technologically feasible.

The time domain approach proposed in this paper is a method that achieves the above objective. It extends the principle of the *reconfiguration with time division multiplexing*

Manuscript received June 30, 1993; revised June 9, 1994. This work was supported in part by the Air Force Office of Scientific Research under Contract AFOSR-89-0469 and in part by the Andrew Mellon Educational Trust in the form of a fellowship to the first author.
C. Qiao is with the Department of Electrical and Computer Engineering, State University of New York at Buffalo, Buffalo, NY 14260.
R. Melheim and D. Chiarulli are with the Department of Computer Science, University of Pittsburgh, Pittsburgh, PA 15260.
S. Levitan is with the Department of Electrical Engineering, University of Pittsburgh, Pittsburgh, PA 15260.
IEEE Log Number 9404870.

Reprinted with permission from *IEEE Journal of Lightwave Technology*, Vol. 12, No. 10, pp. 1854-1862, October 1994.

(RTDM) paradigm [13], [15], [16], to be described in Section II-B. The RTDM resolves the path conflicts by partitioning a set of connections into conflict-free subsets and is thus applicable to dilated networks as well. The basic idea of the proposed time domain approach is to avoid crosstalk in a way similar to avoiding path conflict. More specifically, a set of connections is partitioned into several subsets such that the connections in each subset can be established simultaneously in a network not only conflict-free but also crosstalk-free. As such, the set of connections is established within several *time slots*, one for each subset. Clearly, the connections in some permutations realizable in a dilated network may need two time slots to be established using this time domain approach. The proposed approach can be regarded as "dilating" a network in the time domain, which is described in more details in Section II-C.

In Sections III and IV, we study the connectivity of networks dilated in the space domain and that of networks dilated in the time domain. More specifically, we determine the relationship between the set of permutations realizable in one time slot using the space domain approach and the set of permutations that require two time slots using the proposed time domain approach. In addition, we determine the number of time slots needed to establish arbitrary connections using either approach.

While the space domain approach can be regarded as a way to trade hardware complexity for crosstalk-free communications, the time domain approach trades time complexity, which is equivalent to communication bandwidth. Whether to trade hardware complexity or to trade bandwidth for crosstalk-free communications depends on the specific application being considered. For instance, for multiprocessor interconnections, the computation bandwidth of a processor is much lower than the optical bandwidth. Thus, it is more appropriate to utilize the excessive bandwidth to achieve crosstalk-free communications. In Section V, we discuss the tradeoffs involved in the time domain and space domain approaches. Finally we draw the conclusions in Section VI.

II. Approaches for Supporting Crosstalk-Free Communications

Throughout this study, we consider MIN's with the generalized cube (GC) topology [20] such as the one shown in Fig. 1. This topology is chosen partly because a large GC network can be easily constructed from smaller ones in a recursive way. For example, a 16×16 network is constructed from two 8×8 networks and an additional front stage, as shown in Fig. 2. The recursive characteristics of the topology will facilitate the discussions in later sections. Note that such a GC network is topologically equivalent to many blocking MIN's such as the Omega network [26]. For instance, the GC network in Fig. 1 becomes an 8×8 Omega network after positions of the two middle switches at the middle stage are interchanged. Similarly, by properly interchanging the positions of the switches at the (two) middle stages of the 16×16 GC network shown in Fig. 2, we can obtain a 16×16 Omega network. Based on this topological equivalence,

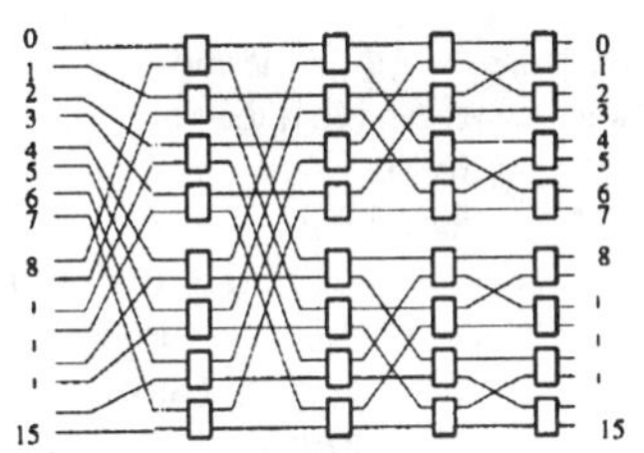

Fig. 2. A 16×16 GC network recursively constructed from 8×8 ones.

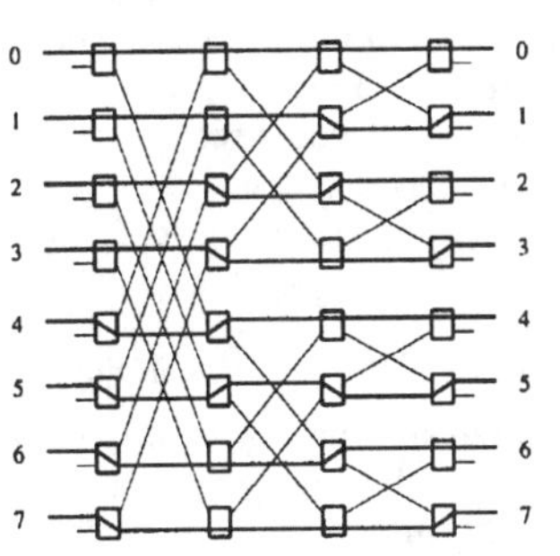

Fig. 3. The realization of the 8×8 identity permutation in a dilated 8×8 GC network.

we conclude that a GC network can realize the same set of permutation as an Omega network. Let this set of permutations be defined as follows.

Definition 1: Let Ω be the set of N by N permutations realizable by an $N \times N$ Omega or GC network.

The next subsection reviews the space domain approach for avoiding crosstalk, followed by a subsection describing RTDM for resolving path conflicts and a subsection describing the proposed time domain approach for avoiding crosstalk and resolving path conflicts as well.

A. Space Domain Dilation for Avoiding Crosstalk

A dilated $N \times N$ network is similar to a $2N \times 2N$ network. A major difference is that in a dilated network, only half of the input and output ports is used. Fig. 3 shows a dilated 8×8 GC network, in which only one of the two input (or output) ports of the switches at the first (or last) stage of the network is actually used. Other than that, this dilated 8×8 network is the same as the 16×16 network shown in Fig. 2. For this reason, we conclude that a dilated GC network is also topologically equivalent to a dilated Omega network. That is, they can realize the same set of permutations without crosstalk.

In a dilated network, a connection between an input and output is established by choosing an appropriate path in the network so that no switch in the network will have both input ports active at the same time. This avoids crosstalk as discussed earlier. By dilating a network, it becomes possible to establish a set of connections, or to realize certain permutations, without crosstalk. In [10], it is shown that a dilated Omega network has the same permutation capability as an Omega network. That is, the set of permutations realizable in a dilated Omega network without crosstalk is also Ω (see

Definition 1). Therefore, based on the topological equivalence mentioned previously, we establish the following lemma.

Lemma 1: Ω is also the set of permutation realizable in a dilated $N \times N$ GC network without crosstalk.

Since GC and Omega networks are blocking networks, Ω will contain some but not all possible permutations. Lemma 1 implies that certain permutations (such as the identity permutation as shown in Fig. 3) can be realized in a dilated GC network without crosstalk while others cannot. That is, dilated GC (or dilated Omega) networks are also blocking networks. This means that the network dilation approach can avoid crosstalks but cannot resolve path conflicts that exist in the original network. In the next section, we describe a time division multiplexing technique that may be used to resolve path conflicts in optically interconnected networks.

B. Reconfiguration with Time Division Multiplexing (RTDM)

The reconfiguration with time division multiplexing (RTDM) was proposed as a solution to the problem of relatively slow network control in resolving path conflicts [13]. The idea in RTDM is a generalization of techniques used in the time-space-time switching networks as in [8], [23], and [24].

With RTDM, a set of connections R is partitioned into several conflict-free subsets, called *mappings*. Thereafter, the connections in each mapping can be established in a network without path conflicts. Based on these mappings, a sequence of network configurations, or in other words, a set of states to which the switches in the network are set, can be determined. Each input and output is then informed of the sequence of configurations, which the network goes through, one configuration during each *time slot*. As a result, the connections in a mapping are established in one time slot and all the connections in R are established within several time slots in a time-division multiplexed fashion. Since no buffering or arbitration is needed, electro-optical switching devices such as lithium niobate directional couplers [2], [8], become suitable for implementing such networks.

If the set of connections R required by an application program is known and does not change during execution, a sequence of network configurations can be determined once at the beginning of execution of the application program. This is referred to as *static reconfiguration*, which is suitable whenever compile time analysis of the connection requirements can be done or a target communication structure is to be embedded in the interconnection network. Static reconfiguration with time-division multiplexing in multistage interconnection networks (MIN's) was studied in [15].

In the cases when the connection requirements change dynamically, the sequence of network configurations also needs to be changed. By time-multiplexing K configurations, K virtual networks are created in the time domain. The amount of control overhead involved in determining and setting the states of switches in these virtual networks can be amortized with concurrent processing of the requests. In addition, a sequence of configurations can capture communication locality in a similar way that a set of pages in a virtual memory system captures memory reference locality [5]. As a result, dynamic reconfiguration with time-multiplexing can effectively reduce control overhead and improve network performance [11], [16].

The principle of the above described RTDM can thus be used to resolve path conflicts in *dilated* networks as well. In doing so, a set of connections is partitioned into subsets that are not only conflict-free but crosstalk-free as well in a dilated network. Such subsets are called crosstalk-free (CF) mappings for the network. We will elaborate on the application of RTDM to path conflict resolution in dilated networks in Section IV. Next, we describe a way to integrate the solutions to crosstalk and path conflicts based on RTDM.

C. Time Domain Dilation

The proposed time domain approach extends the principle of the RTDM paradigm for avoiding crosstalk. More specifically, the same ideas in RTDM can be applied except that a set of connections needs to be partitioned into several subsets to avoid not only path conflicts but crosstalk as well. These subsets are CF-mappings for an undilated network. By establishing the connections in each of these CF-mappings in a separate time slot, crosstalk is avoided without the need for a dilated network and path conflicts are automatically resolved. Note that, these CF-mappings are different from the CF-mappings for a dilated network. One of the differences, for example, is that a CF-mapping for an undilated $N \times N$ network can contain only up to $N/2$ connections, while a CF-mapping for a dilated $N \times N$ network can contain up to N connections.

Assuming that a permutation in Ω needs to be realized, such a time domain approach that avoids crosstalk by using two or more CF-mappings may be considered as "dilating" a network in the *time domain*. In particular, a network dilated in the time domain using two time slots can be regarded as a correspondence to a spatially dilated network. The next two sections study the issues related to the connectivity of the networks that are dilated either in the time or in the space domain.

III. Permutation Capability of the Time and Space Domain Approaches

In this section, we first describe the set of permutations that require just two CF-mappings using the time domain approach. This set is then compared to the set of permutation realizable in one time slot in a dilated network.

In the following discussions, the phase "in one time slot" is usually omitted following the word "realizable" whenever there is no confusion. We first introduce a definition which corresponds to Definition 1 in Section II.

Definition 2: Let Θ be the set of N by N permutations realizable with *two* CF-mappings by an $N \times N$ GC network.

The following theorems state the relationship between the Ω-permutations (see Definition 1) and the Θ-permutations. For the time being, we consider Ω as the set of permutations realizable by a GC network (although it could be considered as the set of permutations realizable by a dilated GC network, as in Lemma 1).

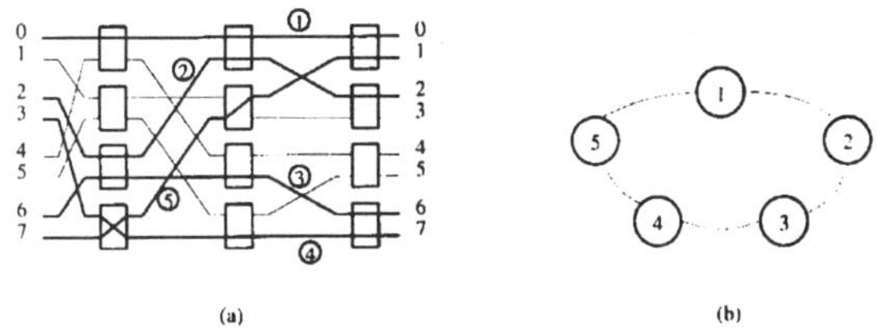

Fig. 4. Possible crosstalk among the five paths.

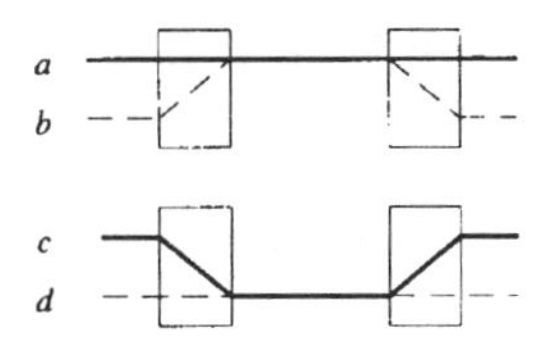

Fig. 5. An example 4×4 Θ-permutation.

Theorem 1: Not every Ω-permutation is a Θ-permutation in a network with $N \geq 8$.

Proof: We first consider the case $N = 8$ and show an example Ω-permutation which cannot be realized with just two CF-mappings. Five paths, numbered from 1 to 5, are drawn in bold lines in Fig. 4(a) as a part of the example Ω-permutation in an 8×8 GC network. Fig. 4(b) shows a graph of five vertices. Each vertex in the graph corresponds to a path in Fig. 4(a). Two vertices are connected by an edge if the two corresponding paths share a switch in the network. For example, vertex 1 and vertex 2 correspond to path 1 and path 2, respectively. Clearly, these two paths cannot be in the same CF-mapping. Since the graph in Fig. 4(b) is a ring of five vertices, it is impossible to establish the five paths in Fig. 4(a) with just two CF-mappings. This proves that the example Ω-permutation including these five paths is not a Θ-permutation. For $N > 8$, an Ω-permutation which includes the above five paths as *partial* paths can be constructed. The resulting Ω-permutation is not a Θ-permutation and therefore the theorem is proved. □

The above theorem indicates that Θ is not a superset of Ω. The following theorem shows the converse.

Theorem 2: Not every Θ-permutation is an Ω-permutation.

To prove this theorem, we first look at an example. Fig. 5 shows a 4×4 GC network and a permutation containing four paths labeled a, b, c, and d. This permutation, which we denote by $\pi(4)$, is not an Ω-permutation since these four paths cannot be established simultaneously without conflict in switch settings. However, since paths a and c can be established in one CF-mapping and paths b and d can be established in another CF-mapping, the permutation is a Θ-permutation.

The proof of the theorem is based on recursive constructions of a permutation $\pi(N)$ that is a Θ-permutation but not an Ω-permutation. $\pi(N)$ will be recursively constructed from the example permutation $\pi(4)$ shown in Fig. 5. Number the paths in $\pi(N)$ from 1 to N in a top-down order based on the position of their originating input ports. Denote the set of odd and even numbered paths by $\pi_{odd}(N)$ and $\pi_{even}(N)$, respectively. The induction hypotheses about $\pi(N)$ are as follows.

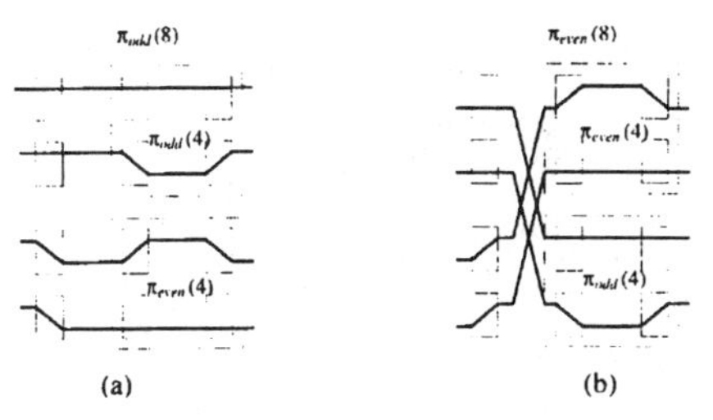

Fig. 6. Recursive construction of the two sets for $N = 8$.

1. Each of the sets, $\pi_{odd}(N)$ and $\pi_{even}(N)$ is a CF-mapping in an $N \times N$ network and their union, $\pi(N)$, is a Θ-permutation.
2. One path from set $\pi_{odd}(N)$ containing a as a partial path conflicts with one path from set $\pi_{even}(N)$ containing b as a partial path. Therefore, the union of the two sets, is not an Ω-permutation.

For $N = 4$, $\pi_{odd}(4)$ contains path 1 (which is a) and path 3 (which is c) while $\pi_{even}(N)$ contains path 2 (which is b) and path 4 (which is d). Clearly, the above hypotheses are true for $N = 4$.

Assume that the hypotheses are true for a network of size N. Based on this assumption, we will prove that the hypotheses are also true for a network of size $2N$. We do that by constructing the two sets, $\pi_{odd}(2N)$ and $\pi_{even}(2N)$, from $\pi_{odd}(N)$ and $\pi_{even}(N)$, in the following way.

To construct $\pi_{odd}(2N)$ in a network of size $2N \times 2N$, the top subnetwork (of size $N \times N$) is set to establish the paths in $\pi_{odd}(N)$ while the bottom subnetwork is set to establish the paths in $\pi_{even}(N)$. After the two subnetworks are set properly, the switches of the upper half of the first stage are then set to the "straight" state while those of the lower half are set to the "cross" state. This completes the construction of set $\pi_{odd}(2N)$. Fig. 6(a) shows the construction of $\pi_{odd}(8)$ from $\pi_{odd}(4)$ and $\pi_{even}(4)$.

Since only one input (an odd numbered one) of each switch at the first stage is used, there will be no crosstalk at that stage when the connections in set $\pi_{odd}(2N)$ are established. Due to the induction hypotheses 1), no crosstalk will be present at later stages (of either subnetwork). Therefore, set $\pi_{odd}(2N)$ is a CF-mapping.

The set $\pi_{even}(2N)$ can be similarly constructed by reversing the roles of the two subnetworks in the above procedure. Fig. 6(b) shows the construction of $\pi_{even}(8)$. The resulting set $\pi_{even}(2N)$ is also a CF-mapping. Since different inputs and outputs are active in set $\pi_{odd}(2N)$ and set $\pi_{even}(2N)$, and each set contains N connections, the union of the two sets is thus a $2N$ by $2N$ permutation. Therefore, the hypotheses 1) is also true for networks of size $2N$.

Note that when the set $\pi_{odd}(2N)$ is constructed, the upper 4×4 subnetwork is set to establish a as a partial path. Similarly, when the set $\pi_{even}(2N)$ is constructed, the same 4×4 subnetwork is set to establish b as a partial path. Since paths a and b conflict in the subnetwork, the path from $\pi_{odd}(2N)$ containing a as a partial path conflicts with the path from $\pi_{even}(2N)$ containing b as a partial path. Thus, hypotheses (2) is also true for networks of size $2N$.

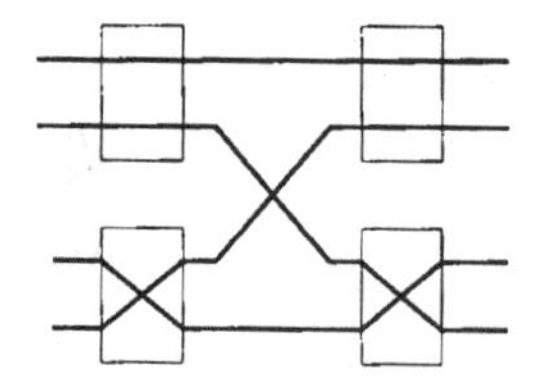

Fig. 7. An example 4×4 permutation in both Ω and Θ.

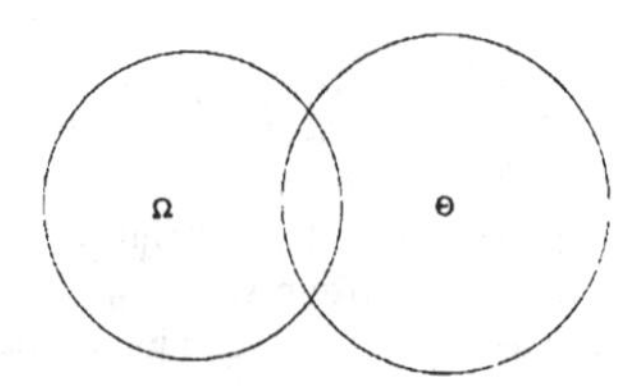

Fig. 8. Relationship between the set Ω and the set Θ.

Since both hypotheses are true, we have proved the theorem by induction. □

Having found from the above two theorems that the Θ-permutations and the Ω-permutations are not the same, we also observe the following.

Theorem 3: Some Ω-permutations are also Θ-permutations (and vice versa).

We will sketch the proof of the theorem by examining an example shown in Fig. 7. Note that the permutation in this figure, which is different from the one shown previously in Fig. 5, belongs to both Ω and Θ. Therefore, the theorem holds for a network with $N = 4$.

For networks with $N > 4$, a recursive construction procedure similar to the one used in Theorem 2 can be carried out and its description is thus omitted. It is worth noting, however, that the recursive construction is now based on the permutation in Fig. 7 instead of Fig. 5. Therefore, the theorem can be proved using a similar induction proof in which the hypotheses (2) becomes "no path in set $\pi_{odd}(2N)$ conflicts with paths in set $\pi_{even}(2N)$ and therefore the union of the two sets is an Ω-permutation." □

So far, we have found that Ω and Θ are two different sets with a nonempty common subset. Fig. 8 summarizes the relationship between the two sets. The figure also shows that Θ is larger (that is, contains more permutations) than Ω. This is because there is a one-to-one (but not onto) mapping from Ω to Θ, as proved below.

Denote by f a perfect shuffle on set $\{0, 1, \cdots, N-1\}$. That is, given any number D $(0 \leq D \leq N-1)$ and its binary representation $d_1 d_2 \cdots d_n$, a perfect shuffle f maps $D = d_1 d_2 \cdots d_n$ to its image $D' = d_2 \cdots d_n d_1$.

Let each connection be represented by an input–output pair (S, D) where $0 \leq S, D \leq N-1$. We may represent a permutation P by N pairs of input–output ports. That is, $P = \{(S_i, D_i) | i = 1, 2, \cdots, N\}$. Denote by F a mapping which maps P to P' by shuffling all output ports of the connections in P. More specifically, we have $P' = F(P) = \{(S_i, f(D_i)) | i = 1, 2, \cdots, N\}$. Fig. 9 illustrates the mapping from P to P' by F where $N = 8$ and $D_i = i - 1$.

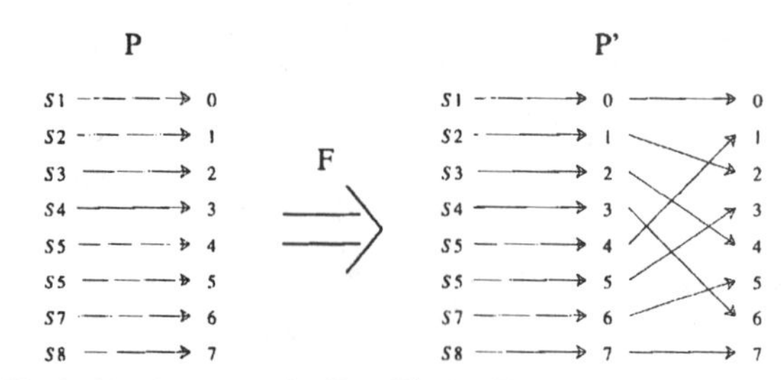

Fig. 9. Mapping permutation P to P' using F.

Lemma 2: F is a one-to-one mapping from Ω to Θ.

Proof: If P_1 and P_2 are two different permutations, then $F(P_1)$ and $F(P_2)$ are also different permutations. This is because function f is a one-to-one mapping between integers. To prove the lemma, we need to show that if P is an Ω-permutation, then its image $P' = F(P)$ is a Θ-permutation. So far, we have considered Ω as a set of permutations realizable by a GC-network as in Definition 1. In proving this lemma, we consider Ω as a set of permutations realizable by a *dilated* GC network as in Lemma 1. That is, we will show that if P can be realized by a dilated GC network with one CF-mapping, then P' can be realized by a GC network with two CF-mappings.

While presenting the general proof for any permutation P in an $N \times N$ network, we demonstrate the ideas used in the proof by applying them to the identity permutation realized by the 8×8 dilated GC network shown in Fig. 3. (Note that the identity permutation in an Ω-permutation according to Lemma 1.)

Divide the connections in P into two subsets, p and q, based on their destinations. More specifically, let p contain the connections whose destinations are $0, 1, \cdots, N/2-1$ and q contain the connections whose destinations are from $N/2$ to $N-1$. For example, the 8×8 identity permutation is divided into

$$p = \{(0, 0), (1, 1), (2, 2), (3, 3)\} \tag{1a}$$

$$q = \{(4, 4), (5, 5), (6, 6), (7, 7)\}. \tag{1b}$$

When P is realized in a dilated network, the connections in p will use the upper subnetwork while the connections in q will use the lower subnetwork. The realization of p and q for the example identity permutation is shown in the left hand of Fig. 10.

Given that P is partitioned into p and q, the permutation $P' = F(P)$ is the union of the two subsets, p' and q', which correspond to p and q, respectively. While the connections in p' have the same sources as those in p, their destinations are $f(0), f(1), \cdots, f(N/2-1)$, that is, $0, 2, \cdots, N-2$. Similarly, the connections in q' have the same sources as those in q but their destinations are $1, 3, \cdots, N-1$. For example, for the 8×8 identity permutation, we have

$$p' = \{(0, 0), (1, 2), (2, 4), (3, 6)\} \tag{2a}$$

$$q' = \{(4, 1), (5, 3), (6, 5), (7, 7)\}. \tag{2b}$$

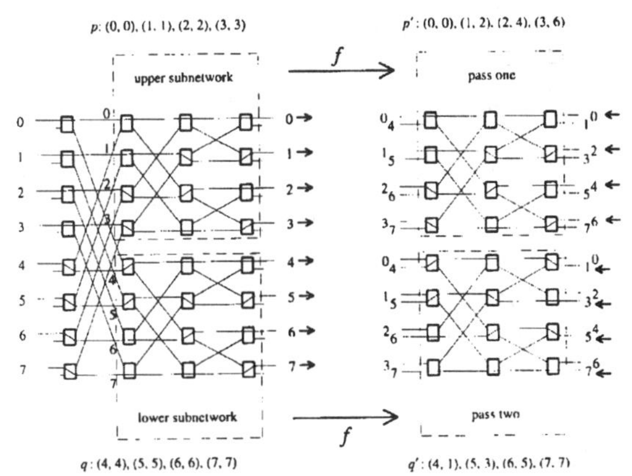

Fig. 10. Emulating the two subnetworks of a dilated network by an undilated network in two passes.

In order to show that P' is a Θ-permutation, it is sufficient to show that both p' and q' are CF-mappings in an undilated GC network.

We first observe that a dilated $N \times N$ GC network (e.g., the 8×8 dilated network in Fig. 3) consists of a first stage and two subnetworks. Each subnetwork is similar to an undilated $N \times N$ GC network (e.g., the 8×8 network in Fig. 1). The input ports of each subnetwork may be numbered in the same way as the undilated network while the output ports are numbered differently. For example, Fig. 10 shows that the output ports of the upper subnetwork of the 8×8 dilated network are numbered 0, 1, 2, and 3, respectively, while those of the undilated network (shown at top right) are numbered 0, 2, 4, and 6, respectively. The difference between these two numberings can be represented by the shuffle function f. Similarly, as shown in Fig. 10, by applying f to output ports 4, 5, 6, and 7 of the lower subnetwork, we obtain output ports 1, 3, 5, and 7 of the undilated network (shown at bottom right), respectively.

In general, by applying f to the numbering of the output ports of the upper subnetwork of an $N \times N$ *dilated* network, we get the numbering of the corresponding output ports of an $N \times N$ *undilated* network. Since p' is obtained by applying f to the destination of the connections in p, we may establish the connections p' in the undilated network in the same way that the connections in p are established in the dilated network. Similarly, we may establish the connections in q' in the undilated network in a way similar to that the connections in q are established in the lower subnetwork of the dilated network. Since both p' and q' are CF-mappings in an undilated $N \times N$ network, P' is a Θ-permutation.

For example, as shown in Fig. 10, by setting the switches of an undilated network (shown at the top right in Fig. 10) in the same way as the upper subnetwork, the four paths in p' (see (2a)) can be established without crosstalk in the same way that the paths in p (see (1a)) are established in the upper subnetwork. Similarly, we may set an undilated network (as shown at the bottom right) according to the lower subnetwork, except that the switches at the last stage should be set differently. This permits the four paths in q' (see (2b)) to be established in the network without crosstalk in a similar way that the four paths in q (see (1b)) are established in the lower subnetwork. This shows that both p' in (2a) and q' in (2b) are a CF-mapping and thus the union of the two subsets is a Θ-permutation. □

In order to show that Θ actually contains more permutations than Ω, we first note that a reverse mapping of F exists. Denote such a mapping by F^-. This mapping can be accomplished by a reverse shuffle of all output ports. More specifically, let a reverse shuffle be denoted by f^-, where $f^-(D = d_1 \cdots d_{n-1} d_n) = d_n d_1 \cdots d_{n-1}$. Given a permutation $P = \{(S_i, D_i) | i = 1, 2, \cdots, N\}$, its image under the reverse mapping F^- is $P' = F^-(P) = \{(S_i, f^-(D_i)) | i = 1, 2, \cdots, N\}$. Note that both $F(F^-)$ and $F^-(F)$ are identity mappings. That is, given any P, $F(F^-(P)) = F^-(F(P)) = P$.

Lemma 3: F is not an *onto* mapping from Ω to Θ.

To prove the lemma, it is sufficient to show that there exists a Θ-permutation π, such that $F^-(\pi)$ is not an Ω-permutation. Note that Ω will again be considered in the context of an undilated network according to Definition 1.

Let π be the Θ-permutation constructed recursively in Theorem 2. π contains connection $(0, 0)$ in $\pi_{odd}(N)$ and connection $(N/2, N/2)$ in $\pi_{even}(N)$. Thus, $F^-(\pi)$ contains $(0, 0)$ and $(N/2, N/4)$. These two connections share the same switch (for input 0 and input $N/2$) and their destinations are both in the upper subnetwork, thus resulting in a conflict in setting that switch. The more formal techniques used in [17] may be applied to show that these two connections in $F^-(Q)$ conflict with each other. Therefore, $F^-(\pi)$ is not an Ω-permutation. For example, when $N = 8$, π contains $(0, 0)$ and $(4, 4)$. Thus, $F^-(\pi)$ contains two conflicting connections, $(0, 0)$ and $(4, 2)$, and thus is not an Ω-permutation.

Since F is an one-to-one mapping from Ω to Θ and F^- is its reverse mapping, the proof that F is not an onto mapping is thus completed. □

Based on the above two lemmas, the following is established.

Theorem 4: There are more Θ-permutations than Ω-permutations.

Note that this theorem implies that the time domain "dilation" is more powerful than the space domain dilation in realizing permutations in blocking networks. As will be shown next, the time domain approach enjoys a similar property as it needs less than twice as many time slots as the space domain approach when establishing an arbitrary set of connections.

IV. Establishing Arbitrary Connections with CF-Mappings

We now consider sets of arbitrary connections, which are not necessarily permutations. Such a set of connections may need to be established in the MIN if the MIN is used as a centralized switching hub. In [24], a dilated slipped banyan (DSB) architecture is proposed which emulates an $N \times N$ completely-connected network by repeatedly realizing N different permutations. Each of these permutations is realized in one CF-mapping for the duration of a time slot, and every input is connected to a different output each time a different permutation is realized. Therefore, a completely-connected network is emulated with N CF-mappings.

In the previous section, it was shown that not every permutation can be realized with two CF-mappings by an undilated network. Thus, it is not clear that an undilated GC network can emulate an $N \times N$ completely-connected network using just $2N$ CF-mappings. However, we establish the following theorem.

Theorem 5: An $N \times N$ completely-connected network can be emulated in $2N$ CF-mappings in an undilated GC network.

Proof: Let $P_1, P_2, \cdots, P_N$ be the N Ω-permutations used when a dilated network emulates a completely-connected network. Since all possible N^2 connections are contained in these permutations, we have $\bigcup_{i=1}^{N} P_i = I \times O$, where I denotes the set of N inputs and O denotes the set of N outputs.

To prove the theorem, we apply the F mapping in Lemma 2 to the above Ω-permutations and consider the permutations: $F(P_1), F(P_2), \cdots, F(P_N)$. Since each of these permutations is a Θ-permutation, it can be realized with two CF-mappings by an undilated network. Therefore, all these N Θ-permutations can be realized with $2N$ CF-mappings. It remains to be shown that any connection in $I \times O$ is in $\bigcup_{i=1}^{N} F(P_i)$.

Given any connection $(S, D) \in I \times O$. Let $D' = f^{-}(D)$. Since (S, D') is also a connection in $I \times O$, it belongs to P_k for some k $(1 \leq k \leq N)$. That is, $(S, D') \in P_k$. According to the definition of mapping F, we have $(S, f(D')) \in F(P_k)$. That is, $(S, f(f^{-}(D))) \in F(P_k)$ and thus $(S, D) \in F(P_k)$. □

For applications that may not require all possible connections at all times, using $2N$ CF-mappings results in low bandwidth utilization. The static and dynamic RTDM summarized in Section II-B are shown to be effective in achieving high communication efficiency [15], [16]. Although the results in [15], [16] do not consider crosstalk, the same ideas can be applied except that a set of arbitrary connections needs to be partitioned into several CF-mappings to avoid not only path conflicts but crosstalk as well. Note that, to establish a set of arbitrary connections R, more than one CF-mapping is needed even in a *dilated* network. This is because, as mentioned earlier, that space domain dilation does not resolve path conflicts. More specifically, a dilated blocking network is still a blocking network. Even in a dilated nonblocking (or rearrangeably nonblocking) network, path conflicts exist between connections with the same source or destination.

We are interested in determining the number of time slots (or CF-mappings) needed to establish a set of arbitrary connections using either the time domain or the space domain approach. To do so, let K_u and K_d be the number of CF-mappings resulted from the partitioning of R in an undilated and a dilated network, respectively. Clearly, $K_u \geq K_d$. Based on Theorem 5, $K_u = 2 \times K_d$ when $R = I \times O$.

When $R \subseteq I \times O$, simulations have been carried out to determine K_d and K_u. A set of random, distinct connection requests is generated from all possible N^2 connection requests. A greedy algorithm is used to partition this set into CF-mappings in either an undilated or a dilated network. Fig. 11

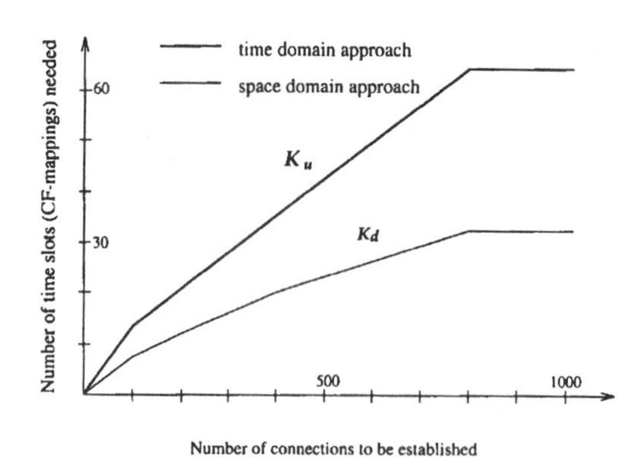

Fig. 11. Simulation results when $N = 32$.

shows both K_d and K_u as a function of the number of requests when $N = 32$. From this figure, we see that $K_u < 2 \times K_d$. This result may be explained by the fact that the time domain approach integrates the solutions to both crosstalk and path conflict problems, and thus is more efficient than the space domain approach which deals with crosstalk avoidance only and still relies on the time multiplexing techniques to resolve path conflicts. For example, when establishing a set of connections that contains two connections with identical inputs or outputs, at least two CF-mappings are needed in either an undilated or a dilated network. As such, the space domain approach may lose its advantage of being able to save twice as much time as the time domain approach in such cases.

V. Tradeoff Analysis

Ordinarily, a network having N input and output ports can support communications between N inputs and outputs. That is, up to N connections may be established in the network.

In order to support crosstalk-free communications among N inputs and outputs, the space domain approach increases the hardware complexity by dilating an $N \times N$ network into one that is essentially equivalent to a $2N \times 2N$ network. Although there may be other factors, we assume that the hardware complexity of a network is proportional to the number of switches, links and electronic driver circuits used in the network. As such, the hardware complexity of a dilated $N \times N$ (or a $2N \times 2N$) network can be considered as twice the hardware complexity of an (undilated) $N \times N$ network. Note that, however, the actual *cost* of a dilated $N \times N$ photonic switching network made of lithium niobate may be less than twice the cost of an $N \times N$ network, as long as constructing such a network is technologically and/or economically *feasible* [9], [25].

The time domain approach, on the other hand, increases time complexity by using more than one time slot to support crosstalk-free communications among N inputs and outputs. Note that this increase in the time complexity also represents a decrease in throughput, or usable bandwidth of the network. For example, in an 8×8 network with 2.5 Gb/s lasers, the time domain approach can only provide a (crosstalk-free) bandwidth of 10 Gb/s. If crosstalk had not been of any concern, an 8×8 configuration with 2.5 Gb/s lasers would have provided a bandwidth of 20 Gb/s. The usable bandwidth for crosstalk-free communications can only be increased if more hardware is used as in the space domain approach.

We note that to achieve crosstalk-free communications with about the same amount of hardware (e.g., a time-dilated $N \times N$ network or a space-dilated $N/2 \times N/2$ network), the space domain approach would have been able to connect only $N/2$ inputs to $N/2$ outputs. But if each of these $N/2$ inputs is time multiplexed with two input sources, and each of these $N/2$ outputs is time multiplexed with two output destinations, then the (crosstalk-free) bandwidth connecting the N inputs to N outputs in the space domain approach is the same as with the time domain approach.

If the future technology allows the transmission rate to scale up faster than the network size, the proposed time domain approach will be a useful way to support crosstalk-free communications. That is, if the cost of increasing the bandwidth of each connection should become as "cheap" as (or even cheaper than) the cost of constructing a network whose size is twice its original size, then one may consider using the time domain approach instead of the space domain approach. For example, let us assume that a goal is to provide 32 channels between 32 input and output pairs with each channel having 2 Gb/s bandwidth. Using the proposed time domain approach, one can achieve the goal by using a 32×32 *undilated* network and lasers each operating at 4 Gb/s. This is appropriate *when* (and if) building a 32×32 *dilated* network which needs lasers that operate only at 2 Gb/s, is *less feasible* than building the smaller (undilated) network that needs lasers at 4 Gb/s. Of course, it would be ideal to have a 32×32 dilated network and use these high speed lasers (each operating at 4 Gb/s) to provide higher bandwidth. In summary, whenever the limit on the network size is reached before the limit on the available bandwidth, the time domain approach may be used as a way to trade the bandwidth for the desired connectivity, as many other multiplexing techniques would do.

We also note that so far, we have assumed that each switch in a network, whether dilated or undilated, is controlled individually. A network control algorithm will be responsible for processing a given set of connection requests and determining the state of each switch in the network. The time, as well as memory, needed for such an algorithm to determine a sequence of network configurations that can establish a given set of requested connections is on the same order whether the network is dilated or not (although, as can be seen from Fig. 11 for example, the number of configurations, i.e., the number of time slots, needed will be different). If column control (or stage control) [3] is used in either the dilated or the undilated network, the connectivity of the network in terms of its capability in realizing permutations will be reduced and the network control will be simplified. It would be interesting to investigate if results that are similar to those theorems and simulation results previously obtained under individual switch control can also be obtained under column control.

In [14], a space–time complexity measure is defined as the product of the number of switches (and links) used in the network and the number of time slots needed to establish a set of crosstalk-free connections. This is similar to the measure used in VLSI design [4], [6], [18], [22]. Simulation results presented in [14] show that the time domain approach improves the space-time tradeoffs over the network dilation approach when used to establish a set of arbitrary connections. An intuitive explanation is that the time domain approach integrates the solutions to the crosstalk and path conflict problems while space domain dilation only deals with crosstalk avoidance and still relies on time multiplexing techniques to resolve path conflicts. Note that although only blocking networks are simulated, these results are expected to be applicable to nonblocking networks as well since path conflicts exist among the connections having the same source or destination. This is in spite of the fact that the set of permutations realizable with two time slots by a nonblocking network would be the same as the set of permutations realizable by a *dilated* nonblocking network (both consist of all possible permutations).

VI. Summary

In this paper, a time domain approach which is an extension of RTDM is proposed. Using this approach, a set of connections to be established is partitioned into several subsets, so that the connections in each subset can be established simultaneously in the network without crosstalk. Such an approach may be regarded as a way to "dilate" a network in the time domain as opposed to the space domain network dilation approach.

The relationship between Θ, which is the set of permutations realizable with two CF-mappings by a GC network (or its equivalence) and Ω, which is the set of permutations realizable by a dilated GC network (or its equivalence), is studied. In particular, it is shown that there is a one-to-one but not onto mapping from Ω to Θ. In other words, the set Θ contains more permutations than the set Ω. This implies that the time domain "dilation" is more powerful in realizing permutations than space domain "dilation."

Note that since dilating a network usually increases the number of stages in the network as well, the time domain approach for avoiding crosstalk will result in less propagation delay and optical path loss than the network dilation approach. It is also more flexible since the same architecture may be used for applications with or without crosstalk budget problems. Finally, it is worth noting that the proposed time domain approach is a useful approach for overcoming the limits imposed on the network size, especially if future technology allows optical bandwidth to scale up faster than network sizes. On the other hand, if the technology allows the feasible construction of large networks and the bandwidth does not scale up accordingly, then the network dilation approach should be used.

Acknowledgment

The authors would like to thank Dr. M. Karol and the anonymous reviewers for their comments which helped us to improve this paper.

REFERENCES

[1] N. K. Ailawadi, R. C. Alferness, G. D. Bergland, and R. A. Thompson, "Broadband photonic switching using guided-wave fabrics," *IEEE LTS Mag.*, pp. 38–42, May 1991.

[2] R. C. Alferness, "Guided-wave devices for optical communication," *IEEE J. Quantum Electron.*, vol. QE-17, pp. 946–957, 1981.

[3] K. E. Batcher, "The multi-dimensional access in STARAN," *IEEE Trans. Comput. (Special Issue on Parallel Processing)*, pp. 174–177, Feb. 1977.

[4] R. P. Brent and H. T. Kung, "The area-time complexity of binary multiplication," *J. ACM*, vol. 28, no. 3, pp. 521–534, 1981.

[5] D. Chiarulli, S. Levitan, R. Melhem, and C. Qiao, "Locality based control algorithms for reconfigurable optical interconnection networks," *Appl. Opt.*, vol. 33, pp. 1528–1537, Mar. 1994.

[6] H. Habelson and P. Andreae, "Information transfer and area-time tradeoffs for VLSI multiplication," *Commun. ACM*, vol. 23, no. 1, pp. 20–23, 1980.

[7] H. Hinton, "A non-blocking optical interconnection network using directional couplers," in *Proc. IEEE Global Telecommun. Conf.*, Nov. 1984, pp. 885–889.

[8] S. Korotky *et al.*, "An experimental synchronized optical network using a high-speed time-multiplexed $Ti:LiNbO_3$ switch," presented at the Optical Fiber Communication Conf., 1989.

[9] T. O. Murphy, C. T. Kemmerer, and D. T. Moser, "A 16×16 $Ti:LiNbO_3$ dilated benes photonic switch module," presented at the Topical Meeting on Photonic Switching, Postdeadline Paper PD3, Salt Lake City, Ut, Mar. 1991.

[10] K. Padmanabhan and A. Netravali, "Dilated network for photonic switching," *IEEE Trans. Commun.*, vol. COM-35, no. 12, pp. 1357–1365, Dec. 1987.

[11] C. Qiao, R. Melhem, D. Chiarulli, and S. Levitan, "Efficient channel allocation for routing in optically interconnected multiprocessor systems," in *SPIE Proc. Conf. Advances in Optical Information Processing V*, vol. 1704, Apr. 1992, pp. 428–439.

[12] C. Qiao and R. Melhem, "Time-division optical communications in multiprocessor arrays," *IEEE Trans. Comput.*, vol. 42, no. 5, pp. 577–590, 1993.

[13] C. Qiao, "A high speed interconnection paradigm for multiprocessors and its applications to optical interconnection networks," Ph.D. dissertation, Dept. Comput. Sci., Univ. Pittsburgh, 1993.

[14] C. Qiao, R. Melhem, D. Chiarulli, and S. Levitan, "A time domain approach for avoiding crosstalk," in *OSA Proc. Photonics in Switching*, vol. 16, Mar. 1993, pp. 133–137.

[15] C. Qiao and R. Melhem, "Reconfiguration with time-division multiplexed MINs for multiprocessor communications," *IEEE Trans. Parallel Distributed Syst.*, vol. 5, no. 4, pp. 337–352, 1994.

[16] C. Qiao, R. Melhem, D. Chiarulli, and S. Levitan, "Dynamic reconfiguration of optically interconnected networks with time division multiplexing," *J. Parallel Distributed Comput.*, vol. 22, pp. 268–278, 1994.

[17] C. Raghavendra and A. Varma, "Fault-tolerant multiprocessors with redundant-path interconnection networks," *IEEE Trans. Comput.*, vol. C-35, no. 4, pp. 307–316, Apr. 1986.

[18] J. E. Savage and S. Swamy, "Space-time tradeoffs for oblivious sorting and integer multiplication," Brown Univ., Tech. Rep. CS-37, 1979.

[19] J. Sharony, K. W. Cheung, and T. E. Stern, "The wavelength dilation concept in lightwave networks—Implementation and system considerations," *IEEE J. Lightwave Technol.*, vol. 11, no. 5/6, pp. 900–907, 1993.

[20] H. J. Siegel and S. D. Smith, "Study of multistage SIMD interconnection networks," in *Proc. Fifth Annu. Symp. Computer Architecture*, April 1978, pp. 223–229.

[21] R. Spanke, "Architectures for large non-blocking optical switches," *IEEE J. Quantum Electron.*, vol. QE-22, pp. 885–889, Aug. 1986.

[22] C. D. Thompson, "Area-time complexity for VLSI," in *Proc. 11th Annu. ACM Symp. Theory of Computing*, May 1979, pp. 81–88.

[23] R. A. Thompson, R. V. Anderson, J. V. Camlet, and P. P. Giordano, "Experimental modular switching system with a time-multiplexed photonic center stage," in *Proc. OSA Topical Meeting Photonic Switching*, Mar. 1989, pp. 212–218.

[24] R. A. Thompson, "The dilated slipped banyan switching network architecture for use in an all-optical local area network," *IEEE J. Lightwave Technol.*, vol. 9, no. 12, pp. 1780–1787, Dec. 1991.

[25] J. E. Watson *et al.*, "A low-voltage 8×8 $Ti:LiNbO_3$ switch with a dilated Benes architecture," *J. Lightwave Technol.*, vol. 8, no. 5, pp. 794–800, May 1990.

[26] C.-L. Wu and T. Y. Feng, "On a class of multistage interconnection networks," *IEEE Trans. Comput.*, vol. C-29, pp. 694–702, Aug. 1980.

Tree-Type Photonic Switching Networks

ANDRZEJ JAJSZCZYK AND H.T. MOUFTAH

The enormous bandwidth offered by optical systems makes photonic switching a very attractive solution for broadband communications. Tree-type architectures play an important role in the design of photonic switching networks. The authors present and discuss all known tree-type architectures in a unified framework, and propose a number of new solutions.

This article presents and compares various architectures of tree-type space-division photonic switching networks and proposes some new solutions as well. The discussed networks can be implemented with guided-wave-based switching elements, or laser diodes and passive splitters and combiners. The following network types are considered: conventional, simplified, and two-active-stage networks. Techniques for improving SNR as well as waveguide crossover minimization are presented and discussed.

Optics has achieved a spectacular success in transmission systems, where a large percentage of new facilities are now making use of optical fiber. However, the full potential of optical technology cannot be used in systems containing electronic nodes.

Among various solutions of photonic switches, those based on the guided-wave technology are the most mature. Guided-wave switches are capable of switching light from one channel to another. This channel guide can be an optical fiber or a waveguide formed in an appropriate substrate. Most guided-wave switches are based on the application of the refractive index change with the amplitude of the electric field.

To obtain switching networks of a higher capacity than that of a single guided-wave switch, the switches can be arranged and interconnected in various ways. Among possible solutions, the following architectures are best known [1, 2]: crossbar, double crossbar, N-stage planar, Benes, dilated Benes, and tree-type. Crossbar and double-crossbar as well as the tree-type switching networks are nonblocking, while Benes and N-stage planar architectures are rearrangeable. The crossbar and N-stage planar switches have poor loss and SNR performance, in contrast to significantly better double crossbar, dilated Benes or tree-type topologies. One of the disadvantages of the double crossbar, Benes, and dilated Benes guided-wave networks is the considerable number of required waveguide crossovers that increase loss, lower SNR, and complicate manufacturing.

Tree-type architectures play an important role in the design of photonic switching networks. Examples of such architectures were proposed by Habara and Kikuchi [3], Spanke [4], Okayama *et al.* [5], Jajszczyk [6, 7], as well as Jajszczyk and Mouftah [8]. Tree-type networks can be applied in both circuit and fast-packet switching systems [8, 9].

Although various space-division photonic networks were surveyed in the past, only the basic, conventional tree architectures were considered [2, 10, 11]. The purpose of this article is to present and discuss all known tree-type architectures in a unified framework, as well as to propose some new solutions.

First, we present tree-type networks, beginning with conventional architectures. We also discuss the influence of an increase of the number of active stages on network performance. Then simplified tree and two-active-stage architectures are discussed. The next section deals with some important implementation issues for both single- and multiple-substrate networks. The application of laser diodes in tree-type architectures is also discussed. In a separate section we compare various characteristics of the tree-type networks.

Tree Architectures

Generic Architecture

A generic architecture of an $N \times N$ tree-structured photonic network is shown in Fig. 1. The network is composed of a set of N optical splitters $1{:}N/2$ and a set of N optical combiners $N/2{:}1$. Splitters and combiners can be passive or active [12]. These splitters and combiners are usually implemented as binary tree structures of 1:2 and 2:1 elements, respectively [4]. Central module and patterns of interconnecting links are specific for each

ANDRZEJ JAJSZCZYK is a professor at EFP — The Franco-Polish School of New Information and Communication Technologies in Poznań.

H. T. MOUFTAH is a professor with the Department of Electrical Engineering, Queen's University at Kingston.

Reprinted with permission from *IEEE Network*, Vol. 9, No. 1, pp. 10-16, January/February 1995.

kind of tree-type network. The module can contain one or more switching stages. All proposed tree-structured networks are nonblocking in the strict sense for point-to-point connections, and some of them have also this capability for multicast connections.

Conventional Tree-Type Networks

A conventional tree architecture is presented in Fig. 2 [3, 4]. The central module in this network is formed by an additional stage of optical splitters and an additional stage of optical combiners, as well as a simple interconnection pattern between these stages. In particular designs, splitters and combiners can be active or passive. Therefore, we have the following three variations: AS/AC, PS/AC, and AS/PC, where AS and AC stand for active splitters and active combiners, respectively. PS denotes passive splitters, while PC is related to passive combiners. The PS/AC version allows multicast connections where multiple outputs (all, in the limiting case) can listen to the same input. One of the important advantages of the conventional tree architecture is the fact that all active switching elements in each vertical column of a splitter and a combiner can have a common driver.

The principal characteristics of tree-structured networks are summarized in a following section. Here, we will discuss some aspects of the SNR performance for the AS/PC architecture. In the case where a common driver is associated with each column of an active splitter and for point-to-point connections, a first order approximation for the SNR is X -10log$_{10}$(log$_2N$), where X is the extinction ratio of a switching element. However, if we assume either separate drivers for each switching element or multipoint-to-one connections are allowed, it can be easily shown that the worst-case SNR is increased to X -10log$_{10}$(N -1). We can note that the assumption of multipoint-to-one connections makes sense for photonic fast packet switching, although the simultaneous transmission of multiple packets (over the same wavelength) to a common output would mean an erroneous state leading to a packet collision.

Enhanced Tree-Type Networks

The SNR performance of the AS/AC conventional tree-type network is excellent. The SNR is significantly lower when either splitters or combiners are passive. The application of passive devices, however, reduces the cost of the network and simplifies its control. The question arises whether it is possible to improve the SNR characteristics by replacing some passive elements in either splitters or combiners by active devices. Let us consider, first, the PS/AC architecture. If splitters are completely passive, noise signals attenuated by only one extinction ratio (first order approximation) can be added to a required signal at log$_2N$ active combiners. Therefore, the SNR is X-10log$_{10}$(log$_2N$). If, however, only p splitter stages are passive and remaining log$_2N$ – p stages contain active 1 x 2 switches, the number of noise signals (which affect the desired signal), encountering only one extinction ratio, is at most 2^p – 1. Therefore, the resultant SNR is

$$\mathrm{SNR} = X - 10\log_{10}(\min\{2^p - 1, \log_2 N\}), p \geq 1, \quad (1)$$

where p is the number of passive stages. It can be easily seen that the improvement of the SNR is rather moderate. The same result holds for the architectures where some passive combiners have been replaced by active devices. We can note that the exact location of passive and active stages within splitters or combiners has no impact on the SNR. However, to preserve switching properties, either splitters or combiners should be completely active (i.e., mixed passive-active stages are not allowed in both splitters and combiners).

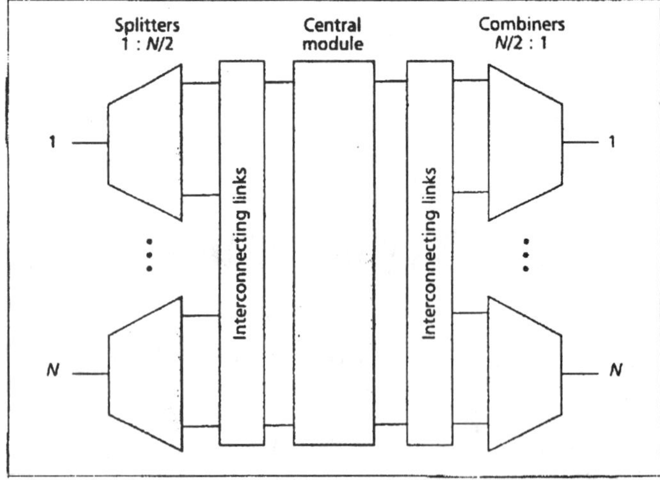

■ Figure 1. *Generic tree-type network.*

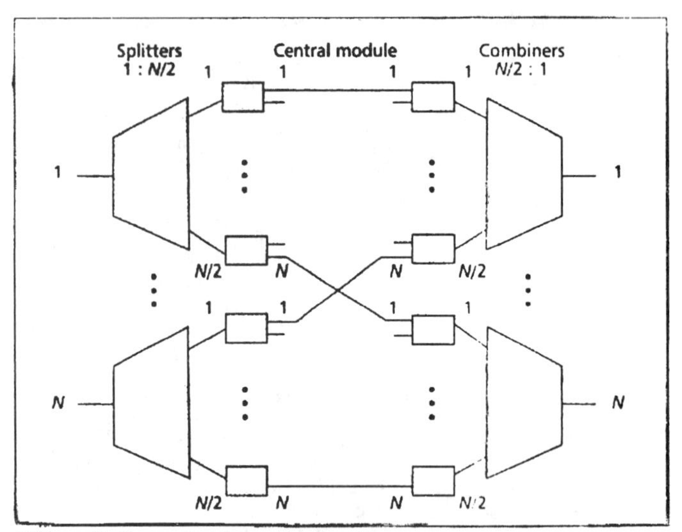

■ Figure 2. *Conventional tree-type network.*

Although the AS/AC architecture has an excellent SNR, it is applicable for point-to-point connections only. To allow broadcasting, passive splitters have to be used. However, the SNR for PS/AC is inferior to that of the AS/AC architecture. To improve the SNR we can add a single on/off switch (e.g., a directional coupler) for each line in the very center of the network. For such a structure we have the following second order approximation

$$\mathrm{SNR} = 2X - 10\log_{10}(\log_2 N) \quad (2)$$

We should note that active switching elements

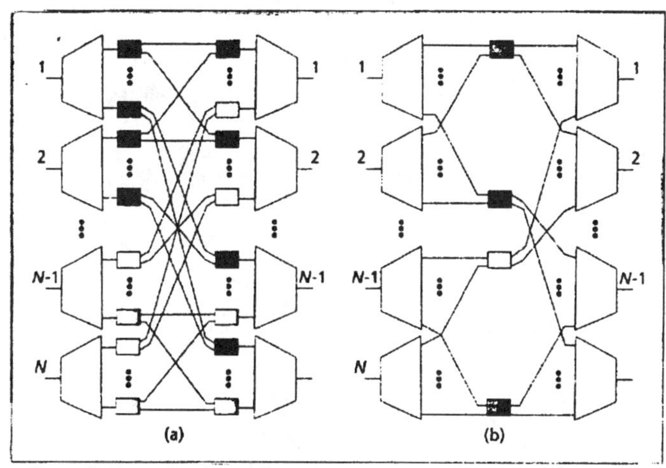

■ Figure 3. *Principle of simplification: a) conventional tree-type network; b) simplified tree-type network.*

in the middle have to be controlled separately (i.e., they need separate drivers), which increases complexity of the network. The discussed structure can also have both splitters and combiners passive (PS/PC). In such a case it contains only one active stage in its middle. The main disadvantage of this approach is poor loss and SNR performance. The latter can be improved, for example, by replacing a single on/off switch with two such switches connected in series. The obtained pair of switches can be controlled by a single driver.

Simplified Tree-Type Networks

Tree-type networks can be condensed as shown in Fig. 3 [5]. The condensation procedure is based on replacing groups of two 1 x 2 and two 2 x 1 elements in a conventional tree-type network by 2 x 2 active switches. Similarly, as in the case of the conventional networks, we can distinguish the following three types of the simplified architectures: AS/AC, PS/AC, AS/PC. The central 2 x 2 switch is always active. The simplified networks are nonblocking in the strict sense and can be used for point-to-point connections only.

Two-Active-Stage Architectures

In some applications, e.g., in fast packet switching, photonic networks containing a small number of active (i.e., controlled) switching stages can be of some advantage, especially when fully photonic control is used. In such a case, control circuitry (associated only with active elements) can be considerably limited [8]. One of the possible solutions is presented in the previous section on enhanced networks. But better characteristics, in terms of loss and SNR, can be achieved in architectures in which a connecting path passes through two active elements [6-8].

A basic architecture of an $N \times N$ two-active-stage switching network is shown in Fig. 4. The network consists of $k (k = \lceil N/2 \rceil$, where the symbol $\lceil x \rceil$ denotes the smallest integer greater than or equal to x) active modules each containing k basic elements, N passive splitters 1:k and N passive combiners k:1. The basic elements can be composed, for example, of three directional couplers (Fig. 4b), for point-to-point connections, or four-directional couplers (Fig. 4c) when multicasting is required.

Two-active-stage networks can be simplified in a similar way as the conventional tree-type systems [7]. An alternative, inverse structure of the two-active-stage network is also possible [8]. Such a configuration facilitates self-routing (as explained in [8]), which is of primary interest in fast packet switching applications.

Implementation Issues

Multiple-Substrate Implementations

Tree architectures are best fitted for multiple-substrate implementations. In such cases most of the networks do not contain any waveguide crossovers. In conventional tree-type networks every 1:N splitter and N:1 combiner can be fabricated on separate substrates. All passive 1:N splitters exhibit the splitting loss of $10\log_{10}N$[dB], under the assumption that the optical power is split equally. Most single-mode passive combiners suffer from the similar combining loss. However, this loss can be eliminated if the combiner devices are made from waveguides with a multimode state in the lateral direction, while maintaining a single-mode state in the vertical direction [13]. Sometimes passive splitters or combiners are implemented as fiber devices to avoid the waveguide-to-fiber coupling loss, which is roughly 1 – 2 dB at each interface [4]. The active 1 x 2 and 2 x 1 switching

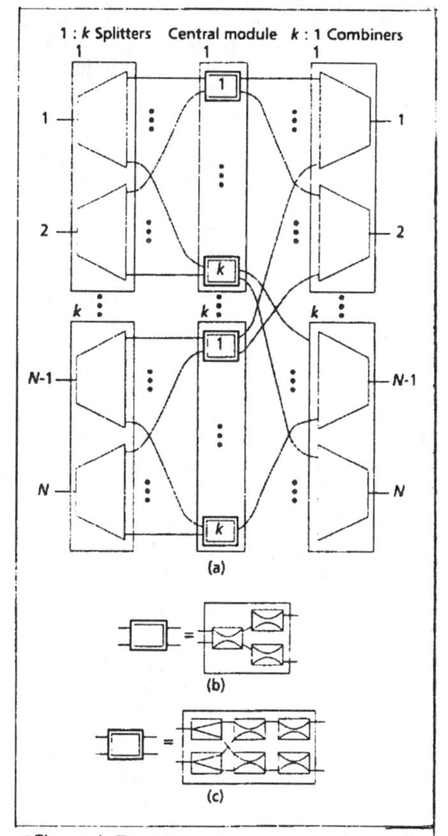

■ Figure 4. *Two-active-stage network: a) general structure; b) active point-to-point switch; c) active multicasting switch.*

elements can be directional couplers, X-switches, or branching switches [14]. Interferometric, Mach-Zehnder, 1 x 16 and 1 x 32 matrices fabricated on lithium niobate substrates have also been recently reported [15, 16].

If only some columns of 1 x 2 or 2 x 1 devices in a splitter or combiner, respectively, are passive, it is always desirable to locate them as close to the center of the complete tree-type network, thus minimizing the number of active components.

In the case of the enhanced multicast network, the central on/off switches can be integrated on the substrates containing active 2 x 1 devices. We should note here that in such a solution, driving circuit configurations are different for the active combiners and the on/off switches. The latter require separate drivers, while the former needs a single driver for each column.

When a two-active-stage network is implemented on multiple substrates, active switches forming the central module (Fig. 4) can be fabricated on a single crystal. If the number of fibers is too large to attach them to the single substrate, the central module can be partitioned, for example, into k separate devices. A similar approach can be used for multiple-substrate implementations of the simplified tree-type networks of Fig. 3b.

Single-Substrate Implementations

The major problem associated with single-substrate implementations of tree-type architectures is the large number of waveguide crossovers. Such crossovers cause crosstalk, signal loss, and increase the manufacturing complexity. These adverse effects can be minimized if the intersection angles are kept above a certain minimum value (usually 7° to 8°, for lithium niobate devices). The manufacturing complexity is expressed by the total number of crossovers on a substrate, while the worst-case insertion loss is related to the maximum number of crossovers between an inlet and an outlet.

The formulas for the maximum and the total number of crossovers for various tree-type architectures are given in a following section. To decrease the number of crossovers, an alternative, nested structure can be used for conventional tree, simplified tree and two-active-stage networks [7]. For example, the central-stage switching elements of Figs. 3b or 4 are replaced by complete networks of Figs. 3b or 4, respectively, or by other nested structures. An 8 x 8 nested and a 4 x 4 matrix with a simplified tree structure, implemented on a single substrate, have been reported [17-19].

Application of Laser Diodes

Tree-type architectures can also be used in networks employing laser diode switches. Application of such diodes in photonic switching has been studied in recent years [13, 20-23]. The use of laser diodes as on/off switches results in additional losses associated with coupling into and out of the laser diode. These losses as well as the losses caused by passive splitting and combining in tree-type networks can be compensated by the gain of laser diodes that operate as optical amplifiers. We should note, however, that because of spontaneous emission associated with each laser diode, successive amplification in cascaded architectures is of a major concern [22]. Since it has been shown, based on spontaneous emission considerations, that it is better to compensate small losses with small gain amplifiers than large losses with large gain amplifiers, laser-diode-based networks can contain more laser diodes (some operating simply as optical amplifiers) than the minimum needed to realize all switching states. For example, it can be justifiable to locate a laser amplifier after every splitter and every combiner in a tree-type network rather than to locate all laser diodes in the center of a tree-type network. Such a solution, however, considerably increases the overall cost of the network.

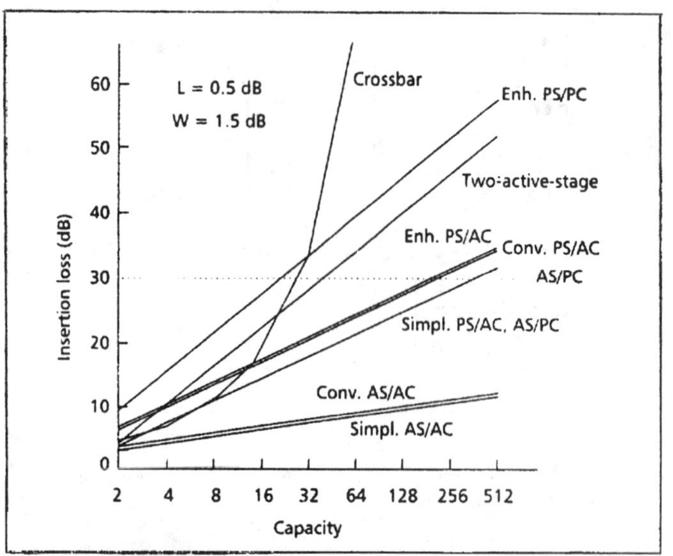

■ Figure 5. *Insertion loss for various architectures.*

Architecture	Insertion loss in dB
Conventional AS/AC	$2L\log_2 N + 2W$
Conventional PS/AC, AS/PC	$(3 + L)\log_2 N + 2W$
Enhanced PS/AC	$(3 + L)\log_2 N + L + 2W$
Enhanced PS/PC	$6\log_2 N + L + 2W$
Simplified AS/AC	$(2\log_2 N - 1)L + 2W$
Simplified PS/AC, AS/PC	$(3 + L)\log_2 N - 3 + 2W$
Two-active-stage (point-to-point)	$6(\log_2 N - 1) + 2L + 2W$
Two-active-stage (multicast)	$6(\log_2 N - 1) + 3 - 2L + 2W$

■ Table 1. *Worst case insertion loss in tree-type architectures.*

Characteristics of Tree-Type Networks

Many characteristics should be considered while designing or comparing various space-division optical switching networks. To select an optimum network structure it is necessary to find a compromise solution, depending on a particular application, accessible technology, etc. The most

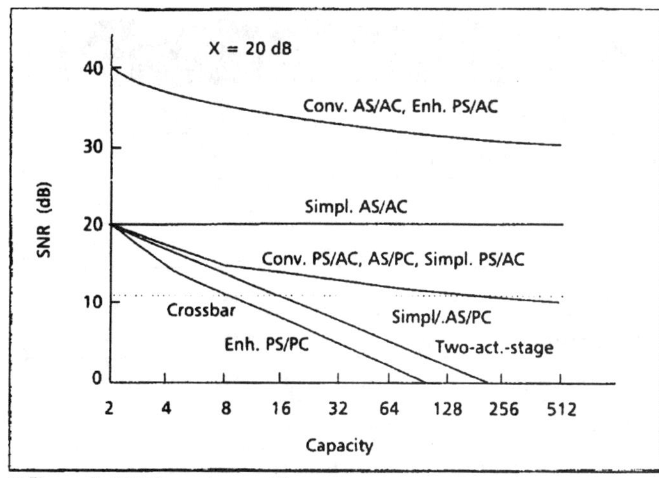

■ **Figure 6.** *SNR for various architectures.*

Architecture	SNR in dB
Conventional AS/AC	$2X - 10\log_{10}(\log_2 N)$
Conventional PS/AC, AS/PC	$X - 10\log_{10}(\log_2 N)$
Enhanced PS/AC	$2X - 10\log_{10}(\log_2 N)$
Enhanced PS/PC	$X - 10\log_{10}(N - 1)$
Simplified AS/AC	X
Simplified PS/AC	$X - 10\log_{10}(\log_2 N)$
Simplified AS/PC	$X - 10\log_{10}(N/2)$
Two-active-stage	$X - 10\log_{10}(N/2)$

■ **Table 2.** *Worst case SNR for tree-type architectures.*

important characteristics can be listed as follows: loss; crosstalk; number of switching elements; number of active stages; number of drivers; number of waveguide crossovers; number of substrates; number of optical fibers; blocking properties; possibility of multicasting; voltage requirements; control complexity; bandwidth; switching speed; stability of parameters; reliability, testability and maintainability; modularity and growability; mass manufacturability; and overall cost.

Here, we limit ourselves to the first six characteristics of tree-type networks composed of guided-wave switches. Structures with center on/off switches are referred to as enhanced networks in all tables and figures.

Loss characteristics are presented in Fig. 5 (optical crossbar loss is given for comparison) and summarized in Table 1, where: N is the number of inputs and outputs, L is the insertion loss of a single active switch, and W is the waveguide-to-fiber coupling loss. The excess losses and the crossover losses are assumed to be equal to zero. We also assumed that all passive 1:2 splitters and 2:1 combiners exhibit 3-dB splitting and combining loss, respectively. The formulas are valid for single-substrate solutions, although they can be easily modified for multiple-substrate networks.

The formulas for the signal-to-noise ratio are shown in Table 2, where X is the extinction ratio of a single switching elements in decibels, and illustrated by Fig. 6. We can note that the formula for the conventional AS/PC architecture is valid under the assumption that a common driver is associated with each column of an active splitter and point-to-point connections are only allowed. A first order approximation of the SNR for the simplified AS/AC architecture is X. The formula given in [5] is, in fact, a second order approximation, and for practical values of X it gives almost identical results. The last expression of Table 2 is valid for both point-to-point and multicast versions of the two-active-stage architecture, as well as for their inverse counterparts. However, if an inverse version allows multipoint-to-point connections, the formula changes to $X - 10\log_{10}(N-1)$ [8].

The formulas for the numbers of active switching elements are contained in Table 3, while the numbers of active stages are listed in Table 4. We can see that two architectures require a fixed number of active stages independently of a network capacity. This feature can be of a particular interest in fast packet switching applications.

Table 5 presents the number of drivers. The conventional tree-type networks are definitely the best as far as this point is concerned, since many of the switching elements can be electrically tied together and driven from the same driver. Other

Architecture	Number of active elements
Conventional AS/AC	$2N(N-1)$
Conventional PS/AC, AS/PC	$N(N-1)$
Enhanced PS/AC	$N(2N-1)$
Enhanced PS/PC	N^2
Simplified AS/AC	$N(5N/4-2)$
Simplified PS/AC, AS/PC	$N(3N/4-1)$
Two-active-stage (point-to-point)	$3N^2/4$
Two-active-stage (multicast)	N^2

■ **Table 3.** *Number of active switching elements in tree-type architectures.*

Architecture	Number of active stages
Conventional AS/AC	$2\log_2 N$
Conventional PS/AC, AS/PC	$\log_2 N$
Enhanced PS/AC	$\log_2 N + 1$
Enhanced PS/PC	1
Simplified AS/AC	$2\log_2 N - 1$
Simplified PS/AC, AS/PC	$\log_2 N$
Two-active-stage	2

■ **Table 4.** *Number of active stages in tree-type architectures.*

Architecture	Number of drivers
Conventional AS/AC	$2N\log_2 N$
Conventional PS/AC, AS/PC	$N\log_2 N$
Enhanced PS/AC	$N\log_2 N + N^2$
Enhanced PS/PC	N^2
Simplified AS/AC	$N(5N/4 - 2)$
Simplified PS/AC, AS/PC	$N(3N/4 - 1)$
Two-active-stage (point-to-point)	$3N^2/4$
Two-active-stage (multicast)	N^2

■ Table 5. *Number of drivers in tree-type architectures.*

Architecture	Total number of crossovers
Conventional, enhanced (non-nested)	$N^2(N-1)^2/4$
Conventional, enhanced (nested; $k = 2$)	$3N^2(\log_2 N)/4 - N^2 + 4$
Simplified (non-nested)	$N^3(N-2)/16$
Simplified (nested; $k = 2$)	$3N^2(\log_2 N)/4 - 5N^2/4 + N$
Two-active-stage (non-nested, point-to-point)	$N^3(N-2)/16$
Two-active-stage (non-nested, multicast)	$N^3(N-2)/16 + N^2/4$
Two-active-stage (nested; $k = 2$, point-to-point)	$3N^2(\log_2 N)/4 - 5N^2/4 + N$
Two-active-stage (nested; $k = 2$, multicast)	$3N^2(\log_2 N)/4 - N^2 + 4$

■ Table 6. *Total number of crossovers in tree-type architectures.*

architectures require separate drivers for each switching element, except the enhanced PS/AC architecture, where active combiners have shared drivers and the switching elements of the central stage are driven separately.

Tables 6 and 7 contain formulas for the total and the maximum number of waveguide crossovers, respectively. In all cases we assumed a single-substrate implementation. Application of multiple substrates reduces these numbers to zero in most architectures. For the conventional, simplified, and two-active-stage networks we gave formulas for two limiting cases, i.e., non-nested as well as nested networks with $k = 2$ (Fig. 4) for subnetworks on all levels of nesting hierarchy. For a general k, the total number of crossovers for a simplified or a point-to-point two-active-stage network can be expressed as follows:

$$C_{N,k} = kN(k-1)(N + N/k - 2)/4 + k^2 C_{N/k,k'} \quad (3)$$

where k' is the parameter of the splitters and combiners in the network nested inside the network of capacity N, and $C_{2,k'} = 0$.

The total number of crossovers for the conventional or two-active-stage multicast networks is increased by one additional crossover for each switching element, i.e., by $(N/2)^2$. The maximum number of crossovers for a simplified or a point-to-point two-active-stage network is [7]:

$$M_{N,k} = (k-1)(N + N/k - 2) + M_{N/k,k'} \quad (4)$$

where $M_{2,k'} = 0$. One additional crossover should be added for conventional and two-active-stage multicast networks.

Conclusion

Tree-type architectures are among the most attractive options for space-division photonic networks. All kinds of networks discussed in this article are nonblocking in the strict sense, some of them even for multicast connections (conventional PS/AC, enhanced, two-active-stage). For multiple substrate implementations the considered networks do not virtually require waveguide crossovers. Appropriate design also allows single-substrate implementations. In such cases, the nested versions of the networks (with $k = 2$) have the lowest number of crossovers.

Architecture	Maximum number of crossovers
Conventional, enhanced (non-nested)	$(N-1)^2$
Conventional, enhanced (nested; $k = 2$)	$3N - 2\log_2 N - 3$
Simplified (non-nested)	$N^2/2 - N$
Simplified (nested; $k = 2$)	$3N - 2\log_2 N - 4$
Two-active-stage (non-nested, point-to-point)	$N^2/2 - N$
Two-active-stage (non-nested, multicast)	$N^2/2 - N + 1$
Two-active-stage (nested; $k = 2$, point-to-point)	$3N - 2\log_2 N - 4$
Two-active-stage (nested; $k = 2$, multicast)	$3N - 2\log_2 N - 3$

■ Table 7. *Maximum number of crossovers in tree-type architectures.*

The maximum capacity of most network types is limited mainly by the total insertion loss and the worst-case SNR. If we assume the limiting value for the insertion loss to be 30 dB (Fig. 5) and the lowest acceptable SNR to be 11 dB (to achieve a 10^{-9} bit error rate [4]; Fig. 6) as well as $L = 0.5$ dB, $W = 1.5$ dB, and $X = 20$ dB, the maximum network capacity is 128 x 128 for the conventional PS/AC and AS/PC, the enhanced PS/AC, and the simplified PS/AC architectures. The simplified AS/PC and the two-active-stage networks can achieve the capacity of 16 x 16, while the enhanced PS/PC architecture, only 8 x 8. The capacity of the conventional AS/AC and the simplified AS/AC networks easily exceeds 2,048 x 2,048, and is limited by other factors rather than the total insertion loss and the SNR. If the smallest possible number of active stages is of a major concern, the two-active-stage and the enhanced PS/PC architectures are the most attractive. The latter seems to be especially viable when optical amplifiers are used as on/off switches.

References

[1] H. S. Hinton, "Photonic switching using directional couplers," *IEEE Commun. Mag.*, vol. 25, no. 5, pp. 16-26, May 1987.

[2] R. A. Spanke, "Architectures for guided-wave optical space switching systems," *IEEE Commun. Mag.*, vol. 25, no. 5, pp. 42-48, May 1987.

[3] K. Habara and K. Kikuchi, "Optical time-division space switches using tree-structured directional couplers," *Electron. Lett.*, vol. 21, pp. 631-632, July 1985.

Tree-type architectures are among the most attractive options for space-division photonic networks.

[4] R. A. Spanke, "Architectures for large nonblocking optical space switches," *IEEE J. Quantum Electron.*, vol. QE-22, pp. 964-967, June 1986.

[5] H. Okayama, *et. al.*, "Optical switch matrix with simplified *N* x *N* tree structure," J. Lightwave Technol., vol. 7, pp. 1023-1028, July 1989.

[6] A. Jajszczyk, "Application of single-stage electronic networks concepts to photonic switching," 1990 Int. Topical Meeting on Photonic Switching Tech. Dig., Kobe, Japan, April 1990, pp. 77-79.

[7] A. Jajszczyk, "A class of directional-coupler-based photonic switching networks," *IEEE Trans. Commun.*, vol. 41, pp. 599-603, April 1993.

[8] A. Jajszczyk and H. T. Mouftah, "An architecture for a photonic fast packet switching fabric," Proc. IEEE Global Telecommun. Conf., GLOBECOM '91, Phoenix, Arizona, Dec. 1991, pp. 1219-1223.

[9] K. K. Goel *et al.*, "Demonstration of packet switching through an integrated-optic tree switch using photo-conductive logic gates," Electron. Lett., vol. 26, pp. 287-289, March 1990.

[10] R. J. Reason, "Optical space switch architectures based upon lithium niobate crosspoints," *Br. Telecommun. Technol. J.*, vol. 7, Jan. 1989.

[11] N. K. Ailawadi, "Photonic switching architectures and their comparison," in S. K. Tewksbury (ed.), Frontiers of Computing Systems Research, vol. I, pp. 129-186, (Plenum, 1990).

[12] S. E. Miller and I. P. Kaminow (eds.), Optical Fiber Telecommunications II, (Academic Press, 1988).

[13] A. Himeno, H. Terui, and M. Kobayashi, "Guided-wave optical matrix switch," J. Lightwave Technol., vol. 6, pp. 30-35, Jan. 1988.

[14] J. E. Watson, M. A. Milbrodt, and T. C. Rice, "A polarization independent 1 x 16 guided-wave optical switch integrated on lithium niobate," *J. Lightwave Technol.*, vol. LT-4, pp. 1717-1721, Nov. 1986.

[15] A. C. O'Donnell, "Polarisation independent 1 x 16 and 1 x 32 lithium niobate optical switch matrices," *Electron. Lett.*, vol. 27, pp. 2349-2350, Dec. 1991.

[16] A. C. O'Donnell and N. J. Parsons, "1 x 16 lithium niobate optical switch matrix with integral TTL compatible drive electronics," *Electron. Lett.*, vol. 27, pp. 2367-2368, Dec. 1991.

[17] A. Matoba, *et. al.*, "Low-drive-voltage 8 x 8 Ti:LiNbO3 switch with simplified tree structure," *Electron. Lett.*, vol. 25, pp. 165-166, Jan. 1989.

[18] S. Suzuki *et al.*, "Photonic space-division switching system for broadband services," in Proc. XIII Int. Switching Symp., vol. I, pp. 153-158, Stockholm, Sweden, May-June 1990.

[19] K. Komatsu *et al.*, "4 x 4 GaAs/AlGaAs optical matrix switch with electro-optic guided-wave directional couplers," in 1990 Int. Topical Meeting on Photonic Switching Tech. Dig., pp. 35-37, Kobe, Japan, April 1990.

[20] M. Ikeda, "Laser diode switch," *Electron Lett.*, vol. 17, pp. 899-900, Nov. 1981.

[21] M. Ikeda, "Switching characteristics of laser diode switch," *J. Quantum Electron.*, vol. QE-19, pp. 157-164, Feb. 1983.

[22] J. D. Evankow, Jr. and R. A. Thompson, "Photonic switching modules designed with laser diode amplifiers," *IEEE JSAC*, vol. 6, pp. 1087-1095, Aug. 1988.

[23] R. Kalman, L. G. Kazovsky, and J. W. Goodman, "Space-division optical switches based on semiconductor optical amplifiers," in *Opt. Fiber Commun. Conf. 1992 Tech. Dig.*, p. 131, San Jose, Calif., Feb. 1992.

Chapter 8

Wavelength-Division Switching

Wavelength-division multiplexing (WDM) has been recognized for a number of years as a promising and applied technology that can be used to increase the aggregate system bit rate when a single fiber is used to convey several messages each on a different wavelength. Such an approach has widely been implemented in the transmission links with system bit rates ranging from 40 Gbit/s to several hundred Gbit/s and beyond. The rapid progress in this area has also been fueled by the discovery of erbium-doped fiber amplifiers (EDFAs) with their large gain, good noise performance, and huge bandwidth that enables a large number of WDM channels to be optically amplified in a transparent fashion.

Recently, more attention has been focused on devices and technologies that can be exploited to enable WDM to move from being a pure transmission technology into a state where it can be applied in transparent all-optical networks. In particular, devices and technologies such as wavelength routing switches, switched sources, tunable sources, and tunable detectors have all been developed and demonstrated.

This chapter gives a description of the technologies, subsystems, and network architectures that rely on multiple wavelength to achieve full-transparent all-optical connectivity joined, in many instances, with features like scalability, modularity, and survivability. The first paper by Brackett et al. gives a good overview of all-optical networks employing multiple wavelengths and is formulated as a proposal that outlines the conceptual approach into multiwavelength networking, the benefits gained, the network architecture, the device technologies used, and proposed applications. The architecture proposed is scalable in the three main dimensions, namely, number of users, network coverage, and network aggregate capacity. The devices and technologies used enable wavelength translation and wavelength reuse. An example is given where eight wavelengths are used enabling up to 100 million users to be interconnected at a possible aggregate network capacity of 10 million Tb/s. The chapter continues with a paper by Smith et al. (paper 8.2) that deals with the acoustooptic wavelength routing switch. The treatment is channeled toward the design of such switches in order to meet the anticipated WDM wavelength routing requirements. The paper sets off with a description of the classic acousto-optic tunable filter (AOTF) and in particular how the passband can be engineered to meet the WDM switching demands. A model is presented that can be utilized in designing the passband shape and in shaping or suppressing the sidebands. Moreover, consideration is given to current research issues including crosstalk in acoustooptic switches, fabrication issues, and the use of the AOTF for gain equalization in optically preamplified WDM systems.

Hildebrand et al. report in paper 8.3 on an extremely versatile device, the so-called "Y-Laser." They describe the basic principle of operation and then go on to give experimental results for the device when used as a tunable source with a continuous-wave (cw) tuning range of more than 50 nm. Other functionalities demonstrated for this device include:

- Wavelength processing; where a data-stream has been transformed from the short-wavelength window (780 or 1300 nm) to the long wavelength window (1550 nm)
- Ultrawide-range wavelength conversion; where conversion has been achieved within the 1550-nm window across ± 20 nm
- Space switching/gating; where gating functions have been verified up to 1 Gbit/s and packet switching up to 5 Gbit/s
- Other functionalities include
  - —Pulse reshaping
  - —Electrically triggerable wavelength bistable device
  - —Mode-locked pulse generator

The paper considers applications in switched networks and in optical communications.

The theme of wavelength processing is continued by Yoo in paper 8.4 where he offers a comprehensive treatment and review of wavelength conversion technologies for WDM networks discussing merits, limitations, and implications for transparent networks. Wavelength conversion is identified as the first potential obstacle in achieving fully transparent WDM networks. It is also argued that wavelength conversion is needed to resolve wavelength contention, thus aiding the realization of strictly transparent optical networks. Different criteria are given that can be used to classify converters. A classification is adopted in which conversion methods are categorized into optoelectronic conversion, optical gating wavelength conversion (employing cross-gain or cross-phase modulation in semiconductor optical amplifiers), and wave-mixing wavelength converters. The last category includes optical-acoustic wave mixing, optical-electric wave mixing, and nonlinear wave mixing. The paper then gives a detailed comparison between the three conversion technologies using signal quality, configuration, and performance (efficiency, bandwidth, transparency) as the main criteria. The paper concludes by giving a useful discussion of transparency, the conversion methods that achieve it and by indicating that wavelength conversion methods with limited transparency are currently available (e.g., optoelectronic), however, fully transparent methods such as wave-mixing are desirable and represent future development directions.

The chapter is concluded by two papers that consider WDM packet networks and some of the limitations of using WDM. Monacos et al. (paper 8.5) describe the components and subsytems that are needed in a self-routing multihop network. In particular, they describe monolithic WDM laser diode and detector arrays, a WDM network interface and an all-optical switching node. Consideration is also given to the packet routing architecture and a four-channel WDM testbed is described. Potential applications and future work are outlined. The last paper in this chapter by Li and Tong deals with crosstalk and interference in WDM networks. It is shown that the interference and crosstalk penalty heavily depends on the linewidth of the lasers used. The authors consider the maximum allowable component crosstalk in optical networks and its accumulated effect on the network size. Using a worst-case scenario and assuming that the laser linewidth is less than the receiver bandwidth, they conclude by giving specifications on the maximum allowable crosstalk for each device in a network of moderate size and in a network of arbitrary size.

A Scalable Multiwavelength Multihop Optical Network: A Proposal for Research on All-Optical Networks

CHARLES A. BRACKETT, FELLOW IEEE, ANTHONY S. ACAMPORA, FELLOW, IEEE,
JOHN SWEITZER, GREGORY TANGONAN, MARK T. SMITH, MEMBER, IEEE,
WILLIAM LENNON, KEH-CHUNG WANG, MEMBER, IEEE,
AND ROBERT H. HOBBS, MEMBER, IEEE

***Abstract*—An architectural approach for very-high-capacity wide-area optical networks is presented, and a proposed program of research to address key system and device issues is described. This paper is a summary of a proposal submitted to DARPA for research on all-optical networks. The proposed network is based upon dense multiwavelength technology and is scalable in terms of the number of networked users, the geographical range of coverage, and the aggregate network capacity. Of paramount importance to the achievement of scalability are the notions of wavelength re-use and wavelength translation. With this approach, each user is attached to the network through a generic access station equipped with a small number of optical transmitters and receivers, each operating on a different wavelength. Using this fixed set of transceivers, a clear optical channel is established by connecting each access station to a number of other access stations through a transparent optical interconnect. This distributed optical interconnect is wavelength-selective and electronically controllable, permitting the same limited set of wavelengths to be re-used among other access stations. Furthermore, by exercising the wavelength-selective switches, the wavelength-routed connectivity between stations can be reconfigured as needed. Finally, a multihop overlay network involving wavelength translation and self-routing fast packet switches permits full connectivity, if desired, among the access stations at the individual virtual circuit level. This architecture and its innovative use of both optical and electronic technologies achieves a total independence between the number of available wavelengths and the number of nodes served, thereby introducing true scalability and modularity to optical networks for the first time. As an example of the power of this approach, it is shown in this proposal that using just eight wavelengths, such a network could in principle interconnect a population of 100 million users over a nationwide geography with an expected delay equal to that of 12 hops. If each such user were to access at a rate of 1 Gb/s, an overall pool of network capacity on the order of 10 million Terabits/s would have been created. The rationale behind the approach, the requisite systems and device technologies, and a planned testbed implementation to demonstrate overall technical feasibility of the approach are discussed.**

Manuscript received January 13, 1993. This work has been supported in part by DARPA.

C. A. Brackett is with Bell Communications Research, Morristown, NJ 07962.

A. S. Acampora is with the Department of Electrical Engineering, Columbia University, New York, NY 10027.

J. Sweitzer is with Northern Telecom, Morristown, NJ 07962.

G. Tangonan is with Hughes Research Laboratories, Malibu, CA 90265.

M. Smith is with Hewlett-Packard Laboratories, Palo Alto, CA 94303-0969.

W. Lennon is with Lawrence Livermore National Laboratories, Livermore, CA 94550.

K. C. Wang is with the Rockwell Science Center, Thousand Oaks, CA 91358.

R. H. Hobbs is with the United Technologies Research Center, East Harford, CT 06108.

IEEE Log Number 9208020.

I. Introduction

THIS paper presents a summary of the principal features of a proposal for research presented to DARPA by the Optical Network Technology Consortium consisting of Bell Communications Research, Columbia University, Hughes Research Laboratories, Northern Telecom, Hewlett-Packard, Lawrence Livermore National Laboratories, Rockwell Science Center, and United Technologies Research Center.

The enormous potential of multiwavelength optical networks to satisfy the emerging needs for broad-band multimedia telecommunications has long been recognized and, over the years, a considerable body of knowledge, technology, and expertise has been developed [1]. In this paper, we propose a multiwavelength network architecture that builds upon this body of knowledge to achieve the goals of scalability and modularity previously missing. This program of research has as its major objectives the development of this architecture, the underlying control algorithms, and the key device technologies required for a demonstration of overall technical feasibility. The proposed network can provide an aggregate capacity measured in Terabits per second and can operate on a LAN, MAN, WAN, or national scale. The network is scalable (one additional user can always be added) and modular (the number of nodes added may be in units of one if desired). Furthermore, the network is dynamically rearrangeable in response to changing user-to-user traffic patterns, equipment status (operational/failed), and the addition or deletion of users. Accordingly, the network is fault tolerant and can provide high quality of service with regard to timeliness and integrity of the delivered information.

The network architectural approach is based on:

1) the use of high-density wavelength division multiplexing (WDM),
2) the use of wavelength to route the signal to its intended destination in the network (wavelength routing),

Reprinted with permission from *IEEE Journal of Lightwave Technology*, Vol. 11, No. 5/6, pp. 736-753, May/June 1993.

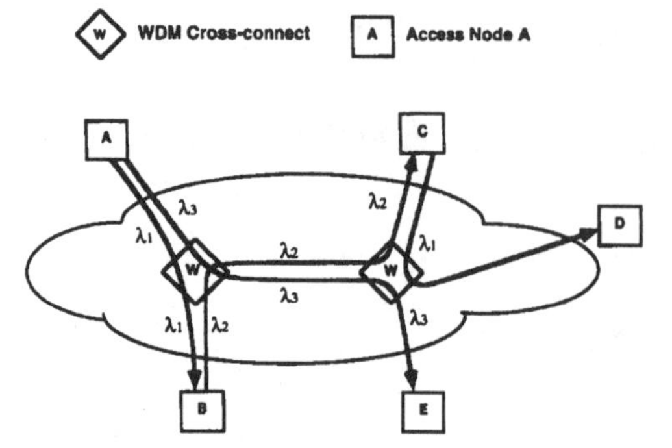

Fig. 1. A proposed multiwavelength network architecture that is both scalable and modular.

3) the use of wavelength-selective switches along with a high degree of wavelength re-use, and
4) the translation of signals from one wavelength to another at selected nodes in the network (wavelength translation) in a multihop arrangement [2]–[5].

These four principles permit networks to be built whose size is essentially unlimited and is independent of the number of wavelengths available. Using these combined techniques, the scalability in total network capacity, performance, and numbers of network nodes becomes commensurate with a national network. The remarkable feature of this architecture is the outstanding performance achievable with a very small number of independent wavelengths. For example, it is shown in this paper that an idealized network of this type, using as few as 8 wavelengths, could in principle support as many as 100 million nodes with a total network capacity on the order of 10 million Terabits/s for peak access rates of 1 Gb/s.

The network we propose (Fig. 1) consists of an all-optical inner portion that contains wavelength-routing cross-connect elements and a common network control to permit their dynamic rearrangeability utilizing the unique properties of acoustooptic wavelength switches. This rearrangeability will allow the dynamic allocation of wavelength and capacity throughout the network to meet changing traffic, service, and performance requirements and to provide a robust, fault-tolerant network. The transparency of this inner all-optical network also accommodates multiple service formats, such as simultaneous digital and analog transport. Each network access node will be able to transparently transmit to and receive from several other nodes by selecting the appropriate wavelength. With an appropriate configuration of the wavelength-routing (WDM) cross-connects, this transparent optical transmission may extend over large distance and through many WDM cross-connects.

Scalability to large numbers of interconnected nodes is achieved by incorporating a wavelength translation function which consists of receiving a signal at one wavelength and retransmitting it on another. This wavelength translation must currently be done under switched control, and so resides in the network access nodes outside the inner all-optical portion. In this way, the wavelength translation can be under the same control as provided by the packet routing headers, and a packet-by-packet wavelength translation results. Besides providing the wavelength translation function, the network access nodes also form the basis for the user interface by providing asynchronous transfer mode (ATM) switching and user access at the desired data rates and signal formats [6]. This is illustrated in Fig. 1 where the diamond-shaped nodes within the all-optical portion at the WDM cross-connects and the squares are the network access nodes. Wavelength λ_3 carries a one-hop communication (one with no intermediate detection or translation) from node A to node E, while a communication from node A to node C is carried in two hops: A to B on λ_1, and B to C on λ_2. Wavelength λ_1 is re-used to carry a signal from C to D. In the general case, all wavelengths are received and retransmitted at each access node simultaneously, implying a new level of integration of WDM components and devices at those nodes.

The research proposed in this program will encompass the device technologies, the network architectures, the network control algorithms associated with those architectures, and a system demonstration designed to prove feasibility of both the device technologies and the architectural concepts.

The device technologies to be investigated will include:

1) the use of acoustooptic tunable filters to form the WDM cross-connects that are the core of the rearrangeable all-optical network,
2) integrated WDM laser and transmitter arrays that are used to simultaneously generate a set of up to eight wavelengths at each access node,
3) integrated OEIC receiver arrays to simultaneously receive all wavelengths at each node, and
4) subcarrier multiplexing technology to demonstrate the simultaneous analog and baseband digital signal transmission and its applications.

The use of broad-band optical amplifiers in WDM systems will be explored to overcome both transmission and component insertion losses on a broad-band basis. Our component and device technologies program is oriented specifically to demonstrating wavelength routing and wavelength translation properties in a cost-effective and functional manner. Our architecture studies will focus on:

1) the control algorithms,
2) the routing algorithms,
3) throughput, availability and performance of the network, and
4) the possible strategies for achieving the broad service potential desired in a survivable network with rapid recovery.

The principle digital data rate for the demonstration portion of this project is expected to be at 2.5 Gb/s, the SONET OC-48 rate, with access interfaces at 2.5 Gb/s, 622 Mb/s, and 155 Mb/s. An initial phase of this work may, however, limit data rates in the packet switching demonstration to 155 Mb/s, the SONET OC-3 rate. In addition to the direct baseband modulation interfaces, sub-carrier multiplexing will allow both analog and digital interfaces for simultaneous transport of more than one channel and format on a single wavelength. Such a capability is needed for certain special-purpose local area networks, as for example, real-time military applications. Among the high-bandwidth applications studies will be network-based 2D/3D real-time rendering, image processing and shared imaging, and the application of the network to such demanding applications as the switched transport of scientific visualization studies.

We have chosen to not consider several technologies that we believe to be either inadequate to meeting the goals, or not sufficiently well developed to be demonstrable at the network level in the foreseeable future. This list includes short-pulse optical time domain techniques such as soliton logic and transmission, femtosecond photonics, CDMA, nonlinear optics and optical logic, and time-division self-routing optical techniques. These technologies may well become important to networks in the future but we choose to limit ourselves to the present challenge of high-density WDM. Passive-star broadcast architectures may be considered where the application makes it desirable (such as in broadcast services or for certain local networking applications), but as a general network structure are not compatible with scalability requirements. We believe that no approach to Terabit networks, including coherent detection or other ultra-dense WDM approaches, can produce a truly scalable and modular network without wavelength re-use achievable through dynamic wavelength routing and wavelength translation. Consequently, coherent detection and optical frequency-division-multiplexing techniques are not part of our proposed research and demonstration, but both could be integrated into our approach as an extension. Neither would, however, alter the architectural conclusions of our approach in a radical way.

Section II describes the concepts and rationale for the approach taken. A description of the WDM network architecture appears in Section III. Network issues are discussed in Section IV and needed device technologies are described in Section V. The proposed test bed is described in Section VI and potential applications for such a network appear in Section VII.

II. Conceptual Approach and Rationale

A. Background

As a discipline for focused research activity the field of optical networks has most distinctly emerged from its infancy and has now entered its early adolescence. The early years were dominated by emphasis on the physical layer: physical topologies, passive optical components and active electrooptic and optoelectronic devices which could be arranged in such a manner as to permit the creation of multiple distinguishable communication channels, all sharing a common fiber-optic medium. These channels would consume a proportionately larger fraction of the enormous bandwidth potential of the medium (and thereby support a correspondingly higher capacity) as compared against a single-channel point-to-point transmission link operating at a data rate equal to any one of the "optical network" channels. The two basic techniques considered for the creation of such multiple channels were wavelength division multiplexing (using either wavelength division multiplexers/demultiplexers or, for finer line spacing, coherent optics) and time division multiplexing (using time-multiplexed streams of the narrow pulses created, for example, by a mode-locked or gain-switched laser; code-division multiplexing is a variant of this time-division approach). Rapidly tunable lasers and receivers were sought to support fast packet switching [7], [8].

As telecommunications systems researchers became attracted to the field, the focus began to shift toward network-level considerations. The unique opportunities of optical networks (e.g., clear-channel circuit switching, physical/logical topological independence, flexibility and rearrangeability of a "permanent" physical plant) became recognized, along with the architectural constraints presented by device limitations. It became clear that the major issues involved in multi-user optical networks were significantly different from those of point-to-point optical transmission systems, and were even different from those encountered in the earliest attempts to use optical fiber as the shared medium of a multi-access network.

Today, the focus has decidedly extended beyond the creation of multiple channels to the allocation of these channels among a multiplicity of high-speed users, each offering multimedia traffic to the network and each demanding some guaranteed quality-of-service. Issues involving signal detectability, power budgets, fiber dispersion, coupling loss, and so forth, although still significant from a physical-layer perspective, were joined by new and challenging networking-related issues. These included:

1) The need for **scalability** to very large networks, and the need for simple **modular growth** capability and **survivability**;
2) User **connectivity** in a multi-channel environment, where many channels are created on a shared optical medium, with each user node having access to only a limited number of such channels;
3) Provisioning of fast, self-routing **packet switching** capability in such an environment;
4) Determination of a practical way for **assigning the channels** among the access nodes, along with algorithms that enable channels to be electronically reassigned to match the node-to-node connectivity with prevailing traffic patterns, network usage, and equipment failures (the last providing a high degree of fault-tolerance);
5) **Traffic performance and fault management**, including admission control, access control (policing), flow control, buffer and bandwidth management, and alternate routing.

B. Generic Approach

The generic approach that we have chosen is shown schematically in Fig. 2. As drawn, the optical network has three functional elements: the "passive" shared medium, the access nodes, and the user port. The optical network resembles in some sense a local area network (LAN): users are interconnected via access nodes that communicate among themselves in a non hierarchical peer-to-peer transport mode

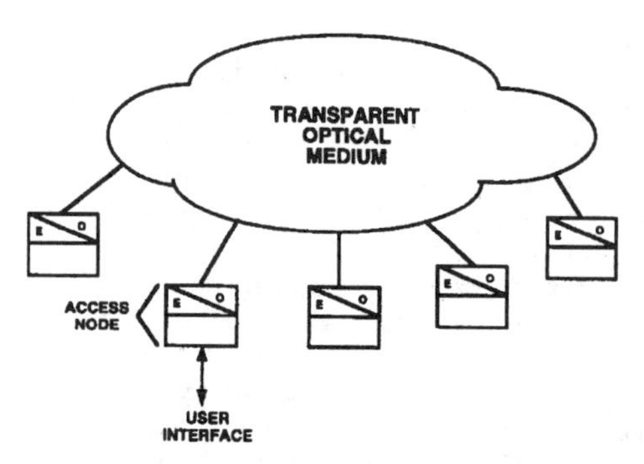

Fig. 2. Functional elements of the network architecture.

over a multi-channel medium. Important differences are as follows:

1) Geographically, the optical transport medium spans a range measured in hundreds or thousands of kilometers, not a fraction of a kilometer.
2) The medium contains a great number of high-speed channels, each operating at a very high speed as appropriate to the network application or service, each providing a direct, non-shared connection between precisely two access nodes (one transmitting on that channel, one receiving from that channel). This is in contrast with shared-media networks that contain only a single high-speed channel operating at a rate set by electronic and/or optical device limitations (e.g., several Gb/s) that is shared among users who access the medium on the basis of a suitable contention-resolving algorithm such as DQDB.
3) The number of access nodes that may be interconnected can grow very large (hundreds of thousands or millions).

Each access node is equipped with some small number (say, m) of optical transmitters and receivers, one for each accessible wavelength channel, where m is much smaller than the number of access nodes N. The multi-channel optical medium, in effect, provides a pool of $m \times N$ directed channels that can be assigned among the various transmitter-receiver pairs, creating a *connectivity* pattern among the access nodes. It is important to note that the connectivity diagram among nodes is independent of the relative physical locations among the nodes and the physical topology of the optical medium.

Channels on the medium are created by a combination of wavelength and space-division switching/multiplexing, thereby permitting a large number of channels to be created by re-using a limited number of wavelengths. Since none of the optical fibers carry more than this limited number of wavelengths, problems associated with very high-density optical frequency division multiplexing are thereby avoided, without compromising the overall network capacity and architectural freedom. Furthermore, by deploying wavelength-selective optical switches, we can dynamically reassign channels to produce some desired connectivity diagrams among access nodes.

The optically transparent portion of this network consists of a distributed arrangement of WDM cross-connects, interconnected in an appropriate fashion with optical fiber, with the losses of both the fiber and the cross-connects compensated at all wavelengths by the generous use of broad-band fiber amplifiers. This "all-optical" portion of the network can be regarded as a large and distributed WDM cross-connect in itself having certain constraints and capabilities, such as a controlled amount of circuit blocking. The properties of such networks will determine the overall capacity and bandwidth available to the various nodes in the network and will be a major portion of the study in this proposal.

For now, we treat the optical medium as transparent; no logical operations upon the information-bearing signals are performed within the medium per se. Although optical amplifiers are to be deployed, their sole function is to linearly amplify wavelength multiplexed signals, thereby maintaining an adequate signal-to-noise ratio. Moreover, although wavelength-selective optical switches are present, these too are transparent; their sole function is to effect the desired connection patterns among access nodes. They are controlled by electronic signals generated by a network configuration manager that is responsive to requests for "clear-channel" connections among the nodes and to equipment status (failures/recoveries).

The time scale of the rearrangeability of the network is, at least initially, considered to be long in comparison to the packet or call transmission or arrival rate, and is therefore considered quasi-static in nature. In telephone parlance, this all-optical portion of the medium would be called a facility-switched network. The packet switching, or call-by-call switching for circuit switched applications, would be achieved in the access nodes by means to be described below. The actual reconfiguration times of the network will be more a function of the reconfiguration algorithms and their implementation than of the switching times of the WDM cross-connect devices.

In effect, from a network perspective, the optical medium provides a means of enabling circuit-switched, broad-band (several gigahertz) transparent connections among the access nodes. These circuit-switched connections become the one-hop paths of the all-optical network, and one key feature of the use of rearrangeable WDM cross-connects is that these one-hop paths can be chosen almost arbitrarily to meet the service and performance requirements of the users. The number of such one-hop connections that can be established by each access node is limited to m, the number of wavelengths generated and received in that access node. The number of virtual circuits that can be established from each access station is unlimited and, if necessary, a virtual circuit can be established from each access station to every other station.

In this approach, the access node serves three functions: providing a user interface, providing means for wavelength translation under switched control, and thirdly the ATM (or other) switching function itself.

First, the access node presents a "universal" interface to the user. We envision two such interfaces. The first is the "clear-channel" interface in which a signal can be generated by a user in any format and is simply carried across the network without being processed. The second is a high-speed (multi-Gb/s) Asynchronous Transfer Mode (ATM) format. Signals presented in the ATM format can be delivered to any other

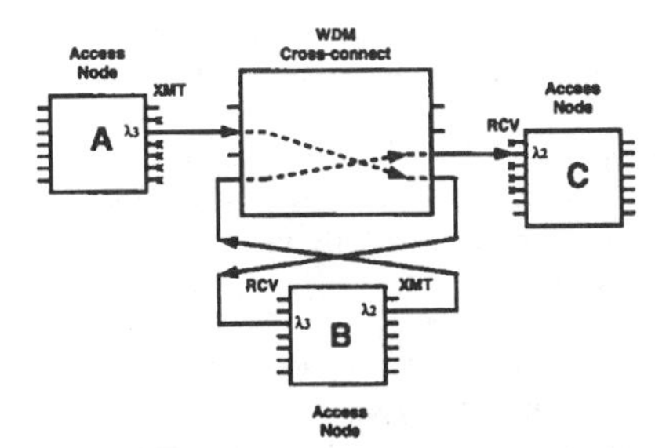

Fig. 3. Indirect connection via wavelength translation.

user, and the number of ATM virtual connections that can be established is essentially unlimited.

The second function provided by the access node is that of wavelength translation, illustrated in Fig. 3, which shows access nodes A, B, and C, and a wavelength division cross-connect (WDM cross-connect) to be described in detail later. Suppose node A wishes to establish a "clear channel" to node C. If node A has no available transmitter wavelengths in common with the available receive wavelengths on node C, then it can transmit to node B on whatever wavelength is available between A and B such as λ_3 as shown in Fig. 3. Internal to node B, the signal is connected to a transmitter on wavelength λ_2 for connection to node B. Since the number of m transmitters and receivers assigned to each node is small, wavelength translation is required to make possible an essentially unlimited number of both connections and network nodes. Clearly, for large N (which may range into the millions), we must insist that the number of wavelengths not be equal to the number of nodes, as would be needed to enable fully connected direct connections. The third function to be implemented in the access node is that of store-and-forward ATM switching. As previously mentioned, each user can, if necessary, establish a virtual connection to every other user. In Fig. 3, we have supposed a direct connection exists between nodes A and B, but not between nodes A and C. We can, nonetheless, route ATM cells from A to C by first sending these cells to B. There, in electronics, the self-routing header (virtual circuit number) will be read, and the cell will be electronically switched to the outbound link from B to C. Since contention may arise for the use of the link from B to C, smoothing buffers are deployed in the ATM switch to permit cells to be delivered on a store-and-forward basis (such buffers must be present in any ATM switch).

The separation of the transparent all-optical transport portion of the network from the logical switching portion of the network allows a very important feature, namely, each wavelength signal may contain packets destined for an unlimited number of destinations, and the required wavelength translation at each node can be accomplished on a packet-by-packet basis. Moreover, this network, by use of standardized networking protocols and transmission and switching formats, can be considered as an evolution of the logical layers of broadband networks, while at the same time being a revolutionary advance in architectural capability. We have chosen the major vehicles for this study to be SONET and ATM.

For cells/packets that are part of communications sessions between two nodes not connected by one-hop paths, multi-hop connections are established. ATM cells can be routed from any user to any other user in multi-hop connections be relaying through intermediate nodes along the links of a suitably chosen path of the connection diagram. While accumulated queueing delay might seem to present an issue, we can easily control this by limiting the number of virtual connections assigned to every link of the connection diagram at call setup time such that the load presented to each link is sufficiently less than the capacity of this link. This is a function that must be done in every ATM network, however, the flexibility in assigning transmission bandwidth in the network considered here will offer the potential for significantly reducing the congestion and delay that normally occurs. The average number of ATM cells queued behind each link can therefore be maintained at a sufficiently small value to insure low end-to-end average latency. By using such store-and-forward techniques, in conjunction with the proposed wavelength routing and translation architecture, a scalable, modularly growable distributed ATM switch will be effectively created over the optical medium. The electronics needed for the self-routing elements and buffers of a large centralized ATM switch have effectively been decomposed and distributed among the geographically distributed access nodes.

C. Benefits

The combination of a geographically distributed "passive" optical medium and a high-functionality access node represents an "optimum" approach in the following sense: we let the optics move vast amounts of information among widely dispersed ports, while letting electronics process protocols, route, and statistically smooth via buffering. Among the benefits anticipated by this approach are the following:

1) **Enormous aggregate capacity,** easily growing to the tens or hundreds of Tb/s, although no user "sees" an access rate greater than that of an individual channel.
2) **Scalability,** the property of always being able to add more nodes. By deploying more copies of the same equipment, the network can handle an increasing number of users located over an enlarging service region while offering higher aggregate capacity.
3) **Modularity,** the property of being able to add only the desired number of new nodes, and updating the logical connectivity diagram to include these new nodes.
4) **Permanence.** Once the optical medium is installed, it assumes the appearance of "permanence" in that new access nodes, new services, and new network architectures can always be added and/or updated without requiring that the pre-existing medium be retrofitted.
5) **Multiple virtual network integration onto a common physical medium.** Some of the "clear channels" among nodes can be used for an analog network, some for a digital network, some for an ATM network, some for a voice network, some for a video network, etc.
6) **Integration of circuit/packet** switching capabilities.
7) **Distributed ATM fabric,** rearrangeable to optimize performance requirements.

8) **Enhanced reliability.** The failure of an access node affects only the users connected via that node; the failed node is then "bypassed" by changing the connection diagram, thereby preserving all other real and virtual connections; if the failed node serves a large number of users such as a local area network, then dual homing on two access nodes may be used for further reliability enhancement.
9) **Independence of the physical topology from the logical interconnectivity.** The connectivity diagram is easily modified in response to changing traffic patterns, service requests, and equipment failure/restoration, without requiring a physical change to the optical medium.

III. WDM Network Architecture

The networks we are proposing are based on an underlying physical layer consisting of fibers joined by dynamically controllable WDM cross-connects and providing purely optical transport between network access nodes, as shown above in Fig. 1. The fact that the physical layer is based on dynamically reconfigurable WDM cross-connects means that the optical layer acts as a controllable interconnection, capable of establishing and taking down connections as required, and modifying connection configurations to react to network faults and changing traffic conditions. All of the connections at the optical level are essentially "circuit switched," as noted above. The time scales over which connections are expected to be held will vary widely. For example, connections supporting a multihop overlay would be semi-permanent (hours or days), changing only to react to network faults or significant changes in traffic patterns or service requirements. On the other hand, a video teleconference might involve a few clear-channel connections held for several minutes. Dedicated service channels may be held indefinitely, but would be reconfigurable to meet the needs of changing private networks, for example.

In the remainder of this section we examine the physical basis for scalability and modularity in WDM networks and the proposed concepts for network reconfigurability, channel assignment, and network control. We then present a detailed description of the proposed physical layer components such as the WDM cross-connects and access node optoelectronics.

A. Scalability and Modularity in WDM Networks

We define **scalability** to be the property that one more access station may always be added to a network, thereby permitting service to be offered to an arbitrarily large population of users spread over some arbitrarily large service domain. We define **modularity** to be the property that only one more node need be added at a time, or at least no more than is desired. These two definitions allow us to say a great deal about the form and function of multiwavelength architectures.

For example, this definition of scalability implies that any scalable architecture will be required to have the property that the number of wavelengths is independent of the number of nodes in the network. This does not need to be the case for very small networks, especially those for which the number of nodes is less than or equal to the number of wavelengths. But as the number of nodes becomes very large, say a million or more, clearly one cannot continue to increase the number of wavelengths.

The total number of wavelengths available depends upon the technologies used and the information bandwidth of the signals to be transported. As an example, if the defining optical bandwidth is the bandwidth of the erbium-doped fiber amplifier, say 35 nm, and the wavelength spacing is of the order of 1 nm, then 35 wavelengths would be the maximum number that could be used, neglecting the serious problems of gain variation across the spectrum. A 1-nm wavelength spacing is typical of some tunable filter types such as the acousto-optic tunable filter (AOTF) mentioned below, but it is also about the minimum practical spacing for signals carrying bit rates of the order of 5 to 10 Gb/s due to crosstalk limitations in simple optical filters or coherent detection schemes. Finer wavelength spacings can be supported with other filters, such as Fabry–Perot types, or with optical heterodyne detection. The difficulty with these approaches is that of providing rearrangeable switching in an integrated fashion of all the wavelengths. Clearly, if the WDM switching technology advances to the point where more wavelengths are possible, then this network architecture will benefit in proportion. One must keep in mind that each wavelength requires signal processing for the configuration and routing algorithms, and that will become a significant contributor to the complexity of the network as the number of wavelength grows. We choose eight wavelengths here because, as shown below, it is sufficient to provide the performance we desire, and because it is compatible with the scale of optoelectronic integration that is considered practical.

This number is far less than many expectations and serves to emphasize that whatever the maximum number of wavelengths is, it is surely finite and far less than the total number of nodes one would desire in a large network.

A second requirement that this definition of scalability imposes on the architecture is that a very large degree of wavelength re-use be achieved naturally in the architecture. Wavelength re-use, discussed in more detail below, is the principal of using each available wavelength many times throughout the network, either in different spatial locations, or over different paths, in a way that signals at the same wavelength never interfere with each other. A natural way of achieving wavelength re-use by the use of wavelength-routing networks will be discussed further below. One consequence of the wavelength cross-connect approach to wavelength routing is that it is essentially impossible to have two signals of the same wavelength interfere with each other in the same transmission path, except for the issues of crosstalk.

The third requirement for scalability is that of wavelength translation. Wavelength translation is the process of moving a signal from its original wavelength to another wavelength, and it is necessary for implementing the degree of interconnectivity necessary to produce truly scalable network architectures. We have discussed above the highly desirable feature of having the wavelength translation performed under switched network control.

In addition, it is very desirable for a network to have the property of being rearrangeable. Conceivably there are applications for networks for which a permanent interconnection pattern is sufficient for all time, but the added flexibility produced by making the interconnection pattern rearrangeable is probably one of the greatest capabilities of multiwavelength networks and so we assume here its necessity. A rearrangeable WDM (wavelength division multiplexed) network allows the passive routing of wavelengths to be changed to accommodate changing application, service, or performance requirements without disturbing the basic fiber topology.

B. An Example: Recirculating Shuffle Multihop Networks

In order to illustrate the use of wavelength routing and wavelength translation in producing modular and scalable networks, we consider as an example two versions of the original multihop lightwave network architecture using a recirculating shuffle interconnection pattern, known in the literature as ShuffleNet [2]–[5]. The first version, as shown in Fig. 4, is a folded multiwavelength optical bus in which each access node also acts as a switching node for traffic on the bus. Each user access node can transmit at two wavelengths, and can receive at two other wavelengths. All signals share the common bus so they must all be distinct, and $N_\lambda = 2N$ must hold, where N_λ is the number of wavelengths, and N is the number of user access nodes. This linear dependence of the number of wavelengths on the number of nodes means that this is not a scalable network as defined above. Each node may transmit directly to two other nodes by appropriate choice of one of that node's two transmit wavelengths. In general, for any node to communicate with any other node, the signal must go through one or more intermediate nodes. This is therefore a multihop network, with each signal being detected at each intermediate node and retransmitted on another wavelength. The ShuffleNet is so named because the assumed logical interconnection pattern is a folded perfect shuffle as is shown in Fig. 5. It is easily seen how a signal can reach any given node from any other by a multihop connection. The perfect shuffle of Figs. 4 and 5 is mathematically characterized by the parameter k in the relationship for the number of nodes in the network, $N = k2^k$. For 8 nodes, $k = 2$. A feature of such networks is that the regularity of the interconnection pattern must be maintained for the mathematical description of them to remain valid. These regular interconnection patterns must be built in completeness, with no missing nodes, for whatever k is chosen. It is not possible to maintain the perfect shuffle design for intermediate numbers of nodes between those for two successive values of k. N therefore jumps by increasing numbers of nodes as the network is scaled upwards, and is therefore not modular in addition to being not scalable.

The logical interconnection pattern of Fig. 5 is independent from the physical interconnection pattern of Fig. 4 in the following sense. The logical interconnection pattern of Fig. 5 could just as well have been implemented on a passive star as on a passive bus, both of which are shared optical media. This independence of logical and physical interconnection patterns

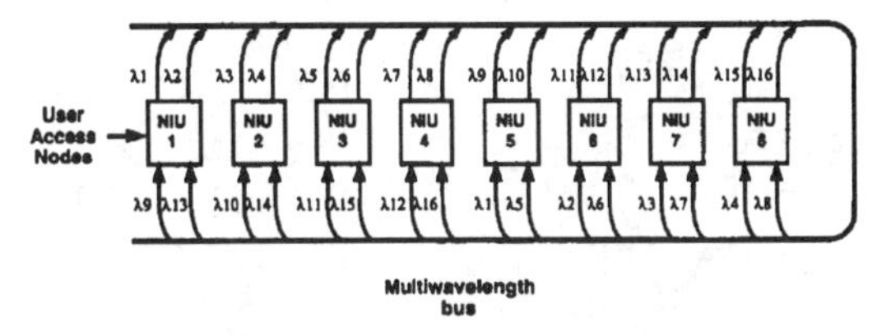

Fig. 4. ShuffleNet, in which each node transmits to and from two distinct nodes, and in which the required number of wavelengths is $N_\lambda = 2N$.

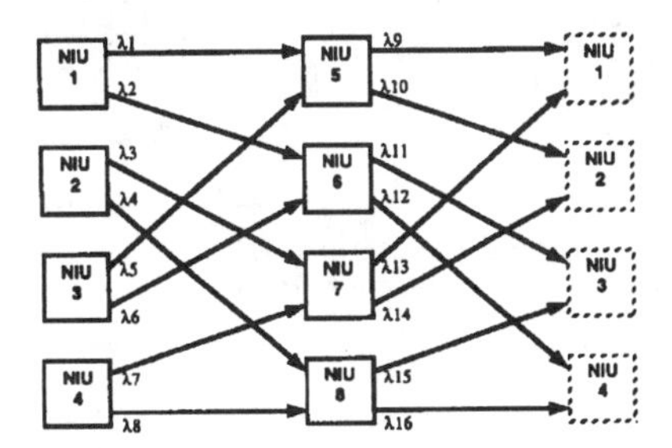

Fig. 5. Perfect shuffle logical interconnection pattern and wavelength assignment.

in a key feature of optical networks in general which derives from the generally passive nature of the optical medium. It is, of course, a property that brings a great deal of flexibility to the network and its application.

The recirculating perfect shuffle logical interconnection was chosen because of its high fan-out efficiency so that closed-form expressions for aggregate capacity could be easily found under the assumption of uniform traffic. The version of ShuffleNet shown in Fig. 4 is noted to not be scalable because of the linear dependence of number of wavelengths and number of nodes, but a second version using wavelength routing, to be discussed below, is.

Another feature of the perfect shuffle is its regular connectivity. This may turn out to be inconsistent with the physical topology of the network nodes and the traffic patterns between them. In one version of ShuffleNet, this connectivity is established on a bus which may be run through every node on the network, difficulties arise when one considers adding a node, or in the event of breaking the fiber. Adding a node requires breaking into the fiber, but it also requires reconfiguring all of the existing transmitter and receiver nodes to obtain a new perfect shuffle pattern (where the above comments of nonmodularity also apply). In the case of a fiber breakage, the network is broken into two halves, one of which can still operate, but the other has no return path so the transmitted signals simply fall into thin space. ShuffleNet can of course be implemented on any shared medium, such as a broadcast star network, with a different set of topological advantages and disadvantages. The star-based ShuffleNet still suffers from the need for too many wavelengths and is not modular for the same reasons as the bus-based version. What is needed is an architecture that will function well on an irregular topology and still achieve the desired capacity and survivability.

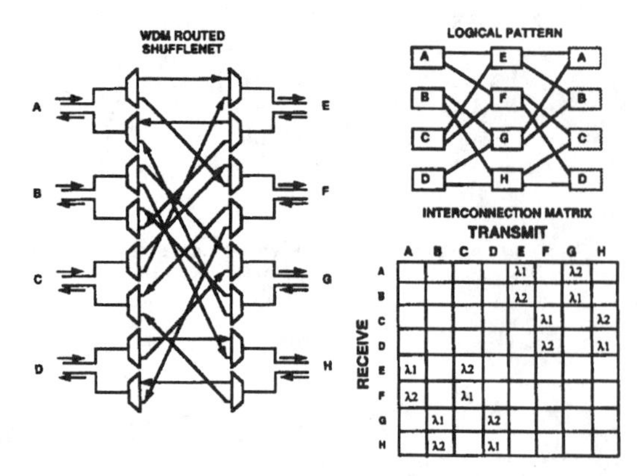

Fig. 6. A wavelength-routed ShuffleNet using a wavelength cross-connect of fixed WDM units. The logical pattern and the wavelength interconnectivity matrix are also shown.

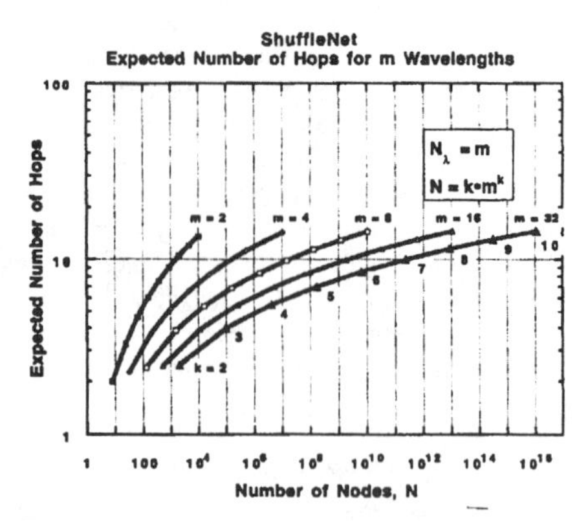

Fig. 7. The expected number of hops, $E(k, m)$, versus the number of nodes N for a wavelength-routed, m-wavelength ShuffleNet.

The lack of scalability in either the bus-based version (Fig. 4) or the star-based version is due to the use of a shared medium, which therefore requires a large number of wavelengths to keep the signals from interfering with each other. Wavelength routing can be used to get around this limitation as is shown in Fig. 6 (version 2). A WDM routed perfect shuffle can, as shown in Fig. 6, be obtained with just two wavelengths. Two wavelengths are required because each node must connect with two other nodes, but the same two wavelengths can be used for every transmitting node in the network. In fact, any size perfect shuffle network can be implemented with just these same two wavelengths. The wavelength-routing concept allows networks to be built where the number of wavelengths and the number of nodes are not related to each other. Of course, aggregate network capacity would suffer because the number of hops would become enormous, causing most of the capacity of each optical link to be consumed by relayed packets. However, as stated above, it is practical to consider the use of up to a few tens of wavelengths. If each network access node can transmit up to m wavelengths, then we can generate an m-connected perfect shuffle with the same form of wavelength routing shown in Fig. 6. The use of m wavelengths, or m-connectedness at each node, has the effect of giving a much larger fanout of accessible destination nodes from each access node, which dramatically decreases the number of hops required to reach any arbitrary destination. It is readily shown that [5] the expected number of hops and the overall capacity, assuming uniformly distributed traffic, are given by:

$$E(k,m) = \frac{km^k(m-1)(3k-1) - 2k(m^k-1)}{2(m-1)(km^k-1)} \tag{1}$$

$$C(k,m) = \frac{B \cdot N \cdot m}{E(k,m)}. \tag{2}$$

where $N = km^k$ is the number of access stations and B is the bit rate being transmitted on each channel. The average number of hops, $E(k,m)$, and the normalized capacity, $C(k,m)$,

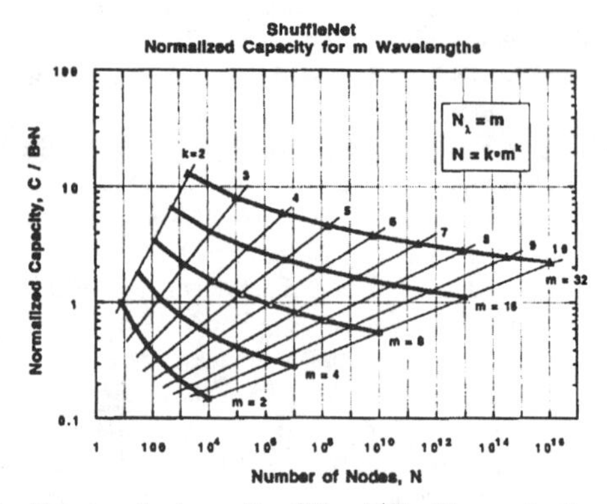

Fig. 8. The normalized capacity, $C(k, m)/B \cdot N$, as a function of the number of nodes for a wavelength-routed, m-wavelength ShuffleNet.

are shown in Figs. 7 and 8, respectively, for numbers of wavelengths, m, ranging from 2 to 32, and for several values of the parameter k.

The normalized capacity, as shown in Fig. 8, gives the average expected capacity for each user, normalized to the bit rate of the channel, B. A normalized capacity $C/B \cdot N = 1$ corresponds to the per user capacity of a multiwavelength broadcast-star network, for reference. We see that for as few as 8 wavelengths, a very significant performance is achieved. For $m = 8$ wavelengths and $k = 8$, the network supports as many as 100 million nodes with an average number of hops between users of about 12. Clearly, more wavelengths provides better performance by decreasing the expected number of hops. If each node were to transmit at about 1.25 Gb/s, such a network for 10^8 nodes would in theory be capable of carrying 10 million Terabits/s total traffic.

The point of this example is not to propose this network as a real solution to real problems, nor to suggest that this architecture would really be practical for serving 100 million users, but instead to demonstrate that the principles of wavelength routing and translation can be used to create truly scalable networks with a very modest number of wavelengths.

It is again emphasized that, although wavelength routing and translation have made the ShuffleNet scalable to dimensions

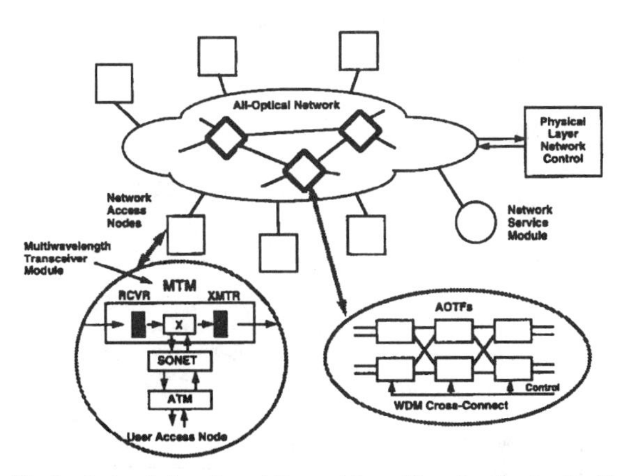

Fig. 9. System design of a scalable, modular multiwavelength network with rearrangeable wavelength routing and switchable wavelength translation.

commensurate with a national network, the resulting network is still not modular nor does its topology necessarily fit any real geographic distribution of network nodes in a practical way. The problem, of course, is the perfect shuffle (PS) interconnection pattern itself. The PS belongs to a class of richly interconnected interconnection networks, and nearly achieves the information-theoretic bound on minimal complexity. This pattern is therefore regular and dense. To add a node, once the network is filled for a given value of m and k, requires the addition of a large number of additional nodes and the rearrangement of the interconnections between all of the earlier nodes.

What is needed is a network topology and interconnection method that allows a nearly arbitrary topology, along with the rules for adding nodes, such that the performance and flexibility of the network are maximized while at the same time the nature of the needs of real networks are accommodated in a natural way. We want to be able to add just one node at a time, wherever desired, without the need to rearrange the rest of the network or to seriously impair its operation.

By using the rearrangeability property of wavelength-routed networks, it becomes possible to modularly add nodes to a multihop network, assign the channels to best accommodate prevailing traffic patterns (which results in fewer relay packets on the optical links and therefore higher capacity, lower delay, and generally superior performance), and route around network failures.

C. Details of the Network Layers

We have thus far outlined elements of a wavelength-routed network that makes possible the spatial re-use of wavelength, and which uses wavelength translation to extend the scalability and flexibility of the network. The concept is illustrated in Fig. 9. The network is conceptually divided into two portions: an inner all-optical portion, and the logical part of the network comprising the network access nodes. Within the all-optical portion are the wavelength cross-connect modules, and within the logical portion are the switching nodes, including the means for performing wavelength translation.

This inner all-optical network is in reality a large wavelength cross-connect, made up of many individual wavelength cross-connect modules that have been interconnected in a manner that provides the physical layer connectivity desired. This inner network carries signals from port-to-port, establishing a set of wavelength-dependent paths through the network. The network controller can determine what those paths should be, based upon an algorithm that attempts to match the network capacity to the topological constraints and traffic requirements of the network users. Reconfiguration can take place when needed, but in many cases, this connectivity will change reasonably slowly with time. Packet switching is performed using these wavelength paths to provide the connectivity between nodes; the packet switching itself is performed in the logical portions of the network access nodes.

An important feature of this design is that two nodes in the network that are connected by any particular wavelength can be as close, or as far away, as the application requires. That is, direct transparent optical connections may extend through as many cross-connects and across as large a physical distance as desired, within the constraints of signal-to-noise ratios, gain variations across the signal spectrum in the amplifiers, and dispersion. The gain variation promises to present serious difficulties for cascaded amplifiers, and will need to be dealt with as part of this program.

This means that a computer in New York might be simultaneously connected to a computer in New Jersey, Washington, D.C., and Berkeley, California, all without intermediate detection and regeneration of signals. And, that interconnection pattern can at any time be changed to provide the same direct transparent interconnection to other locations. Using this property, not only is high performance ensured for those connections with direct end-to-end transparent optical paths, but by routing a significant part of the high-performance circuits over such transparent paths, congestion in the other parts of the network, for other connections, will be reduced.

This property is a feature of wavelength-routed networks, but not of optical star networks or other networks that require digital gateways between them. And it requires the wavelength routing to be rearrangeable and dynamic. This is one of the prime motivations for using acoustooptic tunable filters, to be described later, as the basis for the wavelength cross-connect module. Another is that these filters can independently and simultaneously select and route many tens of wavelengths, considerably simplifying the complexity of building a many-wavelength cross-connect. The acoustooptic tunable filters are configured into WDM cross-connect modules, as shown in Fig. 9, which can be individually reconfigured electronically in less than ten microseconds.

The other portion of the network is the electronic portion, which contains the necessary wavelength-translation devices and resides largely in the network access nodes, as shown in Fig. 9. In this proposal, we take the rather direct, and eminently practical, approach of using multiwavelength receiver and transmitter arrays to provide the wavelength translation function. The receivers and transmitters are configured as arrays, each transmitter and receiver responsible for communication on a single wavelength. Wavelength translation

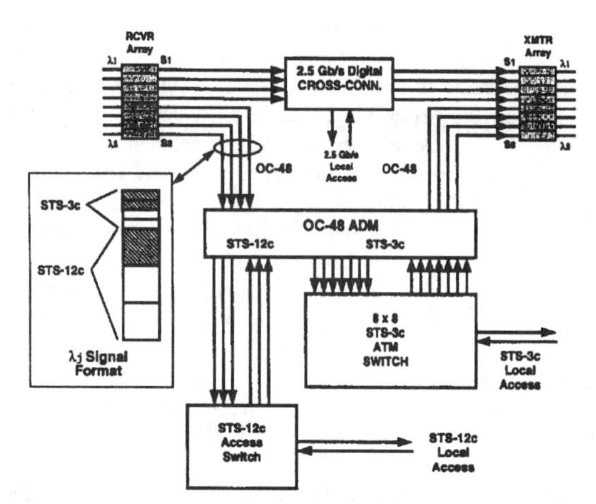

Fig. 10. A generalized switching access node, showing proposed ATM signal interface rates and switching, the multiwavelength transceiver module, and several switching functions. Initial implementations may only contain some of these elements and capabilities.

is performed by detecting at one wavelength and switching that detected signal to a laser of the desired wavelength. This switching can be done either by an electronic cross-connect switch, or in a high-speed packet switch, depending on the service requirements. By including the detector/receiver arrays, the laser/transmitter arrays, and a digital cross-connect chip in a single module we call the Multi-wavelength Transceiver Module, we can have a general network element that is usable with a wide variety of digital formats and which can be reconfigured internally to also allow certain maintenance operations such as fault and error testing.

The use of laser arrays and receiver OEIC arrays as the sources and detectors provides simultaneous access and switching capability to all of the wavelengths in the network at each major access node. This is an essential part of providing the scalable multihop network potential. The laser arrays will be designed to hold their wavelengths accurately and stably, so that these devices as units determine the wavelength channels available in the network. More detail of the access node is shown in Fig. 10. The relative roles of the receiver and transmitter arrays are shown. In an eight-wavelength implementation, four of the wavelengths may be consigned to digital "clear channels" at 2.5 Gb/s, with the digital cross-connect chip providing the required wavelength reconfigurability and local access. Other signal access is performed through a series of high-speed demultiplexing operations, with switching done either in the ATM or circuit domain, as desired.

This structure is very general, and very complex. We propose it here as a test vehicle for the demonstration. In application, some network nodes would have reduced capabilities.

1) *An Example: Network Reconfiguration and Growth:* An example of an interconnected network of WDM cross-connects and network access nodes is shown in Fig. 11. The network access nodes are shown as circles and the WDM cross-connects are shown as diamond shapes and labeled "W." Each network access node has access to m transparent paths through the total network, where m is the number of wavelengths it can

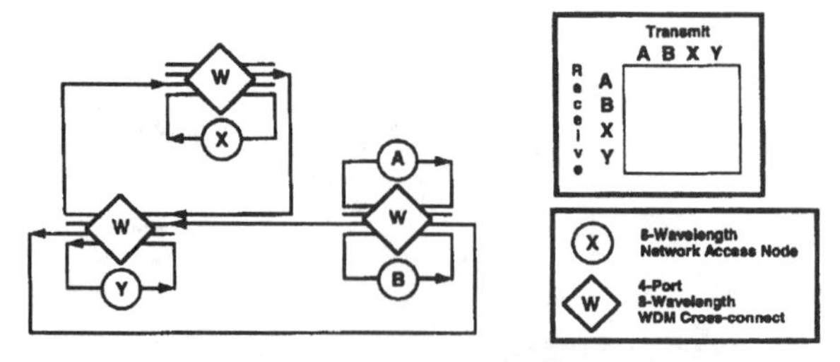

Fig. 11. A WDM cross-connect network example showing the physical connectivity but with no wavelength assignments established.

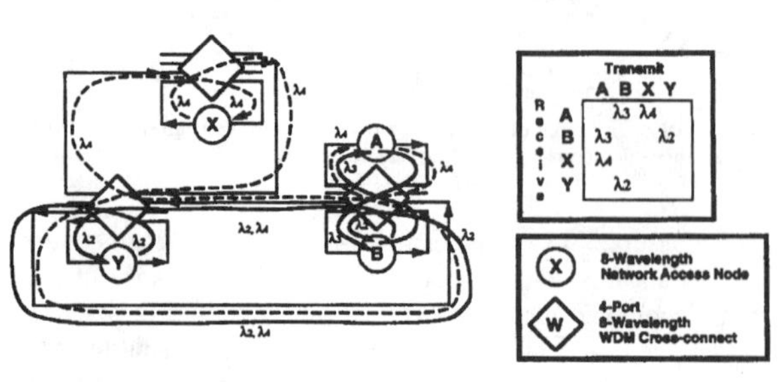

Fig. 12. A symmetrical wavelength routing on the physical interconnection topology of Fig. 11.

generate and detect. On a given cross-connect module, some of the ports will be used for connection to local network access nodes, while others will be used to provide interconnection to other cross-connect modules. Fiber amplifiers are used to compensate both transmission and component insertion losses. The example shown has four nodes, A, B, X, and Y. We have assumed the number of wavelengths available is 8, and that the WDM cross-connects have four ports. A particular interconnection between the cross-connects is shown, however no wavelength-routing paths have been specified, so the wavelength interconnection matrix is empty.

In Fig. 12 we have assumed a particular wavelength routing, in which the interconnections between the four nodes are symmetrical as shown in the interconnection matrix. From access node, "A," we have λ_3 connecting "A" to node "B" on the same module, in both directions. We have assigned λ_4 to connect node "A" to node "X" several modules away, again in both directions. Node "B," however, may also be connected to other nodes, say "Y," across a similar but distinct path on wavelength λ_2. The wavelength interconnection matrix shown in the inset is symmetric because of the reciprocal connections assumed. Three wavelengths are required in this case in order to remove contention or blocking between wavelengths.

Such networks provide the opportunity to construct node-to-node connection diagrams that are independent of the physical topology. For example, in Fig. 13, we have constructed a logical ring on a single wavelength, using the same physical topology, by having A transmit to B, B transmit to X, X transmit to Y, and Y transmit to A to close the ring. Only a single wavelength is required in this case, as is shown in the interconnection matrix. One may, in fact, construct several such rings independently, using the same topology,

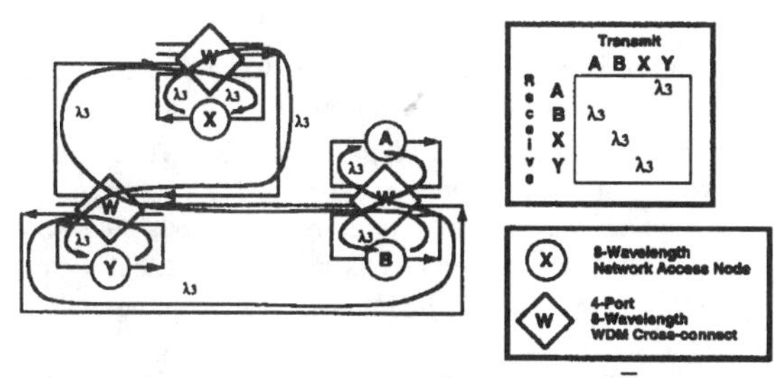

Fig. 13. A logical ring routing upon the physical interconnection topology of Fig. 11.

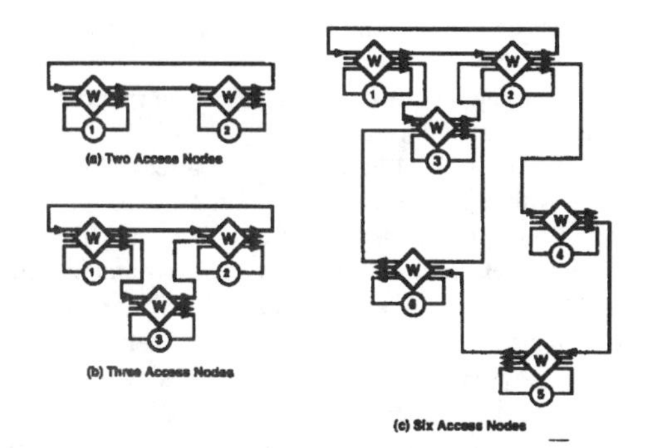

Fig. 14. The addition of nodes to a WDM routed architecture.

although in this case one runs out of nodes. The extension of these techniques to very large networks is not straightforward, as constraints develop such as the avoidance of transparent loops, and the noninterference of wavelengths. To avoid these problems requires a set of engineering rules on the construction techniques to ensure that there are always enough paths and free wavelengths to avoid blocking.

Some interesting constraints become obvious with examples like these. Clearly, any one link between cross-connects must have distinct signals, which is why the B-to-Y and A-to-X must have separate wavelengths, even though they are ultimately going to distinct destinations. This restricts the wavelength re-use and introduces a form of wavelength blocking. We will, in the course of this research, derive the rules that govern the use of wavelength to establish these interconnection patterns as a part of the network architecture and network control algorithm studies.

Another operation of interest is the addition of nodes, as shown in the example of Fig. 14, indicating the ease with which the network may be grown. In addition to adding new nodes to free ports on existing cross-connects, it is also possible to "break" into a fiber span and insert a new node or cluster of nodes, without rearranging the previously established connections.

The network access nodes shown in Fig. 9 must have access (in the most general case) to all wavelengths and be capable of performing cross-connection or switching between those wavelengths to perform the general wavelength-translation function required for scalability.

Each ATM user on the network will have a virtual signaling channel permanently set up to communicate with the network call management process. Call management may be executed in a distributed and coordinated manner by those workstations in charge of ATM switch control, or it may be implemented in a centralized manner for small networks. Whenever a user wants to set up or tear down a virtual channel, it communicates the network call management through the signaling channel using a standard protocol.

D. Reconfigurability, Channel Assignment, and Network Control

Each configuration of the network then is concerned with establishing a connection graph corresponding to the desired connectivity and providing the desired performance. The question then is, how do we create a "good" connection graph? First, we consider the requests for "clear-channel" connections. The number of clear-channel requests can be represented by an $N \times N$ circuit-request matrix, **C**, where element $C(i, j)$ corresponds to the number of clear-channel connections requested from node i to node j. Clearly, every row sum of this matrix (row sum i represents the number of clear-channel connections requested by node i, independent of destination) must be less than m, the number of transmitters deployed at node i. Similarly, every column sum must be less than m (column sum j represents the number of clear-channel connections requested to node j, independent of origin).

We next represent the ATM traffic among nodes by a similar $N \times N$ matrix, **T**, where element $T(i, j)$ represents the number of virtual connections requested from node i to node j, each appropriately weighted by the expected average traffic for each connection.

Both traffic matrices are continually changing as new connections (clear-channel or virtual) are added and existing connections dropped. Also, the dimensionality N of the traffic matrices may change as new users are added, old users are dropped, and as network equipment fails or is restored. Requests for new connections and notification of disconnects are communicated to the network controller using permanent virtual connections of the ATM network. In general, the network controller will be distributed over the network, components of which appear on selected ports of the network. Network controller components communicate among themselves over the ATM network to make collective admission decisions on requests for new connections. In essence, in response to a request for a new connection, the network controller must find a connection diagram (and, if the request is for a virtual connection, also a path over that connection diagram) such that the resulting traffic flow on every channel on the medium is less than the channel capacity. If no such connection diagram/flow can be found, the request must be blocked. The problem can be decomposed into two phases. First, the network controller must attempt to find an acceptable connection diagram. Next, it must attempt to assign wavelengths and set the interconnection pattern of each optical switch to produce that connection diagram.

These problems have been addressed separately. In response to virtual connection requests, a good heuristic has already been developed to find a connection diagram [9], [10]. This heuristic favors the traffic elements of the virtual connection matrix that are largest, that is, the one-hop traffic is maximized. Applications of this heuristic has shown that the traffic handling capability of a large, distributed ATM switch of the type we are proposing is essentially the same as an ideal centralized ATM switch over a range of traffic conditions. In essence, the queueing delay of the centralized switch (lowest possible) is achieved by the distributed switch, and the call-blocking probability of the centralized switch (lowest possible) is achieved by the distributed switch. Fortunately, the problem posed by the heuristic can be readily solved by a linear programming algorithm.

The clear-channel assignment problem is somewhat different. Here, the desired connection diagram is defined by the clear-channel matrix **C**. In response to a request for a "clear-channel" connection, algorithms have previously been developed that seek to find accommodating wavelength assignments and optical switch settings. These algorithms have been shown to provide low call blocking probabilities [11], [12].

It is the intent of the current research to combine all aspects of these problems (integrate clear-channel circuit switching and virtual channel ATM switching). A two-tiered approach is envisioned. Given existing clear-channel and virtual circuit traffic matrices **C** and **T**, suppose a new request for either type of connection is received. First, we seek to find a new connection diagram and virtual circuit flow over the links that accommodate the new request (no channel flow exceeds the channel capacity). Next, we seek to assign wavelengths and set the optical switches to produce the desired connection diagram. If either cannot be found, the new request is blocked. The new algorithms will build upon the separate approaches previously developed, with close attention paid to implementation complexity and feasibility. Our prior experience indicates that a total reconfiguration of either the connection diagram or wavelength/switch setting selection is rarely needed; as old clear-channel or virtual connections are dropped, new connections can be added on the freed-up capacity or with slow migration of the existing connection diagram/physical-level settings.

At the physical layer, network control consists of the allocation of network resources to connection requests. As indicated above, these connections may be quasi-static, supporting logical overlays or long-term services. Or, they may be of short duration, requiring frequent modification. If the list of required connections is known in advance, the allocation problem can be solved and perhaps optimized, without requiring real-time computation. A variant of this problem arises if a fault occurs or traffic demand changes, requiring a modification of the physical connections supporting current services. In that case algorithms are required that can run in real-time. Allocation of dedicated connections or clear channels on demand also requires real-time algorithms. In this case, we consider the resource allocation problem on an "incremental" basis. If a connection is requested at any given time, a physical path, waveband, and channel are allocated to it without rearranging other currently active connections. If that cannot be done, the request is blocked.

IV. Networking Issues

A listing of networking-related issues to be studied as part of the proposed research is an follows:

1) **WDM Network Architectures:** This issue is concerned with both the fundamental physical layer architectures by which wavelength re-use, network scalability, and modularity are implemented, and with potential logical architectures that may be implemented upon those physical layers.

   A number of physical architectures will be analyzed with regard to their suitability for access, inter-office, LAN and WAN environments. The purpose is to find the appropriate size and structure for sub-networks in the different environments. Strategies will then be identified for building end-to-end networks using sub-networks joined by networking gateways.

   For the physical network study, a real network database will be used to ensure a realistic starting point. New services will be added to the traditional service demand forecasts as sensitivity tests for the new architecture.

   On the logical side of the network study, multiple logical sub-networks will be modeled within the same physical fiber network. These sub-networks can be private and public LAN, WAN, T3 and other end-to-end sub-networks. Each sub-network may satisfy a different connectivity requirement such as point-to-point, broadcast, closed-user-group (corporate), etc. The reconfiguration of any of these logical sub-networks should be able to be done independently without affecting others.
2) **Reconfiguration Algorithms** for the optical WDM physical layer: Reconfiguration algorithms will be developed to assign wavelengths to user access station pairs and to control the AOTF switching elements to establish the appropriate interconnections. These algorithms will adapt the interconnections in response to changing traffic patterns, specific requests for circuit-switched connections (both point-to-point and multicast) and equipment failures at the optical level. Both regularly structured AOTF switching matrices and modularly growable physical arrangements will be considered. It is expected that optimum algorithms cannot be produced; this problem is believed to be N-P complete. Good heuristics will be sought, studied, and compared.
3) **Traffic Control Algorithms:** Algorithms responsible for controlling admission of new connections to the network and for flow control over previously established connections will be developed. The goal is to produce the policies that guarantee quality of service for each of the various types of traffic and each of the applications served by existing connections, while maximizing the ability of the network to satisfy future requests for new connections.
4) **Performance Analysis:** Detailed studies of traffic handling capability and throughput/delay performance will be performed. The call-blocking performance for circuit- and virtual-circuit-switched connections, and the delay

versus loading performance for data packets of virtual connections or connectionless service will be found for alternative physical-level architectures.

5) **Target Network Architecture:** The objective here is to produce a target network architecture that embodies all of the knowledge gained in the project, including physical WDM architectures, reconfiguration and traffic control algorithms, and the performance analysis, as well as the limitations found in the technology and device studies. It is intended to answer the fundamental questions concerning optical networks: What are the "right" ways to construct these networks? What are their advantages over conventional approaches? Where should they be used, and for what applications? Are they more reliable, more modular, easier to maintain, in need of less control software, better performing, and of lower cost? Comparisons with alternative approaches will be made.
6) **Network Transition Analysis:** The all-optical network described here has the potential of being scaled up to national size, and capable of carrying high bandwidth future services. One key consideration in the building of any large network is the question of transition. The transition issue is a serious one, as it impacts the business case for the new network. Existing rights of way, fiber routes, and systems, will be used as the basis for transition analysis. Important issues to be considered in all-optical network transition analysis include dispersion compensation techniques, dispersion shifted fiber deployment, central office break-out point and equipment modernization. As the economic scale will be enormous, the migration issue will be a key consideration for the public network operators.

V. Device Technologies

The supporting device technologies that are critical for this architecture, and the demonstration that we will perform, are the acousto-optic tunable filter, integrated laser array transmitters and integrated detector array receivers. We will also depend heavily on erbium-doped fiber optical amplifiers to compensate both transmission and component loss. The key issues associated with each are listed below:

A. Acoustooptic Tunable Filters

At the core of our network proposal is the concept outlined above of an all-optical network of wavelength-routing elements using a network of interconnected wavelength cross-connects, each cross-connect being constructed of acoustooptic tunable filters. We will build research prototype 2 × 2 AOTF tunable filters and fabricate them into 2 × 2 and 4 × 4 WDM cross-connects. These cross-connects will be modular and will contain the required temperature control and rf generators for controlling the modules on 4 to 8 wavelengths. Research on the AOTFs will be conducted to lower the cross-talk and losses and to introduce a single-process fabrication technology.

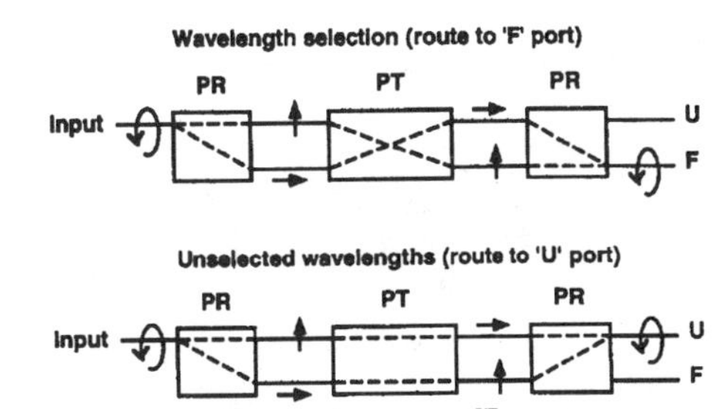

Fig. 15. In the acoustooptic filter, a polarization router (PR) decomposes input light into vertical and horizontal component polarizations, then a wavelength-selective polarization transformer (PT) flips the selected wavelength-channel polarization (top), while not disturbing the unselected wavelength bands (bottom). The output polarization components are recombined in the output polarization router into filtered (F) and unfiltered (U) ports.

1) *Principle of Operation:* The acoustooptic tunable filter (AOTF) [13] consists of three functional units (Fig. 15). At the core of the device is a highly wavelength-selective polarization transformer (PT). Flanking this polarization transformer, a pair of polarization-routing elements (PR) sort the input beam according to polarization and recombine the output beam, in such a way that each channel is either routed to a filtered wavelength port (F) or to an unfiltered port (U). In the form described above, the AOTF functions as a powerful acoustooptic 2 × 2 exchange-bypass switch (AOS). A control command is applied to the polarization-transformer element in the form of an electronically excited sound wave. This acoustic grating is only effective at flipping the polarization state of the specific wavelength for which the optical frequency is an exact multiple of the applied sound frequency. This wavelength specificity allows the incoming optical spectrum to be partitioned into independently sorted wavelength channels. This sound-to-light frequency conversion scale factor is about one-million-to-one, allowing one to leverage large wavelength changes with modest changes in radio-frequency drive. In the 1550-nm optical communication band, the scale factor is 9 nm per MHz about a center frequency of 175 MHz. Typical AOS channel spacings are 1 nm, with tuning ranges from 100–300 nm.

2) *Comparison to Other Filters:* The acoustooptic filter is unique among the candidates for wavelength-selective routing hardware in WDM networks because it is the only filter capable of simultaneously routing many wavelength channels, selected at random. When multiple acoustic frequencies are superimposed on the AOTF crystal, all of the corresponding optical frequencies are simultaneously and independently selected. There is no constraint on the order or adjacency on the selected wavelengths. Other filter types would have to be cascaded in some complex and lossy subnetwork in order to achieve this parallel-processing capability, which comes without added complexity or added loss in the basic AOTF device.

3) *Limitations of Commercially-Available AOTF's:* The AOTF is currently available in bulk-crystal versions, but suffers from high-power consumption (making multiple-wavelength operation impossible), appreciable sidelobe levels

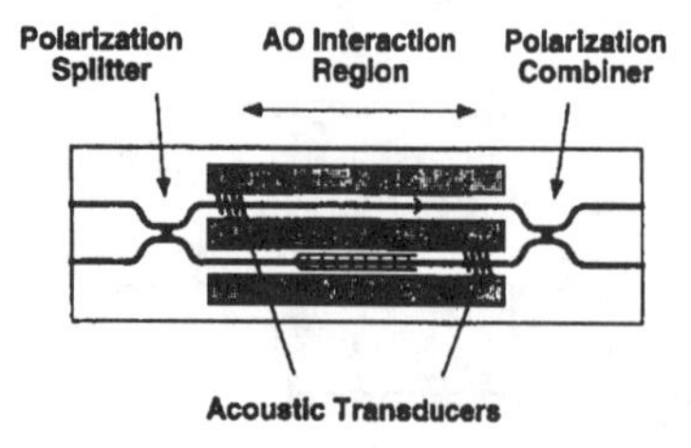

Fig. 16. Schematic diagram of the proposed polarization independent integrated acoustooptic 2 × 2 tunable filter.

(resulting in interchannel interference in WDM system applications), and the requirement of a fixed input polarization (resulting in a potentially severe polarization-dependent insertion loss). An integrated AOTF has the capability to resolve all of the above problems. Power consumption drops from a prohibitive few watts per channel down to a few milliwatts per channel as the optical and acoustic beams are confined to single-mode channel waveguides, in which the optical and acoustic overlap can be made intense. Polarization independence can be achieved in the acoustooptic switch by integrating polarization splitters and polarization combiners on a common substrate, so as to allow any input polarization state to be decomposed, independently transformed by a common acoustic wave and then to be reconstructed in the output combiner. This "polarization diversity" configuration creates a polarization-independent AOS. The problem of sidelobe suppression can be controlled, using techniques of acoustic wave engineering to control the onset and cutoff of the acousto-optic interaction.

4) *Research Prototype: First and Second Generations:* A schematic diagram of the proposed acoustooptic tunable filter is shown in Fig. 16. The polarization conversion takes place in the acoustooptic (AO) interaction region for those wavelengths that have been selected by applying their associated acoustic frequency signal to the acoustic transducers. This device is a fully-integrated 4-port polarization-independent acoustooptic switch module. Fiber V-groove holders etched from silicon provide a precise simultaneous alignment of two singlemode fibers to a pair of titanium-indiffused optical waveguides, both at the input and output sections. The optical waveguides merge into a polarization splitter formed by a hybrid titanium-indiffusion/proton-exchange fabrication technique, and then separate into an interaction region in which each optical waveguide is embedded in its own acoustic waveguide excited by opposing transducers (to eliminate the frequency subspectrum caused by the acoustooptic interaction).

Laboratory experiments have separately demonstrated polarization splitters, polarization-independent operation of an integrated AOTF, the lower-power polarization converter section, the waveguide bends, and the V-groove alignment. The first generation device will be to integrate all of these subcomponents into a low-loss, high-performance whole. The second generation device is to be a refinement of the first, improving the performance of the beamsplitter loss and extinction, reducing bend loss, and reducing the sidelobe intensity by perfecting the apodized acoustic-wave interaction.

As part of the proposed work, six four-by-four acoustooptic switch elements will be packaged in a Benes configuration in which any four input fibers can be connected to any four outputs by suitable selection of cross and bar states for the six switching modules. Each wavelength channel can be simultaneously and independently configured. Crosstalk and resolution limits to such arrangements will be studied and remedies for these problems will be sought. If the crosstalk necessitates it, the switching elements can be dilated to further reduce its effects.

B. Multiwavelength Laser and Receiver Arrays

An array approach to providing multi-wavelength optical detection and signal generation will be used. The laser array [14] will be an integrated array with 8 to 16 lasers on a single chip, depending on the number of wavelengths to actually be realized. In an early phase of the work, four wavelengths will be utilized, with 4-nm spacings, in the most uniform gain portion of the erbium-doped fiber amplifier gain spectrum. The laser array for that phase will consist of 8 lasers, with 2 lasers dedicated to each wavelength to provide redundancy and improve yield. This will be an edge-emitting array of InP distributed feedback (DFB) lasers coupled to 8 fibers with an array alignment technique. In a second phase, a more advanced device will be produced that will provide 8 wavelengths, again with 2 lasers per wavelength, for a total of at least 16 lasers per chip. In addition, the chip for this second phase will contain an on-chip star coupler to combine the outputs of the lasers, and a semiconductor laser amplifier to compensate the loss of the star coupler. A significant advantage of this approach is that only one fiber need be aligned, with considerable improvement to the packaging cost and complexity. All lasers will be directly modulated at up to 2.5 Gb/s.

For the receiver, high-performance OEIC receivers will be realized comprising up to eight detectors monolithically integrated with eight preamplifiers using InP based devices. Each receiver circuit will operate at speeds up to 2.5 Gb/s (OC-48) with a sensitivity of better than -20 dBm at a bit error rate of 10^{-9}.

Both the transmitter and receiver circuits will be completed using GaAs HBT integrated circuits. Both the laser drivers and the receiver gain and retiming circuits will be integrated in arrays to match the packaging advantage of the single-chip detector/preamp and laser chips and will be packaged in receiver and transmitter submodules.

The use of OEIC receiver arrays and integrated laser arrays is essential to the practical realization of WDM multi-channel communications. WDM is essentially a parallel technology, and the use of array receivers and transmitters is intended to overcome the cost factors associated with the use of discrete devices, one per channel, in a natural and practical way. Array technology will also provide improved performance and control of wavelength specifications over discrete single-channel devices. These are additional reasons for preferring an architecture that can achieve the large size and performance with only a small number of operating wavelengths.

VI. Demonstration System

To demonstrate the concepts and technical feasibility of our proposed network, a test system will be implemented. There will be two phases to this demonstration. In the first phase, the networking principles will be demonstrated operating at 155-Mb/s speeds using 4 wavelengths. The network will consist of 4 access nodes and up to $5 - 2 \times 2$ WDM cross-connects, all configured in various experimental topologies to demonstrate and test the switching characteristics using this WDM technology and multihop network architecture. The first phase will include the use of SONET as the carrier and the switching will be done with ATM switches at the network access nodes, both to provide the local access, and to provide the packet-by-packet wavelength translation desired for this architecture. Reconfiguration of the all-optical portion of the network (the WDM cross-connects) will be performed with a workstation providing centralized network control.

The second phase will provide for upgrading the technologies to the OC-48 rate of 2.5 Gb/s, to 8 wavelengths, and to 4×4 WDM cross-connects. These components will be built into a limited number of more integrated modules, and the principal tests to be performed will be of the functional aspects of the technologies, including the ability of the optical amplifiers to operate under the extreme requirements of many channels when cascaded through several stages. SONET channels at STS-48 will be transmitted over the demonstrator network, and the possibility of interfacing to an ATM switch at that rate will be explored.

1) **Network and Technology Demonstration:** The purpose of this demonstration is to prove the architectural concepts of scalability, flexible rearrangeability, modularity, transparency and service flexibility, and fault tolerance. The demonstration will be performed in laboratory facilities at Bellcore in New Jersey.

   The demonstration will be performed with the objectives of:

   a) Demonstrating the functional aspects of the technology, that it works, to uncover technology problems that must be overcome before commercial application is possible, and that the technology can perform in typical environments.
   b) Demonstrating that the network concepts of scalability and modularity through the use of flexible wavelength routing and switched wavelength translation are valid and that the technologies and architectures can be extended to more general configurations.

   The demonstration will consist of several alternate configurations of the access nodes and wavelength-routing cross-connects, under packet-switched operation. The principles of wavelength routing, wavelength translation, and multihop operation will be demonstrated to provide the necessary functionality for scalable, modular networks. It will also include a demonstration of the simultaneous use of analog modulation and subcarrier multiplexing.
2) **Optical Amplifier Performance:** This work will seek techniques to equalize the passband of cascaded erbium-doped fiber amplifiers. The objective is to gain a theoretical understanding from gain spectral modeling of the effectiveness of passive gain spectral equalization and the dependence on pump power, input signal level, and the number of amplifiers in the cascade. The objective is sufficient flatness to permit amplifiers to be cascaded while providing acceptable signal-to-noise ratio at all wavelengths in use.
3) **Integrated WDM Component Technology:** This will be pursued for inclusion in the multiwavelength access node, including research on integrated grating-based demultiplexers in InP, which have the potential of accommodating as many as 50 or more channels in a small device that may be compatible with optoelectronics integration.
4) **Sub-Carrier Multiplexing:** This will include a multiplexing demonstrator involving several satellite video channels (wide deviation FM) and digitally modulated subcarriers (QPSK, FSK, or ASK) at the standard 100–200-Mb/s rates used in the demonstration program. The components used in the demonstrator will include both the optoelectronic components and the subcarrier-multiplexing electronics for the demonstration. The transmitter technology will include an electrooptic $LiNbO_3$ modulator for enhanced linearity. The optical receiver will be constructed using hybrid components and commercial parts.

VII. Applications

As part of this proposed project, the use of PC's and powerful workstations in distributed applications over the network will be explored, with heavy emphasis on cooperative work applications that require the sharing of images, graphics, and video. An important class of these involves storing, retrieving, and sharing images. Another is that of real-time multimedia communications. The focus of this task will be to derive an understanding of the network performance for such applications, and in carrying a range of commercial, industrial, and scientific traffic.

Because the WDM network offers an enormous data bandwidth, applications can become very sophisticated by the use of multiple forms of information-carrying media, both auditory and visual. The more forms of media an application exploits will add complexity to the user nodes themselves. For example, both audio and video media can originate in either analog or digital form. The source of the media can be magnetically or optically stored, or can be real-time source. The data itself can be uncompressed or compressed using a variety of methods, and compression ratios. Furthermore, visual media can be of low or high resolution, and can be still images or full motion video. Applications can use graphics and images intended for a single user, but it is very desirable to be able to interactively share images among many users.

In order for multimedia applications to use PC's and workstations in an optical network environment correctly, issues connected with media manipulation and network utilization need to be resolved. First, the applications must work with different media sources and destinations in a node-independent way and without requiring great amounts of node complexity. Second, they need to be able to effectively use network capabilities. To address these topics we will study how the optical network is used with applications employing graphics, images, and video, and how through the use of network-based visual media servers the complexity of the nodes can be reduced.

The potential bandwidth per user offered by the optical network makes multimedia-based applications very attractive, especially for graphics and other visually based forms of media. Visual media can be digital or analog, and can come from a variety of stored or real-time sources. Destination devices are usually user nodes, and they can be very diverse, ranging from simple PC's or laptops with a relatively-low-resolution display, up to sophisticated scientific workstations or business computers with high screen resolutions. In addition, many destinations will also be sources of multimedia data, for example several engineers at different workstations viewing and manipulating the same image simultaneously. The optical network is an ideal medium to support multimedia data, as it is intended to provide independence of the type of services and data formats that can be sent. For user nodes to be able to directly support many forms of different media from potentially many different sources onto and off the optical network would require complex nodes. We will address this problem by investigating the design and use of modular, scalable network-based visual media servers specifically targeted to support the use of PC's and workstations in distributed applications involving imaging, graphics, and video. A visual media server can reduce the complexity of networked nodes by providing to them graphics rendering, image processing, format translation, and high-bandwidth network access to different media sources. Sharing of images in real time among several users on physically different nodes is an area where a visual media server can greatly assist.

Some specific applications are in the following areas:

1) **Network-based 2D/3D real-time rendering.** The application is to take computer-generated graphics databases, render them into screen space on a networked visual media server, and send the resulting pixel values to network clients. One goal of this is to render to clients at animation rates, which are about 30 frames/s. The advantage of this is that a variety of clients, including those without special graphics hardware, can be used with applications requiring high performance 2-D or 3-D graphics. Doing this at animation rates is not possible without the high-speed optical network.
2) **Network-based image processing.** The application is to take scientific data, apply appropriate image processing algorithms to the data set, and send the resulting image data to network clients. The advantage of using a server to do this is that high-performance digital signal processing for visualization can be done on the server itself, making the application available to a wide range of clients, including those not having special DSP hardware. Our specific application will be to apply Fourier Volume Rendering to medical image data. This involves transferring via the network large amounts of data to the server, and transferring pixel data to the client. The optical network is necessary as it provides enough bandwidth to allow interactive viewing and manipulation of the data set.
3) **Sharing of images.** Another aspect of the network-based Fourier Volume Rendering application is the ability to share images among several network clients. The high bandwidth of the network gives the potential to allow collaboration among several individuals as they interactively examine the same data set. The server reduces the complexity of such sharing as it focuses operations on the data set in one place (the data is not distributed over the clients), and effectively uses the network to distribute the image to the clients in a coherent fashion (all the collaborators are seeing the same thing at the same time).

Bandwidth requirements will vary based on the application, and on the user machines being used. Applications that employ video or some other form of full motion animation will require higher rates. Standardized PC formats of 640×480 (SVGA) and 1024×760 (XGA) will require data bandwidths of 73.7 Mb/s and 188.7 Mb/s respectively to support full motion video at 30 frames/s. The applications running on workstations will require similar bandwidths, but can also support windowing environments that will allow multiple applications to run in different windows. The size and number of these windows will determine the required bandwidth. For example, a single 155.52-Mb/s ATM connection between the server and the user could support two full motion 640×480 windows at 8 bits/pixel. Applications requiring 24-bits/pixel color resolution will require higher bandwidths. A single 622.08-Mb/s ATM connection can support the requirements of the Fourier Volume Rendering application up to animation rates.

Using network capabilities effectively means that a form of media will use network resources in the best way to accomplish what the application needs to do. For this, an understanding of network performance and how media-specific data can be sent on it is necessary. In an asynchronous system such as an ATM-based transmission mode, there exists a need to support isochronous data communications for real-time applications such as those using video sources. The methods for doing this may be different from those for effectively transferring audio data, or high-resolution still images. Our work will attempt to characterize network performance with respect to the requirements of images, graphics and real-time video. We will do this by using our visual media server to drive the network with different media types using a variety of packetization and coding methods. Media types will include full-motion video and still images of various pixel densities and color resolution.

Data will be collected that reflects the performance measured at different user nodes, where the nodes are either PC-class machines or computer workstations.

4) **Analog Networking.** One important feature of our all-optical network design is the independence of network performance from signal format. To demonstrate this powerful capability of our network, we have chosen to demonstrate broadband networking based on subcarrier multiplexing (SCM). By resorting to microwave subcarriers we emphasize one inherent property of our network design—insensitivity to modulation format and service type. Moreover, the use of analog subcarrier multiplexing avoids a major problem with baseband digital Gigabit/s networks, namely the need for each network node in a LAN or MAN to perform digital processing and control at GHz rates.

The rationale for exploiting SCM is further bolstered by two important points. Broad-band networks based on analog subcarrier multiplexing (up to 500 MHz) presently service a significant fraction of today's LAN traffic and are dominant in multiservice environments (data, voice, video conferencing, live video, and office management). We can provide a direct interconnection between distant LAN's by sending over the network the complete spectrum on a single optical carrier. This has significant implications in the building of large LAN's and WAN's. Secondly, we strongly believe that combining the extremely wide bandwidth of analog systems and the ability to handle digital and analog signals on the same optical carrier provides a powerful networking option. In practical terms, the SCM option offers network designers the very real choice of broadband FM satellite channels plus multiple Gigabit/s channels to the desktop without resorting to very expensive 10-Gb/s digital electronics at each node.

To this end we have structured two demonstration channels that will exhibit the power of SCM. The analog channels that will be demonstrated exemplify two types of connections—first, ultra-high speed channels of digitally encoded signals (several SONET signals at 155 Mb/s, 622 Mb/s, and 2.5 Gb/s) and second, 1–2 GHz subcarriers with real-time satellite video (wide-deviation FM-modulated subcarrier) and 100–500 MHz digitally modulated subcarriers (FDDI connections and broad-band coaxial SCM interconnections across the optical network), and HIPPI standard channels. Our purpose will be to compare the performance and complexity of the SCM approach with that of digital TDM in diverse applications.

VIII. Conclusion

In the above, we have outlined a proposed program of research on the subject of multiwavelength all-optical networks with the purpose of integrating together in one study the elements of network architecture, device technology, and a laboratory demonstration, which will test and demonstrate the advantages of these all-optical technologies for network application. We believe that the approach we have chosen is practical and can offer significant opportunity for integrated multimedia networking among large user populations over large service regions.

The attributes to be studied and demonstrated include scalability, modularity, dynamic rearrangeability of the interconnection pattern to meet applications requirements such as connectivity, traffic load, survivability, and multiple-service capability, low latency and timely delivery of information, survivability and quick recovery from faults, and evolvability and growth.

We will produce the device technologies required to implement such a network concept, the network architectural principles that allow the design of large scale-networks, and test algorithms for the control of such a network. Operation of a small network using these technologies will be demonstrated in a laboratory setting.

Acknowledgment

The authors wish to acknowledge the many contributions that have been made by our friends and colleagues, including Bellcore: E. Golstein, C. Guenzer, J. Hayes, P. Kaiser, T. P. Lee, R. Leheny, D. Smith; Columbia University: T. E. Stern; Hughes Research Labs.: R. Walden; Bell Northern Research: L. Coathup, J. M. Glinski, P. Jay, J. McEachern, J. Sitch, M. To; Hewlett-Packard: J. Limb; Rockwell Science Center: K. Pedrotti, C. P. Lee, and many others.

References

[1] C. A. Brackett, "Dense wavelength division multiplexing networks: Principles and applications," *IEEE J. Select. Areas Commun.*, vol. 8, pp. 948–964, 1990.

[2] A. S. Acampora, "A multichannel multihop local lightwave network," in *Proc. IEEE GLOBECOM '87* (Tokyo), Nov. 1987, pp. 1459–1467.

[3] A. S. Acampora, M. J. Karol, and M. G. Hluchyj, "Terabit lightwave networks: The multihop approach," *AT&T Tech. J.*, vol. 66, Nov./Dec. 1987.

[4] A. S. Acampora and M. J. Karol, "An overview of lightwave packet networks," *IEEE Network Mag.*, vol. 3, Jan. 1989.

[5] M. G. Hluchyj and M. J. Karol, "ShuffleNet: An application of generalized perfect shuffles to multihop lightwave networks," *J. Lightwave Technol.*, vol. 9, pp. 1386–1397, 1991.

[6] M. DePrycker, *Asynchronous Transfer Mode: Solution for Broadband ISDN*, New York: Prentice Hall, Nov. 1990.

[7] E. Arthurs, *et al.*, "Multiwavelength optical crossconnect for parallel processing computers," *Electron. Lett.*, vol. 24, pp. 119–120, 1986.

[8] J. Cooper, *et al.*, "Nanosecond wavelength switching with a double-section distributed feedback laser," in *Conf. Proc., CLEO '88* (Anaheim, CA), 1988, paper WA4.

[9] J. F. Labourdette and A. S. Acampora, "Logically rearrangeable multihop lightwave networks," *IEEE Trans. Comm.*, Aug. 1991.

[10] A. S. Acampora and J. F. Labourdette, "A traffic handling comparison of centralized and distributed ATM switching systems," *IEEE INFOCOM '92*, (Florence), May 1992.

[11] K. Bala, T. E. Stern, and Kavita Bala, "Algorithms for routing in a linear lightwave network," in *Proc. INFOCOM '91* (Miami), Apr. 1991.

[12] K. Bala and T. E. Stern, "Topologies for linear lightwave networks," in *Proc. SPIE '91* (Boston), 1991.

[13] D. A. Smith, J. E. Baran, J. J. Johnson, and K. W. Cheung, "Integrated-optic acoustically tunable filters for WDM networks," *IEEE J. Select. Areas Commun.*, vol. 8, pp. 1151–1159, 1990.

[14] C. E. Zah, *et al.*, "Monolithic integration of multiwavelength compressive-strained multi-quantum-well distributed-feedback laser array with star coupler and optical amplifiers," *Electron. Lett.*, vol. 28, pp. 2361–2362, Dec. 1992.

Evolution of the Acousto-Optic Wavelength Routing Switch

David A. Smith, *Senior Member, IEEE,* Rohini S. Chakravarthy, *Senior Member, IEEE,* Zhuoyu Bao, Jane E. Baran, Janet L. Jackel, Antonio d'Alessandro, Daniel J. Fritz, S. H. Huang, X. Y. Zou, S.-M. Hwang, Alan E. Willner, *Senior Member, IEEE,* and Kathryn D. Li

***Abstract*— Through the efforts of many research groups and consortia over the last several years, the acousto-optic tunable filter has evolved into a device capable of high-performance wavelength-selective optical switching and wavelength routing in dense WDM systems. The distinguishing feature of the AO switch is its ability to sustain many independent coexisting passbands, thus allowing in a simple integrated-optic device, the parallel processing capability of much more complex designs. The AOTF has also found a role in active gain equalization of optically amplified networks. In this paper, we review the design of both hybrid and fully integrated AO switches. The theory of operation is reviewed and recent advances in passband engineering are described which have made low-crosstalk, wavelength misalignment-tolerant switches to be possible. Advanced issues such as mechanisms of interchannel crosstalk and its reduction are also discussed. Both device and system issues are covered.**

I. Introduction

A. The Role of the Acousto-Optic Switch in WDM Networks

THE TREND in modern advanced optical networks is to break up the low-loss wavelength spectrum of silica fiber into a densely packed spectrum of independent wavelength channels which are manipulated as separate carriers by transparent wavelength-selective optical filters or wavelength-channel routers. The result is a network, consisting of a number of WDM channels, each of which contains independent time-division-multiplexed data. In this paper we restrict ourselves to wavelength channel isolation by optical filtering, and in particular, by acousto-optic switching and routing of WDM channels.

Manuscript received June 19, 1995; revised November 28, 1995. This work was supported in part by an Advanced Research Projects Agency Optical Technology Consortium Grant (MDA-972-92-H-0010) and by an AFOSR Grant (F30602-91-1-001).

D. A. Smith, R. S. Chakravarthy, and Z. Bao are with the Department of Electrical Engineering and Applied Physics, Case Western Reserve University, Cleveland, OH 44106 USA.

J. E. Baran and J. L. Jackel are with Bellcore, Red Bank, NJ 07701 USA.

A. d'Alessandro is with the Dipartimento di Ingegneria Elettronica, Universita' degli Studi "La Sapienza", 00184 Rome, Italy.

D. J. Fritz is with the United Technologies Research Center, East Hartford, CT 06108 USA.

S. H. Huang, X. Y. Zou, S.-M. Hwang, and A. E. Willner are with the Department of Electrical Engineering, University of Southern California, Los Angeles, CA 90089 USA.

K. D. Li is with the Focused Research, Inc., Sunnyvale, CA 94089 USA.

Publisher Item Identifier S 0733-8724(96)04080-7.

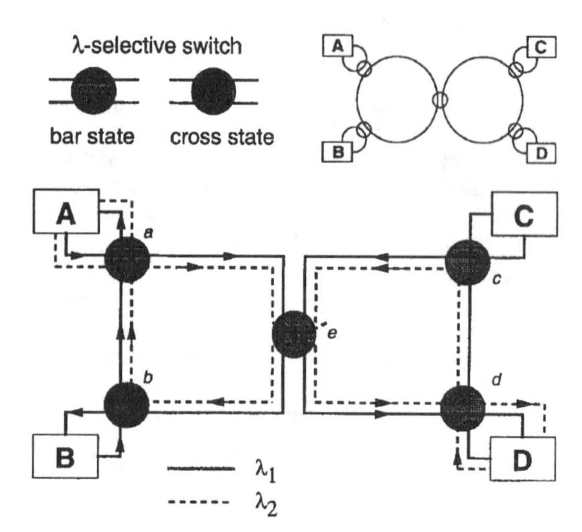

Fig. 1. Example of a wavelength-routed network showing wavelength reuse for $\lambda_1(A \leftrightarrow B$ and $C \leftrightarrow D)$ and independent switch operation (bar or cross state) within each wavelength channel.

Wavelength routed networks have a single physical (fiber) layout but wavelength switches connect access nodes and interconnect fiber rings in such a way as to allow a different connection diagram for each wavelength. The basic element required for such parallel processing is the 2×2 wavelength-selective switch of Fig. 1. Switches *a–d* act as add/drop multiplexers for ports *A–D*, respectively. Switch *e* connects the two fiber rings. The switch can be configured in either the bar (or bypass) state or the cross (or exchange) state for each wavelength channel simultaneously and independently. In the example of Fig. 1, port A is connected to port B and C is connected to D by wavelength λ_1 (solid line), demonstrating wavelength reuse in disjoint portions of the network, as will be required for scalability of networks to vastly many more users than the number of available wavelength channels. The required switch states for switches *abcde*, respectively, are XXXX=. A two-way connection between A and D is achieved using λ_2 (dashed line) using the switch states X==XX. In general, larger wavelength-selective crossconnects can be constructed from arrays of 2×2 switches to achieve dynamic wavelength-channel routing in larger networks.

The use of the acousto-optic tunable filter (AOTF) for wavelength-selective switching has been investigated in

Reprinted with permission from *IEEE Journal of Lightwave Technology*, Vol. 14, No. 6, pp. 1005-1019, June 1996.

the ARPA-funded Optical Network Technology Consortium (ONTC), a WDM demonstration network which uses wavelength-routing to achieve a dynamically interconnected system of nodes [1]. In the ONTC architecture, a given node has all the system wavelengths at its disposal and makes connections through wavelength routing and wavelength translation in one or more hops between multiwavelength transceiver nodes in the system. Each node consists of a WDM wavelength-routing add/drop multiplexer which allows signals passing through to proceed without O/E/O conversion, while channels which are to be extracted or wavelength-translated are dropped into the transceiver. The transparency provided by a wavelength router allows arbitrary signal formats to coexist in a network. If the WDM node function was assumed by the transceiver alone, transparency would be sacrificed. In the ONTC network, for example, subcarrier multiplexed analog signals can be routed trough the network because they are isolated from the digital traffic which penetrates to the transceiver level. The array transceiver has a WDM demultiplexer to sort incoming wavelength channels, followed by a detector array which is connected to an electronic crossconnect. The crossconnect provides arbitrary connection to the array of transmitters, followed by a multiplexer to aggregate the wavelength channels onto a single output fiber. This design allows regenerative wavelength translation. The electronic crossconnect is tapped so as to allow adding or dropping content of some or all of the TDM content of each incoming WDM channel.

The acousto-optic switch (AOS) was chosen as the WDM access node of the ONTC program because it has the unique capability to function as a *multistate* WDM *switch*, a single 2×2 switching element which can be set into the cross state for an *arbitrary subset* of WDM channels while remaining in the bar state for the rest of the channels. Furthermore, the locations of the WDM channels are arbitrary in the AO-switch-based WDM router. Most alternative technologies are either multiport passive wavelength demultiplexers such as the arrayed waveguide grating [2] with fixed channel spacing or can only switch a single WDM channel or comb of channels, such as Fabry-Perot-based filters. Passive WDM router-based architectures rely on agile transmitters or wavelength translators to achieve wavelength routing. WDM crossconnects can be made from wavelength-insensitive optical switches, but WDM crossconnects based on single-state switches must be broken up into as many subcross-connects as are available wavelength bands in the system, and then each wavelength is separately switched. Finally, the re-wired wavelength channels are multiplexed in an output WDM element. An example of a WDM demonstration system using this demux/switch/mux cross-connect design is the so-called MWTN (Multi-Wavelength Transport Network) project in the RACE program [3]. This architecture is N times as complex as the AOTF crossconnect, where N is the number of system wavelengths. Such divide-and-conquer schemes are also specifically locked into the WDM system channel allocation, unless the demultiplexer passbands are agile, whereas the AO switch can support an arbitrary number of arbitrary wavelength band locations, as long as they are sparse enough to be well isolated by the underlying filter passband width.

B. Organization of This Paper

This paper is a study in evolution of integrated acousto-optic technology toward meeting anticipated WDM wavelength routing requirements as well as those unexpected requirements exposed during the interplay between device development and system demonstration of prototype devices. The presentation starts, in Section II-A, with a description of the classical AOTF as a polarization-flipping filter, reviewing how integrated optic and integrated acoustic techniques have provided the means to obtain low-power, polarization-independent, passband-engineered devices to meet WDM switching demands. Hybrid (external-polarizer-based) and fully integrated switch designs are compared in Section II-B. Although optimizing the filter passband shape provides the first line of defense against interchannel crosstalk, space dilation of AO switches and crossconnects provides the extra margin to allow AO devices to achieve extinction below -30 dB. Section III presents a simple but comprehensive AOTF model which provides the conceptual foundation for passband engineering techniques such as sidelobe shaping, sidelobe suppression and passband flattening. Examples of device performance for passband-engineered filters and switches are provided to support this theoretical model. The optimum passband shape of a WDM filter or switch is the robust rectangular transmission profile and the degree to which various designs achieve this goal is investigated. Section IV deals with current research issues including the problem of interchannel crosstalk in AO switches, dense packing of WDM channels by wavelength dilation, and recent progress toward fabricating lossless crossconnects using erbium-doped substrates. Finally, the use of the AOTF as a gain equalizer for optically amplified WDM networks is briefly discussed.

II. Requirements of the Acousto-Optic WDM Switch

A. Comparison of Attributes and Requirements

1) Performance Requirements of a WDM Switch: WDM switching places stringent demands on the acousto-optic switch, demands which at first seem inappropriate for an *analog*, polarization-dependent, moderate-tuning-speed, moderate-width-passband filter with high sidelobes. The WDM access node switch must be *insensitive* to polarization state; it must have a high-extinction *digital* switching characteristic, and it must either possess nanosecond switching time for packet-switching applications or it can be rather slow (millisecond response) for circuit switching. The routing node must have a passband which allows dense WDM with 1–2 nm spacing, and must possess a transmission passband shape which is nearly perfectly rectangular. A rectangular passband is a transmission window which is flat and very high in extinction. Less than -30 dB crosstalk is the anticipated need both in the filter transmission window (to allow wavelength misalignment tolerance) and deep out-of-band rejection (in order to reduce interchannel crosstalk) [4]. Other desirable

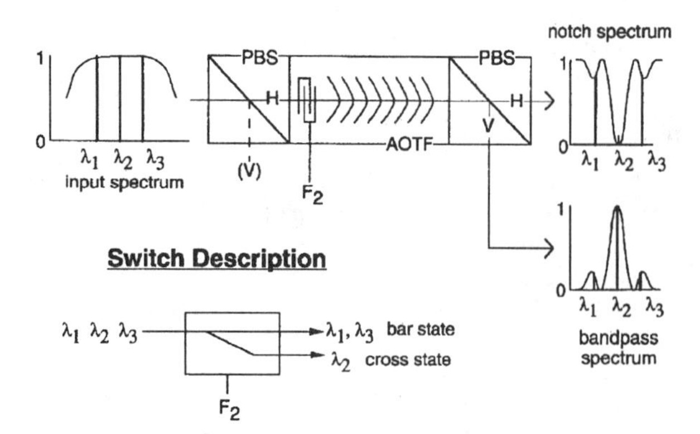

Fig. 2. The AO polarization converting filter operating on either a continuous or discrete input optical spectrum. Wavelength channel 2 has been diverted by the activation of the corresponding resonant photoelastic grating followed by polarization splitting.

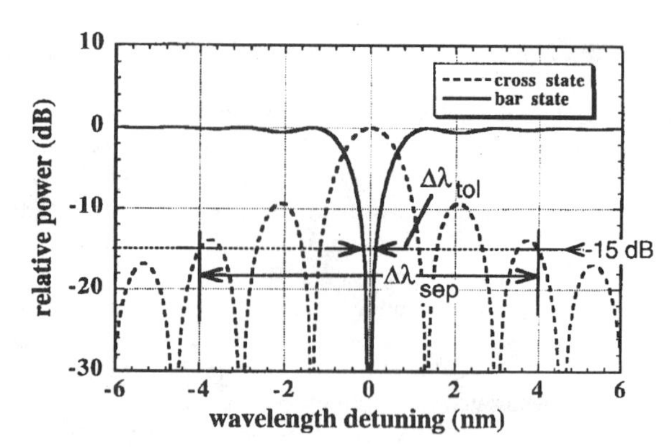

Fig. 3. Bar state (solid line) and cross state (dashed line) transmission spectra for a 20 mm-long AO switch at 1550 nm using lithium niobate parameters. The wavelength misalignment tolerance and interchannel separation at −15 dB crosstalk is identified.

features in a WDM switch include low loss, low power consumption, fail-safe operation so that the crossconnect goes into a manageable state if power is lost, and monolithic fabrication technology for low potential cost. The rest of this paper will be devoted to describing the evolution of the AOTF to meet the requirements of requirements of deployable, scalable and affordable WDM networks.

2) Principle of Operation: The acousto-optic tunable filter uses an acoustic wave to generate a birefringence grating seen by an incoming polarized beam of light. In a collinear AOTF, this birefringence acts to cause a polarization transformation between TE and TM polarization, and if the interaction strength, integrated over the length of the interaction, is just strong enough, a perfect polarization flip (TE–TM interconversion) is achieved. The process is resonant and narrow in spectral width. Because the two polarization states propagate at very different velocities, coupling only can be achieved when the phase-matching condition is met, i.e., when the sound wave momentum just compensates the TE and TM momentum mismatch. Analogous to polarization-preserving fiber, the device will preserve the input polarization TE/TM ratio rather than flip it, unless the polarization beat length $L_b = \lambda/\Delta n$, closely matches the acoustic period $\Lambda = V_s/f_s$. The resonant polarization conversion bandwidth is $\Delta\lambda/\lambda = L_b/L$. In the above, L is the device interaction length, V_s is the SAW velocity, Δn is the waveguide effective index birefringence and f_s is the SAW frequency. Fig. 2 shows the action of an AOTF on either a white light spectrum or a discrete WDM-type input spectrum. From the phase-matching requirement $\Lambda = L_b$, the acousto-optic scale factor A, relating the frequency of light to the resonant sound frequency, is obtained:

$$A = f_s/f_o = V_s\Delta n/c \tag{1}$$

where c is the speed of light in vacuum. The switching speed is equal to the acoustic transit time, i.e. the time τ it takes to establish the birefringence grating, $\tau = L/V_s$. Typical values of interest for lithium niobate-based AO switches for WDM systems centered at 1.55 μm, are V_s = 3.7 km/s, Δn = 0.08, L_b = 20 μm, f_s = 175 MHz, $A = 10^{-6}, L = 2$ cm, τ = 6 ms and $\Delta\lambda$ = 1.5 nm. It should be pointed out that the polarization conversion is accompanied by an imposed optical frequency shift equal in magnitude to the acoustic frequency, but of a sign determined by whether a phonon was absorbed or emitted during the polarization flip. For example, since $n_z < n_x$ in $LiNbO_3$ (i.e. $n_{\rm TE} < n_{\rm TM}$ in x-cut y-propagating samples used for AOTF's), a light beam co-propagating with the SAW beam must gain momentum in TE-to-TM conversion, therefore absorb a phonon, therefore in turn have its frequency upshifted. Similarly, during TM-to-TE conversion, the frequency is decreased by f_s. The signs of the frequency shifts are just the opposite for the case of sound and light counterpropagating.

3) Interchannel Crosstalk and Passband Shape The classical collinear acousto-optic filter transmission function is a sinc-squared function of detuning from resonance, calculated and shown on a log scale in Fig. 3 for the case of a 20 mm-long lithium-niobate-based AOTF centered at 1550 nm. If one's system crosstalk requirement for the 2 × 2 elemental switch of a WDM network were −15 dB, then interchannel spacing and wavelength misalignment tolerance can easily be read from the log transmission-vs-wavelength detuning curve by reading across the −15 dB, line. The −15 dB, threshold passes the cross-state transmission curve I_x at ±4 nm and passes the bar-state transmission plot $I_=(1-I_x)$ at ±0.14 nm. A parameter which clearly defines the desired flat shape of the transmission function is the rectangularity r at a given crosstalk level c as described for the sinc-squared-shaped AOTF passband shape in Fig. 3. At a given crosstalk level the cross state filter width for which the crosstalk is less than c is defined as $\Delta\Lambda_{\rm cross}(c)$ and the equivelant parameter for the bar state is $\Delta\lambda_{\rm bar}(c)$. The value of $\Delta\lambda_{\rm cross}(c)$ is primarily determined by the sidelobe skirt envelope of the cross state transmission function. In the bar state, we define a misalignment crosstalk tolerance width $\Delta\lambda_{\rm tol}(c) = \Delta\lambda_{\rm bar}(c)$. The minimum interchannel separation for a given crosstalk level c is given by $\Delta\lambda_{\rm sep}(c) = \frac{1}{2}[\Delta\lambda_{\rm bar}(c) + \Delta\lambda_{\rm cross}(c)]$. The ratio of wavelength misalignment tolerance to minimum channel separation might aptly be called the *rectangularity* r because it defines the degree to which the transmission function is "rectangular":

$$r(c) = [\Delta\lambda_{\text{tol}}(c)/\Delta\lambda_{\text{sep}}(c)]. \tag{2}$$

The ideal value of r is unity, independent of the crosstalk criterion c in the sense that the ideal rectangular transmission function would allow all wavelengths to pass right up to the wall of the neighboring channel passband. In general, r is quite a bit less than one and becomes smaller as the crosstalk threshold becomes more severe because $\Delta\lambda_{\text{tol}}$ shrinks due to the sharply peaked passband and $\Delta\lambda_{\text{sep}}$ expands since we have to go out ever further until the sidelobe skirt drops to the desired level. As an extreme but crucial example, for the sinc-squared passband shape, we can calculate that r $(-10 \text{ dB}) = 0.20$ but r $(-15 \text{ dB}) = 0.068$ and r $(-20 \text{ dB}) = 0.022$. Since r is a measure of packing density and we always are in need of maximum dense packing due to the limited single-channel transmission capacity and available optical amplifier bandwidth, we must find a way to reduce sidelobes as well as to flatten the passband in order to obtain the maximum possible rectangularity.

4) Passband Engineering: The sinc-squared AOTF passband is a consequence of the smooth, constant-amplitude polarization transformation along the active length of the device. The detailed AOTF passband shape will be shown in Section III to be more generally dependent on the integrated effect of the local phase mismatch and the local acousto-optic interaction strength. The local phase mismatch, proportional to $(\Lambda - L_b)$, depends on the degree of match between the local acoustic period and the local polarization beat length which is in turn dependent on the local acoustic wavelength and birefringence (which has an underlying material dependence as well as a temperature dependence) . The local acousto-optic interaction strength depends on the acoustic intensity which decays with length, but can also be altered by the effects of acoustic waveguiding, acoustic diffraction, and interference with other leaky acoustic modes. The concept of passband engineering is to control the acousto-optic phase and amplitude profile along the device in order to achieve the desirable passband characteristics which depend on the local polarization conversion properties of the device. The achievement of sidelobe suppression for the purpose of reducing interchannel coupling has been a success story of the AOTF. Sidelobe suppression is performed by a gradual tapering of the onset and cutoff of the acousto-optic interaction strength. This technique, known as apodization, has its parallels in many areas from classical beam optics (for reducing diffraction rings) to electronic signal processing (for pulse shaping). In the AOTF, a number of apodization techniques have been demonstrated [5]–[10], but we will concentrate on the use of coupled acoustic waveguides between which an acoustic wave can be made to oscillate, thus producing a sinusoidal interaction strength having the tapered onset and cutoff required for sidelobe suppression. The details of controlling the acousto-optic interaction will be discussed, after the theory underlying the full capabilities of passband engineering has been described in Section III.

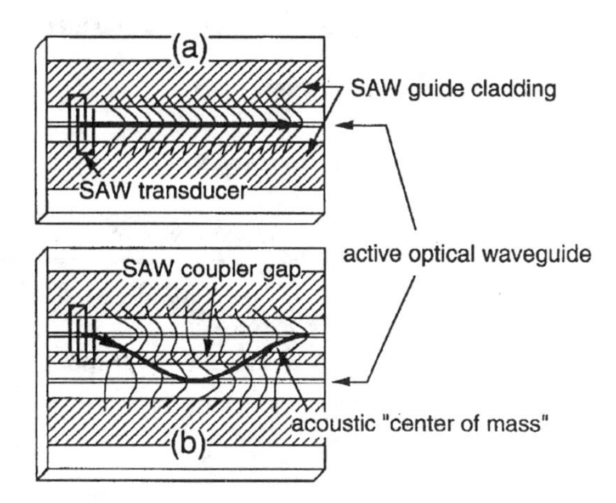

Fig. 4. Design of the integrated AOTF showing acoustic cladding to define SAW waveguides, SAW-generating transducers and the imbedded active optical waveguide. The SAW intensity profile is depicted for (a) the uniform filter and (b) the SAW-coupler-apodized filter.

B. Current AO Switch Technology

1) Monolithic Device Technology: The collinear acousto-optic filter has been developed on x-cut y-propagating lithium niobate because that substrate possesses the combined advantages of a mature integrated-optic fabrication technology and good acousto-optic and piezoelectric figures of merit [11]. Optical signals are coupled in through single-mode fibers which are aligned to single-mode titanium-indiffused optical waveguides, formed by indiffusion of a Ti metal stripe ($\approx$8 μm wide and $\approx$10 nm thick) for about 10 hrs at approximately 1050 °C. The acoustic wave is generated using interdigital transducers (IDT's), usually, but not always, confined to an acoustic waveguide, which is formed by defining cladding barriers with higher surface acoustic wave (SAW) velocity [12]. Acoustic barriers can be fabricated using deep Ti indiffusion, typically by subjecting a 160 nm Ti stripe to more than 20 h of diffusion heating under identical conditions to subsequent optical waveguide metallization and diffusion. Fig. 4 shows an IDT spanning the 100 μ width of a single-mode SAW waveguide. For devices with the 20 mm interaction length required for nanometer passband width, a typical drive power of 10 mW is required for 100% polarization conversion in the 1.55 micron optical band [13]. Sidelobe suppression, as discussed in Section II-A4) can be achieved by making a SAW directional coupler [Fig. 4(b)] from a pair of SAW waveguides separated by a narrow barrier, about 20 μ in width, which evanescently couples the SAW modes, resulting in the required oscillatory SAW amplitude between the SAW waveguides. The SAW beam is generated in the upper SAW guide, while the acousto-optic interaction occurs in the optical waveguide imbedded in the adjacent SAW waveguide.

A number of techniques have been applied to integrate polarizers into the polarization conversion filter, including metal overlays, optically contacted superstrate crystals, all-titanium-waveguide polarization beamsplitters and proton exchange waveguides. Although one can place orthogonal integrated polarizers at opposite ends of the AOTF in order to make a self-contained bandpass filter, the more practical extension to a

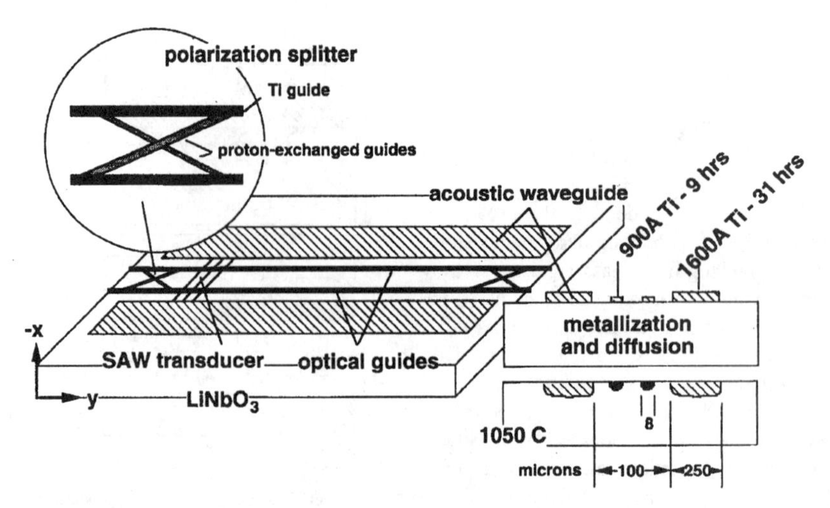

Fig. 5. Design and fabrication parameters for a fully integrated AO switch showing input and output polarization splitters flanking the two active waveguides of this polarization-diversity structure.

polarization-independent filter design is obtained using somewhat more sophisticated TE–TM waveguide beamsplitters instead of simple polarizers. In the standard polarization-diversity configuration used to achieve polarization insensitivity, the polarization converter is flanked by TE–TM splitters on either side, achieving polarization independence and switching in one step. In all such switches, the TE–TM splitter decomposes incoming light into orthogonal polarizations (TE and TM in the local crystal coordinates), separately filters them in the same or adjacent AO interaction regions, and then recombines the two processed beams in an output TE-TM X-junction. Non-resonant light appears at the bar-state or bypass port of the device while resonant, filtered light recombines at the cross-state. The polarization-diversity AO filter of Fig. 5 is in fact the desired 2×2 exchange/bypass switch needed for wavelength-routed WDM systems. The wavelength channel exchange is achieved by applying a SAW frequency to perform 100% polarization interconversion in the wavelength band or bands chosen to crossover to opposite switch ports. All-titanium polarizing beamsplitters (PBS's) have been employed in monolithic AO switches, either using a straightforward directional coupler design [14] or the related bow-tie-type waveguide crossing polarization splitter [15]. Also, a mixed Ti-indiffused/proton-exchanged beamsplitter has been used [16]. In Fig. 5, X-junction Ti/PE waveguide branches have been depicted [17] in an AOTF, shown without apodization for simplicity. Recent progress on fully integrated AO switches has involved the combination of polarization-diversity switch design with deep sidelobe-suppressed acousto-optic geometries [18].

2) Hybrid Acousto-Optic Switches: As is often the case, isolating components in a hybrid structure, though complex and bound to be expensive, allows one to make an overall better-performing switch during the immature stages of monolithic device fabrication technology. The hybrid AO switch currently under development in the ONTC program, is sketched in Fig. 6. The hybrid AO switch connects external polarization splitters to an pair of common-substrate apodized polarization converters by means of polarization-preserving fiber (PPF). The PPF allows one to rotate one of the input polarizations in such a way as to make what is overall a polarization-diversity structure appear to be single-polarization, as far as the AO interaction is concerned. This all-TE device is preferred because the orthogonal signal components are treated exactly the same way, including the potentially troublesome frequency shift on polarization conversion, which is opposite for TE–TM and TM–TE conversions and would cause self-beating in subsequent areas of the network where TE and TM can be mixed, as in another polarization splitter. The opposite frequency shift content of orthogonal polarizations causes problems in a network with multiple polarization splitters as shown in the two-AOTF sequence of Fig. 7 in which a first stage causes a differential TE-TM frequency shift, and the second device decomposes this signal into four terms so that now there is a coherent mixing of what were formerly orthogonal TE and TM states. Finally, the pair produces beats at twice the SAW frequency and of very large magnitude since the second device can redefine TE and TM arbitrarily in its local coordinate system. Differential frequency shift suppression can be obtained by driving the two legs of the polarization-diversity filter from opposite ends. Performance of the hybrid AOTF has been excellent in single and cascaded filter arrangements. The best hybrid device in the ONTC demonstration network was optimized for the wavelength set {1546 nm, 1550, 1554, 1558}, had an average fiber-to-fiber insertion loss of 4 dB for the complete assembly, a switching power of 17.5 dBm/channel and was designed with partial absorber passband flattening over two 17 mm-long sections (see Section III-C) and counterpropagating acoustic drives to achieve alternate-channel wavelength dilation (see Section IV-B). The polarization converter chip of Fig. 6 had heat-sink pads on the outsides of each transducer to guarantee uni-directional SAW propagation and for thermal stability, SAW guide cladding heat absorbers were placed along the interaction path. Slots in the heat sink block provide access for wire-bonding to the two IDT's and access to the partial

TABLE I
SWITCHING CROSSTALK MATRIX FOR A FOUR nm-SPACED AO SINGLE-STAGE HYBRID AO SWITCH. SHADED ENTRIES CORRESPOND TO RESIDUAL (BAR-STATE) POWER (NEGATIVE dB'S) AND CLEAR ENTRIES ARE LEAKAGE LEVELS (NEGATIVE dB'S) INTO THE CROSS PORT

	Crosstalk level (-dB)		cross state	
Channel (nm)	1546	1550	1554	1558
frequency (MHz)	176.17	175.68	175.16	174.65
0000	21	21	18	18
1000		17.6	16	17
0100	20		17.5	17.5
0010	20	19		16.5
0001	20	20	17.5	
0011	18.5	18.5		
0101	18		17	
1001		17	16	
0110	17.5			15.3
1010		16		15
1100			15.5	15.5
0111	17			
1011		15		
1101			14	
1110				13.5
1111				

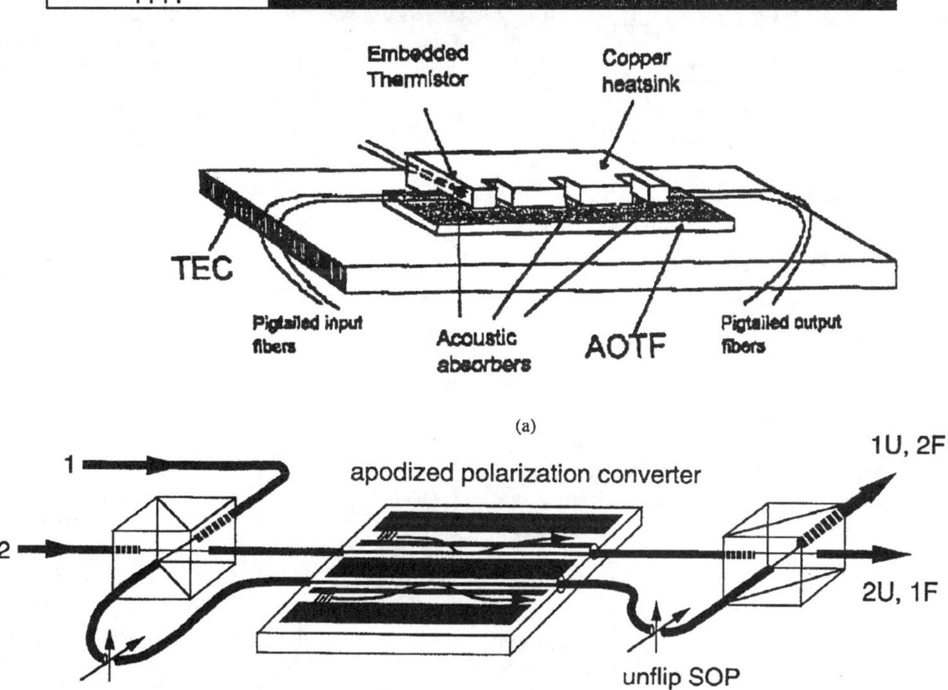

Fig. 6. (a) Detail of the heat-sinking and TE cooler mounting for a polarization-diversity polarization converter which (b) is packaged with external polarization splitters in order to achieve polarization diversity. The term 1U refers to unfiltered light from port 1, 2F is filtered light from port 2, etc.

absorber. This assembly was mounted on a thermoelectric cooler which maintained 20 ± 0.1 °C stability so that the required RF drive (acoustic switching) frequency could be preset. The acousto-optic tuning scalefactor was 127 kHz/nm about a center frequency of 175 MHz for the 1550 nm wavelength region. The switching matrix is shown in Table I. Consider the switch state denoted [1101]) where wavelength channels 1, 2, and 4 are in the cross state. The residual optical powers in the bar port (shaded table entries) was measured to be −15 dB, −12.5 dB, and −13.5 dB for channels 1, 2, and 4, respectively. The leakage of channel 3 into the cross port, though channel 3 was dormant, was −14 dB. The filter transmission function is shown in Fig. 8, showing sidelobe suppression to −17 dB. This optical spectrum analyzer trace

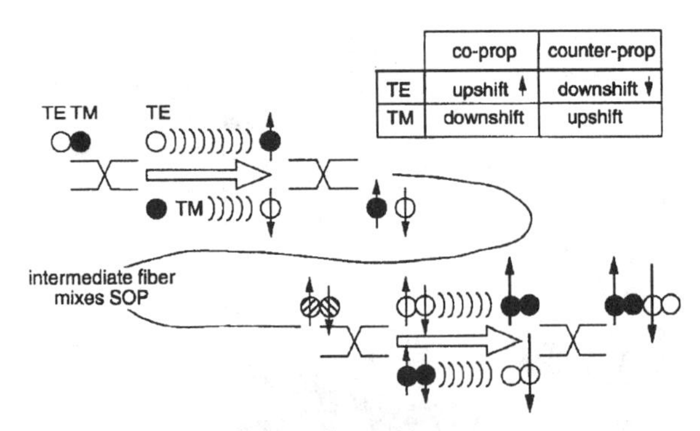

Fig. 7. A series of two polarization-independent filters can impose coherent interference due to mixing of upshifted and downshifted polarization components, according to the induced AO frequency shift table (inset). This is a consequence of the polarization mixing caused by the polarization splitters of the second AO switch.

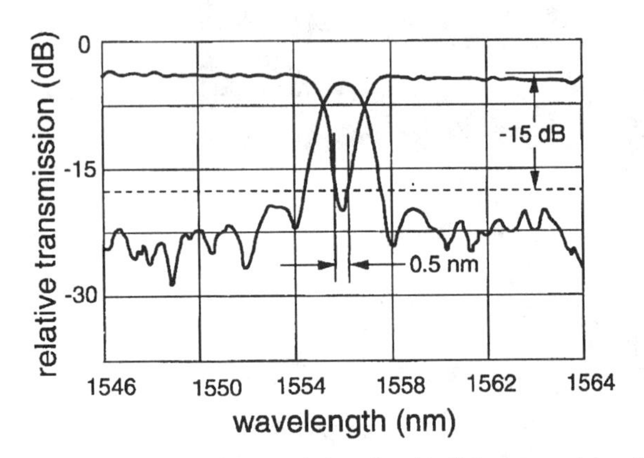

Fig. 8. Cross and bar-state transmission of a white-light spectrum injected into the hybrid filter of Fig. 6, showing sidelobe structure and misalignment tolerance (0.5 nm window) at the −15 dB crosstalk level (dashed line).

was obtained using an EDFA "white-light" source applied to a hybrid single-stage filter with single 17.5 dB RF drive. The partial passband flattening achieved in this switch opened a wavelength misalignment tolerance window of 0.5 nm width, defined by a −15 dB bar-port isolation requirement (95% switching).

Even though the crosstalk obtained for hybrid devices is close to theoretical performance, the requirement for WDM access nodes in a multi-node network can be too stringent to be achieved with practical single-stage AO switches. The crosstalk demands on an optical switch can be estimated as follows. Suppose one requires an intensity signal-to-noise level of −15 dB for adequate bit-error-rate in the system receiver. If a single device has intensity crosstalk of ϵ, the self-beat from this signal is coherent and so the signal-to-noise becomes $S/N = 2\sqrt{(P_o \epsilon P_o)}/P_o = 2\sqrt{\epsilon}$ which implies e must be at least −36 dB to obtain −15 dB relative intensity noise at the receiver. In order to achieve this level of crosstalk, the method of switch space dilation has to be employed in practical demonstration networks [19]. Dilation of the 2 × 2 AO switch, as shown in Fig. 9, requires four 2 × 2 subswitches interconnected so that crosstalk due to sources

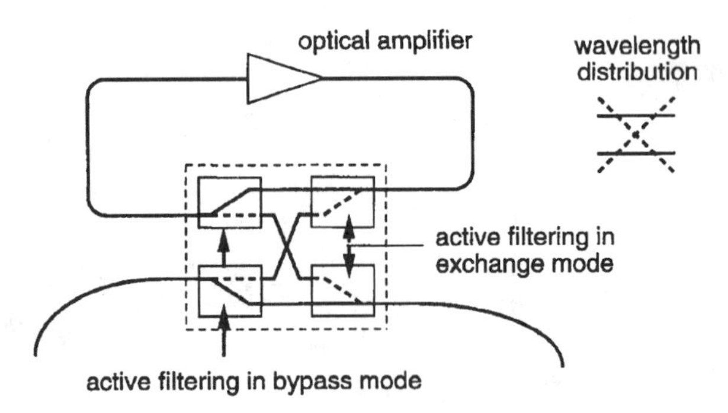

Fig. 9. Dilated AO WDM access node design which treats bar and cross states as active, thus preventing a closed feedback path for the optical amplifier.

such as incomplete polarization conversion and beamsplitter leakage is itself filtered out and dumped to an unused port of the dilated array. A 2 × 2 array consisting of four hybrid apodized 2 × 2 switch elements, has shown −30 dB crosstalk or better for all possible AO switch states for four-nm-spaced WDM channels [20]. The dilated 2 × 2 access node or WDM switch has some significant advantages. First, all crosstalk is reduced to second order, so that a −17 dB leakage subswitch supports a dilated node with almost −35 dB extinction. If the internal wiring of the switch is polarization-preserving, TE-TM frequency shift effects are eliminated altogether. Alternatively, the access node shown in Fig. 9 allows nodes to treat the bar-state wavelengths on an equal footing with the cross-state wavelength channels, with the effect that any wavelength passing through must be filtered, eliminating the potential for unexpected oscillating loops in optically amplified fiber networks. The dilated WDM access node prevents oscillating loops since the path containing the optical amplifier is broken except for wavelength channels specifically excited in the upper left subswitch.

Dilation takes its toll in increased network complexity and cost, but as WDM crossconnects grow, the dilation penalty drops toward a factor of two. A dilated 2 × 2 switch takes 4 elements, but a dilated 4 × 4 requires only 12 elements [21]. Various 4 × 4 crossconnect geometries are possible with leakage terms of order ϵ^2 where ϵ is the intensity leakage per subswitch. Again, values of ϵ^2 of −30 dB should be obtained for larger dilated crossconnects. It is certain that *strictly* nonblocking designs need to be employed, since otherwise active high capacity wavelength channel paths will be interrupted for periods on the order of a switching time [22]. It is hoped that with the maturing of monolithic switches, dilated switches and arrays may be formed as columns on a common substrate, interconnected by fiber ribbons in a way that achieves the required switch hardwiring pattern, but which allows the economy and high yield of common-substrate switch array elements.

III. Theory

A. Modeling the Optical Transmission Function

The collinear acousto-optic filter consists of a narrowband polarization converter placed between crossed polarizers (in the cross state) or between parallel polarizers (in the bar state). The filter operation can be modeled as a directional coupling process between the TE and TM normal waveguide modes which are coupled by stress-induced birefringence. The birefringence is made periodic, in the form of a surface acoustic wave (SAW) grating, as is required for phase-matched coupling between the momentum-mismatched normal modes. As discussed in Section II-A2), perfect phase match is achieved when the polarization beat length, $L_b = \lambda/\Delta n$ exactly matches the acoustic period $\Lambda = V_s/f_s$. In the above, Δn is the waveguide *effective* index birefringence which depends on the detailed waveguide shape, metallization and diffusion history. More generally, we require that the *local values* of $L_b = \Lambda_s$ must hold if we are to have constructive interference of polarization-converted light throughout the length of the device. Phase mismatch can be obtained through change in effective index, which in turn depends on waveguide width, depth and dopant concentration and profile, as well as on local temperature and crystal inhomogeneities, and the local perturbation of sound velocity. Thus, phase match is a function of position, in general. We define a local mismatch between polarization beat length and acoustic period as a phase mismatch per unit length of $\Delta\phi = 2\pi(\Delta n/\lambda - f_s/v_s)$. Defining a reference wavelength λ_o and a mean sound velocity V_{so} we then write

$$\Delta\phi = -2\pi\left(\frac{\Delta\lambda}{L_{b0}\lambda_0} - \frac{\Delta V_s}{\Lambda_{s0}V_{s0}}\right). \tag{3}$$

If we ignore SAW velocity inhomogeneity, although its treatment remains clear from the above equation, we can define a detuning parameter

$$2\delta = -\frac{2\pi}{\Lambda}\left(\frac{\Delta\lambda}{\lambda_0} - \frac{\Delta n - \Delta n_0}{\Delta n_0}\right) \tag{4}$$

where $\Delta\lambda = \lambda - \lambda_o$ is the wavelength detuning from the reference value, which is usually taken to be the peak polarization conversion wavelength, and we have assumed that $|\Delta\lambda|/\lambda_o \ll 1$. The possibility of nonuniform birefringence is included as well in the above equation.

Mathematical modeling of the AOTF transmission function is strictly analogous to the coupled mode theory used in describing the coupling between modes in optical directional couplers. Two parameters describe the AO interaction as a function of position: the acousto-optic interaction amplitude κ and the detuning parameter Δ described above. The acoustic amplitude term is proportional to the product of the acoustic field, the optical and acoustic overlap integral, and the photoelastic coupling factor (which includes a stress-induced electro-optic contribution). If the polarization state of the optical beam is described using the Jones calculus, in which the normalized TE and TM electric field amplitude components take the upper and lower values of a complex vector $\boldsymbol{x} = (x, y)$, then the polarization transformation matrix $\boldsymbol{M}(\kappa, \delta)$ per unit length can be written

$$M(\kappa, \delta) = \frac{1}{\mu}\begin{bmatrix} \mu\cos\mu + j\delta\sin\mu & -j\kappa\sin\mu \\ -j\kappa\sin\mu & \mu\cos\mu - j\delta\sin\mu \end{bmatrix} \tag{5}$$

where $\mu = \sqrt{(\kappa^2 + d^2)}$. For a given interaction length l we substitute $\delta \rightarrow \delta l, \kappa \rightarrow \kappa l$ and $\mu \rightarrow \mu l$ in (5) in order to make it dimensionless. The output polarization Jones vector is found by matrix multiplication: $x_{\text{out}} = Mx_{\text{in}}$. As an example, if the input vector is TE aligned, denoted as (1, 0), the output polarization $(\boldsymbol{x}', \boldsymbol{y}')$, represents bar and cross state powers of $|x'|^2$ and $|y'|^2$, respectively. A uniformly weighted AOTF, subjected to a TE input state of polarization (SOP), has the output cross state (TM) transmission function intensity as a function of position l along the device as

$$P_x(l) = \frac{\kappa^2}{\kappa^2 + \delta^2}\sin^2(l\sqrt{\kappa^2 + \delta^2}). \tag{6}$$

After a device interaction length L, the polarization is perfectly flipped on resonance ($\delta = 0$), and the filter therefore at 100% transmission, when $\kappa L = \pi/2$. The above equation allows us to plot the classical AOTF transmission function in the cross (dashed line) and bar state (solid line) in Fig. 3.

Treatment of devices for which the acoustic amplitude and detuning vary along the device, intentionally or not, requires breaking up the interaction region into a number of subsections of lengths $L_i$which is short enough so that the local κ_i and δ_i can be considered uniform. Then the problem reduces to matrix multiplication

$$\begin{pmatrix} x \\ y \end{pmatrix}_{\text{out}} = \left\{\prod_{i=1}^{N} M(\kappa_i L_i, \delta_i L_i)\right\}\begin{pmatrix} x \\ y \end{pmatrix}_{\text{in}} \tag{7}$$

from which bar and cross state powers can be readily calculated. In the following two sections, control of the optical transmission function, referred to as passband engineering, is described.

B. Apodization

The intentional suppression of sidelobe levels by contouring the acousto-optic interaction along the device interaction length is called apodization, in analogy with a similar technique employed in lens optics, the suppression of diffraction rings by tapering the optical transmission near the lens edges [23]. Sidelobe suppression of the AOTF on either the long or short wavelength side of resonance, often a result of fabrication nonuniformities, has been explained by means of birefringence tapering [24], [25]. Fig. 10 shows two cases of sinusoidal variation in birefringence, specifically, $\Delta n - \Delta n_0 \pm \alpha\sin(\pi z/L)$ where $\alpha = 6 \times 10^{-9}$ has been modeled. Systematic birefringence variations can be obtained in practice by designing-in waveguide width variations which alter the waveguide mode indices along the device, or by nonuniform heating. Indeed, a pair of AOTF's in series with alternately long- and short-wavelength sidelobe suppression was fabricated to yield a two-stage filter with a sidelobe-suppressed composite passband [26]. Unlike birefringence apodization, coupling strength amplitude (κ) tapering can suppress sidelobes *on both sides* of resonance, as is required for high-extinction filtering and wavelength-channel routing. Sinusoidal coupling-strength apodization is simply and elegantly obtained by imbedding the active optical waveguide in the passive arm of a SAW coupler [Fig. 2(b)]. A uniformly weighted SAW directional coupler has a coupling amplitude $\kappa = \pi\kappa_o/2\sin(\pi z/L)$. Fig 11 shows the calculated 10-dB sidelobe reduction achieved by sinusoidal tapering of acoustic amplitude relative to the case of flat coupling amplitude κ_o. The interaction length L was 20 mm and the center wavelength was 1550 nm in this calculation. Since the effective interaction length is shortened, one observes a small degree of filter broadening. All current AO switch designs use SAW-coupler-based apodization with or without additional tapering which allows one to suppress the sidelobe skirt even further. More optimal tapers include the Gaussian taper with

$$\kappa = \kappa_o e^{-(z^2/2\sigma^2)}. \tag{8}$$

Herrmann *et al.* [18] have shown deep sidelobe suppression in a zero-gap SAW coupler with a weighting similar to that of (8) where modulation of coupling strength was obtained by tapering the SAW coupler gap. The SAW coupling between SAW guides depends exponentially on gap and can be used to create a systematic variation of acoustic coupling along the length of the device. Defining a coupling length L_c for complete power transfer between two SAW waveguides, experiments on a series of different-gap couplers have provided a fit to the function

$$L_c(g) = L_0 e^{\beta g} \tag{9}$$

where $L_o = 6.6$ mm, $\beta = 0.021\ \text{mm}^{-1}$ and the gap g is in microns, measured from waveguide edge to waveguide edge [27]. Other authors have obtained similar fitting parameters [28]. In all of these designs, the full switching requirement on resonance is that

$$\int_0^L \kappa(z)\, dz = \frac{\pi}{2}. \tag{10}$$

Other passband-altering consequences can be derived from a simple analysis of polarization evolution along nonuniform acoustic interaction strength devices. For example, a simple acoustic decay along the length of the device broadens the filter and softens the filter nulls significantly. Typical AOTF's have wider passbands than one would expect from a simple model (6) if one neglects the inevitable decaying acoustic field strength [29].

C. Passband Flattening

The relationship between a given optical transmission profile and the acousto-optic interaction profile is reminiscent of the Fourier transform (which applies for $\kappa L \ll \pi/2$, but the general transform is more complex and has been treated in detail by Song [30]. Fig. 12 describes the improvement in passband characteristics derived from successive improvements in acoustic amplitude control toward the general goal of obtaining a rectangular transmission profile. The uniform acoustic interaction strength yields the sinc-squared transmission of the classical AOTF. SAW-coupler apodization suppresses the sidelobes considerably but maintains the severe

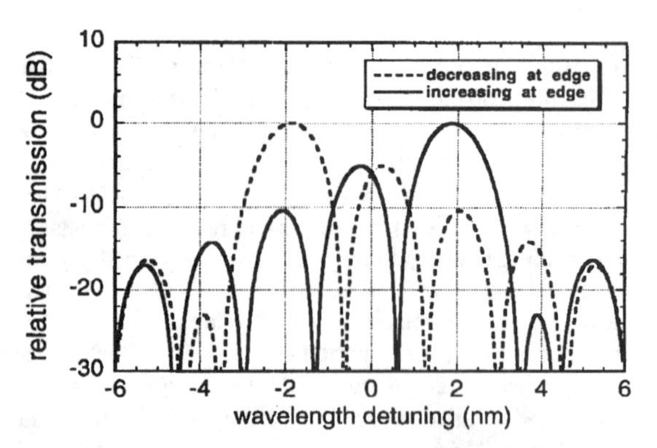

Fig. 10. Transmission spectra of an AOTF with a sinusoidally varying birefringence which produces sidelobe suppression on the short-wavelength side if the birefringence peaks at the device center (dashed line) and suppression of the long-wavelength side with minimum birefringence at device center (solid line).

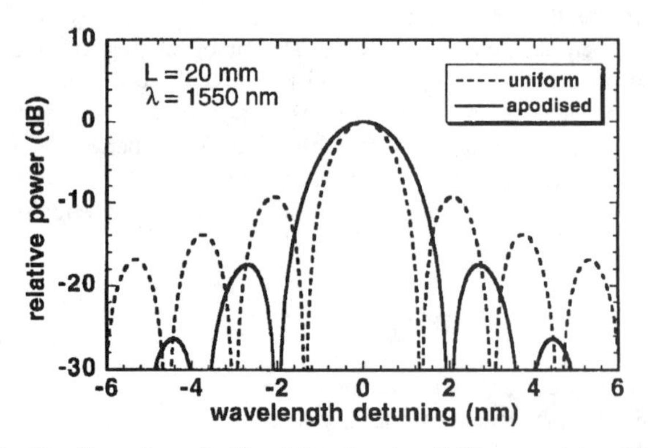

Fig. 11. Comparison of uniform interaction strength filter cross-state spectra (dashed line) with that of a sinusoidally weighted AOTF (solid line) showing significant suppression of the sidelobe skirt.

channel alignment requirement. In the third row, a fourth-order Butterworth coupling coefficient along the device length is shown to yield a rectangular profile [30] and experiments to date as well as subsequent designs are intended to achieve the combined flattened-filter coupling strength characteristics of tapered-onset, oscillating amplitude and decaying envelope. Recent tapered SAW coupler designs [31] strive to achieve the combined characteristics of the Butterworth potential, namely, a tapered onset and a sinusoidally varying acoustic coupling with an overall decaying envelope. Using the exponential dependence of κ on SAW coupler gap, all of these features can be achieved at once. Fig. 13 presents a design, which we call a dual-linear-tapered SAW coupler, which uses minimal crystal real estate to achieve high rectangularity and shows the predicted SAW coupling amplitude as a function of position along the device. By using linear-tapered SAW source waveguides leading into a zero-gap SAW coupler, the desired gradual onset of κ is achieved without the excessive length of the erfc-shaped tapers proposed in [30]. The flattened filter is symmetric about the center so that it can be driven from either direction as described in Section IV-B. The exponentially decaying envelope requires distributed absorbers to flank or overlay the acoustic waveguide. Either acoustically absorbing

TABLE II
PREDICTED RECTANGULARITY MEASURE OF THE THREE FILTER DESIGNS OF FIG. 12, COMPARED TO THE IDEAL RECTANGULAR FILTER

	r (-10 dbB)	r (-15 dB)	r (-20 dB)
Uniform	0.11	0.05	0.02
Apodized	0.25	0.13	0.05
Flattened	0.58	0.46	0.24
Ideal	1.00	1.00	1.00

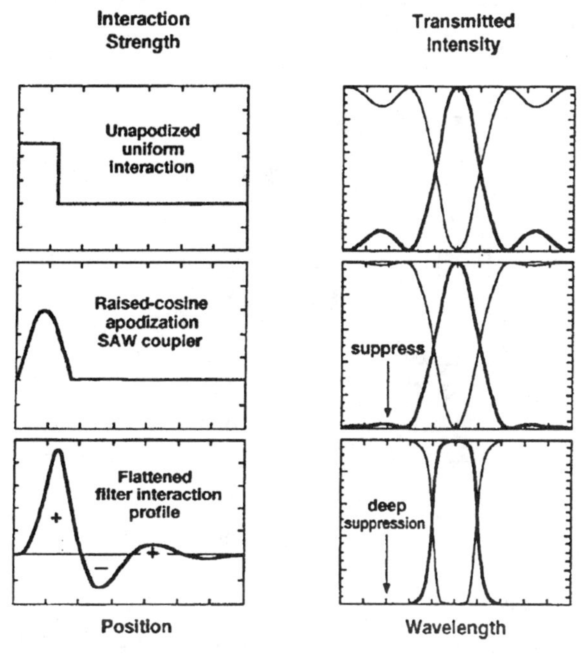

Fig. 12. The progression of passband-engineered devices from the uniform interaction strength (top rows) to the apodized filter (center rows) to the passband-flattened interaction (bottom rows) from the point of view of AO interaction amplitude profile (left columns) and predicted passbands (right columns).

media or electrically lossy attenuators are applicable for the piezoelectric traveling wave in this orientation of $LiNbO_3$. Table II compares the calculated rectangularity (2) at various crosstalk levels for the three cases of progressively more sophisticated passband-engineered AOTF designs: the uniform AOTF, the sinusoidal-apodized AOTF and the above tapered-onset zero-gap coupler with a 60 μ gap tapering down to zero gap over 10 mm and a 6 dB/cm acoustic power absorber [31]. The predicted transmission function for this dual-linear-taper design is plotted in Fig. 14.

In order to demonstrate passband flattening, a preliminary experiment was performed by Jackel *et al.* [32] who predicted that two cycles of a SAW coupler, with the second cycle present but strongly attenuated, would be an easily implemented first approximation to a decaying, oscillating Butterworth amplitude profile. Two cycles of an 17 mm-period apodized SAW coupler were used, and an acoustic absorber was placed at the first acoustic null to reduce the SAW *amplitude* to 10–15% of its value in the previous section (corresponding to a few percent acoustic power transmission), as depicted in Fig. 15. The resulting flattened transmission

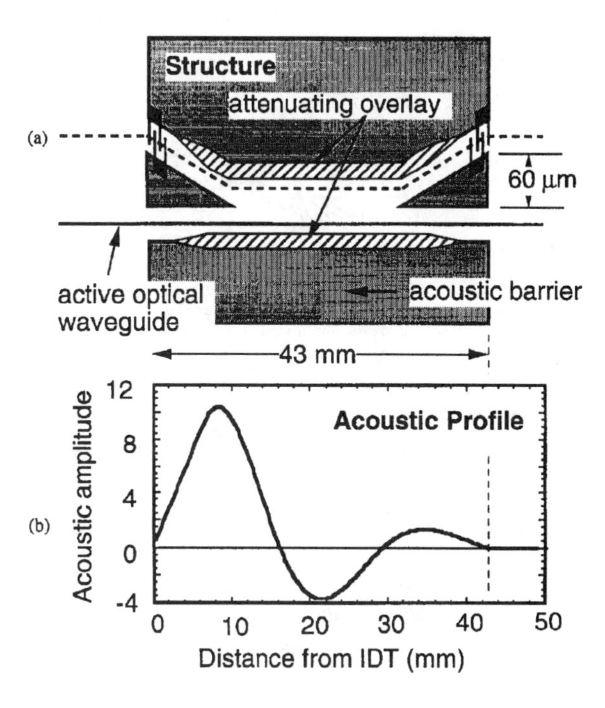

Fig. 13. (a) Dual linear tapered zero-gap coupler with an attenuating overlay designed for passband flattening; (b) Calculated SAW amplitude profiles for right- and left-going SAW beams.

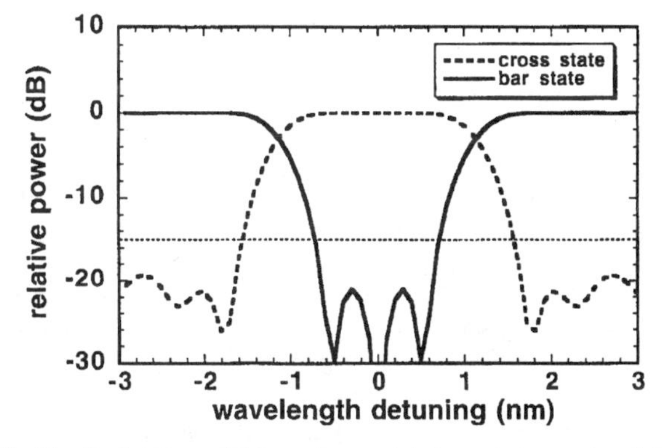

Fig. 14. Predicted very flat bar and cross state transmission spectra of the dual linear-tapered filter of Fig. 13.

peak is easily deduced from the model described above and agrees well with experimental data.

IV. ADVANCED TOPICS IN AO SWITCHES ON WDM APPLICATIONS

A. WDM Channel Intermodulation

Because many gratings can coexist in a single collinear interaction region, the AOTF can excite an arbitrary manifold of passbands. The AO filter has been particularly promising in wavelength switching and spectroscopic applications because of this parallel processing capability. These putatively independent passbands interact in a number of ways, including

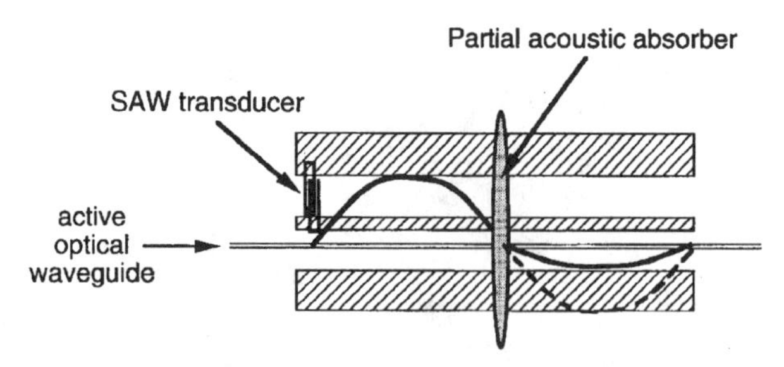

Fig. 15. Double-length SAW coupler [32] with slight leakage across a partial absorber in order to approximate the decaying oscillating amplitude profile required [30] for passband flattening. The dotted line shows SAW amplitude without the partial absorber.

mutual attraction and time-dependence at multiples of the source acoustic beat frequencies. The interchannel coupling arises because the composite acoustic grating is a traveling beat pattern with a characteristic spatial period equal to the sound beat period: $T = V_s/\Delta F$ where V_s is the speed of sound and ΔF is the RF drive difference frequency. Consider the simple case of two passband excitation at sound frequencies $\omega 1$ and $\omega 2$. The composite grating can be written

$$A(t) = 2A\cos(\Delta\omega t/2)\cos(\omega t) \tag{11}$$

where ω is the mean frequency and $\Delta\omega$ is the difference angular frequency. Similar relations can be written for arbitrary excitation spectra if one allows a more complex envelope function. If the resulting acoustic grating beat pattern varies over a distance long compared to the device interaction length, then an incoming photon sees a uniform grating but the degree of polarization conversion depends on when in the overall modulation envelope that photon arrives. Hence the filter is overdriven at some times in its cycle and underdriven at others [33][34]. More important is the case when the beat period is of the order of the device length, which corresponds to typical dense WDM channel switching. Then the AO interaction strength varies markedly along the device length and this pattern changes dramatically with time. The consequence is that the multi-channel passband shape varies in time with a characteristic frequency of ΔF. The detailed time dependence derives from treating the excitation, whatever the acoustic difference frequency is, as a single-period grating at the mean frequency and with amplitude modulated in space and time along the device. Then one can directly use the mathematical formalism for apodized filters in order to study the filter behavior, obtain expected optical beat spectra and deduce the resulting bit error rate degradation. This formalism applies even for a pair of distantly spaced gratings for which the modulation envelope is repeated many times along the device so the time-dependent effects are minimal. A number of authors have used a coupled-mode method to calculate bit-error-rate degradation due to intermodulation [35], [36]. Fig. 16 shows the calculated passband shapes with the dark regions identifying the range of modulation for both a uniform-interaction-strength AOTF and a sinusoidally apodized AOTF for two 4 nm spaced drive frequencies. Note that the suppressed sidelobes help reduce the intermodulation significantly.

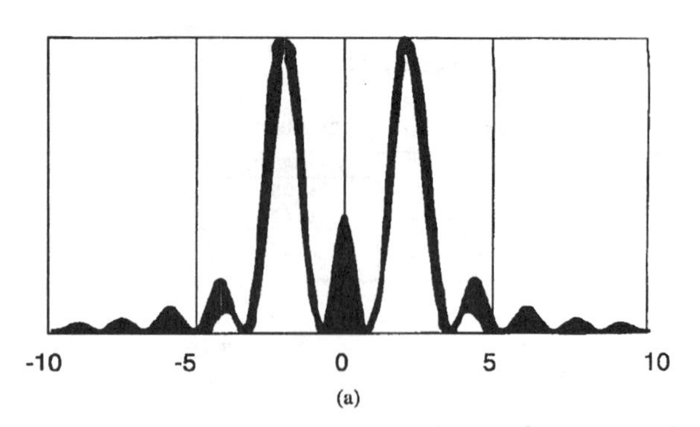

(a)

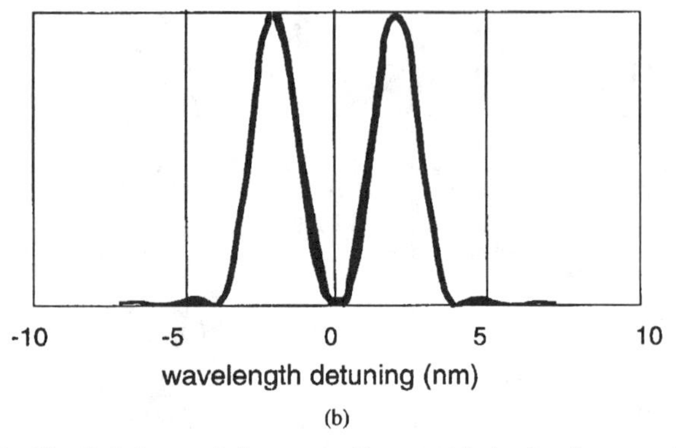

(b)

Fig. 16. Optical transmission spectra (linear scale) showing the range of intermodulation for (a) a uniformly weighted SAW interaction strength and (b) compared to that of a sinusoidally apodized filter.

B. Maximum Dense Packing of Channels

Intermodulation is only low for channel spacings of several filter widths, suggesting a hard limit to ultra-dense packing of WDM channels in AO switched systems. The technique of wavelength dilation [37] provides a means to keep coupling low even at high densities, but the price paid is subdivision of the WDM spectrum into interleaved subsets of wider-spaced wavelength channels and dilation of the switch architecture into subswitches for these wider-spaced WDM subsets. The first effective treatment of wavelength dilation of the AOTF was performed in a fully integrated filter by Jackel *et al.* [38] by dividing a long filter into two weakly coupled halves (which achieved some rudimentary passband flattening as a byproduct—see Section III-C). In this 4 nm-spaced WDM demonstration, the effective channel spacing was actually 8 nm because alternating channels were driven from opposite ends of the device. The combined effects of counterpropagating-drive dilation and partial-absorber flattening resulted in a multichannel operation bit error rate penalty of under 1 dB, even when all four channels were active. Fig. 17 shows the device operation and notch spectra for two 4 nm-spaced channels taken before and after counter-propagating wavelength dilation was employed. In this design, the partial acoustic transmission between device halves makes the wavelength dilation incomplete, though still effective. Also, unlike strict wavelength dilation in which the alternating optical channels are directed to separate AO switches, in the present design-wavelength channels follow the same path through the device,

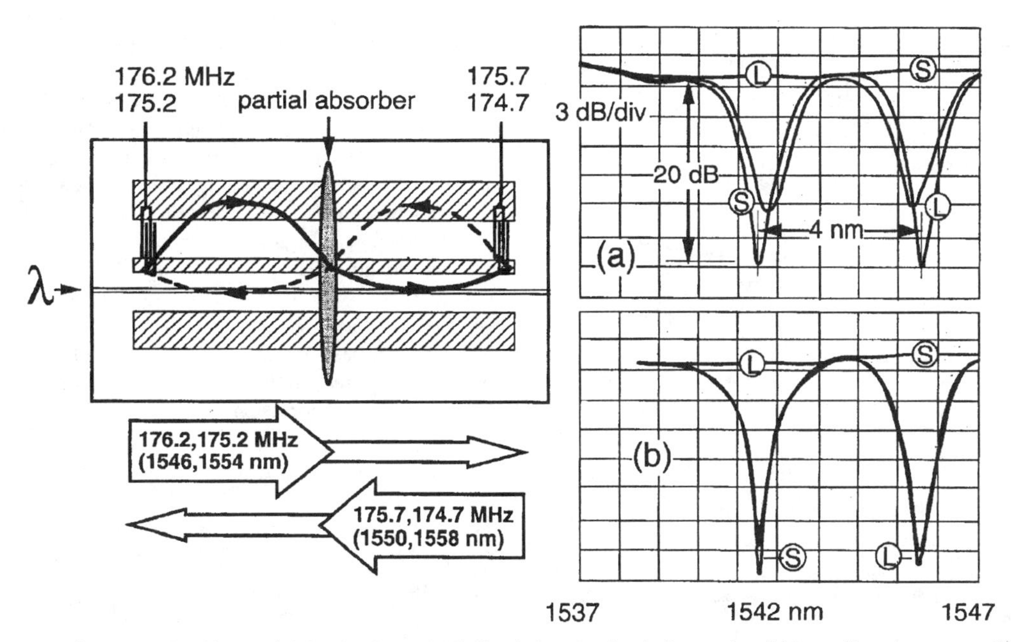

Fig. 17. (Left) Counter-propagating drive, partial absorber flattened AO filter design, showing the frequencies of drive and how the corresponding wavelength channels are addressed. (Right) Bar-state notches for the long (L) and short (S) wavelength channels at 4 nm spacing without counterpropagating drives (trace a) and with them (trace b). Solid and dashed lines depict acoustic amplitude in the co- and counterpropagating directions, respectively.

being alternately resonant in the first and second halves. Therefore, while interactions due to the simultaneous presence of nearest-neighbor acoustic frequencies is eliminated, light of a given wavelength is still affected by the sidelobes of its nearest neighbor.

C. Technology Trends and Issues

Various improvements in technology will continue the drive toward practical WDM crossconnects based on AO filters. One area of development needed for lower cost and component reliability and stability is improvement in integrated subcomponent technology, particularly small footprint high extinction polarization splitters, efficient RF transducers and compact acoustic waveguides for low drive power. Monolithic switch integration (even of dilated switch arrays) is crucial to the penetration of this technology in larger scale demonstration systems. A recent demonstration of a sidelobe-suppressed fully integrated polarization-diversity switch using all-titanium integrated optic and acoustic technology represents the state-of-the-art [19]. Further improvements in passband engineering are expected as optimum rectangularity designs are developed and the acoustic waveguide technology is developed to implement them. A better understanding of the relationship between passband shape and the magnitude of intermodulation at close channel spacings will dictate the range of applicability of AOTF's in ultradense WDM systems. The switching speed of the AOTF is too slow for packet-by-packet routing, but electro-optic switching on a nanosecond time scale has been recently demonstrated over a 10-20 nm subsection of the AOTF tuning range [39]. A traveling-wave electrode provides a birefringence change at electro-optic rates, thus swtching the passband center, but the required drive voltages are prohibitive for large duty-cycle switching applications.

For the 1550 nm communication band AO switches with gain have been demonstrated. It remains to be seen whether these devices are preferable to systems where gain is provided in optical amplifiers specifically devoted for loss compensation, but progress in erbium-doped, 1480 nm-pumped lithium niobate structures is proceeding very rapidly [40]. A typical Er-doped AOTF is formed by pre-doping the substrate with about 10 nm of Erbium metal and diffusing for about 100 hrs at temperatures extremely close to the Curie point. A typical diffusion condition is 100 hour at 1035 °C. The surface, after adequate diffusion time, is free of Er micro-crystallites which indicate residual surface metal. After doping the substrate, standard titanium metallization and other fabrication steps for the AO filter can proceed.

D. AOTF Gain Equalization in Optically Amplified Networks

Applications of the acousto-optic filter in WDM systems extends beyond the role of a digital multichannel wavelength switch. Another area where AOTF's are being used in WDM systems is as analog spectral filters to provide the required wavelength-dependent loss profile to compensate for accumulated system loss and spectral gain nonuniformity, whether due to optical amplifier gain "tilt", spectral dependence of component loss or even source channel intensity differences [41]. In large amplified networks with potentially hundreds of dB of average loss between users, the anticipated gain equalization problem is a growing concern. Fig. 18 shows the design of a four-channel polarization-independent AOTF-based gain-equalizing node following an EDFA or EDFA

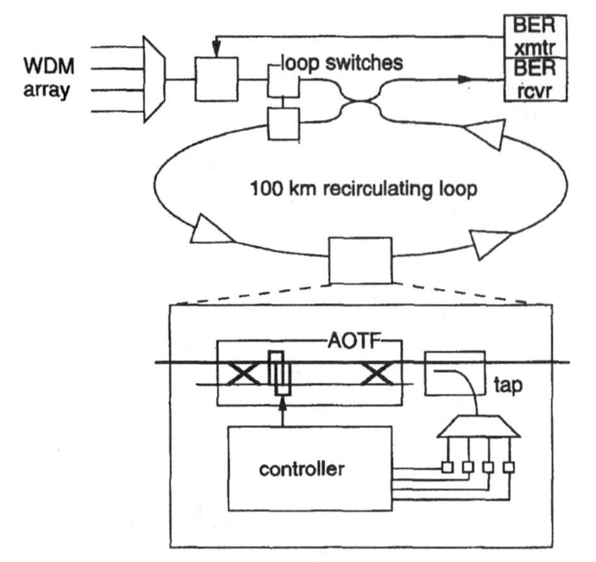

Fig. 18. Recirculating loop experiment to simulate very long amplifier chains in which a four-channel AOTF-based gain equalizer (inset) is placed to maintain uniform S/N over great distances.

chain. The filter output is sampled and the power in each spectral channel is isolated by a demux/detector array and compared to the desired levels. A control circuit adjusts the AOTF drive power in each channel so as to maintain the output levels at a predetermined value, thus actively compensating for whatever mechanisms that may unbalance the optical spectrum [42]. The imposed excess loss in a given channel is determined by the channel transmission

$$T = 1 - \frac{P_x}{P_0} = \cos^2\left(\frac{\pi}{2}\sqrt{\frac{P_{\mathrm{RF}}}{P_{100}}}\right) \quad (12)$$

where P_{RF} is the applied RF drive power in the chosen channel while P_{100} is the drive power required for 100% polarization conversion. The capability of such a periodic servo-control scheme to sustain a uniform signal-to-noise ratio through a long chain of amplifiers is under study in a recirculating loop experiment which contains a number of amplifiers followed by an AOTF gain equalizer [43].

V. Conclusion

A great deal of progress has been made in the development of integrated acousto-optic filters for the demanding application of WDM switching. Originally the AOTF was described as an ideal candidate for WDM switching because it was capable of simultaneous and independent multiple-channel switching, but evolving WDM demonstrations have tended toward systems with fixed wavelength spacings with channel separations which are multiples of 100 GHz (0.8 nm in the 1.5 μm communication band), making arbitrary spacing of questionable merit. Further, the claim of independent channel operation has been shown to be only an approximation since intermodulation forces one to choose channel spacings of two to four times the filter width. Since long substrates are needed to achieve the passband engineering benefits of passband flattening AOTF's seem constrained to channel separations of about 4 nm. Recent 2 nm separated acousto-optic switch array crossconnects have demonstrated −25 dB crosstalk in wavelengh-space dilated configurations [44], but the idea of single-device switches becomes replaced with the reality of a single 2 × 2 AOTF functioning more like a single-transistor filter which must be used in multitransistor switch structures in order to obtain high-fidelity operation. The use of space and wavelength dilation does make AOTF's attractive as moderate-sized WDM crossconnects down to 1.6 nm spacing and possibly closer in future work. A lot effort must go into the study of passband engineering as a means to achieve the somewhat competing demands of passband flattening and intermodulation suppression.

References

[1] C. A. Brackett *et al.*, "A scalable multiwavelength multihop optical network: A proposal for research on all-optical networks," *J. Lightwave Technol.*, vol. 11, pp. 736–753, 1993.
[2] C. Dragone, C. A. Edwards and R. C. Kistler, "Integrated $N \times N$ multiplexer on silicon," *IEEE Photon. Technol. Lett.*, vol. 3, pp. 896–899, 1991.
[3] P. J. Chidgey, "Multi-wavelength transport networks," *IEEE Commun. Mag.*, pp. 28–35, Dec. 1994.
[4] E. L. Goldstein, L. Eskildsen, and A. F. Elrefaie, "Performance implications of component crosstalk in transparent lightwave networks," *IEEE Photon. Technol. Lett.*, vol. 6, pp. 657–660 (1994).
[5] A. Kar-Roy and C. S. Tsai, "Low sidelobe weighted-coupling integrated acoutso-topic tunable filter using focused surface acoustic waves," *IEEE Photon. Technol. Lett.*, vol. 4, pp. 1132–1135, 1992.
[6] ——, "Integrated acousto-optic tunable filters using weighted coupling," *IEEE J. Quantum Electron.*, vol. 30, pp. 1574–1586, July 1994.
[7] Y. Yamamoto *et al.*, "Suppression of sidelobe levels for guided-wave acousto-otpic tunable filters using weighted coupling," *IEEE Trans. Ultrason., Ferroelect. Freq. Contr.*, vol. 40, pp. 814–818, 1993.
[8] H. Herrmann and St. Schmid, "Integrated acousto-optical mode converters with weighted couplng using surface acoustic wave directiona couplers," *Electron. Lett.* vol. 28, pp. 979–980, 1992.
[9] D. A. Smith and J. J. Johnson, "Sideobe suppression in an acousto-optic tunable filter with raised-cosine interaction strength," *Appl. Phys. Lett.*, vol. 61, pp. 1025–1027, 1992.
[10] A. Kar-Roy and C. S. Tsai, "Ultralow sidelobe-level integrated acoutso-optic tunable filters using tapered-gap surface acoustic wave directional couplers," *J. Lightwave Technol.* vol. 12, pp. 977–982, June 1994.
[11] B. L. Heffner, D. A. Smith, J. E. Baran, and K. W. Cheung, "Integrated-optic acoustically-tunable infrared optical filter," *Electron. Lett.* vol. 24, pp. 1562–1563, 1988.
[12] J. Frangen, H. Herrmann, R. Ricken, H. Siebert, W. Sohler, and E. Strake, "Integrated optical acoustically tunable wavelength filter," *Electron. Lett.*, vol. 25, pp. 1583–1584, 1989.
[13] D. A. Smith and J. J. Johnson, "Low drive-power integrated acousto-optic fiolter on X-cut Y-propagating $LiNbO_3$," *IEEE Photon. Technol. Lett.* vol 3, pp. 923–925, 1992.
[14] D. A. Smith, J. E. Baran, K. W. Cheung and J. J. Johnson, "Polarization-independent acoustically-tunable optical filter," *Appl. Phys. Lett.*, vol. 56, pp. 290–292, 1990.
[15] T. Pohlmann, A. Neyer, and A Voges, "Polarization independent $Ti:LiNbO_3$ switches and fiilters," *IEEE J. Quantum Electron.*, vol. 27, pp. 602–607, 1991.
[16] J. E. Baran and D. A. Smith, "Adiabatic 2 × 2 polarization splitter on $LiNbO_3$," *IEEE Photon. Technol. Lett.*, vol. 4, pp. 39–40, 1989.
[17] A. d'Alessandro, D. A. Smith, and J. E. Baran, "Polarization-independent low power acousto-optic tunable filter/switch using APE/Ti polarization splitters on lithium niobate", *Electron. Lett.*, vol. 20, pp. 1767–1769, 1993.
[18] H. Herrmann, K. Schafer, and W. Sohler, "Polarization independent, integrated optical, acoustically tunable wavelength filters/switches with tapered acoustical directional coupler," *IEEE Photon. Technol. Lett.*, vol. 6, pp. 1335, 1994. See also, F. Wehrmann *et al.*, "Fully packaged, integrated optical, acoustically tunable add/drop multiplexers in $LiNbO_3$,"

in *Proc. 7th Europ. Conf. Integr. Opt.*, Delft, The Netherlands, 1995, pp. 487–490.

[19] D. A. Smith, A. d'Alessandro, J. E. Baran, D. J. Fritz, and R. H. Hobbs, "Reduction of crosstalk in an acousto-optic switch by means of dilation," *Opt. Lett.* vol. 19, pp. 99–101, 1994.

[20] D. A. Smith, A. d'Alessandro, J. E. Baran, D. J. Fritz, J. L. Jackel, and R. S. Chakravarthy, "Multi-wavelength performance of an apodized acousto-optic switch," *J. Lightwave Technol.*, to be published.

[21] K. Padmanabhan and A. N. Netravali, "Dilated network for photonic switching," *IEEE Trans. Commun.*, vol. COM-35, pp. 1357–1365, 1987.

[22] P. Winkler, "Control of acousto-optical tunable filters in a 4 × 4 Benes network," Bell Commun. Res. Tech. Memo. Series, May 1994.

[23] G. R. Fowles, *Introduction to Modern Optics*. New York: Holt, Rinehart and Winston, 1975, 2nd ed., pp. 138–139.

[24] D. A. Smith, A. d'Alessandro, J. E. Baran, and H. Herrmann, "Source of sidelobe asymmetry in integrated acousto-optic filters," *Appl. Phys. Lett.* vol. 62, pp. 814–816, 1993.

[25] W. R. Trutna, D. W. Dolfi, and C. A. Flory, "Anomalous sidelobes and birefringence apodization in acousto-optic tunable filters," *Opt. Lett.* vol. 18, pp. 28–30 , 1993.

[26] L. B. Aronson, G. Rankin, and W. R. Trutna, "Reduced-sidelobe integrated acousto-optic filter with birefringence apodization," *Opt. Lett.*, vol. 18, pp. 1721, 1993.

[27] D. A. Smith and J. J. Johnson, "Surface-Acoustic-Wave directional coupler for apodization of integrated acousto-optic filters," *IEEE Trans. Ultrason., Ferroelect. Freq. Contr.*, vol. 40, pp. 22–25 (1993).

[28] H. Herrmann *et al.*, "Tapered acoustical directional couplers for integrated acousto-optical mode converters with weighted coupling," *J. Lightwave Technol.*, vol. 13, pp. 364–374, Mar. 1995.

[29] A. Yariv and P. Yeh, *Optical Waves in Crystals*. New York: Wiley, 1984, pp. 387–396.

[30] G. H. Song, "Toward the ideal codirectional Bragg filter with an acousto-optic filter design," *J. Lightwave Technol.*, vol. 13, pp. 470–480, 1995.

[31] R. S. Chakravarthy, D. A. Smith, A. d'Alessandro, J. E. Baran and J. L. Jackel, "Passband Engineering of Acousto-optic tunable filters," in *Proc. 7th European Conf. Integr. Opt.*, Delft, The Netherlands, Apr. 3–6, 1995.

[32] J. L. Jackel, J. E. Baran, A. d'Alessandro, and D. A. Smith, "A passband-flattened acousto-optic filter," *Photon. Technol. Lett.*, vol. 7, pp. 318–320, 1995.

[33] D. A. Smith, R. S. Chakravarthy, L. Troilo, and A. d'Alessandro, "Passband Collisions and multi-channel crosstalk in acousto-optic filters and switches, " in *Proc. 7th Europ. Conf. Integr. Opt.*, Delft, The Netherlands, Apr. 3–6, 1995.

[34] J. L. Jackel, J. E. Baran, D.A. Smith. R. S. Chakravarthy, and D. J. Fritz, "Observation of modulated crosstalk in multi-channel acousto-optic switches," in *Proc. Integr. Photon. Res. Conf.*, postdeadline paper PD5, Dana Pt., CA, Feb. 23–25, 1995.

[35] F. Tian and H. Herrmann, "Interchannel interference in multiwavelength operation of integrated acousto-optical filters and switches," *J. Lightwave Technol.*, vol. 13, pp. 1146–1154, June 1995.

[36] M. Fukutoku, K. Oda, and H. Toba, "Optical beat-indeuced crosstalk of an acousto-optic tunable filter for WDM network application," to be published.

[37] J. Sharony, K. W. Cheung, and T. E. Stern, "Wavelength dilated switches (WDS)—A new class of suppressed crosstalk dynamic wavelength routing crossconnects," presented at OFC'92, San Jose, CA, 1992.

[38] J. L. Jackel, J. E. Baran, G. K. Chang, M. Z. Iqbal, G. H. Song, W. J. Tomlinson, and R. Ade, "Multi-channel operation of AOTF switches: Reducing channel-to-channel interaction," *Photon. Technol. Lett.*, vol. 7, pp. 370–372, 1995.

[39] L. B. Aronson, "Electro-optic tuning and sidelobe control in acousto-optic tunable filters," *Opt. Lett.*, vol. 20, pp. 46-48, 1995.

[40] R. Brinkmann, M. Dinaud, I. Baumann, Ch. Harizi, W. Sohler, and H. Suche, "Acoustically-tunable wavelength filter with gain," *IEEE Photon. Technol. Lett.*, vol. 6, pp. 519–521, 1994.

[41] S. F. Su, R. Olshansky, G. Joyce, D. A. Smith, and J. E. Baran, "Gain equalization in multi-wavelength lightwave systems using acousto-optic tunable filters," *IEEE Photon. Technol. Lett.*, vol. 4, pp. 269–271, 1992.

[42] S.-M. Hwang, A. E. Willner, Z. Bao, R. S. Chakravarthy, and D. A. Smith, "Active equalization of nonuniform EDFA gain by using multiple AOTF passbands for megameter WDM transmission," submitted for publication.

[43] S. M. Huang and A. E. Willner, "Active equalization of nonuniform EDFA gain by using multiple AOTF passbands for megameter WDM," in *Proc. Conf. Lasers Electro-Opt.*, paper CTUS3, Baltimore, MD, May 21–26, 1995.

[44] J. L. Jackel, M. Goodman, J. Gamelin, W. J. Tomlinson, J. Baran, G. H. Song, C. A. Brackett, D. J. Fritz, R, Hobbs, K. Kissa, R. Ade, and D. A. Smith, "Demonstration of simultaneous and independent switching of eight wavelength channels with 2 nm spacing using a wavelength-dilated acousto-optic switch," in *LEOS'95*, postdeadline paper, San Francisco, CA, Oct. 30–Nov. 2, 1995.

The Y-Laser: A Multifunctional Device for Optical Communication Systems and Switching Networks

Olaf Hildebrand, Michael Schilling, *Member, IEEE,* Dieter Baums, Wilfried Idler, Kaspar Dütting, Gert Laube, and Klaus Wünstel

INVITED PAPER

Abstract—**This paper gives a comprehensive review of the recent progress obtained with various operation modes of a *Y*-shaped all active waveguide device with three optical input / output ports, the so called Y-laser. Used as a monolithic semiconductor light source, single-mode emission with an extremely large cw tuning range of more than 50 nm is accessed. Single current step tuning spans up to 8-nm range. Several-GHz bandwidth AM and FM response is achieved. Operation as a wavelength processing device is demonstrated with data-stream transformation from short wavelength fiber windows (around 780 nm or 1310 nm) to the long wavelength window (1550 nm), as well as with ultra wide range wavelength conversion within the 1550-nm region, across +/− 20 nm (blue and red shift). Optical high-speed space switching / gating functions are verified up to 1 Gb / s, and packet switching of 5-Gb / s data streams is demonstrated. Further features as a pulse reshaping device, as an electrically triggerable wavelength bistable device and as a mode locked pulse generator are addressed. Possible applications of this multifunctional device in future optical communication systems and switching networks are discussed.**

I. Introduction

FIBER-BASED optical communication systems are evolving from simple point-to-point links to versatile system architectures, including subscriber access network structures, and rapid progress is being made towards application of optics not only in transmission but also in switching systems. An optically transparent trans-switching network might become feasible in the future. The evolution of InP-based optoelectronic components is governed by the above system trends, and the number of required components modifications and diversifications is apparently ever increasing [1].

Basically, on the way towards future communication system architectures, we need optoelectronic (OE) components that are able to process optical information in the time, space, and frequency (or wavelength) domains [2]–[4]. Favorably, these new components should be derived from well-known OE device technologies, and they should be easily manufacturable to achieve the cost targets. This demand also means that we have to find new concepts in order to realize the needed components functions with a minimum number of component diversifications.

A high-speed, wavelength-tunable light source is an indispensable key component for future trans-switching networks. The key figure of merit is the maximum accessible wavelength range, or the number of addressable wavelength channels at fixed channel separation. In fact, our first aim in developing the Y-laser was to overcome the limitations of conventional linear multisegment tunable DFB/DBR lasers, which allow for a wavelength tuning range of about 10 nm only [5]. Instead of DFB/DBR Bragg gratings, the Y-laser uses the interferometric principle for wavelength filtering and therefore provides broadband tuning across the whole gain spectrum of the semiconductor material.

A second, required key component is a wavelength converter that changes the wavelength of a high-speed optical data stream without any electronic processing. Again, the accessible wavelength range is of great importance, and red *and* blue shift function of the same component is needed for cascadibility. The Y-laser is excellent for these purposes, and, compared to other approaches [4], it provides very wide range conversion at high speed.

In addition to the above wavelength processing functions we need high-speed space switches to control the route of optical data streams. Different principles and technologies of the switch elements have been proposed [6]. As discussed further below, the Y-'laser' can be operated as a basic and cascadable 1:2 switch element, and, since its segments can be used for optical on/off gating functions, it also allows for manipulations in time.

Apart from the above mentioned key functions, further possible operation modes are addressed below, demonstrating the multifunctionality of the Y-laser. This multifunctionality, achieved with one basic and simple device and technology concept, makes the Y-laser a very attrac-

Manuscript received April 12, 1993. This work was supported in part by the German Minister for Research and Technology within two projects: TK0440 (tunable light source) as part of an R & D project of the Heinrich-Hertz-Institute für Nachrichtentechnik GmbH, Berlin; and TK0581, as part of the current German Joint Research Program on Optical Signal Processing. This work was also partly supported by the RACE 2039 ATMOS project of the European Community.

The authors are with the Optoelectronic Components Division, Research Center, ALCATEL SEL AG, D-7000 Stuttgart 40, Germany.

IEEE Log Number 9211161.

Reprinted with permission from *IEEE Journal of Lightwave Technology*, Vol. 11, No. 12, pp. 2066-2075, December 1993.

tive candidate for application in future complex communication systems: the number of needed components specializations and diversifications can be kept low.

In the following, we first describe the device concept and the basic operation principles of the Y-laser, and we address the device technology. Then, we review the experimental results obtained with the Y-laser in its different operation modes, and we include some new conceptual ideas and suggested system applications.

II. Basic Principles of Operation

The concept of the Y-laser as a tunable, single-mode light source is based on the interferometric principle [7]–[9]: The device can be understood (in a simplified manner) as two standard Fabry–Perot (FP) lasers, each with three electrodes, put together to obtain a Y-laser: The middle segments form a Y-branch with one electrical contact, and one outer segment of each laser is now in common. Due to interference of the FP resonator modes, the emission spectrum of the Y-laser is single mode (supermode operation), and continuous tuning is achieved by synchronous shift of both Fabry–Perot mode sets, whereas step tuning is accomplished by jumping to another supermode.

With full equivalence, the Y-laser can also be regarded as (half a) Mach–Zehnder (MZ) interferometer. In this view, the longitudinal modes of the Y-shaped composite cavity are superimposed by the MZ filter function, and laser oscillation occurs at that longitudinal mode with the lowest overall loss.

A basic theoretical discussion of interferometric lasers, including MZ-type lasers, is given in [7]. For practical purposes we need to understand the wavelength-tuning behaviour of the Y-laser in dependence of geometry and drive currents. We introduce a length difference ΔL of the two branches 3 and 4 (Fig. 1), and we calculate the total effective gain profile and the effective refractive index as a function of the four injection currents. For simplicity, we use the parabolic approximation for the MQW-gain profile, and we assume constant differential gain dg/dN [10] and differential index of refraction dn/dN.

In Fig. 2 a typical loss spectrum is shown demonstrating the resulting superposition of the MZ filter spectrum with the effective gain curve. Laser oscillation will occur at a longitudinal mode with minimum loss, i.e. close to the lowest minimum in the loss spectrum. A change of either the loss spectrum or the gain curve results in wavelength tuning. The following operation modes are of special importance: 1) A shift of the MZ filter spectrum using current I_3 and/or I_4 leads to wavelength jumps (step tuning) between subsequent longitudinal modes which approximately are separated by $\Delta\lambda_{FP} = \lambda^2/n(2L - \Delta L)$ (with L being the total device length). Caused by the effective gain curve, the loss of the MZ resonance changes with wavelength. If the loss becomes too high, the next MZ mode may overcome and a sawtooth-like wavelength-current characteristic results; and 2) A spectral shift of the gain maximum using the gain dominating currents I_1 or I_2 leads to step tuning between (fixed) MZ resonances. In this case, the steps are approximately given by $\Delta\lambda_{MZ} = \lambda^2/2n\Delta L$, i.e. the steps are much larger than for case 1), and the tuning range is limited by the achievable gain curve shift.

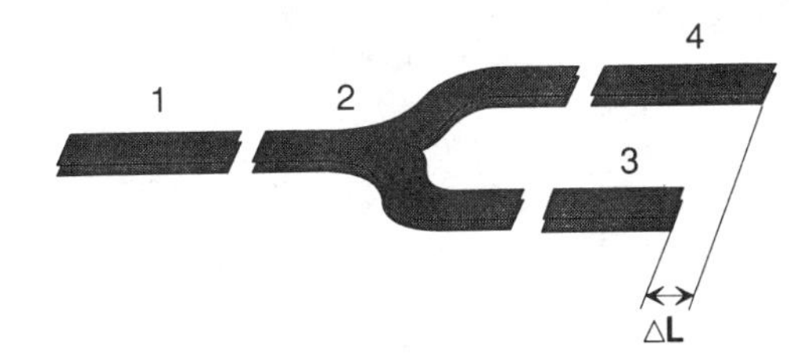

Fig. 1. Schematic drawing of the Y-laser.

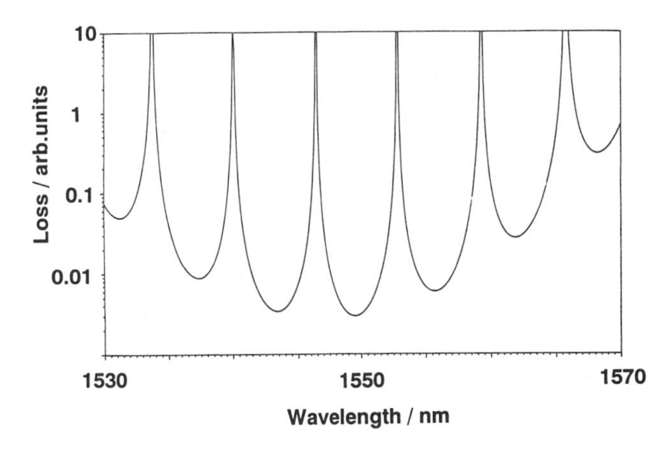

Fig. 2. Typical filter characteristic of the Y-laser. The curve shows the reflectivity of the MZ interferometer multiplied by the gain curve of the material. The closely spaced FP modes are not shown.

From the above it is obvious that the Y-laser can easily be tailored for specific tuning requirements, simply by proper geometry design. Further, step tuning using one *single control current* is easily accomplished, although the remaining currents can be favourably applied, e.g., for additional (continuous) fine tuning or for modulation, or for power control. If all four currents are used for tuning, it is possible deliberately to address any desired wavelength out of the full tuning range, limited by the gain curve only.

The correlation of the theoretically expected single current tuning behaviour and first experimentally achieved device characteristics is shown in Figs. 3 and 4 and is discussed in more detail below [11].

It should be noted that the above discussion in general holds for those operation modes of the Y-laser where active laser oscillation is involved, including, e.g., wavelength conversion or translation. Another basic principle of operation is seen if we consider the Y-'laser' as a 3-dB coupler or splitter with three optical input/output ports, similar to the loss-compensated Y-branch switch reported in [12]. Then, the four injection currents can be used to achieve transparency (or gain) in the respective sections,

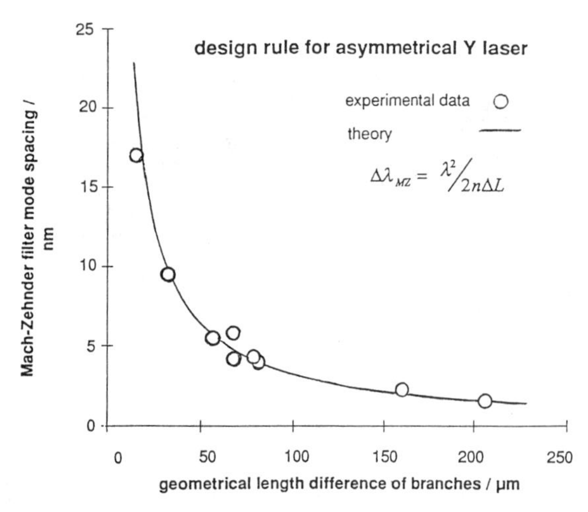

Fig. 3. Relationship between MZ mode spacing and arm length difference. Symbols refer to measurements and the hyperbola represents the formula given in Section II.

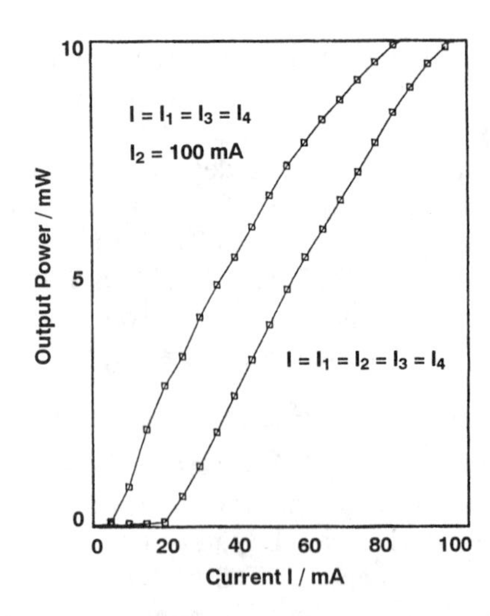

Fig. 5. Optical characteristic of a Y-laser: Power coupled out of segment number 1 versus current per segment.

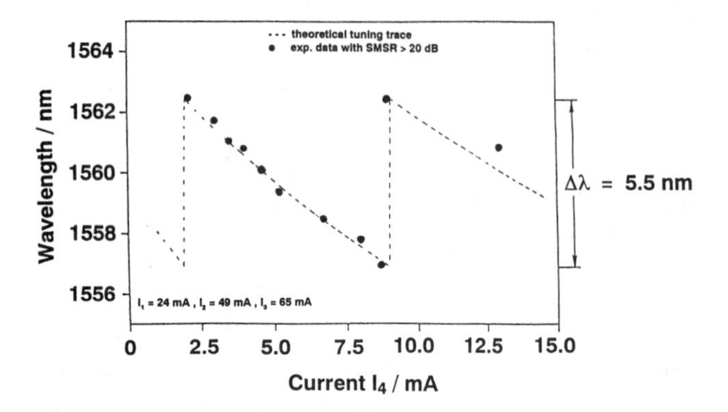

Fig. 4. Tuning of a Y-laser by variation of the current injected into one of the parallel arms. Points are experimental data with SMSR > 20 dB, the line corresponds to the theoretical prediction and serves as a guide to the eye.

and each segment can be switched on/off individually, providing a variety of space/time switch operations to be discussed below. Obviously, in this operation mode the device has to be prevented from laser oscillation so that antireflection coating of the facets is appropriate.

III. Device Technology

The device technology of the Y-laser is specifically simple since, basically, it is derived from standard Fabry–Perot technology, without any DFB or DBR technology needed. The device is built up of a Y-shaped waveguide that includes the optically active layer [13] consisting of InGaAsP bulk or multiple-quantum-well (MQW) material and InP confinement layers, thus leading to a buried heterostructure (BH) configuration. The complete epitaxial growth is done by LPMOVPE [14], [15]. The all-active Y-waveguide structure is segmented into three straight sections and the Y-branching section (see Fig. 1) by separating the contact metallization [9]. Finally, the waveguide end facets are defined by cleaving.

Future evolutions of this device will include modifications adapted to the intended system application as discussed above. Such modifications might involve, for example, optimizations of geometry, replacement of active waveguide segments by passive ones or high/low reflectivity coatings of facets.

IV. Experimental Results With the Y-Laser as a Tunable Light Source

The Y-laser does not have a conventional threshold current since four drive currents are involved. If the same currents are applied to all segments, the typical threshold current per segment is $I_{th} = 20$ mA (see Fig. 5); if a segment is prebiassed (e.g. with 100 mA, see Fig. 5), the 'threshold current' of the other segments is drastically reduced, say, down to 5 mA. Typically, 10 mW optical output power is reached at a bias of 4*50 mA. 30-dB sidemode suppression and a DC linewidth lower than 30 MHz (8 MHz best) have been achieved. 1300-nm and 1550-nm lasers have been realized with comparable data.

An ultra-wide wavelength-tuning range up to 51 nm (e.g. 1528 nm to 1579 nm, see Fig. 6) is realized by proper adjustment of the four electrical injection currents, i.e. by proper setting of the respective refractive indices of the active waveguide segments [16], [17]. Discrete tuning, combined with continuous subnanometer fine tuning, allows operation at arbitrarily defined wavelengths. For instance, 16 channels with 2-nm wavelength separation are successively addressable (Fig. 7) only by variation of the electrical drive currents [18].

We have fabricated Y-laser devices with different values of ΔL, ranging from $\Delta L = 0$ μm (symmetrical devices) to $\Delta L = 200$ μm. In Fig. 3, the experimentally determined MZ channel spacing is plotted versus ΔL, showing the expected relation. This figure represents a simple design rule for asymmetrical Y-lasers: the desired MZ channel spacing is adjusted by the length difference of the two branches of the device.

As an example, the tuning response of a device with $\Delta L = 56$ μm is shown in Fig. 4 together with the theoret-

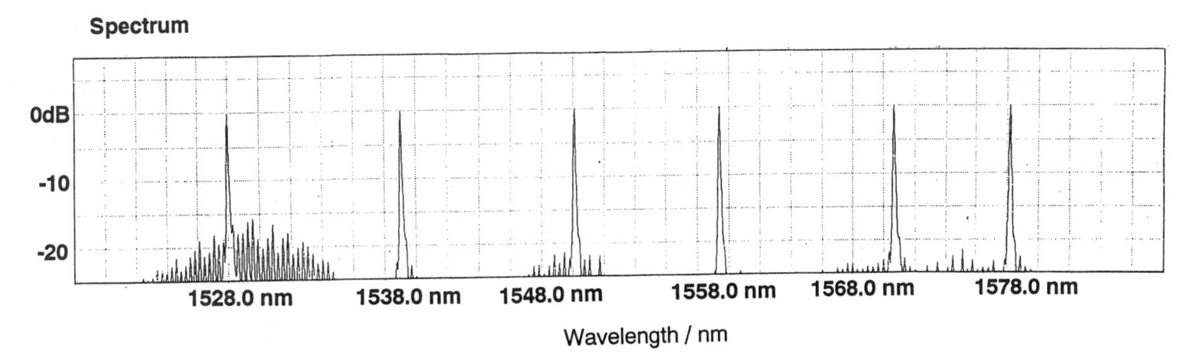

Fig. 6. 51-nm tuning span of a symmetrical Y-laser.

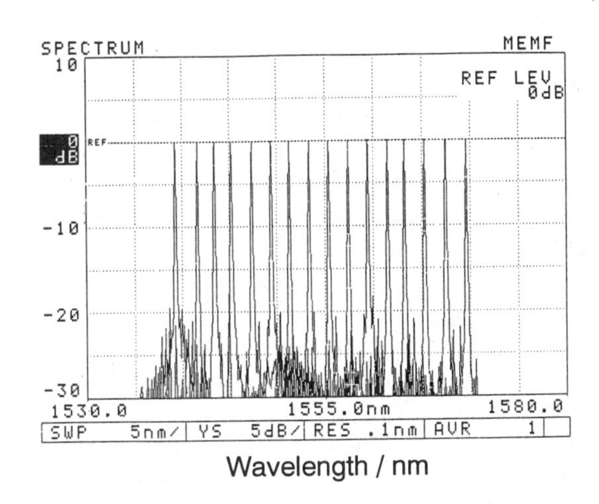

Fig. 7. 30-nm tuning covered by 16 wavelength channels that are spaced by 2 nm.

ical tuning curve. With the change of only one of the four injection currents (here I_4), the MZ filter is tuned across closely spaced FP cavity modes with a full range of 5.5 nm, which corresponds to the MZ filter width.

In an other operation mode (not shown in the figure), the gain curve is shifted by changing one single current (here I_1), while the MZ modes of the asymmetrical device are kept fixed. With this mode, we achieve step tuning between the MZ modes. As an example, for a Y-laser with $\Delta L = 81$ μm (corresponding to a MZ spacing of 4 nm), we observed a 4-nm step tuning across three MZ modes (whole range 8 nm).

The described features of the Y-laser, operated as a tunable light source, makes this device very attractive, especially for system applications where a very large wavelength range of several ten nanometers is to be addressed. For direct detection WDM systems the Y-laser is specifically suited since, for a given number of required channels, the Y-laser allows the use of rather wide channel spacings that reduce effort and cost for channel filters at the receiver side; the design of the Y-laser can be adapted to the desired channel spacing, and a single current channel selection is feasible. For coherent detection systems, the large tuning range of the Y-laser could be favorably applied for the local oscillator, e.g. for coherent multichannel distribution networks, depending on the achievable spectral linewidth improvements over the entire tuning range. Using the FM modulation capability of up to 300 MHz/mA at 2.8 GHz, the Y-laser is also suited, e.g., as an FSK transmitter laser with presettable emission wavelength.

The speed of wavelength setting is an important parameter for the above mentioned applications, but it is even more important for future optical switching systems. With a single injection current modulated, the Y-laser can be switched from one emission wavelength to another with up to 2-Gb/s speed and 9-nm wavelength separation [9], [19], [20].

V. The Y-Laser as a Wavelength Processing Device

The discussion so far was restricted to an optical output of the device as a response to *electrical* inputs. Response to *optical* input signals and the resulting wavelength processing are further key features of the Y-device, as described in the following.

A. Wavelength Transformation (Between Optical Fiber Windows)

Photon injection can be used to manipulate the optical output of Y-laser devices. If short wavelength light is injected to one facet, the free carriers generated due to absorption push the prebiassed device from below to above laser threshold (optical pumping).

First experiments on wavelength transformation are demonstrated in Fig. 8: When a short wavelength optical data stream (here 140 Mb/s at $\lambda_{in} = 1310$ nm) is injected into one facet of the Y-laser, each bit pushes the device above threshold, and the same data stream appears at the output, but is transformed to long wavelength ($\lambda_{out} = 1561$ nm in Fig. 8). An important feature is that the output wavelength λ_{out} of the device is electrically presettable within the tuning range of the Y-laser. Further, since the incoming light is absorbed, wavelength, spectrum, and polarization are not critical. Therefore, wavelength transformation from the first (~ 780 nm) or second (~ 1310 nm) optical fiber window to the third one (~ 1550 nm) is feasible, with a single-mode output spectrum, irrespective of whether input is multimode (Fig. 9).

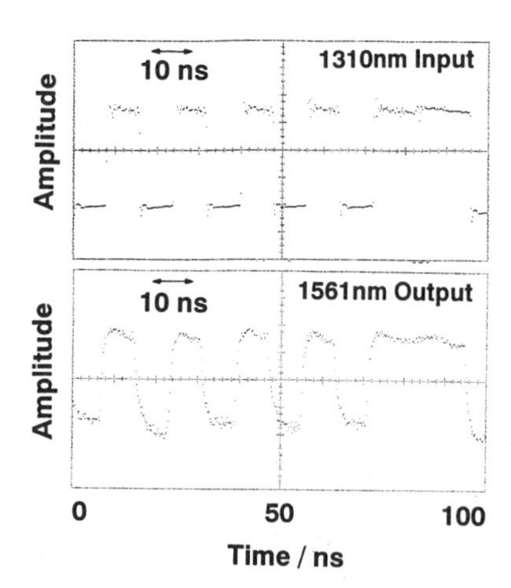

Fig. 8. Dynamic wavelength transformation from λ_{in} = 1310 nm to λ_{out} = 1561 nm. Upper trace is a 140-Mb/s data stream injected into the Y-laser, lower trace is the transformed data stream emitted by the Y-laser.

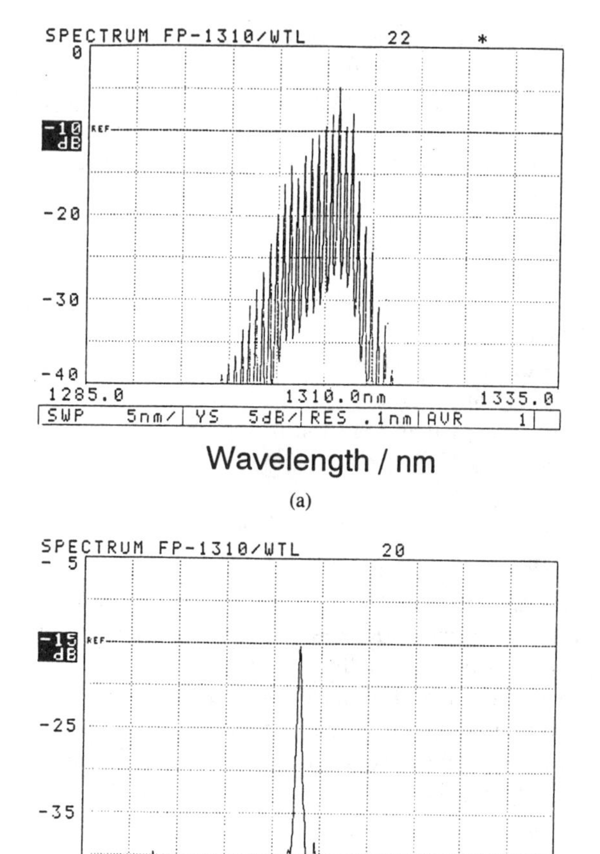

Fig. 9. Wavelength transformation: 1310-nm Fabry–Perot spectrum of the injecting laser (top), and single-mode emission spectrum of the Y-laser at 1561 nm (bottom).

Since both branches of the Y-laser can be used as optical input ports, optical logic functions such as OR or AND become feasible in addition, depending on the chosen bias of the device.

Applications of the described wavelength transformation could be, e.g., where local networks operating at short wavelength shall be connected with 1550 nm transmission systems, or where transformation of 1300 nm FP multimode to 1550 nm single-mode signal is needed.

B. Wavelength Conversion (Within Optical Fiber Window)

Other physical effects occur if the input wavelength λ_{in} is within the gain curve of the Y-laser. The device is preset to operate at the desired λ_{out}, and one (or two) of the facets are used as optical input port(s) [21], [22]. The Y-laser does not react on an optical input at λ_{in} unless the input power exceeds a certain threshold level; below this level, the Y-laser operates at the electrically preset wavelength λ_{out}; above this value, the Y-laser locks to the wavelength of the injected light so that the output wavelength equals λ_{in} (see Fig. 10).

This principle of operation does not only work for red shift ($\lambda_{out} > \lambda_{in}$), but also for blue shift ($\lambda_{out} < \lambda_{in}$), as demonstrated in Fig. 10, where a wavelength conversion by 18 nm (red shift) and 21 nm (blue shift) are shown in the static mode of operation.

The dynamic behavior of wavelength conversion is shown in Fig. 11 where a 2.5-Gb/s optical input signal from a DFB laser operating at 1525 nm is converted (and inverted) to an output data stream at a wavelength of 1570 nm, corresponding to the enormous wavelength shift of 45 nm in this example [16], [17].

C. Pulse Reshaping

The operation modes described in A. and B. involve an optical power dependent nonlinear (or thresholding) behavior so that (with appropriate filters) these effects might be used to clip off power peaks of a data stream or to clip off low power (noise) levels. These features are basic functions for pulse reshaping; if in addition the Y-laser is electrically on/off switched by a proper clock signal, pulse shaping and jitter rejection become feasible. Preliminary pulse shaping experiments are shown in Fig. 12 where an optical sine wave signal at 1300 nm is shaped into a 'more digital' signal form at 1550 nm, with low power levels cut off.

VI. Wavelength Bistability

The Y-laser is able to perform a wavelength latch operation with two possible wavelengths emitted under the same bias conditions. The actual wavelength is determined by the direction of the current changes when approaching the said bias conditions. The corresponding bistable loop is shown in Fig. 13 with 5.5-nm wavelength separation [23]. Control is through the drive current of the branching section 2 (Fig. 1), with 17-mA loop width. These values characterizing the bistable loop are stable under

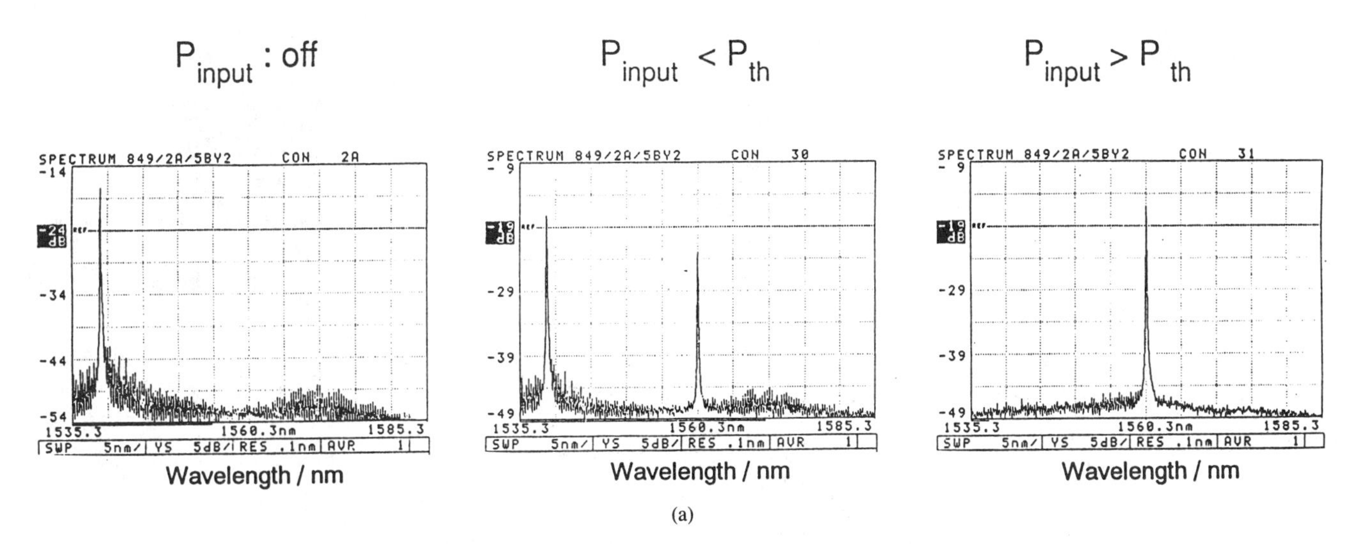

(a)

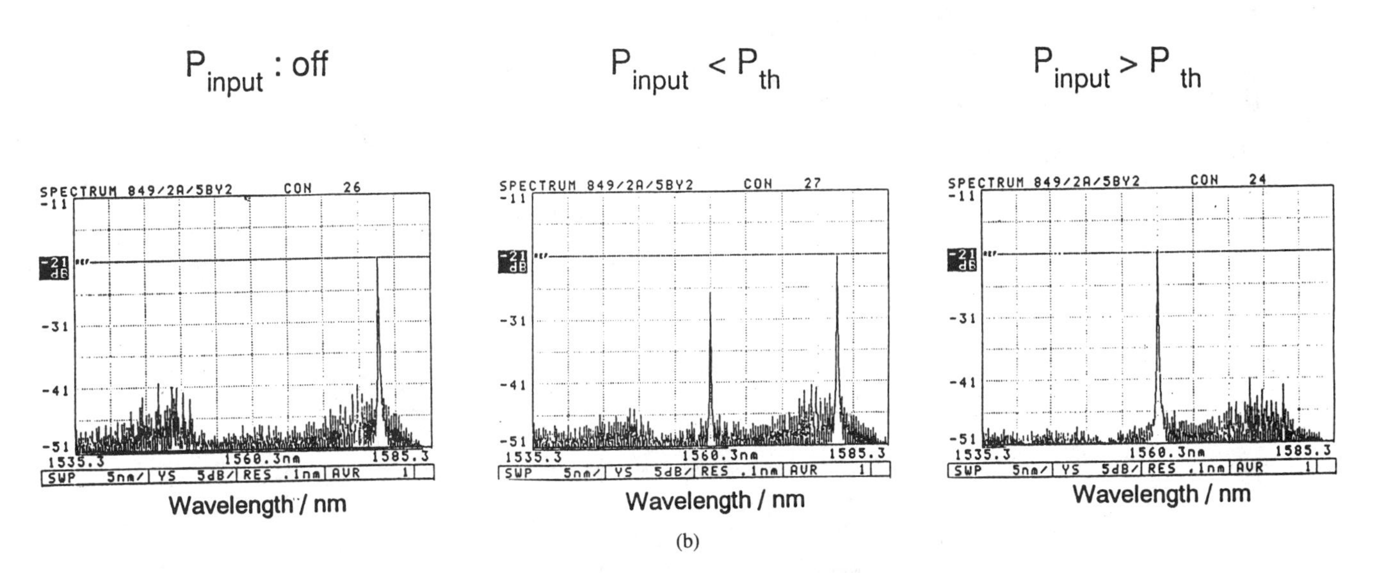

(b)

Fig. 10. Wavelength conversion: Output spectra of the Y-laser in dependence of the injected input power P_{input}. Figures on top: 21 nm blue shift (upconversion), figures at bottom: 18-nm red shift (downconversion).

fixed temperature conditions. The peak wavelength exhibits the typical temperature sensitivity of about 0.1 nm/K within a temperature interval of ~10 K before a mode jump occurs. The bistable looping operation was measured up to 565 Mb/s without speed limitations. Furthermore, the effect is not accompanied by a change in intensity because it is of refractive origin. This mode of operation of the Y-laser needs further investigation and it might be useful for future optical switching applications.

VII. The Y-Laser as an Optical Space Switch

If operated below threshold, the Y-'laser' can be regarded as a 3-dB coupler/splitter, with four electrically controlled absorber/amplifier sections which can be switched on/off individually by the bandfilling effect. Fig. 14 shows a 1:2 space switch experiment where external light is injected into the input port of the device and appears at any one of the two output ports depending on the control currents applied [20], [22]. With the inverse operation a 2:1 optical combiner is obtained, with gain/absorption control features.

The spatial switching operation shows an extinction ratio up to 50 dB; dynamic operation at 1 Gb/s was demonstrated with zero insertion loss due to internal amplification. The required square wave modulation current per output segment was only 30 mA.

This high-speed dynamic operation can be used, for example, for routing or manipulating of optical ATM

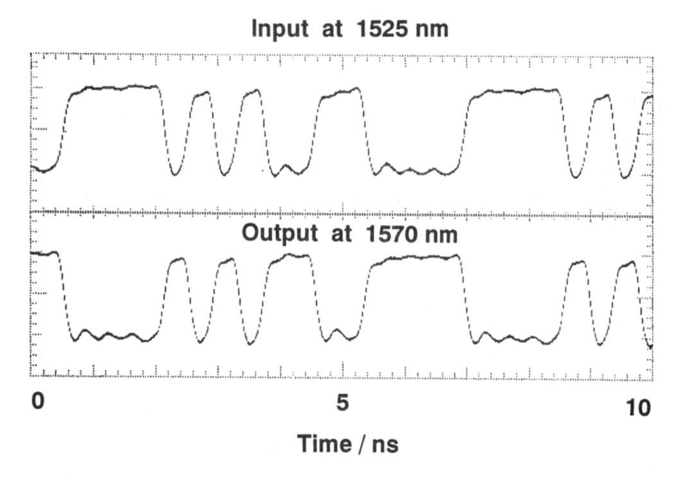

Fig. 11. Dynamic wavelength conversion at 2.5 Gb/s with 45 nm wavelength offset.

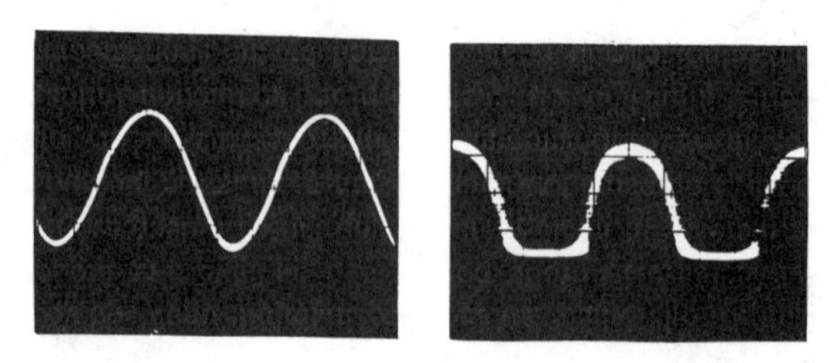

Fig. 12. Reshaping of a signal: picture at left is a sinusoidal optical input signal to the Y-laser; picture at right is the output signal with low power levels cut off.

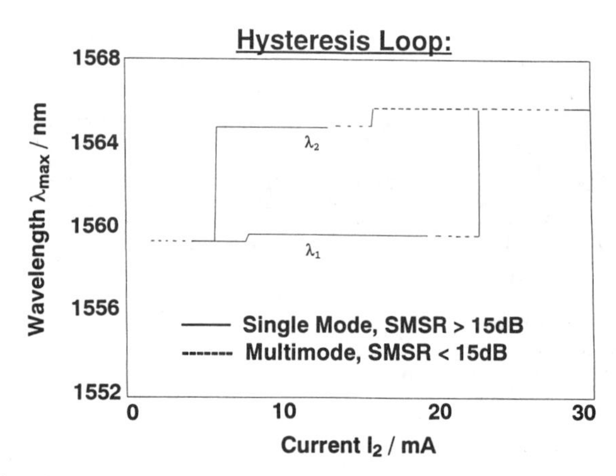

Fig. 13. Hysteresis loop of the emitted wavelength versus current into segment number 2. The other currents are held constant at $I_1 = 59$ mA, $I_3 = 28$ mA, $I_4 = 44$ mA.

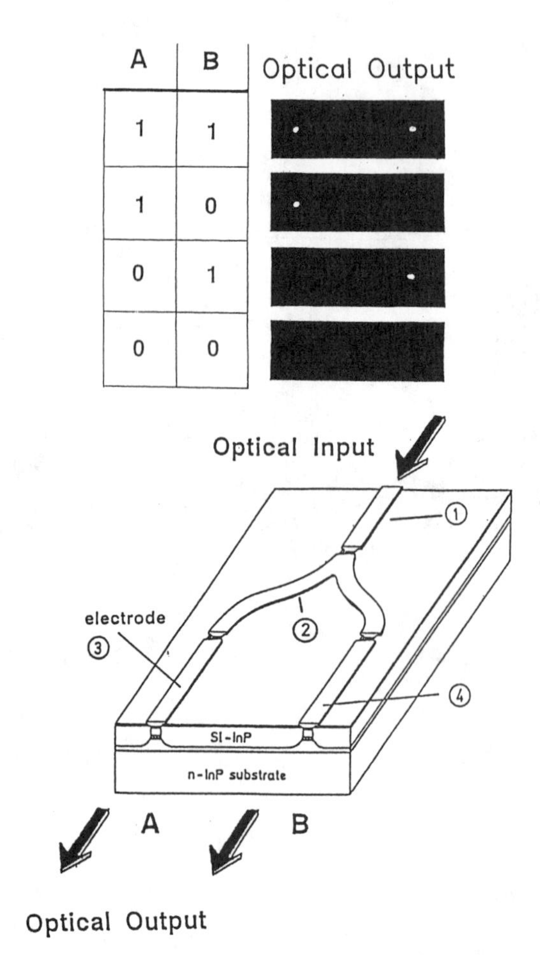

Fig. 14. Spatial switching operation of light injected into segment number 1. The light is on/off switched to either one of the two output ports.

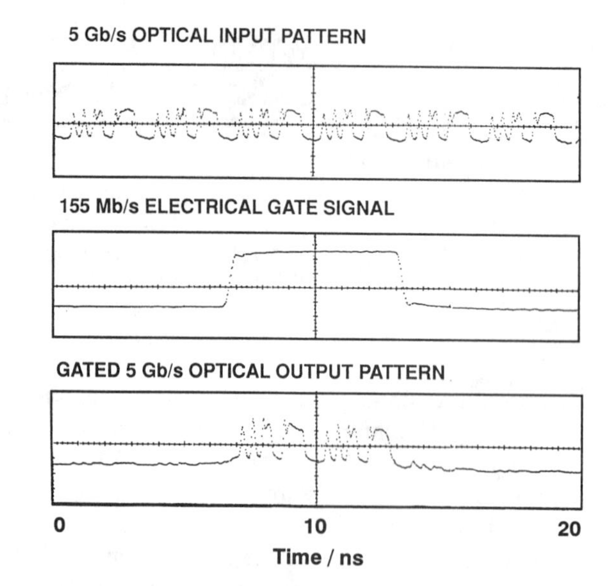

Fig. 15. Packet switching of a 5-Gb/s data stream with a rate of 155 Mb/s selecting one packet.

packets. In Fig. 15, spatial switching of 5-Gb/s data packets is shown at a switching "cell" rate of 155 Mb/s. From the dynamics we estimate a required guard band width of only two bits for complete suppression of bit losses.

Since for space switch functions the Y-'laser' is bit rate transparent there is no *a priori* speed limitation for the traversing data stream, so that the device works even at 10 Gb/s. Together with the aforementioned loss compensation feature, the Y-'laser' as it is, or in low scale integrated (cascaded) versions, is an interesting candidate for 'simple' switch functions such as data drop/insert or listen functions, or fibre delay loop drivers, or even lossless small scale switch matrices. Also, OTDM transmitters and receivers are feasible using cascaded 2:1 combiners (transmitter side), and cascaded 1:2 switches (receiver side) [24].

VIII. The Y-Laser as a Modelocked Pulse Generator

The Y-laser can also be used to generate mode-locked short pulses. Each of the four active segments can be used as an amplifier, absorber, modulator or transparent active waveguide. This gives rise to an interesting experimentalist's playground with a variety of biasing and modulation schemes. Up to now, only a few schemes have been exploited [25].

The experiments were performed with Y-lasers which were antireflection coated to 10^{-4} on the facet of segment 1 (cf. Fig. 1). A 0.5-m-long external resonator, corresponding to 300-MHz repetition rate, was used to couple the light back into the laser. The observed pulse width was 18 ps at 1300 nm, and 23 ps at 1550 nm, resp. (Fig. 16). With a slightly longer pulse of 27 ps, the lowest time-bandwidth-product 0.97 was achieved.

These first experiments demonstrate that principally the Y-laser structure is also suited for short pulse generation via mode-locking, and even shorter pulses are expected from monolithic integrated cavities with shorter length and higher repetition frequencies.

IX. Summary

We have reviewed the recent status of the multifunctional 'Y-laser' device, which is not only suited as an ultra-wide range tunable monolithic light source but also offers features towards wavelength/frequency and space/time processing. It is a technologically simple device, derived from standard, high-speed SIBH-lasers, and modifications/optimizations towards specialized system applications are open.

Used as a light source, single-mode emission with extremely wide tuning range up to 51 nm is achieved under cw operation. Single-current tuning control is accomplished, according to the predictions of a simplified theory that also gives design rules for specific applications.

Wavelength processing was reported, with data-stream transformation from short wavelength fiber windows (~780 nm or ~1310 nm) to long wavelength (~1550 nm), and with high-speed wavelength conversion within the 1550-nm range, including red as well as blue shift over the enormous span of +/− 20 nm. First features as a pulse reshaping device were described, and bistable wavelength switching caused by a refractive index effect were observed and controlled through one current.

Optical space/time switch functions were demonstrated with switch speed up to 1 Gb/s and 0-dB fiber-to-fiber insertion loss and, additionally, optical switching of 5-Gb/s data packets. Possible applications for WDM and OTDM systems, as well as for optical switching networks were discussed. First results on mode-locking of Y-lasers were reported showing the application as a spikes laser producing pulses as short as 18 ps.

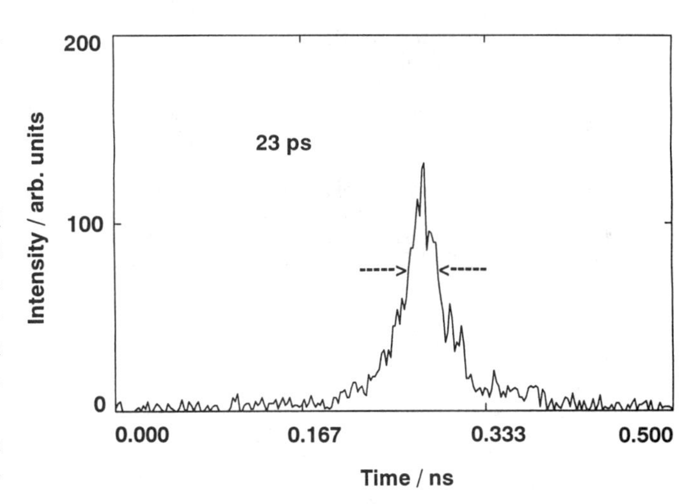

Fig. 16. 1550-nm short pulses generated by the Y-laser by mode-locking. The pulse width is 23 ps.

X. Conclusion

The Y-laser device, originally designed as a large range tunable light source, has evolved to a multifunctional device with a variety of possible applications, especially for processing of optical information in the wavelength/frequency, space and time domains. Its versatility and specific features originate from the basic principle of the Y-arrangement of active waveguide segments: the Y-shape not only involves the interferometric Mach–Zehnder filter function which allows for very wide tuning, but the Y-shape also is the basic geometry needed for any cascading in space. It also introduces a third facet which, e.g., allows to manipulate an optical output in dependence of logic combinations of two optical inputs. The present technological realization of the Y-'laser' device principle probably is not fully optimized, and future evolutions of the device may deviate from the active waveguide structure with four electrical segments, and also standard laser technologies other than the SIBH technology might be applied. Nevertheless, as an essential outcome of the work reviewed in this paper, we feel that the system driven demands for new optoelectronic functions with diversified performance requirements does not necessarily lead to a further diversification of optoelectronic components.

Acknowledgment

The authors would like to thank all of their colleagues at Alcatel-SEL Research Center who contributed to the work. Expert technical support from A. Nowitzki and K. Daub is especially acknowledged. They also thank the group for Solid State Electronics of the University of Marburg for allowing them to do mode-locking experiments at their facility.

References

[1] O. Hildebrand, "Recent trends in InP based optoelectronic components," *Proc. ESSDERC'90*, Nottingham, pp. 505–514, 1990.

[2] H. S. Hinton, "Architectural considerations for photonic switching networks," *IEEE J. Select. Ar. Comm.*, vol. 6, pp. 1209–1226, 1988.

[3] E.-J. Bachus, "Photonic switching—a keystone of multiplexed broadband communication systems," *Proc. 16th ECOC*, vol. 2, Amsterdam, pp. 739–747, 1990.

[4] S. Yamakoshi, "Optically triggered functional device—potential applications of bistable laser diodes." *Proc. 7th IOOC*, Kobe, paper 21A3-3, 1989.

[5] K. Kobayashi and I. Mito, "Single frequency and tunable laser diodes," *J. Lightwave Technol.*, vol. 6, pp. 1623–1633, 1988. S. Murata, I. Mito, and K. Kobayashi, "Tuning ranges for 1.5 μm wavelength tunable DBR lasers," *Electron. Lett.*, vol. 24, pp. 577–579, 1988.

[6] M. Erman, "InP Optoelectronic devices and photonic integrated circuits for high speed packet switching," presented at the Topical Meeting on Photonic Switching '92, Minsk, 1992, *SPIE Proc.*, vol. 1807, 1993.

[7] S. Wang, H. K. Choi, and I. H. A. Fattah, "Studies of semiconductor lasers of the interferometric and ring types," *IEEE J. Quant. Electron.*, vol. 18, pp. 610–617, 1982. I. H. A. Fattah and S. Wang, "Semiconductor interferometric laser," *Appl. Phys. Lett.*, vol. 41, pp. 112–114, 1982.

[8] J. Salzmann, J. S. Osinski, R. Bhat, K. Cummings, and L. Harriot, "Cross coupled cavity semiconductor laser," *Appl. Phys. Lett.*, vol. 52, pp. 767–769, 1988.

[9] M. Schilling *et al.*, "Integrated interferometric injection laser: Novel fast and broad-band tunable monolithic light source," *IEEE J. Quantum Electron.*, vol. 27, pp. 1616–1624, 1991.

[10] E. Zielinski *et al.*, "Optical gain and loss processes in GaInAs/InP MQW laser structures," *IEEE J. Quant. Electron.*, vol. 25, pp. 1407–1416, 1989.

[11] M. Schilling *et al.*, "Asymmetrical Y-laser with simple single current tuning response," *Electron. Lett.*, vol. 28, pp. 1698–1699, 1992.

[12] S. Lindgren, "Loss-compensated optical Y-branch switch in InGaAsP-InP," *J. Lightwave Technol.*, vol. 8, pp. 1591–1594, 1990.

[13] M. Schilling *et al.*, "Widely tunable Y-coupled cavity integrated interferometric injection laser," *Electron. Lett.*, vol. 26, pp. 243–244, 1990.

[14] P. Wiedemann *et al.*, "MOVPE of In(GaAs)P/InGaAs MQW-structures," *J. Cryst. Growth*, vol. 107, pp. 561–566, 1991.

[15] P. Speier *et al.*, "10 Gbit/s MQW-DFB-SIBH lasers entirely grown by LPMOVPE," *Electron. Lett.*, vol. 27, pp. 863–864, 1991.

[16] M. Schilling *et al.*, "6 THz range frequency conversion of 2.5 Gb/s signals by a 1.55 mm MQW based widely tunable Y-laser," in *Proc. 13th IEEE Semiconductor Laser Conf.*, Takamatsu, pp. 272–273, 1992.

[17] W. Idler *et al.*, "Wide range 2.5 Gbit/s wavelength conversion with a tunable Y-laser," in *Proc. 18th ECOC*, Berlin, vol. 1, pp. 449–452, 1992.

[18] K. Wünstel *et al.*, "Y-shaped semiconductor device as a basis for various photonic switching applications," in *Proc. OFC '92*, San José, p. 125, 1992.

[19] W. Idler *et al.*, "High speed integrated interferometric injection laser with 22 nm tuning range," in *Proc. 16th ECOC*, Amsterdam, paper WeF2.1, 1990.

[20] W. Idler *et al.*, "High speed wavelength and spatial switching with a YCCL," presented at the Topical Meeting on Photonic Switching, Salt Lake City, paper FC1, 1991.

[21] O. Hildebrand *et al.*, "The integrated interferometric injection laser (Y-laser): one device concept for various system applications," in *Proc. 17th ECOC/IOOC '91*, vol. 2, Paris, pp. 39–46, 1991.

[22] M. Schilling *et al.*, "Multifunctional photonic switching operation of 1500 nm Y-coupled cavity laser (YCCL) with 28 nm tuning capability," *IEEE Photon. Technol. Lett.*, vol. 3, pp. 1054–1057, 1991.

[23] D. Baums, M. Schilling, W. Idler, G. Laube, and K. Wünstel, "Observation of wavelength-bistability in the interferometric Y-laser," presented at the Topical Meeting on Photonic Switching '92, Minsk, 1992, *SPIE Proc.*, vol. 1807, pp. 494–499, 1993.

[24] E. Zielinski, W. Kuebart, M. Schilling, and O. Hildebrand, "Very high speed and very wide tuning range optoelectronic components," in *Proc. Telecom '91*, Geneva, 1991.

[25] M. Hofmann *et al.*, "Modelocking of interferometric Y-lasers in an external cavity," to be published *Photon. Technol. Lett.*, vol. 5, no. 10, Oct. 1993.

Wavelength Conversion Technologies for WDM Network Applications

S.J.B. YOO, SENIOR MEMBER, IEEE

(Invited Paper)

Abstract—**WDM networks make a very effective utilization of the fiber bandwidth and offer flexible interconnections based on wavelength routing. In high capacity, dynamic WDM networks, blocking due to wavelength contention can be reduced by wavelength conversion. Wavelength conversion addresses a number of key issues in WDM networks including transparency, interoperability, and network capacity. Strictly transparent networks offer seamless interconnections with full reconfigurability and interoperability. Wavelength conversion may be the first obstacle in realizing a transparent WDM network. Among numerous wavelength conversion techniques reported to date, only a few techniques offer strict transparency. Optoelectronic conversion (O/E–E/O) techniques achieve limited transparency, yet their mature technologies allow deployment in the near future. The majority of all-optical wavelength conversion techniques also offer limited transparency but they have a potential advantage over the optoelectronic counterpart in realizing lower packaging costs and crosstalk when multiple wavelength array configurations are considered. Wavelength conversion by difference-frequency-generation offers a full range of transparency while adding no excess noise to the signal. Recent experiments showed promising results including a spectral inversion and a 90 nm conversion bandwidth. This paper reviews various wavelength conversion techniques, discusses the advantages and shortcomings of each technique, and addresses their implications for transparent networks.**

I. Introduction

TELECOMMUNICATIONS is currently undergoing a large-scale transformation. Multimedia services, HDTV, and computer links in the national information highway will undoubtedly benefit us by improving the way of life and increasing efficiency. Such new aspects of telecommunications will rapidly increase the communication traffic, which will in turn demand a national network that can accommodate the entire traffic in a cost effective manner. Single-mode fibers deployed in the public telecommunications network today can potentially accommodate more than 1 Tb/s traffic. Such enormous bandwidth is by far underutilized by even a top-of-the-line telecommunication system (2.5 Gb/s) commercially available today. Higher bit-rate systems at 10 Gb/s are becoming available, however, the dispersion and nonlinearities in the fiber severely limit the transmission distance. Wavelength division multiplexing (WDM) techniques offer a very effective utilization of the fiber bandwidth directly in the wavelength domain, rather than in the time domain. In addition, wavelength can be used to perform functions as routing and switching [1], which becomes an important consideration for realization of an all-optical transparent network layer in the network [2].

Manuscript received April 15, 1995. This work was supported in part by the Advanced Research Project Agency under Contract MDA972-95-30027.

The author is with Bell Communications Research, Red Bank, NJ 07701 USA.

Publisher Item Identifier S 0733-8724(96)04674-9.

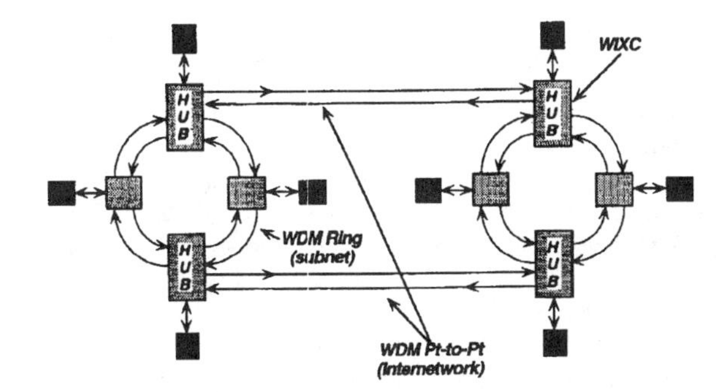

Fig. 1. WDM networks where wavelength interchanging cross-connects (WIXC) are utilized. The two WDM ring subnetworks are connected to each other by four HUB's which include WIXC's (courtesy of [6]).

The number of wavelengths in WDM networks determines the number of independent wavelength addresses, or paths. Although this number may be large enough to fulfill the required information capacity, it often is not large enough to support a large number of nodes. In such cases, the blocking probability rises due to possible wavelength contention when two channels at the same wavelength are to be routed at the same output. One method of overcoming this limitation is to convert signals from one wavelength to another [3]. Fig. 1 illustrates one arrangement for using wavelength conversion for interconnecting WDM networks. This example contains four wavelength interchanging cross-connects (WIXC) and a common network control to permit their dynamic rearrangeability. The WIXC contains space switches and wavelength converters to route signals from any wavelength of any input port to any wavelength of any output port.

The benefit of wavelength conversion varies with network architectures and traffic patterns [4]–[6]. A "small" network with a fixed traffic pattern may not need wavelength conversion and maintain a low blocking rate. A "large" network with dynamic traffic patterns will greatly benefit from wavelength conversion. It is generally accepted that the benefit of wavelength conversion increases with increased traffic.

Reprinted with permission from *IEEE Journal of Lightwave Technology*, Vol. 14, No. 6, pp. 955-966, June 1996.

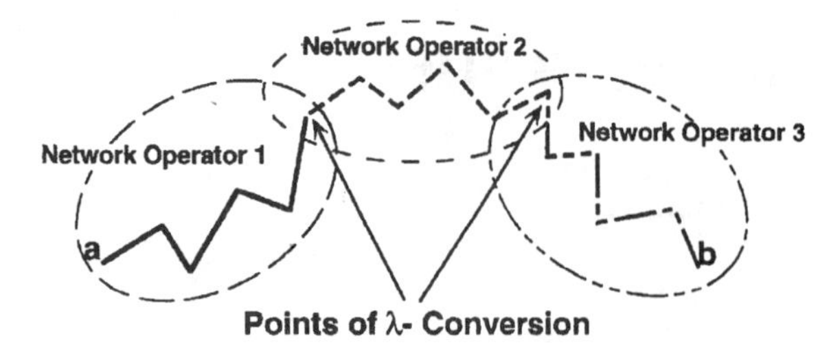

Fig. 2. Distribution of network control and management into smaller subnetworks utilizing wavelength conversion. Network operators 1, 2, and 3 are responsible for their own subnetworks and wavelength assignments within the subnetworks are independent of each other (courtesy of [6]).

Wavelength conversion also allows distributing the network control and management into smaller subnetworks and allows flexible wavelength assignments within the subnetwork. Fig. 2 illustrates this effect. Network operators 1, 2, and 3 are required to manage their own subnetworks, and wavelength conversion may be needed for communication between the subnetworks.

For WDM networks to accommodate vast users and evolve in the future, maximum degrees of transparency and interoperability [7] are desired. In addition, transparent networks facilitate implementations of sophisticated security architectures. The first obstacle in achieving transparency in WDM networks lies in wavelength conversion. The majority of wavelength conversion methods offer only limited transparency. This paper reviews various wavelength conversion techniques, discusses the advantages and shortcomings of each technique, and addresses their implications for transparent networks.

II. Wavelength Converters-Comparison Issues

Wavelength converters reside in two sectors of the network: WIXC's and wavelength adapters. A WIXC is a network element which cross-connects nodes with wavelength conversion when needed [8]. A Wavelength Adapter is an element that changes a noncompliant wavelength to a compliant wavelength. Here, compliant wavelengths are defined as a specific set of wavelengths used in the multiple wavelength optical network. The need for wavelength adaptation disappears once all network elements are based on a set of standard compliant wavelengths. We will limit our discussions to wavelength conversion in WIXC.

There are countless issues to be considered in comparing wavelength conversion techniques. Table I lists some of the key comparison issues. These issues are grouped in three large categories: signal quality, configuration, and performance. *Signal quality* includes signal-to-noise ratio, chirp, amplitude distortion, and extinction ratio, and largely determines the bit-error-rate and the cascadability of wavelength converters. *Configuration* is related to the actual implementation of the wavelength conversion in the WIXC, and is closely linked to the mapping function of the wavelength converter and the resulting WIXC architecture. This category includes control requirements, dynamic ranges of input signals, polarization dependence, filtering requirements, and wall-plug power requirements. Lastly, the *performance* includes conversion

TABLE I
Wavelength Converter Comparison Issues

	COMPARISON ISSUES	COMMON ISSUE
signal quality	s/n ratio chirp amplitude distortion extinction ratio	wavelength registration
configuration	WIXC architecture control and stability polarization dependence optical filtering dynamic range of input signal wall-plug power requirement	
performance	conversion efficiency conversion bandwidth bit-rate-limit and dependence transparency	

efficiencies, conversion bandwidths, and bit-rate (or signal bandwidth) limits. In contrast to these comparison issues, all wavelength converters, regardless of what mechanism is used, must accurately register wavelengths. As we will discuss later in this paper, some of the above comparison parameters are inter-related with each other. For instance, a semiconductor optical amplifier wavelength conversion bandwidth limit is 1 Gb/s for probe power levels at -10 dBm [9], but 20 Gb/s is possible for probe power levels at 4 dBm [10], [11], [60]. Another example is conversion efficiency and noise figure. Some wavelength conversion methods offer a high conversion efficiency at the expense of a high noise figure and high wall-plug power. Wall-plug power can be an important criterion in a large WIXC where total power requirements can add to a very high value and back-up power becomes critical.

Characteristics of an ideal wavelength converter are obvious. In absence of such an ideal element, it is important to balance out the advantages and disadvantages according to the application and extent of the network. Among the issues discussed above, the required level of transparency, the minimum signal-to-noise ratio, and the preferred WIXC architecture are significantly different depending on the network applications.

It is the objective of this paper to consider wavelength converters as black boxes, list the parameters and discuss the tradeoffs. The black box includes all essential elements for wavelength conversion. The black box will have two optical ports for input and output signals, a power connection, and a control interface. Components such as pump or probe lasers will be included in the box, but no regenerators will be included. Filtering requirements will be included and we will make qualitative comparisons. The wall plug power is defined as a total dc power required to operate the entire box. Conversion efficiency is defined as a ratio of the output signal power with respect to the input. Table II summarizes the comparison of wavelength conversion techniques in light of the comparison issues discussed above. Section III overviews three categories of wavelength conversion mechanisms and

TABLE II
COMPARISON OF VARIOUS WAVELENGTH CONVERSION TECHNIQUES

	Optoelectronic Conversion	Cross-gain-modulation in semiconductor amplifiers	Cross-phase-modulation in semiconductor amplifiers	Four-wave-mixing in semiconductor amplifiers	Difference-frequency-generation in semiconductor waveguides
mapping function between input, output, and probe	$\omega_{output} = \omega_{transmitter}$	$\omega_{output} = \omega_{probe}$	$\omega_{output} = \omega_{probe}$	$\omega_{output} = 2\omega_{pump} - \omega_{in}$	$\omega_{output} = \omega_{pump} - \omega_{in}$
signaling category	variable-input-fixed-output	variable-input-fixed-output	variable-input-fixed-output	variable-input-fixed-output	variable-input-fixed-output
transparency	limited	limited	limited	strict	strict
s/n ratio	noise Fig. 7–9 dB s/n > 40 dB if regen.	noise Fig. 7–9 dB	noise figure ~3 dB	s/n < 20 dB	same as input signal s/n ratio
extinction ratio	~10 dB for direct ~20 dB for ext.mod	~8 dB	~15 dB	approximately same as input	same as input
amplitude distortion	mostly digital response	large distortion due to carrier density fluctuation	mostly digital response	negligible for detuning >10 GHz	none
chirp parameter	~3 for direct modulation ~0.3 for electro absorption -1 for Mach–Zehnder mod.	4 for bulk amplifier (for polarization insens. operation) 2.5 for QW amp. (pol. sens. operation)	−0.7~0.4 (depends on the interferometer splitting ratio and inversion of logic)	chirp reversal	chirp reversal
polarization sensitivity	insensitive	sensitive unless critical design is incorporated	sensitive unless critical design is incorporated	sensitive insensitive for multiple pumps	insensitive, polarization diversity possible
optical filtering	not needed	must filter strong input signal	not needed for counter-propagating geometry	must filter strong pump and satellite wavelength	pump filtering trivial. input signal must be filtered.
dynamic range of input signal	17 dB (−17 dBm to 0 dBm for 10 Gb/s)	input power > 0 dBm for 10 Gb/s	3 dB for Extinction Ration > 12 dB	> 10 dB	> 40 dB
wall-plug power	2 W for directly modulated OC-192 6 W for externally modulated OC-192	~200 mW	~200 mW (for pol. sens. QW) ~500 mW (for pol. insens. bulk)	~150 mW	~300 mW
conversion efficiency	17 dB for OC-192	8 dB for OC-192	-2 dB for OC-192	-7 dB	-4 dB (theory) -17 dB (experiment)
conversion bandwidth (3 dB)	extremely broad (limited by the detector)	$\lvert\omega_{input} - \omega_{center}\rvert < 2*\pi*$ (3 THz	$\lvert\omega_{input} - \omega_{center}\rvert < 2*\pi*$ (2 THz)	$\lvert\omega_{input} - \omega_{center}\rvert < 2*\pi*$ (1 THz)	$\lvert\omega_{input} - \omega_{center}\rvert < 2*\pi*$ (12 THz)
bit-rate-limit	~10 Gb/s	~10 Gb/s	~10 Gb/s	> 10 Gb/s	> 10 Tb/s
advantages	• ready for deployment. • regen. improves s/n ratio	• simple configuration • gain in conv. eff.	• reduced chirp, distortion	• chirp reversal • transparent	• chirp reversal • transparent • no excess noise •broad bandwidth
disadvantages	• cost increases with bit-rate and with number of elements • limited transparency	• high noise figure, distortion, and chirp. • limited transparency	• narrow dynamic range of input power • limited transparency	• large spontaneous noise • narrow conversion bandwidth	• phasematching has to be achieved by careful fabrication.
References	21, 22 (components only)	9, 10, 11, 35, 36, 37, 38, 60	39, 40, 41, 42, 43, 44	47, 48, 49, 50, 51, 53	15, 54, 55, 56

Section IV makes detailed comparisons of techniques from each category.

III. WAVELENGTH CONVERSION MECHANISMS

This section briefly overviews three categories of wavelength conversion mechanisms. Inside the black-box, the core of the wavelength converter is in general a three terminal device consisting of input, output, and control terminals. Fig. 3(a) is a functional block diagram of a general wavelength converter. Depending on the mapping functions and the form of control signals, wavelength converters can be classified into three categories: optoelectronic, optical gating, and wave-mixing. Fig. 3(b)–(d) shows functional block diagrams for the three types of wavelength converters. There are also other ways of classifying wavelength converters. Depending on the signal routing mechanisms, they are classified as optoelectronic wavelength converters and all-optical wavelength converters. They can also be classified according to the signal properties: a variable-input-fixed-output converter, a variable-input-variable-output converter, and a fixed-input-variable-output converter. The following subsections describe

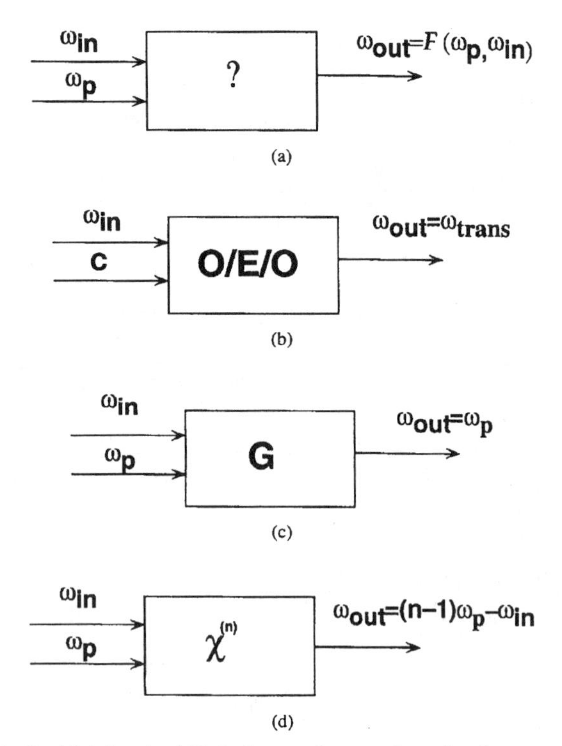

Fig. 3. (a) A functional block diagram of a general wavelength converter. Functional block diagrams of (b) optoelectronic, (c) optical gating, and (d) wave-mixing (e.g., difference-frequency-generation) wavelength converters.

the three categories according to the first classification method, and Section IV compares wavelength conversion mechanisms in more detail.

A. *Optoelectronic (O/E–E/O) Wavelength Conversion*

Most straightforward of all wavelength conversion techniques is a detection of the optical signal and a retransmission of the signal (O/E–E/O). For example, a commercially available SONET repeater with or without a regenerator is an O/E–E/O wavelength converter, provided that the output wavelength is a compliant wavelength. This technique involves an electrical routing of the signal during the wavelength conversion process. This is a variable-input-fixed-output wavelength converter, unless a tunable laser or a laser array is used.

B. *Optical Gating Wavelength Conversion*

A large number of wavelength converters fall into this category. This type of wavelength converter employs an optical device which changes its characteristics depending on the intensity of the input signal. This change is monitored by a cw signal called probe and this probe signal will contain the information in the input signal. There are countless wavelength conversion methods that fall into this category, resulting from the fact that there are numerous optical gating mechanisms available. This category includes semiconductor optical amplifier cross-gain modulation, semiconductor optical amplifier cross-phase modulation, semiconductor lasers with saturable absorption, and nonlinear optical loop mirrors. This is a classical example of variable-input-fixed-output wavelength converters, and we will see more of these wavelength conversion methods as more optical gating methods are utilized.

C. *Wave-Mixing Wavelength Converters*

Perhaps the least explored but offering the highest degree of transparency is wavelength conversion based on wave-mixing. In a broad sense, this category includes optical-acoustic wave-mixing [12], optical-electrical wave-mixing [13], as well as nonlinear optical wave-mixing. The first two can be quite effective frequency translation methods for dense FDM, however, they are unlikely to provide larger frequency translation required for WDM. The nonlinear optical wave-mixing results from nonlinear interactions among the optical waves present in a nonlinear optical material. This mechanism is sensitive to both amplitude and phase information, and is the only category of wavelength conversion methods that offers strict transparency. So far, four-wave-mixing based on the third-order optical nonlinearity and difference-frequency-generation based on the second-order optical nonlinearity have been demonstrated. The utilized mixing functions are parametric, and the mapping functions allow one-to-one mapping of an input wavelength to an output wavelength. Therefore, the conversion process is variable-input-variable-output. One unique feature common to this category of wavelength converters is that they allow simultaneous conversions of multiple input wavelengths to multiple output wavelengths. This property is an outcome of the superposition properties of optical fields. The details of this process are discussed in [14] and [15], and are briefly summarized in the later section.

IV. Comparison of Wavelength Converters

This section compares wavelength conversion techniques in more detail.

A. *Optoelectronic (O/E–E/O) Wavelength Conversion*

Optoelectronic wavelength conversion can be achieved in an optical repeater with or without signal regeneration. It is even possible to undertake the entire WIXC functionality by combining an electronic cross-connect with an optoelectronic repeater. The optoelectronic wavelength conversion technology is more mature and is more readily applicable to field deployment when compared to all-optical counterparts. In addition, the possibility of regeneration of signal and the added capability of network control and management make this approach extremely attractive. The O/E–E/O operation also accommodates wide optical power levels and requires no filtering or polarization control. On the other hand, there are shortcomings associated with this technique, and we will discuss them below.

Currently, full 3R-regeneration (retiming, reshaping, reclocking) is typically performed after every 40 km. However, optical fiber amplifiers are rapidly replacing the regenerators that are already in place today. Transoceanic distance (>9000 km) transmission without regeneration has already been demonstrated by numerous research groups. Due to high cost, low reliability, and poor upgradability, most network operators hope to avoid regeneration unless it is really necessary. For consistent comparisons of wavelength conversion techniques, we will continue our discussions with optoelectronic wavelength conversion without regeneration.

The first limitation of this technology lies in limited transparency. The information in the form of phase, frequency, and analog amplitude is lost during the conversion process. The highest degree of transparency that the optoelectronic conversion can achieve is digital transparency, where digital signals of any bit-rates up to a certain limit are accommodated. The operation bandwidth must cover those of all anticipated applications. The gain-bandwidth limitations, on the other hand, dictates that the performance of the optoelectronic repeater can not be optimized for all bit-rates. Usually, optimizations are made for the high end bit-rate, and the sensitivities at lower bit-rates will deviate from their optimum values.

An optoelectronic wavelength converter consists of a receiver and a transmitter. At the physical level, this is equivalent to a detector, an RF amplifier, and a laser. It is worth noting that RF amplifiers increase the conversion efficiency (defined as a power ratio between the converted signal and the input signal) at the expense of the signal-to-noise ratio. The ratio of signal to noise ratio before and after amplification is defined as a noise-figure [16]. Numerous wavelength converters employ either electrical or optical amplification during the conversion process. Some amplifiers outperform others in terms of noise figures and power requirements. Noise figures as low as 3.1 dB [17] are demonstrated in 980 nm pumped Erbium-doped fiber amplifiers (EDFA), whereas those of broadband RF and semiconductor optical amplifiers lie in the 7–9 dB range [18]–[20]. It is important to note that a combination of an EDFA with a noise figure of 4–5 dB and an all-optical wavelength converter with no excess noise can outperform a typical optoelectronic wavelength converter in numerous categories, including the signal-to-noise ratio and the conversion efficiency.

In terms of wall plug power, high-speed electronics require a constant power supply at high levels to achieve high speed and high gain. The wall plug power required for the 2.5 Gb/s repeater consisting of a receiver front-end, an RF amplifier, and a transmitter is approximately 2 W [21], and is anticipated to be at least double for 10 Gb/s. The power requirement is more severe for a system with an external modulator. This power requirement is higher than those for the majority of all-optical wavelength conversion technologies. For instance, the semiconductor optical amplifier cross-gain modulation wavelength converter which requires a −5 dBm probe signal and a 100 mA constant current translates to approximately 200 mW wall-plug power requirement. However, one should also consider the fact that the optoelectronics system achieves much higher conversion efficiency (gain). The amplification factor or conversion efficiency of an optoelectronic converter is typically 30 dB for 2.5 Gb/s systems and 17 dB for 10 Gb/s systems, whereas the cross-gain modulation wavelength converter offers 10 dB at 2.5 Gb/s and 5 dB at 10 Gb/s.

Another key consideration in comparing optoelectronic versus all-optical wavelength conversion methods is cost. An 2.5 Gb/s optoelectronic wavelength converter realized by combining a receiver and a transmitter costs approximately U.S. $9000 [22]. This is approximately the same as the cost of a polarization insensitive semiconductor amplifier capable

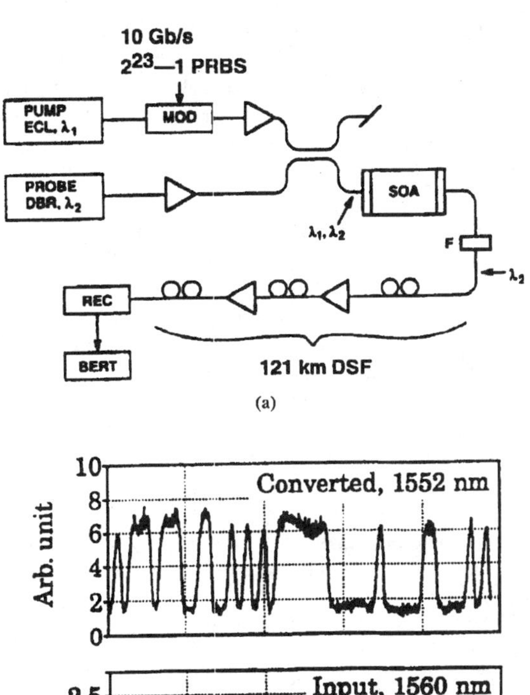

(a)

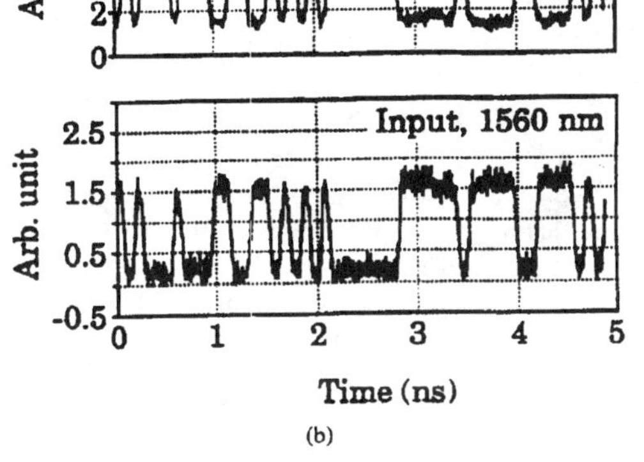

(b)

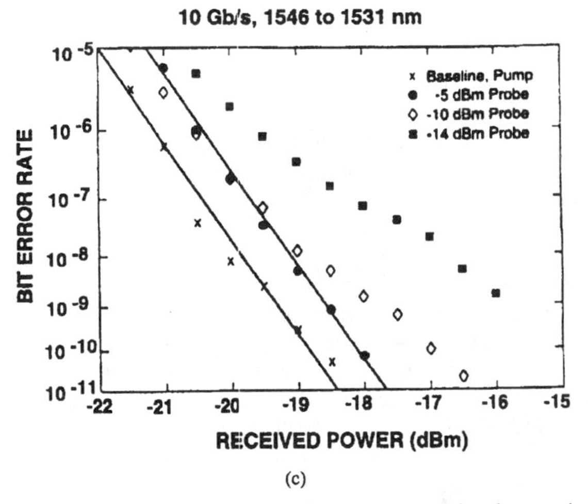

(c)

Fig. 4. (a) A configuration for a cross-gain modulation semiconductor optical amplifier wavelength converter (courtesy of [10]), (b) bit patterns of input and output signals (courtesy of [11]), and (c) bit error rate for 10 Gb/s operation (courtesy of [10]).

of wavelength conversion at 10 Gb/s (~U.S. $10 000) [23]. Although the unit costs are similar in two cases, there are two factors that can eventually make the optoelectronic wavelength converter more costly. The first factor is the market. Since the optoelectronic market is far greater than the other, the cost of the semiconductor amplifier may significantly drop if its market grows to become similar in size. The second is packaging cost. A WIXC will require multiple wavelength converters, and packaging them in an array configuration can significantly reduce the packaging cost per unit converter. This packaging is not as trivial in optoelectronic wavelength converters as in all-optical counterparts because of a strong RF coupling which causes crosstalk. High-speed electronics require sophisticated packaging to avoid crosstalk from other channels. There are a number of groups making effort toward multichannel optoelectronic packaging with low crosstalk and we look forward to good system results.

The robust operation and mature technologies associated with this technology make the optoelectronic wavelength converter attractive for nontransparent WDM network applications.

B. Optical Gating Wavelength Conversion

This category of wavelength converters is very similar to the optoelectronic wavelength converter in terms of its limited transparency and thresholding characteristics [24]. The technologies discussed in this section are far less mature than the optoelectronic method discussed in the previous section. If this category of converters are favored over the optoelectronic ones, it would be due to cost and packaging considerations as discussed above.

The optical gating function is achieved predominantly by the third-order optical nonlinearity of the material. The magnitude of the optical nonlinearity determines the input signal power required to gate the probe signal for wavelength conversion. The materials available today have extremely small third order nonlinearities unless resonant enhancement is used. The problem of resonant nonlinearity is that it is accompanied by resonant absorption and that the resonant transition lifetime limits bit rates. On the other hand, a long interaction length must be used for relatively weak nonresonant nonlinear effects to accumulate.

Saturable absorption provides a relatively simple and compact realization of wavelength conversion [25]–[29]. The input signal saturates the absorption of exciton transitions near the bandgap and allows the probe beam to transmit. This technique showed a bandwidth limit of 1 GHz due to carrier recombinations, and suffered from a bistable behavior when integrated with lasers. Similar optical gating wavelength conversion can be obtained by utilizing gain-suppression mechanism in semicondutor lasers such as tunable DFB lasers [30], Y-Lasers [31], and T-Gate Lasers [32]. The gain suppression mechanism can achieve higher bit-rates than saturable absorption owing to stimulated emission inside the gain material. A recent work reported 10 Gb/s wavelength conversion in a DBR structure [33]. Similarly, cross-gain modulation in semiconductor optical amplifier can also achieve 10 Gbit/s or even higher bit rates [10], [60]. The mechanisms discussed so far in this section suffer relatively large phase modulation in the signal during the gain or absorption modulation. This phase modulation adversely affect transmission capabilities through the conventional single mode fibers. One of the most successful conversion demonstrations is achieved in a semiconductor optical amplifier utilizing cross-phase modulation. Typically, a Mach–Zehnder interferometer is used for converting the phase modulation within one of the arms to an intensity modulation at the output. This is discussed in detail in Section IV-B2). Last, a recent demonstration of a nonlinear-optical-loop-mirror is an example of accumulating a weak nonresonant nonlinear optical effect over a long interaction length. This was made possible in a silica fiber with a propagation loss below 0.3 dB/km. The following sections discuss the details of selected devices.

1) Cross-Gain Modulation in Semiconductor Optical Amplifiers: The gain in a semiconductor optical amplifier saturates as the optical power level increases. Therefore, it is possible to modulate the amplifier gain with an input signal, and, in turn, encode this gain modulation on a separate cw probe signal traveling through the amplifier at another wavelength. Fig. 4(a) illustrates this configuration. As in saturable absorber type wavelength converters, the saturation is never complete, and the extinction ratio of the converted signal is typically below 8 dB. Fig. 4(b) shows the bit patterns of input and output signals. The advantage over the saturable absorption technique is its capability to operate beyond the spontaneous recombination rate of carriers. The presence of the optical probe signal induces stimulated emission which results in reduction of the carrier lifetime. These wavelength converters are polarization dependent due to polarization dependent gain and confinement factors. By utilizing semiconductor amplifiers with bulk or strain-compensated active regions [34], the polarization dependence is greatly reduced. References [35] and [36] discuss polarization independent wavelength conversion at 10 and 20 Gb/s for probe signal powers up to -5 dBm and 3 dBm. Fig. 4(c) shows 10 Gb/s bit-error-rate measurements obtained in [10]. The conversion experiment show a 1 dB penalty for -5 dBm probe power. One of the key shortcomings of this method is a signal-to-noise deterioration due to a large spontaneous emission background. Typical noise figures of semiconductor amplifier are 7–8 dB. Usually, the conversion efficiency is lower than the gain, and the noise figure for the conversion process is even higher than the intrinsic noise figure of the amplifier [20]. In addition, the signal quality further deteriorates with chirping and amplitude distortion caused by carrier modulations. Experiments discussed in [37] show nearly 2 dB dispersion penalty at 6 Gb/s over 20 km of conventional single mode fiber. Two stage cascaded wavelength conversion showed approximately 5 dB power penalty and pulse reshaping [38]. Despite some of shortcomings in signal quality, this is one of the simplest wavelength conversion methods which utilizes a component available today.

2) Cross-Phase Modulation in Semiconductor Optical Amplifiers: Optical signals traveling through semiconductor optical amplifiers undergo a relatively large phase modulation compared to the gain modulation. This is manifested

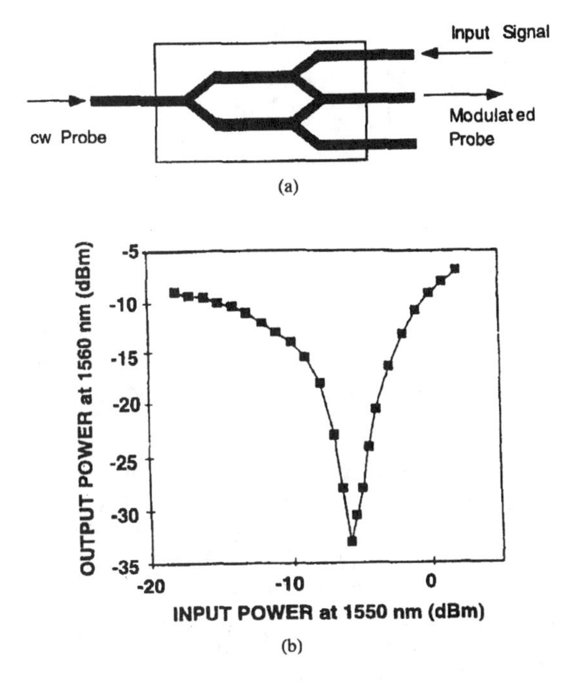

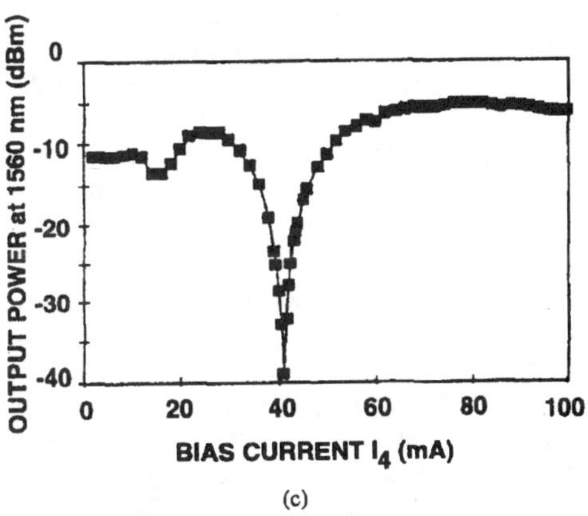

Fig. 5. (a) A Mach–Zehnder configuration for a cross-phase modulation semiconductor optical amplifier wavelength converter, (b) the output power dependence on the input power for a cross-phase modulation semiconductor optical amplifier wavelength converter (courtesy of [43]), and (c) the output power dependence on the input current into one of the arms of a cross-phase modulation semiconductor optical amplifier wavelength converter (courtesy of [43]).

by the large chirp parameters in the cross-gain modulated wavelength converters discussed in the previous section. The cross-phase modulation effect is utilized in an interferometer configuration, typically in a Mach–Zehnder interferometer [39]–[41]. Fig. 5(a) shows an example of a Mach–Zehnder cross-phase modulation wavelength converter. Semiconductor optical amplifiers are incorporated in both arms and electrical currents are injected into both amplifiers. An input optical signal passes through one of the arms and modulates the phase of the arm. The interferometric nature of the device converts this phase modulation to an amplitude modulation in the probe signal. Similar interferometric conversion principles

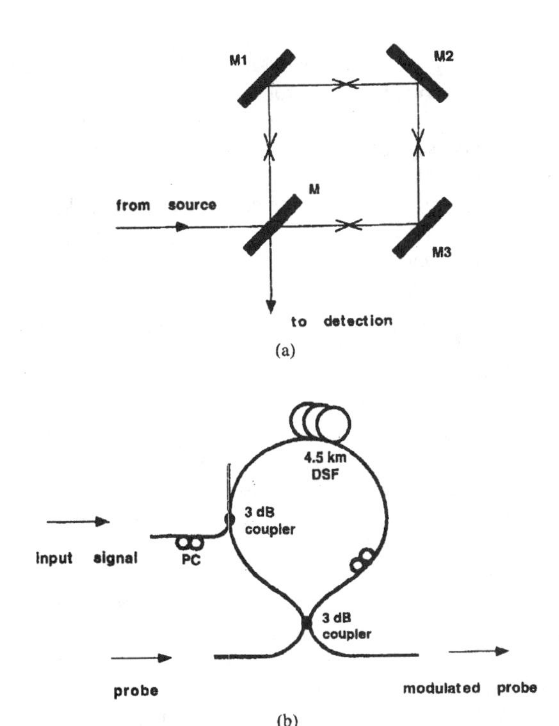

Fig. 6. (a) A classical Sagnac Interferometer consisting of a half silvered mirror M, and three full mirrors M_1, M_2 and M_3 and (b) a nonlinear optical loop mirror using a 4.5 km dispersion shifted fiber as a nonlinear optical element.

apply to a number of variations in waveguide configurations. The interferometer can operate in two different modes, a noninverting mode where an increase in input signal power causes an increase in probe power, and an inverting mode where an increase in input signal power causes a decrease in probe power. Compared to the cross-gain modulation method, the use of an interferometer greatly improves the quality of the converted signal in terms of chirp and extinction ratios. The signal-to-noise ratio is also affected by the spontaneous emission of the amplifier, but high power optical probe reduces the spontaneous noise background. This chirp reduction and the signal-to-noise improvement are more pronounced in a noninverting mode of operation [42]. Theoretically calculated relative intensity noise (RIN) is in the range between -140 and -120 dB/Hz, comparable with that of typical semiconductor laser diodes. In addition, the thresholding characteristics of the interferometer can increase the extinction ratio, and a negative power penalties can be obtained in the conversion when input signals of poor extinction ratios are used [39]. These thresholding characteristics, on the other hand, restrict the dynamic range of the input power and add extra difficulty in controlling input power levels. Fig. 5(b) and (c) shows experimental results from [43]. The figures indicate that a contrast ratio of 15 dB can be obtained for input optical power levels between -7 and -8 dBm, and bias currents between 39–42 mA. Hence, the input power level must be monitored and adjusted to better than 3 dB. Since the input power levels of WIXC are likely to be monitored and controlled accurately in any case, this power accuracy requirement may add negligible complexity to the system. The mechanism which limit the bit rate is identical to the case of cross-gain modulated semiconductor optical amplifier, and higher intensity probe beam allow the bit rate to extend to 10 Gb/s [44]. Relatively high quality of the converted signal make this approach extremely attractive.

3) Nonlinear-Optical-Loop Mirror: A nonlinear-optical-loop-mirror (NOLM) is essentially a Sagnac Interferometer implemented in optical fibers with a nonlinear optical medium. The basic operation principle of a NOLM resembles that of the classical Sagnac interferometer shown in Fig. 6(a). In 1925, Michelson and Gale used the Sagnac interferometer to detect rotation of the earth. A beam of light from a source S is divided into two beams by means of a half-silvered mirror M. The two beams are caused to traverse opposing paths around a loop formed by mirrors M_1, M_2, and M_3, as shown. The beams recombine at M and are reflected into an observing detector in which interference fringes are seen. In the absence of asymmetry in the two paths, i.e., no rotation, the fringes appear dark (destructive interference). As the rotation rate increases, the fringes will evolve bright and dark, alternately. Sagnac interferometers can be easily implemented by means of optical fibers and fiber couplers. Such fiber interferometers are widely used as high precision gyroscopes in aircrafts today. The NOLM is a fiber Sagnac interferometer with a nonlinear optical medium inserted in the loop. Fig. 6(b) is a nonlinear-optical-loop-mirror using an optical fiber as a nonlinear medium [45]. A probe beam is split into two by a 50:50 fiber coupler and propagates in both directions. In the absence of nonlinear interaction, the output port sees no probe beam. An input signal is coupled into the loop via a fiber coupler and propagates in a counter-clockwise direction. This signal modulates the optical index of the nonlinear optical fiber owing to an optical Kerr effect, and causes the phase of the probe beam propagating counter-clockwise to increase relative to that of the clockwise beam. Due to this asymmetry, the output port sees the probe beam. Due to a finite propagation time through the nonlinear element, the probe signal is pulsed (clocked) and needs synchronization with the input signal. All-fiber systems require more than 2 km of optical fibers, and unstable output can be caused due to local index variation in the fiber. Recently, a nonlinear optical loop mirror is implemented by insertion of a semiconductor optical amplifier as a nonlinear optical medium. The demonstrated system is far more compact compared to the all-fiber system. These semiconductor amplifier-based wavelength converters share characteristics discussed in Section III-B2), cross-phase-modulated semiconductor optical amplifier. Reference [46] discusses a polarization insensitive wavelength conversion in a NOLM at 10 Gb/s.

C. Wave-Mixing Wavelength Converters

Wave-mixing rises from a nonlinear optical res ponse of a medium when more than one wave is present. The outcome of wave-mixing effect is a generation of another wave whose intensity is proportional to the product of the interacting wave

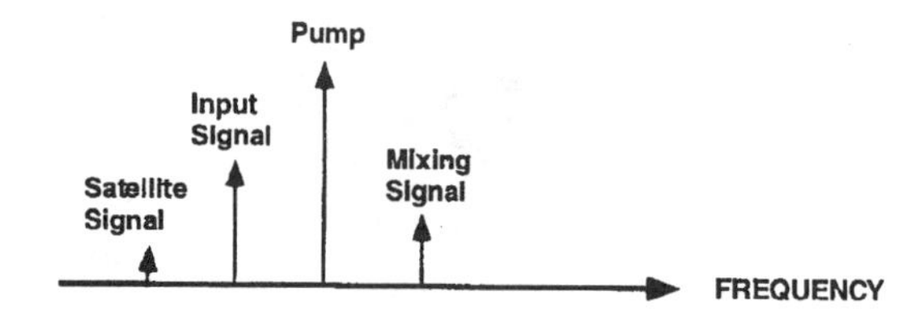

Fig. 7. A four-wave-mixing diagram shown in the frequency domain.

intensities. The phase and frequency of the generated wave is a linear combination of those of the interacting waves. Therefore the wave-mixing preserves both phase and amplitude information, and this is the only category of wavelength conversion that offers strict transparency. It is also the only method that allows simultaneous conversion of a set of multiple input wavelengths to another set of multiple output wavelengths. Last, this method can potentially accommodate signals with extremely high bit-rates exceeding 100 Gbit/s.

Depending on the number of interacting waves involved, they are called three-wave-mixing (two input waves and one output wave), four-wave-mixing (three input waves and one output wave), and so on. Four-wave-mixing is based on one order higher nonlinear optical effects (third order) as compared to three-wave-mixing (second order). Consequently, conversion efficiencies are in general much higher for the three-wave-mixing than for the four-wave-mixing. The wave-mixing effects considered for WDM applications are parametric, and the chirp is reversed during the conversion process. The parametric conversion process is classically free of excess noise unless an active material is used. Optical wave-mixing is typically far less efficient compared to microwave mixing unless high intensities are used. To enhance nonlinearities, a number of conversion techniques involve optical amplifiers, in which case the signal-to-noise ratio is degraded. Comparison of this category of wavelength converters with others should again take the black box approach considering both noise and conversion efficiency. The following two sections discuss four-wave-mixing and difference-frequency-generation in waveguides.

1) Four-Wave-Mixing in Passive Waveguides: Parametric four-wave-mixing in fibers or semiconductor waveguides has been utilized for wavelength conversion applications. A semi-classical picture of four-wave-mixing includes formation of a grating and scattering of a wave off of the grating. Two optical waves form a grating due to intensity patterns in the nonlinear medium. This can be a standing wave grating if the two waves have identical frequency and no relative phase jitters. The nonlinear material responds to this intensity grating by forming a refractive index grating (Kerr grating) or by forming a population grating. The latter occurs if the material changes its energy state due to optical intensity. If the two waves differ in frequency, the grating pattern sweeps in space at a rate corresponding to the frequency difference. If the material response is fast enough to follow this sweep, the grating efficiency is unchanged. However, if the grating sweeps at a faster rate compared to the material response, the grating efficiency will decrease. The third beam present in the

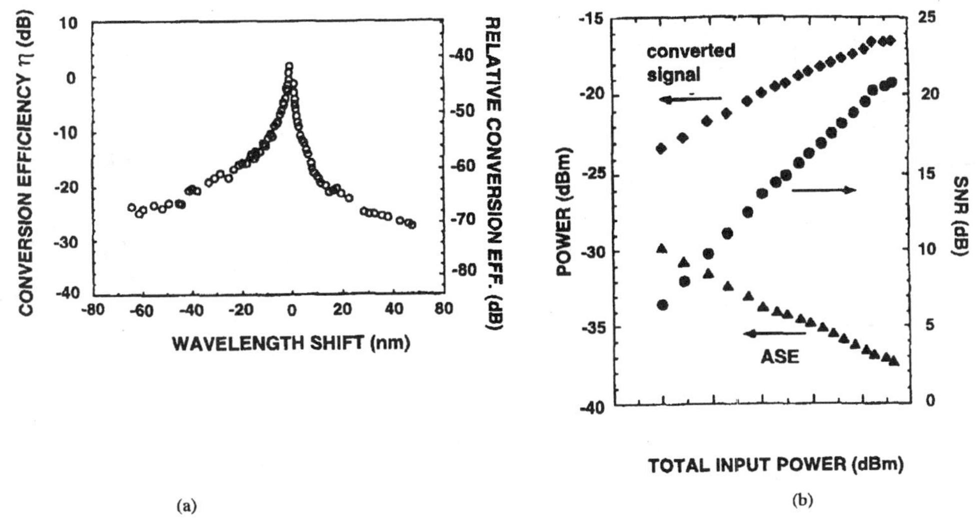

Fig. 8. (a) The conversion efficiency as a function of detuning for four-wave-mixing wavelength conversion in a semiconductor amplifier (courtesy of [53]) and (b) the conversion efficiency, gain, and signal-to-noise ratio as a function of the power level for four-wave-mixing wavelength conversion in a semiconductor amplifier (courtesy of [53]).

material is scattered by the grating. This scattered wave is the wave generated as a result of a four-wave-mixing interaction of the three incident waves. The frequency of the generated wave is offset from that of the third wave by the frequency difference of the first two waves. If one of the three incident waves contain information (amplitude, phase, or frequency), and the others are constant, then the generated wave will contain the same information (except for a possibility of phase conjugation). It is worth noting that there is nothing that distinguishes amongst the three waves unless there is a selection rule imposed by the material and the polarizations. In other words, if two optical waves of two frequencies are present, there can be two possible combinations of four-wave-mixing, which produces two waves. Fig. 7 illustrates this example using a diagram in the frequency domain. Two input optical waves, a pump wave at ω_p and an input signal wave at ω_s are present in a nonlinear medium. The generated wave at $2\omega_p - \omega_s$ is a result of the pump wave scattered by a grating formed by the pump wave and the signal wave. There is another generated wave at $2\omega_s - \omega_p$, which is a result of the signal wave scattered by a grating formed by the pump and the signal waves. The intensity ratio of the first and the latter generated waves equals that of the pump and the signal waves. Typically this ratio is above 20 dB, and for this reason, the first will be called a converted wave and the latter a satellite wave. As the satellite signal wave and the pump wave can overlap in wavelength with another WDM channel, filtering has to be carefully executed to avoid crosstalk noise. System penalties beyond 3 dB can rise in high capacity WDM networks for a crosstalk level as low as −35 dB. Reference [47] discusses a relatively efficient method of reducing the pump signal and a amplifier background. By employing a four-wave-mixing medium within a fiber Sagnac interferometer, only the input and the converted signals are expected to appear on the input port. In practice, imperfection in the 50:50 splitter and birefringence along the fiber loop result in an incomplete nulling of the pump wave.

The conversion efficiency and the bandwidth of this process bear a strong relation to the waveguide dispersion and length. A longer interaction length allows more accumulation of generated wave, but the absorption and the dispersion may limit the useful length. In the case of semiconductor waveguides [48], the losses on the order of 2 dB/cm limits the useful device length to less than 1 cm. In this case, the waveguide dispersion is negligible, and the conversion bandwidth typically exceeds 20 nm. For optical fibers [49], losses are on the order of 0.3 dB/km, and fiber lengths exceeding 1 km are typically used. Dispersion, however, limits the product of the tuning range and the fiber length. Even for dispersion shifted fibers, a 3 dB reduction in conversion efficiency occurs at the product 10 nm $*$ 2 km. Typical conversion efficiencies are in the range of −20 dB for 17 dBm pump power for both semiconductor and fiber waveguides.

2) Four-Wave-Mixing in Semiconductor Optical Amplifiers: Four-wave-mixing in an active medium, such as semiconductor optical amplifiers [50] allows relatively efficient conversion owing to increased optical intensities and resonantly enhanced optical nonlinearities. In such gain enhanced four-wave-mixing, a population grating is formed unless special geometries are used [51], [52]. Population gratings are generally much stronger than phase gratings, and their scattering efficiencies decrease more steeply with detuning of the signal wavelength relative to the pump wavelength. In semiconductor amplifiers, at least three physical mechanisms form gratings. They are carrier density modulation, dynamic carrier heating, and spectral hole burning. These mechanisms have different lifetimes and scattering strengths, and the four-wave-mixing conversion includes contribution from all these effects. Fig. 8(a) shows the conversion efficiency as a function of detuning (courtesy of [53]). At a small detuning, a slow pop-

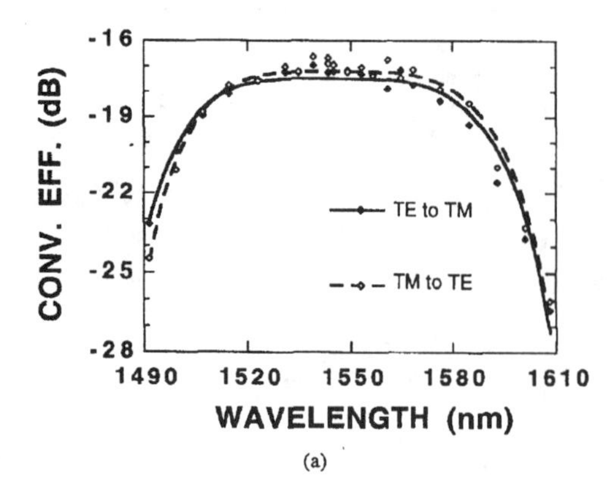

(a)

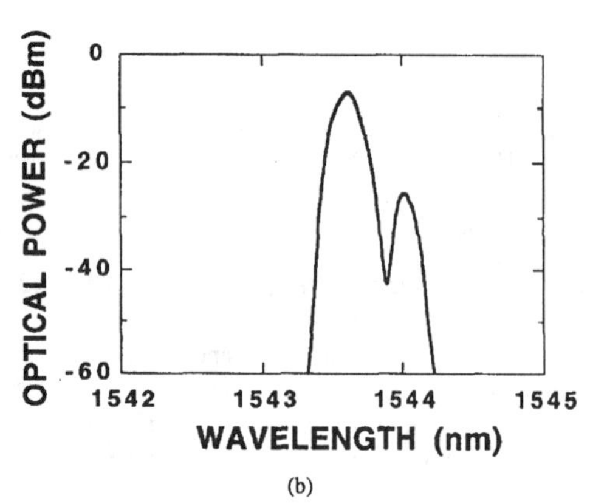

(b)

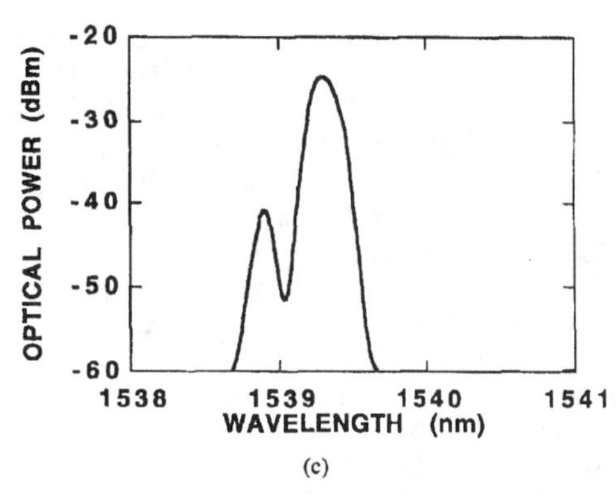

(c)

Fig. 9. Experimental results obtained in wavelength conversion by difference-frequency-generation. (a) Conversion efficiency curves for two input polarizations as a function of input wavelength. Also shown are the spectral inversion effects between, (b) the input wave spectrum, and (c) the output wave spectrum.

ulation grating effect dominates, and shows a high conversion efficiency. At a large detuning, a fast phase grating dominates but the conversion efficiency is significantly lower than the peak value. The asymmetry in the spectra is a result of the asymmetric gain, interband-intraband dynamics, and coherence effects of various contributing mechanisms. Although the gain in the amplifier helps the conversion efficiency to approach 0 dB, it is difficult to achieve a 20 dB signal-to-noise ratio even with the use of a relatively narrow (0.08 nm) spectral filter. Fig. 8(b) shows the conversion efficiency, gain, and signal to noise ratio as a function of the power level.

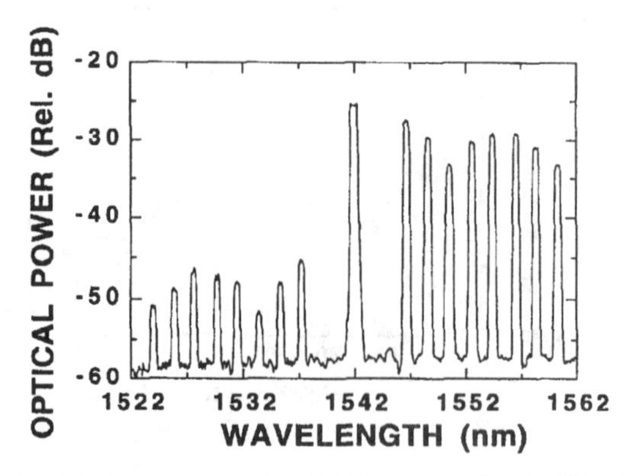

Fig. 10. Simultaneous conversion of eight input wavelengths (1546, 1548, 1550, 1552, 1554, 1556, 1558, 1560 nm) [59] to a set of eight output wavelengths (1538, 1536, 1534, 1532, 1530, 1528, 1526, 1524 nm) achieved by difference-frequency-generation. The structure at 1542 nm is a second-order diffraction of a pump wave at 771 nm.

3) Difference Frequency Generation: Difference-frequency-generation (DFG) is a consequence of nonlinear interaction of the material with two optical waves: a pump wave and a signal wave. DFG like four-wave-mixing in passive waveguides, offers a transparent wavelength conversion with quantum-noise limited operation. It is also capable of chirp-reversal and multiwavelength conversions. DFG wavelength converters based on a $LiNbO_3$ waveguide have been demonstrated [54], [55]. The normalized conversion efficiency is 41%/W-cm^2 which corresponds to a -6 dB conversion efficiency for a 2 cm waveguide and a 100 mW pump. Recently, DFG wavelength conversion in AlGaAs waveguides has been demonstrated [56], [58]. The main difficulty in realizing such a semiconductor wavelength converter lies in the phase-matching of interacting waves. Reference [57] discusses utilization of wafer-bonding and OMCVD growth to achieve periodic domain inversion for quasiphase-matching. Due to a large scattering loss in the waveguide the conversion efficiency was limited to -17 dB which is far below theoretically predicted -4 dB. Fig. 9(a) shows the conversion efficiency as a function of the input wavelength. The figure shows extremely wide conversion bandwidths exceeding 90 nm and polarization diversified operation. In the case of arbitrary polarization input, the conversion efficiency deviates by less than 0.4 dB from the value obtained with a TE or TM input polarization. This method also shows spectral inversion as in the case of four-wave-mixing. Fig. 9(b) and (c) show the measured spectra of the input and output waves. One key advantage of this conversion method is a capability to simultaneously convert multiple input wavelengths. Fig. 10 shows simultaneous conversion of eight input wavelengths (1546, 1548, 1550, 1552, 1554, 1556, 1558, 1560 nm) [59] to a set of eight output wavelengths (1538, 1536, 1534, 1532, 1530, 1528, 1526, 1524 nm). The structure at 1542 nm is a second order diffraction of a pump wave at 771 nm. Although four-wave-mixing can achieve similar multichannel conversion, DFG is free from satellite signals which appear in FWM (see Fig. 7). Difficulty in fabricating a low loss waveguide for high conversion efficiency is the main drawback of this method. On the other hand, strict transparency, the wide conversion bandwidth (90 nm), polarization diversity, polarization independent conversion efficiencies and spectrum inversion capabilities are some of the promising qualities of the semiconductor DFG conversion technique.

V. Conclusion

Wavelength conversion is a key function in providing full scalability in WDM networks. Wave-mixing converters are the only category of wavelength converters that offer a full range of transparency. Opto-electronic converters and optical gating converters offer limited digital transparency. The discussions are two-fold. What is the advantage of strictly transparent wavelength conversion over nontransparent wavelength conversion? Provided that we discard transparent wavelength conversion, what is the advantage of all-optical methods over O/E–E/O conversion? These discussions are similar to those on fiber-amplifiers versus optoelectronic repeaters. Fiber amplifiers are considered all-optical (no electrical signal routing) and strictly transparent. In reality, fiber amplifiers are rapidly replacing the optoelectronic regenerators that are already in place today. Deployment of fiber amplifiers in the trunk does not limit network users' access to signals of any specific protocols or formats. The upgrade of network capacity is simple in networks with such transparent elements. In addition, transparent networks can accommodate complex encrypting methods for enhancing the network security. The advantage of all-optical over optoelectronic methods becomes clear as multiple wavelengths are involved and as cost-effective management of crosstalk must be achieved. As WDM technology becomes viable, transparency becomes more important for seamless evolution of the network with full interoperability. Wavelength conversion methods with limited digital transparency are available today for deployment. On the other hand, wave-mixing wavelength conversion show its potential for strictly transparent WDM network applications.

Acknowledgment

The author would like to thank C. A. Brackett and K. Bala for enlightening discussions. The author is indebted to R. Bhat, C. Caneau, M. A. Koza, A. Rajhel, and N. Antoniades for their efforts in making possible the experimental results discussed in Section IV-C3).

References

[1] C. A. Brackett, "Dense wavelength division multiplexing networks: Principles and applications," *IEEE J. Select. Areas Commun.*, vol. 8, p. 948, 1990.

[2] S. Johansson, "Transparent optical multicarrier networks," in *Proc. ECOC'92*, 1992, vol. We A9.1.

[3] C. A. Brackett, A. S. Acampora, J. Sweitzer, G. Tangonan, M. T. Smith, W. Lennon, K.-C. Wang, and R. H. Hobbs, "A scalable multiwavelength multihop optical network: A proposal for research on all-optical networks," *J. Lightwave Technology*, vol. 11, p. 736, 1993.

[4] R. Ramaswami and K. N. Sivarajan, "Routing and wavelength assignment in all-optical networks," *IEEE/ACM Trans. Networking*, p. 489, 1995.

[5] K.-C. Lee and V. O. K. Li, "A wavelength-convertible optical network," *J. Lightwave Technol.*, vol. 11, p. 962, 1993.

[6] K. Bala *et al.*, "Candidate network architecture for the MONET local exchange Network," 1995.

[7] A. A. M. Saleh, "Transparent optical networks for the next generation information infrastructure,"*OFC'95*, paper ThEl, 1995.

[8] WIXC is formally defined as a network element that accepts multi-wavelength signals from a transport facility (and may accept single wavelength signals at the client network interfaces), and cross-connects individual wavelengths using wavelength interchange (when necessary).

[9] B. Glance, J. M. Wiesenfeld, U. Koren, A. H. Gnauck, H. M. Presby, and A. Jourdan, "High performance optical wavelength shifter," *Electron. Lett.*, vol. 28, p. 1715, 1992.

[10] J. M. Wiesenfeld, J. S. Perino, A. H. Gnauck, and B. Glance, "Bit error rate performance for wavelength conversion at 20 Gbit/s," *Electron. Lett.*, vol. 30, p. 720, 1994.

[11] B. Mikkelsen, M. Vaa, R. J. Pedersen, T. Durhuus, C. Joergensen, C. Braagaard, N. Storkfelt, K. E. StuLkjaer, p. Doussiere, G. Garabedian, C. Graverm, E. Derouin, T. Fillion, and M. Klenk, "20 Gb/s polarization insensitive wavelength conversion in semiconductor optical amplifiers," in *ECOC'93*, postdeadline paper, Thp 12.6, 1993.

[12] K. Shimizu, T. Horiguchi, and Y. Koyamada, "Technique for translating light-wave frequency by using an optical ring circuit containing a frequency shifter," *Optic. Lett.*, vol. 17, p. 1307, 1992.

[13] F. Heismann and R. Ulrich, "Integrated-optical frequency translator with stripe waveguide," *Appl. Phys. Lett.*, vol. 45, p. 490, 1984.

[14] S. J. B. Yoo, R. Bhat, C. Caneau, M. Amersfoort, A. Rajhel, N. Antoniades, and M. A. Koza, "Quasi-phase-matched semiconductor nonlinear optical waveguides,"*J. Lightwave Technol.*, to be submitted.

[15] S. J. B. Yoo, "All-Optical Wavelength Converter for WDM Networks," U.S. Patent 5,434,700, 1993.

[16] N. A. Olsson, "Lightwave systems with optical amplifiers," *J. Lightwave Technol.*, vol. 7, p. 1071, 1989.

[17] R. I. Laming, A. H. Gnauck, C. R. Giles, M. N. Zervas, and D. N. Payne, "High sensitivity optical pre-amplifier at 10 Gbit/s employing a low noise composite EDFA with 46 dB gain," post-deadline papers, in *Optic. Amplifiers Appli. Topic. Meet., 1992*, Washington, DC, 1992, paper PD 13.

[18] Veritech Microwave Datasheet.

[19] B&H electronics Datasheet.

[20] The actual noise figure for semiconductor optical amplifier wavelength converter is higher than that of the amplifier itself. See discussions in: Nobuhiko Kikuchi and Shinya Sasaki, "Noise analysis for optical frequency conversion using nearly degenerate four wave mixing in semiconductor amplifier," *J. Lightwave Technol.*, vol. 11, p. 819, 1993.

[21] Lucent Technologies Datasheet.

[22] Lucent Technologies Preliminary Datasheet.

[23] BT&D (HP) Datasheet.

[24] K. Inoue and K. Oda, "Noise suppression in wavelength conversion using a light injected laser diode," *IEEE Photon. Technol. Lett.*, vol. 7, p. 500, 1995.

[25] H. Kawaguchi, K. Oe, H. Yasaka, K. Magari, and M. Fukuda, "Tunable optical wavelength conversion using a multielectrode distributed-feedback diode with a saturable absorber," *Electron. Lett.*, vols. 23/24, p. 1088, 1987.

[26] H. Kawaguchi, K. Magari, H. Yasaka, M. Fukuda, and K. Oe, "Tunable optical-wavelength conversion using an optically triggerable mutielectrode distributed feedback laser diode," *IEEE J. Quantum Electron.*, vol. 24, p. 2153, 1988.

[27] H. Nobuhara, K. Kondo, S. Yamakoshi, and K. Wakao, "Optical exclusive-or operation using tunable wavelength conversion laser diode," *Electron. Lett.*, vol. 25, p. 1485, 1989.

[28] K. Kondo, H. Nobuhara, S. Yamakoshi, and K. Wakao, "Giga-bit operation of wavelength conversion laser," in *Topic. Meet. Photon. Switch*, Kobe, Japan, 1990, paper 13D9, vol. 1, p. 99.

[29] P. E. Barnsley and P. J. Fiddyment, "Wavelength conversion from 1.3–1.55 μm using split contact optical amplifiers," *IEEE Photon. Technol. Lett.*, vol. 3, p. 256, 1991.

[30] P. Ottolenghi, A. Jourdan, and J. Jaquet, "All-optical wavelength conversion with extinction ratio enhancement using a tunable DBR laser," in *Topic. Meet. Photon. Switch*, Kobe, Japan, 1990, paper 13D9, p. 199.

[31] M. Schilling, W. Idler, D. Baums, K. Dutting, G. Laube, K. Wunstel, and O. Hildebrand, "6 THz range tunable 2.5 Gb/s frequency conversion with a MQW Y-laser," *IEEE J. Quantum Electron.*, vol. 29, 1993.

[32] C.-C. Lu, S. Jiang, P. S. Yeh, P. J. S. Heim, C. E. C. Wood, and M. Dagenais, "Wavelength conversion using a T-gate laser," *IEEE Photon. Technol. Lett.*, vol. 8, p. 52, 1996.

[33] H. Yasaka, K. Takahata, K. Kasaya, and K. Oe, "Frequency response of a unidirectional-output optical frequency conversion device with an

asymmetrical-κ DBR laser structure," *IEEE Photon. Technol. Lett.*, vol. 5, p. 1306, 1993.

[34] S. Dubovitsky, P. D. Dapkus, A. Mathur, and W. H. Steier, "Wavelength conversion in a quantum well polarization insensitive amplifier," *IEEE Photon. Tech. Lett.*, vol. 6, p. 804, 1994.

[35] J. M. Wiesenfeld, B. Glance, J. S. Perino, and A. H. Gnauck, "Wavelength conversion at 10 Gb/s using a semiconductor optical amplifier," *IEEE Photon. Technol. Lett.*, vol. 5, p. 1300, 1993.

[36] B. Mikkelsen M. Vaa, R. J. Pedersen, T. Durhuus, C. Joergensen, C. Braagaard, N. Storkfelt, K. E. Stubkjaer, P. Doussiere, G. Garabedian, C. Graver, E. Derouin, T. Fillion, and M. Klenk, "20 Gbit/s polarization insensitive wavelength conversion in semiconductor optical amplifiers, " *ECOC '93*, Paper ThP 12.6.

[37] J. S. Perino, J. M. Wiesenfeld, and B. Glance, "Fiber transmission of 10 Gbit/s signals following wavelength conversion using a traveling-wave semiconductor optical amplifier," *Electron. Lett.*, vol. 30, p. 256, 1994.

[38] J. M. Wiesenfeld and B. Glance, "Cascadability and fan out of semiconductor optical amplifier wavelength shifter," *IEEE Photon. Technol. Lett.*, vol. 5, p. 1168, 1992.

[39] T. Durhuus, C. Joergensen, B. Mikkelsen, R. J. S. Pedersen, and K. E. Stubkjaer, "All optical wavelength conversion by SOA's in a Mach–Zehnder configuration," *IEEE Photon. Technol. Lett.*, vol. 6, p. 53, 1994.

[40] F. Ratovelomanana, N. Vodjdani, A. Enard, G. Glastre, D. Rondi, R. Blondeau, C. Joergensen, T. Durhuus, B. Mikkelsen, K. E. Stubkjaer, A. Jourdan, and G. Soulage, "An all-optical wavelength-converter with semiconductor optical amplifiers monolithically integrated in an asymmetric passive Mach–Zehnder interferometer," *IEEE Photon. Technol. Lett.*, vol. 7, p. 992, 1995.

[41] X. Pan, J. M. Weisenfeld, J. S. Perino, T. L. Koch, G. Raybon, U. Koren, M. Chien, M. Young, B. I. Miller, and C. A. Burrus, "Dynamic operation of a three-port, integrated Mach–Zehnder wavelength converter," *IEEE Photon. Technol. Lett.*, vol. 7, p. 995, 1995.

[42] X. Pan and T. L. Koch, "Intensity noise characteristics of a Mach–Zehnder wavelength converter," *IEEE Photon. Technol. Lett.*, vol. 7, p. 1276, 1995.

[43] M. Schilling, K. Daub, W. Idler, D. Baums, U. Koerner, E. Lach, G. Laube, and K. Wunsterl, "Wavelength converter based on integrated all-active three-port Mach–Zehnder interferometer," *Electron. Lett.*, vol. 30, p. 2128, 1994.

[44] M. Schilling, W. Idler, G. Laubem, K. Daub, K. Dutting, E. Lach, and K. Wunstel, "10 Gb/s monolithic MQW based wavelength converter in Michelson interferometer configuration," in *OFC'96*, San Jose, CA, 1996, paper WG2.

[45] K. A. Rauschenbach, K. L. Hall, J. C. Livas, and G. Raybon, "All-optical pulse width and wavelength conversion at 10 Gb/s using a nonlinear optical loop mirror," *IEEE Photon. Technol. Lett.*, vol. 6, p. 1130, 1994.

[46] D. A. O. Davies, A.D. Ellis, T. Widdowson, and G. Sherlock, "10 Gb/s data switched semiconductor laser amplifier nonlinear optical loop mirror," *Electron Lett.*, vol. 31, p. 111, 1995.

[47] E. A. Swanson and J. D. Moores, "A fiber frequency shifter with broad bandwidth, high conversion efficiency, pump, and pump ASK cancellation, and rapid tunability for WDM optical networks," *IEEE Photon. Technol. Lett.*, vol. 6, p. 1341, 1994.

[48] H. Q. Le and S. Di Cecca, "Ultrafast, multi-THz detuning, third-order frequency conversion in semiconductor quantum well waveguides," *IEEE Photon. Technol. Lett.*, vol. 4, p. 878, 1992.

[49] K. Inoue, "Tunable and selective wavelength conversion using fiber four-wave mixing with two pump lights," *IEEE Photon. Technol. Lett.*, vol. 6, p. 1451, 1994.

[50] G. P. Agrawal, "Population pulsations and nondegenerate four-wave-mixing in semiconductor lasers and amplifiers," *J. Opt. Soc. Amer. B*, vol. 5, p. 147, 1988.

[51] R. Schnabel, U. Hilbk, Th. Hermes, P. Meibner, Cv. Helmolt, K. Magari, F. Raub, W. Pieper, F. J. Westphal, R. Ludwig, L. Kuller, and H. G. Weber, "Polarization insensitive frequency conversion of a 10 channel OFDM signal using four-wave-mixing in a semiconductor laser amplifier," *IEEE Photon. Technol. Lett.*, vol. 6, p. 56, 1994.

[52] Two counter-propagating pump waves with orthogonal polarizations is one example.

[53] J. Zhou, N. Park, K. J. Vahala, M. Newkirk, and B. I. Miller, "Four-wave mixing wavelength conversion efficiency in semiconductor traveling-wave amplifiers measured to 65 nm of wavelength shift," "*IEEE Photon. Technol. Lett.*, vol. 6, p. 984, 1994.

[54] C. Q. Xu, H. Okayama, and M. Kawahara, "1.5 μm band efficient broadband wavelength conversion by difference frequency generation in a periodically domain-inverted $LiNbO_3$ channel waveguide," *Appl. Phy. Lett.*, vol. 63, p. 3559, 1993.

[55] M. L. Bortz, D. Serkland, M. M. Fejer, and S. J. B. Yoo, "Near-degenerate difference frequency generation at 1.3 μm in $LiNbO_3$ waveguides for application as an all-optical channel shifter,"*CLEO*, 1994, paper CTHD6.

[56] S. J. B. Yoo, C. Caneau, R. Bhat, and M. A. Koza, "Wavelength conversion by quasi-phasematched difference frequency generation in AlGaAs waveguides," *OFC '95, paper, PD-13.*

[57] S. J. B. Yoo, C. Caneau, R. Bhat, M. A. Koza, A. Rajhel, N. Antoniades, "All optical wavelength conversion by quasi-phasematched difference frequency generation in AlGaAs waveguides," *Appl. Phys. Lett.*, vol. 68, p. 2609, 1966.

[58] ——, Quasi-phasematched second-harmonic generation in AlGaAs waveguides with periodic domain inversion achieved by wafer-bonding," *Appl. Phys. Lett.*, vol. 66, p. 3410, 1995.

[59] A Network Access Module (NAM) implemented as part of the ONTC-II program partially supported by Advance Research Project Agency.

[60] S. L. Danielsen, C. Joergensen, M. Vaa, K. E. Stubkjaer, P. Doussiere, F Pommerau, L. Goldstein, R. Ngo, M. Goix, "Bit error rate assessment of a 40 Gbit/s all-optical polarization independent wavelength converter," *OFC'96*, paper PD-12.

All-Optical WDM Packet Networks

STEVE P. MONACOS, JOHN M. MOROOKIAN, LARRY DAVIS,
LARRY A. BERGMAN, MEMBER, IEEE, SIAMAK FOROUHAR, AND JON R. SAUER

Abstract— **This paper describes the components and subsystems for the implementation of a multi-GHz optoelectronic data transport network using self-routing packets in a multi-hop network. The short packet payloads are compressed using optical wavelength division multiplexing techniques, and remain optical from source to destination while traversing the switching nodes. The routing is done with a lean, self-routing *hot potato* protocol in order to avoid the need for data storage at the switching nodes and to provide a fixed node latency equivalent to a few meters of fiber. Sustainable throughput both into and out of the electronic host at each node should exceed 10 Gb/s. Some technical details of the switching nodes and interfaces of the recirculating shuffle network, and the stepped wavelength laser arrays and testbed will be given.**

I. Introduction

OPTOELECTRONIC TECHNOLOGY is now yielding useful new devices and techniques yet to be exploited by practical systems. The limitations of conventional electronics are increasingly severe in terms of achievable data rates and system extent, especially in the interconnection networks of high performance systems. The natural next extension of optoelectronic technology is then into more sophisticated and physically compact interconnection systems. The requirements for the highest possible speeds with only limited logic functions are a good potential match to current optoelectronic technology. To date, however, no high-performance photonic system is able to duplicate the routing capability and minimal source-to-destination delay for word size data transfers. This paper outlines a proposed optical distributed interconnection architecture with unprecedented capacity that is also able to incorporate these desired features [1]. The system would be a multihop, high-capacity network transporting short packets optically from source to destination. The geographical scale is primarily from backplane to campus scale environments, although continental scales are not hard to accommodate. The short packets do not need to be synchronized and result in maximal flexibility for network users with packet sizes of one to several words. To develop such a network will require innovations in optoelectronic component design to meet network architecture requirements, but its existence would open up opportunities in computer and telecommunications system design that are unreachable today.

The development of low-loss, low dispersion silica-based fiber and high speed semiconductor optical sources and photodetectors has resulted in photonics dominating long haul point-to-point transmission systems. The tremendous bandwidth capacity of optical fibers, however, is largely untapped by today's networks due to the speed limitations of electronics in the data path [2]. Present long haul communications are dominated by transmission at 1.55 μm; silica fibers have a low loss region around this wavelength of $\sim$200 nm (THz) [3]. Wavelength division multiplexing (WDM) is a powerful technique to access this tremendous bandwidth by simultaneously transmitting two or more signals at different optical wavelengths over the same fiber [4]. By utilizing multiple channels, each at a lower bit rate and on a separate wavelength, a very large aggregate bit rate is achieved (i.e. number of wavelengths $\times$ bit rate per wavelength) [2]. The use of erbium doped fiber amplifiers (EDFA's) will limit the available bandwidth of the fibers in the 1.55 μm region to the gain bandwidth of the amplifier ($\sim$47 nm or 5.9 THz) [5], which still supports >20 channels at 2.5 Gb/s per channel (assuming 100 GHz/channel allocation [6]). Thus, WDM allows one to access the tremendous bandwidth of optical fibers while still allowing transmitters and receivers to operate at the single-channel transmission rate [4].

WDM systems have a number of important applications related to interoffice and local area networks (LAN) as well as wide area networks (WAN) for long haul communications [7], [8], including high rate communication architectures, supercomputer links [9], [10], separation of control signals from data signals, phased arrays, and concurrent (parallel) processors [11]–[13]. This multiwavelength format can dramatically increase the capacity of present transmission systems without requiring significant technological developments in the modulation bandwidth of high speed electronics, transmitters and receivers.

Current lightwave communications systems are limited by the use of electronics for signal regeneration, packet routing and buffering at intermediate switching nodes. If a WDM data format is used with electronics in the data path, the packets must be optically demultiplexed and converted into separate electronic channels. Each optical channel requires separate hardware to buffer and route the data. Finally, the individual channels are converted back to optical form and multiplexed back into one fiber. By avoiding the use of optoelectronic conversions at intermediate nodes, WDM techniques can further leverage today's technology of intensity-modulated, direct

Manuscript received April 18, 1995; revised November 27, 1995. This work was supported by the BMDO/IST. The research described in this paper was carried out by the Jet Propulsion Laboratory, California Institute of Technology, and was supported by the Ballistic Missile Defense Organization, Innovative Science and Technology Office, and the National Aeronautics and Space Administration. This work was performed as part of JPL's Center for Space Microelectronics Technology.

The authors are with the Jet Propulsion Laboratory, Pasadena, CA 91190 USA.

Publisher Item Identifier S 0733-8724(96)04088-1.

detection systems to realize much higher capacity networks without modification of the network fabric due to the WDM format.

The components needed to realize the full benefit of a WDM network are monolithic WDM laser diode and detector arrays, a WDM network interface, and an all-optical data path switching node. In this paper, we describe ongoing work in these areas. Section II discusses a new type of recirculating network topology well suited to routing of optical packets. In this section, we further describe a switching node implementation for this network and a conceptual interface design to interconnect this network to existing electronic networks. In Section III, we present a simplified version of this interface used to construct a four-channel WDM HIPPI link testbed. In this section, we also discuss progress on a four-element single chip stepped wavelength DFB laser diode array for high-data rate WDM communication systems. Section IV discusses potential applications for multigigabit/sec communications networks. Future work is presented in Section V, and conclusions are given in Section VI.

II. Multicylinder ShuffleNet (MCSN)

To take full advantage of the benefits of WDM technology, we desire a network capable of routing optical packets. The fundamental problem in routing optical packets is that there is currently no good way of storing data in optical form to handle contention problems. As such, a new topology and control structure was developed which places no timing constraints on arrival times of data packets at switching nodes and uses fixed set up latencies for making routing decisions at the switching nodes of the network. The end result is the MCSN architecture with a short packet deflection protocol (hot potato) [1] which can be easily interfaced to one or more industry standard protocols for connection to existing networks.

The goal of the MCSN is to provide a methodology well suited to the routing of optical WDM packets, without header modification or optoelectronic conversions and with very low routing delay for fine-grain distributed supercomputing. Detailed simulations of this network exhibit crossbar like routing characteristics with near 100% availability of the network input ports by significantly reducing the number of packet deflections [1].

One consequence of using a deflection routing protocol is that packet deflections can occur that results in delivering packets out of sequence. A possible solution to resequencing packets is to add a source host tag and packet tag to each packet for use by the receiving host to reorder packets after transmission. Current work focuses on developing a proof-of-concept all-electronic MCSN prototype switching node as a precursor to building an electronic 8-node MCSN. In this section we describe the basic architecture of the MCSN and the functional requirements for the switching nodes and interfaces for this network.

A. MCSN Architecture

The MCSN topology described in [1] is a network which uses a recirculating shuffle network (SN) with an all-optical data path from source to destination. The difficulty with this network architecture is that packet deflections result even at very low network loading [1]. To improve network performance we augment the basic SN topology with multiple parallel copies of the original topology called *routing cylinders*. Additionally, the switching nodes and links of the network dynamically store packets and provide congestion control within the network by routing blocked packets onto alternative routing cylinders. The architecture is easily scalable and uses the hot potato protocol in Ref. [14] but without age/priority information. The MCSN architecture is also designed for *packet asynchronous* traffic, thus avoiding the need to synchronize packets entering the network.

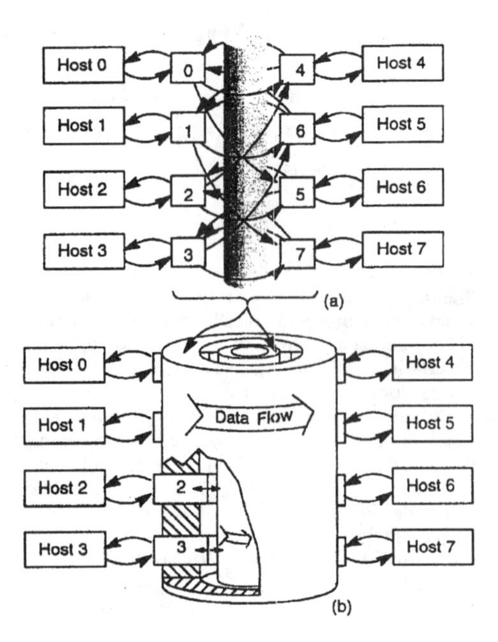

Fig. 1. An eight-node ShuffleNet. (a) Illustrates the single-cylinder ShuffleNet topology [1] and (b) shows a two-cylinder topology where each cylinder performs a perfect shuffle interconnection between stages of switching nodes. The cylinders in Fig. 1(b) are also connected to each other at each switching node to allow packets to circulate through the network on any cylinder.

Fig. 1(a) is an example of an 8-node single-cylinder SN topology. Fig. 1(b) is an example of an 8-node two-cylinder SN topology. An R-cylinder MCSN topology is topologically equivalent to the generic SN topology but has R parallel perfect shuffle interconnections between stages of SN nodes (i.e., we expand each node-to-node link to R *data paths* in the R-cylinder SN) [1]. We accommodate this augmentation to the generic SN topology by expanding the number of ports at each node to R times that of the basic SN switching node shown in Fig. 2. Additionally, some of the R data paths to the host may instead be used for local recirculation links of blocked packets [1].

B. MCSN Node

The MCSN switching nodes internally use a *Permutation Engine* (PE), which is described in detail in [15]. The PE is a simple distributed routing control mechanism for routing

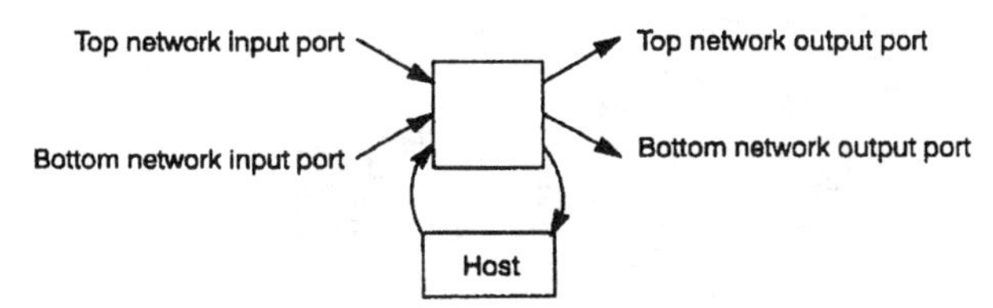

Fig. 2. Single-cylinder SN node with three ports at each node [1].

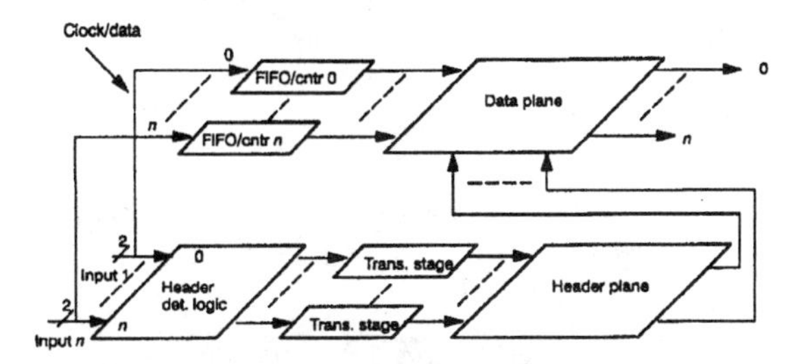

Fig. 3. MCSN switching node configuration.

packet-asynchronous data using only local traffic information to make routing decisions. For the R-cylinder configuration, the PE switching nodes also provide dynamic routing of packets between SN cylinders.

Fig. 3 is a block diagram representation of a MCSN switching node. This node consists of the header detection logic, the header translation stage, the header plane, the first-in–first-out (FIFO) buffers, and the data plane. The header detection logic is used to detect an n bit-serial packet header and convert it to an n-bit parallel header. The translation stage is used to convert a global packet header to a local node output port number. The header plane routes packet headers to establish the input port to output port connection based on the PE routing algorithm. The FIFO buffers are used to delay incoming packets for a fixed time interval based on the set up times of the header detection logic, the translation stage, and the header plane. Finally, the data plane is used to route packets based on the control signals from the header plane.

1) Prototype Switching Node: A prototype all-electronic implementation of the switching node in Fig. 3 is currently in development at JPL. The basic configuration is a 12 input to 12 output PE on a 6U × 220 mm VME wire wrap card [16]. The number of input/output ports was selected based on the MCSN design requirements for an eight-node system [1]. The header translation stage shown in Fig. 3 was omitted to simplify testing and verification of the node design. For an eight-node MCSN demonstration system, a small amount of translation logic can be incorporated into the header detection logic of each node. This logic must be customized for each node to implement the deflection routing protocol similar to that in [14] but without age/priority information. For large scale MCSN's (>100 nodes) a simple memory look up table can be used to implement the translation stage. The basic layout used for this prototype node is shown in Fig. 4. It separates the header plane logic from the data plane logic to allow for ease of scaling in the number of input/output ports. In addition to the node design shown in Fig. 3, Fig. 4 also

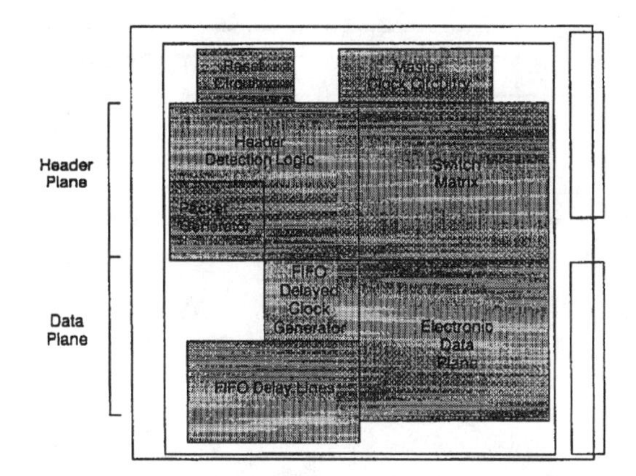

Fig. 4. 12 × 12 MCSN switching node wire wrap board layout.

shows a packet generator. This component is used to generate packet headers for injection into the node to verify routing of the headers and assess packet flow through the node.

C. MCSN Network Interface

A *message* contains the total communication from a source host to a receiving host and is defined as a sequence of packets [1]. A data packet is the indivisible unit of information which may exist as a serial time signal or wavelength-parallel signal as shown in Fig. 5. A serial packet shown at the top of Fig. 5 consists of an n-bit binary packet header; a d-bit payload or data for this header; and a t-bit trailer marking the end of the packet [1]. The bottom of Fig. 5 shows how sequential segments of the serial packet are mapped onto different optical frequencies to result in a wavelength-parallel format. The header contains the address of the desired destination host and marks the beginning of a packet. It is used to modulated the optical frequency λ_H in Fig. 5. The d-bit payload is segmented into N data words which modulate the optical frequencies from λ_1 to λ_N. The start bits denote the beginning of a packet and the location of the packet header at wavelength λ_H. The stop bit marks the end of a packet. These bits modulate λ_C and denote the packet boundaries of the wavelength-parallel packet.

The serial packet format shown at the top of Fig. 5 is suitable for packet switched networks which can handle long duration packets. The performance of such networks is significantly enhanced by using buffers to avoid packet deflection from output port contention [17], [18]. For the proposed all-optical data path switching node, such buffering capability is not currently available. In order to reduce the probability of packet contention at the output ports of a switching node, it is desirable to reduce the packet duration by parallelizing the header, data and trailer information as shown in Fig. 5. For an all-optical data path network, this scheme is realized by using the wavelength division multiplexed (WDM) format shown at the bottom of Fig. 5.

The basic problem is that existing protocols are structured to utilize the strengths of electronics, namely good component functionality with data rates in the 100's of Mb/s range, static storage, and μs path configuration times. These properties

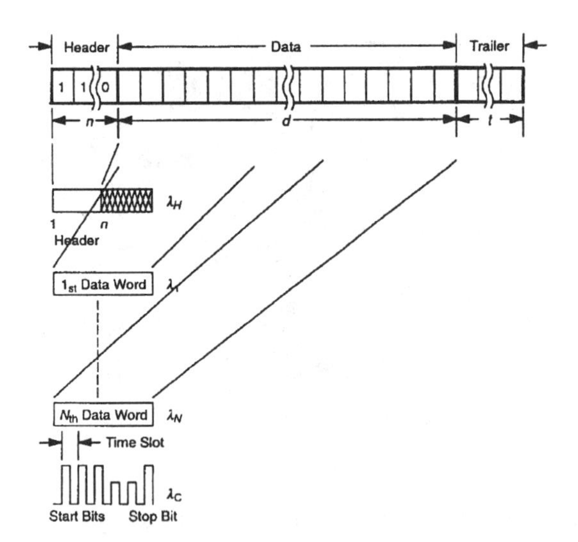

Fig. 5. Packets in serial and wavelength parallel signal formats [1].

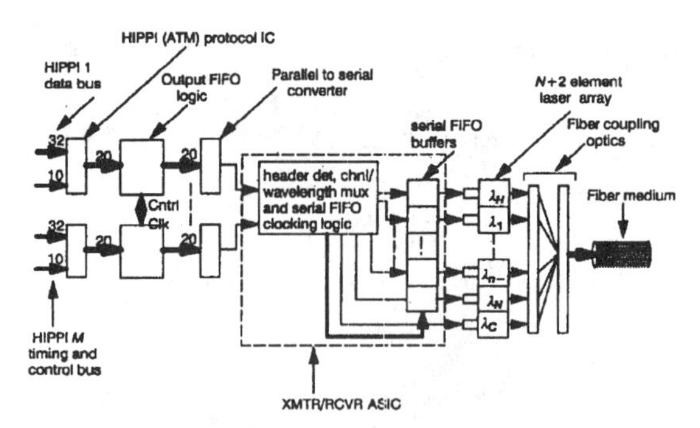

Fig. 6. N wire $\times M$ HIPPI's transmitter.

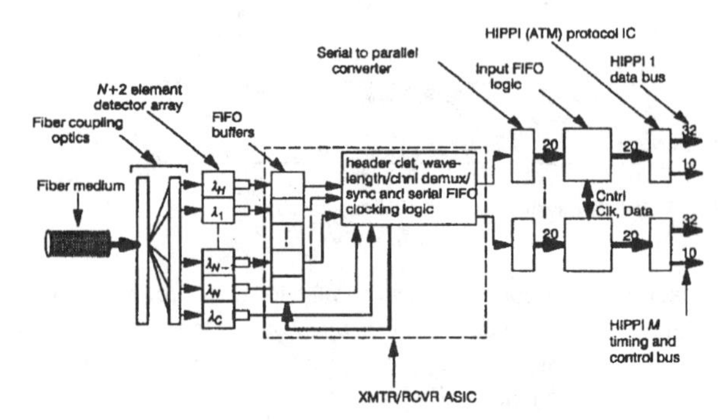

Fig. 7. N wire $\times M$ HIPPI's receiver.

have resulted in a variety of circuit switched architectures where the overhead for path set up is reduced by sending long packets. This construct is similar to a burst transfer meaning a large block of data is sent once the bus—data path—has been acquired. The problem with this approach is that the path set up time grows to 10's of ms for WAN's and is unacceptable for low routing delay applications.

An alternative scheme is to build a datagram or packet switched network. In this protocol a header is prefixed to a packet and is used to dynamically route a packet through the network. In the electronic domain such schemes utilize a store-and-forward approach with electronic buffers for packet storage until the data path of a switching node can be established. This type of network is better suited to WAN's because the path set up is done locally at each switching node so that a global request/acknowledge cycle is avoided.

1) Optoelectronic Space Time-Multiplexer: In looking at the requirements for an all-optical data path network and comparing them to the operation of existing electronic networks, we see a dichotomy in terms of the required functionality of each type of network. The all-optical data path network requires switching nodes capable of handling small asynchronous packets with fixed routing latency. The hot potato deflection protocol provides this functionality by misrouting/deflecting a packet instead of storing it in a buffer for an arbitrary period of time until the desired routing path is achieved. Host devices connected to this network, however, use industry standard protocols such as HIPPI to transmit information. Protocols such as HIPPI are store-and-forward in nature and allow a header or packet to sit at an intermediate switching node for arbitrary time periods as a result.

To connect host devices using industry standard protocols to an all-optical data path network requires developing an interface which converts from existing protocols well suited to electronics to a hot potato style scheme well suited to optics. Such an interface was conceptualized and is called a *space-time multiplexer* (STM) due to its ability to multiplex M multiple electronic sources in a time division multiplexed (TDM) fashion onto a multi-wire link. The space multiplexing function occurs by mapping one electronic source onto an N-bit wide link. Figs. 6 and 7 show the transmitter and receiver sections of this interface respectively to realize an N-bit wide link, where each bit corresponds to one wavelength in the WDM data format in Fig. 5.

The three functional elements of the STM interface are segmentation of large messages into small packets, flow control, and header insertion/detection. The first element looks to break up large messages into many smaller packets. Many electronic protocols allow for large messages to reduce the overhead incurred in setting up the data path. For the envisioned optical data path network, however, large messages pose a serious blocking problem in that optical messages can not be statically stored and will be misrouted if a given message ties up a routing path for an extended period of time. Thus, the efficiency of a deflection routing network is improved by using a small packet size. The third element is a consequence of this packetizing process and entails inserting a header before each packet at the transmitter side and removing this header at the receiver. The second element addresses the global request/acknowledge issue by using local flow control between the local host device and the STM interface. This scheme allows for source throttling as an indirect global flow control mechanism without incurring the long time of flight penalty for a global path set up.

III. WDM Network Components

In this section we describe recent progress in developing the necessary components needed to build the MCSN. These com-

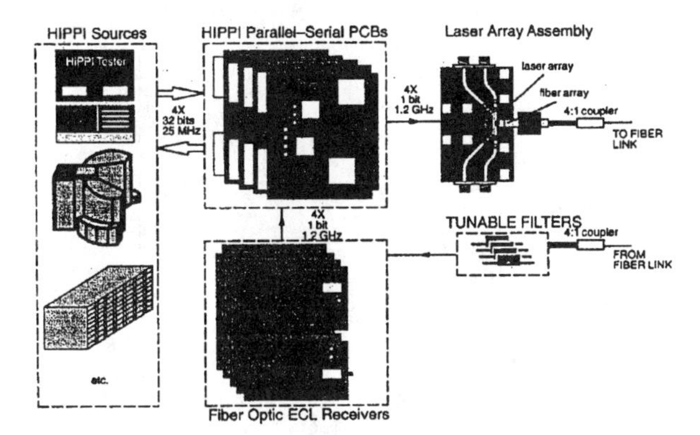

Fig. 8. WDM HIPPI extender block diagram.

ponents consist of a four-wavelength WDM HIPPI link and four-element stepped wavelength laser arrays under development at the Microdevice Laboratory (MDL). Future directions concerning these components are presented in Section V.

A. WDM HIPPI Link

A four channel, multi-gigabit/s, full duplex WDM HIPPI extender link is currently in development at JPL. A block diagram of one of the two identical interfaces is shown in Fig. 8. HIPPI devices at one end of the link provide four parallel HIPPI inputs (32 b wide $\times 25$ MHz $= 800$ Mb/s each) which are routed to the transmit side of the HIPPI parallel-serial printed circuit board, which multiplexes the parallel data down to a serial stream at a 1.2 Gb/s. The four serial HIPPI lines are used to modulate a four-element distributed feedback (DFB) laser diode array which was designed and built at JPL. The optical output from the DFB array is coupled to an array of four optical fibers which are combined to a single optical fiber using a fused coupler. The single fiber output is routed to a 10 km optical fiber link. At the receive end, a 1:4 fused coupler provides four identical copies of the WDM optical signal to a bank of tunable optical bandpass filters. Each filter is manually tuned to pass a single WDM channel and reject the others. Four commercial fiber optic receivers provide ECL-compatible outputs corresponding to the four 1.2 Gb/s serial HIPPI streams. The receiver section of the HIPPI parallel-serial board demultiplexes the serial streams back to 32 b wide HIPPI, which is returned to the destination ports on the HIPPI devices.

B. Four-Element WDM Laser Diode Array

The core of the system described above is a monolithic laser diode array developed at JPL's Microdevices Lab (MDL). The transmitters have four side-by-side single mode DFB lasers made on a single substrate, with each laser emitting light at a slightly different wavelength in the 1.55 μm region. The laser design is an InP-based ridge waveguide laser. Due to the simplicity of fabrication and less stringent fabrication tolerances compared to buried heterostructure lasers, ridge waveguide lasers are seen to have a strong potential for commercial use [19].

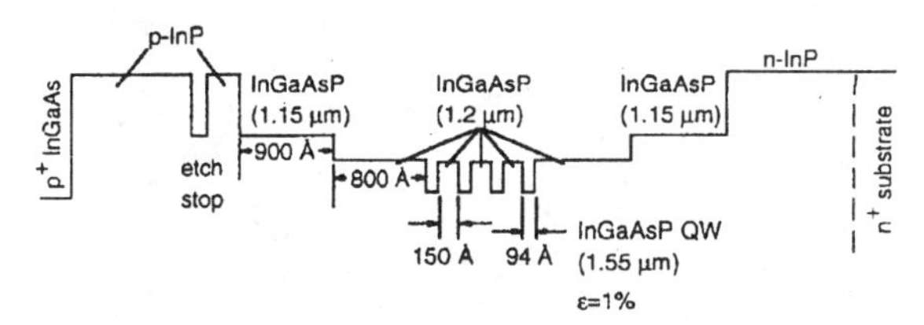

Fig. 9. Layer structure schematic of the InP-based ridge waveguide laser.

1) Laser Array Growth and Fabrication: The laser wafers were prepared by atmospheric pressure metal-organic chemical vapor deposition (MOCVD) on (100)-oriented n^+ InP substrates. The active region consists of four compressively strained ($\varepsilon = 1\%$) InGaAsP quantum wells, each 94 Å wide, with 150 Å barriers of InGaAsP ($\lambda = 1.2$ μm). The optical confinement is provided by a stepped separate confinement heterostructure (SCH) region consisting of 900 Å InGaAsP ($\lambda = 1.2$ μm) and 800 Å InGaAsP ($\lambda = 1.15$ μm), with InP as the top and bottom cladding material. The conduction band profile of the complete laser structure is shown in Fig. 9. Broad area lasers were fabricated to evaluate the quality of the material; measurement of the threshold current and slope efficiency versus cavity length allowed the extraction of the internal quantum efficiency (60%) and the internal loss (17.4/cm).

Fabrication of this material into four-element DFB laser arrays requires e-beam writing of the diffraction gratings, an MOCVD regrowth, and the fabrication of the ridge waveguide structure. The top 4 layers of the laser structure (contact, 2 InP layers, and etch stop in Fig. 9) are removed in order to define the distributed feedback grating in the SCH region. The pitch of the grating for the individual lasers is determined by the modal index and the design criteria of four wavelengths in the range from 1.54–1.56 μm (to be compatible with erbium doped fiber amplifiers). This leads to four grating pitches in the range from 2375–2400 Å. The gratings are e-beam defined in PMMA, and etched into the InGaAsP (1.15 μm) layer using an aqueous solution of HBr and HNO_3. MOCVD is then used to regrow the same 4 layers back onto the structure. Ridge waveguide lasers are then fabricated from this regrown structure. First, the p contact (Ti/Pt/Au) is deposited and annealed; each contact is nominally 3.5 μm wide. A self-aligned wet chemical etch is used to define the ridge waveguide structure. Use of an etch stop allows for reproducible waveguide definition with a pre-determined amount of index-guiding. The amount of index-guiding is dictated by the InP spacer thickness. After the ridge definition etch, polyimide is applied to the wafer and then cured. Oxygen-based reactive ion etching (RIE) is then used to open the polyimide to the p contact. The final top side processing is the lithography and evaporation for the contact metal (Cr/Au). The wafer is then lapped to a thickness of $\sim$100 μm, and then a back contact metal is evaporated (AuGe/Ni/Au). A final anneal completes the laser fabrication, and the devices are then scribed and cleaved.

The devices are soldered to a silicon submount as shown in Fig. 10 and run CW. The submount can be fit into a variety

Fig. 10. Four-element DFB laser submount.

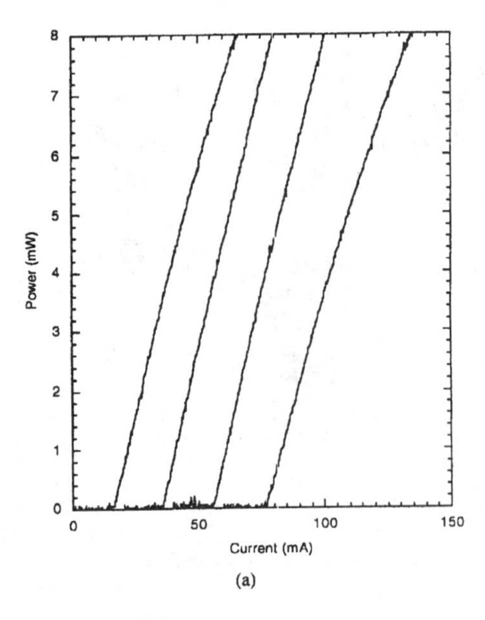

(a)

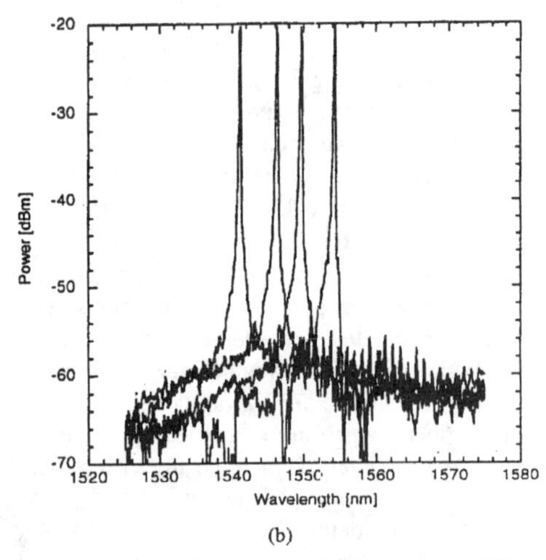

(b)

Fig. 11. (a) L-I characteristics of a typical four-element DFB array; the ordinate is shifted by 20 mA between devices. (b) Spectral characteristics of the four-element DFB array.

of packages and the laser spacing (250 μm) is designed to be compatible with silicon v-groove based fiber arrays. The light vs. current characteristics of a 300 μm long, four-element laser array is shown in Fig. 11(a), showing the uniformity of the threshold and slope efficiency of the devices. Fig 11(b) shows the spectral characteristics of this same array for a drive current of 50 mA, displaying a side mode suppression ratio greater than 20 dB. The finished laser arrays have wavelength separations of approximately 5 nm, very uniform threshold currents as low as 15 mA, output power of several mW, and excellent sidemode suppression ratios.

An important aspect of the WDM laser arrays is the reproducibility of the absolute wavelength and the wavelength spacing. Implementation of WDM systems requires a wavelength reference and definition of the required wavelength spacing in order to build the proper demultiplexing components. However, one finds that the absolute wavelength of the laser emission is directly proportional to the modal index, which can be affected by a number of process variations. Seemingly minor variations in the ridge width and etch depth (less than 1000 Å) can significantly affect the emission wavelength of a DFB laser. It is interesting to note that ridge waveguide structures are less affected by such processing variations than buried heterostructure devices. Figure 12 shows a calculation of the change in emission wavelength with variations in the ridge width—for a ridge width of 3.5 μm, a 1000 Å in the ridge width will change the emission wavelength by approximately 0.5 Å. Buried heterostructure lasers show a much larger variation in the emission wavelength with changes in the active region width [20]. The flexibility of the ridge waveguide structure with respect to process variations allows for good control over the wavelength emission of the laser array. Fig. 13 shows the emission wavelength of the different elements in several laser arrays, displaying the high degree of uniformity in the absolute wavelength and wavelength spacing achieved in our devices. The output wavelength spacing is approximately 5 nm, with a variation ± 1 nm. This accuracy should be adequate for the first generation of WDM devices; however, specifications for future WDM systems may require wavelength spacing of 0.8 nm (100 GHz) [20].

2) Device Speed Performance Tests: Before integration into the WDM HIPPI extender, individual laser arrays are tested to insure that they meet the bandwidth requirement for use in a gigabit/sec system. First, the laser diodes are modulated with a sharp electrical step function. A block diagram of the test setup is shown in Fig. 14. A time-domain relflectometer (TDR) plug-in module for the Tektronix CSA-803 oscilloscope provides a short electrical step function (rise time < 100 ps). This signal is fed from the instrument via coaxial cable to a test jig in which the laser array has been mounted. The 'bias-T'

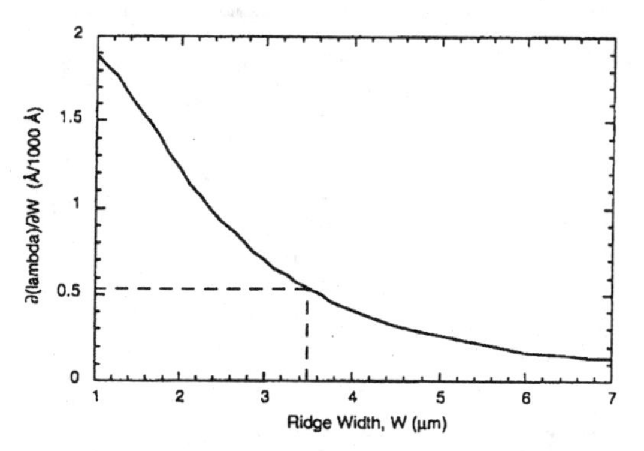

Fig. 12. Variation in the emission wavelength with a 1000 Å change in the ridge width; the calculation was performed using the effective index method. Note that the wider ridges have a reduced sensitivity to this type of processing variation. For our devices with a 3.5 μm ridge width, the wavelength will change by ~0.5 Å with a 1000 Å change in the ridge width.

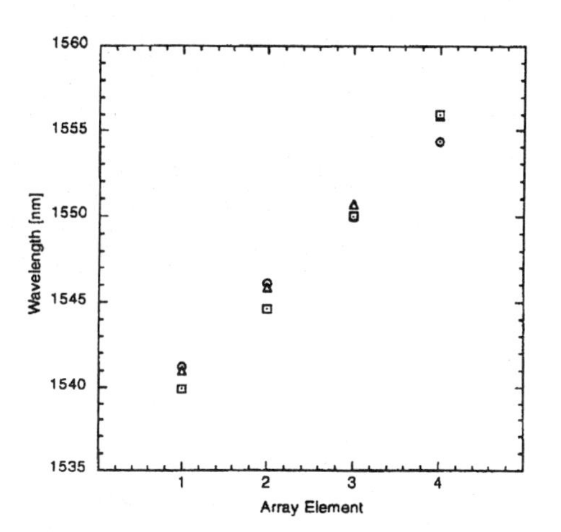

Fig. 13. Emission wavelength vs. array element for a few DFB laser arrays.

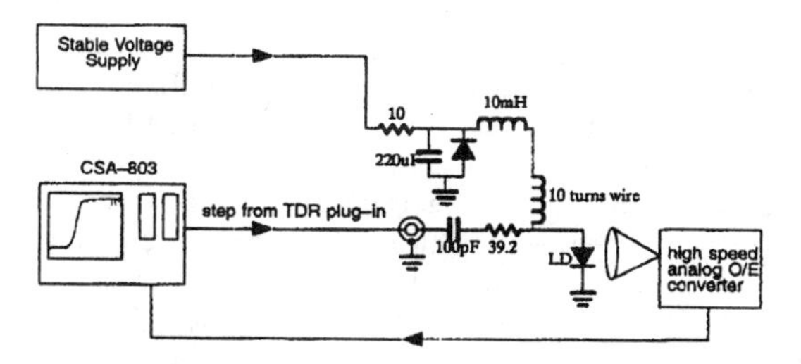

Fig. 14. Step response test setup.

circuit shown in this figure is built using a microstrip design incorporating leadless chip components. The light output is directed to a high-speed analog detector which converts the optical response to an electrical signal which is fed back to the CSA-803 oscilloscope for viewing.

The optical output generally exhibits a degree of cyclical overshoot and undershoot called the relaxation oscillation which gradually dies down as shown in Fig. 15. The frequency of the oscillation places an upper bound on the modulation rate possible under particular bias and drive conditions. For frequencies beyond the relaxation oscillation, the laser response drops off sharply. Therefore, the frequency content of the modulating signal should be kept below the relaxation resonant frequency [21]. It is also preferable for the damping of the oscillations to be high. As the current bias of a laser diode is increased in the linear regime of the L-I curve, the resonant relaxation frequency and the degree of damping increase. In the oscillogram of Fig. 15 the pronounced oscillation is due to the current bias being close to the threshold current. Fig. 16 plots the relaxation resonance against the current bias. It can be seen that this oscillation approaches 3 GHz at higher bias currents. It should be noted that the results of this test are affected by parasitics in both the laser module itself and the surrounding test setup. Thus, the observed resonance is likely limited by the chip components, cables, connectors, and other elements of the test apparatus, but it does indicate that modulation up to the desired rate of 1.2 Gb/s is possible.

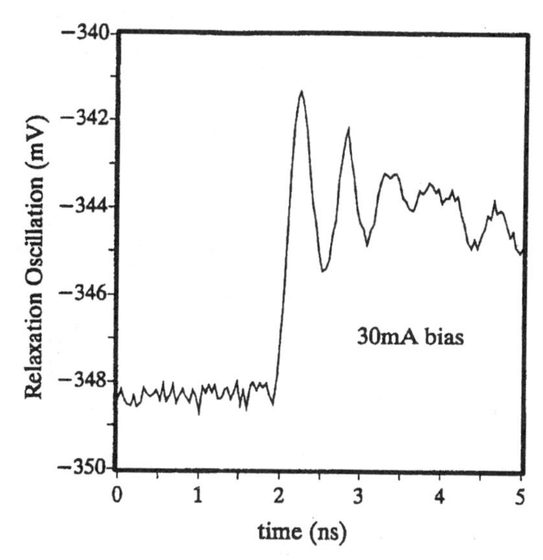

Fig. 15. Relaxation oscillation transient response.

In a second test, the rise and fall times produced in the optical waveform are observed by using a square wave input to each laser. The sum of these should be less than the period of one bit for the desired modulation rate, which is approximately 800 ps for serial HIPPI rates. In Fig. 17, the rise and fall times are limited by the speed of the electrical input to the laser diode (and not the laser response itself), which in this case is a high-speed silicon ECL-level driver.

C. WDM Link Components and System Testing

After completing the testing of individual laser array elements, the next step is to integrate the devices into the WDM HIPPI link for testing and performance characterization of the system. In this section, we describe the details of interfacing the laser array to the driver electronics and coupling of the laser elements to V-groove fiber array. Details concerning system testing are also presented.

1) Laser Diode Module: The DFB array is mounted on a block measuring roughly 250 × 300 × 100 mils. As shown in

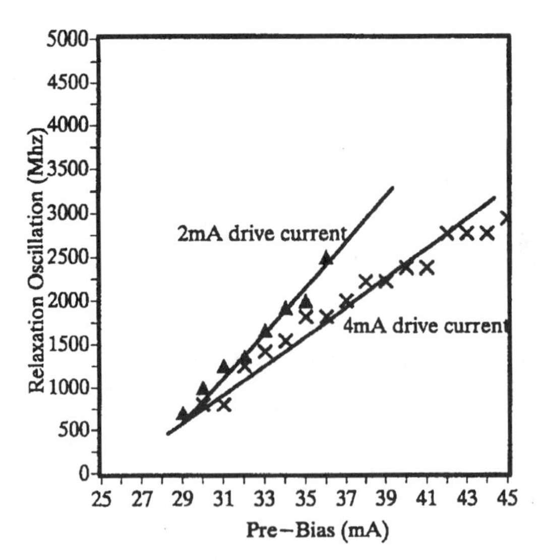

Fig. 16. Relaxation oscillation versus bias.

Fig. 18 the submount is bolted to a custom aluminum block. To the top of this block is mounted a microstrip design printed circuit board on a PTFE substrate which contains the required bias TEE circuitry. The signal input to the bias TEE is routed to the edge of the board. The connector tabs are soldered to these traces and the bodies of the connectors are bolted to the aluminum block for mechanical stability. Mini-clip type connectors are used to bring the bias currents to the inductors in the dc path of the bias TEE.

The optical fiber array to which the light output of the DFB's are coupled, consists of four optical fibers cemented between two silicon v-grooves with an inter-groove spacing of 250 μm to match the physical DFB spacing. The front surface of the fiber array is polished flush to the v-groove endface. The fiber v-groove assembly is positioned in front of the DFB array using a precision 6-axis translation/tilt stage. Both the aluminum block and translation stage may be bolted to an optical bench or to the bottom of a more portable chassis.

Additional printed circuit boards mounted in a card cage along with the HIPPI parallel-serial boards provide a stable bias current to the DFB elements. The bias circuitry provides from 0 to 80 mA of bias current and incorporates a "slow start" feature which gradually ramps up the current to eliminate transient effects. The same printed circuit boards also contain the fiber optic receivers and related circuity.

2) HIPPI Protocol Operation: A HIPPI Tester, made by Input Output Systems Corporation, was used to send and receive parallel HIPPI data in the testbed. Fig. 19 shows the connections between a source and destination device connected by a HIPPI interface. It performs a number of signaling tests and data loopback tests which provide a measurement of the word error rate of the link. Various types of data (all 1's, all 0's, walking 1's, pseudorandom, etc.) may be sent under various signaling conditions.

The following discussion of the HIPPI specification provides background for later discussions on link operation [22], [23]. There are 32 data lines, 4 lines of parity, CLOCK, several handshake lines (REQUEST, CONNECT, READY, PACKET, and BURST), and two interconnect lines (source-to-destination and destination-to-source). The two interconnect lines verify the "hard" connection between the source and destination. If one or more of the interconnect lines are false, then either the HIPPI cables are not connected or one of the HIPPI devices is not powered on. No other signaling is allowed until both interconnect lines are true.

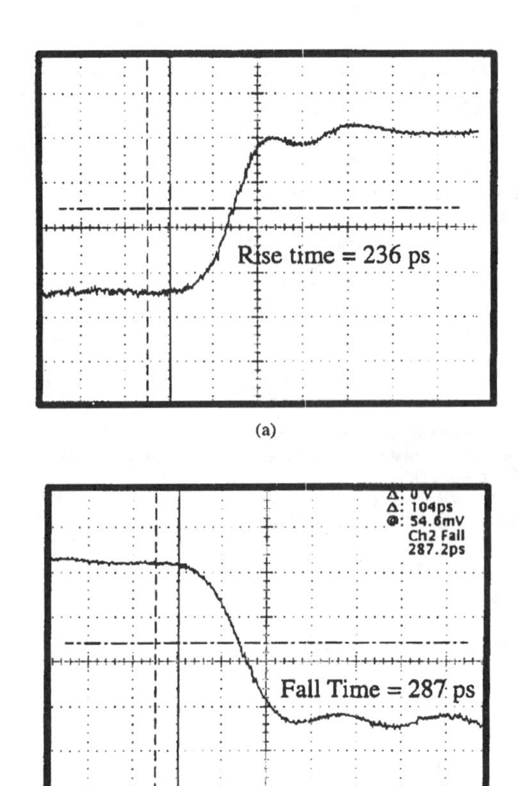

Fig. 17. Rise and fall times.

The operation of the remaining control lines during a typical data transfer is shown in Fig. 20. When the source desires to make a connection, it will assert the REQUEST line and place the I-field (which contains information to select the desired destination) on the 32 b data bus. The destination will respond with CONNECT to make the connection. At this point, the source may assert PACKET to indicate a packet of data is ready to send. The destination send READY indications after CONNECT is asserted for four clock cycles. Each READY indication (at least four clock cycles long) indicates that the destination is ready to receive a burst (256 words of 32 b each). The source will respond by asserting BURST, indicating that the burst is being transferred at a rate of one word per 25 MHz clock cycle. If the REQUEST line is dropped, the destination responds by releasing CONNECT, and the connection is broken. Normally this occurs after BURST and PACKET have been deasserted.

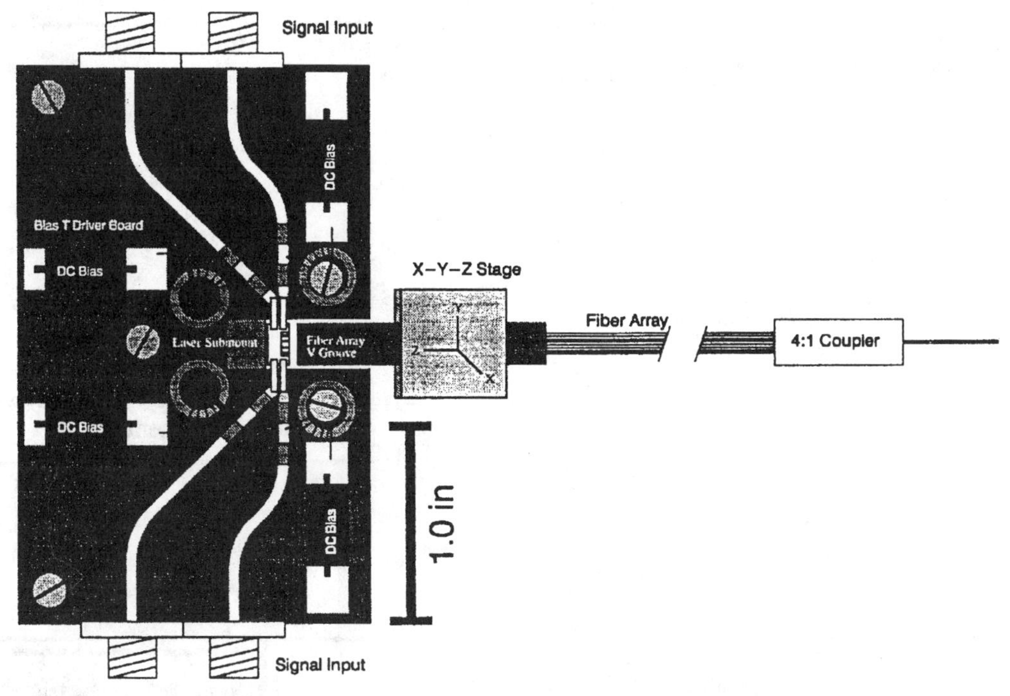

Fig. 18. Laser module and coupling diagram.

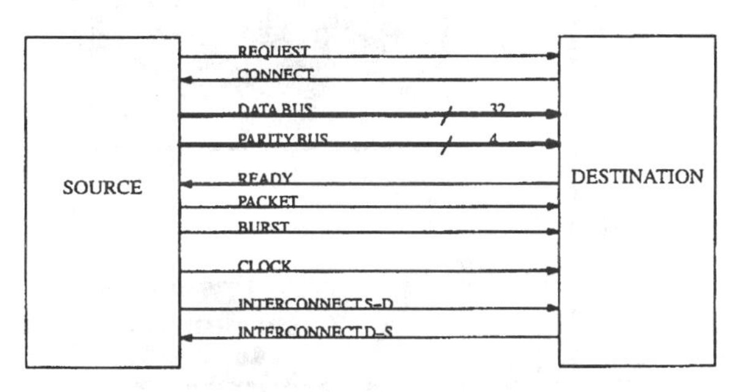

Fig. 19. Parallel HIPPI connections.

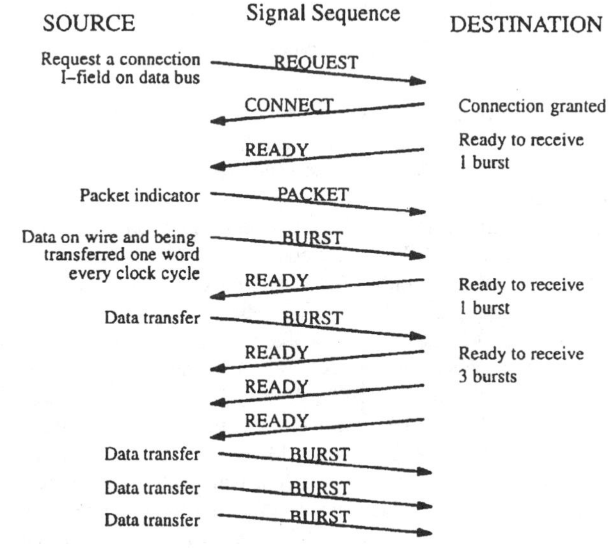

Fig. 20. HIPPI signalling sequence.

3) HIPPI Electronic MUX (Parallel-Serial PCB): To achieve long haul fiber optic transmission of HIPPI data feasibly, the parallel data, parity and control must be reduced to a single serial line. This is accomplished by a JPL-designed HIPPI parallel-serial printed circuit board (PCB). The HIPPI parallel-serial PCB is composed of two major multiplex/demultiplex (MUX/DMUX) chip sets and their supporting components. The first MUX/ DMUX pair translates between the conventional parallel HIPPI format of 8 control and 32 data lines at 25 MHz to a 20 bit-wide form at 50 MHz. A second MUX/DMUX pair converts to/from the 20 parallel lines from/to a serial line which is clocked at a 1.2 GHz rate. MUX and DMUX operations are performed on the same multilayer PCB. An option is included for a local loopback mode in which serial output of the MUX is sent directly to the DMUX input (which is useful for verifying that the PCB is operating correctly).

The second MUX/DMUX pair is responsible for setting up the 1.2 Gb/s link before actual HIPPI data can be sent. Functionally, this MUX/DMUX pair contains a transmitter, receiver, and state machine [24]. The state machine controls the status of the link and has three possible states: frequency acquisition (state 0), waiting for peer (state 1), and sending data (state 2). The state machine decides what state it should be in based on its memory and the type of frame currently

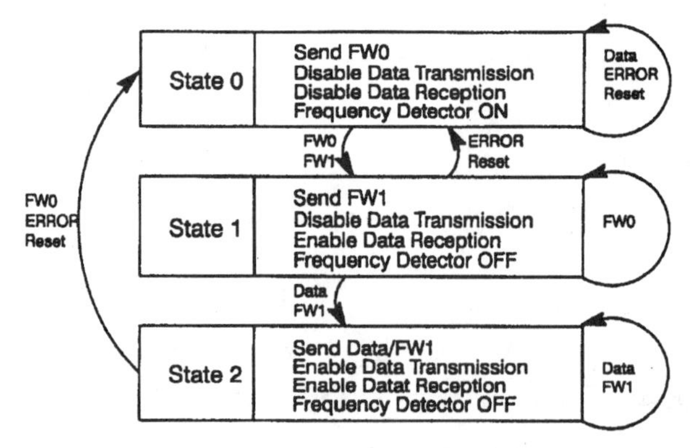

Fig. 21. State machine diagram.

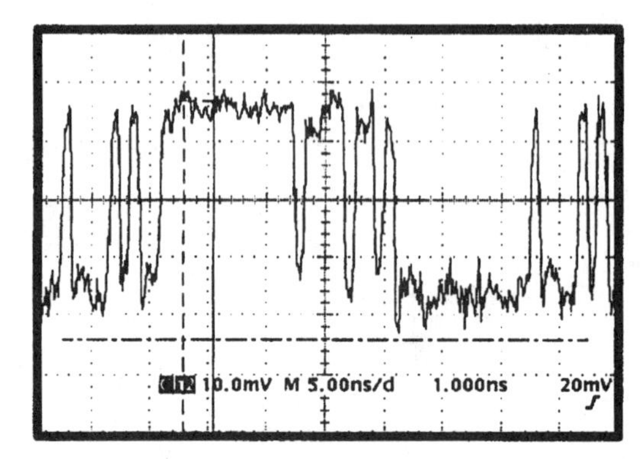

Fig. 22. Optical serial HIPPI 'rest' pattern.

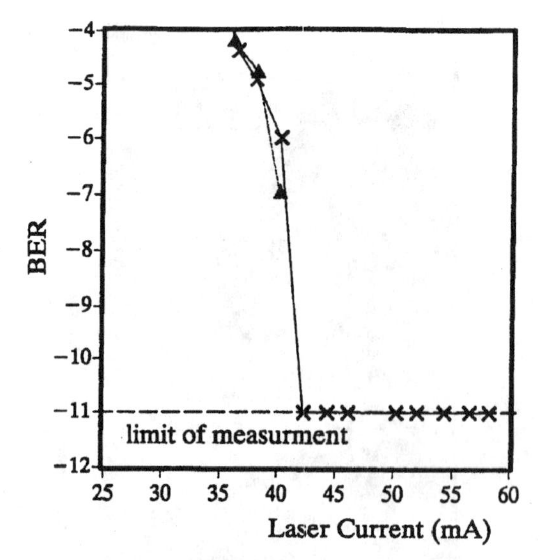

Fig. 23. BER versus laser current.

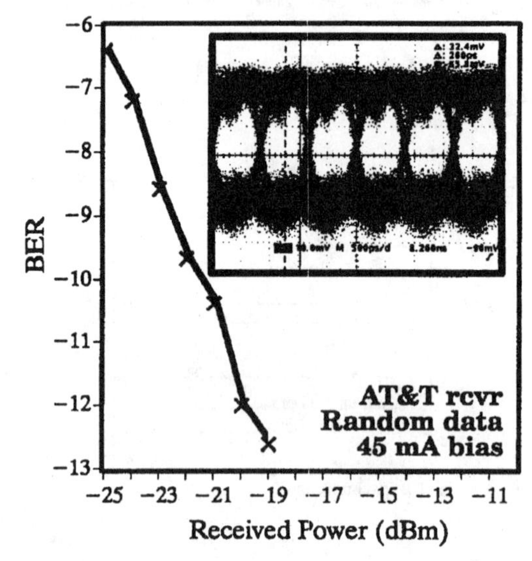

Fig. 24. BER versus loss characteristic.

being received: fill word 0 (FW0), fill word 1 low or high (FW1L or FW1H), data/control word, or an error frame (i.e. the frame is invalid). When the state machine is in state 0, it is in the reset state.

The transmit chip sends FW0 continuously, and the receiver phase locked loop (PLL) is in frequency detection mode. When the receiver detects either FW0 or FW1, the state machine is advanced to state 1. At this stage, the receive phase-locked loop (PLL) is phase-locked and the transmit chip sends FW1 (L/H). If the receiver detects FW0, it remains in state 1. If FW1 or a data word is detected, the state machine advances to state 2. Now data transmission and reception are enabled for the parallel interface. This will result in the local HIPPI interconnect lines being asserted. When all HIPPI interconnect lines are asserted (which means the local and remote state machines are both in state 2), the link has been established and HIPPI data communications can proceed. A state machine diagram appears in Fig. 21.

Fig. 22 shows a sample oscilloscope trace of the laser output of the serial HIPPI data stream. In this case, the link is transmitting a stable data word that corresponds to the link being established, but no actual data being transferred. In order to determine the minimum laser current bias required for error-free transmission, the bias current was varied over a range from around threshold to 60 mA, and the corresponding bit error rate (BER) measured (for random data pattern). The power at the receiver was fixed at -10 dBm. The result in Fig. 23 indicate that for current biases below 42 mA, the BER is markedly increased due to relaxation oscillation like that shown previously in Fig. 15. Fig. 24 shows the results of a test in which the laser current bias was fixed at 45 mA and the received optical power varied via attenuation in the fiber link. It can be seen that for a received optical power greater than approximately -20.7 dBm the BER is at least 10^{-11}.

For a BER of 10^{-11}, this link configuration gives the requirement that the power available to the receiver be greater than -20.7 dBm. Fig. 25 shows the link power budget analysis for the final link using the leased telecom line. This link is approximately 14 Km in length, but has many lossy splices with an estimated loss of 10 dB. This fact, combined with coupler and laser-to-fiber coupling losses, necessitate the use

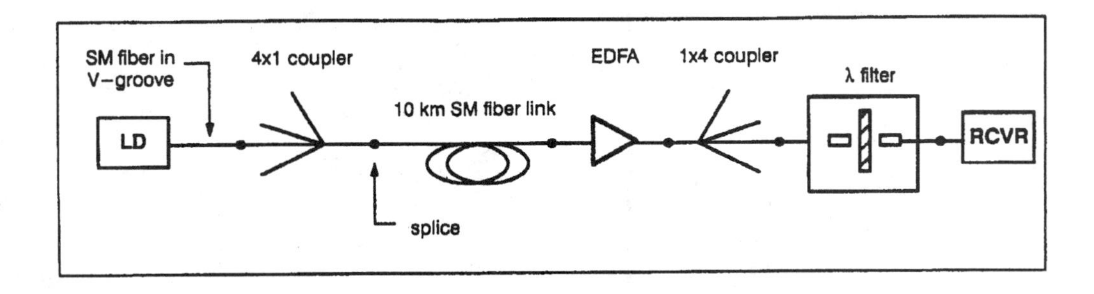

Loss Mechanism	Attenuation	Optical Power
LD to fiber coupling loss		−6.0dBm
4x1 coupler loss	6.0 dB	−12.0dBm
Leased Fiber Link Loss	10.0 dB	−22.0dBm
Erbium Doped Fiber Amplifier (EDFA)	−25dB	+3.0dBm
Splice losses (worst case)	5.0dB	−2.0dBm
1x4 coupler loss	6.0dB	−8.0dBm
Wavelength filter loss	3.5 dB	−11.5dBm

Fig. 25. Link power budget analysis.

of an optical amplifier to boost the signal up to a level the receiver can detect.

IV. Applications

The *Global Grid Data Fusion* program looks to develop and apply state of the art information system technologies to the *real-time* remote theatre defense [25]. Specifically, key technologies such as all-optic terabit networks and teraflop parallel supercomputers, will be used to collect, analyze and correlate multiple image sensor inputs from the battlefield theatre via the Global Grid, locate and classify targets (either ground or aerial), assess defensive strategies through fast battlefield simulators, and finally project solutions back to field commanders in the form of high-resolution 3-D terrain visualizations. By definition, the response time of this complete system must be less than the reaction time of the weapons systems at the remote theatre, typically a few seconds or less.

A. Background

Global Grid represents a new Department of Defense operational doctrine. This doctrine postulates a defense future in which DoD people and resources are highly inter-connected and shared, often by non-DoD users. Past distinctions between tactical and strategic, local and global, defense and nondefense break down and become blurred. Mission planning and situation assessment via teleconferencing become global. Tactical real-time military actions are backed up by computing and analysis capabilities remote from the field of action. Sensor platforms are used to support disaster relief as well as for mobile launcher detection.

Lessons learned from the scud missile strikes during the Gulf War and the more recent humanitarian relief efforts in Africa, clearly indicate satellite imagery, if analyzed and understood *quickly* enough, can reverse a military tactical advantage, or in the case of disaster mitigation, stabilize or even reverse a disease epidemic. Such high resolution satellite imagery (e.g., radar, infrared, or hyperspectral) is typically widely dispersed geographically today, making it difficult to collect quickly. Furthermore, the computational resources needed to process it in real-time often involve large highly specialized supercomputers at various national laboratories. Typically, these already existing resources are not used during crisis management because of inadequate communications infrastructure.

The Global Grid Data Fusion Testbed will apply information and communication system technologies in the following areas: TeraFLOP MPP supercomputer technologies, high-speed low-latency fiber optic networks (100 Gbit/s – 1 Tb/s), WAN based meta-supercomputers, advanced data base management (DBM) and battlefield management (BM) simulators, multi-spectral satellite and ground-based imagers and radar Gigabit relay satellites, neural and fuzzy logic tactical situation assessment algorithms, force structure characterization, force elements geolocation/prediction, and tactical intelligence knowledge engineering (Intelligence Template Mappers).

B. Andromeda Project

The *Andromeda Project* proposes to *glue* together these high performance data archive and processing centers with a high performance multi-gigabit network to create a nationwide *meta-supercomputer* that can be tapped into or exported to distant continents by the Global Grid networks. Much like the civilian air transport reserve was used to transport troops during the Gulf crisis, this vast network of data imagery and highest possible performance supercomputers, always being

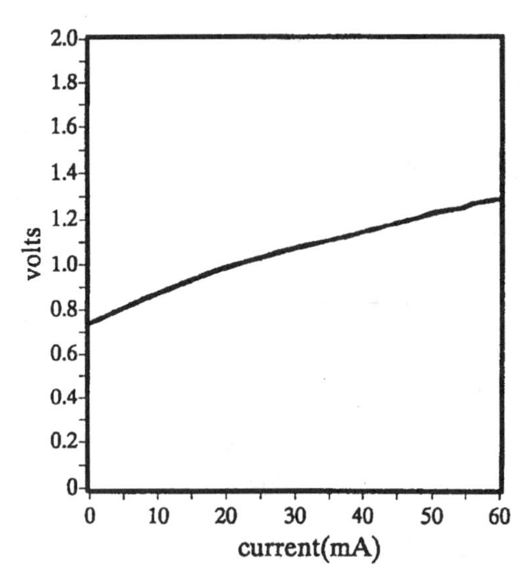

Fig. 26. V/I laser characteristic.

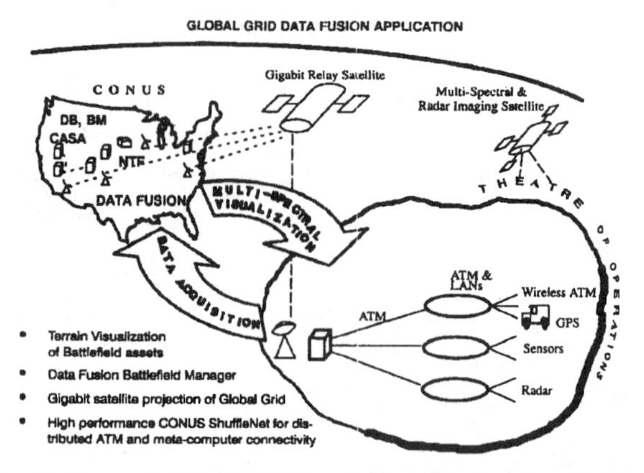

Fig. 27. In a typical Global Grid Data Fusion application, CONUS assets are projected to a remote theatre. First, remote sensor data is collected on a remote theatre of operations, concentrated, and transmitted via a gigabit relay satellite to CONUS where a national meta-supercomputer network processes the data, performs multi-modality correlations between image data sets and existing data bases. Targets and positional coordinates are then supplied to sophisticated battlefield managers and simulators (also meta-supercomputer based) that deliver tactical planning data in the form of 3-D visualizations back to field commanders over the return path of the Global Grid network—all in a few seconds.

updated, could be applied on a moment's notice in a global crisis. This would provide commanders in the field rapid (in minutes) analysis of unfolding crisis, such as population migration, whereabouts of scud missile launchers, spread of brush wildfires, and flood prediction, to name but a few.

An illustration of how CONUS supercomputer assets might be projected to remote corners of the world via the global grid is shown in Fig. 27. Here, remote sensor data first is collected on a remote theatre of operations. This includes on-site sensor data (GPS, radar, IR, IMINT, ELINT, and others) delivered over tactical battlefield C^3 networks as well as satellite image data transmitted back to CONUS over traditional SATCOM channels. Remote ground sensor data is collected, concentrated, and then transmitted via a gigabit relay satellite back to CONUS where a national meta-supercomputer network processes the data (noise removal, map registration, etc), and then performs the *data fusion* operation itself—multi-modality correlations between image data sets and existing data bases. Targets and positional coordinates are then supplied to sophisticated battlefield managers and simulators (also meta-supercomputer based) that deliver tactical planning data in the form of 3-D visualizations back to field commanders over the return path of the Global Grid network—all in a few seconds. It is our goal to make the entire response time short enough that it might even be able to operate as machine in the loop, as part of, for example, a Patriot advanced early warning guidance system, or in providing target location data to F15 fighter pilots during the progress of missions.

An underlying premise of this effort is that some of the highest performance computation and communication systems being developed today can be used to transport, process, and disseminate remote theatre sensor image and radar data *as fast as* it is being collected. Specific technology examples include: massively parallel processor (MPP) supercomputers to process C^3 data in real-time, ubiquitous acquisition and dissemination of large data sets via the Global Grid, creation of powerful, highly reliable and available *meta-supercomputer* networks within CONUS with all-optic terabit networks. In brief, this effort will provide an experimental testbed to explore high performance systems technologies for next generation *WarBreaker* and the resulting new enabling applications for theatre defense.

V. Future Work

Future work focuses on continued development of the components and subsystems discussed above. Specifically, the areas of interest are construction of an 8-node MCSN testbed, addressing issues related to using an optical data plane with the electronic MCSN node for optical packet routing, completion and demonstration of the 4-wavelength WDM HIPPI link, and integration of the WDM laser array with a fiber star coupler.

A. MCSN Development

The prototype MCSN node design provides the foundation for an eight-node MCSN demonstration system. The current switching node design will be converted to a PCB version to allow for ease of replication of the node. The purpose of this system is to validate the simulation results presented in [1]. Additional work to build the MCSN demonstrator system is needed to replace the on board packet generator shown in Fig. 4, with a more sophisticated off board generator to allow for different types of traffic patterns and collection of routing statistics. This generator is currently envisioned as a peripheral card which can be plugged into a PC. With this configuration, we can network up to eight PC's for system performance evaluation in addition to realizing a generalized interconnection for parallel computing applications.

While the MCSN demonstrator system uses electronic packets for rapid system evaluation, this system can also route optical packets due to the short packet deflection protocol described in [1]. To realize a multi-gigabit/sec system requires replacing the electronic data plane in Fig. 3 with an optical equivalent. The primary limitation of current commercially available optical switches is the millisecond set up time, which is not well suited for routing small packets with a deflection protocol. Integrated switching fabrics provide nano second switching speeds, but at expense of increased noise due to crosstalk and optical amplifier noise [26]–[33]. This technology is not yet mature enough for application to the data plane of an MCSN node but the implementations in Refs. [26], [27], [31], [32], [33] look quite promising and warrant further investigation.

B. WDM Components

The main focus of the four-wavelength WDM HIPPI link to this point has been the design and testing of the electronics for the parallel-serial HIPPI PCB, interfacing this board to the four-element stepped wavelength array and characterization of the array elements. The next task is to build the requisite number components to build the four-wavelength link for full-duplex operation in a laboratory setting. This link will be tested to verify and characterize the operation of the four channels before installation of the link between the supercomputing facilities of JPL and the California Institute of Technology (CIT). The purpose of the laboratory test is to validate link operation before introducing additional variables due to transmission line characteristics of the 10 Km link between the two sites.

The primary difficulty in using the MDL laser arrays comes from the labor intensive approach needed to couple the four laser elements into a single fiber. The alignment of the fibers to the array requires three-dimensional positioning and results in low, nonuniform coupling efficiencies. To alleviate this problem, a monolithically integrated laser array and star coupler are presently being developed by the MDL to improve the overall laser-fiber coupling loss. This WDM transmitter will include four DFB lasers and a 4×1 star coupler, thus requiring only a single fiber pigtail and potentially reducing coupling loss and coupling nonuniformity across the array. This star coupler-based monolithically integrated WDM transmitter is also scalable to an N-element laser transmitter with an integrated $N \times 1$ star coupler—a 21-element monolithically integrated transmitter has already been demonstrated that requires coupling to a single fiber [34].

VI. Conclusions

Significant progress has been made in defining the basic components and subsystems needed to realize an all-optical data path network. A network architecture called a multicylinder ShuffleNet was conceptualized and simulated to verify datagram routing of asynchronous optical packets. This simulation model incorporated PE switching nodes to handle short, asynchronous packets and an interface to convert long, streaming packets into short packets well suited to the MCSN concept with a WDM data format.

As a complement to the theoretical and simulation work done, various components and subsystems are under development to validate the proposed WDM networking concept. These efforts include a prototype MCSN switching node for routing asynchronous datagram packets, a HIPPI serial-parallel PCB as the first step in realizing the desired WDM network interface, and the four-element stepped wavelength laser arrays. The last two elements are being integrated into a four channel HIPPI extender to ultimately demonstrate the viability of WDM.

These components and subsystems are viewed as intermediate steps to the final optical network as an enabling technology to multi-gigabit networking applications. Additional progress in both the size of optical switching fabrics and the data path characteristics is needed before these devices can be inserted into the proposed network. We further look to WDM techniques to utilize the THz bandwidth of optical fiber. This technology, however, requires further integration of the WDM laser array (and a detector array) with an integrated optic coupler to simplify system construction and mitigate coupling losses. Encouraged by the current results, we can begin to see the formation for a true multigigabit/second interconnection network.

Acknowledgment

Reference herein to any specific commercial product, process, or service by trade name, trademark, manufacturer, or otherwise, does not constitute or imply its endorsement by the United States Government, the Jet Propulsion Laboratory, or the California Institute of Technology.

References

[1] S. P. Monacos and A. A. Sawchuk, "*A Scalable Recirculating Shuffle Network with Deflection Routing*," submitted to the Special Issue on Optical Networks, *IEEE J. Select. Areas Commun.*

[2] C. A. Brackett, "Dense wavelength division multiplexing networks: Principles and applications," *IEEE J. Select. Areas Commun.*, vol. 8, pp. 948–964, 1990.

[3] G. P. Agrawal and N. K. Dutta, *Semiconductor Lasers.* New York: Van Nostrand Reinhold, 1993, 2nd ed.

[4] T. L. Koch and U. Koren, "Photonic integrated circuits," *AT&T Tech. J.*, pp. 63–74, Jan./Feb. 1992.

[5] B. J. Ainslie, "A review of the fabrication and properties of erbium-doped fibers for optical amplifiers," *J. Lightwave Technol.*, vol. 9, pp. 220–227, Feb. 1991.

[6] T. Koch, "Laser sources for wavelength division multiplexing," in *Conf. Optic. Fiber Commun. (OFC)*, San Diego, CA, 1995.

[7] L. A. Wang, "Medium-density WDM system with Fabry-Perot laser diodes for subscriber loop applications," *IEEE Photon. Technol. Lett.*, vol. 3, p. 554, June 1991.

[8] J. C. Feggler, D. G. Duff, and G. F. Valvo, "10 Gb/s WDM transmission measurement on an installed optical amplifier undersea cable system," *Electron. Lett.*, vol. 31, no. 19, p. 1676, Sept. 1995.

[9] R. G. Blom, L. A. Bergman, R. E. Crippen, E. G. Frost, K. J. Hussey, P. M. Lyster, D. A. Okaya, and D. F. Stanfill, "Interactive, regional-scale geological data exploration and analysis across a gigabit computing network: A part of the casa gigabit network testbed," in *9th Thematic Conf. Geologic Remote Sensing*, Pasadena, CA, Feb. 1–8, 1993.

[10] P. Messina, "CASA gigabit network," *Optic. Fiber Commun. (OFC) Conf. 1992*, San Jose, CA, Feb 4, 1992.

[11] D. J. Blumenthal and J. R. Sauer, "Multiwavelength information processing architectures," in *Topic. Meeting Optic. Comput.*, Palm Springs, CA, Mar. 1993.

[12] D. J. Blumenthal, R. J. Feuerstein, and J. R. Sauer, "First demonstration of multihop all-optical packet switching," *IEEE Photon. Technol. Lett.*, vol. 6, pp. 457–460, Mar. 1994.

[13] A. V. Ramana, H. F. Jordan, J. R. Sauer, and D. J. Blumenthal, "An extended fiber-optic backplane for multiprocessors," in *Proc. 27th Hawaii Int. Conf. Syst. Sci.*, vol. 1, Maui, HI, Jan 1994, pp. 462–470.

[14] J. R. Sauer, "An optoelectronic multi-Gb/s packet switching network," OCS Tech. Rep. 89-06, Feb. 1989.
[15] S. P. Monacos and A. A. Sawchuk, "A permutation engine switching node," submitted to the *J. Parallel Distrib. Comput.*
[16] MUPAC System Packaging Products, MUPAC Corp., 1992.
[17] N. F. Maxemchuk, "Comparison of deflection and store-and-forward techniques in the Manhattan street and shuffle-exchange networks," in *IEEE INFOCOM '89*, Apr. 1989, pp. 800–809.
[18] A. K. Gupta and N. D. Georganas, "Analysis of a packet switch with input and output buffers and speed constraints," in *IEEE INFOCOM '91*, 1991, pp. 694–700.
[19] M. Aoki, T. Tsuchiya, K. Nakahara, M. Komori, and K. Uomi, "High-power and wide-temperature-range operations of InGaAsP-InP strained MQW lasers with reverse-mesa ridge waveguide structure," *IEEE Photon. Technol. Lett.*, vol. 7, pp. 13–15, Jan. 1995.
[20] T. Koch, "Lasers sources for wavelength division multiplexing," in *Conf. Optic. Fiber Commun.*, paper WF, Feb. 1995.
[21] E. E. B. Basch, Ed., *Optical Fiber Transmission.* Howard W. Sams & Co., 1987, pp. 309–313.
[22] T. Russel, *HIPPI and the Issues of HIPPI Data Networking*, Ultra Network Technol., 1991.
[23] Draft AMERICAN NATIONAL STANDARD X3.183-199x
[24] *Preliminary Specification for the Gigabit Rate Transmit Receive Chip Set*," Hewlett–Packard, 1992.
[25] C. Graff, F. Halloran, and C. Lockhart, "Tactical battlefield ATM," in *Proc. IEEE MILCOM*, vol. 2, 1994, pp. 473–478.
[26] W. H. Nelson, A. N. M. Masum Choudhury, M. Abdalla, R. Bryant, E. Meland, W. Niland, and W. Powazinik, "Large-angle 1.3 μm InP/InGaAsP Digital Optical Switches with Extinction Ratios Exceeding 20 dB," OFC 1994 Tech. Dig. Series, pp. 53–54.
[27] T. Kirihara, M. Ogawa, S. Tsuji, and H. Inoue, "High-speed signal-transmission performance in a lossless 4×4 optical switch for photonic switching," OFC 1994 Tech. Dig. Series, p. 55.
[28] T. Kirihara, M. Ogawa, H. Inoue, and K. Ishida, "Lossless and low-crosstalk characteristics in an InP-Based 2×2 optical switch," *IEEE Photon. Technol. Lett.*, vol. 5, pp. 1059–1061, Sept. 1993.
[29] P. Granestrand, B. Langerstrom, P. Svensson, H. Olofsson, J. E. Falk, and B. Stoltz, "Pigtailed Tree-Structured 8×8 $LiNbO_3$ switch matrix with 112 digital optical switches," *IEEE Photon. Technol. Lett.*, vol. 6, pp. 71–73, Jan. 1994.
[30] W. H. Nelson, private communication.
[31] T. Kirihara, M. Ogawa, H. Inoue, H. Kodera, and K. Ishida, "Lossless and low-crosstalk characteristics in an InP-based 4×4 optical switch with integrated single-stage optical amplifiers," *IEEE Photon. Technol. Lett.*, vol. 6, pp. 218–221, Feb. 1994.
[32] R. Nagase, A. Himeno, M. Okuno, K. Kato, K. Yukimatsu, and M. Kawachi, "*Silica-Based* 8×8 *Optical Matrix Switch Module with Hybrid Integrated Driving Circuits and its System Application*," J. of Lightwave Technology, vol. 12, no. 9, pp. 1631–1639, Sept. 1994.
[33] W. H. Nelson, A. N. M. Masum Choudhury, M. Abdalla, R. Bryant, E. Meland, and W. Niland, "Wavelength- and polarization-independent large angle InP/InGaAsP digital optical switches with extinction ratios exceeding 20 dB," *IEEE Photon. Technol. Lett.*, vol. 6, pp. 1332–1334, Nov. 1994.
[34] C. E. Zah and T. P. Lee, "Monolithically integrated multi-wavelength DFB laser arrays and star couplers for WDM lightwave systems," *Opt. Photon. News*, pp. 24–27, Mar. 1993.

Crosstalk and Interference Penalty in All-Optical Networks Using Static Wavelength Routers

CHUNG-SHENG LI, SENIOR MEMBER, IEEE, AND FRANK TONG, SENIOR MEMBER, IEEE

Abstract— **The crosstalk and interference penalty in an all-optical network using static wavelength routers is analyzed in this paper. A worst case methodology is used to derive the upper bound of the penalty. We show that the penalty strongly depends on the linewidth of the laser source. Up to −20 dB in crosstalk can be tolerated in a moderate-size network (≈10^5 nodes), with the ratio of the laser linewidth to the electrical bandwidth less than or equal to unity. Larger linewidth has the advantage of reducing the power penalty incurred by phase-to-amplitude noise conversion. However, the number of wavelength channels will be reduced as well. The maximum tolerable component crosstalk for a network with arbitrary size is reduced to −30 dB.**

I. INTRODUCTION

WAVELENGTH routing for all-optical networks using WDMA has received increasing attention recently [1]–[4]. In a wavelength-routing network, shown in Fig. 1, wavelength-selective elements are used to route different wavelengths to their corresponding destinations. Compared to a network using only star couplers, a network with wavelength routing capability can avoid the splitting loss incurred by the broadcasting nature of a star coupler [5]. Furthermore, the same wavelength can be used simultaneously on different links of the same network and reduce the total number of required wavelengths [1].

The routing mechanism in a wavelength router can either be static, in which the wavelengths are routed using a fixed configuration [6], or dynamic, in which the wavelength paths can be reconfigured [7]. The common feature of these multiport devices is that different wavelengths from each individual input port are spatially resolved and permuted before they are recombined with wavelengths from other input ports. These wavelength routers, however, have imperfections and nonideal filtering characteristics which give rise to signal distortion and crosstalk.

Crosstalk phenomena in wavelength routers have previously been studied [8]–[11]. It was shown in [8] that the maximum allowable crosstalk in each grating (grating as optical demultiplexers and multiplexers in the wavelength router) is −15 dB in an all-optical network with moderate size (say 20 wavelengths and 10 routers in cascade). The results are based on using a 1 dB power penalty criterion and only considering the power addition effect of the crosstalk. Crosstalk can also arise from beating between the data signal and the leakage signal (from imperfect filtering) at the same output channel. The beating of these uncorrelated signals converts the phase noise of the laser sources into the amplitude noise and corrupts the received signals [12] when the linewidths of the laser sources are smaller than the electrical bandwidth of the receiver. Coherent beating, in which the data signal beats with itself, can occur as a result of the beatings among the signals from multiple paths or loops caused by the leakage in the wavelength routers in the system. It was shown in [10] that the component crosstalk has to be less than −20, −30, and −40 dB in order to achieve satisfactory performance for a system consisting of a single, ten, and hundred leakage sources, respectively. However, the authors only considered the case for narrow linewidth sources (using external modulation) and with destructive interference. In [11], the authors simulated the theoretical results reported in [12] with experiments, again, using externally modulated source of 40 MHz linewidth. The situation in which the dominant noise arises from phase-to-intensity noise conversion for different laser linewidths is yet to be studied.

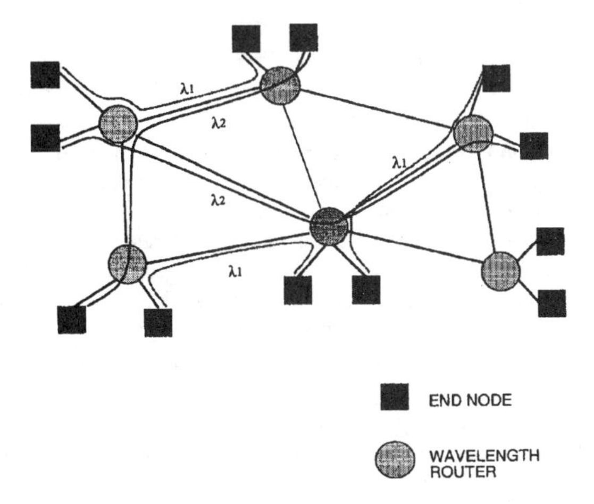

Fig. 1. Topology of a wavelength routing network.

In this paper, we investigate the maximum allowable component crosstalks in optical networks using static wavelength routers. The component crosstalk is a major system issue as the network size can be limited by the crosstalk accumulated through these routers. Apart from what had previously been published, we concentrate our study on crosstalk arising from phase-to-intensity-noise conversion due to the beating between the interfering channels and the data channel. A worst-case methodology, in which the data pattern sent in the crosstalk channels is chosen so that the system penalty is

Manuscript received March 23, 1995; revised October 13, 1995.
The authors are with the IBM Research, T. J. Watson Research Center, Yorktown Heights, NY 10598 USA.
Publisher Item Identifier S 0733-8724(96)04072-8.

Reprinted with permission from *IEEE Journal of Lightwave Technology*, Vol. 14, No. 6, pp. 1120-1126, June 1996.

maximized, is used to evaluate the upper bound of the system penalty for a given crosstalk level. Using this methodology, we found that the crosstalk penalty heavily depends on the linewidth and the difference in deviation from the assigned wavelength of the laser sources. We found that the maximum tolerable component crosstalk is −20 dB (or −40 dB when a demultiplexer/multiplexer pair is considered) for a network of moderate size when both the laser linewidth and the difference in deviation from the assigned wavelength are less than or comparable to the electrical bandwidth of the receiver, since most of the noise power falls within of the receiver bandwidth. Conversely, the system penalty becomes negligible when either of the laser linewidth or this difference in deviation from the assigned wavelength is much larger than the electrical bandwidth of the receiver. This implies that the crosstalk penalty is reduced when the frequency chirping from direct modulation of the laser is also considered. For a small number of wavelengths (say, 4–6), our model predicts that the maximum tolerable component crosstalk is −16.75 dB (or −33.5 dB for a demultiplexer/multiplexer pair) to achieve a power penalty less than 1 dB. This number matches the experimental data reported in [11].

The organization of the rest of the paper is as follows. Section II presents the crosstalk formulation for a static wavelength router. Power penalty of single- and multiple-stage optical networks using static wavelength routers are evaluated in Sections III and IV, respectively. Several numerical examples are given in Section V. This paper is summarized in Section VI.

II. Crosstalk Formulation

A. Wavelength Router Structure

Fig. 2 shows the structure of a static wavelength router which consists of K optical demultiplexers and multiplexers. Each input fiber to an optical demultiplexer is assumed to contain up to M different wavelengths where $M \leq K$. However, we only consider the case where $M = K$. The optical demultiplexer spatially separates the incoming wavelengths into M paths. Each of these paths is then combined at an optical multiplexer with the outputs from the other $M - 1$ optical demultiplexers.

The wavelength routing configuration in Fig. 2 is fixed permanently. The optical data at wavelength λ_j entering the ith demultiplexer exit at the $[(j - i) \bmod M]$th output of that demultiplexer. That output is connected to the ith input of the $[(j - i) \bmod M]$th multiplexer.

Because of the imperfections and nonideal filtering characteristics of the optical multiplexers and demultiplexers, crosstalks occurs in the wavelength routers. On the demultiplexer side, each output contains both the signals from the desired wavelength and that from the other $M - 1$ crosstalk wavelengths. From reciprocity, both the desired wavelength and the crosstalk signals exit at the output on the multiplexer side. Thus, each wavelength at every multiplexer contains $M - 1$ crosstalk signals originating from all demultiplexers.

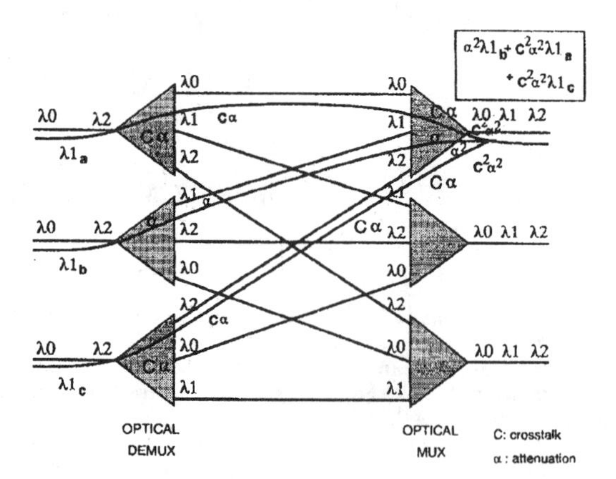

Fig. 2. Structure of a static wavelength routing mode using 3 wavelengths as an example. Calculation of total crosstalk for a wavelength router is also indicated in the figure.

B. Crosstalk Evaluation

The crosstalk effects can be formulated as follows. We assume each end node in Fig. 1 has a fixed-tuned or tunable single-mode laser transmitter (such as a distributed feedback laser) and a fixed-tuned or tunable optical receiver. The optical power for the logical ONE's and ZERO's at the output of the transmitter are P_{ON} and P_{OFF}, respectively, and can be related to the average output power, P_{av}, and extinction ratio, r, as

$$P_{\mathrm{ON}} = \frac{2r}{r+1} P_{\mathrm{av}}$$
$$P_{\mathrm{OFF}} = \frac{2}{r+1} P_{\mathrm{av}} \quad (1)$$

for nonreturn-to-zero (NRZ) data. The electric field of a laser output during a bit can be expressed as

$$E(t) = \sqrt{2P} \cos(\omega t + \Phi(t) + \Theta) \quad (2)$$

where P can be either P_{ON} or P_{OFF}, ω is the angular laser frequency, $\Phi(t)$ is the phase noise process, and Θ is the initial phase of the laser. We only consider the phase-to-intensity noise as all other noise sources (such as the relative intensity noise and the mode-partition noise) are negligible as compared to the phase-to-intensity noise conversion. We denote the input electric field of the jth wavelength at the ith optical demultiplexer in Fig. 2 as $E^{\mathrm{in}}_{i,\lambda_j}$; the power attenuation of wavelength λ_j at the ith output channel of the lth optical demultiplexer as α_{l,i,λ_j}; and the power attenuation of wavelength λ_j at the lth input channel of the ith optical multiplexer as β_{i,l,λ_j}. The output electric field of the jth wavelength at the ith optical multiplexer $E^{\mathrm{out}}_{i,\lambda_j}$ thus equals

$$E^{\mathrm{out}}_{i,\lambda_j}(t) = \sum_{l=0}^{M-1} \sqrt{\beta_{i,l,\lambda_j}} \sqrt{\alpha_{l,i,\lambda_j}} E^{\mathrm{in}}_{l,\lambda_j}(t). \quad (3)$$

It can be found from Fig. 2 that for an ideal demultiplexer, we have $\alpha_{l,i,\lambda_{(i-l) \bmod M}} = 1$ (no attenuation) and $\alpha_{l,i,\lambda_j} = 0$ for $j \neq (i - l) \bmod M$ (full attenuation). Similarly, for an ideal multiplexer, we have $\beta_{i,l,\lambda_{(i-l) \bmod M}} = 1$ and $\beta_{i,l,\lambda_j} = 0$ for $j \neq (l - i) \bmod M$. For nonideal demultiplexers and multiplexers, we have $0 < \alpha_{l,i,\lambda_j} < 1$ and $0 < \beta_{l,i,\lambda_j} < 1$.

To evaluate the worst case scenario (upper-bound in system penalty), we assume that all of the optical multiplexers and demultiplexers have identical characteristics, and that the crosstalk introduced for each channel is also identical. That is, we have $\alpha_{l,i,\lambda_{(i-l)\text{mod}M}} = \alpha, \alpha_{l,i,\lambda_j} = C\alpha$ for $j \neq (i-l) \bmod M$ where C is the ratio of the intensity of a crosstalk channel to that of the signal channel. Typically, a flat −30 dB crosstalk across the bandwidth can be observed in planar waveguide demultiplexer devices [13]. From reciprocity, we have $\beta_{i,l,\lambda_{(l-i)\text{mod}M}} = \beta = \alpha, \beta_{i,l,\lambda_j} = C\beta = C\alpha$ for $j \neq (l-i) \bmod M$. Equation (3) can then be simplified to

$$E_{i,\lambda_j}^{\text{out}} = \alpha\left(E_{(j-i)\text{mod}M,\lambda_j}^{\text{in}} + C\sum_{l\in X_{ij}} E_{l,\lambda_j}^{\text{in}}\right) \quad (4)$$

where $X_{ij} = \{0,\cdots,M-1\} - \{(j-i) \bmod M\}$. The output intensity of wavelength λ_j at the output of the ith wavelength demultiplexer can then be computed:

$$\begin{aligned} P_{i,\lambda_j}^{\text{out}}(t) = \alpha^2\Bigg(& P_{(j-i)\text{mod}M,\lambda_j}^{\text{in}} + C^2\sum_{l\in X_{il}} P_{l,\lambda_j}^{\text{in}} \\ &+ 2C\sum_{l\in X_{il}} \gamma_{il}(t)\sqrt{P_{i,\lambda_j}^{\text{in}} P_{l,\lambda_j}^{\text{in}}}\Bigg) \\ &+ \alpha^2 C^2 \sum_{k\in X_{ik}}\sum_{l\in X_{il}} \gamma_{kl}(t)\sqrt{P_{k,\lambda_j}^{\text{in}} P_{l,\lambda_j}^{\text{in}}} \end{aligned} \quad (5)$$

In (5), $\gamma_{kl}(t)$ is defined as

$$\gamma_{kl}(t) = \cos(\Delta\omega_{kl}t + \Phi_k(t) - \Phi_l(t) + \Theta_k - \Theta_l) \quad (6)$$

where $\Delta\omega_{kl} = \omega_k - \omega_l$ is the frequency difference between the kth and the lth laser sources. The first, second, third and fourth terms on the right-hand side of (5) represents the signal, the dc-crosstalk, the signal-crosstalk beating and crosstalk-crosstalk beating, respectively. While the dc crosstalk adds directly to the signal, the beatings arising from the signal-crosstalk and crosstalk-crosstalk generate the phase-to-intensity noise.

The phase-to-intensity noise at the fixed-tuned or tunable receiver (only a single wavelength is selected) can be computed as follows. Assuming that a transimpedance preamplifier with transimpedance R_F, the output voltages for the ONE's and ZERO's from the preamplifier are

$$\begin{aligned} V_{\text{ON}} &= \frac{\eta_e}{h\nu} S_{\text{ON}} R_F \\ V_{\text{OFF}} &= \frac{\eta_e}{h\nu} S_{\text{OFF}} R_F \end{aligned} \quad (7)$$

where η is the photodetector quantum efficiency, ν is the optical frequency, and h is the Planck's constant. S_{ON} and S_{OFF} are the respective expectation values of (5) for receiving a ONE and ZERO

$$\begin{aligned} S_{\text{ON},xtalk} &= \alpha^2\left(P_{\text{ON}} + C^2\sum_{l\in X_{ij}} P_l\right) \\ S_{\text{OFF},xtalk} &= \alpha^2\left(P_{\text{OFF}} + C^2\sum_{l\in X_{ij}} P_l\right) \end{aligned} \quad (8)$$

where P_l instead of P_{l,λ_j} is used since only one wavelength is considered here. The subscript $xtalk$ denotes the bit pattern $(b_0, b_1, \cdots, b_{M-1})$ sent on the crosstalk channels: $P_l = P_{\text{ON}}$ if $b_l = 1$ and $P_l = P_{\text{OFF}}$ if $b_l = 0$. The total noise power of the received logical ONE's and ZERO's equals [14]

$$\begin{aligned} V_{N,i,xtalk}^2 = \Bigg[&(V_A^*)^2\left(\left(1+\frac{R_F}{R}\right)^2 + \frac{4\pi^2}{3}B_e^2 C_{in}^2 R_F^2\right) \\ &+ R_F^2\left(\frac{4kT}{R} + (I_A^*)^2\right) + 4kTR_F\Bigg]B_e \\ &+ R_F^2\int_{-B_e}^{B_e} S_{i,xtalk}^I(f) \qquad i\in\{\text{ON},\text{OFF}\} \end{aligned} \quad (9)$$

where R, and C_{in} are the preamplifier input resistance, and input capacitance, respectively, k is the Boltzmann's constant, and T is the temperature. $S_{i,xtalk}^I(f)$ is the two-sided noise spectral density of the photodetector current, given as

$$S_{i,xtalk}^I(f) = \frac{\eta e^2}{h\nu} S_{i,xtalk} + \left(\frac{\eta e}{h\nu}\right)^2 S_{i,xtalk}^P(f) \qquad i\in\{\text{ON},\text{OFF}\} \quad (10)$$

where $S_{i,xtalk}^P(f)$ is the optical intensity noise spectral density (see next section). This intensity-related noise comes from the signal and the dc crosstalks (first term), and the beatings between the signal and crosstalks as well as among the crosstalks (second term). The latter is greatly affected by the laser linewidth and the difference in the deviation from the assigned wavelength, as will be fully discussed in the next section.

Assuming Gaussian receiver noise, the bit-error rate at a receiver output is

$$\begin{aligned} P_e = \frac{1}{4}\sum_{j=(0,\cdots,0)}^{(1,\cdots,1)} &\text{Prob}[xtalk = j] \\ &\cdot\left(\text{erfc}\left(\frac{th_{\text{opt}} - V_{\text{OFF},xtalk}}{\sqrt{2}V_{N,\text{OFF},xtalk}}\right)\right. \\ &\left.+ \text{erfc}\left(\frac{V_{\text{ON},xtalk} - th_{\text{opt}}}{\sqrt{2}V_{N,\text{ON},xtalk}}\right)\right) \end{aligned} \quad (11)$$

where th_{opt} is the optimal decision threshold that minimizes the bit-error rate.

III. System Penalty in Single-Stage Configuration

In this section, the system penalty caused by the crosstalk in a single-stage wavelength router will be derived for the following cases.

1) Noncoinciding uncorrelated sources, in which the independent laser sources have slightly different wavelengths of ≤10 GHz,
2) Coinciding uncorrelated sources, in which the independent laser sources have identical wavelengths. Note that this is a special case of (1).

The coinciding correlated case (such as that arises from loops in the network) is not considered here, as it is network-structure-dependent and has to be addressed individually for each specific configuration.

A. Noncoinciding Uncorrelated Sources

Using the approach similar to [14], the noise spectral density $S^P_{i,xtalk}(f)$ can be derived from the autocovariance, $L_P(\tau)$, of (5)

$$\begin{aligned}L_P(\tau) &= E[(P(t+\tau) - E[P(t+\tau)])(P(t) - E[P(t)])]\\ &= 4C^2 \sum_{m\in X_{im}} \sum_{n\in X_{in}} E[\gamma_{im}(t+\tau)\gamma_{in}(t)]\\ &\quad \cdot \sqrt{P_iP_m}\sqrt{P_iP_n}\\ &\quad + 2C^3 \sum_{l\in X_{il}} \sum_{m\in X_{im}} \sum_{n\in X_{in}} E[\gamma_{il}(t+\tau)\gamma_{mn}(t)]\\ &\quad \cdot \sqrt{P_iP_l}\sqrt{P_mP_n}\\ &\quad + 2C^3 \sum_{l\in X_{il}} \sum_{m\in X_{im}} \sum_{n\in X_{in}} E[\gamma_{mn}(t+\tau)\gamma_{il}(t)]\\ &\quad \cdot \sqrt{P_iP_l}\sqrt{P_mP_n}\\ &\quad + C^4 \sum_{k\in X_{ik}} \sum_{l\in X_{il}} \sum_{m\in X_{im}} \sum_{n\in X_{in}} E[\gamma_{kl}(t+\tau)\gamma_{mn}(t)]\\ &\quad \cdot \sqrt{P_kP_l}\sqrt{P_mP_n}. \end{aligned} \tag{12}$$

This equation can further be simplified by keeping only the C^2 term (i.e., ignoring those terms arising from the beatings among the crosstalks) since $C \ll 1$ in most multiplexers and demultiplexers reported. That is,

$$\begin{aligned}L_P(\tau) &\approx 4C^2 \sum_{m\in X_{im}} \sum_{n\in X_{in}} E[\gamma_{mn}(t+\tau)\gamma_{mn}(t)]\\ &\quad \cdot \sqrt{P_iP_m}\sqrt{P_iP_n}\\ &= 4C^2\Bigg(\sum_{m\in X_{im}} E[\gamma_{im}(t+\tau)\gamma_{im}(t)]P_iP_m\\ &\quad + \sum_{\substack{m\in X_{im}\\ m\neq n}} \sum_{n\in X_{in}} E[\gamma_{im}(t+\tau)\gamma_{in}(t)]P_i\sqrt{P_mP_n}\Bigg). \end{aligned} \tag{13}$$

By substituting (6) in (13), the first term can be expanded into

$$\begin{aligned}&E[\gamma_{im}(t+\tau)\gamma_{im}(t)]\\ &\quad = E[\cos(\Delta\omega_{im}(t+\tau) + \Phi_i(t+\tau) - \Phi_m(t+\tau)\\ &\qquad + \Theta_i - \Theta_m)\cos(\Delta\omega_{im}(t) + \Phi_i(t) - \Phi_m(t) + \Phi_i\\ &\qquad - \Theta_m)] = \tfrac{1}{2}\cos(\Delta\omega_{i,m}\tau)e^{-2\pi\Delta\nu|\tau|} \end{aligned} \tag{14}$$

while the second term in (13) can be shown to reduce to zero [14]. Therefore,

$$L_P(\tau) \approx 2C^2P_i \sum_{m\in X_{im}} \cos(\Delta\omega_{i,m}\tau)e^{-2\pi\Delta\nu|\tau|}P_m \qquad i \in \{\mathrm{ON}, \mathrm{OFF}\} \tag{15}$$

and the spectral density is the Fourier transformation of (15):

$$S^P_i(f) = \frac{2\Delta\nu C^2 P_i}{\pi} \sum_{m\in X_{im}} \frac{P_m}{(\Delta\nu)^2 + (f - \Delta f_{i,m})^2} \qquad i \in \{\mathrm{ON}, \mathrm{OFF}\} \tag{16}$$

where $\Delta\nu$ is the linewidth of the laser, and $\Delta f_{j,m}$ is the frequency separation between the the mth crosstalk and the signal.

Total noise power due to phase-to-intensity noise conversion can then be found

$$\begin{aligned}N_{\mathrm{total},i,xtalk} &= \int_{-B_e}^{B_e} S^P_i(f)\,df\\ &= \frac{2C^2}{\pi}P_i \sum_{m\in X_{jm}} P_m\Bigg(\tan^{-1}\frac{B_e - \Delta f_{j,m}}{\Delta\nu}\\ &\quad + \tan^{-1}\frac{B_e + \Delta f_{j,m}}{\Delta\nu}\Bigg) \qquad i \in \{\mathrm{ON}, \mathrm{OFF}\} \end{aligned} \tag{17}$$

where B_e is the electrical bandwidth of the receiver. This equation implies that the phase-to-intensity noise conversion can be ignored when $\Delta f_{j,m} \gg B_e$ or $\Delta\nu \gg |B_e \pm \Delta f_{j,m}|$.

B. Coinciding Uncorrelated Sources

For the coinciding-uncorrelated-sources case, we have $\Delta f_{j,m} = 0$ for all $m \neq j$. Equation (17) can thus be further reduced to

$$\begin{aligned}N_{\mathrm{total},i,xtalk} &= \int_{-B_e}^{B_e} S^P_{i,xtalk}(f)\,df\\ &= \frac{4C^2}{\pi}P_i(L_{i,\mathrm{OFF}}P_{\mathrm{OFF}} + L_{i,\mathrm{ON}}P_{\mathrm{ON}})\\ &\quad \cdot \tan^{-1}\frac{B_e}{\Delta\nu} \qquad i \in \{\mathrm{ON}, \mathrm{OFF}\} \end{aligned} \tag{18}$$

and (8) is reduced to

$$S_{i,xtalk} = P_i + C^2(L_{i,\mathrm{OFF}}P_{\mathrm{OFF}} + L_{i,\mathrm{ON}}P_{\mathrm{ON}}) \qquad i \in \{\mathrm{ON}, \mathrm{OFF}\}. \tag{19}$$

The parameter $L_{i,j}$ describes the number of crosstalk channels that are transmitting the symbol j while the signal channel is transmitting the symbol i. Note that $L_{i,\mathrm{OFF}} + L_{i,\mathrm{ON}} = M - 1$.

The power penalty can be computed by taking the logarithm of the ratio of the required signal level to achieve a specific bit error rate with a crosstalk level of $C, P_{\mathrm{av},C}$, to that of the required signal level to achieve the same bit error rate with zero crosstalk, P_{av}, i.e., $\log_{10}(P_{\mathrm{av},C}/P_{\mathrm{av}})$. A worst-case methodology, in which the crosstalk channels are assumed to transmit bit patterns that maximize the bit error rate, is used to evaluate the system penalty. In this methodology, the crosstalk channels are assumed to transmit a bit pattern that maximizes the system penalty or effectively, minimizes the eye-opening in the eye-diagram analysis. Due to our pessimistic assumptions, the derived results yield the upper bound of the maximum allowable component crosstalk in the system.

The worst case crosstalk can be analyzed in two categories. First, when the data channel is transmitting a ZERO, the worst case degradation on the signal occurs when all crosstalk channels are transmitting ONE's. Note that the total noise is also maximized [see (18)], resulting in the worst case power penalty. When the data channel is transmitting a ONE, the worst case power penalty could occur at $L_{\mathrm{ON,OFF}} = 0, 0 < L_{\mathrm{ON,OFF}} < M - 1$, or $L_{\mathrm{ON,OFF}} = M - 1$, as illustrated

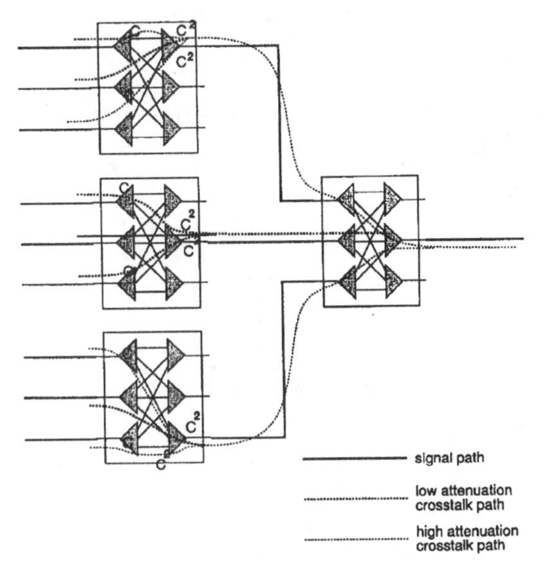

Fig. 3. Structure of cascaded wavelenth routers. Calculation of total crosstalk for a wavelength router is also indicated in the figure.

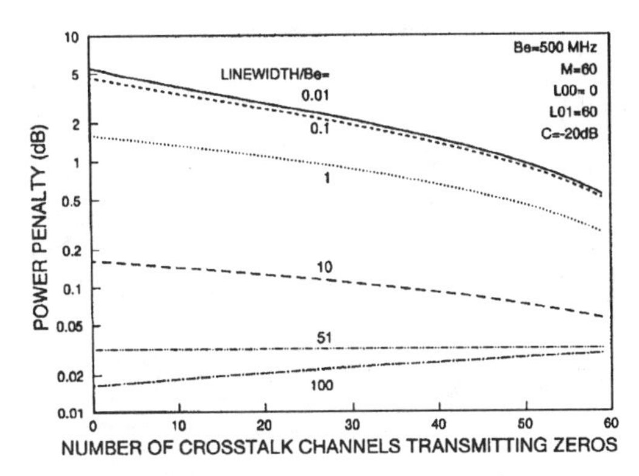

Fig. 4. System penalty as a function of crosstalk channels and are transmitting ZERO's.

in Fig. 4. In this figure, we choose $M = 60$ to illustrate a broad range of possible worst case scenarios. Here, the power penalty is plotted as a function of $L_{\mathrm{ON,OFF}}$ with zero $L_{\mathrm{OFF,OFF}}$. When $\Delta\nu/B_e < 50$, the noise contributed by phase-to-amplitude conversion is the dominant source and the worst case occurs where all of the crosstalk channels are transmitting ONE's. The power penalty becomes almost independent of the number of crosstalk channels transmitting ONE's or ZERO's when $\Delta\nu/B_e \sim 50$. When $\Delta\nu/B_e > 50$, the noise contributed by phase-to-amplitude conversion becomes insignificant and the worst case occurs where all of the crosstalk channels are transmitting ZERO's.

The phase-to-amplitude noise contributed by noncoinciding uncorrelated sources is always smaller than coinciding uncorrelated sources, as shown in (17) and (18). Therefore, we restrict our attention to those situations with coinciding uncorrelated sources for the rest of the paper.

IV. System Penalty in Multistage Configuration

Crosstalk power can accumulate as wavelength routers are cascaded, as shown in Fig. 3. In this section, we evaluate the system penalty in a multistage configuration consisting of coinciding uncorrelated sources as this case represents an upper bound on the system penalty. The constraint set by the accumulated signal-to-noise ratio degradation caused by the crosstalk limits the size of the network to M^H, assuming a fully populated M-ary tree configuration.

Assuming all of the wavelength channels of each fiber are fully populated, we can categorize signal paths as low-attenuation and high-attenuation crosstalk paths. Each stage of wavelength router only allows one low-attenuation path (which is the intended signal path) and $M-1$ high-attenuation paths. In a network consisting of multiple stages of wavelength routers, we only need to consider crosstalk signals going through low-attenuation crosstalk path, as the relatively weak contribution from the high-attenuation paths can be ignored. Each additional stage of wavelength router will increase the number of low-attenuation paths by $M-1$. Therefore, the total number of crosstalk paths after H stages of wavelength routers is $H(M-1)$.

Taking into the consideration of multiple stages, (19) is modified to

$$\begin{aligned} S_{i,xtalk} &= P_i + C^2(L_{i,\mathrm{OFF}}P_{\mathrm{OFF}} + L_{i,\mathrm{ON}}P_{\mathrm{ON}}) \\ &= P_i + C^2(L_{i,\mathrm{OFF}}P_{\mathrm{OFF}} + (H(M-1) - L_{i,\mathrm{OFF}})P_{\mathrm{ON}}) \\ &\qquad i \in \{\mathrm{ON}, \mathrm{OFF}\} \end{aligned} \tag{20}$$

where $L_{i,\mathrm{ON}}$ and $L_{i,\mathrm{OFF}}$ are constrained by $H(M-1)$, the total number of possible crosstalk paths. That is, $L_{i,\mathrm{OFF}} + L_{i,\mathrm{ON}} = H(M-1)$. Equation (18) is modified to

$$\begin{aligned} N_{\mathrm{total},i,xtalk} &= \int_{-B_e}^{B_e} S_{i,xtalk}^P(f)\,df \\ &= \frac{4C^2}{\pi}P_i(L_{i,\mathrm{OFF}}P_{\mathrm{OFF}} + L_{i,\mathrm{ON}}P_{\mathrm{ON}})\tan^{-1}\frac{B_e}{\Delta\nu} \\ &= \frac{4C^2}{\pi}P_i(L_{i,\mathrm{OFF}}P_{\mathrm{OFF}} + (H(M-1) \\ &\quad - L_{i,\mathrm{OFF}})P_{\mathrm{ON}}) \\ &\quad \cdot \tan^{-1}\frac{B_e}{\Delta\nu} \qquad i \in \{\mathrm{OFF}, \mathrm{ON}\}. \end{aligned} \tag{21}$$

The bit error rate at a receiver output of the entire network can be obtained by evaluating (11).

For a specific (H, M, C), the total noise power can be calculated from (20) and (21). The bit error rate can be computed by substituting the signal and noise power into (11). The power penalty can then be obtained by using the same procedure as outlined at the end of the previous section.

V. Numerical Results

Fig. 5 shows crosstalk penalty as a function of crosstalk level C for various values of network size parameters $H(M-1)$. The maximum tolerable component crosstalk level at -16.75 dB (or -33.5 dB when a demultiplexer/multiplexer pair is considered) for $H(M-1) = 5$) in this figure matches the experimental results obtained in [11]. System penalty increases rapidly with the increase of C, and linearly increases

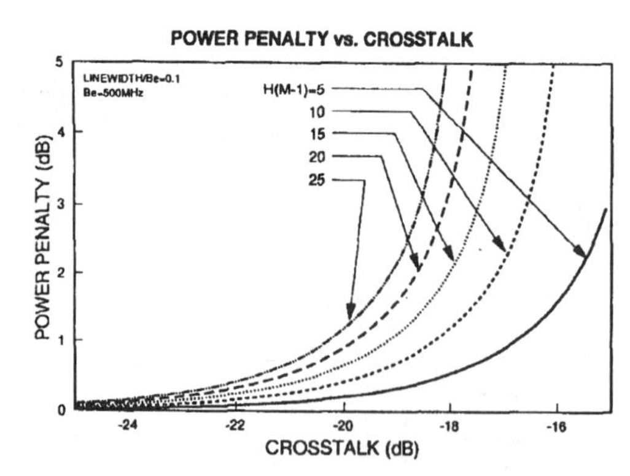

Fig. 5. System penalty as a function of crosstalk.

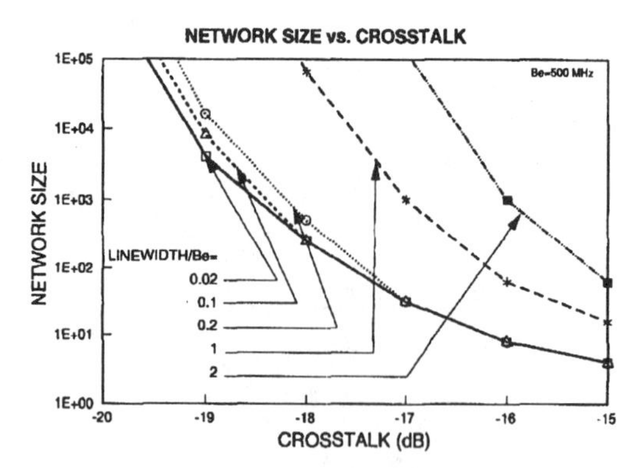

Fig. 7. Network size as a function of crosstalk.

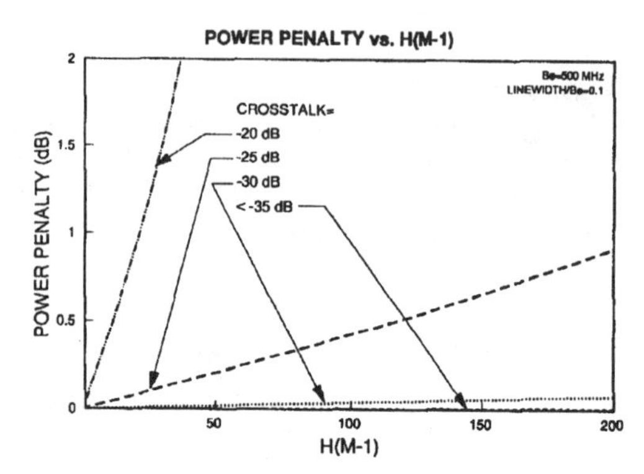

Fig. 6. System penalty as a function of $H(M-1)$.

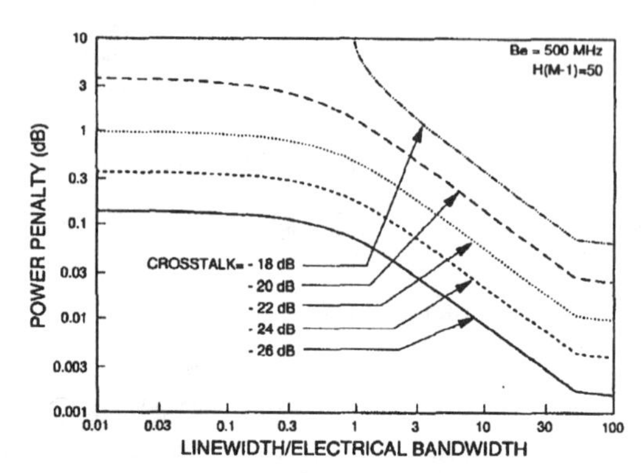

Fig. 8. System penalty as a function of linewidth.

with $H(M-1)$ (see Fig. 6). As shown in Fig. 6 the network can tolerate a crosstalk level C of -30 dB, independent the network size. This assumes a typical laser linewidth of 50 MHz and using a bit error rate of 10^{-9}. The network size is severely limited ($H(M-1) < 100$) as the crosstalk level C approaches ~ -20 dB. Using a 1-dB crosstalk penalty criterion, the maximum achievable network size (M^H) is plotted in Fig. 7 as a function of crosstalk level C for various values of linewidth $\Delta\nu$. For each given crosstalk level, the values in (M, H) have been chosen so that M^H is maximized. Note that the tolerable crosstalk level C is -20 dB for a network size of 10^5. The difference in this value as opposed to the results (-30 dB) generated earlier is because of the optimum network tree configuration, allowing more crosstalk to the same constraint of $H(M-1)$.

The dependency of the crosstalk penalty on the linewidth of a laser sources is shown in Fig. 8 for $H(M-1) = 50$ and $B_e = 500$ MHz. For a given crosstalk level C, there exists a dc crosstalk at large laser-linewidth regime ($\Delta\nu/B_e > 50$) and phase-to-amplitude crosstalk at narrow laser-linewidth regime ($\Delta\nu/B_e < 0.5$). The apparent saturation in the crosstalk penalty in these regimes arises from (18), in which $N_{\mathrm{total},i}$ is determined by the $\tan^{-1}(B_e/\Delta\nu)$. The crosstalk penalty is significantly reduced for systems using lasers with larger linewidth, which could result from the chirp introduced by directly modulating DFB lasers [15]. However, increasing the laser linewidth reduces the number of wavelength channels that can be accommodated by the 30 nm (~3.7 THz) transmission window of the Er-doped fiber amplifier. Furthermore, as a result of the increased laser linewidth, additional power penalty is introduced by the intersymbol interference caused by the fiber dispersion. A trade-off exists and the laser linewidth should be controlled such that the total power penalty is minimized. This is illustrated in Fig. 9, in which the dispersion penalties are plotted as a function of $\Delta\nu/B_e$ for propagation distances of 100, 200, and 500 km, respectively. The dispersion penalty is calculated based on the analytical model described in [16]. For comparison, the power penalty from phase-to-amplitude noise conversion at a crosstalk level of -20 dB with $H(M-1) = 50$ is also shown in the same figure. As an example in Fig. 9, the power penalty from dispersion (propagation distance of 500 km) and from crosstalk intersects at $\Delta\nu/B_e \sim 5$, suggesting an optimal linewidth of 2.5 GHz for the system. In this calculation, the crosstalk power penalty from adjacent channels (due to increased linewidth) is not considered as it has been shown in [15] that the crosstalk with a channel separation of 0.6 nm (at a bit rate less than 1 Gb/s) is negligible.

VI. Summary and Discussion

In this paper, the crosstalk penalties in an all-optical network consisting of single or multiple stages of static wavelength routers are evaluated. A worst-case methodology is adopted in which all of the crosstalk channels are assumed to transmit

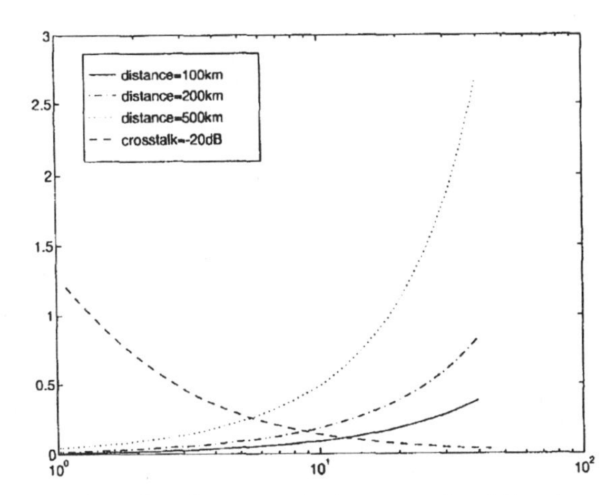

Fig. 9. Comparison of system penalty due to crosstalk and dispersion as a function of linewidth.

data patterns that maximize system penalty. This methodology allows us to derive an upper bound of the crosstalk penalty. Using this methodology and assuming that the laser linewidth is smaller than the receiver bandwidth, we show that the maximum allowable crosstalk for each device is ≈ -20 dB in order to achieve a network with moderate size, and -30 dB for a network with arbitrary size.

REFERENCES

[1] C. A. Brackett, "The principle of scalability and modularity in multi-wavelength optical networks," in *Proc. OFC Access Network*, 1993, p. 44.

[2] S. B. Alexander *et al.*, "A precompetitive consortium on wide-band all-optical network," *J. Lightwave Technol.*, vol. 11, pp. 714–735, May/June 1993.

[3] I. Chlamtac, A. Ganz, and G. Karmi, "Lightpath communications: An approach to high-band-width optical WAN's," *IEEE Trans. Commun.*, vol. 40, pp. 1171–1182, July 1992.

[4] G. R. Hill, "A wavelength routing approach to optical communication networks," in *Proc. INFOCOM*, 1988, pp. 354–362.

[5] R. Ramaswami, "Multiwavelength lightwave networks for computer communication," *IEEE Commun. Mag.*, vol. 31, no. 2, pp. 78–88, Feb. 1993.

[6] M. Zirnigibl, C. H. Joyner, and B. Glance, "Digitally tunable channel dropping filter/equalizer based on waveguide grating router and optical amplifier integration," *IEEE Photon. Technol. Lett.*, vol. 6, pp. 513–515, Apr. 1994.

[7] A d'Alessandro, D. A. Smith, and J. E. Baran, "Multichannel operation for an integrated acousto-optic wavelength routing switch for WDM systems," *IEEE Photon. Technol. Lett.*, vol. 6, pp. 390–393, Mar. 1994.

[8] C.-S. Li, F. Tong, and C. J. Georgiou, "Crosstalk penalty in an all-optical network using static wavelength routers," in *Proc. LEOS Annu. Meeting*, 1993.

[9] C.-S. Li and F. Tong, "Crosstalk penalty in an all-optical network using dynamic wavelength routers," in *Proc. OFC'94*, 1994.

[10] E. L. Goldstein, L. Eskildsen, and A. F. Elrefaie, "Performance implications of component crosstalk in transparent lightwave networks," *IEEE Photon. Technol. Lett.*, vol. 6, pp. 657–660, May 1994.

[11] E. L. Goldstein and L. Eskildsen, "Scaling limitations in transparent optical networks due to low-level crosstalk," *IEEE Photon. Technol. Lett.*, vol. 7, pp. 93–94, Jan. 1995.

[12] J. Gimlett and N. K. Cheung, "Effects of phase-to-intensity noise conversion by multiple reflections on gigabit-per-second DFB laser transmission systems," *J. Lightwave Technol.*, vol. 7, pp. 888–895, June 1989.

[13] K. Okamoto and Y. Inoue, "Silica-based planar lightwave circuits for WDM systems," in *Proc. OFC*, pp. 224–225, Feb. 1995.

[14] C.-S. Li, C. M. Olsen, and D. G. Messerschmitt, "Analysis of crosstalk penalty in dense optical chip interconnects using single-mode waveguides," *J. Lightwave Technol.*, vol. 9, pp. 1693–1701, Dec. 1991.

[15] C.-S. Li, F. Tong, F. Liu, and D. G. Messerchmitt, "Channel capacity optimization of chirp-limited dense WDM/WDMA systems using OOK/FSK modulation and optical filters," *J. Lightwave Technol.*, vol. 10, pp. 1148–1161, Aug. 1992.

[16] S. Yamamoto, M. Kuwazuru, H. Wakabayashi, and Y. Iwamoto, "Analysis of chirp power penalty in 1.55 μm DFB-LD high-speed optical fiber transmission systems," *J. Lightwave Technol.*, vol. 5, pp. 1518–1524, Oct. 1987.

Chapter 9

Optical Cross-Connects

Broadband integrated service digital networks (BISDN) are envisioned to provide high-speed integrated services such as high-speed data communications, video distribution, and multimedia services. To realize these, asychronous transfer mode (ATM) switching systems should have a very large switching capacity with an aggregate throughput of over a terabit per second. With their very large bandwidth and fast switching speed, optical cross-connects will play an important role when such switching targets are approached. As switching cores in switching systems, optical cross-connects have many advantages over their electronic counterparts, such as ultrahigh internal line rate, lack of electronic bottlenecks, and easy control. Optical cross-connects can also be used as digital cross-connects in multigigabit transport networks to reconfigure network interconnections so as to meet traffic and restoration requirements.

Optical cross-connection techniques can be classified under three categories: wavelength-division multiplexing (WDM), space-division multiplexing (SDM), and time-division multiplexing (TDM). Some recent proposed systems are based on a combination of two of these techniques. TDM technique may be appropriate to cross-connects at relatively low data rates. For multigigabit data signals, TDM becomes difficult to implement since it needs ultrashort optical pulse sources and ultrahigh speed synchronization systems. Also, TDM dictates certain limitations on signal formats. The other two multiplexing techniques do not have limitations on signal data rates and formats. The WDM technique can provide a simple cross-connect structure to implement a broadband cross-connect system, however, not without some challenges that restrict the number of available wavelengths for WDM cross-connects (see Chapter 8). The SDM-based optical cross-connects use optical space-division switching networks, such as directional coupler-based $LiNbO_3$ switching arrays, switching arrays based on the gating principle (splitter-combiner), and semiconductor optical amplifiers (SOA)-based switching arrays. Also, combined WDM/SDM based optical cross-connects have been recently introduced. Here, SOA-based space switches are utilized to interconnect WDM active routing cross-connects among which the wavelengths are reused.

Johansson et al. describe in paper 9.1, an optical cross-connect system incorporating a new concept, transparent optical multiwavelength networks, for flexible and high-capacity transport networks. The key technologies used are space switching devices and wavelength-selective devices that make it possible to obtain enhanced flexibility in a transparent wavelength-routed network. Two optical cross-connect demonstrators set up within the Research and Development in Advanced Communications in Europe (RACE) R1003 OSCAR program are described. The first one demonstrates protection switching and restoration of traffic in a future transport network, while the other demonstrates routing of subscriber signals to different service switches in a local exchange.

Multiwavelength cross-connects are considered by Zhong et al. (paper 9.2) where a class of wavelength interchange devices is proposed and, in conjunction with arrayed-waveguide grating multiplexers, are used as building blocks for high-capacity strictly nonblocking multiwavelength optical cross-connects. Three different structures that can be constructed using these devices are described and their blocking performance, complexity, modularity, and wavelength channel spacing are examined.

A WDM-based supercomputer interconnect is presented in paper 9.3. In this paper Kleinrock et al. describe a new WDM network which supports both stream and datagram service and extends reach and functionality of low-latency, high-bandwidth workstation clusters to a campus and metropolitan area network (MAN) setting. The concept is based on combining the rich interconnect structure of WDM with the fast, low-latency mesh of crossbar switches developed for workstation groups. The scheme interconnects asynchronous wormhole routing switches used in parallel supercomputers via multichannel WDM links embedded into an optical star (or tree) topology.

The chapter continues by giving attention to optical cross-connects for transport networks. Jourdan et al. describe in paper 9.4 the optical devices suitable for cross-connect node implementation and present results for experimental demonstrations at data rates up to 10 Gb/s. First they define the functional and physical architecture for a generic all-optical crossconnection node based on both space and wavelength switching techniques, as well as the broadcast-and-select principle. Then they present the devices that could be used for the realization of the four major functions required in the all-optical cross-connect node, namely, multiwavelength space-selection, tunable wavelength selection, wavelength conversion, and multiwavelength optical amplification. This experimental work has highlighted the main physical degradations introduced by all-optical routing devices, and the requirements to overcome these degradations.

In the last paper of this chapter, Okamoto et al. further explore the cross-connect node architectures that can be used in transport networks. They describe the optical path cross-connect (OPXC) node architectures that are essential components of the optical path network. They highlight the significance of the wavelength path (WP) and the virtual wavelength path (VWP) technologies and their potential in greatly enhancing the path layer capability and efficiency of network failure restoration.

Optical Cross-Connect System in Broad-Band Networks: System Concept and Demonstrators Description

SONNY JOHANSSON, MEMBER, IEEE, MATS LINDBLOM, PER GRANESTRAND, BO LAGERSTRÖM, AND LARS THYLÉN, MEMBER, IEEE

(Invited Paper)

***Abstract*—A future-proof transport network, robust to future evolution in network topologies or transmission formats and bit rates, would be achieved by introducing an all-optical transparent layer in the transport network hierarchy. The transparency would permit usage of physically common fiber lines and nodes for different transmission hierarchies and/or formats. A transparent network could be achieved by combining photonic switching with electronic switching technology in the network nodes. A network node would be designed as an optical cross-connect (OXC) in an optical layer, interfacing the transmission links and as a digital cross-connect (DXC) in a electrical layer. A combination of wavelength routing and space division switching in the optical layer would increase the capacity, as well as the flexibility in a network. The latter is especially important, since it allows routing with higher granularity within the optical layer. Two optical cross-connect demonstrators have been set up. The first one demonstrates protection switching and restoration of traffic in a future transport network, while the other one demonstrates routing of subscriber signals to different service switches in a local exchange. Space switches, tunable lasers and filters are key technologies used to obtain enhanced flexibility in a transparent wavelength-routed network. Optical amplifiers are needed in order to maintain the signal level through an optical node. With new technology, e.g., InP, larger monolithic matrices can be made, because both the elementary switches are smaller and because loss can be compensated for by integrating optical amplifiers. Optical building practises is an area that deserves more attention, in order to put it all together.**

I. Introduction

THE capacity needs of the transport network are increasing rapidly. This growth is foreseen even when only considering the existing telecommunication services. In addition to that, we are facing the prospect of broad-band integrated services digital network (B-ISDN) introduction. It is obvious that the capacity requirement of the future transport network will be demanding. In the development of future transport networks, it is important to avoid bottlenecks which would cause severe restrictions on the network utilization. However, the potential bottlenecks will not be associated with the fiber lines, but will be found in the network nodes. Therefore, new technologies are needed to route the increasing traffic capacity through the nodes more efficiently.

Photonic switching, in space and wavelength, is expected to be one key technology in increasing the switching capacity in future broad-band communications systems. It is well known that photonics are good for transmission, but presently not well suited for processing operations in the temporal domain, storage etc. Consequently, it appears that the most efficient near term use of photonic switching will have to rely on the frequency and code transparency of space and wavelength switching devices [1]. Hence, it is likely that the first use of photonic switching will appear in the transport network, where it will be used for the routing of entire multigigabit wavelength channels, without breaking into the data streams. In this way, it is possible to increase the capacity, flexibility, and reliability of the total network.

There is no immediate demand for transparent networks. The present technology for transmission is the synchronous digital hierarchy/synchronous optical network (SDH/SONET). These will form the transport mechanism for many years, providing higher capacity, flexibility, and reliability than earlier technology. However, with introduction of a multi-wavelength based network layer the flexibility can be extended even more. New transmissions format can be introduced on different wavelengths but in the same fibre network.

Around the world, governments are initiating reforms to enhance the efficiency of the telecommunication networks. Over several years, the market will slowly be transformed from a monopolistic market to a more global and competitive market. As a result of the new regulations, the telecommunication's market has already started to diverge into two separate specialized areas: one for bridging distance, provided mainly by network operators and an other area for adding value (i.e., adding extra services) on the network, where new service providers come in. Network operators will then sell capacity on their networks to service providers, who in their part improve the network facilities, which entice more subscribers to use the network.

New network operators will also play an important role. The railway companies are constructing fiber lines along their tracks, the power suppliers are putting fibers into their power

Manuscript received July 27, 1992; revised December 9, 1992. Part of this work was performed within the RACE I project OSCAR (R1033) and the RACE II project multi-wavelength transport network (MWTN) (R2028).

S. Johansson and M. Lindblom are with ELLEMTEL AB, Stockholm, Sweden.

P. Granestrand, B. Lagerström, and L. Thylén are with Fiber Optics Research Centre, Ericsson Telecom AB, Stockholm, Sweden.

IEEE Log Number 9207755.

Reprinted with permission from *IEEE Journal of Lightwave Technology*, Vol. 11, No. 5/6, pp. 688-694, May/June 1993.

cables, and cable TV suppliers are putting fibers into the ground. There will be lots of fibers crossing our countries in the future. Therefore, a successful network operator will be one who can utilize his network most efficiently and, with a flexible network, will be able to accommodate various kinds of service providers or other big leased enterprises networks with different requirements in network topology and, possibly, transmission format. In view of this scenario, where competition is playing an important role, there will be a future demand for transparent and flexible transport networks.

In this paper, we describe a new concept, transparent optical multi-wavelength networks, for flexible and high-capacity transport networks by introducing an optical network layer in the transport hierarchy. Key technologies used are photonic space switching devices in conjunction with wavelength selective devices. This achieves a flexible, wavelength routing network, transparent to code format and bit rates. Issues regarding the surveillance of a transparent optical layer are also elaborated.

Two demonstrators have been set up, primarily to evaluate the use of photonic space switches. One demonstrator shows protection switching and restoration in a meshed transport network of four nodes. The other system demonstrator, an Optical Access Cross-Connect, is used for connecting subscriber lines to appropriate service switches in a local exchange. This demonstrator was the first photonic switching demonstrator accomplished within the European RACE R1003 "OSCAR" project.

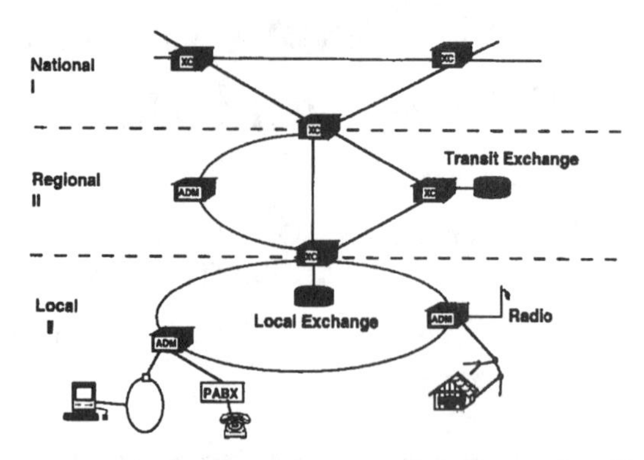

Fig. 1. Network nodes in all the transport network levels, is a future application area for new technologies, e.g., photonic switching.

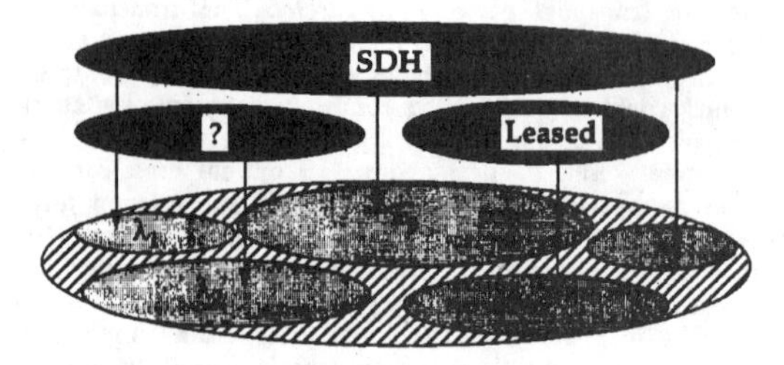

Fig. 2. Introduction of an underlying optical transparent network, divided into a number of wavelength sub-networks, which carries different overlaying networks.

II. A Transparent Optical Network Layer

The fast development of optical transmission systems provides for the capacity required in the future transport networks. However, the increasing capacity will have an impact on the throughput at the network nodes. The nodes will be more likely to become bottlenecks than the fiber lines (Fig. 1). A new technology would be a useful complement to present switching technology in solving this problem. Instead of very large cross-connect matrices for a large number of connections, smaller-sized cross-connects which serve links with much higher capacity would be useful. Photonic switching technology can be utilized in such a manner and therefore will be a strong candidate. A new concept for the transport network is developed by RACE OSCAR, at BT Labs [2] and other references e.g., [3]–[5], based on the Wavelength Division Multiplexing (WDM) technique and/or use of Optical Cross-Connects (OXC's) in the network nodes.

A. Implementation

A new approach can be used for designing transport networks with considerably improved properties compared to presently available networks. An-all optical layer can form a common underlying network for different overlying networks. The transparency would permit usage of physically common fiber lines and nodes for different transmission hierarchies or different code formats. This makes the network "future-proof" and robust to evolution, e.g. during the upgrading of the SDH network, the introduction of new transmission formats, the hosting of service providers with particular requirements, or the setting up of leased lines with different transmission formats. Only the end terminals would have to be modified while the optical transmission path basically remains unchanged.

Wavelength division multiplexing (WDM) technique increases the capacity within the network. However, this is not the only reason for introducing the WDM technique. More important is the fact that the flexibility of the network would also be enhanced, when routing with higher granularity. In addition, several multigigabit streams, even with different transmission formats, can be multiplexed and coexist in the same physical fiber network. Thus, the high throughput capacity demand on the electronics relaxes, since much of the routing can be performed within the optical layer.

An optical network layer which employs multi-wavelength technique can be divided into separate sub-networks, each able to carry any kind of overlying network (Fig. 2) even with different transmission formats. The wavelength channels can be used to extend the capacity for a specific network, e.g., the SDH network, by allocating more wavelength channels to it. Other wavelengths may be used for other networks we like to separate from the ordinary SDH-network. Why not leased networks, for big enterprises that are able to chose their own transmission format, e.g., pure ATM, if that is more efficient in their application.

B. Network Nodes

A network node with the properties given above could be implemented as a combination of an optical cross-connect

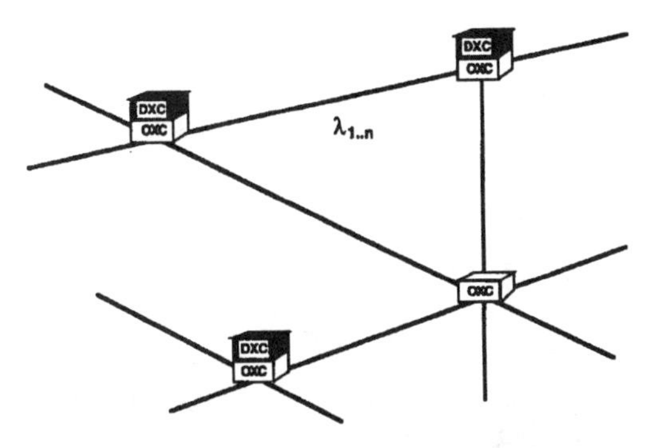

Fig. 3. A transparent and flexible transport network with OXC/DXC nodes.

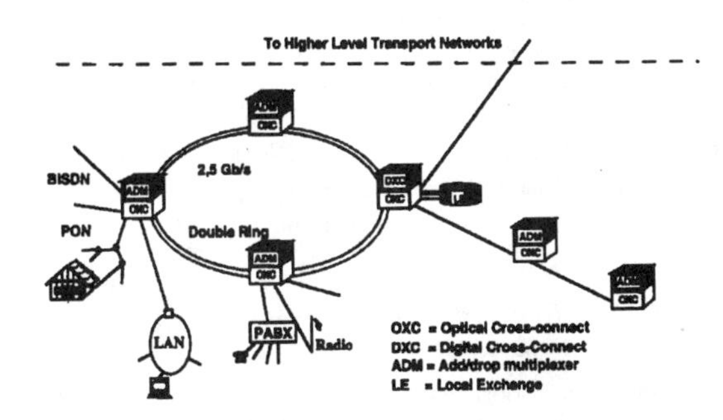

Fig. 4. OXC's in the local transport network.

(OXC) in an optical layer, interfacing with the transmission links, and a digital cross-connect (DXC), in a electrical layer above the OXC (Fig. 3). The OXC accomplishes individual routing in the optical layer of separate wavelength channels consisting of whole bit streams. The OXC can be utilized for protection switching, network restoration or reconfiguration to adapt the network to structural changes. However, not all functionality required in the node can be expected to be handled by an OXC. Hence, the electrical part, the DXC is still needed when breaking in to the bitstreams. The DXC accomplishes grooming and sorting functions, add/drop between different network levels, regeneration of poor signals and resolve wavelength contention. This is accomplished by dropping off the signals in need of more advanced processing from the optical layer to the DXC. Compared to a "stand alone" DXC approach, the OXC/DXC concept decreases the required size of the DXC. The OXC provides for routing of large gigabit traffic streams, so that much of the passing traffic is just bypassed without utilizing the DXC.

OXC's may also be very useful in the local transport network. The local transport network architecture may consist of stars, buses or rings, where ADM's are distributed along the buses or around the rings in order to concentrate the traffic up to the local exchange. Restoration of self-healing rings in case of cable breaks or node failures, would be easily accomplished by the OXC's (Fig. 4) [6].

Traffic through each ADM is dependent on the number of ADM's on the ring/bus. Consequently, a large number of ADM's on a ring generate a lot of traffic, which has to

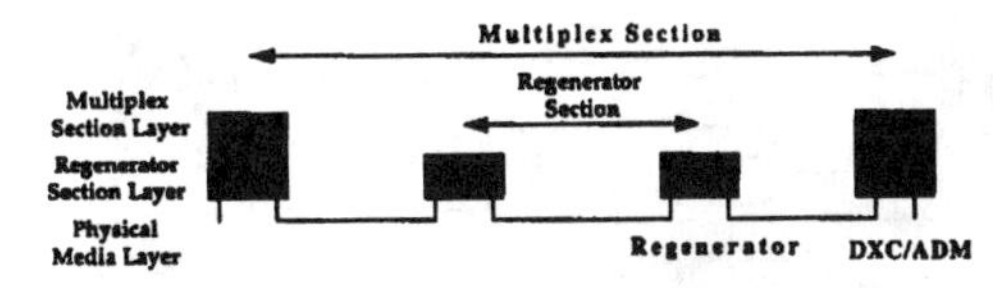

Fig. 5. A SDH path is sub-divided into multiplex sections and regenerator sections. Over each section is surveillance implemented.

be processed in each node. Therefore, much can be gained by implementing OXC's at each ADM node, so that major through traffic can be bypassed without loading the ADM's. The main part of the traffic destined for other ADM's on different wavelength channels is simply bypassed by the OXC. The ADM's add and drop traffic only from selected wavelength channels.

By combining photonic switching technology with electronic switching to take advantage of both technologies, is the most feasible way of introducing photonic switching technology into operating telecommunication networks. The photonic switching should be considered as a complement and support to electronic switching, instead of a competitor in the same application area. It is not a question of an all optical solution by this approach. Although, we do take advantage of the optical transparency. As long as no breakthrough in temporal optics occurs, the main advantages of optics will remain within transmission systems. Therefore, OXC is one of the most promising near-term applications for photonic switching technology. Hence, we are bringing photonics one step closer into telecommunication network. From have being just a point to point transmission technique, photonics will by this be able to form a base for the whole network.

C. Surveillance in a Transparent Network Layer

When introducing an optical transparent network layer, it has to be integrated to the telecommunication management network (TMN) to provide the user a powerful tool to administrate the extended network flexibility. Regarding the surveillance strategy, compared to that in SDH networks, it requires a completely different technique.

SDH is characterized with a comprehensive surveillance. There signals are supervised over each section along the path. The path from end-to-end is divided into shorter sections, i.e., multiplex sections and regenerator sections. A regenerator section is the distance between two regenerators or a regenerator and a multiplexer. A multiplex section reaches between two multiplexers and usually over a number of regenerator sections (see Fig. 5). Bit error rate (BER) measurements are made continuously over a regenerator section to validate the section's performance. In addition, there are service channels allocated for maintenance and control, e.g., for communication between the operation system and a distance regenerator. A multiplex section contains an even more comprehensive BER measurement, and more channels for maintenance and control. Alarm channels used for automatic protection switching are also embedded in the overhead. A multiplexing node can send alarms directly to the previous node when errors are detected, so that protection switching to a preselected alternative link can be made quickly.

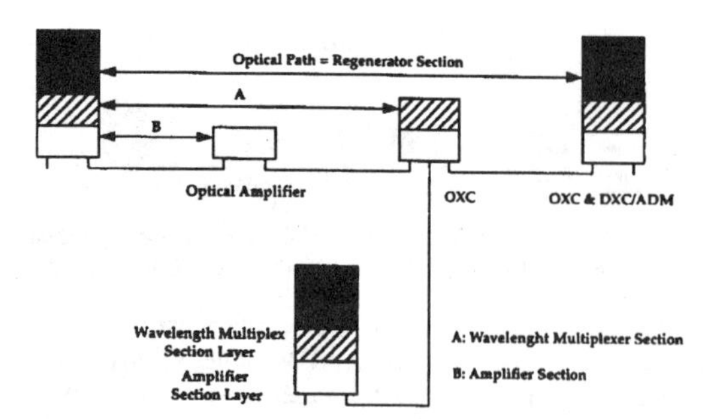

Fig. 6. A new surveillance strategy has to be developed for a transparent wavelength routed optical network. Possible by defining wavelength multiplex sections and amplifier sections.

However, a transparent network with fiber amplifiers along the lines and OXC's within the nodes will affect the surveillance strategy. An optical path, which is the same as a regenerator section, will, in a transparent network layer reach over several nodes. This means that, when a poor signal is detected at the remote end, that there is no means for fault isolation in the network with present technique. Thus, there is no information available which indicates in which node protection switching should take place. Hence, we are not able to accomplish network restoration. The SDH section overhead can only be utilized over a complete optical path (i.e., a regenerator section), since its use requires termination of the signal. The code format and bit rate will be defined when terminating the signal at the intermediate nodes. This, will cause loss of the network transparency. Therefore, a new strategy for surveillance has to be identified, possibly by defining a wavelength multiplex section layer and an amplifier section layer (Fig. 6). Each separate wavelength channel should be surveyed in all optical network nodes. In the optical layer, despite digital transmission, the signal has to be considered as analog since the time domain in the optical layer has not been defined. Accordingly, we are dealing with surveillance of analog systems.

D. Photonic Network Modeling

A CAD tool for photonic systems would be very helpful to a system designer in modelling the transmission capabilities. In this way, critical parameters in photonic network systems would be very easily identified. The advantages and drawbacks of different approaches in a network design could quickly be obtained. Development of such a tool has already begun, based on a semi-classical expression of a spectral density matrix [7]. The tool would be capable of handling effects such as amplifier spontaneous emission (ASE), saturation effects, polarization states, multiple reflection etc. in multi-wavelength systems.

III. Systems Demonstrators

To evaluate the new approach for designing transport networks, two optical cross-connect demonstrators have been set up. The first one, demonstrates protection switching and flexibility switching in a future transport network, while the other one, demonstrates automatic connection of subscriber lines to various service switches in a local exchange. The development of these demonstrators have given us essential experience for our further work in the area of transparent networks.

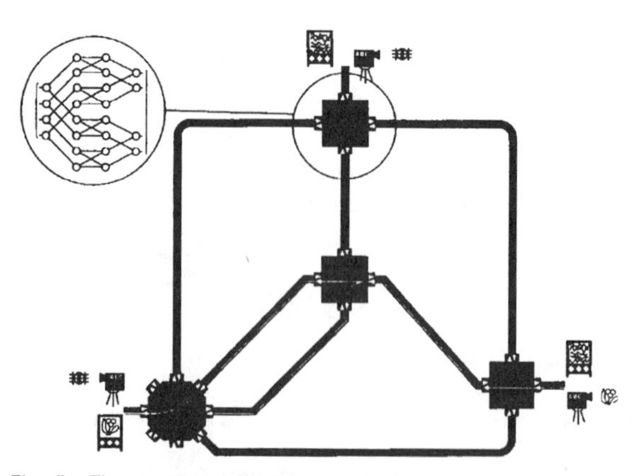

Fig. 7. The network demonstrator architecture with four OXC nodes for protection switching, routing and rearrangement.

A. The Optical Transport Network Cross-Connect Demonstrator

The optical transport network cross-connect demonstrator, demonstrating an application in a future transport network, consists of a single mode fiber-optic network with four OXC nodes (shown at the Telecom '91 exhibition in Geneva, see Fig. 7). The OXC nodes, arranged in a mesh shape, are implemented by photonic space switches. The demonstrator, representing the optical level in a synchronous network, performs optical routing of video signals (transmission rate 140 Mb/s), thus illustrating the broadband capabilities. The network is a purely optical medium in which the operator can demonstrate, e.g., fast protection switching, routing and rearrangement. In this case of fast protection switching, the operator will activate alarm signals on inputs of the node and preprogram the node on what to do if loss of signal occurs. The alarms are also relayed from the node to the operator and displayed on the screen for information and further evaluation. The operator's computer presents a graphic view of the overall network status as well as detailed information on the specific nodes. The activated routes are displayed on the screen together with the transmitters and receivers involved.

The management of the system is performed from a control computer and the management system is developed in the programming language ERLANG [8], a programming language for building robust real-time systems. Erlang is especially suited for rapid prototyping due to its high design efficiency.

The local intelligence in the nodes is capable of handling a number of configuration commands and sending signalling to the operator control computer on, e.g., alarms.

The optical hardware is based on polarization independent packaged $LiNbO_3$ 4×4 space switch matrices as well as on packaged $LiNbO_3$ 1×2 switches as basic building blocks. Each matrix is a strictly non-blocking switch comprised of 24 switch elements [9], each element being a 1×2 switching

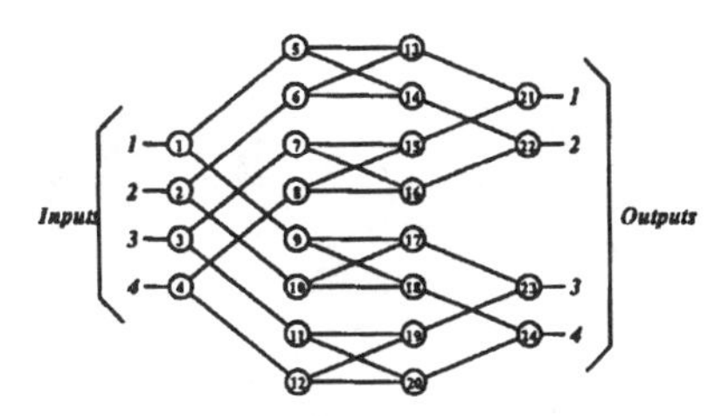

Fig. 8. The internal tree structure of the strictly non-blocking 4 × 4 $LiNbO_3$ switch matrix.

element with a digital response and controlled by *one* voltage (Fig. 8). Other characteristics are: crosstalk < −20 dB (optical), insertion loss ≈ 10 dB, switching voltage ±60 V and 3 dB splitting at 0 V. The 8 × 8 node has also a strictly nonblocking structure built up by cascading sixteen 1 × 2 modules and four 4 × 4 modules. On each input there are repeaters for two reasons: detection of loss of signal and compensation for insertion loss.

B. Optical Access Cross-Connect Demonstrator

The optical access cross-connect demonstrator, developed within the RACE I project OSCAR, was set up in order to demonstrate the application of photonic switching in a future broadband local exchange and to evaluate the performance of photonic switches in a switching fabric [10]. The demonstrator now constitutes an environment for exploring new system concepts utilizing the latest technology in the field of photonics.

The function of the optical access cross-connect is to set up bi-directional optical connections between subscribers and core switches, which maintain different services within the local exchange. Examples of possible core switches are: An ATM switch for all types of ATM services, a communicative services (CS) STM switch for various kinds of STM-carried communicative services or a distributive services (DS) STM switch, for example, which is used for switched STM-carried TV distribution.

As a demonstrator, its complexity is reduced and the subscriber network is considerably simplified. The demonstrator is basically comprised of an optical access cross-connect, to which three broad-band subscribers and a service provider are connected. Each of the three broadband subscribers is equipped with a broad-band communication terminal, here represented by a video telephone and a TV monitor. The service provider supplies the DS-STM switch with distributive services, here represented by TV channels. The transmission rate for the video signals is 140 Mb/s. The demonstrator configuration is shown in Fig. 9.

The optical access cross-connect demonstrator has the capacity of terminating eight optical subscriber lines. It is implemented using six 4 × 4 photonic switch matrices (of the same type as those used in the demonstrator described above) with corresponding control electronics, eight bidirectional WDM's and two passive optical power splitters. The management system is a similar workstation-based tool that is utilized in the demonstrator described above, which is developed in the programming language ERLANG.

Fig. 9. The optical access cross-connect demonstrator system configuration.

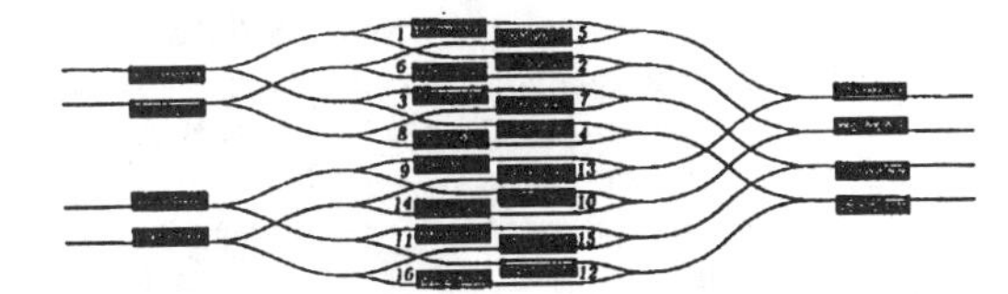

Fig. 10. The 4 × 4 InP based switching matrix structure, with integrated optical amplifiers. The figure illustrates waveguides, integrated optical amplifiers and the bond pads.

In order to maintain bidirectional cross-connection, the optical access cross-connect is divided in two planes, one for each transmission direction. The wavelength 1.55 μm is used upstream and the wavelength 1.3 μm is used downstream. The transmission directions are separated at the subscriber line interface using the bidirectional WDM's.

The incoming traffic from the subscriber lines first enters the bidirectional WDM's, which separates the incoming signals from the outgoing signals. Then the traffic is concentrated towards the CS-STM switch in two stages of optical 1.55-μm switch matrices. Similarly the traffic from the CS-STM switch passes through two stages of photonic 1.3-μm switch matrices before the bidirectional WDM's merges the transmission directions at the subscriber line interface. The procedure is slightly different for traffic from the DS-STM switch. Here the signals first enter one stage of passive optical power splitters before they enter the second stage of photonic 1.3- μm switch matrices.

IV. Key Technology

A. Optical Amplifiers

Fiber amplifiers in long distance transmission systems have become a very important component. A limiting factor in achieving a high bandwidth distance product is the ASE noise build-up. The effect of the ASE can be held back some by maintaining a rather high signal level through the system. The same argument can be applied to amplifiers in photonic switching systems, for compensation of insertion losses. It is not feasible to drop the signal level too much when the signal passes an optical node. Ideally, if the insertion losses were lower than the corresponding amplification over the section, it would be enough to adjust the fibre line amplifier span in a section where an OXC is sited in order to compensate the node loss. However, the insertion loss of a more complicated OXC will exceed allowable losses for the amplification section. Therefore, additional amplification has to be implemented inside the OXC. This has been shown by NEC, where a five stage photonic switch has been developed, by integrating of two intermediate laser amplifier stages [11].

B. Space Switches

The demonstrators described above, are both constructed in $LiNbO_3$ switching technology. At the present time, this is the only technology mature for system demonstrator experiments. In order to implement larger-scale systems of the kind reported here, upgrades in a number of technological areas are necessary. As mentioned above, the signal level has to be maintained through the switching network. The first 4 × 4 InP switch of its kind incorporating integrated optical amplifiers as routing elements has been reported [13]. The chip size of the strictly nonblocking tree structure device is only 3 × 7 mm^2, and contains 24 integrated laser amplifiers which compensate for the losses. A 0-dB insertion loss performance is obtained [14].

Switch sizes can be further increased by using optical interconnects, e.g., attach optics on Si, to cascade switch chips in a hybrid fashion. Here, losses could be compensated for by using Er-doped fiber amplifiers, or possibly, by Er-doping the interconnect medium, so that it performs amplification. As far as crosstalk and losses are concerned, large ($\gg 10 \times 10$) switch sizes are feasible [12], [13]. There is also the possibility of utilizing 3-D free space switch arrays, such as liquid crystal switches [16].

C. Wavelength Selective Devices

Important key technologies for multi-wavelength routing system are practical lasers and filters, tunable or switchable over the Er fibre amplifier window (i.e., $\approx$ 25 nm) or at least a sizable fraction of this. In many cases, when less flexibility is acceptable but the cost is a issue, also passive filters or WDM's can be used. Such technology is already on the market today.

Several tunable wavelength filter technologies are under development, whereas others are already mature. Among the most interesting are acousto-optic tunable filter (AOTF) and active semiconductors. The ATOF has a very large tuning range, which allows access to 100s of channels. Moreover, the ATOF has an appropriate feature with a dynamic selection of a number of non-adjacent channels.

Active semiconductor filters, e.g., distributed feed back (DFB) or distributed Bragg reflection (DBR), have different characteristics. They have a very narrow optical bandwidth, of about 0.5 Å, useful for high dense wavelength division multiplexing (HDWDM) systems. However, the tuning range is limited, so that the Er fibre amplifier window cannot be covered, and the very narrow bandwidth has a disadvantages in cascaded systems. An advantage with this type of filter is a net gain of the selected channel [15].

Wavelength converters would have an important role to fill in a multi-wavelength network. Such components would allow not only routing but also switching in the wavelength domain. Although research is being done in the area, a technological break through is necessary before we will see any wavelength converters in the network.

However, the lack of mature wavelength converters puts certain limitations on the network flexibility. Wavelength contention would occur at a node when two channels at same the wavelength are routed to the same output. The only way to resolve this contention is to convert the wavelength of one of the channels. Without optical wavelength converters, this has to be accomplished electronically. This puts certain limitations of the transparency of the network. Introduction of wavelength converters would merge the separated wavelength sub-networks together in a common and more flexible optical network, where allocation of wavelengths can be more channel-oriented rather than just separate sub-networks.

D. Optical Building Practices

Building practices for both optical and electrical interfaces is one very important area that deserves more attention. There are activities, but still only in small separate areas either in the micro-scale or in the macro-scale, e.g., [17]. However, in order to obtain a "future-proof" solution more coordinating activities which consider the issue as a whole from micro to macro with optics and electronics in conjunction are needed.

V. RACE Involvment

The RACE II project multi-wavelength transport network (MWTN) (R2028), is aiming to study, demonstrate and evaluate an extension of the network hierarchy with a transparent optical multi-wavelength network layer. A demonstrator will be constructed, and relevant technology developed. More about this is found in the adjacent paper in this issue [18].

VI. Conclusions

A new promising system concept to accomplish a flexible high-capacity optical network has been described. A fully-transparent optical layer, divided into a number of wavelength sub-networks, would form an underlying network for various overlying networks. The transport network nodes would be implemented as a combination of a wavelength selective optical cross-connect (OXC) in an optical layer and a digital cross-connect (DXC) in an electrical layer. The OXC is used for routing multigigabit signals, in protection switching, network restoration or reconfiguration of individual wavelength sub-networks. This is one of the most promising near-term applications for photonic switching technology, in space and wavelength. It is not an all optical solution. However, the advantage of the optical transparency is utilized. From being just a point-to-point transmission technology, photonics can by this form a base for the whole network. This brings photonics one step closer into the telecommunication network. The technology is close to mature already today to accomplish a network along this lines. However, there is also room for novel technology in a later stage to improve the functionality in the optical layer, e.g., wavelength converters etc.

Two optical cross-connect demonstrators have been constructed. The implementation of the demonstrators gives valuable experience for further system developments. These results point to the basic feasibility of the approach but also indicate the areas where upgrades are required in order that reasonable network specifications are met. Based on this experience, adequate requirements could also be set for the development of new technology in the field of photonic switching.

REFERENCES

[1] L. Thylén, "Guided-wave switching devices," *Proc. ECOC'91*, vol. 2, pp. 109–116.

[2] G. R. Hill *et al.*, "Lightwave Transport Network," *5th Int. Network Planning Symp.* (Kobe, Japan), May 18–22 1992.

[3] J. A. McEachern, "Gigabit networking on public transmission network," *IEEE Commun. Mag.*, vol. 30, no. 4, pp. 70–78, 1992.

[4] J. Sharony, K. W. Cheung, and T. E. Stern, "Wavelength dilated switches—A new class of suppressed crosstalk, dynamic wavelength-routing cross connects," *OFC'92*, vol. 5, 1992, pp. 88–89.

[5] M. Fujiwara, "Studies on optical digital cross-connect system using photonic switching matrices and optical amplifiers," *Proc. ECOC'91*, vol. 1, pp. 97–100.

[6] A. F. Elrefaie, "Self-healing WDM ring networks with all-optical protection path," *OFC'92 Tech. Dig.*, vol. 5, 1992, pp. 255–256.

[7] G. Berglind and L. Gillner, "A simple method for noise analysis of a chain of linearised optical two-ports," *3'de topical meeting on Optical Amplifiers and Their Applications* (Santa Fe, NM), June 1992.

[8] J. Armstrong and R. Virding, "Erlang—An experimental telephony programming language," *International Switching Symp.*, 1990.

[9] P. Granestrand *et al.*, "Integrated optics 4 x 4 switch matrix with digital optical switches," *Electron. Lett.*, vol. 26, pp. 4–5, 1990.

[10] M. Lindblom, P. Granestrand, and L. Thylén, "Optical access switch—First photonic switching demonstrator within the RACE program," *Proc. Photon. Switching '91*, Postdeadline Papers, pp. 3–6.

[11] C. Burk, M. Fujiwara, M. Yamaguchi, H. Nishimoto, and H. Honmou, "128 line photonic switching system using $LiNbO_3$ switch matrices and semiconductor traveling wave amplifiers," *J. Lightwave Technol.*, vol. 10, no. 5, pp. 610–615, 1992.

[12] M. Fujiwara, S. Suzuki, and H. Nishimoto, "Line capacity consideration for a photonic space-division switching system with switch matrices and optical amplifiers," *Tech. Dig. Photon. Switching* 1990 (Kobe, Japan), paper 13C-3.

[13] P. Granestrand, L. Thylén, B. Stoltz, and J.-E. Falk, "Systems experiment with a packaged 4 x 4 polarization independent switch matrix," *Proc. Photon. Switching* '89, pp. 159–161.

[14] M. Gustavsson *et al.*, "Monolithically integrated 4 x 4 InGaAsP/InP laser aplifier gate switch arrays," *Topical Meet. Optical Amplifiers and Theirs Application*, postdeadline paper, 1992.

[15] O. Sahlén, M. Öberg, and S. Nilsson, "Performance of DBR active filters in 2.4 Gb/s systems," *ECOC/IOOC '91*, vol. 1, part 2, 1991, pp. 445–448.

[16] K. Noguchi, T. Sakano, and T. Matsumoto, "A 128 x 128-channel free-space optical switch using polarization multiplexing technique," *ECOC/IOOC '91* vol. 1, part 1, 1992, pp. 165–168.

[17] Y. Satoh, M. Kurisaka, and T. Sawano, "Orthogonal packaging for a photonic switching system," *OFC'92, Tech. Dig.*, vol. 5, 1992, p. 162.

[18] G. Hill *et al.*, "A transport network layer based on optical network elements," *J. Lightwave Technol.*, this issue, 1993.

Multiwavelength Cross-Connects for Optical Transport Networks

Wen De Zhong, *Member, IEEE,* Jonathan P. R. Lacey, *Member, IEEE,* and Rodney S. Tucker, *Fellow, IEEE*

***Abstract*—Multiwavelength cross-connects (M–XC's) will play a key role in future optical multiwavelength transport networks. In this paper, we propose a class of optical wavelength interchange devices that can be used as basic building-blocks for multiwavelength optical cross-connects. We describe three different multiwavelength cross-connect structures that can be constructed using these building blocks. We investigate their blocking performance and examine issues such as complexity, modularity, and wavelength channel spacing associated with the proposed cross-connect structures.**

I. Introduction

CONTINUOUS advances in photonic technologies such as wavelength division multiplexing (WDM), all-optical wavelength conversion, tunable lasers and filters, optical semiconductor and fiber amplifiers offer the possibility of building novel all-optical transport networks [1]–[3]. To integrate different existing transmission systems (such as SDH and PDH) and different transfer modes (such as STM and ATM) in a single optical transport network [4], and to improve optical path network performance (such as path connectivity and restoration), the optical virtual wavelength path (VWP) concept is a useful approach [4]. Using the VWP scheme, the wavelength assigned to a path in the optical path layer is determined on a link-by-link basis rather than a global basis, resulting in significantly improved path connectivity and reduced number of required wavelengths [4].

In the optical path layer, multiwavelength cross-connects (M–XC's) capable of wavelength translation (or conversion) are crucial. An optical cross-connect is called an M–XC if each input/output fiber carries multiple wavelength channels and if these wavelengths are interchangeable. To elucidate the role of an M–XC, Fig. 1 illustrates a layered optical network consisting of an optical path layer, a switching layer and an access layer. The optical path layer is comprised of M–XC's and long haul transmission fibers. The switching layer consists of electrical or optical switches. These switches can operate either in synchronous transfer mode (STM) or in asynchronous transfer mode (ATM). In the example shown in Fig. 1, a group of switches under an M–XC are looked upon as the users of that M–XC. This grouping can significantly reduces the complexity of path reconfiguration and restoration associated with large networks. The requirement for an M–XC is to transport a large amount of traffic between any two switches

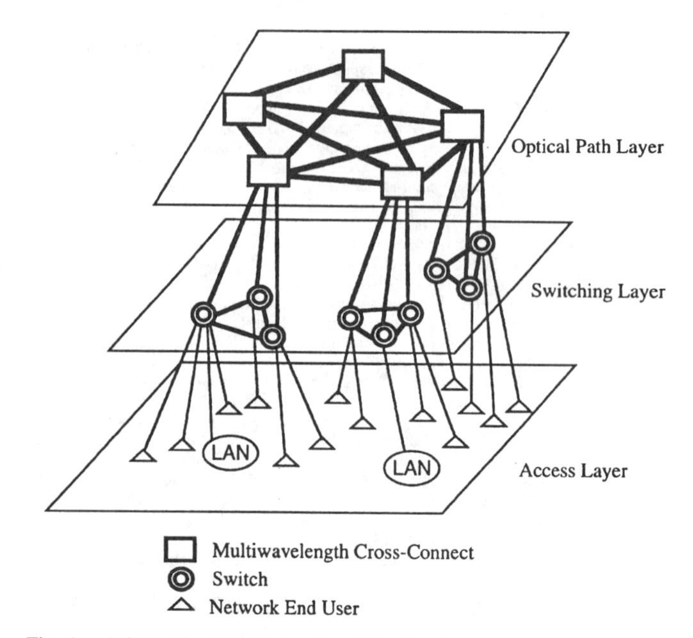

Fig. 1. A layered optical network architecture.

in the switching layer. That is, an M–XC needs to provide a number of wavelength paths between any two switches. Within a wavelength path, electrical paths (such ATM virtual paths and STM paths) can be established. To fully utilize optical bandwidth (in terms of wavelength channels) over each of the fiber links between any two switches and to flexibly configure wavelength paths, it is desirable that a wavelength channel is able to be changed from one wavelength to another when it is switched from an input fiber to an output fiber at an M–XC. This is the essence of the VWP scheme [4].

From the viewpoint of blocking performance, M–XC's can be classified into nonblocking networks and blocking networks [5], [6]. Non-blocking networks can be further divided into strictly nonblocking networks [5] and rearrangeably nonblocking networks [6]. If an M–XC is strictly nonblocking, any idle input wavelength channel on any input fiber can always be connected to any unused output wavelength channel on any output fiber without disturbing any existing wavelength connections [5]. If it is rearrangeably nonblocking, any idle input wavelength channel on any input fiber can always be connected to any unused output wavelength channel on any output fiber by rerouting the existing wavelength connections if necessary [6]. In general, it is desirable that an M–XC is strictly nonblocking because some wavelength channels on a fiber link cannot be used in a blocking network, and a large amount of traffic may be lost during the period of rerouting in

Manuscript received December 18, 1995.

The authors are with the Australian Photonics Cooperative Research Centre, the Photonics Research Laboratory, Department of Electrical and Electronic Engineering, The University of Melbourne, Parkville, Victoria 3052, Australia.

Publisher Item Identifier S 0733-8724(96)05432-1.

Reprinted with permission from *IEEE Journal of Lightwave Technology*, Vol. 14, No. 7, pp. 1613-1620, July 1996.

a rearrangeable nonblocking network. Thus, the objective of this paper is to investigate strictly nonblocking M–XC's.

There have been two major approaches to realize M–XC's. The first approach is to employ space-division (SD) switching followed by wavelength conversion [4]. That is, the wavelength channels carried on each input fiber are first demultiplexed before SD switching. Switching a wavelength channel from an input fiber to a desired output fiber is then carried out by optical SD switches. Following the SD switching, wavelength conversion is performed for individual channels before they are multiplexed and launched to an output fiber. This approach is essentially SD switching and is therefore referred to as the SD approach in this paper. Although the optical SD switch technology is already practical, the SD approach may encounter problems such as a rapid growth in the number of cross-points when the total number of switched wavelength channels is increased [7].

The second approach is to employ wavelength routing in conjunction with wavelength conversion [8]. Since switching a channel from an input fiber to an output fiber is conducted in the wavelength domain, this approach is referred to as the WD (wavelength-division) approach. With the WD approach, wavelength conversion is inevitable because the switching of a channel is determined by the wavelength of that channel. The WD approach is very promising in the sense that it has the potential of providing a large number of wavelength channels with a relatively low complexity when all-optical wavelength conversion technologies are well developed [8]. This paper focuses on the WD approach.

Suzuki *et al.* [8] were the first to propose a wavelength switch (called the λ switch) capable of wavelength interchanging, and have experimentally demonstrated the λ switch based on optical-to-electrical and electrical-to-optical conversions. They have also proposed WDM switching systems using λ switches [8]. One practical difficulty with Suzuki's switching systems is that they need a large number of optical components in terms of separate wavelength multiplexers and demultiplexed to interconnect two adjacent stages. In addition, their WDM switching networks are rearrangeably nonblocking and are not strictly nonblocking.

With the motivation of investigating novel strictly nonblocking M–XC's that have a simple structure and use a minimum number of components, this paper proposes a class of wavelength interchange devices (WID's) as basic building-blocks for construction of M–XC's [22]. The paper further proposes and investigates three M–XC structures, using WID's in conjunction with arrayed-waveguide grating multiplexers (AWGM's) [9], [10]. The conditions under which the proposed M–XC's are strictly nonblocking or rearrangeably nonblocking are derived in a straightforward way. Some key issues associated with the proposed M–XC structures such as complexity, switching capacity expansion (or modularity), and wavelength channel spacing are also investigated.

II. Wavelength Interchange Devices

In this section, we propose four types of WID's, whose structures depend on the number of fibers and wavelengths carried on each input/output fiber [22]. Their application as basic building-blocks for M–XC's is presented in Section III. In general, we denote a WID as $i \times j$ WID(N, K). This device has i input fibers, each having N possible input wavelengths, and j output fibers, each having K possible output wavelengths. In this paper, the basic WID's that are of interest are 1×1 WID(N, K), 1×2 WID(N, K), 2×1 WID(N, K) and 2×2 WID(N, K), since they can be built in a relatively simple way. The reason that N may be different from K will be clear in Section III. These WID's are required to be strictly nonblocking, as our aim is to use them to build strictly nonblocking M–XC's. The examples of WID's presented here use cross-gain modulation (XGM) in semiconductor optical amplifiers (SOA's) [11], [12] to perform wavelength conversion, since this scheme can achieve relatively high conversion efficiency [11] . In addition, the XGM wavelength conversion is relatively tolerant to the wavelength and the power level of the signal, and it is polarization insensitive [11]. However, other wavelength conversion mechanisms such as four-wave mixing in SOA's [13], cross-phase modulation in SOA's [14] and carrier depletion in DBR lasers [15] could be also used.

A 1×2 WID(N, K) has one input fiber carrying M data channels on M of N possible input wavelengths, and two output fibers, which between them carry the same M data channels on M of K possible output wavelengths where $M \leq \text{Min}\{N, K\}$. Fig. 2 shows a block diagram of a 1×2 WID(N, K) and its wavelength interchange characteristics. In the example shown in Fig. 2, the input fiber carries channels A, B, C, G, and M at wavelengths λ_1, λ_3, λ_4, λ_{N-1}, and λ_N, respectively. Channels A and G are switched to the upper output fiber at wavelengths λ_1 and λ_3, respectively; while channels B, C, and M are switched to the lower output fiber at wavelengths λ_1, λ_2, and λ_K, respectively. Fig. 3 shows the schematic configuration of a 1×2 WID(N, K). The input WDM signal is amplified by an EDFA before being demultiplexed. Each individual wavelength signal is fed to a fixed-input tunable-output wavelength converter (FTWC), where it is converted to another wavelength out of K possible output wavelengths. In the FTWC, the input signal of a fixed wavelength and the CW light of wavelength λ_{out} generated from a tunable laser are injected into an SOA in opposite directions [11]. This facilitates the separation of the desired output wavelength λ_{out} from the input wavelength. Also a tunable bandpass filter at the output of the FTWC is used to select the desired wavelength λ_{out} and to block the amplified spontaneous emission (ASE) from the SOA. The individual signals from the FTWC's are then switched by the 1×2 space switches toward one output fiber or the other.

A 2×1 WID(N, K) has two input fibers carrying a total of M WDM channels on M of N possible input wavelengths, and one output fiber carrying the same M WDM channels on M of K possible different output wavelengths where $M \leq \text{Min}\{N, K\}$. Fig. 4 shows a block diagram of a 2×1 WID(N, K) and illustrates its wavelength interchange characteristics. In the example shown in Fig. 4, the upper input fiber carries channels A, C, and M at wavelengths λ_1, λ_3, and λ_N, respectively; the lower input fiber carries channels B and

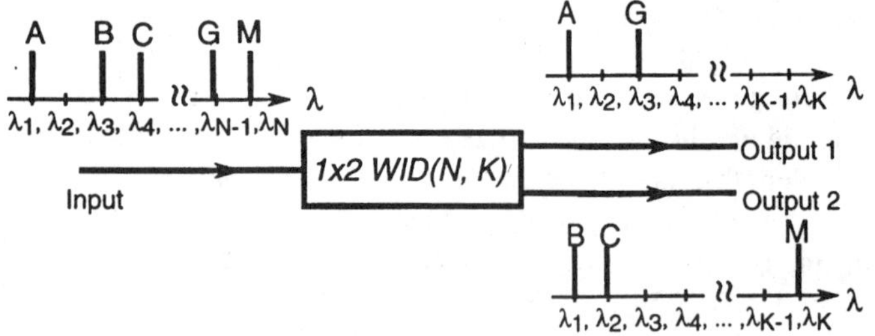

Fig. 2. Block diagram of a 1 × 2 WID(N, K).

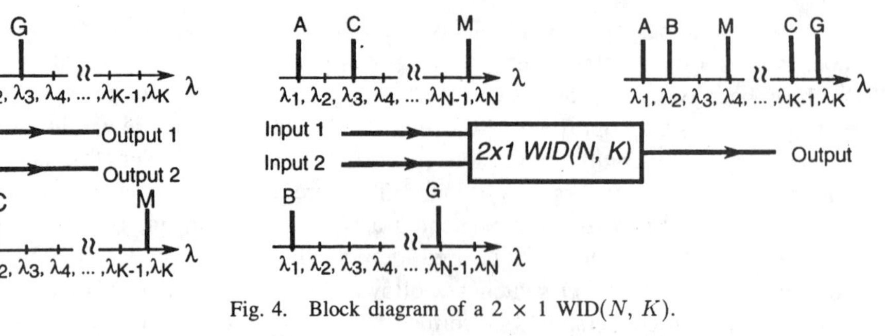

Fig. 4. Block diagram of a 2 × 1 WID(N, K).

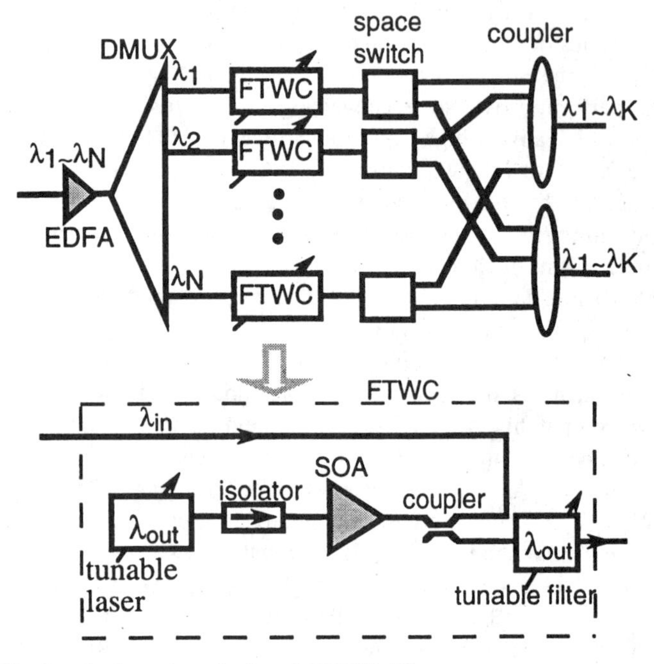

Fig. 3. Configuration of a 1 × 2 WID(N, K).

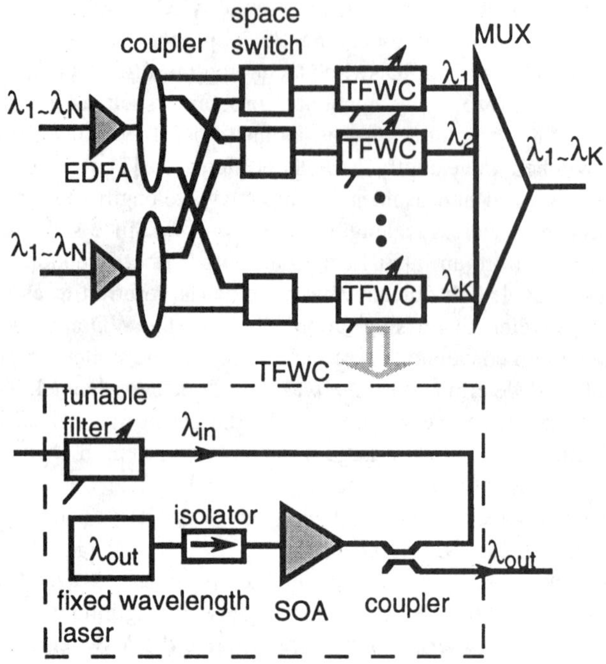

Fig. 5. Configuration of a 2 × 1 WID(N, K).

G at wavelengths λ_1, λ_{N-1} respectively. These channels A, B, C, G, and M emerge from the output fiber at wavelengths λ_1, λ_2, λ_{K-1}, λ_K, and λ_4, respectively. Fig. 5 shows the schematic configuration of a 2 × 1 WID(N, K). Each of the two input WDM signals is amplified and then divided into K equal parts. Each 2 × 1 space switch selects one of the input WDM signals. The selected WDM signal is fed to a tunable-input fixed-output wavelength converter (TFWC), where a specific wavelength out of N possible input wavelengths is selected by a tunable filter and then converted to a preassigned output wavelength. Since the individual signals from the TFWC's have fixed wavelengths, they are multiplexed by a wavelength multiplexer. Note that no bandpass filter is incorporated at the output of the TFWC. The ASE from the SOA is removed at the wavelength multiplexer.

Note that the 1 × 2 WID(N, K) and the 2 × 1 WID(N, K) both become a 1 × 1 WID(N, K) when the space switches in Figs. 3 and 5, one of the two output fibers and the corresponding coupler in Fig. 3, and one of the two input fibers and the corresponding coupler in Fig. 5 are removed. Thus, two configurations of 1 × 1 WID(N, K) are available. Note that the λ switch proposed in [8] is essentially a 1 × 1 WID(N, K) when $N = K = M$. Likewise, a 2 × 2 WID(N, K) can be made by including 2 × 2 space switches. Note also that each wavelength converter in Figs. 3 and 5 is disabled by simply turning off the electrical current injected into the SOA if there is no corresponding input wavelength signal to be converted.

III. Multiwavelength Cross-Connects

We now show how the WID's presented in Section II can be used in conjunction with arrayed-waveguide grating multiplexers (AWGM's) [9], [10] to build M–XC's. Three different M–XC structures are proposed. The conditions under which the proposed M–XC's are strictly nonblocking or rearrangeably nonblocking are derived in a straightforward way. Comparisons between each of the structures are presented in Section IV-A.

A. Structure I

The first M–XC structure (Structure I) is shown in Fig. 6. It has K input/output fibers, each carrying n wavelength channels. It features three columns of 1 × 1 WID's, each column representing a stage. Unlike the WDM cross-connects proposed in [8], the M–XC in Fig. 6 uses an AWGM to

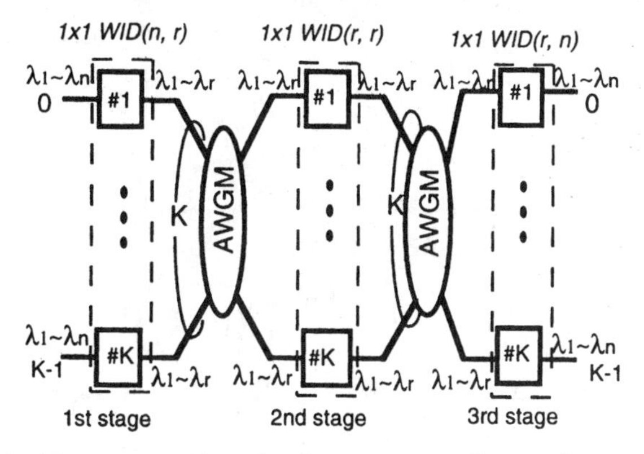

Fig. 6. Three-stage multiwavelength cross-connect: Structure I.

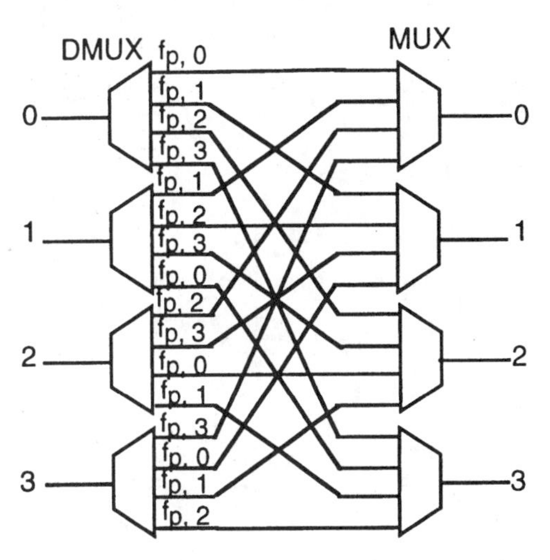

Fig. 7. The equivalent circuit of a 4 × 4 AWGM.

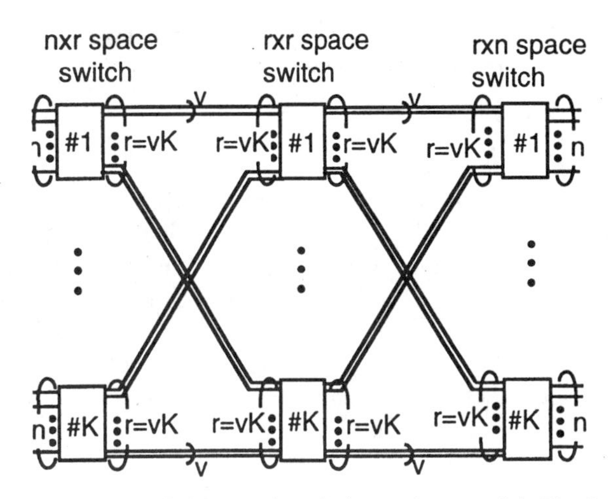

Fig. 8. Three-stage SDM network equivalent to the network in Fig. 6.

interconnect two adjacent stages instead of using separate wavelength multiplexers and demultiplexers. This is done because a single AWGM can perform both wavelength multiplexing and demultiplexing [9], [10]. Another important property of a $K \times K$ AWGM is that it has a free spectral range (FSR) or a frequency period [9], [10]. By making use of this property, a number of wavelength (frequency) channels, each separated by a FSR, can be simultaneously set up through the AWGM between two WID's in two adjacent stages. For simplicity of description, the terms wavelength and frequency will be used interchangeably in the remainder of this paper.

In order to explore the operation of the M–XC in Fig. 6, we now consider models of the AWGM and the overall M–XC structure. Fig. 7 illustrates the equivalent circuit of a 4 × 4 AWGM, in which a connection between a demultiplexer (DMUX) and a multiplexer (MUX) represents a set of frequencies each separated by a FSR [16]. Let $f_0, f_1, \cdots, f_{K-1}$ be the K consecutive operating frequencies within a FSR of a $K \times K$ AWGM. The set of frequencies $f_{p,q}$ that connect the ith input to the jth output of a $K \times K$ AWGM can be expressed in terms of f_q and FSR, i.e., $f_{p,q} = p \times \text{FSR} + f_q$, where q is determined by $q = (i + j)$ modulo K and p is an integer. Note that the operating frequencies $f_0, f_1, \cdots, f_{K-1}$ can be made to be either equally spaced [9], [10] or unequally spaced [17]. The frequency spacing issue will be discussed in Section IV-C. In the remainder of this section, all the frequencies are assumed to be equally spaced and they are all operating frequencies of the AWGM's used.

If a wavelength connection between any two 1 × 1 WID's in adjacent stages in Fig. 6 is looked upon as a space-division (SD) connection, the network in Fig. 6 can be translated into an equivalent three-stage SD network shown in Fig. 8. This is because a 1 × 1 WID(n, r) is equivalent to an $n \times r$ SD switch, and an AWGM can provide multiple wavelength connections between any two WID's in adjacent stages. The network in Fig. 8 resembles a three-stage Clos network [5], except that there exist v links between two SD switches in adjacent stages. These links correspond to v wavelengths, each separated by a FSR of the AWGM, between two WID's in adjacent stages in Fig. 6. According to the strictly nonblocking principle discussed in [5] and the equivalence between the two networks in Figs. 6 and 8, it can be shown in a straightforward way that the network in Fig. 6 is strictly nonblocking if there is an equal number of wavelength connections between any two WID's in adjacent stages and if the following equation is satisfied:

$$r \geq \left\lceil \frac{2n-1}{K} \right\rceil \times K \tag{1}$$

where the notation $\lceil x \rceil$ represents the least integer greater than or equal to x.

It can also be shown in a straightforward way that the network in Fig. 6 is rearrangeably nonblocking [6] if there is an equal number of wavelength connections between any two WID's in adjacent stages and if the following equation is satisfied:

$$r \geq \left\lceil \frac{n}{K} \right\rceil \times K. \tag{2}$$

From (1), it can be seen that, in order to make the network in Fig. 6 strictly nonblocking, the number of wavelengths r required in the WID's is about twice the number of wavelength channels n carried on an input/output fiber. This means that about 50% of the wavelengths used in the WID's are redundant. It is desirable to have the value of r equal to or close to the value of n, since the tuning range of tunable lasers

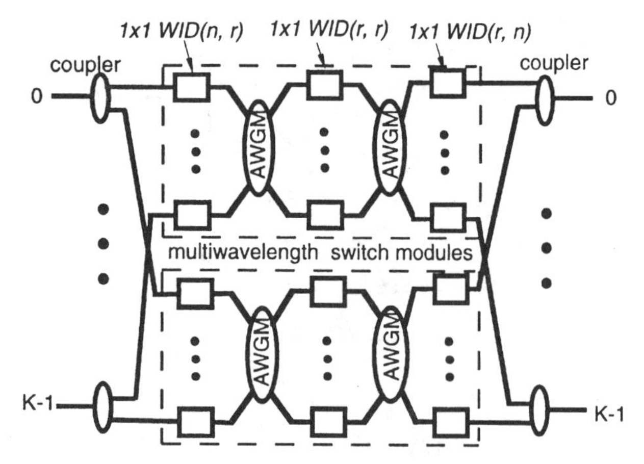

Fig. 9. Three-stage multiwavelength cross-connect: Structure II.

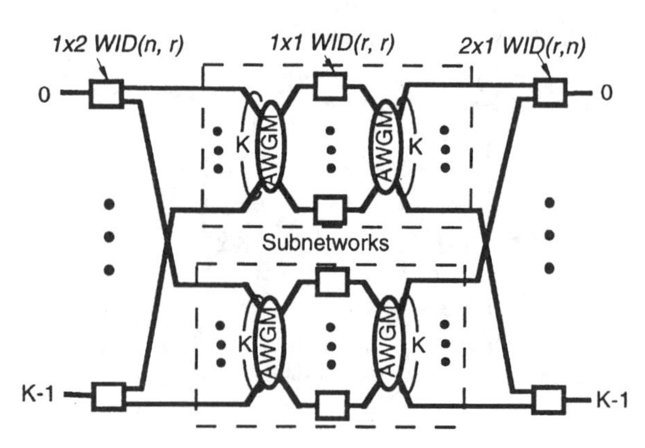

Fig. 10. Three-stage multiwavelength cross-connect: Structure III.

and filters used in WID's is limited [18], [19]. Two modified structures that reduce the number of wavelengths required in the WID's are proposed below.

B. Structure II

The second M–XC structure (Structure II) for strictly nonblocking operation is shown in Fig. 9. Unlike Structure I, the number of output wavelengths from tunable lasers and filters used in the WID's does not need to be a factor of 2 larger than the number of wavelength channels carried on an input/output fiber. However, this reduction is achieved at the expense of adding more components to double connections in space between input and output fibers (see Section IV-A for details).

In Fig. 9, the WDM signal from each input fiber is split by a coupler into two equal parts, each fed to an input of a succeeding $K \times K$ multiwavelength switch module which has exactly the same structure as the network shown in Fig. 6. Therefore, Structure II in Fig. 9 provides doubled connections in space compared with Structure I in Fig. 6. Accordingly, to be strictly nonblocking as is Structure I, Structure II does not need to use tunable lasers and filters that have a tuning range twice the number of wavelength channels carried on an input/output fiber. However, this reduced tuning range comes at the expense of increasing the number of wavelength converters by about 50% (see Section IV-A). Again, based on the strictly nonblocking principle discussed in [5], it can be shown in a straightforward way that Structure II is strictly nonblocking if there is an equal number of wavelength connections between any two WID's in adjacent stages and (2) is satisfied. Thus, the required number of output wavelengths from tunable lasers and filters used in WID's is reduced from $\lceil (2n-1)/K \rceil K$ to $\lceil n/K \rceil K$ (i.e., $r = \lceil n/K \rceil K$).

C. Structure III

As explained above, although Structure II in Fig. 9 reduces the number of wavelengths by about 50%, it requires 50% more wavelength converters than Structure I in Fig. 6. We now propose a third structure (Structure III) that uses the same number of wavelengths as Structure II, but it does not require an increase in the number of wavelength converters. The number of wavelength converters required by Structure III is almost the same as that of Structure I. These reductions in number of wavelengths and wavelength converters are achieved by using wavelength switching in combination with space switching.

Structure III is shown in Fig. 10. It uses three different kinds of WID's, i.e., 1×2 WID(n, r) in the first stage, 1×1 WID(r, r) in the second stage, and 2×1 WID(r, n) in the last stage. The middle section in dashed lines of Fig. 10 includes two identical subnetworks. If one of the subnetworks is removed and the 1×2 WID(n, r)'s in the first stage and the 2×1 WID(r, n)'s in the last stage are made 1×1 WID's, Structure III becomes exactly the same as Structure I. Thus, Structure III provides doubled connections in space between an input fiber and an output fiber, compared with Structure I. Consequently, to be strictly nonblocking, Structure III does not need to use tunable lasers and filters that require twice as many output wavelengths as wavelength channels carried on an input/output fiber. The strictly nonblocking conditions for Structure III are exactly the same as that for Structure II. Compared with Structure II, Structure III substantially reduces the number of wavelength converters.

IV. Discussion

In this section, some key issues associated with the proposed M–XC structures such as complexity, switching capacity expansion (or modularity), and wavelength channel spacing are presented.

A. Comparison

The complexity of the three proposed M–XC structures are compared in terms of number of output wavelengths from tunable lasers and filters, wavelength converters, optical SD switches, and AWGM's. All the M–XC structures are compared under strictly nonblocking operation.

As explained in Section II, there are two available types of 1×1 WID's, of which one is made up of tunable-input fixed-output wavelength converters (TFWC's), and the other is composed of fixed-input tunable-output wavelength converters (FTWC's). The former is referred to as the tunable-input type

TABLE I
COMPARISON OF THREE PROPOSED MULTIWAVELENGTH CROSS-CONNECTS

Structure \ Number	Output wavelength from tunable lasers and filters	Wavelength converters	space switches	AWGM's
Structure I	$\lceil(2n-1)/K\rceil K$	$2nK+\lceil(2n-1)/K\rceil K^2$	0	2
Structure II	$\lceil n/K\rceil K$	$4nK+2\lceil n/K\rceil K^2$	0	4
Structure III	$\lceil n/K\rceil K$	$2nK+2\lceil n/K\rceil K^2$	2nK	4

1×1 WID and the latter as the tunable-output type 1×1 WID. In general, a TFWC is simpler than a FTWC, since the latter uses one more tunable component than the former (see Figs. 3 and 5). However, in most of the following discussion, the cost and complexity of a 1×1 WID is determined simply on the basis of the number of wavelength converters required. In other words, the cost and complexity of a TFWC is assumed to be about the same as that of a FTWC. In comparison, it is always assumed that the type of 1×1 WID that uses fewer wavelength converters is chosen. That is, the tunable-output type of 1×1 WID (i, j) is chosen when $i < j$; whereas the tunable-input type of 1×1 WID(i, j) is adopted when $i > j$. For $i = j$, it may be preferable to use the tunable-input type of 1×1 WID(i, j) since it needs fewer tunable components than the tunable-output type. For a network of the kind shown Fig. 6, it is more economic for the 1×1 WID(n, r) in the first stage to use the tunable-output type and the 1×1 WID(r, n) in the third stage to use the tunable-input type, since r is about twice n in order to make the network strictly nonblocking.

Let K be the number of input/output fibers and n be the number of wavelength channels carried on each input/output fiber. Table I presents and compares the required components for the three proposed M–XC structures. With reference to Table I, it can be seen that Structures I and III both require wavelength converters about to $4nK$, while Structure II needs about $6nK$ wavelength converters, 50% more than Structures I and III. The number of output wavelengths from tunable lasers and filters for Structure I is almost twice that for Structures II and III. The advantage of Structure II is that it does not need space switches.

B. Switching Capacity Expansion

In principle, the total switching capacity (defined as the total number of wavelength channels switched by a M–XC) can be expanded by increasing the number of wavelength channels carried on an input/output fiber, i.e., n, or by increasing the number of ports of the AWGM's, i.e., K, or by increasing both of them. However an increase in n is limited by several factors, such as the tuning range of tunable lasers or filters, channel spacing, and the gain bandwidth of SOA's and EDFA's used in WID's. On the other hand, the available number of input/output ports of an AWGM may be restricted by arrayed-waveguide grating technology [9], [10]. Furthermore, an increase in K requires an increase in the tuning range of tunable lasers and filters, as the number of wavelengths

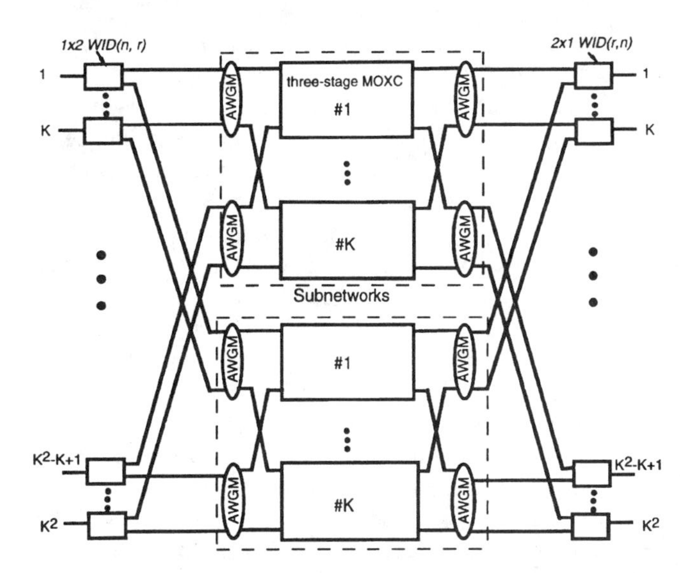

Fig. 11. Five-stage multiwavelength cross-connect adopting Structure III.

required in WID's increases as K grows. In the evolution of a network from small to large scale, it is desirable that the total switching capacity can be increased in a modular fashion without any change for the basic building blocks.

Thus, the most promising way to increase the total switching capacity is to add more WID stages and AWGM's for interconnections between stages, while keeping the values of n and K unchanged. In general, the total switching capacity can be increased by a factor of K for every two stages (one at the input and one at the output) added [5], [6], where K is the number of ports of the AWGM. However, for strictly nonblocking M–XC's with Structure I, the number of wavelengths required in the WID's at the central stage has to be increased by a factor of about two for every two stages added. In contrast, for Structures II and III, the number of wavelengths required in the WID's is independent of the number of stages. Furthermore, Structure III is the most attractive because it needs fewer wavelength converters than Structure II.

Fig. 11 shows a five-stage M–XC adopting Structure III of Fig. 10. As can be seen, there are two identical subnetworks. Each subnetwork includes K three-stage M–XC modules, a set of K $K \times K$ AWGM's connecting to the 1×2 WID(n, r)'s at the first stage, and another set of K $K \times K$ AWGM's connecting to the 2×1 WID(n, r)'s at the last stage. The three-stage M–XC modules are each the same as the network in Fig. 10, which is equivalent to a strictly nonblocking $Kr \times Kr$ SD network if a wavelength connection is looked upon as a SD channel. Thus, the network in Fig. 11 is equivalent to an $nK^2 \times nK^2$ five-stage SD network resembling a five-stage Clos network [5]. Its total switching capacity is nK^2. Its nonblocking conditions are the same as those for the network in Fig. 10.

C. Wavelength Channel Spacing

The question of wavelength channel spacing is an important issue in long haul WDM transmission systems [20], [21]. When wavelength channels are equally spaced, the product terms generated by four-wave mixing (FWM) between channels fall exactly at the channel wavelengths and hence cause crosstalk [20], [21]. This is particular severe when low-dispersion fibers are used [20], [21]. An effective technique to circumvent the FWM crosstalk is to use unequal channel spacing so that the FWM waves do not coincide with the channel wavelengths [20], [21]. For this reason, it may be necessary to use unequally spaced wavelength channels in M–XC's rather than to use equally spaced wavelength channels.

Unfortunately, the design of an M–XC with unequally spaced channels is much more complicated than that of an M–XC with equally spaced channels. This is because an M–XC with unequally spaced channels would require specially designed AWGM's [17] that have unequally spaced operating wavelengths as well as continuously tunable lasers rather than discretely tunable lasers. It is critical to precisely control the wavelength of continuously tunable lasers. In contrast, an M–XC with equally spaced channels can use discretely tunable lasers. A discretely tunable laser provides a set of discrete wavelengths equally spaced by its FSR [18], [19]. Tuning a discretely tunable laser from one wavelength to another is relatively simpler than tuning a continuously tunable laser, since the injection current applied for the discretely tunable laser does not need to be extremely precise and a small injection current disturbance does not change the wavelength [18], [19]. For this reason, it is desirable to have equal channel spacing in M–XC's.

There is a straightforward technique to deal with both the FWM crosstalk generated over long haul transmission fibers and the wavelength control problem arising from tunable lasers used in WID's. Since the cause of FWM crosstalk does not arise from the M–XC but the long haul transmission fibers, the wavelength channels inside the M–XC can be equally spaced as long as the wavelength channels entering and leaving the M–XC are unequally spaced. This means that only the input wavelengths of WID's at the first stage of the M–XC's and the output wavelengths of WID's at the last stage need to be unequally spaced. As already explained, in order to reduce the number of wavelength converters, the WID's for the first stage are made up of fixed-input tunable-output wavelength converters (Fig. 3), and the WID's for the last stage are made up of tunable-input fixed-output wavelength converters (Fig. 5). With reference to Fig. 3, only the wavelength demultiplexer needs to be adapted to unequally spaced output wavelengths, while the tunable lasers can have discrete equally spaced output wavelengths. With reference to Fig. 5, only the fixed-wavelength lasers and the multiplexer need to be adjusted to the same wavelengths that enter the M–XC and are unequally spaced. By this means, all the tunable lasers and AWGM's can have equally spaced operating wavelengths.

V. Conclusion

We have discussed the general concept of wavelength interchange devices (WID's), and have shown how these devices, in conjunction with arrayed-waveguide grating multiplexers, are very powerful building blocks for high-capacity strictly nonblocking multiwavelength optical cross-connects (M–XC's). Three M–XC structures have been proposed and investigated. It has been shown that, by combining wavelength switching and space switching, it is possible to reduce the number of wavelengths required in the WID's and to use the minimum number of wavelength converters. When the number of stages in this cross-connect is increased to expand the total switching capacity, the number of wavelengths required is independent of the number of stages. We have also described a straightforward technique to deal with the FWM crosstalk arising from long haul transmission fibers as well as the wavelength control problem of tunable lasers.

References

[1] Special Issue on Dense Wavelength Division Multiplexing Techniques for High Capacity and Multiple Access Communication Systems, *IEEE J. Select. Areas Commun.*, vol. 8, Aug. 1990.

[2] Special Issue on Broad-Band Optical Networks, *J. Lightwave Technol.* vol. 11, May/June 1993.

[3] T. Miki, "The potential of photonic networks," *IEEE Commun. Mag.*, vol. 32, pp. 23–27, Dec. 1994.

[4] K. Sato *et al.*, "Network performance and integrity enhancement with optical path layer technologies," *IEEE J. Select. Areas Commun.*, vol. 12, pp. 159–171, Jan. 1994.

[5] C. Clos, "A study of nonblocking switching networks," *Bell Syst. Tech. J.*, pp. 407–425, Mar. 1953.

[6] V. E. Benes, "On rearrangeable three-stage connecting networks," *Bell Syst. Tech. J.*, pp. 1481–1492, Sept. 1962.

[7] R. A. Spanke, "Architectures for guided-wave optical space switching systems," *IEEE Commun. Mag.*, vol. 25, pp. 42–48, May 1987.

[8] S. Suzuki *et al.*, "A photonic wavelength-division switching system using tunable laser diode filters," *J. Lightwave Technol.*, vol. 8, pp. 660–666, May 1990.

[9] H. Takahashi *et al.*, "Arrayed-waveguide grating for wavelength division multi/demultiplexer with nanometer resolution," *Electron. Lett.*, vol. 26, pp. 87–88, 1990.

[10] C. Dragone, "An $N \times N$ optical multiplexer using a planar arrangement of two star couplers," *IEEE Photon. Technol. Lett.*, vol. 3, pp. 896–899, Sept. 1991.

[11] K. Stubkjaer *et al.*, "Semiconductor optical amplifiers as linear amplifiers, gates and wavelength converters," in *Proc. ECOC'93*, vol. 1, 1993, pp. 60–67.

[12] J. Wiesenfeld *et al.*, "Wavelength conversion at 10 Gb/s using a semiconductor optical amplifier," *IEEE Photon. Technol. Lett.*, vol. 5, no. 11, pp. 1300–1303, Nov. 1993.

[13] M. A. Summerfield and R. S. Tucker, "Noise figure and conversion efficiency of four-wave mixing in semiconductor optical amplifiers," *Electron. Lett.*, vol. 31, pp. 1159–1160, 1995.

[14] T. Durhuus *et al.*, "All optical wavelength conversion by SOA's in a Mach–Zehnder configuration," *IEEE Photon. Technol. Lett.*, vol. 6, pp. 53–55, Jan. 1994.

[15] ——, "Optical wavelength conversion over 18 nm at 2.5 Gb/s by DBR-Laser," *IEEE Photon. Technol. Lett.*, vol. 5, pp. 86–88, Jan. 1993.

[16] R. A. Barry and P. A. Humblet, "Latin routers, design and implementation," *J. Lightwave Technol.*, vol. 13, pp. 891–899, May/June 1993.

[17] K. Okamoto *et al.*, "Fabrication of unequal channel spacing arrayed-waveguide grating multiplexer modules," *Electron. Lett.*, vol. 31, pp. 1464–1465, 1995.

[18] N. K. Shankaranarayanan *et al.*, "Two-section DBR laser transmitters with accurate channel spacing and fast arbitrary-sequence tuning for optical FDMA networks," in *OFC'94, Tech. Dig.*, vol. Tul2, pp. 36–37, 1994.

[19] B. S. Glance *et al.*, "New advances on optical components needed for FDM optical networks," *J. Lightwave Technol.*, vol. 11, pp. 882–889, May/June 1993.

[20] F. Forghieri, R. Tkach, and A. R. Chraplyvy, "Reduction of four-wave mixing cross-talk in WDM systems using unequally spaced channels," *IEEE Photon. Technol. Lett.*, vol. 6, pp. 754–756, 1994.

[21] R. Tkach *et al.*, "Four-photon mixing and high-speed WDM systems," *J. Lightwave Technol.*, vol. 13, pp. 841–849, May 1995.

[22] W. D. Zhong, J. P. R. Lacey, and R. S. Tucker, "An optical cross-connect," Australian Provisional Patent PN6511/95, Nov. 10, 1995.

The Supercomputer Supernet Testbed: A WDM-Based Supercomputer Interconnect

Leonard Kleinrock, *Fellow, IEEE,* Mario Gerla, *Member, IEEE,* Nicholas Bambos, Jason Cong, Eli Gafni, Larry Bergman, *Member, IEEE,* Joseph Bannister, *Senior Member, IEEE,* Steve P. Monacos, Theodore Bujewski, Po-Chi Hu, B. Kannan, Bruce Kwan, Emilio Leonardi, John Peck, Prasasth Palnati, and Simon Walton

***Abstract*—Current fiber optic networks effectively provide local connectivity among end user computing devices, and can serve as backbone fabric between LAN subnets across campus and metropolitan areas. However, combining both stream service (in which ATM excels) and low latency datagram service (in which cluster networks like Myrinet and POLO excel) has been difficult to realize. This paper describes a new wavelength division multiplexed (WDM) fiber optic network that supports both stream and datagram service and extends reach and functionality of low-latency, high bandwidth workstation clusters to a campus and MAN setting. The novel concept is based on combining the rich interconnect structure of WDM fiber optics with the fast, low-latency mesh of crossbar switches recently developed for workstation groups. This system, called the Supercomputer Supernet (SSN) achieves a high level of performance by replacing the point-to-point copper wire links with a parallel channel (WDM) fiber optic interconnect system. The novel scheme interconnects asynchronous wormhole routing switches used in parallel supercomputers via multi-channel WDM fiber optic links embedded in to an optical star (or tree) "physical" topology. WDM will be used to subdivide the very large fiber bandwidth into several channels, each of Gb/s bandwidth. WDM channels (supporting also time division multiplexing) will be established between modules, thus defining a dense "virtual" interconnection topology, which is dynamically reconfigurable and responds to changing traffic patterns. A pool of channels will be set aside for direct, end-to-end connections between crossbars, providing circuit-switched service for real-time traffic applications.**

I. Introduction

THE Supercomputer Supernet (SSN) currently being developed at UCLA, JPL and Aerospace under ARPA support is a novel, high-performance, scalable optical interconnection network for supercomputers and workstation clusters based on asynchronous wormhole routing crossbar switches.

Manuscript received April 18, 1995. This work was supported by the Advanced Research Projects Agency (ARPA) of the U.S. Department of Defense under Contract DABT63-93-C-0055 and the University of California through an agreement with the National Aeronautics and Space Administration. . This paper was presented in part under the title "The Distributed Supercomputer Supernet—A Multi-Service Optical Intelligent Network." This work was performed by the Center for Space Microelecronics Technology at the Jet Propulsion Laboratory, California Institute of Technology, Pasadena, and the University of California, Los Angeles.

L. Kleinrock, M. Gerla, N. Bambos, J. Cong, E. Gafni, P.-C. Hu, B. Kannan, B. Kwan, E. Leonardi, J. Peck, P. Palnati, and S. Walton are with the University of California, Los Angeles, CA 90024 USA.

L. Bergman and S. P. Monacos are with the Jet Propulsion Laboratory, Pasadena, CA 91109 USA.

J. Bannister and T. Bujewski are with the Aerospace Corporation, El Segundo, CA 90245 USA.

Publisher Item Identifier S 0733-8724(96)04553-7.

The WDM fiber optics extends the geographic coverage range from interdepartmental to campus and even to metropolitan areas. The network provides very high-speed multiple services, supporting hybrid circuit-switched and datagram traffic, and direct or multi-hop connections that are dynamically reconfigurable. At a first networking level, the crossbars locally interconnect workstations, supercomputers, peripheral devices, mass memory, etc. through host interfaces. At a higher networking level, the crossbars are fully interconnected with optical fibers supporting multiple wavelength division multiplexed channels, allowing communication between devices connected to distinct crossbars.

The resulting distributed SSN will be very fast—up to one gigabit per second (Gb/s) per channel—and possess a low latency approaching the physical media propagation time. It will scale up in the number of hosts connected and in geographical coverage for LAN and MAN situations. Using today's technologies, and being guided by emerging ones, the network design integrates the high throughput and parallelism of optics with the high intelligence of electronic processing, being clearly in line with, as well as at the front of modern networking trends.

A. New Enabling Applications

Just as conventional LAN technology enabled such present day applications as network file system (NFS), low-latency, high bandwidth <1 μs, >60 Mbytes/s [MB/s]) workstation cluster networks will likely enable new applications as well. These include the following:

- low cost arrayable video servers,
- distributed memory parallel supercomputers,
- network based memory management,
- real-time data acquisition and processing systems (e.g., radar), and
- video display walls.

One common characteristic of these applications is that they depend on extending the RAM memory of any one workstation in a cluster to any of the others in the group. Such network based distributed memory permit large applications to run beyond the confines of any one machine's memory space on a demand basis.

These functions have long been accepted as a minimum requirement to implement efficient message passing commu-

Reprinted with permission from *IEEE Journal of Lightwave Technology*, Vol. 14, No. 6, pp. 1388-1399, June 1996.

nications on MPP supercomputers, but only recently have been made possible over network interconnected workstations using a new breed of low latency (<1 μs) high bandwidth (>60 MB/s) networks that match the memory bandwidth and responsiveness of workstations.

Adding WDM fiber optics enhances this system in two ways. First, these cluster networks are extended from a machine room (100's m) to a campus setting (LAN), and secondly, the multiple channels of WDM fiber optics make it possible to support circuit switched type information services (e.g., voice, video) concurrently with conventional datagram services with maximum isolation.

In high end supercomputer networks, SSN will also provide the underlying low latency network fabric for interconnecting meta-supercomputer machines with scalable I/O; that is, I/O which is dynamically scalable in degree of parallelism from the host interface, network fabric, tertiary storage, and the application itself. This is particularly important for large data flow applications, such as radar or image processing, where the memory contents of MPP machines must be exchanged or updated on short time intervals commensurate with the real-time data source frame rate. The CASA gigabit testbed demonstrated the power of this meta-computing method using a single channel HIPPI network, but was not capable of setting up multiple channels between MPP machines on a dynamic basis over long distances. Hence, finer grain parallel applications with massive I/O requirements could not be attempted. Likewise, conventional telecom services, such as ATM, cannot efficiently provide this high bandwidth on demand without long setup delay.

B. Optical WDM Network Alternatives

Supercomputer networking and high-speed optical communications are active areas of research. Several optical networks have been proposed [2], and a few have been or are being implemented. Optical WDM network testbeds include LAMBDANET [9], Rainbow [6], the All Optical Network (AON) [8], and Lightning [7]. Two main alternatives have emerged in the design of WDM networks: single-hop and multihop networks.

Instead of using a direct path from source to destination, multihop networks [16] may require some packets to travel across several hops. For this reason, multihop networks are not suitable for high-throughput, real-time, delay-sensitive traffic. In fact, high data rates require loose flow control, which on the other hand gives limited protection against congestion. Alleviating congestion by dropping or deflecting messages is not an acceptable solution. Dropping messages is problematic in supercomputing since, at the high data rates involved, losing the contents of even a single buffer (which can exceed 64 kilobytes) is potentially disastrous. Deflection routing, on the other hand, introduces unpredictable delay and out-of-order delivery, which is intolerable given the high data rates used. Finally, multihop networks do not naturally support broadcast and multicast.

Single-hop networks [15] provide a dedicated, switchless path between each communicating pair of nodes. Each two-party communication requires one party to be aware of the other's request to communicate and to find a free virtual channel over which to communicate. This requires frequency-agile lasers and detectors over a broad range of the optical spectrum. The devices must also be capable of nanosecond reaction times. Furthermore, single-hop networks require substantial control and coordination overhead e.g. rendezvous control and dedicated out-of-band control channels). With current technology single-hop networks cannot readily accommodate bursty, short-lived communications.

Single-hop and multihop networks both suffer from limitations. The major single limitation of single-hop networks is the "complexity" of scaling up to large user populations and therefore high throughputs. If a single wavelength is used, then the throughput is limited by the maximum data rate achievable with affordable digital circuit technology. Capacity can be enhanced by using multiple wavelengths and implementing time and frequency division access schemes as in LAMBDANET and Rainbow. However, to achieve good efficiency in bursty traffic environments, these schemes require frequency-agile, rapidly tuned lasers and detectors over broad ranges of the optical spectrum. Such devices are not yet commercially available, although rapid progress of the technology in this direction has been reported [3]. Still, a major challenge is the production of components with both high tuning speed and broad wavelength range. Furthermore, the coordination of transceivers for short burst exchanges introduces considerable control overhead.

Like earlier testbeds, SSN employs WDM technology and high-speed transmission. However, the key feature that distinguishes the proposed SSN testbed from other approaches is the combination of multiple single-hop on-demand circuits with a multihop virtual embedded network to realize a hybrid architecture. This way, both stream traffic and low latency datagram traffic can be efficiently supported using the appropriate transport mechasnism. The SSN also allows for the dynamic reconfiguration of the virtual topology of its multihop component by slow retuning of its transceivers.

C. Outline of the Paper

In Section II, the SSN architecture will be described in more detail, beginning with the Myrinet high speed low latency LAN that has been developed by Myricom Inc. for workstation clustering application. SSN builds on this basic switching fabric by adding the WDM fiber links and an optical channel interface (OCI) controller for extending the connectivity from a building to a campus setting. In Section III, the overall SSN network protocol suite will be described for datagram and stream services, including datagram transfer, flow control, routing, multicasting, deadlock prevention, and dynamic reallocation of wavelengths for different services on the Myrinet system and the WDM optical backbone. In Section IV, performance studies will be described that analyze the topology tradeoffs, the scaling problem associated with wormhole routing (and a deadlock free routing technique) and the performance of multihop virtual topologies. Finally, in Section V, the SSN testbed for the LAN and MAN setting will be described.

II. Architecture

A. Myrinet

The Myrinet [24] high-speed LAN is an integral part of the SSN. Myrinet, manufactured by Myricom, Inc., is a high-speed, switch-based LAN intended to provide access over a limited geographical area. Myrinet has its roots in the multicomputer world [25], [4], where it was used as the interconnection network for a prototype parallel computer. It uses eight-bit–wide data paths between LAN elements, operating at a data rate of 640 megabits/second. The data channel is a full-duplex point-to-point link from a host interface to a switching node or between switching nodes. The Myrinet LAN transmits nine-bit symbols, eight of which carry data and one of which carries control information. Thus, in addition to data octets, several other nondata symbols are possible. The topology is arbitrary, being any configuration of interconnected host interfaces and switching nodes. Each switching node can have up to 16 ports. The Myrinet LAN has a limited spatial coverage, since the maximum link length is 25 m.

The Myrinet switches are simple, nonblocking switches that make switching decisions for a message by examining the source-supplied routing information in the message header. The complete route from the source node to a destination node is supplied to the requesting source node by a special route-manager software entity. Myrinet uses a form of cut-through routing called wormhole routing, in which the head of the message may arrive at its destination node before the tail has even left source node. This keeps latency very low. If an in-transit message is blocked at a switch, then the progress of the entire message is halted by backpressure. To reduce message latency these switches can switch a message in less than 600 ns.

The full-duplex channels use symbol-by-symbol stop-and-go flow control. Special STOP, GO, and IDLE symbols are available for controlling the flow of messages. Every Myrinet host and switch interface has a so-called slack buffer, which holds a small number of in-transit symbols. The size of the slack buffer is enough to hold twice as many symbols as can propagate simultaneously on a maximum-length link (27 symbols). When the receiving slack buffer has filled beyond a threshold, the receiver sends a STOP symbol to halt the incoming flow. Since it could take up to a full link-propagation delay for the STOP to arrive, the buffer must be able to absorb at least two link's worth of symbols beyond the threshold. When the sender receives the STOP, it immediately throttles its flow. When the switch's port unblocks and the slack buffer has drained below the threshold, the receiver sends a GO symbol to restart the flow.

A Myrinet message consists of a maximum of 5 600 000 symbols. Multicast is not supported in Myrinet. A source node would have to transmit to each destination node a copy of the multicast message.

B. OPTIMIC WDM Fabric

In the two level network architecture of the Supercomputer SuperNet (SSN) project, the second level is an optical backbone network that interconnects several high speed Myrinet LAN's. The optical channel interface (OCI) acts as an interface between the optical backbone and the high speed Myrinet LAN's. The physical optical backbone network can be any architecture–a single passive star coupler, a tree or a star of stars [11]. Space division multiplexing (using multiple fibers) is employed too.

The optical backbone network is viewed as a collection of a pool of wavelengths. By employing wavelength division multiplexing (WDM), the optical backbone network is designed to provide support for circuit switching, packet switching, multicasting and broadcasting.

Circuit switched service can be provided by dedicating a wavelength between two OCI's after an arbitration protocol. Packet switched service is provided by configuring the network as a multihop network [16]. Any scalable multihop virtual topology (like the traditional shufflenet [12] or the bidirectional shufflenet [18]) can be configured on the physical topology.

Since one of the goals of the SSN project is to extend the low-latency, high bandwidth supporting protocols of the high speed Myrinet LAN to the optical backbone network, the optical backbone should be able to support wormhole routing and the backpressure hop-by-hop flow control mechanism. The optical backbone can be configured into any topology that satisfies these requirements. Wormhole routing has a potential for deadlocks. The issue of deadlock prevention in SSN is addressed later.

C. OCI Design

An important contribution of the SSN project is the *optical channel interface* card. The goal of the OCI card is a modular design which focuses on the optical backbone protocol by using off-the-shelf Myrinet and optical components where possible. The OCI card extends the slack buffer concept of the Myrinet scheme to allow for long distance optical links between Myrinet clusters. Additionally, the OCI is responsible for arbitration and routing functions over the optical backbone.

The OCI card is a three port device. Two ports are used to connect one Myrinet port to one optical backbone port. The third port is a standard 9U VME connection for OCI configuration and system monitoring functions. A SPARC 32 b CPU card and one or more OCI boards will be packaged in a single VME enclosure to realize an integrated package for interfacing multiple Myrinet ports to the SSN optical backbone. Use of a VME based system allows for ease of integration by using an industry standard backplane for OCI integration. The OCI chassis is a 9U VME card cage used to house multiple OCI boards and a SPARC based CPU card for monitor and control. Fig. 1 shows the basic configuration of the OCI chassis.

The detailed OCI design provides the basis for the SSN optical backbone. The OCI consists of the dual LANai circuitry used as a Myrinet destination, the low-level data path monitoring logic, the flow control and routing logic, the VME interface, the worm buffers and the WDM fiber optic transceiver. By using *field programmable gate arrays* (FPGA)

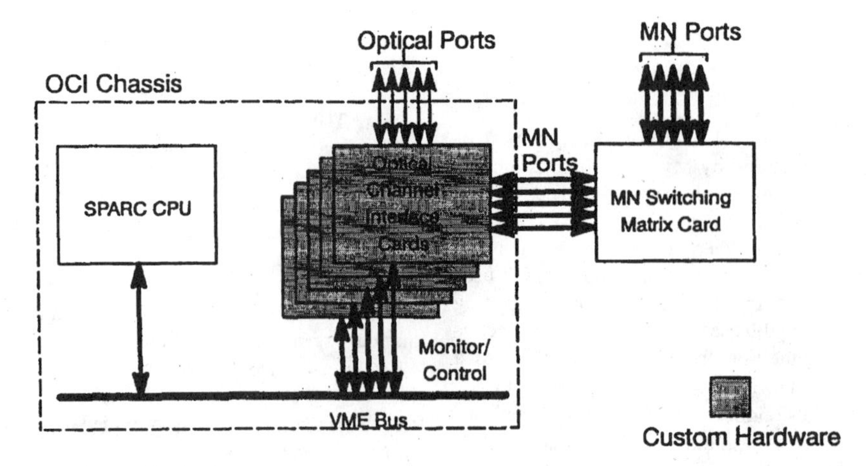

Fig. 1. Optical channel interface (OCI) chassis.

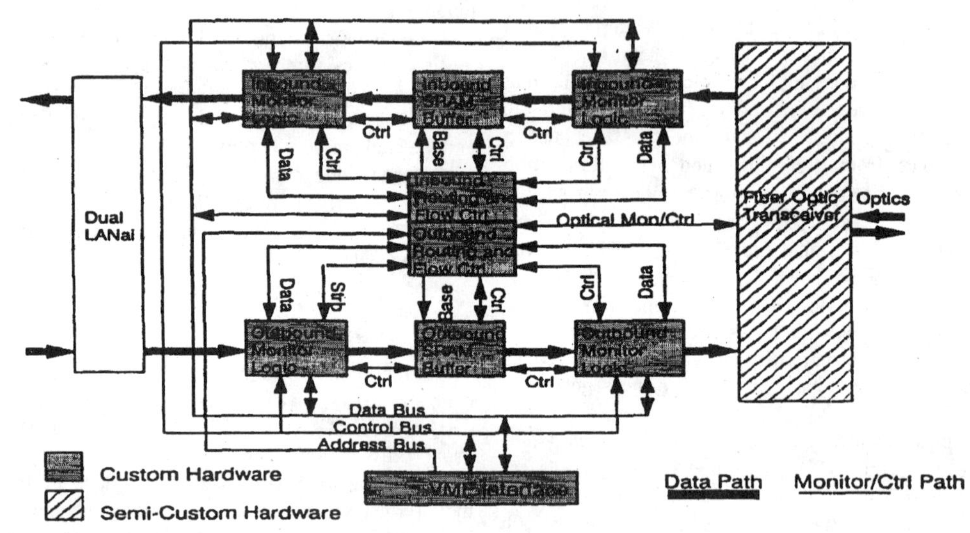

Fig. 2. OCI board block diagram.

for the monitor and control logic of the OCI, we allow for evolution of the design with the existing platform. The detailed block diagram of the OCI is shown in Fig. 2.

The dual LANai circuitry is a two port board which uses a pair of LANai routing processors to emulate a Myrinet source/destination while passing worms to/from the OCI board in a Myrinet format. One of the LANai processors of this circuitry provides the standard repertoire of Myrinet data and control bytes used to transmit and receive worms as part of a Myrinet [23]. The backend of this LANai looks like a DMA engine [23]. To simplify the interface to the backend of this LANai, a second LANai is used to generate a Myrinet like port [23] for transmission of worms to/from an OCI card.

The VME interface circuitry implements the VME protocol to communicate with the SPARC card. This circuitry also provides the data, address and interrupt information path ways from the various OCI logic blocks to a SPARC VME card for configuration of an OCI board and statistic collection operations.

The OCI board also contains worm buffers which are extensions of the slack buffers in the LANai. These buffers are needed due to the long propagation delay of the fiber optic links of the optical backbone. The OCI card also has the capability to handle worm priorities. This functionality is achieved by using random access memory (RAM) to allow for selection of worms based on worm priority. For a simple first-in-first-out buffering scheme, the delay through these buffers is in the neighborhood of one micro second.

The final component of the OCI is the WDM fiber optic transceiver. This hardware is similar to a standard transceiver but also provides the capability to transmit or receive at different optical wavelengths. Tuning at the transmitter will use tunable lasers or stepped wavelength laser arrays, while receiver tuning is accomplished with tunable filters. By providing the flow control and routing logic with the capability to use different wavelengths, we allow for multiple simultaneous communications over the same fiber optic media for improved optical backbone throughput and functionality.

III. Protocols

The key services provided by SSN are as follows.

1) low latency datagram service, to support fine grain distributed supercomputing. Variable size datagrams are allowed (as opposed to fixed size) in order to avoid segmentation/reassembly delay and overhead in origin and destination hosts.
2) high bandwidth, connection oriented service to support scientific visualization, large file transfers and more generally, time critical stream transmissions.

In the SSN project, protocols have been developed both in the Myrinet and in the optical backbone, in order to support the above basic services as well as additional services. The main function of the optical backbone is to extend geographically the reach of a Myrinet, and to permit the "transparent interconnection" of a large number of Myrinet islands.

In the remainder of the section, we briefly review the Myrinet protocols, and then focus on the optical WDM network protocols and network control and management procedures.

A. Myrinet Protocols

The commercial Myrinet already comes equipped with protocols for the support of datagram service. Source routing and backpressure flow control allow efficient transfer of datagrams in the Myrinet. In the SSN program, additional protocols are being implemented to provide:

1) integrated packet and circuit switched service support
2) bandwidth allocation to circuit switched traffic
3) "intelligent", alternate routing to minimize blocking and reduce latency (more generally "congestion management")
4) multicasting
5) priority support and QoS enforcement for different traffic classes.

Some of these protocols have been extensively evaluated via analysis and simulation. Performance results (along with more detailed description of the protocols) are reported in Section IV.

B. WDM Optical Backbone Protocols

Optical backbone protocols must extend transparently the Myrinet services end-to-end while achieving efficient utilization (and reallocation) of expensive backbone resources, efficient scaling, congestion protection and fault tolerance. The WDM optical star/tree architecture is ideally suited to this set of requirements in that it allows high bandwidth interconnection, efficient integration of multiple services and flexible reallocation of channel/bandwidth resources. The optical fabric will be initially a combination of space and wavelength division multiplexing; later in the project, it will be enriched to support also T/WDMA (time and wavelength division multiple access).

The following are the key protocols supported in the optical backbone.

1) **C/S Protocol:** Initially, a separate wavelength/fiber will be allocated to each C/S connection. When T/WDMA will be available, multiple connections with possibly different data rates will be carried in each WDM channel. An efficient technique for supporting multirate connections (based on receiver pipelining and slot retuning) was reported in [13].
2) **Datagram Transfer Protocol:** Two basic options are available here: namely, the single hop scheme (with wavelength retuning at transmitter and receiver on a datagram by datagram basis) and the multihop scheme. In our testbed, we will pursue the multihop scheme, which is less demanding in terms of transmitter/receiver tunability. Later, we will also consider single hop schemes. Section IV presents a comparison of these alternatives.
3) **Flow Control:** The Myrinet backpressure type flow control is extended to the optical backbone by using large slack buffers in the OCI's. Furthermore, virtual channels have been defined on each individual link of the multihop network, so that a single backpressured worm does not clog the link.
4) **Routing Protocol:** Two options are available for routing: the "flat" source routing option, which is an end-to-end extension of the Myrinet routing scheme, and the two-level routing option, where separate routing schemes are used for Myrinet and optical backbone. The latter scheme is more scalable, and offers better flexibility in backbone routing, at the expense of additional implementation complexity (in the OCI). As part of the latter scheme, deflection routing will also be explored.
5) **Multicasting:** is supported both for C/S connections and P/S transfers. Recall that C/S connections are single hop ("broadcast and select"). Thus, the signal can be received by all the OCI's in the multi-cast group, by simply tuning to the transmission wavelength at the proper slot. In the multihop network, multicasting can be achieved by define a proper multicast tree (embedded in the virtual multihop topology). Alternatively, an hamiltonian loop, visiting all the OCI's in the multicast group, can be used.
6) **Priorities:** are implemented in order to handle datagram traffic with different QoS. Furthermore, priority is given to transit traffic (over entry traffic) in the multihop network, in order to maintain the backbone clear of congestion, and stop the overload at the entry points.
7) **Deadlock Prevention:** Deadlocks may occur in the transfer of datagrams in the backbone network. These would bring the entire network to a halt, and therefore must be prevented. Several schemes are now being considered, and are discussed and evaluated in more detail in Section IV.
8) **Dynamic Reallocation of Wavelengths:** to different services. The WDM architecture offers the unique op-

portunity to reallocate bandwidth resources to different services (in our case, C/S and P/S service) based on user requirements. Furthermore, the multihop virtual topology can be dynamically "tuned" to obtain the best match with the current traffic pattern. Previous studies have shown that topology readjustments can lead to significant throughput and delay improvements. Initially, the dynamic reallocation and topology tuning will be in terms of actual wavelength. With the introduction of T/WDMA, finer grain, more efficient reallocation can be achieved. Protocols for dynamic reallocation are currently under development.

IV. Protocol and Performance Studies

There are several research issues that are brought about by the unique two level architecture of SSN. To start, performance models and simulation studies have been developed for improving the performance of the low-latency, high-bandwidth electronic network (Myrinet). We have developed several algorithms that extend the operating region of the Myrinet network to better match the large aggregate throughput of the attached WDM optical network. The algorithms involve refinement of the the timeout scheme to help increase throughput.

Another important component of research is the development of protocols for implementation in the optical WDM fabric. The issue of providing deadlock-free routes in the optical backbone as well as across the whole network has been studied [17]. Further, the properties of the bidirectional shufflenet multihop topology (a derivative of the traditional shufflenet with bidirectional links) have been studied [19], [18]. Bidirectional shufflenet allows for easy extension of the hop-by-hop backpressure flow control mechanism of the electronic LAN to the optical WDM fabric besides increased throughput and shorter average hops.

In this section, we present a few results from the studies on the performance of Myrinet and enhancements to improve its performance. Then, the issue of deadlock-free routing in SSN is addressed. Finally, some results for virtual multihop topologies embedded in the WDM optical backbone are presented. All performance results were obtained using a simulator written in Maisie [20], [1].

A. Enhancing Performance of the Myrinet Network

We have been developing algorithms to enhance the performance of the low-latency, high speed network that supports cluster computing. This must be done in order to adequately harness the large aggregate throughput of the attached WDM optical network. Myrinet implements wormhole switching. Due to low latency, Myrinet provides an effective interface for the hosts and supercomputers to the optical portion of the network. The ideal operating region for wormhole switching networks is in the low load traffic region. Blocking is minimal and thus latency remains low. However, as the traffic load increases, more worms block. A feedback blocking effect occurs and message delay rises rapidly.

To increase the network's effective operating region, we have developed different algorithms to handle the timeout

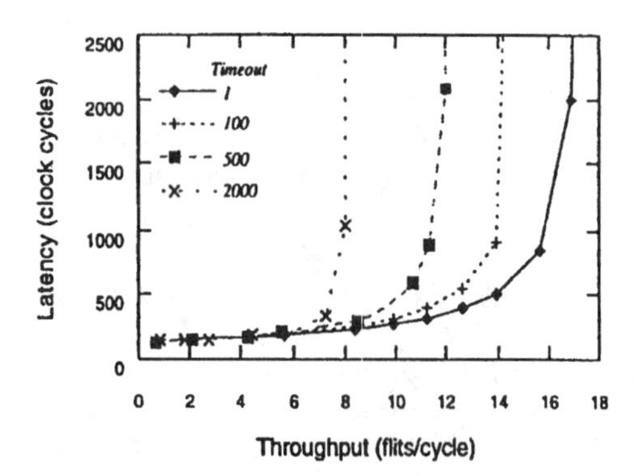

Fig. 3. Performance of a 3×3 torus network with different timeout values (worm length = 100 flits).

parameter. In Myrinet, timeout is used to break deadlocks that may occur in the system due to spurious network errors resulting in misrouted worms. Whenever a worm arrives at a switch, a counter is activated if its desired output port is busy. After some timeout value, a reset signal is sent out that resets the entire network. Currently, the timeout value is set to a very large value (50ms) and the reset mechanism clears all worms in the network. Studies have been done on a refined timeout mechanism where smaller timeout values are used to selectively clear only the worm that has been blocked for a time longer than the timeout value of the system ([10]). The results show that small timeout values better optimize the system (see Fig. 3).

In a different timeout parameter refinement, we implement a scheme using switch state information to determine the timeout value for a blocked worm [14]. When a worm arrives at a switch and the output port it desires is busy with a worm that is itself blocked, the newly arriving worm times out and is retransmitted. Otherwise, if the arriving worm sees that the worm currently using the output port is flowing, it blocks and waits for the flowing worm to be completely transmitted. The motivation behind switch state dependent timeouts (SSD TO) comes from the observation of the feedback blocking effect of a wormhole switching network under medium to high traffic loads. As the traffic load rises, more worms encounter blocked output ports. After a certain threshold, several worms clog the network and prevent other worms from flowing. Using the SSD TO algorithm, only worms that are waiting behind other worms that are making forward progress are allowed to block and wait. All other worms are forced to timeout. The SSD TO improves delay performance over the original large timeout scheme where all worms must block and wait unless a deadlock occurs (see Fig. 4).

B. Deadlock Free Routing in SSN

Wormhole routing can cause deadlocks to happen in the network if a cycle of worms blocking each other exists. Deadlock resolution using timeouts is a costly procedure requiring a network reset. Deadlock free routing is thus desirable for SSN.

Deadlock free routing can be done on any bidirectional topology by forming a spanning tree of the topology and then

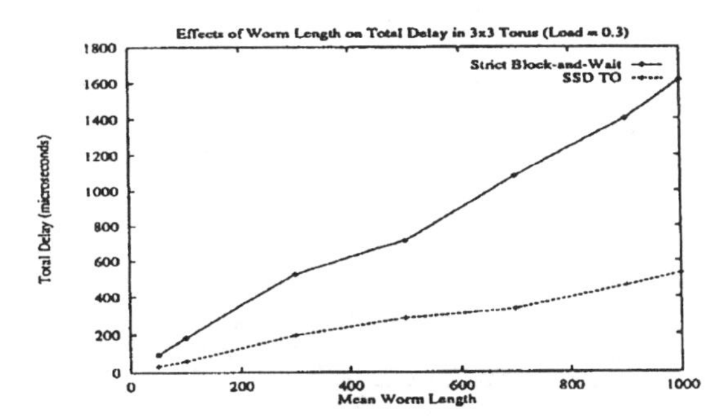

Fig. 4. Delay performance of the 3 × 3 torus network with Switch State Dependent Timeout with varying worm lengths.

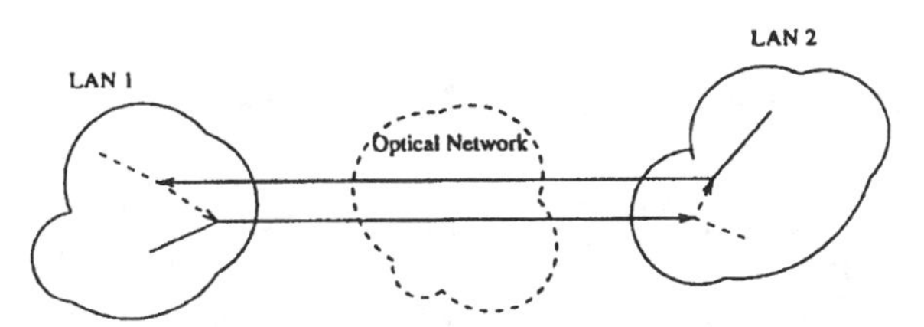

Fig. 5. An example showing that deadlock free routing in each part of the network does not imply a deadlock free network. The solid lines represent a worm destined to a remote LAN that are in a remote LAN; the dashed lines represent local/remote traffic still in the origin LAN. The arrows indicate where blocking occurs.

routing worms on this spanning tree such that the worm first travels zero or more links in the Up direction (toward the root) and then zero or more links in the Down direction (away from the root) toward the destination. This technique is called Up/Down routing.

Another technique for deadlock free routing, on unidirectional as well as bidirectional topologies, is the use of virtual channels [5]. In this approach, each link is decomposed into several virtual channels (all sharing the same bandwidth but each having its own slack buffer). These virtual channels are then connected together to form several levels of virtual networks. After eliminating cycles from each level, a fixed ordering is made among these virtual networks. This fixed ordering determines the routing. An advantage of using virtual channels is that shortest path routing is possible unlike in the Up/Down routing approach.

For the multihop virtual topologies of the traditional shufflenet and the bidirectional shufflenet (in which the links of the shufflenet are bidirectional), deadlock free routing techniques employing Up/Down routing (as in Autonet [22], [21]) and virtual channels have been described elsewhere [17]. Here we present the approach used to perform deadlock free routing in the two level architecture of the SSN project, assuming that the virtual multihop topology implemented in the optical backbone network is the bidirectional shufflenet.

For the high speed electronic mesh network, deadlock free routing can be achieved by employing the technique of the Up/Down routing on a spanning tree of the network. For the optical network, we can use Up/Down routing or virtual channels to achieve deadlock free routing in the bidirectional shufflenet. However, having deadlock free routing schemes in the different parts of the network, does not guarantee deadlock freedom for the network as a whole. An example is shown in Fig. 5.

Several alternatives for having deadlock free routing across the network exist. One way is to use Up/Down routing on a single spanning tree for the whole network. In this approach, either one of the OCI's or one of the stations on the net can behave as the root of the spanning tree. This alternative works correctly since we have bidirectional links in SSN. However, this approach has the drawback that shortest paths are not guaranteed.

Second, we could extend the virtual channel mapping method across the whole network. Namely, we achieve deadlock free routing in the electronic LAN's by dedicating separate sets of virtual channels to origin and remote LAN traffic.[1] Thus, origin LAN traffic does not interfere with remote LAN traffic providing deadlock free routing across the whole SSN network. Unfortunately, Myrinet LAN's currently do not support virtual channels with separate slack buffers for each virtual channel.

A third solution consists of requiring that the OCI in each subnet be the root of the spanning tree used to define the up/down links for that subnet. Then, deadlock prevention can be achieved for the subnet with up/down routing as discussed earlier. For the bidirectional shufflenet deadlock prevention can be achieved using either the Up/Down approach or the virtual channel approach. For deadlock free routing across the whole network, it suffices to require that remote traffic (i.e. a connection between two distinct subnets) be routed UP in the origin LAN and DOWN in the destination LAN. Of course, choosing the OCI as the root of the spanning tree for the subnet is key to this solution. Deadlock prevention is evident from inspection of Fig. 5 where conflict between local and remote worms could not exist with this new rule. Note that this deadlock free routing scheme can be easily extended to the case in which an electronic LAN is connected to the optical backbone via multiple OCI's. In this case, one of the OCI's is elected as the root of the spanning tree (using, for example, a distributed procedure similar to the IEEE 802.1 spanning tree bridge algorithm). Each OCI can then handle remote traffic to and from sources and destinations in its subtree. Of course, the 'root' OCI could handle all remote traffic by itself. However, hosts will be assigned to different OCI's using appropriate criteria e.g. shortest distance) in order to balance the traffic across the electronic LAN, OCI and the bidirectional shufflenet.

One important advantage of the last solution (over the previous two) is that it allows the independent choice of the most cost effective deadlock free routing scheme for each level of the SSN architecture.

C. Performance of Multihop Topologies

Since fast tunable receivers are not yet widely available, studies of the performance of multihop topologies of shuf-

[1] Here, traffic originating within a LAN destined to a host in the same LAN is called local traffic. If the destination is a host on a remote LAN then the traffic is called remote traffic. If the worm is traveling through the LAN to which the origin host belongs then it is said to be in the origin LAN otherwise it is said to be in a remote LAN.

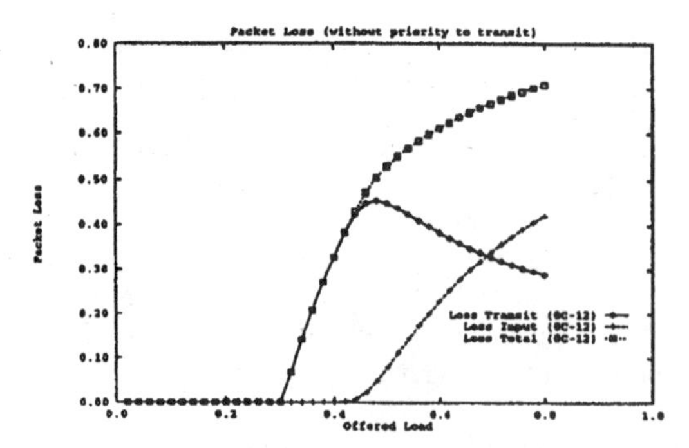

Fig. 6. Packet loss vs Offered load in the case when no priority is given to transit traffic. An 8 node shufflenet topology was used.

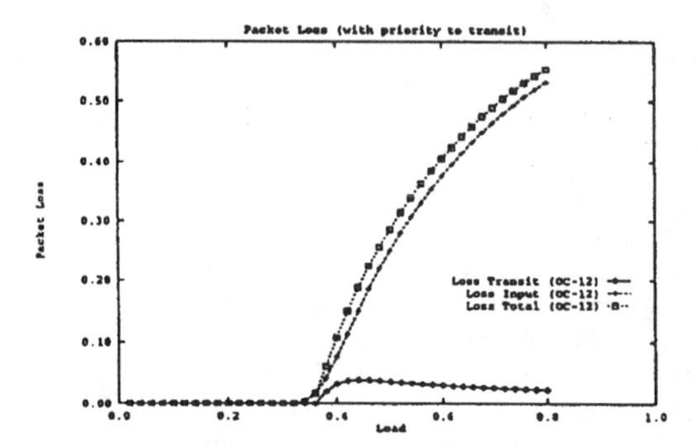

Fig. 7. Packet loss vs Offered load in the case when priority is given to transit traffic. An 8 node shufflenet topology was used.

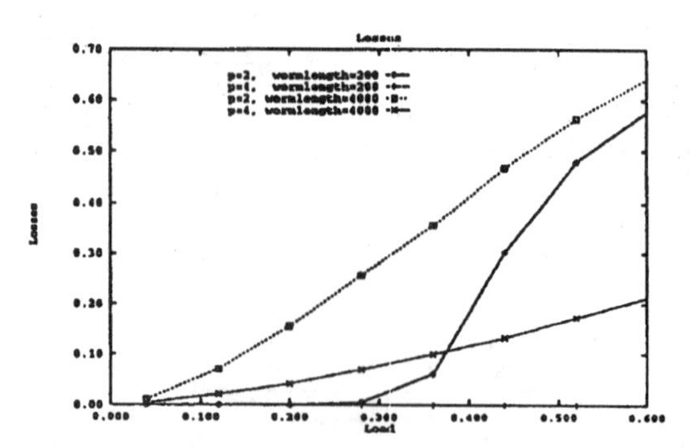

Fig. 8. Worm Loss Percentage versus Offered Load for a 24 node shufflenet and a 32 node shufflenet. The loss is shown for worms of average length 200 and 4000 bytes.

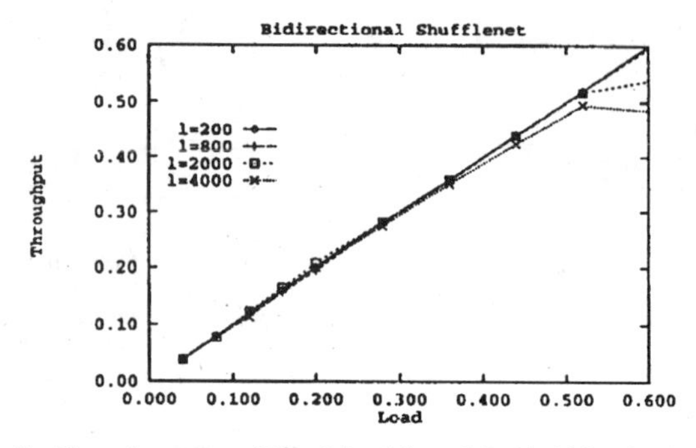

Fig. 9. Throughput versus Offered Load for a 24 node bidirectional shufflenet. Average worm lengths range from 200 to 4000 bytes.

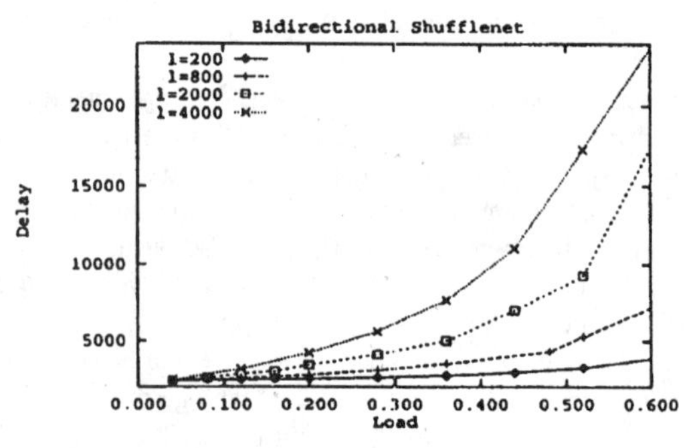

Fig. 10. Delay versus Offered Load for a 24 node bidirectional shufflenet. Average worm lengths range from 200 to 4000 bytes.

flenet and bidirectional shufflenet have been carried out via simulation and analysis as well. Though the WDM optical backbone network can be configured as any virtual multihop topology, we have only studied the traditional shufflenet and the bidirectional shufflenet topologies.

1) Performance of Shufflenet: We studied the effect of giving priority to transit traffic (i.e., traffic which is passing through an intermediate station) over input traffic (i.e, traffic entering the station from the Myrinet LAN. In this study, no flow control was assumed, thus, there was packet loss due to buffer overflow. The speed for the optical link was assumed to be the OC-12 rate (622 Mb/s). In Figs. 6 and 7, we plot the results for the two cases. Fig. 6 shows that in the absence of priority to transit traffic, packet loss starts at about 0.3 offered load and the transit packets are dropped first. However, when priority is given to transit traffic, the loss starts at about 0.35 offered load and almost all the loss is at the input. Since the hop-by-hop backpressure flow control mechanism employed by Myrinet is being extended to the WDM optical backbone fabric, there would be no loss at the input. Thus, the significant result of this study is that giving priority to transit traffic in the traditional shufflenet gives higher throughputs and lower losses.

2) Performance of Bidirectional Shufflenet: A bidirectional shufflenet is a traditional shufflenet with bidirectional links. Each station in the bidirectional shufflenet has twice the number of fixed tuned transmitters and receivers than a station in the traditional shufflenet. The bidirectional nature of this multihop virtual topology makes it easy to extend the hop-by-hop backpressure flow control mechanism to the WDM optical backbone. Also, this topology has a shorter average hop length. In the absence of flow control, the traditional shufflenet shows worm loss (see Fig. 8). In the bidirectional shufflenet, worm loss does not occur because of the flow control mechanism. In Fig. 6, we show the throughput obtained for a 24 node bidirectional shufflenet topology for different average worm lengths. In Fig. 9, we show the delay performance for different worm lengths. The significant results of this study are that the use of bidirectional shufflenet's natural support for flow control eliminates loss and the throughput and delay performances are quite good.

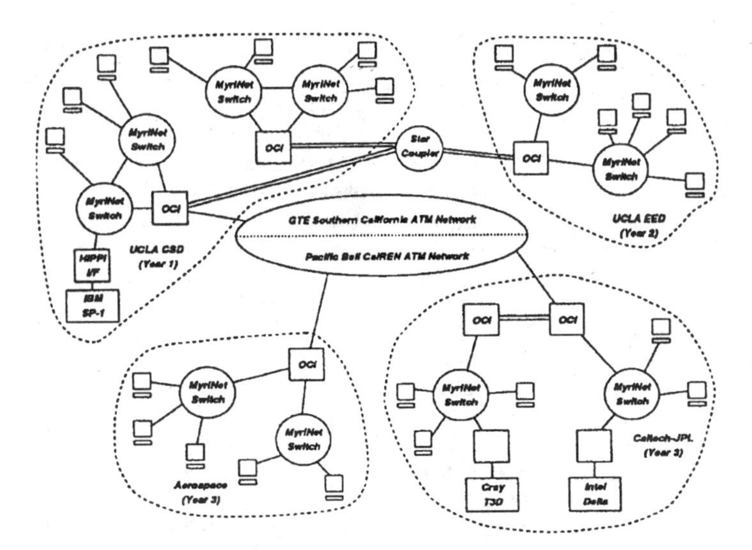

Fig. 11. The SSN testbed consists of four clusters of Myrinet switches: four at the UCLA computer science department, two at the electrical engineering department, two in Pasadena (one at JPL and one at Caltech), and two at Aerospace Corporation.

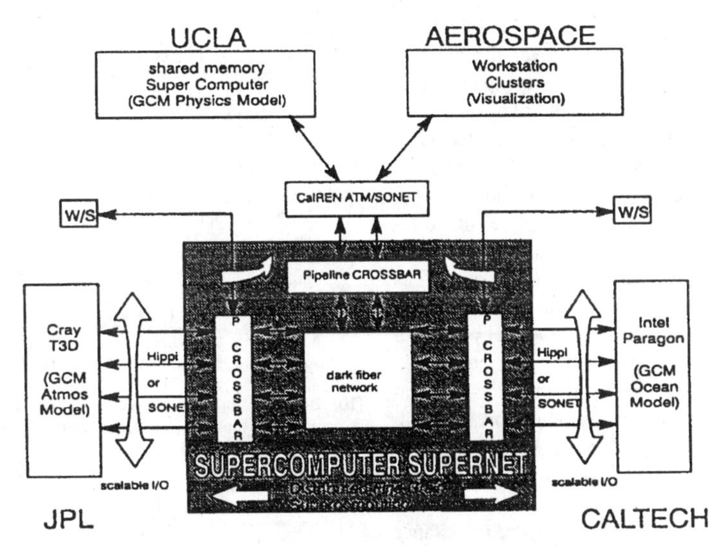

Fig. 12. SSN supports a fine-grain distributed supercomputing and visualization GCM application using Myrinet on one segment of the CASA testbed between JPL and Caltech.

V. Applications and Testbeds

A. Applications

The low-latency, dynamic reconfigurability, and scalability of SSN are expected to enable several new applications in the area of distributed supercomputing and visualization:

1) *Fine Grain Meta-Supercomputer:* The SSN attributes would accelerate the evolution of a network-based operating system with precise synchronization of dispersed processes, fine grain process management on 100's–1000's of processor elements, distributed checkpointing of jobs, and dynamic entry of new hosts.
2) *Real Time Distributed Network Operating System:* Low and predictable (bounded) latency makes SSN well suited for wide area network control and data acquisition applications. Examples include missile tracking, radar telemetry, remote robot control, and in the commercial arena, oil refinery and power plant control, avionics and spacecraft control systems, control of electrical power distribution systems, and factory automation.
3) *Distributed Image Data Base Perusal:* Scientific image-based data-base archival and perusal systems are now being developed in several efforts, such as the UC Sequoia effort and the MAGIC testbed. NASA applications, such as EOS, will require the capability of perusing through terabytes of data very quickly and interactively. A low latency high throughput network will be essential for responding quickly to interactive

control from the user (datagram) and sending image bursts back to the user (streams/circuit switched).

B. Testbeds

1) Campus: The basic SSN testbed topology is shown in Fig. 11. The Myrinet switches are placed in three clusters: a group of four in the UCLA Computer Science department building, a group of two in the UCLA Electrical Engineering department building, two at JPL/Caltech (between two supercomputers), and finally, two at the Aerospace Corporation. OCI's interconnect the clusters as well as selected ports within the largest cluster at the UCLA Computer Science Department.

One fiber optic link segment (14 km) of the CASA gigabit network between JPL and Caltech in the Pasadena area is proposed as the target SSN testbed demonstration site using scalable I/O supercomputers (see Fig. 12). The proposed SSN application that combines elements of (1) and (2) above is the UCLA Global Climate Model (GCM) being developed by R. Mechoso for the CASA project. On the present CASA network, a single HIPPI channel only permits a coarse-grain coupling of the ocean/atmosphere model between the Caltech Intel DELTA (running the ocean model) and JPL Cray YMP (running the atmospheric model). Running over the existing dark fiber, SSN would provide four times the capacity (3.2 Gbit/s) and lower latency routing between the two supercomputers than the present single HIPPI channel with Crossbar Interfaces. This would provide a foundation for a finer grain decomposition of the GCM application. Simultaneously, high performance workstations can interactively capture image results of the running GCM model and peruse through new data sets that would be staged for later GCM runs. The SSN network dynamically allocates/deallocates optical channel bandwidth as workstations or massively parallel processor (MPP) nodes enter/leave the network. The Myrinet APCS network node also accommodates instantaneous reconfiguration of the MPP I/O channels from asynchronous I/O for separate partitioned jobs (e.g., one per quadrant of the MPP) to coherently striped I/O for one large single job.

2) MAN: Between UCLA and Pasadena (a distance of about 30 miles), and between UCLA and Aerospace (a distance of about 15 miles) the network fabric will consist of a single ATM/SONET OC-3 channel provided by the Pacific Bell CalREN (California Research and Education Network) and GTE consortiums (see Fig. 11). At each location, an OCI will be configured to provide a gateway function by incorporating a SONET/ATM network interface with the OCI SPARC CPU controller. A higher performance solution is being explored with a leading routing vendor. Initially, permanent virtual circuits (PVC) will be used between the sites. Striped channel performance, which SSN provides via WDM in a LAN campus setting, can be provided by setting up multiple PVC's and/or SONET OC-3 channels.

VI. Conclusion

As fine grain, closely coupled real-time distributed system applications begin to mature for cluster workstation computing and networking of metamassively parallel processor supercomputers, low-latency rapidly reconfigurable networks with high Gb/s per channel capacity will be required. SSN provides one such network fabric for binding these systems together that is easily scalable in both physical size and number of ports per host. It is also adaptable to a variety of optical transmission techniques, providing multiple growth paths as WDM and spatial optical multiplexing optoelectronics becomes commercially available. Such networks also raise a host of new issues in network management, flow and congestion control, and error recovery that will be the subject of future work.

References

[1] R. Bagrodia, Y. Chen, M. Gerla, B. Kwan, J. Martin, P. Palnati, and S. Walton, "Scalable simulation of a high-speed wormhole network in a parallel language," to be published.

[2] J. Bannister, M. Gerla, and M. Kovačvić, "Routing in optical networks," in *Routing in Communications Networks*, M. Steenstrup, Ed. Englewood Cliffs, NJ: Prentice-Hall, 1995, pp. 187–225.

[3] C. A. Bracket, "Dense wavelength division multiplexing networks: Principles and applications," *IEEE J. Select. Areas Commun.*, vol. 8, pp. 948–964, Aug. 1990.

[4] D. Cohen, G. Finn, R. Felderman, and A. DeSchon, "The ATOMIC LAN," in *IEEE Workshop High Perform. Commun. Subsyst.*, Tucson, AZ, Feb. 1992.

[5] W. J. Dally and C. L. Seitz, "Deadlock-free message routing in multiprocessor interconnection networks," *IEEE Trans. Comput.*, vol. C-36, pp. 547–553, May 1987.

[6] N. R. Dono, P. E. Green, K. Liu, R. Ramaswami, and F. F. Tong, "A wavelength division multiple access network for computer communication," *IEEE J. Select. Areas Commun.*, vol. 8, pp. 983–994, Aug. 1990.

[7] P. W. Dowd, K. Bogineni, K. A. Aly, and J. Perreault, "Hierarchical scalable photonic architectures for high-performance processor interconnection," *IEEE Trans. Comput.*, vol. 42, pp. 1105–1120, Sept. 1993.

[8] S. B. Alexander *et al.* "A precompetitive consortium on wide-band all-optical networks,"*J. Lightwave Technol.*, vol. 11, pp. 714–735, May/June 1993.

[9] M. S. Goodman, H. Kobrinski, M. Vecchi, R. M. Bulley, and J. L. Gimlett, "The lambdanet multiwavelength network: Architecture, applications, and demonstrations," *IEEE J. Select. Areas Commun.*, vol. 8, pp. 995–1004, Aug. 1990.

[10] P.-C. Hu and L. Kleinrock, "A queueing model for wormhole routing with timeout," 1995, presented at the Fourth Int. Conf. Comput. Commun. Networks.

[11] B. Kannan, S. Fotedar, and M. Gerla, "A two level optical star WDM metropolitan area network," in *Proc. GLOBECOM 94 Conf.*, June 1994.

[12] M. Karol and S. Shaikh, "A simple adaptive routing scheme for schufflenet multihop lightwave networks," in *Proc. GLOBECOM 88 Conf.*, 1988, pp. 1640–1647.

[13] M. Kovacevic, M. Gerla, and J. Bannister, "Time and wavelength division multiple access with acoustooptic tunable filters," *Fiber and Integr. Opt.*, vol. 12, no. 2, pp. 113–132, 1993.

[14] B. Kwan and N. Bambos, "Performance of a switch state dependent timeout scheme in a wormhole switching lan," Univ. California, Los Angeles, Tech. Rep. UCLA ENG-95-121, Apr. 1995.

[15] B. Mukherjee, "Wdm-based Local lightwave networks—Part I: Single-hop systems," *IEEE Network*, vol. 6, pp. 12–27, May 1992.

[16] ——, "Wdm-based Local lightwave networks—Part II: Multi-hop systems," *IEEE Network*, vol. 6, pp. 20–32, July 1992.

[17] P. Palnati, M. Gerla, and E. Leonardi, "Deadlock-free Routing in a Hierarchical Supercomputer Interconnection Network," 1995. Submitted for publication.

[18] P. Palnati, E. Leonardi, B. Kannan, and M. Gerla, "Bidirectional Shufflenet: A Multihop Topology for Backpressure Flow Control," to be presented at the Fourth International Conf. on Computer Communications and Networks, 1995.

[19] ——, "Performance analysis of bidirectional shufflenet: A multihop topology for backpressure flow control," to be published.

[20] R. Bagrodia, K. M. Chandy, and J. A Misra, "Message-based approach to discrete-event simulation," *IEEE Trans. Software Eng.*, vol. 13, no. 6, June 1987.

[21] T. L. Rodeheffer, "Experience with Autonet," *Comput. Networks ISDN Syst.*, pp. 623–629, 1993.

[22] M. D. Schroeder, A. D. Birrell, M. Burrows, H. Murray, R. M. Needham, T. L. Rodeheffer, E. H. Satterthwaite, and C. P. Thacker, "Autonet: A high-speed self-configuring local area network using point-to-point links," *IEEE J. Select. Areas Commun.*, vol. 9, pp. 1318–1335, Oct. 1991.
[23] C. Seitz, Private Communication.
[24] C. Seitz, D. Cohen, and R. Felderman, "Myrinet—A gigabit-per-second local-area network," *IEEE Micro*, vol. 15, no. 1, pp. 29–36, Feb. 1995.
[25] C. L. Seitz, J. Seizovic, and W.-K. Su, "The design of the Caltech Mosaic C multicomputer," in *Proc. Symp. Integr. Syst.*, Mar. 1993.

Design and Implementation of a Fully Reconfigurable All-Optical Crossconnect for High Capacity Multiwavelength Transport Networks

AMAURY JOURDAN, FRANCESCO MASETTI, MATTHIEU GARNOT, GUY SOULAGE, AND MICHEL SOTOM

Abstract— **The motivations and application framework for the introduction of all-optical wavelength division multiplexing (WDM) transmission and routing techniques in the transport network are presented. The requirements and functionalities of all-optical transparent routing nodes are discussed, and the physical architecture of a crossconnection node is proposed, to meet these requirements. Optical devices suitable for the node implementation are compared, and first demonstrations of crossconnection function at data rates up to 10 Gb/s are given. These results bring experimental evidence of the high potential of all-optical routing nodes for actual implementation of multiwavelength transport networks.**

I. Introduction

THE current telecommunication environment is evolving toward increasingly heterogeneous but interconnected networks. This evolution, together with a large increase of traffic expected by the end of this decade [1], will add significant complexity to the planning, implementation, and operation of the transport network. In order to simplify these tasks, a layered structure of the transport network was proposed [2], namely, into transmission, path, and circuit layers. As regards the transmission layer, optics have already been widely spread for long distance high bit-rate transmission applications. It is already well accepted that Wavelength-Division-Multiplexing (WDM) techniques [3] have a high potential to upgrade further the capacity of the existing transmission links in a cost effective way, but they also open the door to new, very efficient all-optical routing schemes, replacing what is today performed using complex electronics. All-optical WDM networking could provide an effective answer, in terms of functionalities, performances, and cost, to the requirements of the transport network at the transmission and path layers level, and may also pave the way to less hierarchical network structures. The capabilities of these techniques to meet the requirements of high capacity transport networks will be analyzed in Section II.

Section III includes functionalities and main characteristics needed in the crossconnection nodes, derived from the former network requirements. An Optical Crossconnection Node (OXCN) architecture of the broadcast-and-select type will be proposed, designed to meet these external specifications, and based on both space and wavelength routing functions, including wavelength conversion.

Section IV is first devoted to the presentation of the principle and state-of-the-art performance of the technological solutions to perform the routing functions needed in the OXCN, namely space-switches, tunable wavelength selectors, wavelength converters, and multiwavelength optical amplifiers. In particular, the most suitable and promising devices are highlighted.

Finally, we present in Section V, first experimental results, at data-rates up to 10 Gb/s, aiming at assessing the feasibility of such OXCN's, and their compatibility with transmission constraints. In particular, issues related to crosstalk (coherent or incoherent), noise accumulation and wavelength conversion efficiency are addressed.

II. WDM Techniques in the Transport Network

Notwithstanding the high degree of heterogeneity of the existing network infrastructures (PDH, SDH, ATM in the near future), the transport network should provide global connectivity, high capacity, easy upgradability, high dependability, and flexibility, with a low transport and management cost. A layered structure of the transport network has been introduced to simplify its design and planning [2]. An important feature of a layered architecture is that the properties of any layer can be defined independently from others, and it is possible to add or change a layer without affecting others from a topological point of view. Furthermore, each layer can be partitioned on a different basis; for instance, geographical arrangement, services, administration, or operator. Different types of network partitions can therefore coexist and even experience independent evolution with time.

This paper aims to demonstrate the great potential of optical routing techniques at the network path layer level. This optical path sublayer (Fig. 1) can provide a transparent common infrastructure for all heterogeneous existing networks capable of interworking with the different existing electrical path layers/sublayers.

The introduction of this all-optical path sublayer may of course follow different scenarios depending on the strategies of the network operators, based on national evolution of the traffic demand. For instance, a reference European country with a

Manuscript received April 21, 1995.
The authors are with Alcatel Alsthom Recherche, Route de Nozay, 91460 Marcoussis, France.
Publisher Item Identifier S 0733-8724(96)04675-0.

Reprinted with permission from *IEEE Journal of Lightwave Technology*, Vol. 14, No. 6, pp. 1198-1206, June 1996.

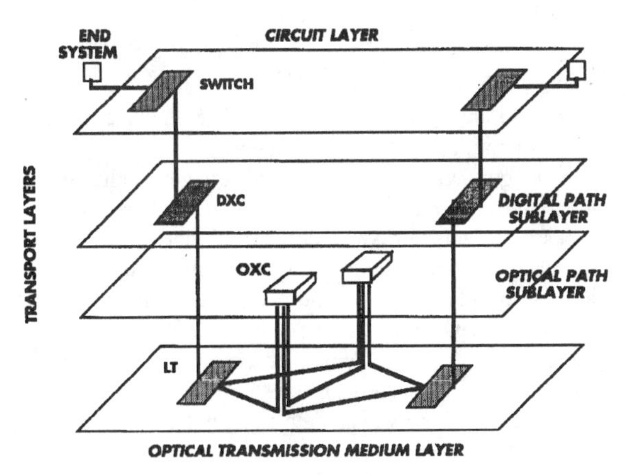

Fig. 1. Network model and relationships to the transport network.

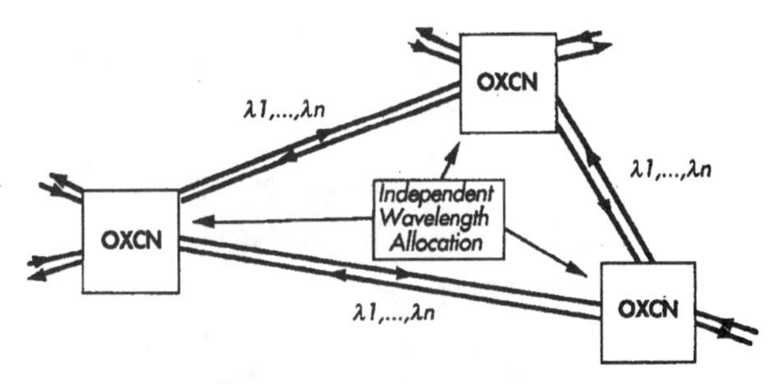

Fig. 2. Link-by-link wavelength allocation.

well developed market for nonswitched services would need about 10 major routing nodes, which could evolve into all-optical routing nodes by the beginning of the next decade, with throughput around 1 Tb/s. Such national infrastructures could then be interconnected to make a relatively well connected (for flexibility and protection purposes) meshed network, the nodes of which would be linked by bidirectional optical highways. For instance, in Europe this could lead to a core transport network of several tens of nodes, with a mean network radius of about 1500 km, with bit-rates per channel in the order of 2.5 to 10 Gb/s. For protection purposes as well as for future proof of such a European backbone, bridging (information transported simultaneously on two different physical paths) and multicasting (transmission to different end-users) capabilities must also be supported by the optical routing nodes.

The introduction of WDM techniques in this core all-optical transport network would imply the following advantages on the network planning and operation:

- to perform simple routing functions in a cost effective manner, through the efficient use of both space and wavelength domains [4]
- to introduce a certain degree of transparency, leading to easier transport of information coming from different preexisting infrastructures and telecommunication standards without major modifications, to a less hierarchical network by allowing simultaneous transport of different bit-rates and existing digital modulation standards (PDH, SDH, ATM), and finally to high tolerance to future change of the traffic demand and telecommunication standards [5]
- to devote cost effective facilities to dynamic and flexible rerouting and path protection, which is all the more important that the total and link-by-link traffic load of the network increases [6]
- to facilitate network scalability; in particular, more wavelengths can be added at low incremental costs [7]
- to increase smoothly the transmission capacity of existing fiber links without reaching the physical limits imposed by current optical technologies [8]

In an all-optical WDM core network, the wavelength channel is considered as a transport resource allocated to a given data stream. An end-to-end connection can be either supported by one wavelength (Wavelength Path) or can physically be transported over different wavelengths through the transmission layer (Virtual Wavelength Path) [6]. In the latter case, the wavelength allocation is performed on a link-by-link basis, and wavelength conversion must be, in principle, possible at all crossconnection nodes.

The wavelength path (WP) technique suffers from a limitation of the network extension, flexibility, scalability and performances, imposed by the total number of wavelengths available, which will certainly be restricted to a few tens for technological reasons (amplification window, limitations on the channel spacing). To cope with this limitation, the same set of available wavelengths must be reused, but only on fully disjointed parts of the network. In this case, the path set up leads to a still critical blocking probability, especially for long connections that cross a large number of common links. Besides, rerouting and protection, as well as introduction of new connections, may imply changes in the wavelength or link global allocation, adding a high complexity to the network management, since the wavelength assignment to light path is a NP-complete problem [9].

On the other hand, the virtual wavelength path (VWP) technique introduces greater flexibility in the wavelength allocation scheme, since the same wavelength is now reusable on all links (Fig. 2). Besides, wavelength conversion is the only practical way to solve blocking situations, allowing to build strictly non-blocking crossconnects. With such nodes, as long as the transmission capacity is available on all links of this connection, any change (for reconfiguration or protection purposes) or addition of one connection is possible without disrupting the other connections throughout the network. This leads to a drastic reduction of the management complexity, as well as improved scalability and flexibility [10]. In particular, a completely reconfigurable path layer topology, where wavelength conversion can be performed at all crossconnection nodes, is capable of accommodating any combination of different types of network partitions. Other transport network organizations, based on geographically independent partitions, have been already identified [11], aimed at limiting wavelength conversion at the partition boundaries. A solution based on VWP crossconnects is obviously also capable of supporting any of these concepts, while remaining open to future network evolutions (Fig. 3).

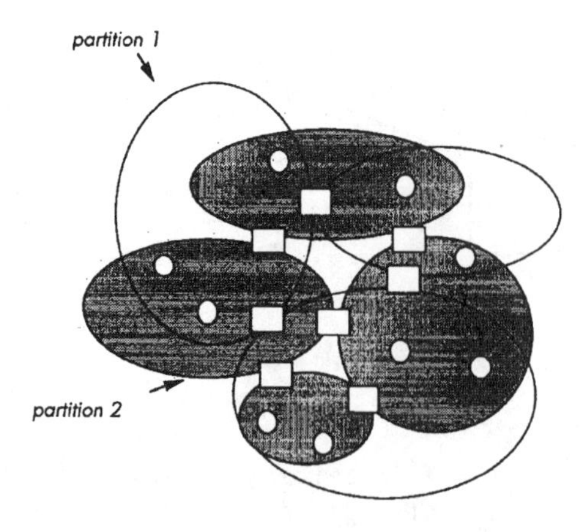

Fig. 3. Example of network with two simultaneous partitions, with fully reconfigurable OXCN's incorporating wavelength converters at the interfaces.

Another major advantage of wavelength conversion techniques lies in their ability to accommodate physical limitations. Since we are considering a network with connections over some thousands of kilometers, the signal degradation coming from noise accumulation, transmission impairments, and crosstalk issues in the routing node might raise severe limitations. Therefore, some signal regeneration might be required in identified nodes of the network. State-of-the-art results about all-optical wavelength converters indicate that they could in the near future perform some signal reshaping.

In conclusion, for all these reasons, we have chosen to design and implement the more flexible and powerful solution, namely a VWP crossconnect.

III. Crossconnection Node Architecture for All-Optical WDM Transport Network

Based on the assumptions of previous sections, namely high network capacity, fully independent wavelength allocation on all links, non-blocking operation as long as spare capacity is available, bridging and multicasting capabilities, and so on ..., we have derived the following main characteristics the all-optical WDM routing nodes should feature:

- the node should crossconnect n ($n \approx 4-8$) m-wavelengths multiplex ($m \approx 4-16$) from n input links to the same number of output links, including bridging and multicasting capabilities, with operating bit-rate per channel in the range of 2.5–10 Gb/s
- the same set of wavelengths can be reused on all input links.
- wavelength conversion should occur on all input channels
- the node operation should be strictly non blocking, even under bridging and multicasting operation

Besides, for fast restoration in case of line failure, the node reconfiguration time should be in the order of 1 ms or below.

Two main different classes of functional architectures are usually distinguished, so called Wavelength-Routing (WR) and Broadcast-and-Select (BS) [3]. They both can be built exploiting the following points:

- space-switching, either at the input or at the output, to perform routing in the space domain
- wavelength-switching, to take advantage of the WDM techniques introduced in the network. There are two types of wavelength switching, wavelength selection using passive devices and wavelength conversion. The latter allows to solve possible blocking situations, since two input channels with the same wavelength may aim to the same destination. Tunability either at the transmitter (or wavelength converter) side, or on the filter side is required, since the wavelength must not be associated to a fixed transmitter-to-receiver (or wavelength converter-to-wavelength converter) path

The two classes of architectures are obtained combining differently the previous points. In some cases, space and wavelength routing can be performed at the same time, which does not affect the functional description [12], [13]. In particular, depending on the side where tunability is introduced, we have an architecture which performs:

- WR: The information is emitted on a wavelength (tunable) which corresponds to a given optical receiver (fixed), as shown in Fig. 4(a);
- BS: The information is broadcast on a given (fixed) wavelength toward all receivers (tunable), which are capable of selecting the proper wavelength aiming at them [Fig. 4(b)].

The two architectures, comparable in terms of number of components, can be both strictly non-blocking in point-to-point configuration but differ in two major features. First, the BS architecture is inherently suitable for multicasting configurations. This is due to the fact that in the former case, a given information can be routed simultaneously on different output links with different wavelengths. Therefore, if some capacity is available on the links, you can always address it. On the other hand, the WR solution implies that the multicast information has the same wavelength on all selected output links. If for some reason no free wavelength is available at the same time on all links, multicasting is impossible, even if spare capacity is available.

Second, for actual implementation of such a routing node, one has to compare the technological complexity and cost efficiency of tunable filters and widely tunable transmitters. Today, commercial or near term solutions are more readily available for tunable filtering. Of course, being closely related to the improvement of the technology, this situation might change very rapidly.

For these two reasons, the BS arrangement is so far our preferred approach.

Fig. 4(c) describes an alternative to Fig. 4(b), where the wavelength selection is performed before the space switching block. Even by using a simple demultiplexer on each input port, tunability is still granted thanks to the space-switching stage. In that case, the space-switching stage operates in a single-wavelength-per-port configuration, which is an advantage, but has also a number of input ports multiplied by

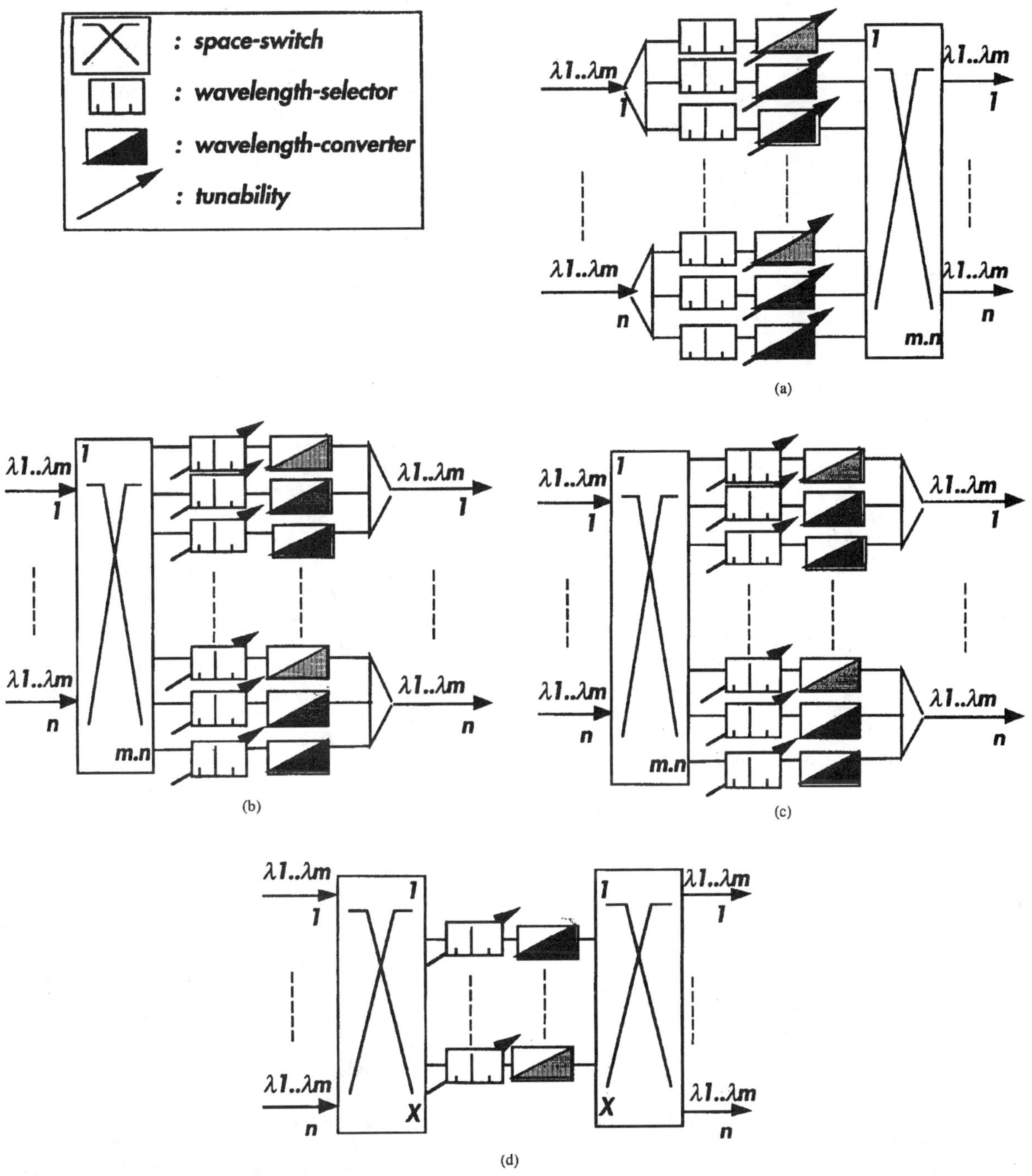

Fig. 4. All-optical crossconnection node architectures: (a) wavelength-routing architecture, (b) broadcast-and-select architecture using tunable filters, (c) broadcast-and-select architecture with input demultiplexing, and (d) broadcast-and-select architecture with sharing of the wavelength converter stage between all input channels.

the number of wavelengths. Therefore, for cost optimization reasons, we have chosen the case of Fig. 4(b), which means that the space-switching stage has to be capable of routing multiwavelength signals.

The solution schematically described on Fig. 4(b) allows us to meet all the requirements stated in Section III, and this is the solution we have chosen to implement. We must nevertheless keep in mind that, if the requirement on strictly non-blocking operation were loosened, one could imagine to decrease the number of wavelength selectors and wavelength converters (so far proportional to the number of output ports and number of wavelengths) by simply sharing a pool of them between

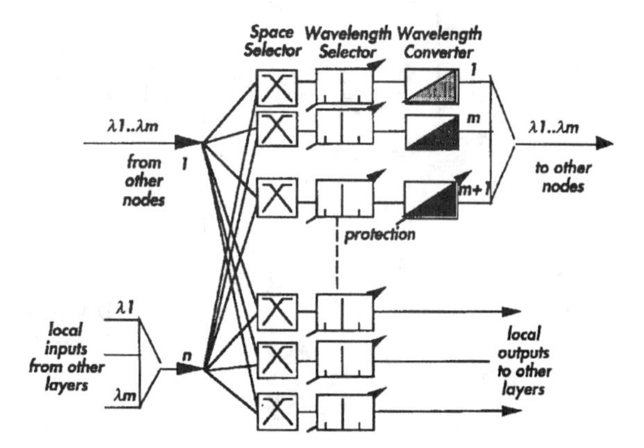

Fig. 5. All-optical crossconnection node featuring local interfaces to other layers and protection using tunable wavelength conversion.

all I/O ports. This solution, presented in Fig. 4(d), would imply to have an extra switching stage after the wavelength routing block. The total throughput of the node will, in this case, depend on the following ratio: (nb of wlgth selectors & converters)/(nb input ports × nb input wlgths).

The local interface with other layers, presented in Fig. 5, is defined in a very straightforward way: one input port is used for local inputs, and one output port for local outputs. Protection of the OXCN in case of failure of one of the devices is possible by adding one routing line per output port. For that application, tunability of the wavelength converter is also required. Fig. 5 will be the basis of our experimental work and physical implementation of the crossconnect.

IV. Physical Implementation of the Crossconnection Node

To implement a routing node according to our architecture, four major optical functions are required. They are:

- space-selection; polarization insensitivity, high path isolation, low insertion loss, low path unfairness and multiwavelength operation capability are the major requirements for this stage,
- tunable wavelength-selection, with low polarization sensitivity, high rejection and low insertion loss,
- wavelength-conversion with fixed output wavelength but varying input wavelength, possibly in the 1530–1560 nm amplifier window, insensitive to input state of polarization, requiring moderate control power, and, if possible, capable of signal reshaping,
- multiwavelength amplification with minimum gain excursion over the whole bandwidth to compensate for the losses during the routing process. In the following, present commercially available or advanced optoelectronic devices are reviewed and the most attractive solution for each function is highlighted.

A. Space-Selectors:

Commercially available solutions exist, such as electromechanical switches, that have very good switching performances, but are quite slow. In addition, they are not compatible with mass production, and will therefore stay very expensive, and $LiNbO_3$ switches, and may not reach the reliability required for actual products. Two advanced solutions are currently under investigation:

1) Integrated Digital-Optical-Switch Matrices on InP [14]: Based on carrier depletion effects and digital behavior to allow polarization insensitive operation, such switches are reliable and compatible with mass production. Main challenges rely in the optimization of the insertion losses and the path isolation. At this stage, a 4 × 4 packaged matrix has been mounted, with the following performance:

- losses below 20 dB (including 6 dB splitting losses)
- mean path isolation of 20 dB
- mean polarization sensitivity of 4 dB
- improved device compactness thanks to the use of a mirror technology in the interconnection stage.

2) Semiconductor Optical Amplifier (SOA) Used as Optical Gates [15]: By controlling the current injected into the active section of a SOA, very high on–off ratios can be reached (in the order of 50 dB), leading to very good path isolation. Besides, SOA's can provide gain to compensate for the splitting and combining losses. The main limitation is the cross-modulation arising from gain saturation effects experienced when operated with several channels simultaneously. To overcome this limitation, a weakly coupled Bragg grating can be superimposed to the active waveguide, leading to laser oscillation, and suppressed gain variations [16]. Such clamped-gain SOA's (CG-SOA's) are very promising, and a 2 × 1 switch has been already demonstrated using this technique with up to four channels at 2.5 Gb/s with a negligible signal degradation. The next step will be the integration of SOA's in arrays, with the interconnection structures for easier packaging.

B. Wavelength Selectors

This function can be performed using conventional tunable fiber Fabry–Perot filters, which have good rejection capabilities. Nevertheless, the stability is limited, leading to complex control electronics.

Another very promising candidate is the Acoustooptic Tunable Filter, based on TE/TM mode conversion in $LiNbO_3$ through the wavelength-dependent interaction with an acoustic wave. Polarization insensitivity is reached by using a polarization diversity scheme [11]. Such devices are very stable, if thermally controlled, and have reached rejections higher than 20 dB. The main challenge is to integrate the polarization splitters and combiners to reach full polarization insensitivity.

C. Wavelength Converters

The wavelength converter can be realized using two different approaches. The most straightforward approach is to use a hybrid solution, consisting of a high speed optoelectronic converter, an electrical regenerator, and a high speed optical transmitter. This solution would, of course, give the best performance in terms of power efficiency, signal regeneration, and wavelength and polarization insensitivity. But such an arrangement has to be designed for a given bit-rate, therefore

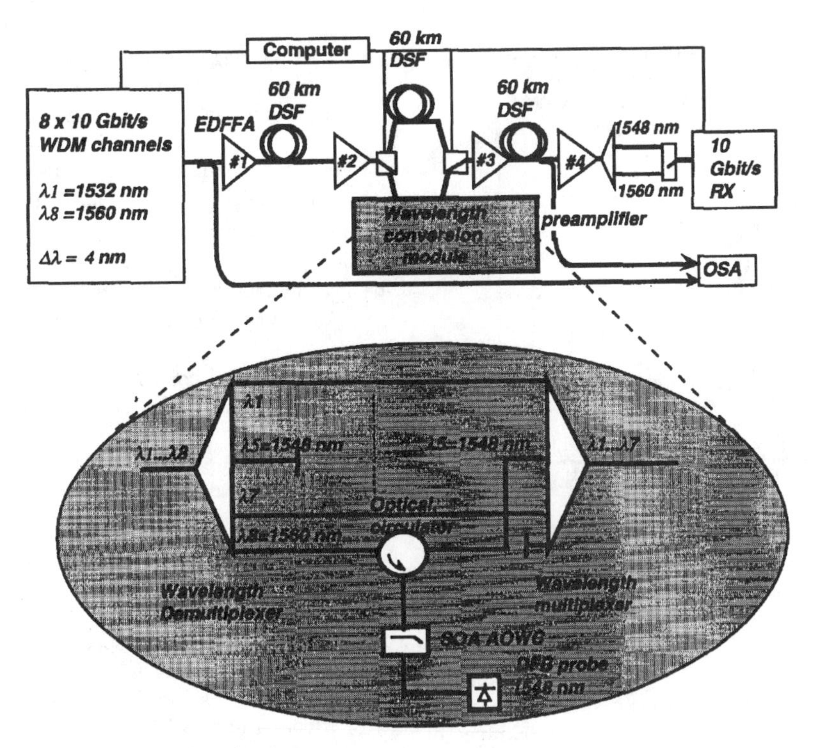

Fig. 6. 10 Gb/s 8-channel WDM transport and routing demonstrator set up.

limiting the network transparency, and is all the more expensive and power consuming as the bit-rate increases, since all parts of the converter become challenging.

We are also following all-optical approaches, which would not break the network transparency, and would be more cost-effective thanks to integration and mass production facilities. Extensive studies have been performed worldwide, and we have particularly focused on two techniques:

1) Carrier Depletion in DBR/DFB Lasers [17]: Optical injection in the active section of a laser oscillator leads to carrier depletion, and therefore to output power modulation. If the optical input power is high enough to switch off the oscillation, high output extinction ratios are reached. Thanks to this nonlinear response, this technique has signal reshaping capabilities. However, a compromise has to be found between good reshaping and speed. At 2.5 Gb/s, extinction ratio improvements have already been demonstrated (from 5 to 10 dB). The main challenges of these techniques are the realization of polarization insensitive active sections, and the optimization of the modulation bandwidth.

2) Carrier Depletion in SOAs: A strong injection into a SOA leads to gain saturation. This gain saturation can be probed by another signal at the output wavelength, and will receive the modulation with inverted polarity. This technique has been demonstrated at more than 10 Gb/s [15], but, as such, is unfortunately not as capable of signal reshaping.

Main results obtained in our lab are the following:

- speed: up to 10 Gb/s with output extinction ratio of 8 dB and penalty <1 dB
- polarization sensitivity <1 dB
- input power: <0 dBm in the fiber
- output power: >+5 dBm
- input wavelength tolerance: >30 nm within 3 dB input power drift (identical input and output wavelengths are tolerated, by using counter directional interaction and anti-reflection coatings below 10^{-4})

A way to obtain this property is to use the refractive index modulation instead of the gain modulation [15]. The information is converted to a phase modulation format, which has to be converted back to amplitude modulation by an interferometric filter (Mach–Zehnder Interferometer). Among other advantages, signal reshaping is possible, thanks to the nonlinear response of the interferometer, and has already been demonstrated. First samples including the SOA's monolithically integrated with the interferometric structure have already been designed and demonstrated at bit rates higher than 5 Gb/s [18]. This device is surely the most promising solution for the near future, because of its signal reshaping properties and also because of the narrow spectrum broadening, which is very attractive for transmission purposes.

D. Optical Amplification

Multiwavelength optical amplification is required at least on the transmission links, and at the interfaces with the OXCN's. Silica-based amplifiers suffer from limitations in multichannel operation, due to their highly wavelength dependent response in the 1530–1560 nm window. Equalization can nevertheless be performed in the 1545–1560 nm region for acceptable flatness [18]. To enlarge this to the full amplification window

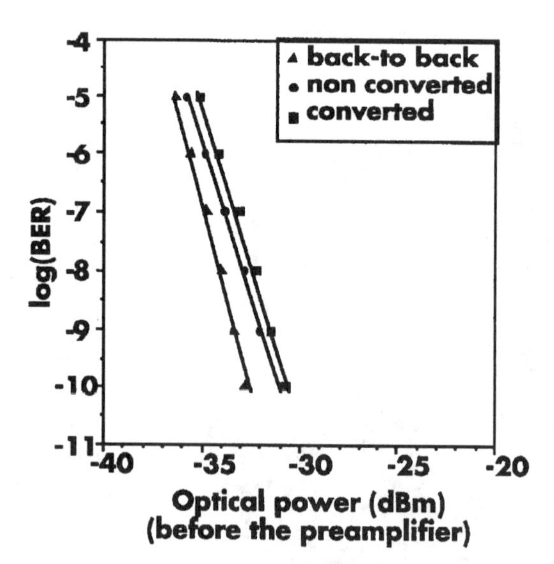

Fig. 7. Bit-error-rate performance of the WDM transport and routing demonstrator, back to back, after transmission, after transmission and wavelength conversion. (PRBS: $2^{23} - 1$).

of 30 nm, we have focused on fluoride-based amplifiers that exhibit without equalization a gain excursion in the range of 1.5 dB in the 1530–1560 nm region, with output signal-to-noise ratios higher than 30 dB, and gain larger than 25 dB [19].

V. Experimental Results

Using the devices described in Section IV, we have already performed physical experiments intended for the assessment of the feasibility of all-optical networks based on the OXCN architecture we have proposed.

A first rack-mounted demonstrator has been built, partially sponsored by the two RACE II "GAIN" and "ATMOS" European programs, including WDM transmission of eight channels, spaced by 4 nm and modulated at 10 Gb/s on 180 km dispersion shifted fiber, through the cascade of four flat gain fluoride-fiber amplifiers [8]. A non- tunable wavelength reallocation module including wavelength demultiplexing and 10 Gb/s wavelength conversion (from 1560 to 1548 nm) using cross-gain modulation in a polarization insensitive SOA was inserted in the middle of the transmission link (Fig. 6). Performances of this demonstrator were quite good: the power and SNR excursion after transmission through 3 in-line amplifiers were, respectively, 4.3 and 1.6 dB only. SNR's were in all cases higher than 26.5 dB, leading to a sensitivity penalty of about 1.5 dB at 1548 nm. The wavelength conversion introduced an extra penalty of only about 0.5 dB, due to the degradation of the output extinction ratio from 10 to 8 dB (Fig. 7).

From this demonstration, we can draw the following conclusions:

- flat-gain fluoride fiber amplifiers are the enabling factor for full use of the 30 nm amplification window for transmission and routing purposes.
- by introducing some routing functionalities in this demonstrator, even if not fully representative yet of the OXCN we have proposed, we have made a first step toward assessing the compatibility between transmission and routing techniques, which is the most critical issue for actual implementation of all-optical WDM networks.

On the other hand, we have built an experimental testbed simulating an OXCN in 4 × 4, 4-wavelength, 4 nm channel spacing configuration, with 10 Gb/s bit-rate per wavelength [21]. For that purpose, we have used advanced routing devices, including a 4 × 1 space-selector based on the digital-optical-switch technique, a polarization insensitive acoustooptic tunable filter, a wavelength ,converter, once again based on cross-gain modulation in a SOA, and fluoride-fiber amplifiers, as depicted in Fig. 8. We have analyzed the sensitivity penalty introduced by the different elements of the OXCN, for different routing configurations (input port, input wavelength, output wavelength). The overall penalty ranged between 1.5–3 dB, as shown in Fig. 9, and the part of each routing block can be split as follows:

1) 0.5 dB came from the space-selector. The path isolation of this selector was in the order of 23 dB, which is high enough to cope with usual heterowavelength crosstalk effects. But based on our assumptions on the network, we have to deal with the same wavelengths coming from different input ports. If, in the worst case, the States of Polarization (SOP) are matched, which is the case simulated in our setup, we are in the conditions of "coherent in-band crosstalk," namely, crosstalk leading to coherent beating between the two channels which are closer than the detection bandwidth: if $E1$ and $E2$ are the electromagnetic field amplitudes of the two input channels under interest, usual crosstalk imposed by $E2$ on $E1$, when the two channels are widely spaced, is proportional to $|E2|^2/|E1|^2$. But in the case of close spacing, it becomes roughly proportional to $2|E2|/|E1|$. Therefore, a 10 dB path isolation, tolerable in the case of wide spacing will raise about as much penalty as 26 dB in the case of close spacing. Therefore, for our application, the path isolation required for minimum penalty in the space-switching stage should be at least in the order of 30 dB, depending on the number of input ports.
2) Between 0.5–1.3 dB came from the wavelength selector. This is simply due to not quite sufficient rejection of the filter (about 15 dB), for 4-wavelength operation. This rejection should be in the order of 18 dB for minimal penalty with 4 channels, to be further increased to tolerate more wavelengths.
3) Between 0.8–1.2 dB came from the wavelength converter. This penalty is due, on one hand, to extinction ratio degradation (from 10 to 8 dB at 1548 nm output wavelength), and was significantly higher for higher output wavelengths. This is due to the fact that the gain peak of the SOA depends as well as the gain on the carrier density, and therefore on the input pump power. Due to this wavelength shift of the gain peak, gain modulation

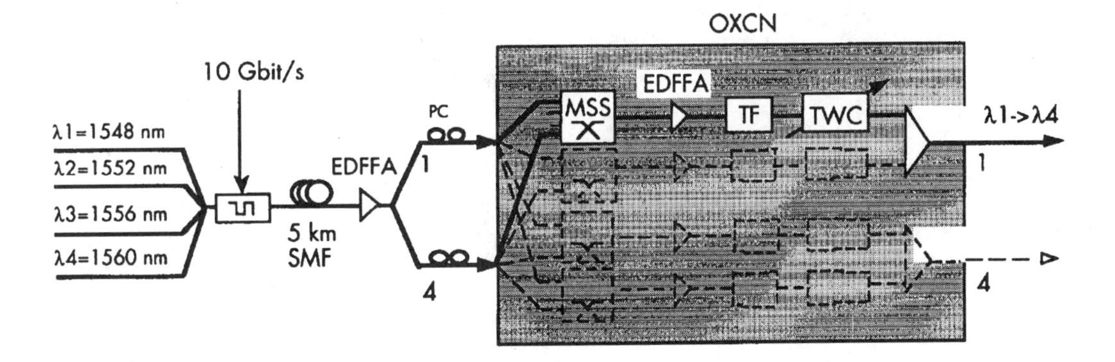

Fig. 8. 4 × 4 4-wavelength, 10 Gb/s data rate all-optical crossconnect experimental testbed.

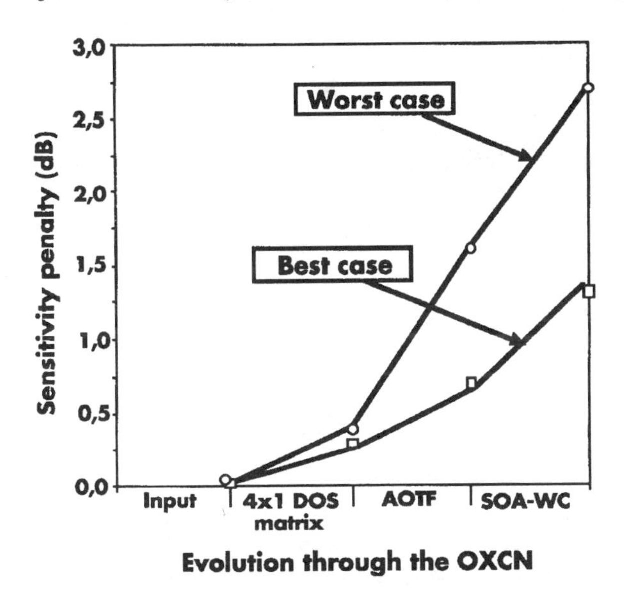

Fig. 9. Evolution of the detection sensitivity power penalty through the crossconnect. (PRBS $2^{23} - 1$).

is more efficient on lower output wavelengths than on higher output wavelengths. In the future, we would like this already attractive wavelength converter to improve on two main parameters:

a. no more dependence on the output wavelength
b. signal reshaping capabilities to compensate for residual degradation in the other routing devices.

In a former OXCN experimental setup [22], we have already demonstrated all-optical signal reshaping at 2.5 Gb/s using the injection locking technique in one of the Fabry–Perot modes of a DFB laser. We demonstrated that it was possible using an all-optical technique to overcome the penalty brought by crosstalk issues. We believe that in the near future such penalty compensation will also be possible using SOA's under cross-phase modulation in an integrated interferometric filter.

These two experiments have shown that optical techniques were becoming usable for routing purposes, according to our OXCN architecture, and with possible compatibility with transmission issues. Besides, we have been able to derive more accurate requirements for the routing components, in order to improve further the performances of the OXCN.

VI. Conclusions

We have indicated the major benefits that all-optical WDM techniques could provide to transport networks, including capacity and connectivity upgrading, transparent interconnection of heterogeneous networks in a cost-effective and future proof manner. Within the framework of the layered network concept we have highlighted the advantages of wavelength conversion to allow network flexible planning and scalability without increasing the management complexity.

We have derived from our network assumptions external specifications for all-optical crossconnection nodes, and we have proposed a functional and physical node architecture meeting these requirements, based on both space and wavelength switching techniques, as well as on the broadcast-and-select principle.

We have also presented the devices that could be used for realization of the four major functions required in the all-optical crossconnection node, namely multiwavelength space-selection, tunable wavelength selection, wavelength conversion and multiwavelength optical amplification.

Finally, we have presented experimental results at 10 Gb/s as first assessments of the feasibility of OXCN's and compatibility with transmission. In particular, operation in 4 × 4, 4-wavelength configuration has been demonstrated with a penalty in all cases lower than 3 dB. Besides, this experimental work has highlighted the main physical degradations introduced by all-optical routing devices, and the requirements to overcome these degradations. These very encouraging results bring experimental evidence of the potential of all-optical routing nodes for actual implementation in multiwavelength transport networks.

References

[1] M. H. Lyons, K. O. Jensen, and I. Hawker, "Traffic scenarios for the 21st century," *BT Technol. J.*, vol. 11, no. 4, pp. 73–83, Oct. 1993.

[2] CCITT, "Architectures of transport network based on the Synchronous Digital Hierarchy (SDH)," Geneva, Switzerland, Recommendation G.803, 1993.

[3] C. A. Brackett, "Dense wavelength division multiplexing networks: Principles and applications," *IEEE J. Select. Areas Commun.*, vol. 8, pp. 948–964, Aug. 1990.

[4] I. Hawker, "Evolution of digital optical transmission network," *BT Technol. J.*, vol. 9, no. 4, pp. 43–56, Oct. 1991.

[5] A. Fioretti, F. Masetti, and M. Sotom, "Transparent routing: The enabling factor toward all-optical networking," in *Proc. European Conf. Opt. Commun.*, Florence, Italy, 1994, pp. 503–509.

[6] K. I. Sato, S. Okamoto, and H. Hadama, "Network performance and integrity enhancement with optical path technologies," *IEEE J. Select. Areas Commun.*, vol. 12, pp. 159–170, Jan. 1994.

[7] C. A. Brackett *et al.*, "A scalable multiwavelength multihop optical network: A proposal for research on all optical network," *J. Lightwave Technol.*, vol. 11, May/June 1993.

[8] B. Clesca, A. Jourdan, S. Artigaud, G. Da Loura, L. Hamon, G. Soulage, J. C. Jacquinot, P. Doussière, P. de Vivie de Régie, D. Bayart, J. L. Beylat, M. Sotom, C. Joergensen, T. Durhuus, K. E. Stubjkaer, M. Semenkoff, and M. Guibert, "A joint RACE experiment: 80 Gbit/s WDM transmission combining wideband optical fiber amplifiers and in-line 10 Gbit/s wavelength converter," in *Proc. European Conf. Opt. Commun.*, Florence, Italy, 1994, vol. 4, pp. 27–31.

[9] I. Chlamtac, A. Ganz, and G. Karmi, "Light path communication: An approach to high bandwidth optical WAN's," *IEEE Trans. Commun.*, vol. 40, pp. 1171–1182, July 1992.

[10] S. Okamoto and K. I. Sato, "Optical path crossconnect systems for photonic transport networks," in *Proc. IEEE GLOBECOM '93*, USA, 1993, pp. 474–480.

[11] M. J. O'Mahony, "The potential of multiwavelength transmission," in *Proc. European Conf. Opt. Commun.*, Florence, Italy, 1994, pp. 907–913.

[12] D. A. Smith, R. S. Chakravarthy, A. d'Alessandro, J. L. Jackel, J. E. Baran, and D. J. Fritz, "Acoustooptic tuned optical filters and switches for WDM systems," in *Tech. Dig. Opt. Fiber Conf.*, San Diego, 1995, paper TuO1.

[13] B. Glance, "A tunable optical add-drop filter," in *Tech. Dig. Opt. Fiber Conf.*, San Diego, 1995, paper TuQ3.

[14] J. F. Vinchant, A. Jourdan, J. Le Bris, G. Soulage, T. Fillion, and E. Grard, "InP 4 × 1 digital optical switch module for multiwavelength cross-connect applications," in *Tech. Dig. Opt. Fiber Conf.*, San Diego, 1995, paper ThK2.

[15] K. Stubkjaer, B. Mikkelsen, T. Durhuus, C. G. Joergensen, C. Joergensen, T. N. Nielsen, B. Fernier, and P. Doussière, "Semiconductor optical amplifiers as linear amplifiers, gates and wavelength converters," in *Proc. European Conf. Opt. Commun.*, Montreux, Switzerland, 1993, paper TuC5.1.

[16] G. Soulage, P. Doussière, A. Jourdan, and M. Sotom, "Clamped gain travelling wave semiconductor optical amplifier as a large dynamic range optical gate," in *Proc. European Conf. Opt. Commun.*, Florence, Italy, 1994, pp. 451–454.

[17] P. Ottolenghi, A. Jourdan, and J. Jacquet, "All-optical wavelength conversion with extinction ratio enhancement using a tunable DBR laser," in *Proc. European Conf. Opt. Commun.*, Montreux, Switzerland, 1993, paper TuC5.5.

[18] T. Durhuus, C. Joergensen, B. Mikkelsen, K. E. Stubkjaer, F. Ratovelomanana, A. Enard, G. Glastre, and N. Vodjdani, "Monolithic integrated Mach–Zehnder wavelength converter: Conversion and transmission experiments at 5 Gbit/s," in *Tech. Dig. Opt. Fiber Conf.*, San Diego, CA, 1995, paper TuO6.

[19] R. A. Betts, S. J. Frisken, and D. Wong, "The split-beam Fourier filter and its application in a gain flattened EDFA," in *Tech. Dig. Opt. Fiber Conf.*, San Diego, CA, 1995, paper TuP4.

[20] D. Bayart, J. Hervo, and F. Chiquet, "Impact of fluoride-based EDFA's gain flatness on design of WDM amplifier cascade," in *Tech. Dig. Opt. Fiber Conf.*, San Diego, CA, 1995, paper TuP2.

[21] A. Jourdan, G. Soulage, G. da Loura, B. Clesca, P. Doussière, C. Duchet, J. F. Vinchant, and M. Sotom, "Experimental assessment of a 4 × 4 4-wavelength all-optical cross-connect at 10 Gbit/s line-rate," in *Tech. Dig. Opt. Fiber Conf.*, San Diego, CA, 1995, paper ThI7.

[22] A. Jourdan, G. Soulage, B. Clesca, M. Sotom, G. Da Loura, J. L. Beylat, P. Doussière, S. Gurib, and J. F. Vinchant, "All-optical cross-connect for transparent multiwavelength transport network," in *Proc. European Conf. Opt. Commun.*, Florence, Italy, 1994, pp. 563–566.

Optical Path Cross-Connect Node Architectures for Photonic Transport Network

SATORU OKAMOTO, MEMBER, IEEE, ATSUSHI WATANABE, MEMBER, IEEE, AND KEN-ICHI SATO, MEMBER, IEEE

Abstract—**This paper explores the optical path cross-connect (OPXC) node architectures that are essential components of the optical path network. Optical path technologies will play a key role in the development of the platform on which the future bandwidth abundant B-ISDN should be created. This paper highlights the wavelength path (WP) and the virtual wavelength path (VWP) technologies, both of which can greatly enhance the path layer capability and the efficiency of network failure restoration. The OPXC, which handles optical paths, is constructed with an optical switching network. Various WDM-based switching networks, which are aimed at LAN applications, have been reported. On the other hand, few WDM-based switching networks for OPXC systems, which are applicable to the nationwide transport network, have been proposed. In this paper, we elucidate the functional conditions required to construct OPXC nodes for WP and VWP global networks. Next, we assess switching network architectures for their applicability to the transport network. It is shown that the OPXC architecture based on DC-switches (delivery and coupling switches) is superior to the other OPXC architectures in terms of optical losses, modularity, and upgradability. Finally, detailed evaluations of the DC-switch-based OPXC node are presented that confirm its feasibility.**

I. Introduction

OPTICAL path technologies based on WDM (wavelength division multiplexing) techniques will be the key to create a powerful and reliable future B-ISDN that allows graceful network evolution [1]–[4]. The optical path network is based on wavelength routing [5]. Each optical path is recognized by its wavelength. Therefore, optical path processing based on wavelength routing can be performed in the optical domain. Because of this optical signal processing, optical path technologies provide an optical platform which has several important merits such as independence from bit rate and transmission mode (PDH, SDH and ATM). For realizing the optical path network, the wavelength path (WP) concept, the virtual wavelength path (VWP) concept and the combination of WP and VWP have been proposed [1], [6]. These optical path schemes will enhance the transport network such as increasing transmission capacity and cross-connect system throughput simultaneously, simplifying network restoration, and improving network upgradability [1]–[4].

To realize the optical path network, establishing the optical path cross-connect (OPXC) node architecture is essential. The OPXC node has two main functions, routing of optical paths and optical path termination. The electrical paths accommodated within the terminated optical paths are fed to the electrical path cross-connect (EPXC). The routing function in the OPXC is performed with an optical switching network. The optical switching network is the key component of the OPXC node. A variety of optical switching networks based on WDM technologies have been proposed. Most of the proposed switching networks were targeted at LAN applications, hence, very few reported optical switching networks are applicable to OPXC.

This paper elucidates the OPXC node architectures to realize WP and VWP networks. First, we propose the functional conditions required to construct OPXC nodes for WP and VWP networks. Next, we investigate existing optical switching network architectures aimed at transport networks and assess their applicability to WP and VWP networks in detail. It is demonstrated that the DC-switch (delivery and coupling switch)-based OPXC architecture, which we newly propose, is superior to all other OPXC architectures in terms of realizing the proposed functional conditions. Finally, detailed evaluations of the DC-switch-based OPXC node are presented and, as a result, the practicability and the effectiveness of the DC-switch are confirmed.

II. Structure of Optical Path Network

Before going on to evaluate OPXC architectures, the key features of WP and VWP networks related to the assessment are summarized in this section.

A. Multiwavelength Transport Network Structure

The transport network can divided into three layers [7] from the functional viewpoint, namely circuit layer, path layer and transmission media layer. The layered architecture simplifies the design, development and operation of the network, and allows smooth network evolution in pace with user demands and technology advances specific to each layer. WDM technologies can enhance transmission capacity, however, the negative impact of even a single transmission link/node failure is increased due to the very high transmission capacity. Therefore, failure immunization techniques and rapid network restoration are required. Several network restoration techniques that act on the path layer have been developed [8], [9]. It has been reported that the optical path layer can be created [1], [2] when WDM technologies are introduced to the transport network. The layered architecture of the transport network with the optical path layer is schematically illustrated

Manuscript received March 25, 1995.

The authors are with NTT Optical Network Systems Laboratories, Kanagawa 238-03, Japan.

Publisher Item Identifier S 0733-8724(96)04617-8.

Reprinted with permission from *IEEE Journal of Lightwave Technology*, Vol. 14, No. 6, pp. 1410-1422, June 1996.

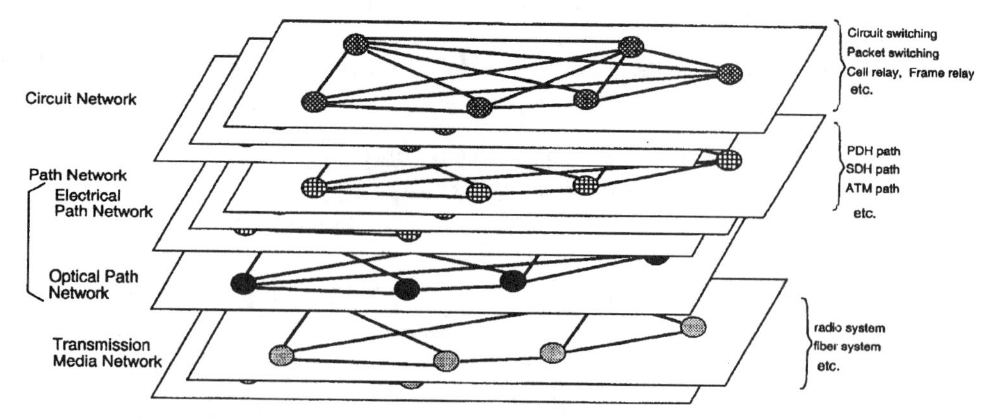

Fig. 1. Layered architecture of transport network for public telecommunication network.

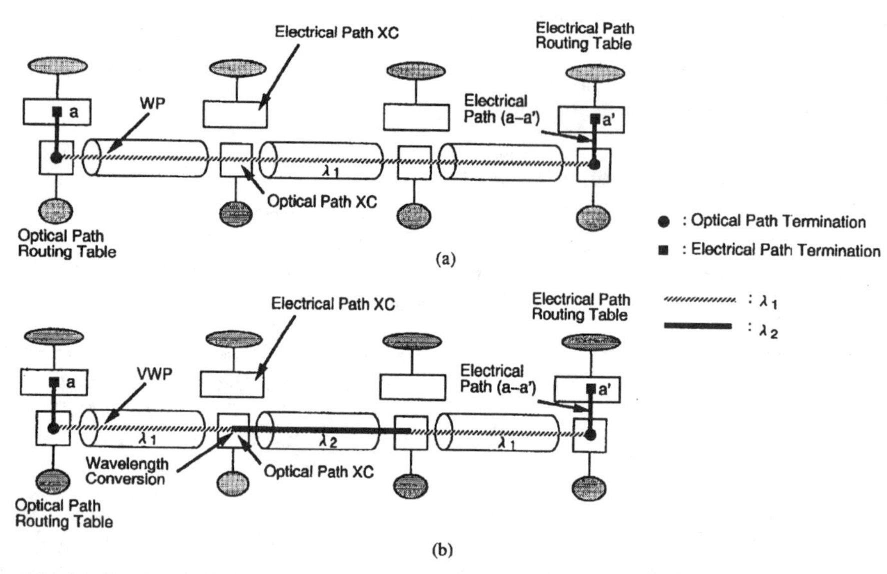

Fig. 2. Path setting of (a) wavelength path (WP) and (b) virtual wavelength path (VWP).

in Fig. 1. If the optical path layer is utilized for network restoration, the major part of the network restoration system would be used in common by all transmission mode networks [1]–[4]. This would more effectively utilize the network's capacity reserved for restoration.

B. WP and VWP

Various lightwave network technologies have been proposed [10]–[13] and different optical path realization techniques have been investigated. Among them, WP and VWP schemes are the most important for trunk transport network application [1], [2]. These optical paths are used to accommodate electrical paths such as PDH paths, SDH paths and VP's. They introduce a new optical path sublayer as depicted in Fig. 1.

A WP is established between two nodes (or node facilities) according to demand by allocating one wavelength for the path [see Fig. 2(a)]. The intermediate nodes along the WP perform WP routing according to the wavelength. WP's are accommodated within a physical network, and in the accommodation process the wavelength assignment problem must be solved simultaneously so that different wavelengths are assigned to all WP's in every physical link throughout the network. Therefore, the WP scheme limits wavelength reuse in the network.

In the VWP scheme, on the other hand, path wavelength is allocated link-by-link to each VWP [see Fig. 2(b)], thus the wavelength of each VWP on a link has local significance instead of global significance as in the case of WP's. For this reason, this scheme is called Virtual WP. VWP termination is done at both end nodes and intermediate nodes perform VWP routing according to a routing table and wavelength conversion as necessary. Therefore, the degree of wavelength reuse in the network is maximized so that fewer wavelengths are needed with the VWP scheme than with the WP scheme. The extent of the wavelength reduction achieved with the VWP scheme depends on various conditions. It has been reported that the difference in required number of wavelengths between the WP scheme and the VWP scheme was little [14], [15] when optical path restoration was not considered.

TABLE I
WP–VWP COMPARISON

Item	WP(Wavelength Path)	VWP(Virtual Wavelength Path)
Wavelength Conversion at OPXC	Unnecessary	Necessary
Wavelength Assignment Problem	Exist	None
Restriction for Network and Path Expansion	Significant	Insignificant
Required Wavelength Number in the Network (Routing Freedom)	Larger (Smaller)	Smaller (Larger)
Path Wavelength Administration	Requires Centralized Control in the Network	Possible with Link-by-Link Distributed Control
Wavelength Preciseness	Absolute Preciseness throughout the Network	Link-by-Link Relative Preciseness

C. Network Restoration for WP and VWP Networks

When a link failure or node trouble occurs, network restoration techniques that act on the path layer are very effective. Failure restoration of a WP/VWP network is performed by rerouting WP's/VWP's [4], [16], [17]. In the case of WP rerouting, the wavelength collision problem must be solved. The effective way is to determine the restoration path route and to assign wavelengths in advance so that wavelength collision will never occur with path rerouting irrespective of which transmission line fails. When failure restoration is considered, it was found that the WP scheme requires more wavelengths than the VWP scheme [4], [16], [17]. References [16]–[18] report that the WP scheme requires 30∼40% more wavelengths.

D. WP and VWP Comparison

The WP and VWP schemes are compared in Table I. The required number of wavelengths for a WP network is always larger than or equal to that for an equivalent VWP network, as mentioned in above. Moreover, the expansibility of the VWP network (links and nodes) is larger than that of the WP network. In particular, considering network restoration, the avoidance of the wavelength allocation problem is a major advantage of the VWP scheme.

On the other hand, concerning OPXC structure, an important point is the wavelength conversion function required in VWP–OPXC's. This function may be performed directly in the optical domain (O-O type) in the future with bistable LD's [19], the four-wave mixing technique [20], [21] and so on. Wavelength conversion via the electrical level (O/E–E/O type), the combination of O/E converter and E/O converter with tunable wavelength LD or the combination of wavelength selective O/E converter and E/O converter with fixed wavelength LD, is very practical. Integrated O/E and E/O circuits offering this function have been reported [22]. This type of wavelength conversion may limit the merit of optical path technologies, such as bit rate and transmission format restriction free. However, this restriction can be minimized by the development of an OPXC system architecture designed to support different transmission signals by simply replacing the O/E and E/O packages, as will be elaborated later. O–O type wavelength converters do not impose any such limit, however, these technologies are still immature.

The limits imposed by optical devices are minimized in the WP scheme, and the hardware requirements are much less than that for VWP, as will be elaborated later. The WP scheme requires more wavelengths than the VWP scheme as mentioned above. This indicates that the capacity of the WP network is smaller than that of the VWP equivalent under the condition of using the same physical network configuration (network topology and OPXC port number at each node) with the same number of wavelengths. In the early stages of optical path introduction, the traffic demand will not be so large to put any reservations on the introduction of WP's, rather the hardware simplicity of the WP–OPXC offsets any limitation present. Therefore, WP's will be very attractive. However, with the enlargement of traffic demand and further progress in device technologies, it is desirable to upgrade the WP network, if installed, to the VWP network to maximize optical path benefits. This upgradability, evolving WP–OPXC to VWP–OPXC, is one of the most important parameters for evaluating OPXC node structure.

III. OPTICAL PATH CROSS-CONNECT NODE ARCHITECTURE

A. System Requirements for OPXC

In this section we propose major five functional conditions for applying the OPXC system to the transport network.

1) Usable Wavelengths: The OPXC system requires optical amplifiers (OA's) because it must compensate the loss of long-haul transmission and the loss of the optical switching networks in the OPXC. An optical fiber amplifier (OFA) e.g., EDFA in the 1.5 μm band is suitable, however, the gain flat region of commercially available OFA's, about 1544 to 1560 nm, restricts the usable wavelength region of WP and VWP networks. Therefore, the usable bandwidth is only about 16 nm when commercially available economical techniques are exploited.

To create a cost-effective and practical system, it is very effective to exploit existing direct intensity modulation for transmission. Thus, channel spacing greater than 1nm is required mainly because of the chirped LD spectrum. If 2 nm channel spacing is utilized, only 8 channels are available with

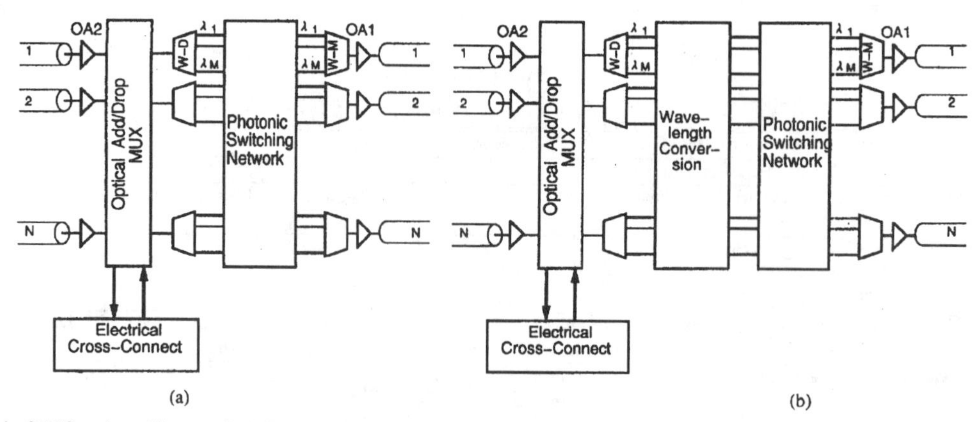

Fig. 3. Generic OPXC node architecture for WP (a) and VWP (b). W–D denotes a wavelength demultiplexer, W–M denotes a wavelength multiplexer, and OA denotes an optical amplifier.

the above EDFA bandwidth. If 1 nm channel spacing becomes possible, 16 channels are usable in the 16 nm range. Therefore, in the early introduction stage, we set the system target at 8∼16 WDM channels. The assessment of OPXC's shown later was based on 8 WDM channels.

In the future, coherent transmission techniques will become practical and over 100 channels will become usable [23]. However, at present, this kind of technique is immature and we can not fully utilize its potential cost-effectively.

2) Transmission Signal Bit Rate: SDH-based optical fiber transmission systems offering up to 2.5 Gb/s (STM-16 [24]) have been introduced, and the same transmission signal bit rate will be used in ATM where ATM cells are mapped into VC-4-16c [25]. Therefore, WP and VWP networks will be required to support 2.5 Gb/s, and in the future 10 Gb/s or so. This paper discusses the support of transmission signal bit rates of 2.5 Gb/s as the first step.

3) OPXC Port Number: Concerning the electrical path layer cross-connect system, e.g., ATM XC, XC systems with total capacity of 20 Gb/s (2.5 Gb/s line speed × 8 ports, or the equivalent capacity systems with lower-speed interfaces) have been developed [26]. To realize an over 8 port system, the M-J XC system architecture can be adapted. Several 8 port XC systems (each system is called "main frame") are connected with a "junction frame."

To enhance the transmission capacity and create an effective network, the cross-connect node processing capability must also be expanded. The OPXC system should support at least 10 times higher throughput than the ATM XC and it is desirable for each OPXC node to handle over 1 Tb/s capacity to enable future bandwidth abundant B-ISDN where end-to-end high-speed path connections are provided very cost-effectively. In the following, considering state-of-the-art WDM and WDM-transmission technologies, 8 wavelength WDM is adopted. A 16 port main frame structure, which supports a total throughput of 320 Gb/s, assuming 8 wavelengths in each fiber, is the first step system target.

4) Nonblocking Characteristics: As explained in Section II, network restoration is performed in the optical path layer. OPXC's that have strictly nonblocking characteristics are necessary in terms of network restoration. This is because the restoration or establishment of new optical paths should have no effect on preestablished active optical paths. Switching times of the order of milliseconds are sufficient for optical switching networks for restoration or new path establishment. This is the major difference from optical packet switching networks. For example, optical ATM switch application requires nanosecond order (or much less than one cell time) switching time.

5) Modularity and Upgradability: In the future B-ISDN, early traffic demands will be low. To enable cost-effective OPXC introduction, the initial versions must offer adequate flexibility and minimum investment but still support later growth and incremental investment as traffic demands increases. Therefore, the modularity of optical switching networks and OPXC node architectures is important. Two kinds of modularity allow these characteristics: link modularity and wavelength modularity. Considering state-of-the-art technologies, however, available wavelengths must be held to about 10 if we are to realize cost-effective and practical OPXC's as mentioned before, so link modularity becomes more important.

As mentioned in Section II, we must be able to upgrade the OPXC architecture from WP's to VWP's as device technologies progress further and traffic demands increases.

B. Optical Switching Network Structure

This section investigates optical switching network structures aimed at OPXC's and evaluates whether they can adapt to WP/VWP networks.

1) Generic OPXC Node Architecture: The generic OPXC node architecture with WP's/VWP's is shown in Fig. 3. At the optical ADM (Add/Drop Multiplexer), the optical path is terminated, and an interface between OPXC and EPXC is created.

The functional difference between WP and VWP–OPXC's is the wavelength conversion function required in the VWP–OPXC. This function is performed with O/E–E/O type wavelength converters as explained in Section II-D. In Fig. 3, wavelength converters are allocated just before the optical switching network. We define this configuration as the prefix type OPXC architecture. In this case, E/O converters with tunable LD's are suitable. On the other hand, it is possible to place the wavelength converters just after the optical switching network. This is defined as the postfix type OPXC architecture. The postfix type OPXC architecture can use fixed wavelength LD's. However, the very high optical loss from the input fibers to O/E converters requires the use of OA's in the optical switching network for loss compensation.

Concerning the WP scheme, there are two categories, the regeneration type WP scheme which uses O/E–E/O type optical regeneration circuits in OPXC nodes like a wavelength converter of the VWP–OPXC, and the nonregeneration type WP scheme which uses in-line OA's to compensate fiber transmission loss and optical switching loss. With the nonregeneration type WP scheme, crosstalk from the switching network, ASE noise from OA's and gain deflection of OA's accumulate with each pass through the OPXC's. These undesired signals deteriorate transmitted signal quality and determine the system application distances. This problem becomes much more significant if the optical power levels among channels are very different. Such a difference can be created by paths traversing different numbers of links and nodes where optical components exhibit wavelength dependent losses. Therefore, an optical regeneration technique which also permits optical power level adjustment is quite useful even for the WP scheme. Considering optical regeneration at the electrical level, the OPXC architectures for WP's and VWP's are basically the same.

Concerning the optical ADM, it is possible to substitute a part of the optical switching network for the optical ADM. This will be mentioned in a later section.

2) Switch Fabric Using SD Switches: The space division (SD) switch is adaptable for the optical switching network shown in Fig. 3. Several SD-switch-based OPXC architectures have been proposed [4], [27]. These optical switching networks are shown in Fig. 4. It is assumed that N links (N incoming and outgoing links) are connected to the node and M wavelengths are multiplexed into each link. If parameters $N = 16$ and $M = 8$ are selected, required SD-switch size of the VWP switch will be 128 × 128, and it must be strictly nonblocking. To construct such a large scale SD-switch, a nonblocking multistage switching network architecture, like the Clos switching network [28], is required. This is shown in Fig. 4(a) and we define this architecture as VWP-1. VWP-1 is a postfix type OPXC, so, fixed wavelength LD's can be used for the OS's. If tunable wavelength LD's can be used as OS's, the multistage switching network can be simplified as shown in Fig. 4(b). We define this architecture as VWP-2.

On the other hand, the WP switch requires no wavelength conversion, and optical path cross-connection is performed while keeping the wavelength constant. Therefore, an $N \times N$ SD switch is sufficient for output port selection, as shown in Fig. 4(c) (WP-1). We evaluated the hardware amounts needed to construct a large SD-switch for WP-1 OPXC and VWP-1/VWP-2 OPXC's, and the results are shown in Fig. 5. In Fig. 5, it is assumed that we can use up to 8 × 8 unit SD switches [29] and the hardware amount of each unit SD switch is assumed to be order n^2 and $n \log_2 n$ where n is the switch size of each unit SD switch. Order n^2 corresponds to the state-of-the-art optical switch technology for example the cross-bar matrix switch, and order $n \log_2 n$ corresponds to the more complex optical switch such as the benes switching network. There are many ways of constructing a multistage Clos switching network with small SD switches. The best combination, as evaluated by the minimum hardware amount, was investigated. Fig. 5 shows that the ratio of VWP hardware to WP hardware, VWP/WP, is around 2.5~5 if $N \geq 16$. This is a significant difference when N is large where total SD switch hardware becomes extremely large. Therefore, VWP-1 and VWP-2 architectures are not applicable for a large scale ($N \geq 16$) OPXC system although it can be realized with fixed wavelength LD's. The required hardware ratios of VWP-2 to WP-1, obviously, are lower than that of VWP-1 to WP-1. However, the VWP-2 architecture requires tunable wavelength LD's, this is a trade-off of the SD-switch-based VWP–OPXC architectures.

The WP-2 architecture shown in Fig. 4(d) [27] is for nonregeneration type WP networks. This OPXC node has tunable transmitters (LD's) and tunable filters. This switching network does not have strictly nonblocking characteristics, but rearrangeable nonblocking characteristics. Therefore, this type of OPXC architecture is not suitable for transport network application where network restoration will be done in the optical path layer.

It is clear that these SD-switch-based switching network architectures have significant disadvantages in terms of poor modularity and upgradability as discussed in Section III-A.

3) Switch Fabric Using Power-Splitters and Filters: The SD-switches shown in Fig. 4 can be replaced with a combination of optical power-splitters and (tunable) wavelength filters. We proposed switch architectures which are composed of MWSF's (multiwavelength selective filters), which can select any combination of wavelengths, WF's (single wavelength filters) and power-splitters [4]. Various types of MWSF's using acousto-optic interaction have been reported. They include the integrated-optic acousto-optic tunable filter (AOTF) [30], those used for protection switches [31], and those for 2 × 2 unit switches [32].

The WP switch architecture is shown in Fig. 6. Each input optical path is routed to its appropriate destination port by controlling the MWSF. Fig. 7 shows the VWP switch architecture. The first MWSF stage leads input optical path signals to the 'appropriate' wavelength converters, and the last MWSF stage routes converted signals to the destination ports. The word 'appropriate' means that signals of the same wavelength are never led to the same star-coupler. The switch shown in Fig. 7 is a type of rearrangeable nonblocking switch. A strictly nonblocking switch architecture composed of optical star-couplers, tunable WF's and gate switches has also reported [33]. This switch architecture is called the "parallel λ-switch," and its configuration is shown in Fig. 8. This architecture is the postfix type OPXC, so, we must compensate the large loss of the power-splitters before the optical receivers; however, it offers the superior characteristics of modularity and upgradability.

For these switching networks, the large loss, temperature stability, high crosstalk level, polarization sensitivity and broad FWHM (a few nanometers or more) of MWSF/tunable filters

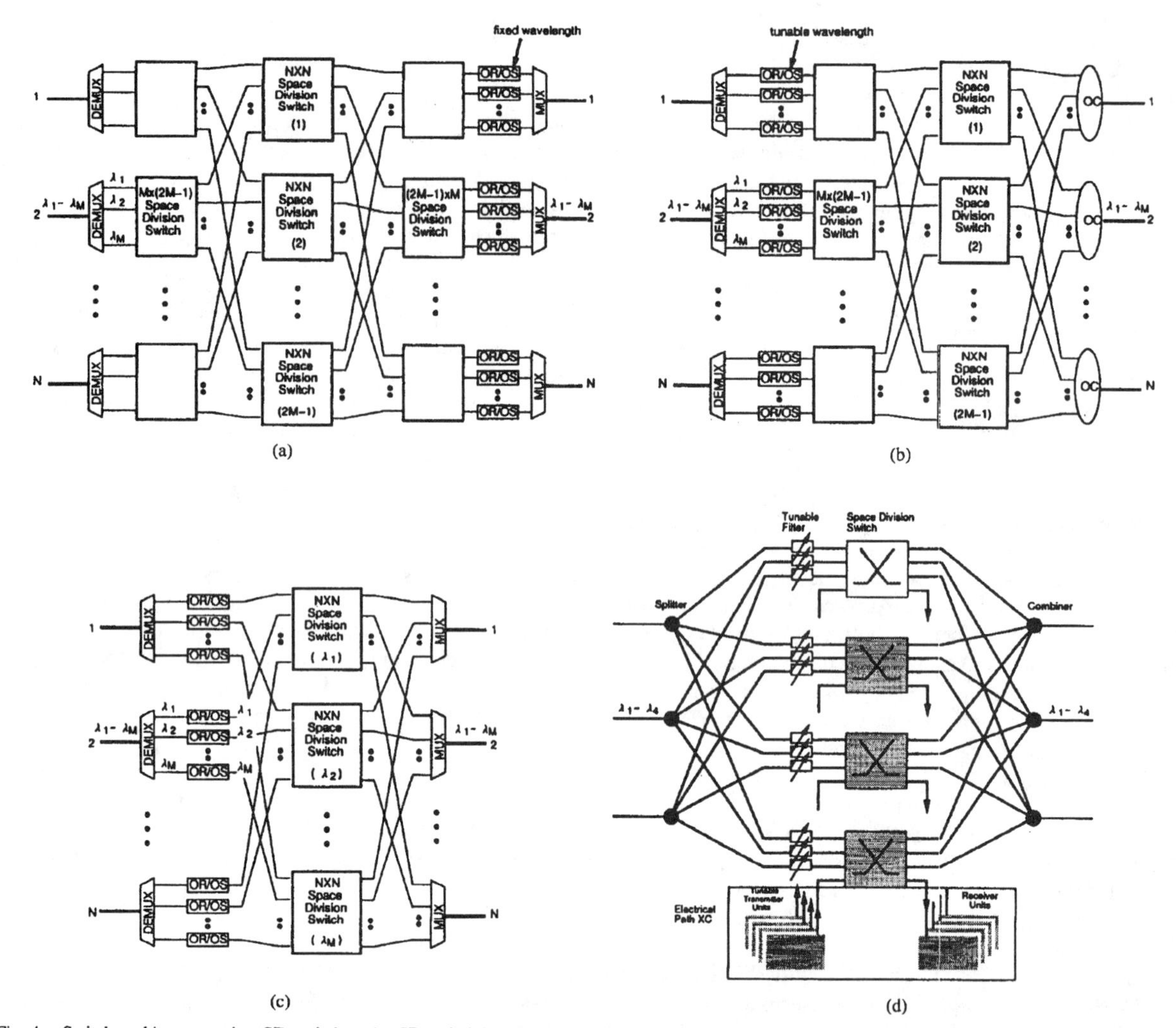

Fig. 4. Switch architecture using SD switches. (a) SD-switch-based OPXC architecture (VWP-1). (b) SD-switch-based OPXC architecture (VWP-2). (c) SD-switch-based OPXC architecture (WP-1). (d) SD-switch-based OPXC architecture (WP-2). DEMUX denotes a wavelength demultiplexer, MUX denotes a wavelength multiplexer, and OC denotes an optical coupler. OR/OS—WP: Optical Signal Regenerator; VWP: Wavelength Converter.

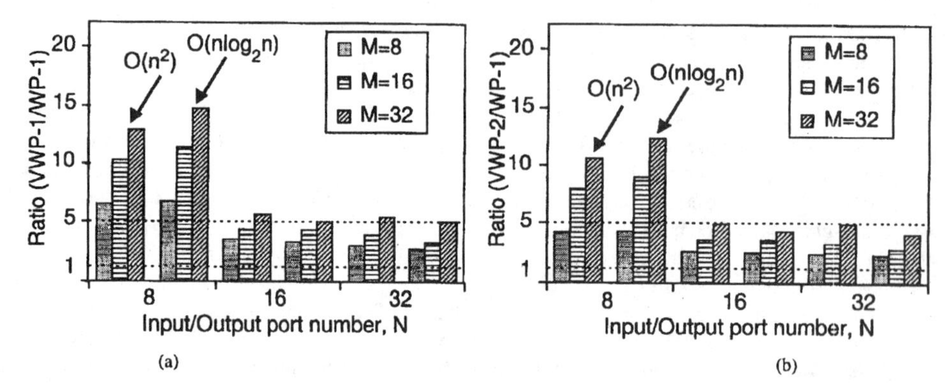

Fig. 5. Required hardware comparison for SD-switch-based OPXC's. (a) VWP-1 versus WP-1 and (b) VWP-2 versus WP-1.

are the main problems in the development of large scale OPXC's, since at least $N \times M$ of them are required. This architecture, therefore, is deemed to lie far beyond the SD-based switch architecture, from the practical point of view.

4) Switch Fabric Using DC-Switches: In 1994, we proposed the delivery and coupling switches (DC-switches)-based OPXC architecture [34], [35] which has superior modularity and upgradability/compatibility characteristics. The points

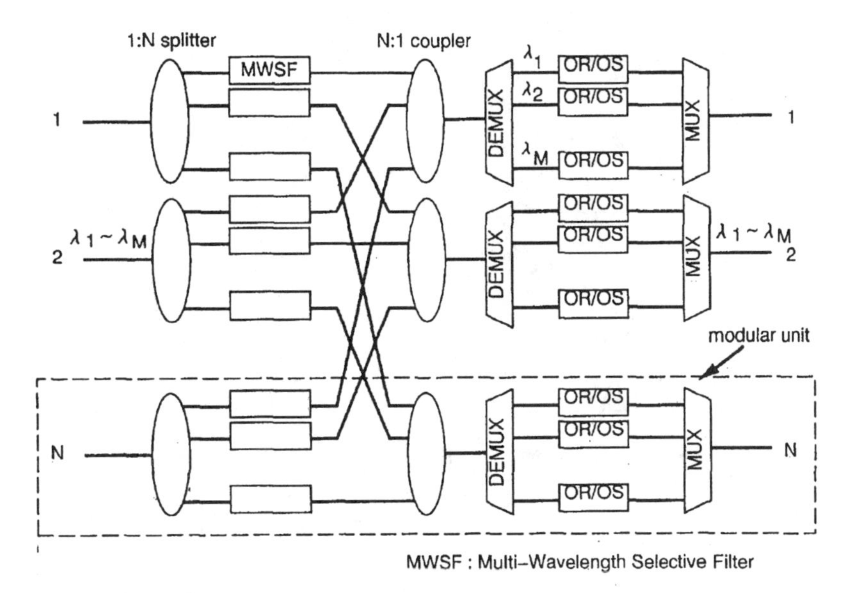

Fig. 6. WP switch architecture using MWSF.

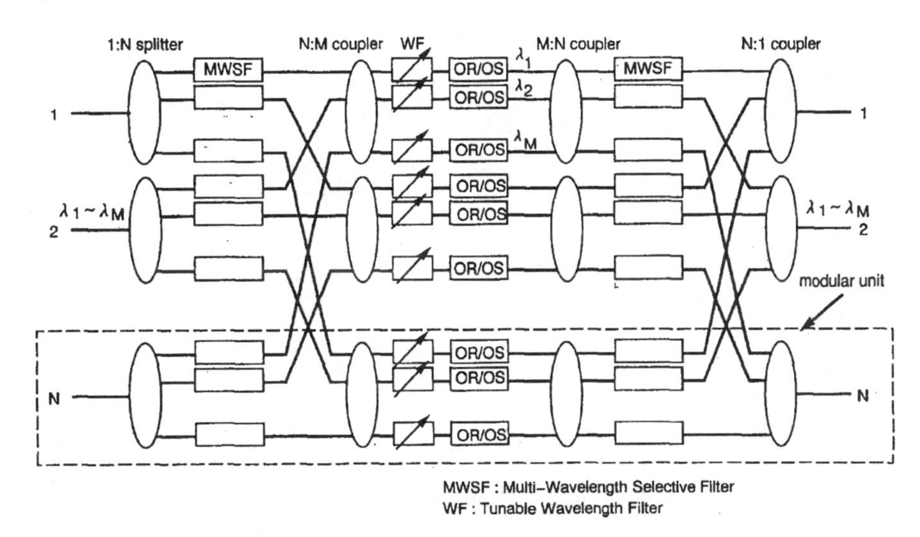

Fig. 7. VWP switch architecture using MWSF.

are summarized below. Fig. 9(a) shows the schematic configuration of the DC-switch-based OPXC, and Fig. 9(b) shows the $M \times N$ delivery and coupling switch configuration. The main feature of the DC-switch is that it allows any of the M incoming optical signals to be connected to any of the N outgoing ports. This is not possible with an SD switch. Of course, optical signals with the same wavelength must be connected to different outgoing ports.

The DC-switch-based OPXC requires fewer switch components and has superior modularity to the SD-switch-based OPXC's. Moreover, when OPXC's are upgraded from WP–OPXC to VWP–OPXC, this OPXC architecture allows easy evolution by simply replacing the OS modules with fixed wavelength LD's to those with tunable wavelength LD's, and leaving other major portions unchanged. Therefore, the compatibility of WP–OPXC and VWP–OPXC is very high. These details are provided in [34], [35].

C. Evaluation of OPXC Architectures

Considering the qualitative evaluation given above, we assessed the OPXC architectures quantitatively from the viewpoint of transport network applicability. Evaluation items are optical loss, modularity, and upgradability of OPXC's.

- **Optical Loss:** The optical loss of each OPXC architecture is shown in Fig. 10. In this evaluation, it is assumed that N (number of links) is 16, M (number of wavelengths) is 8, as discussed in Section III-A and most of the passive components of switching networks are fabricated with PLC (planer lightwave circuit) technologies [36]. PLC technologies are very suitable for mass-production and

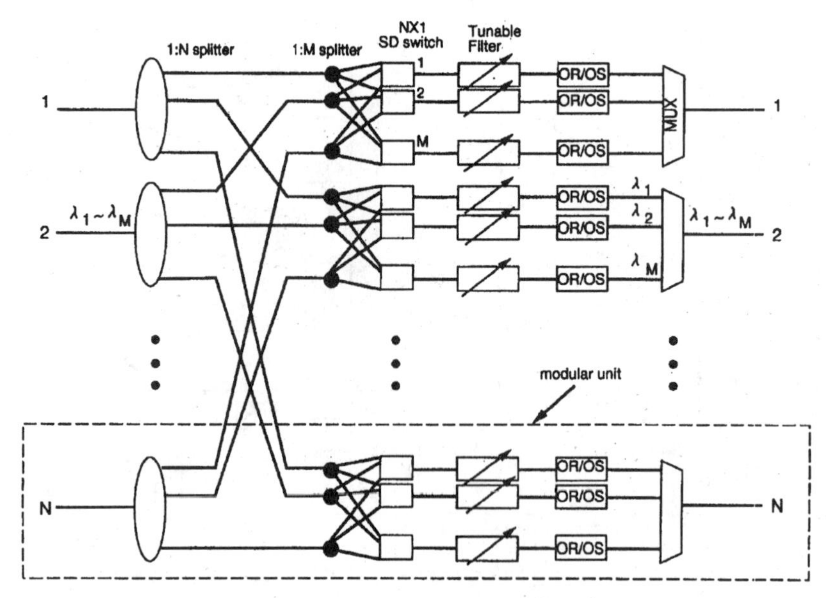

Fig. 8. OPXC architecture using parallel λ-switch.

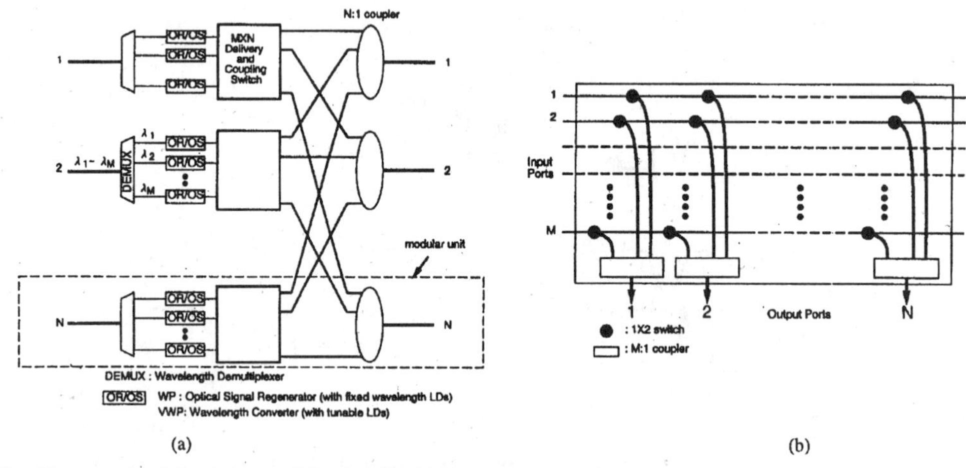

(a) (b)

Fig. 9. OPXC architecture using DC-switches. (a) DC-switch-based OPXC architecture. (b) Example of delivery and coupling switch.

enable significant cost reductions. PLC SD switches have a number of benefits, such as low loss, low crosstalk, polarization insensitivity, long term high stability and high mass productivity.

Evaluated loss values are optical components from OS to OA1 (loss-1) and from OA2 to OR (loss-2), as indicated in Fig. 10. Here, OR and OS are as indicated in Fig. 4 and Figs. 6–9. All the necessary splicing and optical connector losses determined by board and assembly configuration are included in the loss evaluations. The evaluation assumes the worst expected loss value of each optical component and the losses are simply added; no statistical treatment is applied. For a DC-switch-based OPXC, experimental values [37] are also plotted in Fig. 10. The difference between the expected loss value and the experimental one is reasonable, since the expected value simulates the worst case. This is conservative but the small difference confirms the soundness of the loss evaluation of PLC implemented switches. Details of the experimentally obtained performances of the DC-switch-based OPXC will be presented soon [37].

Optical losses must be compensated with OA's. Small loss is naturally desirable, and from the transmission characteristic viewpoint, a smaller difference between loss-1 and loss-2 is desired. The imbalance between loss-1 and loss-2 is reflected in the required characteristics of OA1 and OA2, and the imbalance decreases the possible transmission distance. This is discussed in Section IV-B.

From the viewpoint of optical loss, for the WP scheme, the SD-switch-based OPXC (WP-1) is the best and for the VWP scheme, the parallel λ-switch-based OPXC and the DC-switch-based OPXC are superior.

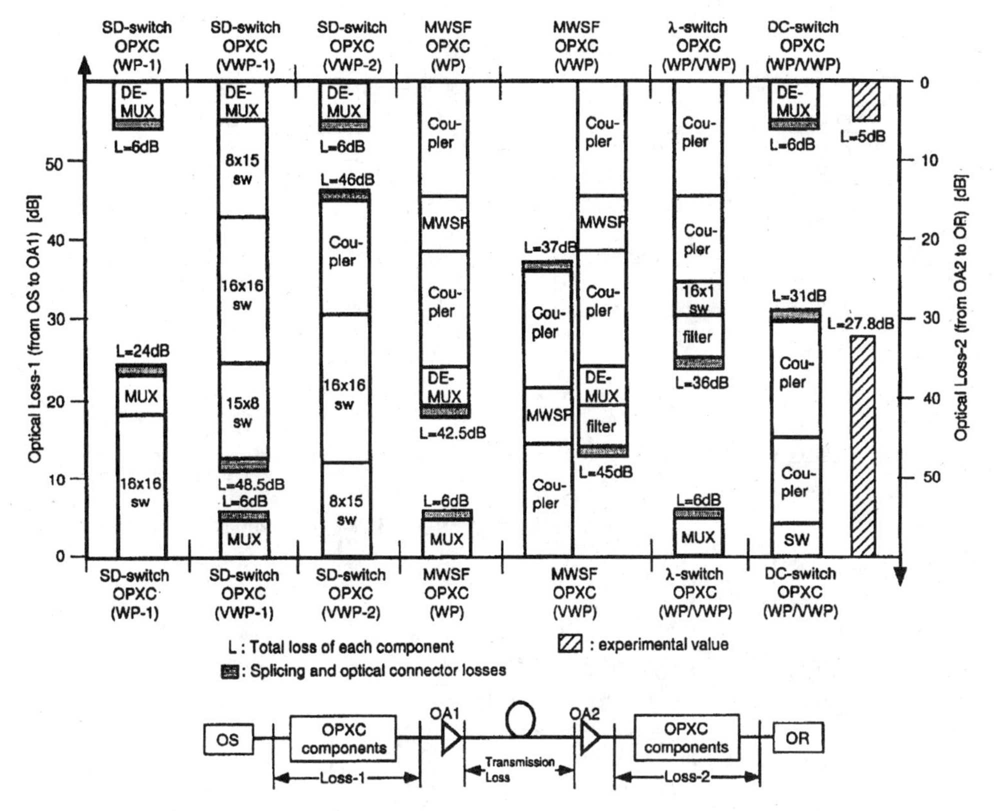

Fig. 10. Optical loss estimation ($N = 16, M = 8$).

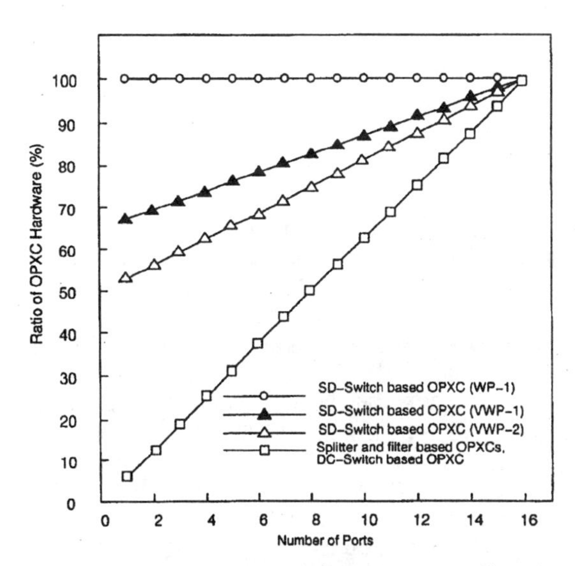

Fig. 11. Modular growth characteristics ($M = 8$, maximum $N = 16$).

- **Modularity:** Modular growth characteristics are evaluated in Fig. 11. Fig. 11 assumes eight wavelengths ($M = 8$) and maximum 16 ports ($N = 16$) and plots ratios of required switch hardware amounts. It can be easily understood that the SD-switch-based WP–OPXC (WP-1) has no modularity against input and output fiber pairs. The DC-switch-based and the parallel λ-switch-based OPXC's are shown to be the best in terms of modularity.
- **Upgradability:** With further progress in device technologies, cost-effective and practical tunable LD's will become feasible. Upgrading the DC-switch-based OPXC's from WP's to VWP's can be done by simply replacing fixed wavelength OS's with tunable wavelength OS's. This indicates that the simple replacement of the OR/OS card (package) allows a WP–OPXC evolve into a VWP–OPXC.

  The parallel λ-switch-based OPXC has similar upgradability. For the WP-OPXC, it is obvious that the tunable filters shown in Fig. 8 can be replaced with fixed wavelength filters. Therefore, upgrading the parallel λ-switch-based OPXC's from WP's to VWP's can be done by simply replacing fixed wavelength filters with tunable wavelength filters.

  Other OPXC architectures have no upgradability, so the commonality between WP and VWP–OPXC is very little. Of course, the VWP–OPXC can be used as a WP–OPXC, but the serious problem of large optical loss remains.

  The evaluation results derived thus far are summarized in Table II. Considering nonblocking capability, optical loss, modularity, upgradability and practicability of OPXC architectures, and their required optical devices,

TABLE II
COMPARISON OF OPXC'S

OPXC type	nonblocking	modularity	upgradability*[1]	loss
SD-switch (WP-1)	strictly	low	no	small
SD-switch (WP-2)	rearrangeable	low	no	small
SD-switch (VWP-1)	strictly	medium	poor	large
SD-switch (VWP-2)	strictly	medium	poor	large
MWSF (WP)	strictly	high	no	large
MWSF (VWP)	rearrangeable	high	poor	extremely large
λ-switch (WP/VWP)	strictly	high	good	medium
DC-switch (WP/VWP)	strictly	high	good	medium

*[1] The compatibility of optical components when upgrading from WP–OPXC to VWP–OPXC.

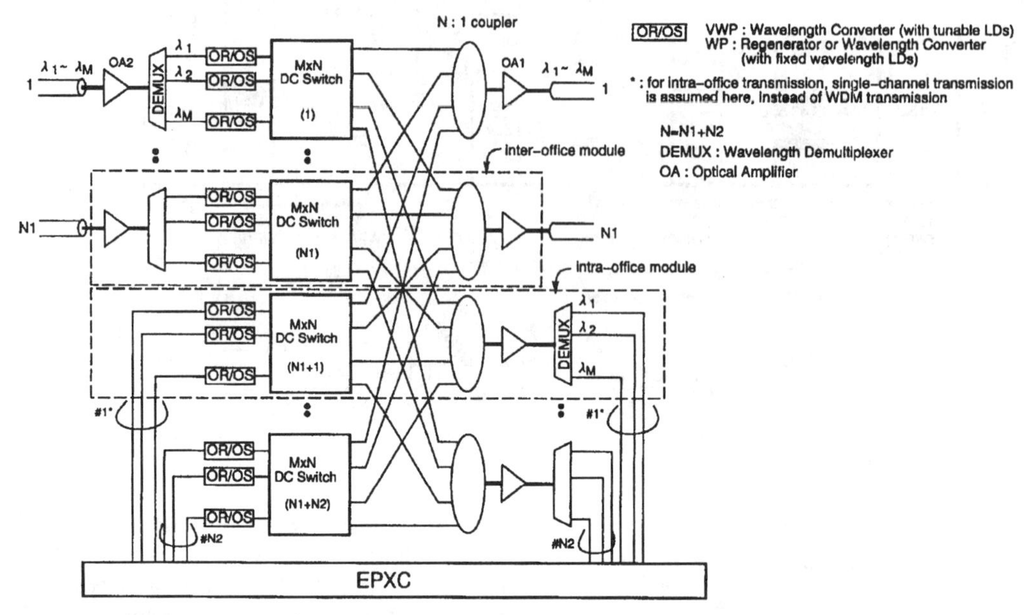

Fig. 12. OPXC node architecture using DC-switches.

the DC-switch-based OPXC architecture is the most suitable for transport network application. In the following section, we discuss the detailed configuration of the DC-switch-based OPXC node and provide transmission performance design results.

IV. OPXC NODE ARCHITECTURE BASED ON DC-SWITCHES

The detailed DC-switch-based OPXC node architecture, which includes the interface structure between OPXC and EPXC, and the transmission characteristics evaluations for global area networks are discussed in this section.

A. Interface Structure Between OPXC and EPXC

Fig. 12 shows the WP/VWP compatible OPXC node architecture based on DC-switches. We discuss here the OPXC-EPXC interface. When the WP scheme is used, the number of add/drop ports (denoted as $N2$ in Fig. 12, where each add/drop port consists of M fibers when WDM transmission was not used for intra-office transmission as shown here) might be larger than with the VWP scheme even when the number of terminating optical paths is the same. For example, two WP's are terminated at the OPXC node. The VWP scheme requires only one intraoffice module ($N2 = 1$), since two terminating VWP's can always be accommodated by one port. However, with the WP scheme, if the assigned wavelengths to the terminating WP's are the same, two intraoffice modules are required ($N2 = 2$). If we permit the arbitrary assignment of wavelengths, in the worst case, $N2$ is the same as $N1$ (because maximum $N1$ incoming WP's can be assigned the same wavelength). The ratio of WP to VWP in terms of $N2$ is plotted in Fig. 13, where arbitrary wavelength assignment for the terminating optical paths is assumed. In Fig. 13, the ratio varies with the number of terminated optical paths and multiplexed wavelengths M. In the case of $M = 8$, the WP scheme requires 50~100% more intra-office modules if special attention is not given to the terminating optical path wavelengths. This ratio can be dropped to about 10~20% by applying the optimized WP accommodation design algorithm [18], [38]. It should be noted that the discussion made above

TABLE III
SYSTEM PARAMETERS

Items	Parameters
OS	0dBm, 2.488Gb/s, NRZ, ER=13dB
OR	APD, BER=10^{-11}@−32dBm (Back-to-back)
N (links)	16
M (channels)	8
OA1	NF=7dB(DC-switch),5dB(parallel λ-switch), output power=0dBm/ch
OA2	NF=5dB(DC-switch),7dB(parallel λ-switch)
DEMUX,filter	FWHM=0.6nm、 crosstalk 25dB(2ch)

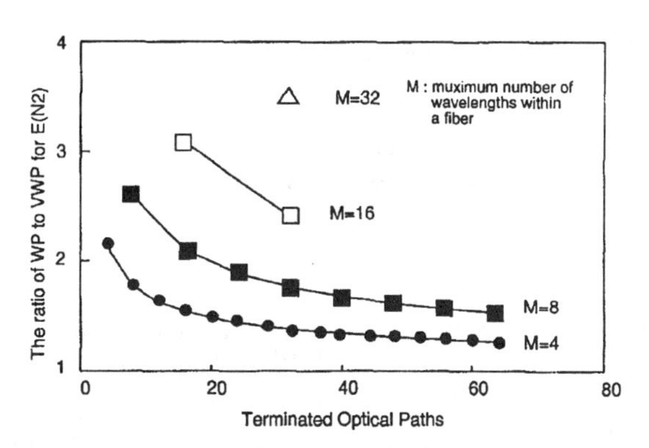

Fig. 13. Expected ratio of $N2$.

holds true to all OPXC's that have nonblocking characteristics for the OPXC-EPXC interface.

B. Transmission Characteristics

To construct a photonic transport network for global area networks offering long-haul transmission and to enable the smooth introduction of OPXC's, it is important that the optical path system is compatible with existing transmission systems in terms of transmission distance. In this section, we theoretically analyze the transmission characteristics to determine whether WP and VWP-OPXC's are compatible with the FTM-2.4G systems [39] of Japan in terms of transmission loss budget. The DC-switch-based OPXC and the parallel λ-switch-based OPXC were evaluated, since they exhibit superior features as shown in Table II. Evaluation results regarding other OPXC's are shown in [40].

Parameters used in the evaluation are summarized in Table III. The evaluated optical path transmission system is OS ⇒ switch components (loss-1) ⇒ OA1 ⇒ optical fiber (transmission loss) ⇒ OA2 ⇒ switch components (loss-2) ⇒ OR (see Fig. 10). The losses of the switch components of each OPXC used in this evaluation are those shown in Fig. 10. The output power of OA1 is limited to 0 dBm/ch, because $M = 8$ means that the total fiber input power is +9 dBm and high input powers cause strong fiber nonlinearity effects. In this evaluation, fiber nonlinearity effects are neglected, since four-wave mixing, the major cause of signal deterioration at this power level, can be avoided by differentiating fiber zero-dispersion wavelength from WDM channels.

While varying the transmission loss from 25 to 40 dB, we calculated the minimum gain and output power of OA2 needed to achieve the SNR (signal-to-noise ratio) of 24.0 dB (BER $\simeq 10^{-16}$) and 22.6 dB (BER $\simeq 10^{-11}$) at the OR. Results are plotted in Fig. 14. In the FTM-2.4G system, we must satisfy the maximum transmission loss of 32 dB [39] which corresponds to the system's transmission range of 80 km and includes all connector/splice losses and system margin. Given this transmission loss (32 dB), the OA2 of DC-switch-based OPXC requires 9 dB gain and −14 dBm output power. For the parallel λ-switch-based OPXC, the required gain is 38 dB and output power is +15 dBm. The parallel λ-switch-based OPXC requires extremely high gain and high output power OA's, the production of which is expensive. This feature is generally applied to postfix type OPXC's such as the parallel λ-switch. Therefore, it is required to put many OA's in the switching network to allow OA2's gain to be moderated, which imposes significant hardware complexity.

If the transmission loss increases, the SNR of the optical signal degrades, so the required minimum gain of OA2 increases abruptly, as seen in Fig. 14. In other words, no matter how large the OA2 gain is, the allowable transmission loss is limited given the specific BER. It was found that WP/VWP network with a transmission distance corresponding to 38.0 dB transmission loss (BER $\leq 10^{-16}$, $N = 16$, $M = 8$) can be realized with the DC-switch-based OPXC. The transmission characteristics analyzed here have been confirmed by experiments which will be presented soon [37]. Thus, the DC-switch-based OPXC systems are more suitable for transport network application than the other switch architectures discussed in this paper.

V. CONCLUSION

This paper has investigated WP and VWP-OPXC system architectures. They can be based on SD-switches, MWSF's, parallel λ-switches and DC-switches. System requirements for the application of the OPXC system to the transport network were elucidated. The system target of eight wavelengths, 16 ports and the transmission signal bit rate 2.5 Gb/s, was defined for the "main frame." OPXC architectures were evaluated in terms of required hardware amount, optical loss, modularity, and transmission characteristics to determine whether they could satisfy system requirements. The modularity of the OPXC node architecture and the upgradability of the OPXC architecture from WP's to VWP's to enable maximum

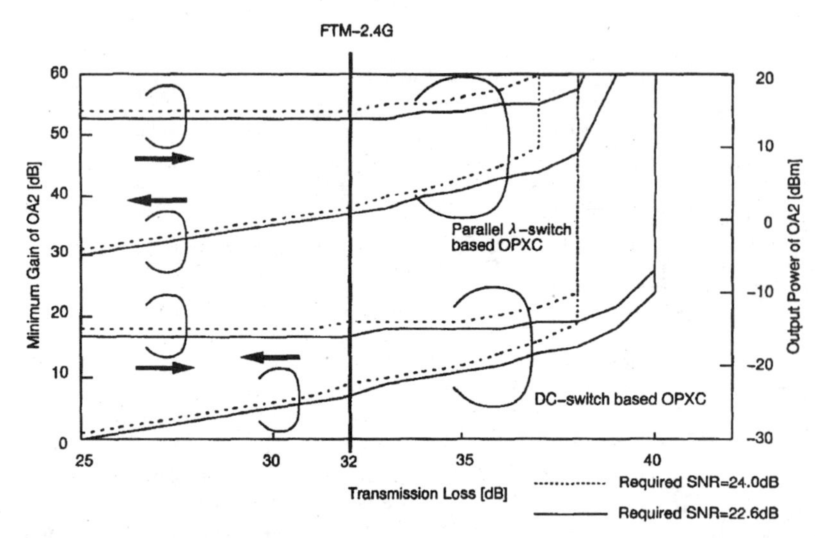

Fig. 14. Required gain and output power of OA2 ($N = 16, M = 8$).

network flexibility are some of the most important evaluation parameters for the OPXC node structure. As a result of our evaluation, it was found that the DC-switch-based OPXC architecture is the most suitable type for the transport network. The DC-switch-based OPXC system architecture provides the maximum commonality between WP and VWP–OPXC, and provides modular growth capability. This architecture allows easy fabrication with state-of-the-art optical device technologies based on PLC. Thus, only technologies suitable for massive production needed be utilized for developing a WP network. The VWP network requires improved OS modules. The DC-switch-based OPXC architecture is, thus, expected to play a key role in constructing optical path (WP and VWP) networks for the future bandwidth abundant B-ISDN.

ACKNOWLEDGMENT

The authors are grateful to T. Miki, H. Yamaguchi, and T. Matsumoto, Executive Managers of NTT, for their support of this work.

REFERENCES

[1] K. Sato, S. Okamoto, and H. Hadama, "Optical path layer technologies to enhance B-ISDN performance," in *Proc. IEEE ICC'93*, Geneva, Switzerland, May 1993, pp. 1300–1307.
[2] ——, "Network performance and integrity enhancement with optical path layer technologies," *IEEE J. Select. Areas Commun.*, vol. 12, pp. 159–170, Jan. 1994.
[3] K. Sato, S. Okamoto, and A. Watanabe, "Optical path and realization technologies," in *Proc. IEEE GLOBECOM'94*, San Francisco, CA, Nov. 1994, pp. 1513–1520.
[4] S. Okamoto and K. Sato, "Optical path cross-connect systems for photonic transport networks," in *Proc. IEEE GLOBECOM'93*, Houston, TX, Nov. 1993, pp. 474–480.
[5] G. R. Hill, "A wavelength routing approach to optical communication networks," in *Proc. IEEE INFOCOM'88*, 1988, pp. 354–362.
[6] K. C. Lee and V. O. K. Li, "A wavelength-convertible optical networks," *J. Lightwave Technology*, vol. 11, pp. 962–970, May/June 1993.
[7] ITU-T Recommendation G.803, "Architectures of transport networks based on the synchronous digital hierarchy (SDH)," Mar. 1993.
[8] S. Hasegawa, A. Kanemasa, H. Sakaguchi, and R. Maruta, "Dynamic reconfiguration of digital cross-connect systems with network control and management," in *Proc. GLOBECOM'87*, 1987, pp. 28.3.1–28.3.5.
[9] T.-H. Wu, *Fiber Network Service Survivability*, Boston, MA: Artech House, 1992.
[10] Special Issue on Broadband Optical Networks, *J. Lightwave Technol.*, vol. 11, May/June 1993.
[11] Special Issue on Dense Wavelength Division Multiplexing Techniques for High Capacity and Multiple Access Communication Systems, *IEEE J. Select. Areas. Commun.*, vol. 8, Aug. 1990.
[12] G. D. Khoe, "Coherent techniques for flexible capacity traffic and access," in *OFC/IOOC'93*, Tutorial Sessions, ThE, San Jose, CA, Feb. 25, 1993, pp. 176–225.
[13] I. Chlamtac, A. Ganz, and G. Karmi, "Lightpath communications: An approach to high bandwidth optical WAN's," *IEEE Trans. Commun.*, vol. 40, pp. 1171–1182, July 1992.
[14] N. Nagatsu, Y. Hamazumi, and K. Sato, "Optical path accommodation design," *IEICE Japan*, vol. CS93-137, Tech. Rep., Nov. 1993.
[15] N. Wauters and P. Demeester, "Wavelength requirements and survivability in WDM cross-connected networks," in *Proc. ECOC'94*, Firenze, Italy, Sept. 25–29, 1994, pp. 589–592.
[16] Y. Hamazumi, N. Nagatsu, and K. Sato, "Number of wavelengths required for optical networks considering failure restoration," in *Proc. OFC'94*, San Jose, CA, Feb. 21–26, 1994, TuE2.
[17] Y. Hamazumi, N. Nagatsu, S. Okamoto, and K. Sato, "Number of wavelengths required for constructing optical path network considering failure restoration," *Trans. IEICE Japan*, vol. J77-B-I, no. 5, pp. 275–284, 1994.
[18] N. Nagatsu and Y. Hamazumi, "WP accommodation design considering failure restoration," in *Proc. IEICE Fall Conf.*, Sendai, Sept. 26–29, 1994, vol. B-736.
[19] H. Kawaguchi, K. Oe, H. Yasaka, K. Magari, M. Fukada, and Y. Itaya, "Tunable optical-wavelength conversion using a multielectrode distributed feedback laser diode with a saturable absorber," *Electron. Lett.*, vol. 23, pp. 1088–1089, 1987.
[20] K. Inoue, "Observation of crosstalk due to four-wave mixing in a laser amplifier for FDM transmission," *Electron. Lett.*, vol. 24, pp. 1293–1295, 1987.
[21] K. Inoue and H. Toba, "Wavelength conversion experiment using fiber four-wave mixing," *IEEE Photon. Technol. Lett.*, vol. 4, pp. 69–72, Jan. 1992.
[22] S. Hata, M. Ikeda, Y. Noguchi, and S. Kondo, "Monolithic integration of an InGaAs PIN photodiode, two InGaAs column gate FET's and InGaAsP laser for optical regeneration," in *Proc. 17th Conf. Solid State Devices and Materials*, 1985, pp. 79–82.
[23] K. Nosu, H. Toba, K. Inoue, and K. Oda, "100 channel optical FDM technology and its applications to optical FDM channel-based networks," *J. Lightwave Technol.*, vol. 11, pp. 764–776, May/June 1993.
[24] ITU-T Recommendation G.707, "Synchronous digital hierarchy bit rates," Mar. 1993.

[25] ITU-T Recommendation G.709, "Synchronous multiplexing structure," Mar. 1993.
[26] I. Tokizawa, H. Ueda, and K. Kikuchi, "ATM transport system architecture and field trial," in *Proc. IEEE GLOBECOM'93*, Houston, TX, Nov. 1993, pp. 1449–1453.
[27] G. R. Hill *et al.*, "A transport network layer based on optical network elements," *J. Lightwave Technol.*, vol. 11, pp. 667–679, May/June 1993.
[28] C. Clos, "A study of nonblocking switching networks," *Bell Syst. Tech. J.*, vol. 32, no. 2, pp. 406–424, 1953.
[29] M. Okuno, K. Kato, Y. Ohmori, and T. Matsunaga, "Improved 8 × 8 integrated optic matrix switch using silica-based planar lightwave circuit," in *Tech. Dig. OEC'92*, vol. 17A4-2, 1992, pp. 264–265.
[30] D. A. Smith, J. J. Johnson, J. E. Baran, K. W. Cheung, and S. C. Liew, "Integrated acoustically tunable optical filters: Devices and applications," in *Proc. OFC'91*, 1991, p. 142.
[31] K. Aida, S. Nishi, and K. Nakagawa, "Optical protection switches for trunk transmission systems," in *Proc. ICC'88*, 1988, pp. 1–5.
[32] J. Sharony, T. E. Stern, and K. W. Cheung, "The wavelength dilation concept—Implementation and system considerations," in *Proc. ICC'92*, 1992, pp. 829–836.
[33] M. Nishio and S. Suzuki, "Photonic wavelength-division switching network using parallel λ-switch," in *Proc. PS'90*, 1990, vol. PD14B-9.
[34] S. Okamoto, A. Watanabe, and K. Sato, "A new optical patch cross-connect system architecture using delivery and coupling matrix switch," *Trans. IEICE Japan*, vol. E77-B, no. 10, pp. 1272–1274, Oct. 1994.
[35] A. Watanabe, S. Okamoto, and K. Sato, "Optical path cross-connect node architecture with high modularity for photonic transport network," *Trans. IEICE Japan*, vol. E77-B, no. 10, pp. 1220–1229, 1994.
[36] M. Kawachi, "Planer lightwave circuits for optical FDM," in *Proc. OEC'92 Tech. Dig.*, Chiba, Japan, July 1992, pp. 302–303.
[37] M. Koga *et al.*, "Design and performance of an optical path cross-connect system based on wavelength path concept," *J. Lightwave Technol.*, this issue, pp. 1106-1119.
[38] N. Nagatsu and K. Sato, "Optical path accommodation design and required cross-connect system scale evaluation," *Trans. IEICE Japan*, vol. E78-B, pp. 1339–1343, Sept. 1995.
[39] H. Tsuji, T. Tsuboi, and H. Arai, "2.4 Gb/s optical transmission system for synchronous digital hierarchy," *NTT R&D*, vol. 40, no. 5, pp. 667–678, May 1989.
[40] S. Okamoto, "Photonic transport network design utilizing optical path cross-connect system," in *Proc. IEICE Fall Conf.*, Sendai, Sept. 26–29, 1994, vol. B-738.

Chapter 10

All-Optical Pilot Systems

In the early and late 1980s, research and development work in the field of optical communications had mainly concentrated on the development of components and at most systems that can be used for high-capacity point-to-point links. The advances thus made on the components front and the maturity of the associated technologies together with the growing demand for multiuser applications has dictated a migration in recent years (starting late 1980s with a great thrust into the 1990s) toward networking. The developments on the systems front had indicated earlier on that one of the main limiting factors that hinders the development of high-speed transparent systems and networks is the optoelectronic conversion and the so-called "electronic bottleneck."

It is due to such a realization that the major players in the telecommunication and networking fields have invested heavily in the development of systems that would avoid any unnecessary electrooptic or optoelectronic conversion. This has marked the start of the "all-optical" systems and networks. It is envisaged that such systems will offer great capacity, networking flexibility, improved functionality, transparency together with ease in the upgrading and reconfiguring phases. Such networks, once implemented can have a significant economic and commercial impact.

Several major consortia have been formed to investigate such all-optical scenarios and indeed a number of pilot systems have emerged out of the efforts of these consortia. It is the aim of this chapter to examine such systems and networks and to illustrate and highlight the challenges and achievements worldwide in the field.

In the United States, the Advanced Research Projects Agency (ARPA) has funded a number of programs including the multiwavelength optical network (MONET) program, and the wideband all-optical networks consortium together with the optical networks technology consortium (ONTC). The MONET project is mainly concerned with WDM networks, the all-optical networks consortium has investigated time-division multiplexing (TDM) networks and wavelength-division multiplexing (WDM) networks, while the ONTC consortium has explored scenarios that encompass WDM, ATM, and SCM.

In Europe, the main thrust into all-optical networking comes in the form of a RACE-funded project given the title COBRA (Coherent Optical Systems Implemented for Business Traffic Routing and Access). The project represents a collaboration between several key technology providers, users, and academic institutions. Other major pilot projects in this area are also reported in this chapter. These include the multiwavelength optical network (LAMBDANET) pioneered by Bellcore, STARNET which is led by Stanford University, and LIGHTNING which represents a collaborative effort between several U.S. universities and research institutions.

The chapter starts by giving an account of the MONET program as reported by some of its key researchers in paper 10.1. The MONET program aims to integrate network architecture, advanced technology, network management, and business drives to achieve high-capacity high-performance, cost-effective reliable transparent multiwavelength optical networks. The program activities are in three parts: network architecture and economics, networking demonstrations, and supporting technologies. This first paper gives an overall view of the program and serves as a good starting point to explore other multiwavelength all-optical networks.

The chapter continues with a paper from the ONTC consortium Phase I Program (paper 10.2) which discusses the use of both wavelength-routing optical cross-connects and virtual path fast-packet ATM switches. Experimental results are reported on reconfigurable WDM networks suited for the transport and switching of diverse types of services demanding large bandwidths. Key network components such as optical amplifiers, optical switches, and multiwavelength transmitters and receivers are described. Attention is also given to the network signaling, control, and management. The paper finally highlights applications.

A report on the Research and Development in Advanced Communications in Europe (RACE) COBRA project is given by Bachus et al. in paper 10.3. The COBRA project has carried out four field trials to verify the suitability of dense WDM combined with heterodyne detection. The use of heterodyne detection has enabled this project to define optical filtering with the ability to resolve closely spaced channels, resulting in a dense WDM network possessing good multiplexing and routing functionalities. The COBRA demonstrators included a reconfigurable-node cross-connect that supports the STM-16 data rate operated by Philips Research in Eindhoven and a demonstrator at the BBC in Kingswood UK, supporting uncompressed high-definition television (HDTV) delivery within a studio environment.

The work carried out by Bellcore in the form of LAMBDANET is reported in paper 10.4 by Goodman et al. The basic network architecture is described including the three main features of LAMBDANET: the star broadcast topology, the use of a unique wavelength by each transmitting node, and the use of wavelength demultiplexing to identify the transmitting node. LAMBDANET can be used for point-to-point applications as well as point-to-multipoint applications. Several experimental configurations and results are reported which mostly are record-breaking figures. For example, a configuration is reported encompassing 18 transmitters at 1.5 Gbit/s per transmitter and 16 receivers, a bandwidth distance product of 1.56 Tbit s^{-1} km. A point-to-point WDM experiment is also reported that achieves a bandwidth distance product of 2.07 Tbit km/s.

The chapter then continues with a paper from the ARPA-funded All-Optical Networks Consortium dealing with time-division multiplexing (TDM) and wideband WDM networks. Barry et al. in paper 10.5 describe an ultrafast (100 Gbit/s) TDM network. They point out that although such networks do provide a total capacity that is the same as that of lower-rate WDM networks, the TDM approach can offer improvements in network performance in terms of user access time, delay, and throughput all depending on the user rates and statistics. In addition to describing the networking issues, the paper also gives a description of various components that have been developed and that are needed in such a system. The components described include a 100 Gbit/s soliton compression laser source, picosecond-fiber laser, picosecond optical bit-phase sensing and clock recovery circuit, together with several other optical devices having various processing and wavelength handling functionalities.

The chapter is finally concluded with a paper by Chiang et al. describing a WDM local area network (LAN) based on a physical star topology. STARNET offers the users two options; a high-speed reconfigurable packet-switched data subnetwork and a moderate-speed fixed-tuned packet-switched subnetwork. The work carried out by the optical communications research laboratory (OCRL) of Stanford University has evaluated the packet switching capabilities of STARNET when used as a fast interface and networking medium interconnecting fast computers. Test results have achieved throughputs of 680 and 571 Mbit/s, respectively, out of the 800 Mbit/s theoretical maximum of the host computer.

MONET: Multiwavelength Optical Networking

RICHARD E. WAGNER, ROD C. ALFERNESS, FELLOW, IEEE, A.A.M. SALEH, FELLOW, IEEE, AND MATTHEW S. GOODMAN, MEMBER, IEEE

(Invited Paper)

Abstract—**This paper presents an overview of the Multiwavelength Optical Networking (MONET) program and summarizes its vision. The program objective is to advance, demonstrate, and integrate network architecture and economics, advanced multiwavelength technology, and network management and control to achieve high capacity, reconfigurable, high performance, reliable multiwavelength optical networks, with scalability to national scale, for both commercial and specialized government applications. The paper describes the major research thrusts of the program including network elements, networking demonstration plans, network control and management, and architecture and economics.**

I. Introduction

THE multiwavelength optical networking (MONET) program was established to define, demonstrate, and help drive industry consensus of how best to achieve multiwavelength optical networking of national scale that serves commercial and specialized government applications. The MONET program participation includes a consortium of partners: AT&T, Bellcore, Lucent Technologies, Bell Atlantic, BellSouth, Pacific Telesis, and SBC, with participation by the National Security Agency and the Naval Research Laboratory.

The primary objective is to integrate network architecture, advanced technology, network management, and business drivers to achieve high capacity, high performance, cost effective, reliable, transparent multiwavelength optical networks that meet DOD needs in the context of systems that are commercially viable. This paper presents a brief overview of the multi-year MONET program, its goals and objectives, its networking demonstrations and its current technological directions.

The MONET program [1] has some concepts and elements that are similar to other programs [2]–[7] and are in some cases derived from those programs. The MONET program concentrates on the interoffice and long distance transport networks and assumes a hierarchical network architecture in order to achieve scalability to national size.

II. Program Objectives and Motivation

The program activities consist of three major parts: Network Architecture and Economics, Networking Demonstrations, and Supporting Technology. Taken together, these aspects of the program will help serve a dual role. The primary purpose is for commercial viability of the WDM networking technology, while providing the necessary leading edge WDM systems to meet specific DOD needs.

The focii of the networking capabilities include expanding, reconfigurable network capacity, interoperability of network management and control between administrative domains, and exploration of the extent of data rate and format transparency. These are all potential strengths of multiwavelength networking. Reconfigurable, multiwavelength networks also offer the potential to gracefully upgrade and extend present networks. The issues related to format transparency over a national scale may have important implications for both military and commercial applications: that is, the ability to transmit and route optical channels largely independent of the detailed transmission format.

There are many degrees of transparency. Current networks are opaque; that is they always regenerate the individual signals between network nodes. The simplest degree of transparency is to digital signals, i.e., transparency to intensity modulated digital signals of arbitrary bit-rate, format and protocol. A more difficult degree of transparency is to intensity modulated signals (both analog and digital) and this is considered as amplitude transparency. Full transparency would require that a network be transparent to any optical signal with amplitude, phase, or frequency modulation. Of course, there are physical limitations to transparent transmission depending on the degree of transparency, the physical transmission distance, and a number of other issues. A critical issue related to transparent transmission is the mechanism to monitor and control such transmissions through a large scale network.

The MONET program recognizes the rapid evolution in switching and transmission technologies for telecommunications networks. Today, connections are often based on circuit switching with local routing of calls. The switched telecommunications trunks today are often based on DS-3 (45 Mbit/s) digital streams, or higher data rate streams which are demultiplexed to DS-3 level for crossconnection. The majority of the routing takes place at DS-3 (or lower) rates in the deployed telecommunications network. There is a rapid evolution for long distance and intra-LATA fiber trunks to SONET data rates with electronic SONET multiplexing. The central switching offices will have SONET crossconnects that can take both asynchronous transfer mode (ATM) and synchronous time division multiplexing (STM) data streams and provide crossconnection between the various data streams.

Manuscript received April 15, 1995. This work is supported in part by the US Defense Advanced Research Projects Agency (DARPA).

R. E. Wagner and M. S. Goodman are with Bellcore, Red Bank, NJ 07701 USA.

R. C. Alferness is with Bell Laboratories, Lucent Technologies, Holmdel, NJ 07733 USA.

A. A. M. Saleh is with AT&T Research, Holmdel, NJ 07733 USA.

Publisher Item Identifier S 0733-8724(96)04696-8.

Reprinted with permission from *IEEE Journal of Lightwave Technology*, Vol. 14, No. 6, pp. 1349-1355, June 1996.

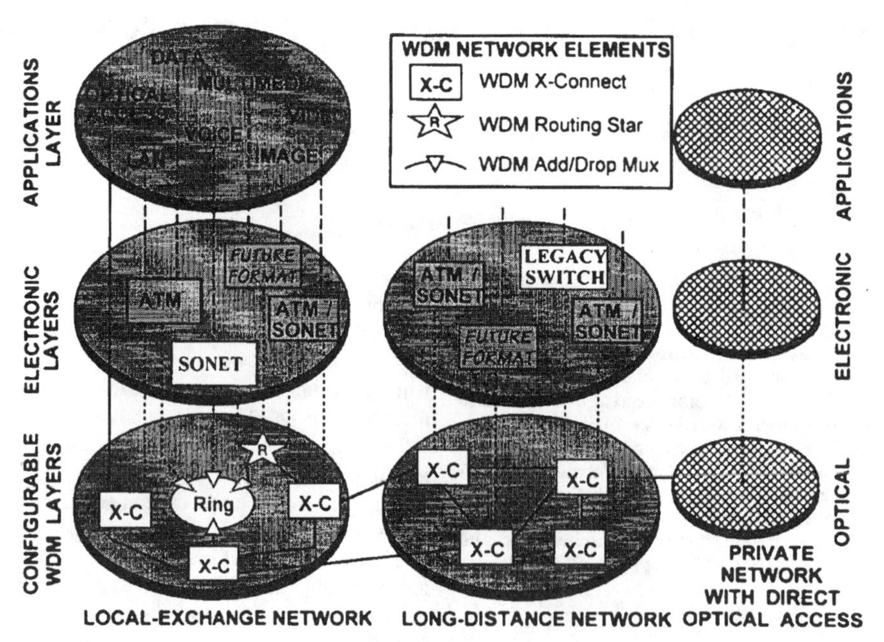

Fig. 1. The MONET perspective (MONET, electronic, and application layers).

In the future, we see an additional layer being added, with wavelength multiplexers and crossconnects (in analogy to the SONET layers). These WDM layers will have the ability to interconnect the data streams in a format independent way, providing a "virtual fiber" for connections. In addition, because of the higher tributary data rates of the WDM multiplexers and crossconnects, the amount of electronic demultiplexing and crossconnection equipment will be reduced.

A high level perspective of the MONET goal is shown in Fig. 1. This figure depicts application layers in several different administrative domains, which are interconnected (electronically) to electronic layers in the hierarchy. These electronic layers, supporting various differing formats, are in turn supported by a reconfigurable optical network layer. Note that this MONET perspective explicitly allows the use of direct optical access from the application layer to the optical layer.

The *transparency* of the MONET layer allows direct optical access and the support of differing transmission formats. Modern telecommunications infrastructures are usually composed of several generations of equipment supporting differing transmission bit rates and formats, since the evolution of these networks occurs gradually. Transparency allows several generations of equipment to coexist simultaneously using the same fiber infrastructure. Transparency thus enables graceful evolution as the formats change in the future.

The MONET program features *reconfigurability* at the multiwavelength optical level. This is important for several of the following reasons. First, it enables restoration at the optical layer for fiber cuts and node failures. Reconfigurability also allows automated provisioning, including rapid provisioning of transparent wavelength channels to build, for example, high capacity configurable private networks for government applications. Reconfigurability of the aggregate capacity is particularly important in telecommunications infrastructures, since the provisioning of significant additional capacity might otherwise be expensive and time consuming. For example, the MONET layer could provide additional capacity to an area after a natural disaster, or in the case of a short term high telecommunications demand event such as a Worlds Fair or Olympic games. At any point in time, there may be several different transmission formats/data rates in use in the network. The MONET approach allows these to all be in use *and routed* over the same fiber infrastructure. In particular, the electronic layer need not be SONET, although a SONET layer is one of particular interest to the MONET program.

Critical to the success of multiwavelength optical networking is the *interoperability* of WDM networks in different administrative domains. These might be, for example, domains operated by different telecommunications operating companies, local exchange carriers, interexchange carriers, or different nations.

III. The MONET Networking Demonstrations and Field Experiment

The MONET networking concepts and technology will be demonstrated in a set of experimental networks consisting of a New Jersey Network, a long distance networking link, and a Washington area network (Fig. 2). Three interconnected MONET testbeds in New Jersey will form the MONET New Jersey Network. The MONET Washington area network, interconnecting Bell Atlantic's Silver Spring laboratory, the Naval Research Lab (Washington, DC) and the National Security Agency (Ft. Meade, MD), is designed to provide a multiwavelength platform in another local exchange area. The Washington area network will be interconnected to the New Jersey Network to provide links for multiwavelength experiments from a local exchange network (in the Wash-

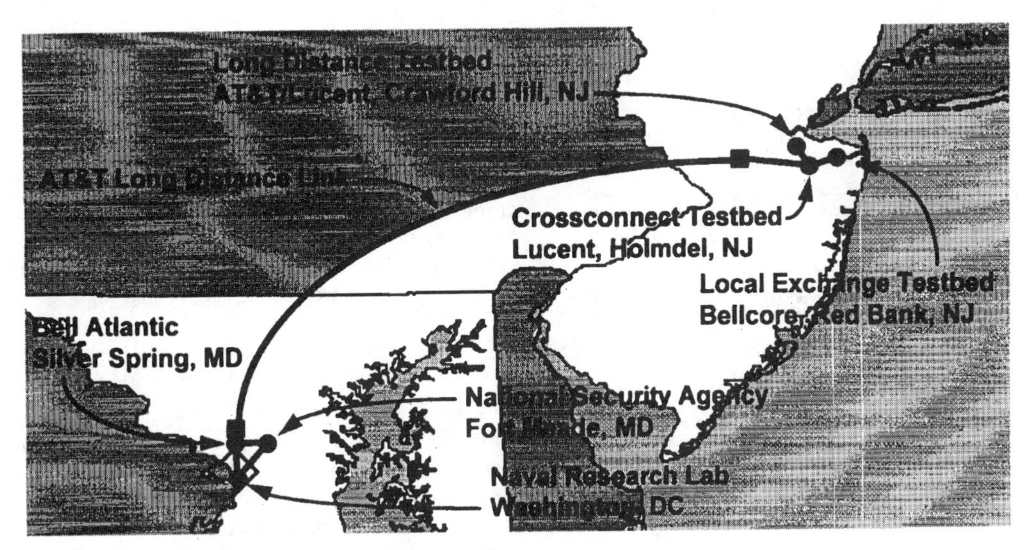

Fig. 2. The MONET field experiment network.

ington area) to a local exchange network (in NJ) over a long distance multiwavelength connection. This will provide a flexible testbed for experiments on network reconfigurability and interoperability over multiple administrative domains.

IV. The MONET Network Elements

To make optical networking cost effective, it is important to determine a minimal set of WDM network building blocks which would provide the functionality and flexibility needed in a WDM network infrastructure. This functionality includes multiwavelength sources and destinations, wavelength adding and dropping, multiwavelength optical gain and gain compensation, multiwavelength routing and crossconnection functions, and wavelength adaptation and interchange functions. In addition to this functionality, the network building blocks also need to have optical monitoring and in some cases optical protection switching.

In the MONET program the network building blocks which implement this functionality are called the MONET network elements. A network element in the multiwavelength optical layer network accepts a multiwavelength signal on the transport interface (TI) and interfaces with the client network (if appropriate) at the MONET client interfaces. In the MONET program, the transport interfaces support eight wavelengths per fiber and the client interfaces support one wavelength per fiber. Importantly, all network elements include bidirectional multiwavelength transmission interfaces and support the network management and control system functions and hence can be monitored and controlled. Thus, network elements are not "bare optical devices" but rather are designed with the appropriate monitoring and control that can be used to construct networks.

In the MONET program, there are currently definitions for seven different types of network elements. These are as follows.

1) A Wavelength Terminal Multiplexer (WTM): The WTM is a network element that accepts a multiwavelength signal from a transport facility and demultiplexes it into single wavelength signals. The WTM also multiplexes the single wavelength signals from the client network and creates a multiwavelength signal that interfaces with the transport facility.

2) A Wavelength Amplifier (WAMP): A network element whose primary function is to provide optical gain. It accepts bi-directional multiwavelength signals and amplifies all the multiwavelength signals.

3) A Wavelength Add/Drop Multiplexer (WADM): A network element that accepts multiwavelength signals from a transport facility and is capable of selectively adding, removing (dropping) or passing through the constituent single wavelength signals.

4) A Fiber Crossconnect (FXC): Which accepts multiwavelength signals and provides fiber-by-fiber crossconnection of these. It also accepts single wavelength signals and provides single wavelength crossconnection, although single wavelength and multiwavelength signals cannot be intermixed.

5) A Wavelength Selective Crossconnect (WSXC): Which is a network element that accepts multiwavelength signals from a transport interface, as well as single wavelength signals at the client interfaces, and crossconnects individual wavelengths. This network element is nonblocking in any wavelength layer, but it is blocking from one wavelength layer to another.

6) A Wavelength Interchanging Crossconnect (WIXC): Which is a network element that accepts multiwavelength signals from a transport interface and can crossconnect signals with wavelength interchange when necessary. This network element is nonblocking between fibers and between wavelength layers.

7) A Wavelength Router (WR): A network element with multiple transport interfaces that internally routes multiwavelength signals among specific fixed transport interfaces. It is equivalent to a WSXC with a fixed routing table.

Fig. 3 illustrates the functionality of one of the MONET network elements, the WADM. Note that the multiwavelength signals are indicated on the MONET Transport Interfaces.

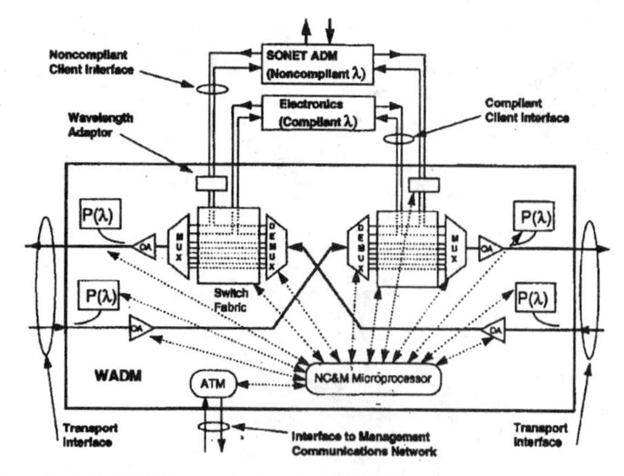

Fig. 3. A MONET network element (WADM).

TABLE I
THE MONET WAVELENGTHS

MONET Channel Number	Absolute Frequency (THz)	Wavelength nm
1	193.500	1549.32
2	193.300	1550.92
3	193.100	1552.52
4	192.900	1554.13
5	192.700	1555.75
6	192.500	1557.36
7	192.300	1558.98
8	192.100	1560.61

The interface to the network management and control system is also indicated. This network element has the capability to interface to existing equipment, which does not have wavelengths matching the MONET set, through an interface called a noncompliant client interface (NCI). This is necessary to provide an evolutionary path for introducing the MONET layer. Alternately, the network element has the capability to interface to equipment that has wavelengths in the MONET defined set through a compliant client interface (CCI).

A. MONET Compliant Wavelengths

The MONET program has defined a set of eight wavelengths that are called the MONET compliant wavelengths. They are shown in Table I. These wavelengths were selected for several reasons. First, they are consistent with general trends in the ITU standards bodies. Second, they are at a frequency spacing of 200 GHz (about 1.6 nm) which allows for dense wavelength packing, while challenging the WDM technologies for sources and demultiplexing filters. The set includes the ITU Reference Frequency near 1552 nm. Within the MONET optical layer, these are the only optical wavelengths that are supported for transmission by the network elements. This set of wavelengths was chosen due to its position in the Er-doped fiber amplifier window and the availability of sources and filters.

B. Wavelength Interchange and Adaptation

Other wavelengths (noncompliant wavelengths) are accommodated through wavelength adaptation in the noncompliant client interfaces. We make a distinction between wavelength interchange (or conversion) and wavelength adaptation. In particular, by wavelength interchange, we mean interchanging an information signal from one MONET compliant wavelength to another compliant wavelength in a way that offers the maximum reasonable transparency to signal format. Wavelength adaptation, on the other hand, involves moving a signal from a noncompliant wavelength (in the 1300 nm band, for example) to a MONET compliant wavelength. The level of format transparency in this operation and the need for the inverse operation are actively being investigated.

V. NETWORK CONTROL AND MANAGEMENT

This program incorporates some management and control functions in each of the network elements. It also incorporates overall network management and control systems for each of the testbeds. Since the NC&M is an essential and often costly portion of the network, it is necessary to understand the new flexibility that is offered by WDM approaches as well as the additional complexity that is required to manage it. The network management communications itself forms a communications hierarchy. This includes a component layer which provides NC&M communications within a network element, a network element layer which provides NC&M communications between network elements, and a network management layer which provides NC&M between architectural units such as rings, or crossconnect meshes.

The network control and management system needs to address three types of issues. The first involves *configuration management*, namely the initial provisioning of the facilities and the reconfiguration of the facilities at the multiwavelength optical layer. The second involves *performance management*, which involves the monitoring of network element hardware performance, and the monitoring of the connection performance. Third, the NC&M needs to be able to perform *fault management*. Fault management is required to maintain network integrity and robustness and to provide for restoration, for example, in the case of a fiber cut.

Two similar approaches to NC&M systems based on different platforms are being developed for the MONET testbeds. These NC&M systems will enable wavelength connections to be setup and released over multiple administrative domains. This is critical, since national scale telecommunications networks usually require interworking among multiple administrative domains. An administrative domain usually does not provide its internal structure to be known outside of its domain; often, details such as network topology, link capacities, and network element capabilities are considered as proprietary. As a result, the capabilities of a particular administrative domain must be exposed to adjacent domains which are different from those available internally. Thus the network management and control problem for the MONET multiwavelength network is significant both in the conceptual complexity that is being addressed, as well as the ability of these systems to scale to national size.

The MONET NC&M system provides for network management within each administrative domain, as well as network management based connection set-up and release over two administrative domains. This level of network management

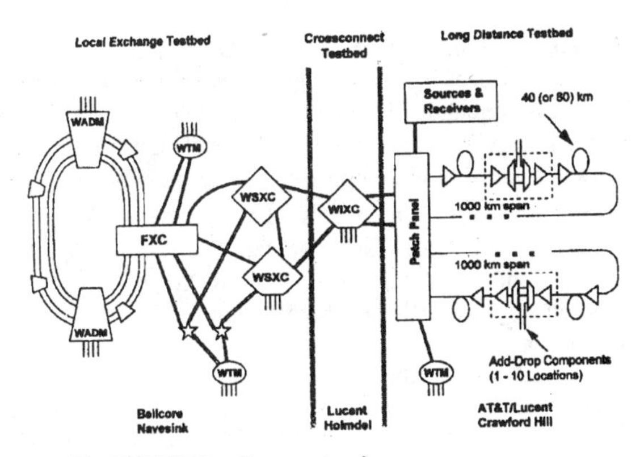

Fig. 4. The MONET New Jersey network.

will include intra-domain network configuration and connection setup, as well as fault isolation and network performance management. The network management and control will be demonstrated for various physical topologies including ring, crossconnect mesh and star networks in the New Jersey Network.

In the Washington area network, basic signaling and limited customer control over the network will also be demonstrated. Finally, the MONET network will interwork with the existing SONET-based ATDNet [8] in the Washington area.

VI. The MONET New Jersey Network

The MONET program includes three testbeds in the New Jersey area. There is a Long Distance Testbed at Bell Laboratories, Lucent Technologies/AT&T in Crawford Hill, NJ; a multiwavelength optical crossconnect testbed at Bell Laboratories, Lucent Technologies in Holmdel, NJ, and a Local Exchange Testbed at Bellcore's Navesink Research and Engineering Center in Red Bank, NJ. All three of these testbeds are currently under construction. The three independent testbeds will be interconnected by Bell Atlantic optical fibers to form the MONET New Jersey Network (Fig. 4).

The long distance testbed will provide a 2000 km dispersion-managed transmission path over which to study multiwavelength networking issues, such as the impact of fiber nonlinearities and dispersion [9], or format transparency over such distances. With various configurations of this testbed, it will be possible to perform experiments ranging from fundamental investigations of the physical limits of WDM transmission, to tests of the viability of various schemes for management of an extended network. By originating a signal in the local exchange testbed, routing it through the long distance testbed and back again to the local exchange testbed, the long distance testbed will provide the ability to extrapolate networking research results to networks of national scale.

The crossconnect testbed will provide crossconnection between the local exchange testbed (LEC) and the long distance testbeds. The crossconnect testbed will be where wavelength interchanging crossconnection will be studied, where components from the enabling technologies effort can be evaluated and where network management and control software can be integrated and tested with the crossconnect and WADM network elements.

The local exchange testbed will test multiwavelength technologies and networking concepts using a number of different local exchange network topologies, including multiwavelength ring experiments, redundant wavelength routed star experiments, and multiwavelength mesh experiments. It will demonstrate the interoperability of these different topologies, and explore the size limitations and interconnection of them. Further, it will explore the issues associated with reconfiguration of the network, and the network management and control of a variety of network architectures.

The New Jersey network will provide a large scale experiment using the testbeds described above, and demonstrating multiwavelength connections over two different NC&M administrative domains. This provides an opportunity to demonstrate real telecommunications networks which must interoperate over multiple domains. The experiments will address and expose potential new limitations imposed by various networking architectures. Several examples of such limitations are already becoming apparent: Component manufacturing tolerances have been studied by demonstrating an eight-node WDM ring [10]. Fast power transients that occur in cascaded network elements during reconfiguration have been identified [11]. Filter magnitude and phase characteristics stemming from cascaded WDM multiplexers and demultiplexers have been shown to limit the allowable source absolute frequency misalignments [12]. The identification of such effects, along with other as yet undiscovered issues, is crucial for the successful design and deployment of a WDM networking layer infrastructure.

VII. Multiwavelength Technology Research

The functionality of the network elements provides the direction for the technology research within the MONET program. The specific network elements requirements provide the technological targets and helps determine the technology choices. Driven by the network elements, the MONET program is investigating a number of multiwavelength technology research directions. These include:

- Four and eight wavelength 2.5 Gb/s InP laser arrays [13],
- InP photodetector/preamplifier arrays [14],
- Wavelength selectable laser sources [15],
- $LiNbO_3$ polarization independent switch arrays [16],
- Wavelength routing space switches [17],
- InP switch arrays [18],
- Wavelength interchanging technology [19], [20],
- Silica-on-silicon multiplexers and routers [21],
- Gain flattened Er-doped fiber amplifiers.

Fig. 5 summarizes how the associated enabling subsystems and technologies are expected to be used in the various network elements.

Research in these technological directions will be used to select WDM components for use in the network elements described above. In particular, optical monitoring will be incorporated in the network elements, along with network management and control interfaces; thus these supporting

Network Elements					Enabling:	
Transport Elements		Networking Elements				
Terminal	Repeater	Add/Drop	Router	Cross-Connect	Subsystem	Technology
✓		✓		✓	Multi-λ Tx	InP
✓		✓		✓	Receiver Arrays	InP
✓		✓		✓	λ Mux/Dmux	SIOB / InP
✓	✓	✓		✓	Amplifier	Fiber / InP
	✓			✓	Dispersion Comp.	Fiber
✓		✓	✓	✓	λ Monitor	SIOB / InP
	✓			✓	Gain Equal.	Passive / InP
			✓		Router	SIOB / InP
		✓		✓	Space Switch	LN / InP / LC
				✓	λ Converter	InP / Fiber

Fig. 5. Enabling subsystems and technologies for MONET network elements.

technologies will not exist as laboratory demonstrations alone, but will be integrated into the network elements which will be part of the networking demonstration infrastructure.

VIII. Network Technology Simulations

In a multiwavelength network, the lightwave signal passes through a number of concatenated optical components such as WDM multiplexers, demultiplexers, amplifiers, and noise limiting bandpass filters all of which can impact the signal by either filtering it or adding noise, or both. The concatenation of optical elements makes the system susceptible to passband misalignments arising from device imperfections, temperature variations and environmental changes. The spectral lines from the laser sources may also have misalignments with the optical components due to manufacturing tolerances or operations specific conditions. In the local exchange portion of the network an optical path may traverse two or three dozen such filtering elements. In the Long Distance network, there may be as many as 50 amplifiers to span a significant portion of a national scale network.

Thus, the MONET program includes activities on the detailed simulation of WDM component performance, network element performance, and end-to-end optical network path performance for various networking topologies such as ring and mesh networks. These simulations will guide understanding of the impairments due to filter misalignments, amplifier noise accumulation and dispersion.

IX. Economics and Architecture Studies

Fundamental to the commercial viability of WDM networking is the understanding of the appropriate architectures and where they are economically suited—deployment of a new infrastructure is certainly driven by economic issues. Thus, MONET has architectural studies addressing both long haul and local exchange areas.

The key research questions for the architectural studies are related to the primary program thrusts:

- *Transparency:* What are the limits, and how much transparency is needed?
- *Scalability:* Can these networks grow gracefully?
- *Reconfigurability:* Where is crossconnection and reconfigurability needed? How large do crossconnects need to be? Is wavelength interchange required?
- *Interoperability:* How do these networks interoperate and how difficult is the complexity management?

Questions addressed in these studies include the structure of long distance and local exchange networks, and the optimal mix of current technology (OC-48), WDM point-to-point systems, and multiwavelength networking topologies such as rings, stars, and meshes.

Initial results from the local exchange carrier economic studies are encouraging. Using a modified version of the commercially available SONET toolkit [22], three metropolitan areas were studied under various scenarios for the growth of bandwidth demand. Under all these scenarios, and with relatively conservative assumptions, WDM networking provided an advantage over non-WDM approaches, ranging from around 15% to around 35%.

Studies of economic benefits prior to MONET indicated that savings from systems employing WDM resulted primarily from savings in the SONET multiplexing equipment. Importantly, the new studies show that by adding the reconfigurable multiwavelength optical layer, there can be savings at the physical fiber layer and from the electronic multiplexing and demultiplexing equipment. That is, these studies indicate that by spending modest resources to incorporate a multiwavelength layer, network operators can lock in savings now, as well as be positioned for further savings as the networks evolve to accommodate larger capacity demands.

Understanding the economic advantages of WDM in the network infrastructure is a major goal of this program. Along with the supporting technology and networking demonstrations, these studies will help clarify the appropriate path for technology adoption and deployment.

X. Challenges for Multiwavelength Optical Networking

The MONET program is an attempt to address simultaneously multiwavelength networking technology, the demonstration of a set of WDM network element building blocks, and networking testbeds, in order to gain an understanding of national scale WDM networks. The key elements addressed in this program are the scalability of the network, reconfigurability at the WDM optical level, exploitation of the format transparency of the WADM's and crossconnects, and interoperability of the network control and management across different administrative domains.

In order for multiwavelength optical networking to be commercially viable, it must provide an advantage to all the major stake holders, namely the network operators, the equipment manufacturers, and the network management operating system software vendors. Our collaboration has partners that specifically combine all these diverse networking perspectives. Thus, MONET has the opportunity to drive WDM from the laboratory environment into a large scale field experiment, involving high bandwidth users, enabling an important step toward commercial viability.

ACKNOWLEDGMENT

This program represents the work of researchers from each of our Consortium partners and we gratefully acknowledge the strong participation of many workers from each of them in this work.

REFERENCES

[1] A. A. M. Saleh, "Overview of the MONET Multiwavelength Optical NETworking Program," in *Proc. Optic. Fiber Conf. (OFC'96)*, San Jose, CA, 1996, ThI3.

[2] C. A. Brackett, A. S. Acampora, J. Sweitzer, G. Tangonan, M. T. Smith, W. Lennon, K. C. Wang, and R. W. Wilson, "A scalable multiwavelength multihop, optical network: A proposal for research on all-optical networks," *J. Lightwave Technol.*, vol. 11, p. 736, 1993.

[3] G. R. Hill, P. J. Chidgey, F. Kaufhold, T. Lynch, O. Sahlen, M. Gustavsson, M. Janson, B. Lagerstrom, G. Grasso, F. Meli, S. Johansson, J. Ingers, L. Gernandez, S. Rotolo, A. Antonielli, S. Tebaldini, E. Vezzoni, R. Caddedu, N. Caponio, F. Testa, A. Scavennec, M. J. O'Mahony, J. Zhou, A. Yu, W. Sohler, U. Rust, and H. Herrmann, "A transport network layer based on optical network elements," *J. Lightwave Technol.*, vol. 11, pp. 667–679, 1993.

[4] S. B. Alexander *et al.*, "A precompetitive consortium on wide-band all optical networks," *J. Lightwave Technol.*, vol. 11, 1993.

[5] M. S. Chen, N. R. Dono, and R. Ramaswami, "A Media access protocol for packet switched wavelength division multiaccess metropolitan area networks," *IEEE J. Select. Areas Commun.*, vol. 8, pp. 1048–1057, 1990.

[6] A. Watanabe, S. Okamoto, and K. Sato, "Optical path crossconnect node architecture with high modularity for photonic transport network," *Trans. IEICE Japan*, vol. E77-B, no. 10, pp. 1220–1229, 1994.

[7] J. Gamelin, M. S. Goodman, J. Jackel, B. Pathak, G. K. Chang, W. J. Tomlinson, R. Cordell, C. E. Zah, T. P. Lee, C. A. Brackett, C. Dreze, D. Pollex, J. Sitch, H. Willemsen, K. Pedrotti, K. C. Wang, R. Walden, W. Stanchina, D. Fritz, R. Ade, R. Hobbs, R. Haigh, K. McKammon, and W. Lennon, "8-Channel reconfigurable WDM networking demonstration with wavelength translation and electronic multicasting at 2.5 Gbit/s," in *Proc. Optic. Fiber Conf. (OFC'96)*, San Jose, CA, 1996, post-deadline paper PD-29.

[8] D. L. Endicott, Jr., "The advanced technology demonstration network (ATDnet): An executive summary," unpublished, 1995.

[9] F. Forghieri, R. Tkach, A. R. Chraplyvy, and A. M. Vengsarkar, "Dispersion compensating fiber: Is there merit in the figure of merit?," in *Proc. Optic. Fiber Conf. (OFC'96)*, San Jose, CA, 1996, paper ThM5.

[10] R. S. Vodhanel, F. Shehadeh, J. C. Chiao, G. K. Chang, C. Gibbons, and T. Suzaki, "Performance of an 8-wavelength 8-Node WDM ring network experiment with 80 Gb/s capacity," in *Proc. Optic. Fiber Conf. (OFC'96)*, San Jose, CA, 1996, post-deadline paper PD28.

[11] J. L. Zyskind, Y. Sun, A. K. Srivastava, J. W. Sulhoff, A. J. Lucero, C. Wolf, and R. W. Tkach, "Fast power transients in optically amplified multiwavelength optical networks," in *Proc. Optic. Fiber Conf. (OFC'96)*, San Jose, CA, 1996, post-deadline paper PD-31.

[12] N. N. Khrais, F. Shehadeh, J. C. Chiao, R. S. Vodhanel, and R. E. Wagner, "Multiplexer eye-closure penalties for 10 Gb/s signals in WDM networks," in *Proc. Optic. Fiber Conf. (OFC'96)*, San Jose, CA, 1996, Post-deadline paper PD-33.

[13] C. E. Zah, M. Amersfoort, F. Faivre, A. Rajhel, P. S. D. Lin, N. C. Andreadakis, R. Bhat, C. Caneau, M. A. Koza, B. Pathak, J. Gamelin, and T. P. Lee, "InP-based multiwavelength laser arrays with integrated combiners for WDM systems," in *Proc. LEOS Annu. Meet.*, San Francisco, CA, 1995, pp. 239–240.

[14] S. Chandrasekhar, L. D. Garrett, L. M. Lunardi, A. G. Dentai, C. A. Burrus, and E. C. Burrows, "Investigation of crosstalk performance of eight-channel p-i-n/HBT OEIC photoreceiver array modules," *IEEE Photon. Technol. Lett.*, to be published.

[15] M. G. Young, T. L. Koch, U. Koren, G. Raybon, A. H. Gnauck, B. I. Miller, M. Chien, K. Dreyer, R. E. Behringer, D. M. Tennant, and K. Feder, "Six-channel WDM transmitter module with ultra-low chirp and stable wavelength selection," in *Proc. 21 European Conf. Opt. Commun., (ECOC'95)*, Brussels, 1995, post-deadline paper Th.B.3.4.

[16] E. J. Murphy, T. O. Murphy, A. F. Ambrose, R. W. Irvin, B. H. Lee, P. Peng, G. W. Richards, and A. Yorinks, "16 × 16 Strictly nonblocking guided-wave optical switching system," *J. Lightwave Technol.*, vol. 14, pp. 352–358, Mar. 1996.

[17] J. S. Patel and Y. Silberberg, "Liquid crystal and grating based multi-wavelength crossconnect switches," *Photon. Technol. Lett.*, vol. 7, pp. 514–516, 1995.

[18] J. E. Zucker, "Advanced semiconductor switches," in *Proc. 7th European Conf. Integr. Opt. (ECIO'95)*, 1995, pp. 37–40.

[19] S. J. B. Yoo, C. Caneau, R. Bhat, and M. A. Koza, "Transparent wavelength conversion by difference frequency generation in AlGaAs waveguides," in *Proc. Optic. Fiber Conf. (OFC'96)*, San Jose, CA, 1996, pp. 129–131.

[20] X. Pan, J. M. Wiesenfeld, J. S. Perino, T. L. Koch, G. Raybon, U. Koren, M. Chien, M. Young, B. I. Miller, and C. A. Burrus, "Dynamic operation of a three-port, integrated Mach–Zehnder wavelength converter," *IEEE Photon. Tech. Lett.*, vol. 7, pp. 995–997, 1995.

[21] C. Dragone, C. A. Edwards, and R. C. Kistler, "Integrated optics $N \times N$ multiplexer on silicon," in *IEEE Photon. Technol. Lett*, vol. 3, pp. 896–899, Oct. 1991.

[22] S. Cosares, D. N. Deutsch, I. Saniee, and O. J. Wasem, "SONET toolkit: A decision support system for designing robust and cost-effective fiber-optic networks," *Interfaces*, vol. 25, pp. 20–41, 1995.

Multiwavelength Reconfigurable WDM/ATM/SONET Network Testbed

GEE-KUNG CHANG, SENIOR MEMBER, IEEE, GEORGIOS ELLINAS, JOHN K. GAMELIN, M.Z. IQBAL, AND CHARLES A. BRACKETT, FELLOW, IEEE

(Invited Paper)

***Abstract*—This paper describes the multiwavelength reconfigurable all-optical network testbed designed and constructed as a part of the Optical Networks Technology Consortium (ONTC) Phase I program with a group of working members consisting of Bellcore, Columbia University, Hughes Research Laboratories, Northern Telecom, Rockwell Science Center, Case Western Reserve University, United Technology Research Center, Uniphase Telecommunications Product, and Lawrence Livermore National Laboratories. The architecture of the testbed and its nodes is presented together with a description of the key component technologies. Network control and management is discussed along with applications and services.**

I. INTRODUCTION

RECENTLY, there has been a surge in research and commercialization activities for wavelength-division-multiplexed (WDM) systems. In 1995, both AT&T and MCI announced plans to deploy WDM systems to enhance the transmission capacity of long-haul networks. This was followed by announcements from several companies including Pirelli and Lucent Technologies of 4–8 channel dense-WDM telecommunication equipment for broadband networks. These early WDM systems provide simple point-to-point connectivity with enhanced transmission bandwidth at lower cost. With the exception of optical line amplifiers, these point-to-point systems all require expensive electronics for grooming of individual trunk channels. This approach becomes increasingly inefficient (and costly) as the network size and capacity scale.

Future high-speed, high capacity optical communication systems will have to handle two particular types of user services: multimedia services to multiple users, and "burst mode" select-cast data transport from user-to-user or from region-to-region. Most of the traffic in this kind of network will be in the "by-pass" mode[1] [1]. We believe that a dynamic reconfigurable WDM channel add/drop function at the user nodes can efficiently process the information for these types of user services, with minimum electronics at the access node, at lower system cost.

We report here results on experimental reconfigurable WDM networks which are well suited to transport diverse types of information services and switch large bandwidth on demand [6]–[8]. We have demonstrated an optical network that exploits WDM optical switching and transport device technologies to create a network which is rearrangeable at the optical layer and transparent to data formats and protocols. This report documents the design and implementation of a reconfigurable, all-optical demonstration network testbed for the Optical Network Technology Consortium (ONTC) Phase I program[2] using both wavelength-routing optical cross-connects and virtual path fast-packet ATM switches.

The demonstration network was designed with five main considerations in mind:

1) to define a simple network to carry out a wide variety of network experiments;
2) to test key network components such as multiwavelength laser arrays, optical cross-connects and multiwavelength optical amplifiers;
3) to be accessible to all competing device technologies;
4) to study multiwavelength network operations, test network control and management techniques applicable in an ATM/WDM network, and study the characteristics of network rearrangeability, control and admission algorithms through wavelength reuse and wavelength translation; and
5) to provide a platform to test applications such as subcarrier multiplexed (SCM) analog-video multicast and multiparty video teleconferencing.

Sections II and III describe the general network testbed and network access node architectures, respectively. Key network components such as optical amplifiers, optical switches, multiwavelength receivers and transmitters are described in Section IV and the network signaling and control is discussed in Section V. This is followed in Section VI by a description of the network management schemes. Network applications are presented in Section VII and conclusions follow in Section VIII.

Manuscript received April 15, 1995. This work was performed as a part of the ONTC consortium under ARPA Contract MDA972-92-H-0010.

G.-K. Chang, J. K. Gamelin, M. Z. Iqbal, and C. A. Brackett are with Bellcore, Red Bank, NJ 07701 USA.

G. Ellinas is with the Center for Telecommunications Research, Department of Electrical Engineering, Columbia University, New York, NY 10027 USA.

Publisher Item Identifier S 0733-8724(96)04695-6.

[1] Through traffic not needed by the local access node.

[2] The Optical Network Technolgy Consortium (ONTC) was formed in October 1992 for the duration of three years and comprised of Phases I and II. Phase I concentrated on network concepts and testbed demonstration. Phase II focused on WDM technology advancement.

Reprinted with permission from *IEEE Journal of Lightwave Technology*, Vol. 14, No. 6, pp. 1320-1340, June 1996.

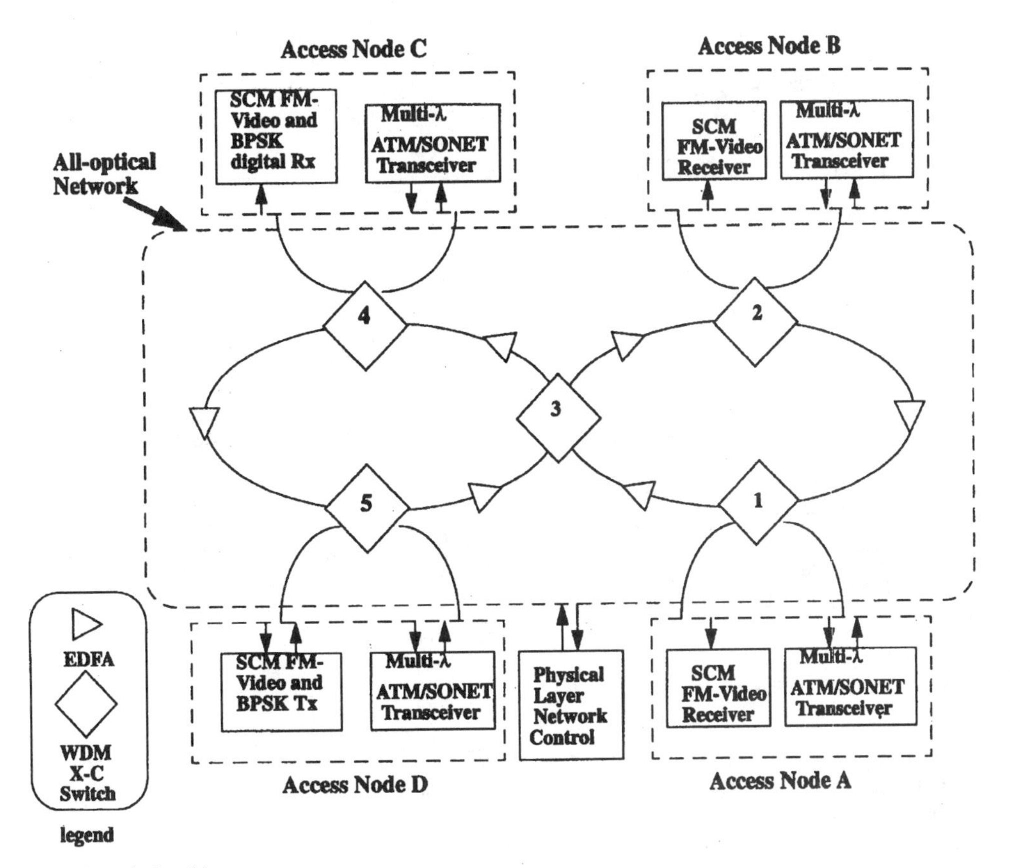

Fig. 1. ONTC network testbed architecture.

II. Network Testbed Architecture

The demonstration system architecture consists of four network access nodes, A–D, distributed in two WDM optical fiber rings (totaling 150 Km of single mode fiber) joined by a 2 × 2 WDM cross-connect module [2], [4] as shown in Fig. 1. Five 2 × 2 WDM switches are used in this system. Four of them are used to connect the access nodes to the fiber rings and the fifth is used to interconnect the two fiber rings together. This cross-connect switch was implemented both by a hybrid WDM optical switch [5] and an acoustooptic tunable filter (AOTF) switch [11], [13] module. Six erbium-doped fiber amplifiers (EDFA's) are also used in our network to compensate for the insertion loss of the optical switches and the transmission loss in the fiber loop. The user terminals connected to an access node, can gain access to the fiber ring through the multiwavelength transceiver array, the ATM switch, and a WDM add/drop switch.

Four optical channels, at wavelengths of 1546, 1550, 1554, and 1558 nm were chosen to establish data communication links between the access nodes. These wavelengths are located in the flat gain spectrum region of the EDFA's. Three of the wavelength channels are modulated at the SONET OC-3 rate and carry both data and control signals between the access nodes. The remaining wavelength (1554 nm) is used to carry four subcarrier multiplexed analog video channels [3], [24], [25]. Subcarrier multiplexed (SCM) transmitters and receivers are installed in various access nodes as shown in Fig. 1.

The system configuration, depicted in Fig. 1, was chosen for this network demonstration because the four access nodes in the two optical fiber rings cannot be fully connected by single-hop optical paths with just four wavelengths, hence additional multihop paths are needed for establishing connections among all access nodes. The multihop connections can be accomplished by utilizing the wavelength selection function of the WDM switches, and the wavelength channel translation capability of the ATM switches located in the access nodes [9], [12], [21]. The rule of connectivity is as follows: in any optical path, each wavelength carries information originating from a single source. The wavelength selective optical cross-connect modules used in this network are capable of independently and simultaneously selecting and switching each wavelength channel to enforce this WDM channel connectivity rule.

III. Network Access Node Architecture

A network access node serves two main purposes:

1) it injects and extracts optical signals to and from the fiber loop, interfacing with the ATM switch for data grooming and information distribution to local user groups, and
2) acts as the optoelectronic wavelength translator (detect and retransmit) in multihop connections.

The architecture of the network access node is shown in Fig. 2. It consists of optical demultiplexing and receiving circuits on one side, and optical multiplexing and transmitting

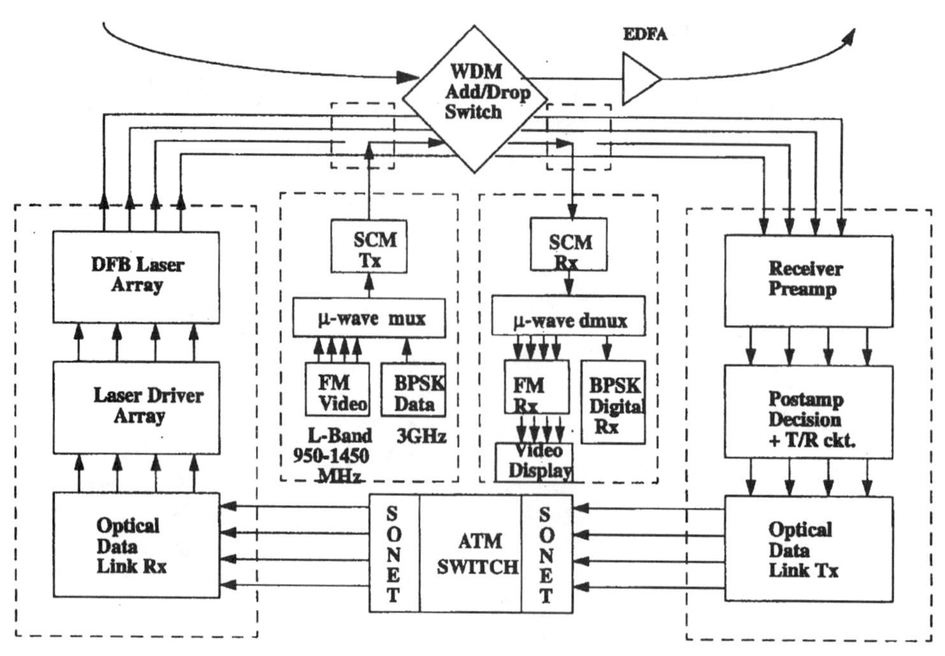

Fig. 2. Key building blocks of a network access node.

circuits on the other side, surrounding an ATM switch in the center. Fig. 2 depicts access node D which is the only access node equipped with an SCM transmitter in addition to an SCM receiver. Each network access node is also equipped with a workstation used for generation of client applications as well as local control of the optical network elements.

The ATM switches on the periphery of the network process mainly local access traffic. They are capable of providing flexible rearrangement of the logical connectivity on a wavelength-by-wavelength basis. In particular, the ATM switches in our network perform two main functions:

1) wavelength translation and routing, channel add/drop, multicast and dynamic reconfiguration and
2) network control and management functions such as traffic routing, connection monitoring, and service provisioning. By carrying out wavelength translation through the E/O and O/E domains, a large number of access nodes can be interconnected with a limited set of wavelength channels.

In the network access node, the system module interfaces are linked together through optical fiber interconnects. The ATM switch is designed to process the four wavelength channels (each at OC-3c, 155 Mb/s data rate) coming from the transceiver module which connects to the outside optical fiber ring through the WDM switch module. The ATM switch also communicates with the user terminals using either TAXI (100 Mb/s) format or OC-3c (155 Mb/s) SONET format depending on the throughput of the user terminal and the required bandwidth for user applications. For our network, the OC-3c transmission rate at 155 Mb/s was adopted [2], [4]. We did not implement this demonstration at higher bit rates because of the limitation and availability of computer and ATM hardware. However, the choice of the system data rate was not a limiting factor in studying network rearrangeability and control algorithms.

IV. Key Network Components

A. *Erbium-Doped Fiber Amplifiers*

Six erbium doped fiber amplifiers (EDFA's) are used in the ONTC phase I network testbed to compensate for the component insertion losses and the fiber transmission losses in the network. In the present testbed architecture, these EDFA's are placed at the outputs of the WDM ring cross-connect and the WDM add/drop switches as shown in Figs. 1 and 2. Rejection of the amplifier spontaneous emission (ASE) peak at 1530 nm can be achieved by using the hybrid optical switches, implemented in this testbed and discussed in Section IV-B-2), which automatically filter out the 1530 nm peak to an energy dumping port (corresponding to the tapping port shown in Fig. 14). Similar rejection can be achieved in the AOTF switches in dilated mode where the wavelength selection is attained using a combination of *bar* and *cross* states so that the 1530 nm ASE noise can always be dumped in an unused port.

The EDFA's integrated in the network testbed were tested and their performance was evaluated in a series of experiments. Our goal was to determine whether we can span a certain number of EDFA's in cascade and still get an acceptable error performance. Fig. 3 shows the experimental setup. Six EDFA's and six WDM switches are put in cascade and a signal is sent through and examined at various stages. In this experiment, the optical power was adjusted to be −20 dBm at every input stage of the EDFA's and the observed gain was about 20 dB for every input channel. After every amplifier we took measurements of the power and the bit-error rate (BER). A

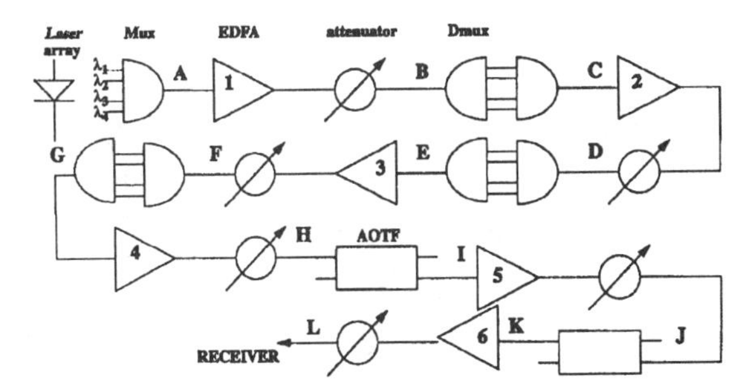

Fig. 3. Cascade EDFA experiment.

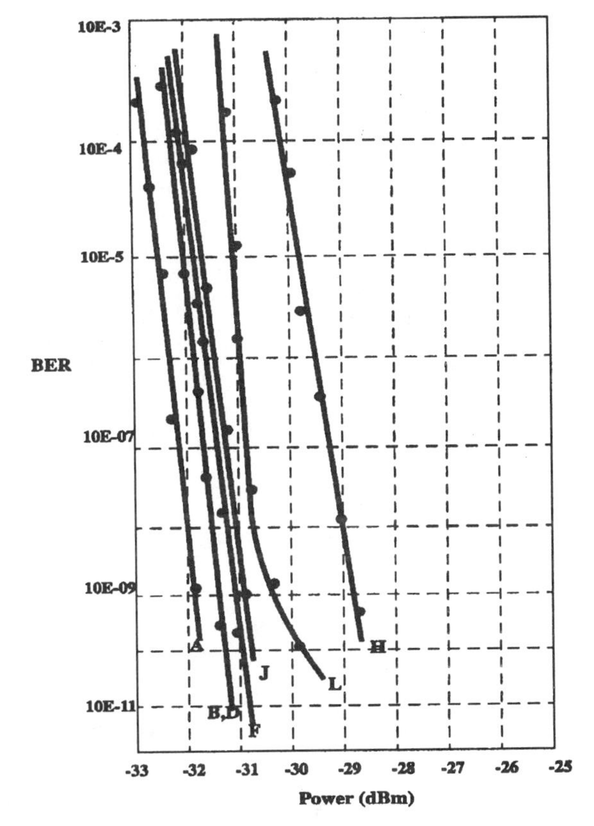

Fig. 4. BER versus power for cascade EDFA experiment.

graph of BER versus Power (dBm) appears in Fig. 4. Even after six amplifiers we observe that for BER 10^{-9} the power penalty is less than 2 dB. At one particular wavelength (1546 nm), the receiver sensitivity was degraded by about 0.5 dB after the first amplifier where the laser array signal to noise ratio (SNR) dropped from 38 to 28 dB. In successive stages, the degradation was smaller as the laser SNR degraded in a much smaller rate. Thus, we concluded that in our network testbed, where in the worst case a signal will span six EDFA's, we can still get an acceptable error performance.

Another point of concern as multiwavelength signals traverse multiple EDFA's before regeneration is the gain equalization. Because the EDFA passband is not flat and the EDFA's operate in cascade the stronger wavelength "grabs" most of

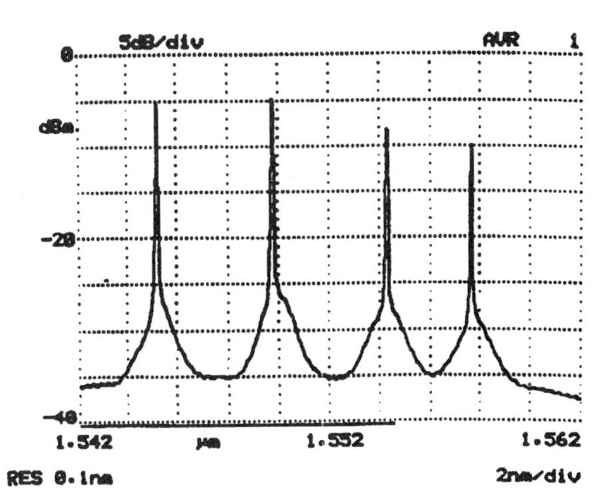

Fig. 5. Gain variation for cascade EDFA experiments.

the power gain, leading to signal gain variations among the wavelengths [18]–[20], [23]. Fig. 5 shows a total of 5 dB amplitude variation in the four wavelengths' power after six EDFA's. Without any signal compensation or equalization this will consume the receiver dynamic range and limit the size of the testbed.

B. High Performance Optical Switches

Optical networks of the future are preferred to be reconfigurable in a short time scale in order to take full advantage of the aggregate bandwidth provided by such systems since the user traffic demands and type of services change rapidly with time. Thus, wavelength selective, wavelength routing WDM switches are key components for any rearrangeable, scalable optical network. In the testbed we explored two different kinds of technology. The first one was based on acoustooptic tunable filters [13] which can independently select and switch wavelength channels, and the second was based on a hybrid switch design using multilayer dielectric thin film interference filters in cascade, to function as an optical multiplexer and demultiplexer, along with discrete 2 × 2 cross-bar optical switches [5].

1) Acoustooptic Tunable Filters: AOTF is a key device technology for performing wavelength-selection, wavelength-routing functions in all-optical networks. In an AOTF device, RF frequencies are launched onto the AOTF device to generate surface acoustic waves to switch wavelengths. RF controllers operated manually or remotely (via an RS-232 interface) are used to configure the wavelengths in the network. AOTF's have the advantages of compact device size, low cost and fast switching speed. However, when we tested the phase I AOTF modules in the network environment we identified various problems.[3] Most of the problems were solved in clever ways through device engineering or packaging improvements and we were able to use this device in our system experiment in a limited way.

[3]This section focuses only on experiments performed with the phase I AOTF modules. Later devices used in the phase II ONTC program and the NTONC program were greatly improved.

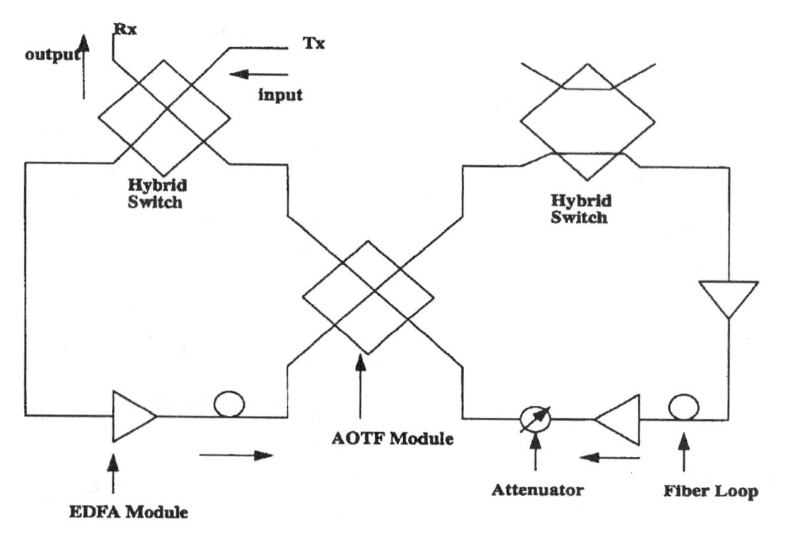

Fig. 6. Experimental set-up of system crosstalk measurement with an AOTF cross-connect module.

System evaluation tests were carried out in order to study the switching and crosstalk requirements of the AOTF cross-connect module as well as characteristics like passband width and phenomena like phase noise to intensity noise conversion.

a) AOTF switching feedthrough/oscillations: While the AOTF has the advantages of independently selecting and switching wavelength channels, it also has the drawback of high intrachannel crosstalk due to switching leakage of signal channels with the same nominal wavelengths. The tolerable limit of leakage crosstalk in an optical network is very stringent due to the interferometric noise, and as we showed, was not easily achievable in a single stage phase I AOTF module.

Experimental results showed that a particular wavelength channel filter can be turned on by applying corresponding RF signals to the acoustic waveguide on the AOTF substrate. A typical channel switching leakage with one RF turned on was about -15 dB. However, when all four RF control signals were turned on, the single channel switching leakage became -9 dB. The leakage is mainly due to imperfect splitting and combining (leakage from polarization beam splitters, fusion splices etc.), denoted interport crosstalk effect, and due to the fact that when we are switching many wavelengths, switching one wavelength can cause partial polarization conversion at a wavelength close to the one we switched (interchannel crosstalk effect).

BER experiments at 155 Mbps data rate were carried out to study the system performance of the AOTF module due to switching leakage. The experimental set-up used is shown in Fig. 6. The AOTF module was used to interconnect the two fiber rings. When it was set in the *cross* state for one of the wavelengths, the signal leaked in the left ring and it interfered with itself after going around the *long loop* (∞ shaped). This resulted in very poor bit error rates. Fig. 7 shows a graph of BER versus Power for increasing leakage. The results showed degradation in terms of receiver sensitivity penalties and BER floors were observed. The BER floors indicated the presence of interferometric noise.

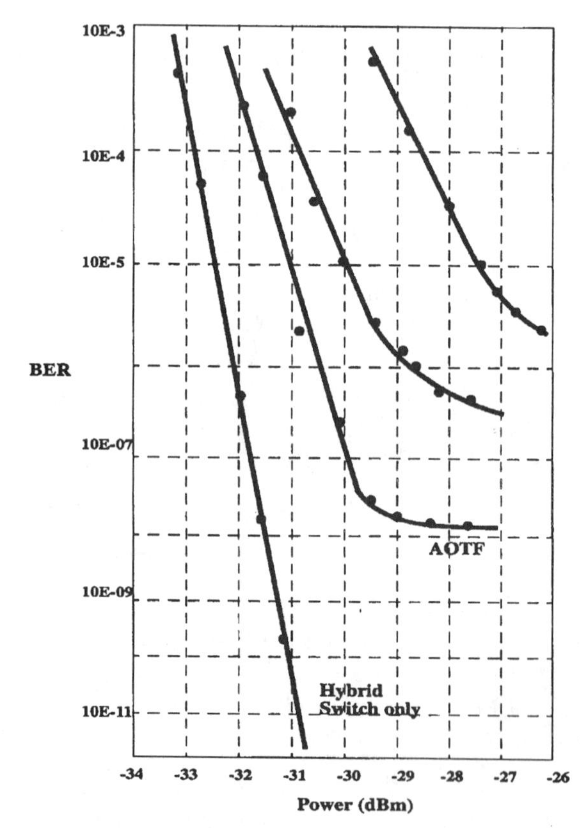

Fig. 7. BER versus power for AOTF system crosstalk measurements.

Switching leakage also caused oscillations in our testbed. Oscillations can occur when the signal circulates in a closed cycle where optical amplifiers are present [22]. Fig. 8 shows the set-up of an experiment performed in order to detect these oscillations. Because of the switching leakage in the AOTF module that interconnects the two fiber rings, we observed leakages that circulated around in the fiber ring on the right.

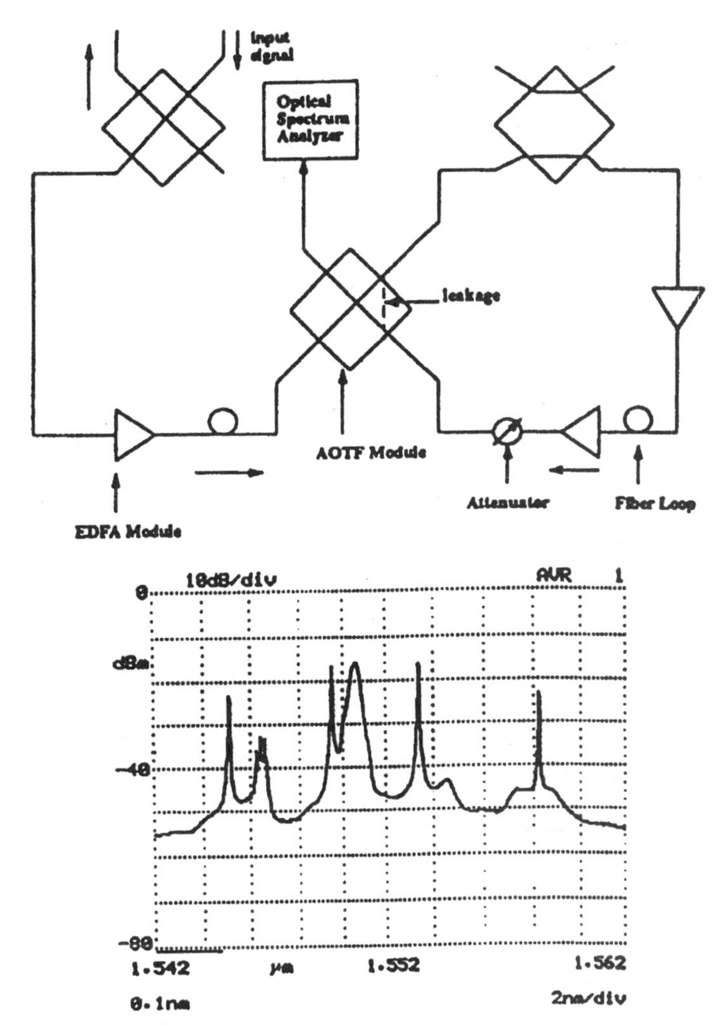

Fig. 8. Experimental setup and optical spectrum for system oscillations with an AOTF cross-connect module.

Thus, oscillations occurred, the noise was amplified, and finally surpassed the original signal (Fig. 8). One way to avoid oscillations is to "cross-connect" the AOTF ("frog" it) [22]. An example of this procedure is shown in Fig. 9. With this scheme we eliminated any closed cycles, thus the noise was considerably reduced.

b) Frequency shifts of the optical carrier: We performed a system experiment to measure the frequency shifts of the optical carrier in the transmitted lightwave signals when traversing through an AOTF module twice. Fig. 10 shows the experimental set-up.

The measured RF spectrum from the experiment showed a 175 MHz and a 350 MHz beat components generated when the signal passed through the AOTF module twice (Fig. 10). These two components appear because of the shifts in the optical frequency induced by the acoustic wave for both the TM and TE states (a + 175 MHz shift for light in the TM state and a − 175 MHz shift for light in the TE state). The shifted and unshifted crosstalk signals coherently beat between themselves and create considerable distortion of the original signal. Fig. 11 shows a "clean" signal received when a 155 Mbps receiver was used in the experimental set-up. The tolerable performance of high intrachannel crosstalk, observed when studying the switching leakage, can be attributed to the small bandwidth of our SONET OC-3 receiver (100 MHz) which rejected the major noise components generated in the AOTF at higher frequencies. When a 2.5 Gb/s SONET OC-48 receiver was used in the same experiment, severe "eye" closures were observed in the BER experiment (Fig. 11). This is directly associated with the frequency shifts. Any receiver with bandwidth less than 175 MHz will not detect the signal interference because the optical carrier frequencies are shifted by the RF modulation frequency of the surface acoustic wave while passing through the AOTF switch.

c) Dilated AOTF cross-connect switch: It was obvious from our experimental results that a single AOTF module could not provide sufficiently high channel isolation during

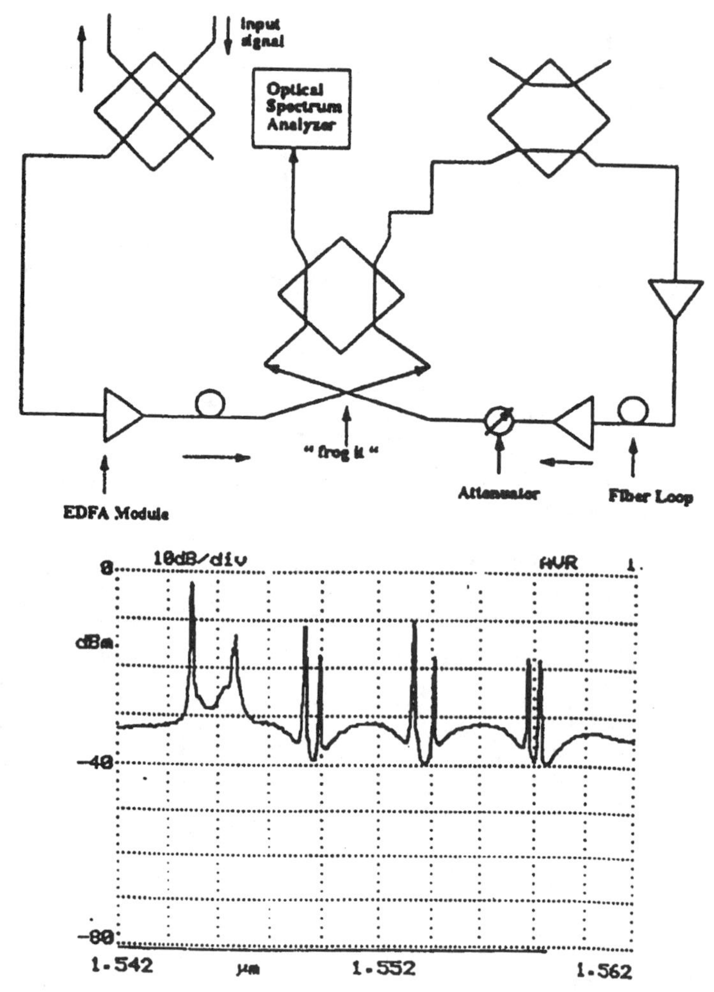

Fig. 9. Experimental setup and optical spectrum for a system with an AOTF cross-connect module with "frog," connectivity.

switching, and interferometric noise would arise to overwhelm the lightwave signals modulated at high frequencies. Thus, we had to build dilated 2 × 2 AOTF's comprising of four individual 2 × 2 AOTF modules, to create energy dumping ports for switching leakage (Fig. 12). By using dilated switches, the leakage is diverted to open-ended ports and does not enter the network. The second order leakage will enter the network but its impact on the actual signal is minimal. Apart from "cleaning up" the leakage signal, dilated switches will also filter out the optical amplifier's ASE noise.

The same configuration as in Fig. 6 was used to test for intrachannel switching leakage but this time instead of a single AOTF module we used a dilated AOTF. The typical intrachannel leakage for the *bar-bar* combination in the dilated switch was about 30 dB and the leakage for "cross-cross" combination was about 26 dB.

The most severe problem encountered by the dilated AOTF was the conversion of laser phase noise into intensity noise. This effect was evident in the dilated case in the absence of any switching. Fig. 13 shows the experimental set-up and the frequency spectrum of the signal at the destination port. Very high interference noise is present at the frequency of the signal. The reason for this phenomenon was traced to the conversion of laser phase noise (chirping) into intensity noise. This can be attributed to the imperfect polarization splitting and the optical path differences inside the AOTF module. The AOTF acts as an interferometer because of the substantial optical path differences in length that exist in the device, thus converting the phase noise in the laser into intensity noise at a frequency characteristic of the path difference. This phenomenon is weak when light traverses one AOTF and increases considerably for a dilated device. An intentional path difference of about 50 cm was eliminated in later devices and the interferometric noise effect was pushed out to a few GHz in RF spectrum. This will improve low noise operation for 2.5 Gb/s OC-48 data transmission.

2) Hybrid Wavelength Selective Cross-Connect Switch: To address the problems presented by the AOTF switch, we also designed and built a hybrid WDM cross-connect switch. The hybrid WDM cross-connect switch was built using multilayer dielectric thin film interference filters in cascade along with discrete 2 × 2 cross-bar, relay-actuated optical switches [5].

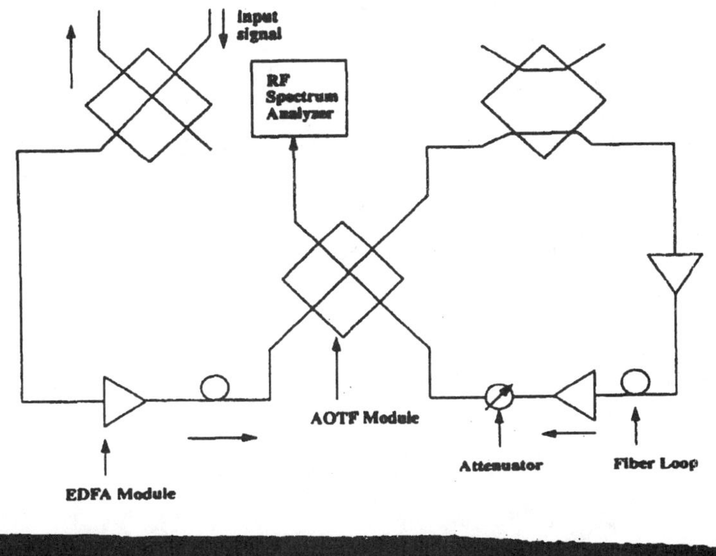

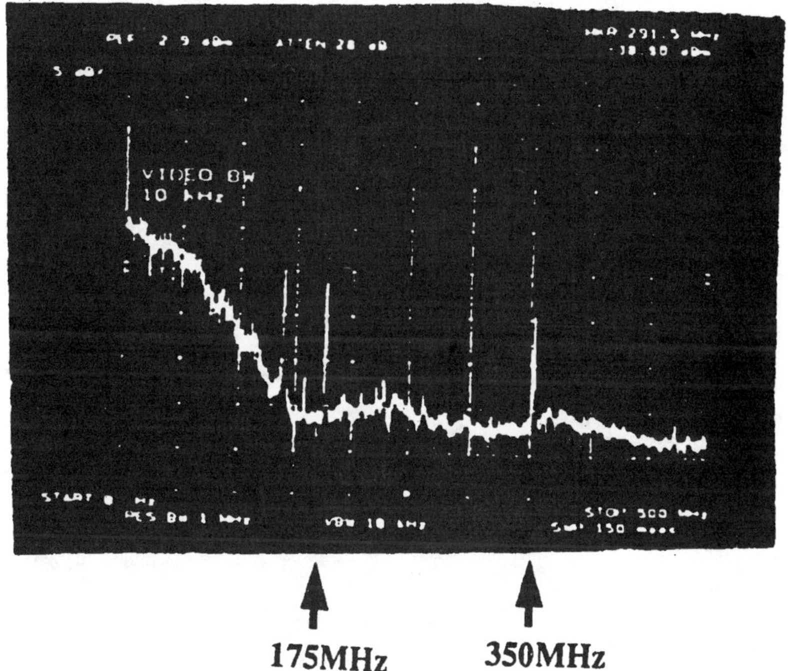

Fig. 10. Experimental setup and RF spectrum for frequency shifts of the optical carrier in the transmitted lightwave signals when traversing through an AOTF module twice.

Optical multiplexers and demultiplexers were made[4] with those kinds of filters preselected for the four wavelength channels used in our testbed. These filters select one wavelength at a time while reflecting the rest. Fig. 14 shows an example of how the wavelengths are separated. This design results in a hybrid, fiber-optic based, four channel (at 1546, 1550, 1554, and 1558 nm) WDM switch where input wavelengths are demultiplexed and wavelengths are switched before multiplexing again to the output port. Fig. 15 shows an example of how this device is used to switch wavelengths λ_2 and λ_3. Because of the internal symmetry of the switch, the inputs and outputs can be used interchangeably and it is even possible to launch wavelengths in bi-directional mode. In addition, there is a build-in reflection tapping port which collects all the rejected light from each wavelength filter and is available for coarse system power monitoring and in-site data processing. This WDM cross-connect switch was deployed as the interconnect module between the two fiber rings in our testbed. Control of the switch can be performed manually or remotely (via an RS-232 interface). We can thus control

[4] By JDS/FIDEL Inc.

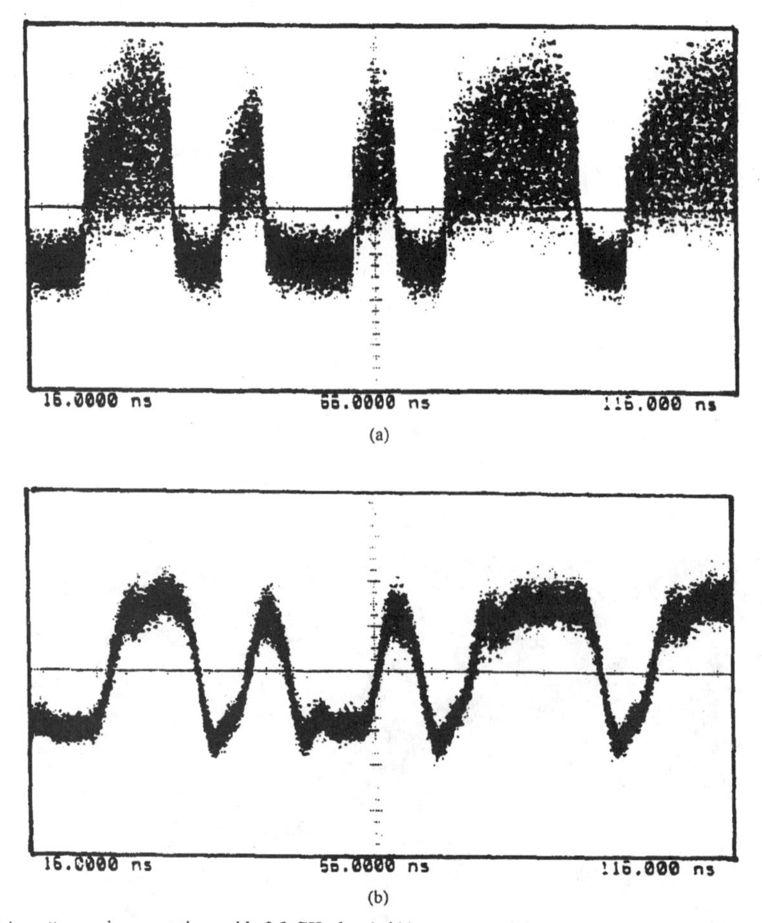

Fig. 11. (a) Transmitted data pattern using a receiver with 2.5 GHz bandwidth at the receiving station. (b) Transmitted data pattern using a receiver with 150 MHz bandwidth at the receiving station.

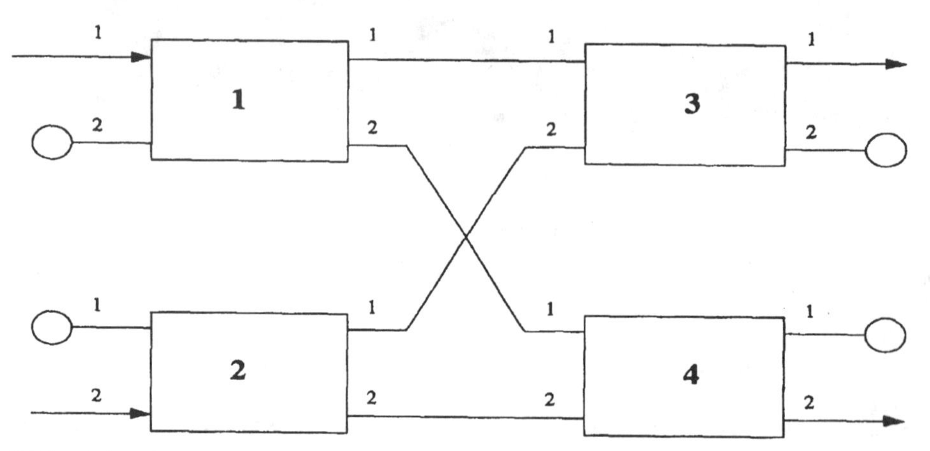

Fig. 12. Dilated AOTF switch.

the switching states for each wavelength channel for system operation and network management.

The hybrid WDM cross-connect switch has wide flat passbands and low insertion loss due to the multilayer interference filter technology. The typical widths of the channel filter passband are 1 nm at $\pm$ 0.1 dB and 2 nm at $\pm$1 dB. Since the passbands are so flat and wide, the laser wavelength registration requirements and the frequency chirping sensitivity due to high speed direct modulation of the lasers are alleviated. This device also has high channel isolation because of the

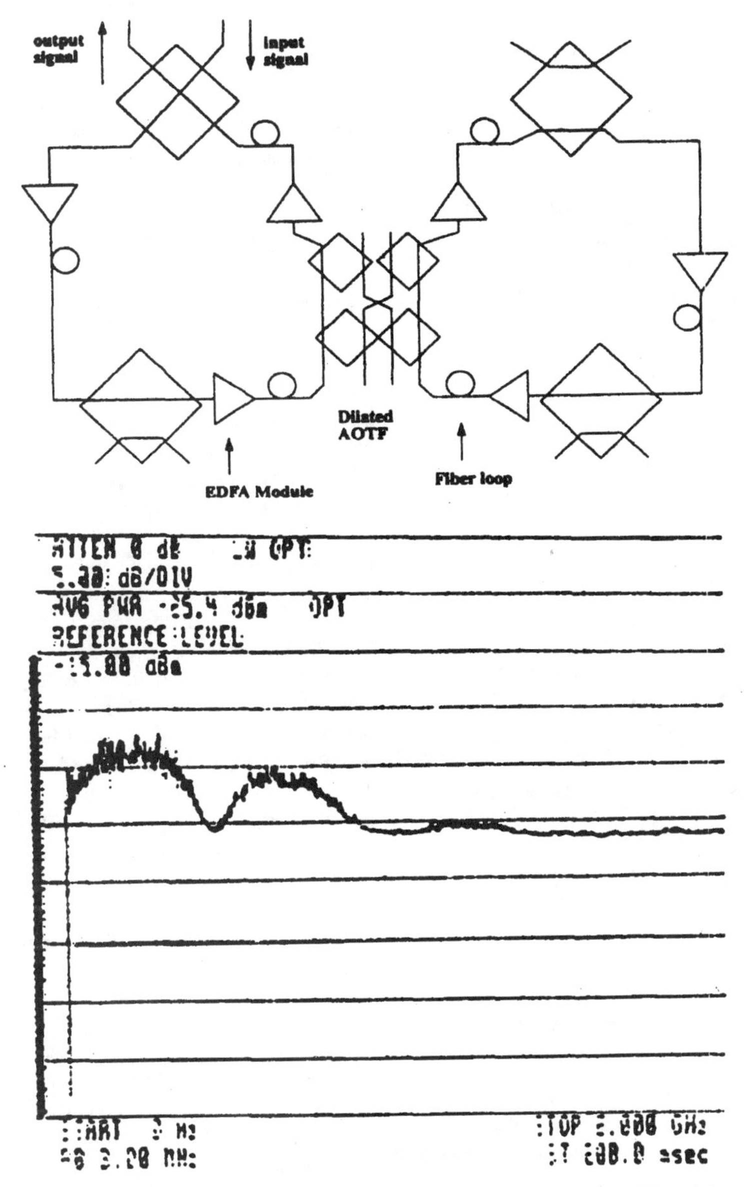

Fig. 13. Experimental set-up and RF spectrum for phase noise to intensity noise conversion with dilated AOTF module.

space division switching design. Specifically, the interchannel crosstalk is less than −30 dB, the intrachannel crosstalk is less than −60 dB and the fiber-to-fiber insertion loss is less than 5 dB. The switch module's temperature sensitivity is 0.03 nm/C, and the switching time is about 10 ms [5]. The low insertion loss of this device enabled us to use only one EDFA, after each switch, for signal amplification. The low crosstalk and wide flat passbands are very critical physical design features of this switch. These features are essential if transparent WDM networks of this kind are to be scaled to large sizes without serious degradation from WDM component imperfections.

Fig. 16(a)–(d) shows the switching performance of the switch. Fig. 16(a) and (b) shows the four wavelengths of two different laser arrays at the input ports, and Fig. 16(c) and (d) shows the four wavelengths at the output ports with wavelengths λ_2 and λ_3 interchanged. The filter passbands defined by optical fiber amplifier spontaneous emission noise are clearly visible in the output spectrum of Fig. 16(c) and (d). The switching leakage is so low that only one wavelength is observed in any passband.

3) Wavelength Add/Drop Multiplexer (WADM): In an all optical network the WDM switches at the access nodes have to deliver and accept individual wavelengths (add/drop feature). A WADM switch requires only one multiplexer/demultiplexer pair at the network side. At the access station side the four wavelengths are launched or received directly at the I/O ports

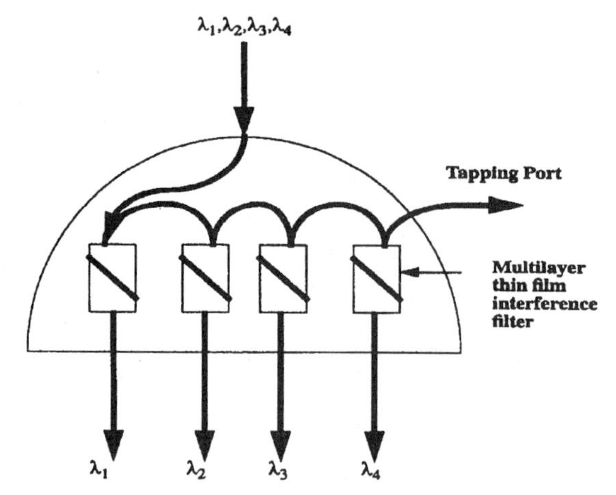

Fig. 14. The four wavelength optical demultiplexer with tapping port.

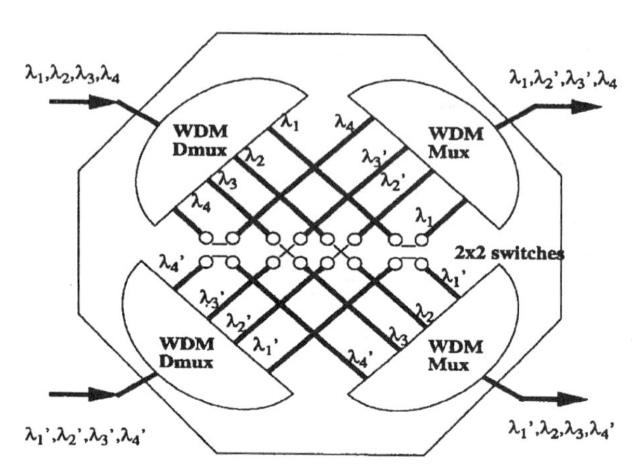

Fig. 15. 2 × 2 WDM cross-connect in a configuration to switch λ_2 and λ_3.

of 2 × 2 optical switches. Fig. 17 shows an example of a WADM architecture and Fig. 18 shows a complete access node switch module in 9U VME format, which was integrated with other network elements in a VME cage.

C. Multiwavelength Transmitter Module

This section discusses the performance of the four-channel WDM transmitters employed in each access node. Fig. 19 depicts the schematics of the transmitter unit featuring the interfaces to the optical and electronic network elements. The module features integrated driver and laser arrays [14], [15], [17] to reduce package size and cost for commercial applications. Functionally, the module translates the 1300 nm SONET/ATM switch output to 1550 nm band signals. Optical input from the SONET/ATM switches undergoes optical-to-electronic conversion using the data link receivers. The ECL-compatible data link outputs modulate the laser driver array which in turn modulates the laser optical output. The laser outputs are optically multiplexed before they are introduced into the network.

The transmitter module must satisfy three system operation requirements:

1) accurate wavelength registration in each channel must be assured for uniform, low-penalty switching through the network cross-connects;
2) sufficient modulated power must be provided into each output to assure a sufficient system loss-budget margin; and
3) the incoming data streams must be converted with low channel crosstalk.

We designed, fabricated and tested the transmitter motherboard. This transmitter motherboard is designed to accommodate and operate a multiwavelength DFB laser array [14], [17] and a GaAs HBT IC laser driver array. A temperature controller, power supplies, digital multimeters, optical data link receivers for OC-3 data transmissions as well as other related electronics were incorporated on the board. Four multimeter displays on the front panel can be switched to show the digital readout of sixteen tunable potentiometers for controlling the operation of each of the four wavelength lasers in the laser array. The laser threshold current, the ac modulation depth, the wavelength-tuning current, and ECL power supply voltages to control the GaAs driver IC are provided and monitored by this motherboard. It was build on an 8-layer, double width VME card for plugging into a standard equipment cabinet. Fig. 20 shows a transmitter unit complete with power supplies, bias adjustments and thermoelectric control in a VME cage format.

1) System Performance Evaluation: Fig. 21 shows the measured optical-optical response of a typical multiwavelength transmitter module on a motherboard. Excellent flatness with a bandwidth of 120 MHz was observed. To demonstrate the transmitter performance through the network, Fig. 22 shows the OC-3 eye diagrams and BER versus power characteristics through three distinct optical paths. Referring to the testbed diagram in Fig. 1, the longest optical path, denoted *long loop*, is shaped as a "∞," and it traverses a total of 150 Km of single mode fiber and five WDM switches in addition to six in-line EDFA's. The *long loop* starts from access node D across switch 5 and switch 3, through switch 2 and switch 1 in the *bar* state, across switch 3, through switch 4 in the *bar* state, and finally across switch 5 back into a receiver in access node D. The *short loop* starts from access node D across switch 5, through switch 3 and switch 4 in the *bar* state, and across switch 5 back to a receiver in access node D. It is shaped as an "O" and covers 80 Km of single mode fiber, three WDM switches, and three EDFA's. The shortest-path is the *back-to-back* loop through switch 5 in the *bar* state in access node D.

The back-to-back sensitivity was −32.5 dBm and 2 dB power penalty was observed for the long loop. We believe this is in part due to the combinations of noise generated inside the AOTF cross-connect module (used to interconnect the two fiber rings) and the degradation of the signal through a cascade of six EDFA's for multiwavelength operation. The eye diagrams of the *back-to-back* and *long loop* at 155 Mb/s are also compared in Fig. 22. Both eyes are fully open and only a slight jitter degradation was detected for the *long loop.*

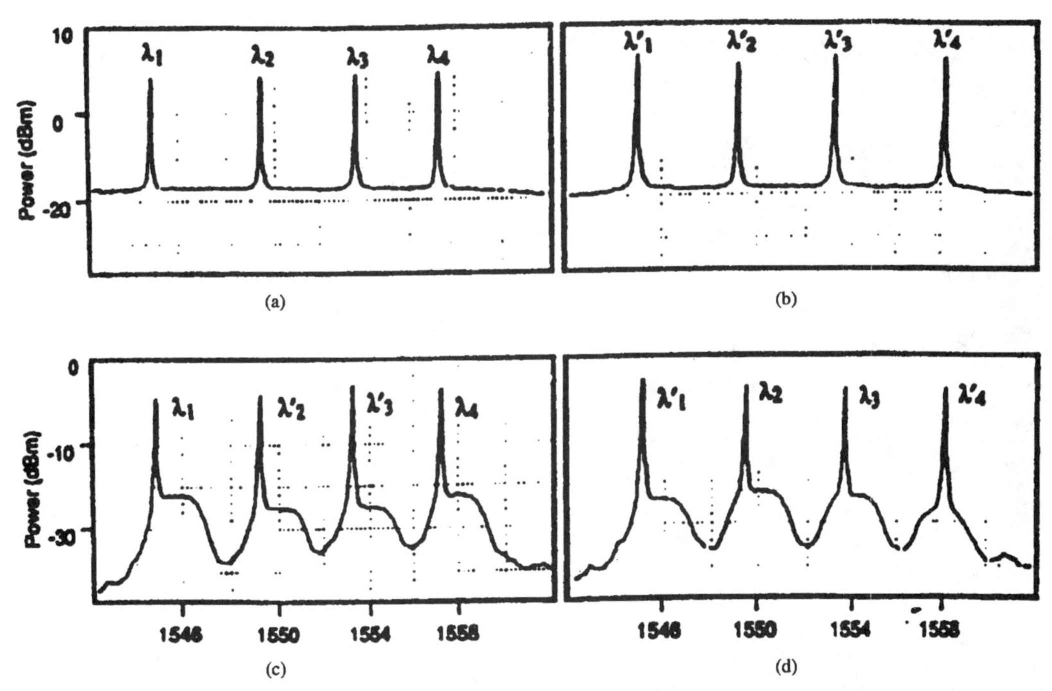

Fig. 16. Optical spectrum of laser array I (a) and II (b) at two input ports. Wavelengths 2 and 3 are switched at the output ports (c) and (d).

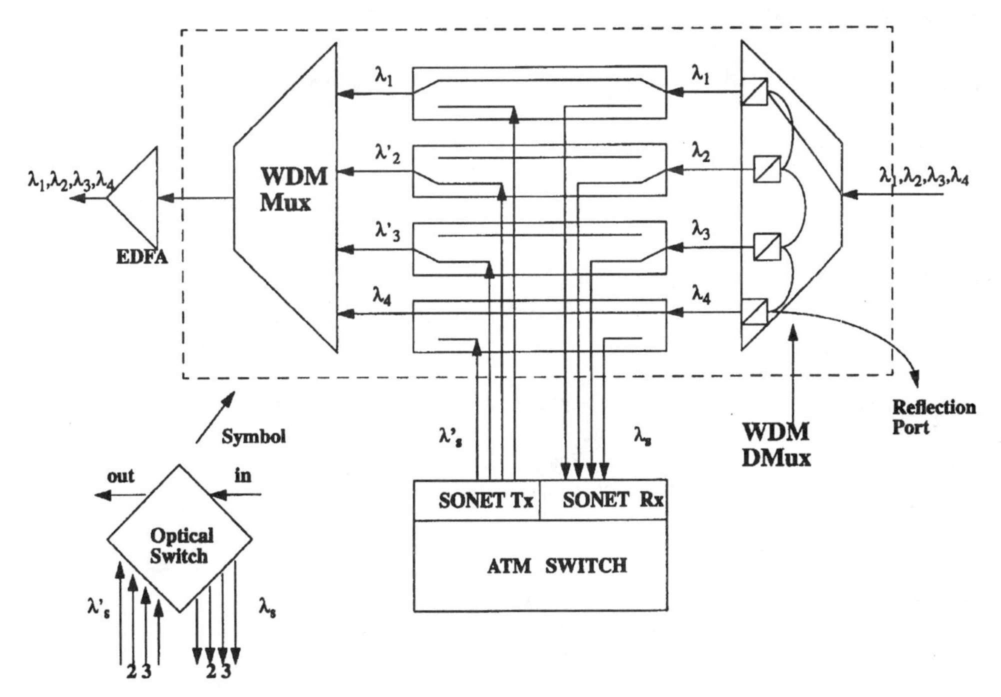

Fig. 17. Wavelength selective fiber optic add/drop switch.

The use of integrated laser driver and laser arrays poses challenges to crosstalk elimination. Interchannel crosstalk of less than −15 dB for frequencies up to 200 MHz were measured. The magnitude of crosstalk was found to be independent of channel number indicating common circuit node coupling. To assess the power penalty, Fig. 23 shows the measured BER of a typical unit under single and four-channel operation. The transmitter was used in a back-to-back configuration and all optical inputs were modulated with independent $2^{15} - 1$ pseudorandom bit streams. Other modules

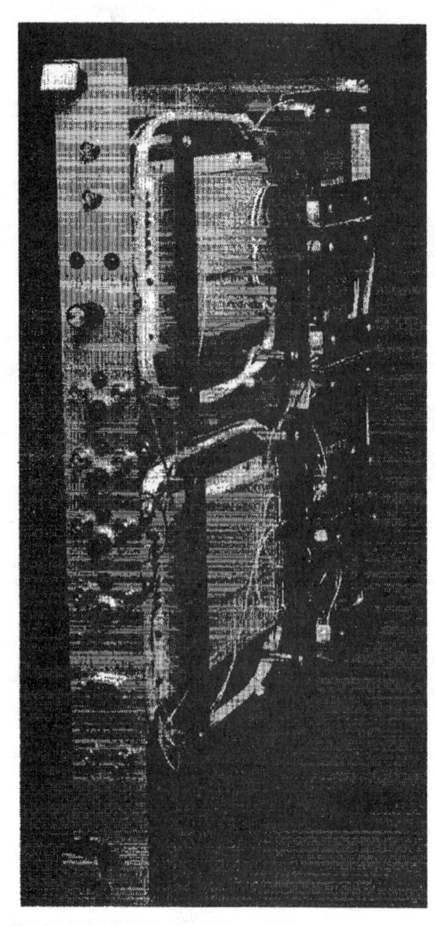

Fig. 18. Hybrid WDM optical switch module.

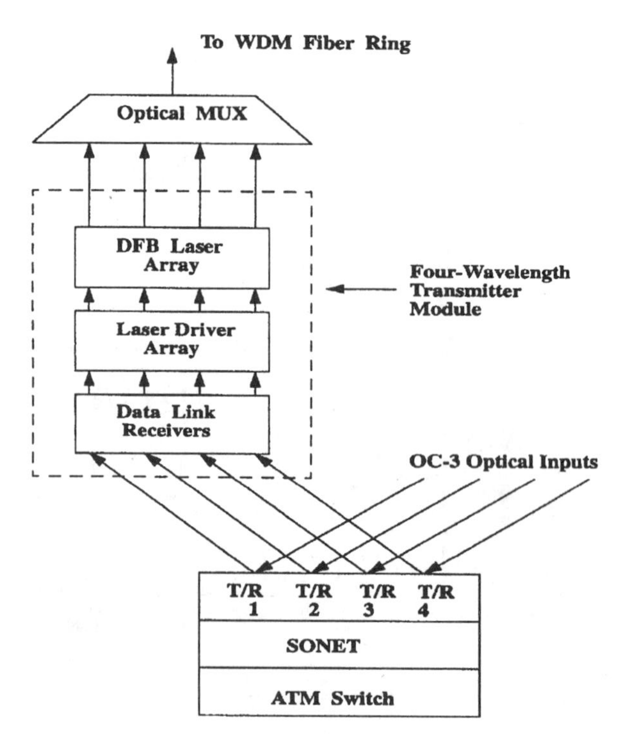

Fig. 19. Schematic of 4-channel WDM transmitter module.

showed similar responses with power penalties of 0.1 dB or less.

D. Multiwavelength Receiver Module

We tested and evaluated a packaged four channel InGaAs–InP PIN-HEMT OEIC receiver front-end fabricated at Bellcore and a packaged AlGaAs–GaAs HBT receiver IC array incorporating four-channel integrated post amplifier, timing recovery (T/R), and decision circuits fabricated at Rockwell Sciences Center. We evaluated these components in the system environment and integrated them onto a prototype receiver motherboard to perform the receiver array function in the network access node. The prototype motherboard carries out O/E, electrical signal processing and E/O functions. A special board on which an automatic voltage ramp control applied to VCO's was implemented by external microprocessors and operational amplifiers. The front panel of the motherboard was designed to provide LED indicators for the phase-locking status of the clock recovery circuits.

1) Receiver Array Testbed Setup—Performance Evaluation: The packaged Rockwell data regeneration circuit was mounted on a daughter board. Two kinds of receiver front-ends were employed to convert the incoming SONET OC-3 signals to electrical signals: the first one employed a data link receiver for 155 Mbps operation while the second employed a four-channel OEIC receiver. Optical data inputs were derived from two sources:

1) directly from the ATM switch with OC-3 network interface carried by 1546, 1550, 1554, and 1558 nm wavelengths
2) DFB laser at 1544 nm modulated by electrical signals from a pseudorandom bit stream generator at 155 Mbps.

The combined receiver module sensitivity at 10^{-9} bit error rate was measured at -32 dBm receiver input power, limited by the commercial receiver front-end. We also evaluated the phase-detect and phase-lock functions of the clock recovery circuit using external loop filters and feedback control voltages. All four channels in this integrated receiver IC array were tested to be fully functional and met the system specifications. Input data interruption and reconnection experiments were carried out successfully to recover the phase-lock clock signals.

The packaged four channel receiver front-end using InGaAs–InP PIN-HEMT OEIC was also evaluated in our testbed. The eye diagram for one of the channels, shown in Fig. 24, was obtained using a DFB laser at 1544 nm modulated at 155 Mbps NRZ with a $2^{15}-1$ pseudorandom bit stream. The eyes were fully open and the receiver sensitivities for the four channels were measured at -26 dBm at 10^{-9} BER. However, the measured receiver sensitivity was 5 dB less than the designed target due to lower unity gain frequency of the HEMT's in OEIC. When the differential outputs of the OEIC receiver were connected to the post-receiver array

Fig. 20. Transmitter module with VME-based motherboard.

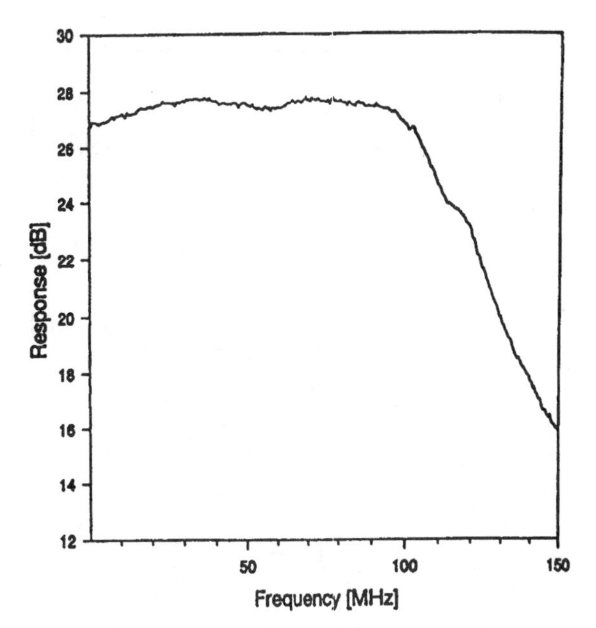

Fig. 21. Small signal optical-optical frequency response of the transmitter unit.

through AC coupling capacitors on the prototype motherboard, receiver sensitivities of about −26 dBm were obtained for the four channels using a post amplifier. The single-ended receiver output exhibited less than 1 dB penalty in receiver sensitivity when compared with the differential operation.

V. Network Signaling and Control

This section presents the system signaling and control in our demonstration network. The control scheme of wavelength-routing switches in the physical platform on which the control signals are transported, is the single most important step in setting up such a network testbed. The objective was to establish channel connectivity between the users and adopt control algorithms that can utilize the network capacity to meet wide variety of data service demands from the users or traffic patterns.

A. Network Signaling

Control packets are generated from a network central controller located in access node C and are optically distributed to the workstations in the other access nodes. The signaling is achieved using in-band isochronous ATM data cells for controlling both SCM and ATM/SONET traffic [2]. A small part of the bandwidth used to facilitate connections among the users in the network is used to carry the signaling packets from a network central controller to local node controllers (local workstations) located in each access node. The control signals are transported by way of the existing optical fiber layer in a manner identical to data transmission without requiring an overlay layer or an out-of-band embedded channel. The objective was to reconfigure the network with the minimum interruption of information flow and maximum utilization of network bandwidth for data transport. In order to be able to send the signaling messages from the central controller to the local controllers our control program ensures that a connection (single-hop or multihop) always exists from the central controller to all the local controllers.

Even though we initially adopted in-band signaling in our network testbed, various schemes including out-of-band signaling and electrical overlay were also performed. Out-of-band signaling in the network testbed was performed using a 1300 nm wavelength. For this scheme, we once again used the extra tapping port of the WDM switches (Fig. 14). For the electrical overlay layer scheme we interconnected all workstations used for control using an Ethernet overlay. By using an extra physical control overlay layer to transmit the network control signals we compromised in network powering and equipment cost but we gained in system reliability.

B. Network Control

A physical layer network central controller is required in this testbed to determine how to assign appropriate communication routes between the users (although in a real implementation, distributed control would be desirable). In the central control station, a software program was designed to monitor the WDM switch configurations according to preselected routing tables. These tables are used to establish system connectivity and to

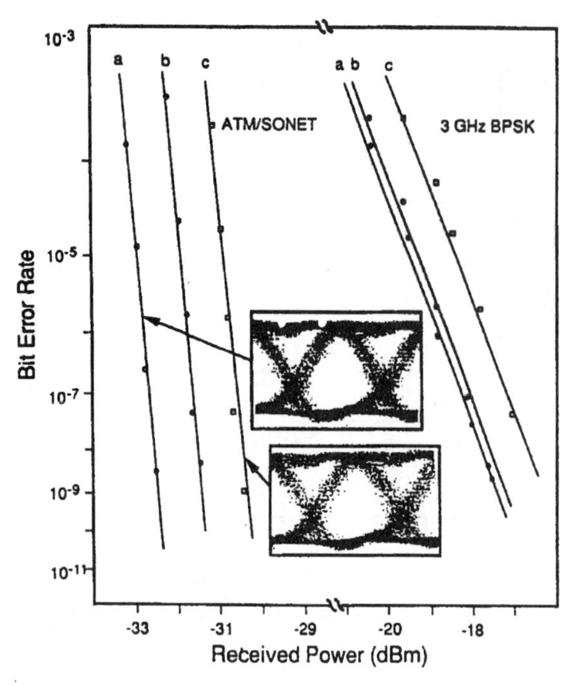

Fig. 22. BER versus Power characteristics and eye diagrams for simultaneous transmission of a SCM channel at 3 GHz BPSK and a ATM/SONET channel modulated at 155 Mbps for (a) back-to-back, (b) short-loop, and (c) long-loop.

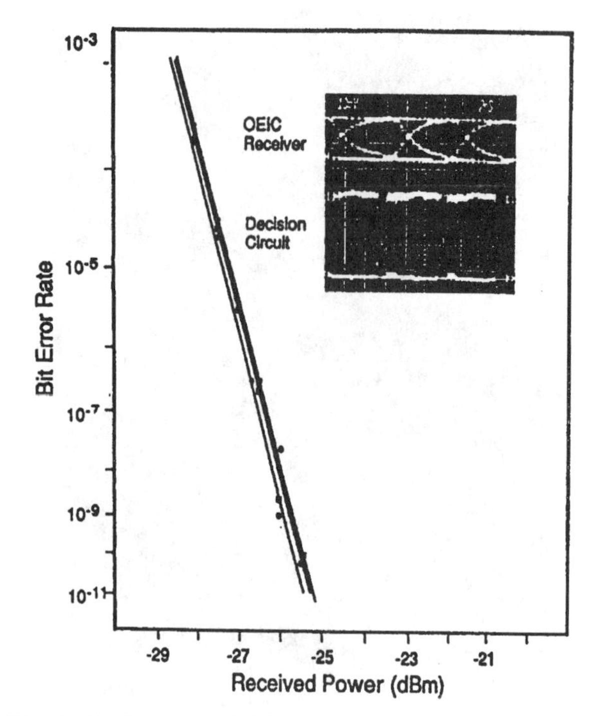

Fig. 24. Eye diagrams and BER plots for a receiver array module.

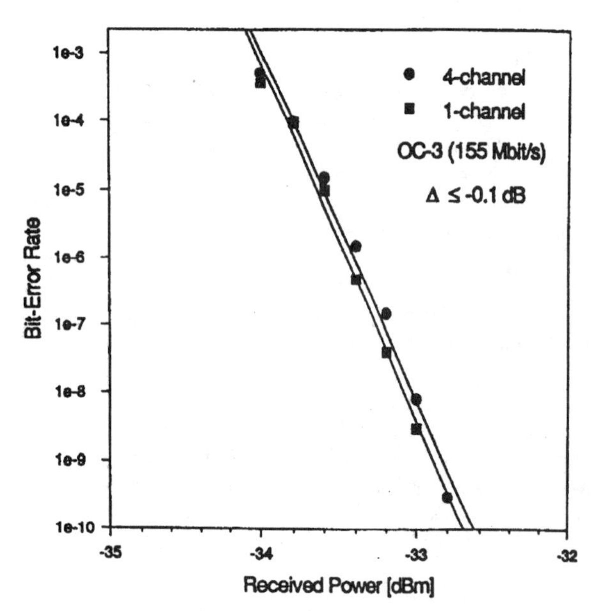

Fig. 23. BER versus power characteristics for a transmitter module in a back-to-back configuration.

test the network rearrangeability. The general rules employed are:

1) maximize the single-hop connections and
2) assigning one wavelength as the information carrier to enter every access node in the testbed.

We can add a new user at a time to this network and maintain the channel connectivity and average bandwidth per user for existing users. The goal is to build as many look up tables as possible to meet the user's demands in varying traffic patterns and new services. Physical constraints and implementation algorithms for setting up those tables have been designed and examined in advance, to ensure the continuity of data transport and service provisioning.

A graphical user interface (GUI) network control program was installed on the workstation located in access node C to serve as the central controller for our reconfigurable all-optical network testbed. The graphical user interface (shown in Fig. 26) enables the network manager to observe the status of the network testbed at any given moment and can be employed to manually reconfigure the network. Hyperlinked icons of the ATM switches and the workstations can give us information on the status of the ports, the amount of data that flows through each port (for the ATM switches) as well as status information on the workstations.

A user initiated control program was also designed for the network testbed. Any user has the capability to request a connection to another user or to a specific service. The request is sent to the central controller using a connection protocol and the central controller decides whether it will accommodate this request. This decision is based on connectivity constraints, resource constraints, and existing connection priorities. If it decides to accommodate the request the testbed is then automatically reconfigured and the GUI display is automatically updated. Additionally, scheduling was also accommodated in our network testbed. Using the GUI we were able to schedule

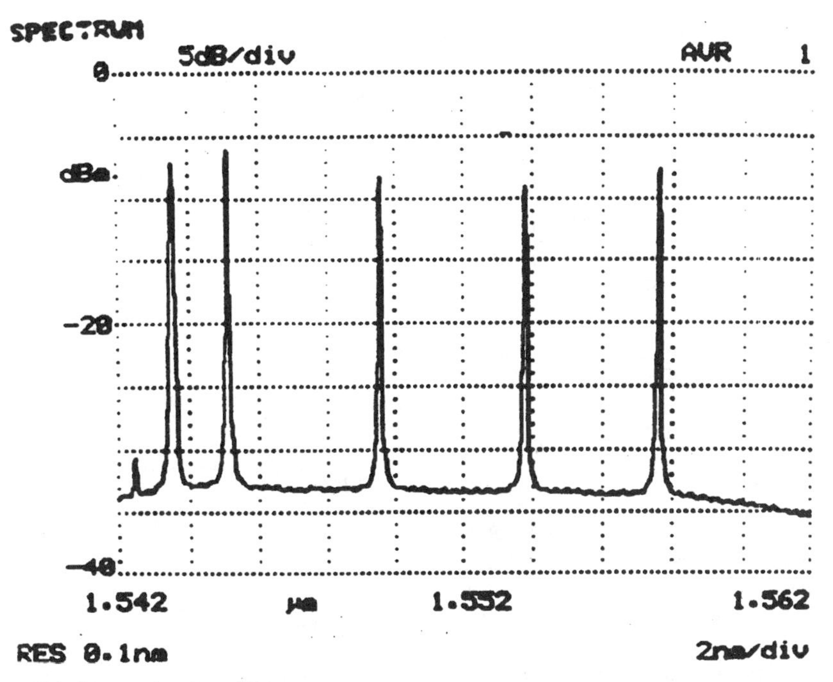

Fig. 25. Optical spectrum including wavelength at 1544 nm.

services to different users at different times. One example we illustrated was the rotation of the SCM analog video service. Two out of three SCM receivers were scheduled to receive the service at the same time. The location of these two receivers changed with time, thus the service was "rotating" around the network as if the events were being scheduled ahead of time.

We designed a network control scheme in which the wavelength routing switches are controlled "locally" by electronic signals applied from the local workstations (user terminals) [2], [4]. This network reconfiguration command is executed via a computer serial port (RS-232) originating from the local workstations through a switch/relay driver controller using TTL logic. In this arrangement, the central controller transmits optical control signals, the local workstations in the access nodes receive and interpret these signals, and in turn apply electronic signals to their respective local WDM switches.

The control of the optical cross-connect module that interconnects the two fiber rings is carried out by the workstation located in the nearest access node (currently we have upgraded the network testbed to use a dedicated workstation for control). Since the cross-connect module that interconnects the two fiber rings may be located hundreds of kilometers away from any of the access nodes, sending an electronic signal to its switch controller may not be easily achieved. One way to accomplish this is by using an extra wavelength dedicated as an embedded channel to transport the control information to the switch controller in order to reconfigure the switch. Using a separate DFB laser at 1544 nm we were able to demonstrate that we can propagate a fifth wavelength through the network while using the same network components (switches, EDFA's, etc.) that are shared by the other four wavelengths in the network. This wavelength was used to control the cross-connect module. Fig. 25 shows the fifth wavelength together with the four wavelengths launched from our laser array (after the multiplexer). We were able to launch and extract the fifth wavelength in our network using the existing components because of the design of the optical switches used. The extra ports that are available at each multiplexer/demultiplexer (Fig. 14) are used to tap a small fraction of the input power for monitoring purposes. These tapping ports are precisely the ones that enable us to launch and extract the fifth wavelength.

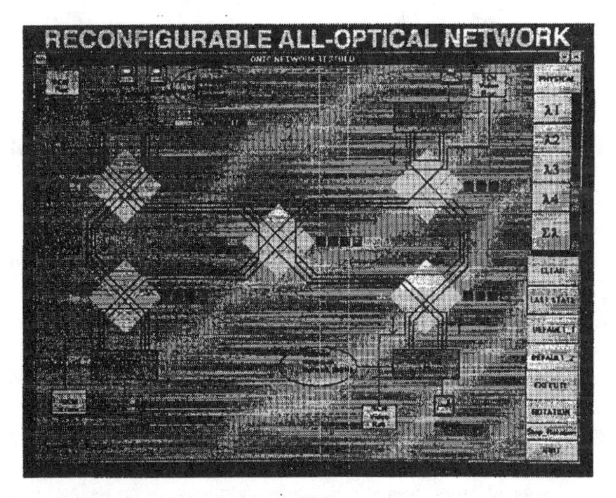

Fig. 26. Graphical User Interface (GUI).

VI. Network Management

Management of large WDM networks involves new issues originating from the superposition of a rearrangeable opti-

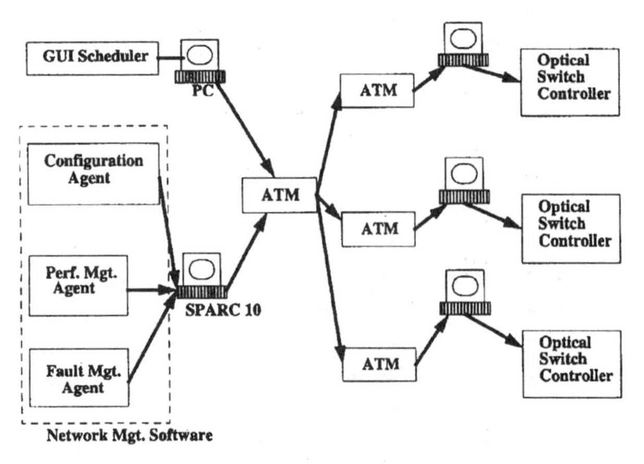

Fig. 27. Management network system.

cal layer supporting virtual channels on the physical fiber plant. Wavelength monitoring, channel power regulation, and optical layer (containing multiple wavelengths) fault detection/recovery are key WDM-specific elements requiring supervision and control from a management system for reliable performance. In this section, we discuss the design and implementation of the network management system used in the ONTC phase I testbed. Fig. 27 shows the high-level schematic of the complete Network Management system.

A. Network Element Management

1) Wavelength Performance Monitoring: The registration and power stability of laser sources provides a key source of concern for the long-term network performance. Wavelength misalignments or drift can cause channels to fall at the edges of the randomly distributed passbands of filters (in multiplexers/demultiplexers) within the network, leading to bandwidth-dependent penalties and error floors in the more extreme cases. Monitoring of the optical characteristics of the laser array sources was thus an integral consideration for network management. For this purpose, we incorporated 10 dB fiber couplers for all channels within the access nodes. Central wavelength, power, and laser temperature/current were tracked using computer-controlled equipment. The monitoring apparatus allowed long-term stability measurements as well as sensitivity investigations with environmental conditions.

2) Channel Power Regulation: Dynamically reconfigurable networks, while offering enhanced flexibility, also introduce potential channel interactions due to transient and nonlinear responses of active optical components as wavelength connections are changed. For example, optical amplifiers possess strong cross-saturation characteristics (channel gain dependent on signal power at other wavelengths) in the saturated operation regime necessitated by network power requirements. Furthermore DFB laser power and wavelength vary with modulation and bias points (temperature, current). Maintenance of stable operation conditions is thus imperative and was managed at several levels. First, all laser channels were continuously modulated for laser stability. To minimize path variations, the attenuators were adjusted to provide link transparency averaged over the network wavelengths. The working power budget for the system demonstration is shown in Table I for the network configuration of Fig. 1. An optical fiber gain of 20 dB was chosen to optimize the in-band signal-to-noise ratio and gain flatness as well as the out-of-band noise suppression ratio.

TABLE I
POWER BUDGET FOR THE DEMONSTRATION TESTBED

1	**Transmitter array power budget**	
	DFB laser array,modulated output per facet	**- 3 dBm**
	Fiber array coupling	**-11 dB**
	Multiplexer insertion loss	**-4 dB**
	Connector loss (2)	**-1 dB**
	Transmitter fiber amplifier gain	**+15 dB**
	Power launched into the ring per channel	**-4 dBm**
2	**Loss in transport path**	
	Optical cross-connect module insertion loss	**-9 dB**
	Fiber transmission loss	**-9 dB**
3	**Receiver array power budget**	
	Receiver fiber amplifier gain	**+15 dB**
	Optical cross-connect insertion loss	**-9 dB**
	WDM Demux (1x4)	**-4 dB**
	Connector loss (2)	**-1 dB**
	Power received by detector array per channel	**-21 dBm**
4	**Receiver Sensitivity**	**-31 dBm**
5	**System power margin**	**10 dB**

Active power equalization to provide transient and relative channel stability was implemented in a prototype node controller incorporated into the central cross-connect. The controller was placed prior to the WDM multiplexers and regulated channel powers to assure nominal −20 dBm input to the output optical amplifier. Channel powers were monitored using 10 dB couplers and electronic feedback circuitry used to control voltage-tunable attenuators for dynamic equalization within 100 ms.

B. Management Software

A commercial network management software called Net-Expert was customized and implemented in the testbed to provide limited network level management functions. The network management system currently supports configuration management of five multiwavelength optical switches and configuration management and fault management of four ATM switches. The network management system supports interfaces to each of the ATM switches, an interface to the GUI, and an interface to an administrative process that enables ATM switch port and optical switch provisioning information to be entered. A hierarchical graphical user interface is again used for this management system. Four main windows are displayed. The "gateway management window" shows the status of the ATM switch gateways and the "alert management window" shows

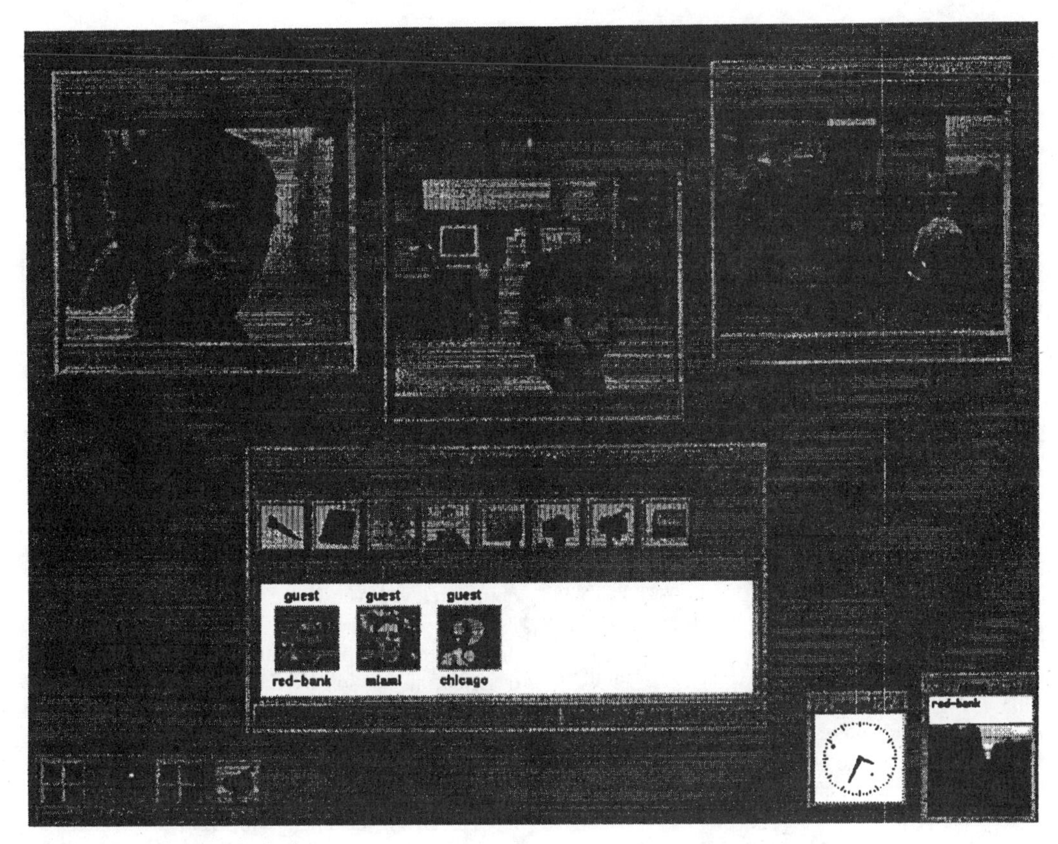

Fig. 28. Multiparty teleconferencing application.

the status of all the elements we monitor in the network (ATM switch ports, optical switches). Filters have been added so that the alert window can contain only faults, only provisioned alert information, or only gateway information. The third window displays the whole network and the fourth window allows us to zoom on any element in the network for more specific information such as the settings of an optical switch or the status of an ATM switch port. The second level graphic for the ATM switches shows which ports for each ATM switch are provisioned and which ports are not.

The management software always checks whether the connection to the ATM switches is currently established. If not, a warning alert is generated. The ATM switches are also polled periodically for switch counts for every provisioned OC-3 port. These counts include bit interleaved parity errors (BIP), loss of signal errors (LOS), far end block errors (FEBE), alarm indication signal errors (AIS) and far end receiver failure errors (FERF). If any errors appear in the switch counts, a major alert is generated for the specific port. These alerts can be cleared manually from the network manager, or automatically if the port counts go back to their normal values at the next polling interval. A log file is kept and updated, detailing the alerts that occur in the network.

VII. Network Applications

Clear channel transmission, i.e., transmission on dedicated optical paths without requiring electrical regeneration, of sub-carrier multiplexed analog and digital modulated signals in all-optical networks [16] offers the advantages of high capacity, low cost, and readily available upgradability of video and data services. One of the applications demonstrated in our reconfigurable network testbed was the distribution of analog video and binary phase shifted keying (BPSK) data signals co-propagated with interactive multimedia digital services transported on ATM/SONET over 150 Km of single mode fiber and a cascade of six EDFA's and five WDM cross-connects [3], [24], [25]. We were thus able to demonstrate the transparency of the network to both analog and digital signals.

Transmission of SONET/ATM digital signals and SCM analog-video and digital signals in the testbed was carried out simultaneously. We used wavelengths λ_1 (1546 nm), λ_2 (1550 nm), and λ_4 (1558 nm) for the multimedia teleconferencing application between access nodes A, C, and D (multicast audio and video signal from any one node to the other two) and wavelength λ_3 (1554 nm) for the FM-video signal distribution from access node D to access nodes A, B, and C. The objective was to study the performance limitations of SCM and SONET/ATM signals traversing through a common set of network elements in the testbed.

A. Multiparty Teleconferencing

We demonstrated transmission of ATM-switched, SONET-transported signals used for multimedia applications in the network testbed. These multimedia applications provide voice,

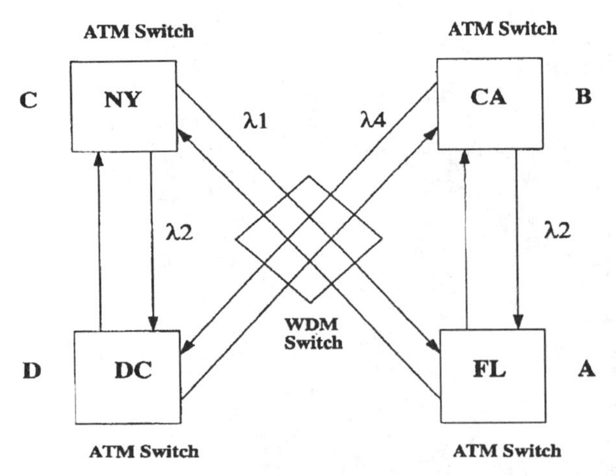

Fig. 29. ATM switch connections in the ONTC testbed.

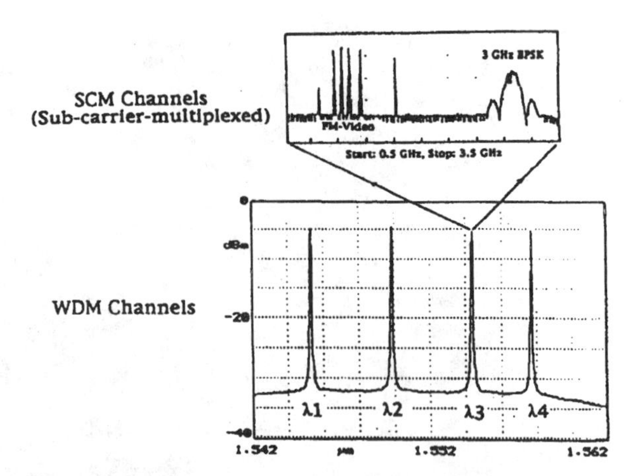

Fig. 30. Subcarrier multiplexed FM-video channels and a 3 GHz BPSK channel modulated at 155 Mbps carried by λ_3 at 1554 nm.

data, still-image and live video/voice services to various users. Fig. 28 shows an example of three users communicating in the network.

The paths from one user to another can be single or multihop [12]. Looking at Fig. 29, a user connected at ATM switch FL communicates with a user connected at ATM switch CA using wavelength λ_2. This is a single hop optical connection. When a user connected at ATM switch FL wants to communicate with a user connected at ATM switch DC; however, a two-hop connection is needed. It uses wavelength λ_1 to make a connection from ATM switch FL to ATM switch NY and wavelength λ_2 to make a connection from ATM switch NY to ATM switch DC. Fig. 29 also shows how we can reuse wavelengths in our testbed. Wavelength λ_2 for example, is used in both rings at the same time to connect ATM switches FL and CA and ATM switches NY and DC, respectively.

B. Transport of Subcarrier Multiplexed Signals

Four analog video channels, FM-modulated between 950–1450 MHz, and a 3 GHz BPSK channel modulated at 155 Mb/s are transmitted by a DFB laser diode at 1554 nm as shown in Fig. 30 [3], [24], [25]. Four VCR's at access node D supplying analog video were used to simulate satellite down linked microwave video channels with a standard spacing of 30 MHz/each in the FM domain. The SCM transmitter up converts the head-end video and injects it into the λ_3 port of the optical add/drop switch at access node D. Since the ATM switch is not required for analog applications, it is bypassed in the clear channel SCM demonstration. The SCM receivers demultiplex the signal and display the four separate analog channels.

The SCM application was designed for head-end distribution and transmission integrity studies. We studied the crosstalk effects of analog signals such as FM-modulated CATV channels transported by an optical wavelength channel sharing a common set of EDFA's with digital modulated WDM channels and traversing through the same set of WDM cross-connects comprising of both AOTF-based and fiber-optical relay-driven based optical switches.

In our clear channel SCM transmission experiment, three different optical paths were used in this testbed (Fig. 1) as described in Section IV-C1). Fig. 22 shows the BER plots of simultaneous transmission of digital SCM and SONET/ATM channels modulated at 155 Mb/s for the three clear channel paths mentioned previously. For both cases an AOTF cross-connect was used to interconnect the two fiber rings. The 3 GHz BPSK channel, with 18.3% modulation depth and a carrier to noise ratio (CNR) larger than 34 dB at the transmitter, shows a receiver sensitivity of -17.2 dBm at 10^{-9} BER for the *back-to-back* loop, with no power penalty observed for the *short* loop, and only a small power penalty of 0.8 dB for the *long loop* (attributed to the noise generated inside the AOTF module due to interference of the signal and leakages from imperfect polarization splitting and not well matched optical paths for TE and TM modes). The modest sensitivity of the 3 GHz BPSK channel is mainly due to the simple receiver front-end made of an InGaAs PIN photodiode with a 50 Ω load. We have improved the sensitivity by 13.7 dB to -30.9 dBm in a later experiment that employed an EDFA as the optical preamplifier followed by an optical filter before illuminating the PIN photodetector.

For comparison, the BER plots of the ATM/SONET channel carried by λ_1 modulated at 155 Mb/s, illustrated in Fig. 22, indicate a receiver sensitivity of -32.5 dBm at 10^{-9} BER for the *back-to-back* loop which is specified as the sensitivity of the SONET OC-3 receiver. As stated before, in Section IV-C-1, receiver power penalties of about 2 dB were observed for the long loop.

C. Selective SCM Multicast

The three optical paths for the clear channel SCM transmission experiment discussed in the previous section all refer to point-to-point connections between the SCM transmitter and one of the SCM receivers.

A new scheme was later adopted that enabled multicast wavelength connections. With this scheme each user can

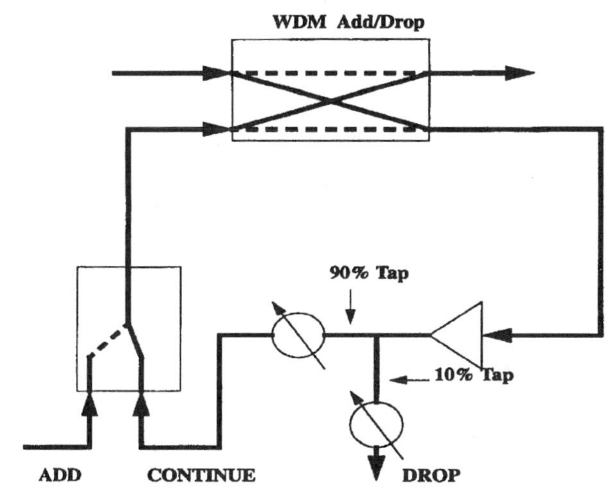

Fig. 31. Drop and continue implementation for wavelength λ_3.

receive the analog-video service regardless of anybody else already receiving the same service. We can multicast the analog-video service to a selective number of users or we can broadcast the service to everybody who is connected to the network. In order to achieve this we used a drop and continue scheme. If a user decides that it wants the SCM service and the central manager grants the request, the WADM switch at that particular node will switch to the *cross* state for wavelength λ_3. We do not however drop all the power to that access node but just 10%. The other 90% continues along the loop. If no other access node chooses to receive this service the remainder of the power for λ_3 will drop at the receiver of access node D. Fig. 31 shows how the drop and continue function is implemented in one access node.

VIII. Conclusion

We were successful in designing, integrating and demonstrating a reconfigurable all-optical network testbed using multiwavelength DFB laser arrays, multichannel OEIC receivers, optical cross-connect switches, erbium doped fiber amplifiers, and ATM/SONET switches. The performance of these network components working together in the testbed was studied and analyzed. We investigated point-to-point and selective multicast transport of multimedia services over ATM/SONET along with analog video and digital signals using subcarrier multiplexing. We also investigated network control, signaling and management schemes for such a network. The innovations of this project included multiwavelength laser arrays, optical cross-connect switches, multiwavelength optical amplifiers, network control and management algorithms, multiwavelength network operations and scalable network architectures.

Our findings suggest that WDM all-optical networks can be deployed in MAN and WAN environments to provide broadband service applications. Some of the results from the study of this network can be summarized as follows:

1) precision WDM laser arrays are achievable;
2) crosstalk in WDM switches is a serious impairment to scalability;
3) EDFA cascades have cost-performance tradeoff;
4) AOTF has serious limitations for dense WDM networks;
5) WDM channel power equalization is necessary; and
6) flexible, low-cost bandwidth is available.

The reconfigurable all-optical network testbed was demonstrated in San Diego for the OFC'95 conference and again in Leesburg, Virginia for Bellcore's Customer Solutions Forum in May, 1995. By taking apart and reassembling the testbed in a very short time we proved the robustness of a network which employed advanced WDM components. Therefore, large scale deployment of multiwavelength reconfigurable optical networks appears to be a realistic goal.

Acknowledgment

The authors would like to thank the program managers from ARPA, B. Hui, A. Yang, and R. Leheny for their continuous support of the ONTC phase I program. They would also like to acknowledge the following people who contributed and supported ONTC phase I testbed activities: T. P. Lee, C. E. Zah, W. C. Young, R. Bhat, R. Cordell, W. J. Tomlinson, C. Lin, P. Kaiser, R. Wagner, R. Standley, E. Goldstein, L. Eskildsen, Y. Silberberg, J. Patel, D. Smith, A. D'Alessandro, M. Goodman, K. Bala, B. Pathak, H. Shirockmann, J. Young, W. Chen, P. Grabbe, H. Song, J. Jackel, L. Curtis, D. Mahoney, F. Favire, P. Lin, N. Andreadakis, M. Koza, C. Caneau, A. Rajhel, J. Baran, J. Ringo, R. Kuchibhotla, C. Chen, S. Chang, V. Shah, E. Mason, and R. Spicer, all from Bellcore; A. Acampora and T. Stern from Columbia University; K. C. Wang, K. Pedrotti, F. Zuca, K. Nary, and P. Zampardi from Rockwell Science Center; G. Tangonan, J. Schaffner, and J. Pikulski from Hughes Research; and R. Hobbs and D. Fritz from United Technology Research Center.

References

[1] A. A. M. Saleh, "Transparent optical networks for the next-generation information infrastructure," in *Tech. Dig. OFC'94*, San Diego, CA, Feb. 1995, paper ThE1.
[2] G. K. Chang, M. Z. Iqbal, K. Bala, G. Ellinas, J. Young, H. Shirockmann, C. E. Zah, L. Curtis, B. Pathak, D. Mahoney, J. Gamelin, E. Goldstein, L. Eskildsen, and C. A. Brackett, "Experimental demonstration of a reconfigurable WDM/ATM/SONET multiwavelength network testbed," in *Tech. Dig. OFC'94*, San Jose, CA, Feb. 1994, post deadline paper PD-9.
[3] G. K. Chang, M. Z. Iqbal, G. Ellinas, H. Shirockmann, J. C. Young, R. R. Cordell, C. A. Brackett, J. H. Schaffner, G. L. Tangonan, and J. L. Pikulski, "Subcarrier multiplexing and ATM/SONET clear channel transmission in a reconfigurable multiwavelength all-optical network testbed," in *Tech. Dig. OFC'95*, San Diego, CA, Feb. 1995, paper ThI1.
[4] G. K. Chang, M. Z. Iqbal, J. Gamelin, and G. Ellinas, "Multiwavelength reconfigurable all-optical network testbed," in *Tech. Dig. NFOEC'95*, Boston, MA, June 1995.
[5] M. Z. Iqbal and G. K. Chang, "High performance optical switches for multiwavelength rearrangeable optical networks," in *Govt. Microcircuit Application Conf.*, San Diego, CA, Nov. 1994, pp. 475–478.
[6] C. A. Brackett, A. S. Acampora, J. Sweitzer, G. Tangonan, M. T. Smith, W. Lennon, K. C. Wang, and R. H. Hobbs, "A scalable multiwavelength multihop optical network: A proposal for research on all-optical networks," *J. Lightwave Technol.*, vol. 11, pp. 736–753, May/June 1993.
[7] G. R. Hill, P. J. Chidgey, F. Kaufhold, T. Lynch, O. Sahlen, M. Gustavsson, M. Janson, B. Lagerstrom, G. Grasso, F. Meli, S. Johansson, J. Ingers, L. Fernandez, S. Rotolo, A. Antonelli, S. Tebaldini, E. Vezzoni, R. Caddedu, N. Caponio, F. Testa, A. Scavennec, M. J. O'Mahony, J. Zhou, A. Yu, W. Sohler, U. Rust, and H. Herrmann,

"A transport network layer based on optical network elements," *J. Lightwave Technol.*, vol. 11, pp. 667–679, May/June 1993.

[8] S. B. Alexander, R. S. Bondurant, D. Byrne, V. W. S. Chan, S. G. Finn, R. Gallager, B. S. Glance, H. A. Haus, P. Humblett, R. Jain, I. Kaminow, M. Karol, R. S. Kennedy, A. Kirby, H. Q. Le, A. A. M. Saleh, B. A. Schofield, J. H. Shapiro, N. K. Shankaranarayanan, R. E. Thomas, R. C. Williamson, and R. W. Wilson, "A precompetitive consortium on wide-band all optical networks," *J. Lightwave Technol.*, vol. 11, pp. 714–735, May/June 1993.

[9] J. F. Labourdette and A. S. Acampora, "Logically rearrangeable multi-hop lightwave networks," *IEEE Trans. Commun.*, vol. 39, Aug. 1991.

[10] R. R. Cordell, W. J. Tomlinson, and G. K. Chang, "All-optical multi-wavelength networks—The optical network technology consortium," in Tech. Dig. NFOEC'94, June 1994.

[11] J. E. Baran, D. A. Smith, and A. D'Alessandro, "Multiwavelength performance of an apodized acousto-optic switch," in *Tech. Dig. OFC'94*, San Jose, CA, Feb. 1994, paper TuM5.

[12] A. S. Acampora, "A multihop local lightwave network," in *Proc. IEEE GLOBECOM'87*, Tokyo, Japan, Nov. 1987, pp. 1459–1467.

[13] D. A. Smith, J. E. Baran, J. J. Johnson, and K. W. Cheung, "Integrated-optic acoustically tunable filters for WDM networks," *IEEE J. Select. Areas Commun.*, vol. 8, pp. 1151–1159, 1990.

[14] C. E. Zah *et al.*, "1.5 mm Compressive-strained multiple-quantum-well 20-wavelength distributed-feedback laser arrays," *IEE Electron. Lett.*, vol. 28, pp. 824–826, 1992.

[15] W. C. Young *et al.*, "4-λ laser array multi-chip transmitter module using conical-microlensed fibers," in *Integrat. Photon. Res. Conf.*, Dana Point, CA, Feb. 1995.

[16] R. Olshansky, V. A. Lanzisera, and P. M. Hill, "Subcarrier multiplexed lightwave systems for broadband distribution," *J. Lightwave Technol.*, vol. 7, pp. 1329–1342, Sept. 1989.

[17] C. E. Zah and T. P. Lee, "Monolithic integrated multi-wavelength DFB laser arrays and star couplers for WDM lightwave systems," *Optics Photon. News*, Mar. 1993.

[18] E. Goldstein, "Multiwavelength fiber-amplifier cascades for networks," in *Tech. Dig. OFC'94*, San Jose, CA, Feb. 1994, pp. 65–66.

[19] E. L. Goldstein, V. da Silva, L. Eskildsen, M. Andrejco, and Y. Silberberg, "Inhomogeneously broadened fiber-amplifier cascade for wavelength-multiplexed systems," *IEEE Photon. Technol. Lett.*, vol. 5, pp. 543–545, May 1993.

[20] E. L. Goldstein, L. Eskildsen, V. da Silva, M. Andrejco, and Y. Silberberg, "Suppression of dynamic cross-saturation in multiwavelength lightwave networks with inhomogeneously broadened fiber amplifiers," *IEEE Photon. Technol. Lett.*, vol. 5, pp. 937–939, Aug. 1993.

[21] K. Bala, F. R. K. Chung, and C. A. Brackett, "Optical wavelength routing, translation and packet/cell switched network," in *Tech. Dig. OFC'95*, San Diego, CA, Feb. 1995, paper Thl5. Also Bellcore TM-24506 and TM-24507, Feb. 1995.

[22] K. Bala and C. A. Brackett, "Cycles in wavelength routed optical networks," in *LEOS Summer Topical Mtg.*, Lake Tahoe, NV, Aug. 1994.

[23] L. Eskildsen, E. L. Goldstein, G. K. Chang, M. Z. Iqbal, and C. Lin, "Self-regulating WDM amplifier module for scalable lightwave networks," *IEEE Photon. Technol. Lett.*, vol. 6, pp. 1321–1323, Nov. 1994.

[24] J. H. Shaffner, G. Tangonan, J. L. Pikulski, G. K. Chang, and M. Z. Iqbal, "Extended utilization of the all-optical network through subcarrier multiplexing," *IEEE Photon. Technol. Lett.*, 1996, to be published.

[25] G. K. Chang, M. Z. Iqbal, J. K. Gamelin, G. Ellinas, J. H. Schaffner, G. L. Tangonan, and J. L. Pikulski, "Clear channel transmission of ATM/SONET and subcarrier multiplexed signals in a reconfigurable multiwavelength network testbed," *IEEE Photon. Technol. Lett.*, to be published.

Coherent Optical Systems Implemented for Business Traffic Routing and Access: The RACE COBRA Project

ERNST-JÜRGEN BACHUS, TERESA ALMEIDA, PIET DEMEESTER, MEMBER, IEEE, GEERT DEPOVERE, ALFRED EBBERG, MARIO RUI FERREIRA, GIOK-DJAN KHOE, FELLOW, IEEE, OSCAR KONING, RICHARD MARSDEN, JOHN RAWSTHORNE, AND NICO WAUTERS

(Invited Paper)

***Abstract*—The RACE COBRA consortium has performed four field trials to verify the suitability of dense WDM combined with heterodyne detection (for which the term CMC, "Coherent Multicarrier" is used) for different network applications. The main advantage in using heterodyne detection is the simple access to individual optical channels by only switching local lasers to the desired channel. Heterodyne detection acts as a perfect optical filter. Thus, a plurality of dense spaced optical carriers can be placed, e.g., within one WDM window. The additional freedom in the channel multiplex provides some advantages which are shown in the demonstrators: A routing flexibility in the RENO (Reconfigurable Node) cross-connect demonstrator supporting STM-16 at Philips Research in Eindhoven. A routing flexibility in the demonstrator at BBC in Kingswood Warren, UK, supporting uncompressed HDTV communication in a studio environment. By shifting switching functions to the terminals, the network is kept completely passive. This is also obtained in the local area videoconference trial at CET in Aveiro, Portugal, using high quality standard TV and in the BCPN trial at KPN Research in Leidschendam, NL, where ATM-streams are supported for multimedia services. The paper outlines the demonstrator concepts as well as some subsystems and key components developed by the consortium.**

I. Introduction

ONE OF the objectives of the RACE program was to promote the competitiveness of the community's telecommunications industry, operators, and service providers in order to make available to final users the services which will sustain the competitiveness of the European economy. The COBRA consortium reflects this objective by including among its partners two network operators and one service provider, as well as three industrial partners and two research institutions. The consortium was therefore in a position to investigate from many aspects the potential of innovative systems for short and medium term IBC (Integrated Broadband Communication) implementation.

Since IBC was first defined as the objective for the RACE program, there has been an important shift in perception. The orientation of the work has shifted from "Exploring Options" under RACE Phase I to "Preparing for Implementation" under Phase II. Work carried out in the COBRA project is aimed toward this latter exercise. Emphasis is placed on the exploitation of the optical frequency domain by coherent techniques, but other technologies are used as well where appropriate. Examples are the use of ATM in the KPN Research field trial, the coexistence of the WTDM system in the TV studio BCPN, and the use of SDH interfaces and optical space switches in the optical RENO cross-connect demonstration.

II. COBRA Consortium and RACE Contributions

The RACE Phase II project R2065 COBRA (Coherent Optical systems implemented for Business traffic Routing and Access) covered the period of 1992–1995. The project participants are Philips (prime contractor), the Heinrich-Hertz-Institut (HHI), GEC-Marconi Materials Technology Limited, Siemens AG, Interuniversity Microelectronics Centre (IMEC), British Broadcasting Corporation (BBC), Centro de Estudos de Telecomunicacoes (CET), and PTT Telecom, NL (and KPN Research as a subcontractor). Since 1994, the Project Leadership has been subcontracted by Philips to the Eindhoven University of Technology.

COBRA was a follow-up of the RACE Phase I project CMC (Subscriber Coherent Multichannel System). In the CMC project, an important highlight was the design and construction of an engineered CMC distribution system. In the COBRA project, the knowledge and techniques developed in the CMC project were further enhanced and four different but

Manuscript received April 28, 1995.

E.-J. Bachus is with the Heinrich-Hertz-Institut, Einsteinufer 37, 10587 Berlin, Germany.

T. Almeida and M. R. Ferreira are with the Centre de Estudos de Telecomunicacoes, Rua Eng. Jose Ferreira Pinto Basto, 3800 Aveiro, Portugal.

P. Demeester and N. Wauters are with the Department of Information Technology, University of Gent-IMEC, Gent, Belgium.

G. Depovere is with Philips Research Laboratories, 5656 AA Eindhoven, The Netherlands.

A. Ebberg was with Siemens AG, Corporate Research and Development, 81730 München, Germany. He is now with the Fachhochschule Westkueste, 25746 Heide, Germany.

G.-D. Khoe is with the University of Technology Eindhoven, 5600 MB Eindhoven, The Netherlands.

O. Koning is with Royal PTT Nederland N.V., KPN Research, 2260 AK Leidschendam, The Netherlands.

R. Marsden is with British Broadcasting Corporation, Research and Development Department, Kingswood Warren, Tadworth, Surrey, KT20 6NP, U.K.

J. Rawsthorne is with GEC-Marconi Materials Technology Limited, Caswell, Towcester, Northants, NN12 8EQ, U.K.

Publisher Item Identifier S 0733-8724(96)04779-2.

Reprinted with permission from *IEEE Journal of Lightwave Technology*, Vol. 14, No. 6, pp. 1309-1319, June 1996.

complementary multiwavelength systems were implemented in field trials.

The activities in COBRA have been split into different working groups: the WG Implementation coordinated the implementation of the demonstrators, the WG Technology specified the components and the link to the overall RACE program was established through the WG Systems. A public deliverable, the "COBRA-Report on Flexible Broadband Networks" was issued annually by this working group [1]–[4]. In these reports, the progress of the COBRA project was outlined and updated each year. Each COBRA Report also contains a summary of worldwide progress and trends in the area of multiwavelength networks and dedicated components.

The Working Group Systems also participated in a number of RACE activities such as the Project Line 1 (PL1) meetings and associated workshops, and Sub Technical Groups (STG's), which have been set up to draft specifications, and to provide essential inputs which are now included in the RACE Common Functional Specifications (CFS's) and in standardization proposals submitted to ETSI (European Telecommunications Standard Institute).

Cooperation with other RACE projects has been set up. The COBRA demonstrator at BBC shared the physical network simultaneously with the demonstrator of the RACE project WTDM. The network of the COBRA CET demonstrator permits the operating status of the demonstrator to be assessed by collecting information on the status of each network element. This specific activity has been performed in close cooperation with RACE project ICM. Also a Techno-economical analysis was carried out with assistance of RACE projects TITAN and INTERACT.

III. Optical Network Trends and COBRA

The aim of this chapter is to point out main features of the COBRA concept for the optical networks in the demonstrators.

A. Trends Toward High Capacity

10 Gb/s transmission systems are already commercially available. The next generation of WDM systems focusing on four or eight wavelength channels within the EDFA band are under development. Impressive advances on the laboratory level have recently been shown: the transmission of 400 Gb/s using optical TDM-processing techniques [5] and the arrival at the 1 Tb/s bound through multiplexing of wavelength channels (55×20 Gb/s [6] and 10×100 Gb/s [7]). The question arises, where the limit will be. In [8] the ultimate Shannon limit for the full band between 1.3 μm and 1.6 μm is estimated at 300 Tb/s. Pushing toward this frontier is of course an attractive approach. A distinct approach is to take an advanced but already engineered technique and to demonstrate various network applications. This was done in the COBRA project and is outlined in the following chapters.

B. The COBRA Concept

COBRA used a multiwavelength technology known as CMC (Coherent Multicarrier) developed in the RACE phase I project R1010 [9] which is based on optical heterodyne detection.

In a heterodyne receiver, the incoming signal is converted into a microwave signal by combining it with the light of a local oscillator laser at a slightly different frequency. The resulting intermediate frequency (IF) signal in the microwave region can be processed by conventional techniques which are well known from radio systems. The local laser frequency is locked to the signal light frequency by an automatic frequency control loop. Up to now this is the usual control scheme and also applied within COBRA.

The first obvious advantage offered by heterodyne detection is the high sensitivity of the receiver (and a corresponding long transmission span). With a sufficient amount of local oscillator power, a heterodyne receiver offers a sensitivity close to the shot-noise-limit. Meanwhile, this advantage has partly been eroded since the introduction of optical fiber amplifiers. The sensitivity of an optical direct detection receiver equipped with an optical fiber preamplifier approaches the sensitivity achieved with a heterodyne receiver.

1) Heterodyne Detection: A Perfect Optical Filter: The major advantages, however, offered by a heterodyne receiver are the selectivity and the tunability. A heterodyne receiver acts as a nearly perfect optical filter with very attractive features. The optical filtering function of the heterodyne receiver is determined by its electrical filter in the IF branch. In contrast to optical filters used in direct detection systems, multipole electrical filters of excellent passband characteristics are easily implemented in a coherent receiver. These nearly ideal optical filtering characteristics will be extremely useful when selecting a single optical channel from a high number of densly spaced channels transmitted on a fiber. A narrow channel spacing is advantageous, especially in the case of systems using a number of concatenated optical fiber amplifiers. As the frequency response of an optical fiber amplifier is not ideally flat, every additional amplifier reduces the available bandwidth leaving only a few nanometers in some cases.

Finally, the heterodyne receiver can be switched electrically to any desired frequency channel by changing the frequency of the local oscillator. Tuning can be performed very quickly compared to thermally or mechanically adjusted optical filters. A tuning range of more than 10 nm (>1 THz) [10] and a passband of around 1 GHz can be realized simultaneously. In contrast to preamplified direct detection systems, heterodyne systems can be operated over the entire wavelength range of the optical fiber (e.g., from 1300 nm to 1600 nm) and are not restricted to the passband of a particular optical amplifier. Note that the tuning range of semiconductor lasers is very large in principle. A continuous tuning of 242 nm ranging from 1.32 μm to 1.562 μm has already been shown in [11]. Some recent advances in CMC technology are reported in [12]–[20].

In our terminology, the term CMC (Coherent Multicarrier) is used for (dense spaced) optical carriers (which all have to fulfill certain coherence-conditions) combined with heterodyne detection. Channel selection (which is a switching function) is performed within the receiver. Thus, in any cases, the optical path is terminated by the corresponding transmitter-receiver pair. (If one wishes to route transparent optical WDM channels through a network, frequency selective all-optical devices like multiplexer, filter, router, and frequency converter are

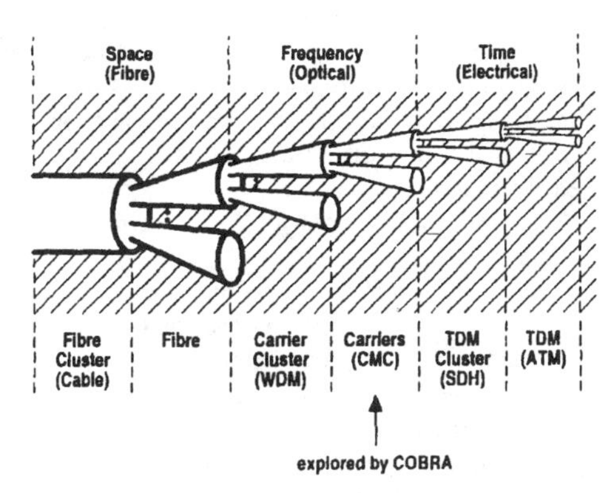

Fig. 1. Hierarchical structure of multiplexing. The COBRA project aims at the potential of the carrier "pipes" and the interaction with the others.

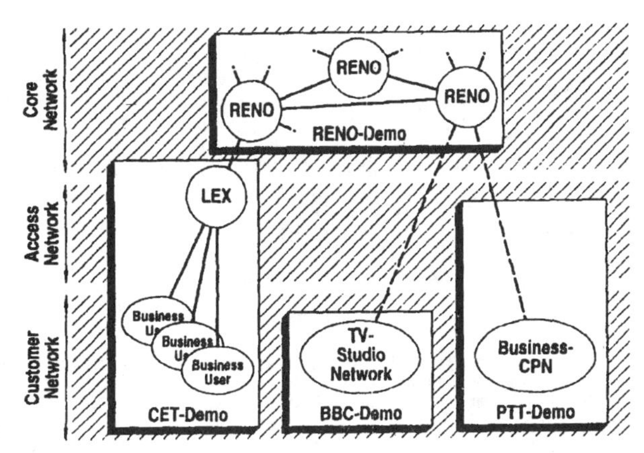

Fig. 2. The four COBRA demonstrators and the network segments covered by these demonstrators.[1]

indispensible). Most of the key technical problems associated with heterodyne detection have already been addressed and solved during recent years [21], [9], [22], [23], [24]. The technical advantages of coherent systems are evident; however, it still remains to be shown that these systems are economically competitive compared to other solutions. A step toward cost reduction is the monolithic integration of a heterodyne receiver front end on a single chip [25], which has also been one activity within the COBRA project.

2) Illustration of the Multiplexing Concept: A layered transport network architecture for SDH has been commonly accepted and will be standardized by ITU [26]. The ITU-T recommendation G.803 also explicitely states, that "the necessity of an explicit description of a WDM layer in the transmission media layer is for further study." The concept can be extended with new layers representing optical multiplexing and cross-connect functionality as proposed in [27], [28], a WDM layer and a CMC layer. The additional CMC multiplexing explored by the COBRA consortium is illustrated in Fig. 1.

The style of this figure differs from that commonly used (showing the layered areas) in order to illustrate the fact that a WDM-channel is a bundle of CMC-channels, each CMC-channels a bundle of TDM-channels which itself may split into the layers of SDH and ATM.

Physical and logical topology of a network need not to be identical, and can be optimized independently. Logical reconfiguration of a network (for example, in the case of varying traffic or service demands), is possible without changing the physical topology. Furthermore, optical multicarrier technology can also attribute to network survivability, by offering a very flexible way of rerouting traffic. For these purposes the RENO's (see Ch. IV-A) act as optical carrier cross-connects providing a means for flexible optical network reconfiguration.

3) Main Features of the COBRA Concept: The main features of the COBRA concept shown in the demonstrators are summarized in brief:

—A fine granularity of wavelength channels. The channel spacing can be performed close to the bandwidth of the channels.

—A dedicated wavelength channel for each application. This is shown for the transport modules of SDH (STM-1, STM-16), for uncompressed TV and HDTV distribution and for ATM-streams.

—COBRA networks are based on the broadcast-and-select principle. "Broadcast" by means of star-couplers, "select" by tuning of receivers.

—Optical paths are always terminated by the corresponding transmitter-receiver pairs. Each receiver performs a fully electrical regeneration of the digital signals (3R).

—No optical filters or multiplexers are used for discriminating individual CMC channels.

—Simple upgrading is possible by adding just more carriers (within the range of tunability and power budget).

—Unlimited cascadability is possible through modular subnetworks. The resulting multihop network is of course not all-optical.

—The optical COBRA networks are circuit switched networks (packet switching with CMC is indeed an important option for further investigations).

—Emphasis is laid on demonstrating practical applications including the management of the demonstrator networks.

IV. The COBRA Trials

The demonstrators have been chosen to illustrate different aspects of an overall communications network. The RENO demonstrator is a part of the core network, forming the switching centres at which SDH traffic is routed [29]. The three other demonstrators address three types of BCPN, each with specific customer requirements (Fig. 2).

The individual demonstrators are now outlined by short descriptions and figures followed by a table summarizing their main parameters.

A. The RENO Core Network

The general structure of a single RENO (Reconfigurable Node) is shown in Fig. 3. The maximum number of fibers and carriers per fiber is assumed to be N and M, respectively. The

[1] The drawn lines in the figure are indicating optical connections, not necessarily single fibers. A connection may consist of a single fiber (duplex mode) or two fibers (simplex mode).

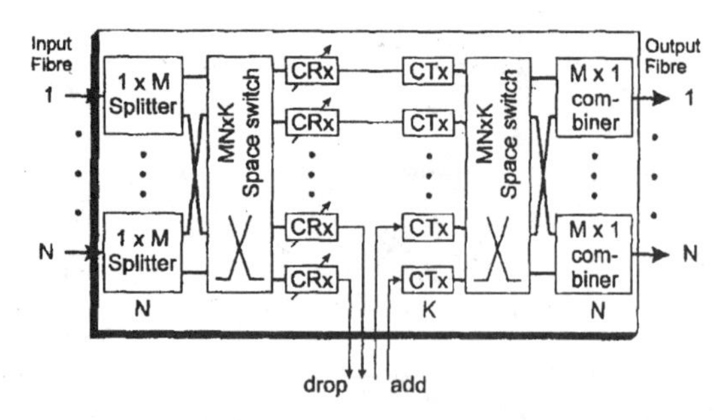

Fig. 3. General structure of the RENO's (Reconfigurable Nodes). CT_x is a Coherent Transmitter at a fixed optical frequency, CR_x is a Coherent Receiver switchable to any of the M carriers ($K \leq MN$).

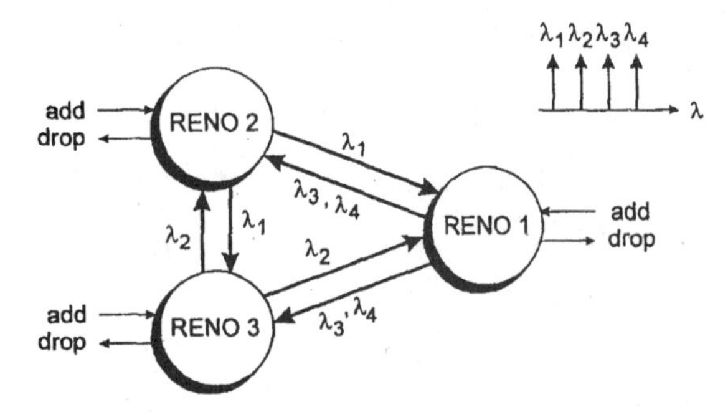

Fig. 4. The RENO core network demonstrator.

incoming carriers are transferred through the RENO to the outgoing fibers via space switch matrices and CR_x-CT_x pairs ("C" stands for coherent) performing the wavelength translation. Also add-drop-functions are implemented. The transmitters have a fixed wavelength, the receivers are tunable (it has to be mentioned here that a plurality of different node-architectures are possible to obtain the same functionalities). A part of a core network based on RENO cross-connects is depicted in Fig. 4. The nodes are interconnected by unidirectional optical fiber links. Generally, such nodes will be located in major cities, also interfacing via dedicated optical links to the local exchanges (LEX's). In the demonstrator all CT_x and CR_x are working at 2.5 Gb/s for transmit and receive STM-16 signals of the SDH [30], [27]. In [31] it is shown that the SDH management facilities can be used to locate errors (monitoring) and to transport network information across the optical layer (signaling). In [28] it is shown how signaling and surveillance facilities can be achieved through the addition of a low bitrate signal to each wavelength channels.

The RENO demonstrator is restricted in terms of available hardware (4 wavelengths for 3 RENO's), but illustrates the full capability of the RENO concept.

A typical feature of the RENO network is the possibility to adapt the traffic per fiber to changes in the actual short-term traffic demand [32]-[35]. Each transmitter can be allocated to any trunk. This is cost effective, because reductions in hardware requirements can be made through statistical traffic analysis, the number of CR_x-CT_x pairs being determined using standard network planning techniques. Spare CR_x-CT_x pairs may also be present to allow extra capacity for special events. These may also be used for replacing any CR_x-CT_x pair that has become unavailable. In the case of a connection failure between two nodes, the total traffic between the two nodes may be easily rerouted via one or more other nodes of the network. This avoids the necessity for extra optical switch hardware and links.

Some nodes will in practice only interface with a small number of other nodes, and will consequently only have a limited number of CR_x-CT_x pairs. Upgrading may be achieved by simply adding extra hardware modules (optical switch, tunable receiver, and transmitter), assuming that large enough optical splitters and couplers were initially installed. Performance studies which calculate the wavelength requirement and the survivability for realistic networks are published in [33], [34].

B. Broadband Video Multi Access—The BBC TV Studio Network Demonstrator

The objective of the BBC Demonstrator was to show an application of CMC technology in broadband routing—the transport of full-band television signals in a studio center. A coherent transmission layer has been specified and developed to determine the potential capacity of a practical network, and to show that coherent techniques are an effective means of extending close-spaced WDM. It is estimated that coherent techniques could increase the capacity of a conventional WDM system by about a factor of eight.

There is a requirement in a TV studio center to interconnect technical equipment on, say, an hour-by-hour or day-by-day basis, to draw together the technical facilities for a particular program production. (This is distinct from the rapid synchronized switching reqired for editing and program "cuts"). The network must be nonblocking, point-to-multipoint, with the capacity for several hundred video, audio, and ancillary signals.

RACE Projects R2001 and 1036 have developed a network to meet these requirements based on a combination of close-spaced wavelength multiplexing and electrical multiplexing. This system, known as WTDM, provides only just enough capacity for a typical studio center. The work in COBRA shows the extra capacity and optical power margin offered by coherent techniques. This can be used for more signals, for HDTV as opposed to conventional TV signals, to increase network flexibility, or a combination of these. It has also been demonstrated that WTDM and CMC can operate on the same optical network. The ability to tune the CMC receivers eliminates some of the high-speed electrical switching that would be needed in a WTDM network.

The typical demonstration is shown in Fig. 5. The coherent layer itself consists of three sets of 2.5 Gb/s transmission hardware (one each from GMMT, Philips and Siemens), and a central controller (developed by HHI and GMMT, and known within the project as the wavelength manager). The nominal channel spacing of the coherent transmitters is 0.25 nm (about 31 GHz), adjustable for experiments, grouped around 1553 nm. The WTDM lasers operate at 2 nm intervals from 1530

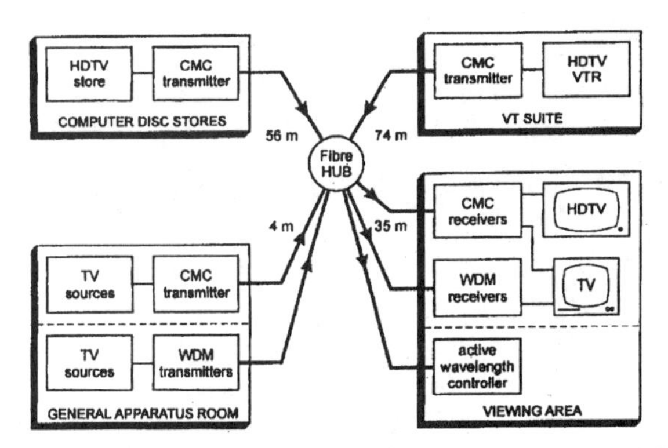

Fig. 5. The BBC Studio demonstrator: A merge of the RACE WTDM and the COBRA project. HDTV studio terminals in the COBRA partare communicating at the STM-16 line bitrate of 2.5 Gb/s (uncompressed digitized video channels in the payload). All the COBRA CMC channels fit into a single WDM slot of the WTDM testbed.

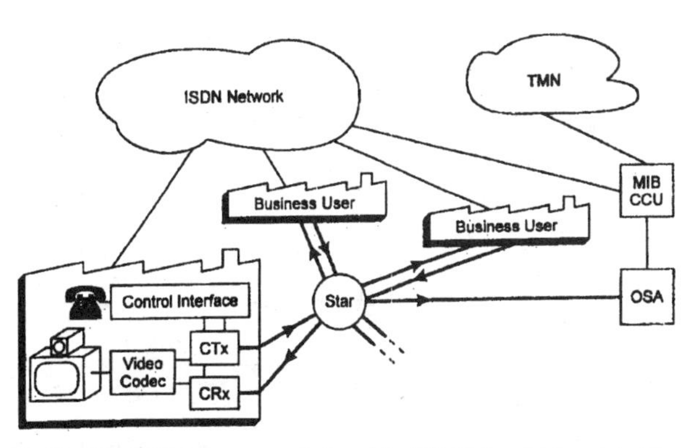

Fig. 6. The CET demonstrator designed for 140 Mb/s point-to-point videoconferences. The network is controlled via the ISDN Network by the Central Control Unit (CCU). Remote management of the demonstrator is performed by the TMN platform which has access to a Managed Information Base (MIB).

nm to 1560 nm. The coherent transmitters and receivers are connected to the WTDM network's central optical star by additional 4 × 4 couplers, and in this arrangement replace the 1552 and 1554 nm channels. Program signals are connected to the network via SDH multiplexing equipment developed in R1036, and audio, TV, and HDTV interfaces developed in R1081. The bit rates of these signals range from 3.072 Mb/s to 1.152 Gb/s.

The system was originally demonstrated during September 1994. This demonstration linked a series of technical areas at the BBC's Research and Development Department in Surrey, UK, and was held as part of the Department's biannual "open days." COBRA equipment was used to convey signals ranging from audio to uncompressed HDTV over an existing optical network. The equipment was reassembled in March 1995 at RAI (Italien, State Television) Technical Research Center Turin, and a further trial was conducted in autumn 1995 at the main studios of NRK (the Norwegian State Broadcaster), Oslo. These last two events were collaborations with RACE 2001, WTDM Pilot.

C. Videoconferencing in Aveiro Town—The CET Demonstrator

The COBRA demonstrator implemented by CET in Aveiro aims at assessing the viability of videoconference services by means of coherent techniques for the transmission of digitized standard video signals over a PON with a star topology. Coherent equipment is used which was developed under the R1010 project and designed for the distribution of high quality video channels at 140 Mb/s [36], [37].

The field trial is based upon a standard singlemode fiber passive optical network (PON) transmitting video signals between three sites (CET, Telecom Shop and Telecoms Institute) Fig. 6. The PON is configured as a star scheme with one duplex fiber cable per user. For the purposes of system control, one output of the central four-by-four star coupler is monitored by an optical spectrum analyzer, allowing measurement of optical power and frequency stability.

The optical network is supported by an N-ISDN network running in copper symmetrical pairs. This network uses the ELDIS Local Digital Exchange developed by CET and interconnects the Control Unit with each of the users premises. There, an ISDN phone set is connected to a PC featuring an ISDN card and through which both digitized voice/sound and control signals are to be passed. The network permits the remote management of the demonstrator by a PON TMN platform which has access to a Managed Information Base (MIB) located in the COBRA Control Unit (CCU). The CCU contains updated information on the status of each network element. The functionalities of this platform include alarm management and configuration monitoring. This activity has been performed in close cooperation with R2059 ICM.

The CET demonstrator was installed in the Aveiro town area using the existing duct and cable network infrastructure from Telecom Portugal. The distance between each of the users and the control center (CET) is not greater than 3 km. Bidirectional interactive videoconference services (both point to point and point to multipoint) are available to business users, allowing telelearning/training and joint business meetings to be held. These define the requirements for the remote control of terminals.

In a realistic environment with a large number of users, a number of questions arise, answers to which are difficult to find from a small size demonstrator. These include spectrum management aspects, arbitration of the videoconferencing session, PON monitoring and alarm processing issues as well as delays introduced by network control. In addition, if less optical carriers are available than the total number of users, congestion is likely to occur and must be accommodated. These issues are studied using a network model with at least 100 users and a smaller number of available optical carriers. The results obtained complement the tests of the demonstrator.

D. Multimedia Communication in Leidschendam—The PTT-BCPN Demonstrator at KPN Research

It is generally expected that a future business customer premises network (BCPN) must deal with a large variety of broadband services, each specified by a different bit rate and traffic characteristic. In future BCPN's several connection types (like point-point, multipoint-multipoint, and point-

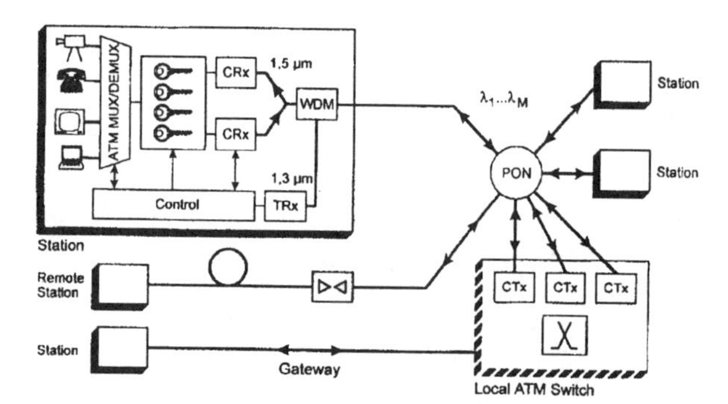

Fig. 7. The concept of the PTT-BCPN demonstrator at KPN Researchin Leidschendam, showing coarse WDM, CMC and ATM multiplexing (TR_x = transmitter and receiver working at $\lambda = 1.3$ μm, CTR_x = coherent transmitter and receiver in the 1.55 μm waveband). The network can be upgraded by alocal ATM switch (dashed lines).

multipoint) must be supported, besides which, the total network capacity is expected to increase significantly. Since it is at present almost impossible to foresee which applications will become prevalent, advanced BCPN's must have great flexibility and must be able to supply a high transport capacity.

Coming back to Fig. 1 showing the multiplex hierarchy, the demonstrators described so far are transporting their services on a corresponding single optical carrier. A higher granularity is obtained in the PTT BCPN demonstrator by introducing ATM multiplexing on each individual carrier [38]–[40]. Thus, the stations in Fig. 7 are individually linked by circuit switching through tuning of corresponding CT_x-CR_x pairs. Each carrier is transporting data at the STM-1 line rate of 155.52 Mb/s. The multimedia terminals of the linked stations are now communicating via additional TDM multiplexing, which is chosen here as ATM multiplexing in order to obtain a flexible granularity and to interface the future B-ISDN standard.

The basic idea of this concept is as follows:

1) By multiplexing a bundle of ATM circuits onto an optical carrier the concept "ATM optical path" introduced by Sato *et al.* [41] is realized as a measure to obtain very high capacity networks based on ATM.
2) Since CMC switching allows for point-to-point as well as for point-to-multipoint connections and ATM multiplexing allows for addressing individual terminals, a high number of circuit configurations can be established without any ATM switch in the network (it has to be noted that an ATM switch is indeed not a part of the demonstrator).
3) By introducing ATM switches (a single centralized switch or many decentralized switches within the stations) the network can easily be upgraded to multipoint-to-multipoint connections without changing the physical topology (the PON) and the optical hardware. In the final test and evaluation phase of the demonstrators these issues will be addressed in more detail.

The hardware configuration of the PTT BCPN trial in Fig. 7 consists of five transmitters and receivers connected to each other by means of a passive optical star. The transmitters and receivers are distributed among four stations. Each station comprises at least one audio/video (de)coder and an ATM (de)multiplexer.

TABLE I

Short name of the trial	RENO demonstrator	BBC demonstrator	CET demonstrator	PTT demonstrator
Location of the trials	Philips Research Eindhoven, NL	Kingswood Warren, UK	Aveiro, P	KPN Research Leidschendam, NL
Application shown	Cross-connected core network	HDTV in a studio	Videoconference	multimedia, medical
Number of nodes in the trial	3 OCCs (potential: unlimited)	1 passive star	1 passive star	1 passive star
Number of terminals connected to the nodes in the trial	(3 for testing only)	5	3	4
Number of optical carriers and line transmitters realized Demonstrator concept allows for	4 in total 16 per link	3 per node 8 per WDM window	3 per node 128 per node	4 per node 128 per node
Number of line receivers realized	4 in total	3	3	4
Number of potential line receivers due to power budget limit (without amplifiers)	16 per link	more than 100	more than 4000	more than 4000
Fibre length in the trials	60 km between OCCs	below 1 km	average 3 km	up to 50 km
OAM and signalling via	separate fibre network	separate electrical network	public ISDN	WDM overlay at λ=1.3µm
Optical center frequency in GHz	193036.5	193040.8	192150.0	192150.0
Carrier spacing in GHz	15	30	20	25
Frequency stabilization implemented	yes	yes	yes	no
Bitrate per carrier in Mbit/s	2488.32	2488.32	140	155.52
Service per carrier shown	routing of STM-16	one uncompressed HDTV in STM-16	one PAL video channel	ATM streams
Modulation scheme	CPFSK	CPFSK	FSK	FSK
Coding scheme	NRZ	NRZ	Manchester	Manchester
Level of demonstration	field trial	application pilot	application pilot	application pilot

The use of a passive optical network implies that all information is sent to all subscribers. Therefore, in many applications, network security is important. Implementation of network security functions is introduced in the demonstrator at three levels: User authentication during call setup, exchange of encryption keys, and high-speed encryption of user data. Since encryption is performed on the ATM adaption layer, services at various bit-rates, multiplexed on a single CMC channel, can be encrypted separately. Dedicated hardware and software is part of the demonstrator in order to make encryption of the ATM data stream possible.

Network control is another important aspect in the BCPN concept. Network control includes three main aspects: Call-control, wavelength management, and security/key control. In the BCPN a direct detection overlay network on the same fiber infrastructure at $\lambda = 1.3$ μm is used for the network control functions [42].

A third important aspect tested in the demonstrator is optical amplification. The optical amplifier (EDFA) will either compensate the splitting loss in the passive optical star or the loss in the length of installed fiber. In the demonstrator two configurations are tested successfully: Unidirectional traffic with a transmissive star and an unidirectional optical amplifier, and bidirectional traffic with a reflective star and a bidirectional optical amplifier [43], [44].

E. COBRA Demonstrators Overview

The table on the following page summarizes main parameters of our demonstrators. Also, some technical details which

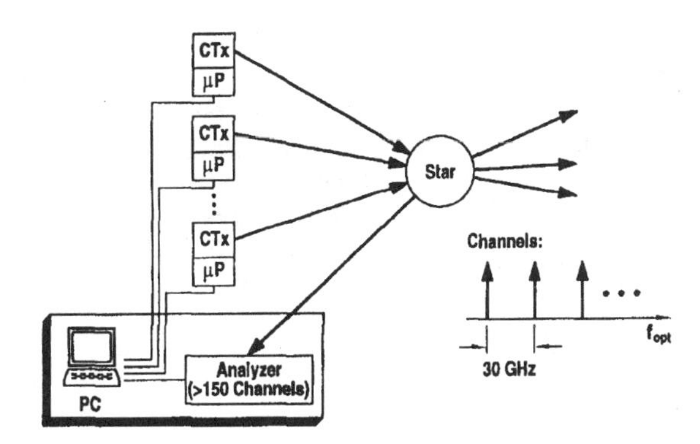

Fig. 8. The Wavelength Manager for controlling coherent transmitters.

are not explicitly mentioned in the foregoing chapters are listed here.

V. SUBSYSTEMS AND COMPONENTS

This chapter summarizes some CMC related hardware developments within the COBRA project.

1) The Wavelength Manager: In WDM and CMC networks a wavelength manager in general is responsible for three functions:

1) Monitoring the status of all WDM/CMC channels
2) Stabilization of the transmitter frequencies
3) Switch-off a transmitter which fails.

Since the temperature stabilized transmitter lasers are operating within their predefined windows and with very low intrinsic frequency drift, all demonstrators have been working during the demonstration periods without the explicit necessity of additional control. However, for large networks with a large number of carriers, and in order to maintain a high reliability, wavelength control is indispensible. The wavelength manager developed in the COBRA project is described in detail in [45]. It consists of a commercial high resolution scanning Fabry–Perot spectrum analyzer featured with a galvatron reference (Kr-line at 1547.825 nm). The total absolute accuracy of transmitter frequency control is within 2.5 GHz, the relative accuracy is below 200 MHz.

The wavelength manager was implemented in the demonstrator at BBC controlling the transmitters as shown schematically in Fig. 8.

2) Laser With 1600 GHz Tuning Range: BH-TTG lasers, fabricated by an all-MOVPE process, have been developed with an active region comprising six strained quaternary quantum-wells. An optimized n-contact geometry (double-sided arrangement) is used. By applying either forward or reserve bias to the tuning region, the continuous tuning range has been extended to the record value of 13.2 nm (1.6 THz). An output power level above 1 mW can be maintained over 12 nm tuning range, with the linewidth remaining below 30–40 MHz [10]. Work has been carried out on TTG-structures with current blocking layers for improved current confinement and therefore enhanced output power.

3) Lasers With Flat FM-Response From 0 to Above 3 GHz: A new type of laser has been developed that solves the problem of the low-frequency dip in the FM-response of conventional lasers. This QCSE (quantum confined Stark effect) TTG (tunable twin guide) laser shows an extremely flat FM-response from dc to 3 to 4 GHz and allows simultaneously wavelength tuning [46]. Two versions—with polyimid layer for lowest capacitance and without polyimid but very small contact pads—have been produced.

4) Common Laser Designs for Transmitter and Local Oscillator: The QCSE-TTG laser has been found to be suitable for both the transmitter and the receiver local oscillator. For all types of single and double-section DFB lasers investigated, it appeared very difficult to meet the linewidth specification in combination with the low and high cut-off frequency specifications for the FM-response. A three-section tunable DBR laser has however been found to be successful, and also able to function as both transmitter and local oscillator [47]. For the transmitter, the FM-response of the modulated Bragg section was flattened to 2.5 GHz by means of an external electronic equalization circuit.

5) Integrated InP-Based Polarization Controller: An integrated InP-based polarization controller having the "semi-insulating embedded ridge structure" has been developed [48], [49]. The structure is fabricated in three epitaxy steps and by a combination of wet and dry chemical etching. A special procedure was developed, resulting in a completely flat regrown structure. Electrical measurements on the regrown structure showed leakage currents to be negligible.

6) Micro-Optical Polarization Hybrid: A novel fully packaged micro-optical polarization diversity hybrid, showing high performance and being extremely fabrication and alignment tolerant was designed. The design employs highly parallel glass plates, which result in parallel and equidistant input and output beams, thereby eliminating critical fabrication and alignment procedures. At the target wavelength of 1550 nm, typical fiber-to-fiber insertion loss is 0.7 dB, balancing is 50 ± 3, polarization extinction ratios are better than >25 dB, and measured back reflection is smaller than < -58 dB. Wavelength insensitive behavior has been observed, resulting in a wide useable spectral operating range in excess of 90 nm. The packaged hybrid is robust, reliable, and allows for low-cost fabrication techniques [22]–[24].

7) Low-IF Receiver Design Implemented: A CPFSK receiver has been developed for 2.5 Gb/s NRZ with an IF of 2 GHz and an IF bandwidth of 4 GHz [18], [50]. This receiver uses a new concept, based on the suppression of the "double-IF" signal which is generated in the frequency demodulation process. The concept is especially attracive when combined with electrical dispersion compensation techniques.

8) Multichannel EDFA Developed for Bidirectional CMC Operation: An erbium-doped bidirectional fiber amplifier has been developed and tested [51]. It was found that the saturation of the amplifier is homogeneous, i.e., only dependent on the sum of the two bidirectional input powers. Besides, the crosstalk from reflections and backscattering, and the influence of gain saturation and noise processes was being examined. From this analysis, it came out that large reflections, even when causing laser action at 1531 nm, only slightly influenced the coherent bidirectional transmission at 1560 nm. It was also found that the crosstalk from Rayleigh backscattering between

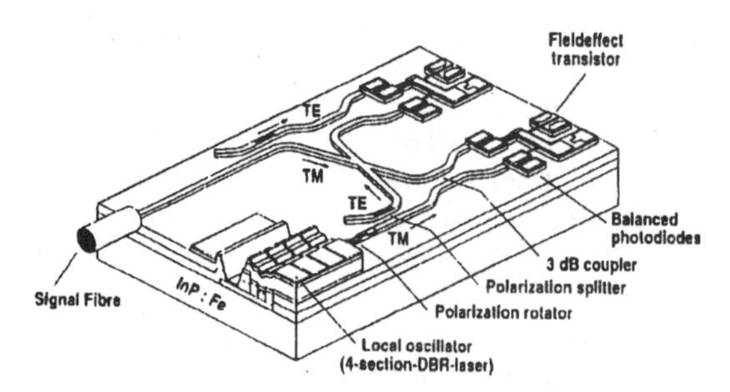

Fig. 9. Schematic view of a fully integrated polarization diversity heterodyne receiver.

counterpropagating signals was increased when the gain was larger than the transmission loss after the EDFA. By a combination of experiment and simulation, it is predicted that up to 10 000 155 Mb/s channels can be amplified simultaneously, which corresponds to a total transmission capacity of over 1 Tb/s. Simulations show that in this case the access shot noise from the amplified channels has a significant influence in the system performance.

9) Fully Integrated Heterodyne Receiver OEIC: Within the COBRA project an OEIC concept has been developed that integrates the basic elements of an isolator-free heterodyne receiver: a four section DBR local laser, a 3 dB coupler, balanced photodiodes, and a JFET. In the meantime, the HHI has upgraded this device to a fully packaged polarization diversity receiver and has carried out first system experiments. Since this is a breakthrough in component technology with the potential of mass production supporting the introduction of coherent multicarrier technology in the future, a brief outline of this device is given here. On a single 2-inch InP-wafer a number of 100 heterodyne receiver chips of size 9×0.6 mm shown schematically in Fig. 9 can be produced. The process based on selective area MOVPE involves 23 lithographic and seven epitaxial growth steps for the integration of 17 different photonic elements [25]. A first packaged device was tested at 140 Mb/s exhibiting a sensitivity of -34 dBm (BER $= 10^{-9}$) and a polarization dependence below 0.5 dB [52].

VI. Conclusion

The implementation and the test of the four COBRA trials has been finished by end of 1995. The consortium has promoted coherent multicarrier technology and has gained expertise ranging from components technology to practical applications on a service level. The trials are covering different network segments for the evaluation of specific practical aspects. The RENO demonstrator shows that the flexibility of today's SDH-based core networks can be increased by CMC technology. The demonstrator in the BBC TV studio has proven the high capacity of an internal passive optical network and the simplicity to access ultra-broadband sources. The videoconference trial at CET demonstrates a fastidious service with real users in an operational environment and the trial at KPN Research a high capacity business customer network supporting ATM based multimedia communication.

Acknowledgment

This paper is written by the members of the Working Group Systems of the COBRA project. However, we wish to acknowledge the work of all our colleagues within the consortium. We also acknowledge the support of our individual companies and the support by the Commission of the EU within the RACE phase II program. We thank B. Mikulski for performing the manuscript and G. Reise for the drawing the figures.

References

[1] E. J. Bachus, I. Borges, P. Demeester, A. Ebberg, G. Heydt, G. D. Khoe, A. C. Labrujere, B. Mikulski, and J. R. Rawsthorne, "COBRA—Flexible broadband networks," *First R2065 COBRA Report*, Sept. 1992.

[2] E. J. Bachus, C. R. Batchellor, I. Borges, P. Demeester, G. Depovere, A. Ebberg, M. B. Ferreira, G. D. Khoe, A. C. Labrujere, B. Mikulski, J. R. Rawsthorne, G. Reise, and N. Wauters, "COBRA—Flexible broadband networks," *Second R2065 COBRA Report*, Aug. 1993.

[3] E. J. Bachus, I. Borges, P. Demeester, G. Depovere, A. Ebberg, G. D. Khoe, O. J. Koning, and J. R. Rawsthorne, "COBRA—Flexible broadband networks," *Third R2065 COBRA Report*, Sept. 1994.

[4] E. J. Bachus, T. Almeida, P. Demeester, G. Depovere, A. Ebberg, M. Ferreira, G. D. Khoe, O. J. Koning, J. R. Rawsthorne, and N. Wauters, "COBRA—Flexible broadband networks," *Fourth R2065 COBRA Report*, Sept. 1995.

[5] S. Kawanishi, H. Takara, T. Morioka, O. Kamatani, K. Takiguchi, T. Kitoh, and M. Saruwatari, "400 Gbit/s TDM transmission of 0.98 ps pulses over 40 km employing dispersion slope compensation," *OFC '96*, San José, CA, vol. PD24, Feb. 26–Mar. 1, 1996.

[6] H. Onaka, H. Miyata, G. Ishikawa, K. Otsuka, H. Ooi, Y. Kai, S. Kinoshita, M. Seino, H. Nishimoto, and T. Chikama, "1.1 Tb/s WDM transmission over a 150 km 1.3 μm zero-dispersion single-mode fiber," *OFC '96*, San José, CA, vol. PD19, Feb. 26–Mar. 1, 1996.

[7] T. Morioka, H. Takara, S. Kawanishi, O. Kamatani, K. Takiguchi, K. Uchiyama, M. Saruwatari, H. Takahashi, M. Yamada, T. Kanamori, and H. Ono, "100 Gbit/s $\times 10$ channel OTDM/WDM transmission using a single supercontinuum WDM source," *OFC'96*, San José, CA, vol. PD21, Feb. 26–Mar. 1, 1996.

[8] E. J. Bachus, "Transparency in optical networks," *EFOC&N '93*, The Hague, Proc. Papers on networks, pp. 219–224, June 30–July 2, 1993.

[9] G. D. Khoe, G. Heydt, I. Morgan, P. Demeester, A. Ebberg, A. Labrujere, and J. Rawthorne, "Coherent multicarrier technology for implementation in the customer access," *J. Lightwave Technol.*, vol. 11, pp. 695–713, May 1993.

[10] T. Wolf, S. Illek, J. Rieger, B. Borchert, and M.-C. Amann, "Tunable twin-guide (TTG) distributed feedback laser with over 10 nm continuous range," *Electron. Lett.*, vol. 29, pp. 2142–2125, 1993.

[11] M. Bagley, R. Wyatt, D. J. Elton, H. J. Wickes, P. C. Spurdens, C. P. Seltzer, D. M. Cooper, and W. J. Devlin, "242 nm continuous tuning from a grin-SC-MQW-BH InGaAsP laser in an extended cavity," *Electron. Lett.*, vol. 26, no. 4, pp. 267–269, 1990.

[12] P. W. Hooijmans and M. T. Tomesen, "Analytical analysis of IF-penalties for different FSK and CPFSK detection schemes," *J. Lightwave Technol.*, vol. 10, pp. 649–659, May 1992.

[13] P. W. Hooijmans, M. T. Tomesen, G. F. G. Depovere, R. A. J. M. van Gils, F. A. J. Dumont, and P. H. G. M. Thijssen, "1 Gbit/s CPFSK phase diversity optical transmission for high band width efficiency," in *Proc. 18th ECOC*, Berlin, Sept. 27–Oct. 1, 1992, paper We-A8.3.

[14] C. R. Batchellor, M. G. Holliday, and A. M. Thorley, "A novel coherent communications transceiver," in *Proc. 19th ECOC*, Montreux, Sept. 12–16, 1993, paper WeP9.4.

[15] P. Hooijmans, M. Tomesen, G. Depovere, R. van Gils, F. Dumont, and P. Thijssen, "1 Gbit/s CPFSK phase diversity optical transmission for high bandwidth efficiency," in *Proc. 18th ECOC*, Berlin, Sept. 27–Oct. 1, 1992, paper WeA8.3, pp. 357–360.

[16] M. Tomesen, P. Hooijmans, C. Wong, G. Depovere, L. Tiemeijer, H. van Tongeren, and J. Kokkelink, "Novel single laser coherent transceiver with a semiconductor optical amplifier as signal booster," in *Proc. 18th ECOC*, Berlin, Sept. 27–Oct. 1, 1992, paper WeA8.6, pp. 369–372.

[17] P. W. Hooijmans and M. T. Tomesen, "Analytical analysis of IF-Penalties for different FSK and CPFSK detection schemes," *J. Lightwave Technol.*, vol. 10. no. 5, pp. 649–659, 1992.

[18] M. Tomesen, G. Depovere, R. van Gils, R. Witlox, D. Schouten, E. Pennings, and A. Staring, "Novel heterodyne CPFSK receiver allowing dispersion equalization in a narrow IF bandwidth starting from nearly DC," in *Proc. 20th ECOC*, Firenze, Sept. 25–29, 1994, vol. 1, pp. 73–76.

[19] E. Meissner, H. Rodler, and M. Lades, "Pattern independent 2.5 Gbit/s AMI-CPFSK transmission with −44 dBm receiver sensitivity," *Electron. Lett.*, vol. 30, no. 4, pp. 345–346, 1994.

[20] E. Meissner, M. Lades, and R. Rodler, "Optical 2.5 Gbit/s heterodyne receiver with an IF of only 3.0 Ghz and a reveiver sensitivity of 46.6 dBm," in *Proc. 20th ECOC*, Firenze, Sept. 25–29, 1994, vol. 2, pp. 781–784.

[21] R. Noé, H. J. Rodler, A. Ebberg, G. Gaukel, B. Noll, J. Wittmann, and F. Auracher, "Comparison of polarization handling methods in coherent optical systems," *J. Lightwave Technol.*, vol. 9, no. 10, pp. 1353–1366, 1991.

[22] E. C. M. Pennings, D. Schouten, and G. D. Khoe, "Ultra fabrication tolerant micro-optical polarization-diversity hybrid," in *Proc. 20th ECOC*, Firenze, Sept. 25–29, 1994, vol. 1, pp. 217–220.

[23] E. C. M. Pennings, D. Schouten, R. A. J. C. M. van Gils, G. F. G. Depovere, and G. D. Khoe, "Fully-packaged ultra fabrication-tolerant micro-optical polarization-diversity hybrid," in *Proc. of OFC'95*, San Diego, CA, Feb. 26–Mar. 3, 1995.

[24] E. C. M. Pennings, D. Schouten, G. D. Khoe, R. A. J. C. M. van Gils, and G. F. G. Depovere, "Ultra fabrication tolerant fully packaged micro-optical polarization diversity hybrid," submitted at JLT.

[25] R. Kaiser, D. Trommer, F. Fidorra, H. Heidrich, S. Malchow, D. Franke, W. Passenberg, W. Rehbein, H. Schroeter-Janßen, R. Stenzel, and G. Unterbörsch, "Monolithically integrated polarization diversity heterodyne receivers on GaInAsP/InP," *Electron. Lett.*, vol. 30, pp. 1446–1447, 1994.

[26] *ITU-T Recommendation G.803*, "Architectures of transport networks based on the synchronous digital hierarchie (SDH)," Mar. 1993.

[27] N. Wauters, P. Demeester, and G. F. G. Depovere, "GTNA representation of optical cross-connected networks," in *Proc. EFOC&N '95 Conf. European Fiber Optic Commun. Networks*, Brighton, England, June 27–30, 1995.

[28] ——, "Functional representation of optical transport networks and surveillance monitoring techniques," in *10th Int. Conf. Integrat. Optics Optical Fiber Commun.*, IOOC '95, Hong Kong, June 26–30, 1995.

[29] G. Depovere, M. Tomesen, P. Hooijmans, N. Wauters, and P. Demeester, "A flexible cross-connect network using multiple optical carriers," in *Proc. 19th ECOC*, Montreux, paper TuP4.5, Sept. 12–16, 1993.

[30] G. Depovere, M. Tomesen, R. v. Gils, N. Wauters, D. Vercauteren, and P. Demeester, "Laboratory demonstration of a 2.5 Gbit/s SDH-compatible optical cross-connect network," in *Proc. 20th ECOC*, Firenze, Sept. 25–29, 1994, vol. 2, pp. 571–574.

[31] N. Wauters, C. Vercauteren, P. Demeester, P. Lagasse, G. Depovere, and P. Hooijmans, "Introduction in SDH-networks of reconfigurable nodes using optical frequency division multiplexing," in *Proc. 19th ECOC*, Montreux, Sept. 12–16, 1993.

[32] G. D. Khoe, Tutorial, "Coherent techniques for flexible capacity traffic and access," in *Proc. OFC/IOOC'93*, San Jose, CA, Tutorial Digest, paper ThE, Feb. 21–26, 1993, pp. 177–225.

[33] N. Wauters and P. Demeester, "Wavelength requirement and survivability in WDM cross-connected networks," in *Proc. 20th ECOC*, Firenze, vol. 2, Sept. 25–29, pp. 589–592, 1994.

[34] N. Wauters and P. Demeester, "Influence of wavelength translation in optical frequency multiplexed networks," in *IEEE/LEOS 1994, Summer Topic. Meet. Optic. Networks Enabling Technol.*, Lake Tahoe, NV, July 11–13, paper M4.3, 1994.

[35] G. D. Khoe, "Coherent multicarrier lightwave technology for flexible capacity networks," *IEEE Commun. Mag.*, pp. 40–51, Mar. 1994.

[36] C. R. Batchellor, B. T. Debney, A. M. Thorley, G. D. Khoe, C. v. Helmolt, F. Auracher, and P. Lagasse, "System assessment of the RACE 1010 multi-channel demonstrator," *Optic. Quantum Electron.*, Special Issue on Advances in Optoelectronics in Europe, vol. 26, pp. S517–S528, 1994.

[37] C. R. Batchellor, B. T. Debney, A. M. Thorley, T. J. B. Swanenburg, G. Heydt, F. Auracher, and P. Lagasse, "A coherent multichannel demonstrator," *IEE Colloq. RACE Optic. Syst. Demonstrators*, London, Digest no. 1993/115, May 17, 1993.

[38] O. J. Koning, R. J. Brinkman, M. O. van Deventer, P. H. A. Venemans, L. Wennekes, J. P. Boly, M. K. de Lange, J. Vanderwege, B. Meuris, and A. Ebberg, "A multi-carrier business customer premises network," in *Proc. The European Symp. Advan. Networks Services*, Amsterdam, Mar. 20–23, 1995.

[39] A. C. Labrujere, M. O. van Deventer, and O. J. Koning, "Coherent multi-channel technology in the local loop," *Fiber Optic Networks Video-Commun. Conf. '93*, Berlin, Apr. 1993.

[40] O. J. Koning, A. C. Labrujere, P. H. A. Venemans, J. P. Boly, R. J. Brinkman, and M. K. de Lange, "A flexible broadband coherent optical system for application in a business customer premises network," *RACE Open Workshop Broadband Access*, Nijmegen, The Netherlands, June 1993.

[41] K. Sato, S. Okamoto, and H. Hadama, "Optical path layer technologies to enhance B-ISDN performance," in *ICC '93, Proc.*, pp. 1300–1307.

[42] B. Meuris, B. Staelens, J. Zhou, J. Vandewege, O. J. Koning, R. J. Brinkman, and P. H. A. Venemans, "Implementation of a control network and wavelength management system for a coherent multi-carrier business customer premises network," in *Proc. EFOC&N '95 Conf. European Fiber Optic Commun. Networks*, Brighton, England, June 27–30, 1995.

[43] M. O. van Deventer, O. J. Koning, and A. J. Lous, "Interaction between channels in a bidirectionally amplified multicarrier system," in *Proc. 20th ECOC*, Firenze, Sept. 25–29, 1994, vol. 1, pp. 411–414.

[44] M. O. van Deventer and O. J. Koning, "Crosstalk and channel spacing in a coherent multichannel system using erbium doped fiber amplifiers," *IEEE Photon. Technol. Lett.*, vol. 6, pp. 260–262, 1994.

[45] U. Krüger, K. Krüger, C. R. Batchellor, U. Fischer, C. v. Helmolt, and U. Nagengast, "Wavelength manager covering the EDFA-band and capable of controlling hundreds of transmitters," *20th ECOC*, Florence, Italy, vol. TuP 28, Sept. 25–29, 1994.

[46] T. Wolf, K. Droegenmueller, B. Borchert, H. Westermeier, E. Veuhoff, and H. Baumeister, "Tunable twin-guided lasers with flat frequency modulation response by quantum confined stark effect," *Appl. Phys. Lett.*, no. 60, pp. 2472–2474, 1992.

[47] A. A. M. Staring, J. J. M. Binsma, P. I. Kuindersma, E. J. Jansen, P. J. A. Thijs, T. van Dongen, and G. F. G. Depovere, "Wavelength-independent output power from an injection-tunable DBR laser," *IEEE Photon. Technol. Lett.*, vol. 6, pp. 147–149, Feb. 1994.

[48] P. Kaczmarski, G. M. Morthier, R. Baets, B. H. Verbeek, and E. C. M. Pennings, "Numerical analysis of electrooptic polarization rotation in In-based embedded waveguides," in *Proc. ECIO'93*, Neuchatel, Apr. 18–22, 1993, pp. 13-6/13-7.

[49] P. Kaczmarski, G. M. Morthier, R. Baets, B. H. Verbeek, and E. C. M. Pennings, "Analysis of InP-based guided-wave electrooptically rotatable waveplates for polarization transformers," in *Proc. OFC/IOOC '93*, San Jose, CA, Feb. 1993, paper WH7.

[50] R. A. J. C. M. van Gils, M. T. Tomesen, and G. F. G. Depovere, "2.5 GBit/s CPFSK system with low IF-frequency," *Electron. Lett.*, vol. 30, no. 14, pp. 1153–1154, 1994.

[51] O. J. Koning, "An EDFA, suitable for bidirectional multi-channel signal amplification," *RACE PL1 Workshop Optic. Ampl. Applicat.*, Brussels, July 6, 1994.

[52] U. Hilbk, Th. Hermes, P. Meißner, F.-J. Westphal, G. Jacumeit, R. Stenzel, and G. Unterbörsch, "Stable operation of a monolithically integrated InP-heterodyne polarization diversity receiver module including a tunable laser in an experimental OFDM system," in *Proc. 20th ECOC*, Florence, Italy, Sept. 25–29, 1994, vol. 1, pp. 77–81.

The LAMBDANET Multiwavelength Network: Architecture, Applications, and Demonstrations

MATTHEW S. GOODMAN, HAIM KOBRINSKI, MEMBER, IEEE, MARIO P. VECCHI, MEMBER, IEEE, RAY M. BULLEY, AND JAMES L. GIMLETT

***Abstract*—This paper summarizes work conducted at Bellcore over the last few years concerning the LAMBDANET™ multiwavelength optical network. We describe the basic network architecture and some variations, including discussion of several applications based on these architectures. Finally, we present experimental results demonstrating the technological feasibility of this approach. The LAMBDANET design incorporates three basic ingredients: 1) it associates a unique optical wavelength with each transmitting node in a cluster of nodes; 2) the physical topology is that of a broadcast star; 3) each receiving node identifies transmitting nodes based on the transmission wavelength through wavelength demultiplexing. This network design features full connectivity among the nodes, large nonblocking throughput, data format transparency, and flexible control. The LAMBDANET network may be used for both point-to-point and point-to-multipoint applications. The paper discusses possible applications for this basic structure including broadcast video, private virtual networks, and packet switching. Experiments to verify the salient optical characteristics demonstrated the largest multicast WDM transmission capacity measured to date. With 18 transmitters at 1.5 Gb/s per transmitter and 16 receiving nodes, the largest point-to-point bandwidth-distance product is 1.56 Tb · s^{-1} · km, and the point-to-multipoint figure of merit is 21.5 Tb · s^{-1} · km · node. A point-to-point WDM experiment (no star coupler) is also described that achieved a bandwidth-distance product of 2.07 Tb · km/s and 36 Gb/s total transmitted bandwidth.**

I. Introduction

THE advent of single-mode optical fibers in public telecommunication networks has presented an enormous, potentially useable frequency bandwidth. To date, the deployment of optical-fiber public telecommunication networks has been largely as a replacement for metallic transmission lines, without using the potential capabilities unique to fiber-based systems. This paper explores the use of a portion of the available bandwidth through a *multiwavelength optical network* design as a backbone network to support broad-band services. We identify this approach as the LAMBDANET design. This backbone network has the virtue that it starts to unload some of the distribution and routing functions traditionally allocated to electronics into the optical domain. Many variations on the same basic structure are possible, and we discuss, as examples, a number of applications to support both existing services and new broad-band services. The LAMBDANET design has a number of desirable characteristics including large network throughput, point-to-point and point-to-multipoint transmission capability, and a physical design that is consistent with the likely future developments of optoelectronics and optoelectronic integrated circuit technology.

The large transmission capacities available from a single-mode optical fiber may vastly exceed the switching capacities currently available. Applications to help alleviate this potential switching bottleneck are also presented. Results from a series of LAMBDANET experiments have confirmed the technological feasibility of this network architecture. Furthermore, these experimental results indicated large throughput for point-to-point traffic and even larger throughput for point-to-multipoint distribution applications.

II. Basic LAMBDANET Architecture

The LAMBDANET architecture [1], [2] (Fig. 1) is composed of a cluster of N communication nodes [e.g., telephone company central offices (CO)] connected by single-mode optical fibers to a hub location. An essential feature of this architecture is that each node within a cluster transmits its information on a unique wavelength using single-frequency laser diodes (e.g., distributed feedback (DFB) lasers). The optical fibers are passively coupled at the hub location using an $N \times N$ transmissive star coupler. Each network node requires at least two optical fibers: one for transmitting to the star coupler and one for receiving information from the star coupler. Each output fiber from the star coupler thus carries *all* of the wavelength channels to its corresponding node. In order to separate the received set of wavelengths, several approaches and technologies have been considered depending on the application. These approaches include wavelength tunable filters [3] and receivers [4], and multiple fixed receivers. For a network in which individual nodes are distributors of information (as opposed to the end users), it is likely that each node may simultaneously utilize information from several transmitting nodes. Therefore, we consider a design in which each node has a diffraction-grating wavelength demultiplexer followed by as many as N optical receivers. Thus, each node may receive and process, asynchronously and in parallel, transmissions from

Manuscript received September 18, 1989; revised March 25, 1990.

M. S. Goodman, H. Kobrinski, M. P. Vecchi, and R. M. Bulley are with Bell Communications Research, Morristown, NJ 07960.

J. L. Gimlett is with Bell Communications Research, Red Bank, NJ 07701.

IEEE Log Number 9034455.

™LAMBDANET is a trademark of Bell Communications Research.

Reprinted with permission from *IEEE Journal on Selected Areas in Communications,* Vol. 8, No. 6, pp. 995-1004, August 1990.

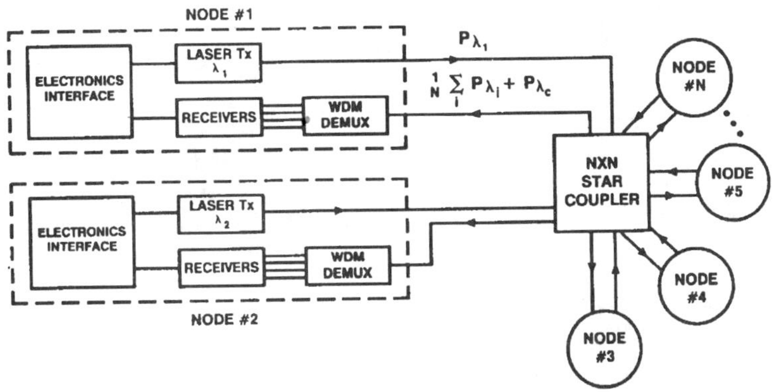

Fig. 1. Block diagram of a LAMBDANET star network.

all the other nodes on the cluster.[1] Effectively, each LAMBDANET node has a dedicated private line to all the other nodes on the cluster—there is no contention for the use of its unique wavelength.

The hub location within a hypothetical LAMBDANET cluster of telephone company central offices, for example, would no longer use high-speed electronic switching, but would use passive optical distribution to provide connectivity between the CO's. Switching (after demultiplexing the information) would be performed in a distributed manner at each central office for distribution to the subscribers.

The network features large cluster throughputs (our experiments demonstrated up to 36 Gb/s). However, no electronic system in the network needs to process data at higher than the individual transmitter rate (~2 Gb/s, and time multiplexing and demultiplexing are only performed at this rate). The specific design of time multiplexing in the LAMBDANET is application dependent. Several potential designs are possible in accordance with the foreseen traffic and service requirements of different applications.

Each node may also have a control transmitter tuned to a wavelength which is common to all the nodes on the cluster (λ_c), as indicated in Fig. 1. This user–user channel may be used for all process (call) setup/takedown and all user–user signaling functions. This common control channel could, for example, have a fixed, time-slotted structure, thereby avoiding contention for its use.

Thus, the LAMBDANET architecture exhibits the following features:

- large cluster throughput;
- point-to-point and point-to-multipoint communications possible;
- independence of wavelength channels;
- transparency to data form and format;
- ability to use both analog and digital data streams in the same system;
- common user–user signaling channel;
- system is robust under optical fiber link failure.

[1]Each node may receive its own wavelength as well, to be used for system error detection, etc.

The LAMBDANET design requires N receivers at each node, which could be viewed as a disadvantage of this design. However, recent advances in OEIC (optoelectronic integrated circuit) receiver-array technology [5] may well mitigate this difficulty. Since the LAMBDANET design uses optical power splitting to distribute the information, the available power budget is another potentially serious limitation. We address these considerations below.

III. Optical Power Budget Considerations

Device characteristics of currently available optical components limit, through optical power budget constraints, the number of nodes and the physical span in the LAMBDANET design. The star topology is one of the most power-efficient designs for distribution networks. Fig. 2 shows a computed power budget for a LAMBDANET system indicating tradeoffs between network span, number of nodes, and transmission bit-rates for reasonable assumptions about near-term device characteristics. The parameters used in these calculations included 10 log N star-coupler splitting loss, 0.1 dB splices and coupler excess losses, 4 dB grating demultiplexer loss, 0.3 dB/km optical fiber loss, and 3 dB system margin. The horizontal lines correspond to the available optical power budgets (launch power - receiver sensitivity) at the specified bit rates assuming 0 dBm launch power. The figure shows the actual optical loss as a function of the number of nodes on a cluster for node–node distances of 5 and 40 km. For example, a cluster of 16 nodes, representing telephone company central offices, supporting a data rate of 2.5 Gb/s per node could be spaced with a maximum CO–CO span of 30 km. These limits were experimentally verified, with parameters similar to this example, and will be described later in this paper. Corresponding to these 16 central offices, 16 different wavelengths, with a minimum channel separation of approximately 2 nm, could be used in the spectral region around 1.55 μm. There are several reasons to design multiwavelength systems with near-minimum channel separations, including minimizing the required spectral range of the laser transmitters and

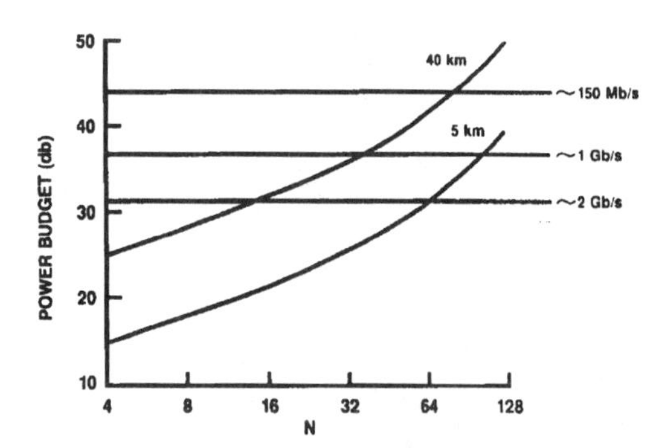

Fig. 2. Computed LAMBDANET power budget.

the requirements for optical amplification. For applications with a limited number of channels, the channel spacing could be increased above these minimum separations (e.g., 5 nm), with concommitantly decreased tolerances on device characteristics and grating design.

Although the 30 km CO–CO span is sufficient for cluster designs in most metropolitan areas, the use of optical amplifiers with even modest gain (about 10 dB net gain) will effectively remove the distance restriction and (or) allow an increase in the number of nodes. In addition, the alleviation of power budget constraints, using optical amplification, will enable implementation of network designs more complex than simple star networks in the appropriate applications. The issues associated with multiwavelength optical amplification of broad spectral range are currently being investigated in many laboratories. Recent demonstrations involving semiconductor laser amplifiers and doped optical-fiber amplifiers have shown encouraging results.

IV. LAMBDANET System Applications

We describe several potential network applications of the LAMBDANET architecture for video distribution, intra-LATA distribution of point-to-point traffic, and finally as the fabric for an optoelectronic packet switch.

A. Voice Traffic Transport and Virtual Network Applications

A typical application of the LAMBDANET architecture is shown in Fig. 3, which depicts an intra-LATA interoffice transport network composed of LAMBDANET clusters. In each cluster, there is a separate master node which is used for cluster synchronization and monitoring functions, node restart/stop, and may be used for administrative functions. In addition, each cluster has a gateway node which provides connectivity from this cluster to other clusters. The gateway nodes could be connected in a variety of ways, depending on the nature and extent of the intercluster traffic patterns. Each cluster could have, in addition, a centralized database node. This node validates the database requests and enters and retrieves the information from the database. The database node also distributes the relevant control information to each affected node.

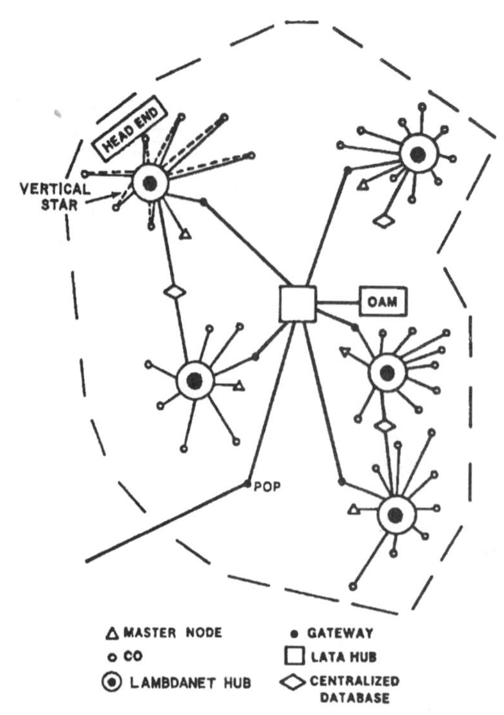

Fig. 3. LAMBDANET LATA (hypothetical).

Fig. 4 shows a hypothetical communications structure within a LAMBDANET cluster for a voice-traffic application. In a related application, the LAMBDANET architecture allows the implementation of a virtual private network for large business customers, as described elsewhere [6]. The hardware architecture associated with these applications is schematically indicated in the figure. For example, for the virtual network application, a business customer defines its virtual network as a set of customer points; each point has a number of directly connected *T1* lines (or their ≈ 1.5 Mb/s equivalent via fiber) to the nearest CO node on the nearest LAMBDANET cluster. In the central database, there could be records indicating the customer-point virtual address, the local node (complete with cluster identifier), the priority of service provision requested, authorization codes (for incoming and outgoing *T1* line screening), time of day/week routing information, and other features selected. A portion of this centralized information, relevant to call processing at each node, is retained locally and is accessible by the user–user channel processor at that node. For example, in a 16-node LAMBDANET system, the incoming *T1* traffic at each central office is groomed and multiplexed in the time domain to form, for example, 155 Mb/s broad-band data channels. Each 155 Mb/s channel represents up to ≈ 90 *T1* equivalent channels originating at one central office and destined for a specific central office. Sixteen of these channels are multiplexed with bit interleaving to form the 2.5 Gb/s transmission data stream leaving a particular node at its unique wavelength. There are, thus, approxi-

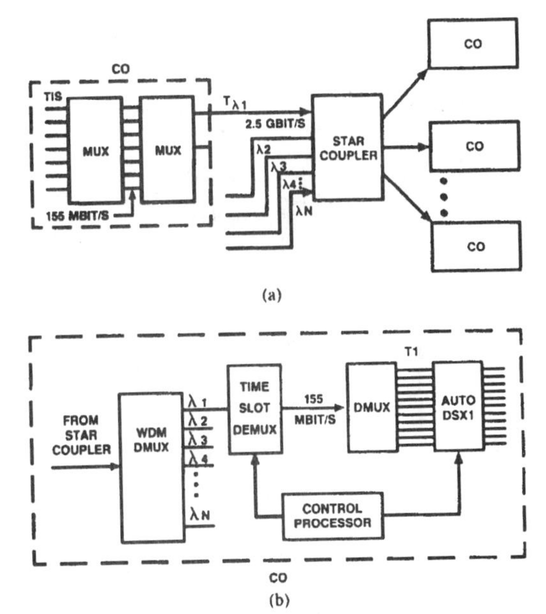

Fig. 4. Voice traffic in a LAMBDANET. (a) Details of the transmitting portion of the network and (b) details of the receiving portion at a CO are shown schematically.

mately 1400 *T1* equivalent lines that may be transmitted from a particular CO destined for all other CO's [Fig. 4(a)]. This number could be increased by either raising the transmission data rate or by using multilayer LAMBDANET stars. Fig. 4(b) shows the distribution of the information to each of the nodes on the local LAMBDANET cluster.

B. Broadcast Video

The LAMBDANET architecture can also be used to support distribution of video programs from a central location to several remote video-switching centers. In this variation, as shown in Fig. 5, all the transmitters are colocated in the central node and each remote switching center has access, continuously, to all the broadcast programs. The figure details a video center (head-end) station which receives video information from both satellite broadcast and from alternate video sources (for example, a videotape library and studios). For example, up to 256 such video channels (at 155 Mb/s data rate for high quality) are time multiplexed to form 16 high-speed data streams (transmission rate = 2.5 Gb/s) which are transmitted on 16 different wavelengths at the video center. These wavelengths are distributed by the star coupler (located at the video center) to all connected nodes. Wavelength demultiplexing and time demultiplexing thus separate the individual video channels at each node. Specific programs may then be switched, based on request, to the subscribers. This application takes advantage of the inherent broadcast nature of the LAMBDANET architecture. An experiment designed to demonstrate this application is described below.

Although the example of Fig. 5 depicts a digital time-multiplexed system, significant variations are possible. In particular, analog broadcast systems (with or without subcarrier multiplexing) could be implemented instead, or in combination with, the TDM system. Due to the data transparency and independence of the wavelength channels, only the endpoint devices, not the network structure, need to be modified to incorporate different data forms and formats.

C. Multilayer LAMBDANET Stars

Fig. 6 shows a multilayer LAMBDANET star application. Three independent multiwavelength stars are shown, along with an enlargement of a hypothetical central office. In this design, the two LAMBDANET layers that provide voice-traffic transport allow for routing redundancy and multiple connectivity (the star couplers may be in different locations and the connecting fibers to each coupler may be over alternate routes) along with additional capacity for future growth. (All interconnections between the stars are accomplished using electronic processing.) Incoming calls to the central switching office are presented at the line side of the local switch (near the top of the figure). After suitable grooming and time multiplexing, these trunk groups are transmitted over this office's unique wavelength to either star *A* or *B*. This node also receives information from other nodes through stars *A* and *B* which, after suitable wavelength and time demultiplexing, and data filtering, are presented to the trunk side of the local switch. In addition, this node may receive high-speed broadcast video information from a video center (head-end) station which it demultiplexes (first in wavelength, then in time) and switches locally to its appropriate customers.

D. The LAMBDANET as a Packet-Switch Fabric

The LAMBDANET design has other potential applications, including using it as a fabric for high-performance optoelectronic packet switching. In this application, there are *N* input ports and *N* output ports on the packet switch (Fig. 7). It is the job of the packet-switch fabric to route and transport packets from the input ports to their desired output port or ports. In the LAMBDANET packet-switch fabric, each *input port* has an associated unique wavelength. Each *output port* has *N* receivers (a broadcast and select network). In this switch fabric, the packets arriving at a given input port are transmitted at the input port wavelength. This information is broadcast to all the switch output ports. Each output port buffers the information received at all wavelengths, and then determines if the received packet is destined for this output port. The information received on each wavelength destined for a particular output port is then multiplexed into a single output-data stream, while the information not destined for this output is discarded.

There are several potential advantages to using a LAMBDANET fabric or similar structures for packet switches. First, there is a very large throughput possi-

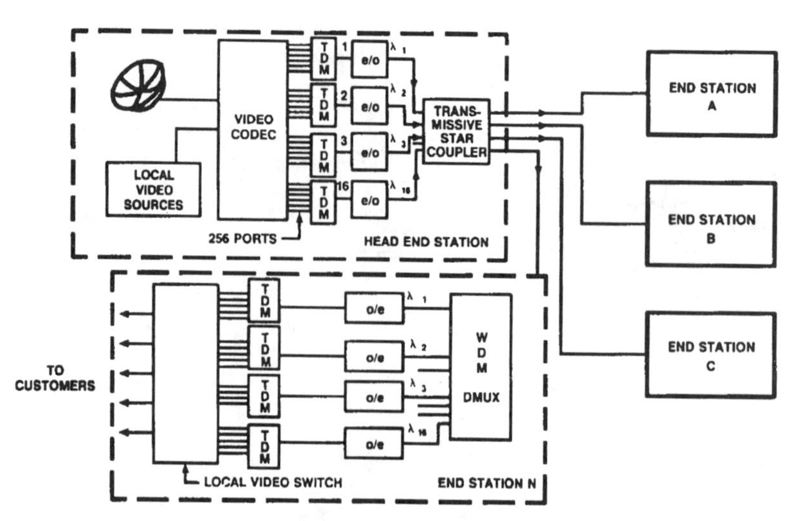

Fig. 5. LAMBDANET video distribution. Example shows the distribution of video services from a head-end station to a number of remotely located switching nodes.

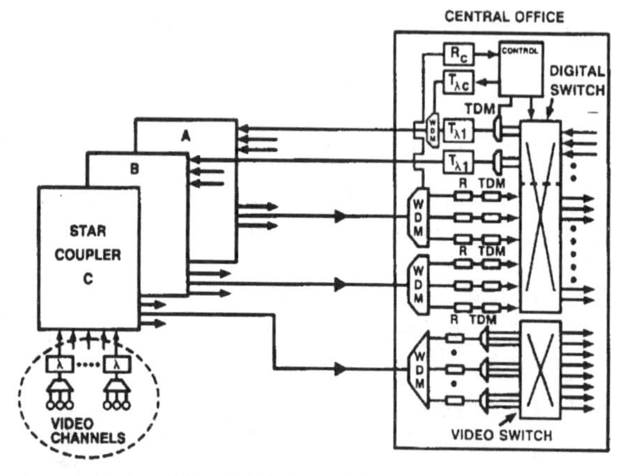

Fig. 6. Multilayer LAMBDANET: a multistar network for distribution of broad-band services.

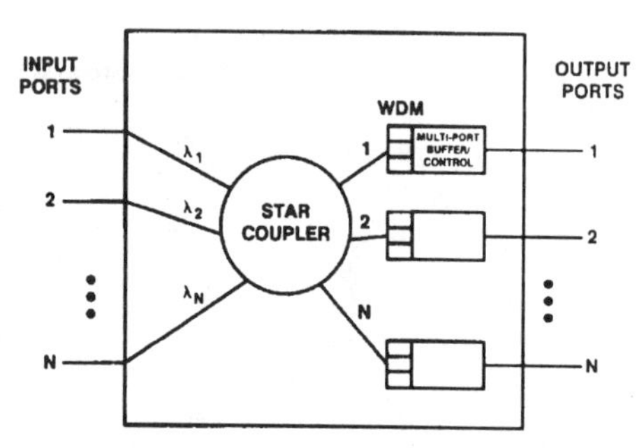

Fig. 7. LAMBDANET as a packet-switch fabric.

ble—with 64 nodes each transmitting at 2 Gb/s, peak throughputs greater than 100 Gb/s are possible. Since this switch fabric is output buffered (rather than input buffered), there is no head-of-the-line blocking, which will yield efficient transport. Furthermore, point-to-point, selective multicast transport, and broadcast packets are all possible. The LAMBDANET switch also features a single-stage interconnection—no intermediate stages of comparisons (where errors could propagate) are needed. However, this approach has the disadvantage that many optoelectronic receivers are required at each output port.

For a packet switch based on this architecture, typical switch parameters include probable packet sizes of 512-4096 b/packet. Thus, packet transmission times, assuming 2 Gb/s transmission rates, are in the neighborhood of 0.25-2 μs. Multiported buffer memories, as indicated in the figure, are required to multiplex at ≈ 2 Gb/s, within the range of currently available technology.

A number of variations of this theme have been independently developed. These include the AT&T Photonic Knockout Switch [7] and the Bellcore StarTrack Multicast Switch [8], [9], which use wavelength-tunable filters at the receiving ports to reduce the number of receivers required.

V. Experimental Demonstration System

Experiments were performed to establish the feasibility of the optical component requirements in a LAMBDANET network of realistic size. The experimental demonstrations had several objectives. First, we wanted to test the feasibility of using many closely spaced wavelengths in a realistic system. Thus, these experiments were based on a modular, fully connectorized system (Fig. 8). Second, we wanted to demonstrate a design as similar as possible to one of the architectures described above. We chose to implement the video broadcast architecture, since that one features the optical backbone network with minimal control requirements. Finally, within the context of

Fig. 8. LAMBDANET overall modular design.

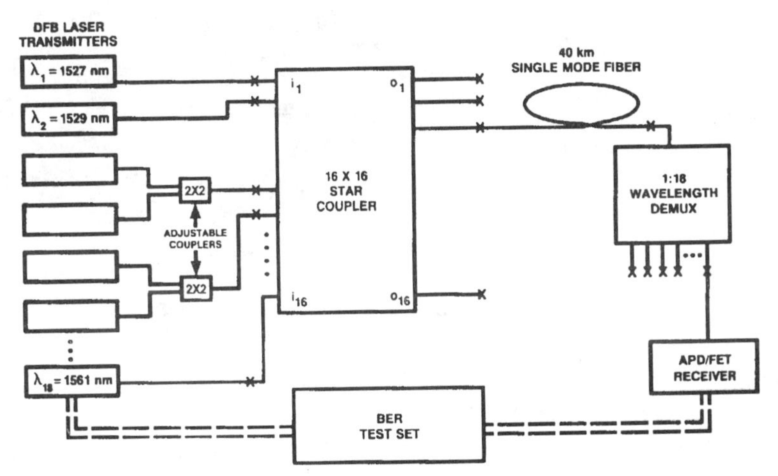

Fig. 9. LAMBDANET experimental setup.

the video broadcast simulation, we wanted to demonstrate the large throughput capability of this network architecture using commercially available components.

The demonstration system is indicated in Fig. 9 and it resembles Fig. 5, illustrating the video broadcast application. The LAMBDANET experimental network incorporates elements constructed from commercially available components. The 16 × 16 single-mode star coupler (Gould, Inc.) was constructed from wavelength-flattened 3 dB directional couplers. It had a 14.4 dB average loss at 1.5 μm with up to 6 dB nonuniformity. DFB laser diodes (Hitachi, Ltd.) were selected with center wavelengths spaced by ≈2 nm in the 1527–1561 nm range. The DFB lasers had "pigtailed" fibers and were operated without isolators. Modular transmitter units were built for these lasers and, for precise wavelength tuning, the temperature was adjusted and then stabilized to within ±0.01°C.

Fig. 10 shows the spectral distribution of the 18 wavelengths as seen at one of the outputs of the star coupler. A grating wavelength demultiplexer was constructed for this experiment (Stimax design, Instruments S.A.). This device had one single-mode fiber input and 18 multimode-fiber outputs with center wavelengths spaced ≈2 nm apart over the 1527–1561 nm range. WDM insertion losses were <3.5 dB. The demultiplexer 3 dB channel bandwidths were <0.85 nm and the nearest neighbor crosstalk level was < −35 dB. The high-impedance receiver, with

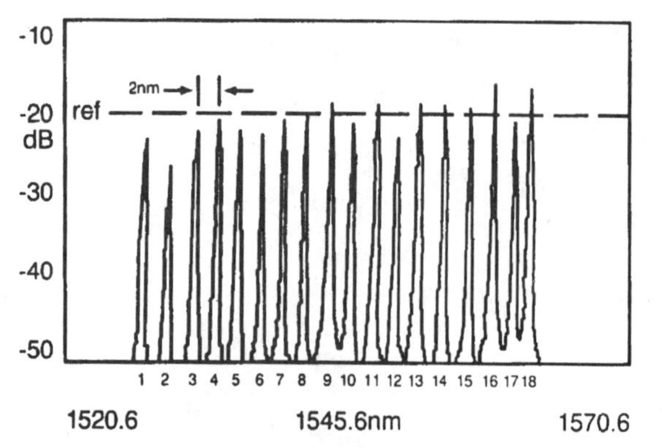

Fig. 10. LAMBDANET experimental power spectrum: power spectral distribution for 18 channels spaced by 2 nm, as measured at the output of the star coupler.

a GaAs FET front-end and 2 GHz bandwidth, was based on a commercial InGaAs APD (Fujitsu) detector with a multimode-fiber pigtail. The receiver sensitivity for 10^{-9} BER was -34.5 dBm at 1.5 Gb/s and -32.0 dBm at 2 Gb/s.

A single-mode fiber with zero dispersion at 1.3 μm was used in our measurements. The system span was determined by the channel with the most limited power budget, which was determined to be primarily dependent on the nonuniformity of the star coupler. Therefore, we took advantage of the rather large nonuniformity of the star coupler to use 18 transmitters (Fig. 9) without reducing the minimum span of the system.

VI. Experimental Results

BER measurements were sequentially taken for each laser, with NRZ $(2^{15} - 1)$ pseudo-random modulation. At 1.5 Gb/s, 18 channels were transmitted (by adding two 2 × 2 directional couplers before the star coupler) over 57.8 km of the single-mode fiber [2]. The average fiber loss was 0.28 dB/km, including connector and splice losses. The received power levels ranged between -28.1 and -32.5 dBm [Fig. 11(a)] for BER = 10^{-9}, indicating an up to 6 dB system power penalty due to fiber dispersion, reflection noise, and extinction ratio penalties. Fig. 11(b) shows the eye patterns for the input and the received signals after the star coupler, long optical fiber, and demultiplexing system.

Experiments [11] at 2 Gb/s have demonstrated the transmission of 16 wavelength-multiplexed channels with 32 Gb/s transmitted bandwidth over 40 km of fiber. Variations in received power between the 16 channels were observed, and the measurements indicate required power levels between -25.4 and -28.1 dBm for BER = 10^{-9}. Fig. 12 shows the BER curves for the received signals for the best (channel 10) and the worst (channel 9) channels. The overall system power penalty ranged from ≈2 dB to 7 dB, due to the same sources as cited above for the 1.5 Gb/s measurements.

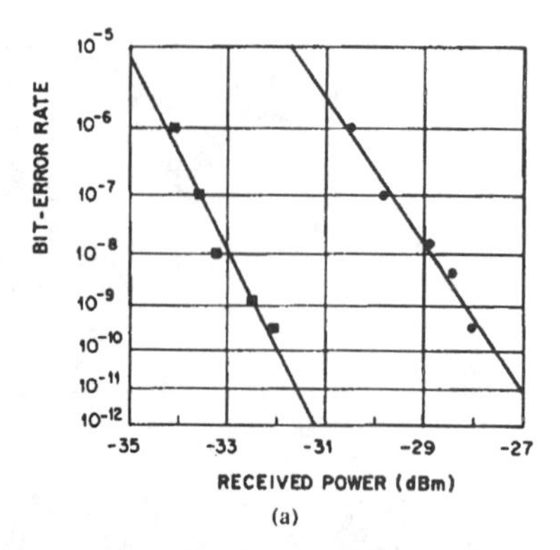

(a)

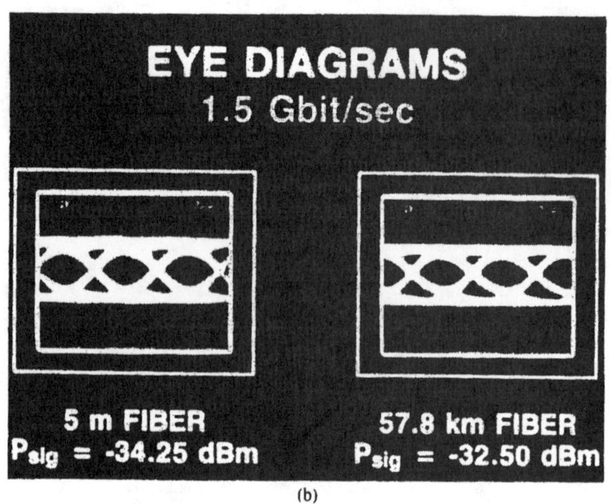

(b)

Fig. 11. LAMBDANET bit-error-rate: (a) curves and (b) eye diagrams at 1.5 Gb/s.

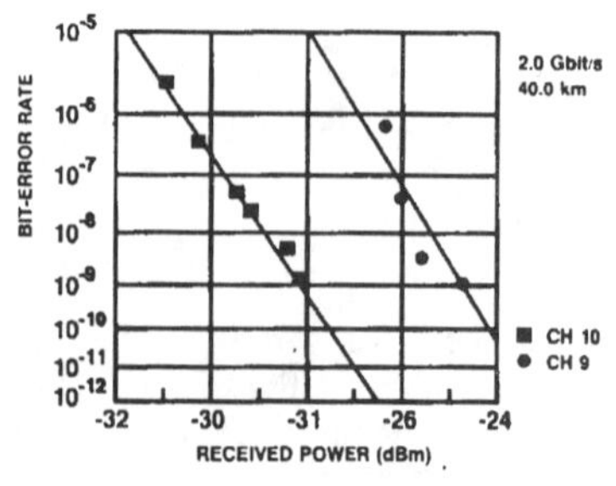

Fig. 12. LAMBDANET bit-error-rate curves 2 Gb/s. The channels corresponding to the lowest and highest sensitivity are indicated.

Crosstalk isolation, due to the cumulative optical power from all other channels on each individual channel, was measured to be < -27 dB, and no BER degradation was observed due to this crosstalk level. The point-to-point bandwidth-distance product thus achieved was 1.56 Tb/s · km for 1.5 Gb/s data and 1.28 Tb/s · km for 2 Gb/s data. The reduced value at 2 Gb/s resulted from increased

TABLE I
SUMMARY OF HIGH BIT-RATE RESULTS

	THROUGHPUT Gbit/s	POINT-TO-POINT Tbit-km/sec	POINT-TO-MULTIPOINT (Tbit-km/sec) • node
LAMBDANET 1.5 Gbit/s, 18 λ's, 57.8 km	27.0	1.56	21.5
LAMBDANET 2.0 Gbit/s, 16 λ's, 40.0 km	32.0	1.28	18.0
WDM 2.0 Gbit/s, 18 λ's, 57.5 km	36.0	2.07	****

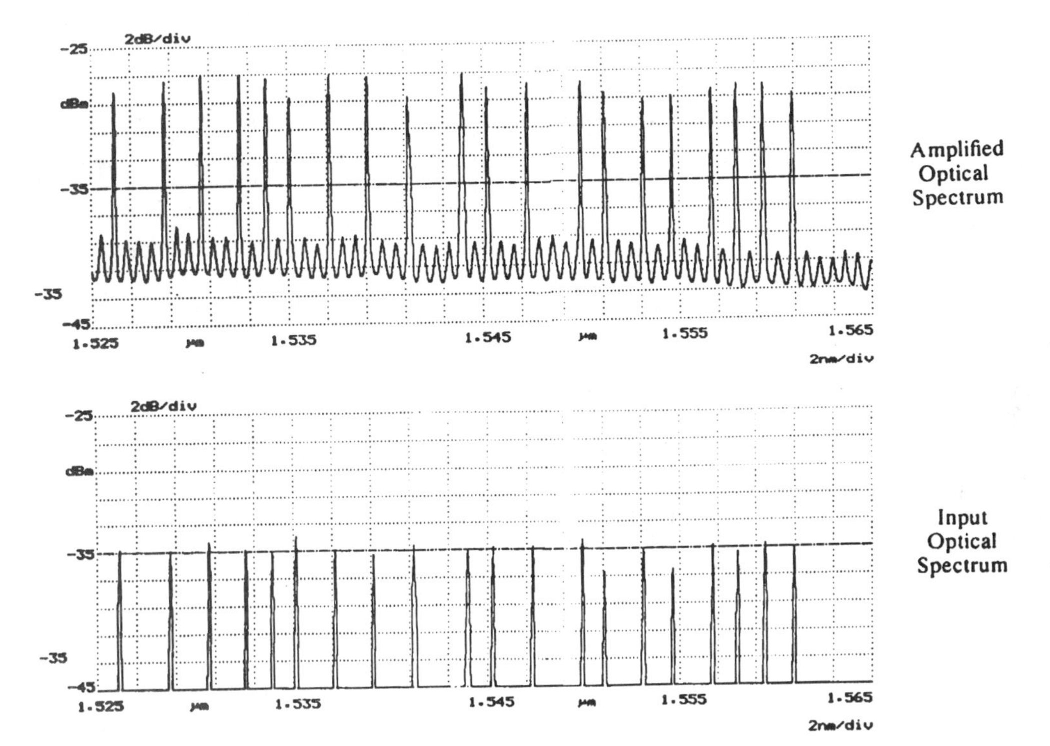

Fig. 13. Multiwavelength 20-channel amplification experiment: 20 input wavelength channels (bottom) and the corresponding amplified spectrum (top).

system power penalties and the use of 16 (as opposed to 18) lasers at the higher data rate.

The LAMBDANET structure allows the simultaneous broadcasting of information from each input port to all 16 of the output ports. The total broadcasting point-to-multipoint capacity of our system can be summarized in a figure of merit of 21.5 (Tb/s) · (km · node) for 18 wavelengths at 1.5 Gb/s and of 18 (Tb/s) · (km · node) for 16 wavelengths at 2 Gb/s. Notice that these multicasting figures of merit have taken into account the nonuniformity of the star coupler and transmitters to obtain a true aggregate throughput over all the output ports of the star coupler.

A measurement of the point-to-point transmission (not a LAMBDANET architecture) was also performed, as a reference, to compare the performance of our system to that of other WDM experiments. Eighteen wavelengths, each modulated at 2 Gb/s, were multiplexed onto a single fiber using a grating-based multiplexer (single-mode in and single-mode out, insertion loss 6.5 dB) instead of the star coupler. After a transmission of over 57.5 km of standard single-mode fiber, the 18 channels were demultiplexed and detected using the same equipment as in the LAMBDANET measurements. Thus, the point-to-point WDM measurement demonstrated a throughput of 36 Gb/s on a single-fiber strand, with a (bit rate) · (distance) product of 2.07 (Tb/s) · km. The results, in terms of bit-rate · distance products from all these experiments, is summarized in Table I.

As previously mentioned, simultaneous optical amplification of WDM signals would greatly simplify equivalent electronic amplification and would allow alleviation of the power-budget constraints in multiwavelength networks and the LAMBDANET architecture, in particular. In an experiment that used the LAMBDANET hardware and a semiconductor laser amplifier, 20 channels were simultaneously amplified with 6–8 dB net gain [12]. The spectrum of the input and amplified channels is shown in Fig. 13. These results correspond to a four-fold increase in the number of nodes in a LAMBDANET cluster or a 20 km increase in its span.

VII. Conclusions

We have presented a general, multiwavelength optical-backbone network design featuring large nonblocking multicast throughput, data transparency, and the ability to take advantage of both analog and digital transmission. The multiwavelength optical network is shown to have usefulness and functionality beyond simple point-to-point transmission. This design has a number of potential applications including transmission of voice and data in the interoffice public network, distribution of video services, and as a fabric for a high-performance electrooptic switching system.

Transmission measurements for the LAMBDANET architecture have been successfully demonstrated for bit rates up to 2 Gb/s. At 1.5 Gb/s, the bandwidth-distance product is $1.56 \text{ Tb} \cdot \text{s}^{-1} \cdot \text{km}$, and since the same information is transmitted on each of the 16 output ports, the network capacity for *broadcasting* is $\approx 21.5 \text{ Tb} \cdot \text{s}^{-1} \cdot \text{km} \cdot \text{node}$. At 2 Gb/s, the measurements have already indicated, respectively, $1.28 \text{ Tb} \cdot \text{s}^{-1} \cdot \text{km}$ and $\approx 18 \text{ Tb} \cdot \text{s}^{-1} \cdot \text{km} \cdot \text{node}$. These results represent the largest multicast (not using repeaters) transmission capacities measured to date.

Acknowledgment

The architectures and experiments described would not have been possible without the dedicated effort of many people. In particular, the authors thank C. A. Brackett, S. S. Cheng, G. Coquin, L. Curtis, J. Georges, T. P. Lee, C. N. Lo, K. W. Loh, and P. W. Shumate of Bellcore for their contributions to this work. The authors also thank H. Matsumura and Y. Koga of Hitachi, Ltd. for assistance in obtaining the DFB lasers used in this work.

References

[1] M. S. Goodman, H. Kobrinski, and K. W. Lo, "Application of wavelength division multiplexing to communication network architectures," in *Proc. Int. Conf. Commun.*, Toronto, Canada, 1986, vol. 2, pp. 931–934.

[2] H. Kobrinski, R. M. Bulley, M. S. Goodman, M. P. Vecchi, C. A. Brackett, L. Curtis, and J. L. Gimlett, "Demonstration of high capacity in the LAMBDANET architecture: A multiwavelength optical network," *Electron. Lett.*, vol. 23, p. 824, 1987.

[3] D. B. Payne and J. R. Stern, in *Proc. ECOC/IOOC 85*, Venice, 1985.

[4] H. M. Foisel, "Ten-channel coherent HDTV distribution system," in *13th European Conf. Opt. Commun.*, Helsinki, Finland, 1987, p. 287.

[5] W. S. Lee, S. W. Bland, and A. J. Robinson, "Monolithic GaInAs/InP photodetector arrays for high density wavelength division multiplexing," *Electron. Lett.*, vol. 24, pp. 1143–1145, 1988.

[6] M.S. Goodman, H. Kobrinski, M. P. Vecchi, and K. D. Cooley, "An application of the LAMBDANET multiwavelength network: Wideband virtual networks," in *Proc. Globecom 87*, Tokyo, Japan, 1987.

[7] K. Y. Eng, "A photonic knockout switch for high speed packet networks," in *Proc. IEEE/IEICE Global Telecomm. Conf.*, Tokyo, Japan, Dec. 1987, pp. 47.2.1–47.2.5.

[8] T. T. Lee, M. S. Goodman, and E. Arthurs, "Star-track: A broadband optical multicast switch," to be published.

[9] M. S. Goodman, "Multiwavelength networks: New approaches to switching," in *Proc. Opt. Fiber Commun. Conf.*, Houston, TX, Feb. 1989, p. 152.

[10] K. Kobayashi and I. Mito, "Progress in narrow-linewidth tunable laser sources," in *Opt. Fiber Commun. Conf.*, Reno, NV, Jan. 1987, paper WC1, p. 148.

[11] M. P. Vecchi, R. M. Bulley, M. S. Goodman, H. Kobrinski, and C. A. Brackett, "High-bit-rate measurements in the LAMBDANET multiwavelength optical star network," in *Opt. Fiber Commun. Conf.*, New Orleans, LA, Jan. 1988, paper WO2, p. 95.

[12] G. Coquin, H. Kobrinski, C. E. Zah, F. K. Shokoohi, C. Caneau, and S. G. Menocal, "Simultaneous amplification of 20 channels in a multiwavelength distribution system," *Photon. Technol. Lett.*, vol. 1, pp. 176–178, 1989.

All-Optical Network Consortium—Ultrafast TDM Networks

RICHARD A. BARRY, MEMBER, IEEE, VINCENT W.S. CHAN, FELLOW, IEEE,
KATHERINE L. HALL, MEMBER, IEEE, EMILY S. KINTZER, J.D. MOORES, MEMBER, IEEE,
KRISTIN A. RAUSCHENBACH, MEMBER, IEEE, ERIC A. SWANSON, MEMBER, IEEE,
LAURA E. ADAMS, C.R. DOERR, STEVEN G. FINN, MEMBER, IEEE,
HERMANN A. HAUS, LIFE FELLOW, IEEE, ERICH P. IPPEN, FELLOW, IEEE,
WILLIAM S. WONG, STUDENT MEMBER, IEEE, AND M. HANER, MEMBER, IEEE

(Invited Paper)

***Abstract*—We describe recent results of the Advanced Research Projects Agency (ARPA) sponsored Consortium on Wideband All-Optical Networks which is developing architectures, technology components, and applications for ultrafast 100 Gb/s time-division multiplexing (TDM) optical networks. The shared-media ultrafast networks we envision are appropriate for providing low-access-delay bandwidth on demand to both future high-burst rate (100 Gb/s) users as well aggregates of lower-rate users (i.e., a heterogeneous user population). To realize these goals we are developing ultrafast network architectures such as HLAN, described here, that operate well in high-latency environments and require only limited processing capability at the ultrafast bit rates. We also describe results on 80-Gb/s, 90-km soliton transmission, 100-Gb/s soliton compression laser source technology, picosecond short-pulse fiber ring lasers, picosecond-accuracy optical bit-phase sensing and clock recovery, all-optical injection-locked fiber figure-eight laser clock recovery, short-pulse fiber loop storage, and all-optical pulse width and wavelength conversion.**

I. Introduction

In 1993, an Advanced Research Projects Agency (ARPA) sponsored Consortium made up of AT&T Bell Laboratories, Digital Equipment Corporation, and the Massachusetts Institute of Technology was formed to develop architectures and technologies for future high-speed high-capacity wavelength-division multiplexing (WDM) optical networks and to develop architecture and technology for ultrafast, 100 Gb/s, time-division multiplexing (TDM) networks. This paper describes our optical TDM architecture and details key technology milestones toward realization of this system. A second separate paper in this special issue addresses Consortium work on WDM networks.

Manuscript received April 6, 1995; revised August 8, 1995. This work was conducted under the auspices of the Consortium on Wideband All-Optical Networks with full support by the Advanced Research Projects Agency. The work at Massachusetts Institute of Technology in Cambridge was also supported by JSEP and AFOSR.

R. A. Barry, V. W. S. Chan, K. L. Hall, E. S. Kintzer, J. D. Moores, K. A. Rauschenbach, and E. A. Swanson are with the Massachusetts Institute of Technology, Lincoln Laboratory, Lexington, MA 02173–9108 USA.

L. E. Adams, C. R. Doerr, S. G. Finn, H. A. Haus, E. P. Ippen, and W. S. Wong are with the Massachusetts Institute of Technology, Cambridge, MA 02139 USA.

M. Haner is with AT&T Bell Laboratories, Holmdel, NJ 07974 USA.

Publisher Item Identifier S 0733-8716(96)03686-4.

II. Ultrafast Time-Domain Network Architecture

Ultrafast optical TDM networks have the potential to provide truly flexible bandwidth on demand at burst rates of 100 Gb/s. The total capacity of these single-channel local and metropolitan area networks (LAN's and MAN's) may be the same as a number of WDM lower-rate channels, but they provide potential improvements in network performance in terms of user access time, delay and throughput, depending on the user rates and statistics. In addition, end node equipment is conceptually simpler for single-channel versus multi-channel approaches. Recently, impressive network subsystems for these types of systems have been demonstrated [1]–[4]. To date, these subsystems have been for bit-interleaved multiplexing. In this paper, we describe architecture and technology for single-stream, 100-Gb/s local/metropolitan area shared media area networks (LAN/MAN) with slotted TDM. We will argue that for a network servicing a combination of high-speed and low-speed users, a 100-Gb/s slotted TDM network has several important operating advantages over a bit-interleaved multiplexing approach.

As shown in the strawman architecture of Fig. 1, we envision high-end single users, such as high-speed video servers, terabyte media banks and supercomputers that may operate at speeds from 10–100 Gb/s, as well as aggregates of lower speed users, communicating via a network with an operating speed of ≈100 Gb/s. Bandwidth on demand packet service and guaranteed bandwidth are desired. As described further below, gateways will permit the interconnection of the 100-Gb/s LAN/MAN's to each other, e.g., through a nationwide all-optical backbone network. In this section, we describe the general requirements of a such a network and motivate the choice of slotted TDM. We then describe the medium access node, the gateway node and a candidate architecture for such a network.

A. General Architectural Requirements

The network we are focusing on serves a heterogeneous population of users. While the optical medium operates at an aggregate rate of 100 Gb/s, user requirements vary from 1–100 Gb/s. The most important functions of these networks

Reprinted with permission from *IEEE Journal on Selected Areas in Communications,* Vol. 14, No. 5, pp. 999-1013, June 1996.

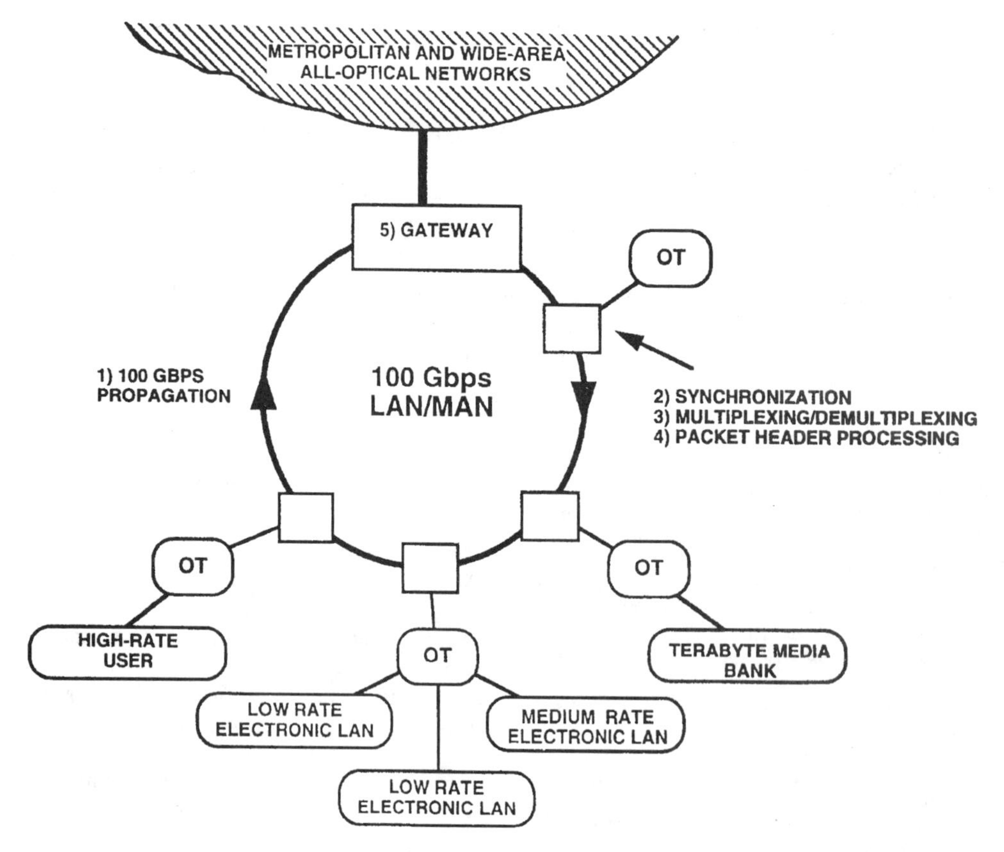

Fig. 1. Strawman architecture for 100 Gb/s optical LAN/MAN.

are to 1) provide a backbone to interconnect high-speed networks, e.g., Gb/s LAN's, 2) quickly transfer very large data blocks, 3) switch large aggregations of traffic, and 4) provide flexible lower-rate access to high-speed devices, e.g., video servers. Although we are designing a 100-Gb/s network, the architecture will be scalable in principle to faster media rates.

We assume the heterogeneous user population requires both guaranteed bandwidth (GBW) connections and bandwidth-on-demand (BOD) services of differing average information rates and burstiness. To serve these users, the network provides point-to-point and point-to-multipoint services between the access stations. The BOD services should, in addition to supporting a variety of average rates, be fair and make efficient use of the network resources.

Relying on technology which is available today, the network stations will operate with 1–10-Gb/s electronics and an all-optical "front-end." However, in the not too distant future, we expect technological advances to permit the addition of stations operating with up to 40-Gb/s electronics [5]. Furthermore, in the more distant future, electronic and optical devices should permit the addition of stations capable of processing data at 100 Gb/s. In the face of this technology evolution, the network architecture must be evolvable, allowing fast stations to co-exist with slower stations. As a general principle, the network architecture should seamlessly support the addition of a 100-Gb/s station.

B. Multiplexing Alternatives

The multiplexing scheme is a network parameter that greatly influences the design, performance, and evolvability of the network. We now briefly describe some implications of the multiplexing choice. For 100-Gb/s shared media network, candidate multiplexing schemes include WDM, bit-interleaved TDM, and time-slotted TDM. At some level of mathematical abstraction, these three systems are identical. However, from a practical point of view, the three schemes are vastly different; each technology provides its own set of devices and "black boxes" to access the shared media. In WDM and bit interleaved TDM, many small channels are shared by the access nodes operating at a peak rate which is a small fraction of the media rate, e.g., 1 Gb/s or 100 Mb/s. In slotted TDM, one fast channel is shared by access nodes capable of bursting at 100 Gb/s rates. While WDM technology is currently more mature than very fast TDM, as TDM technology matures the choice of a multiplexing scheme will become more a function of the system and networking implications than the availability of the underlying transmission technology.

If we look at existing lower-rate shared media networks such as Ethernet at 10 Mb/s, token ring at 10 Mb/s, or FDDI at 100 Mb/s, it is clear the most common choice for multiplexing is a single stream rather than many smaller channels. In these networks, the media access nodes electronically process the single channel at the full media rate, but current limitations make 100-Gb/s processing of the single stream infeasible. In the slotted systems we are considering here, however, we only require electronic processing to implement simple media access protocols at the slot rate, 10–100 Mslots/s. As all-optical logic technology matures, many functions of these protocols, e.g., address recognition, can be implemented optically which eventually allows for even faster media rates.

It is an obvious choice to supply a channel rate of 100 Gb/s when users (all users, or some users in a heterogeneous population) are capable of bursting data out at that rate. However, if all users operate at lower rates and one considers BOD service, the media access protocol is simpler in a slotted TDM system than in a WDM or bit interleaved TDM system. The reason is the existence of optical buffers which make it possible for access nodes to accept data at the media rate for a short period of time. In the switching literature, this capability is known as speed-up [6].

C. Medium Access Control (MAC)

The medium access control (MAC) layer regulates access to the physical channel(s). One key MAC layer function in a shared medium network is to avoid or recover from channel collisions when two transmitters access a channel simultaneously. We envision a slotted system with empty slot markers that indicate when a node can write data into a slot.

In WDM systems, the MAC node can accept data from media at a rate limited by transceiver technology and the number of transceivers per station. This greatly complicates the MAC protocol [7]. In the 100-Gb/s single channel LAN/MAN, an all-optical front end will allow the node to accept data at the medium rate. The length of time a node can accept data at the medium rate is determined by the traffic statistics and the amount of memory in the node. The longer the period, the simpler the MAC protocol.

A transmitting and receiving medium access node is shown in Fig. 2(a) and (b), respectively. A slow protocol logic unit is used to interface to the next layer in the protocol stack and to perform such functions as start-up, failure recovery, by-pass control in the case of component failure, and to provide an interface to the fast protocol logic unit. The slow protocol logic unit operates at a rate much slower than the slot rate. The fast protocol logic unit operates at the slot rate and regulates access to the media. In addition this unit controls the first in first out (FIFO) buffers, including the data interface control (i.e., keeping track of data-ready information). Clock recovery is used to recover slot clock, and possibly bit-clock, information from the network. The slot processor operates at or near the 100 Gb/s bit rate, and performs empty slot recognition (transmitter) and data address recognition (receiver). The interface driver provides an interface between the next protocol layer and the FIFO.

The optoelectronic FIFO buffers data, performing speed-up and buffering for the transmitter, and buffering high rate data for lower speed readout in the receiver. Blanking can be used to remove slotted data from the bus, and is currently an option in the architecture. In order to reduce the required amount of optical memory, we have adopted a two stage design for the FIFO, shown in Fig. 2. Here the first stage consists of optical buffers which can accept data at 100 Gb/s. The buffers are then emptied by high-speed electronics, 10–20 Gb/s, into electrical memory. This memory is then emptied at the station rate, e.g., 1 Gb/s. We are currently envisioning 10–20 optical buffers per receiver node to accommodate traffic with reasonable (1e-9) packet loss rates.

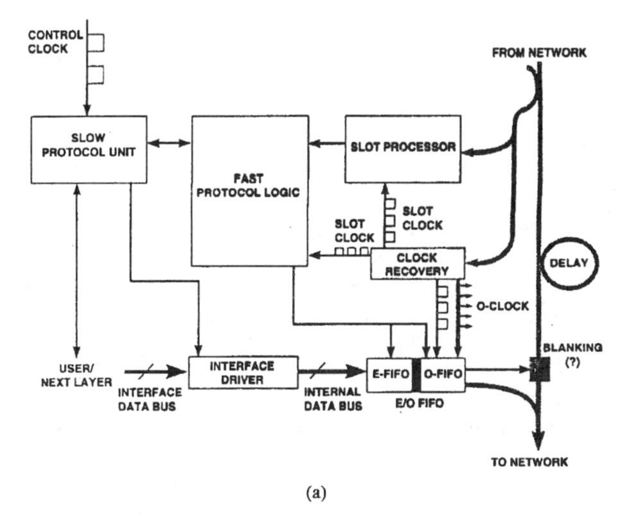

(a)

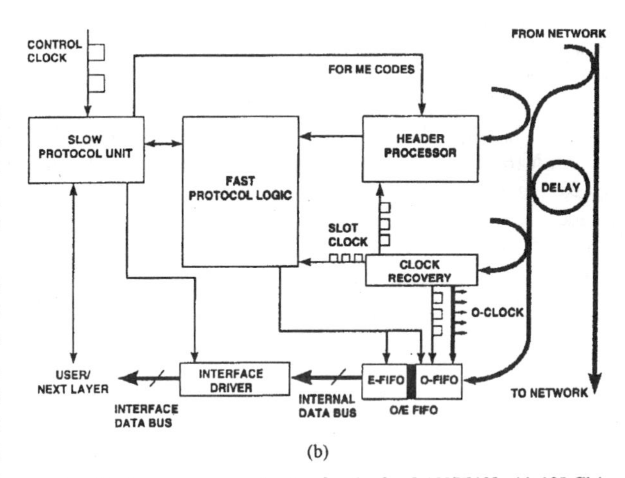

(b)

Fig. 2. Medium access node diagram for ultrafast LAN/MAN with 100 Gb/s burst rate access, (a) transmitter node and (b) receiver node. O-CLOCK: optical bit clock, 100 Gb/s rate; Slot clock: on the order of 10–100 Mslot/s rate.

D. Gateway Node

Internetworking between geographically separated LAN/MAN network segments is an important capability for a complete network architecture. In the 100-Gb/s networks we are considering, internetworking is complicated by the fact that 100-Gb/s signals cannot typically propagate over very long distances. A device which adapts 100-Gb/s data streams to the capabilities of a wide area network (WAN) network transmission system is required. We call such a device a Gateway node.

To interconnect 100-Gb/s networks over a wide area, a WAN network with very high bandwidth is required, such as the all-optical wideband network (AON) [8]. The AON is a multichannel WDM optical network capable of carrying many 10-Gb/s channels across thousands of kilometers. The traffic could be carried over, for example, ten 10-Gb/s AON channels if the full bandwidth of the LAN/MAN requires wide area connectivity. More likely, a smaller fraction (say 10–20%) would require wide area access, and would utilize fewer WDM channels. To transmit through the AON, the gateway must perform three functions. First it must demultiplex the 100-Gb/s stream into multiple 10-Gb/s streams. Second, the short pulses in each of the demultiplexed streams must be broadened for transmission over wide areas. Third, the gateway needs to be able to assign specific wavelengths to each stream to be carried through the WDM network. To receive data from a remote TDM network the gateway must essentially perform the inverse of the transmit process. It must receive multiple streams through the AON and convert them back to the TDM network wavelength. It then must retime the streams to achieve proper bit phasing between each stream. Next, the individual data pulses must be shortened and multiplexed together again to form a 100-Gb/s data packet. Finally, the packet needs to be synchronized with the main timing of the TDM network. The optical buffer technology we are developing is one alternative to address to the synchronization problem. Later we describe a pulse-width and wavelength converter to convert data pulse formats within the gateway.

E. HLAN Architecture

There are two characteristics of these 100-Gb/s slotted TDM networks that significantly impact the architectural design. First, they operate with long propagation delays, i.e., there are many data packets in flight at one time in the network—a high-latency environment. Second, there is limited processing that can be performed at the 100-Gb/s rates. Many discussions have occurred in the literature concerning the architecture of very high-speed optical multi-access networks in the presence of long propagation delays [9], [10]. There are several difficulties in creating such networks. One difficulty is to design an architecture that simultaneously provides GBW and random access BOD services. A second problem is to provide efficient and "fair" bandwidth sharing among BOD users in the presence of both moderately loaded and overloaded network traffic conditions. A third difficulty is to develop algorithms simple enough to execute at the rates required by optical networks. There have been many articles in the literature addressing these problems (for example, see [11] and [12]) but they do not appear to simultaneously satisfy all the above criteria.

HLAN is a frame-based slotted architecture implemented on a helical unidirectional bus which we believe satisfies these criteria. Fig. 3 illustrates the HLAN structure for a helical ring physical structure. The architecture can also be implemented in a linear structure which may be more appropriate for MAN distances and topologies. A headend generates frames of empty slots and puts them on the bus. The first slot of each frame is coded uniquely. GBW traffic is transmitted on the GBW

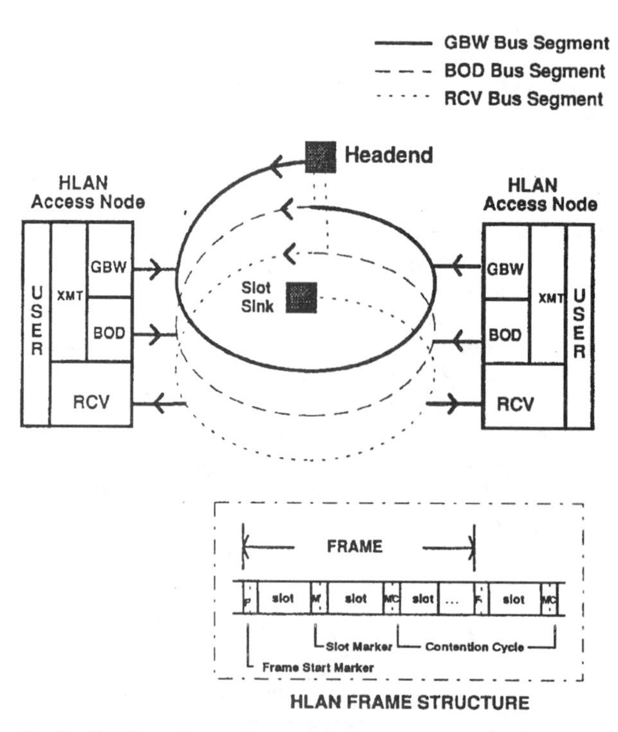

Fig. 3. HLAN topology and protocol.

Segment, BOD traffic is transmitted on the BOD Segment and data is received on the RCV Segment. Note that since all users receive traffic on the third segment, only one physical helical unidirectional bus is required—all receivers are downstream of the transmitters.

A node acquires GBW services by requesting them from the headend over the BOD service or some other facility. If bandwidth is available, the headend allocates reserved slots to the node on the GBW segment. The control of access to the GBW segment is relatively simple. Users can access the HLAN at their guaranteed rate using only a counter, a flip-flop, a few gates and slot marker detection logic [13]. Using existing gallium arsenide logic, HLAN slot rates of 10^7 to 10^8 slots per second can be supported. Optical detection of slot markers and bit timing permits network data rates of 100 Gb/s [14].

Fair and efficient BOD service is provided via the BOD segment. The headend creates credit allocations for BOD traffic using C slot markers. Intervals between credit allocations contain a number of unallocated slots. Users receive M credits for each C slot marker received when they have data to send. Credits are reduced by one for each slot that is used. When there is no data to send, a node's credit count is reset to M. Note that unused GBW slots are available for use by BOD users. To prevent lockouts from high traffic upstream nodes, the headend monitors free slots at the end of the bus. If no free slots are observed for a specified period of time the length of the credit allocation interval is increased, thus decreasing the bandwidth available to individual nodes. If multiple free slots are observed, the length of the interval is reduced, returning more bandwidth to individual nodes. The system is "fair" in that all users will get equal throughput when they have data to send. The algorithms are simple to implement and require only a few logic elements in addition to slot marker detection.

High-speed optical multi-access networks have two principal performance measures—efficiency and delay. It can be shown that there is a lower bound to the average queueing access delay (W_Q) for multi-access slotted unidirectional bus networks under Poisson arrival traffic statistics of $W_Q = 1/(2(1-\rho))$, where the slot time is normalized to 1 and ρ is the total traffic load. Since in a unidirectional slotted system the service time and the propagation time between two nodes *is* fixed, minimizing average queue delay minimizes the overall system delay. An interesting result is that this lower bound can be achieved for traffic with stationary statistics with a "Greedy" slot scheduling algorithm. Utilizing a Greedy algorithm, any node that has data to send will transmit it in any empty slot. The Greedy algorithm is also maximally efficient in that no slot that could be used is ever left empty.

The Greedy algorithm achieves its optimal average delay performance at the expense of fairness among the nodes on the network. Fig. 4(a) shows a plot comparing the optimal delay of the Greedy algorithm to a "fair" TDM algorithm for a 10-node network with uniform traffic. As can be seen, the performance improvement of the Greedy algorithm under heavier loads is significant. However while the average performance is greatly improved, the relative performance between nodes is quite unfair. In Fig. 4(b), the average delay in queue for the first node and the last node on a unidirectional bus are compared for a Greedy scheduling algorithm. Using classical results of queuing [15], the ratio of the delay of the last node to the first node on a unidirectional bus can be shown to approach a limit of $(1-\rho)^{-2}$ for large networks. Under overloaded conditions ($\rho > 1$) the last node's delay becomes infinited. This implies that for large networks under heavy loads, a Greedy algorithm would not be appropriate.

Much of the work in the literature for high-speed networks [16]–[18] has been devoted to developing algorithms that behave like a greedy algorithm under light load and like a "fair" TDM algorithm under heavy load or overload ($\rho > 1$) traffic conditions. Many of these algorithms work well but require operations which are not well suited to very high-speed optical networks. The proposed algorithms usually fall into two categories. Either they are not suited to networks with very high propagation times relative to average message transmission time (e.g., Ethernet, FDDI, DQDB, Expressnet, S), or they are too complicated or require operations such as bit level read-modify-write operations that are not suitable for 100-Gb/s optical communications (e.g., [12]).

HLAN is an attempt to develop a protocol to achieve integrated GBW and BOD service with efficiency and fairness, which is simple to implement, and requires no bit level read-modify-write operations. We believe that the aspects of the protocol that need to operate close to the bit rate, such as slot synchronization and header processing, can be implemented in optical logic so that the protocol will scale to bit rates above of 100 Gb/s. The headend monitors the network load and adjusts the credit allocation interval to adapt dynamically to changes in traffic demand. If the credit allocation interval is set to "one" (i.e., every slot has a C slot marker), then the

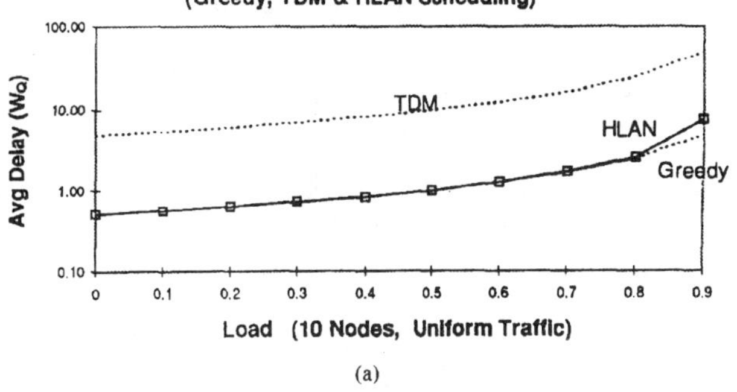

(a)

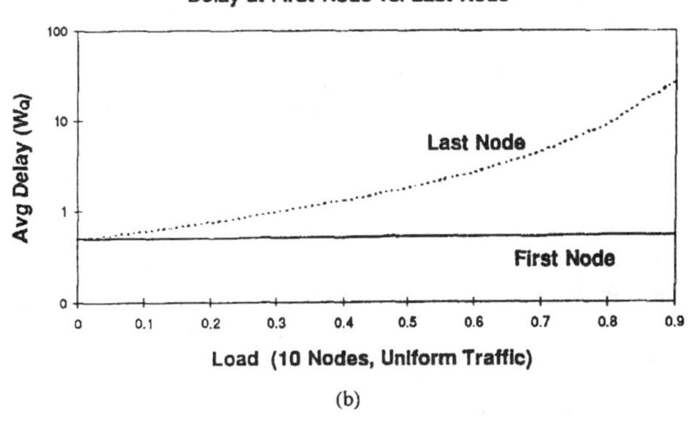

(b)

Fig. 4. (a) Average delay at a node as a function of load, comparing optimal system with bit-interleaved TDM scheduling. (b) Average queueing delay comparing first and last node for Greedy algorithm.

HLAN protocol emulates the Greedy algorithm. For low loads the Greedy algorithm is appropriate. As the load increases the credit allocation interval is also increased making the system behave more like a TDM system which is "fairer" under heavy load. Fig. 4(a) shows simulation results for average node delay for a 10-node system under uniform load using the HLAN algorithm (with $M = 2$) compared with the optimal (Greedy) performance and TDM performance curves. Enhancements of the HLAN algorithm are also being explored which give individual nodes different credit allocations based upon their traffic requirements. Our goal is to try to achieve "fair" and near optimal performance while still keeping the HLAN algorithm simple.

III. Ultrafast Optical Time-Domain Technology

In this section, we present several key technologies that support the subsystem functions (shown in Fig. 1) including soliton transmission, short pulse sources, clock recovery, optical buffers and pulse-width and wavelength converters.

A. Transmission

We have performed ultrafast soliton transmission experiments to demonstrate stable high bit rate soliton transmission. A baseband data rate of 10 Gb/s was used, followed by optical

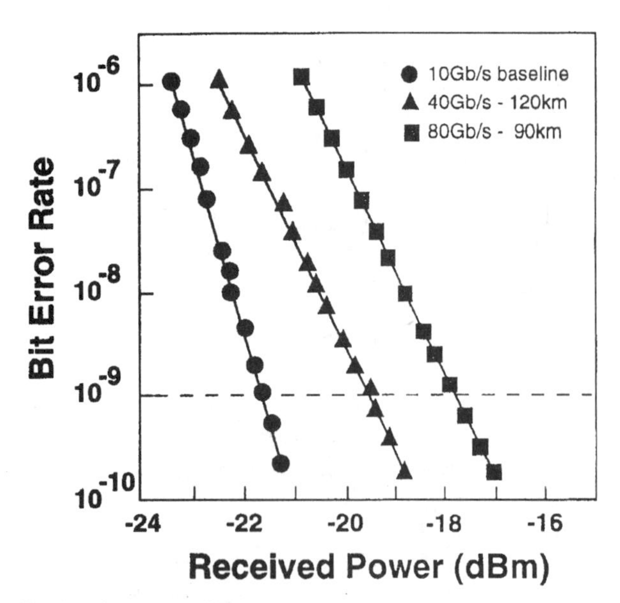

Fig. 5. Bit-error rate (BER) measurements for 10 Gb/s, 40 Gb/s, and 80 Gb/s soliton transmission.

multiplexing to form 40-Gb/s and 80-Gb/s soliton streams. The data signal was optically demultiplexed and recovered to show transmission performance. The pulse source used was a grating-coupled, mode-locked, external cavity semiconductor laser. Transform limited pulses at 1539 nm of 4.6-ps duration were modulated at 10 GHz. An external $LiNbO_3$ waveguide modulator was used to generated $2^{23} - 1$ PRBS sequences or framed data packets to simulate real asynchronous mode traffic. The data streams were optically multiplexed using a series of 3 dB optical couplers, to form 40 Gb/s and 80 Gb/s aggregate data rates. Bidirectionally pumped EDFA's were used to preamplify the modulated soliton stream and then to periodically amplify the soliton streams over the transmission span. The transmission spans consisted of multiple 18 km sections of dispersion shifted fiber (DSF), $D = 0.26$ ps/nm/km, combined with EDFA's. Fig. 5 shows the experimental bit error rate (BER) results for transmission of 40 Gb/s at 120 km and 80 Gb/s at 90 km total spans. In each case aggregate data sequences were demultiplexed to the baseband data rate of 10 Gb/s, then error rates were measured. The BER traces show no error floor down to a BER of 10^{-10} and penalties of 2 dB at 40 Gb/s and 3.5 dB at 80 Gb/s compared with the baseline transmission at 10 Gb/s. The bit error rates reported here are for the worst case channels, (four and eight channels for 40 Gb/s and 80 Gb/s, respectively) in the demultiplexed data. A variety of signaling tests were performed on the demultiplexed data, including characterization of systematic and random errors, noise source analysis, and intersymbol interference measurements.

To gain better understanding of these results we have numerically simulated soliton propagation at 80 Gb/s over 80 km. In soliton transmission, there is a tradeoff in selecting pulse widths. If the pulses are too broad, and the propagation distance sufficiently large, the interaction between pulses can result in a loss of timing information [19], [20]. If the mean dispersion is very low, choosing broader pulses could lead to amplitude fluctuations and errors in detection. On the other hand, as the pulses are narrowed, ASE-induced timing jitter increases, more power is required at the transmitter, and the minimum amplifier spacing to avoid soliton degradation decreases with the soliton period [21]–[23].

We have performed simulations to observe the coarse effects of interaction between solitons and soliton degradation/continuum interaction which result from choosing wider amplifier spacings. Although amplifier noise is included in the simulations, the associated timing jitter and energy fluctuations are small. Fig. 6(a) shows an initial 80-Gb/s data pattern and Fig. 6(b) shows the output after simulated 80 km propagation. A split-step Fourier algorithm with periodic boundary conditions is used. The pulse width is 4 ps and a path-averaged soliton power is launched. The fiber dispersion is 0.4 ps/nm/km, and loss is 0.14 dB/km. The amplifier spacing is 20 km, and the amplifier noise figure used is three. There is 1 dB of excess loss at each amplifier (splice losses, etc.). It is assumed that the components in each amplifier module, including the erbium-doped fiber itself, act as a filter of 30-nm bandwidth. All of the initial pulses are chosen to be in phase to enhance the soliton–soliton attraction, making this a "worst" case. Nevertheless, the output is acceptable-the timing and energy fluctuations are not severe.

To demonstrate the effect of changing the initial pulse separation, a similar simulation was run at the higher bit rate of 100 Gb/s. In this case, the input bit pattern is the same as shown in Fig. 6(a), but there is more overlap between the tails of neighboring pulses than in the 80-Gb/s case. The output after propagating through 80 km of fiber with the same parameters as in the previous simulation is shown in Fig. 6(c). Increasing the bit rate, and therefore reducing the ratio of the bit interval to the pulse width, has led to severe degradation of the data. Complicated pulse interactions have occurred, with one pulse compressing by more than a factor of two, and with the apparent generation of a new pulse.

We have also simulated propagation of 1.8-ps pulses at 111 Gb/s over 1000 km, by utilizing compensation techniques of filtering and phase modulation at every amplifier [23]. Through the action of dispersion, phase modulation provides potential wells for pulses (ONE's), and filtering damps out the oscillations of pulses within these wells. Modulation also spreads the spectra of the noise and continuum in the bit intervals containing ZERO's, and filtering dissipates those frequency components outside the center of the passband. With compensation, the packet is well-preserved over a much greater distance, though with potentially increased complexity, cost, and reliability issues.

B. Ultrashort-Pulse Fiber Laser

Short pulse optical sources support several of the key subsystems in ultrafast TDM systems, including sources for optical modulation, demultiplexing, optical clocks and optical logic. We require sources for optical modulation and demultiplexing that operate at lower bit rates, 10 Gb/s and lower, as well as sources for synchronization and clocks for optical

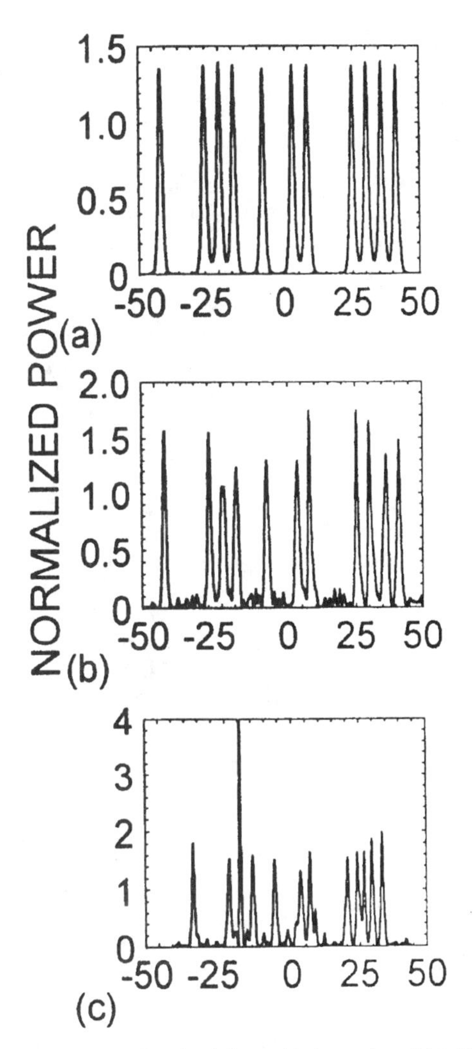

Fig. 6. 80 Gb/s propagation simultaions with 4 ps pulses: (a) Initial condition, (b) after 80 km, and (c) 100 Gb/s propagation simulations with 4 ps pulses after 80 km. Note that the time parameter is normalized in these figures.

logic and optical buffers which must operate with the bit rate of the system.

Erbium-doped fiber lasers are potentially attractive sources for high bit-rate optical networks because they produce picosecond, and even sub-picosecond, solitons in the 1550 nm wavelength band, they are low-noise, and they can be diode-pumped. The shortest pulses are obtained from passively mode-locked systems [24]–[26] which work best when only one pulse is circulating in the fiber resonator. With the minimum required length of erbium gain fiber, this limits the repetition rate to the order of 50 MHz.

Higher repetition rates can be achieved by harmonic mode-locking, making use of an active modulator to synchronize multiple pulses in the loop. These systems are susceptible to pulse-to-pulse energy fluctuations because of the long upper-state lifetime of erbium. They have been stabilized with an interferometrically-controlled Fabry–Perot in the loop [27], with an intensity dependent loss mechanism [28], and by using solitons in conjunction with a bandwidth filter. Still, none of these approaches has led to ultrashort pulses. Passive methods for inducing more than one pulse per round-trip have also been reported [29], [30], but they have not led to very high repetition rates.

To achieve ultrashort pulse durations at high repetition rate, one would like to combine the best aspects of both passive and active mode-locking: passive pulse shaping to produce short pulses; and active modulation to define the repetition rate and provide synchronization to an external clock. The problem inherent in such an approach is that when pulses become short compared to the modulation window, the modulator loses its ability to restore timing. Then, either ultra-precise cavity length control or regenerative drive electronics is needed for stability.

Recently, we reported a scheme that attains harmonic mode-locking of ultrashort pulses in an unusual way-with asynchronous modulation [31]. Illustrated in Fig. 7(a), the laser looks like most other actively-mode-locked fiber ring lasers. It incorporates waveplates (for polarization control), an isolator/polarizer (to insure unidirectional operation), a birefringent plate (for bandwidth control) and a $LiNbO_3$ phase modulator resonant at 1 GHz, approximately the 28th harmonic of the round-trip rate of the ring. Without the modulator but with the waveplates adjusted for passive mode-locking (additive-pulse mode-locking [26]) the laser produces sub-picosecond pulses. When the modulator is turned on and the waveplates are adjusted for stabilization by additive-pulse limiting [28], longer 40-ps pulses are generated in synchronism with the modulator. But, when the nonlinearity is adjusted for short pulses and the modulator is on, stable synchronous operation is very difficult to achieve.

The new, stable ultrashort pulse mode is reached by detuning the modulator frequency from the cavity harmonic by 5–20 kHz. Then a steady pulse stream develops at the 28th harmonic. An autocorrelation measurement and corresponding spectrum for the 1.6-ps pulse train attained at 1 GHz is shown in Fig. 7(b). The laser is stabilized by the asynchronous phase modulation in a way not unlike that provided by the sliding-frequency filters proposed for long-distance soliton communications [32] or the continuously frequency-shifted filter for soliton generation [33]. The low intensity background is shifted out of the central filter passband while the solitons are better able to readjust. In the asynchronously mode-locked laser there is the additional benefit that the modulation is close enough to the harmonic frequency to induce phase coherence between pulses. No careful stabilization of either the modulator frequency or the cavity length is necessary. Higher repetition rates and shorter pulses seem feasible and can be expected in the future. Another interesting application may be the use of asynchronous phase modulation to enhance the stability of an optical data storage loop [34].

C. 100-Gb/s Soliton Compression Source

Many techniques for short-pulse train generation, including gain switching, active and passive mode locking in fiber and semiconductor lasers, and electro-absorption and electro-

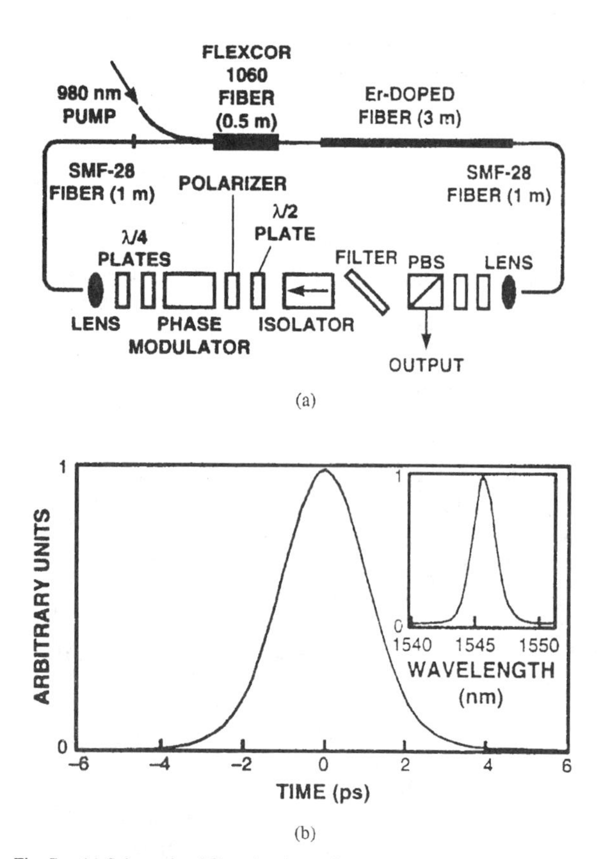

Fig. 7. (a) Schematic of fiber ring laser. (b) An autocorrelation measurement and corresponding spectrum for the 1.6 ps pulse train attained at 1 GHz.

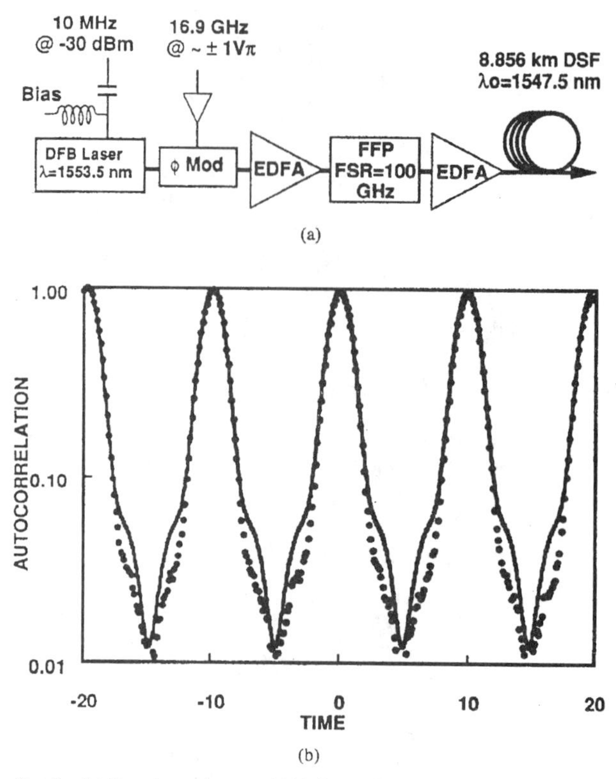

Fig. 8. (a) Experimental setup of 100 GHz soliton compression laser source. (b) Measured (solid line) and simulated (dashed line) background-free autocorrelation function.

optic modulation do not readily lend themselves to frequency agile pulse train generation at 20 GHz and above. Typically, to generate higher repetition frequencies, a lower repetition rate pulse train ($\sim$10 GHz) is used in conjunction with a multiport tapped delay line. Several authors have investigated soliton compression of the beat signal between two optical carriers as an alternative technique [35]–[40]. Some of the attractive features of the latter technique include the ability to generate very high repetition rate pulse trains ($\sim$1 THz) [35], very short pulse widths ($\sim$200 fs) [36], high output power, and electronic tunability. However, as these pulse trains are typically generated using two separate laser sources, timing jitter occurs from the relative phase fluctuations. Typical DFB lasers have a beat signal linewidth greater than 1 MHz which implies that the pulse train will decorrelate in about 1 μs. Even narrow-linewidth external cavity lasers and coupled cavity lasers are not sufficient for most applications where phase locking to external pulse trains is required [38].

We have demonstrated several techniques that use a single DFB laser source, an external modulator, and soliton compression to achieve high frequency near-transform-limited pulse train generation with potentially very low timing jitter, wavelength tunability, and electronic repetition rate tunablility [39], [40]. Because the two optical carriers are generated from a single laser source they have nearly identical frequency noise; thus the resulting timing jitter should be low. One technique utilizes a Mach–Zehnder modulator biased at a transmission null to generate a pulse train at twice the drive frequency. We have measured nearly transform-limited pulses at 20 and 40 GHz. Another technique, with higher repetition rate capability, uses a phase modulator driven at more than V_π to produce multiple optical sidebands and optical filtering to select out two phase side-bands 100 GHz apart. Fig. 8(a) shows the experimental set-up of this latter technique and Fig. 8(b) shows the theoretical and experimental background free autocorrelation traces. We estimate the pulse width is 1.37 ps, the duty cycle is 7.2:1, and the time-bandwidth product is $\sim$0.45. More nearly transform pulses can be obtained using dispersion tailored fiber [40].

D. Clock Recovery

Clock recovery is an essential element in the optical access node to enable the receiver decision circuitry to be synchronized with the incoming data stream. At data rates beyond a few Gbit/s, it becomes increasingly difficult to perform the clock recovery electronically. Optical clock recovery techniques that can be scaled to data rates of many tens of Gb/s or even 100 Gb/s will be essential to high rate communication links of the future. We have demonstrated two clock recovery schemes that are scalable to high rates, one based on an electro-optical phase-lock loop (PLL), and one based on optical mode-locking.

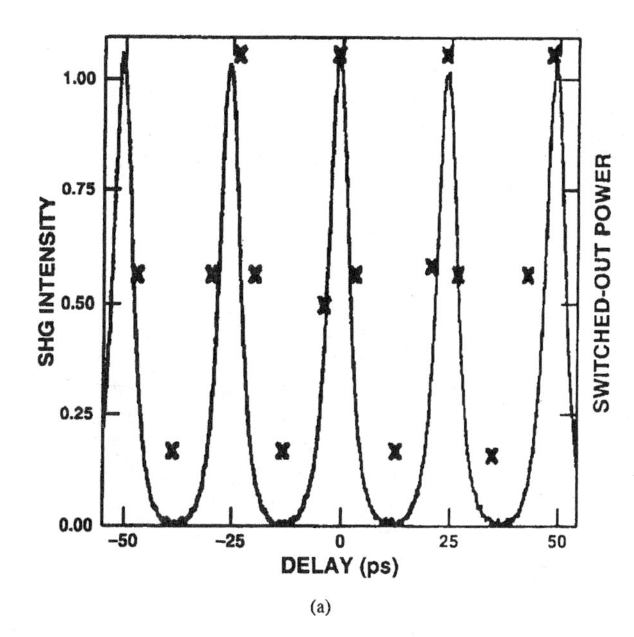

(a)

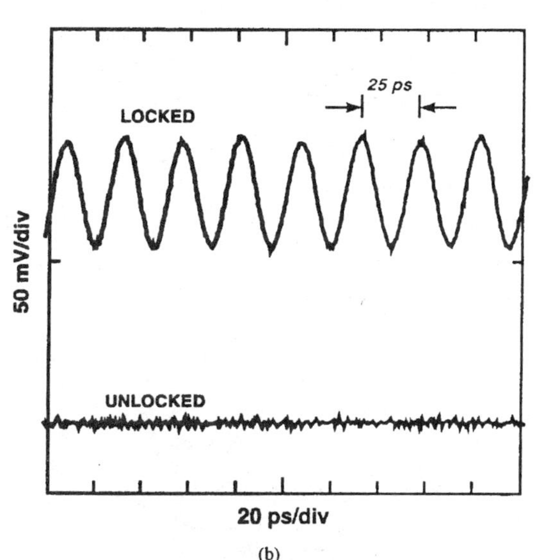

(b)

Fig. 9. (a) Measured average output power from the NOLM as a function of delay between the two pulse streams (crosses) and an autocorrelation trace of the 40 GHz pulse stream (solid line). (b) Sampling oscilloscope traces the 40 GHz optical pulse stream when the two laser sources are synchronized (to trace) and not triggered (bottom trace).

For recovering local slot and bit clocks from the slot clock which is distributed over the LAN/MAN, electro-optical PLL clock recovery is appropriate. In this scheme, a local optical clock whose repetition rate is controlled by some RF drive source is locked to the optical stream derived from the network. Such schemes have been demonstrated previously using four-wave mixing [41] and gain switching [42] in diode amplifiers and second harmonic generation in a $LiIO_3$ crystal [43] as the all-optical bit phase sensors. We have demonstrated the use of a nonlinear optical loop mirror (NOLM) as an all-optical bit phase sensor [44]–[46]. The NOLM consists of a 3-dB fiber coupler with two of the ports joined through a length of fiber. Optical clock pulses are split at the coupler into two counterpropagating pulses which acquire identical phase shifts as they traverse the loop. These pulses interfere constructively at the coupler and are reflected back through their input port. The loop can be imbalanced by introducing a high power data pulse that copropagates with one of the clock pulses, nonlinearly shifting its phase. The two clock pulses are no longer in phase when they arrive back at the 3-dB coupler and some fraction of the input signal is transmitted through the output port.

Rather than making the NOLM relatively insensitive to timing jitter by utilizing the pulse walkthrough associated with polarization dispersion [47], [48] and wavelength dispersion [49]–[52], the NOLM bit phase sensor requires minimum pulse walkthrough and is sensitive to timing errors. In this application, the center wavelengths for clock and data pulse streams are equally spaced around the zero dispersion wavelength of the fiber in the loop, and their group velocity is the same [53]. Any mismatch in timing between the two pulse streams will result in less power being switched out from the loop. The demultiplexing or modulation is more sensitive to timing jitter but the NOLM can be used to perform these functions with minimal pulse walkthrough [2]. Thus, a NOLM may be used simultaneously for clock recovery and optical demultiplexing or modulation.

We have used the NOLM bit phase sensor to synchronize a 10-GHz harmonically mode-locked external cavity laser and a 40-GHz soliton compression laser described earlier. The 10-GHz pulse stream of 8-ps pulses had a center wavelength of 1541.6 nm and the 40-GHz stream of 3-ps pulses had a center wavelength of 1552.4 nm. A tunable filter at the NOLM transmission port was used to distinguish the two pulse streams. The NOLM used in this demonstration had a zero dispersion wavelength of 1547 nm. Initially, the RF drives to the two sources were externally locked to a common 10-MHz clock and an electrical phase delay was used to walk the two pulse streams through each other in time. The crosses in Fig. 9(a) show the measured average output power from the transmission port of the NOLM as a function of delay between the two pulse streams. For comparison, the autocorrelation of the 40 GHz pulse stream is also shown (solid line). Notice that there is excellent contrast between the measured NOLM output power when the pulse streams are aligned in time and when they are not. This NOLM output power provides the error signal in the electrical portion of the PLL.

We demonstrate the synchronization scheme by triggering a digital sampling oscilloscope with a portion of the electrical drive to the 10 GHz source, and measuring the 40 GHz optical pulse stream from, the soliton source. Here, the RF drive sources generate 10-GHz and 20-GHz sinusoids, but they are no longer locked to a common 10-MHz clock. When the electro-optic PLL is enabled the 40-GHz optical pulse stream (detected using a 45-GHz photodiode), Fig. 9(b), is synchronous with the 10-GHz drive to the mode-locked diode laser, and therefore to the trigger for the sampling scope. However, when the PLL is disabled, there is no correlation between the two sources and the signal on the scope are not

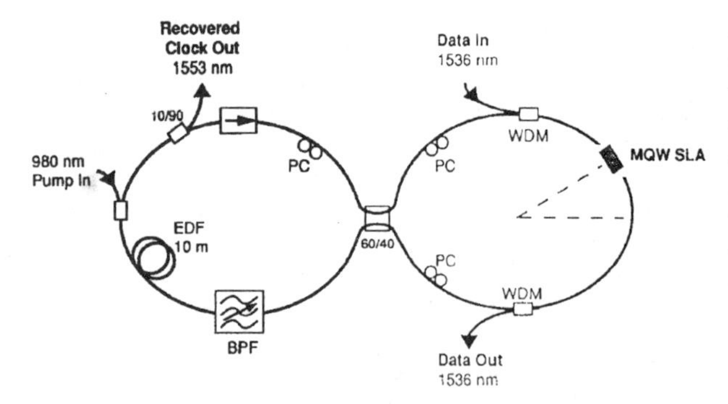

Fig. 10. Fiber figure-eight laser clock recovery system.

triggered (Fig. 9(b), bottom trace). Note that the traces shown in Fig. 9 have been averaged over many samples and are not a snapshot in time.

The holding range for this opto-electronic circuit was measured by enabling the PLL and manually detuning the RF drive frequency to the mode-locked diode laser. The measured holding range was ±20 MHz. We believe this detuning is limited because the optical pulses produced by the mode-locked diode laser broaden as the RF drive is tuned away from resonance which results in reduced phase accuracy. Larger holding ranges are expected for nonresonant short pulse sources. The clock acquisition time for this clock recovery scheme will probably be limited by propagation time through the NOLM. We envision the NOLM as a bit phase sensor when multiple functions are required such as simultaneous modulation or demultiplexing and where relatively low bandwidth (kHz) PLL's are required (e.g., clock smoothing).

All-optical clock recovery, in which no electrical components, or opto-electronic conversion are utilized, not only plays a useful role at the receiver but it permits all-optical data regeneration to be performed in the transmission system to improve the signal-to-noise ratio. Optical clock recovery techniques have in common the use of some component with an optical nonlinearity. This nonlinear component can be an optical fiber, but the weak nonlinearity of fiber requires that very long lengths be used, typically many kilometers [54]. The long propagation time through the fiber implies these clock recovery systems will require a long time (and thus, many data bits) to lock up and may have a narrow range of clock frequency over which the system will lock. While these systems can be shortened using fiber with higher nonlinearity [55], the use of a semiconductor waveguide, where the nonlinearity is three to four orders of magnitude higher than standard fiber, provides the potential to realize much shorter cavities.

Semiconductor waveguides have been employed in a variety of different clock recovery schemes. Electro-optical phase locked loop clock recovery has been demonstrated at high data rates using semiconductor waveguides to derive a lower-rate prescaled clock [41]. Semiconductor nonlinearity has also been used to demonstrate all-optical clock recovery in a fiber ring laser [56].

We have been working on all-optical clock recovery using a semiconductor amplifier in a fiber laser with a figure-eight geometry (Fig. 10). One loop of the figure eight is a NOLM containing the semiconductor amplifier, located off-center in the NOLM [57]. The other loop contains gain and output coupling for the fiber laser, an isolator to force unidirectional operation and a bandpass filter. An advantage of the figure eight geometry for optical mode-locking, as opposed to a ring laser geometry, is that relatively small phase changes can be translated to large amplitude changes in the NOLM. A data pulse acting on the SLA induces a time-dependent phase shift which, in turn, modifies the transmissivity of the NOLM. An advantage of the figure-eight geometry, as opposed to a ring geometry, for optical mode locking is that the NOLM provides a direct conversion of phase shift to amplitude modulation.

The offset of the semiconductor amplifier from the middle of the NOLM, combined with the temporal response of the semiconductor medium to the data pulse, determines the duration of the transmission change of the NOLM. This permits the fiber figure-eight laser to mode lock to the data rate via the modulation of the NOLM transmission. Transform limited clock pulses that are shorter in duration than the data pulses have been demonstrated.

The overall cavity length can be kept relatively short (a few tens of meters in this case) so that the fiber laser can lock to a large range of incoming data rates [58]. To date, clock recovery has been demonstrated at rates up to 2.5 Gb/s in this system. The data stream consisted of 30-ps gain switched diode laser pulses externally modulated with a $2^{23}-1$ PRBS. The pulses were amplified to $\sim$1 pJ per pulse. The recovered clock pulse-width was 20 ps. The clock output had low timing jitter, with the jitter remaining below 100 fs as the input data rate was tuned over a 50 kHz range. Other work has shown that gain and index nonlinearities can be made very fast [59], indicating that this system should be scalable to higher rates.

An additional advantage of the figure-eight geometry is that locking time and stability in the presence of a long string of zeros can be enhanced by taking advantage of self phase modulation effects that result from using a coupler for the NOLM with a split ratio different from 50/50. With a 60/40 coupler, we have measured locking time and stability by gating a stream of all ones at the data input port. The peak power of the clock output pulses has been observed to build up from cw to 70% of its steady state value in $\sim$1 μs (or nine roundtrips of the figure-eight cavity). When the data stream is gated off, the peak power of the clock output pulses remains within 70% of its steady state value for 22 roundtrips of the figure-eight cavity (or 7000 zeros at 2.5 Gb/s). These numbers may improve further by increasing the coupling to the SLA.

E. Optical Short Pulse Storage Loop

It is clear that a slotted TDM network requires buffering. Furthermore, it would be desirable to maintain the data in its optical form and to buffer the data at the bus data rate. With data rates in the 100 Gb/s range, electronics does not provide a simple solution, and we seek optical implementations. Furthermore, we anticipate that most end-users will be incapable of processing data at the bus rate, and this creates a need for rate conversion, which can be achieved with a storage loop in conjunction with an optical switch. Packet lengths within an order of magnitude of 10 kb are anticipated for our 100-Gb/s LAN/MAN. At 100 Gb/s, the duration of a 10 kb packet is 100 ns. Based upon anticipated tolerable latencies, storage time requirements are likely to be no more than hundreds of microseconds for these applications. Several storage loop designs have been proposed and demonstrated [60]–[65], and recirculating-loop transmission experiments have demonstrated pulse storage as well [66]–[68]. We have demonstrated storage of 1.76 kb of data at 20 Gb/s. Stable operation is achieved using amplitude modulation, filtering, and artificial fast saturable absorption.

The layout of our experiment is shown in Fig. 11(a). The storage ring consists of 2 m of erbium-doped fiber with a 980/1550 WDM coupler with 4-m Flexcor pigtails, a fiber-pigtailed polarization-insensitive isolator, 12 m of SMF-28 fiber, a polarization controller (PC), a 90/10 output coupler, a $LiNbO_3$ amplitude modulator of 12 GHz 3 dB bandwidth, and an air gap. The air gap contains two interference filters, two half-wave plates (HWP), one quarter-wave plate (QWP), and one polarizing beamsplitter (PBS). The pump for the erbium-doped fiber is a MOPA (master oscillator power amplifier) supplying up to 400 mW of 980-nm pump power in the ring.

Figure 11(b) shows a portion of a stored pattern, generated from noise. A 45 GHz photodiode detects the optical pulse stream. The pulses are displayed on a digital sampling oscilloscope that is triggered at the cavity fundamental frequency of 7.78 MHz. The loop stores 1760 b at 20 Gb/s, for several minutes. The number of ONE's stored can be adjusted by changing the pump power or waveplate settings. With sufficient gain, the loop is an harmonically mode-locked laser. The loading and unloading of these loops has been demonstrated at lower rates [34].

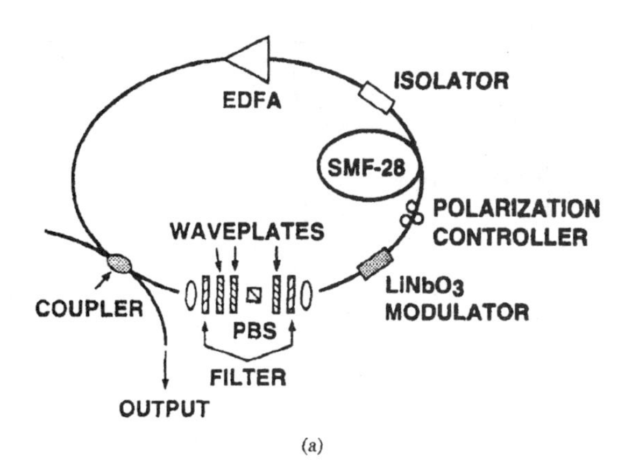

(a)

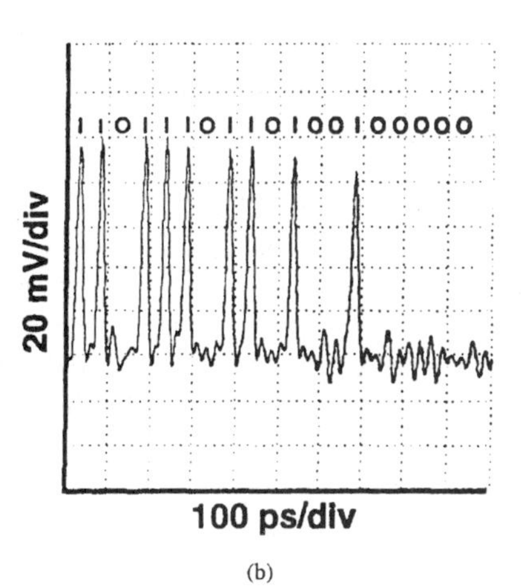

(b)

Fig. 11. (a) Experimental setup of fiber storage loop. (b) Sampling oscilloscope output showing a portion of a 20 Gb/s stored data pattern (11 011 101 101 001 000 000).

F. Pulse-Width and Wavelength Converter

As described above, the optical gateway converts some portion of the local 100-Gb/s data to data compatible with a wider area, WDM network. These wide area connections will utilize lower-bit-rate multiple wavelength channels [18] and will require long pulses (20–40 ps) if soliton transmission is used [69]. This conversion of signals from a high-rate TDM stream to a lower-rate WDM stream will require both rate conversion or demultiplexing and pulse-width and wavelength conversion. Pulse-width and wavelength conversion can be achieved using a NOLM [70] configured as an AND gate [52].

This NOLM may be configured as a pulse-width and wavelength converter by utilizing pulse walkthrough of the data and clock pulses in the device. In this set-up, wavelength dispersion [49]–[52] causes the short optical data pulse at one wavelength to walkthrough the long optical clock pulse at another wavelength [71]. The required pulse walkthrough is any amount greater than the longer pulse width and less than the bit period. Too little walkthrough decreases the switching efficiency and too much walkthrough causes intersymbol interference. Depending on the parameters of the fiber used in the loop mirror and on the data rate, a variety of pulse-widths and wavelengths can be accommodated. For example, using 2.2 km of dispersion shifted fiber with a zero dispersion wavelength of 1564 nm we realize a 20-nm range over which the data and clock pulse walkthrough is 20–50 ps. This allows pulse-width conversion to 20-ps pulses, at bit rates less than 20 GHz, over this range.

We have demonstrated this tunability of the pulse-width and wavelength converter. Fig. 12(a) shows the measured intensity autocorrelation functions of the data pulse (solid line) and the data-encoded clock output pulse (dashed line). Fig. 12(b) shows the measured BER's for the system. The BER for the short pulse (8 ps) data stream before it is launched into the NOLM is shown by the circles. The BER for the encoded clock pulse (20 ps) stream is shown for two clock center wavelengths, 1537 nm (squares) and 1557 nm (triangles). The BER for the encoded clock pulse stream shows a small penalty when compared to the data pulse stream, presumably due to imperfect switching in the NOLM. However, the BER's for encoded clock streams at various wavelengths are identical.

IV. Summary

This paper has reviewed some of the architecture and technology required to enable multiple access communication

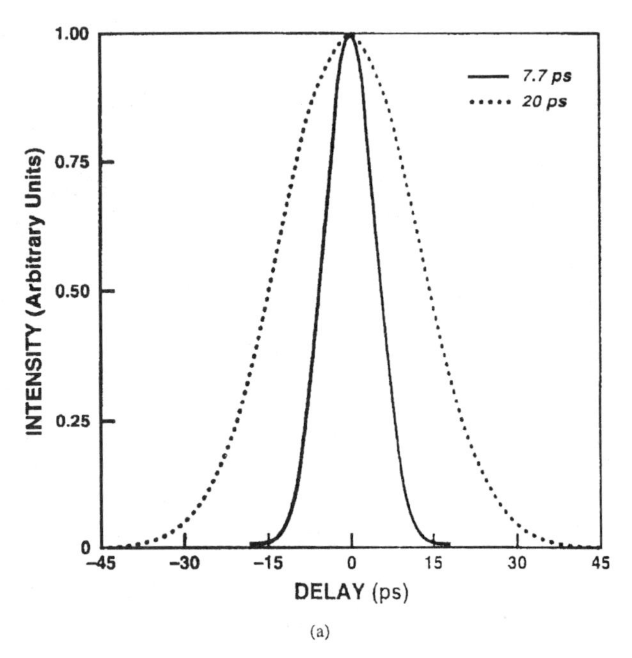

(a)

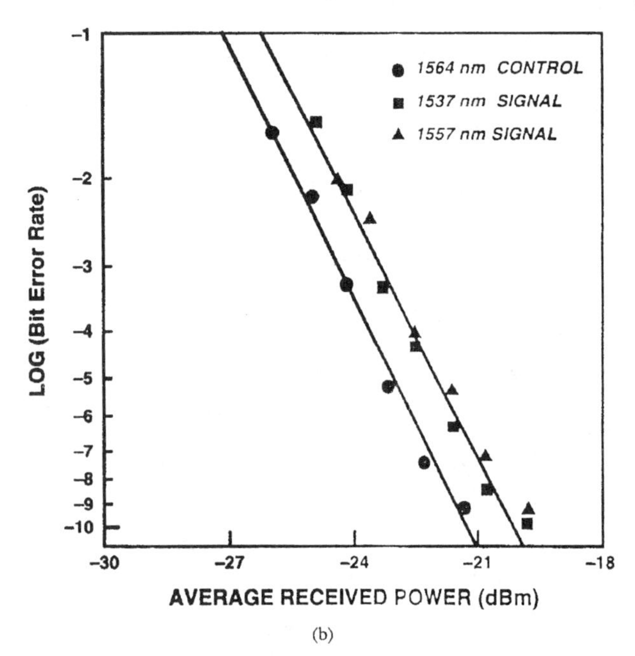

(b)

Fig. 12. (a) Measured intensity autocorrelation functions for the data pulses (solid line) and the encoded clock pulses (dashed line). (b) BER as a function of average received power for the data pulse stream (circles), the encoded clock stream at 1537 nm (squares) and the encoded clock stream at 1557 nm (triangles).

systems that provide truly flexible bandwidth on demand with simpler network control, efficient resource allocation and short access delay. These systems are anticipated to provide service to future high-end users that are capable of high-speed burst rates, but also to support aggregates of lower speed users. These networks will rely on new evolving technologies, which must mature to become cost-effective when compared to other network solutions. However, assuming the technology matures sufficiently, the network performance advantages of TDM systems like the one described here will make it, we believe, a natural choice, particularly for providing bandwidth on demand service.

Our candidate architecture, HLAN, is a single-stream 100-Gb/s local/metropolitan area network (LAN/MAN) based on a slotted system in a unidirectional bus topology. The key technologies we have demonstrated include 80 Gb/s, 90-km soliton transmission, 100-GHz short pulse clock sources and short-pulse high repetition rate fiber ring lasers, 40-Gb/s clock recovery, 20-Gb/s optical short-pulse fiber loop buffers and an all-optical pulse-width and wavelength converter. These demonstrations, and represent an important step toward eventual system demonstrations.

Acknowledgment

The authors gratefully acknowledge a number of people who have made valuable contributions to this program. R. Bondurant has provided invaluable insight and support. S. Lepage, C. Fennelly, and B. Lucia for assistance. Also, the authors thank A. Chiu for her valuable contributions to the HLAN protocol.

References

[1] A. D. Ellis, T. Widdowson, X. Shan, and D. G. Moody, "Three-node, 40 Gbit/s OTDM network experiment using electro-optic switches," *Electron. Lett.*, vol. 30, p. 1333, 1994.

[2] K. Suzuki, K. Iwatsuki, S. Nishi, and M. Saruwatari, "Error-free demultiplexing of 160 Gbit/s pulse signal using optical loop mirror including semiconductor laser amplifier," *Electron. Lett.*, vol. 30, p. 1501, 1994.

[3] D. Cotter, K. Smith, M. Shabeer, D. C. Rogers, and D. Nesset, "Ultrafast self-routing packet networks," in *Tech. Dig. Opt. Fiber Commun. Conf., OFC'95*, San Diego, CA, paper WJ1.

[4] T. Morioka, S. Kawanishi, H. Takara, and M. Saruwatari, "Multiple-output, 100 Gbit/s all-optical demultiplexer based on multichannel four-wave mixing pumped by a linearly chirped square pulse," *Electron. Lett.*, vol. 30, p. 1959, 1994.

[5] K. A. Rauschenbach *et al.*, "Optical time-domain technology," *LEOS'94*, Boston, MA, 1994, WD4.

[6] J. Y. Hui, *Switching and Traffic Theory for Integrated Broadband Networks*. Boston, MA: Kluwer, 1994, p. 171.

[7] B. Mukherjee, "WDM-based local lightwave networks, Part 1: Single-hop systems," *IEEE Networks Mag.*, vol. 1, pp. 12–27, 1992.

[8] S. B. Alexander *et al.*, "A precompetitive consortium on wideband all-optical networks," *J. Lightwave Technol.*, vol. 11, pp. 714–735, 1993.

[9] S. R. Sachs, "Alternative local area access protocols," *IEEE Commun. Magazine.*, vol. 26, no. 3, p. 25, 1988.

[10] *IEEE Project 802.6*, Proposed Standard, Doc. no. P806.2.6/D12, 1990.

[11] C. Yeh, M. Lin, M. Gerla, and P. Rodrigues, "RATO-net: A random-access protocol for unidirectional ultra-high-speed optical fiber network," *J. Lightwave Technol.*, vol 8, no. 1, 1990.

[12] G. C. Watson and S. Tohme, "S++—A new MAC protocol for Gb/s local area networks," *IEEE J. Select. Areas Commun.*, vol. 11, no. 4, 1993.

[13] S. J. Finn, "HLAN-an architecture for optical multi-access networks," in *IEEE 1995 Dig. LEOS Summer Top. Mtg.*, 1995, p. 45–47.

[14] J. R. Sauer, M. N. Islam, and S. P. Dijaili, "A soliton ring network," *J. Lightwave Technol.*, vol. 11, no. 12, pp. 2182–2187, 1993.

[15] D. Bertsekas and R. Gallager, *Data Networks*, 2nd ed. Englewood Cliffs, NJ: Prentice-Hall, 1992, pp. 221–240.

[16] H. R. Muller, M. M. Nassehi, J. W. Wong, E. Zurfluh, W. Bux, and P. Zafiropulo, "DQMA and CRMA: New access schemes for Gbit/s LAN's and MAN's," in *Proc. IEEE INFOCOM'90*, 1990, pp. 185–191.

[17] G. Watson, S. Ooi, D. Skellern, and D. Cunningham, "HANGMAN Gbit/s network," *IEEE Network Mag.*, 1992.

[18] G. Watson and S. Tohme, "S++—A new MAC protocol for Gb/s local area networks," *IEEE J. Select. Areas Commun.*, vol. 11. no. 4, May 1993.

[19] J. P. Gordon, "Interaction forces among solitons in optical fibers," *Opt. Lett.*, vol. 8, no. 11, pp. 596–598, Nov. 1983.

[20] K. Smith and L. F. Mollenauer, "Experimental observation of soliton interaction over long fiber paths: Discovery of a long-range interaction," *Opt. Lett.*, vol. 14, no. 22, pp. 1284–1286, 1989.

[21] J. P. Gordon and H. A. Haus, "Random walk of coherently amplified solutions in optical fiber transmission," *Opt. Lett.*, vol. 11, pp. 665–667, 1986.

[22] J. P. Gordon and L. F. Mollenauer, "Effects of fiber nonlinearities and amplifier spacing on ultra-long distance transmission," *J. Lightwave Technol.*, vol. 9, no. 2, pp. 170–173, 1991.

[23] J. D. Moores, W. S. Wong, and H. A. Haus, "Stability and timing maintenance in soliton transmission and storage rings," *Opt. Commun.*, vol. 113, nos. 1, 2, 3, pp. 153–175, 1994.

[24] I. N. Duling III, "Subpicosecond all-fiber erbium laser," *Electron. Lett.*, vol. 27, pp. 544–545, 1991.

[25] V. J. Matsas, T. P. Newton, D. J. Richardson, and D. N. Payne, "Self-starting passively mode-locked fiber ring soliton laser exploiting nonlinear polarization rotation," *Electron. Lett.*, vol. 28, pp. 1391–1393, 1992.

[26] K. Tamura, H. A. Haus, and E. P. Ippen, "Self-starting additive pulse mode-locked erbium fiber ring laser," *Electron. Lett.*, vol. 28, pp. 2226–2228, 1992.

[27] G. T. Harvey and L. F. Mollenauer, "Harmonically mode-locked fiber ring last with an internal Fabry-Perot stabilizer for soliton transmission," *Opt. Lett.*, vol. 18, pp. 107–109, 1993.

[28] C. Doerr, H. A. Haus, E. P. Ippen, M. Shirasaki, and K. Tamura, "Additive-pulse limiting," *Opt. Lett.*, vol. 19, pp. 31–33, 1994.

[29] E. Yoshida, Y. Kimura, and M. Nakazawa, "Laser diode-pumped femtosecond erbium-doped fiber laser with a sub-ring cavity for repetition rate control," *Appl. Phys. Lett.*, vol. 60, pp. 932–934, 1992.

[30] M. L. Dennis and I. N. Duling III, "High repetition rate figure eight laser with extracavity feedback," *Electron. Lett.*, vol. 28, pp. 1894–1896, 1992.

[31] C. R. Doerr, H. A. Haus, and E. P. Ippen, "Asynchronous soliton mode locking," *Opt. Lett.*, vol. 19, pp. 1958–1960, 1994.

[32] L. F. Mollenauer, J. P. Gordon, and S. G. Evangelides, "The sliding-frequency guiding filter: An improved form of soliton jitter control," *Opt. Lett.*, vol. 17, pp. 1575–1577, 1992.

[33] F. Fontana, P. Franco, M. Midrio, M. Romagnoli, S. Wabnitz, and Y. Kodama, "Stable soliton generation in continuously frequency-shifted erbium fiber lasers," in *Tech. Dig., CLEO'94*. Optical Society of America, Washington, DC, 1994, p. 50.

[34] C. R. Doerr, W. S. Wong, H. A. Haus, and E. P. Ippen, "Additive-pulse mode-locking/limiting storage ring," *Opt. Lett.*, vol. 19, pp. 1747–1749, 1994.

[35] S. V. Chernikov, J. R. Taylor, and R. Kashyap, "1-THz optical pulse train generation at 1.5 μm," in *Proc. Conf. Lasers Electro-Optics*, Paper CTuH2, Anaheim, CA, 1994, pp. 73–74.

[36] S. V. Chernikov, E. M Dianov, D. J. Richardson, R. I. Laming, and D. N. Payne, "144 Gbit/s soliton train generation through raman self-scattering of a dual frequency beat signal in dispersion decreasing optical fiber," *Appl. Phys. Lett.*, vol. 63, pp. 293–295, 1993.

[37] E. A. Swanson and S. R. Chinn, "23 GHz and 123 GHz soliton pulse generation using two CW lasers and standard single-mode fiber," *IEEE Photon. Technol. Lett.*, vol. 6/7, pp. 796–798, 1994.

[38] S. B. Chernikov, J. R. Taylor, and R. Kashyap, "Integrated all optical fiber source of multigigahertz soliton pulse train," *Electron. Lett.*, vol. 29, pp. 1788–1789, 1993.

[39] E. A. Swanson, S. R. Chinn, K. Hall, K. A. Rauschenbach, R. S. Bondurant, and J. W. Miller, "100-GHz soliton pulse train generation using soliton compression of two phase side bands from a single DFB laser," *IEEE Photon. Technol. Lett.*, vol. 6, no. 10, pp. 1194–1196, 1994.

[40] E. A. Swanson and S. R. Chinn, "40-GHz pulse train generation using soliton compression of a Mach-Zehnder modulator output," *IEEE Photon. Technol. Lett.*, vol. 7, no. 1, pp. 114–116, 1995.

[41] O. Kamatani, S. Kawanishi, and M. Saruwatari, "Prescaled 6.3 GHz clock recovery from 50 Gb/s TDM optical signal with 50 GHz PLL using four-wave-mixing in a travelling wave laser diode amplifier," *Electron. Lett.*, vol. 30, pp. 807–809, 1994.

[42] S. Kawanishi and M. Saruwatari, "Ultra-high-speed PLL-type clock recovery circuit based on all-optical gain modulation in travelling-wave laser diode amplifier," *J. Lightwave Technol.*, vol. 11, pp. 2123–2129, 1993.

[43] S. P. Dijaili, J. S. Smith, and A. Dienes, "Timing synchronization of a passively mode-locked dye laser using a pulsed optical phase lock loop," *Appl. Phys. Lett.*, vol. 55, pp. 418–420, 1989.

[44] K. L. Hall, K. A. Rauschenbach, E. A. Swanson, S. R. Chinn, and G. Raybon, "Picosecond-accuracy all-optical bit phase sensing using a nonlinear optical loop mirror," to be published: *IEEE Photon. Technol. Lett.*, 1995.

[45] P. Nelson and N. J. Doran, "Optical sampling scope using nonlinear fiber loop mirror," *Electron. Lett.*, vol. 27, pp. 204–205, 1991.

[46] L. F. Mollenauer, P. V. Mamyshev, and M. J. Neubelt, "Measurement of timing jitter in filter-guided soliton transmission at 10 Gbps and achievement of 375 Gb/s-μm error free, at 12.5 and 15 Gb/s," *Opt. Lett.*, vol. 19, pp. 704–706, 1994.

[47] J. D. Moores, K. Bergman, H. A. Haus, and E. P. Ippen, "Demonstration of optical switching by means of solitary wave collisions in a fiber ring reflector," *Opt. Lett.*, vol. 16, pp. 138–140, 1991.

[48] N. A. Whitaker, Jr. *et al.*, "All-optical arbitrary demultiplexing at 2.5 Gb/s with tolerance to timing jitter," *Opt. Lett.*, vol. 16, pp. 1838–1840, 1991.

[49] M. Jinno and M. Abe, "All-optical regenerator based on the nonlinear fiber sagnac interferometer," *Electron. Lett.*, vol. 28, pp. 1350–1351, 1992.

[50] K. J. Blow, N. J. Doran, B. K. Nayar, and B. P. Nelson, "Two-wavelength operation of the nonlinear fiber loop mirror," *Opt. Lett.*, vol. 15, pp. 248–250, 1990.

[51] K. Uchiyama, H. Takara, T. Morioka, S. Kawanishi, and M. Saruwatari, "Effects of the control pulse walk-off on BER performance of non-linear optical loop mirror demultiplexer," *Electron. Lett.*, vol. 29, pp. 1313–1314, 1993.

[52] M. Jinno and T. Matsumoto, "Nonlinear Sagnac interferometer switch and its applications," *IEEE J. Quantum Electron.*, vol. 28, pp. 875–882, 1992.

[53] H. Bulow, Th. Pfeiffer, and G. Veith, "Polarization-insensitive all-optical demultiplexing over a wide wavelength range in a short-fiber nonlinear optical loop mirror," *Electron. Lett.*, vol. 30, pp. 68–69, 1994.

[54] K. Smith and J. Lucek, "All-optical clock recovery using a mode-locked laser," *Electron. Lett.*, vol. 28, pp. 1814–1816, 1992.

[55] D. Williams, M. Holmes, J. Devaney, R. Manning, J. Lucek, K. Smith, and B. Ainslie, "Optical mode locking of an erbium-doped fiber laser at 10 GHz by using 20 m of highly nonlinear silica-based fiber," in ,*Optical Fiber Communication*, 1995 Technical Digest Series. Washington, DC: Optical Society of America, 1995.

[56] D. Patrick and R. Manning, "20 Gbit/s all-optical clock recovery using semiconductor nonlinearity," *Electron. Lett.*, vol. 30, pp. 151–152, 1994.

[57] Sokolov *et al.*, "A terahertz optical asymmetric demultiplexer (TOAD)," in *OSA Conf. Photonic Switching*, Paper PD-4, Palm Springs, CA, 1993.

[58] L. E. Adams, E. S. Kintzer, and J. G. Fujimoto, "All-optical clock recovery using a mode-locked figure eight laser with a semiconductor nonlinearity," *Electron. Lett.*, vol. 30, pp. 1696–1697, 1994.

[59] R. Manning, D. Davies, D. Cotter, and J. Lucek, "Enhanced recovery rates in semiconductor laser amplifiers using optical pumping," *Electron. Lett.*, vol. 30, pp. 787–788, 1994.

[60] H. A. Haus and A. Mecozzi, "Long-term storage of a bit stream of solitons," *Opt. Lett.*, vol. 17, no. 21, pp. 1500–1502, 1992.

[61] V. I. Belotitskii, E. A. Kuzin, M. P. Petrov, and V. V. Spirin, "Demonstration of over 100 million round trips in recirculating fiber loop with all-optical regeneration," *Electron. Lett.*, vol. 29, no. 1, pp. 49–50, 1993.

[62] H. Avramopoulos and N. A. Whitaker, Jr., "Addressable fiber-loop memory," *Opt. Lett.*, vol. 18, no. 1, pp. 22–24, 1993.

[63] J. D. Moores, "All-optical soliton communications: Devices and limitations," Ph.D. dissertation, MIT, Cambridge, MA, 1994.

[64] C. R. Doerr, W. S. Wong, H. A. Haus, and E. P. Ippen, "Additive-pulse mode-locking/limiting storage ring," *Opt. Lett.*, vol. 19, no. 21, pp. 1747–1749, 1994.

[65] J. D. Moores, W. S. Wong, and H. A. Haus, "Stability and timing maintenance in soliton transmission and storage rings," *Opt. Commun.*, vol. 113, no. 1/2/3, pp. 153–175, 1994.

[66] M. Nakazawa, H. Kubota, E. Yamada, and K. Suzuki, "Infinite-distance soliton transmission with soliton controls in time and frequency domains," *Electron. Lett.*, vol. 28, no. 12, pp. 1099–1100, 1992.

[67] M. Eiselt, W. Pieper, G. Grosskopf, R. Ludwig, and H. G. Weber, "One million pulse circulations in a fiber ring using a SLALOM for pulse shaping and noise reduction," *IEEE Photon. Technol. Lett.*, vol. 4, p. 422, 1993.

[68] L. F. Mollenauer, E. Lichtman, G. T. Harvey, M. J. Neubelt, and B. M. Nyman, "Demonstration of error-free soliton transmission over more than 15000 km at 5 Gbit/s, single-channel, and over more than 11000 km at 10 Gbit/s in two-channel WDM," *Electron. Lett.*, vol. 28, no. 8, pp. 792–794, 1992.

[69] L. F. Mollenauer, E. Lichtman, G. T. Harvey, M. J. Neubelt, and B. M. Nyman, "Demonstration of error-free soliton transmission at 2.5 Gb/s over more than 14000 km," *Electron. Lett.*, vol. 27, p. 2055, 1991.
[70] K. J. Blow and N. J. Doran, "Nonlinear optical loop mirror," *Opt. Lett.*, vol. 13, p. 56, 1988.
[71] K. A. Rauschenbach, K. L. Hall, J. C. Livas, and G. Raybon, "All-optical pulse width and wavelength conversion at 10 Gb/s using a nonlinear optical loop mirror," *IEEE Photon. Technol. Lett.*, vol. 6, p. 1130, 1994.

Implementation of STARNET: A WDM Computer Communications Network

Ting-Kuang Chiang, Sanjay K. Agrawal, Derek T. Mayweather, Dan Sadot, Charles F. Barry, *Member, IEEE*, Michael Hickey, and Leonid G. Kazovsky, *Fellow, IEEE*

***Abstract*— STARNET is a broadband backbone optical wavelength-division multiplexing (WDM) local area network (LAN). Based on a physical passive star topology, STARNET offers all users two logical subnetworks: a high-speed reconfigurable packet-switched data subnetwork and a moderate-speed fix-tuned packet-switched control subnetwork. Thus, STARNET supports traffic with a wide range of speed and continuity characteristics. We report the analysis and implementation of an entire STARNET two-node network, from the optical to the computer layer, at the Optical Communications Research Laboratory (OCRL) of Stanford University. To implement the two logical subnetworks, we designed and implemented two different techniques: combined modulation and multichannel subcarrier multiplexing (MSCM). OCRL has already demonstrated several combined modulation techniques such as phase shift-keyed and amplitude shift-keyed (PSK/ASK), and differential phase shift-keyed and amplitude shift-keyed (DPSK/ASK), yielding combined ASK/DPSK modulation receiver sensitivities better than -32 dBm.**

OCRL has designed and implemented a high-speed high-performance packet-switched STARNET computer interface which enables high-throughput transfer to/from host computer, low latency switching, traffic prioritization, and capability of multicasting and broadcasting. With this interface board, OCRL has achieved average transmit and receive throughputs of 685 Mb/s and 571 Mb/s, respectively, out of the 800 Mb/s theoretical maximum of the host computer bus. The incurred packet latency due to the interface for a specified multihop network configuration has been simulated to be 24 μs. Using simulation and experimental results, it is shown that STARNET is highly suitable for high-speed multimedia network applications.

I. Introduction

NEXT generation communication networks will be required to handle exciting new applications such as desktop video-conferencing, interactive TV, supercomputer interconnection, and tele-medicine applications. The increasing number of users who will use these applications will require ultra-high total network throughputs ranging from several hundreds of Gb/s to perhaps even several Tb/s. To meet these requirements, advanced optical communications networks, capable of transmitting, receiving, switching, multiplexing, and demultiplexing this tremendous amount of traffic must be developed. Although optical fiber is well suited for high-speed traffic transport, the bursty nature of computer traffic, restrictive latency requirements (particularly for video-conferencing applications), and the large number of users makes it difficult to fully utilize the fiber's capacity in optical networks. In addition to providing all these services, future networks should also be economical, scalable, modular, reliable, fault-tolerant, and interoperable (i.e., support heterogeneous architectures, protocols, and data formats).

Data communications networks capable of supporting Tb/s applications over both the local area and wide area must overcome key bottlenecks imposed by technological limitations. Electronic bottlenecks are imposed by the components needed for serialization, deserialization, and clock recovery. Photonics bottleneck arises from the photodetector bandwidth, limited wavelength tuning range, relatively slow tuning times, insufficient wavelength stability, and lack of optical buffering. The protocol bottleneck is the perceived need to make every system interoperable with, and backward compatible to pre-existing protocols. To solve the first two (electronic and photonic) technological bottlenecks, and to advance the communication link toward exploiting the huge optical fiber capacity, a multichannel transmission scheme is needed, with each channel running at about the capacity of electronics, i.e., several to few tens of Gb/s.

Multichannel networks [1]–[2] can be implemented using optical fiber through wavelength-division multiplexing (WDM), subcarrier multiplexing (SCM), time-division multiplexing (TDM), code-division multiple access (CDMA), or a combination of these techniques. With WDM, each channel is transmitted on a unique optical carrier wavelength. These carriers can then be combined with $N \times N$ couplers for passive star topologies, with couplers and taps for bus and ring topologies, or with more sophisticated components for mesh configurations.

Transmission of both payload and network control data is essential in multichannel networks. Different network control techniques and methods to transmit both payload and control data are briefly described below (Section II). Two promising methods to transmit two data streams on the same optical carrier are combined modulation [3] and multichannel subcarrier multiplexing (MSCM) [4]. Combined modulation involves the transmission of independent data streams using a different modulation format for each stream (e.g., amplitude shift keying, ASK, together with phase shift keying, PSK), thus using one laser for two independent data streams. MSCM utilizes transmission of the payload data at baseband and

Manuscript received April 6, 1995; revised August 10, 1995. This work was supported in part by the BMDO under Grant DASG60-93-C-0054, the National Science Foundation under Grant ECS-9111766, and Digital Equipment Corporation.

The authors are with the Department of Electrical Engineering, Stanford University, Stanford, CA 94305-4055 USA.

Publisher Item Identifier S 0733-8716(96)03683-9.

Reprinted with permission from *IEEE Journal on Selected Areas in Communications,* Vol. 14, No. 5, pp. 824-839, June 1996.

control data on a dedicated subcarrier frequency over the same optical carrier, again requiring only one laser for two data streams. STARNET, a WDM LAN computer communications network breadboard developed at the Optical Communications Research Laboratory (OCRL) of Stanford University, is the first optical network utilizing combined modulation; the second version of STARNET is designed to utilize MSCM. In this paper we will present the design, analysis, and progress of STARNET. The paper is organized as follows. In Section II, we discuss the two key issues in high-speed optical networks that STARNET is aimed to solve: network control and multimedia traffic requirements. In Section III, the STARNET network architecture is described and the basic node structure is introduced. Sections IV and V describe two different STARNET optical layer implementations utilizing combined modulation and MSCM, respectively. Section VI describes the architecture, arbitration algorithms, and performance of the STARNET high-speed computer interface. Conclusions are presented in Section VII.

II. Issues in High-Speed Optical Networks

A. Network Control

One of the most challenging problems in high-speed optical networking is the signaling and access control. Techniques of handling the network control issue can be divided into the following categories: a) pre-assigning access at network configuration time; b) dynamically coordinating access on a demand basis; c) random access; and d) a combination of those. One of the first WDM network demonstrations, which utilized pre-assigned access was the Bellcore Lambdanet [5]. The most complete in-band WDM network demonstration to date, utilizing in-band signaling with tunable receivers, is the IBM Rainbow network [6]–[7]. The signaling protocol for coordinating the retuning of the optical tunable receiver is an in-band polling procedure. To establish high-speed connections between nodes, for bursty traffic and packet switching applications, an out-of-band control channel is more suited. The control channel can be transmitted on a unique dedicated wavelength [5], or multiplexed on the same wavelength as the payload channel (i.e., same optical carrier) with subcarrier modulation or a combined modulation technique. A hybrid multiplexing scheme that combines WDM with SCM, thus dividing the available optical bandwidth, is used in the Columbia University network Teranet [8], [9]. An example of a wide area network (WAN) with dedicated transceivers is the test bed of the wide-band-all-optical network consortium [10]. Combined modulation offers a good solution for transmitting out-of-band control signaling along with the payload data on the same optical carrier. This method requires only one transmitter per node. Therefore, combined modulation provides the required functionality for networks with dedicated control transceivers, but is simpler and potentially less expensive. An additional method for transmitting an independent control channel is the use of MSCM [11]–[13]. With this technique, nodes transmit a payload channel at baseband and a control channel at a unique microwave frequency on a single optical carrier. The microwave frequencies are selected so that a single photodetector can be used to receive all control channels simultaneously. This technique, like combined modulation, requires a single transmitting laser.

B. Multimedia Traffic Characteristics and Requirements

Furthermore, high-speed networks must support a variety of traffic with highly diverse characteristics. The growth of multimedia applications and the increase in the number of users demanding those services continue to push bandwidth requirements higher. Further, the real-time, interactive nature of these new services require low latency transport.

Data transmission types can be divided into four categories: 1) bursty, low-speed—examples include remote logins, emails, and file transfer; 2) continuous, low-speed—examples include real-time voice transmission; 3) bursty, high-speed—examples include compressed video, images, and parallel computer interconnection; and 4) continuous high-speed—examples include real-time uncompressed video. Ideally, networks must simultaneously support all four categories of data with acceptable efficiency.

III. STARNET Overview

A. STARNET Architecture

To meet the diverse traffic and control demands discussed above, STARNET employs two logical subnetworks: a reconfigurable high-speed subnetwork and a fixed-ring moderate-speed subnetwork. Packet-switching is used in both subnetworks for more efficient use of channel bandwidth. The high-speed logical subnetwork operates at up to 2.5 Gb/s/node. The moderate-speed subnetwork is FDDI compatible and operates at the link rate of 125 Mb/s.

The FDDI-compatible moderate-speed logical subnetwork provides full connectivity between all nodes. This subnetwork supports the transmission of the network control information which manages the reconfiguration of the high-speed subnetwork.

The reconfigurable high-speed logical subnetwork is suitable for transmitting both continuous and bursty data. For high bandwidth, connection-oriented services the high-speed subnetwork is dynamically reconfigured to optimize the bandwidth utilization. For high-bandwidth communication among a cluster of nodes, multinode communication is achieved through electronic multihop.

B. STARNET Physical Layer

The STARNET physical architecture utilizes a passive star, as shown in Fig. 1. Optical power from each node is distributed equally through a passive star coupler to all nodes in the network. Logically, STARNET consists of two separate subnetworks which are realized over the same physical star topology using combined modulation or MSCM on the same lightwave.

Each node contains a single transmitter and two receivers: the main, or payload, receiver and the auxiliary receiver. Each node's transmitter has a unique and fixed wavelength. The

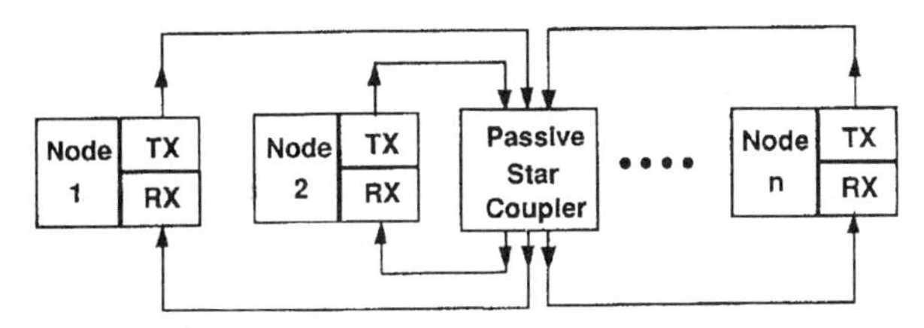

Fig. 1. Physical topology of STARNET.

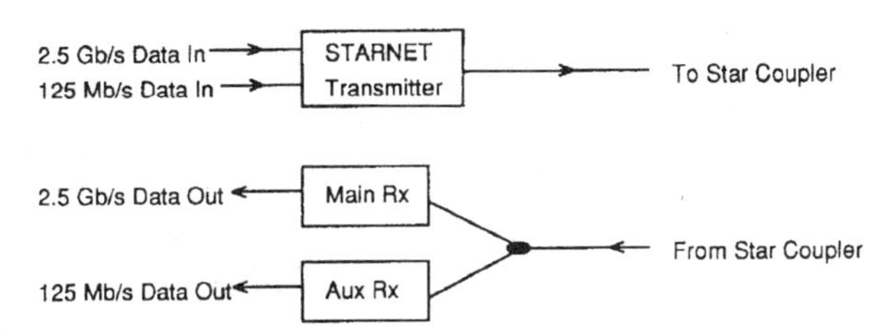

Fig. 2. STARNET transmitter and receivers.

main and auxiliary receivers are separately tunable. The two logical subnetworks are realized by simultaneously transmitting high-speed (2.5 Gb/s) payload data and moderate-speed (125 Mb/s) data to the passive star on a single lightwave using combined modulation (STARNET-I) or MSCM (STARNET-II), as discussed later. The main receiver demodulates the high-speed traffic and the auxiliary receiver detects the moderate-speed traffic and control information (see Fig. 2).

Each node's auxiliary receiver is tuned to the previous node transmitter. The first node in the resulting frequency comb (Fig. 3) is tuned to the last node. Logically, this tuning configuration forms a ring for the moderate-speed subnetwork as shown in Fig. 4. The FDDI ring remains in a fixed configuration throughout the network operation unless there is a node failure.

The main (2.5 Gb/s) receiver can be tuned to any specified node as governed by the network traffic characteristics. Several independently operating high-speed subnetworks can be formed. Electronic multihop is utilized to relax hardware tuning speed requirements, while achieving packet switching in an intrinsically circuit-switched environment.

IV. STARNET-I Optical Layer

The STARNET-I optical transceivers utilize coherent detection technique. The data stream multiplexing is achieved by using combined modulation where the amplitude and phase of the optical carrier represent separate data streams. We designed and experimented with both an external modulation version and direct modulation version of STARNET-I.

A. Physical Implementation

1) Externally Modulated Transceiver: The STARNET-I demonstration network at Stanford University is built using miniature diode-pumped Nd:YAG lasers. The Nd:YAG lasers have low linewidth (less than 100 Hz) and are thus suitable for phase modulation. We have demonstrated both PSK/ASK and DPSK/ASK combined modulation schemes [14]–[16]. These schemes are applicable to low-linewidth semiconductor lasers as well.

Figure 5 shows the block diagram of a two-node STARNET experiment. It includes two DPSK/ASK transceivers. A fraction of the light of the transmitter laser is tapped off to be used as the local oscillator (LO) of the auxiliary ASK receiver. The remaining power is coupled into a $LiNbO_3$ combined phase/amplitude modulator, and the output lightwave is sent to a passive star coupler. Fig. 6 shows the block diagram of the STARNET node transmitter.

The light from the passive star contains signals from all the channels; it is sent to a coupler splitting the power between the ASK and DPSK receivers. For the DPSK receiver, an additional Nd : YAG laser is used as the LO and is combined with the received optical signal. Figs. 7 and 8 show the block diagrams of the STARNET DPSK and ASK node receivers, respectively.

2) Directly Modulated Transceiver: We have demonstrated a direct modulation version of STARNET-I [17], [18] which utilizes semiconductor lasers to enable higher level integration of transmitters and receivers. In this experiment, the transmitter contained a DFB laser integrated with an electro-absorption modulator (EAM) section, as shown in Fig. 9. The bandwidth of the electro-absorption section is a few GHz; this section is directly modulated, yielding 2.5 Gb/s ASK modulated data. The grating section is modulated by slightly varying the bias current generating the 125 Mb/s FSK data. Since the transmitter laser is directly modulated, the two receivers need two separate tunable LO lasers. Fig. 10 shows the block diagram of the ASK and SF-FSK (single-frequency FSK) receivers.

B. Performance Analysis

1) Receiver Sensitivity Penalty: The receiver sensitivity penalty imposed by the multiplexing of the control and data channels using combined modulation is analyzed in [3] and [15]. The modulation indices of the two streams can be optimized according to their bit-rate ratio so that both channels have the same sensitivity. It has been shown [3] that when the bit rate of (D)PSK is 2.5 Gb/s and that of ASK is 125 Mb/s, the penalty is less than 3 dB if the transmitters and receivers are optimized.

2) Power Budget: The maximum number of nodes, $N_{\max}$, that is limited by the power budget can be found by solving the following equation:

$$P_{\min} = \frac{P_l}{B_R} \frac{m}{2N_{\max}} 10^{-(s/10) \log_2 (N_{\max})} \tag{1}$$

where $P_{\min}$ is the receiver sensitivity (photons/bit), P_l is the launched number of photons per second, B_R is the bit rate, m is the system safety margin, and the factor $1/2$ accounts for having two receivers at each node. We assume that the $N \times N$ star coupler consists of $\log_2 N$ stages of 2×2 couplers, and each 2×2 coupler has an excess loss of s dB. The factor $10^{-(s/10) \log_2 N}$ represents the excess loss of the whole star coupler. If we assume the transmitter power is 0 dBm ($\lambda =$ 1550 nm), $B_R = 2.5$ Gb/s, $m = 0.05$ (− 13 dB), $s = 0.5$ dB,

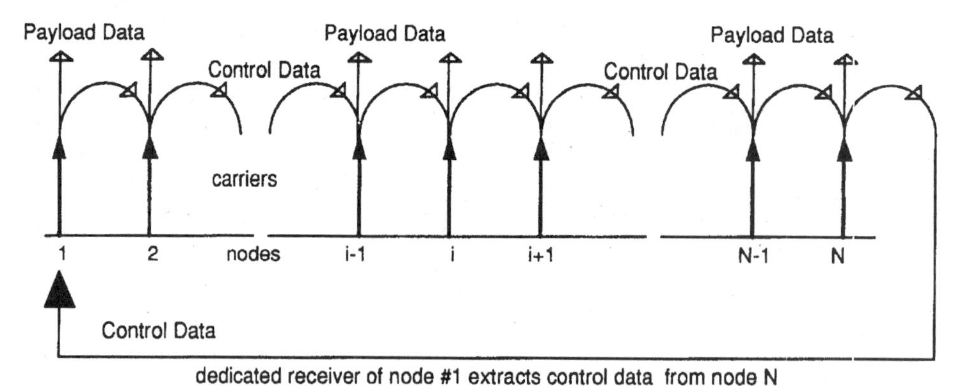

Fig. 3. STARNET frequency comb.

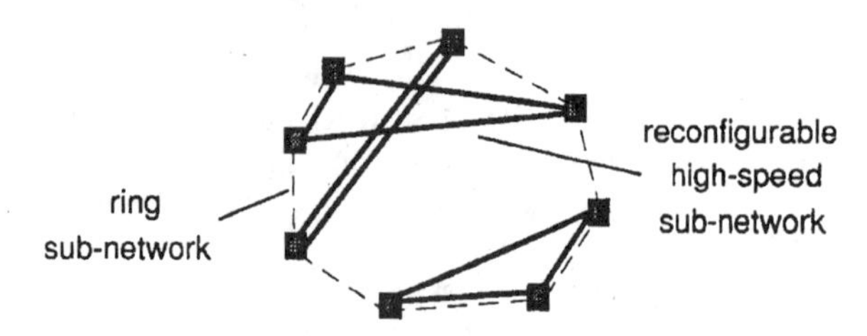

Fig. 4. STARNET logical topology.

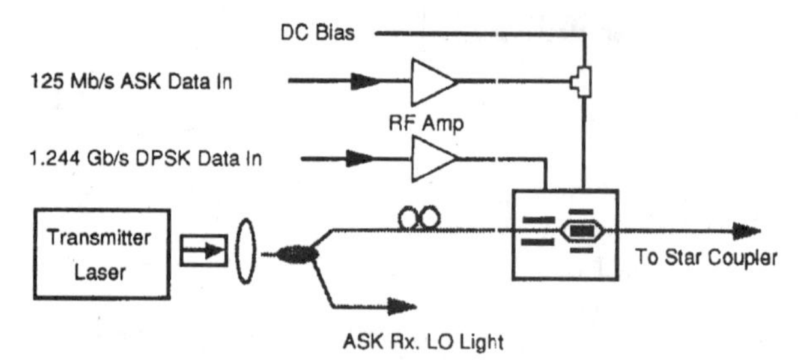

Fig. 6. Block diagram of the STARNET-I transmitter.

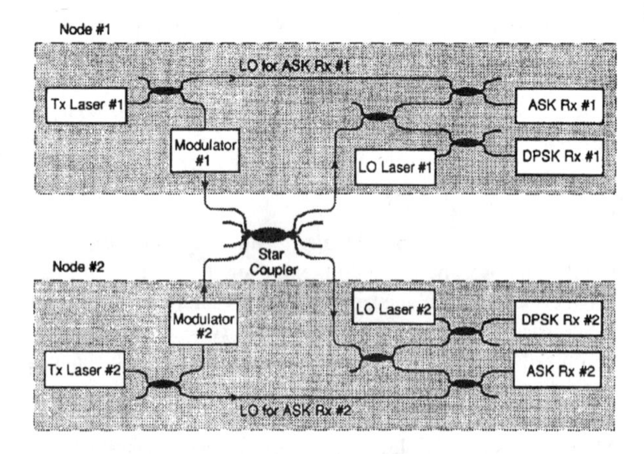

Fig. 5. Block diagram of the two-node STARNET-I experimental setup.

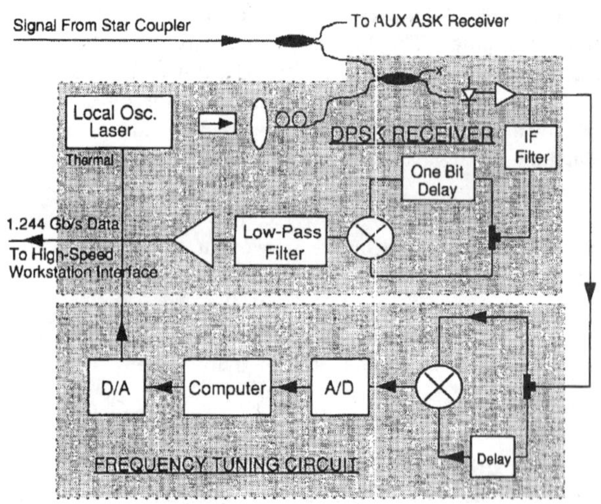

Fig. 7. Block diagram of the STARNET-I DPSK node receiver.

$P_{\min} = 100$ (for a DPSK receiver in DPSK/ASK systems), the power budget limits the maximum number of nodes to 300.

3) Optical Bandwidth Requirement: Since the control and data channels are separately modulated on the amplitude and phase of the optical carrier, the resulting optical spectrum is the convolution of the amplitude-modulated spectrum and the phase-modulated spectrum. The full width at half maximum (FWHM) bandwidth of the optical spectrum, B_o, is approximately equal to twice the sum of the two bit rates

$$B_o \approx 2(B_{RC} + B_{RD}) \tag{2}$$

where B_{RC} and B_{RD} are the bit rates of the control and data channels, respectively.

4) Channel Spacing: To avoid spectrum overlap in the main and auxiliary receivers, unequally spaced optical comb is used in the channel allocation of STARNET-I [3]. This scheme minimizes the optical bandwidth for heterodyne detection [19]. Fig. 11 shows the optical frequency allocation using the minimum bandwidth optical comb. If the minimum channel spacing $\delta/2$ is equal to $2B_o$, then the average channel spacing, f_{SP}, can be expressed by

$$\begin{aligned} f_{SP} &\approx 3B_o \\ &\approx 6(B_{RC} + B_{RD}). \end{aligned} \tag{3}$$

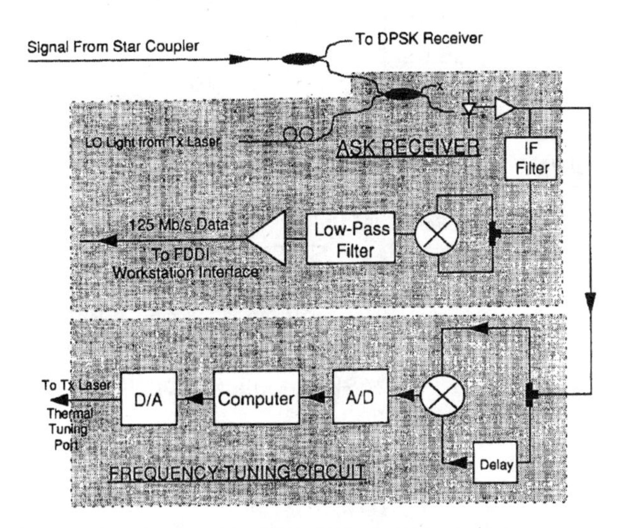

Fig. 8. Block diagram of the STARNET-I ASK node receiver.

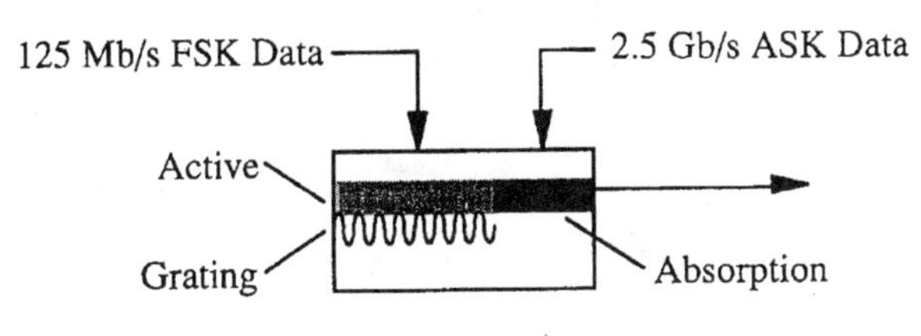

Fig. 9. Diagram of a DFB laser with integrated EAM.

5) Photodetector Bandwidth Requirement: Since the transmitter laser is used to detect the previous node in the optical comb, the IF center frequency of the auxiliary receiver is fixed at the minimum channel spacing $\delta/2$. The required photodetector bandwidth, B_{PD}, is approximately equal to two and a half times the width of the optical spectrum

$$\begin{aligned} B_{PD} &= \tfrac{5}{2} B_o \\ &\approx 5(B_{RC} + B_{RD}). \end{aligned} \tag{4}$$

6) Estimated Number of Channels: The optical heterodyne detection provides excellent channel isolation due to high suppression in electrical filtering; therefore, the receiver sensitivity is independent of the total number of channels. The total number of channels in STARNET-I is limited by: i) power budget and ii) available optical bandwidth. The constraint of power budget is discussed above. The maximum number of channels stemming from bandwidth requirements is bounded by the ratio of the available optical bandwidth to the average channel spacing.

We make the following assumptions: a) the bandwidth of the fiber is 20 THz in the 1550 nm window; b) the bandwidth of the EDFA is 35 nm; c) the laser tuning range is 10 nm; and d) the bit-rate ratio of data and control channels, B_{RD}/B_{RC}, is 20.

The maximum number of nodes that the network can support is plotted against the data channel bit rate in Fig. 12.

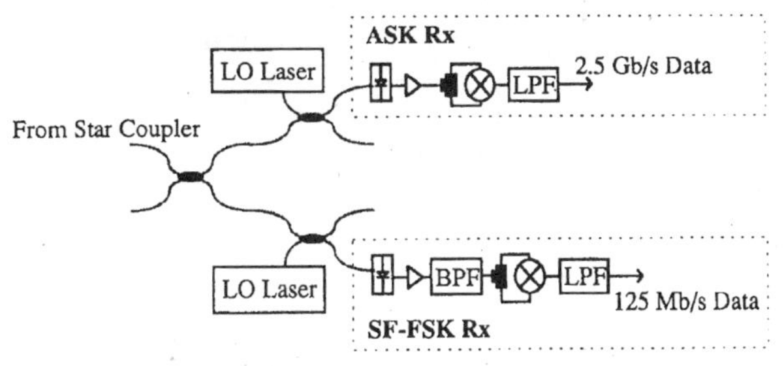

Fig. 10. Block diagram of the ASK and SF-FSK receivers for directly modulated STARNET-I nodes.

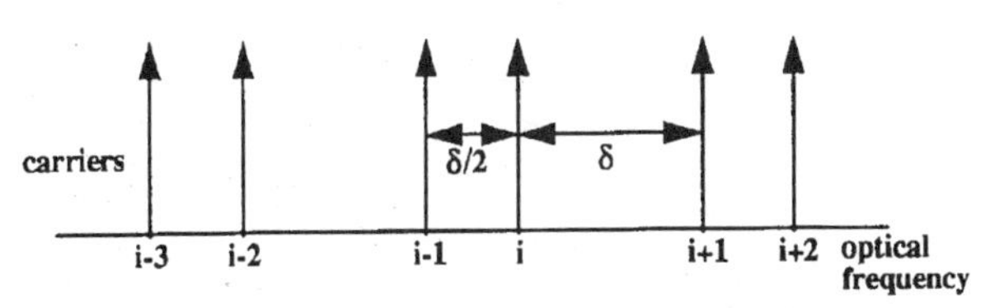

Fig. 11. Minimum bandwidth optical comb used in STARNET-I.

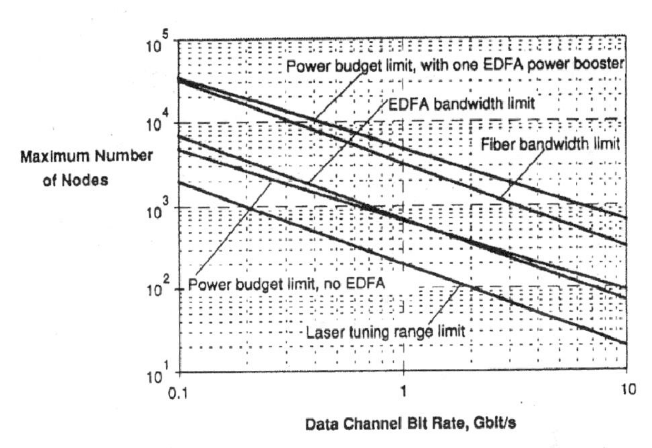

Fig. 12. Maximum number of nodes as a function of data channel bit rate for STARNET-I.

Fig. 12 shows that the fiber bandwidth and the EDFA bandwidth limits are well above the laser tuning range limit. For our experimental STARNET with the bit rate of 2.5 Gb/s, the maximum number of nodes is 80, and is limited by the laser tuning range.

7) Wavelength Stabilization: The wavelength separation among transmitter lasers is automatically maintained by the fixed-tuned auxiliary receivers. If there is a frequency drift in the first node of the optical comb, the rest of the nodes will adjust accordingly through the AFC circuits of the auxiliary receivers.

8) Hardware Complexity: The coherent receivers shown in Figs. 7 and 8 consist of many components and are not polarization independent which is a major drawback of coherent systems. However, it is possible to integrate all the optical components in one chip and achieve polarization diversity at the same time [20]. With the advance in device technology, the coherent transceivers have the potential to be integrated in one chip and thus be compact and robust.

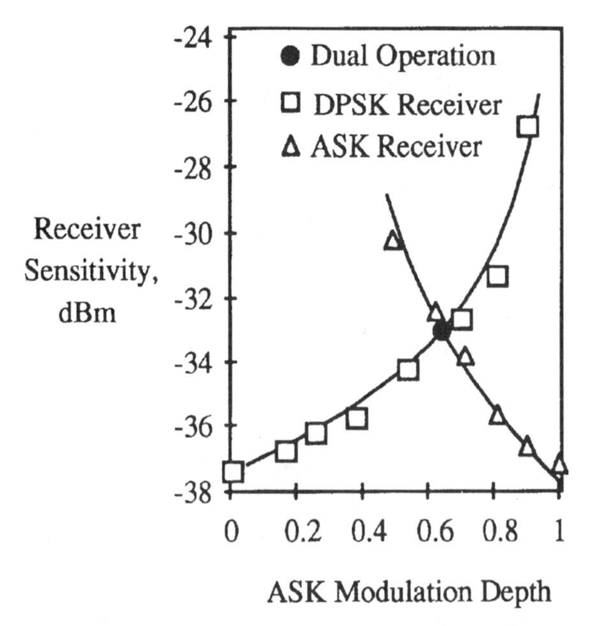

Fig. 13. Experimental optimum ASK modulation depth.

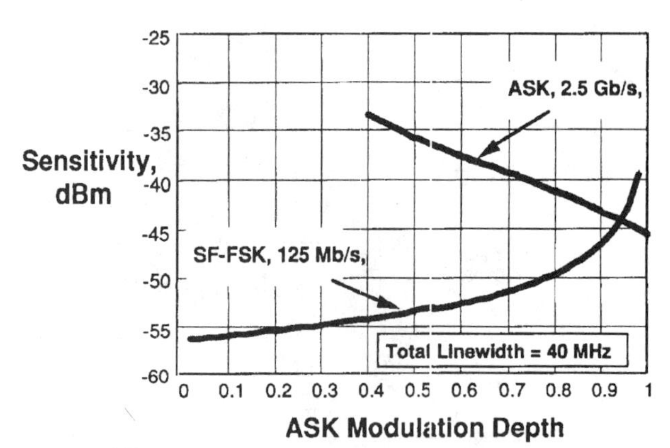

Fig. 14. Theoretical coherent ASK/FSK receiver sensitivity.

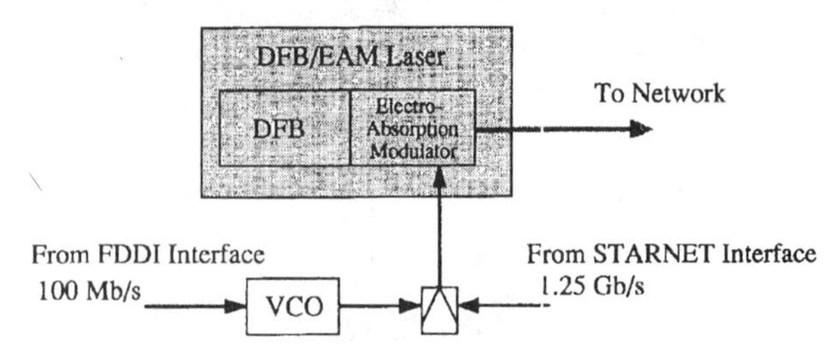

Fig. 15. Block diagram of the STARNET-II transmitter.

C. Experimental Results

1) Externally Modulated Transceiver: To investigate the properties and demonstrate the operation of the combined modulation format, we constructed the DPSK/ASK transceiver shown in Figs. 6–8. Fig. 13 shows the measured experimental ASK and PSK receiver sensitivities versus the ASK modulation depth m. For optimum operation, we adjusted m to a level where the ASK-FDDI receiver *and* the DPSK receiver operate with bit-error rates (BER's) of 10^{-9} simultaneously. This occurs for $m = 0.65$ as shown by the solid black circle in Fig. 13. The resulting optimum receiver sensitivities are -33 dBm, measured at the input to the 3 dB coupler in each of the receivers.

2) Directly Modulated Transceiver: To characterize the performance of a STARNET node using a DFB/EAM, we constructed both square-law ASK and SF–FSK receivers shown in Fig. 10. Fig. 14 shows the theoretical shot-noise-limited receiver sensitivity for simultaneous transmission of coherent ASK and FSK data using the DFB/EAM and coherent receivers [21]. It is seen that the theoretical optimum ASK modulation depth is 0.95 corresponding to a -44 dBm receiver sensitivity for both SF–FSK and ASK receivers. However, in the experimental setup, the received FSK signal is not clean enough to drive the lower-speed FDDI link when the ASK modulation is applied. The main reason is that there is still some chirp induced from the EAM section of the device which interferes with the FSK signal. We expect further developments in DFB/EAM technology will solve this problem.

3) Channel Spacing: To assess STARNET-I channel spacing, a two node experiment was conducted, as described in Fig. 5. We investigated experimentally the impact of the crosstalk due to the second transmitter. Results show that with a transmitted bit rate of 1.25 Gb/s, an optical channel spacing of 10 GHz appears to be satisfactory: the resulting power penalty is less than 1 dB.

V. STARNET-II Optical Layer

The STARNET-II optical transceivers utilize direct detection rather than coherent detection to simplify the nodes' structure and reduce long-term cost. Direct detection does not require an additional local oscillator laser, is polarization independent, and less sensitive to temperature fluctuations. Integrated coherent receivers are not available currently and direct detection receivers are ready to deploy.

In STARNET-II, the payload data channel is intensity-modulated on the optical carrier. Wavelength selection is achieved using passive optical filters. The control channel of STARNET-II is implemented using the MSCM signaling scheme [11]–[13].

A. Physical Implementation

STARNET-II node design is shown in Figs. 15 and 16. The transmitter module (Fig. 15) includes a 3-dBm DFB/EAM laser used to transmit both control and payload information. The control data from FDDI interface is FSK-modulated on a dedicated subcarrier frequency and combined with the payload data from the STARNET computer interface. The resulting signal amplitude modulates the optical carrier; each node transmits on a dedicated wavelength. The node receiver module consists of payload and control receivers (Fig. 16). The integrity of the control data is maintained via MSCM

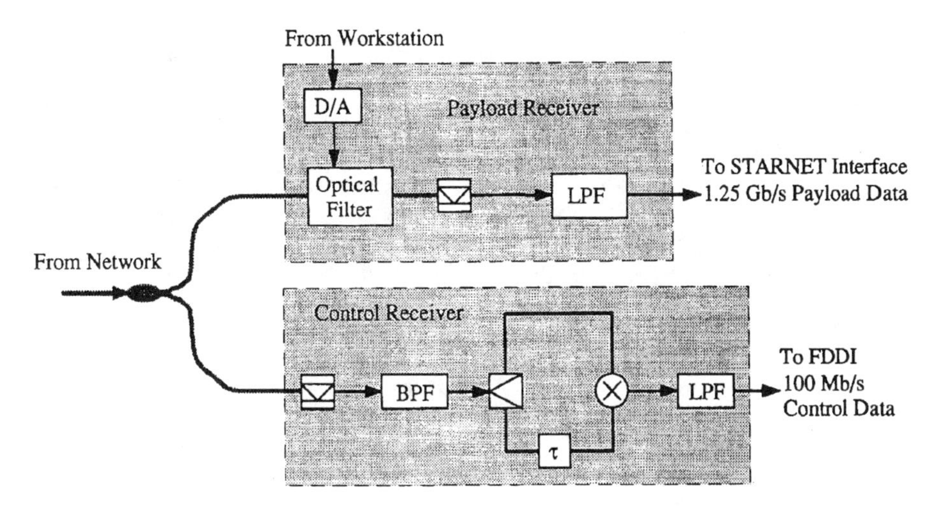

Fig. 16. Block diagram of the STARNET-II receiver.

while the payload data is optically separated with a tunable optical filter.

B. Performance Analysis

1) Receiver Sensitivity Penalty: The multiplexing of the control and data channels using SCM introduces additional receiver sensitivity penalty. The subcarrier control channel consumes a portion of the signal power and does not contribute to the payload reception. Furthermore, the subcarrier signal also introduces an extra crosstalk which degrades the payload reception. The impact of payload data on the subcarrier control data reception is similar.

The modulation depths of the subcarrier and payload signals can be optimized according to their bit-rate ratio so that both channels have the same sensitivity. The sensitivity penalty is minimized in this case. It has been shown [13] that when the bit-rate ratio of the data and control channels is greater than 20, the penalty is less than 3 dB.

2) Power Budget: The maximum number of nodes $N_{\max}$ limited by the power budget can be found by solving (2) similar to the analysis in Section IV-B. If we assume the same system parameters as in Section IV-B, except that $P_{\min} = 2000$ (for a direct detection APD receiver in MSCM systems), and no optical amplifier is used, the power budget limits the maximum number of nodes to 23.

Since the direct detection receiver is thermal-noise limited, the sensitivity can be greatly improved by using an EDFA preamplifier. Use of the EDFA for preamplification is the optimal configuration in direct detection systems to overcome thermal noise. This is in contrast to the coherent case where the received signal is strong, thus the EDFA is used to boost the transmitter power after external modulation [22], [23]. For the direct detection system, if an EDFA preamplifier is used, $P_{\min} = 200$, and the maximum number of nodes can be extended to 165.

3) Optical Bandwidth Requirement: To avoid crosstalk in subcarrier channels, the spacing of subcarrier frequencies is approximately equal to $4B_{RC}$. The data channel is located in the baseband. The total signal bandwidth is approximately $B_{RD} + 4NB_{RC}$. The minimum width of the optical spectrum B_o is approximately equal to twice the total signal bandwidth (the bandwidth can be significantly higher if laser chirp is large)

$$B_o \approx 2(B_{RD} + 4NB_{RC}). \tag{5}$$

4) Channel Spacing: Since tunable optical filter and direct detection are used in STARNET-II, the main concern of channel spacing is the crosstalk through the stopband of the optical filter. Here we roughly assume that the crosstalk through optical filter is negligible outside twice the optical bandwidth; then, the minimum channel spacing, f_{SP}, can be expressed as

$$\begin{aligned} f_{SP} &\approx 2B_o \\ &\approx 4(B_{RD} + 4NB_{RC}). \end{aligned} \tag{6}$$

5) Photodetector Bandwidth Requirement: The required photodetector bandwidth B_{PD} for the direct detection receiver is approximately equal to the total signal bandwidth

$$B_{PD} \approx B_{RD} + 4NB_{RC}. \tag{7}$$

6) Estimated Number of Channels: The MSCM scheme requires wider channel bandwidth as the number of nodes increases. The total number of channels in STARNET-II is limited by: i) power budget; ii) available optical bandwidth; and iii) photodetector bandwidth. The constraint of power budget is discussed above. The number of channels is bounded by the ratio of the available optical bandwidth and the average channel spacing.

We make the following assumptions: a) the bandwidth of the fiber is 20 THz in the 1550 nm window; b) the bandwidth of the EDFA is 35 nm; c) the filter tuning range is 60 nm; and d) the bandwidth of photodetector is 18 GHz.

Figure 17(a) shows the maximum number of nodes that the network can support versus the data channel bit rate, B_{RD}, when $B_{RD}/B_{RC} = 20$. Fig. 17(a) shows that in multi-Gb/s region, the power budget restricts the maximum number of

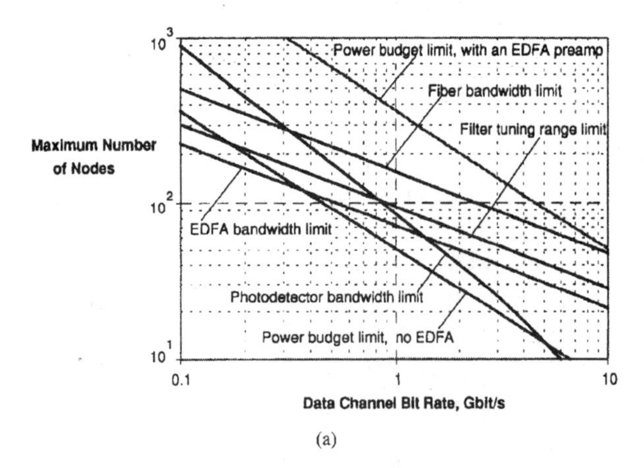

(a)

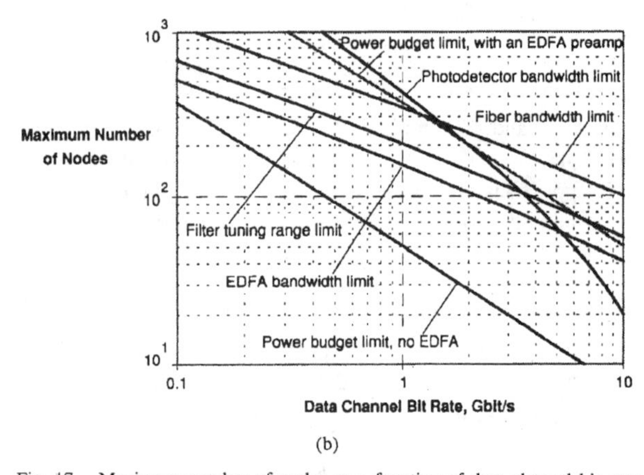

(b)

Fig. 17. Maximum number of nodes as a function of data channel bit rate for STARNET-II. (a) $B_{RD}/B_{RC} = 20$ and (b) $B_{RD}/B_{RC} = 200$.

nodes be 23, when $B_{RD} = 2.5$ Gb/s, if no EDFA is used. If an EDFA preamplifier is used in each node, then the photodetector bandwidth limits the maximum number of nodes to 30, when $B_{RD} = 2.5$ Gb/s. Even if higher-bandwidth photodetector is used, the EDFA bandwidth still limits the network size to 42 nodes.

To further increase the network size, the bit rate of control channel needs to be reduced. Fig. 17(b) shows the results when $B_{RD}/B_{RC} = 200$. In this case, if an EDFA preamplifier is used in each node, the EDFA bandwidth is the limiting factor; it limits the network size to 90 nodes when $B_{RD} = 2.5$ Gb/s.

7) Wavelength Stabilization: While in the coherent detection case the need for frequency stabilization is absolute (so the heterodyned signal will fit into the microwave components' bandwidth), in the incoherent/direct detection case the optical frequency accuracy depends on the *optical* (rather than electronic) filtering bandwidth requirements, which are usually much less restrictive. When dense WDM is used as in STARNET, calibrated frequency sources or Fabry–Perot etalons may be used.

8) Hardware Complexity: The direct detection receivers are simpler and less expensive than coherent receivers. Direct detection does not require an additional local oscillator laser, is polarization independent, and less sensitive to temperature fluctuations, but has inferior sensitivity. EDFA preamplifiers are needed to improve sensitivity. However, EDFA's are hard to integrate with other components.

C. Comparison of STARNET-I and STARNET-II

- STARNET-I utilizes coherent detection and thus provides better sensitivity than direct detection STARNET-II.
- Combined modulation utilizes optical spectrum more efficiently than the MSCM scheme. Thus STARNET-I can incorporate more nodes than STARNET-II.
- The optical bandwidth requirement of STARNET-I is proportional to N, the number of nodes; the optical bandwidth requirement of STARNET-II is proportional to $N + \alpha N^2$ where α depends on the bit-rate ratio of the control and data channels. If the control channel bit rate is much lower than the data channel bit rate, then the optical bandwidth requirement of STARNET-II is proportional to N. Otherwise, the requirement is proportional to N^2.
- The network size of STARNET-I is limited by the tuning range of lasers; the limit is 80 nodes for the 10-nm tuning range. If the tuning range of the laser is extended beyond 35 nm, the size of STARNET-I is limited to 300 by the power budget. In this case, even an EDFA power booster is used to extend the limit of power budget, the network size is still limited to 300 by the bandwidth of EDFA.
- The network size of STARNET-II is limited to 23 by the power budget if no optical amplifiers are used. If an EDFA preamplifier is used in each node, the size of STARNET-II is limited to 30 by the photodetector bandwidth. If the control channel bit rate is reduced to 12.5 Mb/s, the size of STARNET-II can be extended to 90; the EDFA bandwidth is the dominant limiting factor in the latter case.
- Relative wavelength stabilization in STARNET-I is built-in while STARNET-II requires separate stabilization circuits.
- STARNET-I relies on the development of integrated polarization diversity receivers to achieve compactness and robustness. STARNET-II requires EDFA preamplifiers to improve sensitivity; EDFA's are hard to integrate with the rest of the receiver components.

VI. STARNET: WDM Network Interface

A. Goal of the Network Interface

STARNET is intended for a variety of applications from electronic mail (low bandwidth), medical image transfer (high bandwidth, low loss), video retrieval (high bandwidth and bounded latency), and live video conferencing (high bandwidth and low and bounded latency). Requirements for some of these applications are quantified in Fig. 18. STARNET provides two logical subnetworks to serve these applications in the campus area environment: a moderate speed subnetwork and a high-speed subnetwork. The moderate-speed (125 Mb/s) subnetwork is intended for applications with bursty bandwidth requirements and relatively short durations such as file transfer and e-mail. In addition, the moderate-speed subnetworks provides a signaling and control channel for the high-speed

Traffic	Maximum Delay	Average Bandwidth
Voice	250 ms	0.064 Mbps
TV Video	250 ms	100 Mbps
Compressed Video	250 ms	2-10 Mbps
Video Conference	250 ms	0.256-2 Mbps
Image Transfer	1-5 s	2-10 Mbps
Data Transfer	N/A	2-100 Mbps

Fig. 18. Real-time requirements for some broadcast and interactive services.

subnetworks. We choose the moderate-speed subnetwork to be FDDI compatible and use a standard transport and network protocols such as TCP/IP and UDP/IP for communication over the moderate-speed subnetwork.

The high-speed subnetwork is intended for high-bandwidth, real-time, and relatively long duration applications such as multimedia applications, medical image transfer, video retrieval, and video-conferencing. As we can see in Fig. 18, interactive applications such as video-conferencing, together with the large number of expected users, impose challenging bandwidth and latency requirements. Therefore, our design goal is to support these requirements for multiple live video-conferencing sessions in the campus area network. In order to achieve this goal, our interface must provide a large throughput to the network while off-loading from the host processor as much as possible the transport and lower layer protocols. The design goals are listed below:

- high-throughput to the network and to the host per node;
- low latency switching;
- support of simultaneous packet streams(applications) of varying bandwidth;
- support of prioritized traffic;
- fair bandwidth allocation among competing streams;
- minimization of packet loss;
- minimization of host loading;
- elimination of packet reassembling and reordering;
- support of multicast and broadcast connections;
- modularity/upgradeability; and
- economy.

Some of these goals are conflicting; for example, minimizing dropped packets requires larger buffer sizes which can increase latency; large packet size improves the throughput to/from the network at the cost of low-latency switching. With these assumptions, goals and tradeoffs in mind, we now examine the high-speed interface architecture in detail.

B. High-Speed Interface Architecture

The current version of the high-speed interface board provides an interface between an optical network operating at a line/data rate of 1.25/1.00 Gb/s and the TURBOchannel bus of Digital workstations operating at 800 Mb/s. Fig. 19 shows a simplified block diagram of the high-speed subnetwork interface. The key elements of the architecture are the three queues (one each, for transmit, receive, and pass-through traffic) which provide buffers between the host and the network. There are three 32-b wide key datapaths. The first datapath is bidirectional and connects the transmit and receive queues to the host queues through the TURBOchannel I/O bus. The second path accepts incoming network packets in the form of 16 consecutive 32-b words from the deserializer. The third path supplies outgoing word-aligned packets to the serializer.

Having designed the datapath, the control logic was partitioned into three distinct areas and implemented in separate field programmable gate arrays (FPGA's). These functions are performed by two bandwidth arbiters and a routing switch. The first arbiter is responsible for managing access to the TURBOchannel I/O bus (transmit versus receive), and the second manages access to the high-speed output link (transmit versus pass). The routing switch determines if the packet header is valid, and whether the packet should be received, passed-through, or multicast. In case of multicast, packet is both received and passed-through in the interface card.

Although the STARNET architecture is intended to support communication up to 2.5 Gb/s, our design is flexible as to the actual data rate and format. We partitioned our design into two modules: the high-speed packet switch module and the serializer/deserializer module. In the current implementation, serialization and deserialization of the 32-b data words operates at 1.25 Gb/s. The high-speed packet switch module board is operating at 1.25 Gb/s although it is laid out so that it is upgradeable to 2.5 Gb/s.

We now describe transmission and reception of packets over STARNET. The packets are organized as 16, 32-b words. The first word of the packet is a packet header which contains a virtual circuit identifier (VCI), packet sequence number, service class and synchronization information. The remaining 15 words are available for user data. Packets are organized contiguously in host memory in buffers of up to 8 KB (128 packets). Buffers are aligned on 8 KB boundaries to optimize memory access.

The interface board first loads a pointer from a list of 64 transmit buffers[1] in host memory using direct memory access (DMA). The pointer includes the physical base address of the buffer in host memory as well as additional bits that are used to determine the length of the buffer and its relative share of bandwidth. Consecutive packets are taken from each buffer according to its allocated bandwidth and placed in the interface's transmit queue to await access to the network. Each transmit buffer is serviced similarly, in round-robin fashion. This technique provides fair access to the network bandwidth. Moreover, the technique also reduces overall burstiness in the network since it can limit the number of consecutive packets with identical destinations. In retrospect, we note that performance gains could be made by keeping the pointer list on the interface rather than in host memory at the cost of increased interface complexity.

[1] Various size limitations mentioned in the implementation are arbitrarily limited by our implementation and not the architecture itself.

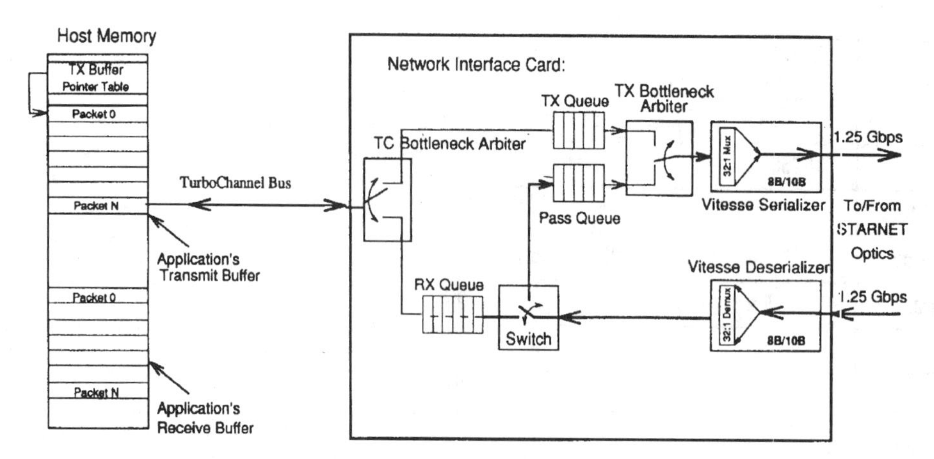

Fig. 19. Block diagram of the high-speed subnetwork interface board.

After deserialization, incoming packets are routed as described later into either the pass and/or the receive queues. In the case of packets destined for the host, the interface uses the VCI and the packet number of the packet header to generate a host memory DMA address. Packets are then placed in the receive queue, and thereafter are transferred directly into an application's buffer space. So, although the node receives packets from one connection interspersed with packets from another connection, all packets on a particular virtual circuit are not only received in order, but also placed into the application's buffer in sequence. Thus, no reassemble or reordering of packets is necessary. This greatly reduces the overhead normally associated with network interfaces and enables an extremely lightweight, low-latency connection.

C. Virtual Circuit Routing

On-the-fly virtual circuit routing is performed using a lookup table in RAM (see Fig. 19). The least significant byte of the packet header is used to address a location in RAM. The data at this location includes bits which specify whether to receive or retransmit a packet (or both, in the multicast case). The RAM data also includes an outgoing virtual circuit number in the retransmit case or a pointer directly to host memory in the receive case. On-the-fly routing is essential to the off-loading of network duties from the host and provides for extremely low switching latencies.

D. Novel Arbitration Scheme of Traffic Streams to Control Network Access and Traffic Latency

A "bottleneck" exists whenever there are two or more streams simultaneously competing for a given network resource. There are two such bottlenecks associated with the high-speed network interface: the first is at the I/O bus; the second is at the 1.0 Gb/s (data rate) serial output link. The bottlenecks exist because the I/O bus bandwidth must be shared between the node's transmit and receive streams, as well as with the host processor and other I/O devices; the serial output link must be shared between the node's transmit stream and the pass-through traffic. The allocation of bandwidth at each bottleneck is performed in hardware by two arbiters, as shown in Fig. 19. The arbiter at the host bus is the TC arbiter; at the network interface is the TX arbiter. In the following, we refer to the bandwidth allocation policy as the STARNET arbitration algorithm.

We are investigating several versions of the STARNET arbitration algorithm. We first describe the basic algorithm and then briefly discuss some of the dynamic capabilities we are investigating.

1) Basic STARNET Arbitration Scheme: The basic STARNET arbitration scheme works by dividing time into a fixed number of time slots which we call a cycle. A slot is defined as the time it takes to transmit one packet. At the beginning of each cycle a number of tokens are granted to each of the competing streams (e.g., PASS stream and TX Stream in the case of the TX ARBITER). The ratio of the number of tokens granted per cycle to the number of slots per cycle yields the average bandwidth allocated to the stream. For example, if there are 32 slots per cycle, 4 TX tokens per cycle and 24 PASS tokens per cycle, then this corresponds at a 1 Gb/s link rate to 125 Mb/s allocated to the TX stream and 750 Mb/s allocated to the PASS stream.

During each time slot of the cycle, the algorithm chooses whether or not to send a TX or PASS packet based upon the state of each of the stream FIFO's and the number of tokens available to the streams. Continuing with the above example, let's say at the beginning of the 32-slot cycle there are 16 packets queued in the TX FIFO and 20 packets in the PASS FIFO. The STARNET arbitration scheme first sends the first four packets of the TX stream, thereby exhausting the TX tokens allocated for the cycle. The arbiter then sends the 20 PASS packets (leaving four PASS tokens). The arbiter then waits until another PASS packet is queued in the PASS FIFO or until the cycle ends. At the beginning of the next cycle, the token counts are reset to their allocated values. Tokens are not accumulated from cycle to cycle.

The scheme is quite similar to the familiar leaky-bucket scheme which fills a bucket with tokens at an allocated fixed rate up to a maximum number of tokens (i.e., bucket depth). The purpose of both algorithms is to control the average transmitted bandwidth as well as the burstiness of the transmitted streams, in essence shaping the distribution of packet traffic in the network. Just as the token rate in the leaky-bucket scheme sets the maximum average bandwidth of the traffic, so does the tokens per cycle parameter in the STARNET arbitration scheme. In the leaky-bucket scheme, the depth of the bucket sets the limit on the burst length for the stream. In the STARNET scheme, the maximum burst length is set by the number of tokens allocated to each stream per cycle. The differences between the two algorithms are quite subtle; the main point to be taken is that both algorithms shape the traffic entering the network by maintaining an average rate and a maximum burst length. The main driving force behind using our algorithm is that it required fewer logic cells in the actual implementation.

2) Dynamic Control of Bandwidth: Given the basic implementation of the algorithm, we now discuss how we dynamically modify the token rates based on the state of the stream FIFO's. The goal is to drop as few packets as possible. In the case of the TX ARBITER, TX packets are competing with PASS packets. Although we would like to always guarantee the TX bandwidth, we make it a priority to not lose PASS packets. Therefore, by monitoring the number of packets in the PASS FIFO, we can temporarily increase the number of tokens allocated to the PASS stream in order to keep the PASS FIFO from overflowing. In doing so, however, great care must be taken when straying off the average allocated bandwidth. For instance, simply increasing the PASS stream's bandwidth without decreasing the TX stream bandwidth would likely result in overloading all the down-stream nodes. We are investigating several heuristics for dynamically trading off of bandwidth between TX and PASS streams. The results in this paper correspond in pseudo-code to

```
if (pass_fifo_almost_full) {
    pass_tokens_per_cycle
            = min(maximum_pass_tokens_per_cycle,
            average_pass_tokens_per_cycle
            + average_tx_tokens_cycle/2);
    tx_packets_per_cycle = average_tx_tokens
            -(pass_tokens_per_cycle
            - average_pass_tokens_per_cycle);
}
else
    pass_tokens_per_cycle
            = average_pass_tokens_per_cycle;
    tx_tokens_per_cycle
            = average_tx_tokens_per_cycle;
}
```

This heuristic maintains the average total bandwidth of traffic flowing through the node while prioritizing the PASS traffic. A similar heuristic is used by the TC ARBITER at the host's I/O bus with the RX stream receiving the priority over the HOST TX stream when the RX stream FIFO is almost full. This makes sense, since it would be more wasteful for a packet to have traversed the entire network only to be dropped at the receiver's end than to delay granting a new packet access to the network. Two drawbacks of our heuristic are: first, a node which does not strictly adhere to the negotiated parameters can easily disrupt the network by flooding the network with its packets which could reduce downstream node's bandwidth in half; and second, since the network is multihop it is naive to assume that TX and PASS bandwidth can be traded off one-for-one. Strictly, only the bandwidth of competing traffic with the same destination node can be fairly traded off; any other trade results in traffic imbalance. However, the simulation and experimental results presented in the sequel of the paper are for uniformly destined traffic are not impacted by these traffic imbalances. Our ongoing work will investigate the above heuristic and others, including the leaky-bucket scheme in order to improve STARNET's performance.

3) Dynamic Control of Latency: When access to the network is unrestricted, packets spend very little time at local queues, and packets arriving in bulk leave in bulk. Therefore, traffic admitted to the network from each node is very bursty. Such bursty traffic will subsequently incur a high queuing latency when multiplexed with other traffic streams as it traverses the network. STARNET arbitration scheme serves to discipline the node's access to the network by controlling the flow of packets out of each of the node's queues. Disciplined admission of traffic to the network reduces the burstiness of the traffic [25], [26]. This reduced burst length traffic will incur smaller latencies through network queues. This flow control results in the node's traffic paying a larger latency penalty to get into the network. Once in the network, the traffic pays a smaller average latency penalty as it travels through queues in the multihop network, reducing the overall latency of the traffic in the network. In TX arbiter, by controlling number of TX slots (out of the total of eight slots), we control the transmit stream's access to the network and control the overall network latency. The above results will be substantiated in Section IV-F, when we evaluate the performance of STARNET.

E. Experiment Status

We have implemented the interface on a modest $6'' \times 9''$ printed circuit board (PCB), as shown in Fig. 20. We have conducted several experiments to measure the maximum transmission and reception performance of the high-speed interface. To measure the maximum transmission performance, we filled the user buffers with pre-made packets and then initiated the data transfer. The CPU polled the transmit buffer pointer queue until all buffered packets were sent. When all the packets had been sent, the time was noted. The header word is not counted as transmitted data. Although the transmit throughput is variable, depending upon host memory loading; we obtained an average throughput of 685 Mb/s out of the 800 Mb/s theoretical maximum.

To measure the maximum receive rate, we first filled the receive queue and then allowed DMA to begin. We measured transaction time with an on-board counter. On average, we obtained a throughput of 571 Mb/s. The maximum receive throughput is less than the transmit throughput because received packets must be handled individually whereas transmit packets are sent in bulk.

Fig. 20. Photograph of the high-speed interface board used in STARNET.

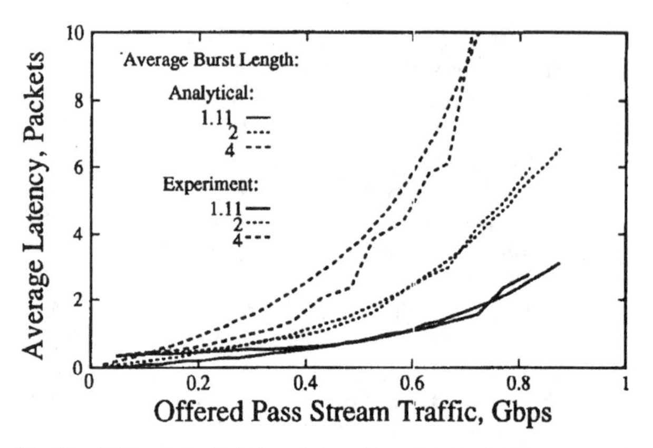

Fig. 21. Minimum bandwidth optical comb used in STARNET-I.

F. STARNET Latency Performance

Consider again multimedia applications listed in Fig. 18: voice, TV video, and video conferencing. The latter imposes the most strict limits on acceptable latency and jitter. We see that a tolerable delay for video-conferencing is some 250 ms. In a campus network, it is expected that a relatively large number of real-time video-conferences will take place concurrently, such as remote viewing of classes. We have performed several simulations and experiments to determine the suitability of the high-speed interface for a campus network environment. We will show below that the STARNET interface meet video-conferencing constraints for a reasonable number of hops between STARNET nodes. This result is obtained by measuring the average latency of packets through each node for various traffic conditions. We also show how the STARNET arbitration scheme is utilized to further reduce the overall network latency. In the following, we describe our network model and present simulation results and experimental measurements.

Our simulation model for the interface follows Fig. 19. The main elements are the routing switch, the three queues, and the mechanisms for arbitrating access to the host and to the network. The three queues are: a) the transmit queue which buffers transmit traffic from local host before it is transmitted to the network, b) the receive queue which buffers traffic from network before it is received at the local host's memory, and c) the pass queue which buffers the traffic from the network which is being forwarded through the node. There are two arbiters, one at each bottleneck. The first is the TX Arbiter at the network, where transmit and pass traffic competes; the second is the TC Arbiter, where transmit and receive traffic compete.

Figure 21 compares analytical results and experimental measurements of the average latency seen by a particular node offering bursty transmit traffic versus the bandwidth of the pass through traffic for three values of transmit traffic burstiness. The data rate before 8B/10B encoding is 1.0 Gb/s and the average rate of the transmit traffic is 100 Mb/s. Burstiness is defined by the average number of consecutive packets that the host generates in a busy period. We used a simple two-state Markov model as a source of bursty transmit traffic. The analytical curves show that the transmit latency increases both with increasing offered pass traffic, and with the burstiness of the data. This effect is expected because the stream is exceeding its allocated share of bandwidth during the burst periods and shows that the STARNET arbitration flow control imposes a latency penalty on bursty transmit traffic. Also plotted are experimental curves which show good agreement with the analytical results. The measurements are obtained by having the first node send data through the second while the second transmits bursty traffic. Packet latency is measured by starting an on-board event timer when a specially tagged packet is placed in the transmit queues and stopping the timer when the packet exits the transmit queue. Discrepancies between the simulation results and the experimental measurements are mostly due to both the approximations made in the simulation model and the difficulties in timing events within UNIX.

Having now shown good correspondence between our experimental results and our simulation data, we now consider the interface's operation in a simulated eight-node network with transmit bandwidth of 200 Mb/s.

Figure 22 shows the latency of packets in each of the system queues versus offered transmit traffic per node for an eight node STARNET subnetwork. Here we chose the burst length to be four, which is true for transmission of MPEG 2 traffic over STARNET because the size of an MPEG2 transport stream packet is 188 bytes which fits four STARNET cells. Latency increases in each of the system queues as the offered transmit traffic per node increases, and the total output link throughput approaches 1.0 Gb/s. Assuming uniform destination distribution, the total offered traffic each node's output link is given by $Tx^*(N+1)/2$, where Tx is the offered transmit bandwidth per node, and N is the number of nodes in the subnetwork. For 200 Mb/s transmit traffic per node and eight nodes, this gives 900 Mb/s total offered traffic at each of the output links and 1.6 Gb/s total transmit traffic on the subnetwork. The transmit latency is always higher than the pass-through latency because (a) the bursty transmit traffic was allocated only one fourth of the

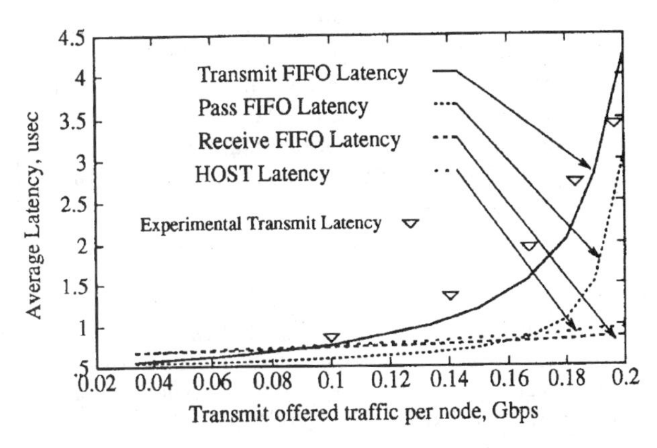

Fig. 22. Latency behavior of traffic passing through the host, transmit, pass, and receive queues of the high-speed interface versus the offered transmit traffic. Results were obtained for an eight-node subnetwork using OPNET.

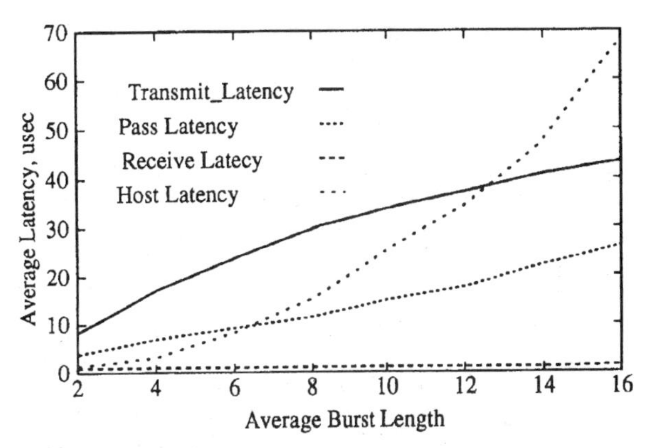

Fig. 23. Latency behavior of system queues as the average burst length of transmit traffic increases.

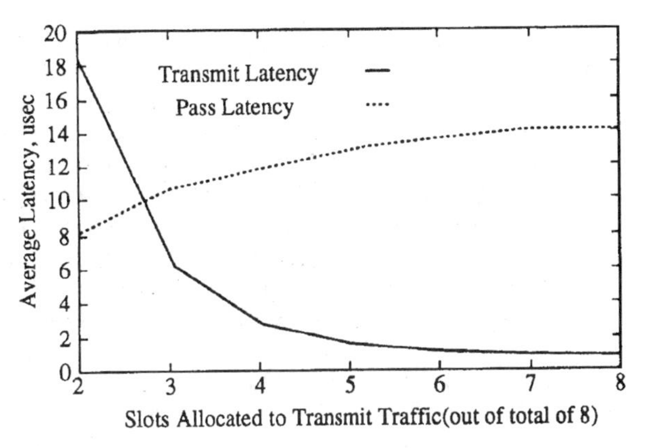

Fig. 24. Latency behavior of STARNET bandwidth arbitration scheme in TX arbiter.

link bandwidth (250 Mb/s) and (b) pass-through traffic coming from the subnetwork is less bursty due to the smoothing effect of the bandwidth arbitration scheme. Packets from the host and packets for the host (receive queue) generally experience the least delay because there is very little congestion across I/O bus, the total traffic is less than half of 800 Mb/s capacity of I/O bus.

Figure 23 shows the latency of packets in each of the system queues for a particular node versus the average burst length of transmit traffic; the average bit rate is 200 Mb/s/node in an eight-node subnetwork. As mentioned earlier, packets experience greater latency in each of the system queues as the burst length of transmit traffic increases. This is because when packets arrive in bulk to a queue, the instantaneous bandwidth of the incoming channel exceeds the outgoing channel bandwidth of the queue. Therefore, packets arriving toward the end of the bulk have to wait in the queue until packets at the head of the bulk are serviced; thus increasing the average latency seen by the packet in the system queue. Also notice that the pass queue and the receive queues are least affected among system queues. This is a result of the STARNET arbitration flow control. By controlling the slot allotment to competing traffic streams, the STARNET arbitration reduces the burstiness of the outgoing traffic. By the time the network traffic comes to receive and pass queue, it has gone through several queue arbitrations; therefore its burstiness has been significantly reduced, resulting in smaller latency.

The STARNET arbitration scheme allows the number of slots allocated to competing streams to be adjusted in order to control latencies in the system queues. We seek to investigate the latency behavior as we vary the slot allocation in TX and TC arbiters.

Figure 24 shows simulation results of the average latency seen by the transmit stream and the pass stream in an eight-node network offering 200 Mb/s transmit traffic with an average burst length of four versus the maximum number slots allocated to the transmit stream in a TX arbiter out of total of eight slots (e.g., four slots out of eight corresponds to 500 Mb/s out of 1.0 Gb/s). Fig. 24 shows that the latency seen by the transmit stream decreases significantly as we increase the number of slots allocated to the transmit queue, while pass latency increases because its slots are allocated for the transmit stream, leaving the pass-through stream with fewer slots. As we decrease the number of slots allocated to transmit stream, transmit latency increases much faster than pass latency decreases. It is desirable to keep the number of slots allocated to the transmit stream low so that the traffic pays a high latency price to gain admission to the network and a smaller latency in hopping through a node in the multihop network reducing the overall network latency.

Figure 25 shows the average latency seen by the receive stream and the host stream (traffic coming from host computer) versus the number of slots allocated to the transmit stream in the TC arbiter out of total of eight slots. In this case, it is desirable to allocate more slots to the receive queue to minimize the receive packet drops since the receive queue is only 512 packets deep while the host queue, being the main memory of the workstation, is many magnitudes deeper.

The foregoing performance analysis verifies our arbitration scheme in controlling the latency as a result of shaping the traffic and shows how the STARNET will perform under different traffic conditions. Based on these results, one can optimize slot allocations to minimize the overall latency in the network.

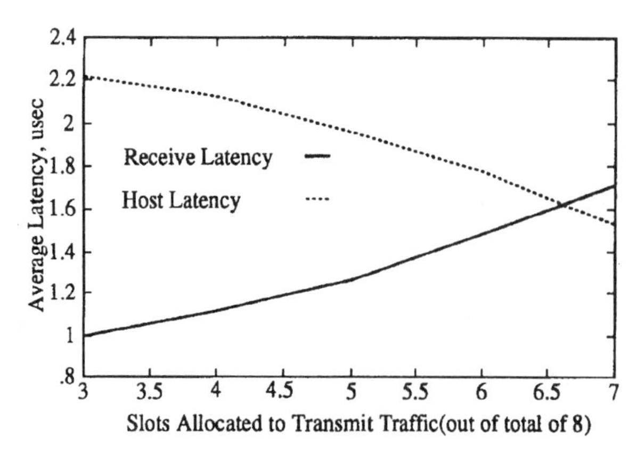

Fig. 25. Latency behavior of STARNET bandwidth arbitration scheme in TC arbiter.

We now show how the above behavior relates to multimedia service requirements. A broadcast quality MPEG-II video conference requires about 6 Mb/s of guaranteed, low-latency bandwidth. Thirty two such MPEG-II sessions will generate approximately 200 Mb/s of traffic from each node. Calculation of the average latency that the stream will incur on its path from the host to the destination requires summing the latency in each queue that it passes through. For an eight-node subnetwork, the maximum total latency is the sum of the time spent in the host queue, the transmit queue, a maximum of six hops through pass queues and finally in the receive queue. For 200 Mb/s total transmit traffic, we can obtain these latencies from Fig. 22. Fig. 22, and other latency plots, have been obtained for the burstlength of four STARNET cells which had been chosen earlier to accommodate MPEG2 traffic which has a fixed transport stream packet size of 188 bytes including 4 bytes of header and 184 bytes of payload. To pack 188 bytes of payload data, it takes four STARNET cells which are 64 bytes each. MPEG2 traffic over STARNET will always transmitted in clumps of four STARNET cells, therefore a burstlength of four was chosen for the latency measurements.

The latencies are, respectively: host, 1.0 μs; transmit, 4.0 μs; pass, 3.0 μs; receive, 1.0 μs. This yields a total latency of $(1.0 + 4.5 + 6 * 3.0 + 1.0)\ \mu s = 24.5\ \mu s$. If we assume 10 km (50 μs) radius passive star network, the total latency through the network is still only about 724.5 μs. Note that the inherent propagation delay in the fiber is in fact far greater than the delay in the queues. Therefore, a moderate-sized STARNET can easily support both the throughput and the delay requirements for video-conferencing, even for 32 simultaneous video conferencing sessions per node. This corresponds to a total of 256 MPEG-II broadcast quality video conferencing sessions per eight-node high-speed subnetwork virtual LAN. In case of common MPEG I video conferences ($\sim$2 Mb/s traffic), $64 * 8 = 512$ such conferences can be supported. Furthermore, for 200 Mb/s traffic per node in the eight node network with the burst length of four (required to accommodate 188 byte MPEG II transport stream packet) we experimentally found the packet loss ratio to 1.95×10^{-7}. In this experiment, we used two experimental nodes A and B, where B generated the network traffic. We had set up virtual circuit tables on both nodes such that traffic from B hops back and forth eight times before it is received to generate eight-node network traffic in the pass queue of B. Then, we transmitted stream from A to B and received it back in A and measured the packet loss. We sent 1.024×10^8 packets and lost only 20 packets. Thus, we have shown STARNET to be suitable for interactive applications in campus networks.

VII. Conclusion

STARNET is a broad-band optical WDM LAN based on a physical passive star topology. It is under investigation in the Optical Communication Research Laboratory at Stanford University. STARNET provides both a high-speed reconfigurable packet-switched data subnetwork and a moderate-speed fixed-ring packet-switched control subnetwork over the same physical network. The high-speed subnetwork operates at the bit rate of up to 1.25 Gb/s/node and the moderate-speed subnetwork operates at the bit rate of 125 Mb/s.

The reconfigurable high-speed data subnetwork is suitable for transmitting both continuous and bursty data. For high-bandwidth, connection-oriented services, the high-speed subnetwork is dynamically reconfigured to optimally support the current service demands. For high-bandwidth communication between a cluster of nodes, multinode communication is achieved through electronic multihop by buffering and dynamically re-routing packets at each node.

The FDDI-compatible moderate-speed subnetwork is designed to support both continuous and bursty moderate-speed payload data with relaxed packet latency requirements and to support services requiring error-free transmission, such as file transfer. In addition, the moderate-speed logical subnetwork provides full connectivity between all nodes. This subnetwork supports the transmission of the network control information including data managing the reconfiguration of the high-speed subnetwork. Thus, STARNET can effectively support diverse traffic with a wide range of speed and continuity characteristics.

The data corresponding to the two subnetworks is multiplexed onto the same optical carrier by using combined modulation in STARNET-I and MSCM in STARNET-II. The two schemes are demonstrated by experimental optical transceivers. The performance of these two schemes have been analyzed. The results show that the combined modulation uses the optical bandwidth more efficiently and can support more nodes than the MSCM scheme. However, the combined modulation scheme is, at the moment, more difficult to implement than the MSCM scheme.

We designed, implemented, and investigated a high-speed high-performance STARNET computer interface which features high-throughput direct memory access to/from host computer, low latency switching, traffic prioritization, and capability of multicasting and broadcasting. The high-speed low-latency feature of the STARNET interface enables efficient electronic multihop through the STARNET high-speed subnetwork. The interface is implemented for DEC5000 workstations and achieves average transmit and receive throughputs of 685 Mb/s and 571 Mb/s (out of the 800 Mb/s TURBOchannel bus capacity), respectively. The total latency of the interface board is very small since the CPU is not involved in the decision making. The total latency incurred for an eight-node subnetwork with uniform destination distribution due to the STARNET interface is measured to be 24 μs. For a 10 km radius subnetwork, the total latency including propagation delay is approximately 724 μs, which is well below the requirements for low latency applications such as video conferencing. The above features make STARNET highly suitable for high-speed multimedia network applications.

ACKNOWLEDGMENT

The authors would like to thank Dr. T. K. Fong and Dr. P. T. Poggiolini for their valuable contributions to this work. The authors would also like to thank the reviewers of this paper for their valuable comments.

REFERENCES

[1] B. Mukherjee, "WDM-based local lightwave networks part I: Single-hop systems," *IEEE Networks*, pp. 12–17, May 1992.

[2] ——, "WDM-based local lightwave networks part II: Multihop systems," *IEEE Networks*, pp. 20–32, July 1992.

[3] L. G. Kazovsky and P. T. Poggiolini, "STARNET: A multi-gigabit-per-second optical LAN utilizing a passive WDM star," *J. Lightwave Technol.*, vol. 11, no. 5/6, pp. 1009–1027, May/June 1993.

[4] M. Cerisola, T. K. Fong, R. T. Hofmeister, L. G. Kazovsky, C. L. Lu, P. Poggiolini, and D. J. M. Sabido IX, "Subcarrier multiplexing of packet headers in a WDM optical network and a novel ultrafast header clock-recovery technique," in *OFC'95*, San Diego, CA, 1995, pp. 273–274.

[5] M. S. Goodman, H. Kobrinski, M. P. Vecchi, R. M. Bulley, and J. L. Gimlett, "The LAMBDANET multiwavelength network: Architecture, applications, and demonstrations," *IEEE J. Select. Areas Commun.*, vol. 8, no. 6, pp. 995–1004, Aug. 1990.

[6] N. R. Dono, P. E. Green, K. Liu, R. Ramaswami, and F. F. Tong, "A wavelength division multiple access network for computer communications," *IEEE J. Select. Areas Commun.*, vol. 8, no. 6, pp. 983–993, Aug. 1990.

[7] F. J. Janniello, R. Ramaswami, and D. G. Steinberg, "A prototype circuit-switched multi-wavelength optical metropolitan-area network," in *IEEE Int. Commun. Conf.*, 1992.

[8] R. Gidron and A. Temple, "Teranet, a multihop multichannel ATM lightwave network," in *IEEE Int. Commun. Conf.*, 1991.

[9] R. Gidron, "Teranet: A multi gigabit per second hybrid/packet switched lightwave network," in *Proc. SPIE 1579, Adv. Fiber Commun. Technol.*, Boston, Sept. 3–4, 1991, pp. 40–48.

[10] S. Alexander *et al.*, "A precompetitive consortium on wide-band all-optical networks," *J. Lightwave Technol.*, vol. 11, no. 5/6, pp. 714–735, May/June 1993.

[11] S. F. Su and R. Olshansky, "Performance of multiple access WDM networks with subcarrier multiplexed control channels," *J. Lightwave Technol.*, vol. 11, no. 5/6, pp. 1028–1033, May/June 1993.

[12] P. Poggiolini and S. Benedetto, "Theory of subcarrier encoding of packet headers in quasiall-optical broadband WDM networks," *J. Lightwave Technol.*, vol. 12, no. 10, pp. 1869–1881, Oct. 1994.

[13] C.-L. Lu, D. J. M. Sabido IX, P. Poggiolini, R. T. Hofmeister, and L. G. Kazovsky, "CORD—a WDMA optical network: Subcarrier-based signaling and control scheme," *IEEE Photon. Technol. Lett.*, vol. 7, no. 5, pp. 555–557, May 1995.

[14] M. J. Hickey, C. F. Barry, C. A. Noronha, and L. G. Kazovsky, "Experimental PSK/ASK transceiver for the STARNET WDM computer communication network," *IEEE Photon. Technol. Lett.*, vol. 5, no. 5, pp. 568–571, May 1993.

[15] M. J. Hickey and L. G. Kazovsky, "The STARNET coherent WDM computer communication network: Experimental transceiver employing a novel modulation format," *J. Lightwave Technol.*, vol. 12, no. 5, pp. 876–884, May 1994.

[16] F.-J. Westphal, M. Hickey, and L. Kazovsky, "An experimental DPSK/ASK transceiver for the STARNET coherent WDM network," in *CLEO'94*, Anaheim, CA, May 1994, pp. 337–338.

[17] M. Hickey, F.-J. Westphal, J. Fan, G. Jacobsen, and L. Kazovsky, "Combined ASK/FSK modulation for the STARNET WDM computer network," in *IEEE/LEOS '94, Summer Top. Mtg. Dig., Opt. Networks, Enabling Technol.*, Lake Tahoe, NV, July 1994, pp. 38–39.

[18] M. Hickey and L. G. Kazovsky, "Combined frequency and amplitude modulation for the STARNET coherent WDM network," *IEEE Photon. Technol. Lett.*, vol. 6, no. 12, pp. 1473–1475, 1994.

[19] V. O'Byrne, "A method for reducing the channel spacing in a coherent optical heterodyne system," *IEEE Photon. Technol. Lett.*, vol. 2, no. 7, July 1990.

[20] F. Ghirardi, J. Brandon, F. Huet, M. Carre, J. Thomas, A. Bruno, and A. Carenco, "InP-based 10-GHz bandwidth polarization diversity heterodyne photoreceiver with electrooptical adjustability," *IEEE Photon. Technol. Lett.*, vol. 6, no. 7, pp. 814–816, July 1994.

[21] L. G. Kazovsky and O. K. Tonguz, "ASK and FSK coherent lightwave systems: A simplified approximate analysis," *J. Lightwave Technol.*, vol. 8, no. 3, pp. 338–352, Mar. 1990.

[22] L. G. Kazovksy, D. J. M. Sabido IX, and T. K. Fong, "Dynamic range of analog optical links: Do optical amplifiers help?" in *LEOS '95 Summer Topical Meeting*, 1995.

[23] D. J. M. Sabido IX, M. Tabara, T. K. Fong, and L. G. Kazovsky, "Experimental investigation of the impact of EDFA's on coherent AM analog optical links," in *OFC'94*, San Jose, CA, 1994, pp. 264–266.

[24] A. Demers, S. Keshav, and S. Shenker, "Analysis and simulation of a fair queuing algorithm," *Internetworking: Res. Experience*, vol. 1, pp. 3–26, 1990.

[25] I. Cidon, R. Guerin, and A. Khamisy, "On protective buffer policies," in *LEOS '93 Summer Top. Mtg., Gigabit Networks*, Santa Barbara, CA, July 1993, pp. 36–37.

[26] V. Anantharam *et al.*, "Burst reduction properties of the leaky bucket flow control scheme in ATM networks," *IEEE Trans. Commun.*, vol. 42, no. 12, pp. 3085–3089, Dec. 1994.

Author Index

Subject Index

A

P

R

S

About the Editors

Hussein T. Mouftah is a professor at Queen's University in Kingston, Ontario, and is principal investigator for the Telecommunications Research Institute of Ontario (TRIO). He is also director of the IEEE Communications Society magazines and served as editor-in-chief for the IEEE Communications Society (1995-97). He has consulted for government and industry in the areas of computer networks, digital systems, and fault tolerant computing. Dr. Mouftah holds a number of patents and has published widely in the areas of computer networks, digital systems, and system reliability. He is the author of *Microprocessors and Microcomputers: Principles, Design, and Applications* (Jackson Press, 1985). In 1989, he received the Engineering Medal for Research and Development from the Association of Professional Engineers of Ontario (APEO).

Jaafar M. H. Elmirghani is a principal lecturer and BT Reader in Telecommunication Systems at the University of Northumbria, U.K. He is editor of the *IEEE Communications Magazine* and also serves as technical editor for the *Journal of Optical Communications* (JOC). He currently acts as secretary for both the IEEE UK&RI Communications Chapter and the IEEE COMSOC Signal Processing and Communication Electronics (SPCE) Technical Committee. Dr. Elmirghani has published over 100 technical papers in optical communication systems, signal processing, and statistical communication theory.